ASTRONOMY AND ASTROPHYSICS ABSTRACTS

A Publication of the Astronomisches Rechen-Institut Heidelberg
Member of the Abstracting Board of the International
Council of Scientific Unions

Volume 8
Literature 1972, Part 2

Edited by
S. Böhme · W. Fricke · U. Güntzel-Lingner
F. Henn · D. Krahn · U. Scheffer · G. Zech

Springer-Verlag Berlin Heidelberg GmbH 1973

Astronomisches Rechen-Institut
Heidelberg
Director: Prof. Dr. W. Fricke

Astronomy and Astrophysics Abstracts
Editor-in-Chief: F. Henn

Astronomy and Astrophysics Abstracts
is prepared under the auspices
of the International Astronomical Union

ISBN 978-3-662-12286-0   ISBN 978-3-662-12284-6 (eBook)
DOI 10.1007/978-3-662-12284-6

Preface

Astronomy and Astrophysics Abstracts, which has appeared in semi-annual volumes since 1969, is devoted to the recording, summarizing and indexing of astronomical publications throughout the world. It is prepared under the auspices of the International Astronomical Union (according to a resolution adopted at the 14th General Assembly in 1970).

Astronomy and Astrophysics Abstracts aims to present a comprehensive documentation of literature in all fields of astronomy and astrophysics. Every effort will be made to ensure that the average time interval between the date of receipt of the original literature and publication of the abstracts will not exceed eight months. This time interval is near to that achieved by monthly abstracting journals, compared to which our system of accumulating abstracts for about six months offers the advantage of greater convenience for the user.

Volume 8 contains literature published in 1972 and received before March 15, 1973; some older literature which was received late and which is not recorded in earlier volumes is also included.

The authors of papers who have sent us abstracts on request have effectively contributed to the success of our service. We should like to express our gratitude to them. We acknowledge with thanks contributions to this volume by Dr. J. Bouška, who surveyed journals and publications in the Czech language and supplied us with abstracts in English, and by the Commonwealth Scientific and Industrial Research Organization (C.S.I.R.O.), Sydney, for providing titles and abstracts of papers on radio astronomy.

We also extend our warmest thanks to Miss Helga Ballmann, Mrs Monika Betz, and Mrs Karola Gudé, who typed the text of this volume on IBM 72 Composers and compiled the pages from abstract slips in a perfect form for offset reproduction, to Miss Gisela Nollert, for punching material for the author index and the subject index.

Heidelberg, April 1973

Siegfried Böhme
Walter Fricke
Ulrich Güntzel-Lingner
Frieda Henn
Dietlinde Krahn
Ute Scheffer
Gert Zech

Contents

Stellar Systems

Introduction

Astronomical bibliographies

Astronomy and Astrophysics Abstracts begins documentation and abstracting as from the year 1969. For information on astronomical literature before this date consultation of one of the following bibliographies is suggested:

(1) J. J. de Lalande, Bibliographie Astronomique, Paris 1803 (this work covers the time from 480 B. C. to the year 1803, VIII + 966 pages).

(2) J. C. Houzeau, A. Lancaster, Bibliographie générale de l'astronomie, Volume I (in two parts), Bruxelles 1882, 1887, Volume II, Bruxelles 1889. The complete title of Volume II is "Bibliographie générale de l'astronomie ou catalogue méthodique des ouvrages, des mémoires et des observations astronomiques, publiés depuis l'origine de l'imprimerie jusqu'en 1880". A new edition of these volumes was prepared by D. W. Dewhirst in 1964.

(3) Bibliography of Astronomy, 1881 - 1898. The literature of this period was recorded on standard slips by the Observatoire Royal de Belgique. From the material (some 52.000 items) a microfilm version was produced by University Microfilms Limited, Tylers Green, High Wycombe, Buckinghamshire, England, in 1970.

(4) Astronomischer Jahresbericht, 1899 gegründet von Walter Wislicenus, herausgegeben vom Astronomischen Rechen-Institut in Heidelberg (formerly in Berlin), Verlag W. de Gruyter, Berlin. For the period from 1899 to 1968 sixty-eight volumes were published, each of which, in general, covers the literature of one year.

(5) Bulletin Signalétique – Section, Astronomie, Physique Spatiale, Géophysique. Published by Centre de Documentation du Centre National de la Recherche Scientifique, Paris. This publication is a continuation of "Bibliographie Mensuelle de l'Astronomie" founded in 1933 by the Société Astronomique de France. The publication is continued.

(6) Referativnyj Zhurnal. Founded in 1953 and published by Vsesoyuznyj Institut Näuchnoj i Tekhnicheskoj Informatsii, Akademiya Nauk, Moskva. The publication is continued.

Concept of Astronomy and Astrophysics Abstracts

This abstracting service aims to present a comprehensive documentation of the literature in all fields of astronomy and astrophysics. It appears in semi-annual volumes, two of which cover the literature of a calendar year. The half-yearly period of issue is regarded as an optimal period of time for summarizing papers into subject categories and for the presentation of abstracts as quickly as possible after the publication of the original literature. The time limits at which the documentation begins and ends for a volume are not sharply defined, except in the sense that all literature will be covered which was received by the editors within these limits.

Vol. 8 is devoted to the recording summarizing and indexing of astronomical publications of the year 1972 received from August 16, 1972 to March 15, 1973; it also records a number of papers issued before 1972 but received within the given period of time.

The main characteristics of the concept of Astronomy and Astrophysics Abstracts may be summarized briefly.

(1) Titles of papers are given in the language of their authors whenever possible. If they are not in English but supplied with English translations they will be given in English. Abstracts are presented in English, French or German. Titles of papers in Russian are given in English.

(2) Authors' abstracts are used whenever possible. As a rule, popular articles were not abstracted; however their titles are usually given with the notation "Popular article".

(3) As a rule, each paper has been classified into one of 108 numbered subject categories and allocated a serial number within the category. In this way each item is numbered by six figures, the first three of which indicate the number of the category. Three further figures indicate the serial number within the category, which was allocated in the order of the receipt of the abstract. Reference to an abstract in Volume 1 is indicated by "01" before the number of the category; for example, 01.074.028, denotes Volume 1, category 074, abstract 028. Vol. 2 is indicated by "02", etc., Vol. 8 by "08".

A paper may have been classified into more than one category. Then its abstract has been allocated a number in one of the categories involved, and in the other category (or categories) the paper has been indicated by the title and a reference to the abstract number.

Papers whose authors are not named were treated like those with authors' names, with one exception: reports from correspondents of journals whose names were unknown were not numbered.

(4) There are categories which suggest the presentation of the material in subject groups. For instance, a subject group may be formed by all information received on the same solar eclipse, comet, nova, etc. The unsorted presentation of such material in a subject category would be inconvenient for the user, even if the individual comet, etc. were included in the subject index.

The following subject categories are subdivided into subject groups:

008 Observatories, Institutes. The publications of observatories and astronomical institutes are listed in alphabetical order of the towns of the institutions, each town forming a numbered subject group. For each publication a reference to an abstract number is made.

010 Societies, Associations, Organizations. The publications of each one form a subject group. The groups are presented in alphabetical order.

079 Solar eclipses. All publications related to one solar eclipse form a subject group.

103 Comets: Listed Objects. All publications related to the same comet form a numbered group.

124 Novae. All publications related to one nova form a subject group.

125 Supernovae. All publications related to one supernova form a subject group.

(5) Border fields of astronomy and astrophysics have been taken into account by presenting titles of papers occasionally without abstracts. The selection of papers for inclusion has been made according to the degree of relevance to astronomical research.

Transliteration of the Russian alphabet

The transliteration of the Russian alphabet in use in Astronomy and Astrophysics Abstracts is presented here.

А	а	a	Р	р	r
Б	б	b	С	с	s
В	в	v	Т	т	t
Г	г	g	У	у	u
Д	д	d	Ф	ф	f
Е	е	e	Х	х	kh
Ё	ё	e	Ц	ц	ts
Ж	ж	zh	Ч	ч	ch
З	з	z	Ш	ш	sh
И	и	i	Щ	щ	shch
Й	й	j	Ъ	ъ	"
К	к	k	Ы	ы	y
Л	л	l	Ь	ь	'
М	м	m	Э	э	eh
Н	н	n	Ю	ю	yu
О	о	o	Я	я	ya
П	п	p			

This transliteration was recommended by the Abstracting Board of the International Council of Scientific Unions in 1969. It is essentially the same as the transliteration proposed by the Academy of Sciences, Moscow, and used by the Referativnyj Zhurnal (see Referativnyj Zhurnal, 51. Astronomiya, 1969 No. 1). It may be noted that the letters can be read and printed by usual data processing machines.

In the literature however the names of Russian authors can be found transliterated in different ways. We present the names in the form in which they are given in the references cited.

Sources of information

The majority of sources of information for this volume are given in section **001 Periodicals** and in section **008 Observatories, Institutes.** The term "periodical" has been used in its widest sense for publications in a sequence of undetermined duration, even if the intervals of appearance are not regular. Section 001 records 291 periodicals with their full titles and with abbreviations which are in use in Astronomy and Astrophysics Abstracts. It may be noted that the titles of the periodicals are given in their original languages, and that Russian titles have been transliterated applying the transliteration given above. Section 008 records 162 periodicals; these are publication series of observatories and astronomical institutes which have not been included in section 001. The abbreviations of the titles of the periodicals have been given so that in most cases they permit recognition of the full title without recourse to the key in section 001. The steadily growing number of periodicals makes it necessary to use more extensive abbreviations and to abandon the use of very condensed ones.

Other abstracting journals have been consulted in order to examine the degree of completeness of our service. Occasionally, in particular in Physics Abstracts, Referativnyj Zhurnal, and Bulletin Signalétique abstracts of papers were found which had not come to our attention. In such cases Astronomy and Astrophysics Abstracts cites these papers, but also gives reference to the abstracting service which acted as the source.

Classification into a scheme of subject categories

The subdivision of astronomy and its border fields into sub-ject categories is facilitated by the fact that the astronomical objects appear to be particularly well suited for the formation of categories. Sun, moon, earth, planets, comets, and meteorites, the various kinds of stars, galaxies, radio sources, quasars, and pulsars etc. suggest natural subdivisions. It may be assumed that such subdivisions can be maintained for long periods of time. Experience shows, however, that progress in research may imply changes in the classification scheme, in particular, in fields where the expansion of knowledge is explosive.

A few explanatory remarks may be in order on some of the subject categories. Section 002 includes short news notes whose titles and authors are given, but the authors of the notes have not been included in the author index. In section 003 books on astronomy and astrophysics and its border fields are listed which came to our notice from August 1972 to March 1973. References to book reviews are given if the review appeared quickly.

For completeness of documentation, personal notes (section 006) and obituaries (section 007) are listed. In section 012 (Proceedings of Colloquia, Congresses, Meetings, and Symposia) the proceedings etc. are listed with titles and editors. The individual papers are classified into their corresponding subject categories, but not included in the subject index. The main subjects of these symposia are cited in the index under section 012.

Errata to papers communicated by the authors are listed at the end of the corresponding subject categories.

Author index and subject index

The subject category and the serial number forming six figures for each abstract have been used as a means of reference in the author index and the subject index. These references are more precise than page references. They offer considerable advantages in indexing by means of data processing machines, and they are more convenient for the user.

The author index of this volume contains 7758 names. A complete reference comprises six figures, three for the subject category and three for the serial number within the category. In the case of more than one reference to abstracts in one category, the number of the category is given only once and not repeated in the immediately following references. The total number of papers (some do not give names of authors) recorded in this volume is about 7100.

We consider the subject index as only a first approximation to an optimal index covering all fields of astronomy and astrophysics and their border fields. Several iterative steps appear to be necessary until an index has been compiled for one of the subsequent volumes which may then serve as a kind of standard for the near future. The assigning of one or more key words to a paper is undoubtedly a difficult task. Some journals have started giving key words together with the titles of papers. These key words are chosen by the authors themselves and are in many cases identical with our designations of subject categories with no additional specification. In fact, in some cases it may be more useful to refer to a subject category as a whole than to an item number, in particular, if the total number of abstracts in a category is very small, and if more specific key words do not provide a proper description of the paper.

While each volume is scheduled to contain an author index and a subject index, the magnetic tapes containing the index information will be used to produce separate index volumes (authors and subjects) at intervals of a few years.

The text of the publication was typed on IBM 72 Composers in the editorial office, and it was given to the printer in a form ready for offset reproduction. The author index and the subject index were compiled and printed by means of electronic computer (Siemens 2002).

Abbreviations

AAS	American Astronomical Society
AAVSO	American Association of Variable Star Observers
Abh.	Abhandlungen
Abstr.	Abstract
Abt.	Abteilung
Acad.	Academy, etc.
Accad.	Accademia
Adv.	Advances
AG	Astronomische Gesellschaft
AIAA	American Institute of Aeronautics and Astronautics
AJB	Astronomischer Jahresbericht
Akad.	Akademie
An.	Anales, etc.
Ann.	Annals, etc.
Arch.	Archiv, etc.
Ark.	Arkiv
ASA	Astronomical Society of Australia
Asoc.	Asociación
ASP	Astronomical Society of the Pacific
Ass.	Association
ASSA	Astronomical Society of Southern Africa
Astrofis.	Astrofisica, etc.
Astrofiz.	Astrofizika, etc.
Astron.	Astronomy, etc.
Astronaut.	Astronautics, etc.
Astrophys.	Astrophysics, etc.
ASV	Astronomical Society of Victoria
ASWA	Astronomical Society of Western Australia
Atmosph.	Atmosphere, etc.
BA	Bulletin Astronomique
BAA	British Astronomical Association
BAN	Bulletin of the Astronomical Institutes of the Netherlands
Ber.	Berichte
BIH	Bureau International de l'Heure (Paris)
Bol.	Boletin
Boll.	Bolletino
Bull.	Bulletin
Byull.	Byulleten' (Bulletin)
Circ.	Circular
Cl.	Classe
Coll.	Collection
Commun.	Communication
Comun.	Comunicazioni
Contr.	Contributions, etc.
COSPAR	Committee on Space Research
C.S.I.R.O.	Commonwealth Scientific Industrial Research Organization
Dep.	Department
Diss.	Dissertation
Div.	Division
Dokl.	Doklady (Reports)
ESO	European Southern Observatory
ESRO	European Space Research Organization
Fis.	Fisica, etc.
Fiz.	Fizika, etc.
Fys.	Fysica, etc.
Géod.	Géodésie, etc.
Geod.	Geodesy, etc.
Geofis.	Geofisica, etc.
Geofiz.	Geofizika, etc.
Geofys.	Geofysik, etc.
Geol.	Geology, etc.
Geogr.	Geography, etc.
Geophys.	Geophysics, etc.
Ges.	Gesellschaft
Glav.	Glavnyj (Main)
Gos.	Gosudarstvennyj (State)
HRD	Herzsprung-Russell diagram
Hydrogr.	Hydrography, etc.
IAF	International Astronautical Federation
IAU	International Astronomical Union
ICSU	International Council of Scientific Unions
IEEE	Institute of Electrical and Electronics Engineers
Industr.	Industry, etc.
Inform.	Information
Inst.	Institute, etc.
Instn.	Institution
Ionosph.	Ionosphere, etc.
Issled.	Issledovaniya (Research)
Ist.	Istituto
Izv.	Izvestiya (News)
Jb.	Jahrbuch
JO	Journal des Observateurs
Journ.	Journal
Kl.	Klasse
Lab.	Laboratory
Mag.	Magazine
Mat.	Matematica, etc.
Math.	Mathematics, etc.
Mech.	Mechanics, etc.
Med.	Mededelingen
Medd.	Meddelande, Meddelser
Mekhan.	Mekhanika, etc.
Mém.	Mémoires
Mem.	Memoirs, Memorandum, etc.
Meteorol.	Meteorology, etc.
MIT	Massachusetts Institute of Technology
Mitt.	Mitteilungen
MVS Sonneberg	Mitteilungen über Veränderliche Sterne, Sonneberg
Nachr.	Nachrichten
NASA	National Aeronautics and Space Administration
Nat.	Naturwissenschaftlich, etc.
Naut.	Nautics, etc.
NBS	National Bureau of Standards
NRAO	National Radio Astronomy Observatory (Green Bank)
NRL	Naval Research Laboratory (Washington)
Obs.	Observatory, etc.
OSA	Optical Society of America
Oss.	Osservatorio, Osservazioni, etc.
Ped.	Pedagogika, etc. (Pedagogics)
Phil.	Philosophical
Phys.	Physics, etc.
Planet.	Planetary
Priklad.	Prikladnoj (Applied)
Proc.	Proceedings
Progr.	Progress, etc.
Pubbl.	Pubblicazioni
Publ.	Publications
Rap.	Raportoj
RAS	Royal Astronomical Society
RAS Canada	Royal Astronomical Society of Canada
Rech.	Recherches
Rend.	Rendiconti

Rep.	Report	Techn.	Technics, etc.
Repr.	Reprint	Tekhn.	Tekhnika, etc.
Res.	Research	Teor.	Teoreticheskij
Rev.	Review, etc.	Terr.	Terrestrial, etc.
Ric.	Ricerche	TH	Technische Hochschule
Roy.	Royal, etc.	Theor.	Theoretical
SAF	Société Astronomique de France	Tidssk.	Tidsskrift
SAI	Società Astronomica Italiana	Trans.	Transactions
SAO	Smithsonian Astrophysical Observatory	Trudy	Trudy (Publications)
SAS	Société Astronomique de Suisse	Tsentr.	Tsentral'nyj (Central)
Sci.	Science, etc.	Tsirk.	Tsirkulyar (Circular)
Sect.	Section	TU	Technical University
Ser.	Series, etc.	Uch. Zap.	Uchenye Zapiski (Treatise)
S. I. R.	Service International Rapide des Latitudes	Univ.	University, etc.
Sitz.-Ber.	Sitzungsberichte	URSI	Union Radio Scientifique Internationale
Soc.	Society	Verh.	Verhandlungen
Soobshch.	Soobshcheniya (Communications)	Veröff.	Veröffentlichungen
Sternw.	Sternwarte	Wet.	Wetenschappen
Stud. Cerc.	Studii şi Cercetari	Wiss.	Wissenschaften, etc.
Supl.	Suplemento	Zeitschr.	Zeitschrift
Suppl.	Supplement	ZfA	Zeitschrift für Astrophysik
SuW	Sterne und Weltraum	Zhurn.	Zhurnal (Journal)

Periodicals, Proceedings, Books, Activities

001 Periodicals

AAS Photo-Bull.
AAS (American Astronomical Society) Photo-Bulletin. Published by the Working Group on Photographic Materials. Produced by Eastman Kodak Co., Rochester, N. Y.

Abh. Deutsch. Akad. Wiss. Berlin
Abhandlungen der Deutschen Akademie der Wissenschaften zu Berlin. Klasse für Mathematik, Physik und Technik. Publisher: Akademie-Verlag, Berlin.

Acad. Roy. Belgique, Bull. Cl. Sci.
Académie Royale de Belgique, Bulletin de la Classe des Sciences (Koninklijke Academie van België, Mededelingen van de Klasse der Wetenschappen). 5ᵉ Série. Palais des Académies, Bruxelles.

Acta Astron.
Acta Astronomica. Publisher: Komitet Astronomii, Polskiej Akademii Nauk, Warszawa - Kraków.

Acta Phys. Austriaca
Acta Physica Austriaca. Publisher: Springer-Verlag, Wien.

Acta Univ. Carolinae Math. Phys.
Acta Universitatis Carolinae, Mathematica et Physica. Administrace: Matematicko-fyzikálni fakulta University Karlovy, Praha.

Actas Acad. Nacional Cienc. Lima
Actas de la Academia Nacional de Ciencias Exactas, Fisicas y Naturales de Lima. Lima - Peru.

Adv. Astron. Astrophys.
Advances in Astronomy and Astrophysics. Publisher: Academic Press, New York — London.

AIAA Journ.
AIAA Journal. A Publication of the American Institute of Aeronautics and Astronautics devoted to Aerospace Research and Development. Published by the American Institute of Aeronautics and Astronautics, New York, N.Y.

Am. Scient.
American Scientist. Society of Sigma Xi, New Haven, Conn.

Ann. d'Astrophys.
Annales d'Astrophysique. Revue internationale bimestrielle publiée par le Centre National de la Recherche Scientifique et éditée par son Service d'Astrophysique, Paris. After Vol. 31 replaced by "Astronomy and Astrophysics".

Ann. Françaises Chronométrie Micromécanique
Annales Françaises de Chronométrie et de Micromécanique, publication annuelle de l'Observatoire de Besançon, du Centre Technique de l'Industrie Horlogère et de la Société Française de Chronométrie et de Micromécanique. Rédaction et administration: Observatoire de Besançon. Publiées avec le concours du Centre National de la Recherche Scientifique et des organismes corporatifs.

Ann. Géophys.
Annales de Géophysique. Revue Internationale trimestrielle, publiée par le Centre National de la Recherche Scientifique, Paris.

Ann. Obs. Astron. Météorol. Toulouse
Annales de l'Observatoire Astronomique et Météorologique de Toulouse. Publisher: Gauthier-Villars, Paris.

Ann. Physics
Annals of Physics. Publisher: Academic Press Inc., New York, N.Y.

Ann. Physik
Annalen der Physik. 7. Folge. Publisher: Johann Ambrosius Barth, Leipzig.

Ann. Physique
Annales de Physique. Publisher: Masson et Cie., Paris.

Ann. Soc. Sci. Bruxelles
Annales de la Société Scientifique de Bruxelles. Série I: Sciences Mathématiques, Astronomiques et Physiques. Published by Institut de Physique, Heverlé-Louvain.

Annual Rep. Astron. Inst. Greece
Annual Reports of the Astronomical Institutes of Greece. Published by the Greek National Committee for Astronomy. Academy of Athens, Research Center for Astronomy and Applied Mathematics.

Annual Rev. Astron. Astrophys.
Annual Review of Astronomy and Astrophysics. Publisher: Annual Reviews Inc., Palo Alto, California.

Ann. Univ.-Sternw. Wien
Annalen der Universitäts-Sternwarte Wien. In Kommission bei Ferd. Dümmlers Verlag, Bonn.

Anzeiger. Österreich. Akad. Wiss. Math.-Nat. Kl.
Anzeiger. Österreichische Akademie der Wissenschaften. Mathematisch-Naturwissenschaftliche Klasse. Publisher: Springer-Verlag, Wien.

Applied Optics
Applied Optics. A monthly publication of the Optical Society of America. Published for the Optical Society of America by the American Institute of Physics, New York, N. Y.

Arch. Sci. Genève
Archives des Sciences, éditées par la Société de Physique et d'Histoire Naturelle de Genève. Publisher: Imprimerie Kundig, Genève. Subscription address: Librairie Payot, Genève.

Ark. Astron.
Arkiv för Astronomi. Utgivet av Kungliga Svenska Vetens-

kapsakademien, Stockholm. Printed by Almqvist & Wiksell, Stockholm.

Ark. Geofys.
Arkiv för Geofysik. Kungliga Svenska Vetenskapsakademien, Stockholm. Printed by Almqvist & Wiksell, Stockholm.

Artificial Satellites
Artificial Satellites. Publication of Polish Scientific Institutions. Polish Academy of Sciences, National Committee of Geophysics and Geodesy, National Committee for Space Research, Warsaw. Publishing Office: Palac Kultury i Nauki, Warszawa.

Asoc. Argentina Astron. Bol.
Asociación Argentina de Astronomía. Boletin. Editor: Instituto Argentino de Radioastronomía, Provincia de Buenos Aires, Argentina. Printer: Talleres Gráficos "Renovación", La Plata, República Argentina.

Astrofizika
Astrofizika. Izdatel'stvo Akademii Nauk Armyanskoj SSR, Erevan. [An English translation is published in "Astrophysics".]

Astrofiz. Issled. Izv. Spets. Astrofiz. Obs.
Astrofizicheskie Issledovaniya. Izvestiya Spetsial'noj Astrofizicheskoj Observatorii. Akademiya Nauk SSSR. Publishers: Izdatel'stvo "Nauka", Leningradskoe Otdelenie, Leningrad.

Astron. Astrophys.
Astronomy and Astrophysics. A European Journal. Published by Springer-Verlag, Berlin — Heidelberg — New York.

Astron. Astrophys. Suppl. Ser.
Astronomy and Astrophysics. Supplement Series. A European Journal. Published by the Astronomical Institute Lausanne and Geneva Observatory, Switzerland, on behalf of the Board of Directors.

Astronaut. Acta
Astronautica Acta. An Archive Journal of the International Academy of Astronautics. Published by Pergamon Press, New York — Oxford.

Astronaut. Aeronaut.
Astronautics & Aeronautics. A Publication of the American Institute of Aeronautics and Astronautics. Published monthly by the American Institute of Aeronautics and Astronautics, Easton, Pennsylvania.

Astron. in der Schule
Astronomie in der Schule. Zeitschrift für die Hand des Astronomielehrers. Herausgegeben vom Verlag Volk und Wissen, Berlin. Redaktion: Sternwarte Bautzen.

Astron. Journ.
The Astronomical Journal. Published for the American Astronomical Society by the American Institute of Physics, New York, N. Y. Editorial Office: Department of Astronomy, Columbia University, New York, N. Y.

Astron. Nachr.
Astronomische Nachrichten. Publisher: Akademie-Verlag, Berlin.

Astron. Soc. Pacific Leaflet
Astronomical Society of the Pacific. Leaflet. Edited by the Astronomical Society of the Pacific, San Francisco, California.

Astron. Tidssk.
Astronomisk Tidsskrift. Edited by Astronomisk Selskab, København; Norsk Astronomisk Selskap, Oslo; Svenska Astronomiska Sällskapet, Stockholm. Printed by John Griegs Boktrykkeri, Bergen.

Astron. Tsirk.
Astronomicheskij Tsirkulyar, izdavaemyj Byuro Astronomicheskikh Soobshchenij Akademii Nauk SSSR. Tipografiya Astrosoveta AN SSSR, Moskva.

Astron. Vestn.
Astronomicheskij Vestnik. Publishers: Izdatel'stvo "Nauka", Moskva.

Astron. Zhurn. Akad. Nauk SSSR
Astronomicheskij Zhurnal. Akademiya Nauk SSSR. Publishers: Izdatel'stvo "Nauka", Moskva. [An English translation is published in "Soviet Astronomy AJ"].

Astrophysics
Astrophysics. The Faraday Press cover-to-cover translation of Astrofizika. The Faraday Press, Inc., New York, N. Y.

Astrophys. Journ.
The Astrophysical Journal. Published in collaboration with the American Astronomical Society by the University of Chicago Press, Chicago, Illinois.

Astrophys. Journ. Suppl. Ser.
The Astrophysical Journal. Supplement Series. Published in collaboration with the American Astronomical Society by the University of Chicago Press, Chicago, Illinois.

Astrophys. Letters
Astrophysical Letters. An International *EXPRESS* Journal. Published monthly by Gordon and Breach Science Publishers Ltd., New York — London — Paris.

Astrophys. Norvegica
Astrophysica Norvegica. Edited by The Institute of Theoretical Astrophysics, University of Oslo (Det Norske Videnskaps-Akademi i Oslo). Universitets-forlaget, Oslo.

Astrophys. Space Sci.
Astrophysics and Space Science. An International Journal of Cosmic Physics. Published by D. Reidel Publishing Company, Dordrecht — Holland.

Atti Accad. Nazionale Lincei. Mem.
Atti della Accademia Nazionale dei Lincei. Serie Ottava. Memorie. Classe di Scienze fisiche, matematiche e naturali. Sezione I: Matematica, Meccanica, Astronomia, Geodesia e Geofisica. Published by Accademia Nazionale dei Lincei, Roma.

Atti Accad. Nazionale Lincei. Rend.
Atti della Accademia Nazionale dei Lincei. Serie Ottava. Rendiconti. Classe di Scienze fisiche, matematiche e naturali. Published by Accademia Nazionale dei Lincei, Roma.

Australian Journ. Phys.
Australian Journal of Physics. Published by the Commonwealth Scientific and Industrial Research Organization, East Melbourne, Victoria.

Australian Journ. Phys. Astrophys. Suppl.
Australian Journal of Physics, Astrophysical Supplement.
Published by Commonwealth Scientific and Industrial
Research Organization, East Melbourne, Victoria.

BAV Rundbrief
BAV Rundbrief. Mitteilungsblatt der Berliner Arbeitsgemeinschaft für Veränderliche Sterne. Editor: BAV Berliner Arbeitsgemeinschaft für Veränderliche Sterne eV.,
Berlin.

BBSAG Bull.
Bedeckungsveränderlichen Beobachter der Schweizerischen Astronomischen Gesellschaft, [Swiss Astronomical
Society's Eclipsing Variable Observers], Bulletin. To be
obtained from R. Diethelm, Winterthur, Switzerland.

Bild der Wiss.
Bild der Wissenschaft. Zeitschrift über die Naturwissen-
schaften und die Technik in unserer Zeit. Publisher:
Deutsche Verlagsanstalt, Stuttgart.

Bol. Inst. Mat., Astron., Fis. Univ. Nacional Córdoba
Boletin del Instituto de Matematica, Astronomia y
Fisica, Universidad Nacional de Córdoba (R. A.).Direc-
ción General de Publicaciones, Córdoba (Argentina).

Bol. Liga Latinoamericana Astron.
Boletin de la Liga Latinoamericana de Astronomia. Publi-
cado por la Asociacion Argentina Amigos de la Astronomia, Buenos Aires, Argentina.

Boll. Geod. Sci. Affini
Bolletino di Geodesia e Scienze Affini. Pubblicazione
dell'Istituto Geografico Militare, Firenze.

Boundary-Layer Meteorology
Boundary-Layer Meteorology. An International Journal
of Physical and Biological Processes in the Atmospheric
Boundary Layer. Published by D. Reidel Publishing Company, Dordrecht—Holland.

British Astron. Ass. Circ.
British Astronomical Association, Circular. Editorial
Office: 97 Hawkswood Drive, Hailsham, Sussex.

Bull. American Astron. Soc.
Bulletin of the American Astronomical Society. Published
for the American Astronomical Society by the American
Institute of Physics Inc., New York, N. Y.

Bull. Astron. Inst. Czechoslovakia (BAC)
Bulletin of the Astronomical Institutes of Czechoslovakia.
Published under the auspices of the Czechoslovak Acade-
my of Sciences by Academia, Praha. Editor: Astronomic-
al Institutes of the Czechoslovak Academy of Sciences,
Praha.

Bull. Astron. Inst. Netherlands (BAN)
Bulletin of the Astronomical Institutes of the Nether-
lands. Publisher: North-Holland Publishing Company,
Amsterdam. After Vol. 20 replaced by "Astronomy and
Astrophysics".

Bull. Géod.
Bulletin Géodésique, being the Journal of the International Association of Geodesy. Nouvelle Série. Publié
par le Bureau Central de l'Association Internationale
de Géodésie, Paris.

Bull. Geograph. Survey Inst.
Bulletin of the Geographical Survey Institute. Published
by the Geographical Survey Institute, Ministry of Construction, Tokyo, Japan.

Bull. Obs. Astron. Beograd
Bulletin de l'Observatoire Astronomique de Béograd.
Editor: Observatoire Astronomique de Béograd. Printed
by Naucna delo, Béograd.

Bull. Sci. Yougoslavie
Bulletin Scientifique. Conseil des Academies des Sciences
et des Arts de la RSF de Yougoslavie. Section A: Sciences
Naturelles, Techniques et Médicales. Redaction et Administration: Opatička ul. 18/II, Zagreb (Yougoslavie).

Bull. Signal.
Bulletin Signalétique. Section 120: Astronomie, Physique
spatiale, Géophysique. Centre de Documentation du
Centre Nationale de la Recherche Scientifique, Paris.

Bull. Signal.
Bulletin Signalétique. Bibliographie des Sciences de la
Terre. Section 220, Cahier A: Minéralogie, Géochimie,
Géologie extraterrestre. Centre de Documentation du
C.N.R.S., Paris; Département Documentation du B.R.
G.M., Orléans.

Bull. Soc. Roy. Sci. Liège
Bulletin de la Société Royale des Sciences de Liège.
L'Université, Liège.

Byull. Abastuman. Astrofiz. Obs.
Abastumanskaya Astrofizicheskaya Observatoriya, Gora
Kanobili. Byulleten'. Akademiya Nauk Gruzinskoj SSR.
Publishers: Izdatel'stvo "Metsniereba", Tbilisi.

Byull. Stantsij Optichesk. Nablyud. Iskusstv. Sputnikov Zemli
Byulleten' Stantsij Opticheskogo Nablyudeniya Iskusstvennykh Sputnikov Zemli. Published by Astronomicheskij Sovet Akademii Nauk SSSR, Moskva.
Beginning with number 60 (1971) the title of the publication changed in Nablyudeniya Iskusstvennykh Nebesnykh Tel.

Canadian Journ. Phys.
Canadian Journal of Physics. Published by the National
Research Council of Canada, Ottawa. Printed in Canada
by the University of Toronto Press, Toronto, Ont.

Celestial Mechanics
Celestial Mechanics. An International Journal of Space
Dynamics. Publishers: D. Reidel Publishing Company,
Dordrecht—Holland.

Ciel et Terre
Ciel et Terre. Bulletin de la Société Belge d'Astronomie,
de Météorologie et de Physique du Globe. Administration: Avenue Circulaire, 3, Bruxelles. Printed by Imprimerie R. Louis, Bruxelles.

Circ. d'Information
Circulaire d'Information. Union Astronomique Internationale. Commission des Etoiles Doubles. Address: Observatoire de Meudon, Meudon, France.

Coelum
Coelum. Periodico bimestrale per la Divulgazione dell'
Astronomia. Editor: Osservatorio Astronomico Univer-
sitario di Bologna.

Comments Astrophys. Space Phys.
Comments on Astrophysics and Space Physics. A Journal of Critical Discussion of the Current Literature. Comments on Modern Physics: Part C. Publishers: Gordon and Breach Science Publishers, Inc., New York — London

Comptes Rendus Acad. Bulg. Sci.
Comptes Rendus de l'Académie bulgare des Sciences. (Doklady Bolgarskoj Akademii Nauk). Sofia.

Comptes Rendus Acad. Sci. Paris
Comptes Rendus hebdomadaires des Séances de l'Académie des Sciences, publié avec le concours du Centre National de la Recherche Scientifique. Imprimerie: Gauthier-Villars, Paris.

Contr. Atmosph. Phys.
Contributions to Atmospheric Physics — Beiträge zur Physik der Atmosphäre. Publisher: Friedrich Vieweg & Sohn, Braunschweig.

Cosmic Electrodynamics
Cosmic Electrodynamics. An International Journal devoted to Geophysical and Astrophysical Plasmas. Printed in The Netherlands by D. Reidel Publishing Company, Dordrecht—Holland.

COSPAR Inform. Bull.
COSPAR. Information Bulletin. Address: COSPAR Secretariat, Paris.

Deutsche Geod. Kommission Bayer. Akad. Wiss.
Deutsche Geodätische Kommission bei der Bayerischen Akademie der Wissenschaften. Reihe A: Höhere Geodäsie; Reihe B: Angewandte Geodäsie; Reihe C: Dissertationen; Reihe D: Tafelwerke; Reihe E: Geschichte und Entwicklung der Geodäsie. Published by Verlag der Bayerischen Akademie der Wissenschaften, München.

Documentat. Observateurs
Documentation des Observateurs. Rédaction: Station d'Astrophysique de Forcalquier.

Documentat. Observateurs Circ.
Documentation des Observateurs. Circulaire. Rédaction: Station d'Astrophysique de Forcalquier.

Dokl. Akad. Nauk
Doklady Akademii Nauk SSSR. Seriya Matematika, Fizika. Publishers: Izdatel'stvo "Nauka", Moskva.

Dunsink Obs. Publ.
Dunsink Observatory Publications. The Observatory of the School of Cosmic Physics, Dublin Institute for Advanced Studies, Dublin.

Earth Extraterr. Sci.
Earth and Extraterrestrial Sciences. Published by Gordon and Breach Science Publishers, London.

Earth Planet. Sci. Letters
Earth and Planetary Science Letters. A Letter Journal devoted to the Development in Time of the Earth and Planetary System. Publisher: North-Holland Publishing Company, Amsterdam.

El Universo
El Universo. Organo de la Sociedad Astronomica de Mexico, Mexico, D. F.

Endeavour
Endeavour. A review of the progress of science, published in four languages by Imperial Chemical Industries Limited, London.

ESO Bull.
European Southern Observatory, Bulletin. Edited by European Southern Observatory. Office of the Director: Hamburg.

Fortschritte Phys.
Fortschritte der Physik. Publisher: Akademie-Verlag, Berlin.

Gaz. Astron. Mém.
Gazette Astronomique. Mémoires van het Sterrenkundig Genootschap van Antwerpen, (de la Société d'Astronomie d'Anvers), Antwerpen. Printer: «De Voorzorg», A. Van Leuvenhaege, Antwerpen.

Geochim. Cosmochim. Acta
Geochimica et Cosmochimica Acta. Journal of the Geochemical Society. Publishing House: Pergamon Press, Ltd., Oxford.

Geodezja Kartografia
Geodezja i Kartografia. Komitet Geodezji Polskiej Akademii Nauk. Publisher: Państwowe Wydawnictwo Naukowe, Warszawa.

Geomagn. Aeronom.
Geomagnetizm i Aehronomiya. Akademiya Nauk SSSR. Izdatel'stvo "Nauka", Moskva [An English translation is published in "Geomagnetism and Aeronomy".]

Geophys. Journ.
The Geophysical Journal of the Royal Astronomical Society. Published for the Royal Astronomical Society by Blackwell Scientific Publications, Oxford — Edinburgh.

Gerlands Beiträge Geophys.
Gerlands Beiträge zur Geophysik. Publisher: Akademische Verlagsgesellschaft Geest & Portig K.-G., Leipzig.

Glasnik Mat.
Glasnik Matematicki. Published by the Society of Mathematicians and Physicists of the S. R. of Croatia. Publisher: Drustvo Matematicara i Fizicara S. R. Hrvatske, Zagreb.

Helvetica Phys. Acta
Helvetica Physica Acta. Schweizerische Physikalische Gesellschaft. Publisher: E. Birkhäuser, Basel.

Hemel en Dampkring
Maandblad van de Nederlandse Vereniging voor Weer-en Sterrenkunde en van de Vereniging voor Sterrenkunde, Meteorologie, Geophysica en Aanverwante Wetenschappen in Belgie. Publisher: Wolters-Noordhoff N. V., Groningen.

IAU Circ.
International Astronomical Union, Circular. Central Bureau for Astronomical Telegrams, Smithsonian Astrophysical Observatory, Cambridge, Mass.

IBM Journ. Res. Development
IBM Journal of Research and Development. Published bimonthly by International Business Machines Corporation, Armonk, New York.

Icarus
Icarus. International Journal of Solar System Studies. Publisher: Academic Press, New York – London.

ICSU Bull.
ICSU Bulletin. International Council of Scientific Unions. Secretariat: 7, Via Cornelio Celso, Rome, Italy.

IEEE Spectrum
IEEE Spectrum. Published monthly by the Institute of Electrical and Electronics Engineers, Inc., New York, N. Y.

Inform. Bull. Southern Hemisphere
Information Bulletin of the Southern Hemisphere. Editorial Office: Observatorio Astronómico, La Plata, Argentina.

Inform. Bull. Variable Stars
Commission 27 of the I.A.U. Information Bulletin on Variable Stars. Konkoly Observatory, Budapest.

Infrared Physics
An International Research Journal. Publisher: Pergamon Press Ltd., Oxford – London – New York.

International Journ. Theor. Phys.
International Journal of Theoretical Physics. Publisher: Plenum Publishing Company, Donington House, London.

Irish Astron. Journ.
The Irish Astronomical Journal. A Quarterly Publication under the auspices of the Observatories of Armagh and Dunsink. Subscription address: Managing Editor, Irish Astronomical Journal, Armagh Observatory, Northern Ireland.

Izv. Akad. Nauk Armyan. SSR
Izvestiya Akademii Nauk Armyanskoj SSR. Fizika. Publisher: Izdatel'stvo AN Armyanskoj SSR, Erevan.

Izv. Glav. Astron. Obs. Pulkovo
Izvestiya Glavnoj Astronomicheskoj Observatorii v Pulkove. Akademiya Nauk SSSR. Izdanie Glavnoj astronomicheskoj observatorii v Pulkove, Leningrad.

Izv. Komissii Fiz. Planet
Izvestiya Komissii po Fizike Planet. Akademiya Nauk SSSR. Astronomicheskij Sovet. Moskva.

Izv. Krymskoj Astrofiz. Obs.
Izvestiya Krymskoj Astrofizicheskoj Observatorii. Akademiya Nauk SSR. Publishers: Izdatel'stvo "Nauka", Moskva.

Jenaer Rundschau (Jena Review)
Jenaer Rundschau (Jena Review). Publisher: VEB Verlag Technik, Berlin.

JETP Letters
JETP Letters. A translation of JETP Pis'ma v Redaktsiyu of the Academy of Sciences in the USSR. Published semimonthly by the American Institute of Physics, Lancaster, Pennsylvania.

Journ. Astronaut. Sci.
The Journal of the Astronautical Sciences. Published by the American Astronautical Society Inc., Baltimore, Md.

Journ. Astron. Soc. Victoria
The Journal of the Astronomical Society of Victoria.

Printed by D. Buscombe Printers, Glen Waverley, Victoria.

Journ. Astron. Soc. Western Australia
The Journal of the Astronomical Society of Western Australia. Edited by the Astronomical Society of Western Australia, Perth, W. A.

Journ. Atmosph. Sci.
Journal of the Atmospheric Sciences. Published by the American Meteorological Society, Boston, Mass.

Journ. Atmosph. Terr. Phys.
Journal of Atmospheric and Terrestrial Physics. Publishers: Pergamon Press, Oxford – London – New York.

Journ. British Astron. Ass.
Journal of the British Astronomical Association. Subscription address: British Astronomical Association, Burlington House, Piccadilly, London.

Journ. British Interplanet. Soc.
Journal of the British Interplanetary Society. Printed in Great Britain by Unwin Brothers Ltd., The Gresham Press, Old Woking, Surrey, and published by The British Interplanetary Society, London.

Journ. Fluid Mechanics
Journal of Fluid Mechanics. Published by Cambridge University Press, London – New York.

Journ. Geophys. Res.
Journal of Geophysical Research. An International Scientific Publication. Published three times a month by the American Geophysical Union, Washington, D. C. First section: Space Physics; Second section: Physics and chemistry of the solid earth, planetology, geodesy; Third section: Oceans and atmospheres.

Journ. History Astron.
Journal for the History of Astronomy. Publisher: Science History Publications Ltd., Cambridge, England. American Representative: Neale Watson Academic Publications, Inc., New York City, U.S.A.

Journ. Math. Phys.
Journal of Mathematical Physics. Published by the American Institute of Physics, New York, N. Y.

Journ. Navigation
The Journal of Navigation. Published quarterly by The Royal Institute of Navigation at the Royal Geographical Society, London.

Journ. Optical Soc. America
Journal of the Optical Society of America. Publisher: American Institute of Physics, New York.

Journ. Phys. A. General Phys.
Journal of Physics A. General Physics. Europhysics Journal. Published by the Institute of Physics and the Physical Society, London, England, in association with the American Institute of Physics, New York.

Journ. Physique
Journal de Physique. Publication de la Société Française de Physique, Paris.

Journ. Plasma Phys.
Journal of Plasma Physics. Publishers: Cambridge University Press, London.

Journ. Proc. Roy. Soc. New South Wales
Journal and Proceedings of the Royal Society of New South Wales. Published by the Society, Science House, Sydney.

Journ. Quant. Spectrosc. Radiat. Transfer
Journal of Quantitative Spectroscopy & Radiative Transfer. Publisher: Pergamon Press, Oxford – New York.

Journ. Roy. Astron. Soc. Canada
The Journal of the Royal Astronomical Society of Canada, devoted to the advancement of astronomy and allied sciences. Printed by the University of Toronto Press, Toronto, Ontario, Canada.

Kometn. Tsirk. *Kiev*
Kometnyj Tsirkulyar. Gruppa po Issledovaniyu Komet Astrosoveta i Mezhduvedomstvennyj Geofizicheskij Komitet Akademii Nauk SSSR. Kievskij Universitet im. T. G. Shevchenko.

Komety i Meteory
Komety i Meteory. Akademiya Nauk Tadzhikskoj SSR. Astronomicheskij Sovet Akademii Nauk SSSR. Publishers: Izdatel'stvo "Donish", Dushanbe.

Kosmich. Issled.
Kosmicheskie Issledovaniya. Akademiya Nauk SSSR. Publishers: Izdatel'stvo "Nauka", Moskva.

Kozmos
Kozmos. Popular Astronomical Journal of the Slovak Central Observatory in Hurbanovo. Publisher: Slovenská ústredná hvezdáren v Hurbanove.

L'Astronomie
L'Astronomie et Bulletin de la Société Astronomique de France. Revue mensuelle. Rédaction: Société Astronomique de France, Paris.

L'Universo
L'Universo. Rivista dell'Instituto Geografico Militare. Direzione, Redazione e Amministrazione: Istituto Geografico Militare, Firenze.

Magnitnye Polya Solnech. Pyaten
Magnitnye Polya Solnechnykh Pyaten. (Supplements to Solnechnye Dannye. Byulleten' (*Solar Data*)). Publishers: Izdatel'stvo "Nauka", Leningrad.

Math. Rev.
Mathematical Reviews. Published by the American Mathematical Society, Providence, R. I.

Mem. Fac. Sci. Kyoto Univ.
Memoirs of the Faculty of Science, Kyoto University. Series of Physics, Astrophysics, Geophysics, and Chemistry. Printed by Yamashiro Printing Publishing Co. Ltd., Kamigyo, Kyoto.

Mem. Roy. Astron. Soc.
Memoirs of the Royal Astronomical Society. Published for the Royal Astronomical Society by Blackwell Scientific Publications, Oxford – Edinburgh.

Mem. Soc. Astron. Italiana
Memorie della Società Astronomica Italiana. Nuova Serie. Pubblicate sotto gli auspici del Consiglio Nazionale dell Ricerche. Publisher: Tipografia Baccini & Chiappi, Firenze.

Mercury
Mercury. The Journal of the Astronomical Society of the Pacific. Published by the Astronomical Society of the Pacific, San Francisco, California.

Messtechnik
Messtechnik (Zeitschrift für Instrumentenkunde). Publishers: Verlag Friedrich Vieweg & Sohn GmbH, Braunschweig.

Meteoritics
Meteoritics. The Journal of the Meteoritical Society. Published quarterly by The Meteoritical Society and Arizona State University Bureau of Publications. Editorial address: Center for Meteorite Studies, The Arizona State University, Tempe, Arizona.

Meteoritika
Akademiya Nauk SSSR. Komitet po Meteoritam. Publishers: Izdatel'stvo "Nauka", Moskva.

Mitt. Astron. Ges.
Mitteilungen der Astronomischen Gesellschaft, Hamburg. Printed by G. Braun, GmbH, Karlsruhe.

Monatsber. Deutsch. Akad. Wiss. Berlin
Monatsberichte der Deutschen Akademie der Wissenschaften zu Berlin. Mitteilungen aus Mathematik, Naturwissenschaft, Medizin und Technik. Publisher: Akademie-Verlag, Berlin.

Monthly Notes Astron. Soc. Southern Africa
Monthly Notes of the Royal Astronomical Society of Southern Africa. Published by the Astronomical Society of Southern Africa, Royal Observatory, Cape Province, South Africa.

Monthly Notes International Polar Motion Service
Monthly Notes of the International Polar Motion Service. Published by the Central Bureau, International Latitude Observatory of Mizusawa, Mizusawa-shi, Iwate-ken, Japan.

Monthly Notices Roy. Astron. Soc.
Monthly Notices of the Royal Astronomical Society. Published for the Royal Astronomical Society by Blackwell Scientific Publications, Oxford – Edinburgh.

Moon
The Moon. An International Journal of Lunar Studies. Publisher: D. Reidel Publishing Company, Dordrecht – Holland.

MVS Sonneberg
Mitteilungen über Veränderliche Sterne. Edited by Sternwarte Sonneberg (Zentralinstitut für Astrophysik, Bereich Sternphysik) der Deutschen Akademie der Wissenschaften zu Berlin.

Nablyud. Iskusstv. Nebesn. Tel
Nablyudeniya Iskusstvennykh Nebesnykh Tel. Published by Astronomicheskij Sovet Akademii Nauk SSSR, Moskva.

Nachr. Akad. Wiss. Göttingen
Nachrichten der Akademie der Wissenschaften in Göttingen. II. Mathematisch-Physikalische Klasse. Vandenhoeck & Ruprecht, Göttingen.

Nachr. Karten-, Vermessungswesen
Nachrichten aus dem Karten- und Vermessungswesen. Editor: Institut für Angewandte Geodäsie (Abt. II des

Deutschen Geodätischen Forschungsinstituts). Published by Verlag des Instituts für Angewandte Geodäsie, Frankfurt a. M.

Nature
Nature. Editorial and Publishing Offices: Macmillan Journals Limited, 4 Little Essex Street, London; 711 National Press Building, Washington, D. C.

Nature, Phys. Sci.
Nature, Physical Science. Editorial and Publishing Offices: Macmillan Journals Limited, London – Washington.

Naturwissenschaften
Die Naturwissenschaften. Publisher: Springer-Verlag, Berlin – Heidelberg – New York.

Nauchn. Informatsii
Nauchnye Informatsii. Astronomicheskij Sovet Akademii Nauk SSSR, Moskva.

Nuovo Cimento
Il Nuovo Cimento. Rivista Internazionale e Organo della Società Italiana di Fisica, Series A, B. Publisher: Nicola Zanichelli, Editore, Bologna.

Nuovo Cimento Lettere
Lettere al Nuovo Cimento, a Cura della Società Italiana di Fisica. Editrice Compositori, Bologna.

Nuovo Cimento Rivista
Rivista del Nuovo Cimento a cura della Società Italiana di Fisica. Editrice Compositori, Bologna.

Nuovo Cimento Suppl.
Supplemento al Nuovo Cimento. Publisher: Nicola Zanichelli, Editore, Bologna.

Observations Artificial Earth Satellites
Observations of Artificial Satellites of the Earth (Nablyudeniya Iskusstvennykh Sputnikov Zemli). Magyar Tudományos Akadémia Csillagvizsgáló Intézete, Budapest.

Observatory
The Observatory. A Review of Astronomy. Publishers: The Editors of "The Observatory", Royal Greenwich Observatory, Herstmonceaux Castle, Hailsham, Sussex, England.

Optik
Optik. Zeitschrift für das gesamte Gebiet der Licht- und Elektronenoptik. Publishers: Wissenschaftliche Verlagsgesellschaft mbH., Stuttgart.

Orion Schaffhausen
Orion. Zeitschrift der Schweizerischen Astronomischen Gesellschaft (SAG). Bulletin de la Société Astronomique de Suisse (SAS). Administration: Generalsekretariat der SAG, Schaffhausen.

Österreich. Zeitschr. Vermessungswesen
Österreichische Zeitschrift für Vermessungswesen. Editor and Publisher: Österreichischer Verein für Vermessungswesen, Wien.

Peremennye Zvezdy, Byull.
Peremennye Zvezdy, Byulleten', izdavaemyj Astronomicheskim Sovetom Akademii Nauk SSSR. Published by Astronomicheskij Sovet Akademii Nauk SSSR, Moskva.

Peremennye Zvezdy, Prilozhenie
Peremennye Zvezdy, Prilozhenie (The Variable Stars, Supplement). Astronomicheskij Sovet Akademii Nauk SSSR, Moskva.

Phil. Mag.
The Philosophical Magazine. A Journal of Theoretical, Experimental and Applied Physics. Eighth Series. Publisher: Taylor & Francis, Ltd., London.

Phil. Trans. Roy. Soc. London
Philosophical Transactions of the Royal Society of London. Series A, Mathematical and Physical Sciences. Published by the Royal Society, London.

Phys. Abstr.
Physics Abstracts. Science Abstracts, Series A. An INSPEC Publication, published by The Institution of Electrical Engineers, London.

Phys. Ber.
Physikalische Berichte. Herausgegeben von der Deutschen Physikalischen Gesellschaft e. V.und von der Deutschen Akademie der Wissenschaften zu Berlin. Friedrich Vieweg & Sohn, Braunschweig.

Phys. Blätter
Physikalische Blätter. Physik-Verlag, Mosbach/Baden.

Phys. Bull.
Physics Bulletin. Published by the Institute of Physics and the Physical Society, London, England.

Phys. Earth Planet. Interiors
Physics of the Earth and Planetary Interiors. A journal devoted to observational and experimental studies of the Earth and Planetary interiors and their theoretical interpretation by the physical sciences. Publisher: North-Holland Publishing Company, Amsterdam, Netherlands.

Phys. Fluids
The Physics of Fluids. Published by the American Institute of Physics, New York, N.Y.

Phys. Letters
Physics Letters. Volumes A and B. Publisher: North-Holland Publishing Company, Amsterdam.

Phys. Rev. A
Physical Review A, General Physics. Published for the American Physical Society by the American Institute of Physics, Lancaster, Pa., and New York, N.Y.

Phys. Rev. B
Physical Review B, Solid State. Published for the American Physical Society by the American Institute of Physics, Lancaster, Pa., and New York, N. Y.

Phys. Rev. C
Physical Review C, Nuclear Physics. Published for the American Physical Society by the American Institute of Physics, Lancaster, Pa., and New York, N.Y.

Phys. Rev. D
Physical Review D, Particles and Fields. Published for the American Physical Society by the American Institute of Physics, Lancaster, Pa., and New York, N.Y.

Phys. Rev. Letters
Physical Review Letters. Published weekly by The Amer-

ican Physical Society, New York, N. Y.

Phys. Today
Physics Today. Published by the American Institute of
Physics, New York, N.Y.

Physica
Physica. Publishers: North-Holland Publishing Company,
Amsterdam, The Netherlands, on request of the Founda-
tion "Physica", Utrecht.

Physica Scripta
Physica Scripta. (Formerly Arkiv för Fysik). Published
by the Royal Swedish Academy of Sciences, Stockholm.

Planet. Space Sci.
Planetary and Space Science. Pergamon Press, Oxford –
London – New York.

Plasma Physics
Plasma Physics. Publisher: Pergamon Press, Oxford,
England.

Pokroky
Pokroky matematiky, fyziky a astronomie. Editor:
Jednota čs. matematiků a fyziků. Publisher: Academia,
Praha.

Postępy Astron.
Postępy Astronomii. Czasopismo Poświecone Upowszech-
nianiu Wiedzy Astronomicznej. Polskie Towarzystwo
Astronomiczne, Warszawa. Printed in Poland by Pánstwo-
we Wydawnictwo Naukowe, Lódź.

Priroda
Priroda. Publishers: Izdatel'stvo "Nauka", Moskva.

Proc. Astron. Soc. Australia
Proceedings of the Astronomical Society of Australia.
Published for the Society by Sydney University Press,
Sydney.

Proc. Cambridge Phil. Soc.
Proceedings of the Cambridge Philosophical Society
(Mathematical and Physical Sciences). Publishers: Cam-
bridge University Press, London.

Proc. IEEE
Proceedings of the IEEE. Published monthly by the In-
stitute of Electrical and Electronics Engineers, Inc., New
York, N. Y.

Proc. Koninkl. Nederl. Akad. Wet.
Koninklijke Nederlandse Akademie van Wetenschappen.
Proceedings. Series B, Physical Sciences. Publishers:
North-Holland Publishing Company, Amsterdam.

Proc. National Acad. Sci. U.S.A.
Proceedings of the National Academy of Sciences of the
United States of America. Published monthly by the Na-
tional Academy of Sciences, Washington, D.C.

Proc. Roy. Soc. London
Proceedings of the Royal Society of London. Series A:
Mathematical and Physical Sciences. Published by the
Royal Society, London.

Progr. Theor. Phys. Japan
Progress of Theoretical Physics. Published for the Research Institute for Fundamental Physics and the Physic-

al Society of Japan. Publication Office: Progress of
Theoretical Physics, Yukawa Hall, Kyoto University,
Kyoto, Japan.

Progr. Theor. Phys. Suppl.
Supplement of the Progress of Theoretical Physics. Published for the Research Institute for Fundamental Physics and The Physical Society of Japan. Publication Office: Progress of Theoretical Physics, Yukawa Hall, Kyoto University, Kyoto, Japan.

PTB Mitt.
PTB Mitteilungen. Amts- und Mitteilungsblatt der Physi-
kalisch-Technischen Bundesanstalt, Braunschweig –
Berlin.

Publ. Astron. Soc. Japan
Publications of the Astronomical Society of Japan. Pub-
lished by the Astronomical Society of Japan. Office of
the Society: Tokyo Astronomical Observatory, Mitaka,
Tokyo. Agent: Maruzen Co. Ltd. (Export Department),
Nihonbashi, Tokyo, Japan.

Publ. Astron. Soc. Pacific
Publications of the Astronomical Society of the Pacific.
Published in Provo, Utah, by the Astronomical Society
of the Pacific, San Francisco, California. Printed by
Brigham Young University Press, Provo, Utah.

Publ. Roy. Obs. Edinburgh
Publications of the Royal Observatory, Edinburgh.
Published by The Royal Observatory, Edinburgh,
Scotland.

Publ. Tartu Astrofiz. Obs.
W. Struve nimelise Tartu Astrofüüsika Observatooriumi,
Publikatsioonid. Eesti NSV Teaduste Akadeemia, Tartu.

Quarterly Journ. Roy. Astron. Soc.
Quarterly Journal of the Royal Astronomical Society.
Published for the Royal Astronomical Society by Black-
well Scientific Publications, Oxford.

Radio Sci.
Radio Science. Published by the American Geophysical
Union, Richmond, Virginia.

Referativ. Zhurn. 51. Astron.
Referativnyj Zhurnal. 51. Astronomiya. Vsesoyuznyj
Institut Nachnoj i Tekhnicheskoj Informatsii. Moskva.

Referativ. Zhurn. 52. Geod. i Aehros"emka
Referativnyj Zhurnal. 52. Geodeziya i Aehros"emka.
Vsesoyuznyj Institut Nauchnoj i Tekhnicheskoj Infor-
matsii. Moskva.

Referativ. Zhurn. 62. Issled. kosm. prostranstva
Referativnyj Zhurnal. 62. Issledovanie Kosmicheskogo
Prostranstva. Vsesoyuznyj Institut Nauchnoj i Tekhni-
cheskoj Informatsii. Moskva.

Rep. Progr. Phys.
Reports on Progress in Physics. Published by The Institute of Physics and the Physical Society, London.

Rev. Geophys. Space Phys.
Reviews of Geophysics and Space Physics (formerly Reviews of Geophysics). Published by the American Geophysical Union, Richmond, Virginia.

Revista Astron.
Revista Astronomica. Organo de la Asociación Argentina Amigos de la Astronomia, Buenos Aires.

Rev. Modern Phys.
Reviews of Modern Physics. Published for The American Physical Society by the American Institute of Physics, Lancaster, Pa., and New York, N.Y.

Rev. Sci. Instruments
Reviews of Scientific Instruments. Published by the American Institute of Physics, Lancaster, Pa., and New York, N.Y.

Rezul'taty Nablyud. Sovet. Iskusstv. Sputnikov Zemli
Rezul'taty Nablyudenij Sovetskikh Iskusstvennykh Sputnikov Zemli. Published by Astronomicheskij Sovet Akademii Nauk SSSR, Moskva. Replaced after No. 140 by Rezul'taty Nablyudenij Iskusstvennykh Sputnikov Zemli.

Rezul'taty Nablyud. Iskusstv. Sputnikov Zemli
Rezul'taty Nablyudenij Iskusstvennykh Sputnikov Zemli. Published by Astronomicheskij Sovet Akademii Nauk SSSR, Ryazanskij Gosudarstvennyj Pedagogicheskij Institut, Ryazan'.

Ric. Sci.
La Ricerca Scientifica. Serie Seconda. Rivista del Consiglio Nazionale delle Ricerche. Consiglio Nazionale delle Ricerche, Roma.

Říše hvězd
Říše hvězd. Czechoslovak popular astronomical journal. Publisher: Orbis, Praha.

Roy. Astron. Soc. New Zealand Circ.
Royal Astronomical Society of New Zealand, Variable Star Section, Circular. Publication Office: Greerton, Tauranga, New Zealand.

Roy. Astron. Soc. New Zealand Variable Star Sect. Repr.
Royal Astronomical Society of New Zealand, Variable Star Section, Reprint. Publication Office: Greerton, Tauranga, New Zealand.

Rumanian Sci. Abstr.
Rumanian Scientific Abstracts. Natural Sciences. Publishers: The Scientific Documentation Centre of the Academy of the Socialist Republic of Romania, Bucureşti.

Sci. American
Scientific American. Published monthly by Scientific American, Inc., New York, N.Y.

Science
Science. American Association for the Advancement of Science, Washington, D.C.

Sci. Progr. Découverte
Science Progrès Découverte (formerly Science Progrès, La Nature). Revue publiée avec la participation du Palais de la Découverte. Published by Dunod, Editeur, Paris. Imprimerie Bayeusaine, Bayeux.

Sci. Rep. Tôhoku Univ.
The Science Reports of the Tôhoku University. First Series (Physics, Chemistry, Astronomy). Published by the Faculty of Science, Tôhoku University, Sendai, Japan.

Sitz.-Ber. Bayer. Akad. Wiss.
Bayerische Akademie der Wissenschaften. Mathematisch-Naturwissenschaftliche Klasse. Sitzungsberichte. Publisher: Verlag der Bayerischen Akademie der Wissenschaften, München.

Sitz.-Ber. Deutsch. Akad. Wiss. Berlin
Sitzungsberichte der Deutschen Akademie der Wissenschaften zu Berlin. Klasse für Mathematik, Physik und Technik. Publisher: Akademie-Verlag, Berlin.

Sitz.-Ber. Heidelberger Akad. Wiss.
Sitzungsberichte der Heidelberger Akademie der Wissenschaften. Mathematisch-Naturwissenschaftliche Klasse. Publisher: Springer-Verlag, Heidelberg.

Sitz.-Ber. Österreich. Akad. Wiss.
Sitzungsberichte. Österreichische Akademie der Wissenschaften. Mathematisch-Naturwissenschaftliche Klasse. Abteilung II: Mathematik, Astronomie, Meteorologie und Technik. Publisher: Springer-Verlag, Wien.

Sky Telescope
Sky and Telescope. Published by Sky Publishing Corporation, Cambridge, Mass.

Smithsonian Contr. Astrophys.
Smithsonian Contributions to Astrophysics. Smithsonian Institution Astrophysical Observatory, Cambridge, Mass. Printed by Smithsonian Institution Press, City of Washington. For sale by the Superintendent of Documents, U. S. Government Printing Office, Washington, D. C.

Smithsonian Year
Smithsonian Year. Annual Report of the Smithsonian Institution, including the financial report of the Executive Committee of the Boards of Regents. Published by the Smithsonian Institution, Washington, D.C.

Solar Physics
Solar Physics. A Journal for Solar Research and the Study of Solar Terrestrial Physics. Publishers: D. Reidel Publishing Company, Dordrecht—Holland.

Solnechnye Dannye Byull.
Solnechnye Dannye. Byulleten'. *(Solar Data)*. Publishers: Izdatel'stvo "Nauka", Leningradskoe Otdelenie, Leningrad.

Soobshch. Byurakan. Obs.
Soobshcheniya Byurakanskoj Observatorii. Akademiya Nauk Armyanskoj SSR, Erevan.

Soobshch. Gos. Astron. Inst. Shternberg
Soobshcheniya Gosudarstvennogo Astronomicheskogo Instituta im P.K. Shternberga. Publishers: Izdatel'stvo Moskovskogo Universiteta, Moskva.

Southern Stars
Southern Stars. The Journal of the Royal Astronomical Society of New Zealand (Inc.). Address of the Society: P.O. Box 3181, Wellington C1, New Zealand.

Soviet Astron. AJ
Soviet Astronomy AJ. A translation of the Astronomical Journal of the Academy of Sciences of the USSR. Published by the American Institute of Physics, Inc., New York, N.Y.

Spaceflight
Spaceflight. A Publication of the British Interplanetary

Society. Printed by Eyre & Spottiswoode Limited at Grosvenor Press, Portsmouth, and published by the British Interplanetary Society, London.

Space Science Rev.
Space Science Reviews. Publishers: D. Reidel Publishing Company, Dordrecht—Holland.

Springer Tracts Modern Phys.
Springer Tracts in Modern Physics. (Ergebnisse der exakten Naturwissenschaften). Springer-Verlag, Berlin—Heidelberg—New York.

Sterne
Die Sterne. Zeitschrift für alle Gebiete der Himmelskunde. Johann Ambrosius Barth, Leipzig.

Sternenbote
Sternenbote. Monatsschrift für Österreichs Amateurastronomen. Publisher: Astronomisches Büro, Hermann Mucke, Wien.

Stockholms Obs. Ann.
Stockholms Observatoriums Annaler. Printed by Almquist & Wiksell, Stockholm.

Strolling Astronomer
The Strolling Astronomer. The Journal of The Association of Lunar and Planetary Observers, Publication Office: The Strolling Astronomer, Box 3 AZ, University Park, New Mexico.

Stud. Cerc. Astron.
Studii şi Cercetări de Astronomie. Editura Academiei Republicii Socialiste România. Editorial Office: Observatorul Astronomic, Bucureşti.

Stud. Geophys. Geod.
Studia geophysica et geodaetica. Published for the Geophysical Institute of the Czechoslovak Academy of Sciences by Academia, Praha.

Stud. Soc. Sci. Torunensis
Studia Societatis Scientiarum Torunensis, Toruń – Polonia. Sectio F (Astronomia).

Stud. Univ. Babeş-Bolyai
Studia Universitatis Babeş-Bolyai. Series Mathematica-Physica. Publishers: Intreprinderea Poligrafica, Cluj.

SuW
Sterne und Weltraum. Astronomische Monatsschrift. Publisher: Verlag Sterne und Weltraum Dr. Vehrenberg, Düsseldorf, Germany.

Tellus
Tellus, a bi-monthly Journal of Geophysics. Svenska Geofysiska Foreningen. Printed in Sweden by Almqvist & Wiksells Boktryckeri AB, Uppsala.

Trans. Astron. Obs. Yale Univ.
Transactions of the Astronomical Observatory of Yale University. Published by the Observatory, New Haven.

Trans. Roy. Soc. Canada
Transactions of the Royal Society of Canada. Published by the Royal Society of Canada, National Research Building, Ottawa.

Trudy Astrofiz. Inst. Alma-Ata
Trudy Astrofizicheskogo Instituta, Alma-Ata. Akademiya Nauk Kazakhskoj SSR. Publishers: Izdatel'stvo "Nauka" Kazakhskoj SSR, Alma-Ata.

Trudy Glav. Astron. Obs. Pulkovo
Trudy Glavnoj Astronomicheskoj Observatorii v Pulkove. Akademiya Nauk SSSR. Izdanie Glavnoj astronomicheskoj observatorii v Pulkove, Leningrad.

Trudy Inst. Teor. Astron.,*Leningrad*
Trudy Instituta Teoreticheskoj Astronomii. Akademiya Nauk SSSR. Publishers: Izdatel'stvo "Nauka", Leningrad.

Trudy Tashkent. Astron. Obs.
Trudy Tashkentskoj Astronomicheskoj Observatorii. Akademiya Nauk Uzbekskoj SSR. Publishers: Izdatel'stvo "FAN" Uzbekskoj SSR, Tashkent.

Tsirk. Astron. Inst. Tashkent
Tsirkulyar Astronomicheskogo Instituta. Akademiya Nauk Uzbekskoj SSR. Izdatel'stvo "FAN" Uzbekskoj SSR, Tashkent.

Tsirk. Astron. Obs. L'vov
Tsirkulyar. Astronomicheskaya Observatoriya. L'vovskij Ordena Lenina Gosudarstvennyj Universitet imeni Ivana Franko. Publisher: Izdatel'stvo L'vovskogo Universiteta, L'vov.

Umschau
Umschau in Wissenschaft und Technik. Umschau-Verlag Frankfurt a. M.

Urania Barcelona
Urania. Revista de Astronomia y Ciencias Afines. Organo de la Sociedad Astronómica de España y América, Barcelona; Unión Nacional de Astronomia y Ciencias Afines, Madrid.

Urania Kraków
Urania. Miesiecznik Polskiego Towarzystwa Milośników Astronomii, Kraków. Publisher: Krakowska Drukarnia Prasowa, Kraków.

Vasiona
Vasiona. Revue d'Astronomie et d'Astronautique. Bulletin de la Société Astronomique "R. Bosković", Beograd.

VdS Nachrichtenblatt
Nachrichtenblatt der Vereinigung der Sternfreunde e.V. After Vol. 18, No. 3 published in combination with "Sterne und Weltraum". Verlag Sterne und Weltraum Dr. Vehrenberg, Düsseldorf, Germany.

Veröff. Astron. Rechen-Inst. Heidelberg
Veröffentlichungen des Astronomischen Rechen-Instituts Heidelberg. Verlag G. Braun, Karlsruhe.

Veröff. Sternw. Sonneberg
Deutsche Akademie der Wissenschaften zu Berlin. Institut für Sternphysik. Veröffentlichungen der Sternwarte in Sonneberg. Publisher: Akademie-Verlag, Berlin.

Vesmír
Vesmír. Přírodovědecky časopis Čs. akadmie věd. Publisher: Academia, Praha.

Vestn. Khar'kov. Univ.
Vestnik Khar'kovskogo Universiteta. Seriya Astronomicheskaya. Publishers: Izdatel'stvo Khar'kovskogo Universiteta, Khar'kov.

Vestn. Kiev. Univ.
Vestnik Kievskogo Universiteta. Seriya Astronomii.
Publishers: Izdatel'stvo Kievskogo Universiteta, Kiev.

VJS Naturforsch. Ges. Zürich
Vierteljahresschrift der Naturforschenden Gesellschaft
in Zürich. Printer and Publisher: Leeman AG, Zürich.

Weltraumfahrt
Weltraumfahrt, Raketentechnik. Publisher: Umschau-
Verlag, Frankfurt a/Main.

Wiss. Zeitschr. Friedrich-Schiller Univ. Jena
Wissenschaftliche Zeitschrift der Friedrich-Schiller-Uni-
versität. Jena. Mathematisch-Naturwissenschaftliche
Reihe. Edited by the Rektor der Friedrich-Schiller-Uni-
versität Jena.

Wiss. Zeitschr. Humboldt-Univ. Berlin
Wissenschaftliche Zeitschrift der Humboldt-Universität
zu Berlin. Mathematisch-Naturwissenschaftliche Reihe.
Edited by the Rektor der Humboldt-Universität, Berlin.

Yamamoto Circ.
Yamamoto Circular. Published by the Yamamoto Observatory, Kamitanakami – Kiryutyo, Otu, Siga-ken, Japan.

Zeitschr. Angew. Physik
Zeitschrift für Angewandte Physik. Publisher: Springer-
Verlag, Berlin–Heidelberg–New York.

Zeitschr. Astrophys. (ZfA)

Zeitschrift für Astrophysik. Publisher: Springer-Verlag,
Berlin–Heidelberg–New York. After Vol. 69 (1968)
replaced by "Astronomy and Astrophysics".

Zeitschr. Geophys.
Zeitschrift für Geophysik. Publisher: Physica-Verlag,
Würzburg, Germany.

Zeitschr. Naturforschung
Zeitschrift für Naturforschung. Europhysics Journal.
Teil a: Astrophysik, Physik, Physikalische Chemie.
Published by Verlag der Zeitschrift für Naturforschung,
Tübingen, Germany.

Zeitschr. Physik
Zeitschrift für Physik. Publisher: Springer-Verlag, Berlin–
Heidelberg–New York.

Zemlya i Vselennaya
Zemlya i Vselennaya. Astronomiya, Geofizika, Issledovaniya Kosmicheskogo Prostranstva. Nauchno-Populyar-
nyj Zhurnal Akademii Nauk SSSR. Publishers: Izdatel'-
stvo "Nauka", Moskva.

Zentralblatt Math. Grenzgebiete
Zentralblatt für Mathematik und ihre Grenzgebiete. Publisher: Springer-Verlag, Berlin–Heidelberg–New York.

Zvaigžņota Debess
Latvijas PSR Zinātņu Akadēmijas Radioastrofizikas
Observatorijas Populārzinatniks Gadalaiku Izdevums.
Izdevnieciba "Zinātne", Riga.

002 Bibliographical Publications

002.001 Science news.
Priroda, No. 7.72, p. 104 - 113; No. 8.72, p. 101 - 110; No. 9.72, p. 103 - 114 (1972). In Russian.
(7) Content of antinuclei in primary cosmic rays; Regular revision of the age of the universe; Water on the moon? Radar investigations of the Martian topography. (8) "MAC"[1972 - 25B] and Intercosmos 66 (*S. A. Nikitin*); Expedition to the crater Descartes; Remnants of microorganisms in a meteorite? New catalogues of supernovae and quasars (*L. V. Samsonenko*); Diameter of a star – 15 km! X-ray flare on the sun on December 10, 1970. (9) Transmission from the Venus surface; The gravitational field on the moon; The dislocation mechanism of the lunar ground; News of Sirius B; The carbon core of stars and formation of pulsars; Diamonds in a gigantic Siberian meteoritic crater.

002.002 On the books in physics and astrophysics being published by "Nauka" Publishing House (Glavnaya Redaktsiya Fiziko-matematicheskoj Literatury) in 1972.
V. V. Vlasov.
Uspekhi fiz. nauk., Vol. 106, 744 - 746 (1972). In Russian.
Abstr. in Referativ. Zhurn. 51. Astron., 9.51.59 (1972).

002.003 News from science and other informations.
Zemlya i Vselennaya, 1972, No. 4. In Russian.
The motion of Toro (*A. N. Simonenko*), p. 17; The riddle of compact galaxies satellites (*B. V. Komberg*), p. 24 - 25; The activity of the nuclei of galaxies – result of the evolution of stars? p. 25; New neighbouring galaxies, p. 25; New minor planets, p. 35; Radio radiation of stars, p. 35; Cyg XR-1 – double star with heavy component, p. 35; Flight of Apollo 16, p. 45; Emission of the earth's magnetosphere, p. 65; The nature of the Martian polar caps, p. 73; Very unusual X-ray sources, p. 74; Exchange of lunar soil, p. 74.

002.004 Annotations on papers on geomagnetism and aeronomy published in "News of higher educational establishments. Radiophysics", 1971, Vol. 14, Nos. 6, 8–10.
Geomagn. Aeronom., Vol. 12, 964 - 965 (1972). In Russian.

002.005 Index to Special Reports.
Smithsonian Astrophys. Obs., *Cambridge, Mass.*, Special Reports 251 - 325, 3 + 43 pp. (1972).
This index covers the materials published in the SAO Special Reports Numbers 251 through 325, issued between October 13, 1967, and October 15, 1970. The index includes four sections: 1) list of individual reports by number, date, authors, title, and abstract; 2) list of authors; 3) titles of papers; and 4) significant references to particular subjects.

002.006 Science news.
Priroda, No. 10.72, p. 104 - 113 (1972). In Russian.
Record brightness of a supernova of our century? (*Yu. P. Pskovskij*); Again no quarks; Radio radiation of Antares; The evolution of Mars has only begun; Fall of a large meteorite on the moon; News of motions in the earth's mantle.

002.007 News from science and other informations.
Zemlya i Vselennaya, 1972, No. 5. In Russian.
Centaurus XR-3 – a double system, p. 16; Lunar seismic experiment, p. 16; New meteorite crater on the moon, p. 16; Is the neutrino stable? p. 26; Observations of radio sources with record resolution on terrestrial conditions, p. 27; Phobos and Deimos (*A. N. Simonenko*), p. 59; Names on lunar maps (*V. A. Shishakov*), p. 60 - 61; Paleomap of the moon, p. 70; Memorial tablet of the first Astronomical Society of Russia (*E. G. Demidovich*), p. 76.

002.008 Science news.
Priroda, No. 11.72, p. 90 - 99 (1972). In Russian.
Satellite jubilee (*S. A. Nikitin*); Dry ice on Mars; Lithium in magnetic stars (*N. S. Polosukhina*).

002.009 News and views.
Nature, Vol. 238, 127 - 134, 187 - 194, 244 - 250, 307 - 312, 429 - 434 (1972).
High energy gamma rays from the galactic centre, p. 127; First data from TD-1A, p. 129; Is β Persei a transient X-ray star? p. 133; Position of the pulsar PSR 1749–28, p. 190 - 191; Atomic clocks coming and going, p. 244 - 245; Solar system: Assembling the planets, p. 250; New data on interstellar gas, p. 311 - 312; Variable stars: Smokescreen stars, p. 431.

002.010 News.
Nature, Phys. Sci., Vol. 238, 33, 49, 81 - 82, 97, 113 - 114, 129 (1972). – Gravitation: G results refuted, p. 33; Halley's comet and Chinese records, p. 49; A place in the sun, p. 81; Atmosphere: Drifts and temperatures, p. 81; NP 0532: Random walk, p. 82; Gamma rays from Crab nebula, p. 97; Are quasars and galaxies associated?, p. 97; Cosmic rays: Energy dependence, p. 114; X-ray astronomy: Candidates for Her X-1, p. 129.

002.011 News and views.
Nature, Vol. 239, 7 - 12, 65 - 72, 129 - 136, 250 - 256, 305 - 312, 365 - 372, 426 - 434, 487 - 494 (1972). – Phyllosilicates and the solar system, p. 7; Pulsations in white dwarf stars, p. 10; Measuring optical frequencies and the speed of light, p. 65 - 66; Supernovae: Insight into type I, p. 66; Magnetic field structure of 3C 273A, p. 66; Cosmochemistry: Abundance discussed, p. 68; Faraday rotation in radio galaxies and quasars, p. 71 - 72; Cyg X-3 blows up, p. 130; Recognition of gravitational radiation, p. 134 - 135; Still no quarks, p. 136; Moon-making in three dimensions, p. 251; Selenology: More Luna 20 analyses, p. 256; Is Cen X-3 a neutron star? p. 256; Reversed geomagnetic events in the Brunhes epoch, p. 305 - 306; Cen X-3 a white dwarf? p. 307 - 308; Inside cool interstellar clouds, p. 311; Apollo clock corrections, p. 312; Difficulties in telling QSOs from blue stars, p. 367 - 368; Outburst of Cyg X-3, p. 426 - 427; First observations from Copernicus, p. 487; Southern hemisphere astronomy: Schmidt survey plans, p. 489 - 490; Measuring Schwarzschild coordinates, p. 492 - 493.

002.012 News.
Nature, Phys. Sci., Vol. 239, 33 - 34, 49 - 50, 81 - 82, 97 - 98, 113, 137 - 138 (1972).
High resolution map of M31, p. 33; Interstellar H_2S, p. 33 - 34; Motion faster than light, p. 49; Solar neutrinos: Escape from a paradox, p. 49 - 50; Cosmogony: Origin of elements, p. 81 - 82; Position of Cyg X-3, p. 82; Soft X-ray survey of the galactic plane, p. 97; Relativity: Dirac's equation, p. 98; Modulation collimators for X-ray astronomy, p. 113; Is continental drift sacrosanct? , p. 137 - 138; Selenology: More on impact craters, p. 138.

002.013 Astronomical notebook. J. S. Griffith.
Journ. British Interplanet. Soc., Vol. 25, 669 - 681, 739 - 745 (1972).
The magnetism of the earth, p. 669 - 670; The thermal evolution of the moon, p. 670 - 671; The atmosphere of Venus, p. 671 - 672; Deimos, p. 672; Occultations—Io, Jupiter and Beta Scorpii, p. 672 - 673; A supernova close to the sun? p. 673; The solar wind, p. 673 - 674; Star formation, p. 674 - 675;

Red giants and novae, p. 675 - 676; Cepheids, binaries and white dwarfs, p. 676 - 677; Quasars, p. 677; Cosmology, p. 678; Galaxies, p. 678; Interstellar molecules, p. 678 - 679; X-ray pulsations, supernovae, pulsars and neutron stars, p. 679-681; Jupiter's radiation, p. 739; Surface conditions on Venus, p. 739 - 740; Titan's atmosphere, p. 740; Cosmology, p. 740 - 741; Stellar formation and binary fission, p. 741 - 742; A binary with a postcataclysmic nova component? p. 742; Pulsar timings and black holes, p. 742 - 743; A new class of astronomical objects? p. 743; Quasars, p. 743 - 744.

002.014 **Pêle-mêle.** J. Lequeux.
     Sci. Progrès Découverte, 99e année, No. 3429, p. 25 - 30; No. 3430, p. 24 - 26; No. 3431, p. 30 - 31; No. 3432, p. 28 - 30; No. 3433, p. 32 - 33; No. 3435, p. 28 - 30; No. 3436, p. 28; No. 3437, p. 23 - 24; No. 3439, p. 18 - 21 (1971). – Le diamètre des quasars, No. 3429, p. 25 - 26; La radioastronomie au service de l'astronomie, p. 26 - 27. Deux nouvelles voisines: Des galaxies elliptique et spirale, No. 3430, p. 24. Des X mous dans la Boucle du Cygne, No. 3431, p. 30 - 31. Les quasars: vraiment à de grandes distances, No. 3432, p. 28. Des nouvelles de Maffei 2, No. 3433, p. 33. Les découvertes d'«Uhuru», No. 3435, p. 30; Prouver la relativité générale..., p. 30. Des quasars muets, No. 3436, p. 28; Du centre de la Galaxie un rayonnement infrarouge, p. 28. Encore des molécules interstellaires, No. 3437, p. 24. A 20000 années-lumière: JP 1933, No. 3439, p. 18; Le rayonnement venu du fond des âges..., p. 18 - 19.

002.015 **News notes.**
     Sky Telescope, Vol. 44, 86 - 87, 157 - 158, 225, 297-298, 367 - 368 (1972). – (2) A radio problem at Cambridge; World's largest optical telescope nears completion; Variable star observing from space; Orbits of Phobos and Deimos; Dutch optical designer; Ultraviolet astronomy satellite; X-rays from rich clusters of galaxies. (3) Giacobinid meteor shower; Two notable eclipsing binaries; Radio observations of a supernova outburst; Another interstellar molecule; More about the supernova in NGC 5253; Messier 13 as viewed in a giant telescope; New Astronomer Royal; Asteroid names. (4) Interstellar hydrogen sulfide; Giant radio outburst of Cygnus X-3; Cepheid of longest period; Cometary handbooks; New AAVSO periodical. (5) Nova in the LMC; Halley's comet and a hypothetical new planet; Progress on the Anglo-Australian telescope. (6) Unusual new asteroids; Astrophysicists meet in New York; Radcliffe Observatory bicentennial; Origin and evolution of the earth-moon system; Plans for a joint American-Soviet space effort; 1,309 light curves.

002.016 **Bibliography.**
     Z. Kopal, M. Moutsoulas, J. W. Salisbury (Editors). The Moon, Vol. 5, 233 - 250, 457 - 481 (1972). – Current critical bibliography of the entire field of lunar studies.

002.017 **News and comments.** E. J. Öpik.
     Irish Astron. Journ., Vol. 10, 153 - 170 (1971). Prehistoric astronomy; Dating the technical advances of civilization; Interstellar dust; The Parkes hydrogen-line survey of the Milky Way; The frequency of double stars; Proper motion survey; White dwarfs; Proper motions with reference to galaxies; The masses of quasars; The RR-Lyrae variables. Stars in the making: The young clusters; "Stellar evolution before the main sequence" (IAU Colloquium). Solar granulation; Precambrian meteorite impact; The exact diameter of Io. Satellites and double stars.

002.018 **Forschung und Technik.**
     Phys. Blätter, 28. Jahrgang, p. 421 - 423, 473 - 475, 518 - 520, 568 - 570 (1972).
Nukleon-Antinukleon-Systeme, p. 421 - 422; Neutrinos mit nichtverschwindender Ruhmasse? , p. 422 - 423; Mehrfache Materieverluste des galaktischen Kerns, p. 473; Molekülwolken bilden galaktischen „ Rauchring", p. 520; Bestimmung von c mit nur ±18 m/s Meßfehler, p. 568 - 569.

002.019 **Short scientific communications.**
     Priroda (NRB), Vol. 21, No. 3, p. 89 - 97 (1972). In Bulgarian. – Abstr. in Referativ. Zhurn. 51. Astron., 12.51.43 (1972).

002.020 **Science news.**
     Priroda, No. 12.72, p. 101 - 110 (1972). In Russian. Flights of the automatic stations "Prognoz" (*S. A. Nikitin*); Mars 3 investigates the planet; The atmosphere of neutron stars.

002.021 **News from science and other informations.**
     Zemlya i Vselennaya, 1972, No. 6. In Russian. Bright supernova in the galaxy NGC 5253, p. 32; Unknown superheavy element? p. 32; Measurement of the density of the lunar atmosphere, p. 32; Mars 2 and Mars 3 carried out the program, p. 52 - 53; Impact of a micrometeorite, p. 60.

002.022 **News and views.**
     Nature, Vol. 240, 9 - 16, 69 - 76, 125 - 132, 179 - 186, 251 - 258, 376 - 384, 438 - 444, 520 - 526 (1972). Closed time and black body radiation, p. 15; Lunar alkali metal poverty, p. 70 - 71; Identification of a southern sky Uhuru source, p. 76; Gravitational waves? p. 125 - 126; Identification of GX 17 + 2, p. 131; Do all supernovae beget pulsars? p. 179; Black hole energy extraction problems, p. 184; The hunting of the S.N.U., p. 251; Partons and the origin of the universe, p. 257; Isotopes in meteorites, p. 378 - 379; Gravitation: Quantization pursued, p. 382, 383; Continental drift: Pre-permian probing, p. 383, 384; Steady state obituary? p. 439; More about solar neutrinos, p. 443; Neutrinos: Double beta decay? p. 444; What controls cosmic ray modulation? p. 444; Ancient astronomy at the Royal Society, p. 522, 523; Rate of occurrence of supernovae, p. 526.

002.023 **News.**
     Nature, Phys. Sci., Vol. 240, 1 - 2, 25 - 26, 49 - 50, 73, 97 - 98, 121, 145, 169 (1972).
Fluctuations in Cyg X-1, p. 1 - 2; Interkosmos looks at cosmic rays and VLF, p. 2; Tenerife beckons infrared observers, p. 25 - 26; Instrumentation: Speeding-up galaxy, p. 50; Quasar angular diameters and cosmology, p. 73; How the solar wind interacts with non-magnetic planets, p. 97; ESRO IV launched, p. 98; The moon's occasional atmosphere, p. 121; Infrared astronomy: Clues in variability of extragalactic sources, p. 145; Optical pulsar in Hercules, p. 169; Jovian frost, p. 169; D-region lunar influence, p. 169.

002.024 **News: search & discovery.**
     Physics Today, Vol. 25, No. 1, p. 17 - 20; No. 2, p. 17 - 20; No. 3, p. 17 - 20; No. 4, p. 17 - 20; No. 5, p. 17 - 20 No. 6, p. 17 - 20; No. 7, p. 17 - 20; No. 8, p. 17 - 20; No. 9, p. 17 - 20; No. 10, p. 17 - 20; No. 11, p. 17 - 20; No. 12, p. 17 - 20 (1972).
(1) Neutron stars grow magnetic whiskers, p. 17 - 18; More power and smoother surface for Arecibo dish, p. 20. (2) Redshift problem intensifies, says Burbidge, p. 17 - 19. (3) More support for the big bang, p. 17 - 18. (4) Absolute laser frequency measurements, p. 17 - 18. (5) New Mexico site for radio telescope, p. 17 - 18; Alfvén on cosmic rays, sunspots, antimatter, p. 20. (6) Gravity waves attract theories and experiments, p. 17, 18 - 20; Multimirror infrared telescope planned, p. 17 - 18. (7) Infrared interferometer to measure size and shape of stars, p. 17, 19 - 20; Fine-structure components of Hα line resolved, p. 20. (8) The case of the missing solar neutrinos, p. 17 - 18. (10) Cave drawing is evidence of Crab nebula supernova, p. 20. (11) Neutron star cores: solid or

liquid? , p. 17 - 19; Ultraviolet and X-ray observatory in orbit, p. 19.

002.025 Astronomijas jaunumi.
Zvaigžņotā debess, 1971. gada rudens, p. 17 - 21.
Pulsārs – gamma staru avots (*A. Balklavs*), p. 17; Saules uzliesmojumi jaunā skatījumā (*V. Kasinskis*), p. 17 - 19; Jauni radioteleskopi (*A. Spektors*), p. 19 - 21; Cianoacetilēns un skudrskābe starpzvaigžņu telpā (*A. Alksnis*), p. 21.

002.026 Astronomijas jaunumi.
Zvaigžņotā debess, 1971./72. gada ziema, p. 12 - 19.
Jaunas atziņas par pulsāriem (*A. Balklavs*), p. 12 - 15; Gulbja V 1057 zvaigznes straujā pārvēršanās (*A. Alksnis*), p. 15 - 18; Metīlspirts un citas jaunatklātas starpzvaigžņu telpas molekulas (*A. Alksnis*), p. 18; Antaresa radiostarojums (*A. Alksnis*), p. 19.

002.027 Astronomijas jaunumi.
Zvaigžņotā debess, 1972. gada pavasaris, p. 6 - 11.
Lodveida zibens un Saules aktivitāte (*A. Balklavs*), p. 6 - 8; Aminoskābes Marčisonas meteorītā (*E. Cielēns*), p. 8 - 9; Jauna hipotēze par kvazāru starojuma mehānismu (*A. Balklavs*) p. 9; Orbitālās automātiskās stacijas novēro komētas (*J. Francmanis*), p. 10; Nova Cefeja zvaigznājā (*L. Duncāns*), p. 10 - 11.

002.028 Astronomijas jaunumi.
Zvaigžņotā debess, 1972. gada vasara, p. 23 - 29.
Kosmologijas jaunumi (*A. Balklavs*), p. 23 - 25; Ziemeļu Vainaga ϑ ir dubultzvaigzne (*J. Francmanis*), p. 25; Kvazāri uzdod jaunu mīklu (*I. Šmelds*), p. 25 - 26; Kolapsārs aptumsuma dubultzvaigznē (*U. Dzērvītis*), p. 26 - 29.

002.029 Rassegna delle riviste e notizie brevi.   P. Maffei.
Coelum, Vol. 40, 151 - 154, 189 - 195, 228 - 234 (1972).

002.030 Mitteilungen aus Wissenschaft und Literatur.
Sterne, 48. Jahrgang, p. 183 - 188, 246 (1972).
(3) Zerplatzte Neutrinos (*H. Lambrecht*); Die Deutung der Rotverschiebung in den Spektren extragalaktischer Systeme (*F. Schmeidler*); Die optische Variabilität von 3C 273 (*F. Schmeidler*); Wasser auf dem Mond (*J. Classen*); Nachtrag zu dem Bericht "Wasser auf dem Mond" (*J. Classen*); Neue Beobachtung einer lunaren Leuchterscheinung (*G. Loibl*); (4) Aufschlag eines grossen Meteoriten auf dem Mond (*L. Kühn*).

002.031 Nouvelles brèves.
Ciel et Terre, Vol. 88, 314 - 318, 491 - 496 (1972).
(4) Le mouvement des satellites de Mars; Redécouverte de la comète périodique Tempel 1; Découverte de nouvelles étoiles doubles; Nouvelles autour de Sirius B; Les supernovae découvertes depuis 1885; Le premier jour de la semaine; (6) L'accroissement du diamètre de l'ombre de la terre lors des éclipses de lune; L'occultation partielle de Vénus du 17 avril 1972; Occultations d'étoiles par Jupiter; Le mouvement des satellites de Jupiter; Notes cométaires; Les Puppides: L'averse prédite n'a pas eu lieu; Mesures de diamètres stellaires; L'orbite de Pageos 1.

002.032 Noted in the current journals.
D. Morrison, N. D. Morrison.
Mercury, (Journ. Astron. Soc. Pacific), Vol. 1, No. 5, p. 10 - 11; No. 6, p. 14 - 16 (1972).
(5) The atmosphere of Titan; Third lunar science conference; Observational cosmology; Speckle interferometry; Discovery of a bright supernova. (6) Are the redshifts of galaxies and quasars of cosmological origin; Gamma rays from the Crab nebula; The surface of Venus.

002.033 Kleine Kepler-Bibliographie.   M. List.

Kepler Festschrift 1971, (see 003.010), p. 266 - 273 (1971).

002.034 Annual index of authors, to Astrophysical Journal, volumes 171 - 178, parts 1 and 2 and to the Supplement Series, volume 24, pages 167 - 492.
Astrophys. Journ., Vol. 178, No. 3, Part 3, 63 pp. (1972).

002.035 Chronicle.
Urania Kraków, Vol. 43, 203 - 204, 245 - 249, 274 - 278, 305 - 308 (1972). In Polish.
(1) Radio observations of supernova remnants, *M. Panków;* Discovery of two nearby galaxies, *A. Marks;* Have the Trojans ever been Jupiter's satellites?, *A. Marks;* New meteorite craters in Africa, *A. Marks.* (2) Peculiar properties of the X ray pulsar Cyg X-1, *B. Kuchowicz;* Another quasar in the vicinity of a galaxy, *B. Kuchowicz;* Helium abundance anomalies on the surface of the peculiar star CU Vir, *T. Szymczak;* Atmospheric pressure on Mars, *T. Szymczak;* Will the change of direction of flow of the Siberian rivers influence the rotation of earth? *T. Szymczak.* (3) Properties of the nucleus of the NGC 5128 galaxy, *B. Kuchowicz;* Pioneer 10, *S. R. Brzostkiewicz;* A giant meteorite falls on the moon, *S. R. Brzostkiewicz;* (4) The Universe is bigger and older, *K. Ziołkowski;* Project of the world's largest radiotelescope, *B. Kuchowicz;* Diameters of the largest minor planets, *K. Ziołkowski;* New orbital elements of the hypothetical transplutonian planet, *S. R. Brzostkiewicz.*

002.036 Kurzberichte aus der Forschung.
SuW, Vol. 11, 233 - 236, 279 - 281, 307 - 309, 343 - 345 (1972).
(1) Erste UV-Sternspektren von TD-1A; Ein transplutonischer Planet? ; Gravitations-Rotverschiebung einer solaren Spektrallinie; Röntgenpolarisation des Crab-Nebels; Temperaturmessungen am Mars; Die Mondkruste ist 65 km dick; Simulierter Meteoritenaufprall; Hellste Supernova seit 35 Jahren; Transmissionsgitter statt Objektivprismen; Simulierte Supernova-Ausbrüche. (2) Über die Entstehung des Saturnringes; Quasare: Massenverlust durch Strahlungsdruck; Umweltverschmutzung als Wiege des Lebens; Sieben Antiprotonen; Infrarot-Exzesse; Werkmontage des 2.2-m Teleskops begonnen. (3) Radioausbruch von Cyg X-3; Aktive Galaxien mit Radio-Schweifen; Weitere Ergebnisse von Mars 2 und 3; Das amerikanische Sechs-Spiegel-Teleskop; Ergebnisse der Venus 8-Sonde. (4) Suche nach Transpluto; Der Durchmesser von Io; Präzisionsmessungen der Lichtgeschwindigkeit; Zum C^{12}/C^{13}-Verhältnis der interstellaren Materie; Übereinstimmung von optischer- und Radio-Spiralstruktur; Cen X-3 – Neutronenstern oder Weißer Zwerg?

002.037 Astronomy and Astrophysics Abstracts. Vol. 7, Literature 1972, Part I.
S. Böhme, W. Fricke, U. Güntzel-Lingner, F. Henn, D. Krahn, U. Scheffer, G. Zech (Editors).
Published for Astronomisches Rechen-Institut, Heidelberg by Springer-Verlag, Berlin – Heidelberg – New York. 10 + 526 pp. Price DM 72.00 [Subscription price per volume DM 57.60] (1972).

002.038 Bibliographie générale des marées terrestres, 1800– 1971.   P. Melchior (Editor).
Ass. Internationale Géod., Centre International des Marées Terrestres, Obs. Roy. Belgique, Bruxelles. 5 + 115 pp. Price fb 350.00 (1972).

002.039 Commission 41 (History of Astronomy) of the IAU: Bibliography of books and papers published in 1969–1971 on the history of astronomy.
Prepared by N. Lavrova, edited by Z. K. Sokolovskaya, L. M. Vasil'eva.
Tipografiya Astronomicheskogo Soveta Akad. Nauk SSSR, Moskva. 11th and 12th issues, 25 + 26 pp. (1972).

002.040 **Annotations on astrophysical papers published in the journal "Radiofizika", Vol. 14, Nos. 1 - 12 for the year 1971.**
Astron. Zhurn. Akad. Nauk SSSR, Vol. 49, 908 - 909 (1972). In Russian.

002.041 **AFCRL bibliography for 1970 October - 1971 June.** J. W. Salisbury (Editor).
Icarus, Vol. 16, 581 - 637; Vol. 17, 234 - 264, 265 - 288 (1972).
Presented is a bibliography on lunar and planetary subjects furnished by the Air Force Cambridge Research Laboratory, Laurence G. Hanscom Field, Bedford, Mass.

002.042 **Astronomical notebook.** J. S. Griffith.
Spaceflight, Vol. 14, 311 - 314, 353 - 356 (1972).
(8) Constants of the solar system; Astronomical constants and relativity; Venus; The circumterrestrial dust cloud; The triplet system, earth-moon-Toro; Tektites — terrestrial or lunar? ; Mars; The formation of the solar system; The moon and the earth; Capture of satellites by planets; Planetary quarantine orbits; Surface of Mercury; (9) The origin of the moon; Lunar latitude and longitude; Transient lunar phenomena; Toro — an interloper in the earth/moon system; Carbon dioxide frost on the Martian polar caps; Martian surface and wind conditions; Gravitational fields of the giant planets; Radio-echo observations of meteor-streams; The early solar system; Comets as interplanetary probes? ; Comets and non-gravitational forces; Stability and Saturn's rings; Satellites of the outer planets; Electromagnetic surface waves; The surface of Deimos.

002.043 **Solar physics. A journal for solar research and the study of solar terrestrial physics. Index to volumes 11 – 25 (1970–1972).**
D. Reidel Publishing Company, Dordrecht — Holland/Boston — U. S. A. 29 pp. (1972).

003 Books (Astronomy and Astrophysics)

003.001 **Astrometry and astrophysics.** V. I. Voroshilov (Editor).
Respublikanskij mezhvedomstvennyj sbornik. Ser. Astrometriya i Astrofizika, No. 15. Akademiya Nauk Ukrainskoj SSR. Izdatel'stvo "Naukova Dumka", Kiev. 104 pp. Price 98 Kop. (1972). In Russian. — The papers included are abstracted in their corresponding subject categories — see 031.024, 041.012, 061.023, 063.013, 074.046, 097.068, 102.054, 114.066 - 114.068, 121.042, 123.022, 123.023.

003.002 **Xth conference of the USSR on the propagation of radio waves. Theses of reports. Section I. Ionospheric propagation of radio waves.**
AN SSSR. Nauch. sovet po probl. "Rasprostr. radiovoln", Sib. in-t zemn. magn. ionosfery i rasprostr. radiovoln Sib. otd. AN SSSR, Irkutsk. un-t. Nauka, Moskva. 560 pp. (1972). In Russian.

003.003 **Cosmology, fusion & other matters. George Gamow Memorial Volume.** F. Reines (Editor).
Adam Hilger Ltd., London. 14 + 320 pp. Price $ 15.00 (1972). — The individual contributions within the subject scope of Astronomy and Astrophysics Abstracts are included in their corresponding categories — see abstracts 005.008 - 005.010, 061.033 - 061.035, 062.034, 066.045, 143.027, 162.031 - 162.036.

003.004 **Éléments de cosmologie philosophique et scientifique.** S. Mazierski.
Księgarnia ŚW. Wojciecha, Poznań — Warszawa — Lublin. 414 pp. Price zł 205.00 (1972). In Polish.

003.005 **Annual Review of Astronomy and Astrophysics, Vol. 10.**
L. Goldberg, D. Layzer, J. G. Phillips (Editors).
Annual Reviews Inc., Palo Alto, California. 9 + 513 pp. Price $ 10.50 (1972). — The individual contributions are included in their corresponding subject categories — see abstracts 041.022, 061.042, 065.069, 065.070, 066.052, 066.053, 073.083, 073.084, 077.041, 080.037, 125.020, 131.072, 141.521, 161.006.

003.006 **Radiation measurement in space.**
Yu. I. Gal'perin, L. S. Gorn, B. I. Khazanov.
Atomizdat, Moskva, 344 pp. Price 3 Rbl. 21 Kop. (1972). In Russian. — Review in Referativ. Zhurn. 62. Issled. kosmich. prostranstva, 11.62.325 (1972).

003.007 **Young stellar complexes. Astroclimate.**
Astron. in-t AN UzSSR, FAN, Tashkent. 180 pp. Price 1 Rbl. 2 Kop. (1972). In Russian.

003.008 **Vistas in astronomy, Vol. 14.**
A. Beer (Editor).
Pergamon Press, Oxford — New York — Toronto — Sydney — Braunschweig. 7 + 300 pp. Price £ 10.00 (1972). — The individual contributions are included in their corresponding subject categories — see abstracts 064.036, 095.002, 113.002, 155.044, 158.096.

003.009 **Encyclopedia of physics. (Handbuch der Physik).**
S. Flügge (Chief editor). Vol. 49/4: Geophysics III (Part IV). K. Rawer (Editor).
Springer-Verlag, Berlin — Heidelberg — New York. 6 + 579 pp. Price DM 175.00 (1972). — The individual contributions to this volume are included in their corresponding subject categories — see abstracts 062.054, 084.317, 084.318, 084.421.

003.010 **Kepler Festschrift 1971. Zur Erinnerung an seinen Geburtstag vor 400 Jahren.**
Editor: Naturwissenschaftlicher Verein Regensburg, under the redaction of E. Preuss. Acta Albertina Ratisbonensia, Vol. 32, 273 pp. Price DM 24,50 (1971). — The individual contributions within the subject scope of Astronomy and Astrophysics Abstracts are included in their corresponding categories — see abstracts 002.033, 004.044 - 004.050, 005.018, 005.019, 07.042.061, 08.042.072.

003.011 **Tables on the scattering of light of a polydisperse system of spherical particles.**
Eh. G. Yanovitskij, Z. O. Dumanskij.
Akademiya Nauk Ukrainskoj SSR. Glavnaya Astronomicheskaya Observatoriya. Izdatel'stvo "Naukova Dumka", Kiev. 124 pp. Price 30 Kop. (1972). In Russian.

003.012 **Astrometriya i Astrofizika**, Vyp. (No.) 17.
E. P. Fedorov (Editor).
Respublikanskij Mezhvedomstvennyj Sbornik. Akademiya Nauk Ukrainskij SSR. Izdatel'stvo "Naukova Dumka", Kiev. 128 pp. Price 1 Rbl. 15 Kop. (1972). In Russian. – The papers included are abstracted in their corresponding subject categories – see abstracts 031.058, 071.060, 071.061, 073.100, 074.092, 082.205 - 082.207, 126.021.

003.013 **Problems of galactic and extragalactic astronomy.**
G. M. Idlis (Editor).
Akademiya Nauk Kazakhskoj SSR; Trudy Astrofizicheskogo Instituta. Vol. 19. Izdatel'stvo "Nauka" Kazakhskoj SSR, Alma-Ata. 136 pp. Price 1 Rbl. 26 Kop. (1972). In Russian. The individual contributions are included in their corresponding subject categories – see abstracts 007.000, 042.073, 066.089, 066.090, 114.136, 114.137, 132.029 - 132.031, 151.032 - 151.035,155.056 - 155.058, 158.113, 158.114, 160.020, 160.021.

003.014 **Catalogue of selected compact galaxies and of post-eruptive galaxies.**
Prepared by F. Zwicky, with the collaboration of M. A. Zwicky. Published by F. Zwicky, Guemligen, Switzerland. Printed in Switzerland by Offsetdruck L. Speich, Zürich. 32 + 388 pp. Price Sfr. 205.00 (1971).
After an introduction on origin, scope and purpose of the catalogue data on more than 3700 objects are listed including over 250 spectra and symbolic velocities of recession.

003.015 **The cosmic ν.** B. Kuchowicz.
Nuclear Energy Information Center Warsaw. NEIC-RR-47, 130 pp. (1972).
All the major developments in cosmic neutrino research, from astrophysical to cosmic-ray and gravitational aspects, for the periods from 1965 up to the beginnings of 1972 are referenced in this booklet.

003.016 **Collected papers presented at the session of the Scientific Council on the complex problem "Radio astronomy".** IZMIRAN, 13–16 October, 1970.
AN SSSR. In-t zemn. magn., ionosfery i rasprostr. radiovoln, Moskva. 179 pp. Price 1 Rbl. (1972). In Russian. – Review in Referativ. Zhurn. 51. Astron., 1.51.68 (1973).

003.017 **New entertaining astronomy.** V. N. Komarov.
Nauka, Moskva. 280 pp. Price 52 Kop. (1972). In Russian.

003.018 **The revolution in astronomy.** P. W. Hodge.
Translated from the English edition. Mir, Moskva. 152 pp. Price 37 Kop. (1972). In Russian.

003.019 **Apparatus and methods of reduction of radio astronomical observations.**
A. Eh. Balklavs (Editor).
Izdatel'stvo "Zinatne"; Trudy Radioastrophys. Obs. Akad. Nauk Latvijskoj SSR, *Riga*, Vol. 14, 127 pp. Price 38 Kop. (1972). In Russian. – The papers included are abstracted in their corresponding subject categories – see abstracts 033.057 - 033.064, 074.096, 077.063, 077.064, 079.104.

003.020 **Physics of solar continuum radio bursts.**
A. Krüger.
Akademie-Verlag, Berlin. 206 pp. Price DM 39.00 (1972). Review in Nature, Phys. Sci., Vol. 240, 72; 1972 (*P. A. Sweet*) Contents: 1.) Introduction; 2.) Basic phenomenological results; 3.) Theoretical foundations of the emission of radio waves in a plasma; 4.) The interpretation of radio bursts; 5.) Summary and conclusions.

003.021 **Space physics and space astronomy.**
M. D. Papagiannis.
Gordon and Breach, Science Publishers, New York – Paris – London. 14 + 293 pp. Price $ 14.50 (1972). – Contents: 1.) Planetary atmospheres; 2.) The ionosphere; 3.) The magnetosphere; 4.) The active sun; 5.) The interplanetary space; 6.) Solar-terrestrial relations; 7.) Solar and planetary space astronomy; 8.) Galactic space astronomy.

003.022 **The physics of pulsars.** A. M. Lenchek (Editor).
Gordon and Breach, Science Publishers, New York – London – Paris. 9 + 173 pp. Price $ 14.50 (1972). – The individual contributions are included in their corresponding subject categories – see abstracts 065.129, 065.130, 141.539 - 141.551.

003.023 **Perturbation methods in non-linear systems.**
G. E. O. Giacaglia.
Applied Math. Sci., Vol. 8. Springer-Verlag, New York – Heidelberg – Berlin. 8 + 369 pp. Price DM 27.80 (1972). – Contents: (1) Canonical transformation theory and generalizations; (2) Perturbation methods for Hamiltonian systems. Generalizations; (3) Perturbations of integrable systems; (4) Perturbations of area preserving mappings; (5) Resonance; Appendix: Remarks, some open questions and research topics.

003.024 **Cosmic rays.** A. M. Hillas.
Selected Readings in Physics. Pergamon Press, Oxford – New York – Toronto – Sydney – Braunschweig. 8 + 297 pp. Price DM 44.40 (1972).
Contents: (1) Discovery; (2) The nature of the radiation; (3) Particles produced by the cosmic rays; (4) The primary cosmic radiation; (5) Radio waves from the Galaxy; (6) Extensive air showers; (7) The origin of cosmic rays; (8) Local peculiarities; (9) Other cosmic radiations. After presentation of this outline of the more recent work on the subject the volume contains reprints of 16 key papers which mark some turning points in the study of cosmic rays.

003.025 **Tables K11: Two-star fix without use of altitude difference method. Vol. IV N (Lat. 30° – 39° 30' N).**
S. M. Kotlarić.
Edited by Hidrografskog Instituta Jugoslavenske Ratne Mornarice, Split. 29 + 367 pp. (1972). In Serbo-Croatian.

003.026 **Astronomy in Toruń, Nicholas Copernicus' native town.** C. Iwaniszewska (Editor).
Toruń Scientific Society, Toruń. Popular Science Ser. No. 21 = Copernican Publ. No. 1, 75 pp. Price zł 15.00 (1972). – Contents: Nicholas Copernicus, founder of modern astronomy; History of the Toruń Observatory; Domes and telescopes; Research work; General education activities; Publications.

003.027 **Dictionary of scientific biography, Vol. V: Emil Fischer – Gottlieb Haberlandt.**
C. C. Gillispie (Editor-in-chief).
Charles Scribner's Sons, New York. 13 + 624 pp. Price $ 35.00 (1972). – Review in Sky Telescope, Vol. 44, 187 (1972).

003.028 **Problems of cosmic physics. Vypusk 7.**
S. K. Vsekhsvyatskij (Editor).
Mezhvedomstvennyj Nauchnyj Sbornik. Izdatel'stvo Kievskogo Universiteta, Kiev. 180 pp. Price 1 Rbl. 20 Kop. (1972). In Russian. – The papers included are abstracted in their corresponding subject categories – see 062.076, 065.152, 073.113, 084.356, 102.074, 104.055, 105.129, 106.039, 114.175, 124.105, 132.036, 158.131, 158.132.

003.029 **Vltavíny a tektity.** R. Rost.
Academia, Praha. 244 + 16 pp. Price Kčs. 15.00 (1972).

003.030 **Mimozemské civilizace – Problémy mezihvězdného spojení.** S. A. Kaplan (Editor).
Academia, Praha. 312 pp. Price Kčs. 36.00 (1972).

003.031 **Koperníkův svět.** H. Bietkowski, W. Zonn.
Orbis, Praha. 172 pp. Price Kčs. 50.00 (1972).

003.032 **The Pulkovo Observatory. Review of the history and scientific activities.** A. N. Dadaev.
Akademiya Nauk SSSR, Glavnaya Astronomicheskaya Observatoriya. Izdatel'stvo "Nauka", Leningradskoe Otdelenie, Leningrad. 149 pp. Price 49 Kop. (1972). In Russian.

003.033 **Selections from the scientific correspondence of Elihu Thomson.**
H. J. Abrahams, M. B. Savin (Editors).
The MIT Press, Cambridge, Mass., London. 14 + 569 pp. Price $ 17.50 (1971). – Review in Journ. Optical Soc. America, Vol. 62, 1014 (1972).

003.034 **Solar-terrestrial physics.** An account of the wave and particle radiations from the quiet and active sun, and of the consequent terrestrial phenomena.
S.-I. Akasofu, S. Chapman.
The International Series of Monographs on Physics. At the Clarendon Press, Oxford, 23 + 901 pp. Price £ 25.00 (1972). Contents: (1) The sun and interplanetary space; (2) The internal structure and magnetic field of the earth; (3) The terrestrial atmosphere: Photochemistry; (4) Dynamics of the upper atmosphere and dynamo action; (5) The formation of the magnetosphere; (6) Energetic particles, plasma, and electromagnetic waves in the magnetosphere; (7) Solar storms and their extension into interplanetary space; (8) Magnetospheric storms; Appendices 1–11.

003.035 **Problems of modern cosmogony.**
V. A. Ambartsumyan (Editor).
Nauka, Moskva. 2nd revised and enlarged edition. 470 pp. Price 2 Rbl. 53 Kop. (1972). In Russian. – Review in Referativ. Zhurn. 51. Astron., 1.51.74 (1973).

003.036 **Problèmes de cosmogonie contemporaine.**
V. Ambartsumyan, L. Mirzoyan, G. Sahakian, S. Vsekhsvyatskii, V. Kazutinski.
Éditions Mir, Moskva. Translation from Russian into French. 370 pp. Price 98 F belges. (1971). – Reviews in Ciel et Terre, Vol. 88, 498 - 499; 1972 (*J. Meeus*); Sky Telescope, Vol. 44, 118 (1972).

003.037 **Theory and analysis of phased array antennas.**
N. Amitay, V. Galindo, C. P. Wu.
John Wiley & Sons,Ltd., Chichester. 480 pp. Price £ 8.85 (1972). – Review in Journ. British Interplanet. Soc., Vol. 25, 621 (1972).

003.038 **Radiation transport in spectral lines.**
R. G. Athay.
Geophysics and Astrophysics Monographs (an international series of fundamental textbooks), Vol. 1. D. Reidel Publishing Company, Dordrecht – Holland. 13 + 263 pp. Price hfl. 41.00 (1972). – Contents: (1) Introduction; (2) The line source function; (3) The two-level case: one spectral line; (4) The multilevel case: two or more lines; (5) Line profiles; (6) Total intensities of lines; (7) The line blanketing effect; (8) Numerical methods.

003.039 **L'astrophysique nucléaire.**
J. Audouze, S. Vauclair.
Presses Universitaires de France, Paris. "Que sais-je" No. 1473, 126 pp. (1972).

003.040 **The optical transfer function.** K. R. Barnes.
Monographs on Applied Optics, No. 3. Adam Hilger, Ltd., London. 78 pp. Price £ 5.00 (1971). – Review in Journ. British Astron. Ass., Vol. 83, 71; 1972 (*H. E. Dall*).

003.041 **Outlines of the motion of cosmic bodies.**
V. V. Beletskij.
Nauka, Moskva. 360 pp. Price 1 Rbl. 81 Kop. (1972). In Russian. – Review in Referativ. Zhurn. 62. Issled. kosmich. prostranstva, 12.62.197 (1972).

003.042 **Course of spherical astronomy. (For high school students in the field of engineering geodesy).**
N. A. Belova.
Nedra, Moskva. 182 pp. Price 42 Kop. (1971). In Russian.

003.043 **Annuaire 1973 du Bureau des Longitudes.** Encyclopédie Physique et Spatiale. Gauthier-Villars, Éditeur, Paris, 14 + 849 + A 22 + B 10 + C 80 + D 34 pp. (1972). Part 1: Éphémérides astronomiques; Part 2: Hydrosphère; atmosphère; Part 3: Astrophysique; Part 4: Électromagnétisme; particules; radiations; Part 5: Géographie mondiale; Part 6: Supplément pour l'année 1974; Part 7: Notices; Part 8: Tables analytiques.

003.044 **Astronomie, Lehrbuch für Klasse 10.**
H. Bernhard, O. Günther, K. Lindner, J. Stier.
Volk und Wissen, Volkseigener Verlag, Berlin. Price MDM 5.00 (1971). – Review in Astron. Tidsskr., Årg. 5, p. 148; 1972 (*V. Hegvad*).

003.045 **Nicolaus Copernicus and the heliocentric theory in Polish culture till the end of the eighteenth century.**
B. Bienkowska.
Ossolineum, Wrocław. 295 pp. Price Zł. 70 (1971). In Polish. Review in Journ. History Astron., Vol. 3, 220; 1972 (*J. R. Ravetz*).

003.046 **Johannes Kepler. Beiträge zu seinem Lebensbild.**
E. Bindel.
Verlag Freies Geistesleben, Stuttgart. 133 pp. Price DM 15.00 (1971). – Review in Zeitschr. Geophys., Vol. 37, 755; 1971 (*H. Schwentek*).

003.047 **Études sur l'astronomie Indienne et sur l'astronomie Chinoise.** J. B. Biot.
Albert Blanchard, Paris. 398 pp. Price F 25.00 (1969). Review in Naturwissenschaften, 59. Jahrgang, p. 319; 1972 (*W. Hartner*).

003.048 **Precision measurement and calibration, Vol. 5. Selected NBS papers on frequency and time.**
B. E. Blair, A. H. Morgan (Editors).
National Bureau of Standards (NBS), U. S. Government Printing Office, Washington D. C. 555 pp. Price $ 6.00 (1972).

003.049 **Einstein's theory of relativity.** M. Born.
Mir, Moskva. Translated from the English edition. 2nd edition. 368 pp. Price 1 Rbl. 72 Kop. (1972). In Russian. Review in Priroda, No. 7.72, p. 122 (1972).

003.050 **Topics in modern physics.**
W. E. Britten, H. Odabasi (Editors).
Adam Hilger Ltd., London. 353 pp. Price £ 10.00 (1971). Review in Observatory, Vol. 92, 106; 1972 (*V. P. Myerscough*).

003.051 **Astronomy in colour.** P. Lancaster Brown.
Blandford Press Ltd., London. 263 pp. Price £ 1.65 (1972). – Review in Spaceflight, Vol. 14, 397; 1972 (*M. J. Anslow*).

003.052 **Relativistic celestial mechanics.** V. A. Brumberg.
Izdatel'stvo "Nauka". Glavnaya Redaktsiya Fiziko-
Matematicheskoj Literatury, Moskva. 384 pp. Price 1 Rbl. 94
Kop. (1972). In Russian. – Review in Referativ. Zhurn. 51.
Astron., 11.51.680 (1972).

003.053 **High energy physics. Vol. 5.**
E. H. S. Burhop (Editor).
Pure and Applied Physics, Vol. 25–5. Academic Press Inc.,
New York. 282 pp. Price $ 16.50 (1972).

003.054 **Collected scientific papers of Meghnad Saha.**
S. Chatterjee (Editor).
Sree Saraswaty Press, Calcutta, India. 414 pp. Price $ 18.00
(1970). – Review in Phys. Today, Vol. 25, No. 2, p. 55 - 56;
1972 (*S. G. Brush*).

003.055 **Stellar evolution.** Lectures at 3rd Summer Institute
for Astronomy and Astrophysics, SUNY at Stony
Brook, 18 June – 16 July 1969.
H.-Y. Chiu, A. Muriel (Editors).
The MIT Press, Cambridge, Mass.–London. 14 + 812 pp.
Price $ 15.00, £ 6.75 respectively (1972). – Review in Sky
Telescope, Vol. 44, 256 (1972).

003.056 **Einstein: The life and times.** R. W. Clark.
World Publishing Company, New York, N.Y. 718 pp.
Price $ 15.00 (1971). – Review in Phys. Today, Vol. 25, No.
11, p. 51 - 53; 1972 (*N. Balazs*).

003.057 **The Allende, Mexico, meteorite shower.**
R. S. Clarke, Jr., E. Jarosewich, B. Mason, J. Nelen,
M. Gomez, J. R. Hyde.
Smithsonian Contributions to the Earth Sciences, No. 5.
Smithsonian Institution Press, Washington, D.C. 4 + 53 pp.
Price $ 1.25 (1971). – Review in Journ. British Astron. Ass.,
Vol. 82, 390 - 391; 1972 (*K. B. Hindley*).

003.058 **Introducere în astronomie.** I. Curea.
Tipografia Universității din Timişoara, 399 pp.
(1971). – Review in Stud. Cerc. Astron., Vol. 17, 287; 1972
(*C. Drâmbă*).

003.059 **Scientific instruments of the seventeenth and
eighteenth centuries and their makers.**
M. Daumas.
B. T. Batsford Ltd., London; Praeger Publishers, Inc., New
York. 361 pp. + 142 plates. Price £ 10.00, $ 38.50 respectively
(1972). – Reviews in Nature, Vol. 239, 294; 1972 (*G. L'E.
Turner*); Sky Telescope, Vol. 44, 325 (1972).

003.060 **Methods in nonlinear plasma theory.**
R. C. Davidson.
Academic Press, New York – London. 285 pp. Price $ 18.50
(1971).

003.061 **Turbulence phenomena.** An introduction to the
eddy transfer of momentum, mass, and heat,
particularly at interfaces. J. T. Davies.
Academic Press, Inc., New York. 412 pp. Price $ 19.50 (1972).

003.062 **The view from space.** M. E. Davies, B. C. Murray.
Columbia University Press, New York. 12 + 163 pp.
Price $ 14.95 (1972). – Review in Space Science Rev., Vol.
14, 176; 1972 (*E. L. G. Bowell*).

003.063 **En bok om universum.** D. Dietz.
C. W. K. Gleerups Bokförlag, Lund. 149 pp. Price
Skr. 15.00 (1971). – Review in Astron. Tidssk., Årg. 5, p.
193 - 194; 1972 (*S. Wramdemark*).

003.064 **Atmospheric Optics. Vol. 2.**
N. B. Divari (Editor). Translated from the Russian
edition (Moscow, 1970) by F. L. Sinclair.
Consultants Bureau, New York. 6 + 164 pp. Price $ 27.50
(1972).

003.065 **The earth: Our physical environment.**
W. L. Donn.
John Wiley & Sons Ltd., Chichester. 512 pp. Price £ 5.70
(1972). – Review in Journ. British Interplanet. Soc., Vol. 25,
560 (1972).

003.066 **Robots in the sky.** D. Dwiggins.
Golden Gate Junior Books, San Carlos, Calif. 80 pp.
Price $ 4.95 (1972). – Review in Sky Telescope, Vol. 44,
325 (1972).

003.067 **Theory of fully ionized plasmas.** G. H. Ecker.
Academic Press, New York – London. 344 pp.
Price $ 19.50 (1972).

003.068 **The theory of altitudes in the earth's gravitational
field.** V. F. Eremeev, M. I. Yurkina.
Nedra, Moskva. 144 pp. Price 1 Rbl. 60 Kop. (1972).
In Russian. – Review in Referativ. Zhurn. 52. Geod.
Aehros"emka, 8.52.60 (1972).

003.069 **Motion of the earth's pole from 1890.0 to 1969.0.**
E. P. Fedorov, A. A. Korsun', S. P. Major, N. I.
Panchenko, V. K. Taradij, Ya. S. Yatskiv.
Naukova dumka, Kiev. 264 pp. Price 86 Kop. (1972).
In Russian. – Review in Referativ. Zhurn. 51. Astron., 2.51.
161 (1973).

003.070 **Johannes Kepler, Opera omnia in 8 Bänden. Vol. 1.**
C. von Frisch (Editor).
Verlag Dr. H. A. Gerstenberg, Hildesheim. 16 + 679 pp.
(1971). – Reprint of the first volume of the Kepler–Edition,
(Frankfurt/M. – Erlangen), 1858/71.

003.071 **An introduction to modern optics.**
A. K. Ghatak.
McGraw-Hill Book Company, New York. 368 pp. Price
$ 9.95 (1971).

003.072 **Dictionary of scientific biography, Vol. VI: J. Ha-
chette–J. Hyrtl.** C. C. Gillispie (Editor-in-chief).
Charles Scribner's Sons, New York. 13 + 619 pp. Price $ 35.00
(1972).

003.073 **Atmospheres.** R. M. Goody, J. C. G. Walker.
Prentice-Hall Inc., Englewood Cliffs, N.J., USA.
10 + 150 pp. Price $ 7.75, cloth; $ 2.95, paper, respectively
(1972). – Review in Sky Telescope, Vol. 44, 325 (1972).

003.074 **Astronomy and astrophysics for the 1970's. Vol. 1:
Report of the Astronomy Survey Committee.**
J. L. Greenstein.
National Academy of Sciences, Washington. 136 pp. Price
$ 4.75 (1972). – Reviews in Mercury (Journ. Astron. Soc.
Pacific), Vol. 1, No. 6, p. 8 (1972); Sky Telescope, Vol. 44,
391 - 396; 1972 (*J. B. Irwin*).

003.075 **Planetary quarantine: Principles, methods and
problems.** L. B. Hall (Editor).
Gordon and Breach, Science Publishers, Inc., New York –
London. 173 pp. Price £ 6.05 (1971).

003.076 **Atomspektren.** W. R. Hindmarsh.
Translated from English by R. Sube, and edited by
K. Meyer.

Wissenschaftliche Taschenbücher, WTB Vol. 76. Akademie-Verlag, Berlin; Pergamon Press, Oxford; Vieweg & Sohn, Braunschweig. 457 pp. Price DM 29.50 (1972).

003.077 Coronal expansion and solar wind.
A. J. Hundhausen.
Physics and Chemistry in Space, Vol. 5. Springer-Verlag, Berlin – Heidelberg – New York. 12 + 238 pp. Price DM 68.00 (1972). – Contents: 1.) History and background; 2.) The identification and classification of some important solar wind phenomena; 3.) The dynamics of a structureless coronal expansion; 4.) Chemical composition of the expanding coronal and interplanetary plasma; 5.) High-speed plasma streams and magnetic sectors; 6.) Flare-produced interplanetary shock waves; 7.) Concluding remarks.

003.078 The UFO experience: A scientific inquiry.
J. A. Hynek.
Henry Regnery Co., Chicago, Ill. 12 + 276 pp. Price $ 6.95 (1972). – Review in Nature, Vol. 239, 529; 1972 (*D. G. King-Hele*).

003.079 The sun and the ionosphere. Short-wave solar radiation and its effect on the ionosphere.
G. S. Ivanov-Kholodnyi, G. M. Nikol'skii.
Translated from Russian. Israel Program for Scientific Translations, Jerusalem. 5 + 366 pp. Price £ 13.15 (1972).

003.080 Optimization techniques in lens design.
T. H. Jamieson.
Monographs on Applied Optics No. 5. Adam Hilger Ltd., London. 106 pp. Price £ 5.00 (1971). – Review in Journ. British Astron. Ass., Vol. 83, 72; 1972 (*H. E. Dall*).

003.081 Point to the stars.
J. M. Joseph, S. L. Lippincott.
McGraw-Hill Book Company, New York. 96 pp. Price $ 4.72 (1972). – Review in Sky Telescope, Vol. 44, 398 (1972).

003.082 Elements of the information theory of atmospheric visibility. V. K. Kagan, K. Ya. Kondrat'ev.
Israel Program for Scientific Translations, Jerusalem. 9 + 140 pp. Price £ 5.00 (1971). – Review in Journ. British Astron. Ass., Vol. 83, 65; 1972 (*V. Barocas*).

003.083 Plasma astrophysics.
S. A. Kaplan, V. N. Tsytovich.
Nauka, Moskva. 440 pp. Price 2 Rbl. 23 Kop. (1972). In Russian. – Review in Referativ. Zhurn. 51. Astron., 8.51. 199 (1972).

003.084 Laboratory exercises for general astronomy.
L. Kelsey, D. Hoff, J. Neff.
University of Iowa, Department of Physics and Astronomy, Iowa City [Available from University of Iowa Campus Stores, Iowa City]. Variously paged, (168 pp.). Price $ 3.00 (1971). Review in Sky Telescope, Vol. 44, 325 (1972).

003.085 Taschenatlas der Sternbilder.
J. Klepešta, A. Rükl.
Verlag Dausien, Hanau/Main. 2. Edition, 292 pp. Price DM 5.80 (1972). – Review in Sternbote, 15. Jahrgang, p. 191 - 192; 1972 (*H. Mucke*).

003.086 Problems and methods of solar observations.
M. A. Klyakotko.
Nauka, Moskva. 59 pp. Price 32 Kop. (1971). In Russian.

003.087 La conquête des planètes (The conquest of the planets). P. Kohler.
Éditions Albin Michel, Paris. 288 pp. Price F 18.00 (1972).

Review in Spaceflight, Vol. 14, 436; 1972 (*J. Tiziou*).

003.088 Space research: Physical and technical principles.
M. G. Kroshkin.
Israel Program for Scientific Translations, Jerusalem. 314 pp. Price $ 6.00 (1972). – Review in Sky Telescope, Vol. 44, 398 (1972).

003.089 Dimensional analysis and group theory in astrophysics. R. Kurth.
Pergamon Press, Oxford – New York – Toronto – Sydney – Braunschweig. 14 + 235 pp. Price DM 83.25 (1972).

003.090 Naked eye stars. R. H. Lampkin.
Gall & Inglis, London. 2nd edition. 46 pp. Price 75 p. (1972). – Review in Journ. British Astron. Ass., Vol. 82, 391; 1972 (*P. Moore*).

003.091 Advances in geophysics, volume 15.
H. E. Landsberg, J. Van Mieghem.
Academic Press, London. 350 pp. Price £ 8.40 (1972). – Review in Journ. British Interplanet. Soc., Vol. 25, 622 (1972)

003.092 Specular reflection. E. P. Lavin.
Monographs on Applied Optics No. 2. Adam Hilger Ltd., London. 105 pp. Price £ 5.00 (1971). – Review in Journ. British Astron. Ass., Vol. 83, 70 - 71; 1972 (*H. E. Dall*).

003.093 Atomic and molecular structure: The development of our concepts. W. J. Lehman.
John Wiley & Sons Ltd., Chichester. 464 pp. Price £ 5.00 (1972). – Review in Journ. British Interplanet. Soc., Vol. 25, 621 (1972).

003.094 Moon rocks and minerals.
A. A. Levinson, S. R. Taylor.
Pergamon Press, Oxford – London. 222 pp. Price £ 5.00 (1972). – Review in Journ. British Interplanet. Soc., Vol. 25, 619 - 620; 1972 (*C. A. Cross*).

003.095 Die Sonnenuhren, Kunstwerke der Zeitmessung und ihre Geheimnisse. L. M. Loske.
Springer-Verlag, Berlin – Heidelberg – New York. 102 pp. Price DM 7.80 (1971). – Review in Sterne, 48. Jahrgang, p. 189 - 190; 1972 (*S. Marx*).

003.096 Metal dielectric multilayers. J. Macdonald.
Monographs on Applied Optics No. 4. Adam Hilger Ltd., London. 78 pp. Price £ 5.00 (1971). – Review in Journ. British Astron. Ass., Vol. 83, 71; 1972 (*H. E. Dall*).

003.097 Proceedings of the second conference on the origin of life. L. Margulis (Editor).
Gordon and Breach, Science Publishers, Inc., New York – London – Paris. 238 pp. Price £ 2.30 (1971). – Review in SuW, Vol. 11, 322 (1972).

003.098 Allgemeine Astronomie. Eine Einführung in die Wissenschaft von den großen Räumen. J. Meurers.
Verlag Rombach + Co. GmbH, Freiburg (Germany). 260 pp. Price DM 29.00 (1972).

003.099 1973 Yearbook of astronomy.
P. Moore (Editor).
W. W. Norton & Company, Inc., New York, N.Y. 247 pp. Price $ 5.95 (1972). – Review in Sky Telescope, Vol. 44, 399 (1972).

003.100 The new guide to the planets. P. Moore.
W. W. Norton & Company, Inc., New York, N.Y. 224 pp. Price $ 7.95 (1972).

003.101 **Planeterna och universum.** P. Moore.
Det Bästa, Stockholm. 192 pp. Price Skr. 139.50
(1972). – Review in Astron. Tidssk., Årg. 5, p. 152 - 153;
1972 (*G. Lyngå*).

003.102 **Challenge of the stars.** P. Moore, D. A. Hardy.
Mitchell Beazley Ltd., London. 63 pp. Price
Australian RMP $ 5.95; £ 2.25 respectively (1972). – Reviews
in Journ. Astron. Soc. Victoria, Vol. 25, 73; 1972 (*J. L. Perdrix*); Journ. British Astron. Ass., Vol. 83, 64; 1972 (*C. A. Ronan*); Spaceflight, Vol. 14, 397; 1972 (*A. T. Lawton*).

003.103 **Mars: The red world.**
P. Moore, with illustrations by D. A. Hardy.
World's Work Ltd., Tadworth. 48 pp. Price £ 1.20 (1971).
Review in Journ. British Astron. Ass., Vol. 82, 477; 1972
(*C. S. J. H. Daniel*).

003.104 **Astronomy from a space platform.**
G. W. Morgenthaler, H. D. Greyber (Editors).
Science and Technology Series, Vol. 28. American Astronautical Society, Publications Office, Tarzana, Calif. 15 + 398 pp.
Price $ 17.50 (1972). – Review in Sky Telescope, Vol. 44,
399 (1972).

003.105 **Planning challenges of the 70's in space.**
G. W. Morgenthaler, R. Morra (Editors).
Advances in the Astronautical Sciences, Vol. 26. American
Astronautical Society, Publications Office, Tarzana, Calif.
445 pp. Price $ 16.75 (1970). – Review in Icarus, Vol. 17,
231 - 232; 1972 (*L. H. Wasserman*).

003.106 **Essentials of astronomy.** L. Motz, A. Duveen.
Columbia University Press, New York – London.
8 + 712 pp. Price £ 6.50 (1971). – Review in Journ. British
Astron. Ass., Vol. 82, 477; 1972 (*C. A. Ronan*).

003.107 **Sonne, Mond und Sterne über dem Reich der Inka.**
R. Müller.
Verständliche Wissenschaft, Vol. 110. Springer-Verlag, Berlin –
Heidelberg – New York. 8 + 85 pp. Price DM 9.80 (1972).

003.108 **Egyptian astronomical texts, III: Decans, planets,
constellations and zodiacs.**
O. Neugebauer, R. A. Parker.
Brown University Press, Providence, R. I. 10 + 273 pp. Price
$ 45.00 (1969). – Review in Journ. History Astron., Vol. 3,
217; 1972 (*O. Gingerich*).

003.109 **Medieval chronicles and the rotation of the earth.**
R. R. Newton.
The Johns Hopkins University Press, Baltimore, MD. 10 +
826 pp. Price $ 15.00 (1972).

003.110 **The Titius-Bode law of planetary distances.**
Its history and theory. M. M. Nieto.
International Series of Monographs in Natural Philosophy,
Vol. 47. Pergamon Press, Oxford – New York – Toronto –
Sydney – Braunschweig. 12 + 161 pp. Price DM 40.70 (1972).

003.111 **Analytical theory of optimization in gravitational
fields.** V. S. Novoselov.
Leningr. un-t, Leningrad. 318 pp. Price 2 Rbl. 8 Kop. (1972).
In Russian.

003.112 **Fernrohr-Selbstbau.** H. Oberndorfer.
Verlag Sterne und Weltraum, Düsseldorf. SuW
Taschenbuch 1. 2nd revised edition. 175 pp. Price DM 11.60
(1972).

003.113 **Advances in space science and technology, Vol. 11.**
F. I. Ordway, III (Editor).
Academic Press, New York – London. 22 + 502 pp. Price
$ 29.50 (1972).

003.114 **An introduction to the ionosphere and magneto-
sphere.** J. A. Ratcliffe.
Cambridge University Press, London – New York. 10 + 256
pp. Price £ 4.00, $ 14.50 respectively (1972). – Reviews in
Journ. British Astron. Ass., Vol. 83, 65 - 66; 1972 (*V. Barocas*); Journ. British Interplanet. Soc., Vol. 25, 624 (1972);
Nature, Phys. Sci., Vol. 240, 72; 1972 (*N. M. Brice, W. E.
Swartz*).

003.115 **Black holes, gravitational waves and cosmology.**
M. Rees, R. Ruffini, J. Wheeler.
Gordon and Breach, Science Publishers, Inc., London – New
York. 330 pp. (1972).

003.116 **Russian in space.** E. Riabchikov (Editor).
George Weidenfeld & Nicolsons Ltd., London. 300
pp. Price £ 3.50 (1972). Review in Spaceflight, Vol. 14, 320;
1972 (*G. E. Perry*).

003.117 **New optics.** V. Ronchi.
L. S. Olschki Editore, Firenze. 141 pp. Price Lire
2500 (1971). – Review in Applied Optics, Vol. 11, 2390;
1972 (*S. Tolansky*).

003.118 **Earth sciences.**
S. K. Runcorn (Editor).
Applied Science Publishers Ltd., London. Vol. 1, 20 + 502,
Vol. 2, 16 + 539, Vol. 3, 16 + 499 pp. Price £ 20.00 the three
volumes (1971). – Review in Nature, Vol. 238, 417 - 418;
1972 (*V. A. Eyles*).

003.119 **Interplanetary flight and communication.** Vol. 3,
No. 8: Theory of space flight; No. 9: Astronaviga-
tion. Theory, annals, bibliography. N. A. Rynin.
Translated from the Russian edition. [Available from National
Technical Information Service, Springfield, Va.]. No. 8, 6 +
340 pp., N72 - 22809; No. 9, 4 + 236 pp., N72 - 22645, Price
each vol. paper $3.00, microfiche $95.00 (1971).

003.120 **Invisible astronomy.** C. A. Ronan.
J. B. Lippincott Company, Philadelphia, Pa. 173 pp.
Price $7.95 (1972).

003.121 **Zonal spectrophotometric standards. A study of
the energy distribution in absolute units in the
spectra of 109 stars.** V. M. Tereshchenko, A. V. Kharitonov.
Trudy Astrofiz. in-ta. AN KazSSR, Vol. 21, 185 pp. (1972).
In Russian.

003.122 **Radiation measurement of the temperature of
slightly heated bodies.**
V. G. Vafiadi, M. M. Miroshnikov (Editors).
Lenin Belorussian State University Press, Minsk. 195 pp.
Price 1 Rbl. 48 Kop. (1969). - Review in Applied Optics, Vol.
11, A18; 1972 (*C. W. Erickson*).

003.123 **Lunokhod I, a rolling laboratory on the moon.**
A. P. Vinogradov (Editor).
Nauka, Moskva. 128 pp. (1971). In Russian.

003.124 **Gravitation and cosmology: Principles and applications of the general theory of relativity.**
S. Weinberg.
John Wiley & Sons Inc., New York, 30 + 658 pp. Price $18.95
(1972). - Reviews in Phys. Today, Vol. 25, No. 3, p. 74

(1972); Science, Vol. 178, 786 (1972); Sky Telescope, Vol. 44, 325 (1972).

003.125 **Widmann-Schütte: Welcher Stern ist das?**
Franckh'sche Verlagshandlung/Kosmos-Verlag, Stuttgart. 18.enlarged edition. 180 pp. Price DM 14.80 (1972). - Reviews in Orion Schaffhausen, 30. Jahrgang, p. 191 (1972); SuW, Vol. 11, 322; 1972 (*H. Vehrenberg*).

003.126 **Passion to know: The world's scientists.**
M. Wilson.
Doubleday & Company, Inc., New York. Price $10.00 (1972). Review in Sci. American, Vol. 227, No.4 p. 121 - 122; 1972 (*P. Morrison*).

003.127 **The legacy of George Ellery Hale: Evolution of astronomy and scientific institutions, in pictures and documents.**
H. Wright, J. N. Warnow, C. Weiner (Editors).
The MIT Press, Cambridge, Mass.– London. 7 + 293 pp. Price $17.50 (1972). - Reviews in Journ. Roy. Astron. Soc. Canada, Vol. 66, 322 - 323; 1972 (*R. P. Broughton*); Phys. Abstr., Vol. 75, No. 84946 (1972); Sky Telescope, Vol. 44, 119 (1972).

003.128 **Origin and history of the earth. (Collection of articles).** M. Zh. Zhandaev (Editor).
Kazakhskij universitet, Alma-Ata. 293 pp. Price 1 Rbl. (1972). In Russian.

003.129 **Astrophysics of high energies.** T. C. Weekes.
Translated from the English edition. Mir, Moskva. 248 pp. Price 1 Rbl. 68 Kop. (1972). In Russian. – Review in Referativ. Zhurn. 51. Astron., 9. 51. 39 (1972).

003.130 **Physics of the solar system.** S. I. Rasool (Editor).
National Aeronautics and Space Administration. 523 pp. Price $3.75 (1972). – Review in Sky Telescope, Vol. 44, 324 - 325 (1972).

003.131 **Hamlet's mill. An essay on myth and the frame of time.** G. de Santillana, H. von Dechend.
Gambit, Inc., Boston, Ma. 25 + 505 pp. Price $10.00 (1969). Essay review in Journ. History Astron., Vol. 3, 206 - 211; 1972 (*C. Payne-Gaposchkin*).

003.132 **Galaxies.** H. Shapley, revised by P. W. Hodge.
Harvard University Press, Cambridge, Mass. 3. ed., 12 + 232 pp. Price $10.00 (1972). – Review in Sky Telescope, Vol. 44, 119 (1972).

003.133 **Statistical antenna theory.** Y. S. Shriftin.
The Golem Press, Boulder, Col. 370 pp. Price $16.00 (1971). – Review in IEEE Spectrum, Vol. 9, No. 7, p. 84 (1972).

003.134 **Dynamics of cometary atmospheres. Neutral gas.**
L. M. Shul'man.
Naukova dumka, Kiev. 244 pp. Price 1 Rbl. 47 Kop. (1972). In Russian. – Review in Referativ. Zhurn. 51. Astron., 8. 51. 303 (1972).

003.135 **On the moon with Apollo 16.** G. Simmons.
National Aeronautics and Space Administration.
EP - 95, Superintendent of Documents, U. S. Government Printing Office, Washington, D. C. Price $1.00, 1972. – Review in Sky Telescope, Vol. 44, 119 (1972).

003.136 **Lectures in scattering theory.** A. G. Sitenko.
Translated into English by P. J. Shepherd.
Vol. 39 of the Internat. Series of Monographs in Natural Phi-

losophy. Pergamon Press, Oxford - New York - Toronto - Sydney - Braunschweig. 269 pp. Price £5.00; $13.50 respectively (1971). – Review in Jenaer Rundschau, (Jena Rev.), 17. Jahrgang, p. 258; 1972 (*C. Hofmann*).

003.137 **Physics of the space environment.**
R. E. Smith, S. T. Wu (Editors).
National Aeronautics and Space Administration, Washington D.C. NASA SP - 305. [To be obtained from National Technical Information Service, Springfield, Va.], 171 pp. Price $3.00 (1972). – Review in Sky Telescope, Vol. 44, 398 - 399 (1972).

003.138 **Light scattering in planetary atmospheres.**
V. V. Sobolev.
Nauka, Moskva. 336 pp. Price 1 Rbl. 99 Kop. (1972). In Russian. – Review in Referativ. Zhurn. 51. Astron., 9. 51. 195 (1972).

003.139 **The ethereal aether.** L. S. Swenson, Jr.
University of Texas Press, Austin, Tx. 361 pp. Price $10.00 (1972). – Review in Sky Telescope, Vol. 44, 186 - 187; 1972 (*L. N. Cooper*).

003.140 **Aurorae and processes in the magnetosphere.**
S. I. Isaev, M. I. Pudovkin.
Nauka, Leningrad. 244 pp. Price 1 Rbl. 85 Kop. (1972). In Russian. – Review in Referativ. Zhurn. 62. Issled. kosmich. prostranstva, 2. 62. 184 (1973).

003.141 **Jahressternkarten.** K. Schütte.
Franckh'sche Verlagshandlung/Kosmos–Verlag, Stuttgart. 60 plates. Preis DM 7.80 (1972). – Review in Orion Schaffhausen, 30. Jahrgang, p. 192; 1972 (*K. Locher*); SuW, Vol. 11, 322 (1972).

003.142 **Mikołaj Kopernik.** S. Grzybowski.
Książkaj Wiedza, Warszawa, Poland. 332 pp. Price zł 15.00 (1972).

003.143 **Communication with extraterrestrial intelligence.**
Proceedings of the first international conference on extraterrestrial civilizations, 1971 Sept., Soviet Armenia.
C. Sagan (Editor).
Massachusetts Institute of Technology, Cambridge, Mass. Price $ 12.50, $ 3.95 (1972). – Review in Sci. American, Vol. 227, No. 3, p. 194 (1972).

003.144 **Physics of the moon and planets.** International symposium. Kiev, 15–22 October, 1968.
Astronomical Council of the USSR Academy of Sciences. Nauka, Moskva. 472 pp. Price 3 Rbl. 16 Kop. (1972). In Russian.

003.145 **Modern ideas on the moon.**
Institut kosmicheskikh issledovanij AN SSSR.
Nauka, Moskva. 127 pp. Price 1 Rbl. 61 Kop. (1972). In Russian.

003.146 **Selected works of H. Poincaré. Vol. 2. New methods of celestial mechanics. Topology. Number theory.**
Nauka, Moskva. 1000 pp. Price 4 Rbl. 51 Kop. (1972). In Russian.

003.147 **Space research 1969–70.**
Science Research Council, H.M.S.O. 122 pp. Price £ 1.20 (1972). – Review in Journ. British Interplanet. Soc., Vol. 25, 686; 1972 (*A. D. Farmer*).

003.148 **Guidelines for employment opportunities in astronomy.** Committee on Manpower and Employment.

American Astronomical Society, Princeton, N.J. 36 pp. (1972).

003.149 New technique in astronomy. Proceedings of the Conference of the Commission of Instrument making of the USSR Academy of Sciences. Sverdlovsk, 1970, July 1–3. Vypusk (No.) 4.
AN SSSR. Astronomicheskij sovet. Nauka, Leningrad. 84 pp. Price 70 Kop. (1972). In Russian.

003.150 Atlas & gazetteer of the near side of the moon. NASA, Washington, D.C. 538 pp., 404 Lunar Orbiter IV photographs, Sp-241, Price $ 15.00 (1971). — Review in Journ. British Interplanet. Soc., Vol. 25, 684; 1972 (*A. E. Roy*).

003.151 Annual report 1971 of the center for short-lived phenomena.
Smithsonian Institution, Office of Environmental Sciences. 310 pp. Price $ 5.00 (1972). — Review in Sky Telescope, Vol. 44, 256 (1972).

003.152 Apparatus for cosmic investigations. In-t kosmich issled. AN SSSR.
Nauka, Moskva. 267 pp. Price 1 Rbl. 83 Kop. (1972). In Russian.

003.153 Our world in space and time. Kingsland Camp, Vantage, 230 pp. Price $ 6.00 (1971). — Review in Sky Telescope, Vol. 44, 256 (1972).

003.154 Astronomy and astrophysics for the 1970'2, Vol. 1. Report of the Astronomy Survey Committee.
To be obtained from Printing and Publishing Office, National Academy of Sciences, Washington, D.C. (1972). — Review in Phys. Today, Vol. 25, No. 8, p. 69 - 70 (1972).

003.155 J. C. Poggendorff, Biographisch-literarisches Handwörterbuch der exakten Naturwissenschaften.
Vol. VIIb, Part 4, 2nd and 3rd number.
H. Salié (Editor).
Published by Akademie-Verlag, Berlin. 160 + 160 pp. Price DM 48.00 (1972).

003.156 The Milky Way. An elusive road for science. S. L. Jaki.
Science History Publications, New York. 368 pp. Price $14.95 (1972).

003.157 Galileo's intellectual revolution. The middle period, 1610 – 1632. W. R. Shea.
Science History Publications, New York. 200 pp. Price $ 15.95 (1972).

003.158 Zur vorläufigen Bahnbestimmung künstlicher Erdsatelliten. K. H. Ilk.

Zentralstelle f. Luft- u. Raumfahrtdokumentation u. -information, ZLDI, d. Dt. Forschungs- u. Versuchsanst. f. Luft- u. Raumfahrt e.V., München. 103 pp. (1972).

003.159 Tabellen von Mie-Streufunktionen. R. H. Giese.
Zentralstelle f. Luft- u. Raumfahrtkokumentation u. -information, ZLDI, d. Dt. Forschungs- u. Versuchsanst. f. Luft- u. Raumfahrt e.V., München. 149 pp. (1972).

003.160 Sonnenbeobachtung und -forschung mit Raketen. Zentralstelle f. Luft- u. Raumfahrtdokumentation u. -information, ZLDI, d. Dt. Forschungs- u. Versuchsanst. f. Luft- u. Raumfahrt e.V., München. 124 pp. (1972).

003.161 Der große Augenblick in der Astronomie. H.-W. Gaebert.
Loewes Verlag Ferdinand Carl KG, Bayreuth. 374 pp. (1972).

003.162 Deutsche und niederländische astronomische Instrumente des 11.–18. (elften bis achtzehnten) Jahrhunderts. E. Zinner.
Verlag C. H. Beck, München. 10 + 688 + 80 pp. (1972).

003.163 Das Taschenbuch vom Sternenhimmel. P. von Eynern.
Wilhelm Heyne Verlag, München. 236 pp. (1972).

003.164 Gravitation. E. G. Valens.
Umschau-Verlag, Frankfurt/Main. 164 pp. (1972).

003.165 Nomenklatur zur Beschreibung von Strahlungsmessungen und -rechnungen. F. Kasten.
Zentralstelle f. Luft- u. Raumfahrtdokumentation u. -information, ZLDI, d. Dt. Forschungs- u. Versuchsanst. f. Luft- u. Raumfahrt e.V., München. 34 pp. (1972).

003.166 Das Fischer-Lexikon: Astronomie. Fischer-Taschenbuch-Verlag, Frankfurt/Main. 363 pp. (1972).

003.167 Untersuchung der Emissionskontinua von Krypton und Xenon an Stosswellenplasmen. W. Weber.
Zentralstelle f. Atomkernenergie-Dokumentation, ZAED, Leopoldshafen. 110 pp. (1972).

003.168 Möglichkeiten der Streulichtanalyse zur Dichte- und Temperaturmessung in nichtstrahlenden Gasen. G. Schweiger.
Zentralstelle f. Luft- u. Raumfahrtdokumentation u. -information, ZLDI, d. Dt. Forschungs- u. Versuchsanst. f. Luft- u. Raumfahrt e.V., München. 32 pp. (1972).

003.169 Gestirnkunde in der Steinzeit. H. W. Pfannenmueller.
An der Bastei 6: [Selbstverlag], Neustadt (a.d.Aisch). 16 pp. (1972).

004 History of Astronomy, Chronology

004.001 **The uses of the alignments at Le Menec carnac.**
A. Thom, A. S. Thom.
Journ. History Astron., Vol. 3, 151 - 164 (1972).

We show that the layout of Le Menec alignments was such that they provided, for four of the backsights, a means of extrapolating to the position on the ground corresponding to the declination maximum from observations made on three successive nights. We cannot be certain that we have interpreted exactly the builders' intentions, but we show a method which could have given a sufficiently close approximation to the ideal extrapolation distance.

004.002 **Thomas A. Edison and infra-red astronomy.**
J. A. Eddy.
Journ. History Astron., Vol. 3, 165 - 187 (1972).

004.003 **The 'Abd al-A'imma astrolabe forgeries.**
O. Gingerich, D. King, G. Saliba.
Journ. History Astron., Vol. 3, 188 - 198 (1972).

004.004 **The Milky Way from Galileo to Wright.**
S. L. Jaki.
Journ. History Astron., Vol. 3, 199 - 204 (1972).

004.005 **Britain's first observatory?** D. J. Bryden.
Journ. History Astron., Vol. 3, 205 (1972). —Note.

004.006 **L'astrologie de Kepler: Le sens d'une réforme.**
G. Simon.
L'Astronomie, 86ᵉ année, p. 325 - 336 (1972).

004.007 **King Edward VI's defence of astronomy.**
B. Hellyer, H. Hellyer.
Journ. British Astron. Ass., Vol. 82, 362 - 366 (1972).

004.008 **Commentary of Khady-zade ar-Roumy on the astronomical treatise of Chagmini.** (A brief survey).
Z. A. Pashaev.
Trudy Samarkand. un-ta, 1972, vyp. (No.) 202, p. 202 - 204. In Russian. — Abstr. in Referativ. Zhurn. 51. Astron., 8.51.6 (1972).

004.009 **John Tebbutt, his observatory, and a probable nova.**
J. Ashbrook.
Sky Telescope, Vol. 44, 236, 240 (1972).

004.010 **De geschiedenis van de sterrenkunde (3).**
G. W. E. Beekman.
Hemel en Dampkring, Vol. 70, 216 - 218 (1972).

004.011 **History of the use of balloons in scientific experiments.** G. Pfotzer.
Space Science Rev., Vol. 13, 199 - 242 (1972). — Invited paper (see 012.002).

004.012 **Johannes Kepler and the new astronomy.**
O. Gingerich.
Quarterly Journ. Roy. Astron. Soc., Vol. 13, 346 - 373 (1972). George Darwin Lecture delivered on 1971 December 10.

004.013 **Heavenly harmony and earthly harmonics.**
D. G. King-Hele.
Quarterly Journ. Roy. Astron. Soc., Vol. 13, 374 - 395 (1972). Harold Jeffreys Lecture delivered on 10 December 1971.

004.014 **Zur Geschichte der Astronomie in Berlin im 16. bis 18. Jahrhundert. I. Eine Quellenübersicht.**
D. Wattenberg.
Sterne, 48. Jahrgang, p. 161 - 172 (1972).

004.015 **Vor 50 Jahren: Die IAU und die Frage der Kalenderreform.** P. Aufgebauer.
Sterne, 48. Jahrgang, p. 173 - 176 (1972).

004.016 **Die bedeutendsten Sternwarten vor 100 Jahren.**
J. Classen.
Sterne, 48. Jahrgang, p. 177 - 181 (1972).

004.017 **Le système copernicien jusqu'à Kepler.**
A. Hayli.
L'Astronomie, 86ᵉ année, p. 384 - 394 (1972).

004.018 **A propos de quel jour était-ce?** A. Hamon.
L'Astronomie, 86ᵉ année, p. 429 - 432 (1972).

004.019 **The practical application of astronomy in ancient and medieval Armenia.** B. E. Tumanyan.
Uch. zap. Erevan. un-t. Estestv. n., Vol. 3 (118), 43 - 52 (1971). In Armenian. — Abstr. in Referativ. Zhurn. 51. Astron., 9.51.4 (1972).

004.020 **A telescope with a history for Exeter.**
A. R. Hutchings.
Journ. British Astron. Ass., Vol. 82, 427 - 430 (1972).

004.021 **A northern California pictograph that may be another record of the Crab nebula supernova explosion.**
J. C. Brandt, S. P. Maran, M. M. Kennedy, R. S. Harrington.
Bull. American Astron. Soc., Vol. 4, 319 (1972). – Abstr. AAS.

004.022 **Possible use of Stonehenge.**
R. R. Newton, R. E. Jenkins.
Nature, Vol. 239, 511 - 512 (1972).

004.023 **De geschiedenis van de sterrenkunde (4).**
G. W. E. Beekman.
Hemel en Dampkring, Vol. 70, 252 - 255 (1972).

004.024 **About the observatory of Copernicus.**
J. Classen.
Sky Telescope, Vol. 44, 307, 312 (1972).

004.025 **De Kepler à Newton.** J. Lévy.
L'Astronomie, 86ᵉ année, p. 465 - 471 (1972).

004.026 **Giovanni Keplero nel quarto centenario della nascita.** F. Zagar.
Atti XV Riunione Soc. Astron. Italiana, Bologna 1971, (see 012.013), p. 5 - 27 (1972).

004.027 **The laws of motion.** G. J. Whitrow.
British Journ. History Sci., Vol. 5, 217 - 234 (1971). – Presidential address to the British Society for the History of Science.

004.028 **Georg Christoph Lichtenberg and the Opera inedita of Tobias Mayer.** E. G. Forbes.
Ann. Sci. (*GB*), Vol. 28, 31 - 42 (1972).

004.029 **De geschiedenis van de sterrenkunde (5).**
G. W. E. Beekman.
Hemel en Dampkring, Vol. 70, 291 - 294 (1972).

004.030 **An old Chinese way of finding the volume of a**

sphere.   T. Kiang.
Math. Gaz., Journ. Math. Ass., Vol. 56, (No. 396), 88 - 91 = Dunsink Obs. Repr., No. 64 (1972).

004.031   The visitation book of Dunsink observatory, 1791 to 1924.   P. A. Wayman.
Irish Astron. Journ., Vol. 10, 135 - 141 (1971).

004.032   Centenaries 1472–1972.   C. M. Botley.
Journ. British Astron. Ass., Vol. 83, 28 - 30 (1972).

004.033   Tycho's star.   P. G. A. Smith.
Journ. British Astron. Ass., Vol. 83, 50 - 52 (1972).
Report of the Historical Section of the British Astron. Ass.

004.034   Johann Bayer and his star atlas.   B. Hetherington.
Journ. British Astron. Ass., Vol. 83, 52 - 53 (1972).
Report of the Historical Section of the British Astron. Ass.

004.035   Uzbekistan's astronomy in the past 50 years.   V. P. Shcheglov.
Priroda, No. 12.72, p. 89 - 93 (1972). In Russian.

004.036   The astronomical establishment in 1570.   A. Romer.
Phys. Today, Vol. 25, No. 3, p. 9 (1972). – Letter.

004.037   Keplero e Galileo.   L. Rosino.
Accad. Nazionale Lincei, Celebrazioni Lincee 54: 'Giovanni Keplero' « Giornata Lincea », indetta nella ricorrenza del IV centenario della nascita, Roma, 20 novembre 1971 (*Estratto*), p. 63 - 73 = Oss. Astron. Padova, Comun. Rassegne, No. 97 (1972).

004.038   Copernicus – half a millenium in retrospect.   K. Hujer.
Publ. Astron. Soc. Pacific, Vol. 84, 640 - 641 (1972). – Abstr. Astron. Soc. Pacific.

004.039   Astronomische Datierung von Kunstwerken. I.   A. Beer.
Sterne, 48. Jahrgang, p. 227 - 245 (1972).

004.040   Le monument solaire de Bagneux.   L. Janin.
L'Astronomie, 86ᵉ année, p. 521 - 529 (1972).

004.041   Corrección del calendario solar en Xochicalco.   L. E. Arochi.
El Universo, Vol. 26, 112 - 113 (1972).

004.042   Le cadran « aux étoiles ».   L. Janin.
Orion Schaffhausen, 30. Jahrgang, p. 171 - 175 (1972).

004.043   De geschiedenis van de sterrenkunde (6).   G. W. E. Beekman.
Hemel en Dampkring, Vol. 70, 324 - 327 (1972).

004.044   Kepler und die „Kopernikanische Wende".   W. Gerlach.
Kepler Festschrift 1971, (see 003.010), p. 11 - 20 (1971).

004.045   Die betrachtende Kreatur im trinitarischen Kosmos.   W. Petri.
Kepler Festschrift 1971, (see 003.010), p. 64 - 98 (1971).
Free and selective translation of Books I and IV of Keplers "Epitome Astronomiae Copernicanae".

004.046   Die quantitative Beschreibung der Planetenbewegung von Johannes Kepler in seinem handschriftlichen Nachlaß.   V. Bialas.
Kepler Festschrift 1971, (see 003.010), p. 99 - 140 (1971).

004.047   Das Kepler-Gedächtnishaus in Regensburg.   W. Boll.
Kepler Festschrift 1971, (see 003.010), p. 166 - 174 (1971).

004.048   Über astronomische Beobachtung in Regensburg.   A. Menath.
Kepler Festschrift 1971, (see 003.010), p. 186 - 190 (1971).

004.049   Geschichte und Ergebnisse der meteorologischen Forschung in Regensburg.   K. Rocznik.
Kepler Festschrift 1971, (see 003.010), p. 191 - 243 (1971).

004.050   Ropé und Nutus in Keplers Astronomie.   H. M. Nobis.
Kepler Festschrift 1971, (see 003.010), p. 244 - 265 (1971).

004.051   Något om äldre tiders komettänkande.   U. R. Johansson.
Astron. Tidssk., Årg. 5, p. 157 - 167 (1972).

004.052   On the Senshū-Goyomi.   T. Watanabe.
Mem. Japan Astron. Study Ass., No. 17, Vol. 5, 1 - 4 (1971). In Japanese.

004.053   On Ochikochi Dooin.   S. Yokoyama.
Mem. Japan Astron. Study Ass., No. 17, Vol. 5, 65 - 89 (1971).   In Japanese.

004.054   Pre-analysis of the astronomical chart drawn on the wall of Takamatsu funeral mound found near Nara in Japan most recently.   S. Imoto.
Mem. Japan Astron. Study Ass., No. 18, Vol. 5, 97 - 104 (1972).   In Japanese.

004.055   On the Sensyū-Goyomi, again.   T. Watanabe.
Mem. Japan Astron. Study Ass., No. 18, Vol. 5, 105 - 108 (1972).   In Japanese.

004.056   Japanese investigations of "Juji-reki" in Kammon period.   A. Kodama.
Mem. Japan Astron. Study Ass., No. 18, Vol. 5, 165 - 174 (1972).   In Japanese.

004.057   History of astronomy.   Z. Paprotny.
Urania Kraków, Vol. 43, 281 - 284 (1972).
In Polish.

004.058   Die Goldhörner von Gallehus.   W. Hartner.
Bild der Wiss., [Deutsche Verlags-Anstalt (Stuttgart)], 9. Jahrgang, p. 1210 - 1216 (1972).
In 1639 and 1734 two gold horns dating from the early North Germanic era were found in Gallehus, Denmark. The author succeeded in interpreting the symbolic meaning of the figures and in deciphering the inscriptions. The horns had been made because of the solar eclipse of the year 413. The magic device was intended to prevent the imminent end of the world.

004.059   Sundials—the Common Vertical in N. W. Kent.   C. S. J. H. Daniel.
Reprinted from the Conference Handbook of The Institute of Craft Education and The College of Craft Education, 1972, p. 90 - 109.
A brief sketch history of the science and art of gnomonics of the art of dialling with particular reference to the Common Vertical Sundial, illustrated by examples in North West Kent/South East London.

004.060   On some mathematical and astronomical manuscripts of Nasir at-Din at-Tusi from the collection

of the Institute of Oriental Studies of the Uzbek Academy of Sciences.   Kh. Tllashev.
Izv. AN UzSSR. Ser. fiz.-mat. n., 1972, No. 4, p. 63 - 65. In Russian. – Abstr. in Referativ. Zhurn. 51. Astron., 1.51.4 (1973).

004.061   On a comment of Ansari on the astronomical treatise of Ali Kushchi "Risalaj dar falakiyat".
A. U. Usmanov.
Trudy Samarkand. un-ta imeni Alishera Navoi, 1972, vyp. (No.) 203, p. 47 - 52. In Russian. – Abstr. in Referativ. Zhurn. 51. Astron., 1.51.6 (1973).

004.062   Halley and the Traité de la lumière of Huygens: New light on Halley's relationship with Newton.
W. R. Albury.
Isis, [Smithsonian Institution, Washington, D.C.], Vol. 62, 445 - 468 (1971). – Abstr. in Zentralblatt Math. Grenzgebiete, Vol. 236, No. 01003 (1972).

004.063   Tobias Mayer's method for calculating the circumstances of a solar eclipse.   E. G. Forbes.
Ann. Sci., (GB), Vol. 28, 177 - 189 (1972).

004.064   Kepler and the comets.   N. Richter.
Wiss. Fortschritt, Vol. 21, 536 - 539 (1971) = Mitt. Karl-Schwarzschild-Obs. Tautenburg, No. 60 (1972). In German.

004.065   Das heliozentrische System in der griechischen, persischen und indischen Astronomie.
B. L. van der Waerden.
Neujahrsblatt Naturforsch. Ges. Zürich, [Kommissionsverlag Leemann AG, Zürich], No. 172, 55 pp. Price sFr 13.00 (1970). – Abstr. in Math. Rev., Vol. 42, No. 12 (1971).

004.066   Astronomical information of Zakharia Sarkavag.
B. E. Tumanjan, G. G. Georgobiani.
Byull. Abastumansk. Astrofiz. Obs., No. 43, p. 270 - 272 (1972). In Georgian.
   The astronomical information of an Armenian historian Z. Sarkavag (XVII century), inserted in his composition "Description of Armenian History", are commented. In particular the date of a bolide fall, which the historian witnessed himself, is refined. The appearance of a comet at the end of May 1679 is supposed to be a new observation, thus far unknown in the history of comet astronomy.

004.067   Application of the knowledge of Kepler's work to lunar astronomy.   D. V. Trifunović, Yu. A. Belyj.
Vasiona, Vol. 20, 49 - 55 (1972). In Serbo-Croatian.

004.068   Kopernikus und die Erfindung des Projektionsplanetariums.   H. Letsch.
Jenaer Rundschau, (Jena Rev.), 17. Jahrgang, p. 315 - 318 (1972).

004.069   Los nombres de Orión y sus estrellas.
A. Paluzíe Borrell.
Urania Barcelona, Año 56, No. 274, p. 185 - 196 (1971).

004.070   J. Kepler und die Harmonie der Welt.
A. Erck, P. Wengler.
Wiss. Zeitschr. Techn. Hochschule Ilmenau, Vol. 18, 123 - 138 (1972).

004.071   Die geistigen Vorläufer des heliozentrischen Planetensystems des Copernicus.   J. Hoppe.
Astron. in der Schule, 9. Jahrgang, p. 129 - 131 (1972).

004.072   Charakterisierung der astronomischen Leistungen des Copernicus.   O. Günther.
Astron. in der Schule, 9. Jahrgang, p. 131 - 133 (1972).

004.073   Der Kampf um die wissenschaftliche Beweisführung zur copernicanischen Lehre.   K.-G. Steinert.
Astron. in der Schule, 9. Jahrgang, p. 134 - 137 (1972).

004.074   Der Beitrag der copernicanischen Ideen zur Entwicklung eines wissenschaftlichen Weltbildes.
R. Wahsner.
Astron. in der Schule, 9. Jahrgang, p. 138 - 140 (1972).

004.075   Observing tower of the former Prague observatory.
J. Klepešta.
Říše hvězd, Vol. 53, 226 - 227 (1972).   In Czech.

004.076   The prehistoric stargazer.   K. Menzel.
Journ. Astron. Soc. Western Australia, 1972 October, p. 4 - 8, 1972 November, p. 2 - 7.

Lockyer's telescope.
Nature, Vol. 240, 176 (1972).

Astronomy in Toruń, Nicholas Copernicus' native town.   See Abstr. 003.026.

Nicolaus Copernicus and the heliocentric theory in Polish culture till the end of the eighteenth century. See Abstr. 003.045.

Études sur l'astronomie Indienne et sur l'astronomie Chinoise.   See Abstr. 003.047.

Scientific instruments of the seventeenth and eighteenth centuries and their makers. See Abstr. 003.059.

Hamlet's mill. An essay on myth and the frame of time.   See Abstr. 003.131.

The Milky Way. An elusive road for science. See Abstr. 003.156.

Galileo's intellectual revolution. The middle period, 1610 – 1632.   See Abstr. 003.157.

Deutsche und niederländische astronomische Instrumente des 11.–18. (elften bis achtzehnten) Jahrhunderts. See Abstr. 003.162.

Gestirnkunde in der Steinzeit.
See Abstr. 003.169.

Errata

004.901   Erratum: 'Heavenly harmony and earthly harmonics' [Quarterly Journ. Roy. Astron. Soc., Vol. 13, 374 - 395 (1972)].   D. G. King-Hele.
Quarterly Journ. Roy. Astron. Soc., Vol. 13, 594 (1972).

005 Biography

005.001 **Gerhard Herzberg–Nobel Laureate, 1971.**
H. L. Welsh.
Journ. Roy. Astron. Soc. Canada, Vol. 66, 183 - 188 (1972).

005.002 **Zukor, pintor del cielo.**   E. Lastra.
El Universo, No. 99, Vol. 26, 63 - 64 (1972).

005.003 **Flammarión.**   J. Rubi.
El Universo, No. 99, Vol. 26, 76 (1972).

005.004 **To Poland with love: U. S. brass, scientists warm to Copernicus fete.**   D. Shapley.
Science, Vol. 177, 683 - 684 (1972).

005.005 **Some centenaries of 1971.**   C. A. Ronan.
Quarterly Journ. Roy. Astron. Soc., Vol. 13, 396 - 402 (1972).
The purpose of this note is to draw attention to three Fellows of the Society who died a century ago: Augustus de Morgan (1792–1871), Sir John Herschel (1792–1871), Charles Babbage (1791–1871).

005.006 **Kepler, mathématicien et physicien.**
P. Costabel.
L'Astronomie, 86e année, p. 395 - 405 (1972).

005.007 **K. Eh. Tsiolkovskij and world science.**
N. G. Belova.
Zemlya i Vselennaya, 1972, No. 5, p. 62 - 65. In Russian.

005.008 **Gamow – and mathematics.**   S. M. Ulam.
G. Gamow Memorial Volume, (see 003.003), p. 272 - 279 (1972).

005.009 **George Gamow – an appreciation.**
M. M. Shapiro.
G. Gamow Memorial Volume, (see 003.003), p. 300 - 303 (1972).

005.010 **Memories of Gamow.**
R. A. Alpher, R. Herman.
G. Gamow Memorial Volume, (see 003.003), p. 304 - 313 (1972).

005.011 **Harlow Shapley – cosmographer and humanitarian.**
B. J. Bok.
Sky Telescope, Vol. 44, 354 - 357 (1972).

005.012 **E. P. Mason and the nebulae.**   J. Ashbrook.
Sky Telescope, Vol. 44, 366 - 367 (1972).

005.013 **Professor K. Kozieł – 40 years of research.**
K. Rudnicki.
Postępy Astron., Vol. 20, 365 - 366 (1972). In Polish.

005.014 **Profesors Kārlis Šteins – jubilārs.**   L. Roze.
Zvaigžņotā debess, 1971. gada rudens, p. 1 - 9.

005.015 **Pirmoreiz par Johanu Kepleru latviešu valodā.**
Z. Cirse.
Zvaigžņotā debess, 1971. gada rudens, p. 30.

005.016 **Ivans Žongolovičs.**   G. Chebotarev.
Zvaigžņotā debess, 1972. gada pavasaris, p. 32 - 35.

005.017 **Teodors Grothuss un viņa devums zinātnei.**
J. Stradiņš, I. Daube.

Zvaigžņotā debess, 1972. gada pavasaris, p. 39 - 47.

005.018 **Johannes Kepler als theologischer Denker.**
J. Hübner.
Kepler Festschrift 1971, (see 003.010), p. 21 - 44 (1971).

005.019 **Kepler und die Gegenreformation.**   M. List.
Kepler Festschrift 1971, (see 003.010), p. 45 - 63 (1971).

005.020 **Johannes Kepler zum 400. Geburtstag.**
W. Gerlach.
Verlag der Bayerischen Akademie der Wissenschaften, München. 24 pp. (1972). – Address given at the annual meeting of the Bayerische Akademie der Wissenschaften, Munich, 1971 December 4.

005.021 **Chronological table of the works of M. Harada.**
M. Harada.
Mem. Japan Astron. Study Ass., No. 17, Vol. 5, 90 - 92 (1971).   In Japanese.

005.022 **Nicolaus Copernicus. (7), (8), (9), (10).**
S. R. Brzostkiewicz.
Urania Kraków, Vol. 43, 229 - 238, 262 - 269, 294 - 301, 332 - 338 (1972). In Polish.

005.023 **"Nicolaus Copernicus" – the anniversary exposition in the Chorzów Planetarium and Observatory.**
L. Zajdler.
Urania Kraków, Vol. 43, 269 - 274 (1972). In Polish.

005.024 **Zum 400. Geburtstag von Johannes Kepler.**
F. Schmeidler.
Mitt. Sternw. München, Vol. 2, No. 1, 8 pp. (1972).

005.025 **Nicolaus Copernicus – Zeit, Leben und Werk.**
D. B. Herrmann.
Astron. in der Schule, 9. Jahrgang, p. 125 - 128 (1972).

005.026 **Gustav Gruss (1854 - 1922).**   V. Guth.
Říše hvězd, Vol. 53, 161 - 162 (1972). In Czech.

005.027 **Special Issue, dedicated to Ernst Julius Öpik to mark his 75th birthday.**
E. M. Lindsay, S. Grew (Editors).
Irish Astron. Journ., Vol. 10, Special Issue, 92 pp. (1972).
Contents: Ernst Julius Öpik, *E. M. Lindsay*; Classified list of astronomical publications; Collisions of meteoric atoms, *D. R. Bates*; Öpik's contributions to planetary physics, *S. F. Dermott*; News and comments by a great teacher, *D. Hoffleit*; Öpik's contributions to lunar studies, *Z. Kopal*; Öpik and multiple colour photometry, *G. E. Kron*; Ernst Öpik and meteoritics, *P. M. Millman*; Ernst Öpik's double star researches, *D. J. K. O'Connell*; A review of Öpik's contributions in the field of exospheres and interplanetary gas, *S. F. Singer*; Stellar statistics, *A. G. Velghe*; The phenomenon of the ice ages, *P. A. Wayman*; Ernst Öpik's research on comets, *F. L. Whipple*; Ernst Öpik and the structure of red giants, *M. H. Wrubel*; E. Öpik–composer, *M. de Vermond*; A hymn to humanity, autographed, *E. Öpik*.

005.028 **M. P. Barabashov (Life and work).**
V. J. Ehzers'kij, K. N. Kuz'menko, V. Kh. Pluzhnikov.
Visn. Kharkiv. Univ. No. 82, (Ser. Astron., No. 7), p. 5 - 11 (1972). In Ukrainian.

Astronomy in Toruń, Nicholas Copernicus' native town. See Abstr. 003.026.

Johannes Kepler. Beiträge zu seinem Lebensbild. See Abstr. 003.046.

Einstein: The life and times. See Abstr. 003.056.

Mikołaj Kopernik. See Abstr. 003.142.

A lunar crater dedicated to Luis Enrique Erro. See Abstr. 094.220.

006 Personal Notes

P. Ahnert, 75th birthday. H. Lambrecht. Sterne, 48. Jahrgang, p. 193 (1972).

H. Alfvén received the Franklin Medal. Phys. Today, Vol. 25, No. 1, p. 91 (1972).

H. Alfvén received the Kepler Gold Medal. Phys. Today, Vol. 25, No. 3, p. 93, 95 (1972).

M. Beyer received the Comet Medal of the Astronomical Society of the Pacific. G. Perkins. Mercury, (Journ. Astron. Soc. Pacific), Vol. 1, No. 4, p. 18 (1972).

M. Beyer received the Comet Medal of the Astronomical Society of the Pacific. SuW, Vol. 11, 263, 286 (1972).

B. J. Bok, president of the American Astronomical Society. Phys. Today. Vol. 25, No. 9, p. 72 (1972).

E. M. Burbidge, vice-president of the American Astronomical Society. Phys. Today, Vol. 25, No. 9, p. 72 (1972).

T. J. Deeming received the McIntyre Award. Monthly Notes Astron. Soc. Southern Africa, Vol. 31, 88 - 89 (1972).

B. H. Evans received the McIntyre Award. Monthly Notes Astron. Soc. Southern Africa, Vol. 31, 88 - 89 (1972).

D. S. Evans received the McIntyre Award. Monthly Notes Astron. Soc. Southern Africa, Vol. 31, 88 - 89 (1972).

G. B. Field, director of the Harvard College Observatory. Phys. Today, Vol. 25, No. 3, p. 95 (1972).

G. B. Field, director of Smithsonian Astrophysical Observatory. Sky Telescope, Vol. 44, 365 (1972).

H. Friedman received the Michelson Medal. Phys. Today, Vol. 25, No. 11, p. 81 (1972).

S. Goldfarb received the McIntyre Award.

Monthly Notes Astron. Soc. Southern Africa, Vol. 31, 88 - 89 (1972).

G. Haro, director of the National Institute of Astrophysics, Optics, and Electronics, Tonantzintla, Mexico. Mercury, (Journ. Astron. Soc. Pacific), Vol. 1, No. 4, p. 9 (1972).

P. Jordan, 70th birthday. J. Ehlers. Phys. Blätter, 28. Jahrgang, p. 468 - 469 (1972).

B. F. Kinahan received the Gold Medal of the Royal Astronomical Society of Canada for 1971. Journ. Roy. Astron. Soc. Canada, Vol. 66, 316 (1972).

G. Kuiper received the Kepler Gold Medal. Phys. Today, Vol. 25, No. 3, p. 93, 95 (1972).

P. Ledoux received the Eddington Medal. F. Hoyle. Observatory, Vol. 92, 78 - 79 (1972).

P. Ledoux received the Eddington Medal. F. Hoyle. Quarterly Journ. Roy. Astron. Soc., Vol. 13, 485 (1972).

Kam-Ching Leung, director of the Behlen Observatory, University of Nebraska, Lincoln. Mercury, (Journ. Astron. Soc. Pacific), Vol. 1, No. 4, p. 8 (1972).

B. Y. Levin received the Kepler Gold Medal. Phys. Today, Vol. 25, No. 3, p. 93, 95 (1972).

R. Lüst, president of Max-Planck-Gesellschaft. Phys. Blätter, 28. Jahrgang, p. 335 (1972).

R. Lüst, president of Max-Planck-Gesellschaft. SuW, Vol. 11, 227 (1972).

R. Lüst received the Guggenheim-Price. SuW, Vol. 11, 296 (1972).

D. Malacara, technical director of the National Institute of Astrophysics, Optics, and Electronics, Tonantzintla, Mexico. Mercury, (Journ. Astron. Soc. Pacific), Vol. 1, No. 4, p. 9 (1972).

E. J. Öpik received the Kepler Gold Medal.

D. E. Osterbrock, director of Lick Observatory.
Science, Vol. 177, 1179 (1972).

L. Perek, received la dixième médaille de l'ADION.
J.-C. Pecker.
L'Astronomie, 86ᵉ année, p. 366 - 367 (1972).

H.-U. Sandig, 60th birthday. K.-G. Steinert.
Wiss. Zeitschr. Techn. Univ. Dresden, Vol. 19, 87 - 88 (1970).

J. S. Shklovsky received the Bruce Gold Medal.
H. Weaver.
Mercury, (Journ. Astron. Soc. Pacific), Vol. 1, No. 4, p. 6 - 7 (1972).

H.I.S. Thirlaway received the Gold Medal of the Royal Astronomical Society. F. Hoyle.
Observatory, Vol. 92, 78 (1972).

H. I. S. Thirlaway received the Gold Medal of the Royal Astronomical Society. F. Hoyle.
Quarterly Journ. Roy. Astron. Soc., Vol. 13, 484 - 485 (1972).

H. Urey received the Kepler Gold Medal.
Phys. Today, Vol. 25, No. 3, p. 93, 95 (1972).

Phys. Today, Vol. 25, No. 3, p. 93, 95 (1972).

C. F. von Weizsäcker, 60th birthday.
W. Heisenberg.
Phys. Blätter, 28. Jahrgang, p. 319 - 321 (1972).

W. J. Welch, director of the Radio Astronomy Laboratory.
Mercury, (Journ. Astron. Soc. Pacific), Vol. 1, No. 5, p. 13 (1972).

W. J. Welch, director of the Radio Astronomy Laboratory at the University of California, Berkeley.
Phys. Today, Vol. 25, No. 11, p. 81, 83 (1972).

F. Whipple received the Kepler Gold Medal.
Phys. Today, Vol. 25, No. 3, p. 93, 95 (1972).

F. Zwicky received the Gold Medal of the Royal Astronomical Society.
Mercury, (Journ. Astron. Soc. Pacific), Vol. 1, No. 4, p. 8 (1972).

F. Zwicky received the Gold Medal of the Royal Astronomical Society. F. Hoyle.
Observatory, Vol. 92, 77 - 78 (1972).

F. Zwicky received the Gold Medal of the Royal Astronomical Society. F. Hoyle.
Quarterly Journ. Roy. Astron. Soc., Vol. 13, 483 - 484 (1972).

007 Obituaries

W. Anderfuhren died 1972 July 18. M. Roud.
Orion Schaffhausen, 30. Jahrgang, p. 187 (1972).

L. Arbey, 1908 - 1972 March 27. V. Maitre.
Ann. Françaises Chronométrie Micromécanique, Année 1972, p. 5.

Louis Arbey, 1908–1972 March 27.
V. Maître.
L'Astronomie, 86ᵉ année, p. 478 - 479 (1972).

G. S. Badalyan, 1908 - 1972 April 6.
Astron. Tsirk., No. 720, p. 7 - 8 (1972). In Russian.

M. P. Barabashov, 1894 March 30 – 1971 April 29.
Visn. Kharkiv. Univ. No. 82, (Ser. Astron., No. 7), p. 3 - 4 (1972). In Ukrainian.

W. M. Baxter, 1896 – 1971 December 9.
H. Hill.
Quarterly Journ. Roy. Astron. Soc., Vol. 13, 477 (1972).

E. Bazzi died 1972 November 4. M. Schürer.
Orion Schaffhausen, 30. Jahrgang, p. 188 (1972).

L. A. Brown, 1903 March 10 – 1971 Oct. 26.
J. L. White.
Quarterly Journ. Roy. Astron. Soc., Vol. 13, 478 (1972).

S. Chapman, 1888 January 29 - 1970 June 16.

G. Fanselau.
Gerlands Beiträge Geophys., Vol. 80, 273 - 276 (1971).

S. Chapman, 1888 Jan. 29 – 1970 June 14.
T. G. Cowling, V. C. A. Ferraro.
Quarterly Journ. Roy. Astron. Soc., Vol. 13, 464 - 476 (1972).

R. T. Cullen, 1886 April 12 - 1972 January 29.
L. S. T. Symms.
Quarterly Journ. Roy. Astron. Soc., Vol. 13, 592 (1972).

V. A. Dombrovskij died 1972 February 1.
Astron. Tsirk., No. 719, p. 6 - 7 (1972). In Russian.

A. B. F. Duncan died 1972 August 29.
C. R. Tolbert.
Phys. Today, Vol. 25, No. 11, p. 85 (1972).

A. B. F. Duncan died 1972 August 29.
Science, Vol. 178, 650 (1972).

W. J. Eckert, 1902 - 1971.
Celestial Mechanics, Vol. 6, 3 (1972).

F. J. Escalante died 1972 March 3.
A. G. Solis.
El Universo, No. 99, Vol. 26, 67 - 69 (1972).

V. G. Fesenkov died on 1972 March 12.
G. F. Sitnik.

Astron. Tsirk., No. 713, p. 4 - 7 (1972). In Russian.

V. G. Fesenkov, 1889 Jan. 13 - 1972 March 12.
Meteoritika, vyp. (No.) 31, p. 162 - 163 (1972). In Russian.

V. G. Fesenkov died 1972 March 12.
Trudy Astrofiz. Inst., *Alma-Ata*, Vol. 19, 129 (1972). In Russian.

V. G. Fesenkov, 1889 January 13 — 1972 March 12.
Vestn. AN KazSSR, 1972, No. 4, p. 67. In Russian.

K. A. Grigoryan, 1928 June 2 — 1971 August 29.
Astron. Tsirk., No. 719, p. 7 - 8 (1972). In Russian.

M. Humason, 1891 - 1972, June 18.
Mercury, (Journ. Astron. Soc. Pacific), Vol. 1, No. 5, p. 12 (1972).

M. L. Humason died 1972 June 18.
Phys. Today, Vol. 25, No. 10, p. 59 (1972).

M. L. Humason, 1891 Aug. 19 - 1972 June 18.
Sky Telescope, Vol. 44, 71, 87 (1972).

M. G. J. Minnaert, 1893 February 12 — 1970 October 26.   C. de Jager.
Utrechtse Sterrekundige Overdrukken, No. 160, 6 pp.(1971).

H. Shapley, 1885 November 2 - 1972 October 20.
Nature, Vol. 240, 429 - 430 (1972).

H. Shapley, 1885 - 1972 October 20.
E. Wiedemann.
Orion Schaffhausen, 30. Jahrgang, p. 186 (1972).

H. Shapley, 1885 Nov. 2 - 1972 Oct. 20.   B. J. Bok.
Sky Telescope, Vol. 44, 354 - 357 (1972).

H. Shapley, 1885 Nov. 2 - 1972.
SuW, Vol. 11, 296 (1972).

J. Q. Stewart died 1972 March 19.
J. W. Stewart.
Phys. Today, Vol. 25, No. 6, p. 75 (1972).

J. Q. Stewart died 1972 March 19.
Science, Vol. 177, 733 (1972).

P. Tardi, 1897 June 4 - 1972 August 5.
Bull. Géod., Nouvelle Sér., Année 1972, No. 106, p. 378 - 381.

P. Tardi, 1897 June 4 - 1972 August 5.
Comptes Rendus Acad. Sci. Paris, Vol. 275, Vie Académique, 122 - 123 (1972).

P. Tardi died 1972 August 5.
B. Morando.
L'Astronomie, 86e année, p. 440 (1972).

Y. Väisälä, 1891 September 6 — 1971 July 21.
L. Oterma.
Procès-Verbaux Comité International des Poids et Mesures, 2e Sér., Vol. 39, 129 - 133 (1971).

Y. Väisälä, 1891 September 6 — 1971 July 21.
L. Oterma.
Sci. Revuo, Beograd, Vol. 22, 133 - 140 = Astron.-Optika Inst. Univ. Turku, Informo No. 37/I (1971). In Esperanto.

S. V. Voroshilova-Romanskaya (1886 - 1969).
L. D. Kostina.
Zemlya i Vselennaya, 1972, No. 4, p. 50 - 51. In Russian.

H. Zanstra, 1894 November 3 - 1972 October 2.
D. Koelbloed.
Hemel en Dampkring, Vol. 70, 313 - 315 (1972).

H. Zanstra died 1972 October 2.   D. S. Evans.
Monthly Notes Astron. Soc. Southern Africa, Vol. 31, 140 - 141 (1972).

Reports, communications and publications of observatories and astronomical institutes are recorded in this section; included are numbered series of reprints. Whenever possible, the numbers of the abstracts referring to the publications are given. Observatories and institutes are listed in alphabetical order of their towns. In some cases observatory publications do not give the name of the town; the following list which gives names and towns of some institutions may serve as an aid in such cases.

Aarne Karjalainen Observatory	Oulu, Finland
Algonquin Radio Observatory	Lake Traverse, Ontario, Canada
Allegheny Observatory	Pittsburgh, Pennsylvania
Archenhold-Sternwarte	Berlin-Treptow, Germany
Arthur J. Dyer Observatory	Nashville, Tennessee
Astronomical Latitude Station, Polish Academy of Sciences	Borowiec, Poland
Bosscha Observatory	Lembang, Indonesia
Boyden Observatory	Bloemfontein, South Africa
Bureau International de l'Heure	Paris, France
Cajigal Observatory	Caracas, Venezuela
California Institute of Technology	Pasadena, California
Cape of Good Hope	Cape Town, South Africa
Carter Observatory	Wellington, New Zealand
Catalina Station	Tucson, Arizona
Cavendish Laboratory	Cambridge, England
Ceskoslovenská Akademie Ved Astronomický Ustav	Praha, Czechoslovakia
Chamberlin Observatory, University of Denver	Denver, Colorado
Commonwealth Observatory	Canberra, Australia
Corralitos Observatory	Las Cruces, New Mexico
David Dunlap Observatory, University of Toronto	Richmond Hill, Ontario
Dearborn Observatory	Evanston, Illinois
Department of Astronomy and Observatory, Univ. California	Los Angeles, California
Department of Astronomy, University of Texas	Austin, Texas
Division Radiophysics, C.S.I.R.O. University Grounds	Sydney, N.S.W., Australia
Dominion Astrophysical Observatory	Victoria, British Columbia
Dominion Observatory	Ottawa, Ontario
Dominion Radio Astrophysical Observatory	Penticton, British Columbia
Dudley Observatory	Albany, New York
Dunsink Observatory	Dublin, Ireland
Engelhardt Observatory	Kazan, R.S.F.S.R.
European Southern Observatory	Hamburg, Federal German Republic
Five College Observatories	Amherst, Massachusetts
Florida State University Radio Observatory	Tallahassee, Florida
Flower and Cook Observatories, University of Pennsylvania	Philadelphia, Pennsylvania
Fraunhofer Institut	Freiburg, Federal German Republic
Georgetown Observatory	Washington, D.C.
Goddard Space Flight Center	Greenbelt, Maryland
Goethe Link Observatory, University of Indiana	Bloomington, Indiana
Hale Observatories	Pasadena, California
Harvard College Observatory	Cambridge, Massachusetts
Harvard Radio Astronomy Station	Cambridge, Massachusetts
Haystack Observatory	Westford, Massachusetts
Heinrich-Hertz-Institut	Berlin, Germany
High Altitude Observatory, University of Colorado	Boulder, Colorado
Institute for Astronomy, University of Hawaii	Honolulu, Hawaii
Institute for Theoretical Astronomy (Institut Teoreticheskoj Astronomii)	Leningrad, R.S.F.S.R.
Institute of Theoretical Astrophysics, Blindern	Oslo, Norway
Inter-American Observatory	Cerro-Tololo, (La Serena), Chile
International Latitude Observatory	Mizusawa, Japan
Joint Institute for Laboratory Astrophysics (JILA)	Boulder, Colorado
Kandilli Observatory	Istanbul, Turkey
Kansas University Observatory	Lawrence, Kansas
Kapteyn Astronomical Laboratory	Groningen, Netherlands
Karl-Schwarzschild-Observatorium	Tautenburg, German Democratic Republic
Kenneth Mees Observatory	Rochester, New York
Kwasan Observatory	Kyoto, Japan
Lamont-Hussey Observatory	Bloemfontein, South Africa
Leander McCormick Observatory University of Virginia	Charlottesville, Virginia
Lee Observatory	Beirut, Lebanon
Leopold-Figl-Observatorium	Wien, Austria
Leuschner Observatory	Berkeley, California
Lick Observatory	Santa Cruz, (Mount Hamilton), California
Lindheimer Astronomical Research Center	Evanston, Illinois
Lockheed Solar Observatory	Saugus, California
Lohrmann-Observatorium für Geodätische Astronomie	Dresden, German Democratic Republic
Louisiana State University Observatory	Baton Rouge, Louisiana
Lowell Observatory	Flagstaff, Arizona
Lunar and Planetary Laboratory	Tucson, Arizona
Max-Planck-Institut für Astronomie	Heidelberg, Federal German Republic
Max-Planck-Institut für Phyik und Astrophysik	München, Federal German Republic
Max-Planck-Institut für Radioastronomie	Bonn, Federal German Republic
McDonald Observatory	Fort Davis, Texas
McMath Hulbert Observatory	Pontiac, Michigan
Michigan State University Observatory	East Lansing, Michigan
Molonglo Radio Observatory, University of Sydney	Sydney, New South Wales
Mount Cuba Observatory	Wilmington, Delaware
Mount John Observatory	Lake Tekapo, New Zealand
Mount Palomar Observatory	Pasadena, California
Mount Wilson Observatory	Pasadena, California
Mullard Radio Astronomy Observatory	Cambridge, England
Narrabri Observatory, University of Sydney	Sydney, New South Wales

National Bureau of Standards   **Washington**, D. C.
National Observatory,USA   **Kitt Peak**, Arizona
National Radio Astronomy   **Charlottesville**, Virginia
   Observatory   **Green Bank**, West Virginia
   Tucson, Arizona
New Mexico State
   University Observatory   **Las Cruces**, New Mexico
Nizamiah Observatory   **Hyderabad**, India
Nuffield Radio Astronomy
   Laboratories, Jodrell Bank
   University of Manchester   **Manchester**, England
Observatoire Royal de Belgique   **Uccle**, Belgium
Observatorio de Cartuja   **Granada**, Spain
Observatorio del Ebro   **Tortosa**, Spain
Observatory Fabra   **Barcelona**, Spain
Observatory, University of
   Michigan   **Ann Arbor**, Michigan
Ohio State University
   Radio Observatory   **Columbus**, Ohio
Ole Roemer-Observatoriet   **Aarhus**, Denmark
Owens Valley Radio   **Pasadena**, California
   Observatory
Perkins Observatory, Ohio State
   and Wesleyan Universities   **Delaware**, Ohio
Purple Mountain Observatory   **Nanking**, China
Radcliffe Observatory   **Pretoria**, South Africa
Remeis-Sternwarte   **Bamberg**,
   Federal German Republic
Republic Observatory   **Johannesburg**, South Africa
Rosemary Hill Observatory   **Gainesville**, Florida
Royal Radar Establishment,
   Radio Astronomy Division   **Malvern**, England

Sagamore Hill Radio Observatory **Bedford**, Massachusetts
Saint-Michel, l'Observatoire   **Haute Provence**, France
San Fernando Observatory   **El Segundo**, California
Smithsonian Astrophysical
   Observatory   **Cambridge**, Massachusetts
Specola Astronomica Vaticana   **Castel Gandolfo**, Italy
Specola di Padova   **Asiago**, Italy
Sproul Observatory   **Swarthmore**, Pennsylvania
Sternberg Observatory   **Moscow**, R.S.F.S.R.
Steward Observatory,
   University of Arizona   **Tucson**, Arizona
United States Naval Observatory   **Washington**, D.C.
University of Florida,
   Radio Observatory   **Gainesville**, Florida
University of Illinois Observatory **Urbana**, Illinois
University of Michigan
   Observatories   **Ann Arbor**, Michigan
University of South Florida
   Observatory   **Tampa**, Florida
Uttar Pradesh State Observatory **Naini Tal**, India
Van Vleck Observatory   **Middletown**, Connecticut
Wallace Observatory   **Cambridge**, Massachusetts
Warner and Swasey Observatory   **Cleveland**, Ohio
Washburn Observatory   **Madison**, Wisconsin
West Melton Observatory   **Christchurch**, New Zealand
Yale University Observatory   **New Haven**, Connecticut
Yerkes Observatory   **Williams Bay**, Wisconsin
Zentralinstitut für Astrophysik,
   Sternwarte Babelsberg, (Fach-
   bereich Kosmische Physik)   **Potsdam-Babelsberg**, German
   Democratic Republic

008.001   Abastumani

     Chronicle.   T. M. Borchkhadze.
Astron. Tsirk., No. 672, p. 8 (1972). In Russian.

     Chronicle (1971).
Byull. Abastumansk. Astrofiz. Obs., No. 43, p. 273 - 275
(1972). In Russian.

     **Abastumanskaya Astrofizicheskaya Observatoriya,
Gora Kanobili, Byulleten'**, Akademiya Nauk Gruzinskoj SSR,
Nos. **41** (08.012.018), **42** (08.012.019; N. M. Martsvaladze,
L. M. Fishkova, 08.082.178; L. M. Fishkova, 08.082.179),
43 (N. L. Magalashvili, J. I. Kumsishvili, 08.122.134; I. F.
Alania, 08.122.135; A. E. Vasilevsky, 08.064.071; V. I.
Voroshilov, N. B. Kalandadze, V. I. Kuznetsov, 08.131.123;
V. I. Voroshilov, N. B. Kalandadze, V. I. Kuznetsov, 08.131.
124; M. S. Kazanasmas, 08.131.125; G. N. Salukvadze, 08.
117.024; R. M. West, 08.114.172; Ts. S. Khetsuriani, 08.073.
112; E. I. Tetruashvili, 08.074.106; Ts. S. Khetsuriani, E. I.
Tetruashvili, R. I. Kiladze, 08.074.107; Ts. S. Khetsuriani, E.
I. Tetruashvili, 08.074.108; V. M. Iskandarova, 08.082.231;
A. Sh. Khatisov, 08.098.082; Sh. A. Sabashvili, 08.063.043;
R. M. Dzigvashvili, 08.151.052; V. T. Khukhunaishvili, 08.042
093; Z. D. Mestiashvili, 08.034.144; Z. D. Mestiashvili, 08.034.
145; B. E. Tumanjan, G. G. Georgobiani, 08.004.066).

008.002   Alma Ata

     Astronomy in the Kazakh SSR.   G. M. Idlis.
Zemlya i Vselennaya, 1972, No. 5, p. 54 - 59. In Russian.

     Akademiya Nauk Kazakhskoj SSR. **Trudy Astrofi-
zicheskogo Instituta**, *Alma-Ata*, Vol. 19 (G. M. Idlis, 08.003.

013), 22 (A. V. Kharitonov, E. A. Glushkova, L. N. Knyazeva,
N. N. Morozova, V. T. Rebristyj, T. V. Solodovnikova, V. M.
Tereshchenko, L. D. Frishberg, 08.114.161).

008.003   Asiago

     **Contributi dell'Osservatorio Astrofisico dell'Univer-
sità di Padova in Asiago**, Nos. 253 (F. Bertola, F. Ciatti,
05.125.104), 254 (F. Ciatti, R. Barbon, 06.125.006), 255
(F. Ciatti, L. Rosino, F. Bertola, 06.125.101), 256 (F. Bertola,
F. Lucchin, E. Nasi, 06.158.130), 257 (S. M. Hassan, 07.153.
017), 258 (M. Perinotto, 06.132.004), 259 (C. Barbieri, F.
Bertola, 07.141.114), 260 (F. Ciatti, A. Mammano, 08.124.
103), 261 (S. M. Hassan, 08.153.003), 262 (R. Barbon, S.
D'Odorico, 08.158.003), 263 (R. Barbon, 08.158.004), 264
(F. Bertola, H. Arp, 04.125.101), 265 (F. Ciatti, A. Mammano,
L. Rosino, 07.114.078), 266 (L. Pigatto, L. Rosino, 07.122.
080), 267 (D. A. Klinglesmith, P. L. Bernacca, H. Frey, 07.
114.081), 268 (C. Barbieri, L. Rosino, 07.113.032), 269
(P. L. Bernacca, F. Ciatti, 08.114.016), 270 (M. Perinotto,
K. Wurm, 06.132.018).

008.004   Auckland

     Auckland Observatory.
Inform. Bull. Southern Hemisph., No. 21, p. 17 (1972). – Current research report.

008.005   Bamberg

     Besuch der Remeis-Sternwarte in Bamberg, BRD.

P. Reinhard.
Sternenbote, 15. Jahrgang, p. 150 - 152 (1972).

Die Dr. Remeis-Sternwarte Bamberg und die "Veränderlichen Sterne" als die Objekte ihrer Forschung.
See Abstr. 120.012.

Veröffentlichungen der Remeis-Sternwarte Bamberg, Astronomisches Institut der Universität Erlangen–Nürnberg, Vol. 8, Nos. 92 (H. Mauder, 07.121.004), 99 (W. Strohmeier, 07.123.012), Vol. 10, Nos. 101 (F. M. Sosna, 07.123.024), 102 (W. Strohmeier, 07.121.055), 103 (D. H. Martins, 08.123.026).

008.006 Beirut

Lee Observatory, American University of Beirut, Lebanon, Monthly Bulletin, Astronomical Section, 1972 April - September (F. Bruin, H. Hourani, N. G. Bustati, 08.075 033).

008.007 Belo Horizonte

Instituto de Ciencias Exatas da Universidade Federal de Minas Gerais. (Institute of Exact Sciences of the Federal University of Minas Gerais).
Inform. Bull. Southern Hemisph., No. 20, p. 13 (1972).
Current research report.

Instituto de Ciencias Exatas da Universidade Federal de Minas Gerais (Institute of Exact Sciences of the University of Minas Gerais).
Inform. Bull. Southern Hemisph., No. 21, p. 8 - 9 (1972).
Current research report.

008.008 Berlin

Heinrich-Hertz-Institut. Solare Beobachtungsergebnisse. Deutsche Akademie der Wissenschaften zu Berlin, Zentralinstitut für Solar-Terrestrische Physik, Berlin-Adlershof. HHI Solar Data, Vol. 23, 1972 March – August (C.-U. Wagner, A. Böhme, F. Fürstenberg, D. Scholz, S. Böhm, 08.075.024).

008.009 Berlin - Treptow

Veröffentlichungen der Wilhelm-Foerster-Sternwarte Berlin, Nos. 28 (H. Heuseler, H. T. Piehl, AJB 68, 84.32), 29 (H. Heuseler, AJB 68, 84.33), 30 (H. Heuseler, 02.097.037).

008.010 Besançon

Annales de l'Observatoire de Besançon, Université de Besançon, Vol. 8, Fasc. 5 (F. Puel, 08.098.037; M. Sauzeat, 05.034.006; F. Barlier, C. Berger-Jaeck, J. L. Falin, J. P. Lespes, 05.082.025; M. Sauzéat, 05.031.011; J. Cottet, B. Prud'homme, 08.031.055; J. P. Lespés, J. L. Falin, M. Ill, 06.082.055; V. Maitre, 08.041.035; G. Hilaire, 07.094.082).

008.011 Bloemfontein

Boyden Observatory.   A. H. Jarrett.
Inform. Bull. Southern Hemisph., No. 20, p. 1 - 4 (1972).

Boyden Observatory.   A. H. Jarrett.
Inform. Bull. Southern Hemisph., No. 20, p. 19 (1972).
Current research report.

Boyden Observatory.   A. H. Jarrett.
Inform. Bull. Southern Hemisph., No. 21, p. 18 (1972). – Current research report.

Boyden Observatory. A brief history of the observatory.   A. H. Jarrett.
Irish Astron. Journ., Vol. 10, 129 - 134 (1971).

008.012 Bogotá

Publicaciones del Observatorio Astronómico Nacional, Universidad Nacional de Colombia, Facultad de Ciencias, *Bogotá,* No. 5 (M. H. Ibañez S., 08.072.058).

008.013 Bologna

Laboratorio di Tecnologia e Studio delle Radiazioni Extraterrestri, Bologna.   D. Brini.
Ricerca Scientifica, Anno 40, p. 269 - 271 (1970).

008.014 Bonn

Mitteilungen der Astronomischen Institute Bonn, Nos. 129 (M. Grewing, H. Heintzmann, 07.143.064), 131 (M. Grewing, U. Mebold, K. Rohlfs, 08.131.128), 139 (C. Wulf-Mathies, 08.082.144), 140 (07.008.026).

Veröffentlichungen der Astronomischen Institute Bonn, No. 85 (M. Römer, 08.082.234).

Max-Planck-Institut für Radioastronomie, Bonn.
O. Hachenberg, P. G. Mezger, R. Wielebinski.
Naturwissenschaften, 59. Jahrgang, p. 627 - 629 (1972).
Report on activities 1970 – 1971.

Max-Planck-Institut für Radioastronomie, Bonn.
Sonderdrucke, Nos. 41 (U. Mebold, 07.131.115), 43 (P. Stumpff, 07.033.027; H. G. Girnstein, W. Voss, 07.033.028; J. Schraml, 07.033.029), 45 (P. Stumpff, 08.103.105), 46 (M. Grewing, M. Walmsley, 07.131.122), 47 (R. Wielebinski, W. E. Wilson, 07.033.025), 48 (O. Hachenberg, P. G. Mezger, R. Wielebinski, 07.008.026), 49 (E. Fürst, 08.074.014), 50 (F. F. Gardner, J. B. Whiteoak, 07.131.131), 51 (M. A. F. Thiel, 08.141.060), 53 (A. v. Kap-herr, H. J. Wendker, 08.131.019), 54 (G. V. Schultz, W. Wiemer, 08.114.024), 55 (T. L. Wilson, 08.131.009), 57 (F. F. Gardner, J. B. Whiteoak, 08.141.053), 58 (A. H. M. Martin, D. Downes, 08.065.015), 59 (J. Barsuhn, 08.022.068), 60 (I. I. K. Pauliny-Toth, K. I. Kellermann, M. M. Davis, E. B. Fomalont, D. B. Shaffer, 07.141.118), 61 (R. Wohlleben, H. Mattes, O. Lochner, 08.033.116), 62 (J. B. Whiteoak, F. F. Gardner, 08.131.031).

008.015 Borowiec

Polish Academy of Sciences, Astronomical Latitude Station, Borowiec, Circular, Nos. 121 - 123 (08.044.030).

008.016 Bruxelles

Astrofysisch Instituut, Vrije Universiteit, Brussel, Nos. 16 (E. P. J. van den Heuvel, AJB 66, 71.30), 17 (E. P. J. van den Heuvel, AJB 66, 71.29), 20 (R. Hendrickx, AJB 66, 64.29), 21 (R. Hendrickx, AJB 66, 64.30), 35 (W. van Rensbergen, AJB 67, 67.61), 36 (W. van Rensbergen, AJB 67, 67.62), 37 (C. de Jager, L. Neven, AJB 68, 64.52), 39 (C. de Jager, AJB 68, 66.58), 40 (C. de Jager, AJB 68, 64.54), 41 (C. de Jager, AJB 68, 23.34), 58 (B. Asselbergh, W. van Rensbergen, 01.133.017), 59 (W. van Rensbergen, 01.022.063), 62 (C.de Loore, C. de Jager, 04.064.020), 63 (C. de Jager, C. de Loore, 04.073.003), 65 (C.de Jager, 08.071.050), 69 (C. de Loore, 05.074.041), 71 (W. van Rensbergen, J. Wuyts, 04.133.009), 72 (W. van Rensbergen, 06.022.155), 73 (W. van Rensbergen, 05.022.077), 74 (C. de Jager, C. de Loore, 05.064.036), 78 (W. van Rensbergen, 07.022.085), 79 (C. de Jager, L. Neven, 03.071.002).

008.017 Bucarest

Observations solaires, Académie de la République Socialiste de Roumanie, Observatoire de Bucarest, Secteur Solaire. Rotations 1570 - 1582 (C. Popovici, E. Ţifrea, V. Dinulescu, A. Dimitriu, S. Nicolescu, G. Mariş, I. Niţă, 08. 075.030).

008.018 Buenos Aires

Departamento de Física, Facultad de Ciencias Exactas y Naturales, Universidad de Buenos Aires. (Department of Physics, School of Exact and Natural Sciences, University of Buenos Aires). Inform. Bull. Southern Hemisph., No. 20, p. 5 (1972). Current research report.

Instituto de Astronomía y Física del Espacio. (Institute of Astronomyand Space Physics). Inform. Bull. Southern Hemisph., No. 20, p. 5 - 6 (1972). Current research report.

Instituto de Astronomía y Física del Espacio (Institute of Astronomy and Space Physics). Inform. Bull. Southern Hemisph., No. 21, p. 2 - 3 (1972). Current research report.

Instituto de Astronomía y Física del Espacio, (IAFE), *Buenos Aires, Argentina.* **Serie Publicaciones de Registros** (Data Report), No. 1 (08.082.232).

Instituto de Astronomía y Física del Espacio, (IAFE), *Buenos Aires, Argentina,* **Tirada Aparte** Nos. 1 (V. N. de Monteagudo, J. Sahade, 06.114.085), 2 (V. S. Niemela, 08.114.005).

008.019 Byurakan

Astrophysics in Soviet Armenia. See Abstr. 013.017.

008.020 Cambridge, Engl.

Cambridge Observatories. Report for the year end-ing 1971 September 30. R. O. Redman. Quarterly Journ. Roy. Astron. Soc., Vol. 13, 436 - 447 (1972).

008.021 Cambridge, Mass.

Smithsonian Astrophysical Observatory. – Annual report for the year ended 30 June 1970. Smithsonian Year 1970 = Smithsonian Publ. 4766, p. 45 - 49 (1970).

Smithsonian Astrophysical Observatory. – Annual report for the year ended 30 June 1971. Smithsonian Year 1971 = Smithsonian Publ. 4767, p. 45 - 50 (1971).

Smithsonian Institution. Astrophysical Observatory. Research in Space Science. SAO Special Reports, Nos. 343 (R. F. C. Vessot, 08.066.029), 344 (G. E. O. Giacaglia, C. A. Lundquist, 08.021.029), 346 (J. R. Cherniack, 08.042.040).

008.022 Canberra

Uppsala Southern Station. Inform. Bull. Southern Hemisph., No. 20, p. 12 (1972). Current research report.

008.023 Cape Town

Department of Astronomy, University of Cape Town. A. P. Fairall. Inform. Bull. Southern Hemisph., No. 20, p. 19 - 20 (1972). Current research report.

National Institute for Telecommunications Research, C. S. I. R., Radio Astronomy Division. G. D. Nicolson. Inform. Bull. Southern Hemisph., No. 20, p. 20 (1972). Current research report.

South African Astronomical Observatory. Inform. Bull. Southern Hemisph., No. 20, p. 22 - 23 (1972). Current research report.

Department of Astronomy, University of Cape Town. Inform. Bull. Southern Hemisph., No. 21, p. 19 (1972). – Current research report.

Royal Observatory Annals, (Joint Publications of the Royal Greenwich and Cape Observatories). See Abstr. 008.041.

Royal Observatory Bulletins, (Joint Publications of the Royal Greenwich Observatory, Herstmonceux, Royal Observatory, Cape of Good Hope). See Abstr. 008.041.

008.024 Castel Gandolfo

Ricerche Astronomiche, Specola Vaticana, Città del Vaticano, Vol. 8, Nos 12 (W. J. Miller, A. A. Wachmann, 08.123.039), 13 (G. V. Coyne, 08.121.071), 14 (D. J. K. O'Connell, 08.121.072), 15 (G. V. Coyne, 08. 115.013), 16

(G. V. Coyne, 08.114.118).

Specola Vaticana, *Castel Gandolfo*, Comunicazione, Nos. 53 (G. V. Coyne, 08.064.043), 54 (P. J. Treanor, E. Salpeter, 08.034.063), 55 (P. J. Treanor, 08.009.001).

008.025 Catania

Osservatorio Astrofisico di Catania, Pubblicazione, No. 148 (G. Godoli, V. Sciuto, M. L. Sturiale, R. A. Zappalà, E. Catinoto, G. Domina, G. Celeani, G. Sapienza, 08.075.011).

008.026 Cincinnati

Minor Planet Circulars (MPC), Nos. 3353 - 3406 (P. Herget, 08.098.087).

008.027 Córdoba

Observatorio Astronómico (Astronomical Observatory, National University of Córdoba). Inform. Bull. Southern Hemisph., No. 21, p. 3 - 5 (1972). Current research report.

008.028 Cracow

Cracow Observatory, Reprints Nos. 90 (M. Winiarski, 06.121.073), 91 (M. Winiarski, 06.096.021), 92 (T. Z. Dworak, M. Winiarski, 08.124.101), 93 (Z. Klimek, 08.117. 009), 94 (I. Toborek, 08.161.003).

008.029 Culgoora

The Culgoora Solar Radio Observatory. N. R. Labrum. Solar Physics, Vol. 27, 496 - 504 (1972). – Report from solar institute.

008.030 Dresden

Mitteilungen des Lohrmann-Observatoriums der Technischen Universität Dresden, Nos. 25 (08.012.021), 25a (J. Dittrich, 08.044.033; K.-G. Steinert, 08.031.084; H. Potthoff, 08.031.085), 26 (K.-G. Steinert, 07.032.018).

Technische Universität Dresden, Lohrmann-Observatorium, Zirkular, Nos. 56 - 60 (08.045.052).

008.031 Dublin

Contributions from the Dunsink Observatory, No. 10 (T. Kiang, 07.103.118).

Dunsink Observatory, Reprints Nos. 60 (C. J. Butler, P. A. Wayman, 02.021.002), 61 (A. D. Andrews, T. W. Rackham, P. A. Wayman, 01.105.035), 63 (C. J. Butler, 08.122. 084), 64 (T. Kiang, 08.004.030), 65 (C. J. Butler, 07.122.075).

Communications of the Dublin Institute for Advanced Studies, Series C. Dunsink Observatory Publications, Vol. 1, No. 6 (C. J. Butler, 08.159.010).

008.032 Edinburgh

Royal Observatory, Edinburgh. Report for the year ending 1972 March 31. H. A. Brück. Quarterly Journ. Roy. Astron. Soc., Vol. 13, 580 - 591 (1972).

Report of the Astronomer Royal for Scotland for the year ending 31st March 1972. H. A. Brück. Science Research Council. The Royal Observatory, Edinburgh, 14 pp. (1972).

The Royal Observatory Edinburgh 1822 - 1972. H. A. Brück. Separate print Royal Obs. Edinburgh. Produced by Edinburgh University Press and printed by Westerham Press Limited, Kent. 38 pp. (1972). – Survey of the Observatory's history and review of its present activities.

Communications from the Royal Observatory, Edinburgh, Nos. 122 (T. J. Lee, 08.022.115), 123 (M. J. Smyth, 08.034.037), 126 (R. J. Dodd, 07.031.019), 129 (J. W. Campbell, 08.034.055), 130 (T. J. Lee, 07.131.114).

Publications of the Royal Observatory Edinburgh, Vol. 7, Nos. 7 (M. T. Brück, 08.113.027), 8 (M. T. Brück, 08.113.033).

008.033 Flagstaff

Lowell Observatory Bulletin, *Flagstaff, Arizona*, No. 159 = Vol. 7, No. 22 (L. A. Riley, J. S. Hall, 08.094.097).

008.034 Fort Davis

A description of the lunar ranging station at McDonald Observatory. See Abstr. 034.101.

008.035 Frankfurt

Veröffentlichungen des Astronomischen Instituts der Universiät Frankfurt (Main), Nos. 36 (R. Henkel, 06.072. 055), 37 (R. Hartmann, 06.072.069), 38 (W. Gleissberg, 06.072.068), 39 (R. Henkel, 08.072.038).

008.036 Freiburg

Fraunhofer Institut, Map of the sun. 1972 July 1 – December 31 (08.075.028).

008.037 Glasgow

Department of Astronomy, University of Glasgow. Report for the period 1969 October to 1972 September. P. A. Sweet. Quarterly Journ. Roy. Astron. Soc., Vol. 13, 538 - 541 (1972).

Department of Geology, University of Glasgow.
Report for the period 1971 March 1 to 1972 February 29.
A. C. McLean.
Quarterly Journ. Roy. Astron. Soc., Vol. 13, 542 - 544 (1972).

008.038 Gothenburg

**Research Laboratory of Electronics and Onsala
Space Observatory**, Chalmers University of Technology, Gothenburg, Sweden. **Research Report** No. 101 (B. O. Rönnäng, 08.131.079).

008.039 Graz

Mitteilungen der Universitätssternwarte Graz, No. 10 (H. Haupt, 07.098.013).

008.040 Green Bank

National Radio Astronomy Observatory, *Green Bank*, **Reprints**, Series A, Nos. 243 (A. H. Bridle, M. M. Davis, E. B. Fomalont, J. Lequeux, 07.141.054), 244 (W. J. Wilson, A. H. Barrett, 07.114.048), 245 (K. W. Riegel, R. M. Crutcher, 07.155.025), 246 (G. S. Downs, A. R. Thompson, 07.141.066), 247 (R. M. Hjellming, C. M. Wade, E. Webster, 07.121.091), 248 (D. F. Dickinson, B. E. Turner, 07.131.076), 249 (K. I. Kellermann, 08.033.035), 250 (G. L. Verschuur, T. Cram, R. Giovanelli, 07.131.095), 251 (W. C. Saslaw, D. S. De Young, 07.151.069), 252 (A. H. Bridle, M. J. L. Kesteven, 07.141.081), 253 (W. R. Burns, 07.033.021), 254 (W. B. Burton, 07.155.061), 255 (F. J. Kerr, G. R. Knapp, 07.131.132), 256 (I. I. K. Pauliny-Toth, K. I. Kellermann, M. M. Davis, E. B. Fomalont, D. B. Shaffer, 07.141.118), 257 (F. Biraud, 07.141.160), 258 (M. R. Kundu, R. H. Becker, 08.125.003), 259 (A. H. Bridle, M. M. Davis, E. B. Fomalont, J. Lequeux, 08.141.012), 260 (N. J. Evans II, R. E. Hills, O. E. H. Rydbeck, E. Kollberg, 08.034.107).

National Radio Astronomy Observatory, *Green Bank*, **Reprints**, Ser. B, Nos. 315 (R. L. Brown, 07.131.081), 316 (D. B. Shaffer, M. H. Cohen, D. L. Jauncey, D. I. Kellermann, 07.158.112), 317 (L. E. Snyder, D. Buhl, 08.131.074), 318 (A. A. Penzias, P. M. Solomon, K. B. Jefferts, R. W. Wilson, 07.131.092), 319 (B. Zuckerman, M. Morris, P. Palmer, B. E. Turner, 07.153.019), 320 (D. C. Backer, 07.141.550), 321 (R. N. Whitehurst, M. S. Roberts, 08.158.008), 322 (J. W. Findlay, 07.141.176), 323 (S. T. Gottesman, J. J. Broderick, R. L. Brown, B. Balick, P. Palmer, 07.125.101), 324 (M. A. Gordon, 07.125.025), 325 (R. L. Brown, 07.131.134), 326 (A. C. E. Sinclair, J. P. Basart, D. Buhl, W. A. Gale, 08.093.004), 327 (W. J. Webster, Jr., A. C. Webster, G. T. Webster, 07.101.005), 328 (P. Thaddeus, M. L. Kutner, A. A. Penzias, R. W. Wilson, K. B. Jefferts, 08.131.029), 329 (R. W. Wilson, A. A. Penzias, K. B. Jefferts, P. Thaddeus, M. L. Kutner, 08.132.015), 330 (R. L. Brown, 07.061.040).

008.041 Greenwich

Royal Observatory Annals, (Joint Publications of the Royal Greenwich and Cape Observatories), Nos. 6 (08.075.003), 7 (A. W. J. Cousins, 08.113.037).

Royal Observatory Bulletins, (Joint Publications of

the Royal Greenwich Observatory, Herstmonceux, Royal Observatory, Cape of Good Hope), Nos. 171 (A. L. T. Powell, 08.114.087), 172 (G. A. Harding, R. S. Harbour, K. P. Tritton, 08.113.025), 173 (V. A. French, A. L. T. Powell, 08.114.093), 174 (J. B. Alexander, 08.122.124), 175 (J. B. Alexander, 08.115.023), 176 (E. A. Epps, 08.113.060).

008.042 Groningen

Nederlandse Vereiniging voor Weer- en Sterrenkunde. Observations of Variable Stars. Report (Kapteyn Astronomical Laboratory, Groningen – Netherlands), No. 22 (L. Plaut, H. Feijth, 08.123.024).

008.043 Hamburg

Hamburger Sternwarte. Sonderdruck, Serie A, Nos. 6 (E. Pansch, C. de Vegt, 06.096.031), 7 (D. Lauterborn, S. Refsdal, M. L. Roth, 05.065.118), 8 (L. Kohoutek, 07.133.004), 9 (C. de Vegt, H. Ebner, 07.041.003), 10 (H. Bossen, 07.121.010), 11 (D. Lauterborn, A. Weigert, 07.117.010), 12 (E. Høg, 07.032.031), 13 (A. Günther, H. Kox, 07.041.054), 14 (U. K. Gehlich, J. Prölss, R. Wehmeyer, 08.121.015), 15 (H. Kähler, 08.065.021), 16 (M. L. Roth, A. Weigert, 08.065.019).

European Southern Observatory. Annual report 1971. A. Blaauw.
Printed in the Federal Republic of Germany by Bergedorfer Buchdruckerei von E. Wagner, Hamburg-Bergedorf, 73 pp. (1972).

25./26. Jahresbericht 1970/71. H. U. Roll.
Published by Deutsches Hydrographisches Institut, Hamburg. 138 pp. (1972).

Deutsches Hydrographisches Institut, Hamburg. **Astronomische Zeit- und Breitenbestimmungen, Empfangszeiten von Zeitsignalen**, 1972 January - September (08.044.036).

008.044 Hannover

Technische Universität Hannover, Astronomische Station des Instituts für Theoretische Geodäsie, No. 8 (K. Pilowski, R. Winter, 08.032.035).

008.045 Heidelberg

Astronomy and Astrophysics Abstracts, Vol. 7 (S. Böhme, W. Fricke, U. Güntzel-Lingner, F. Henn, D. Krahn, U. Scheffer, G. Zech, 08.002.037).

Astronomisches Rechen-Institut in Heidelberg, Mitteilungen, Serie A, Nos. 56 (W. Fricke, 07.013.007), 57 (P. Brosche, J. Sündermann, 08.081.016), 58 (W. Fricke, 08.043.001), 59 (J. Schubart, 07.098.012), 60 (J. Schubart, 07.098.034), 61 (W. Gliese, 07.111.002), 62 (W. Fricke, 08.041.022).

Astronomisches Rechen-Institut in Heidelberg, Mitteilungen, Serie B, Nos. 30 (W. Lohmann, 07.153.023), 31 (R. E. Laubscher, 08.042.021), 32 (W. Gliese, 08.155.045).

Max-Planck-Institut für Astronomie, Heidelberg-Königstuhl.   H. Elsässer.
Naturwissenschaften, 59. Jahrgang, p. 548 - 549 (1972).
Report on activities 1970 - 1971.

Calar Alto-Verträge perfekt.
SuW, Vol. 11, 210 (1972).

008.046 Helsinki

Contributions from Observatory and Astrophysics Laboratory, University of Helsinki, (E. Anttila, 08.065.071; H. Oja, 08.042.062; I. V. Tuominen, 08.065.072; I. V. Tuominen, 08.065.073; I. V. Tuominen, 08.065.074; O. Vilhu, 08.064.030).

008.047 Ioannina

University of Ioannina, Contributions from the Laboratory of Astronomy, Ioannina, Greece, No. 6 (S. N. Svolopoulos, S. Kapranidis, 08.121.088).

008.048 Johannesburg

Republic Observatory.   J. Hers.
Inform. Bull. Southern Hemisph., No. 20, p. 20 - 22 (1972). Current research report.

Republic Observatory, Johannesburg. Report for the year ending 1971 December 31.   J. Hers.
Quarterly Journ. Roy. Astron. Soc., Vol. 13, 460 - 462 (1972).

008.049 Kharkov

Visnik Kharkivs'kogo Universitetu. Vidavnitstvo Kharkivs'kogo Ordena Trudovogo Chervonogo Prapora Derzhavnogo Universitetu imeni O. M. Gor'kogo, Kharkiv, 1972, No. 82, [Ser.] Astronomiya, Vipusk (No.) 7 (08.007.000; V. J. Ehzers'kij, K. N. Kuz'menko, V. Kh. Pluzhnikov, 08.005. 028; M. P. Barabashov, V. O. Ehzers'ka, V. J. Ehzers'kij, I. I. Latinina, 08.094.253; M. P. Barabashov, V. O. Ehzers'ka, V. J. Ehzers'kij, N. P. Stadnikova, 08.094.254; O. M. Starodubtseva, 08.097.122; V. I. Bistrits'kij, V. P. Vasil'ehv, 08.031.088; V. I. Bistrits'kij, V. P. Vasil'ehv, 08.074.109; K. M. Derkach, M. G. Zuev, 08.041.050; M. F. Khodyachikh, 08.099.087; A. D. Egorov, O. F. Vantsan, 08.031.089; P. P. Pavlenko, 08.055. 024; V. Kh. Pluzhnikov, 08.092.013; M. P. Bondarevs'kij, K. N. Kuz'menko, G. A. Orlov, 08.103.100; M. G. Zuev, 08.034. 148; P. P. Pavlenko, 08.034.149).

008.050 Kiev

Report about the work of the Astronomical Observatory of the Kiev University in 1966 - 1970.
A. F. Bogorodskij.
Vestn. Kiev. Un-ta, Ser. Astron., No. 14, p. 107 - 110 (1972). In Russian.

Solar physics at the Kiev University Observatory.
N. A. Yakovkin.
Solar Physics, Vol. 27, 493 - 495 (1972). – Report from solar institute.

008.051 Kodaikanal

Kodaikanal Observatory, Kodaikanal. Report for period January 1970 to March 31 1971.   M. K. V. Bappu.
Quarterly Journ. Roy. Astron. Soc., Vol. 13, 448 - 452 (1972).

Astrometriya i Astrofizika, Kiev, Vyp. (Nos.) 15 (V. I. Voroshilov, 08.003.001), 17 (E. P. Fedorov, 08.003.012).

Kodaikanal Observatory, Bulletins, Series A, No. 208 (V. Natarajan, R. Rajamohan, 08.119.010).

Kodaikanal Observatory, Bulletins, Series B, No. 206 (M. K. V. Bappu, 08.075.009).

Kodaikanal Observatory, Bulletins, Series C, No. 198 (M. K. V. Bappu, 08.075.010).

008.052 Krim

The number of photometric nights and seeing at the Crimean Astrophysical Observatory.   E. A. Vitrichenko.
Astron. Tsirk., No. 677, p. 7 - 8 (1972). In Russian.

Chronicle.
Izv. Krymskoj Astrofiz. Obs., Vol. 44, 174 - 175 (1972). In Russian.

Izvestiya Krymskoj Astrofizicheskoj Observatorii, Akademiya Nauk SSSR, Tom (Vol.) 44 (P. F. Chugainov, 08.122.027; R. E. Gershberg, S. A. Kaplan, 08.122.028; M. E. Boyarchuk, 08.114.028; V. V. Golovatyj, V. I. Pronik, 08.134.003; I. I. Pronik, K. K. Chuvaev, 08.158.022; S. I. Gopasyuk, T. T. Tsap, 08.071.014; S. I. Gopasyuk, B. Kalman, 08.071.015; D. N. Rachkovsky, 08.071.016; M. J. Guseynov, 08.072.012; V. A. Kotov, 08.034.007; E. A. Baranovsky, 08.072.013; A. N. Koval, 08.073.021; N. N. Stepanyan, 08.073.022; Nguen-Ngan, 08.072.014; N. N. Eruchev, L. I. Tsvetkov, 08.077.010; A. F. Lagutin, 08.031.010; A. K. Dabakhov, 08.031.011; V. A. Efanov, A. G. Kislyakov, G. V. Lebskij, I. G. Moiseev, A. J. Naumov, 08.077.011), 45 (S. I. Gopasyuk, T. T. Tsap, 08.080.055; S. I. Gopasyuk, T. T. Tsap, 08.080.056; E. E. Dubov, 08.073.110; L. I. Yurovskaya, 08. 077.067; G. F. Eliseev, 08.071.073; N. A. Dimov, A. B. Severny, 08.082.228; N. A. Dimov, A. M. Zvereva, A. B. Severny, 08.082.229; N. M. Shakhovskoy, Yu. S. Efimov, 08.131.122; Yu. S. Efimov, N. M. Shakhovskoy, 08.122.131; R. E. Gershberg, 08.122.132; N. I. Shakhovskaya, 08.122.133; P. F. Chugainov, 08.115.027; M. Yu. Skulsky, 08.121.106; T. A. Ryabchikova, 08.114.171; V. V. Golovatyi, V. I. Pronik, 08. 134.013; I. I. Pronik, 08.158.130; V. A. Efanov, V. I. Zagatin, I. G. Moiseev, H. M. Tovmassian, V. B. Shteinshleger, 08.141. 129; A. G. Gorshkov, I. G. Moiseev, V. A. Soglasnov, 08.141. 130; N. A. Dimov, 08.034.139; G. M. Popov, 08.031.080; Z. N. Kuteva, Yu. A. Sabinin, 08.034.140).

008.053 Kyoto

Contributions from the Kwasan and Hida Observatories, University of Kyoto, Nos. 201 (K. Nakayama, 07.073. 041), 202 (J. Kubota, T. Tamenaga, Y. Funakoshi, T. Kureizumi, 07.073.042), 203 (T. Tsubaki, H. Kurokawa, M. Kanno, 07.074.006), 204 (M. Kanno, T. Tsubaki, H. Kurokawa, 07.074.007), 205 (J. Kubota, T. Tamenaga, K. Yoshikawa, 07.073.068), 206 (S. Miyamoto, 08.097.081), 207 (S. Miyamoto, 08.097.082).

008.054 **La Plata**

Observatorio Astronómico. (Astronomical Observatory, National University of La Plata).
Inform. Bull. Southern Hemisph., No. 20, p. 7 (1972).
Current research report.

Observatorio Astronómico (Astronomical Observatory, National University of La Plata).
Inform. Bull. Southern Hemisph., No. 21, p. 5 - 6 (1972).
Current research report.

008.055 **Las Cruces**

Blue Mesa Observatory.   J. Cuffey.
Contr. Obs. New Mexico State Univ., Las Cruces, Vol. 1, 70 - 72 (1972).

Contributions of the Observatory of New Mexico State University, Las Cruces, New Mexico, Vol. 1, No. 3, (J. Cuffey, 08.008.055; J. Cuffey, 08.082.230; J. R. Gallivan, 08.031.081; D. R. Hollars, T. B. Kirby, 08.034.141; E. J. Reese, 08.099.086; H. A. Beebe, D. R. Hollars, 08.063.042; W. L. Reitmeyer, 08.034.142; M. L. Davis, 08.035.003; C. D. Houghton, 08.034.143).

008.056 **Lembang**

Report for the years 1970 - 1971.   B. Hidajat.
Separate print Bosscha Obs., Dep. Sci., Bandung Inst. Technol., 5 pp. (1972).

Contributions from the Bosscha Observatory, Bandung Institute of Technology, Department of Science, Nos. 43 (B. Hidajat, 05.123.028; B. Hidajat, M. U. Akyol, 07.122. 095), 44 (B. Hidajat, W. Sutantyo, 08.153.027), 45 (S. M. Larson, 08.099.078).

Bandung Institute of Technology, Department of Science. **Publications of the Bosscha Observatory,** No. 5 (W. Sutantyo, 08.061.064).

008.057 **Leningrad**

Byulletin' Instituta Teoreticheskoj Astronomii, Akademiya Nauk SSSR, Vol. 13, Nos. 3 (N. S. Samoilova-Yakhontova, 08.098.002; G. A. Chebotarev, 08.101.001; T. K. Nikolskaya, 08.021.003; M. S. Petrovskaya, 08.042.010; E. P. Filianskaya, 08.042.011; F. B. Khanina, 08.098.003; B. S. Vozdvizhensky, 08.103.100; J. Židů, 08.098.004; L. I. Chernykh, 08.098.005), 4 (V. A. Izvekov, 08.093.025; V. A. Izvekov, 08.041.031; S. I. Nikolaev, 08.042.063; T. K. Nikolskaya, 08.042.064; M. S. Petrovskaya, 08.081.037; M. L. Sveshnikov, 08.042.065; L. L. Filenko, 08.052.023; L. I. Chernykh, 08.098.028), 5 (M. A. Dirikis, 08.098.029; A. M. Zhandarov, V. G. Kiselev, 08.042.066; B. I. Lokhin, 08.052.024; E. N. Polyakhova, 08.052.025; V. I. Voronenko, F. F. Kalihevich, 08.098.030; L. I. Chernykh, 08.098.031; V. L. Brumberg, A. V. Egorova, 08.042.901).

Ephemerides of minor planets for 1973, (G. A. Chebotarev, 08.098.026).

008.058 **London**

University of London Observatory. – Report for the year ending 1971 December 31.   C. W. Allen.
Quarterly Journ. Roy. Astron. Soc., Vol. 13, 545 - 548 (1972).

008.059 **Madrid**

Boletín Astronómico del Observatorio de Madrid, Instituto Geografico y Catastral, Sección 2ª, Astronomia, Vol. 8, No. 1 (J. Pensado, 08.075.037; M. M. Lorón, J. Pensado, J. Claver, 08.092.014; R. Carrasco, M. M. Lorón, J. Pensado, M. de Pascual, 08.079.100).

Universidad de Madrid – Facultad de Ciencias. Seminario de Astronomia y Geodesia, Publicación, Nos. 69 (J. M. Torroja, 06.008.061), 70 (R. Vieira, R. Ortiz, 08.046. 040).

008.060 **Malvern**

Royal Aircraft Establishment: Space Department. Report for year ending 1971.   D. G. King-Hele.
Quarterly Journ. Roy. Astron. Soc., Vol. 13, 453 - 459 (1972).

008.061 **Manchester**

Astronomical Contributions from the University of Manchester, Series III, Nos. 226 (E. Graham, 05.065.074), 234 (E. Budding, 04.094.031), 236 (K. H. Elliott, 03.132. 037), 237 (Z. Kopal, 04.117.012), 238 (A. C. Danks, 04.132. 016), 239 (J. Meaburn, 04.132.015), 240 (J. Meaburn, 04. 034.059), 241 (E. Budding, Z. Kopal, 04.121.040), 242 (J. Meaburn, 04.131.112), 243 (Z. Kopal, 05.117.011), 244 (Z. Kopal, 05.121.011), 245 (C. G. T. Haslam, F. D. Kahn, J. Meaburn, 05.157.006), 246 (J. E. Dyson, J. Meaburn, 05.132.017), 247 (R. A. James, F. D. Kahn, 05.065.106), 248 (A. C. Danks, J. Meaburn, 05.132.021), 249 (Z. Kopal, A. K. M. S. Ali, 05.042.033), 250 (J. D. Mihalov, C. P. Sonett, J. H. Binsack, M. D. Moutsoulas, 05.094.017).

008.062 **Minneapolis**

University of Minnesota, Minneapolis, Minnesota, Separate prints (W. J. Luyten, 08.112.009; W. J. Luyten, 08.112.010; W. J. Luyten, A. E. La Bonte, 08.112.011).

008.063 **Mizusawa**

Annual Report of the International Polar Motion Service 1970, (S. Yumi, 08.045.031).

Monthly Notes of the International Polar Motion Service, 1972 Nos. 5–10 (08.045.046).

008.064 **Mons**

Université de Mons. Département d'Astrophysique. Communications. Nos. 27 (L. Houziaux , A. Ringuelet, 08.

041.033), 30 (A. Delcroix, 08.046.028), 31 (G. Houziaux, 08. 041.034).

008.065 Moskva

Chronicle.   L. N. Bondarenko.
Astron. Tsirk., No. 680, p. 6 - 7 (1972). In Russian.

Soobshcheniya Gosudarstvennogo Astronomicheskogo Instituta im. P. K. Shternberga. Izdatel'stvo Moskovskogo Universiteta, Nos. 178 (V. V. Nesterov, 08.045.034; L. M. Khommik, 08.041.023; A. P. Gulyaev, 08.041.024; V. G. Shamaev, 08.041.025; L. M. Khommik, O. A. Kozina, N. N. Kabaeva, A. P. Gulyaev, 08.041.026), 179 (L. V. Rykhlova, 08.045.035; V. V. Nesterov, 08.045.036, 08.045.037; I. M. Kalinina, 08.041.027; A. P. Gulyaev, 08.041.028; N. N. Kabaeva, 08.041.029; L. P. Panteleeva, 08.041.030), 180 (N. I. Kozhevnikov, G. F. Sitnik, A. I. Khlystov, 08.082.123; A. I. Khlystov, 08.071.051; V. S. Shevchenko, V. E. Slutsky, V. I. Kardopolov, V. G. Khetselius, 08.082.124).

Trudy Gosudarstvennogo Astronomicheskogo Instituta im. P. K. Shternberga. Izdatel'stvo Moskovskogo Universiteta, Vol. 43, vyp. 2 (A. L. Kogan, P. A. Stroev, 08.081.038; V. L. Panteleev, 08.081.039; L. A. Savrov, 08.081.040; A. A. Orlov, 08.042.067; N. A. Solovaya, 08.042.068; E. P. Aksenov, L. M. Domozhilova, 08.052.026; E. P. Aksenov, L. M. Domozhilova, 08.052.027; E. E. Mukin, 08.052.028; N. E. Kurochkin, 08.141.081.

008.066 Mount Stromlo

Mount Stromlo and Siding Spring Observatory. Research School of Physical Sciences, The Australian National University. − Report for the year ending 1971 December 31. O. J. Eggen.
Quarterly Journ. Roy. Astron. Soc., Vol. 13, 550 - 566 (1972).

008.067 München

Mitteilungen der Sternwarte München, Band 2, No. 1 (F. Schmeidler, 08.005.024).

Veröffentlichungen der Sternwarte München, Band 7, Nos. 14 (P. Labitzke, 06.041.032), 15 (H. G. Groth, 08.114.112), 16 (B. Wolf, 08.064.007).

Max-Planck-Institut für Physik und Astrophysik, München: Institut für Astrophysik, L. Biermann; Institut für extraterrestrische Physik, R. Lüst.
Naturwissenschaften, 59. Jahrgang, p. 616 - 619 (1972). Report on activities 1970 - 1971.

Max-Planck-Institut für Physik und Astrophysik, München, Separate print (K. Jockers, Rh. Lüst, 08.103.121).

Max-Planck-Institut für Physik und Astrophysik, Institut für Extraterrestrische Physik, Garching, München, Separate prints (H. A. Mayer-Hasselwander, E. Pfeffermann, K. Pinkau, H. Rothermel, M. Sommer, 07.155.073; V. Schönfelder, A. Hirner, K. Schneider, 08.032.040; K. Pinkau, 08.031.075; H. Billing, 08.131.113).

008.068 Naini Tal

Uttar Pradesh State Observatory, *Naini Tal*, Reprints, Nos. 51 (G. S. D. Babu, 05.116.012), 52 (V. P. Gaur, M. C. Pande, B. M. Tripathi, G. C. Joshi, 06.072.002), 53 (M. C. Pande, B. M. Tripathi, V. P. Gaur, 06.072.003), 54 (M. C. Pande, B. M. Tripathi, C. S. Murthy, V. P. Gaur, 06.122.008), 55 (T. R. Bhatt, S. D. Sinvhal, 07.123.007).

008.069 Napoli

Istituto Universitario Navale, Astronomía Generale e Sferica, Napoli. Seminario Nos. 1 (E. Fichera, 08.046.033), 2 (E. Fichera, 08.046.034), 3 (E. Fichera, 08.046.037).

008.070 Nashville

The Arthur J. Dyer Observatory, Vanderbilt University, Nashville, Tennessee, Reprint Series 1, Nos.58 (E. Ye. Khachikian, D. W. Weedman, 06.158.092), 61 (D. S. Hall, 07.121.034), 62 (D. W. Weedman, 07.158.005), 63 (M. G. Smith, D. W. Weedman, 07.132.008).

The Arthur J. Dyer Observatory, Vanderbilt University, Nashville, Tennessee, Reprint, Series 2, Nos. 6 (D. S. Hall, A. M. Heiser, 07.121.019), 7 (D. W. Weedman, 06.113.020), 8 (D. S. Hall, 07.122.071).

008.071 Odessa

Odesas Observatorija − jubiläre.   I. Daube. Zvaigžņotā debess, 1972. gada pavasaris, p. 26 - 31. − Report on the observatory of Odessa which has been founded a hundred years ago.

008.072 Onsala

Publications from Onsala Space Observatory, Onsala, Sweden, Nos. 63 (B. T. Cato, 08.131.106; M. Cato, T. Cato, P. Landgren, A. Sume, 03.141.174; T. Cato, J. Elldér, B. Höglund, O. E. H. Rydbeck, B. Rönnäng, A. Sume, 07.131.136; M. A. Gordon, T. Cato, 08.157.001), 64 (T. Cato, J. Elldér, B. Höglund, O. E. H. Rydbeck, B. Rönnäng, A. Sume, 07.131. 136), 66 (N. J. Evans II, R. E. Hills, O. E. H. Rydbeck, E. Kollberg, 08.034.107), 73 (M. M. Litvak, A. Sume, 08.131. 107), 75 (B. Rönnäng, 08.033.053).

008.073 Oslo

UBV-system for a 12 inch telescope.
See Abstr. 113.065.

Institute of Theoretical Astrophysics, Blindern− Oslo. Report, Nos. 32 (O. Havnes, 08.112.014), 33 (P. Maltby, L. Staveland, 08.072.066), 34 (S. Sivertsen, 08.113.065), 35 (Ø. Hauge, 08.071.074), 36 (L. Staveland, 08.071.075), 37 (H. Sørli, O. Engvold, 08.071.076).

Institutt for Teoretisk Astrofysikk, Blindern − Oslo. Småtrykk, Nos. 71 (R. Stabell, 08.066.162), 72 (T. Hansen, 08.092.012), 73 (T. Ringnes, 08.094.245), 74 (T. Ringnes,

08.094.246), 75 (O. P. Sveen, 08.079.004), 76 (O. P. Sveen, 07.079.006), 77 (R. Brahde, 07.092.013), 78 (T. Ringnes, 08.094.247).

008.074 Ottawa

Contributions of the National Research Council of Canada, Ottawa, Canada, Nos. 12300 (Y. L. Chow, 08.033. 056), 12424 (D. J. Bradt, L. A. Higgs, S. G. Jones, T. G. O'Neill, J. L. Wolfe, 07.033.015), 12509 (M. B. Bell, E. R. Seaquist, 07.158.057), 12644 (B. H. Andrew, W. J. Medd, G. A. Harvey, J. L. Locke, 07.141.129), 12657 (G. A. Harvey, B. H. Andrew, J. M. MacLeod, W. J. Medd, 07.141.170), 12723 (J. M. MacLeod, L. H. Doherty, 07.158.165), 12819 (L. H. Doherty, L. A. Higgs, J. M. MacLeod, 08.132.020).

Contributions from the Earth Physics Branch, Department of Energy, Mines and Resources, Ottawa, Canada, Nos. 360 (D. R. Auld, B. Caner, 08.084.353), 398 (M. R. Dence, J. Popelar, 08.105.065), 402 (M. R. Dence, 08.105. 066), 403 (R. W. Tanner, 08.081.053).

Publications of the Earth Physics Branch, Department of Energy, Mines and Resources, Ottawa, Canada, Vol. 42, Nos. 4 (E. I. Loomer, G. J. van Beek, 08.084.349), 9 (A. B. Cook, S. J. Sprysak, 08.084.350); Vol. 43, Nos. 1 (G. J. van Beek, H. R. Reny, 08.084.351), 7 (D. R. Auld, B. D. Lowe, 08.084.352).

008.075 Oxford

Department of Astrophysics,University of Oxford. Report for the year ending 1971 December 31. D. E. Blackwell. Quarterly Journ. Roy. Astron. Soc., Vol. 13, 567 - 572 (1972).

008.076 Padova

Osservatorio Astronomico di Padova. Comunicazioni e Rassegne, Nos. 82 (L. Rosino, 05.122.126), 83 (L. Nobili, 05.065.139), 84 (R. Barbon, 05.158.117), 85 (G. Barbaro, 06.065.091), 86 (C. Barbieri, 06.141.182), 87 (F. Bertola, 06.158.096), 88 (F. Ciatti, 06.011.048), 89 (G. Romano, 06.123.072), 90 (L. Rosino, 08.155.051), 91 (C. Barbieri, 06.031.066), 92 (C. Barbieri, 07.141.074), 93 (G. Romano, 07.123.017), 94 (L. Rosino, D. di Martino, 06.123.021), 95 (L. Rosino, 07.124.006), 96 (G. Romano, 08.141.001), 97 (L. Rosino, 08.004.037), 98 (F. Bertola, 08.158.085).

Pubblicazioni dell'Osservatorio Astronomico di Padova, Nos. 161 (G. Romano, 05.122.062), 162 (G. Pinto, G. Romano, 07.123.006), 163 (G. Romano, M. Perissinotto, 08.123.009).

008.077 Paris

Bureau International de l'Heure, (B.I.H.), Circulaires B/C, Nos. 196 - 200 (08.045.049).

Bureau International de l'Heure, (B.I.H.), Circular D 69 - D 73 (08.044.043).

Bureau International de l'Heure, (B.I.H.), Circular E 1 (B. Guinot, 08.044.041), E 2 (B. Guinot, 08.044.042).

Observatoire de Paris, Section d'Astrophysique, à Meudon, **Cartes synoptiques,** Vol. 5, Fasc. 3, années 1970 - 1971 (M. J. Martres, 08.075.022).

008.078 Pasadena

Twenty years astronomy with the 48-inch Schmidt telescope on Palomar Mountain. See Abstr. 013.010.

An interferometric seeing test on Mt. Wilson. See Abstr. 082.014.

008.079 Pereyra Iraola

Instituto Argentino de Radioastronomia. (Argentine Radioastronomy Institute). Inform. Bull. Southern Hemisph., No. 20, p. 7 - 9 (1972). Current research report.

008.080 Porto Alegre

Observatorio Astronômico do Instituto da Universidade Federal do Rio Grande do Sul. (Astronomical Observatory, Institute of Physics of the Federal University of Rio Grande do Sul). Inform. Bull. Southern Hemisph., No. 20, p. 17 (1972). Current research report.

Observatorio Astronómico do Instituto de Física da Universidade Federal do Rio Grande do Sul (Astronomical Observatory, Institute of Physics of the Federal University of Rio Grande do Sul). Inform. Bull. Southern Hemisph., No. 21, p. 10 - 11 (1972). Current research report.

008.081 Praha

Académie Tchécoslovaque des Sciences, Institut Astronomique, **Station de l'Heure à Prague,** Série 5, No. 18; Série 6, Nos. 1 - 2 (L. Webrová, V. Ptáček, 08.044.039).

Publications of the Astronomical Institute of the Charles University, Nos. 65 (J. Bouška, A. Mrkos, 08.103.123), 66 (A. Mrkos, 08.103.007).

008.082 Pretoria

Radcliffe Observatory, Pretoria. – Report for the year ending 1972 March 31. A. D. Thackeray. Quarterly Journ. Roy. Astron. Soc., Vol. 13, 573 - 579 (1972).

Radcliffe Observatory. Inform. Bull. Southern Hemisph., No. 21, p. 19 - 23 (1972). Current research report.

The Radcliffe Observatory 1772–1972. A. D. Thackeray, with an introduction by F. Hoyle. Printed for The Radcliffe Trust by The Riverside Press Ltd., London and Whitstable. 56 pp. (1972).

Two hundred years of the Radcliffe Observatory.
A. D. Thackeray.
Nature, Vol. 239, 313 - 315 (1972).

Communications from the Radcliffe Observatory,
Pretoria, Nos. 111 (P. W. Hill, Abstract of 06.112.001), 114
(J. Menzies, 07.154.008), 115 (R. J. Dickens, 07.154.022).

Radcliffe Observatory, *Pretoria,* **Reprints,** Nos.
105 (A. D. Thackeray, 04.008.084), 106 (M. W. Feast,
07.122.050), 107 (T. Lloyd Evans, 08.119.009), 108 (T.
Lloyd Evans, 08.122.083), 109 (M. W. Feast, 08.122.019),
110 (M. W. Feast, 07.122.151), 111 (A. D. Thackeray, 07.
155.082), 112 (T. Lloyd Evans, 07.115.018), 113 (R. M.
Catchpole, 08.119.005), 116 (R. J. Dickens, 08.154.021),
117 (M. W. Feast, R. Woolley, N. Yilmaz, 08.122.016).

008.083 Pulkovo

**The Pulkovo Observatory. Review of the history
and scientific activities.** See Abstr. 003.032.

**Izvestiya Glavnoj Astronomicheskoj Observatorii
v Pulkove,** No. 188 (08.033.001 - 08.033.032).

Trudy Glavnoj Astronomichskoj Observatorii v
Pulkove, Seriya 2, Vol. 80 (L. F. Gorel, 08.041.001; G. K. Gorel, 08.041.002).

Trudy Glavnoj Astronomicheskoj Observatorii v
Pulkove (R. S. Gnevysheva, 08.075.001).

008.084 Richmond Hill

David Dunlap Observatory, Richmond Hill, Ontario.
D. A. MacRae.
Journ. Roy. Astron. Soc. Canada, Vol. 66, 229 - 231 (1972).

Photometric weather at Toronto.
See Abstr. 082.090.

Communications from the David Dunlap Observa-
tory, University of Toronto, Richmond Hill, Ontario, Canada,
Nos. 301 (W. L. Gorza, 06.119.012), 302 (J. R. Percy, 06.122.
135), 314 (S. van den Bergh, 06.124.107), 315 (J. D. Fernie,
V. Sherwood, D. L. DuPuy, 07.122.015), 316 (S. van den
Bergh, 07.131.068), 317 (S. van den Bergh, 06.113.050),
318 (C. T. Bolton, 07.142.052), 319 (S. van den Bergh, 07.
122.061), 320 (R. G. Lake, R. C. Roeder, 07.141.122), 321
(E. R. Seaquist, 07.062.007), 322 (P. P. Kronberg, C. J.
Pritchet, S. van den Bergh, 07.158.058), 323 (S. van den
Bergh, 06.158.138), 324 (J. D. Fernie, 07.113.011), 325 (S.
van den Bergh, 07.155.044), 326 (N. R. Walborn,
07.114.071), 327 (S. P. S. Anand, 07.062.020), 328 (P. P.
Kronberg, R. C. Conway, J. A. Gilbert, 07.141.072), 329 (J.
R. Percy, 08.153.002), 330 (J. R. Percy, K. Madore, 07.115.
015), 331 (J. D. Fernie, 08.113.020), 332 (M. D. T. Naylor,
S. P. S. Anand, 07.117.026), 333 (C. M. Coutts, 07.122.085),
334 (M. J. Clement, 07.064.046).

Publications of the David Dunlap Observatory,
University of Toronto, *Richmond Hill, Ontario, Canada,* Vol.
3, No. 5 (J. F. Heard, C. Fehrenbach, 08.112.013).

008.085 Riga

**Hronika. Latvijas PSR Zinātņu akadēmijas Radio-
astrofizikas observatorijā.** I. Daube.
Zvaigžņotā debess, 1972. gada pavasaris, p. 61 - 62.

Latvia – centre of investigations on red giants.
A. Balklavs.
Zemlya i Vselennaya, 1972, No. 6, p. 19 - 21. In Russian.
Report on research at Radioastrophys. Obs. Akad. Nauk Latv.
SSR, Riga.

**Akademiya Nauk Latvijskoj SSR, Radioastrofizi-
cheskaya Observatoriya. Trudy** (*Transactions*) **Observatorii,**
Riga, Vol. 14 (A. Eh. Balklavs, 08.003.019).

008.086 Rio de Janeiro

Observatorio Nacional. (National Observatory).
Inform. Bull. Southern Hemisph., No. 20, p. 13 - 14 (1972).
Current research report.

Contribuições do Observatório do Valongo, Universidade Federal do Rio de Janeiro, Série II, No. 14 (08.099.101).

Contribuições do Observatório do Valongo, Universidade Federal do Rio de Janeiro, Série III, No. 22 (08.096.024).

**Universidade Federal do Rio de Janeiro, Centro de
Ciencias Matemáticas e da Natureza** (Federal University of
Rio de Janeiro – Center of Mathematical and Natural Sciences
– Valongo Observatory).
Inform. Bull. Southern Hemisph., No. 21, p. 9 - 10 (1972).
Current research report.

008.087 Rochester

C. E. Kenneth Mees Observatory, University of
Rochester, Rochester, N. Y., **Reprints,** Nos. 32 (J. E. Littleton.
H. M. Van Horn, H. L. Helfer, 07.080.017), 33 (A. Kovetz,
D. Q. Lamb, H. M. Van Horn, 07.061.030), 34 (J. F. W.
Perry, H. L. Helfer, 07.155.066), 35 (C. R. Sturch, S. L. Sharp-
less, 08.114.081), 38 (L. Taff, M. P. Savedoff, 08.131.063).

008.088 Roma

Monthly Bulletin. Osservatorio Astronomico di
Roma, Nos. 169 - 175 (M. Cimino, M. Torelli, A. Cacciani,
V. Croce, R. Flamini, U. Bartolini, 08.075.031).

Photographic Journal of the Sun, Osservatorio
Astronomico di Roma, Nos. 53 - 56, 62 - 65 (M. Cimino,
08.075.032).

008.089 Rosario

Actividades del Observatorio.
Obs. Astron. Municipal Rosario, (*Argentina*), Bol. No. 2, p.
30 - 32 (1972).

**El laboratorio fotográfico del Observatorio Astro-
nómico Municipal de Rosario.** C. F. Sosa.

Obs. Astron. Municipal Rosario, (*Argentina*), Bol. No. 2, p. 25 - 29 (1972).

Observatorio Astronómico Municipal de Rosario, *Argentina,* Boletín No. 2 (J. L. Sersic, 08.158.099; V. Capolongo, 08.013.013; J. A. Gutierrez, 08.075.012; R. R.Barbarroja, 08.033.049; C. F. Sosa, 08.008.089; 08.008.089).

Observatorio Astronómico Municipal, Rosario – Argentina. Departamento de Fisica Solar, Contribuciones, Serie 1, Nos. 4 (J. A. Gutiérrez, V. Capolongo, R. Barbarroja, C. F. Sosa, 08.075.034), 5 (R. Barbarroja, V. Capolongo, J. A. Gutiérrez, O. F. Liesche, L. A. Mansilla, 08.075.035).

008.090 **San Juan**

Observatorio 'Félix Aguilar'. ('Félix Aguilar' Astronomical Observatory, School of Engineering, National University of Cuyo).
Inform. Bull. Southern Hemisph., No. 20, p. 9 - 10 (1972). Current research report.

Yale-Columbia Southern Observatory and U.S. Naval Observatory Southern Hemisphere Expedition.
Inform. Bull. Southern Hemisph., No. 20, p. 6 - 7 (1972). Current research report.

008.091 **San Miguel**

Observatorio Nacional de Física Cósmica (National Observatory of Cosmic Physics).
Inform. Bull. Southern Hemisph., No. 20, p. 11 (1972).–Current research report.

Observatorio Nacional de Física Cósmica (National Observatory of Cosmic Physics).
Inform. Bull. Southern Hemisph., No. 21, p. 7 (1972). – Current research report.

008.092 **Santiago**

Departamento de Astronomía, Universidad de Chile. (Astronomy Department, University of Chile).
Inform. Bull. Southern Hemisph., No. 20, p. 18 (1972). Current research report.

Departamento de Astronomía, Universidad de Chile (Astronomy Department, University of Chile).
Inform. Bull. Southern Hemisph., No. 21, p. 14 - 16 (1972). A) National Astronomical Observatory, Cerro Calán; B) Cerro El Roble Astronomical Station; C) Maipú Radioastronomical Observatory. – Current research reports.

Departamento de Astronomía, Universidad de Chile, Facultad de Ciencias Fisicas y Matematicas, Observatorio Astronomico Nacional, Cerro Calán, Santiago de Chile, **Publicaciones,** Vol. 2, Nos. 1 (F. Noel, K. Czuia, P. Guerra, 08.041.032; J. Stock, A. R. Upgren, 08.031.049; A. Gutiérrez-Moreno, H. Moreno, 08.082.152), 2 (A. Gutiérrez-Moreno, H. Moreno, J. Stock, 08.114.107; C. Contreras, J. Stock, 08.112.008; H. Wroblewski, L. Panaiotov, S. Vásquez, 08.098.032), 3 (J. Stock, H. Wroblewski, 08.114.108).

Universidad de Chile, Departamento de Astronomía, *Santiago,* Separata 8 (T. D. Carr, M. A. Lynch, M. P. Paul,

G. W. Brown, J. May, N. F. Six, V. M. Robinson, W. F. Block, 08.099.083), (H. Moreno, 05.113.037), (A. Gutiérrez-Moreno, H. Moreno, 07.113.003).

008.093 **São José dos Campos**

Observatorio Astronômico do Instituto Tecnológico de Aeronáutica (Astronomical Observatory of the Aeronautical Technical Institute).
Inform. Bull. Southern Hemisph., No. 20, p. 16 - 17 (1972). Current research report.

008.094 **São Paulo**

Centro de Radio Astronomía e Astrofísica da Universidade Mackenzie (C. R. A. A. M.). (Center of Radio Astronomy and Astrophysics, Mackenzie University).
Inform. Bull. Southern Hemisph., No. 20, p. 14 - 15 (1972). Current research report.

Instituto Astronômico e Geofísico da Universidade de São Paulo (Astronomical and Geophysical Institute of São Paulo University).
Inform. Bull. Southern Hemisph., No. 20, p. 15 - 16 (1972). Current research report.

Centro de Radio Astronomia e Astrofísica da Universidade Mackenzie (CRAAM) (Center of Radio Astronomy and Astrophysics, Mackenzie University).
Inform. Bull. Southern Hemisph., No. 21, p. 11 - 12 (1972). Current research report.

Instituto Astronómico e Geofísico da Universidade de São Paulo (Astronomical and Geophysical Institute of São Paulo University).
Inform. Bull. Southern Hemisph., No. 21, p. 12 - 13 (1972). Current research report.

008.095 **Sendai**

Sendai Astronomiaj Raportoj, Nos. 122 (M. Kubo, 08.162.039), 123 (A. Kudô, 08.131.071), 124 (M. Takeuti, Y. Shibata, 08.122.081), 125 (M. Kubo, 08.162.040), 126 (K. Arai, 08.143.032), 128 (S. Kikuchi, 08.042.059).

008.096 **Shemakha**

Astronomy in the Azerbaidzhan SSR.
G. F. Sultanov.
Zemlya i Vselennaya, 1972, No. 5, p. 49 - 53. In Russian.

008.097 **Sonneberg**

Mitteilungen über Veränderliche Sterne, *Sonneberg,* Vol. 6, No. 3,(L. Meinunger, 08.123.062; W. Wenzel, 08.123.063; W. Götz, W. Wenzel, 08.122.126; H. Geßner, 08.122.127; H. Geßner, 08.123.064; C. Thänert, 08.123.065; C. Thänert, 08.123.066; I. Meinunger, 08.123.067; E. Splittgerber, 08.123.068; H.-J. Blasberg, 08.123.069; H.-J. Blasberg, 08.122.129; H.-J. Blasberg, 08.123.070; W. Zschocke, 08.123.071; W. Zschocke, 08.123.072; S. Bratner, 08.123.073; W. Wenzel, H. Geßner, 08.142.148).

008.098 St. Andrews

University Observatory, St. Andrews. Reprints,
Nos. 39 (I. G. van Breda, 05.034.021), 40 (D. W. N. Stibbs,
06.091.002), 41 (P. W. Hill, 06.112.001), 42 (T. R. Carson,
07.065.140), 43 (R. W. Hilditch, 06.123.015), 44 (R. W. Hilditch, 07.121.020), 45 (I. G. van Breda, 07.031.016), 46
(I. G. van Breda, P. W. Hill, 07.034.103), 47 (C. L. Stephens,
I. G. van Breda, 07.031.032).

008.099 Sydney

Sydney Observatory.
Inform. Bull. Southern Hemisph., No. 20, p. 12 (1972).
Current research report.

Sydney Observatory. Report for the year ending
1971 December 31.   H. Wood.
Quarterly Journ. Roy. Astron. Soc., Vol. 13, 463 (1972).

Sydney Observatory Papers, Nos. 64 (K. P. Sims,
08.096.012), 65 (W. H. Robertson, 08.098.027).

Division of Radiophysics, C.S.I.R.O., Sydney
(Epping, New South Wales, Australia), **Separate prints** (A. J.
Shimmins, J. G. Bolton, 08.141.002; O. B. Slee, C. S. Higgins,
06.142.068; F. F. Gardner, R. X. McGee, 05.131.074; J. B.
Whiteoak, F. F. Gardner, 05.158.058; K. J. Johnston, B. J.
Robinson, J. L. Caswell, R. A. Batchelor, 07.131.010; B. A.
Peterson, J. G. Bolton, 07.141.009; D. K. Milne, 08.125.011;
G. A. Dulk, O. B. Slee, 08.125.018; J. L. Caswell, 08.131.057;
J. K. Merkelijn, 08.141.061; J. R. Dickel, D. K. Milne, 08.125.
021; R. T. Schilizzi, I. A. Lockhart, J. V. Wall, 08.141.093;
R. X. McGee, J. W. Brooks, R. A. Batchelor, 08.159.013;
N. W. Broten, 08.159.014; R. X. McGee, J. W. Brooks, R. A.
Batchelor, 08.159.015; R. X. McGee, L. M. Newton, 08.159.
016; D. J. McLean, K. V. Sheridan, R. T. Stewart, J. P. Wild,
06.077.036; V. Radhakrishnan, J. W. Brooks, W. M. Goss, J. D.
Murray, U. J. Schwarz, 07.033.001 (Abstr.); V. Radhakrishnan,
J. D. Murray, P. Lockhart, R. P. J. Whittle, 07.131.002(Abstr.);
V. Radhakrishnan, W. M. Goss, J. D. Murray, J. W. Brooks,
07.131.003 (Abstr.); W. M. Goss, V. Radhakrishnan, J. W.
Brooks, J. D. Murray, 07.131.004 (Abstr.); V. Radhakrishnan,
W. M. Goss, 07.131.005 (Abstr.); E. J. Schmahl, 07.077.033;
A. C. Riddle, 07.077.034; R. T. Stewart, 07.077.035; I. D.
Palmer, R. P. Lin, 07.077.036; I. D. Palmer, S. F. Smerd, A. C.
Riddle, 07.077.037; J. A. Roberts, J. C. Ribes, J. D. Murray,
D. J. Cooke, 07.141.130; J. V. Wall, 08.141.091; J. V. Wall,
08.141.092; G. A. Day, J. L. Caswell, D. J. Cooke, 08.157.008;
A. J. Shimmins, J. G. Bolton, 08.141.131).

008.100 Tampa

**Astronomical Contributions from the University of
South Florida at Tampa,** Nos. 16 (W. D. Googe, H. Eichhorn,
C. F. Luckac, 04.031.002), 22 (S. Sofia, 01.131.107), 23
(S. Sofia, 04.142.031), 29 (R. E. Wilson, 04.121.012), 30
(H. Eichhorn, A. Rust, 03.112.006), 32 (R. E. Wilson, L. W.
Twigg, 03.142.039), 33 (R. E. Wilson, 04.142.035), 36
(R. E. Wilson, E. J. Devinney, 05.121.047), 37 (R. E. Wilson,
05.142.036), 38 (R. E. Wilson, 06.121.065), 39 (J. H. Hunter,
Jr., 04.131.032), 40 (G. Gatewood, H. Eichhorn, 07.112.011),
41 (J. H. Hunter, S. Sofia, 06.133.020), 42 (A. F. Aveni,
J. H. Hunter, 07.065.047), 43 (S. Sofia, 06.142.034), 44
(S. Sofia, 07.125.004), 46 (S. Sofia, 06.065.103), 47 (E. J.
Devinney, Jr., 06.121.045), 48 (S. Sofia, 03.142.002), 49
(H. Eichhorn, G. D. Gatewood, S. Sofia, 04.112.025), 50

(B. Garfinkel, A. Jupp, C. Williams, 05.042.013), 51 (F. W.
Fallon, R. E. Murphy, 07.099.005), 53 (H. Eichhorn, 06.041.
031), 55 (T. B. Horák, 07.121.033), 56 (R. E. Wilson, 06.121.
089), 57 (R. E. Wilson, 07.142.062), 58 (S. Sofia, 07.142.063).

008.101 Tartu

Estonian astronomers investigate galaxies and stars.
L. Luud.
Zemlya i Vselennaya, 1972, No. 6, p. 22 - 24. In Russian.

Tartu Astronoomia Observatoorium, Teated, No. 40
(J. Einasto, 08.155.043).

008.102 Tashkent

Uzbekistan's astronomy in the past 50 years.
V. P. Shcheglov.
Priroda, No. 12.72, p. 89 - 93 (1972). In Russian.

**Astronomical investigations in the Uzbek SSR and
short-range perspectives of their development.**
V. P. Shcheglov.
Zemlya i Vselennaya, 1972, No. 5, p. 42 - 48. In Russian.

008.103 Tautenburg

**Ten years of research with the 134 cm Schmidt
telescope.**   See Abstr. 013.011.

**Fünf Jahre Spektroskopie am Karl-Schwarzschild-
Observatorium Tautenburg.**   See Abstr. 114.174.

**Zentralinstitut für Astrophysik. Mitteilungen des
Karl-Schwarzschild-Observatoriums Tautenburg** der Deutschen
Akademie der Wissenschaften zu Berlin, Nos. 60 (N. Richter,
08.004.064), 61 (08.013.024), 62 (L. Richter, N. B. Richter,
W. Wenzel, 06.113.044), 63 (S. Marx, W. Pfau, N. Richter,
06.031.045).

008.104 Teide

**Estudio de las condiciones del Observatorio del
Teide para la observación astronómica. V. Astronomía in-
frarroja. Cantidad de vapor de agua.**
C. Sánchez-Magro, D. K. Aitken, J. Alvarez, R. E. Jennings,
P. G. Polden, J. J. Tood.
Urania Barcelona, Año 56, No. 274, 215 - 226 = Obs. Astron.
Teide, Tenerife, Publ. No. 15 (1971).

Observatorio Astronomico del Teide, Tenerife
"Islas Canarias". Publicación Nos. 12 (J. M. Torroja, 06.008.
103), 13 (C. Sánchez-Magro, F. Sánchez, 06.105.154), 14
(C. Sánchez-Magro, F. Sánchez, 06.034.112), 15 (C. Sánchez-
Magro, D. K. Aitken, J. Alvarez, R. E. Jennings, P. G. Polden,
J. J. Tood, 08.008.104).

008.105 Thessaloniki

**Université de Thessaloniki, Annuaire de l'Institut
Météorologique et Climatologique,** Nos. 36 - 37 (G. C. Livadas,
08.082.238).

Publications of the Meteorological-Climatological Institute of the University of Thessaloniki, Nos. 1 (G. C. Livadas, A. Arseni-Papadimitriou, 08.082.239), 2 (T. J. Makroyannis, A. Arseni-Papadimitriou, 08.082.240).

008.106 Tokyo

A report on the use of the computer at the Tokyo Astronomical Observatory.
T. Tsumita, T. Hirayama, Y. Nishino.
Univ. Tokyo, Tokyo Astron. Obs. Rep. (No. 60), Vol. 16, 19-24 (1972). In Japanese.

Annals of the Tokyo Astronomical Observatory, University of Tokyo, Second Series, Vol. 13, No. 2 (K. Saito, 08.074.043; S. Hata, A. Tojo, 08.074.044), No. 3 (Y. Yamashita, 08.114.130).

Tokyo Astronomical Bulletin, Tokyo Astronomical Observatory, Second Series, Nos. 219 (M. Saitō, H. Sato, N. Sato, 08.121.066), 220 (M. Kitamura, A. Yamasaki, 08.121.067), 221 (M. Kiyokawa, M. Kitamura, M. Saitō, H. Sato, N. Sato, H. Ogata, 08.121.094), 222 (K. Ichimura, Y. Shimizu, 08.122.123).

Tokyo Astronomical Observatory, Reprints. Nos. 417 (H. Kinoshita, 08.042.043), 418 (K. Nariai, 08.119. 006), 419 (M. Saitō, H. Sato, 08.121.062), 420 (N. Sekiguchi, 08.044.018), 421 (M. Makita, 07.073.056), 422 (S. Okazaki, M. Nasaka, 08.045.011), 423 (S. Iijima, S. Okazaki, 08.044. 003), 424 (N. Sekiguchi, 08.045.029), 425 (Y. Kozai, 08.034. 100), 426 (T. Hirayama, 08.073.009).

University of Tokyo, Tokyo Astronomical Observatory, Report, No. 60, Vol. 16, No. 1 (R. Nagai, 08.042.078; M. Mizohara, S. Ito, N. Kobayashi, 08.021.008; T. Tsumita, T. Hirayama, Y. Nishino, 08.008.106; T. Hirayama, S. Hamana, I. Shimizu, A. Tokuya, H. Miyazaki, Y. Yose, F. Moriyama, 08.032.036; K. Hurukawa, 08.094.209; S. Nishimura, E. Watanabe, 08.031.059; J. Matsumoto, M. Kiyokawa, 08.031. 060; S. Nishimura, 08.031.061; K. Saito, S. Shinozawa, 08.093.041; S. Iijima, 08.031.062; S. Iijima, Y. Niimi, 08.044. 023; M. Yoshinari, K. Fujiwara, 08.034.112; S. Iijima, Y. Adachi, I. Oguma, 08.034.113; H. Kinoshita, 08.042.079; H. Kinoshita, G. Hori, 08.052.034; K. Oki, H. Nakajima, Y. Shiomi, N. Shibuya, H. Sekiguchi, S. Aiba, S. Tachikawa, 08.033.052).

Time and Latitude Bulletins, Tokyo Astronomical Observatory, Vol. 46, Nos. 1 - 8 (M. Huruhata, 08.044.040).

008.107 Tonantzintla

Boletin de los Observatorios Tonantzintla y Tacubaya, Vol. 6, No. 38 (G. Haro, G. González, 08.122.136; G. Haro, E. Chavira, 08.122.137; A. D. Andrews, 08.113.062; A. D. Andrews, 08.122.138; P. Pişmiş, 08.122.139; A. G. D. Philip, J. Stock, 08.114.173; E. E. Mendoza V., 08.113.063; E. E. Mendoza V., J. Luna, T. Gómez, 08.082.233).

008.108 Torino

Il rifrattore fotografico dell'Osservatorio Astronomico di Torino. See Abstr. 032.037.

Contributi dell'Osservatorio Astronomico di Torino, (Pino Torinese), Nos. 60 (G. Cocito, N. Missana, C. Moranzino, 07.032.012), 61 (E. Zanoner, 08.034.114), 62 (C. Moranzino, 08.031.002), 63 (M. Boggio, M. G. Fracastoro, G. Francese, C. Morais, 08.032.037), 64 (M. A. Vogliotti, V. Zappalà, 08.098.006).

Osservatorio Astronomico di Torino, Pino Torinese. Time Service, Bulletin No. 2 (C. Moranzino, 08.044.038).

Pubblicazioni Varie Fuori Serie dell'Osservatorio Astronomico di Torino (Pino Torinese), Nos. 48 (M. G. Fracastoro, 07.008.145), 49 (M. Boggio, M. G. Fracastoro, G. Francese, 08.079.100), 50 (S. Vaghi, 08.010.017), 51 (08.047.032).

008.109 Tortosa

Publicaciones del Observatorio del Ebro, Miscelánea, Nos. 25 (A. Romañá, 08.015.017), 26 (E. Galdón, 06.083. 018).

008.110 Toruń

Bulletin of the Astronomical Observatory of N. Copernicus University in Toruń, No. 49 (W. Iwanowska, 08.155.061; W. Wegner, 08.082.214; L. Adamiak, S. Gąska, K. Wiercioch, 08.104.049; S. Gąska, 08.098.043).

Astronomy in Toruń, Nicholas Copernicus' native town. See Abstr. 003.026.

008.111 Turku

Astronomia-Optika Institucio, Universitato de Turku, Informo No. 37 (L. Oterma, 08.007.000; Y. Väisälä, 08.081.056; Y. Väisälä, L. Oterma, 08.031.086).

008.112 Uccle

Bulletin Astronomique. (Astronomisch Bulletin). Observatoire Royal de Belgique (Koninklijke Sterrenwacht van België), Vol. 8, No. 1 (H. Debehogne, 08.098.044; H. Debehogne, 08.098.045; H. Debehogne, 08.098.046; H. Debehogne, 08.103.129; H. Debehogne, 08.055.013; J. Dommanget, E. Van Dessel, 08.096.019; C. Gonze, R. Gonze, 08.077.012; S. Arend, R. R. de Freitas Mourão, 08.118. 013; J. Dommanget, 08.118.014).

Observatoire Royal de Belgique (Koninklijke Sterrenwacht van Belgie), **Communications** (Mededelingen), Série B, Nos. 71 (P. Melchior, 08.044.002), 72 (P. Melchior, R. Dejaiffe, R. Verbeiren, 08.045.004), 73 (P. Melchior, R. Dejaiffe, 07.041.020), 74 (P. Melchior, R. Dejaiffe, 07.041.019), 76 (P. Melchior, 08.045.038).

008.113 Utrecht

Utrechtse Sterrekundige Overdrukken, Sterrewacht 'Sonnenborgh', Utrecht, Nos. 129 (A. C. Brinkman, W. de Graaff,

M. L. Shaw, 04.073.063),130 (C. de Jager,C. de Loore, 04.073. 003), 131 (C. de Loore, C. de Jager, 04.064.020), 132 (R. C. Canfield, 05.071.001), 133 (R. C. Canfield, 05.064.001), 134 (P. S. Conti, E. P. J. van den Heuvel, 04.153.037), 135 (T. de Graauw, B. P. T. Veltman, 04.031.046), 136 (J. Houtgast, 04.071.053), 137 (A. Schadee, 04.072.047), 138 (C. de Jager, 07.080.029), 139 (C. de Jager, 08.071.050), 140 (M. Kuperus, 07.074.088), 141 (M. A. J. Snijders, A. B. Underhill, 05.064.021), 142 (A. D. Fokker, J. Houtgast, C. de Jager, 05.008.137), 143 (R. Rosenberg, 04.074.019), 144 (H. van de Stadt, 08.034.132), 145 (W. van Rensbergen, J. Wuyts, 04.133.009), 146 (N. Nieuwenhuijzen, 04.022.035), 147 (W. van Rensbergen, 03.071.003), 148 (T. de Groot, 04. 077.014), 149 (H. G. van Bueren, M. Kuperus, 04.077.016), 150 (R. J. Bessey, M. Kuperus, 03.080.011), 151 (H. F. van Beek, G. A. Stevens, 04.078.030), 153 (E. L. J. van Dessel, 04.071.054), 154 (M. Kuperus, 05.064.028), 155 (J. Houtgast, 04.007.000), 156 (C. de Jager, C. de Loore, 05.064. 036), 157 (O. Havnes, 05.143.035), 158 (E. P. J. van den Heuvel, P. S. Conti, 05.065.019), 159 (E. P. J. van den Heuvel, 05.114.027), 160 (C. de Jager, 08.007.000), 161 (C. Zwaan, J. Buurman, 06.072.028), 162 (J. Houtgast,O. Namba, R. J. Rutten, T. de Graauw, 03.071.035), 163 (N. Niewenhuijzen, C. Veth, 08.034.133), 164 (A. Jacobs, 05.022.025), 165 (C. de Jager, A. Hammerschlag, W. Werner, 06.034.072), 166 (O. Havnes, P. S. Conti, 06.065.009), 167 (W.Van Rensbergen, 05.022.077), 168 (C. de Jager, 06.064.049), 169 (L.D. de Feiter, 06.073.017), 170 (O. Havnes, 05.116.009), 171 (H. Rosenberg, 06.074.039), 172 (L. J. Lantwaard, H. van de Stadt, 06.034.134), 173 (A. D. Fokker, 06.077.013), 174 (J. H. Dijkstra, W. Werner, 06.034.055), 175 (A. J. F. den Boggende, H. F. van Beek, A. C. Brinkman, H. T. J. A. Lafleur, 06.034.060), 176 (J. H. Dijkstra, W. de Graaff, L. J. Lantwaard, 06.031.022), 177 (C. de Jager, 07.031.044), 178 (K. Fredga, J. A. Högbom, 06.034.079), 179 (A. Jacobs, 07. 022.054), 180 (A. T. Forrester, H. G. van Bueren, 08.033. 068), 181 (H. J. Lamers, 07.114.023), 182 (D. H. Menzel, M. Minnaert, B. Levin, A. Dollfus, B. Bell, 06.094.002), 183 (R. Mewe, 07.074.023), 184 (R. Mewe, 07.076.007), 185 (C. de Jager, L. Neven, 07.071.004), 186 (M. Kuperus, 07.080. 004), 187 (H. Houtgast, O. Namba, R. J. Rutten, J. W. Wijbenga, 07.071.003), 188 (R. H. Hammerschlag, 08.022. 138), 189 (H. Rosenberg, 08.074.099), 190 (A. Schadee, 06. 022.032), 191 (A. C. Brinkman, M. L. Shaw, 07.076.021).

008.114 Victoria

The Dominion Astrophysical Observatory, Victoria, B. C.   J. B. Hutchings.
Journ. Roy. Astron. Soc. Canada, Vol. 66, 276 (1972).

Dominion Astrophysical Observatory,Victoria, B.C. Report for the year April 1971 to 1972 March 31. K. O. Wright.
Quarterly Journ. Roy. Astron. Soc., Vol. 13, 526 - 537 (1972).

Contributions from the Dominion Astrophysical Observatory, Victoria, B.C., Nos. 133 (J. B. Hutchings, 02. 114.024), 158 (C. L. Morbey, J. B. Hutchings, 05.096.006), 159 (D. Crampton, 05.114.048), 160 (G. Hill, S. C. Morris, G. A. H. Walker, 05.113.013), 162 (J. R. Stilborn, J. M. Fletcher, F. D. A. Hartwick, 07.034.012), 165 (D. Crampton, 06.114.015), 166 (J. B. Hutchings, K. O. Wright, 06.121.057), 167 (E. Brosterhus, E. Pfannenschmidt, F. Younger, 07.082. 040), 169 (D. Crampton, 07.115.003), 171 (J. B. Hutchings, J. R. Auman, A. C. Gower, G. A. H. Walker, 06.114.116), 172 (A. H. Batten, M. Plavec, 06.119.005, 06.121.033), 173 (E. H. Richardson, G. J. Odgers, 07.032.023), 174 (C. D. Scarfe, 06. 119.018), 175 (J. B. Hutchings, R. O. Redman, 07.122.070),

179 (A. H. Batten, R. P. Olowin, 06.117.025), 180 (J. B. Hutchings, 08.114.120), 182 (F. D. A. Hartwick, J. E. Hesser, G. Hill, 07.122.132).

Publications of the Dominion Astrophysical Observatory, Victoria, B.C., Vol. 14, Nos. 3 (C. L. Morbey, 08.096. 011), 4 (J. B. Hutchings, 08.064.072), 5 (A. H. Batten, B. Szeidl, 08.119.014).

008.115 Vilnius

Lithuanian astronomers gave rise to the polychromatic photometry of stars.   V. Straižys.
Zemlya i Vselennaya, 1972, No. 6, p. 16 - 18. In Russian.

Vilniaus Astronomijos Observatorijos Biuletenis (Bulletin of the Vilnius Astronomical Observatory), Nos. 33 (G. Kavaliauskaitė, 08.113.048; A. Bogdanovičius, V. Straižys, 08.113.049; R. Bartkus, 08.114.140; A. Pučinskas, 08.117. 036), 34 (K. Zdanavičius, V. B. Nikonov, J. Sūdžius, V. Straižys, Z. Sviderskienė, R. Kalytis, E. Jodinskienė, E. Meištas, G. Kavaliauskaitė, V. Jasevičius, G. Kakaras, A. Bartkevičius, A. Gurklytė, R. Bartkus, A. Ažusienis, J. Sperauskas, A. Kazlauskas, V. Žitkevičius, 08.113.050; V. Žitkevičius, V. Straižys, 08.113.051).

008.116 Warsaw

Politechnika Warszawska, Obserwatorium Astronomiczno-Geodezyjne w Józefosławiu, (Warsaw Technical University, Astronomic-Geodetical Observatory at Józefosław), Latitude Circular, Nos. 39 - 42 (L. Pieczyński, 08.045.050).

008.117 Washington

Astronomical Papers prepared for the use of the American Ephemeris and Nautical Almanac, Vol. 21, Part I (P. M. Janiczek, 08.098.007).

Publications of the United States Naval Observatory, Washington, Second Series, Vol. 20, Part 1 (K. A. Strand, 08.032.019), Vol. 22, Part 3 (R. S. Harrington, B. F. Mintz, 08.098.018), Part 4 (C. E. Worley, 08.118.017).

United States Naval Observatory, Washington, D.C., Circular, Nos. 138 (G. H. Kaplan, 08.041.045), 139 (G. H. Kaplan, P. K. Seidelmann, E. Smith, 08.041.046).

U. S. Naval Observatory, Washington, D. C., Time Service Publications, Series 4, Nos. 283 - 308 (08.044.046); Series 7, Nos. 236 - 261 (08.044.047).

008.118 Waterloo

Contributions of the University of Waterloo Observatory, Nos. 9 (M. P. FitzGerald, 07.114.079), 10 (G. A. Bakos, 07.121.036), 11 (R. A. Bell, M. P. FitzGerald, 08.155.047), 16 (W. J. F. Wilson, M. P. FitzGerald, 08.155.036).

008.119 Wellington

Astronomical handbook and public programme for

1972. W. J. H. Fisher.
Astron. Bull. Carter Obs. Wellington, New Zealand, No. 76, 27 pp. (1972).

Report of the Carter Observatory Board. For the year ended 1972, March 31. R. P. Gough.
Astron. Bull. Carter Obs. Wellington, New Zealand, No. 77, p. 1 - 2 (1972).

Report of the director. For the year ended 1972, March 31. W. J. H. Fisher.
Astron. Bull. Carter Obs. Wellington, New Zealand, No. 77, p. 3 - 6 (1972).

008.120 Wien

Astronomische Mitteilungen Wien, No. 9 (J. Hopmann, 08.155.042).

008.121 Zürich

Publikationen der Eidgenössischen Sternwarte Zürich, Vol. 14, No. 1 (M. Waldmeier, 08.075.006).

Quarterly Bulletin on Solar Activity (Zürich), Nos. 175 - 176 (M. Waldmeier, R. Howard, R. Michard, G. Olivieri, M. Bernot, 08.075.007).

009 Notes on Observatories, Planetaria, and Exhibitions

009.001 Earthquake risks in the region around Castelgrande. P. J. Treanor.
Mem. Soc. Astron. Italiana, Nuova Ser., Vol. 43, 347 - 364 (1972).
  This report assembles the available data of earthquakes in the region around Castelgrande, Italy, evaluates their completeness and statistical limitations, and uses them to estimate the degree of risk involved for an astronomical observatory in this region.

009.002 The Vanderbilt Planetarium. P. L. Harrison.
Sky Telescope, Vol. 44, 72 - 76 (1972).

009.003 The Newton Observatory in Manitoba. J. B. Newton.
Sky Telescope, Vol. 44, 120 - 124 (1972).

009.004 Two university field stations in the northwest.
Sky Telescope, Vol. 44, 150 (1972).

009.005 The Radcliffe Observatory, Oxford, 1772 - 1929. V. Reade.
Journ. British Astron. Ass., Vol. 82, 377 - 378 (1972).

009.006 The Newton Observatory. J. B. Newton.
Journ. Roy. Astron. Soc. Canada, Vol. 66, 224 - 226 (1972).

009.007 Das Planetarium der Volkssternwarte München. H. Oberndorfer.
SuW, Vol. 11, 250 - 251 (1972).

009.008 Stuttgarter Volkssternwarte 50 Jahre alt. M. Gerstenberger.
SuW, Vol. 11, 251 - 252 (1972).

009.009 Fünfzig Jahre Schulsternwarte Bautzen. H. Bernhard.
Astron. in der Schule, 9. Jahrgang, p. 75 - 77 (1972).

009.010 Besuch in Heluan. W. Lauterbach.
Sterne, 48. Jahrgang, p. 181 - 183 (1972).

009.011 Siding Spring. E. W. Titterton.
Nature, Vol. 239, 417 (1972).

009.012 First people's observatory in Moscow. V. K. Lutskij.
Zemlya i Vselennaya, 1972, No. 4, p. 62 - 64. In Russian.

009.013 Sholinsk people's observatory. I. P. Yartsev.
Zemlya i Vselennaya, 1972, No. 4, p. 70 - 71. In Russian.

009.014 Projet d'un Centre d'Études et de Recherches Géodynamiques et Astronomiques (CERGA).
CERGA, No. 1, 1 + 25 + 12 pp. (1970).
  La création d'un Centre d'Études et de Recherches Géodynamiques et Astronomiques (CERGA) est destinée à rassembler en un Etablissement unique des équipes de chercheurs travaillant dans différents domaines de l'astronomie, de la géophysique, de la géodésie et de la mécanique céleste. Le CERGA sera implanté sur le plateau de Calern au nord de Grasse et dans son voisinage immédiat. Ce choix est issu de l'analyse de la prospection astrométrique menée en France de 1966 à 1970. Les six membres du groupe qui a établi le présent projet s'engagent solidairement à en assurer le développement et la réalisation: F. Barlier, B. Guinot, J. Lévy, (Observatoire de Paris); P. Couteau (Observatoire de Nice); M. Lefebvre (Département de géodésie spatiale, CNES); J. Kovalevsky (Bureau des Longitudes).

009.015 Istituto di Fisica dell'Atmosfera (IFA), Roma. M. Giorgi.
Ricerca Scientifica, Anno 40, p. 469 - 479 (1970).

009.016 Laboratorio di Ricerca e Tecnologia per lo Studio del Plasma nello Spazio, Roma. A. Egidi.
Ricerca Scientifica, Anno 40, p. 631 - 633 (1970).

009.017 A new observatory at Chapel Hill. R. S. Knapp, W. A. Christiansen.
Sky Telescope, Vol. 44, 288 - 289 (1972).

009.018 From the history of amateur astronomy in Odessa. V. A. Smirnov.
Zemlya i Vselennaya, 1972, No. 5, p. 71 - 74. In Russian.

009.019 Hawaii's Mauna Kea Observatory today. D. Morrison, J. T. Jefferies.
Sky Telescope, Vol. 44, 361 - 365 (1972).

009.020 L'osservatorio astronomico "G. Horn d'Arturo" dell'Associazione Astrofili Bolognesi. L. Baldinelli.
Coelum, Vol. 40, 175 - 185 (1972).

009.021 Francijas observatorijās. J. Francmanis.
Zvaigžņotā debess, 1971./72. gada ziema, p. 26 - 38.

009.022 **L'osservatorio astronomico "G. Horn d'Arturo"
dell'Associazione Astrofili Bolognesi.**
L. Baldinelli.
Coelum, Vol. 40, 219 - 227 (1972).

009.023 **From the proceedings of the Commission of
Popular Astronomical Observatories and Planetari-
ums.** P. Rybka.
Urania Kraków, Vol. 43, 249 - 252 (1972). In Polish.

009.024 **Observatories and planetariums in Czechoslovakia.**
O. Obůrka.
Urania Kraków, Vol. 43, 322 - 327 (1972). In Polish.

009.025 **Geophysical Fluid Dynamics Laboratory, Meteorological Office, Bracknell, Berkshire.** – Report for
the year ending 1971 December 31. R. Hide.
Quarterly Journ. Roy. Astron. Soc., Vol. 13, 549 (1972).

009.026 **The Leningrad planetarium.** S. Sadžakov.
Vasiona, Vol. 20, 56 - 59 (1972). In Serbo-Croatian.

009.027 **Kunst und Wissenschaft im Planetarium.**
D. B. Herrmann.

Blick in das Weltall, Archenhold-Sternw. Berlin-Treptow,
1972, No. 11, p. 99 - 102.

009.028 **Drei Jahre Planetarium Bogotá/Kolumbien.**
C. Garavito.
Jenaer Rundschau, (Jena Rev.), 17. Jahrgang, p. 360 - 361, 363
(1972).

009.029 **The Urania observatory in Budapest.** I. Molnár.
Kozmos, Vol. 3, 83 - 84 (1972). In Slovak.

009.030 **El planetario de Chorillos.** N. A. de Chavarry.
Bol. As. Peruana Astron., Vol. 6, (No. 148), 533
(1971).

009.031 **Max-Planck-Institut für Kernphysik Heidelberg.**
Jahresbericht 1971.
B. Martin (Editor).
Published by Max-Planck-Inst. für Kernphysik, Heidelberg.
22 + 250 pp. (1972). – Included is a report of the department
of cosmochemistry.

**Kopernikus und die Erfindung des Projektions-
planetariums.** See Abstr. 004.068.

010 Societies, Associations, Organizations

010.001 **American Association of Variable Star Observers
(AAVSO)**

Notes on the AAVSO annual fall meeting.
Sky Telescope, Vol. 44, 376 - 377 (1972).

010.002 **American Astronomical Society (AAS)**

The TGEA (*Task Group for Education in Astrono-
my*) – a new educational venture of the AAS.
G. Verschuur, D. G. Wentzel.
Bull. American Astron. Soc., Vol. 4, 405 - 406 (1972).

Report of the High Energy Astrophysics Division.
H. Bradt.
Bull. American Astron. Soc., Vol. 4, 407 - 408 (1972). – In-
cluded is a list of invited papers.

**Abstracts of papers presented at the meeting of the
AAS Division of High Energy Astrophysics, held 23–25
October 1972 at Pasadena, California.**
Bull. American Astron. Soc., Vol. 4, 409 - 417 (1972).

**Abstracts of papers presented at the joint meeting of
the AAS Division on Dynamical Astronomy/SIAM, held 19–
20 October 1972 at Austin, Texas.**
Bull. American Astron. Soc., Vol. 4, 417 - 423 (1972).

**Late-paper abstracts from the 138th meeting of the
American Astronomical Society, held 15–18 August 1972 at
East Lansing, Michigan.**
Bull. American Astron. Soc., Vol. 4, 423 - 425 (1972).

010.003 **Association of Lunar and Planetary Observers
(ALPO)**

Announcements.
Strolling Astronomer, Vol. 23, 226 - 228; Vol. 24, 38 - 40
(1972).

**Educational and observational aids offered by the
ALPO.** H. D. Jamieson.
Strolling Astronomer, Vol. 23, 222 - 223 (1972).

010.004 **Astronomical Society of Australia (ASA)**

No publication received.

010.005 **Astronomical Society of Czechoslovakia**

No publication received.

010.006 **Astronomical Society of the Pacific (ASP)**

In focus.
Mercury, (Journ. Astron. Soc. Pacific), Vol. 1, No. 4, p. 8 - 9;
No. 5, p. 12 - 13 (1972). – Personal notes, institutional activi-
ties, and announcements.

Minutes of the meeting of the directors.
Mercury, (Journ. Astron. Soc. Pacific), Vol. 1, No. 5, p. 19 -
20 (1972).

Infrared excesses in stars. Scientific meeting in
Santa Cruz, 1972 June 26 - 28.

L. E. Salanave, W. A. Stein.
Mercury, (Journ. Astron. Soc. Pacific), Vol. 1, No. 5, p. 8 - 9 (1972).

Astronomical Society of the Pacific 83rd annual meeting, May 5, 1972. J. G. Phillips.
Mercury, (Journ. Astron. Soc. Pacific), Vol. 1, No. 5, p. 20 - 22 (1972).

Annual report of the treasurer for the year ending 31 December 1971. P. Davis, J. G. Phillips.
Mercury, (Journ. Astron. Soc. Pacific), Vol. 1, No. 5, p. 22 - 23 (1972).

010.007 Astronomical Society of Southern Africa (ASSA)

Notices.
Monthly Notes Astron. Soc. Southern Africa, Vol. 31, 71, 87, 119, 139, 151 (1972).

Proceedings of the annual general meeting, 1972.
Monthly Notes Astron. Soc. Southern Africa, Vol. 31, 104 - 111 (1972). – Included are: Report of Council for 1971–1972, *A. H. Jarrett, T. W. Russo;* Section reports: Occultation Section, *A. G. F. Morrisby;* Comet and Meteor Section, *J. C. Bennett;* Variable Star Section, *R. P. de Kock.*

Astronomical Society of Southern Africa. Centre reports for 1971-72.
Monthly Notes Astron. Soc. Southern Africa, Vol. 31, 123 - 126 (1972).

010.008 Astronomical Society of Victoria (ASV)

Society notes.
Journ. Astron. Soc. Victoria, Vol. 25, 59 - 60, 75 - 76 (1972).

The Astronomical Society of Victoria – the beginning. J. L. Perdrix.
Journ. Astron. Soc. Victoria, Vol. 25, 30 - 36 (1972).

Auroral Section. T. B. Tregaskis.
Journ. Astron. Soc. Victoria, Vol. 25, 58 (1972).

010.009 Astronomical Society of Western Australia (ASWA)

Reports of proceedings – 236th – 241st ordinary meetings.
Journ. Astron. Soc. Western Australia, 1972 June, August – December.

Annual general meeting.
Journ. Astron. Soc. Western Australia, 1972 July, with the presidential address by R. Lincoln.

010.010 Astronomische Gesellschaft (AG)

La 53ª riunione annuale della Società Astronomica Tedesca. F. Zagar.
Mem. Soc. Astron. Italiana, Nuova Ser., Vol. 43, 569 - 572 (1972).

010.011 Astronomisk Selskab København

No publication received.

010.012 British Astronomical Association (BAA)

Notices.
Journ. British Astron. Ass., Vol. 82, 322 - 325, 426; Vol. 83, 2 - 4 (1972).

Meetings of the Association.
Journ. British Astron. Ass., Vol. 82, 326 - 335; Vol. 83, 5 - 9 (1972).

Lunar Section. P. Moore.
Journ. British Astron. Ass., Vol. 82, 455 - 458; Vol. 83, 33 - 37 (1972).

Solar Section. H. Hill.
Journ. British Astron. Ass., Vol. 83, 57 (1972).

New members elected.
Journ. British Astron. Ass., Vol. 82, 394 - 397; Vol. 83, 75 - 76 (1972).

Report of the exhibition meeting, 1972 May 31.
A. C. Curtis.
Journ. British Astron. Ass., Vol. 82, 335 - 340 (1972).

Report of the council on work during the session 1971 July 1 to 1972 June 30 to be presented to members of the Association at the annual general meeting, 1972 October 25.
Journ. British Astron. Ass., Vol. 82, 402 - 425 (1972). – Included are Sections' reports.

010.013 British Interplanetary Society (BIS)

Society news.
Spaceflight, Vol. 14, 393, 433 - 434, 475 - 477 (1972).

27th annual general meeting.
Spaceflight, Vol. 14, 351 - 352 (1972).

010.014 Committee on Space Research (COSPAR)

Brief report of the XVth meeting of COSPAR.
Astronaut. Acta, Vol. 17, 923 - 924 (1972).

XVᵉ assemblée plénière de COSPAR, qui s'est tenue à Madrid, du 12 au 23 mai 1972. M. Roy.
Comptes Rendus Acad. Sci. Paris, Vol. 275, Vie Académique, p. 59 - 60 (1972).

Upper atmosphere and space investigations in the USSR in 1971. COSPAR report. 15th plenary meeting, Madrid, Spain.
Nauka, Moskva. 3 pp. (1972). In Russian. – Abstr. in Referativ. Zhurn. 62. Issled. kosm. prostranstva, 9.62.24 (1972).

The XVth General Assembly of COSPAR, Madrid, 14–24 May 1972. C. Popovici.
Stud. Cerc. Astron., Vol. 17, 283 - 285 (1972). In Romanian.

010.015 European Space Research Organization (ESRO)

Communiqué ESRO.
Boll. Geod. Sci. Affini, Anno 31, p. 413 - 418 (1972). –
Le conseil du CERS/ESRO approuve la résolution couvrant le programme 1972 – 1980; Le lancement du satellite TD - 1A de l'ESRO reporte au 10 mars 1972; L'ESRO participera au programme commun Canada/U.S. de satellite de télécommunications.

010.016 International Astronautical Federation (IAF)

No publication received.

010.017 International Astronomical Union (IAU)

Urgent need for new programmes of visual double stars observations in the southern hemisphere.
Inform. Bull. Southern Hemisph., No. 21, p. 31 (1972). – Circular letter from I.A.U. Commission 26 to the Southern Astronomical Organizations adhering to the International Astronomical Union.

The meeting of the National Research Council Associate Committee on Astronomy and the National Committee for Canada of the IAU at Montreal, May 11–13, 1972.
G. A. H. Walker.
Journ. Roy. Astron. Soc. Canada, Vol. 66, 213 - 214 (1972).

Il colloquio I.A.U. n. 22: «Asteroidi, comete, materia meteorica». S. Vaghi.
Mem. Soc. Astron. Italiana, Nuova Ser., Vol. 43, 403 - 404 (1972). – Nizza, 1972 April 4–6.

010.018 Meteoritical Society

Minutes of the council meeting of the Meteoritical Society. 34th annual meeting, August 1971, Tübingen, West Germany. T. E. Bunch.
Meteoritics, Vol. 7, 233 - 235 (1972).

Historical records of the Meteoritical Society.
H. G. Fales.
Meteoritics, Vol. 7, 237 - 242 (1972).

010.019 Nederlandse Vereniging voor Weer- en Sterrenkunde.

Verenigingsnieuws.
Hemel en Dampkring, Vol. 70, 228 - 230, 245, 353 - 354 (1972).

Jongerenwerkgroep.
Hemel en Dampkring, Vol. 70, 274 - 276 (1972).

010.020 Polskie Towarzystwo Astronomiczne (PTA)

No publication received.

010.021 Polskie Towarzystwo Miłośników Astronomii (PTMA)

Chronicle of the Polish Amateur Astronomical Society.
Urania Kraków, Vol. 43, 278 - 279, 311 - 313, 345 - 349 (1972). In Polish.

010.022 Royal Astronomical Society (RAS)

Meetings of the Society.
Observatory, Vol. 92, 69 - 76, 77 - 83, 109 - 115, 116 - 124, 153 - 160 (1972).

Meetings of the Society.
Quarterly Journ. Roy. Astron. Soc., Vol. 13, 479 - 482 (1972).

Annual general meeting 1972 March 10.
F. Hoyle.
Quarterly Journ. Roy. Astron. Soc., Vol. 13, 307 - 309 (1972).

Report of the council to the one hundred and fifty-second annual general meeting of the Society.
Quarterly Journ. Roy. Astron. Soc., Vol. 13, 310 - 316 (1972). This report refers to the calendar year 1971.

Report of the honorary auditors for the year 1971.
R. S. Peckover, R. C. Smith.
Quarterly Journ. Roy. Astron. Soc., Vol. 13, 317 (1972).

Treasurer's accounts for the year ending 1971 December 31. A. Hunter.
Quarterly Journ. Roy. Astron. Soc., Vol. 13, 318 - 326 (1972).

Progress and present state of the Society.
Quarterly Journ. Roy. Astron. Soc., Vol. 13, 327 (1972).

Royal Astronomical Society meeting on 'Geophysical plasma dynamics'. P. Kendall, H. Rishbeth.
Quarterly Journ. Roy. Astron. Soc., Vol. 13, 520 - 523 (1972). 1972 June 4.

010.023 Royal Astronomical Society of Canada (RAS Canada)

The Royal Astronomical Society of Canada 1968. New Service Awards.
Journ. Roy. Astron. Soc. Canada, Vol. 66, 227 (1972).

From Saskatoon Centre.
Journ. Roy. Astron. Soc. Canada, Vol. 66, L35 (1972).

Toronto Centre news. D. Ostler.
Journ. Roy. Astron. Soc. Canada, Vol. 66, L36 (1972).

Canadian astronomers meet in Vancouver.
Sky Telescope, Vol. 44, 92 (1972).

010.024 Royal Astronomical Society of New Zealand (RAS New Zealand)

Royal Astronomical Society of New Zealand. Variable Star Section.
Inform. Bull. Southern Hemisph., No. 21, p. 17 (1972). – Current research report.

010.025 **Schweizerische Astronomische Gesellschaft (SAG)**

Aus der SAG und den Sektionen.
Orion Schaffhausen, 30. Jahrgang, p. 164 - 165 (1972).

Bericht über die Generalversammlung 1972 der Schweizerischen Astronomischen Gesellschaft in Zürich.
M. Bornhauser.
Orion, 30. Jahrgang, p. 156 - 157 (1972).

010.026 **Sociedad Astronómica de México**

Actividades de la Sociedad.
El Universo, No. 98 - 99, Vol. 26, 37 - 39, 48 - 49 (1972).

010.027 **Società Astronomica Italiana (SAI)**

Verbale dell'assemblea generale della Società Astronomica Italiana.   C. Bartolini, F. Marchesini, A. Leani.
Atti XV Riunione Soc. Astron. Italiana, Bologna 1971, (see 012.013), p. 29 - 42 (1972). – Bologna, 10 Ottobre 1971.

010.028 **Société Astronomique de France (SAF)**

Les séances de la Société.   B. Clouet.
L'Astronomie, 86e année, p. 367 - 374, 433 - 435, 487 - 490 (1972).

Assemblée générale statutaire du 19 juin 1972.
B. Clouet.
L'Astronomie, 86e année, p. 480 - 486 (1972).

La vie de la Société Astronomique de France.
L. Tartois.
L'Astronomie, 86e année, p. 490 - 493 (1972).

Prix et médailles décernés par la Société.
L'Astronomie, 86e année, p. 494 - 496 (1972).

Les médailles des soixante ans.
L'Astronomie, 86e année, p. 497 (1972).

Commission des instruments et de la photographie astronomique.   A. Hamon.
L'Astronomie, 86e année, p. 501 - 517 (1972).

010.029 **Société Astronomique "R. Bošković"**

No publication received.

010.030 **Société Chronométrique de France**

Communications présentées à la réunion d'automne de la Société Chronométrique de France le 25 Novembre 1971.
Ann. Françaises Chronométrie Micromécanique, Année 1972, p. 7 - 33.

Communications présentées au Congrès de Chronométrie franco-allemand de Fribourg en Brisgau, les 2, 3 et 4 Juin 1972.
Ann. Françaises Chronométrie Micromécanique, Année 1972, p. 225 - 319.

010.031 **Société Belge d'Astronomie, de Méteorologie et de Physique du Globe**

Réunions mensuelles.
Ciel et Terre, Vol. 88, 319 - 320, 417 - 418, 497 (1972).

Assemblée générale statutaire du 25 mars 1972.
M. Bauduin, M. Ducuroir.
Ciel et Terre, Vol. 88, 411 - 416 (1972).

010.032 **Svenska Astronomiska Sällskapet**

Svenska Astronomiska Sällskapet; Astronomiska Sällskapet Tycho Brahe; Göteborgs Astronomiska Klubb. Styrelsens berättelse för år 1971.
E. Holmberg, G. Darsenius, P. Å. Björklund, B. A. Lindblad, N. Ryde, K. Rynefors.
Separate print, Svenska Astronomiska Sällskapet, Stockholm. 8 pp. (1972).

010.033 **VAGO (Astronomical-Geodetical Society of the USSR)**

Astronomy and geodesy in Azerbajdzhan. Plenum of the Central Council of VAGO in Baku. V. A. Bronshtehn.
Zemlya i Vselennaya, 1972, No. 4, p. 52 - 54. In Russian. 1972, March 15 - 18.

010.034 **Vereniging voor Steerenkunde, Belgie**

No publication received.

010.035 **Argentine Astronomical Association**

No publication received.

010.036 **Canadian Astronomical Society**

Second regular meeting of the Canadian Astronomical Society at the Université de Montréal, May 11–13, 1972.
P. M. Millman.
Journ. Roy. Astron. Soc. Canada, Vol. 66, 215 - 216 (1972).

010.037 **Nachrichten der Vereinigung der Sternfreunde e. V.**
SuW, Vol. 11, 246 - 248, 286 - 288, 323 - 325, 352, 354 (1972).

010.038 **Bruno - H. - Bürgel - Kreis Berlin.**   D. Wattenberg.
Blick in das Weltall, Archenhold-Sternw. Berlin-Treptow, 1972, No. 9, p. 81 - 82. – Annual meeting 1972.

010.039 **U. A. I. Unione Astrofili Italiani. Contributi scientifici ed organizzativi.**   A. Leani (Editor).
Segreteria Editoriale U.A.I., Cremona (Italy). Anno 2, No. 2, 36 pp. (1972).

010.040 Il 5° congresso nazionale, (Pescara: 17-18-19 Settembre 1971). A. Leani (Editor).
U.A.I., Unione degli Astrofili Italiani – «Padus» Cremona (Italy). 50 pp. (1972).

010.041 I.U.A.A. International Union of Amateur Astronomers. Contributions.
Edited and directed by A. Leani.
I.U.A.A. Bull., Cremona (Italy), No. 4, 37 pp. (1972).

010.042 Bodas de plata de la Asociación Peruana de Astronomía y la quinta convención Latinoamericana de Astronomía.
Bol. As. Peruana Astron., Vol. 6, (No. 148), 528 - 532 (1971).

010.043 Institut de France. Académie des Sciences. Annuaire pour 1973.
Gauthier-Villars, Paris. 257 pp. (1972).
Cet annuaire a pour but d'exposer l'état de l'Académie des sciences dans le présent et dans le passé. Il contient en outre des indications sur les concours de ses prix et sur ses fondations.

011 Reports on Colloquia, Congresses, Meetings, Symposia, and Expeditions

011.001 Proceedings of the fourth astrometric conference.
F. Zagar.
Mem. Soc. Astron. Italiana, Nuova Ser., Vol. 43, 409 - 411 (1972).

011.002 Report of a conference on "The organic analysis and carbon chemistry of lunar samples", October 26 - 28, 1971, College Park, Maryland. S. Chang, R. S. Young.
Icarus, Vol. 16, 578 - 580 (1972).

011.003 Report on the Cambridge meeting of the A.A.S. Working Group on Photographic Materials in Astronomy – Part II.
AAS Photo-Bull., 1972, No. 2, p. 6 - 17.
The individual contributions are included in their corresponding subject categories. – See abstracts 036.003 - 036.006, 158.026.

011.004 Report of PZT meeting. W. Markowitz.
IAU Symposium No. 48, (see 012.004), p. 241 (1972).

011.005 The problem of connection with extraterrestrial civilizations. Conference in Byurakan.
L. M. Gindilis.
Vestn. AN SSSR, 1972, No. 3, p. 82 - 88. In Russian. – Abstr. in Referativ. Zhurn. 62. Issled. kosm. prostranstva, 8.62.9 (1972). – 1971 September 6–11.

011.006 Royal Astronomical Society meeting on the D and E regions of the ionosphere.
H. Rishbeth, P. C. Kendall.
Quarterly Journ. Roy. Astron. Soc., Vol. 13, 403 - 408 (1972).
Rocket and satellite observations of stratospheric warmings (R. A. Hamilton); D-region N(h) profiles at South Uist and their relevance to winter anomaly and stratospheric warmings (E. R. Williams); Some relations between mid-latitude and high-latitude radio absorption during the winter of 1970 - 1071 (J. K. Hargreaves); E-region phenomena connected with the lower atmosphere (G. M. Brown); Studies of the propagation of 60 kHz signals on the Rugby–Exeter path, which are reflected at about 80 km height by day (I. M. Warman); Results from a number of D-region experiments carried on a single Petrel rocket (J. E. Hall); The importance of negative ion changes in solar flare effects in the D-region (L. Thomas, P. M. Gondhalekar); Short-term phase variability of long-path VLF signals (R. C. Bailey); Energetic electron precipitation as an ionizing source in the mid-latitude D-region (M. P. Gough); Upper atmospheric sodium: A brief review (M. Gadsden); A model ionosphere for summer noon (W. C. Bain).

011.007 Geomagnetic daily variations: Sources, current and induction.
Quarterly Journ. Roy. Astron. Soc., Vol. 13, 409 - 413 (1972). A discussion meeting held at the Scientific Societies Lecture Theatre on 1972 January 28.
Ionospheric conductivities and their day-night variation (H. Rishbeth); The current function associated with solar and lunar daily geomagnetic variations (S. R. C. Malin); Day-to-day variability of Sq (D. M. Schlapp); Electromagnetic induction in the oceans (E. Bullard); Results from transoceanic cables (M. L. Richards, S. K. Runcorn); Observations of sea tidal effects (W. G. V. Rosser); Earth potential studies at Aberystwyth (G. M. Brown, D. R. E. Fraser); A comment on ocean tidal effects at Stonyhurst (J. E. Worthy); Interactions between ionospheric and earth currents (A. T. Price); Electromagnetic induction in thin sheets (B. A. Hobbs); Applications of functional analysis to the solution of geomagnetic integral equations (V. C. L. Hutson, P. C. Kendall, S. R. C. Malin).

011.008 Drittes Mondkolloquium in Houston. J. Classen.
Sterne, 48. Jahrgang, p. 141 - 143 (1972). 1972, Januar 10 - 13.

011.009 Studies on solar activity. Conference in Moscow.
L. A. Vedeshin.
Vestn. AN SSSR, 1972, No. 4, p. 91 - 94. In Russian. – 1971, Nov. 15 - 22. – Abstr. in Referativ. Zhurn. 51. Astron., 9.51.20 (1972).

011.010 Structure and kinematics of the Galaxy. The conference in Rostow-on-Don 27 - 30 Sept. 1971.
E. D. Pavlovskaya.
Vestn. AN SSSR, 1972, No. 3, p. 114 - 115. In Russian.

011.011 IVth colloquium of telescope constructors.
M. M. Shemyakin.
Zemlya i Vselennaya, 1972, No. 4, p. 54 - 56. In Russian. Baku, 1972, March 13 - 14.

011.012 Ground based and space astronomy.
L. Opiela, P. L. Sowerby.
Journ. British Astron. Ass., Vol. 82, 442 - 448 (1972). – A report on a Royal Society Symposium.

011.013 History, philosophy, and sociology of science.
Science, Vol. 177, 1216 - 1217 (1972). − Annual meeting of AAAS, Washington, D.C., 1972 December.

011.014 Annual general meeting of the USSR Academy of Sciences, March 1 - 2, 1972.
Vestn. AN SSSR, 1972, No. 5, p. 3 - 4. In Russian.

011.015 Scientific sessions of the Division of General Physics and Astronomy and of the Division of Nuclear Physics of the USSR Academy of Sciences, September 29 - 30, 1971.
Uspekhi fiz. nauk, Vol. 107, 153 - 159 (1972). In Russian. − Abstr. in Referativ. Zhurn. 51. Astron., 10.51.17 (1972).

011.016 Scientific sessions of the Division of General Physics and Astronomy and of the Division of Nuclear Physics of the USSR Academy of Sciences, November 24 - 25, 1971.
Uspekhi fiz. nauk, Vol. 107, 325 - 328 (1972). In Russian. Abstr. in Referativ. Zhurn. 51. Astron., 10.51.18 (1972).

011.017 3rd International Conference on vacuum ultraviolet radiation physics, Tokyo, 30 August−2 September 1971.   G. L. Weissler.
Applied Optics, Vol. 11, 2395 (1972).

011.018 Two western conventions.   D. Milon.
Sky Telescope, Vol. 44, 290 - 295 (1972).
In August, amateur astronomers from across the country attended two national conventions, whose programs were supplemented with sometimes lengthy field trips to observatories and other institutions of special astronomical interest.

011.019 Space technology: EROS, ERTS, and the space shuttle.
Science, Vol. 178, 187 (1972). − Program for the AAAS Symposium, 1972 December 27.

011.020 Chronicle. VIth conference of the USSR on the physics of comets.   D. A. Andrienko.
Astron. Zhurn. Akad. Nauk SSSR, Vol. 49, 1132 - 1133 (1972). In Russian. English translation in Soviet Astron. AJ, Vol. 16, No. 5. − Kiev, 1971 November 1−4.

011.021 European Space Symposium, Rome 12 - 14 June 1972.   N. Simmons.
Journ. British Interplanet. Soc., Vol. 25, 746 - 748 (1972).

011.022 Eine Astrophotosafari nach Südafrika.
E. Alt, E. Brodkorb, K. Rihm.
SuW, Vol. 11, 310 - 312 (1972).

011.023 4. Internationale Planetariumsleiter-Tagung in Canada und USA.   A. Kunert.
SuW, Vol. 11, 319 - 320 (1972).

011.024 XXIInd Astronautical Congress.   K. B. Serafimov.
Spisanie B"lg. AN, Vol. 18, No. 1, p. 83 - 86 (1972). In Bulgarian.

011.025 Scientific session of the Division of General Physics and Astronomy of the USSR Academy of Sciences, January 19−20, 1972.
Uspekhi fiz. nauk, Vol. 107, 507 - 518 (1972). In Russian.

011.026 Double-star colloquium in Swarthmore.
W. D. Heintz.
Circ. Inform. (U.A.I. Commission des Étoiles Doubles), Obs. Meudon, No. 58, Annexe bis (1972). − 1972, April 12 - 15.

011.027 Il secondo congresso dell'International Union of Amateur Astronomers.   L. Baldinelli.
Coelum, Vol. 40, 200 - 201 (1972).

011.028 The problems of astronomy and geodesy at the ordinary plenary session of the Central Committee of the USSR Astronomical and Geodetical Society.
A. V. Butkevich, E. V. Gromov, L. S. Khrenov.
Geod. i kartografiya, 1972, No. 7, p. 60 - 63. In Russian. − Abstr. in Referativ. Zhurn. 51. Astron., 12.51.20 (1972).

011.029 On the XVth meteorite conference.
A. N. Simonenko.
Zemlya i Vselennaya, 1972, No. 6, p. 58 - 60. In Russian. Kaluga, USSR, 1972, May 29 - June 1.

011.030 The 1972 meeting of the AAS Division for Planetary Sciences, University of Hawaii, Kailua-Kona, Hawaii, March 20 - 24, 1972.   D. P. Cruikshank.
Icarus, Vol. 17, 714 - 719 (1972).

011.031 Astronomi Latvijas Valsts universitātes XXX zinātniskajā konference.   L. Roze.
Zvaigžņotā debess, 1971. gada rudens, p. 38 - 42.

011.032 Samarkandā pulcējas Saules pētnieki.
N. Cimahoviča.
Zvaigžņotā debess, 1971. gada rudens, p. 46 - 51. − Samarkand, 1972 April 7 - 10.

011.033 Astrofiziķu plēnums Rīgā.
Z. Alksne, J. Francmanis.
Zvaigžņotā debess, 1971./72. gada ziema, p. 43 - 53. 1971, May 25 - 29.

011.034 PSRS Zinātņu akadēmijas Astronomiskās padomes prezidija izbraukuma sesija Odesā.   J. Francmanis.
Zvaigžņotā debess, 1972. gada pavasaris, p. 48 - 51.

011.035 Symposium über solar-terrestrische Physik.
H. Lambrecht, F. W. Jäger.
Sterne, 48. Jahrgang, p. 223 - 227 (1972). − Leningrad, 1970 May 12 - 19.

011.036 Tagung 1972 der Internationalen Union der Amateur-Astronomen (IUAA) in Malmö (Schweden).
R. A. Naef.
Orion Schaffhausen, 30. Jahrgang, p. 179 - 180 (1972). − 1972 July 31 − August 4.

011.037 Astronomical conferences and meetings.
Astrophys. Letters, Vol. 11, 228 - 229; Vol. 12, 61 - 62, 123 - 124, 185 - 186, 243 - 244 (1972).

011.038 Astronomy and air pollution.
P. W. Hodge, N. Laulainen, R. J. Charlson.
Science, Vol. 178, 1123 - 1124 (1972).
The University of Washington sponsored a 1-day seminar entitled "The use of astronomical techniques for the study of atmospheric deterioration," in honour of C. G. Abbot on the occasion of his 100th birthday. Two topics were considered: the use of photometry to study the extinction of light by small particles in the atmosphere, and the use of spectroscopic techniques for determining the concentrations of molecular contaminants.

011.039 Die BAV-Tagung 1972.   W. Braune.
BAV Rundbrief, 21. Jahrgang, p. 37 - 41 (1972).

011.040 IUAA−kongressen i Malmö 1972.   P. Linde.
Astron. Tidsskr., Årg. 5, p. 183 - 188 (1972).

011.041 Recollections from the international meeting of amateur astronomers in Bologna 1971.
J. Kazimierowski.
Urania Kraków, Vol. 43, 313 - 315 (1972). In Polish.

011.042 IInd general assembly of the International Union of Amateur Astronomers. K. Ziołkowski.
Urania Kraków, Vol. 43, 342 - 345 (1972). In Polish.

011.043 XVII Meeting of the Argentine Astronomical Association, Córdoba, October 25 - 27, 1971.
Inform. Bull. Southern Hemisph. No. 20, p. 25 - 28 (1972).
In celebration of the Córdoba Observatory's centenary.

011.044 First Latin American astrometric working meeting, Córdoba Observatory, November 4 - 5, 1971.
Inform. Bull. Southern Hemisph. No. 20, p. 28 - 32 (1972).
Part of the Córdoba Observatory's centenary celebrations.

011.045 The sixteenth Herstmonceux conference, 1972 April 5 and 6. Cosmic X-ray sources.
Observatory, Vol. 92, 193 - 216 (1972). – During the conference papers are presented by K. A. Pounds, G. K. Miley, P. Murdin, B. L. Webster, F. Pacini, J. E. Pringle, J. C. Jackson, L. V. Morrison, J. L. Culhane, D. W. Sciama, A. C. Fabian, J. C. Brown, A. H. Gabriel, R. G. Bingham, A. H. Lategan, B. E. J. Pagel.

011.046 Report of a conference on astrophysics, held in honour of Prof. R. O. Redman at The Observatories and Institute of Theoretical Astronomy, Cambridge, 1972 July 18–20.
Observatory, Vol. 92, 217 - 231 (1972).

011.047 Symposium on extraterrestrial civilizations. Translated by M. Mijić.
Vasiona, Vol. 20, 62 - 66 (1972). In Serbo-Croatian.
Byurakan, Sept. 6 - 11, 1971.

011.048 La Scuola estiva "Physics and chemistry of upper atmospheres". R. Pallavicini.
Mem. Soc. Astron. Italiana, Nuova Ser., Vol. 43, 567 - 568 (1972).

011.049 Working seminar on the physics of the interstellar medium. Instituto de Astronomía y Física del Espacio, Buenos Aires, August 3–14, 1972.
Inform. Bull. Southern Hemisph., No. 21, p. 24 - 25 (1972).

011.050 Meeting on dMe and flare stars. ITA Astronomical Observatory, São José dos Campos, SP– Brazil, August 24–25, 1972.
Inform. Bull. Southern Hemisph., No. 21, p. 25 (1972).

011.051 First Latin American astrophysical meeting. La Plata Observatory, October 16–17, 1972.
Inform. Bull. Southern Hemisph., No. 21, p. 26 (1972).

011.052 XVIII meeting of the Argentine Astronomical Association. La Plata Observatory, October 18–20, 1972.
Inform. Bull. Southern Hemisph., No. 21, p. 26 - 29 (1972).

011.053 Regionaltagung der V.d.S. in Bremen. E. Fuchs.
Nachr. Olbers-Ges. Bremen, No. 86, p. 9 - 10 (1972).

011.054 Symposium on the Buckland Park antenna array. Report of a meeting organized by the (Australian Institute of Physics) South Australian Branch on 17th June 1971 at the South Australian Institute of Technology.
Australian Physicist, Vol. 9, No. 3, 41 - 47 (1972).

011.055 Czechoslovak conference on stellar astronomy and astrophysics, Cikháj, Oct. 12 - 14, 1971.
J. Grygar.
Pokroky, Vol. 17, 111 - 112 (1972). In Czech.

011.056 The 1972 W.A.A. – A.L.P.O. convention at Riverside. C. Vaucher.
Strolling Astronomer, Vol. 24, 35 - 38 (1972).
The twentieth annual convention of the Association of Lunar and Planetary Observers was held on August 16–19, 1972, in conjunction with the Convention of the Western Amateur Astronomers at Bannockburn, in Riverside, California.

012 Proceedings of Colloquia, Congresses, Meetings, and Symposia

012.001 **Conference of the working group on analytic methods in celestial mechanics.** Commission of celestial mechanics of the Astronomical Council of the USSR Academy of Sciences, Leningrad, August 3 - 7, 1970.
Compiled by Sh. G. Sharaf.
Nablyud. Iskusstv. Nebesn. Tel, No. 62, 184 pp. (1971). In Russian. − The individual contributions are included in their corresponding subject categories − see abstracts 021.001, 021.002, 042.001 - 042.009, 066.001, 098.001.

012.002 **Technology and utilization of stratospheric balloons.** Proceedings of a Colloquium held on the occasion of the fourteenth plenary meeting of the Committee for Space Research (COSPAR), Seattle, U.S.A. − 26−28 June 1971 and organized by Solar Particles and Radiations Monitoring Organization (SPARMO). Sponsored by International Astronomical Union (IAU) − International Union of Geodesy and Geophysics (IUGG) − Federation of Astronomical and Geophysical Services (FAGS).
L. D. de Feiter, G. Kremser (Editors), with an introduction and a summary of the panel discussions by L. D. de Feiter.
Space Science Rev., Vol. 13, 197 - 365 (1972). − The individual contributions are included in their corresponding subject categories − see abstracts 004.011, 013.001, 031.005, 034.002, 082.011 - 082.013, 084.206, 142.008, 143.004.

012.003 **The motion, evolution of orbits, and origin of comets.** International Astronomical Union, Symposium No. 45, Held in Leningrad, U.S.S.R., August 4 - 11, 1970.
G. A. Chebotarev, E. I. Kazimirchak-Polonskaya, B. G. Marsden (Editors).
D. Reidel Publishing Company, Dordrecht−Holland. 33 + 521 pp. Price hfl. 120.00 (1972). − The individual contributions are included in their corresponding subject categories − see abstracts 021.004, 042.023, 042.024, 091.005, 091.006, 098.009 - 098.012, 099.008, 099.009, 100.005, 102.007 - 102.050, 103.001 - 103.003, 103.101 - 103.111, 104.005 - 104.011, 106.008, 107.004 - 107.007.

012.004 **Rotation of the earth.** International Astronomical Union, Symposium No. 48, held in Morioka, Japan, 9−15 May 1971.
Edited by P. Melchior, S. Yumi, with the cooperation of Lady Jeffreys.
D. Reidel Publishing Company, Dordrecht−Holland. XXII + 244 pp. Price hfl. 60.00 (1972). − The individual contributions are included in their corresponding subject categories − see abstracts 011.004, 032.005, 041.009, 041.010, 043.001, 044.002 - 044.011, 045.002 - 045.030, 054.002, 081.009 - 081.016, 082.041.

012.005 **Infrared detection techniques for space research.** Proceedings of the fifth ESLAB/ESRIN Symposium, held in Noordwijk, The Netherlands, June 8 - 11, 1971.
V. Manno, J. Ring (Editors), with an introduction by E. A. Trendelenburg and a general discussion directed by F. Kneubühl.
D. Reidel Publishing Company, Dordrecht−Holland. Astrophysics and Space Science Library, Vol. 30, 11 + 344 pp. Price hfl. 90.00 (1972). − The individual papers are included in their corresponding subject categories − see abstracts 032.006 - 032.008, 034.013 - 034.042, 082.044, 114.044.

012.006 **Symposium on the morphology and physics of magnetospheric substorms.** XVth General Assembly of IUGG, Moscow, August 1971.

Planet. Space Sci., Vol. 20, 1361 - 1563, with an introduction by S.-I. Akasofu, p. 1361 (1972). − The individual papers presented during the symposium are included in their corresponding subject categories − see abstracts 082.046, 083.013, 084.016, 084.222 - 084.229, 084.407.

012.007 **Conference on theoretical ionospheric models** held at the Pennsylvania State University, University Park, Pennsylvania, from 14 to 16 June 1971.
Journ. Atmosph. Terr. Phys., Vol. 34, 1563 - 1841 (1972). The individual contributions within the subject scope of Astronomy and Astrophysics Abstracts are included in their corresponding subject categories − see abstracts 082.059 - 082.064, 083.018 - 083.022.

012.008 **The moon.** International Astronomical Union Symposium No. 47, held at the University of Newcastle-upon-Tyne, England, 22 - 26 March, 1971.
Edited by S. K. Runcorn, H. C. Urey, with an introduction by G. Fielder.
D. Reidel Publishing Company, Dordrecht − Holland. 16 + 471 pp. Price hfl. 120.00 (1972). − The individual contributions are included in their corresponding subject categories − see abstracts 053.016, 053.017, 094.057 - 094.094.

012.009 **Earth's magnetospheric processes.** Proceedings of a symposium organized by the Summer Advanced Study Institute and ninth ESRO Summer School, held in Cortina, Italy, August 30−September 10, 1971.
B. M. McCormac (Editor).
Astrophysics and Space Science Library, Vol. 32. D. Reidel Publishing Company, Dordrecht − Holland. 8 + 417 pp. Price hfl. 115.00 (1972). − The individual contributions are included in their corresponding subject categories − see abstracts 074.054, 078.022 - 078.023, 083.035, 083.036, 084.029 - 084.036, 084.264 - 084.285, 084.414 - 084.417.

012.010 **Comets. Scientific data and missions.** Proceedings of the Tucson comet conference, Tucson, Arizona, 1970 April 8 - 9. G. P. Kuiper, E. Roemer (Editors).
Published by Lunar and Planetary Laboratory, University of Arizona, Tucson, Arizona. 8 + 222 pp. (1972). − The individual papers are included in their corresponding subject categories − see 051.009 - 051.012, 102.056 - 102.069, 103.100, 103.121, 103.122, 104.025.

012.011 **2. Symposium über Veränderliche.** Darmstadt, 4./5. March 1972, with a preface by H. Zipprich.
BAV Rundbrief, 21. Jahrgang, Sonder-Rundbrief, 56 pp. (1972). − The individual contributions are included in their corresponding subject categories − see abstracts 031.025, 117.013, 120.005 - 120.007, 121.044 - 121.048, 122.056.

012.012 **Ultraviolet and X-ray spectroscopy of astrophysical and laboratory plasmas.** IAU Colloquium No. 14, held at Utrecht 24−26 August, 1971.
A. H. Gabriel (Editor).
Space Sci. Rev., Vol. 13, Nos. 4, 5 and 6, 489 - 889 (1972). − The individual contributions are included in their corresponding subject categories − see abstracts 022.074 - 022.084, 034.091 - 034.093, 062.043 - 062.048, 071.040 - 071.044, 073.067 - 073.075, 074.068 - 074.070, 076.027 - 076.034, 114.092, 142.080.

012.013 **Atti della XV Riunione della Società Astronomica Italiana e Giornate di Studio su 'La rotazione come**

fenomeno e fattore evolutivo nell' universo', Bologna, 8–9 Ottobre 1971, with an introduction by G. Righini.
Printed by Tipografia Baccini & Chiappi, Firenze. 212 pp. (1972). – The individual contributions are included in their corresponding subject categories – see abstracts 004.026, 010.027, 065.063, 066.049, 080.028, 080.029, 116.013, 141.519, 158.084, 158.085.

012.014 **Conference on the role of Schmidt telescopes in astronomy**, Hamburg, March 21–23, 1972. **Proceedings.**
Edited by U. Haug.
To be obtained from Hamburger Sternwarte, Bergedorf.
14 + 160 pp. (1972). – The individual contributions are included in their corresponding subject categories – see abstracts 013.010, 013.011, 032.026 - 032.029, 034.094, 034.095, 041.016 - 041.019, 112.007, 113.028 - 113.030, 114.094 - 114.096, 141.074, 141.075, 155.039, 155.040, 158.088, 158.089.

012.015 **Transactions of the First Republican Scientific-Methodical Conference of lecturers in physics and astronomy at pedagogical colleges of Kazakhstan, Alma-Ata, December 1969.**
Kazakhsk. gos. ped. in-t, Alma-Ata. 182 pp. Price 1 Rbl. 50 Kop. (1972). In Russian.

012.016 **Space Research XII.** Proceedings of open meetings of Working Groups of the Fourteenth Plenary Meeting of COSPAR, Seattle, Washington, USA, 21 June–2 July, 1971 and of 'the symposium on total solar eclipse of 7 March, 1970', Seattle, Washington, USA, 18, 19 and 21 June, 1971 and of 'the symposium on dynamics of the thermosphere and ionosphere above 120 km', Seattle, Washington, USA, 24, 25 and 26 June, 1971 and of 'the symposium on high angular resolution astronomical observations from space', Seattle, Washington, USA, 28, 29, 30 June and 1 July, 1971. Vol. 1, 2.
S. A. Bowhill, L. D. Jaffe, M. J. Rycroft (Editors).
Akademie-Verlag, Berlin. 40 + 1815 pp. Price DM 260.00 (1972). – The individual contributions within the subject scope of Astronomy and Astrophysics Abstracts are included in their corresponding categories – see abstracts 031.042 - 031.048, 032.031 - 032.033, 034.096 - 034.106, 046.026, 046.027, 052.029, 053.021, 053.022, 066.057, 071.052, 073.087, 074.076, 076.035 - 076.037, 078.028 - 078.030, 082.130 - 082.151, 083.046 - 083.054, 084.048 - 084.053, 084.310 - 084.314, 084.419, 091.030, 091.031, 093.026, 093.027, 094.160 - 094.178, 105.044 - 105.054, 106.026, 106.027, 114.104, 114.105, 131.075 - 131.077, 143.034, 143.035. – The papers presented at the 'Symposium on total solar eclipse of 7 March 1970' have appeared in two special issues of the journals: Solar Physics and Journ. Atmosph. Terr. Phys.

012.017 **Symposium on planetary atmospheres and surfaces,** held 10–13 May 1972, in Madrid, Spain, with an introduction by C. Sagan and a foreword by G. H. Pettengill.
Icarus, Vol. 17, 289 - 524, 540 - 542 (1972). – This present issue is devoted almost entirely to papers on Mariner 9 and Mars 2 and 3 missions presented at this symposium. – The individual contributions are included in their corresponding subject categories – see abstracts 097.083 - 097.095, 097.097.

012.018 **Proceedings of the International Symposium on "Optical properties of the upper atmosphere and the circumterrestrial cosmic space",** with a preface by E. K. Kharadze and a discussion of reports and resolutions. Tiflis, 1969, October 2–6.
Byull. Abastumansk. Astrofiz. Obs., No. 41, 143 pp. (1972).
In Russian. – The individual contributions are included in their corresponding subject categories – see abstracts 082.159 - 082.169, 104.043.

012.019 **Reports presented at the Colloquium on aurorae and airglow,** with prefaces by V. I. Krasovskij and E. K. Kharadze and final remarks by V. I. Krasovskij. Abastumani, June 9–10, 1970.
Byull. Abastumansk. Astrofiz. Obs., No. 42, p. 1 - 118 (1972).
In Russian. – The individual contributions are included in their corresponding subject categories – see abstracts 082.170 - 082.177, 084.056, 084.329.

012.020 **Colloquium on supergiant stars.** Proceedings of the Third Colloquium on astrophysics held in Trieste, September 6 - 8, 1971. M. Hack (Editor).
Published by Osservatorio Astronomico, Trieste. 371 pp. Price $ 9.50 (1972). – The individual contributions are included in their corresponding subject categories – see abstracts 064.043, 064.051 - 064.060, 065.105 - 065.113, 113.055 - 113.058, 114.120, 114.146 - 114.154, 115.018 - 115.021, 119.009, 122.083, 122.117 - 122.119, 155.047, 159.017.

012.021 **Geodätische Astronomie. III. Internationales Kolloquium an der TU Dresden, veranstaltet vom Lohrmann-Observatorium der Sektion Geodäsie und Kartographie,** with an introduction by H.-U. Sandig.
Wiss. Zeitschr. Techn. Univ. Dresden, Vol. 21, No. 3, 585 - 638 = Mitt. Lohrmann-Obs. Techn. Univ. Dresden, No. 25 (1972). – The individual papers are included in their corresponding subject categories – see abstracts 031.082, 031.083, 032.045 - 032.047, 041.047, 041.048, 044.031, 044.032, 045.047, 045.048, 046.041, 046.043, 046.044, 055.016 - 055.020, 094.225, 111.007.

012.022 **Publications of the 18th Astrometrical Conference of the USSR (Pulkovo, 2. - 5. June 1969).**
M. S. Zverev (Editor).
Akademiya Nauk SSSR, Glavnaya Astronomicheskaya Observatoriya (Pulkovo). Izdatel'stvo 'Nauka', Leningradskoe Otdelenie, Leningrad. 355 pp. Price 2 Rbl. (1972). In Russian. – The individual papers are included in their corresponding subject categories–see abstracts 031.090, 032.052 - 032.060, 034.150 - 034.157, 035.005, 041.051 - 041.080, 044.044, 045.051, 093.049, 094.255 - 094.265, 112.015, 118.018.

012.023 **High pressure physics and planetary interiors.** Proceedings of the conference held at the Lunar Science Institute, Houston, Texas, March 1–3, 1972.
S. K. Runcorn (Editor).
Phys. Earth Planet. Interiors, (Special Issue), Vol. 6, Nos. 1–3, 18 + 209 pp. = Lunar Sci. Inst. Contr. No. 89 (1972). – This issue is dedicated to R. Wildt, Professor of Astrophysics, Dep. Astron., Yale Univ., on the occasion of his retirement in the spring of 1973. Included is a list of his scientific publications. The individual papers presented during the conference are included in their corresponding subject categories – see abstracts 022.141 - 022.150, 062.079, 081.058 - 081.062, 091.049 - 091.057, 094.267, 094.268, 099.091 - 099.096.

012.024 **Mantle and core in planetary physics.** Proceedings of the International School of Physics «Enrico Fermi», Course 50. Varenna, Italy, 13 th - 25 th July 1970.
J. Coulomb, M. Caputo (Editors).
Academic Press, New York – London. 13 + 277 pp. Price $ 17.50 (1971). – The individual papers within the subject scope of Astronomy and Astrophysics Abstracts are included in their corresponding categories – see abstracts 081.065, 081.066, 084.358, 084.359.

013 Reports on Astronomy in Various Countries and Particular Fields, International Cooperation

013.001 Scientific ballooning services. A. L. Morris.
Space Science Rev., Vol. 13, 243 - 257 (1972).
Conference paper (see 012.002).

013.002 Report on astronomy: A new golden age.
W. D. Metz.
Science, Vol. 177, 247 - 249 (1972).

013.003 British optical astronomy. G. Burbidge.
Nature, Vol. 239, 117 - 118 (1972). – Letter.

013.004 Non-stationary phenomena in the world of stars and galaxies. V. A. Ambartsumyan.
Priroda, No. 7.72, p. 2 - 10 (1972). In Russian.

013.005 Future belongs to astrophysics.
L. A. Artsimovich.
Priroda, No. 9.72, p. 2 - 4 (1972). In Russian.

013.006 Instationary phenomena in the world of stars and galaxies. V. A. Ambartsumyan.
Zemlya i Vselennaya, 1972, No. 4, p. 2 - 12. In Russian.

013.007 The progress in sciences and techniques and the work of institutions belonging to the Department of Sciences of the Kazakh Academy of Sciences on the universe and the earth. I. I. Bok.
Vestn. AN KazSSR, 1972, No. 4, p. 8 - 11. In Russian.

013.008 Non-stationary phenomena in the realm of stars and galaxies. V. A. Ambartsumyan.
Vestn. AN SSSR, 1972, No. 5, p. 33 - 45. In Russian. – Abstr. in Referativ. Zhurn. 51. Astron., 10.51.92 (1972).

013.009 Quelques progrès récents de l'astronomie.
J. Kovalevsky.
L'Astronomie, 86e année, p. 445 - 461 (1972).

013.010 Twenty years astronomy with the 48-inch Schmidt telescope on Palomar Mountain. R. Minkowski.
The role of Schmidt telescopes in astronomy. Conference Hamburg 1972, (see 012.014), p. 5 - 8 (1972).

013.011 Ten years of research with the 134 cm Schmidt telescope. N. Richter.
The role of Schmidt telescopes in astronomy. Conference Hamburg 1972, (see 012.014), p. 141 - 143 (1972).

013.012 The Federation of Astronomical and Geophysical Services (FAGS).
ICSU Bull., No. 27, 8 - 22 (1972).

The Services are concerned with many different aspects of astronomy and geophysics, but their activities follow a fairly uniform general pattern which includes the following elements: the collection of the raw data; the analysis and synthesis of these data leading, in some cases, to conclusions; the publication of the data, in a more concise form, and their distribution to scientists and others who wish to use them.

013.013 Las investigaciones heliofísicas en la Argentina.
V. Capolongo.
Obs. Astron. Municipal Rosario, (*Argentina*), Bol. No. 2, p. 11 - 13 (1972).

We give a report of the actual development of solar physics in Argentina and describe the research work in solar observatories (Observatorio de Física Cósmica de San Miguel, Observatorio Astronómico Municipal de Rosario, and Observatorio Biblioteca "C. C. Vigil" of Rosario). This report was presented at the V° Congreso Latinoamericano de Astronomía, Lima, 1971 August 7 - 15.

013.014 Nichtstationäre Erscheinungen in der Welt der Sterne und Galaxien. V. A. Ambarzumjan.
SuW, Vol. 11, 334 - 338 (1972).

013.015 Astronomy today. N. Nikolov.
Mat. i fizika (NRB), Vol. 15, No. 3, p. 52 - 54 (1972). In Bulgarian.

013.016 Astronomy in the Ukraine. E. P. Fedorov.
Zemlya i Vselennaya, 1972, No. 6, p. 2 - 8.
In Russian. – Remarks on research and instruments at the observatories in Nikolayevsk, Kharkov, Odessa, L'vov and on the Krim.

013.017 Astrophysics in Soviet Armenia. L. V. Mirzoyan.
Zemlya i Vselennaya, 1972, No. 6, p. 9 - 15.
In Russian.

013.018 Astronomy in Siberia and the Far East.
R. B. Teplitskaya.
Zemlya i Vselennaya, 1972, No. 6, p. 25 - 31. In Russian.

013.019 Progress in geomagnetism and aeronomy in the USSR.
Geomagn. Aeronom., Vol. 12, 969 - 973 (1972). In Russian.

013.020 USSR – 50 years and the Soviet astronomy.
Astron. Zhurn. Akad. Nauk SSSR, Vol. 49, 1137 - 1138 (1972). In Russian. English translation in Soviet Astron. AJ, Vol. 16, No. 6.

013.021 A major optical telescope for Canada.
K. O. Wright.
Journ. Roy. Astron. Soc. Canada, Vol. 66, 317 - 321 (1972). Memorandum presented from the Associate Committee on Astronomy.

013.022 Afstandsmålinger i Universet. P. Gammelgaard.
Astron. Tidssk., Årg. 5, p. 174 - 182 (1972).

013.023 Astronomer's luck. W. H. McCrea.
Quarterly Journ. Roy. Astron. Soc., Vol. 13, 506 - 519 (1972).

A shortened version of a talk given in University College London on 1972 June 5. The state of astronomy at any epoch depends to a great extent upon the examples of various phenomena that are available for observational study, upon the instruments that are available for this study, upon the existence of relevant mathematical and physical theories and of serviceable accounts of them, upon the combination of workers who happen to be active and interactive at the time, and so on. These are all matters of 'luck' in the sense in which the word is used in this article.

013.024 Cosmos in sight. Astrophysical research in DDR Tautenburg Centre for collecting facts.
Wiss. Fortschritt, Vol. 21, 551 - 557 (1971). = Mitt. Karl-Schwarzschild-Obs. Tautenburg, No. 61 (1972). In German.

013.025 **Which physical and astrophysical problems seem to be of particular importance and interest at the present? II.** V. L. Ginzburg.
Cesk. Cas. Fis. A, Vol. 22, 349 - 365 (1972). In Czech.

013.026 **L'Astronomie en France.** J. Cantacuzène.

Rapport établi auprès du Ministère des Affaires Etrangères, Paris. 7 + 3 pp. (1972).

Springtime for some European astronomers.
Nature, Vol. 240, 250 (1972).

014 Teaching in Astronomy

014.001 **Do-it-yourself astrophysics.** G. Y. Haig.
Journ. British Astron. Ass., Vol. 82, 357 - 361 (1972).
  Two nomograms are given to determine basic astrophysical data of stars and to construct the Hertzsprung-Russell-diagram.

014.002 **Einführung in die Astronomie − einmal anders.** A. Kunert.
SuW, Vol. 11, 249 (1972).

014.003 **International aspects of astronomical education.** N. S. Nikolov.
Mat. i fizika (NRB), Vol. 14, No. 6, p. 55 - 56 (1971). In Bulgarian. − Abstr. in Referativ. Zhurn. 51. Astron., 8.51.34 (1972).

014.004 **Photoelectric experimental demonstrations during lessons in astronomy.**
F. M. Poroshin, M. L. Aksenov.
Uch. zap. Omsk. gos. ped. in-t, 1971, vyp. (No.) 63, p. 67 - 77. In Russian. − Abstr. in Referativ. Zhurn. 51. Astron., 8.51.35 (1972).

014.005 **Astronomical photographic documents, their preparation and usage at high school lessons in astronomy.**
F. M. Poroshin, L. A. Golovina, E. I. Kovyazin.
Uch. zap. Omsk. gos. ped. in-t, 1971, vyp. (No.) 63, p. 78 - 86. In Russian. − Abstr. in Referativ. Zhurn. 51. Astron., 8.51.36 (1972).

014.006 **Einige Methoden und Ergebnisse der modernen Sonnenforschung.** J. Staude.
Astron. in der Schule, 9. Jahrgang, p. 90 - 94 (1972).

014.007 **Employment in astronomy.**
Journ. Roy. Astron. Soc. Canada, Vol. 66, 279 (1972). − "Guidelines for employment opportunities in astronomy", published by the American Astronomical Society.

014.008 **Interdisciplinary studies: A third career for astronomers.** G. S. Mumford.
Mercury, (Journ. Astron. Soc. Pacific), Vol. 1, No. 4, p. 4 (1972).

014.009 **School telescope BAM-5A.**
G. S. Minasyan, A. M. Gevorkyan.
Zemlya i Vselennaya, 1972, No. 5, p. 74 - 75. In Russian.

014.010 **Astronomin i skolan: Nyt om astronomiundervisningen.** H. E. Jørgensen.
Astron. Tidssk., Årg. 5, p. 132 - 135 (1972).

014.011 **Astronomie und Schule. Bücher für die Unterrichtsvorbereitung und Unterrichtsgestaltung.**
A. Kunert.
SuW, Vol. 11, 320 - 321 (1972).

014.012 **Seminar lessons on astronomy.**
E. Lachkova.
Mat. i fizika (NRB), Vol. 15, No. 3, p. 54 - 56 (1972). In Bulgarian.

014.013 **Der Entwicklungsgedanke und seine Darstellung im Astronomieunterricht.** H. Bernhard.
Astron. in der Schule, 9. Jahrgang, p. 99 - 104 (1972).

014.014 **Wichtige Fortschritte in der astronomischen Forschung im Jahre 1971.** O. Günther.
Astron. in der Schule, 9. Jahrgang, p. 104 - 106 (1972).

014.015 **To teach science: more and better.**
M. Hulin, A. Lagarrigue, J. Lequeux.
Sci. Progrès Découverte, 99e année, No. 3437, p. 4 - 7 (1971). In French.
  The study of the physical sciences appears to be an indispensable counterbalance to a curriculum still too theoretical and abstract − particularly true of the secondary schools. It appears to be already evident that the teaching of the physical sciences should be begun in the first year of secondary school and that eventually all measures should be taken to encourage students to take the initiative. It is also planned to give astronomy a greater role in this program since it will provide examples in all fields of physics.

014.016 **From astronomical clubs at school to collectives of young amateur astronomers.** S. S. Vojnov.
Zemlya i Vselennaya, 1972, No. 6, p. 71 - 74. In Russian.

014.017 **Dažas piezīmes par vidusskolu astronomijas mācību grāmatu.** A. Alksnis.
Zvaigžņotā debess, 1972. gada pavasaris, p. 57 - 60.

014.018 **Tèma "Koperniks".**
Ã. Alksne, I. Rabinovičs.
Zvaigžņotā debess, 1972. gada vasara, p. 49 - 53.

014.019 **Astronomy and aids for teachers.** M. B. Stewart.
Mercury, (Journ. Astron. Soc. Pacific), Vol. 1, No. 4, p. 10 - 12 (1972).
A lecture demonstration apparatus for simulation of eclipses and light variations of eclipsing binary stars; An experiment in the astronomy survey course at a large university.

014.020 **England's astronomical education?**
A. W. Lintern Ball.

Quarterly Journ. Roy. Astron. Soc., Vol. 13, 486 - 505 (1972).
Historical setting; A survey of secondary schools; Public examinations; Non-examination courses; Higher education; Education for the general public; Conclusion–an appeal; Appendix: University of London general certificate of education examination. New syllabus in astronomy at ordinary level for examination in and after June 1974.

014.021 **Amateur observations of the sun.**
M. A. Kljakotka, compiled by D. Pekez.
Vasiona, Vol. 20, 60 - 62 (1972). In Serbo-Croatian.

014.022 **Measurement of distant celestial bodies.**
B. M. Ševarlić.
Vasiona, Vol. 20, 74 - 77 (1972). In Serbo-Croatian.

014.023 **A topical approach to astronomy for nonspecialists.**
F. G. Liming, Jr.
American Journ. Phys., Vol. 40, 1575 - 1584 (1972).

014.024 **An experimental determination of the velocity of light from stellar spectrograms.**
R. B. Culver, R. G. Leisure.
American Journ. Phys., Vol. 40, 1585 - 1587 (1972).

014.025 **Ein Planetarium für die Londoner Schulen.**
P. Richards-Jones.
Jenaer Rundschau, (Jena Rev.), 17. Jahrgang, p. 358 - 360, 363 (1972).

Astronomie, Lehrbuch für Klasse 10.
See Abstr. 003.044.

015 Miscellanea

015.001 **Survival of common terrestrial microorganisms under simulated Jovian conditions.**
P. Molton, C. Ponnamperuma.
Nature, Vol. 238, 217 - 218 (1972).

015.002 **¿ Qué es un buen telescopio?** E. Roel.
El Universo, No. 98, Vol. 26, 3 - 11 (1972).

015.003 **La exploración del espacio. Velocidades y periodos orbitales.** F. Mardus.
El Universo, No. 98, Vol. 26, 25 - 27 (1972).

015.004 **Le phénomène des substances organiques terrestres et son extrapolation à l'univers.** A. Zelenka.
Orion, 30. Jahrgang, p. 142 (1972).

015.005 **Cours d'astronomie de la S.A.F. 7. Les étoiles: Étude générale.** G. Oudenot.
L'Astronomie, 86e année, p. 406 - 428 (1972).

015.006 **Questionnaire of CETI.**
Compiled by T. V. Mavrina.
Zemlya i Vselennaya, 1972, No. 4, p. 57 - 61. In Russian.

015.007 **Finding Venus or Mercury in daylight.**
A. C. Curtis.
Journ. British Astron. Ass., Vol. 82, 438 - 439 (1972).

015.008 **Observatories and city lights – one city fights light pollution.** A. A. Hoag.
Mercury, (Journ. Astron. Soc. Pacific), Vol. 1, No. 5, p. 2 - 3 (1972).

015.009 **Uidentifiserte flygende objekter – "flygende taller-kener"?** O. P. Sveen.
Astron. Tidsskr., Årg. 5, p. 107 - 112 (1972).

015.010 **Non-human artifacts in the solar system.**
G. V. Foster.
Spaceflight, Vol. 14, 447 - 453 (1972).

015.011 **Die Natur ist nicht rein spekulativ zu ergründen.**
E. Verhülsdonk.
SuW, Vol. 11, 332 - 333 (1972).

015.012 **Cours d'astronomie de la S.A.F. 8. Les étoiles: Amas ouverts, étoiles doubles, étoiles variables.**
M. Dumont.
L'Astronomie, 86e année, p. 531 - 545 (1972).

015.013 **El fin del mundo.** R. Compte Porta.
El Universo, Vol. 26, 97 - 100 (1972).

015.014 **Intelligence in the universe.** F. J. Dyson.
Mercury, (Journ. Astron. Soc. Pacific), Vol. 1, No. 6, p. 9 (1972).

015.015 **Modern geocentricism.** J. Mergentaler.
Urania Kraków, Vol. 43, 328 - 332 (1972).
In Polish.

015.016 **Zur Tradition der "Unwettersterne".**
P. Kunitzsch.
Zeitschr. Deutsch. Morgenländ. Ges., [Kommissionsverlag Franz Steiner GmbH, Wiesbaden], Vol. 122, 108 - 117 (1972).

015.017 **El problema de la vida fuera de la tierra.**
A. Romañá.
Publ. Obs. Ebro, Miscelánea, No. 25, 63 pp. (1969). – Discurso inaugural del año academico 1969–1970, Real Academia de Ciencias Exactas, Fisicas y Naturales, Madrid.

015.018 **Thunder: Shock waves in pre-biological organic synthesis.** A. Bar-Nun, M. E. Tauber.
Space Life Sci., Vol. 3, 254 - 259 (1972).

Interstellar flight and intelligence in the universe.
See Abstr. 051.015.

Applied Mathematics, Physics

021 Mathematics, Computing, Machine Programs

021.001 Features of algorithmic language ALGOL 68.
B. K. Martynenko.
Nablyud. Iskusstv. Nebesn. Tel, No. 62, p. 113 - 121 (1971).
In Russian.

021.002 Convergence in the method of a small parameter.
K. V. Kholshevnikov.
Nablyud. Iskusstv. Nebesn. Tel, No. 62, p. 153 - 168 (1971).
In Russian.

**021.003 On ill-conditioned systems of equations in astro-
nomical practice and a method for their solution.**
T. K. Nikolskaya.
Byull. Inst. Teoret. Astron., *Leningrad*, Vol. 13, 148 - 151
(1972). In Russian.
The present paper deals with a method for the solution
of ill-conditioned systems of linear equations. Instead of formal solution a projection of the solution on the eigenvectors
subspace corresponding to «large» eigenvalues is determined.
Two examples for the solution of the equations of condition
obtained by improvement of orbits using this method are
given.

**021.004 The use of the electronic computer for the urgent
publication of astronomical material.**
V. A. Ivakin.
IAU Symposium No. 45, (see 012.003), p. 103 - 104 (1972).

**021.005 A precompiler for the formula manipulation system
TRIGMAN.** W. H. Jefferys.
Celestial Mechanics, Vol. 6, 117 - 124 (1972).
A translator has been written which simplifies the pro-
gramming of problems with the formula manipulation system
TRIGMAN. It allows for the introduction of a new data type,
SERIES, into a FORTRAN program, and translates the user's
program into legal FORTRAN. It should not be difficult to
adapt it for other formula manipulation systems now being
used in celestial mechanics.

**021.006 An exercise in symbolic programming: Computa-
tion of general normalized inclination functions.**
J. A. Campbell.
Celestial Mechanics, Vol. 6, 187 - 197 (1972).
Computation of the general normalized inclination functions $A^k_{lm}(i)$ defined by R. H. Gooding is studied as an
example of a representative problem in symbolic programming
applied to celestial mechanics. The performance of three
different computing systems on the problem is discussed, together with aspects of their operation which are likely to require the user's attention in most such computations. An
Appendix contains the values of the functions through $l=6$.

**021.007 Loss due to missing data in efficiency of a locally
optimal test for homogeneity with respect to very
rare events.** P. S. Puri.
Proc. National Acad. Sci., *U.S.A.*, Vol. 67, 749 - 756 (1970).
With reference to observations of supernovae in galaxies,
a locally optimal test for the hypothesis of homogeneity of the
observational units with respect to occurrence of supernovae
(treated as a very rare event) is obtained for the case where the
data are available only for those galaxies where at least one
supernova is observed. It is shown that the loss in the asymptotic efficiency of the test due to lack of reporting of galaxies
with no supernovae is very heavy and in fact is infinite for the
case of very rare events.

021.008 Computer program for ephemerides.
M. Mizohara, S. Ito, N. Kobayashi.
Univ. Tokyo, Tokyo Astron. Obs. Rep. (No. 60), Vol. 16,
9 - 18 (1972). In Japanese.

**021.009 Multi-off-grid methods in multi-step integration of
ordinary differential equations.** P. R. Beaudet.
Bull. American Astron. Soc., Vol. 4, 417 - 418 (1972). – Abstr.
AAS.

021.010 Higher order Runge-Kutta methods. D. G. Bettis.
Bull. American Astron. Soc., Vol. 4, 418 (1972).
Abstr. AAS.

**021.011 Multiple shooting procedures for nonlinear boun-
dary problems.** R. Bulirsch.
Bull. American Astron. Soc., Vol. 4, 418 (1972). – Abstr. AAS.

021.012 The order of differential equation methods.
J. C. Butcher.
Bull. American Astron. Soc., Vol. 4, 418 (1972). – Abstr. AAS.

**021.013 On stiffness in chemical kinetic transport calcula-
tions.** J. S. Chang.
Bull. American Astron. Soc., Vol. 4, 418 (1972). – Abstr. AAS.

**021.014 On the non-equivalence of maximum polynomial
degree Nordsieck-Gear and classical methods.**
R. Danchick.
Bull. American Astron. Soc., Vol. 4, 418 (1972). – Abstr. AAS.

**021.015 Gauss-Jackson and Runge-Kutta evaluation and
analysis for the interplanetary free-flight predictor
problem.** C. H. Drinnan.
Bull. American Astron. Soc., Vol. 4, 418 - 419 (1972). – Abstr.
AAS.

**021.016 Shooting-splitting method for sensitive two-point
boundary value problems.**
P. J. Firnett, B. A. Troesch.
Bull. American Astron. Soc., Vol. 4, 419 (1972). – Abstr. AAS.

**021.017 A study of numerical integration routines for a
nutrient utilization ecological problem.**
D. B. Frazho, W. F. Powers, R. P. Canale.
Bull. American Astron. Soc., Vol. 4, 419 (1972). – Abstr. AAS.

**021.018 Use of Green's functions in the numerical solution
of two-point boundary value problems.**
L. J. Gallaher, I. E. Perlin.
Bull. American Astron. Soc., Vol. 4, 419 (1972). – Abstr. AAS.

**021.019 A method of numerical integration for trajectories
with variational equations.** W. H. Goodyear.

Bull. American Astron. Soc., Vol. 4, 420 (1972). – Abstr. AAS.

021.020 **Phase space analysis in numerical solution of ordinary differential equations.** B. E. Howard.
Bull. American Astron. Soc., Vol. 4, 420 (1972). – Abstr. AAS.

021.021 **Comparing numerical methods for ordinary differential equations.** T. E. Hull.
Bull. American Astron. Soc., Vol. 4, 420 (1972). – Abstr. AAS.

021.022 **Ellipsoidal bounds for the propagation of uncertainty along trajectories.** W. Kahan.
Bull. American Astron. Soc., Vol. 4, 420 (1972). – Abstr. AAS.

021.023 **Changing stepsize in the integration of differential equations using modified divided differences.**
F. T. Krogh.
Bull. American Astron. Soc., Vol. 4, 420 - 421 (1972). – Abstr. AAS.

021.024 **Local extrapolation in the solution of ordinary differential equations.** L. F. Shampine.
Bull. American Astron. Soc., Vol. 4, 422 (1972). – Abstr. AAS.

021.025 **Convergence and error of the Bubnov-Galerkin method.** F. Stenger.
Bull. American Astron. Soc., Vol. 4, 422 (1972). – Abstr. AAS.

021.026 **Discretization of ordinary differential equations of infinitely long time intervals.** H. J. Stetter.
Bull. American Astron. Soc., Vol. 4, 422 (1972). – Abstr. AAS.

021.027 **Extrapolation methods for the solution of initial value problems and their practical realization.**
J. Stoer.
Bull. American Astron. Soc., Vol. 4, 422 - 423 (1972). – Abstr. AAS.

021.028 **Computer programs in astrophysics.** H.-C. Thomas.
Comput. Phys. Commun., (*Netherlands*), Vol. 3, Suppl., p. 151 - 156 (1972).
The author discusses the relation between computer generated stellar models and reality, the use of computers for analysis of optical and radioastronomical observations and the construction of a theoretical model of the universe. He also considers the use of computer programs in understanding the mechanism which causes pulsations in variable stars.

021.029 **Sampling functions for geophysics.**
G. E. O. Giacaglia, C. A. Lundquist.
Smithsonian Astrophys. Obs., *Cambridge, Mass.*, Special Rep. No. 344, 6 + 93 pp. (1972).
The main objective of the present work is a comprehensive mathematical development of the sampling-function formalism for representing geophysical quantities around a spheri-

cal or a nearly spherical body such as the earth. The discussion here unifies and largely supersedes the mathematical developments in the earlier references. The final section concerns some geophysical applications of the sampling-function formalism that augment the applications discussed in the 1971 references.

On the theory, techniques, and data processing of very long baseline interferometry. See Abstr. 033.053.

A library of standard programmes for constructing numerical theories for studying the motion and evolution of the orbits of the minor bodies of the solar system. See Abstr. 042.024.

Acceleration of the numerical integration of equations of motion in celestial mechanics. See Abstr. 042.076.

On processing of latitude observations by the least squares method. See Abstr. 045.034.

Construction de grilles cartographiques à l'aide d'un ordinateur. See Abstr. 046.028.

Plate reduction for the stellar triangulation. See Abstr. 046.035.

Correction of solar intensity measurements for stray light. See Abstr. 071.075.

Programs for the processing of meteorological data by computers. See Abstr. 082.240.

Method of computations for relative positions on the lunar disk. See Abstr. 094.209.

Calculs d'éphémérides de petites planètes à l'aide d'un ordinateur IBM 1130. See Abstr. 098.037.

The solution of problems of cometary astronomy on electronic computers. See Abstr. 102.017.

A method for absolute calibration of objective-prism plates suitable for computer reduction. See Abstr. 114.022.

Computer programmes for differential curve-of-growth analysis. See Abstr. 114.087.

Computer solution of eclipsing-binary light curves by the method of differential corrections. See Abstr. 121.073.

A new approach to periodogram analyses. See Abstr. 122.053.

022 Physical Papers Related to Astronomy and Astrophysics

022.001 Perturbations in the α-system of the TiO molecule.
J. G. Phillips, S. P. Davis.
Astrophys. Journ., Vol. 175, 583 - 588 (1972).

A breaking-off of branches in bands of the α-system of TiO ($C^3\Delta - X^3\Delta$) has been ascribed to a series of perturbations experienced by excited vibrational states in the upper ($C^3\Delta$) electronic state. Approximate vibrational and rotational constants can be derived for the perturbing electronic state. The excitation energies of the perturbing state do not match with those of any known triplet or singlet electronic state of TiO, nor can convincing extrapolations be made from lower-lying states.

022.002 The cross-section of suprathermal proton bremsstrahlung. E. Haug.
Astrophys. Letters, Vol. 11, 225 - 226 (1972).

The suprathermal proton cross-section is calculated for various values of the proton energy. The results are compared with previous work.

022.003 Spectroscopic studies of molecular structure.
G. Herzberg.
Science, Vol. 177, 123 - 138 (1972).

022.004 Birch's law: Why is it so good? D. H. Chung.
Science, Vol. 177, 261 - 263 (1972).

022.005 Two-photon decay of metastable hydrogenic atoms.
R. Novick.
Science, Vol. 177, 367 (1972).

022.006 Spin alignment of OH molecules by directed infrared radiation. V. V. Burdjuzha, D. A. Varshalovich.
Astron. Zhurn. Akad. Nauk SSSR, Vol. 49, 727 - 736 (1972).
In Russian. English translation in Soviet Astron. AJ, Vol. 16, No. 4.

Pumping of OH molecules by directed radiation at $\lambda \simeq 100\,\mu$ is discussed. Magnetic sublevel populations are calculated for arbitrary angles between the direction of IR-radiation and that of the magnetic field. The dependence of τ on the angle between the direction of the line of sight and that of the magnetic field is determined.

022.007 Interpolation formulae for the electron impact excitation of ions in the H-, He-, Li-, and Ne-sequences.
R. Mewe.
Astron. Astrophys., Vol. 20, 215 - 221 (1972).

The cross sections for electron impact excitation from the ground state of H-, He-, Li-, and Ne-like ions are approximated by interpolation formulae with four parameters that can be integrated analytically over a Maxwellian electron velocity distribution to give the corresponding rate coefficients.

022.008 Broadening and shift of magnesium lines by van der Waals interaction with argon atoms and by microfields. V. Helbig, H. J. Kusch.
Astron. Astrophys., Vol. 20, 299 - 304 (1972).

Line broadening and shift of 3 neutral and 3 ionic lines of magnesium are investigated using a gas — stabilized high — pressure arc running in an argon atmosphere between a carbon cathode and a magnesium anode. From the total line profiles the broadening parameters for van der Waals interaction and for electron impact are obtained. The results are compared with theoretical values obtained from the Lindholm-Foley theory and from quantum-mechanical electron impact calculations.

022.009 Radiative lifetimes for some resonance transitions of Fe I and Fe II in the region between 2300 Å and 3050 Å, and the application to iron abundance determinations in the sun and in the QSO PHL 938.
G. E. Assousa, W. H. Smith.
Astrophys. Journ., Vol. 176, 259 - 264 (1972).

The radiative lifetimes for five Fe I and three Fe II upper states which result in transitions terminating in the ground state were obtained with our electron-beam phase-shift apparatus. We found lifetimes ranging from 8.4 to 2.6 ns for the various transitions. These are converted to absolute oscillator strengths by using relative transition probabilities taken from the literature. The resulting oscillator strengths are applied to problems of the solar iron abundance and the column densities of iron observed in the spectrum of PHL 938.

022.010 The oscillator strengths of the SO $A\,^3\Pi - X\,^3\Sigma^-$ band systems. W. H. Smith.
Astrophys. Journ., Vol. 176, 265 - 266 (1972).

The oscillator strengths for the SO $A\,^3\Pi - X\,^3\Sigma^-$ band systems are calculated using RKR Franck-Condon factors and the radiative lifetime previously derived from electron-beam phase-shift measurements.

022.011 Oscillator strengths for allowed $nd-n'f$ transitions in the helium isoelectronic sequence.
R. T. Brown, J.-L. M. Cortez.
Astrophys. Journ., Vol. 176, 267 - 270 (1972).

Variational wave functions have been used to obtain wavelengths and oscillator strengths for $nd-n'f$ transitions, through n and n' equal to 8, in the helium isoelectronic sequence through S XV. Of the 660 transitions studied, observed wavelengths are available for only 10, all in He I. Differences between observed and calculated values range from 0.07 percent to less than 0.004 percent. Agreement between dipole-length and dipole-velocity oscillator strengths for the great majority of transitions is to within 1 or 2 percent. The accuracy of the wave functions is not great enough to provide reliable values for $\Delta n = 0$ transitions.

022.012 Transient radiative heat transfer in a non-gray medium. D. G. Doornink, R. G. Hering.
Journ. Quant. Spectrosc. Radiat. Transfer, Vol. 12, 1161 - 1174 (1972).

This study is devoted to investigating the influence of spectral effects on the transient temperature and radiative flux distribution within a non-gray medium. The system chosen for study is a stationary, plane layer of non-conducting material enclosed between black walls.

022.013 Check of quantum-mechanical electron broadening calculations for Mg$^+$ and Ca$^+$ resonance lines.
D. E. Roberts, A. J. Barnard.
Journ. Quant. Spectrosc. Radiat. Transfer, Vol. 12, 1205 - 1216 (1972).

An experimental check has been made of the first completely quantum-mechanical calculations of the electron broadening of ion lines. The widths and shifts of the relevant lines (the resonance transitions of Mg$^+$ and Ca$^+$) were measured as functions of electron density and temperature. In addition, calculations were done which remove some of the deficiencies of earlier semi-classical calculations.

022.014 Excited-state cesium photoionization cross sections.
J. C. Weisheit.
Journ. Quant. Spectrosc. Radiat. Transfer, Vol. 12, 1241 -

1248 (1972).

Photoionization cross sections of excited-state cesium atoms $Cs(6^2 P_{1/2})$ and $Cs(6^2 P_{3/2})$ are computed at incident photon wavelengths greater than 1500 Å. Both the spin—orbit perturbation of the valence-electron orbital and the polarization interaction between the valence-electron and the core are included in the calculations. Coefficients for the rate of radiative recombination to each $(6^2 P_j)$ fine-structure level are determined for temperatures below 3500^e K.

022.015 **Vibrational transition probabilities and r-centroids for some diatomic molecular band systems.**
P. D. Singh, M. M. Shukla.
Journ. Quant. Spectrosc. Radiat. Transfer, Vol. 12, 1249 - 1252 (1972).

022.016 **Effect of atomic polarizability on low-energy free—free radiative transitions.** H. A. Hyman.
Journ. Quant. Spectrosc. Radiat. Transfer, Vol. 12, 1253 - 1256 (1972). – Note.

022.017 **Shapes and widths of ammonia lines collision-broadened by hydrogen.** P. Varanasi.
Journ. Quant. Spectrosc. Radiat. Transfer, Vol. 12, 1283 - 1289 (1972).

022.018 **Spectres d'absorption dans l'ultraviolet lointain de Be, B, C, N, Mg, Al et Si.**
J. M. Esteva, G. Mehlman-Balloffet, J. Romand.
Journ. Quant. Spectrosc. Radiat. Transfer, Vol. 12, 1291 - 1303 (1972).

022.019 **Determination des constantes Stark de raies du xenon.** A. Lesage, J. Richou.
Journ. Quant. Spectrosc. Radiat. Transfer, Vol. 12, 1313 - 1318 (1972).

022.020 **Morse Franck-Condon factors and r-centroids for some bands of AlO $(C^2 \Sigma^+ - X^2 \Sigma^+)$, AuBe $(A_{1/2} - X^2 \Sigma^+)$ and BiO $(B^4 \Sigma^+ - X^2 \Pi_{1/2})$ systems.**
G. C. Singh.
Journ. Quant. Spectrosc. Radiat. Transfer, Vol. 12, 1343 - 1346 (1972). – Note.

022.021 **The band spectra of yttrium oxide.**
J. B. Shin, R. W. Nicholls.
Journ. Roy. Astron. Soc. Canada, Vol. 66, 222 - 223 (1972). Abstr. Canadian Astron. Soc.

022.022 **Étude des fondements des systèmes autorégulateurs spontanés.** C.-A. Bogdanski.
Comptes Rendus Acad. Sci. Paris, Sér. B, Vol. 275, 199 - 202 (1972).

022.023 **Laser compression of matter to super-high densities: Thermonuclear (CTR) applications.**
J. Nuckolls, L. Wood, A. Thiessen, G. Zimmerman.
Nature, Vol. 239, 139 - 142 (1972).

Hydrogen may be compressed to more than 10,000 times liquid density by an implosion system energized by a high energy laser. This scheme makes possible efficient thermonuclear burn of small pellets of heavy hydrogen isotopes, and makes feasible fusion power reactors using practical lasers.

022.024 **Compton effect and spectral features of the intensity of scattered emission.** V. M. Charugin.
Astron. Tsirk., No. 686, p. 5 - 7 (1972). In Russian.

022.025 **An SCF-MO-CI calculation of the C_2 molecule using an atomic basis of contracted Gaussian lobe functions.** J. Barsuhn.

Zeitschr. Naturforschung, Vol. 27a, 1031 - 1041 (1972). In German.

022.026 **Theoretical black-body radiation temperature 2.2 K.** C. T. J. Dodson.
Nature, Phys. Sci., Vol. 239, 64 (1972).

022.027 **A shock tube determination of the electronic transition moment of the CN red band system.**
J. O. Arnold, R. W. Nicholls.
Journ. Quant. Spectrosc. Radiat. Transfer, Vol. 12, 1435 - 1452 (1972).

The electronic transition moment of the CN red band system and its variation with internuclear separation have been determined absolutely from spectral emission measurements behind incident shock waves.

022.028 **On the validity of the Dicke approximation for computing collision narrowed profiles of quadrupole lines in hydrogen atmospheres.** G. E. Hunt, J. S. Margolis.
Bull. American Astron. Soc., Vol. 4, 359 (1972). – Abstr. AAS.

022.029 **Model potential calculations of lithium transitions.** T. C. Caves, A. Dalgarno.
Journ. Quant. Spectrosc. Radiat. Transfer, Vol. 12, 1539 - 1552 (1972).

022.030 **Half-life of ^{10}Be.** F. Yiou, G. M. Raisbeck.
Phys. Rev. Letters, Vol. 29, 372 - 375 (1972).

The half-life of ^{10}Be has been measured to be $(1.5 \pm 0.3) \times 10^6$ yr, a value in significant disagreement with the previously accepted value of $(2.7 \pm 0.4) \times 10^6$ yr. We discuss several implications of the revised half-life.

022.031 **Total absorption cross sections of several gases of aeronomic interest at 584 Å.**
W. L. Starr, M. Loewenstein.
Journ. Geophys. Res., Vol. 77, 4790 - 4796 (1972).

022.032 **Electron energy deposition in CO_2.**
T. Sawada, D. J. Strickland, A. E. S. Green.
Journ. Geophys. Res., Vol. 77, 4812 - 4818 (1972).

The study of the molecular properties of CO_2 has taken on new interest and importance in light of recent analyses of the Mariner Mars UV data. Such knowledge is essential for constructing realistic models of the Mars upper atmosphere. In this report we consider the interaction of energetic electrons with CO_2 and predict the distribution of energy among the possible loss channels.

022.033 **Electron impact excitation cross sections and energy degradation in CO.**
T. Sawada, D. L. Sellin, A. E. S. Green.
Journ. Geophys. Res., Vol. 77, 4819 - 4828 (1972).

We determine a comprehensive set of electron impact cross sections for carbon monoxide. If the cross sections are combined with similar studies on CO_2 and H_2O, for example, such calculations should be useful in interpreting observations of dayglow emissions from the upper atmospheres of such planets as Mars and Venus.

022.034 **Radio wave polarization at partial scattering.**
Yu. V. Beresin, N. A. Matiyasevitch, V. I. Smirnov.
Geomagn. Aeronom., Vol. 12, 830 - 835 (1972). In Russian.

022.035 **Structure of the pressure-induced infrared spectrum of hydrogen in the first overtone region.**
A. Watanabe, J. L. Hunt, H. L. Welsh.
Canadian Journ. Phys., Vol. 49, 860 - 863 (1971).

Laboratory measurements on the hydrogen overtone spectra are of interest as a check on the present state of the theory of induced infrared absorption. Such studies have also a practical application in the estimation of the hydrogen content of planetary atmospheres (Herzberg 1952; Welsh 1969; Poll 1970). We therefore present in this communication some measurements on the first overtone under conditions for which a fairly accurate analysis can be made.

022.036 Effective cross sections of excitation of Ne I and Ne II lines in the ultraviolet.
J. M. Smirnov, J. D. Sharonov.
Astron. Zhurn. Akad. Nauk SSSR, Vol. 49, 1102 - 1106 (1972). In Russian. English translation in Soviet Astron. AJ, Vol. 16, No. 5.

Effective cross sections of 35 Ne I and 38 Ne II lines have been measured in the energy region from threshold value up to 350 eV. The lines investigated are situated in the 2700–3700 Å spectral region.

022.037 Branching ratios in photoelectron spectroscopy.
J. A. R. Samson, J. L. Gardner.
Journ. Optical Soc. America, Vol. 62, 856 - 858 (1972).

The problem of measuring branching ratios or transition probabilities using the technique of photoelectron spectroscopy is discussed in terms of the angular distribution of the ejected photoelectrons. Equations are developed that can be used in conjunction with any type of electron-energy analyzer to correct for any discrimination in the analyzer caused by electrons of varying angular distribution. These equations have been applied to determine the branching ratio (or transition-probability ratio) of the A/B electronic states of CO_2^+ produced by the photoionization of CO_2 by the 584-Å He I line. This is an important process in the atmospheres of Mars and Venus. After excitation of these states by solar radiant energy, the states rapidly decay to the ground state of the ion, producing in the process a large amount of fluorescent radiant energy which contributes to the day and night glows of the planet's atmosphere. It is therefore important to know the absolute production rate of these excited states of the CO_2^+ ion.

022.038 Analysis of the spectrum of triply ionized magnesium (Mg IV). M.-C. Artru, V. Kaufman.
Journ. Optical Soc. America, Vol. 62, 949 - 957 (1972).

022.039 Transition probabilities and collision-induced transitions in excited levels of neon. R. A. Lilly.
Journ. Optical Soc. America, Vol. 62, 1023 - 1026 (1972).

022.040 Radiative lifetimes of the $A^2\Pi$ state of CO^+.
R. Anderson, R. Sutherland, N. Frey.
Journ. Optical Soc. America, Vol. 62, 1127 - 1130 (1972).

022.041 Analysis of the spectrum of five-times-ionized zirconium (Zr VI).
J. O. Ekberg, J. E. Hansen, J. Reader.
Journ. Optical Soc. America, Vol. 62, 1134 - 1139 (1972).

022.042 Analysis of the spectrum of six-times-ionized niobium (Nb VII).
J. O. Ekberg, J. E. Hansen, J. Reader.
Journ. Optical Soc. America, Vol. 62, 1139 - 1142 (1972).

022.043 Analysis of the spectrum of seven-times-ionized molybdenum (Mo VIII) and isoelectronic comparison of the spectra Y V–Mo VIII.
J. O. Ekberg, J. E. Hansen, J. Reader.
Journ. Optical Soc. America, Vol. 62, 1143 - 1148 (1972).

022.044 New lines of neon ions in the range 50–200 Å.
H. Hermansdorfer.
Journ. Optical Soc. America, Vol. 62, 1149 - 1152 (1972).

About 180 new spectral lines of Ne V, Ne VI, Ne VII, and Ne VIII have been measured and identified on spectrograms obtained from a large theta-pinch apparatus. The spectrograms covered the range from 50 to 200 Å. The lines were classified by means of isoelectronic extrapolations and interpolations.

022.045 Absolute scale for Si I gf-values from wall-stabilized arc measurements. E. Schulz-Gulde.
Astron. Astrophys., Vol. 21, 313 - 314 (1972). – Research note.

022.046 Measurement of the Stark broadening parameters of some singly ionized argon lines.
N. Konjević, J. Labat, L. Ćirković, J. Purić.
Zeitschr. Physik, Vol. 235, 35 - 43 (1970).

Half-widths of fifteen Stark broadened argon II lines have been measured in argon plasma behind the reflected shock wave. The measured A II linewidths are compared with theoretical and other experimental results. It is shown that a) the broadening of A II lines is in good agreement with the theory, b) line broadening increases linearly with electron density, and c) the Stark broadened lines follow the dispersion profile to the distance of at least three halfwidths from the line center.

022.047 Atomic beam magnetic resonance investigations in the $2p^2$ ^3P ground multiplet of the stable carbon isotopes ^{12}C and ^{13}C.
G. Wolber, H. Figger, R. A. Haberstroh, S. Penselin.
Zeitschr. Physik, Vol. 236, 337 - 351 (1970).

The hyperfine structure separation in the $^3P_2$ state of ^{13}C is of interest to radio astronomers: by means of the line corresponding to this separation it should be possible to check on the concentration of elementary carbon in interstellar space. A preliminary search in July 1969 for this line in absorption in the directions of the Crab nebula and the galactic centre was, however, not sucessful.

022.048 Absorption cross section and oscillator strength of the autoionizing line HgI 1126,6 Å.
R. Lincke, B. Stredele.
Zeitschr. Physik, Vol. 238, 164 - 171 (1970). In German.

The absorption cross section of the strongest autoionizing line of neutral mercury has been measured photoelectrically. The integrated cross section yielded an f-value of 0.53± 0.03. This compares well with the f-values of the corresponding lines in Zn I and Cd I, but is considerably smaller than the value expected from dispersion measurements.

022.049 Photoionization and photoabsorption cross sections of CO_2 at 584 Å.
J. A. R. Samson, J. L. Gardner, J. E. Mentall.
Journ. Geophys. Res., Vol. 77, 5560 - 5566 (1972).

022.050 Electronic excitation of N_2 and dissociative excitation of O_2 by proton impact.
J. H. Moore, Jr.
Journ. Geophys. Res., Vol. 77, 5567 - 5572 (1972).

022.051 Stark-effect calculations in the second Lyman-series of ionized helium. M. Korten, H. F. Berg.
Zeitschr. Physik, Vol. 239, 322 - 330 (1970).

The profiles of some lines of the second Lyman-series of He II, broadened by microscopic Stark-effect in plasmas of given temperatures and densities were calculated and compared to available experimental data. The comparison yielded reasonable agreement as to the resulting halfwidths and lesser agreement in details of the line-centers and -wings.

022.052 Contribution à l'étude du radical OH$^+$ (système de bandes $^3\Pi_i \rightarrow {}^3\Sigma^-$). D. Rakotoarijimy.
Physica, Vol. 49, 360 - 382 (1970).

022.053 Oscillator strengths of the resonance lines of some rare gases. J. P. de Jongh, J. van Eck.
Physica, Vol. 51, 104 - 112 (1971).

The oscillator strengths of several resonance lines of He, Ne, Ar and Kr in the wavelength range between 522 Å and 1165 Å have been determined by measuring the self-absorption of the radiation as a function of the gas pressure at low pressures. The self-absorption of the radiation of the $1^1S - 2^1P$ line of He is used as a reference.

022.054 Transition probabilities and radiative lifetimes for Ne II. B. F. J. Luyken.
Physica, Vol. 51, 445 - 460 (1971).

022.055 Effect of temperature on atomic hydrogen lines excited in the afterglow of water vapour.
S. K. Gupta, R. V. Shukla, S. K. Jain, A. N. Srivastava.
Physica, Vol. 51, 520 - 525 (1971).

Effect of temperature on the excitation processes of the H_α and H_β lines of atomic hydrogen in the afterglow of water vapour has been studied in a dynamical flow system in the temperature range $27-340°C$. It has been found that the intensity of these lines decreases with increasing temperature. It is inferred that the mechanism causing the excitation of these emissions is quite involved. However, the applicability of the present results to the terrestrial exosphere has been demonstrated.

022.056 Emission cross sections for NI and NII multiplets and some molecular bands for electron impact on N_2. J. F. M. Aarts, F. J. De Heer.
Physica, Vol. 52, 45 - 73 (1971).

022.057 A modern version of the Ole Roemer experiment. R. S. McMillan, J. D. Kirszenberg.
Sky Telescope, Vol. 44, 300 - 301 (1972).

022.058 The e.p.r. spectrum of vibrationally excited hydroxyl radicals.
P. N. Clough, A. H. Curran, B. A. Thrush.
Proc. Roy. Soc. London, Ser. A, Vol. 323, 541 - 554 (1971).

The fine energy structure of OH, particularly Λ-doubling, is important in radio astronomy. Interstellar microwave emission by OH in several rotational levels of both ground-state multiplet components has been detected (Yen, Zuckerman, Palmer & Penfield 1969), and mechanisms for its origin have been proposed (Litvak, Zuckerman & Dickinson 1969). The detection of emission from excited vibrational levels, which may be possible with knowledge of the Λ-doubling frequencies given here, could assist the identification of the emission mechanism.

022.059 Electron-ion recombination in a dense molecular gas. D. R. Bates, V. Malaviya, N. A. Young.
Proc. Roy. Soc. London, Ser. A, Vol. 320, 437 - 458 (1971).

A semi-quantal method is developed for treating the recombination of electrons and positive ions in a dense molecular gas. Extensive calculations are carried out relating to recombination in hydrogen, in nitrogen, in carbon dioxide and in damp mixtures of gases.

022.060 Improved apparatus for the measurement of absolute f-values. G. D. Bell, E. F. Tubbs.
Rev. Sci. Instruments, Vol. 41, 435 - 438 (1970).

022.061 Dielectronic satellite spectra for highly-charged helium-like ion lines. A. H. Gabriel.
Monthly Notices Roy. Astron. Soc., Vol. 160, 99 - 119 (1972).

Calculations have been carried out in intermediate coupling of the wavelengths and intensities of the satellite lines situated on the long wavelength side of the helium-like ion resonance lines, recently observed from solar flares. Earlier calculations up to aluminium have been extended up to iron and copper.

022.062 Ionization balance for ions of Na, Al, P, Cl, A, K, Ca, Cr, Mn, Fe and Ni.
M. Landini, B. C. Fossi.
Astron. Astrophys., Suppl. Ser., Vol. 7, 291 - 310 (1972).

The ionization balance is computed for ions of Na, Al, P, Cl, A, K, Ca, Cr, Mn, Fe and Ni for temperatures from 10^4 to 10^8 K. The computations include collisional ionization, autoionization, radiative recombination and dielectronic recombination. Computations for low density plasmas and solar atmosphere conditions are performed.

022.063 Lifetimes of some highly excited levels in the Pb-I spectrum measured by the Hanle method.
S. Garpman, G. Lidö, S. Rydberg, S. Svanberg.
Zeitschr. Physik, Vol. 241, 217 - 235 (1971).

Natural lifetimes of some highly excited levels in the Pb-I spectrum have been measured by the zero field level crossing (Hanle) method. The results are compared with lifetimes derived from oscillator strengths given in the literature.

022.064 Measurement of oscillator strengths in the singlet system of neutral magnesium.
R. Lincke, B. Ziegenbein.
Zeitschr. Physik, Vol. 241, 369 - 379 (1971). In German.

022.065 Measurements of double differential cross sections in electron impact ionization of helium and argon.
H. Ehrhardt, K. H. Hesselbacher, K. Jung, M. Schulz, T. Tekaat, K. Willmann.
Zeitschr. Physik, Vol. 244, 254 - 267 (1971).

Double differential cross sections (angular distributions and energy loss spectra) have been measured of electrons after ionizing electron collisions with helium at primary energies E_0 between 25 eV and about 260 eV and with argon at $E_0 = 75$, 150 and 200 eV.

022.066 Lifetimes and relative initial state populations of some hydrogen atomic states using beam-foil spectroscopy. D. Schürmann, W. Schlagheck, P. H. Heckmann, H. H. Bukow, H. v. Buttlar.
Zeitschr. Physik, Vol. 246, 239 - 247 (1971). In German.

022.067 Populations of excited atoms: Sensitivity to low-energy cross-sections. A. K. Dupree.
Astrophys. Letters, Vol. 12, 125 - 128 (1972).

At low energies two sets of theoretical cross-sections for electron-hydrogen collisions between highly excited atomic levels differ by more than a factor of 100. However, these different cross-sections change the population of high $(n \sim 100)$ atomic levels and the quantity $d \ln b/dn$ by less than a factor of 2 under the conditions of temperature, density, and radiation field expected in H I clouds. It may be possible to infer the appropriate cross-sections from low-frequency observations of radio recombination lines.

022.068 Molecular calculations concerning a new candidate for the unidentified emission line at 89.190 GHz.
J. Barsuhn.
Astrophys. Letters, Vol. 12, 169 - 172 (1972).

SCF-calculations indicate that the electronic ground state of the radical CCH is a $^2\Sigma^+$ state. The computed rigid-rotor, J = 1 → 0, transition-frequency of $^2\Sigma^+$ is 88.9 GHz, which coincides within the error limits of the approximation to the X-ogen frequency at 89.190 GHz.

022.069 Beam-foil studies of neon below 1000 Å.
J. A. Kernahan, A. Denis, R. Drouin.

Physica Scripta, (*Sweden*), Vol. 4, 49 - 51 (1971).

Spectra and mean lives for lines of Ne I to Ne VII have been obtained using the beam-foil technique in the wavelength range 350–920 Å. A new level in Ne VI, $2p^{32}D$, is reported.

022.070 Radiative lifetimes in Sc I–Sc III.
R. Buchta, L. J. Curtis, I. Martinson, J. Brzozowski.
Physica Scripta, (*Sweden*), Vol. 4, 55 - 59 (1971).

We have studied the spectra of scandium (600–6000 Å) with the beam-foil method. We also measured the mean lifetimes of 20 excited Sc I–Sc III levels and generally found good agreement with previous theoretical calculations and emission measurements.

022.071 Lifetimes and oscillator strengths in spectra of Be, B and C. J. Bromander.
Physica Scripta, (*Sweden*), Vol. 4, 61 - 63 (1971).

Lifetimes of 41 excited terms in the spectra of Be I, Be II. B I, B II, B III, C I, C II and C III are presented.

022.072 The transitions $3s^23p^k-3s3p^{k+1}$ of Ti VI and Ti VII.
L. Å. Svensson.
Physica Scripta, (*Sweden*), Vol. 4, 111 - 112 (1971).

022.073 Beam-foil spectroscopy of carbon in the vacuum ultraviolet. M. C. Poulizac, J. P. Buchet.
Physica Scripta, (*Sweden*), Vol. 4, 191 - 194 (1971).

Beam-foil spectra of the different carbon ions C II to C V have been studied between 300 Å and 1300 Å in the energy range 0.4 MeV–2 MeV. Eleven new lines have been identified. The mean lives of numerous excited levels of C II, C III, C IV and C V have been measured and are compared with other experimental and theoretical data.

022.074 The photoionization cross-section of SI.
G. Tondello.
Space Sci. Rev., Vol. 13, 553 (1972). – Abstract of conference paper IAU Colloquium No. 14 (see 012.012).

022.075 Stark broadening of UV nickel lines.
R. D. Bengtson, M. H. Miller, R. A. Roig.
Space Sci. Rev., Vol. 13, 554 (1972). – Abstract of conference paper IAU Colloquium No. 14 (see 012.012).

022.076 Energy levels and classification problems in spectra of highly ionized elements of the fifth period.
M. E. Zohar, B. S. Fraenkel.
Space Sci. Rev., Vol. 13, 555 - 556 (1972). – Conference paper IAU Colloquium No. 14 (see 012.012).

022.077 Theoretical studies on transition wavelengths and transition probabilities. A. Dalgarno.
Space Sci. Rev., Vol. 13, 559 (1972). – Abstract of invited paper IAU Colloquium No. 14 (see 012.012).

022.078 Energy levels and spectra of the Li I and Be I isoelectronic sequences in the fourth row.
S. Goldsmith, U. Feldman, L. Oren, L. Cohen.
Space Sci. Rev., Vol. 13, 560 (1972). – Abstract of conference paper IAU Colloquium No. 14 (see 012.012).

022.079 Absolute intensity calibration at 26 Å by branching ratios to the visible.
F. E. Irons, N. J. Peacock.
Space Sci. Rev., Vol. 13, 561 - 562 (1972). – Conference paper IAU Colloquium No. 14 (see 012.012).

022.080 Observation of argon lines at normal pressure in the vacuum ultraviolet.
D. Müller, Č. Vadla, V. Vujnović.
Space Sci. Rev., Vol. 13, 563 - 564 (1972). – Conference

paper IAU Colloquium No. 14 (see 012.012).

022.081 Ionization equilibrium for ions of Na, Al, P, Cl, A, K, Ca, Cr and Mn.
M. Landini, B. C. Fossi.
Space Sci. Rev., Vol. 13, 586 - 587 (1972). – Conference paper IAU Colloquium No. 14 (see 012.012).

022.082 A universal function for ionization of atoms by structureless charged particles of arbitrary mass and charge. O. Bely, P. Faucher.
Space Sci. Rev., Vol. 13, 588 (1972). – Conference paper IAU Colloquium No. 14 (see 012.012).

022.083 X-ray spectra from highly ionized iron and nickel.
J. L. Schwob, B. S. Fraenkel.
Space Sci. Rev., Vol. 13, 589 - 591 (1972). – Conference paper IAU Colloquium No. 14 (see 012.012).

022.084 The classification of Fe IX to XVI emission lines and isoelectronic lines in laboratory and solar spectra. B. C. Fawcett.
Space Sci. Rev., Vol. 13, 606 - 607 (1972). – Conference paper IAU Colloquium No. 14 (see 012.012).

022.085 The spectrum of FeH: Laboratory and solar identification. P. K. Carroll, P. McCormack.
Astrophys. Journ., (*Letters*), Vol. 177, L33 - L36 (1972).

A new complex molecular spectrum has been observed under high resolution in the blue-green region. The molecule responsible has been identified as FeH. Comparison of the laboratory wavelengths with those in the revised Rowland table shows many coincidences with weak unidentified solar lines, especially lines which either are enhanced in the sunspot spectrum or are observed only in the spot.

022.086 Experimental investigation of radio wave scattering by a plasma cylinder in a waveguide.
N. Z. Derycott, I. V. Bajrachenko.
Vestn. Kiev. Un-ta, Ser. Astron., No. 14, p. 64 - 70 (1972). In Russian.

The results of an experimental study of radio wave scattering by a plasma cylinder in a waveguide are presented: The measurements of the longitudinal component of the electric field have shown that the scattering in the region of main resonance is of a dipole character; in the region of secondary resonances the scattering has more complicated character.

022.087 Oscillator strengths and ground-state photoionization cross-sections for Mg^+ and Ca^+.
J. H. Black, J. C. Weisheit, E. Laviana.
Astrophys. Journ., Vol. 177, 567 - 572 (1972).

022.088 On the absorption spectrum of calcium in solid benzene. D. A. Williams.
Observatory, Vol. 92, 174 - 178 (1972).

022.089 Collisional excitation of the $4\,^2F$ levels in lithiumlike Ne VIII. H.-J. Kunze.
Phys. Rev. A, General Phys., Vol. 4, 111 - 113 (1971).

Collisional excitation rates from the ground state to the $4f$ levels in lithiumlike Ne VIII have been derived from the absolute intensity of the $3\,^2D-4\,^2F$ radiative transition.

022.090 Electron-impact excitation of auto-ionizing levels in cesium. Y. B. Hahn, K. J. Nygaard.
Phys. Rev. A, General Phys., Vol. 4, 125 - 132 (1971).

Auto-ionizing states in Cs between 12 and 20 eV have been studied by electron impact. We have been able to identify about 20 levels, and the agreement with spectroscopic data is excellent.

022.091 **Charge-transfer excitation of Ar$^+$ in low-energy He$^+$+ Ar collisions.** M. Lipeles.
Phys. Rev. A, General Phys., Vol. 4, 140 - 146 (1971).

022.092 **Energy spectra of metastable oxygen atoms produced by electron-impact dissociation of O_2.**
W. L. Borst, E. C. Zipf.
Phys. Rev. A, General Phys., Vol. 4, 153 - 161 (1971).

022.093 **Low-energy elastic and fine-structure excitation scattering of ground-state C$^+$ ions by hydrogen atoms.** J. C. Weisheit, N. F. Lane.
Phys. Rev. A, General Phys., Vol. 4, 171 - 182 (1971).

022.094 **S and P states of the helium isoelectronic sequence up to $Z = 10$.**
Y. Accad, C. L. Pekeris, B. Schiff.
Phys. Rev. A, General Phys., Vol. 4, 516 - 536 (1971).
Calculations have been made of the energy levels and other properties of the states $n\,^1S$, $n\,^3S$, $n\,^1P$, and $n\,^3P$, $n = 2$ to 5, for atoms belonging to the helium isoelectronic sequence up to $Z = 10$, and also for the higher excited S states of helium. The theoretical term values, including the contributions from the mass-polarization correction and the relativistic effects of order α^2 are listed.

022.095 **Vibrorotational excitations of H_2^+ by e^+ impact. I.**
F. H. M. Faisal.
Phys. Rev. A, General Phys., Vol. 4, 596 - 601 (1971).
Analogies between Coulomb excitations of nuclei and ionic molecules by charged projectiles are utilized to calculate vibrorotational excitations of H_2^+ molecular ions by e^+ impact by a semiclassical method developed in the nuclear case.

022.096 **Excitation of the O I (^3S) and N I (^4P) resonance states by electron impact on O and N.**
E. J. Stone, E. C. Zipf.
Phys. Rev. A, General Phys., Vol. 4, 610 - 613 (1971).
The absolute cross sections for the excitation of the O I (^3S) and N I (^4P) resonance states by electron impact on atomic oxygen and nitrogen have been measured over the aeronomically important energy range from threshold to 150 eV.

022.097 **Autoionization in the uv photoabsorption of atomic calcium.** V. L. Carter, R. D. Hudson, E. L. Breig.
Phys. Rev. A, General Phys., Vol. 4, 821 - 825 (1971).
Detailed measurements of the absolute photoabsorption cross section of atomic calcium between 2028 and 1753 Å and 1589 and 1424 Å are presented.

022.098 **f values for transitions between the low-lying S and P states of the helium isoelectronic sequence up to $Z = 10$.** B. Schiff, C. L. Pekeris, Y. Accad.
Phys. Rev. A, General Phys., Vol. 4, 885 - 893 (1971).
f values have been computed for the transitions $m\,^1S - n\,^1P$, $m = 1$-5, $n = 2$-5 and $m\,^3S - n\,^3P$, m, $n = 2$-5 for members of the helium isoelectronic sequence up to $Z = 10$. The values obtained are accurate to within 1% or better for the large majority of the transitions.

022.099 **Close-coupling studies of rotational excitation in H–H_2 collisions.**
E. F. Hayes, C. A. Wells, D. J. Kouri.
Phys. Rev. A, General Phys., Vol. 4, 1017 - 1025 (1971).

022.100 **Radiative lifetime of the 2 $^1S_0$ metastable state of helium.**
R. S. Van Dyck, Jr., C. E. Johnson, H. A. Shugart.
Phys. Rev. A, General Phys., Vol. 4, 1327 - 1336 (1971).

022.101 **Theory of atomic structure including electron correlation. IV. Method for forbidden-transition probabilities with results for [O I], [O II], [O III], [N I], [N II], and [C I].** C. Nicolaides, O. Sinanoğlu, P. Westhaus.
Phys. Rev. A, General Phys., Vol. 4, 1400 - 1410 (1971).
Electric quadrupole transition probabilities for the oxygen atom auroral line $^1S_0$-$^1D_2$ ($\lambda = 5577$ Å) and the ^2P-^2D, ^1S-^1D lines in C I, N I, N II, O II, and O III, which are of atmospheric and astrophysical interest, are calculated.

022.102 **Continuum processes in atomic nitrogen.**
S. Ormonde, M. J. Conneely.
Phys. Rev. A, General Phys., Vol. 4, 1432 - 1445 (1971).

022.103 **Electron capture and loss in collisions of heavy ions with atomic oxygen.**
H. H. Lo, L. Kurzweg, R. T. Brackman, W. L. Fite.
Phys. Rev. A, General Phys., Vol. 4, 1462 - 1476 (1971).
Electron-capture and electron-loss cross sections for various gaseous (N$^+$, O$^+$, Ar$^+$, Kr$^+$, and Xe$^+$) and metallic (Al$^+$, K$^+$, Fe$^+$, Ba$^+$, and Ba^{++}) ions in collisions with atomic oxygen have been measured in the energy range 30 keV to 2 MeV.

022.104 **Differential and integral cross sections for the electron-impact excitation of the $a^1\Delta_g$ and $b^1\Sigma_g^+$ states of O_2.** S. Trajmar, D. C. Cartwright, W. Williams.
Phys. Rev. A, General Phys., Vol. 4, 1482 - 1492 (1971).

022.105 **Absolute transition probabilities of phosphorus.**
M. H. Miller, R. A. Roig, R. D. Bengtson.
Phys. Rev. A, General Phys., Vol. 4, 1709 - 1722 (1971).
A gas-driven shock tube was used to measure the absolute strengths of 21 P I lines and 126 P II lines (3300 Å$<\lambda< 6900$ Å).

022.106 **Frequency dependence of the speed of light in space.** Z. Bay, J. A. White.
Phys. Rev. D, Particles and Fields, Vol. 5, 796 - 799 (1972).
To characterize the possible dispersion of the velocity of light in space (vacuum) a Cauchy-type formula, $n^2 = 1 + A/\nu^2 + B\nu^2$, is used. It is shown that relativity only allows a nonzero A term, independent of the nature of the waves or a quantization thereof. Recent experimental data provide upper bounds for A and B, limiting thereby the dispersion in the microwave, infrared, visible, and ultraviolet regions of the spectrum to less than one part in 10^{20}.

022.107 **Cross sections for the production of excited products in the photoionization of N_2, O_2, CO, and N_2O by 58.4-nm radiation.**
K. Monahan, T. S. Wauchop.
Journ. Geophys. Res., Vol. 77, 6262 - 6265 (1972).
Brief report.

022.108 **Measurement of the photodetachment cross section for O_4^- at high pressure.** J. A. Burt.
Journ. Geophys. Res., Vol. 77, 6280 - 6281 (1972). – Letter.

022.109 **Temperature and pressure dependence of CO_2 extinction coefficients.**
W. B. DeMore, M. Patapoff.
Journ. Geophys. Res., Vol. 77, 6291 - 6293 (1972). – Letter.

022.110 **Properties of photons determined by interferometric spectroscopy.**
H. A. Gebbie, R. A. Bohlander, R. P. Futrelle.
Nature, Vol. 240, 391 - 394 (1972).
Black body spectroscopy has been neglected by experimentalists for fifty years. We propose to apply interferometric methods to black bodies with the prospect of getting temperature values based on frequency standards in the range from 90 K and upwards. When applied to higher temperature sources, the same methods could tell us new things about

photons.

022.111 Radiative-lifetime studies of the emission continua of the hydrogen and deuterium molecules.
W. H. Smith, R. Chevalier.
Astrophys. Journ., Vol. 177, 835 - 839 (1972).

Radiative lifetimes have been measured for states of the hydrogen and deuterium molecules that give rise to continua in the 1400–3500 Å region. We compare our results with the theoretical and experimental results available in the literature. The application of these results to the destruction of the hydrogen and the CH molecules in the interstellar medium is discussed.

022.112 Oscillator strengths of weak Fe I resonance lines measured by combined hook and absorption techniques. M. C. E. Huber, E. F. Tubbs.
Astrophys. Journ., Vol. 177, 847 - 854 (1972).

Oscillator strengths of nine weak Fe I resonance lines determined from nearly simultaneous hook and photoelectric absorption measurements are presented. Comparison of these results with the commonly used oscillator strengths of Corliss and Tech supports the hypothesis of Bell and Upson that the data reflect an intensity-dependent error.

022.113 K-shell photoionization cross-sections.
E. Daltabuit, D. P. Cox.
Astrophys. Journ., Vol. 177, 855 - 859 (1972).

Approximate values for the threshold energies, threshold cross-sections, and the energy dependence of the cross-sections for K-shell photoionization are tabulated for H, He, C, N, O, Ne, Mg, Si, and S in all stages of ionization.

022.114 CH and CH$^+$ formation in ion-molecule reactions.
T. P. Stecher, D. A. Williams.
Astrophys. Journ., (*Letters*), Vol. 177, L141 - L144 (1972).

The ion-molecule reactions between C$^+$ and H$_2$, which are normally endothermic, become exothermic and proceed with high rates when vibrational energy in H$_2$ exceeds the endothermicity. Excitation by ultraviolet photons in the Lyman and Werner bands leaves H$_2$ excited vibrationally. Vibrational relaxation is slow. We suggest that CH$^+$ and CH can be produced by these ion-molecule reactions in about the amounts that are observed in the interstellar medium.

022.115 The condensation of H$_2$ and D$_2$: Astrophysics and vacuum technology. T. J. Lee.
Journ. Vacuum Sci. Technology, Vol. 9, 257 - 261 = Commun. Roy. Obs., Edinburgh, No. 122 (1972).

We have measured the vapor pressure above layers of solid hydrogen and deuterium, grown either directly onto a copper substrate or onto previously condensed argon layers, at temperatures in the range 1.6 to 5.2 K. The results of our experiments are discussed in relation to the design of hydrogen cryopumps, gauge calibration and the conditions required for the condensation of H$_2$ on solid grains in interstellar gas clouds.

022.116 The spectra of highly ionized aluminum (Al VI–X) in the extreme-ultraviolet and soft X-ray regions.
F. P. J. Valero, D. Goorvitch.
Astrophys. Journ., Vol. 178, 271 - 276 (1972).

Spectral lines corresponding to transitions in the five-to-nine-times-ionized Al atom are measured and classified in the spectral range from 48 to 340 Å. The spectra were photographed by using a laser-produced plasma as a light source and a 3-m grazing-incidence spectrograph. Different techniques employed to separate stages of ionization are discussed.

022.117 Absolute Raman scattering cross-section of molecular hydrogen. R. W. Carlson, W. R. Fenner.
Astrophys. Journ., Vol. 178, 551 - 556 (1972).

The absolute Raman scattering cross-section of H$_2$ for the pure rotational lines and the Q-branch of the vibrational transition has been determined at the incident wavelengths 4800 and 5145 Å. Using the measured cross-sections we have determined the molecular polarizabilities.

022.118 Direct measurement of the lifetimes and predissociation probabilities for rotational levels of the OH and OD $A\ ^2\Sigma^+$ states. B. G. Elmergreen, W. H. Smith.
Astrophys. Journ., Vol. 178, 557 - 564 (1972).

Quantitative measurements for the lifetimes and predissociation probabilities for the F_1 and F_2 components for individual rotational lines of the $A\ ^2\Sigma^+$ state for the OH and OD molecules are obtained from absolute phase-shift measurements. Astrophysical implications of preassociation in OH are indicated.

022.119 Observations of shifts of hydrogen lines.
R. D. Bengtson, G. R. Chester.
Astrophys. Journ., Vol. 178, 565 - 569 (1972).

Laboratory measurements of the profiles of Stark-broadened lines emitted from a radiofrequency-generated plasma ($n_e \approx 1.2 \times 10^{13}$ cm^{-3}) show small redshifts for lines arising from odd upper levels (N_{upper} = 13, 15, 17). No shift is noted for the even lines. No asymmetries could be found within experimental error. Shifts of odd Balmer lines were also observed in the spectrum of Sirius.

022.120 Coulomb-Born-Oppenheimer cross-sections for excitation of hydrogenic ions of infinite Z by electron impact. A. D. Parks, D. H. Sampson.
Astrophys. Journ., Vol. 178, 571 - 575 (1972).

Coulomb-Born-Oppenheimer cross-section calculations are made for all the $2l \rightarrow 3l'$ and $1s \rightarrow 3l'$ transitions for hydrogenic ions with $Z = \infty$. The calculations are made for impact electron energies of 1 and 1.5 in threshold units. On the basis of these results some slight modifications of previous semi-empirical cross-section formulae are suggested.

022.121 Lifetimes of excited states in Mg I, Cd I, and Mg II.
T. Andersen, L. Mølhave, G. Sørensen.
Astrophys. Journ., Vol. 178, 577 - 582 (1972).

The beam-foil technique has been used to measure mean lives of some excited states in Mg I, Cd I, and Mg II. The reported Mg I and Cd I lifetimes are considerably shorter than the mean lives recently reported by Schaefer, using the delayed-coincidence technique.

022.122 The probabilities of infrared and radio transitions of molecules OH and CH.
V. V. Burdjuzha, D. A. Varshalovich.
Astron. Zhurn. Akad. Nauk SSSR, Vol. 49, 1211 - 1215 (1972). In Russian. English translation in Soviet Astron. AJ, Vol. 16, No. 6.

The probabilities of all radiative transitions between lower rotation levels and levels of Λ-doublets of molecules OH and CH are calculated.

022.123 Analysis of X-ray fluorescence without dispersion for investigations of cosmic and terrestrial objects.
Yu. A. Surkov, B. M. Andrejchikov, I. K. Akhmetshin, V. K. Khristianov, I. D. Shevaleevskij.
Kosmich. Issled., Vol. 10, 930 - 937 (1972). In Russian.

022.124 Magnetic multipole transition probabilities.
R. H. Garstang.
Journ. Physique, Vol. 31, Suppl. C4, p. C4-189 - C4-190 (1970).

022.125 Further identifications in the Ar IX spectrum.

J. P. Connerade.
Solar Physics, Vol. 27, 130 - 131 (1972). — Research note.

**022.126 Non-X-ray background suppression in wire wall pro-
portional counters with the use of coaxial anodes.**
A. N. Bunner, D. McCammon, W. L. Kraushaar, F. O. Williamson.
Bull. American Astron. Soc., Vol. 4, 411 (1972). — Abstr. AAS.

**022.127 Measurement of an optical frequency and the speed
of light.**   Z. Bay, G. G. Luther, J. A. White.
Phys. Rev. Letters, Vol. 29, 189 - 192 (1972).
We report the measurement of the frequency of the 633-
nm red laser line. Combination of the optical frequency with
the known wavelength yields c to an accuracy higher than
previously known.

022.128 Calculation of photoabsorption processes in helium.
A. Dalgarno, H. Doyle, M. Oppenheimer.
Phys. Rev. Letters, Vol. 29, 1051 - 1052 (1972).
The cross section for photoionization is expressed as an
integral over the dipole response function in a form that yields
a unified procedure for calculating the background and the
resonance contributions. Application is made to absorption by
triplet metastable helium.

**022.129 Speed of light from direct frequency and wavelength
measurements of the methane-stabilized laser.**
K. M. Evenson, J. S. Wells, F. R. Petersen, B. L. Danielson,
G. W. Day, R. L. Barger, J. L. Hall.
Phys. Rev. Letters, Vol. 29, 1346 - 1349 (1972).
The frequency and wavelength of the methane-stabilized
laser at 3.39μm were directly measured against the respective
primary standards. Multiplication yields the speed of light
$c = (299\ 792\ 456.2 \pm 1.1)$ m/sec, in agreement with and 100
times less uncertain than the previously accepted value.

022.130 Relative transition probabilities for krypton.
M. H. Miller, R. A. Roig, R. D. Bengtson.
Journ. Optical Soc. America, Vol. 62, 1027 - 1029 (1972).
Relative transition probabilities of the more prominent
Kr I and Kr II lines have been measured using a gas-driven
shock tube as a spectroscopic source. Results for 22 Kr I lines
$(4200 < \lambda < 8500$ Å) and 33 Kr II lines $(4000 < \lambda < 5300$ Å)
have estimated reliabilities of 8–50%. The data are compared
with other measurements and with theoretical calculations.

022.131 Influence of temperature on the spectrum of water.
G. M. Hale, M. R. Querry, A. N. Rusk, D. Williams.
Journ. Optical Soc. America, Vol. 62, 1103 - 1108 (1972).
The normal-incidence spectral reflectance of water at
5, 27, and 70°C has been measured in the spectral region between 5000 and 350 cm^{-1}. From the measured values of spectral reflectance we have determined the optical constants n_r
and n_i by Kramers-Kronig methods. The band strengths $S_B =
\int n_i(\nu) d\nu$ and bandwidths have been determined for the absorp-
tion bands near 3400, 1640, and 600 cm^{-1} at each temperature. A similar study of deuterium oxide at 27°C has been
conducted for purposes of comparison.

022.132 Lifetime measurements in Ar II–Ar VIII.
A. E. Livingston, D. J. G. Irwin, E. H. Pinnington.
Journ. Optical Soc. America, Vol. 62, 1303 - 1308 (1972).
We have measured radiative decay times from the beam-
foil spectra of Ar II–Ar VIII in the wavelength region 450–
2450 Å. The mean lifetimes for 20 excited levels are presented.
We compare the results with theoretical calculations and other
measurements and discuss the agreement with oscillator
strengths in several isoelectronic sequences.

022.133 Measurement of the velocity of light.   K. M. Baird.

Journ. Optical Soc. America, Vol. 62, 1359 - 1360
(1972). — Abstr. Optical Soc. America.

**022.134 Molecular branching-ratio method for intensity
calibration of optical systems in the vacuum ultraviolet.**   M. J. Mumma.
Journ. Optical Soc. America, Vol. 62, 1459 - 1466 (1972).
A state-of-the-art review is given of the molecular branching-ratio method for intensity calibration in the vacuum
ultraviolet. Ways are described for determining both relative
and quantitative responses in the wavelength range 1000 Å <
$\lambda < 3000$ Å.

022.135 Doubly-excited terms in lithium and beryllium.
H. G. Berry, J. Bromander, I. Martinson, R. Buchta.
Physica Scripta, (Sweden), Vol. 3, 63 - 67 (1971).
We give a summary of the available experimental information about doubly-excited terms in the Li I isoelectronic sequence and discuss some new classifications and lifetime measurements. Some new classifications in Be I are also suggested
and we give the results of lifetime measurements for some
singly-excited levels in Li II, Be I, and Be II.

022.136 Lifetime measurements in Si II, Si III, and Si IV.
H. G. Berry, J. Bromander, L. J. Curtis, R. Buchta.
Physica Scripta, (Sweden), Vol. 3, 125 - 132 (1971).
We have measured radiative decay times in Si II, Si III,
and Si IV in the wavelength region 700–6000 Å using the
beam-foil technique. The lifetimes and transition probabilities
have been evaluated by alternative methods of curve-fitting
and cascade analysis. These results are compared with theoretical transition probabilities, and values in other members of the
isoelectronic sequences. The present estimates of silicon in
astrophysical objects are not changed by our measurements
of transition probabilities in Si II and Si III. The solar photospheric and coronal abundance estimates of silicon relative to
hydrogen thus still differ by a factor of three. We have measured the transition probabilities of most of the silicon lines
observed in the red-shifted quasar spectra.

022.137 Current status of the quest for quarks.
Y. S. Kim, N. Kwak.
Fields and Quanta, (GB), Vol. 3, 1 - 138 (1972).
A review of quark searches in air showers, terrestrial mat-
ter and stellar objects is presented.

**022.138 Jones - Vektoren geschrieben in rechts- und links-
zirkularen Komponenten.**   R. H. Hammerschlag.
Optik, Vol. 34, 595 - 597 = Utrechtse Sterrekundige Overdrukken, No. 188 (1972).

022.139 Towards a new determination of the speed of light.
C. C. Bradley, G. J. Edwards, D. J. E. Knight,
W. R. C. Rowley, P. T. Woods.
Phys. Bull., (GB), Vol. 23, 15 - 18 (1972).
The authors determine the speed of light by using a sta-
bilized CO_2 laser and a Fabry-Perot interferometer. The result
is accurate to 1 part in 10^8.

022.140 Extended analysis of the spectrum of Mg IV.
G.-A. Johannesson, T. Lundström, L. Minnhagen.
Physica Scripta, Vol. 6, 129 - 137 (1972).

**022.141 Isentropic compression of fused quartz and liquid
hydrogen to several Mbar.**
R. S. Hawke, D. E. Duerre, J. G. Huebel, R. N. Keeler, H.
Klapper.
Phys. Earth Planet. Interiors, Vol. 6, (see 012.023), 44 - 47
(1972).

022.142 Correlation of theory and experiment for high-pres-

sure hydrogen.
W. G. Hoover, M. Ross, C. F. Bender, F. J. Rogers, R. J. Olness.
Phys. Earth Planet. Interiors, Vol. 6, (see 012.023), 60 - 64 (1972).

022.143 **Thermodynamics of hydrogen—helium mixtures at high pressure and finite temperature.**
W. B. Hubbard.
Phys. Earth Planet. Interiors, Vol. 6, (see 012.023), 65 - 68 (1972).

022.144 **Phase equilibria in fluid mixtures at high pressures: the He—CH_4 system.**
W. B. Streett, A. L. Erickson, J. L. E. Hill.
Phys. Earth Planet. Interiors, Vol. 6, (see 012.023), 69 - 77 (1972).

022.145 **Equation of state and phase diagram of dense hydrogen.**   G. I. Kerley.
Phys. Earth Planet. Interiors, Vol. 6, (see 012.023), 78 - 82 (1972).

022.146 **Ground state energy of solid molecular hydrogen at high pressure.**   C. Ebner, C. C. Sung.
Phys. Earth Planet. Interiors, Vol. 6, (see 012.023), 83 - 90 (1972).

022.147 **Calculations of electrical transport properties of liquid metals at high pressures.**
R. Evans, A. Jain.
Phys. Earth Planet. Interiors, Vol. 6, (see 012.023), 141 - 145 (1972).

022.148 **Lattice model calculation of elastic and thermodynamic properties at high pressure and temperature.**
H. H. Demarest, Jr.
Phys. Earth Planet. Interiors, Vol. 6, (see 012.023), 146 - 153 = Publ. Inst. Geophys. Planet. Phys., Univ. California, Los Angeles, No. 1087 (1972).

022.149 **Disproportionation of Fe_2SiO_4 to $2FeO + SiO_2$ at pressures up to 250kbar and temperatures up to 3000 °C.**   W. A. Bassett, L.-C. Ming.
Phys. Earth Planet. Interiors, Vol. 6, (see 012.023), 154 - 160 (1972).

022.150 **Transport properties of liquid metal hydrogen under high pressures.**   R. C. Brown, N. H. March.
Phys. Earth Planet. Interiors, Vol. 6, (see 012.023), 206 - 209 (1972).

022.151 **Hartree-Fock calculations of the Mg I spectrum in the extreme ultraviolet.**
M. W. D. Mansfield, J. P. Connerade.
Physica Scripta, Vol. 6, 191 - 194 (1972).
Hartree-Fock calculations have been performed for the seven principal configurations in Mg I. The predicted levels are compared with Newsom's (1971) observations. Agreement is generally good and new identifications are proposed.

022.152 **Electron impact excitation rates for helium.**
R. S. Benson, J. L. Kulander.
Solar Physics, Vol. 27, 305 - 318 (1972).
Electron impact excitation rates are calculated for all transitions in He I and II between individual terms for $n \leqslant 4$. Rates calculated for a large number of different experimental, theoretical and semi-empirical cross sections are compared.

The velocity of light.
Sky Telescope, Vol. 44, 353, 365 (1972).

Errata

022.901 **Errata: 'Lifetime and quenching rates for the H_2 continuum'** [Journ. Quant. Spectrosc. Radiat. Transfer, Vol. 12, 117 - 121 (1972)].
R. T. Thompson, R. G. Fowler.
Journ. Quant. Spectrosc. Radiat. Transfer, Vol. 12, 1367 (1972).

022.902 **Erratum: "The lithium-like spectra of K XVII through Mn XXIII in the extreme-ultraviolet region"** [Astrophys. Journ., Vol. 174, 209 - 214 (1972)].
S. Goldsmith, U. Feldman, L. Oren (Katz), L. Cohen.
Astrophys. Journ., Vol. 175, 589 (1972).

022.903 **Erratum: "Fokker-Planck equations for charged-particle transport in random fields"** [Astrophys. Journ., Vol. 172, 319 - 326 (1972)].   J. R. Jokipii.
Astrophys. Journ., Vol. 176, 557 (1972).

022.904 **Erratum: RKR Franck—Condon factors for blue and ultraviolet transitions of some molecules of astrophysical interest and some comments on the interstellar abundance of CH, CH^+, and SiH^+** [Journ. Quant. Spectrosc. Radiat. Transfer, Vol. 12, 947 - 958 (1972)].   H. S. Liszt, W. H. Smith.
Journ. Quant. Spectrosc. Radiat. Transfer, Vol. 12, 1591 (1972).

Instruments and Astronomical Techniques

031 Optics, Methods of Observation and Reduction

031.001 **I fondamenti matematici del metodo di Bruns.**
E. Fichera.
Mem. Soc. Astron. Italiana, Nuova Ser., Vol. 43, 269 - 272 (1972).
We analyse mathematically the method of Bruns to show its inapplicability to determine the errors in the graduation of a meridian circle.

031.002 **Sulla precisione nelle misure dei passaggi stellari.**
C. Moranzino.
Mem. Soc. Astron. Italiana, Nuova Ser., Vol. 43, 273 - 277 (1972).
This is a comparison between the accuracy obtained by the same observer, when the transit of a star is observed using a zenith telescope or a transit instrument.

031.003 **Random wave-front perturbations and telescopic star images.**   E. H. Linfoot, R. C. Witcomb.
Monthly Notices Roy. Astron. Soc., Vol. 158, 199 - 231 (1972).
A computational model of random perturbations induced in wavefronts in the entry pupil of telescopes of moderate aperture has been constructed and used to predict the instantaneous distribution of intensity in the perturbed stellar images formed by such telescopes. The model and its predictions are compared with the observations.

031.004 **Identification and removal of phase errors in interferometry.**   R. H. T. Bates, P. J. Napier.
Monthly Notices Roy. Astron. Soc., Vol. 158, 405 - 424 (1972).
Using the concept of the complex zeros of the Michelson interferogram it is shown how phase errors can be identified in measured interferograms. Computational procedures are described in detail and are illustrated with examples involving measured interferograms (of Taurus A and radio source PO349-27) and ideal data.

031.005 **The polariscope program.**   T. Gehrels.
Space Science Rev., Vol. 13, 319 (1972). – Conference paper (see 012.002).

031.006 **On the possibility of utilizing a narrow-angle television camera for determination of elements of the moon's rotation.**   V. B. Gurevich.
Astron. Zhurn. Akad. Nauk SSSR, Vol. 49, 853 - 859 (1972). In Russian. English translation in Soviet Astron. AJ, Vol. 16, No. 4.
A method is proposed of utilizing results of measurements of positions of stars, passing the field of view of a narrow-angle television camera, established vertically on the lunar surface, for the determination of elements of the lunar rotation. Considerations and calculations testified to practicability of such an experiment are given.

031.007 **Autocorrelation methods to obtain diffraction-limited resolution with large telescopes.**
C. E. KenKnight.
Astrophys. Journ., (*Letters*), Vol. 176, L43 - L45 (1972).
A new method uses an interferometer to display the Fourier transform of the brightness distribution of an object. A magnitude limit of +9 is easy on large telescopes and may become much fainter. Application includes the infrared.

031.008 **The water cooled secondary mirror for the 450 mm telescope at Wollongong University College.**
R. W. Upfold.
Journ. British Astron. Ass., Vol. 82, 369 - 371 (1972).

031.009 **A short description of the improvements made at the transit instrument of the Wrocław Observatory.**
G. M. Petrov, J. Bem.
Postępy Astron., Vol. 20, 257 - 260 (1972). In Polish.

031.010 **An optical system with steering mirror.**
A. F. Lagutin.
Izv. Krymskoj Astrofiz. Obs., Vol. 44, 122 - 133 (1972).
An optical system is considered in which the transversal shifts of the image in the field of view are due to the displacements of only one mirror (steering mirror). This system is compared with one having usual arrangement of the optical elements. Some examples of the use of the system are made; in particular, the system that was used for the observation of the sun from an airplane is described.

031.011 **A method to obtain composite photographs.**
A. K. Dabakhov.
Izv. Krymskoj Astrofiz. Obs., Vol. 44, 134 - 136 (1972). In Russian.
The design of a special device to obtain composite photographs is described. Examples of composite photographs of Mars and a stellar field are given.

031.012 **A fast optical method of precise angular and linear measurements.**   S. B. Novikov.
Astron. Tsirk., No. 672, p. 4 - 5 (1972). In Russian.

031.013 **A study of reflection coefficients of 122-cm Crimean reflector mirrors.**
E. A. Vitrichenko, K. P. Lianzuridy.
Astron. Tsirk., No. 680, p. 4 - 6 (1972). In Russian.

031.014 **On the accuracy of measurements of the inclination of the horizontal axis of a striding level.**
Ya. V. Naumov.
Trudy TsNII geod., aehros"emki i kartogr., 1972, vyp. (No.) 169, p. 46 - 50. In Russian. – Abstr. in Referativ. Zhurn. 51. Astron., 8.51.174 (1972).

031.015 **Resolution enhancement of astronomical spectra.**
T. J. Ulrych, J. R. Auman, J. A. Eilek, G. A. H. Walker.
Astron. Astrophys., Vol. 21, 125 - 130 (1972).
A technique which may considerably improve the effective resolution of a spectrum, and which is designed to give accurate line positions, is presented and is illustrated by application to synthetic and real spectra. The method also yields estimates of the central line intensity, a line width parameter, and some notion of the line shape.

031.016 **Microanalysis of materials by backscattering spectrometry.**
M.-A. Nicolet, J. W. Mayer, I. V. Mitchell.
Science, Vol. 177, 841 - 849 (1972).

031.017 **Resolution enhancement of atmospherically degraded astronomical photographs by digital computer processing.**   P. H. Richter.
Bull. American Astron. Soc., Vol. 4, 312 - 313 (1972).
Abstr. AAS.

031.018 **The application of computers for photometric processing of images.**   M. F. Shabanov.
Astron. Tsirk., No. 704, p. 1 - 3 (1972). In Russian.

031.019 **The system of automatic processing of images.**
Yu. N. Lipskij, M. F. Shabanov.
Astron. Tsirk., No. 704, p. 3 - 5 (1972). In Russian.

031.020 **New procedure for making Schmidt corrector plates.**
G. Lemaître.
Applied Optics, Vol. 11, 1630 - 1636 (1972).
We describe what we call the dioptric elasticity method of making Schmidt plates. We give the elasticity theory, discuss our shop methods, and show the very satisfactory results.

031.021 **Twyman-Green interferometer to test large aperture optical systems.**   D. G. Kocher.
Applied Optics, Vol. 11, 1872 - 1874 (1972).
This letter describes a simple unequal path laser interferometer that can be used for wavefront testing of large aperture focusing optical systems and components.

031.022 **The reduction of the coma of off-axis guide stars.**
M. V. Penston, C. M. Lowne.
Observatory, Vol. 92, 100 - 101 (1972). – Letter.

031.023 **Die Physiologie der visuellen astronomischen Beobachtung.**   W. Weiss.
Sterne, 48. Jahrgang, p. 146 - 160 (1972).

031.024 **On determining the weights when smoothing observational data.**   A. I. Emets, Ya. S. Yatskiv.
Astrometriya i Astrofiz., *Kiev*, No. 15, (see 003.001), p. 9 - 15 (1972). In Russian.
The question is considered on the determination of weights when smoothing observational data by Whittaker's method. Estimates of the weights and corresponding calculations for the case of dependent errors of latitude observations are given.

031.025 **Untersuchungen zur Argelandermethode.**
U. Bastian, P. Hemmer.
BAV Rundbrief, 21. Jahrgang, Sonder-Rundbrief, (see 012.011), p. 13 - 15 (1972).

031.026 **Recent developments in digital image processing at the Image Processing Laboratory at the Jet Propulsion Laboratory.**   D. A. O'Handley, W. B. Green.
Proc. IEEE, Vol. 60, 821 - 828 (1972).
Image processing of spacecraft images has been carried on at the Jet Propulsion Laboratory since 1964. The most recent advances in removal of geometric distortion and residual image effects along with various types of mapping projections are covered. The recent applications of image processing to the areas of biomedicine, forensic sciences, and astronomy are discussed.

031.027 **Space-variant image motion degradation and restoration.**   A. A. Sawchuk.
Proc. IEEE, Vol. 60, 854 - 861 (1972).

A description of motion degradation in linear incoherent optical systems is presented. Given a mechanical description of the motion, an equivalent linear space-variant system containing all the motion effects is derived, and detailed examples of common types of variant and invariant motion are included. Following a review of restoration techniques for motion blur, a method for image restoration applicable to a large class of space-variant systems is presented. A computer simulation of space-variant restoration is included.

031.028 **Simulated star for testing a guiding telescope.**
N. N. Raimov, V. S. Tataurov, G. R. Pekki, K. Ya. Kutorkina.
Optiko-mekh. prom-st', 1972, No. 5, p. 26 - 27. In Russian.
Abstr. in Referativ. Zhurn. 51. Astron., 10.51.109 (1972).

031.029 **Fresnel-zone-plate spectrometer with central stop.**
P. N. Keating, R. K. Mueller, T. Sawatari.
Journ. Optical Soc. America, Vol. 62, 945 - 948 (1972).
The marked chromatic aberration of a Fresnel-zone-plate lens can be used as a variable narrow-band optical filter, and thus a holographic spectrometer. The filtering properties of the zone plate are analyzed, with particular emphasis on the effect of a central stop over the low-order rings. It is shown that the central stop can markedly improve the rejection of frequencies outside the pass band without significant degradation of the spectrometer resolution.

031.030 **Zone plates and their aberrations.**
M. Young.
Journ. Optical Soc. America, Vol. 62, 972 - 976 (1972).
The zone plate is an optical device that depends on interference, not reflection or refraction, for its image-forming properties. This paper derives the third-order and chromatic aberrations of the zone plate.

031.031 **Restoring with maximum entropy, II: Superresolution of photographs of diffraction-blurred impulses.**
B. R. Frieden, J. J. Burke.
Journ. Optical Soc. America, Vol. 62, 1202 - 1210 (1972).
Can photographic images, e.g., corresponding to star clusters, be superresolved? Or alternatively, can already good (diffraction-limited) images be improved by restoring methods? To test this hypothesis, objects have been prepared that can be resolved only if the bandwidth in the restoration exceeds that of the image data.

031.032 **Pulse risetimes in proportional counters.**
G. R. Ricker, Jr., J. J. Gomes.
Rev. Sci. Instruments, Vol. 40, 227 - 233 (1969).
We present a method of calculating proportional counter pulse shapes for both point and extended ionization tracks. The method is applicable to any counter gas for which the electron drift velocity as a function of the electric field is known.

031.033 **A controlled laser pulse width technique.**
F. Rainer.
Rev. Sci. Instruments, Vol. 40, 368 - 370 (1969). – Note.

031.034 **Les mesures par double image.**   P. Muller.
L'Astronomie, 86e année, p. 472 - 477 (1972).

031.035 **Die Intensitätsverteilung im Beugungsbild von Teleskopspiegeln mit unregelmäßiger Oberflächenwelligkeit.**   G. Schwesinger.
Optik, Vol. 34, 553 - 572 (1972).

031.036 **Improved three-lens field correctors for paraboloids.**
C. G. Wynne.
Monthly Notices Roy. Astron. Soc., Vol. 160, 13P - 18P

(1972).

A form of triple corrector is described giving an image spread within about 1 arc sec over a field of 20 arc min diameter at the prime focus of an f/3.3 paraboloid mirror. Glasses are used having good transmission in the near ultra-violet.

031.037 Some tests of the Vaníček method of spectral analysis. J. Taylor, S. Hamilton.
Astrophys. Space Sci., Vol. 17, 357 - 367 (1972).

Comparisons are made between the Vaníček method and the more usual method consisting of preliminary removal of the systematic noise followed by Fourier analysis. Formulas relating the two methods are developed and a series of comparative plots of simple spectra are presented.

031.038 Device for determination of the mean moment of star transits. M. P. Ogriņš.
Uch. zap. Latv. gos. un-ta imeni Petra Stuchki, Vol. 169, 3 - 17 (1972). In Russian.

031.039 Investigation of the device for determination of the value of the mean moments of star transits.
M. P. Ogriņš.
Uch. zap. Latv. gos. un-ta imeni Petra Stuchki, Vol. 169, 18 - 29 (1972). In Russian.

031.040 On the automatic identification of stars.
J. K. Balodis.
Uch. zap. Latv. gos. un-ta imeni Petra Stuchki, Vol. 169, 63 - 75 (1972). In Russian.

The problem of automatic identification of stars on photographic plates by a computer is discussed.

031.041 Optical arrays. M. J. Disney.
Monthly Notices Roy. Astron. Soc., Vol. 160, 213 - 232 (1972).

The relative effectiveness (speed) of a large telescope for a wide range of observations is compared with that of an array of smaller telescopes of the same total cost. It is concluded that the array is always more effective if the new television type detector systems are to be used.

031.042 Refraction effects due to moving media in Doppler measurements. C. Ferencz, G. Tarcsai.
Space Research XII, (see 012.016), Vol. 1, 595 - 600 (1972).

031.043 High angular resolution from ground-based telescopes: A general outline of the problem.
J. Rösch.
Space Research XII, (see 012.016), Vol. 2, 1633 - 1656 (1972).

031.044 High resolution solar observations. R. B. Dunn.
Space Research XII, (see 012.016), Vol. 2, 1657 - 1669 (1972).

031.045 High resolution obtained by photoelectric scanning techniques. J. S. Hall.
Space Research XII, (see 012.016), Vol. 2, 1689 - 1694 (1972).

031.046 High resolution solar infrared observations.
P. J. Turon, P. J. Léna.
Space Research XII, (see 012.016), Vol. 2, 1695 - 1700 (1972).

031.047 The capabilities of the spin-scan imaging technique.
T. Gehrels, V. E. Suomi, R. J. Krauss.
Space Research XII, (see 012.016), Vol. 2, 1765 - 1769 (1972).

031.048 The data-handling problem with television recording of spectra. A. B. Underhill, D. A. Klinglesmith.
Space Research XII, (see 012.016), Vol. 2, 1771 - 1775 (1972).

031.049 Characteristics of objective prisms.
J. Stock, A. R. Upgren.
Publ. Dep. Astron.,Univ. Chile, Obs. Astron. Nacional, Cerro Calán, Santiago de Chile, Vol. 2, (No. 1), 11 - 21 (1970).

The mounting angle for minimum dispersion, the distortion of the field, the formation of ghost spectra, and the effective transparency of objective prisms are discussed, and some practical applications of the results are considered.

031.050 La scelta di un apparecchio fotografico per l'astrofilo. W. Ferreri.
Coelum, Vol. 40, 186 - 188 (1972).

031.051 Rayleigh's water test: An easy test for optical flats.
D. Pye.
Journ. British Astron. Ass., Vol. 83, 10 - 18 (1972).

031.052 The construction of a 250 mm telescope.
A.Marlow, P. Marlow.
Journ. British Astron. Ass., Vol. 83, 19 - 27 (1972).

031.053 Astronomical applications of differential interferometry.
C. C. Counselman, III, H. F. Hinteregger, I. I. Shapiro.
Science, Vol. 178, 607 - 608 (1972).

Intercomparison of radio signals received simultaneously at several sites from several sources with small mutual angular separation provides a powerful astrometric tool. Applications include tracking the Lunar Rover relative to the Lunar Module, determining the moon's libration, measuring winds in Venus's lower atmosphere, mapping Mars radiometrically, and locating the planetary system in an inertial frame.

031.054 No ieceres līdz īstenībai. (Pirmie astronomiskie kupoli no stiklaplasta mūsu zemē). E. Bervalds.
Zvaigžņotā debess, 1972. gada vasara, p. 14 - 22.

031.055 Analyse critique des moyens de mesures du profil des paliers et tourillons d'une lunette méridienne.
J. Cottet, B. Prud'homme.
Ann. Obs. Besançon, Univ. Besançon, Vol. 8, (Fasc. 5), 205 - 216 (1971).

031.056 Binokulares Sehen im Weltraum. E. Wiedemann.
Orion Schaffhausen, 30. Jahrgang, p. 175 - 176 (1972).

031.057 Grosse optische Teleskopspiegel in Skelett-Bauweise.
A. Hoffmann.
Orion Schaffhausen, 30. Jahrgang, p. 176 - 178 (1972).

031.058 On the estimate of the frequency characteristic of a smoothing operator. Ya. S. Yatskiv.
Astrometriya i Astrofiz., Kiev, Vyp. (No.) 17, (see 003.012), p. 123 - 125 (1972). In Russian.

An attempt is made to estimate the «mean frequency characteristic» of the graphical method of smoothing. The method is illustrated by the analysis of the latitude observations at Washington from 1915.9 to 1941.0.

031.059 Digitization of photoelectric photometry at the Okayama Astrophysical Observatory.
S. Nishimura, E. Watanabe.
Univ. Tokyo, Tokyo Astron. Obs. Rep. (No. 60), Vol. 16, 49 - 55 (1972). In Japanese.

031.060 Digital recording system for photoelectric photometry at the Dodaira Station.
J. Matsumoto, M. Kiyokawa.
Univ. Tokyo, Tokyo Astron. Obs. Rep. (No. 60), Vol. 16, 56 - 63 (1972). In Japanese.

031.061 **Reduction program of digital data in photoelectric photometry.** S. Nishimura.
Univ. Tokyo, Tokyo Astron. Obs. Rep. (No. 60), Vol. 16, 64 - 71 (1972). In Japanese.

031.062 **Re-examination of the reduction method for PZT data.** S. Iijima.
Univ. Tokyo, Tokyo Astron. Obs. Rep. (No. 60), Vol. 16, 163 - 189 (1972). In Japanese.

031.063 **Tips für die Astropraxis.**
SuW, Vol. 11, 240 - 242, 313 - 315 (1972).
Quarzuhren für Amateurastronomen, (*K. Fischer*), p. 240 - 242; Himmelsaufnahmen mit Kamera und Feldstecher, (*A. Langkavel*), p. 313; Unentwickelte Filme für die Sonnenbeobachtung mit bloßem Auge, (*F. Dorst*), p. 313; Zum Thema: Sonnen- und Zenitprisma, (*R. Brandt, A. v. Querfurth*) p. 313 - 314; Schiefspieglermontierung kritisch betrachtet (*M. Bernhardt*), p. 314; Lieferbare Astro-Emulsionen, (*H. Vehrenberg*), p. 314; Das Fernrohr für den Amateur (*R. Brandt*), p. 315.

031.064 **Guide to scientific instruments, 1972–73.**
Compiled by R. G. Sommer (Assistant Editor).
Science, Vol. 178A, No. 4063A, 194 + 138A pp. (1972).

031.065 **Photon-counting system for rapidly scanning low-level optical spectra.** H. D. Pruett.
Applied Optics, Vol. 11, 2529 - 2533 (1972).
A signal-dependent method of photon counting is described that reduces the time required to scan low-level optical spectra but does not sacrifice either the SNR or the dynamic range achieved by conventional methods. This variable-measurement-time technique is particularly applicable to spectra having a wide dynamic range.

031.066 **Real-time computer for monitoring a rapid-scanning Fourier spectrometer.** G. Michel.
Applied Optics, Vol. 11, 2671 - 2675 (1972).
A real-time Fourier computer has been designed and tested as part of the Lunar and Planetary Laboratory's program of airborne infrared astronomy using Fourier spectroscopy. The value and versatility of this device are demonstrated with specific examples of laboratory and in-flight applications.

031.067 **Image quality in telescopes with image motion compensation by secondary mirror control.**
M. Bottema, R. A. Woodruff.
Applied Optics, Vol. 11, 2965 - 2967 (1972).
Image motion compensation by servocontrolled tilt of the secondary mirror is applied in a 40-cm diam $f/7.5$ Cassegrainian balloon-borne telescope. A tilted-aplanatic configuration is used, i.e., the mirrors are aspherized to render the stabilized image free of third-order coma. A spatial resolution of 2 sec of arc is maintained in the presence of pointing errors up to 10 min of arc. The most important remaining aberrations are third-order astigmatism and fifth-order spherical aberration. For the latter, a general expression is presented that can also be applied to other two-mirror telescopes.

031.068 **Evaluation of a real-time, figure control system for a spaceborne telescope mirror.**
H. J. Robertson, G. T. Volpe.
Journ. Optical Soc. America, Vol. 62, 1339 (1972). – Abstr. Optical Soc. America.

031.069 **Some all-reflective nonrotationally symmetric telescope designs.** B. Tatian.
Journ. Optical Soc. America, Vol. 62, 1393 (1972). – Abstr. Optical Soc. America.

031.070 **Interferometer star tracking.** A. B. DeCou.
Journ. Optical Soc. America, Vol. 62, 1401 (1972).
Abstr. Optical Soc. America.

031.071 **Environmental evaluation of protected reflecting coatings for telescope mirrors.**
P. R. Yoder, Jr., W. A. Howland.
Journ. Optical Soc. America, Vol. 62, 1401 (1972). – Abstr. Optical Soc. America.

031.072 **On the mutual error compensation of supporting forces in the systems for axial and lateral support of astronomical mirrors.** L. L. Voskresenskij.
Izv. vyssh. ucheb. zavedenij. Geod. i aehrofotos"emka, 1972, No. 2, p. 119 - 126. In Russian. – Abstr. in Referativ. Zhurn. 51. Astron., 1.51.116 (1973).

031.073 **On increasing the accuracy of the lateral support of an astronomical mirror.** L. L. Voskresenskij.
Izv. vyssh. ucheb. zavedenij. Geod. i aehrofotos"emka, 1972, No. 3, p. 127 - 132. In Russian. – Abstr. in Referativ. Zhurn. 51. Astron., 1.51.117 (1973).

031.074 **Grapho-analytical method of prediction of solar transits across the field of view of electron-optical systems.** K. M. Konstantinovich.
Izv. vyssh. ucheb. zavedenij. Geod. i aehrofotos"emka, 1972, No. 2, p. 69 - 73. In Russian. – Abstr. in Referativ. Zhurn. 51. Astron., 1.51.285 (1973).

031.075 **Analysis procedure of gamma ray astronomy spark chamber data.** K. Pinkau.
Nuclear Instruments and Methods, Vol. 104, 517 - 523 = Separate print Max-Planck Inst. Phys. Astrophys., Inst. Extra-terr. Phys., München (1972).
An analysis procedure is presented for the measurement of gamma ray arrival directions in spark chambers. This method is based on the conception that data are not used for measurement once multiple scattering has overtaken the reading error. The relevant formalism is presented and lower limits of the possible angular resolution are given.

031.076 **Some telescope makers' remarks about large astronomical telescope construction.**
A. Bayle, J. Espiard.
Nouvelle Revue Optique Appl., Vol. 3, No. 2, p. 67 - 73 (1972). In French.
Discusses the production of aspherical deformation surfaces during the grinding stage; surface checking by high precision spherometers; grinding and polishing tools with suitable flexibility; optical testings.

031.077 **An astronomical interface for a small computer.** R. E. Nather.
Rev. Sci. Intruments, Vol. 43, 1012 - 1017 (1972).
Interface logic has been assembled which permits a small computer to record variable star data at high speed, display the accumulating light curve in real time and exercise control over the data acquisition process. The interface design is quite general and has been used successfully to implement several other instrumental systems.

031.078 **A method of reconstructing Mertz shadowgrams.** R. L. F. Boyd, R. F. Booth, R. C. Rawlings.
Journ. Phys. E, Sci. Instruments, Vol. 5, 808 - 811 (1972). See Phys. Abstr., Vol. 75, No. 66129 (1972).

031.079 **Astronomical use of computers in data acquisition and control.** W. M. Ruting.
Proc. Instn. Radio Electron. Engineers Australia, Vol. 33, No. 7, p. 336 (1972). – Abstract.

031.080 **Methods of calculation and testing of Ritchey - Chrétien systems.** G. M. Popov.
Izv. Krymskoj Astrofiz. Obs., Vol. 45, 188 - 195 (1972). In Russian.

A method of calculation of Ritchey - Chrétien systems by electronic computer is described. Two simple methods of testing concave aspherical surfaces are examined.

031.081 **A plate calibration technique.** J. R. Gallivan.
Contr. Obs. New Mexico State Univ., Las Cruces, Vol. 1, 77 - 80 (1972).

A photographic magnitude sequence can be calibrated by using a second surface mirror to form extra images which differ by a given magnitude on a photographic plate. The calibration extends to the limiting magnitude and is obtained without extra telescope time.

031.082 **The numerical identification of stars with the SBG-camera.** I. Synek.
Wiss. Zeitschr. Tech. Univ. Dresden, Vol. 21, No. 3, (see 012.021), 622 - 623 (1972).

031.083 **On automatic angle measurements and a proposition of their application into zenith distance measurements on the surface of the moon.**
B. Kołaczek, H. Z. Kowalski, J. B. Rogowski.
Wiss. Zeitschr. Techn. Univ. Dresden, Vol. 21, No. 3, (see 012.021), 637 - 638 (1972).

031.084 **Verzeichnungsuntersuchung eines Astrographenobjektivs.** K.-G. Steinert.
Vermessungstechnik, 20. Jahrgang, p. 272 - 274 = Mitt. Lohrmann-Obs. Techn. Univ. Dresden, No. 25a/II (1972).

031.085 **Über ein verbessertes Verfahren zur Beobachtung von Sterndurchgängen durch den Meridian.**
H. Potthoff.
Vermessungstechnik, 20. Jahrgang, p. 224 - 226 = Mitt. Lohrmann-Obs. Techn. Univ. Dresden, No. 25a/III (1972).

031.086 **Évolution des objectifs astronomiques et des miroirs pour les télescopes.** Y. Väisälä, L. Oterma.
Sci. Revuo, Beograd, Vol. 22, 145 - 156 = Astron.-Optika Inst. Univ. Turku, Informo No. 37/III (1971). In Esperanto.

031.087 **Betrachtungen zur Herstellung optischer Bauteile für astronomische Geräte.** W. Pfaff.

Jenaer Rundschau, (Jena Rev.), 17. Jahrgang, p. 345 - 348 (1972).

031.088 **Investigation of the photographic congruency in a system of equidensities.**
V. I. Bistrits'kij, V. P. Vasil'ehv.
Visn. Kharkiv. Univ. No. 82, (Ser. Astron., No. 7), p. 53 - 60 (1972). In Ukrainian.

031.089 **Determination of the diameters of star images from observations at the transit instrument at the Kharkov Astronomical Observatory.**
A. D. Egorov, O. F. Vantsan.
Visn. Kharkiv. Univ. No. 82, (Ser. Astron., No. 7), p. 75 - 78 (1972). In Ukrainian.

031.090 **A device for photographic positional observations of the moon.** N. F. Bystrov.
Publ. 18th Astrometrical Conference 1969, (see 012.022), p. 327 - 330 (1972). In Russian.

031.091 **Instrumental polarization concerning magnetographic measurements.** F. W. Jäger.
Solar Physics, Vol. 27, 481 - 488 (1972).

It is shown in detail in which way magnetographic measurements may be affected by instrumental polarization. To eliminate this influence we propose a combined method to partial compensation and computation practicable successfully with only a moderate expenditure of calculations and instrumental equipment.

An introduction to modern optics.
See Abstr. 003.071.

Optimization techniques in lens design.
See Abstr. 003.080.

New technique in astronomy.
See Abstr. 003.149.

Energetic calibration of instruments used for measurements of scattered light in the upper atmosphere.
See Abstr. 082.120.

The role of Schmidt telescopes in the study of galactic structure. (Photometric methods).
See Abstr. 155.039.

032 Astronomical Instruments

032.001 The multiple-mirror telescope project.
R. J. Weymann, N. P. Carlton.
Sky Telescope, Vol. 44, 159 - 163 (1972).

032.002 Un astrografo automatico.   E. Mutti.
Coelum, Vol. 40, 147 - 150 (1972).

032.003 A compact short-focus telescope with spherical optical surfaces.   C. L. Stong.
Sci. American, Vol. 227, No. 2, p. 110 - 113, 117 (1972).

032.004 Die Instrumentation des Figl-Observatoriums für Astrophysik.
SuW, Vol. 11, 211 - 215 (1972).

032.005 On the automatic electronic astrolabe.
I. Tsubokawa, S. Hokugo.
IAU Symposium No. 48, (see 012.004), p. 150 - 159 (1972).

032.006 The NASA 91.5 cm aperture airborne telescope.
M. Bader, F. C. Witteborn.
Astrophys. Space Sci. Library, Vol. 30, (see 012.005), 29 - 31 (1972).

032.007 A 32-cm airborne infrared observatory.
P. Léna, N. Coron, C. Darpentigny, K. Hammal, G. Vanhabost.
Astrophys. Space Sci. Library, Vol. 30, (see 012.005), 32 - 40 (1972).

032.008 Far-infrared observations of celestial objects by balloon-borne telescope.   W. F. Hoffmann.
Astrophys. Space Sci. Library, Vol. 30, (see 012.005), 41 - 52 (1972).

032.009 An investigation of pneumatic mirror support for astronomical instruments.   V. M. Dul'kin.
Trudy Kazan. aviats. in-ta, 1971, vyp. (No.) 138, p. 124 - 130. In Russian. – Abstr. in Referativ. Zhurn. 51. Astron., 8.51.88 (1972).

032.010 A Wisconsin amateur's 20-inch Newtonian reflector.
R. Parmentier.
Sky Telescope, Vol. 44, 258 - 261 (1972).

032.011 Nieuwe grote optische telescopen.   A.Greve.
Hemel en Dampkring, Vol. 70, 219 - 225 (1972).

032.012 Le «Grand Schmidt» de L'Observatoire de Haute-Provence.   A. Heck.
Orion, 30. Jahrgang, p. 138 - 141 (1972).

032.013 Der Faltrefraktor.   H. Treutner.
Orion, 30. Jahrgang, p. 146 - 148 (1972).

032.014 The results of matched patrol telescope systems.
R. G. Hendl, R. F. Carnevale, F. W. Ward, Jr.
Bull. American Astron. Soc., Vol. 4, 384 (1972). – Abstr. AAS.

032.015 Extending the stellar field of view of Ritchey-Chretien telescopes.   S. Rosin, M. Amon.
Applied Optics, Vol. 11, 1623 - 1629 (1972).
The field size of Ritchey-Chretien telescopes is aberrationally limited by undercorrected astigmatism and field curvature. A pair of equally and oppositely oriented flat transmitting plates can neutralize either or both of these aberrations locally in the field to restore the image. These plates may be used to extend the available image area by an order of magnitude.

032.016 Wide-field tilted-component telescope: A Leonard extended Yolo all-reflecting system.
R. A. Buchroeder, A. S. Leonard.
Applied Optics, Vol. 11, 1649 - 1651 (1972).

032.017 Operational tests of the AFCRL 152-cm telescope.
W. E. Carter.
Applied Optics, Vol. 11, 1651 - 1653 (1972).

032.018 Adaptation of the Schupmann medial telescope to a large scale astronomical optical system.   J. J. Villa.
Applied Optics, Vol. 11, 1814 - 1821 (1972).
To circumvent the known limitations, the Schupmann lens was modified by replacing the refractive objective with a spherical mirror producing a new catadioptric lens configuration adaptable to large-scale astronomy. The design parameters and performance data are given for an $f/5.4$, 5.5-m focal length design covering a 2° full field.

032.019 The 61-inch astrometric reflector system.
K. A. Strand.
Publ. United States Naval Obs., *Washington*, Second Ser., Vol. 20, Part 1, 32 pp. (1971).
The design of the telescope was based on the requirements that it should reach a limiting magnitude of 18 in 10 minutes of direct photography, have a focal length in excess of 35 feet, a coma-free field of one-half degree diameter, an easily collimated optical system, and a mounting which would cause minimum deflection of the optical system. Another desirable feature was to obtain low expansion materials for the optical system, to permit observations in the early evening hours.

032.020 Telephoto lenses for X-ray telescopes and neutron cameras.   H. Wolter.
Zeitschr. Angew. Phys., Vol. 31, 152 - 155 (1971). In German.
Among the optics for X-rays several types, called second order X-ray optics, distinguish themselves by their special qualification for the application as X-ray telescopes on satellites and space stations. They contain a combination of collecting and scattering areas, telephoto lenses as it were, the construction length of which from the outside to the picture plane is smaller than their focal length.

032.021 A compact 16-inch Cassegrain reflector.
J. McClure.
Sky Telescope, Vol. 44, 327 - 331 (1972).

032.022 Ombygning af en traditionel aekvatoreal til coudé-system.   P. Darnell.
Astron. Tidssk., Årg. 5, p. 136 - 138 (1972).

032.023 Automation of a photoelectric transit instrument.
K. A. Šteins, A. V. Ivanovs.
Uch. zap. Latv. gos. un-ta imeni Petra Stuchki, Vol. 169, 30 - 39 (1972). In Russian.

032.024 About the error from operating the adjusting screws of the level.   K. A. Šteins.
Uch. zap. Latv. gos. un-ta imeni Petra Stuchki, Vol. 169, 53 - 57 (1972). In Russian.

032.025 About the stability of the horizontal axis of the transit instrument at the Astronomical Observatory

of the Latvian State University.   L. A. Roze.
Uch. zap. Latv. gos. un-ta imeni Petra Stuchki, Vol. 169, 58 -
62 (1972). In Russian.

032.026   AURA Schmidt telescope programme.
         R. Buchroeder, R. Lynds.
The role of Schmidt telescopes in astronomy. Conference
Hamburg 1972, (see 012.014), p. 127 - 134 (1972).

032.027   U. K. 48-inch Schmidt telescope project.
         V. C. Reddish.
The role of Schmidt telescopes in astronomy. Conference
Hamburg 1972, (see 012.014), p. 135 - 136 (1972).

032.028   The ESO Schmidt telescope.   O. Heckmann.
         The role of Schmidt telescopes in astronomy. Con-
ference Hamburg 1972, (see 012.014), p. 137 - 139 (1972).

032.029   The role of Schmidt telescopes in astronomy: Sum-
         mary of the conference.   B. Strömgren.
The role of Schmidt telescopes in astronomy. Conference
Hamburg 1972, (see 012.014), p. 157 - 160 (1972).

032.030   Telescopes that see red.   P. Connes.
         Sci. Progrès Découverte, 99e année, No. 3439, p. 39-
47 (1971). In French.
         The author reminds how and why infrared telescopes
have been built, explaining the reasons for which their diameters have practical limitations : 6 meters. He then goes on to
explain why an increase in this diameter is justified for the
study of infrared. In conclusion, the author describes the infrared telescopes, or light collectors, already built (or planned)
in France and elsewhere.

032.031   High angular resolution solar observation from bal-
         loon borne instruments.   K. O. Kiepenheuer.
Space Research XII, (see 012.016), Vol. 2, 1701 - 1711 (1972).

032.032   High resolution solar observations from space.
         H. Zirin.
Space Research XII, (see 012.016), Vol. 2, 1751 - 1763 (1972).

032.033   Technical problems and plans for high angular reso-
         lution optical telescopes.   M. J. Aucremanne.
Space Research XII, (see 012.016), Vol. 2, 1787 - 1807 (1972).

032.034   Ein kurz gebautes Spektrohelioskop.
         F. N. Veio.
Orion Schaffhausen, 30. Jahrgang, p. 178 - 179 (1972).

032.035   Untersuchung und Erprobung einer optischen Kipp-
         achse aus Spiegel, Doppelkeil und automatischem
Nivellier bei einem transportablen Passageinstrument.
K. Pilowski, R. Winter.
Techn. Univ. Hannover, Astron. Station Inst. Theor. Geod.,
No. 8, 28 pp. (1972).
         Project and test of an "optical axis" consisting of mirror,
double wedge and automatic level with a transportable transit
instrument. The elimination of the irregularities of the pivots
in their influence upon inclination and azimuth of the axis
are as well discussed as the elimination of the inclination itself
without any axis level – and that for the determination of
geodetic longitudes and azimuths and astronomical right ascensions.

032.036   Balloon-borne solar telescope.
         T. Hirayama, S. Hamana, I. Shimizu, A. Tokuya,
H. Miyazaki, Y. Yose, F. Moriyama.
Univ. Tokyo, Tokyo Astron. Obs. Rep. (No. 60), Vol. 16,
25 - 39 (1972). In Japanese.

032.037   Il rifrattore fotografico dell'Osservatorio Astronomi-
         co di Torino.
M. Boggio, M. G. Fracastoro, G. Francese, C. Morais.
Atti Fondaz. G. Ronchi, Anno 27, No. 2, p. 179 - 196 = Contr.
Oss. Astron. Torino, No. 63 (1972).
         Within the development program of instruments of the
Torino Astronomical Observatory, following the astrometric
line of research, the new photographic refracting telescope is
described, and in particular the design of a triplet objective
and the astronomical testing of the instrument, inasmuch as
the field, the magnitude limit and the quality of stellar images
are concerned.

032.038   General analysis of aplanatic Cassegrain, Gregorian,
         and Schwarzschild telescopes.
W. B. Wetherell, M. P. Rimmer.
Applied Optics, Vol. 11, 2817 - 2832 (1972).
         The properties of two-conic reflecting aplanats are ana-
lyzed and discussed on the basis of third order aberration
theory. Techniques for designing infinite conjugate two mirror
aplanats and computing their image properties are developed.
The secondary mirror alignment characteristics of Ritchey-
Chrétien and aplanatic Gregorian telescopes are examined and
neutral point locations defined. Design configurations corrected for a third Seidel aberration (astigmatism, image curvature, or distortion) are identified and their properties discussed. The properties of Ritchey-Chrétien and aplanatic
Gregorian telescopes are compared.

032.039   Proposed 131-cm $f/3.5$ achromatic Schmidt tele-
         scope.   R. A. Buchroeder.
Applied Optics, Vol. 11, 2968 - 2971 (1972).
         An achromatic Schmidt design is proposed for use in
twin instruments at Kitt Peak and Cerro Tololo (Chile). The
use of an achromatic plate, rather than the conventional
Schmidt plate, allows extended spectral coverage by virtue of
improved aberration correction. Some schemes to convert the
Schmidt to a Cassegrainian telescope during bright moon
periods are evaluated.

032.040   A telescope for soft gamma ray astronomy.
         V. Schönfelder, A Hirner, K. Schneider.
Separate print Max-Planck-Inst. Phys. Astrophys., Inst. Extra-
terr. Phys., München. MPI-PAE/Extraterr. 74, 3 + 34 pp.
(1972).
         A gamma-ray telescope using the double Compton proc-
ess is described, which measures extraterrestrial gamma ray
fluxes in the energy range 1–10 MeV.

032.041   Super-reflectors – an appreciation of low scatter.
         G. F. Marshall.
Journ. Optical Soc. America, Vol. 62, 1369 (1972). – Abstr.
Optical Soc. America.

032.042   Large aperture rocket camera.   U. Ludwig.
         Journ. Optical Soc. America, Vol. 62, 1393 - 1394
(1972). – Abstr. Optical Soc. America.

032.043   Beitrag zur Weiterentwicklung des Zirkumzenitals.
         K.-G. Steinert.
Geod. Geophys. Veröff., Nationalkomitee Geod. Geophys.,
Deutsche Demokratische Republik, Deutsche Akad. Wiss.
Berlin, Ser. 3, No. 20, 79 pp. (1970).

032.044   Installazione dell'astrolabio Danjon all'Osservatorio
         di Merate.   F. Mazzoleni.
Mem. Soc. Astron. Italiana, Nuova Ser., Vol. 43, 487 - 499
(1972).

032.045   Ein kleines Modell des Diazenitals Nušl–Frič und
         die Ortsbestimmung.   E. Buchar.

Wiss. Zeitschr. Techn. Univ. Dresden, Vol. 21, No. 3, (see 012.021), 597 - 599 (1972).

032.046 Ergebnisse von Probebeobachtungen mit dem Photo-Zenitteleskop (PZT). M. Meinig.
Wiss. Zeitschr. Techn. Univ. Dresden, Vol. 21, No. 3, (see 012.021), 611 - 613 (1972).

032.047 Genauigkeitsforderungen an ein objektiviertes Astrolab. J. Höpfner.
Wiss. Zeitschr. Techn. Univ. Dresden, Vol. 21, No. 3, (see 012.021), 614 - 616 (1972).

032.048 The Canadian telescope at the Las Campanas Observatory. N. R. Walborn.
Inform. Bull. Southern Hemisph., No. 21, p. 1 (1972).

032.049 Einsatz der Automatischen Kamera für Astrogeodäsie aus Jena im ISAGEX-Programm.
G. Karský, J. Kostelecký, J. Rambousek, I. Synek.
Jenaer Rundschau, (Jena Rev.), 17. Jahrgang, p. 337 - 339 (1972).

032.050 On a telescope stabilization system having an electric drive.
V. A. Strezhnev, A. M. Danilov, E. G. Moskvich.
Trudy Kazan. aviats. in-ta, 1972, vyp. (No.) 146, p. 80 - 88. In Russian. – Abstr. in Referativ. Zhurn. 51. Astron., 2.51.94 (1973).

032.051 Telescopes in the Mediterranean. P. Mayer.
Říše hvězd, Vol. 53, 225 - 226 (1972). In Czech.

032.052 The new Pulkovo transit instrument in the southern hemisphere. V. M. Vasil'ev.
Publ. 18th Astrometrical Conference 1969, (see 012.022), p. 90 - 93 (1972). In Russian.

032.053 On the new transit instrument of the Pulkovo time service. N. N. Pavlov.
Publ. 18th Astrometrical Conference 1969, (see 012.022), p. 151 - 158 (1972). In Russian.

032.054 Preliminary results of an investigation of Sukharev's horizontal meridian circle. G. I. Pinigin.
Publ. 18th Astrometrical Conference 1969, (see 012.022), p.

158 - 165 (1972). In Russian.

032.055 Automation of observations with a transit instrument. V. I. Verkhozin, L. M. Malomyzhev.
Publ. 18th Astrometrical Conference 1969, (see 012.022), p. 173 - 175 (1972). In Russian.

032.056 A probable reason of unexcluded flexure of the Nikolayev vertical circle. V. P. Sibilev.
Publ. 18th Astrometrical Conference 1969, (see 012.022), p. 188 - 190 (1972). In Russian.

032.057 The law of variations of flexure of Wanschaff's vertical circle. A. S. Kharin.
Publ. 18th Astrometrical Conference 1969, (see 012.022), p. 190 - 193 (1972). In Russian.

032.058 Character of temperature deformations of the tube of Wanschaff's vertical circle.
A. A. Korsun, N. T. Mironov, A. S. Kharin.
Publ. 18th Astrometrical Conference 1969, (see 012.022), p. 193 - 198 (1972). In Russian.

032.059 Investigation of the flexure of the Pulkovo vertical circle during day-time. G. S. Kosin, M. Miyatov.
Publ. 18th Astrometrical Conference 1969, (see 012.022), p. 198 - 201 (1972). In Russian.

032.060 Distortion and penetrating power of the Maksutov double-meniscus astrograph installed at Mount Robles (Chile). L. A. Panaiotov.
Publ. 18th Astrometrical Conference 1969, (see 012.022), p. 234 - 239 (1972). In Russian.

Guide to scientific instruments, 1972–73.
See Abstr. 031.064.

Some results from the automatic PZT at Richmond, Florida. See Abstr. 041.010.

The large space telescope program.
See Abstr. 051.017.

Schmidt telescopes and radio astronomy.
See Abstr. 141.075.

033 Radio Telescopes and Equipment

033.001 **The radio telescope RATAN–600.** S. E. Khaikin,
N. L. Kajdanovskij, Y. N. Parijskij, N. A. Esepkina.
Izv. Glav. Astron. Obs. Pulkovo, No. 188, p. 3 - 12 (1972).
In Russian.
  A project of a radio telescope with variable profile an-
tenna with extremely large wavelength region of 0.4–21 cm is
described.

033.002 **The methods of radioastronomical observations with
the RATAN–600.** Y. N. Parijskij, O. N. Shivris.
Izv. Glav. Astron. Obs. Pulkovo, No. 188, p. 13 - 39 (1972).
In Russian.
  A consideration of azimuthal aperture synthesis of a two-
dimensional radio image is made in terms of spectral analysis.
Some possible variants of using variable profile antennas (VPA)
as the system of aperture synthesis, are considered. A comparison of the VPA with other systems of aperture synthesis is
made. All the numerical calculations in the paper are given for
the radio telescope RATAN–600; the experimental data were
obtained with the Pulkovo large radio telescope.

033.003 **Determination of the sizes of reflecting elements and
calculation of the electric parameters of the RATAN-
600.** B. V. Braude, N. A. Esepkina, N. L. Kajdanovskij,
Y. N. Parijskij, O. N. Shivris.
Izv. Glav. Astron. Obs. Pulkovo, No. 188, p. 40 - 53 (1972).
In Russian.
  The vertical and horizontal sizes of the reflecting elements
are determined. An effective area of the radio telescope and its
antenna pattern are calculated, real admissions for construction and deformations being made.

033.004 **On the operation of the variable profile antenna with
a plane periscopic reflector.**
L. M. Gindilis, N. A. Esepkina, N. S. Kardashev.
Izv. Glav. Astron. Obs. Pulkovo, No. 188, p. 54 - 57 (1972).
In Russian.
  The possibility of operating the variable profile antenna
with a plane periscopic reflector is considered. A comparison
with the Kraus system is made.

033.005 **Peculiarities of the operation of variable profile
antennas in observations near the zenith.**
N. A. Esepkina, Y. N. Parijskij.
Izv. Glav. Astron. Obs. Pulkovo, No. 188, p. 58 - 62 (1972).
In Russian.
  The possibility of the operation of variable profile antennas at the zenith and the requirements for a special feed are
considered. It is shown that a variable profile antenna can
operate near the zenith under the condition of parallel aperture synthesis, parasitic circular polarization being absent.

033.006 **Aberrations of the main mirror of the variable
profile antenna and scanning the antenna pattern by
shifts of the primary feed.** A. A. Stotskij.
Izv. Glav. Astron. Obs. Pulkovo, No. 188, p. 63 - 76 (1972).
In Russian.

033.007 **On the phase errors at the aperture of the parabolic
aerial at the feed shifted from the focus.**
V. M. Spitkovskij.
Izv. Glav. Astron. Obs. Pulkovo, No. 188, p. 77 - 82 (1972).
In Russian.

033.008 **Calculations of the noise temperature of the RATAN-
600 antenna.** A. A. Stotskij.

Izv. Glav. Astron. Obs. Pulkovo, No. 188, p. 83 - 88 (1972).
In Russian.

033.009 **The construction of reflecting elements and the
secondary mirror of the RATAN–600.**
A. Z. Amstislavskij, A. I. Kopylov, M. I. Prosmushkin.
Izv. Glav. Astron. Obs. Pulkovo, No. 188, p. 89 - 96 (1972).
In Russian.

033.010 **A double-reduction condenser breaking as used in a
monomotor drive for the reflecting sections of
RATAN-600.** G. S. Golubchin.
Izv. Glav. Astron. Obs. Pulkovo, No. 188, p. 97 - 100 (1972).
In Russian.

033.011 **A reading and setting device for the sections of the
RATAN–600 circular reflector.**
A. I. Kopylov, O. V. Chukanov, O. N. Shivris.
Izv. Glav. Astron. Obs. Pulkovo, No. 188, p. 101 - 106 (1972).
In Russian.

033.012 **A reference reading and setting device for a section
of the RATAN–600 circular reflector.**
Y. K. Zverev, A. M. Contorero, V. T. Martynov, Y. I. Pavlju-
chenkov, E. I. Stepanov, O. V. Chukanov.
Izv. Glav. Astron. Obs. Pulkovo, No. 188, p. 107 - 109 (1972).
In Russian.

033.013 **Errors arising in the RATAN–600 due to thermal
effects.** Y. L. Shakhbazjan.
Izv. Glav. Astron. Obs. Pulkovo, No. 188, p. 110 - 113 (1972).
In Russian.

033.014 **Geodesy in the process of mounting and adjustment
of the RATAN–600.**
A. G. Belevitin, Y. K. Zverev.
Izv. Glav. Astron. Obs. Pulkovo, No. 188, p. 114 - 119 (1972).
In Russian.

033.015 **Laying out of the foundations of the RATAN–600
sections.** Y. K. Zverev, A. P. Glumov.
Izv. Glav. Astron. Obs. Pulkovo, No. 188, p. 120 - 122 (1972).
In Russian.

033.016 **The adjustment of the variable profile antenna.**
G. B. Gelfreikh, Y. K. Zverev, A. A. Stotskij,
O. N. Shivris.
Izv. Glav. Astron. Obs. Pulkovo, No. 188, p. 123 - 128 (1972).
In Russian.

033.017 **Experimental tests of the autocollimation method
for adjustment of the variable profile antenna.**
A. A. Stotskij, N. Khodzhamukhamedov.
Izv. Glav. Astron. Obs. Pulkovo, No. 188, p. 129 - 138 (1972).
In Russian.

033.018 **A radioastronomical method of adjustment of
variable profile antennas.** G. B. Gelfreikh.
Izv. Glav. Astron. Obs. Pulkovo, No. 188, p. 139 - 148 (1972).
In Russian.

033.019 **A control of the antenna parameters by radioastro-
nomical observations.** Y. N. Parijskij.
Izv. Glav. Astron. Obs. Pulkovo, No. 188, p. 149 - 151 (1972).
In Russian.

033.020 **Radiometers of the continuous spectrum. The**

schemes and main parameters.   D. V. Korolkov.
Izv. Glav. Astron. Obs. Pulkovo, No. 188, p. 152 - 167 (1972).
In Russian.

Various circuits of the microwave radiometers and their characteristics are considered. The limits of sensitivity of real constructions are described.

033.021   Radiospectrographs for the RATAN–600.
I. V. Gosachinskij, A. F. Dravskikh, T. M. Egorova, V. A. Prozorov, N. F. Ryzhkov.
Izv. Glav. Astron. Obs. Pulkovo, No. 188, p. 168 - 171 (1972). In Russian.

The main problems of the study of radio lines are described. Starting from these problems, the requirements for the receiving equipment are laid down and the main parameters of radiometers and spectro-analyzers to be attached to the radiotelescope given. The block-scheme and the principle of operation of a modulation balance radiometer are described.

033.022   The radio spectrograph operating at 21 centimeter.
N. F. Ryzhkov.
Izv. Glav. Astron. Obs. Pulkovo, No. 188, p. 172 - 179 (1972). In Russian.

A radio spectrograph with many channels operating at 21 cm attached to the Pulkovo large radio telescope is described. This radio spectrograph is intended for the study of interstellar neutral hydrogen.

033.023   A protection of radiometers from pulse interferences.
V. A. Prozorov.
Izv. Glav. Astron. Obs. Pulkovo, No. 188, p. 180 - 183 (1972). In Russian.

The permissible radio fluxes interfering the radio telescope have been evaluated. The method of protecting the radiometer from radar pulses by means of an amplitude limiter is described.

033.024   Survey of the sky with the RATAN–600.
L. M. Gindilis, N. S. Kardashev.
Izv. Glav. Astron. Obs. Pulkovo, No. 188, p. 184 - 187 (1972). In Russian.

The problems of surveying the sky are discussed as well as the characteristics of the periscopic part of the RATAN–600 in the survey regime.

033.025   The astrophysical problems to be solved with the RATAN–600.
G. B. Gelfreikh, I. V. Gosachinskij, Y. N. Parijskij.
Izv. Glav. Astron. Obs. Pulkovo, No. 188, p. 188 - 194 (1972). In Russian.

033.026   On the possibilities of obtaining radio images of celestial bodies with a resolution more than 10^{-2}
seconds of arc.   Y. N. Parijskij, A. A. Stotskij.
Izv. Glav. Astron. Obs. Pulkovo, No. 188, p. 195 - 212 (1972). In Russian.

033.027   Measurements of the antenna parameters and calibration of the sensitivity of the RATAN–600 at 8 mm
in the radar regime.   V. K. Golovkov.
Izv. Glav. Astron. Obs. Pulkovo, No. 188, p. 213 - 215 (1972). In Russian.

033.028   Computer reductions of the observations of the neutral hydrogen radio line obtained with the
stationary antenna.   Z. A. Alferova, N. V. Bystrova, I. V. Gosachinskij, Z. G. Trunova.
Izv. Glav. Astron. Obs. Pulkovo, No. 188, p. 216 - 220 (1972). In Russian.

033.029   An experimental test of the possibility of scanning the diagram of the variable profile antenna.
V. M. Spitkovskij.
Izv. Glav. Astron. Obs. Pulkovo, No. 188, p. 221 - 225 (1972). In Russian.

033.030   Experimental investigations of illumination of the variable profile reflector of the Pulkovo large radio telescope by radiotechnical methods.   V. N. Borovik, N. G. Peterova, V. M. Spitkovskij, G. M. Timofeyeva.
Izv. Glav. Astron. Obs. Pulkovo, No. 188, p. 226 - 230 (1972). In Russian.

033.031   On the sensitivity of the intensity radiointerferometer.   D. V. Korolkov.
Izv. Glav. Astron. Obs. Pulkovo, No. 188, p. 231 - 237 (1972). In Russian.

033.032   On the precision of coordinates measured by means of the variable profile antenna.
N. M. Lipovka, A. A. Stotskij.
Izv. Glav. Astron. Obs. Pulkovo, No. 188, p. 238 - 243 (1972). In Russian.

The precision of measuring the coordinates of bright radio sources by means of the variable profile antennas is shown to be determinated by the conditions of radio waves propagation through the turbulent atmosphere. The results of measuring the fluctuations of the tilt caused by the atmospheric inhomogeneities for different altitudes are given. The expected precision of the coordinates of cosmic sources measured with the RATAN–600 is evaluated.

033.033   Electrooptical processing for radio astronomy.
H. Stark.
Proc. IEEE, Vol. 60, 1009 - 1010 (1972).

The electrooptical processor can furnish continuously and unambiguously, in the form of an optical image, the distribution of radio emission present in the entire effective scan sector of an array without the necessity of electronic steering.

033.034   On drives and guidance of small radio telescopes.
K. L. Smith.
Observatory, Vol. 92, 136 - 139 (1972).

033.035   Intercontinental radio astronomy.
K. I. Kellermann.
Sci. American, Vol. 226, No. 2, p. 72 - 83 (1972).

The structure of quasars and other radio sources is being examined with interferometers consisting of radio telescopes separated by distances approaching the diameter of the earth.

033.036   A multichannel radio spectrograph for dm-wavelength for solar radio astronomy.
G. Zimmermann, E. Hiss.
Zeitschr. Angew. Phys., Vol. 30, 210 - 215 (1970). In German.

The description of a multichannel radio spectrograph is given. Its purpose is the investigation of finestructures in the spectrum of solar bursts. Some registrations show examples of these structures.

033.037   A redundancy reduction method for the registration of solar radio bursts.   G. Zimmermann.
Zeitschr. Angew. Phys., Vol. 30, 370 - 376 (1971). In German.

A multichannel radio spectrograph for solar observations generates a data quantity too large for conventional registrations. A short survey over the methods of data reduction is given. After a description of the data an improved redundancy reduction method on the basis of least squares approximation is explained. A technical realisation is proposed.

033.038   Noise spectrum measurements from 10 Hz to 1 MHz

using a tunable switching radiometer. I. Instrumentation. H. A. Buckmaster, R. S. Rathie.
Canadian Journ. Phys., Vol. 49, 849 - 852 (1971).

The first application of a switching (Dicke) radiometer to high-precision, noise spectrum measurements from 10 Hz to 1 MHz is described. The solution of the instrumentation problems unique to this frequency range are discussed in detail. The advantage of this radiometer is that it can be employed in a null mode so that it is direct reading and the measurement precision at all frequencies is limited fundamentally by the bandwidth and averaging time or by the calibration of the reference noise source and its attenuator.

033.039 **Noise spectrum measurements from 10 Hz to 1 MHz using a tunable switching radiometer. II. Application to EPR.**   H. A. Buckmaster, R. S. Rathie.
Canadian Journ. Phys., Vol. 49, 853 - 859 (1971).

The $1/f$ noise spectra and conversion gain of 10 GHz point-contact, Schottky barrier and backward microwave diodes are compared with the object of determining the relative merit of their application in electron paramagnetic resonance (EPR) spectrometers.

033.040 **A submillimeter radio telescope with an n-InSb detector.**   A. S. Vardanjan, A. N. Vystavkin, I. A. Iskhakov, V. N. Listvin, N. A. Savich, A. V. Sokolov.
Astron. Zhurn. Akad. Nauk SSSR, Vol. 49, 986 - 989 (1972). In Russian. English translation in Soviet Astron. AJ, Vol. 16, No. 5.

033.041 **Unattended controller for a frequency synthesizer.**   F. V. Cairns, T. H. Shepertycki.
Rev. Sci. Instruments, Vol. 40, 113 - 115 (1969).

An electronic controller, made of integrated circuits and inexpensive silicon controlled rectifiers, has been designed as a coupler of interface between a card reader and a frequency synthesizer to permit sequential selection of one of the 5×10^9 frequencies available from the frequency synthesizer.

033.042 **Beam switching Cassegrain feed system and its applications to microwave and millimeterwave radioastronomical observations.**
S. D. Slobin, W. V. T. Rusch, C. T. Stelzried, T. Sato.
Rev. Sci. Instruments, Vol. 41, 439 - 443 (1970).

033.043 **An analog shift register.**   G. S. F. Orsten.
Rev. Sci. Instruments, Vol. 41, 957 - 959 (1970).

We describe a shift register which is operated by a digital clock as usual, but shifts analog rather than binary signals. This system can be used, among other purposes, as a delay line, in which mode it overcomes many of the limitations of LC delay lines with extremely long delays.

033.044 **The distribution of current on a cylindrical antenna.**   C. C. Harvey.
Proc. Cambridge Philos. Soc., Vol. 70, 351 - 381 (1971).

The iterative solution of Hallen's integral equation for the current distribution on a lossless dipole transmitting antenna is studied from first principles, and a general solution in terms of an expansion function is derived. Three low order solutions are derived in detail: the zero-order solution using both a constant expansion parameter and a simple expansion function, and the first-order solution using a constant expansion parameter. The results are compared and special attention is paid to the very short antenna, although the results are good for lengths up to and including the half-wave dipole. The results are presented in such a way that the first-order admittance of any given antenna may be readily calculated.

033.045 **Multiple-frequency operation of the Culgoora radioheliograph.**

K. V. Sheridan, N. R. Labrum, W. J. Payten.
Nature, Phys. Sci., Vol. 238, 115 - 116 (1972).

Modifications to the Culgoora radioheliograph are being carried out to enable observations to be made at 160, 80 and 43.25 MHz. The observing frequency will be changed second-by-second between pictures in a programmed sequence. The 160 and 80 MHz radio frequency systems are now complete and work is in progress for the 43.25 MHz frequency and on the digital equipment to provide the rapid frequency change-over. Some sample results are presented of nearly-simultaneous 80 and 160 MHz observations of a solar storm centre.

033.046 **The 5-km radio telescope at Cambridge.**   M. Ryle.
Nature, Vol. 239, 435 - 438 (1972).

This instrument is the latest in a series employing the principle of aperture synthesis which have been constructed at Cambridge since 1957. It is designed to provide radio maps with a resolution of 1 arc s which is comparable with that of optical telescopes. The chief purpose of the telescope is to increase our understanding of the physical mechanisms occurring in radio sources, both galactic and extragalactic.

033.047 **Radiotelescope for regular observations of the sun at a frequency of 550 MHz.**   A. T. Nesmyanovich, A. M. Sviridov, V. V. Chmil, V. N. Smelyansky.
Vestn. Kiev. Un-ta, Ser. Astron., No. 14, p. 94 - 96 (1972). In Russian.

033.048 **Listening to the universe.**   E.-J. Blum.
Sci. Progrès Découverte, 99ᵉ année, No. 3435, p. 4 - 14 (1971). In French.

In order to improve the quality of the radioastronomical observations, it has been necessary to continually modify the radiotelescopes. The surfaces of single antenna telescopes have been increased and the radioastronomers have also turned their attention to interferometers. The author then analyses the problem of the sensitivity of radiotelescopes. Finally, the future development of radiotelescopes is discussed briefly.

033.049 **Radioheliografía en el Observatorio Astronómico Municipal de Rosario - I.**   R. R. Barbarroja.
Obs. Astron. Municipal Rosario, (*Argentina*), Bol. No. 2, p. 23 - 24 (1972).

033.050 **A solar radio spectrograph with high time resolution.**   B. L. Gotwols, J. Phipps.
Solar Physics, Vol. 26, 386 - 392 (1972).

Instrumentation for obtaining high time resolution dynamic spectra of solar radio bursts at decimetric wavelengths is described. The spectrograph sweeps the frequency range of 565–1000 MHz at a rate of 100 times per second. All data are recorded both on film and as an analog signal on magnetic tape. The frequency and flux calibrations are discussed. A sampling system which allows the activity at three discrete frequencies to be plotted on a chart recorder is described.

033.051 **The 15 m Cracow radiotelescope. I. Technical description and observational possibilities.**
J. Machalski, J. Masłowski.
Acta Astron., Vol. 22, 419 - 425 (1972).

033.052 **17 GHz interferometer.**
K. Ōki, H. Nakajima, Y. Shiomi, N. Shibuya, H. Sekiguchi, S. Aiba, S. Tachikawa.
Univ. Tokyo, Tokyo Astron. Obs. Rep. (No. 60), Vol. 16, 241 - 258 (1972).   In Japanese.

033.053 **On the theory, techniques, and data processing of very long baseline interferometry.**
B. Rönnäng.
Res. Lab. Electronics, Onsala Space Obs., Chalmers Univ.

Techn., Gothenburg, Sweden, Res. Rep. No. 105, 7 + 83 pp. = Publ. Onsala Space Obs., No. 75 (1971).

The following report gives the basic theoretical background for interferometric observations at radio wavelengths, and describes the receiver system for very long baseline interferometry and the computer programs which process the recorded data to get the fringe amplitude and fringe phase versus frequency. We derive the formulae giving the baseline parameters, the geometrical time delay and fringe rate for given station coordinates and radio source coordinates. The positions of some of the VLBI stations are tabulated and parameters for nine important interferometer baselines are listed. A brief review of the theory of the two-element radio interferometer is given. The relation between the cross-correlation function and the cross-power spectrum for the receiver system at each station is given. The sensitivity formulae for an interferometric observation are derived and the one-bit correlation method of cross-spectral analysis is discussed.

033.054 **On a method of compensating the radiation of the earth's atmosphere in submillimeter radio telescopes.** A. S. Vardanyan, V. D. Shtykov.
Izv. vyssh. ucheb. zavedenij. Radiofizika, Vol. 15, 948 - 949 (1972). In Russian. – Abstr. in Referativ. Zhurn. 51. Astron., 1.51.130 (1973).

033.055 **Equipment for the observation of fast fluctuations of solar radio emission.**
A. A. Bezotosnyj, O. G. Gontarev.
Trudy Sektora ionosfery. AN KazSSR, Vol. 3, 129 - 131 (1972). In Russian. – Abstr. in Referativ. Zhurn. 51. Astron., 1.51.131 (1973).

033.056 **On designing a supersynthesis antenna array.**
Y. L. Chow.
IEEE Trans. Antennas Propagation, Vol. AP-20, 30 - 35 = Contr. NRC Canada, No. 12300 (1972).

An analytical approach to the design of a supersynthesis antenna array is applied in this paper.

033.057 **Some questions of computation of feed systems of parabolic antennae and results of an investigation of a model antenna.** Ya. Ya. Ikaunieks, K. I. Mogil'nikova, G. A. Ozolin'sh, A. A. Pruzhanskaya.
Trudy Radioastrophys. Obs., *Riga,* Vol. 14, (see 003.019), 3 - 22 (1972). In Russian.

This article presents results of the antenna feed system design and an experimental investigation of the basic electric characteristics of the antenna model. The two-wave feed system is adapted for the reception of radio waves of any polarization.

033.058 **Phase-stable system of the heterodyne frequency transmission for a radio interferometer with high resolving power.** G. A. Ozolin'sh.
Trudy Radioastrophys. Obs., *Riga,* Vol. 14, (see 003.019), 23 - 31 (1972). In Russian.

Tolerances to parameters of the phase-coherent oscillator system of a radio interferometer are considered. Balance of the system may be performed by corresponding change of the frequency of the subsidiary generator.

033.059 **Experimental investigation of the phase instability of a parametric frequency multiplier by means of a varactor.** Yu. Ya. Litsis, G. A. Ozolin'sh.
Trudy Radioastrophys. Obs., *Riga,* Vol. 14, (see 003.019), 33 - 38 (1972). In Russian.

The phase instability of a parametric frequency multiplier is investigated in order to explain its application possibility in a heterodyne system of a radio interferometer. The principal cause of phase instability is the instability of the signal ampli-

tude of the multiplied frequency.

033.060 **Variable delay line for a two-antenna radio interferometer.** M. K. Ehliass.
Trudy Radioastrophys. Obs., *Riga,* Vol. 14, (see 003.019), 39 - 52 (1972). In Russian.

A variable delay line is suggested for a two-antenna radio interferometer operated as a system of aperture synthesis. The delay is operated by a generator with automatic frequency change. The errors of the system are considered and the maximum basis of an interferometer is derived, at which the required accuracy can be secured.

033.061 **A decimal impulse frequency converter.**
A. G. Gajlans, M. K. Ehliass.
Trudy Radioastrophys. Obs., *Riga,* Vol. 14, (see 003.019), 53 - 58 (1972). In Russian.

The scheme of a decimal impulse frequency converter is considered, consisting of counters and logical schemes which can be operated with variable division factor. The regularity of the output impulses on the time axis is considered and the maximum error is given relative to a sequence of pulses with "fixed" frequency. The impulse frequency converter has been tested in the frequency range from 200 Hz to 250 kHz.

033.062 **Solar radiometer at 187 MHz.** N. V. Demakov.
Trudy Radioastrophys. Obs., *Riga,* Vol. 14, (see 003.019), 59 - 66 (1972). In Russian.

A radio telescope for integral radio observations of the sun at the frequency of 187 MHz is described. The radiometer is of modulation scheme; its sensitivity is $3°K$.

033.063 **Scientific program and project of the RT-10 radio telescope for solar radio observations.**
A. R. Avotin'sh, A. Eh. Balklavs, Eh. Ya. Bervalds, M. G. Kamenskij, N. P. Tsimakhovich.
Trudy Radioastrophys. Obs., *Riga,* Vol. 14, (see 003.019), 67 - 83 (1972). In Russian.

033.064 **On the solution of the antenna smoothing equation.** A. Eh. Balklavs.
Trudy Radioastrophys. Obs., *Riga,* Vol. 14, (see 003.019), 85 - 93 (1972). In Russian.

Two methods of solving the integral equation of antenna smoothing are presented. The first is based on the filtering property of the function sin x/x, the second on the expansion of the passing curves and power diagrams of a radio telescope or a radio interferometer in Fourier series. Both methods are suitable only for treatment of the results of observations with computers.

033.065 **The M.I.T. K_u-band radio interferometer.**
G. C. Papadopoulos, B. F. Brke.
Radio Sci., Vol. 7, 667 - 674 (1972).

The construction of a K_u-band radio interferometer and some preliminary observations are reported.

033.066 **Extra-high resolution radiointerferometry.**
L. I. Matveyenko.
Journ. Astronaut. Sci., Vol. 19, 73 - 76 (1971).

A discussion of recent radiotelescope interferometer experiments between stations in the United States and the U.S.S.R. is given. Telescopes at Greenbank, Virginia and the Crimean Astrophysical Observatory were used to observe the details of several galaxies.

033.067 **A multi-channel commutated reflection oscillator for a solar radiospectrograph.**
A. F. Dravskikh, Z. V. Dravskikh, A. S. Kutuzov.
Solnechnye Dannye 1972 Byull., No. 9, p. 95 - 98 (1972). In Russian.

033.068 Phase shifts in balanced mixing.   A. T. Forrester, H. G. van Bueren.
Optics Commun., Vol. 4, 224 - 225 = Utrechtse Sterrekundige Overdrukken, No. 180 (1971).

A very simple argument to derive the properties of an optical balanced mixer system is given and the effect of eventual asymmetry is indicated.

033.069 Radiometer at 8.4 mm for radio astronomy.   P. Pampaloni, G. Tofani.
Mem. Soc. Astron. Italiana, Nuova Ser., Vol. 43, 523 - 534 (1972).

A radiometer at 8.4 mm wavelength for radio astronomy is described. The radiometer is an I.F. switched superheterodyne receiver with no R.F. amplification, which is going to be used on a 10-m primary focus parabolic reflector. A feed for this reflector is also described.

033.070 A dual frequency, dual polarized feed for radio-astronomical applications.
M. E. J. Jeuken, M. H. M. Knoben, K. J. Wellington.
Nachrichtentechn. Zeitschr., (*Germany*), Vol. 25, 374 - 376 (1972).

The Dutch radioastronomers have at their disposal an array of twelve reflector antennas. In order to use this system as intensively as possible the system should be suitable for measurements at two frequencies, viz 610 MHz and 4995 MHz, without changing the feed. The solution of this problem consists of a coaxial system.

033.071 The techniques of present-day radioastronomy.   R. D. Davies, H. P. Palmer.
Journ. Phys. E, Sci. Instruments, Vol. 5, 949 - 957 (1972).

The objectives of modern radioastronomical research are outlined in this review and a range of techniques is discussed. The radioastronomer uses both large steerable telescopes and unfilled aperture systems for many differing astronomical programmes.

033.072 RF transistor generator−25 W, 378 MHz.   G. Colla, G. Sinigaglia, G. Tomassetti, A. Righi.
Alta Frequenza, Vol. 41, 542 - 553 (1972). In Italian.

A new solid-state local oscillator for the 'Northern Cross' radio telescope is described. The high degree of stability specified rendered the design of the oscillator exceptionally complicated.

033.073 The quasi-linear intensity interferometer.   R. H. MacPhie.
IEEE Trans. Antennas Propagation, Vol. AP-20, 755 - 763 (1972).

A new type of radio intensity interferometer (fourth-order correlation system) suitable for use at long baselines is described. It provides both amplitude and phase information about the distribution of radio sources which excite its two antennas.

033.074 Radio telescope – in Japan.   K. Akabane.
Butsuri, (*Japan*), Vol. 27, 563 - 571 (1972). In Japanese.

The author describes a 45 m radiotelescope under construction in Japan. Its purpose is: (1) to achieve maximum collimating force near 1 cm; (2) to allow the gravitation distortion due to the inclination of the parabola.

033.075 Gain and noise figure of Ga As transferred-electron amplifiers at 34 GHz.
S. Baskaran, P. N. Robson.
Electronic Letters, Vol. 8, Nr. 5, p. 109 - 110 (1972).

033.076 Noise performance of InP reflection amplifiers in Q band.   S. Baskaran, P. N. Robson.
Electronics Letters, Vol. 8, No. 5, p. 137 - 138 (1972).

Experimental results of reflection gain and noise figure for C.W. circuit-stabilized indium phosphide amplifiers at about 33 GHz are reported. A noise figure of 7.5 db is observed at 23 db gain. − NF

033.077 Approximating true log output at high frequencies.   D. Clifford.
Electronics, Vol. 45, No. 3, p. 70 - 72 (1972).

New way to build high-frequency log amplifiers eliminates interstage phase shift by using twin-gain amplifier blocks that consist of a unity-gain non-limiting amplifier and a high-gain limiting amplifier. − NF

033.078 A digital-to-analog conversion circuit using third-order polynomial interpolation.
W. P. Dotson, J. H. Wilson.
National Aeronautics Space Administration, Greenbelt, NASA Techn. Rep. TR R-382, 52 pp. (1972).

033.079 Impedance circuits imbedding an LC-lattice two-port.   K. H. Haase.
U.S. Air Force Cambridge Res. Labs. Air Force Systems Command Data Sci. Lab. Phys. Sci. Res. Paper No. 455, AFCRL-71-0303, 91 pp. (1971).

033.080 Paraboloidal-reflector illumination with conical scalar horns.   M. S. Narasimhan, Y. B. Malla.
Electronics Letters, Vol. 8, No. 5, p. 111 - 112 (1972).

The authors describe a systematic procedure for calculating the half flare angle of a conical scalar horn, illuminating a paraboloidal reflector of known F/D ratio, with a view to realising a prescribed sidelobe level for the secondary radiation. NF

033.081 Communications systems benefit from monolithic crystal filters.   R. C. Smythe.
Electronics, Vol. 45, No. 3, p. 48 - 51 (1972).

Bandwidths to several tenths of a percent at frequencies above 5 MHz are achieved without inductors, providing lower cost, smaller size, and performance advantages over discrete-element crystal filters. − NF

033.082 Monolithic crystal filter in production after 5 years.   R. C. Smythe.
Electronics, Vol. 45, No. 2, p. 29 - 30 (1972).

Popular, descriptive report of monolithic crystal filters operating at 8 MHz. − NF

033.083 Compensation of spherical reflector aberrations by planar array feeds.   N. Amitay, H. Zucker.
IEEE Trans. Antennas Propagation, Vol. AP-20, 49 - 62 (1972).

The feasibility of correcting phase aberrations of spherical reflector antennas with planar array feeds has been investigated. This type of feed seems to be particularly attractive for applications requiring several closely spaced beams.−JWB

033.084 The locating reflectometer.   P. I. Somlo.
IEEE Trans. Microwave Theory Techn. Vol. MTT-20, 105 - 112 (1972).

A microwave measuring instrument which is capable of producing a visual or graphical record of the reflection coefficient γ along the length of the waveguide component is described. The principle of operation is the evaluation in real time by an analog method of the Fourier transform of γ as a function of frequency resulting in γ as a function of distance.

033.085 Method of conjugate gradients for antenna pattern synthesis.   T. S. Fong, R. A. Birgenheier.

Radio Sci., *(USA)*, Vol. 6, 1123 - 1129 (1971).

033.086 Simple overall absolute calibrator for millimetre-wavelength horn-radiometer systems.
A. C. Gordon-Smith, C. J. Gibbins.
Electronics Letters, Vol. 8, No. 3, p. 59 - 60 (1972).
A low noise calibrator for the 3-mm region using a concave mirror to reflect the radiation from absorber immersed in liquid nitrogen into the radiometer horn. – *DNC*

033.087 Annular resonant structures and their uses as microwave filters. R. T. Irish.
Radio Electronic Engineer, Vol. 42, No. 2, p. 85 - 90 (1972).
Microwave filters using annular resonant elements in printed circuits are described. They may be designed to eliminate recurrent pass-bands at frequencies above the design centre. – *DNC*

033.088 Broad-band parametric amplifier design.
G. R. Branner, E. R. Meyer, P. O. Scheibe.
IEEE Trans. Microwave Theory Techn., Vol. MTT-20, 176 - 178 (1972).
A computerized optimization technique is employed to provide design values for broad-band parametric amplifiers. Some results of the procedure are presented. – *NF*

033.089 A transportable 16-GHz solar telescope for atmospheric transmissivity measurements.
K. C. O'Brien, R. C. Heidt, G. J. Owens.
Radio Sci., *(USA)*, Vol. 7, 215 - 221 (1972).
A transportable 16-GHz solar telescope is now being used to investigate the atmospheric transmission of solar radiation, with special emphasis on the effects of precipitation. Antenna is 40 inches, overall noise temperature of 740 kelvins and I.F. bandwidth of 225 MHz. – *NF*

033.090 Microwave integrated oscillators for broad-band high-performance receivers.
H. C. Okean, E. W. Sard, R. H. Pflieger.
IEEE Trans. Microwave Theory Techn., Vol. MTT-20, 155 - 164 (1972).
The design and development of a variety of microwave integrated circuit oscillators is described. Examples include thin-film microstrip oscillators at C, Ku, X, S and L bands. *NF*

033.091 Microwave integrated tunnel-diode amplifiers for broad-band high-performance receivers.
H. C. Okean, P. J. Meier.
IEEE Trans. Microwave Theory Techn., Vol. MTT-20, 165 - 172 (1972).

033.092 A new project of 8-cm radioheliograph.
M. Ishiguro, S. Énomé, H. Tanaka.
Proc. Res. Inst. Atmosph. Nagoya Univ., Vol. 19, 105 - 108 (1972).
A brief description of the additions planned to the Nagoya University T-array. – *BMT*

033.093 Lateral displacement of log-periodic paraboloid feed.
P. A. McInnes, E. W. Munro, A. J. T. Whitaker.
Electronics Letters, Vol. 8, No. 10, p. 249 - 250 (1972).

033.094 Fourier-subtractive holographic imaging of microwave-antenna apertures.
A. P. Anderson, J. C. Bennett, P. A. McInnes.
Electronics Letters, Vol. 8, No. 11, p. 277 - 278 (1972).

033.095 Broad-band diode phase shifters. R. V. Garver.
IEEE Trans. Microwave Theory Techn., Vol.

MTT-20, 314 - 323 (1972).
Design figures are presented for four types of diode phase shifters: switched line, reflection, loaded line, and a new type using lumped-element high-pass and low-pass circuits. Comparison of their bandwidths shows that most of them can work over an octave bandwidth. – *NF*

033.096 A frequency multiplier for a UHF receiver.
S. L. Cachia.
Australian Electronics Engineering, Vol. 5, No. 6, p. 13 - 15 (1972).

033.097 Evaluation of the illumination correlation function in antenna tolerance theory.
K. K. Dey, P. Khastgir.
International Journ. Electronics, Vol. 32, 537 - 543 (1972).

033.098 Arrays with electrically long elements.
R. W. P. King, S. S. Sandler.
Radio Sci., *(USA)*, Vol. 7, 593 - 602 (1972).

033.099 Low-noise intermediate-frequency amplifiers for 55-GHz radiometers. V. T. Kjartansson.
Mass. Inst. Technology, Res. Lab. Electronics, Quarterly Progr. Rep., No. 105, p. 13 - 15 (1972).
Tentative circuits for broadband amplifiers 10 - 110 MHz and 5 - 500 MHz are discussed. Single frequency noise figures of < 1 db at 60 MHz and 1.6 db at 250 MHz were obtained. *JBS*

033.100 The principles of pulse signal recovery from gravitational antennas.
M. J. Buckingham, E. A. Faulkner.
Radio Electronic Engineer, Vol. 42, 163 - 171 (1972).
The authors relate the problem of the gravitational antenna to engineering problems of pulse signal detection and suggest improvements in current detection practice. – *TWC*

033.101 Corrugations improve monopulse feed horns.
D. Davis.
Microwaves, Vol. 11, No. 4, p. 58 - 63 (1972).
The author describes use of circumferential corrugations in a square waveguide for monopulse feeds. – *BMT*

033.102 Surface waves in the corrugated conical horn.
J. K. M. Jansen, M. Jeuken.
Electronics Letters, Vol. 8, No. 13, p. 342 - 344 (1972).
Surface waves do exist for corrugated conical horns and will propagate if the depth of the grooves is less than one quarter of a wavelength. – *DNC*

033.103 On the relation between modes in rectangular, elliptical, and parabolic waveguides and a mode-classifying system. T. Larsen.
IEEE Trans. Microwave Theory Techn., Vol. MTT-20, 379 - 384 (1972).

033.104 Antenna tolerance theory – A survey of basic methods and recent developments. T. B. Vu.
Proc. Instn. Radio Electronics Engineers Australia, Vol. 33, 268 - 274 (1972).
The author discusses the maximum error and statistical methods of accounting for errors in large reflecting antennas. *DNC*

033.105 The performance and design of 2 : 1 bandwidth log-periodic dipole arrays. B. G. Evans.
Radio Electronic Engineer, Vol. 42, 225 - 232 (1972).

033.106 The M.I.T. K_u-band radio interferometer.

G. D. Papadopoulos, B. F. Burke.
Radio Sci., (USA), Vol. 7, 667 - 674 (1972).

The two 8-ft paraboloids operate at 1.75 cm. The maximum baseline of 100 metres corresponds to resolution of 35 arc sec. PDP-8 computer in the system is used for pointing, tracking, delay compensation and real-time data analysis. System phase stability was better than 10 deg over 2 hours. *ACM*

033.107   Technique for compensating for reflector-antenna-surface errors with long correlation lengths.
E. A. Parker, P. R. Cowles.
Electronics Letters, Vol. 8, No. 14, p. 366 - 367 (1972).

The method makes use of a secondary mirror to compensate for errors with large correlation intervals. Experimental results using a 2.8 m reflector are given. – *BMT*

033.108   Design of Cassegrain antennas employing dielectric cone feeds.
P. J. B. Clarricoats, C. E. R. C. Salema, S. H. Lim.
Electronics Letters, Vol. 8, No. 15, p. 384 - 385 (1972).

Design procedure is given for a Cassegrain antenna using a dielectric cone feed. There is good agreement between measured and predicted radiation patterns for a 1.2 metre paraboloid using the feed at approximately 11 GHz. – *ACM*

033.109   Parametric amplifiers; performance versus operating frequency.   J. Dupraz, M. Creac'h.
Electr. Commun., Vol. 47, No. 2, p. 101 - 107 (1972).

033.110   Mode conversion using circumferentially corrugated cylindrical waveguide.   B. M. Thomas.
Electronics Letters, Vol. 8, No. 15, p. 394 - 396 (1972).

033.111   Radiation characteristics of dielectric cones.
C. E. R. C. Salema, P. J. B. Clarricoats.
Electronics Letters, Vol. 8, No. 16, p. 414 - 416 (1972).

033.112   A postdetection method of measuring predetection RF signal-10-noise ratio.   L. V. Blake.
IEEE Trans. Aerospace Electronic Systems, Vol. AES-8, 168 - 173 (1972).

A method of determining the predetection signal-to-noise power ratio in a radio receiving system by measurement of average postdetection signal-plus-noise and noise-only voltages is described – *JBS*

033.113   Polarization mismatch errors in radio phase interferometers.
E. I. Muehldorf, M. A. Teichman, E. Kramer.
IEEE Trans. Aerospace Electronic Systems, Vol. AES-8, 135 - 140 (1972).

An analysis is presented which deals with the effects of polarization mismatch errors on the accuracy of a phase interferometer used for position location of unknown emitters.

033.114   Design tests of the fully steerable, wideband, decametric array at the Clark Lake Radio Observatory.
J. R. Fisher.
Diss. Fac. Graduate School, Univ. Maryland. 105 pp. (1972).

This thesis gives the performance of the T array consisting of 720 spiral aerials operating between 10 and 110 MHz. *BMT*

033.115   The design of low noise amplifiers.
R. H. S. Riordan.
Proc. Instn. Radio Electronics Engineers Australia, Vol. 33, 382 - 392 (1972).

033.116   Simple small primary feed for large opening angles and high aperture efficiency.

R. Wohlleben, H. Mattes, O. Lochner.
Electronics Letters, Vol. 8, No. 19, p. 474 - 476 (1972).

033.117   Applications of the Josephson effects in the millimetre and submillimetre wavelength regions.
T. G. Blaney.
Radio Electronic Engineer, (GB), Vol. 42, 303 - 308 (1972).

033.118   Statistics of rectified, RC-filtered, band-limited noise.   R. Hurst, P. J. Edwards.
Proc. Instn. Radio Electronics Engineers Australia, Vol. 33, 442 - 444 (1972).

The results of determinations of the probability density function of narrow-band, rectified, RC-filtered electrical noise are given.

033.119   Conductive contacting spheres on the centre of the broad wall of rectangular waveguides.
P. I. Somlo, D. L. Hollway.
Electronics Letters, Vol. 8, No. 20, p. 507 - 508 (1972).

033.120   On the radiation from an open-ended corrugated pipe carrying the He_{II} mode.
C. M. Knop, H. J. Wiesenfarth.
IEEE Trans. Antennas Propagation, Vol. AP-20, 644 - 648 (1972).

033.121   On the efficiency and radiation patterns of mismatched shaped cassegrainian antenna systems.
J. Dijk, E. J. Maanders, J. P. F. Sniekers.
IEEE Trans. Antennas Propagation, Vol. AP-20, 653 - 655 (1972).

033.122   Precision integrator resets as it samples.
D. J. Knowlton.
Electronics, Vol. 45, No. 18, 78 - 79 (1972).

033.123   Radiation properties of conical scalar horns.
M. S. Narasimhan, B. V. Rao.
Proc. Instn. Electr. Engineers (GB), Vol. 119, 1092 - 1094 (1972).

033.124   Measurement of varactor capacitance parameters.
R. B. Smith, B. Bramer.
Radio Electronic Engineer, Vol. 42, 381 - 387 (1972).

033.125   Cross polarizing effects of a water film on a parabolic reflector at microwave frequencies.
P. A. Watson, S. I. Ghobrial.
IEEE Trans. Antennas Propagation, Vol. AP-20, 668 - 671 (1972).

033.126   Ionospheric research – an Australian report.
D. G. Cole.
Australian Physicist, Vol. 9, 169 - 172 (1972).

Includes news that the Adelaide array is to have an ultrasonic image forming system devised by Briggs. – *JAR*

Theory and analysis of phased array antennas.
See Abstr. 003.037.

Statistical antenna theory.   See Abstr. 003.133.

Identification and removal of phase errors in interferometry.   See Abstr. 031.004.

Guide to scientific instruments, 1972–73.
See Abstr. 031.064.

A flux-density scale for microwave frequencies.
See Abstr. 141.072.

034 Astronomical Accessories

034.001 A low-cost pulse amplifier and discriminator for photon counting. D. J. Taylor.
Publ. Astron. Soc. Pacific, Vol. 84, 379 - 381 (1972).

A low-cost pulse amplifier and discriminator with a +5V output pulse and 75 nsec pulse resolution is described and brief instructions for adjustment, cable termination, and elimination of interference are given.

034.002 Instrumentation for balloon and rocket experiments. K. A. Anderson.
Space Science Rev., Vol. 13, 337 - 360 (1972). – Conference paper (see 012.002).

034.003 A variable spacing modulation collimator for X-ray astronomy.
D. J. Adams, A. F. Janes, C. H. Whitford.
Astron. Astrophys., Vol. 20, 121 - 130 (1972).

A modulation collimator instrument for high spatial resolution studies in X-ray astronomy is described. The technique employed is suitable for use on a three axis stabilised vehicle, and produces an output which is a Fourier transform of the field observed. The angular resolution of the instrument is calculated, and its sensitivity is shown to compare favourably with that of conventionally collimated instruments. The instrument can be used to locate pulsating stellar X-ray sources.

034.004 An Irtran-1 reflection filter for the 20 μ atmospheric window. J. J. Wijnbergen.
Astron. Astrophys., Vol. 20, 159 - 160 (1972).

The low temperature reflection of Irtran-1 has been measured to justify its use as a 20 μ filter in astronomical photometers. Calculation of a four times reflection filter shows a mean transmission of 61% between the 30% reflection limits at 16.9 and 22.0 μ. Short wavelengths can be suppressed by a factor of 10^{-5}.

034.005 Les micromètres à fils. P. Couteau.
L'Astronomie, 86e année, p. 355 - 360 (1972).

034.006 Imaging in astronomy.
C. D. Mackay, A. S. Milsom.
Nature, Vol. 239, 116 - 117 (1972). – Review on Spectracon image tube (Instrument Technology Ltd., from about £ 3000).

034.007 The systematic errors of the total vector H of magnetic field measurements with the Crimean Astrophysical Observatory magnetograph. V. A. Kotov.
Izv. Krymskoj Astrofiz. Obs., Vol. 44, 77 - 86 (1972). In Russian.

The influence of some «effective» asymmetry of the spectral line λ5250 on the accuracy of the total vector H of magnetic field is considered for the measurements with the Crimean magnetograph.

034.008 A teaching astronomical spectrophotometer.
K. A. Innanen, W. G. Weller.
Journ. Roy. Astron. Soc. Canada, Vol. 66, 189 - 192 (1972).

A description is given of a compact, relatively inexpensive spectrophotometer for use at the Cassegrain focus of telescopes of moderate aperture for teaching purposes and occasional research.

034.009 Astrospektrographie mit Spiegelteleskopen. II. Einprismenspektrograph mit 60° Flintprisma.
C. Albrecht.
SuW, Vol. 11, 243 - 244 (1972).

034.010 Experimental comparisons of the International Pyrheliometric Scale with the absolute radiation scale. R. C. Willson.
Nature, Vol. 239, 208 - 209 (1972).

A series of radiometer comparison tests has been carried out at the JPL Table Mountain Observatory during the period May 1968 to August 1970. Solar irradiance measurements by instruments reproducing the International Pyrheliometric Scale (IPS) have been compared with measurements made simultaneously by a number of different JPL Active Cavity Radiometers and PACRAD radiometers.

034.011 Investigation of an interference polarization filter for Hα with pass-band of ±16Å.
E. A. Makarova, E. H. Rakhmatulina.
Astron. Tsirk., No. 690, p. 3 - 6 (1972). In Russian.

034.012 Sky background compensation in a phase photoelectric installation.
V. E. Brandt, M. I. Malyshev.
Astron. Tsirk., No. 694, p. 6 - 7 (1972). In Russian.

034.013 An infrared photometer for the balloon-borne telescope Thisbe. D. Lemke.
Astrophys. Space Sci. Library, Vol. 30, (see 012.005), 53 - 59 (1972).

034.014 A balloon-borne helium-cooled interferometer for investigation of the isotropic submillimetre background. J. E. Beckman, E. I. Robson.
Astrophys. Space Sci. Library, Vol. 30, (see 012.005), 63 - 72 (1972).

034.015 Infrared rocket astronomy. K. Shivanandan.
Astrophys. Space Sci. Library, Vol. 30, (see 012.005), 73 - 76 (1972).

034.016 An interference filter radiometer with cooled optics and a cooled PbS detector for rocket application.
W. Bangert, E. Krieg, R. Scheidle.
Astrophys. Space Sci. Library, Vol. 30, (see 012.005), 77 - 83 (1972).

034.017 Infrared interferometry from satellites.
B. J. Conrath, R. A. Hanel.
Astrophys. Space Sci. Library, Vol. 30, (see 012.005), 84 - 92 (1972).

034.018 The Meteosat dual-channel radiometer.
A. L. Peraldi.
Astrophys. Space Sci. Library, Vol. 30, (see 012.005), 93 - 99 (1972).

034.019 Infrared detectors. Survey of the present state of the art. P. Léna.
Astrophys. Space Sci. Library, Vol. 30, (see 012.005), 103 - 113 (1972).

034.020 Liquid helium-cooled bolometers. G. Chanin.
Astrophys. Space Sci. Library, Vol. 30, (see 012.005), 114 - 120 (1972).

034.021 A new type of helium-cooled bolometer.
N. Coron, G. Dambier, J. Leblanc.
Astrophys. Space Sci. Library, Vol. 30, (see 012.005), 121 - 131 (1972).

034.022 **Performance tests of InSb and Ge bolometers.**
P. E. Clegg, J. S. Huizinga.
Astrophys. Space Sci. Library, Vol. 30, (see 012.005), 132 - 140 (1972).

034.023 **An infrared transducer for space applications.**
M. Chatanier, G. Gauffre.
Astrophys. Space Sci. Library, Vol. 30, (see 012.005), 141 - 146 (1972).

034.024 **Josephson detection.** M. Renard.
Astrophys. Space Sci. Library, Vol. 30, (see 012.005), 147 - 153 (1972).

034.025 **Astronomical polarimetry at 5 and 10μ.**
J. C. D. Marsh, R. F. Jameson.
Astrophys. Space Sci. Library, Vol. 30, (see 012.005), 154 - 159 (1972).

034.026 **Problems and design of black-body references.**
H. P. Baltes, P. Stettler.
Astrophys. Space Sci. Library, Vol. 30, (see 012.005), 160 - 167 (1972).

034.027 **Cooling systems for spaceborne infrared experiments.**
R. W. Breckenridge, Jr.
Astrophys. Space Sci. Library, Vol. 30, (see 012.005), 171 - 188 (1972).

034.028 **Balloon-borne infrared cryostats.** G. Chanin.
Astrophys. Space Sci. Library, Vol. 30, (see 012.005), 189 - 195 (1972).

034.029 **Survey of the present state of art of infrared filters.**
S. D. Smith, G. D. Holah, J. S. Seeley, C. Evans, R. Hunneman.
Astrophys. Space Sci. Library, Vol. 30, (see 012.005), 199 - 218 (1972).

034.030 **Metallic grid filters.** G. Chanin.
Astrophys. Space Sci. Library, Vol. 30, (see 012.005), 219 - 224 (1972).

034.031 **New narrow-band metallic mesh filters for the 50μ region.** M. Heidemann.
Astrophys. Space Sci. Library, Vol. 30, (see 012.005), 225 - 233 (1972).

034.032 **High-pass filters for far infrared astrophysical investigation.** M. G. Baldecchi, B. Melchiorri, G. M. Righini.
Astrophys. Space Sci. Library, Vol. 30, (see 012.005), 234 - 242 (1972).

034.033 **Filters for far infrared astronomy.**
J. J. Wijnbergen, W. H. Moolenaar, G. de Groot.
Astrophys. Space Sci. Library, Vol. 30, (see 012.005), 243 - 256 (1972).

034.034 **Submillimetre and millimetre mapping using interference filters.** P. A. Ade, J. A. Bastin.
Astrophys. Space Sci. Library, Vol. 30, (see 012.005), 257 - 264 (1972).

034.035 **Interferometric spectrometry for infrared astronomy.**
D. H. Martin.
Astrophys. Space Sci. Library, Vol. 30, (see 012.005), 267 - 280 (1972).

034.036 **Far infrared broad band interferometry.**
J. P. Baluteau, N. Epchtein, J. Gay, J. P. Verdet.
Astrophys. Space Sci. Library, Vol. 30, (see 012.005), 281 - 283 (1972).

034.037 **Rapid scanning Michelson interferometer.**
M. J. Smyth.
Astrophys. Space Sci. Library, Vol. 30, (see 012.005), 284 - 287 (1972).

034.038 **Field compensated interferometer for stellar spectrometry.** A. Girard.
Astrophys. Space Sci. Library, Vol. 30, (see 012.005), 288 (1972).

034.039 **A SISAM interferometer and a simple Michelson-interferometer with spherical mirrors for space application.** H.-J. Bolle, M. Bottema, W. Völker, A. Zickler.
Astrophys. Space Sci. Library, Vol. 30, (see 012.005), 289 - 305 (1972).

034.040 **Liquid helium-cooled grating monochromator.**
M. Herse.
Astrophys. Space Sci. Library, Vol. 30, (see 012.005), 306 - 312 (1972).

034.041 **One- and two-dimensional computational problems associated with interferometry.** Y. Biraud.
Astrophys. Space Sci. Library, Vol. 30, (see 012.005), 313 - 331 (1972).

034.042 **Remarks on infrared technology and spectroscopy in astrophysics.** F. Kneubühl.
Astrophys. Space Sci. Library, Vol. 30, (see 012.005), 332 - 334 (1972).

034.043 **Apparatus PG-1 for the study of the radiation characteristics in the neighbourhood of the earth by satellite Intercosmos 3.** V. Veselý, M. Šícha, L. Láska, M. Novák, V. Řezáčová, J. Studnička, M. Tichý, S. Fischer, J. Dubinský.
Bull. Astron. Inst. Czechoslovakia, Vol. 23, 293 - 297 (1972).

034.044 **On a design of the slits of a magnetograph photometer.** V. M. Grigoryev, V. E. Stepanov.
Solnechnye Dannye 1972 Byull., No. 4, p. 98 - 104 (1972).
In Russian.

034.045 **The polarization properties of a diffraction spectrograph.** G. M. Nikolsky, Ts. S. Khetsuriani.
Solnechnye Dannye 1972 Byull., No. 5, p. 91 - 94 (1972).
In Russian.
An investigation of the polarization properties of the horizontal solar telescope spectrograph of the Abastumani Observatory is made by means of a photoelectric polarimeter.

034.046 **Eine Justiereinrichtung für transportable parallaktische Instrumente mit Hilfe des Polarsterns.**
H. Blickisdorf.
Orion, 30. Jahrgang, p. 149 - 151 (1972).

034.047 **Three-channel SEC vidicon system for photometric observations of the solar corona.** H. Y. Chiu, A. W. Peterson, L. M. Kieffaber, J. C. Brandt, R. W. Hobbs, S. P. Maran.
Bull. American Astron. Soc., Vol. 4, 309 (1972). – Abstr. AAS.

034.048 **Calibration of coudé spectrograms.**
I. Furenlid, A. G. Millikan.
Bull. American Astron. Soc., Vol. 4, 324 (1972). – Abstr. AAS.

034.049 **A photon counting polarimeter.** J. F. Dolan.
Bull. American Astron. Soc., Vol. 4, 338 - 339

(1972). – Abstr. AAS.

034.050 **Photographic image calibration with a 45° second surface reflector plate.** J. Cuffey.
Bull. American Astron. Soc., Vol. 4, 339 (1972). – Abstr. AAS.

034.051 **New multichannel spectrometer at Sacramento Peak.**
R. B. Dunn, G. L. Epstein, R. W. Hobbs, S. P. Maran.
Bull. American Astron. Soc., Vol. 4, 381 (1972). – Abstr. AAS.

034.052 **The GSFC EUV and X-ray spectroheliograph on OSO-7.** J. H. Underwood, W. M. Neupert.
Bull. American Astron. Soc., Vol. 4, 394 (1972). – Abstr. AAS.

034.053 **Ein Photonen zählendes Photometer.**
K. D. Rakosch.
Meßtechnik 2.72, [Zeitschr. Instrumentenkunde, 80. Jahrgang], p. 45 - 49 = Mitt. Univ.-Sternw. Wien (1972).
The amplifier described here was developed for photoelectric measurements using photomultipliers. It is a single photon counting device and it should be used for measuring very faint light sources or very fast variable light sources. For both purposes it is remarkably superior to any high input impedance direct current amplifier.

034.054 **Infrared heterodyne solar radiometry.**
J. H. McElroy.
Applied Optics, Vol. 11, 1619 - 1622 (1972).
An infrared heterodyne radiometer has sufficient sensitivity for application to high resolution spectrometric studies of the sun and the absorption of sunlight in the atmosphere. In the 10-μm region, the resolution can typically range from less than 6.7×10^{-3} cm^{-1} to 6.7×10^{-2} cm^{-1}. The minimum detectable signal can approach 10^{-24} W/Hz. The infrared heterodyne radiometer (IHR) can also be used for atmospheric propagation studies at the emission frequencies of the CO_2 laser for the examination of the feasibility of space-to-ground laser communication links.

034.055 **Sensitive stellar photometer with a Cassegrainian optical system for use in the 1300–2000-Å region.**
J. W. Campbell.
Applied Optics, Vol. 11, 1801 - 1809 (1972).
A sensitive photoelectric photometer designed for astronomical measurements of early type stars in the 1300–2000-Å region is described. The photometer, which uses a wide aperture Cassegrainian optical system, has produced over ninety-four stellar observations with a limiting magnitude of m_v = 6.5. A number of methods of wavelength isolation have been used and are described, together with details of the method of absolute calibration of the instrument.

034.056 **Low temperature filter for photometry in the 20-μ atmospheric window.** J. J. Wijnbergen.
Applied Optics, Vol. 11, 1810 - 1813 (1972).
In a search for a filter for a 20-μ astronomical photometer the reststrahlen reflection of Irtran-1 and a ZnO pellet has been measured at low temperatures in the 15–50-μ range. With a mean reflection of 82% between 16μ and 22.3μ the Irtran-1 material appears to be an efficient low temperature filter suitable for this spectral window, especially when using several reflections.

034.057 **Fabry lens.** J. Michlovic.
Applied Optics, Vol. 11, 490 - 491 (1972).

034.058 **Comments on: Fabry lens.** M. F. A'Hearn.
Applied Optics, Vol. 11, 1874 (1972).

034.059 **Exit slit mirrors for the Ebert spectrometer.**

W. G. Fastie.
Applied Optics, Vol. 11, 1960 - 1963 (1972).
The use of a very long straight entrance slit in an Ebert grating spectrometer with two plane mirrors at the shorter exit slit to increase the energy density is described. This system has been employed in a far uv rocket spectrometer to provide higher sensitivity than has been achieved previously. The imaging properties and required slit and mirror adjustments are presented. Experimental results are included.

034.060 **Fabry-Perot spectrometer adjustment for the compensation of Doppler shift from rapidly rotating and rapidly flowing sources.** J. T. Trauger, F. L. Roesler.
Applied Optics, Vol. 11, 1964 - 1969 (1972).
This paper discusses the theory of Doppler compensation with a Fabry-Perot spectrometer, using as a practical example the application of the PEPSIOS spectrometer to the study of Jupiter.

034.061 **Pressure-scanned echelle grating plus Fabry-Perot stellar spectrophotometer.** J. Caplan.
Applied Optics, Vol. 11, 1978 - 1985 (1972).
A photoelectric stellar spectrophotometer, with 0.3-Å resolution, for use near Hα (6563 Å) is described. A Fabry-Perot etalon and an echelle grating are simultaneously pressure-scanned. The Fabry-Perot can be bypassed and the grating system used alone to obtain narrow-band (4.89-Å) Hα indices, in which case the pressure system allows the necessary radial velocity corrections to be made conveniently.

034.062 **OSO-5 dim light monitor.**
G. B. Burnett, J. G. Sparrow, E. P. Ney.
Applied Optics, Vol. 11, 2075 - 2081 (1972).
The design, calibration, and operation of the University of Minnesota's experiment on board the satellite OSO-5 are described. The instrument was designed to measure zodiacal light, airglow, and lightning.

034.063 **A portable night-sky photometer.**
P. J. Treanor, E. Salpeter.
Observatory, Vol. 92, 96 - 99 (1972). – Note.

034.064 **An improved chopper for use in infrared photometry.** I. S. Glass.
Observatory, Vol. 92, 140 - 141 (1972). – Note.

034.065 **Einige Probleme der Laser-Spektroskopie.**
W. S. Letochow, S. L. Mandelstam.
Jenaer Rundschau (Jena Review), 17. Jahrgang, p. 192 - 198 (1972).

034.066 **Image motion in the Culgoora solar magnetograph. – The role of vibration.**
J. L. Goldberg, B. Dorien-Brown.
Publ. Astron. Soc. Pacific, Vol. 84, 534 - 540 (1972).
This paper describes part of a broader study directed to the improvement of the quality of solar photographs obtained with the Culgoora magnetograph and is concerned only with the role of mechanical vibration.

034.067 **Optically pumped magnetometers.** L. Engelhard.
Zeitschr. Geophys., Vol. 37, 1 - 37 (1971). In German.
Optically pumped magnetometers can be used for measurements of the earth's and interplanetary magnetic field. The basic principle of operation is the Zeeman effect of free atoms, which must be optically pumped. The different possibilities of measuring the resonant frequency and the properties of the various gases, which are used, are discussed in detail.

034.068 **The influence of temperature on the work of an interference filter.**

A. A. Kmito, V. Ya. Matveev, I. V. Morozova.
Trudy Gl. geofiz. observ., 1972, vyp. (No.) 275, p. 34 - 38.
In Russian. − Abstr. in Referativ. Zhurn. 62. Issled. kosmich.
prostranstva, 10.62.157 (1972).

**034.069 On the dispersion of radiative temperatures measured
with a meteorological earth satellite.**
L. B. Krasil'shchikov, I. V. Morozova, M. G. Rudakov, L. B.
Rudneva.
Trudy Gl. geofiz. observ., 1972, vyp. (No.) 275, p. 31 - 33.
In Russian. − Abstr. in Referativ. Zhurn. 62. Issled. kosmich.
prostranstva, 10.62.159 (1972).

**034.070 All-reflection interferometer for use as a Fourier-
transform spectrometer.**
R. A. Kruger, L. W. Anderson, F. L. Roesler.
Journ. Optical Soc. America, Vol. 62, 938 - 945 (1972).

This paper describes the design and preliminary tests of a
two-beam interferometer that utilizes only reflection optics,
and which can be used as a Fourier-transform spectrometer.
This interferometer can be used in all regions of the electromagnetic spectrum where surfaces with suitable reflectances are
available.

034.071 Performance of S-192 (Hg, Cd) Te arrays.
N. C. Aldrich, J. D. Beck.
Applied Optics, Vol. 11, 2153 - 2156 (1972).

Very high performance (Hg, Cd) Te photoconductive detectors have been fabricated for use on the S-192 experiment,
which is a multispectral scanner being built by Honeywell for
the NASA Manned Space Center's Skylab. The S-192 will scan
the earth from Skylab and record data in twelve near ir spectral bands and one long wavelength band.

**034.072 Nimbus limb radiometer, Apollo fine sun sensor, and
Skylab multispectral scanner.**
J. C. Kollodge, J. R. Thomas, R. A. Weagant.
Applied Optics, Vol. 11, 2169 - 2176 (1972).

Examples of three different types of electrooptical systems developed by the Honeywell Radiation Center for NASA
are described. One is a multichannel infrared ($\sim 15\,\mu$) radiometer that will permit temperature and constituent inferences
over the globe; it carries a one-year supply of cryogenics for
the trimetal infrared detectors. The second is the Apollo telescope mount fine sun sensor, a tracking device making use of
solar radiation and the transmission near critical angle of refraction, that will track within ±2 sec of arc to a designated
point on the sun. The final example is the Skylab S-192 multispectral (thirteen channels from $0.4\,\mu$ to $12\,\mu$) mapper for a
variety of earth resources applications.

**034.073 High resolution solar imaging at 1.65 μ with a 64-
element array.**   P. J. Turon, D. Stefanovitch.
Applied Optics, Vol. 11, 2177 - 2180 (1972).

An imaging system using a 64-element array of PbS detectors has been successfully tested at Kitt Peak. The first
images of the low contrast solar granulation at 1.65 μ have
been obtained with a signal-to-noise ratio better than 200 and
a resolution of the order of 1 sec of arc. The problem of
imaging systems for low contrast objects in the infrared is discussed.

**034.074 Interferometric spectropolarimetry: Alternate ex-
perimental methods.**   A. L. Fymat.
Applied Optics, Vol. 11, 2255 - 2264 (1972).

Three alternate methods of obtaining spectra of the intensity and state of polarization of light are proposed. The
methods make use of a two-beam amplitude division interferometer using the technique of Fourier spectroscopy.

034.075 Calculation of middle ultraviolet radiation detector

response to solar radiation as a function of altitude.
R. G. Johnson.
Applied Optics, Vol. 11, 2372 - 2374 (1972). − Letter.

**034.076 Adaption de la caméra électronique au corono-
graphe.**   J.-P. Rozelot, R. Despiau.
Comptes Rendus Acad. Sci. Paris, Sér. B, Vol. 275, 613 - 616
(1972).

Nous avons voulu mettre à profit les avantages de la
caméra électronique (absence de seuil à l'origine dans la carac-
téristique, linéarité de la réponse en densité permettant une
photométrie précise, gain appréciable du temps de pose, résolution élevée des émulsions, etc.) dans un but précis d'expérience:
l'étude de l'activité coronale dans les régions polaires, afin de
comparer, pour une même raie et pour de faibles valeurs, les
intensités obtenues par les deux méthodes, visuelle et photo-
graphique.

**034.077 A current-to-frequency converter for astronomical
photometry.**   D. J. Taylor.
Rev. Sci. Instruments, Vol. 40, 559 - 562 (1969).

A simple current-to-frequency converter has been
developed which works directly with photomultiplier currents.

**034.078 Degradation of continuous-channel electron multi-
pliers in a laboratory operating environment.**
L. A. Frank, N. K. Henderson, R. L. Swisher.
Rev. Sci. Instruments, Vol. 40, 685 - 689 (1969).

Measurements of the counting rates ($\sim 10^3$ to 10^4 counts/
sec) of continuous-channel electron multipliers mounted in an
electrostatic analyzer responding to a monoenergetic beam of
electrons display a degradation of their gain (fatigue) which
is proportional to the accumulated counts. Our present investigation is directed toward the determination of the useful
lifetime for a particular species of this device, a Bendix model
4013 'Channeltron,' in a typical laboratory application and in
a spaceflight environment.

**034.079 A rocket borne scintillation spectrometer for observ-
ing cosmic X-rays.**
J. Harri, M. McGee, A. Toor.
Rev. Sci. Instruments, Vol. 40, 703 - 708 (1969).

A rocket borne spectrometer system sensitive to 6–80
keV photons is described. Also described is a detector rotation
mechanism that allows the look angle of the detector with
respect to the rocket axis to be changed during flight.

**034.080 An improved collimator for extreme ultraviolet and
X-rays.**   J. H. Underwood.
Rev. Sci. Instruments, Vol. 40, 894 - 896 (1969).

The mechanical collimator for X-rays and extreme ultraviolet described by McGrath may be improved by replacing the
plates pierced with apertures arranged in a regular array with
plates pierced with apertures arranged at random. With this
scheme, fewer plates are required to suppress off-axis radiation.

034.081 The Pioneer 8 cosmic dust experiment.
O. E. Berg, F. F. Richardson.
Rev. Sci. Instruments, Vol. 40, 1333 - 1337 (1969).

A cosmic dust sensor comprising a unique array of
sensors, sensor controls, and in-flight calibrations was placed
in a heliocentric orbit (0.99 to 1.088 a.u.) on 13 December
1967 on board the Pioneer 8 satellite. The ionization and
momentum imparted by the impact of a cosmic dust particle
upon a surface are measured by the sensor and used to determine the particle's direction, speed, and mass.

**034.082 Location of primary cosmic ray particles in photo-
graphic emulsion using spark chambers.**
E. R. Goza, S. Krzywdzinski, E. G. Stafford.

Rev. Sci. Instruments, Vol. 41, 219 - 225 (1970).

Spark chambers have been used in conjunction with photographic emulsion to investigate the accuracy and reliability in the location of primary cosmic rays in emulsion exposed at balloon elevations. The location of these primaries in the emulsion stack has been determined using measurements made on spark chamber photographs.

034.083 A rugged, stable, differential platinum resistance thermometer.
L. Kleven, L. Lofgren, P. Felsenthal.
Rev. Sci. Instruments, Vol. 41, 541 - 544 (1970).

A new stable platinum resistance thermometer, and the methods of construction, calibration, and testing are described. The design provides a differential (±2K) thermometer over the range of 200 to 270 K. After environmental testing, the thermometer has demonstrated differential stability of better than $\pm 1 \times 10^{-3}$ K over one year.

034.084 An electrostatic analyzer with no fringe field for measurements of low energy particles on space vehicles. F. Mariani.
Rev. Sci. Instruments, Vol. 41, 807 - 812 (1970).

An electrostatic analyzer is discussed which has no fringe field effects. The basic idea is to form the equipotential surfaces around an ideal conductor for which the Dirichlet problem can be exactly solved.

034.085 The soft particle spectrometer in the ISIS-I satellite. W. J. Heikkila, J. B. Smith, J. Tarstrup, J. D. Winningham.
Rev. Sci. Instruments, Vol. 41, 1393 - 1402 (1970).

The soft particle spectrometer uses electrostatic deflection for simultaneous measurement of the differential energy spectra of electrons and protons. The energy range is from 10 eV to 10 keV per unit charge with ±40% energy spread. A typical electron spectrogram for a transpolar pass is included as an example of the data reduction procedure used.

034.086 Two spectrographs for use with image intensifier tubes. R. L. Sharpless, R. A. Young.
Rev. Sci. Instruments, Vol. 41, 1628 - 1636 (1970).

This article describes two spectrographs using image intensifier tubes for increased sensitivity. One spectrograph uses an echelle grating to give a resolution of about 0.10 nm and has all-reflecting optics for use in the ultraviolet. It is usable over the range from 250 to >600 nm and covers about 150 nm on a 24 × 36 mm frame. The second spectrograph uses transmission optics of high optical speed and covers roughly the range from 400 to 760 nm on a 24 mm long frame with a resolution of about 2 nm.

034.087 Laser goniometry on satellites. C. Veret.
Bull. Groupe Recherches Géod. Spatiale, Obs. Meudon, No. 1, p. 7 - 23 (1971).

In view of complementing the laser rangefinding to determine the position of a satellite from a single station, ONERA has developed an apparatus with which it is possible to obtain the angular co-ordinates. The angular accuracy reached is 0.5 second of arc, i.e. 4 m at 1500 km.

034.088 Réalisation d'une nouvelle génération de télemètres laser. Étude des possibilités techniques de réalisation. J. Gaignebet.
Bull. Groupe Recherches Géod. Spatiale, Obs. Meudon, No. 1, p. 24 - 74 (1971).

034.089 La station de télémétrie laser de l'Observatoire du Pic du Midi. A. Orszag, O. Calamme.
Bull. Groupe Recherches Géod. Spatiale, Obs. Meudon, No. 1, p. 75 - 85 (1971).

034.090 Controle et essais des réflecteurs laser déposés sur la lune. M. Fournet.
Bull. Groupe Recherches Géod. Spatiale, Obs. Meudon, No. 1, p. 114 - 141 (1971).

034.091 The use of synchrotron radiation in the energy calibration of astronomical apparatus.
G. A. Gurzadyan, J. B. Ohanesyan.
Space Sci. Rev., Vol. 13, 642 - 646 (1972). – Conference paper IAU Colloquium No. 14 (see 012.012).

034.092 Spectroscopic techniques in X-ray astronomy.
L. van Speybroeck.
Space Sci. Rev., Vol. 13, 845 - 869 (1972). – Invited paper IAU Colloquium No. 14 (see 012.012).

034.093 A focusing X-ray telescope monochromator.
A. S. Filler, B. S. Fraenkel.
Space Sci. Rev., Vol. 13, 870 (1972). – Conference paper IAU Colloquium No. 14 (see 012.012).

034.094 A proposal for a flint objective prism for the ESO Schmidt camera. E. H. Geyer.
The role of Schmidt telescopes in astronomy. Conference Hamburg 1972, (see 012.014), p. 61 - 62 (1972).

034.095 A new projecting comparator for large size plates.
S. L. T. J. van Agt.
The role of Schmidt telescopes in astronomy. Conference Hamburg 1972, (see 012.014), p. 97 (1972).

034.096 AFCRL lunar laser instrumentation status report.
W. E. Carter, D. H. Eckhardt, W. G. Robinson.
Space Research XII, (see 012.016), Vol. 1, 177 - 186 (1972).

034.097 A preliminary system of lunar laser ranging.
A. Tachibana, Y. Yamamoto, M. Takatsuji, K. Murasawa, Y. Kozai.
Space Research XII, (see 012.016), Vol. 1, 187 - 195 (1972).

034.098 The SAO lunar laser.
C. G. Lehr, M. R. Pearlman, J. A. Monjes.
Space Research XII, (see 012.016), Vol. 1, 197 - 204 (1972).

034.099 La station de télémétrie laser de l'Observatoire du Pic-du-Midi et l'acquisition des cataphotes français de Luna 17. A. Orszag, J. Rösch, O. Calame.
Space Research XII, (see 012.016), Vol. 1, 205 - 209 (1972).

034.100 Lunar laser ranging experiments in Japan.
Y. Kozai.
Space Research XII, (see 012.016), Vol. 1, 211 - 217 (1972).

034.101 A description of the lunar ranging station at McDonald Observatory.
E. C. Silverberg, D. G. Currie.
Space Research XII, (see 012.016), Vol. 1, 219 - 233 (1972).

034.102 The Apollo retroreflector arrays and a new multi-lensed receiver telescope. J. E. Faller.
Space Research XII, (see 012.016), Vol. 1, 235 - 246 (1972).

034.103 Optical properties of the Apollo laser ranging retroreflector arrays.
R. F. Chang, C. O. Alley, D. G. Currie, J. E. Faller.
Space Research XII, (see 012.016), Vol. 1, 247 - 259 (1972).

034.104 Le réflecteur laser de Lunokhod. M. Fournet.
Space Research XII, (see 012.016), Vol. 1, 261 - 277 (1972).

034.105 **Solar observations from Skylab.** G. K. Oertel.
Space Research XII, (see 012.016), Vol. 2, 1739 - 1749 (1972).

034.106 **A stellar interferometer based on coherent detection.** H. G. van Bueren, H. van de Stadt.
Space Research XII, (see 012.016), Vol. 2, 1777 - 1786 (1972).

034.107 **Statistics of the radiation from astronomical masers.** N. J. Evans II, R. E. Hills, O. E. H. Rydbeck, E. Kollberg.
Phys. Rev. A, General Phys., Vol. 6, 1643 - 1647 = National Radio Astron. Obs., Green Bank, Repr. Ser. A, No. 260 (1972).

The results of an experimental determination of the statistical properties of radiation from OH maser sources are reported and interpreted. The radiation is found to have Gaussian statistics with no deviations greater than 1%.

034.108 **An echelle spectrograph for image tubes.** A. Danielsson, P. Lindblom.
Physica Scripta, Vol. 5, 227 - 231 (1972).

An echelle spectrograph specially developed for image tubes is presented. A Czerny-Turner mount is used and the optimization of the optics in respect to aberrations, order sorters and the focal surface is discussed. The image tube used is an image dissector tube with a photocathode 20 mm in diameter. The wavelength range covered is 1700 − 4000 Å.

034.109 **Transfer function of a T.V. system used in astronomy.** V. V. Prokofieva, T. A. Chuprakova.
Astron. Tsirk., No. 719, p. 1 - 3 (1972). In Russian.

034.110 **A polarimetric method of measuring radial velocities.** K. Serkowski.
Publ. Astron. Soc. Pacific, Vol. 84, 649 - 651 (1972). − Presented at the Santa Cruz meeting of the Astronomical Society of the Pacific, 26–29 June 1972.

Stellar light approaching the spectrometer slit is artificially polarized in such a way that the plane of linear polarization changes rapidly with wavelength. For each resolution element in the spectrum a photon-counting image tube measures the intensity of the light and the position angle of its polarization. Wavelengths determined from these position angles are independent of the position of the stellar image on the spectrometer slit; therefore a wide open slit can be used for radial velocity measurements. An echelle spectrometer on an 80-inch telescope is expected to give a radial velocity accurate to ± 1 km sec^{-1} in about four minutes of observing the profile of a spectral line of a 10th magnitude star.

034.111 **A heliostat for measuring the solar flux spectrum.** D. F. Gray.
Publ. Astron. Soc. Pacific, Vol. 84, 721 - 722 (1972).

In order to use the sun as a spectrophotometric reference for other stars, it is desirable to measure the spectrum of the sun integrated over the disk. A device to facilitate such measurement is described.

034.112 **Accuracy of the timing device of the PZT.** M. Yoshinari, K. Fujiwara.
Univ. Tokyo, Tokyo Astron. Obs. Rep. (No. 60), Vol. 16, 203 - 212 (1972). In Japanese.

034.113 **On a light attenuating plate for the PZT.** S. Iijima, Y. Adachi, I. Oguma.
Univ. Tokyo, Tokyo Astron. Obs. Rep. (No. 60), Vol. 16, 213 - 221 (1972). In Japanese.

034.114 **Studio del misuratore di coordinate Hauser.** E. Zanoner.
Contr. Oss. Astron. Torino, No. 61, 12 pp. (1972).

This is a preliminary study of the two-coordinate measuring machine which has been recently acquired by the Observatory of Turin from Firma Hauser, Bienne, Switzerland. The angle between the coordinates has been deduced, as well as the ratio between the two screw pitches.

034.115 **Mariner 9 Michelson interferometer.** R. Hanel, B. Schlachman, E. Breihan, R. Bywaters, R. Chapman, M. Rhodes, D. Rodgers, D. Vanous.
Applied Optics, Vol. 11, 2625 - 2634 (1972).

The Michelson interferometer on Mariner 9 measures the thermal emission spectrum of Mars between 200 cm^{-1} and 2000 cm^{-1} (between 5 μm and 50 μm) with a spectral resolution of 2.4 cm^{-1} in the apodized mode. A noise equivalent radiance of 0.5×10^{-7} W cm^{-2} sr^{-1}/cm^{-1} is deduced from data recorded in orbit around Mars. The instrument performance is demonstrated by calibration data and samples of Mars spectra.

034.116 **Michelson-type Fourier spectrometer for the far infrared.** K. Sakai.
Applied Optics, Vol. 11, 2894 - 2901 (1972).

This paper describes the construction, performance, and applications of a Michelson-type Fourier spectrometer designed for spectroscopy in the 10-cm^{-1} to 200-cm^{-1} spectral region. The electron spin resonance of some biological substances and the refractive index of solids, for example, have been measured with this spectrometer.

034.117 **Coherence interferometer and astronomical applications.** J. B. Breckinridge.
Applied Optics, Vol. 11, 2996 - 2998 (1972). − Rapid communication.

034.118 **Investigations of the polarization interference Hα filter of the firm "Opton" with a pass-band shifting within the limits of ± 16.0 Å.** E. A. Makarova, E. Kh. Rakhmatulina.
Solnechnye Dannye 1972 Byull., No. 9, p. 67 - 72 (1972). In Russian.

034.119 **The measuring and information system of a solar magnetograph.** G. A. Korkotjan.
Solnechnye Dannye 1972 Byull., No. 9, p. 89 - 94 (1972). In Russian.

The measuring and information system of the Pulkovo six-channel solar magnetograph is described. This system allows to automatize the reduction of the obtained information. The main technical characteristics of the system, the block-scheme of the commutator and a description of its operation are given.

034.120 **A graphite crystal polarimeter for stellar X-ray astronomy.** M. C. Weisskopf, R. Berthelsdorf, G. Epstein, R. Linke, D. Mitchell, R. Novick, R. S. Wolff.
Rev. Sci. Instruments, Vol. 43, 967 - 976 (1972).

The first crystal X-ray polarimeter to be used for X-ray astronomy is described.

034.121 **A high-performance large-aperture window for photography from a space platform.** C. J. Chocol.
Journ. Optical Soc. America, Vol. 62, 1339 (1972). − Abstr. Optical Soc. America.

034.122 **An infrared stellar interferometer using heterodyne detection.** M. A. Johnson.
Journ. Optical Soc. America, Vol. 62, 1350 (1972). − Abstr. Optical Soc. America.

034.123 **Geometric calibration of the Mariner-9 vidicon camera system.** J. B. Seidman, J. E. Kreznar.
Journ. Optical Soc. America, Vol. 62, 1351 (1972). − Abstr.

Optical Soc. America.

034.124 Correction for residual image effects in Mariner–9 television images. P. L. Jepsen, A. A. Schwartz.
Journ. Optical Soc. America, Vol. 62, 1351 (1972). – Abstr. Optical Soc. America.

034.125 Removal of photometric distortion from Mariner–9 television images. W. B. Green, R. M. Ruiz.
Journ. Optical Soc. America, Vol. 62, 1351 - 1352 (1972). Abstr. Optical Soc. America.

034.126 Diffraction of X-rays by shallow blazed ruled gratings. E. G. Loewen, R. J. Speer, D. Turner.
Journ. Optical Soc. America, Vol. 62, 1390 (1972). – Abstr. Optical Soc. America.

034.127 An ultrastable, multiple-pass, plane-mirror, Fabry-Perot interferometer.
C. Roychoudhuri, M. Hercher.
Journ. Optical Soc. America, Vol. 62, 1400 (1972). – Abstr. Optical Soc. America.

034.128 Transfer function of the Williams interferometer used as a Fourier spectrometer. M. Bottema.
Journ. Optical Soc. America, Vol. 62, 1438 - 1443 (1972).
The interferometer consists of a beamsplitter and two spherical mirrors. Each mirror forms an image of the source on the detector. The transfer function is calculated for a point source and various detectors comparable in size to the diffraction images. The path difference is introduced by moving either one or both of the mirrors.

034.129 Design of a celestial Thomson-scattering X-ray polarimeter. P. B. Landecker.
IEEE Trans. Nuclear Sci., Vol. NS-19, 463 - 475 (1972).
See Phys. Abstr., Vol. 75, No. 51795 (1972).

034.130 An electronic 'Brownie' in Martian orbit.
B. Shore.
Optical Spectra, (USA), Vol. 6, No. 5, p. 35 - 38 (1972).

034.131 Stellar interferometer. P. L. Kebabian.
Quarterly Progr. Rep. Res. Lab. Electron., Mass. Inst. Technology, No. 104, p. 84 - 87 (1972).

034.132 An experiment with an optical heterodyne interferometer. H. van de Stadt.
Optics Commun., Vol. 2, 153 - 156 = Utrechtse Sterrekundige Overdrukken, No. 144 (1970).
This paper describes an experimental verification of the theory of a heterodyne interferometer, applied to the measurement of the degree of coherence of a light source.

034.133 Rotating - mirror wavelength shifter.
N. Niewenhuijzen, C. Veth.
Optics and Laser Technol., 1971, p. 89 - 92 = Utrechtse Sterrekundige Overdrukken, No. 163 (1971).
A mechanical method is used to shift the wavelength of a laser light source over a small wavelength range. The device, constructed for use in the field of electronic spectroscopy, uses curved mirrors to give a good wavelength shift stability and a high duty cycle.

034.134 Una nuova elettronica per il fotometro a iride di Becker. G. Canton.
Mem. Soc. Astron. Italiana, Nuova Ser., Vol. 43, 475 - 480 (1972).

034.135 A satellite experiment to measure the intensity and energy spectrum of gamma rays from solar flares in the range 50-500 MeV. G. F. Bignami, C. J. Bland, O. Citterio, A. J. Dean, P. Inzani.
Nuclear Instruments Methods, (Netherlands), Vol. 103, 149 - 156 (1972).
A high energy solar gamma-ray telescope incorporating a lenticular Cherenkov counter for directional measurement and an energy calorimeter is described. The instrument is included in the payload of TD-1 ESRO spacecraft launched into a sun-pointing orbit during 1972. The results of laboratory and accelerator tests are presented and the sensitivity and measurement capability for solar flare gamma rays is discussed.

034.136 A two-element telescope of high collecting efficiency for sub-millimetre astronomy.
J. E. Beckman, J. A. Shaw.
Infrared Phys., Vol. 12, 219 - 234 (1972).
The special problems of the sub-millimetre region of the spectrum for observational astronomy are outlined and followed by a detailed description of the properties of the hot electron InSb bolometer.

034.137 On the calibration system of the amplifying tract of spherical ion trappers. S. K. Chapkunov.
Comptes Rendus Acad. Bulg. Sci., Vol. 25, 1093 - 1096 (1972).

034.138 A television photometer for use in astronomy.
W. M. Ruting.
Proc. Instn. Radio Electron. Engineers Australia, Vol. 33, No. 7, p. 336 (1972). – Abstract.

034.139 A high-resolution photoelectric spectrometer.
N. A. Dimov.
Izv. Krymskoj Astrofiz. Obs., Vol. 45, 182 - 187 (1972). In Russian.
A double-beam photon counting spectrometer at the coudé focus of the 2.6-meter telescope is described. Profiles of $\lambda 4250 \text{ Å}$ FeI and $\lambda 4254 \text{ Å}$ CrI lines have been registered for stars down to 6^m.

034.140 Discrete photoelectric tracking systems with storing of error signal. Z. N. Kuteva, Yu. A. Sabinin.
Izv. Krymskoj Astrofiz. Obs., Vol. 45, 196 - 202 (1972). In Russian.
There have been considered two variants of discrete photoelectric tracking systems with improved threshold sensitivity, providing tracking of a light source one order less bright compared to continuous systems.

034.141 A flexure test of the Blue Mesa spectrograph.
D. R. Hollars, T. B. Kirby.
Contr. Obs. New Mexico State Univ., Las Cruces, Vol. 1, 81 - 82 (1972).

034.142 Cold box design for a photoelectric photometer.
W. L. Reitmeyer.
Contr. Obs. New Mexico State Univ., Las Cruces, Vol. 1, 103 - 104 (1972).

034.143 Electronic shutter for planetary photography.
C. D. Houghton.
Contr. Obs. New Mexico State Univ., Las Cruces, Vol. 1, 107 - 108 (1972).

034.144 An optical-mechanical part of the automatic stellar electrophotometer. Z. D. Mestiashvili.
Byull. Abastumansk. Astrofiz. Obs., No. 43, p. 253 - 256 (1972). In Russian.
The construction of an optical-mechanical part of the automatic stellar electrophotometer is considered. Its replaceable parts are autonomous and move radially with respect to the optical axis. The advantage of such a design compared

with other present-day designs is stated.

034.145 **Intercomparison of the methods of photomultiplier photocurrent registration in measuring weak light signals.** Z. D. Mestiashvili.
Byull. Abastumansk. Astrofiz. Obs., No. 43, p. 257 - 269 (1972). In Russian.

The methods of weak light signals measurement are considered. Three methods of photocurrent measurements are compared. The noise sources and their contribution to the accuracy of measurement in determining photocurrents with a photomultiplier are considered. Experimentally, both in laboratory and during observation, two methods of photocurrent registration are intercompared: pulse counting and direct current measuring methods.

034.146 **Standard deviation of scatterometer measurements from space.** R. E. Fischer.
IEEE Trans. Geoscience Electronics, Vol. GE-10, 106 - 113 (1972).

034.147 **Koordinatenmeßgerät ASCORECORD 3 DP aus Jena.** H. G. Beck, R. Tietsch.
Jenaer Rundschau, (Jena Rev.),17. Jahrgang, p. 349 - 353 (1972).

034.148 **Apparatus for automatic guiding of the wire of a micrometer of the Kharkov meridian circle.**
M. G. Zuev.
Visn. Kharkiv. Univ. No. 82, (Ser. Astron., No. 7), p. 85 - 87 (1972). In Ukrainian.

034.149 **Investigation of the instrument UIM-21.**
P. P. Pavlenko.
Visn. Kharkiv. Univ. No. 82, (Ser. Astron., No. 7), p. 87 - 102 (1972). In Ukrainian.

034.150 **A contact micrometer for the Odessa meridian circle.** B. V. Novopashennyj.
Publ. 18th Astrometrical Conference 1969,(see 012.022), p. 175 - 178 (1972). In Russian.

034.151 **A visual–photoelectric device for astrometrical measurements.** B. K. Bagil'dinskij, V. D. Shkutov.
Publ. 18th Astrometrical Conference 1969, (see 012.022), p. 178 - 184 (1972). In Russian.

034.152 **A new device for determinations of the personal error in observations with the vertical circle.**
G. K. Zimmerman.
Publ. 18th Astrometrical Conference 1969, (see 012.022), p. 185 - 188 (1972). In Russian.

034.153 **Automatic device for measurements of the films with the circle readings.**
A. M. Bobrov, L. L. Vagushchenko, I. N. Nabokov, A. M. Stafeev.
Publ. 18th Astrometrical Conference 1969, (see 012.022), p. 208 - 214 (1972). In Russian.

034.154 **The electronic pendulum level.** H. Sandig.
Publ. 18th Astrometrical Conference 1969, (see 012.022), p. 222 - 226 (1972). In Russian.

034.155 **On temperature influences upon the level readings.**
V. Milovanovich.
Publ. 18th Astrometrical Conference 1969, (see 012.022), p. 226 - 229 (1972). In Russian.

034.156 **A system of automatic setting onto a star image.**
L. M. Zatsiorskij.
Publ. 18th Astrometrical Conference 1969, (see 012.022), p. 239 - 243 (1972). In Russian.

034.157 **Photographic registration of the positional circle and scale readings of the Repsold heliometer.**
A. S. Mamakov.
Publ. 18th Astrometrical Conference 1969, (see 012.022), p. 246 - 250 (1972). In Russian.

034.158 **A comet photometer for the amateur.** R. B. Minton.
Strolling Astronomer, Vol. 24, 14 - 17 (1972).

034.159 **Scanning electrospectrometer with photon counter and digit-printing output.**
S. Z. Omarov, M. S. Gadzhiev, P. N. Shustarev.
Astron. Tsirk., No. 738, p. 5 - 6 (1972). In Russian.

034.160 **A subtractive double pass spectrograph for solar observations.** P. Mein, M. Blondel.
Solar Physics, Vol. 27, 489 - 492 (1972).

Two-dimensional images in locally monochromatic light are obtained, using a subtractive double pass spectrograph. The spatial resolution is high and independent of the spectral bandwidth. The system is very suitable for fine structure observations in strong lines.

Fresnel-zone-plate spectrometer with central stop.
See Abstr. 031.029.

The data-handling problem with television recording of spectra. See Abstr. 031.048.

Guide to scientific instruments, 1972–73.
See Abstr. 031.064.

Die Instrumentation des Figl-Observatoriums für Astrophysik. See Abstr. 032.004.

High angular resolution solar observation from balloon borne instruments. See Abstr. 032.031.

Velocity aberration – application to laser measurements. See Abstr. 055.006.

A non imaging approach to solar oblateness measurements. See Abstr. 080.004.

A dual wavelength ground-based auroral scanner.
See Abstr. 084.041.

The prospects for mineral analysis by remote infrared spectroscopy. See Abstr. 094.015.

The application of objective gratings to faint star photometry. See Abstr. 113.025.

The application of Michelson interferometers to high resolution astronomical spectrometry.
See Abstr. 114.007.

An inexpensive data system for a rapid scanner.
See Abstr. 114.061.

Zur Praxis der photographischen Veränderlichen-Beobachtung. See Abstr. 120.007.

Polarization observations of nonstable stars and extragalactic objects. I. Equipment, method of observation and reduction. See Abstr. 131.122.

SEC vidicon system.
See Abstr. 141.509.

Techniques in balloon X-ray astronomy.
See Abstr. 142.008.

035 Clocks and Frequency Standards

035.001 Automatic starting and setting of a clock.
A. Staniforth.
Rev. Sci. Instruments, Vol. 41, 132 - 134 (1970). – Note.

035.002 A hybrid sundial. A. Ericson.
Sky Telescope, Vol. 44, 299, 305 (1972).

035.003 Advance-retard circuitry for digital clock.
M. L. Davis.
Contr. Obs. New Mexico State Univ., Las Cruces, Vol. 1, 105 - 106 (1972).

035.004 Comparison of precise clocks by the national television network. R. W. Archer.
Proc. Instn. Radio Electron. Engineers Australia, Vol. 33, No. 3, p. 97 - 100 (1972).
Comparison of atomic clocks in Sydney, Melbourne and Canberra using the national television network. – *DNC*

035.005 An electronic punch-chronograph.
V. A. Vyuchkov, A. F. Kunin, A. A. Sorokin, D. A. Shpagin.
Publ. 18th Astrometrical Conference 1969, (see 012.022), p. 230 - 234 (1972). In Russian.

Le monument solaire de Bagneux.
See Abstr. 004.040.

Le cadran « aux étoiles ». See Abstr. 004.042.

036 Photographic Auxiliaries

036.001 Neue Dimensionen der photographischen Technologie. H. Volkmann.
Umschau, 72. Jahrgang, p. 449 - 455 (1972).
A short survey is given of some recent developments which have increased the efficiency of optical photographic techniques and evaluation in the visible, as well as the ultraviolet and infrared, spectral ranges. After a consideration of the advances in the calculation of optical systems brought about by electronic calculators, special attention is given to new developments in astrophotography, microscopy and aerial photography.

036.002 Preliminary results of vacuum hypersensitization applied to several Kodak spectroscopic emulsions.
W. C. Miller.
AAS Photo-Bull., 1972, No. 2, p. 4 - 5, 19.

036.003 Electrography at Yale – A brief progress report.
R. J. Zinn, E. B. Newell.
AAS Photo-Bull., 1972, No. 2, (see 011.003), p. 6 - 7.

036.004 Kodak spectroscopic plate, type IIIa-J, applied to very low dispersion stellar spectra. A. A. Hoag.
AAS Photo-Bull., 1972, No. 2, (see 011.003), p. 12 - 13.

036.005 Response of Kodak spectroscopic plate, type IIIa-J, to baking in various controlled atmospheres.
A. G. Smith, H. W. Schrader, W. W. Richardson.
AAS Photo-Bull., 1972, No. 2, (see 011.003), p. 14 - 15.

036.006 Report of the subgroup on the storage of astronomical plates for archival purposes. W. F. van Altena.
AAS Photo-Bull., 1972, No. 2, (see 011.003), p. 15 - 17.

036.007 Indirekte Astrofarbenphotographie durch subtraktive Farbmischung. E. Brodkorb.
SuW, Vol. 11, 347 - 348 (1972).

Positional Astronomy. Celestial Mechanics

041 Positional Astronomy, Star Catalogues and Atlases

041.001 **Catalogue of proper motions of 12590 faint stars in the +25° to −20° declination zone.** L. F. Gorel.
Trudy Glav. Astron. Obs. Pulkovo, Ser. 2, Vol. 80, 5 - 172 (1972). In Russian.

The proper motions of 12590 stars were derived from the comparison catalogues: AGK3R, KSZ (+25° to −20°) observed with the Repsold meridian circle at Nikolayev Observatory during 1956–1965; AGK2 (+25° to −2°), Yale (−2° to −20°) photographic observations of the zone catalogues; AGK1 – the zone catalogues of the Astronomische Gesellschaft. The mean errors of annual proper motion are: for $\alpha \pm 0^s\!.0002 - 0^s\!.0003$, for $\delta \pm 0''\!.003 - 0''\!.005$. The derived μ_a and μ_δ were compared with the system proper motions of Yale and GC.

041.002 **Observations of Jupiter, Saturn, Uranus and Neptune made with the zonal astrograph at the Nikolayev Observatory during 1964–1967.** G. K. Gorel.
Trudy Glav. Astron. Obs. Pulkovo, Ser. 2, Vol. 80, 173 - 197 (1972). In Russian.

041.003 **Corrections to star catalogues from satellite observations.** C. A. Williams.
Monthly Notices Roy. Astron. Soc., Vol. 158, 125 - 149 (1972).

A theoretical investigation is made to explore the possibility of obtaining systematic corrections to star catalogues from observations of artificial satellites. A model is established to represent the system of equations formed when such corrections are determined simultaneously with corrections to the geocentric position of the observer. An analytical expression for the covariance matrix is set up for the two-dimensional (planar) case. The resulting correlations and weights are discussed.

041.004 **Modern aberration for observations of any date.** R. d'E. Atkinson.
Astron. Journ., Vol. 77, 512 - 517 = Publ. Goethe Link Obs., Indiana Univ., *Bloomington,* No. 138 (1972).

A short computer program is described which delivers modern values of the Besselian C and D for any desired dates and times, without interpolation between 0^h values and without requiring any input from the Ephemerides. The results have been compared with the Ephemeris at 5-day intervals from 1960 Jan. 0 to 1973 Jan. 1.

041.005 **Right ascensions in the FK 4 system by meridian observation with the Döllen method.**
S. Mancuso, L. Milano, E. Proverbio.
Astron. Astrophys., Vol. 19, 393 - 397 (1972).

Using independent values of clock corrections and instrumental azimuths, variations in RA positions of FK 4 stars are investigated. By means of two systems of equations, we have calculated the corrections Δa_a and Δa_δ for FK 4 referring to the local reference system (RS) in the sense RS–FK4. Comparison with other reference systems have been carried out.

041.006 **50 Jahre Geschichte des Fixsternhimmels.**
P. Aufgebauer, L. Brandt.
SuW, Vol. 11, 224 - 226 (1972).

041.007 **Bucharest KSZ catalogue of faint stars for 1950.0, declination zone −11° to +11°.**
Carried out at the meridian circle of Bucharest under the supervision of E. Marcus.
Publishing House of the Academy of the Socialist Republic of Romania, Bucharest. 364 pp. Price Lei 31.00 (1972). – Contents: (1) Bucharest catalogue of faint stars, *E. Marcus*; Connexion of the Bucharest KSZ right ascensions to the FK4 system, *T. Ionescu*; Connexion of the Bucharest KSZ declinations to the FK4 system, *M. Tudor*; Influence of the curvature of the parallel on declination observations at the Bucharest meridian circle, *T. Ionescu*; The collimation constant, *E. Toma*; Seasonal variation of the inclination to the equator of the meridian circle rotation axis, *P. Paraschiv*; Variation of the azimuth of the meridian circle during the period of KSZ observations, *M. Popa*; List of collaborators to the KSZ programme. (2) Catalogue of 3939 faint stars, declination zone −11° to +11°. Mean coordinates for the equinox 1950.0. (3) Mean coordinates 1950.0 for each observed position.

041.008 **Systematic errors $\Delta\alpha_\delta \cos \delta$ and $\Delta\alpha_\alpha \cos \delta$ of FK4 as compared with the General Catalogue of the USSR Time Services.** P. M. Afanasjeva, V. L. Gorshkov.
Astron. Tsirk., No. 684, p. 6 - 8 (1972). In Russian.

041.009 **Corrections to star catalogues from satellite observations.** C. A. Williams.
IAU Symposium No. 48, (see 012.004), p. 131 - 132 (1972).

041.010 **Some results from the automatic PZT at Richmond, Florida.** D. R. Monger.
IAU Symposium No. 48, (see 012.004), p. 136 - 138 (1972).

041.011 **New results of astronomical observations in Chile.** L. A. Panaiotov, K. N. Tavastsherna.
Vestn. AN SSSR, 1972, No. 3, p. 73 - 76. In Russian. – Abstr. in Referativ. Zhurn. 51. Astron., 9.51.141 (1972).

041.012 **On the accuracy of determination of the zero points of a fundamental catalogue from observations of major and minor planets.** D. P. Duma.
Astrometriya i Astrofiz., *Kiev*, No. 15, (see 003.001), p. 3 - 9 (1972). In Russian.

The calculations of weights of the equinox and equator corrections for five major and minor planets are described. It is shown that the weight of the equinox correction decreases essentially when the distance planet–earth increases. Hence, the observations of minor planets are less efficient than the observations of Venus, Mercury and Mars in determining this correction. But for the calculation of the correction to the equator point the observations of minor planets are more favourable than those of the above major ones. The conclusions are illustrated by the results of calculating these corrections.

041.013 **Zur dritten Auflage des Falkauer Atlasses.**
H. Vehrenberg.
SuW, Vol. 11, 282 - 285 (1972).

041.014 **Expansion of the annual aberration in series in terms of the fundamental arguments.**

V. S. Gubanov.
Astron. Zhurn. Akad. Nauk SSSR, Vol. 49, 1112 - 1122
(1972). In Russian. English translation in Soviet Astron. AJ,
Vol. 16, No. 5.

041.015 **On the errors in the circumpolar zone of the cata-
logue AGK2–AGK3.** V. V. Telnyuk-Adamchuk.
Vestn. Kiev. Un-ta, Ser. Astron., No. 14, p. 97 - 102 (1972).
In Russian.

It is shown that the circumpolar zone of AGK2–AGK3
contains essential systematic deviations from the system of
FK4. This conclusion is based on the results of the comparison of AGK2–AGK3 with seven catalogues (epochs from 1855
to 1951) and on the results of comparison of a catalogue of
520 circumpolar stars (system of AGK2–AGK3) with FK4.

041.016 **Astrometry with Schmidt telescopes.**
W. J. Luyten, A. E. La Bonte.
The role of Schmidt telescopes in astronomy. Conference
Hamburg 1972, (see 012.014), p. 33 - 38 (1972).

041.017 **Computational solution for positions on whole
Schmidt-plates.** Report on reduction of coordinates
measured on 22 plates of the Bergedorf Schmidt telescope.
W. Dieckvoss.
The role of Schmidt telescopes in astronomy. Conference
Hamburg 1972, (see 012.014), p. 39 - 43 (1972).

041.018 **Emploi du télescope de Schmidt en astrométrie.**
P. Bru, P. Lacroute.
The role of Schmidt telescopes in astronomy. Conference
Hamburg 1972, (see 012.014), p. 45 - 49 (1972).

041.019 **Astrometry with a Schmidt camera.**
G. van Herk.
The role of Schmidt telescopes in astronomy. Conference
Hamburg 1972, (see 012.014), p. 51 (1972).

041.020 **Contribution à la recherche d'un site d'observatoire
astrométrique.** F. Laclare.
Thesis Fac. Sci. Paris, Univ. Paris. Printed at Obs. Paris. 3 +
134 + 68 pp. (1969).
For the estabilishment of an astrometrical observatory in
France the decision on the site was made on the basis of data
gathered at six stations. The data include informations on
meteorological qualities which are analysed in the first part of
this paper. In the second part the results of astronomical
observations especially the accuracy of astrolabe observations
are presented and compared for the different stations.

041.021 **Conversion tables for standard epochs 1855.0 -
2000.0 matching to Atlas Stellarum 1950.**
H. Vehrenberg.
Treugesell-Verlag KG, Düsseldorf. 10 + 45 pp. (1972).
The "Conversion tables for standard epochs 1855 - 2000"
are part of Atlas Stellarum and are to be used for a quick
determination of approximate positions of stars and celestial
objects for the equinoxes 1855, 1875, 1900, 1925, 1975, and
2000. They should enable the user to determine the co-ordinates without calculation within the limits of accuracy of the
atlas, a task which would otherwise be very difficult and time
consuming.

041.022 **Fundamental systems of positions and proper motions.** W. Fricke.
Annual Rev. Astron. Astrophys., Vol. 10, (see 003.005), 101 -
128 (1972). – Contents: (1) Introduction; (2) Principles of
establishing a fundamental frame of reference; (3) Fundamental
declinations and right ascensions; (4) New instrumental techniques; (5) The Fourth Fundamental Catalogue (FK4); (6)
Comparisons of the FK4 with other systems; (7) Determina-

tions of precession; (8) Suggested precessional corrections;
(9) Contributions to the improvement of the FK4; (10) Photographic reference systems; (11) Proper motions with respect
to galaxies; (12) Positions of faint objects; (13) Absolute positions of radio sources.

041.023 **Rigorous reduction of the catalogue of right ascensions of 589 FKSZ stars, observed at Moscow
(1953 - 1958), to the FK4 system.** L. M. Khommik.
Soobshch. Gos. Astron. Inst. Shternberga, No. 178, p. 18 - 33
(1972). In Russian.

041.024 **The optimum distribution of reference stars for one
case of reduction of observations of right ascensions.**
A. P. Gulyaev.
Soobshch. Gos. Astron. Inst. Shternberga, No. 178, p. 34 - 39
(1972). In Russian.

041.025 **About the determination of the value of one revolution of the R. A. micrometer.** V. G. Shamaev.
Soobshch. Gos. Astron. Inst. Shternberga, No. 178, p. 40 - 44
(1972). In Russian.

041.026 **Observations of right ascensions of the sun and
major planets on the Moscow meridian circle GOMZ.**
L. M. Khommik, O. A. Kozina, N. N. Kabaeva, A. P. Gulyaev.
Soobshch. Gos. Astron. Inst. Shternberga, No. 178, p. 45 - 47
(1972). In Russian.
Observations of the sun, Mercury and Venus in 1969.

041.027 **Some questions of reducing observations of declinations with the ZTL-180.** I. M. Kalinina.
Soobshch. Gos. Astron. Inst. Shternberga, No. 179, p. 40 - 45
(1972). In Russian.
Preliminary analysis of a 2.5-years series of declination
observations at the Moscow zenith telescope.

041.028 **Testing a method of successive approximations in
reducing right ascensions.** A. P. Gulyaev.
Soobshch. Gos. Astron. Inst. Shternberga, No. 179, p. 46 - 51
(1972). In Russian.

041.029 **The system of declinations of the GOMZ meridian
circle of the Moscow Observatory.**
N. N. Kabaeva.
Soobshch. Gos. Astron. Inst. Shternberga, No. 179, p. 52 - 54
(1972). In Russian.

041.030 **Investigation of the magnitude equation from plates
of the 15'' astrograph.** L. P. Panteleeva.
Soobshch. Gos. Astron. Inst. Shternberga, No. 179, p. 55 - 60
(1972). In Russian.

041.031 **Calculation of the precession and nutation in rectangular equatorial coordinates on electronic computers.**
V. A. Izvekov.
Byull. Inst. Teoret. Astron., *Leningrad,* Vol. 13, 210 - 214
(1972). In Russian.
Rigorous formulae of precession and nutation in rectangular coordinates are derived. Displacements of the poles of the
equator and the ecliptic are considered to be known. Approximate formulae representing the elements of matrices of nuta-
tion including third orders in $\Delta\psi$ and $\Delta\epsilon$ are obtained. A short
description is given for the programme to calculate precession
and nutation on the BESM-4 electronic computer with an
accuracy up to 10^{-10}.

041.032 **Results of observations made with the Danjon astro-
labe at Santiago, 1967.** F. Noel, K. Czuia, P. Guerra.
Publ. Dep. Astron., Univ. Chile, Obs. Astron. Nacional, Cerro
Calán, Santiago de Chile, Vol. 2, (No. 1), 1 - 10 (1970).

The results in time, latitude, and altitude obtained during 1967, with the Danjon astrolabe at the Observatorio Astronómico Nacional, Cerro Calán, Santiago de Chile, are given.

041.033 **Ecliptic coordinates of selected hot stars.**
L. Houziaux, A. Ringuelet.
Univ. Mons, Dép. Astrophys., Commun. No. 27, 53 pp. (1972)

This table lists the 1970.0 ecliptic coordinates λ, β of 1789 stars. The latter have been selected according to their expected brightness around 2000 Å. It comprises thus mostly hot objects since the V magnitude limit has been set up at 6.5 for the O type, at 6 for the B type, at 5.5 for the A type, at 5 for the F type, and at 3 for the G type. These criteria correspond roughly to a monochromatic flux of 5×10^{-11} erg cm^{-2} sec^{-1} A^{-1} at 2000 Å, except for the F stars, where the threshold is somewhat lower. 1970.0 ecliptic coordinates of all stars in the Smithsonian Astrophysical Observatory Star Catalog have been computed and rearranged according to increasing longitudes.

041.034 **Finding list for Bayer and Flamsteed stars in the "Ecliptic coordinates of selected hot stars" catalogue.** G. Houziaux.
Univ. Mons. Dép. Astrophys. Commun. No. 31, 2 + 9 pp. (1972).

In order to locate easily bright stars with Bayer or Flamsteed designations in the "Ecliptic coordinates of selected hot stars" catalogue (L. Houziaux, A. Ringuelet, see 041.033), we have rearranged these stars by constellations, given in alphabetical order. For each object, we give the 1970.0 ecliptic longitude. Among the 1789 stars of the list in ecliptic coordinates, 1126 appear in the Bayer or/and Flamsteed catalogues.

041.035 **Positions de 282 étoiles FK3.** V. Maitre.
Ann. Obs. Besançon, Univ. Besançon, Vol. 8, (Fasc. 5), 222 - 231 (1972).

Les positions d'étoiles FK3 données dans un tableau résultent des observations méridiennes effectuées à l'Observatoire de Besançon entre le 6 mars 1946 et le 14 mai 1957. Elles ont servi à l'établissement du deuxième catalogue méridien de Besançon et à quelques observations méridiennes de planètes; ces dernières ont été publiées au Journal des Observateurs.

041.036 **Coordonnées écliptiques de 48 étoiles zodiacales.**
J. Denoyelle, J. Meeus.
Ciel et Terre, Vol. 88, 467 - 473 (1972).

Celestial longitudes and latitudes of 48 bright zodiacal stars, referred to the mean equinox of data, are given from 1600 B.C. to A.D. 2800.

041.037 **A spectral analysis of errors of Δa_a type.**
G. P. Pilnik.
Astron. Zhurn. Akad. Nauk SSSR, Vol. 49, 1328 - 1330 (1972). In Russian. English translation in Soviet Astron. AJ, Vol. 16, No. 6.

Catalogue errors are determined by the chain method using Time Service observations. Spectral functions of Δa_a errors are given for FK3, N30 and FK4.

041.038 **Application of an electronic computer to the deduction of a differential catalogue.** E. V. Khrutskaya.
Astron. Zhurn. Akad. Nauk SSSR, Vol. 49, 1330 - 1331 (1972). In Russian. English translation in Soviet Astron. AJ, Vol. 16, No. 6. - Short note.

041.039 **New results of determining precise declinations of stars by means of a zenith-telescope.**
L. S. Bratoljubova.
Astron. Zhurn. Akad. Nauk SSSR, Vol. 49, 1331 - 1333

(1972). In Russian. English translation in Soviet Astron. AJ, Vol. 16, No. 6.

A catalogue of declinations of 30 circumzenithal stars of the Poltava Observatory is given.

041.040 **De nærmeste stjerner.** O. Havnes.
Astron. Tidssk., Årg. 5, p. 168 - 173 (1972).

041.041 **Observed positions of Jupiter III (Ganymede) and corrections to the tabular place of the star SAO 186800.** T. W. Russo.
Monthly Notes Astron. Soc. Southern Africa, Vol. 31, 157 - 158 (1972). - Note.

041.042 **Über das stochastische Verhalten von photographisch bestimmten Stern- und Satellitenkoordinaten.** G. Seeber.
Deutsche Geod. Kommission, Bayer. Akad. Wiss., München, Ser. C, No. 178, 4 + 116 pp. (1972). - Diss. Rheinisch. Friedrich-Wilhelms-Univ. Bonn.

041.043 **Las rotaciones en astronomía.**
J. M. Correas Dobato.
Urania Barcelona, Año 56, No. 274, p. 159 - 166 (1971).

041.044 **Comparison between MD and GC catalogues and analysis of the proper motion errors.**
E. Proverbio, A. Poma.
Mem. Soc. Astron. Italiana, Nuova Ser., Vol. 43, 413 - 422 (1972).

In comparing the catalogue of Melchior and Dejaiffe with that of Boss, if some working hypotheses are admitted, the proper motion errors of Boss's catalogue can be determined. Statistical analysis of the residuals between the proper motion corrections observed and calculated roughly confirms the hypotheses on which the work is based and points up the existence of a systematic difference between the two catalogues. This difference can be explained by an error of the order of $+ 0\overset{''}{.}03$ in the annual precessional constant in declination.

041.045 **Geocentric solar data.** G. H. Kaplan.
United States Naval Obs., *Washington, D.C.,* Circ. No. 138, 10 pp. (1972).

This Circular contains geocentric solar data for each day in 1978 at 0^h Universal (Greenwich) Time, referred to the equator and equinox of date. These data may be used to determine the position of the sun in the sky at a given date and time in any year, with low accuracy.

041.046 **Astrometric ephemeris of Pluto 1970–1990.**
G. H. Kaplan, P. K. Seidelmann, E. Smith.
United States Naval Obs., *Washington, D.C.,* Circ. No. 139, 45 pp. (1972).

Based on the new integrated ephemeris of Pluto (Cohen, Hubbard, Oesterwinter, 1967), the astrometric ephemeris of Pluto was calculated. The method of calculating the ephemeris differs from that illustrated in the *Explanatory Supplement* in that the aberration corrections are applied in rectangular coordinates, and conversion to spherical coordinates is only made at the end.

041.047 **Die Anwendung der Rechentechnik in der praktischen Astronomie.**
V. S. Gubanov, D. D. Polozhentsev, M. S. Chubej.
Wiss. Zeitschr. Techn. Univ. Dresden, Vol. 21, No. 3, (see 012.021), 589 - 591 (1972).

041.048 **Bestimmung der Richtungskorrektionen und der Meridiandurchgangszeit aus den Zapfenunregelmäßigkeiten mittels Zapfenabweichungen.**

J. Kabeláč, L. Zajíček.
Wiss. Zeitschr. Techn. Univ. Dresden, Vol. 21, No. 3, (see 012.021), 592 - 595 (1972).

041.049 Astronomical determinations with the universal instrument DKM-3A.   B. N. D'yakov.
Geod. i kartografiya, 1972, No. 9, p. 23 - 27. In Russian.
Abstr. in Referativ. Zhurn. 52. Geod. Aehros"emka, 2.52.88 (1973).

041.050 Determination of right ascensions of major planets with the meridian circle of the Kharkov Astronomical Observatory in 1968–1970.
K. M. Derkach, M. G. Zuev.
Visn. Kharkiv. Univ. No. 82, (Ser. Astron., No. 7), p. 66 - 69 (1972). In Ukrainian.

041.051 Report of the astrometric commission of the USSR Astronomical Council.   M. S. Zverev.
Publ. 18th Astrometrical Conference 1969, (see 012.022), p. 5 - 21 (1972). In Russian.

041.052 Present state and perspectives of fundamental astrometry.   A. A. Nemiro.
Publ. 18th Astrometrical Conference 1969, (see 012.022), p. 22 - 29 (1972). In Russian.

041.053 Determination of the zero-points of a star catalogue from observations of artificial earth satellites.
D. P. Duma.
Publ. 18th Astrometrical Conference 1969, (see 012.022), p. 30 - 37 (1972). In Russian.

041.054 On the determination of catalogue errors from observations of minor planets.   V. I. Orelskaya.
Publ. 18th Astrometrical Conference 1969, (see 012.022), p. 43 - 46 (1972). In Russian.

041.055 Positional observations with the 40 - cm astrograph of the Crimean Astrophysical Observatory.
L. I. Chernykh.
Publ. 18th Astrometrical Conference 1969, (see 012.022), p. 46 - 49 (1972). In Russian.

041.056 Twelve-year series of observations (1953 - 1964) with the Pulkovo polar tube.   N. M. Bakhrakh.
Publ. 18th Astrometrical Conference 1969, (see 012.022), p. 49 - 53 (1972). In Russian.

041.057 Some results of day-time observations with the large transit instrument at Pulkovo.   E. Bem.
Publ. 18th Astrometrical Conference 1969, (see 012.022), p. 60 - 65 (1972). In Russian.

041.058 Errors of star catalogues and the TU2 time system.   L. A. Roze.
Publ. 18th Astrometrical Conference 1969, (see 012.022), p. 65 - 67 (1972). In Russian.

041.059 Analytic representation of differences O − C obtained with the Zeiss transit instrument at the Bukarest observatory.   L. Rusu.
Publ. 18th Astrometrical Conference 1969, (see 012.022), p. 68 - 71 (1972). In Russian.

041.060 Systematic errors of the zenith distances of stars obtained with the meridian circle of the Engelhardt Observatory.   A. I. Nefed'eva.
Publ. 18th Astrometrical Conference 1969, (see 012.022), p. 71 - 76 (1972). In Russian.

041.061 Statistical evaluation of the results of latitude observations with a prismatic astrolabe at Pulkovo during 1963.2 - 1968.7.   N. P. Godisov.
Publ. 18th Astrometrical Conference 1969, (see 012.022), p. 76 - 79 (1972). In Russian.

041.062 Some results of absolute determinations of declinations for the southern sky.   K. N. Tavastsherna.
Publ. 18th Astrometrical Conference 1969, (see 012.022), p. 80 - 83 (1972). In Russian.

041.063 Some results of observations with the Zeiss transit instrument at the Cerro-Calan Observatory (Chile).
V. N. Shishkina.
Publ. 18th Astrometrical Conference 1969, (see 012.022), p. 84 - 89 (1972). In Russian.

041.064 On systematic and accidental differences of star catalogues.   A. N. Kur'yanova, E. P. Fedorov.
Publ. 18th Astrometrical Conference 1969, (see 012.022), p. 94 - 107 (1972). In Russian.

041.065 Estimate of the accuracy of a new method of reduction of meridian observations from comparison with the AGK3R catalogue.   A. A. Izvekova.
Publ. 18th Astrometrical Conference 1969, (see 012.022), p. 107 - 118 (1972). In Russian.

041.066 On the principle of systematic homogeneity of instrumental parameters in differential determinations of right ascensions.   L. M. Khommik.
Publ. 18th Astrometrical Conference 1969, (see 012.022), p. 119 - 126 (1972). In Russian.

041.067 Improvement of an initial catalogue by means of coordinates of the stars under determination.
A. P. Gulyaev.
Publ. 18th Astrometrical Conference 1969, (see 012.022), p. 126 - 130 (1972). In Russian.

041.068 Distribution of the accidental errors for the Goloseyevo catalogue of declinations of latitude stars.   A. S. Kharin, Y. S. Yatskiv.
Publ. 18th Astrometrical Conference 1969, (see 012.022), p. 130 - 133 (1972). In Russian.

041.069 Absolute determinations of equatorial star declinations by micrometric measurements.
E. I. Krejnin, S. A. Murry.
Publ. 18th Astrometrical Conference 1969, (see 012.022), p. 133 - 142 (1972). In Russian.

041.070 Methods of determination of star declinations with a zenith telescope.   S. V. Drozdov.
Publ. 18th Astrometrical Conference 1969, (see 012.022), p. 142 - 145 (1972). In Russian.

041.071 On the declinations of the latitude program stars as determined with the zenith telescope.
I. M. Kalinina.
Publ. 18th Astrometrical Conference 1969, (see 012.022), p. 146 - 151 (1972). In Russian.

041.072 Automation of observations of right ascensions of stars.   A. M. Stafeev, L. A. Vagushchenko.
Publ. 18th Astrometrical Conference 1969, (see 012.022), p. 165 - 169 (1972). In Russian.

041.073 Photoelectric registration of observations with the transit instrument.   Ya. Valikhevich.
Publ. 18th Astrometrical Conference 1969, (see 012.022), p.

169 - 172 (1972). In Russian.

041.074   **On the influence of the position of the horizontal axis on the determination of zenith distances by means of a vertical circle.**   G. Teleki, M. Miyatov.
Publ. 18th Astrometrical Conference 1969, (see 012.022), p. 201 - 206 (1972). In Russian.

041.075   **Preliminary system of declinations for the new meridian circle of the Sternberg Institute.**
N. N. Kabaeva.
Publ. 18th Astrometrical Conference 1969, (see 012.022), p. 206 - 207 (1972). In Russian.

041.076   **The present-day state of the problem on calculation of refraction in astronomical observations.**
I. G. Kolchinskij.
Publ. 18th Astrometrical Conference 1969, (see 012.022), p. 250 - 261 (1972). In Russian.

041.077   **The mean refraction and periodic atmospherical oscillations.**   E. Bem.
Publ. 18th Astrometrical Conference 1969, (see 012.022), p. 261 - 263 (1972). In Russian.

041.078   **On the determination of the influence of anomalous refraction.**   G. Teleki.
Publ. 18th Astrometrical Conference 1969, (see 012.022), p. 264 - 270 (1972). In Russian.

041.079   **Investigations of lateral refraction at the Bukarest Observatory.**   E. Marcus.
Publ. 18th Astrometrical Conference 1969, (see 012.022), p. 275 - 280 (1972). In Russian.

041.080   **On the influence of aberrations and decentering of optical systems upon astrometrical observations.**
B. K. Bagil'dinskij, T. S. Belorossova, N. V. Merman.
Publ. 18th Astrometrical Conference 1969, (see 012.022), p. 330 - 346 (1972). In Russian.

041.081   **International Information Bureau on Astronomical Ephemerides.**
Bull. B.I.I.E.A. (IAU–COSPAR), Paris. Information cards, Nos. 30 - 50 (1972).

Astronomical applications of differential interferometry.   See Abstr. 031.053.

Re-examination of the reduction method for PZT data.   See Abstr. 031.062.

Studio del misuratore di coordinate Hauser.
See Abstr. 034.114.

Plate reduction for the stellar triangulation.
See Abstr. 046.035.

On the determination of anomalous refraction out of astrometrical measurements in the zenith zone.
See Abstr. 082.041.

Photographic observations of Mars at the Main Astronomical Observatory of the Ukrainian Academy of Sciences in 1963–1967.   See Abstr. 097.068.

Accurate positions of the planet Pluto in the years 1969–1970.   See Abstr. 101.004.

Richtungsabhängige Genauigkeitsunterschiede bei Eigenbewegungen.   See Abstr. 112.005.

The results of determination of absolute proper motions of stars with respect to galaxies.   See Abstr. 112.015.

A comparison of photoelectric and visual measurements of plates with the double star ADS 7251.
See Abstr. 118.018.

Optical positions for 21 3C objects.
See Abstr. 141.078.

042 Celestial Mechanics

042.001 A method of integration of the equations of planetary motion in rectangular coordinates. The Pluto perturbations from Neptune. L. K. Babadzhanyanz.
Nablyud. Iskusstv. Nebesn. Tel, No. 62, p. 5 - 21 (1971).
In Russian.

042.002 An approximate analytic theory of Trojan motion. G. M. Bazhenov.
Nablyud. Iskusstv. Nebesn. Tel, No. 62, p. 22 - 30 (1971).
In Russian.

042.003 Trigonometric linear second-order theory of the secular perturbations in the motion of the major planets. V. A. Brumberg, A. V. Egorova.
Nablyud. Iskusstv. Nebesn. Tel, No. 62, p. 42 - 72 (1971).
In Russian.

042.004 The determination of the perturbations by Laplace-Newcomb's method using high-speed computers.
N. A. Budnikova.
Nablyud. Iskusstv. Nebesn. Tel, No. 62, p. 73 - 90 (1971).
In Russian.

042.005 Sur le développement du potentiel de la terre et de la lune en fonctions de Lamé.
D. V. Zagrebin.
Nablyud. Iskusstv. Nebesn. Tel, No. 62, p. 91 - 92 (1971).
In Russian.

042.006 Secular perturbations of the major planets. G. A. Krasinsky, L. J. Pius.
Nablyud. Iskusstv. Nebesn. Tel, No. 62, p. 93 - 112 (1971).
In Russian.

042.007 Expansions of the disturbing function. M. S. Petrovskaya.
Nablyud. Iskusstv. Nebesn. Tel, No. 62, p. 122 - 133 (1971).
In Russian.

042.008 Possible applications of Hansen's method of partial anomalies. V. I. Skripnichenko.
Nablyud. Iskusstv. Nebesn. Tel, No. 62, p. 134 - 137 (1971).
In Russian.

042.009 On periodic solutions of the generalised restricted three-body problem near libration solutions.
T. K. Shinkarik.
Nablyud. Iskusstv. Nebesn. Tel, No. 62, p. 169 - 181 (1971).
In Russian.

042.010 On expansions of the disturbing function in the case of close commensurability of the mean motions. M. S. Petrovskaya.
Byull. Inst. Teoret. Astron., *Leningrad*, Vol. 13, 152 - 156 (1972). In Russian.
The expansion of the perturbation function derived previously (Petrovskaya, 1970) has been used to construct special series for the case of close commensurability of the mean motions of the bodies. The series are expected to converge rapidly as compared with the general two-argument expansion.

042.011 On the stability of motion in the restricted three-body problem near the collinear libration points taking into account light pressure. E. P. Filianskaya.
Byull. Inst. Teoret. Astron., *Leningrad*, Vol. 13, 157 - 159 (1972). In Russian.

The paper deals with the problem of motion of an infinitely small body in the gravitation field of the sun and earth, the pressure of radiation being taken into consideration. Collinear libration solutions are proved to be devoid of absolute stability.

042.012 Comparison of the classical and the global solutions of the Ideal Resonance Problem. B. Garfinkel.
Celestial Mechanics, Vol. 5, 451 - 469 (1972).
The Ideal Resonance Problem is defined by the Hamiltonian $F = B(y) + 2\epsilon A(y)\sin^2 x$, $\epsilon \ll 1$. The classical solution of the Problem, expanded in powers of ϵ, carries the derivative B' as a divisor and is, therefore, singular at the zero of B', associated with resonance. With α denoting the resonance parameter, it is shown that the classical solution is valid only for $\alpha^2 \geqslant 0(1/\mu)$. In contrast, the global solution (Garfinkel *et al.*, 1971), expanded in powers of $\mu = \epsilon^{1/2}$, removes the classical singularity at $B' = 0$, and is valid for all α. The two solutions are compared with regard to their general behavior and their accuracy.

042.013 A two-parameter survey of periodic orbits in the restricted problem of three bodies.
P. J. Shelus.
Celestial Mechanics, Vol. 5, 483 - 489 (1972).
Within the context of the restricted problem of three bodies, we wish to show the effects, caused by varying the mass ratio of the primaries and the eccentricity of their orbits, upon periodic orbits of the infinitesimal mass that are numerical continuations of circular orbits in the ordinary problem of two bodies. Seven such families (comprised of a total of more than 2000 orbits) with equally spaced mass ratios from 0.0 to 1.0 and eccentricities of the orbits of the primaries in a range 0.0 to 0.6 are investigated.

042.014 Numerical stabilization of the differential equations of Keplerian motion. J. Baumgarte.
Celestial Mechanics, Vol. 5, 490 - 501 (1972).
A stabilization of the classical equations of two-body motion is offered. It is characterized by the use of the regularizing independent variable (eccentric anomaly) and by the addition of a control-term to the differential equations.

042.015 A nonlinear stability problem in the three-dimensional restricted three-body problem.
K. T. Alfriend.
Celestial Mechanics, Vol. 5, 502 - 511 (1972).
The question of whether or not there is a transfer of energy between the in-plane motion and out-of-plane motion in the neighborhood of L_4 in the restricted problem of three bodies is investigated. The in-plane motion is assumed to be finite and the out-of-plane motion to be infinitesimal.

042.016 Single collision periodic orbits of a new type.
G. Bozis, G. Antonacopoulos.
Astron. Astrophys., Vol. 20, 73 - 77 (1972).
A method is described for establishing collision periodic orbits in the restricted problem for which the velocity of the massless moving point at one of the primaries forms angles different from 0° or 180° with the positive x-axis. Samples of orbit families are shown grafically.

042.017 On the dynamical effects of a distortion of the trajectories of particles in the problem of two bodies with corpuscular radiation. T. B. Omarov.
Astron. Zhurn. Akad. Nauk SSSR, Vol. 49, 872 - 878 (1972).
In Russian. English translation in Soviet Astron. AJ, Vol. 16,

No. 4.

The relative motion of two corpuscularly active bodies with due regard for a gravitational interaction between these bodies and the system of eruptive masses is considered. It is supposed that moderately large speed of a particles emanation from the surface of these bodies is independent of time.

042.018 On the problem of the expansion of the disturbing function. E. I. Timoshkova.
Astron. Zhurn. Akad. Nauk SSSR, Vol. 49, 879 - 885 (1972). In Russian. English translation in Soviet Astron. AJ, Vol. 16, No. 4.

Using the addition theorem for the associated Legendre functions, an expansion of the disturbing function is obtained. Separately the inclination function is considered and a number of properties and equalities is derived.

042.019 Generalization of the sphere of interaction in the restricted problem of four bodies. A. Drożyner.
Postępy Astron., Vol. 20, 261 - 264 (1972). In Polish.

Assuming a four-body system in the form proposed by Su-Shu-Huang the sphere of interaction of the moon relative to the earth and sun has been determined. A physical interpretation of the changes of the dimension of this sphere is also given.

042.020 Celestial mechanics in the twentieth century. J. S. Griffith.
Journ. Roy. Astron. Soc. Canada, Vol. 66, 193 - 200 (1972).

042.021 A determination of the motion of the ecliptic. R. E. Laubscher.
Astron. Astrophys., Vol. 20, 407 - 414 (1972).

This paper provides a discussion of the theoretical motion of the ecliptic based on improved values for the reciprocal planetary masses. Power series in time for the position of the pole of the ecliptic are derived which then yield the corrections $+0\rlap{.}''036 \pm 0\rlap{.}''006$ per century and $-0\rlap{.}''029 \pm 0\rlap{.}''010$ per century at 1850.0 to Newcomb's values of the secular variation of the obliquity and the first-order term in the planetary precession, respectively. The derivation of the cited error estimates is discussed in detail.

042.022 The effects of viscous friction on the precession and nutation of celestial bodies. Z. Kopal.
Astrophys. Space Sci., Vol. 16, 347 - 371 (1972).

The aim of the present study has been to set the system of differential equations which govern the precession and nutation of self-gravitating globes of compressible viscous fluid, due to the attraction exerted on the rotating configuration by its companion; and to construct their approximate solution which are correct to terms of the second order in small dependent variables of the problem. The physical significance of the new results will be discussed. It is demonstrated that the axes of rotation of deformable components in close binary systems are initially inclined to the orbital plane, viscous dissipation produced by dynamical tides will tend secularly to 'rectify' their positions until perpendicularity to the orbital plane has been established. An application of the results of the present study to the dynamics of the earth-moon system discloses that the observed inclination of $1\rlap{.}°5$ of the lunar equator to the ecliptic cannot be regarded as being secularly constant, but representing the present deviations from perpendicularity of oscillatory motion of very long period.

042.023 A numerical method of integration by means of Taylor-Steffensen series and its possible use in the study of the motions of comets and minor planets.
V. F. Myachin, O. A. Sizova.
IAU Symposium No. 45, (see 012.003), p. 83 - 85 (1972).

042.024 A library of standard programmes for constructing

numerical theories for studying the motion and evolution of the orbits of the minor bodies of the solar system.
N. A. Bokhan.
IAU Symposium No. 45, (see 012.003), p. 86 - 89 (1972).

042.025 Capture in the restricted three-body problem. G. Horedt.
Acta Astron., Vol. 22, 55 - 66 (1972).

The paper deals with close single encounters between massless particles and the less massive component of a binary system which revolves on a circular orbit. The Laplace jovicentric method is used to determine the probabilities of collisions and the probabilities that after the encounter the orbit becomes a hyperbola, direct ellipse, or a retrograde ellipse. For the elliptic orbits the mean values and root-mean-square deviations of major semi-axes, eccentricities and inclinations are determined.

042.026 Periodic perturbation of the libration points of the restricted three-body problem due to presence of a resisting medium and both gravitational and radiative fields of a fourth body. V. Matas.
Bull. Astron. Inst. Czechoslovakia, Vol. 23, 262 - 265 (1972).

This paper is concerned with the existence of periodic solutions around the disturbed libration points of the restricted three-body problem. The existence of periodic solutions of an infinitesimal body has been proved provided that a perturbation due to a resisting medium appeared as well as the perturbations owing to both gravitational and radiative effects of a fourth body.

042.027 Application of the restricted hyperbolic three-body problem to a star-sun-comet system.
M. B. Faintich.
Celestial Mechanics, Vol. 6, 22 - 26 (1972).

The equations of motion for a third body of small mass are developed in the problem where the two primary bodies are in hyperbolic orbits about each other. The equations are applied to a hypothetical star-sun-comet system to determine the effect of the stellar encounter on the orbit of the comet.

042.028 About the first integrals of the generalized problem of translatory-rotary motion of rigid bodies.
G. N. Duboshin.
Celestial Mechanics, Vol. 6, 27 - 39 (1972).

The generalized problem of translatory-rotatory motion of rigid bodies, whose elementary particles act upon each other according to arbitrary laws of forces along the straight line joining them, is discussed. It is shown that the first integrals for this general problem, analogous to the integrals of the problem of the translatory-rotatory motion of rigid bodies, whose elementary particles act according to the Newtonian law, exist under certain well known conditions.

042.029 (E, r, α, β) summability in the hyperbolic two-body problem. G. Zelmer.
Celestial Mechanics, Vol. 6, 40 - 43 (1972).

An (E, r, α, β) summation method is a summability procedure which can be used for purposes of analytic continuation. In this paper the solutions of the hyperbolic two-body problem are used as a test case for the methods $(E, r, 0, 1)$ and $(E, r, 0, 2)$ and effective continuation of the solutions is obtained.

042.030 Motor integrals of a generalized Kepler motion. P. Kustaanheimo.
Celestial Mechanics, Vol. 6, 52 - 59 (1972).

It is shown that there appear, also in the non-relativistic case, interesting new integrals as well as a natural generalization for the Kepler motion, when the motor method is applied.

042.031 Critical inclinations in planetary problems.

G. A. Krasinsky.
Celestial Mechanics, Vol. 6, 60 - 83 (1972).

The general conception of the critical inclinations and eccentricities for the N-planet problem is introduced. The connection of this conception with the existence and stability of particular solutions is established. In the restricted circular problem of three bodies the existence of the critical inclinations is proved for any values of the ratio of semi-major axes α. The asymptotic behaviour of the critical inclinations as $\alpha \to 1$ is investigated.

042.032 **Mass effects in the problem of three bodies.**
        V. Szebehely.
Celestial Mechanics, Vol. 6, 84 - 107 (1972).

This is a study of the dynamical behavior of three point masses moving under their mutual gravitational attraction in a plane. The initial positions and velocities are identical for all cases studied and only the masses of the participating bodies change in the series of numerical experiments. In this way the effect of the coupling terms in the differential equations of motion are investigated. The motion in all 125 cases begins with an interplay between the three bodies, followed by temporary ejections or by an eventual escape. The total mass of the system is kept constant while the mass-ratios change from 1 to 5. The initial velocities being zero, the total energy is negative in all cases. The numerical results are tabulated regarding escape time, ejection period, total energy, escape energy, terminal velocity, semi-major axis, and eccentricity.

042.033 **A new proof of the conditions for a canonical transformation.**   J. A. Breves Filho.
Celestial Mechanics, Vol. 6, 108 - 110 (1972).

042.034 **Normality condition in the ideal resonance problem.**
        B. Garfinkel.
Celestial Mechanics, Vol. 6, 151 - 166 (1972).

The ideal resonance problem is defined by the Hamiltonian $F = B(y) + 2\mu^2 A(y) \sin^2 x$, $\mu \ll 1$. The paper derives the normality condition for a first-order solution. The results are applied to the problem of the critical inclination of a satellite of an oblate planet.

042.035 **Possible effects of anisotropy of G on celestial orbits.**   J. P. Vinti.
Celestial Mechanics, Vol. 6, 198 - 207 (1972).

Will (1971) has discussed the possibility of detecting anisotropy in the gravitational constant G, by means of laboratory experiments. In the present note I try to see what astronomical consequences would arise from the special anisotropy that he considers.

042.036 **On a new form for the differential equations of relative motion of the three-body problem.**
C. A. Altavista.
Celestial Mechanics, Vol. 6, 208 - 213 (1972).

A new method for the development of the disturbing function of the three-body problem is outlined in this paper. A special process is devised to get the distance Δ between two planets P_1 and P_2 in terms of their heliocentric distances. It is then shown that the differential equations of relative motion of this problem can be brought in an homogeneous set of differential equations.

042.037 **Le développement du potentiel dans le cas d'une densité lisse.**   C. Kholchevnikov.
Celestial Mechanics, Vol. 6, 214 - 220 (1972).

The potential of a body of revolution is expanded in a series of spherical functions. It is proved that for a body of smooth structure the coefficients of expansion decrease in a power law.

042.038 **A new regularization of the planar problem of three bodies.**   J. Waldvogel.
Celestial Mechanics, Vol. 6, 221 - 231 (1972).

A new method of simultaneously regularizing the three types of binary collisions in the planar problem of three bodies is developed: The coordinates are transformed by means of certain fourth degree polynomials, and a new independent variable is introduced.

042.039 **Single close encounters in the planetary problem.**   G. Horedt.
Celestial Mechanics, Vol. 6, 232 - 241 (1972).

We consider a large mass M and two small masses m_1 and m_2 $(m_1 \approx m_2; m_1, m_2 \ll M)$. The orbit of m_1 is initially circular and the motion of m_2 hyperbolic with respect to M. An approximative method, similar to the theory of stellar encounters, is used to determine the probabilities of collisions, hyperbolas, direct and retrograde ellipses, as well as the mean values of the semimajor axes and their root mean square deviation after the encounter.

042.040 **Computation of Hansen coefficients.**
        J. R. Cherniack.
Smithsonian Astrophys. Obs., *Cambridge, Mass.*, Special Report, No. 346, 3 + 9 + A 16 pp. (1972).

This paper describes some procedures for computer development of Hansen coefficients. The method of von Zeipel and Andoyer is found most efficient. A table extends the method from 7th to 12th order.

042.041 **On elliptic orbits in the problem of two fixed centres.**   F. I. Kiselev.
Mekh. tverd. tela. Resp. mezhved. sb., 1972, vyp. (No.) 4, p. 130 - 133. In Russian. – Abstr. in Referativ. Zhurn. 51. Astron., 10.51.136; 62. Issled. kosmich. prostranstva, 10.62.282 (1972).

042.042 **The secular variations in the "restricted" problem and the stability of the circular orbits.**
G. Antonacopoulos.
Astron. Astrophys., Vol. 21, 265 - 269 (1972).

We discuss the stability of the circular orbits in the restricted problem in the case of the secular variations in relation with the Delaunay variable H, the ratio of the semimajor axes α and the order of the secular part of the expansion of the disturbing function.

042.043 **First-order perturbations of the two finite body problem.**   H. Kinoshita.
Publ. Astron. Soc. Japan, Vol. 24, 423 - 457 = Tokyo Astron. Obs. Repr., No. 417 (1972).

This investigation deals with the motion of a system of two rigid bodies, one spherical and one triaxial, which experience no other forces than mutual gravitational attraction. The first-order perturbations are derived by the Hori-Lie transformation.

042.044 **A tidal period of 1800 years.**   W. de Rop.
Tellus, Vol. 23, 261 - 262 (1971). – Shorter contribution.

042.045 **A numerical investigation of secular terms of the planetary disturbing function.**
G. Antonacopoulos.
Astrophys. Space Sci., Vol. 17, 267 - 276 (1972).

The secular terms of the planetary disturbing function are given, after elimination of short period terms by von Zeipel's transformation. The adequacy of this expansion up to terms of eighth order in the inclination and eccentricity is investigated by numerical processes, as a function of the Keplerian elements a, e and i. The eccentricity e' of the outer planet is taken equal to zero.

042.046 **Stability of the libration points of a rotating triaxial ellipsoid.** S. G. Zhuravlev.
Celestial Mechanics, Vol. 6, 255 - 267 (1972).

The problem of stability of the equilibrium points (the libration points) in the problem of motion of a mass point in the neighbourhood of a rotating triaxial ellipsoid is investigated in the strict sense. In the plane of parameters, depending on the form and dynamical characteristics of the ellipsoids, the regions of stability and instability of the libration points are obtained.

042.047 **An exact analytical solution of Kepler's equation.** C. E. Siewert, E. E. Burniston.
Celestial Mechanics, Vol. 6, 294 - 304 (1972).

Complex-variable analysis is used to develop an exact solution to Kepler's equation for both elliptic and hyperbolic orbits. The method is based on basic properties of canonical solutions to appropriately posed Riemann problems, and the final results are expressed in terms of elementary quadratures.

042.048 **Sur la stabilité des points d'équilibre triangulaires dans le problème restreint elliptique.** N. X. Vinh.
Celestial Mechanics, Vol. 6, 305 - 321 (1972).

La stabilité du mouvement d'un petit corps au voisinage des points triangulaires dans le problème restreint elliptique est discutée. Les courbes de stabilité dans le plan (μ, e) sont obtenues jusqu'au quatrième ordre en e par la méthode du prolongement analytique. Les coefficients des séries obtenues sont données de façon exacte. Les exposants caractéristiques du système des équations aux variations sont obtenus par un procédé d'intégration matricielle.

042.049 **Expansions of the derivatives of the disturbing function in planetary problems.**
M. S. Petrovskaya.
Celestial Mechanics, Vol. 6, 328 - 342 (1972).

A method is suggested to develop literal expansions of derivatives of the disturbing function especially for the case of large values of the major axis ratio λ. The series remain convergent as well if $\lambda = 1$, unless the eccentricities vanish at the same time. The treatment holds true in the case when usual analytical expansions are not valid, that is if the orbits have points equidistant from the primary. The general case is considered too, the intersecting orbits being included.

042.050 **The transition from elliptic to hyperbolic orbits in the two-body problem by slow loss of mass.**
L. van der Laan, F. Verhulst.
Celestial Mechanics, Vol. 6, 343 - 351 (1972).

Transition from elliptic to hyperbolic orbits in the two-body problem with slowly decreasing mass is investigated by means of asymptotic approximations. Analytical results by Verhulst and Eckhaus are extended to construct approximate solutions for the true anomaly and the eccentricity of the osculating orbit if the initial conditions are nearly-parabolic. To illustrate the results quantitatively we calculate the eccentricity as a function of time for Jeans-Eddington functions $n = 0(1)$ 5 and 18 nearly-parabolic initial conditions to find that 93 out of 108 elliptic orbits become hyperbolic.

042.051 **Sufficient conditions for return in the three-body problem.** E. M. Standish, Jr.
Celestial Mechanics, Vol. 6, 352 - 355 (1972).

Sufficient conditions are given, which, if fulfilled, enable one to determine, rigorously, that an ejected particle will not escape from a three-body system.

042.052 **Periodic orbits in the rectilinear restricted three-body problem.** R. Broucke.

Journ. Mécanique, *Paris*, Vol. 10, 449 - 465 (1971).

It is shown that the problem is a non-conservative generalization of the two-fixed centres problem. Numerical results are given when the two masses are equal and also when the ratio of masses is variable. Thirteen isolated periodic orbits are given.

042.053 **Set-theoretical aspects of the restricted problem of three bodies.** R. Kurth.
Arch. Rat. Mech. Analysis, Vol. 41, 108 - 120 (1971). – Abstr. in Zentralbl. Math. Grenzgebiete, Vol. 228, No. 70017 (1972).

042.054 **Anwendung der Matrizenrechnung auf die theoretische Astronomie.** J. Witkowski.
Zeszyty nauk. Akad. Górn.-Hutn. Stanisław Staszic, Kraków 254, Zeszyt spec., Vol. 18, 337 - 353 (1971). In Polish.

042.055 **Les orbites périodiques à trois dimensions autour des points L_1, L_2, L_3 de Lagrange.** D. N. Katsis.
Bull. Soc. math. Grèce, Nouvelle Sér., Vol. 11, No. 2, p. 70 - 114 (1971). In Greek.

042.056 **Improbability of collisions in Newtonian gravitational systems.** D. G. Saari.
Trans. American Math. Soc., Vol. 162, 267 - 271 (1972).

It is proved that the set of initial conditions for an assembly of point masses interacting according to Newton's law of gravitation leading to a collision in finite time has measure zero.

042.057 **Numerical studies of the 3-body problem.** D. Greenspan.
SIAM Journ. Applied Math., Vol. 20, 67 - 78 (1971).

A numerical solution is presented for the general plane non-degenerate 3-body problem. The force is taken as $-(G/r^2) + (H/r^m) - \alpha |v|$, where the addition of a short-range repulsion with $m \geq 2$ simulates collisional effects and α is a viscous damping force. 8 examples of general 3-body motions, with differing masses, initial conditions and dampings, are given.

042.058 **Sur le mouvement d'un triple bâtonnet dans un champ newtonien.** M. Pascal.
Journ. Mécanique, *Paris*, Vol. 11, 147 - 160 (1972).

We consider the problem of the translational-rotational motion of a rigid body of particular shape in the gravitational field of a fixed particle. A particular solution of this problem where the mass center of the body moves on a circular orbit is investigated. Sufficient conditions for the stability of such a motion are discussed by means of Liapounov's direct method. Necessary conditions of stability are obtained by investigating the equations for the perturbations in the first approximation. The results are compared with the study of Beletskii.

042.059 **Auto-mapping and penetration theorem.** S. Kikuchi.
Sci. Rep. Tôhoku Univ., First Ser., Vol. 55, 49 - 62 (1972).

Any orbit of a set of all the orbits belonging to a certain dynamical system can be mapped onto another orbit of the same set by a special transformation called auto-mapping. The auto-mapping is derived by making use of the penetration theorem and parameter transformation. An application is made to Lagrange's theory of variation of constants.

042.060 **On a new form of coordinate expansion in non-perturbed elliptic motion.** Eh. A. Borisov.
Izv. vyssh. ucheb. zavedenij. Geod. i aehrofotos"emka, 1972, No. 1, p. 67 - 76. In Russian. – Abstr. in Referativ. Zhurn. 62. Issled. kosmich. prostranstva, 11.62.209 (1972).

042.061 **Computation of Schwarzschild's periodic solutions in the restricted three-body problem.**
E. Leimanis, B. R. Olund.
Astron. Nachr., Vol. 294, 47 - 64 (1972).

Periodic solutions may be characterized by the mean values and initial values of their orbital elements. Accordingly in Chapter I a method is developed for calculating the mean values and in Chapter II a method is given for calculating the initial values of the orbital elements in the case of Schwarzschild's periodic solutions. The methods are illustrated by numerical calculations for the commensurabilities $2:1$, $3:1$, $3:2$, and $5:3$ of the mean motions of the planetoid and Jupiter.

042.062 **The two-body problem in the theory of gravitation without potential.** H. Oja.
Ann. Acad. Sci. Fennicae, Ser. A, VI.Phys., No. 377, 40 pp. = Contr. Obs. Astrophys. Lab. Univ. Helsinki (1971).

The theory of gravitation without potential is based on eight axioms. The force law deduced from the axioms contains, in the form treated here, eight structure constants, three of which have been determined in connection with the one-body problem. The value of one more constant will be obtained from our results concerning the two-body problem. The theory does not assume constancy of masses. In the two-body problem the masses are found to change periodically. The perturbations caused by this change are of the same type as those due to the other terms in the equation of relative motion, i.e. advance of periastron, acceleration of the centre of mass and dilatation of the sidereal period.

042.063 **On first approximation stability of the libration points of a three-axial ellipsoid under constantly acting perturbations.** S. I. Nikolaev.
Byull. Inst. Teoret. Astron., *Leningrad,* Vol. 13, 215 - 219 (1972). In Russian.

This paper deals with the stability conditions of the libration points L_1, L_3 of a three-axial ellipsoid under constant perturbations on the base of calculation of the characteristic exponents of the system of six linear first-order differential equations with nearly periodical coefficients. The characteristic exponents are presented in the form of asymptotic series in powers of a small parameter. The terms of zero and first approximations are defined.

042.064 **On a method of formation of normal places.**
T. K. Nikolskaya.
Byull. Inst. Teoret. Astron., *Leningrad,* Vol. 13, 220 - 224 (1972). In Russian.

The present paper deals with the formation of normal places by means of approximation of observation series by Chebyshev polynomials. For the choice of the regression curve Fisher's test has been used. The normal place formation is performed according to the minimum variance principle.

042.065 **Some questions on the treatment of observations of major planets.** M. L. Sveshnikov.
Byull. Inst. Teoret. Astron., *Leningrad,* Vol. 13, 231 - 245 (1972). In Russian.

The paper deals with the construction and the analytical investigation of the normal system for the determination of the corrections to the orbital elements of the major planets from optical observations. The coefficients of the normal system are presented in the integral form. Evaluation of the accuracy of the system solution is given. The theory is illustrated by a numerical example for the case of fictitious observations of Mercury.

042.066 **The difference equations for estimate biases in the filtering problem.**
A. M. Zhandarov, V. G. Kiselev.

Byull. Inst. Teoret. Astron., *Leningrad,* Vol. 13, 294 - 299 (1972). In Russian.

The procedure of an improvement of two close orbits is discussed. The a priori probability of belonging of the measurement to one of the given orbits is supposed to be known. An identification of the measurement and the orbits is made by some decision rule. The difference equations are defined which describe the behaviour of the biases of the recurrent orbit estimates computed by Kalman's method.

042.067 **On short-periodic solar perturbations in the motion of planetary satellites.** A. A. Orlov.
Trudy Gos. Astron. Inst. Shternberga, Vol. 43, 30 - 37 (1972). In Russian.

The problem of the motion of a satellite under the influence of solar perturbations is considered. With the help of Zeipel's method the formulas for the computations of the perturbations of short and intermediate periods to second-order terms are obtained.

042.068 **Application of von Zeipel's method to the stellar three-body problem.** N. A. Solovaya.
Trudy Gos. Astron. Inst. Shternberga, Vol. 43, 38 - 51 (1972). In Russian.

Von Zeipel's method is applied to the solution of the stellar three-body problem. The formulas obtained allow to determine the short-period and the intermediate-period terms of the second order.

042.069 **Étude de quelques équations différentielles linéaires et non linéaires avec applications à la mécanique céleste.** Nguyen Xuan Vinh.
Thesis, Sci. Math. Univ. Paris. Centre Documentation C.N.R.S, (1972-06.05), 239 pp. (1972).

042.070 **The perturbation region in the problem Su Shu Huang.** A. Drożyner.
Postępy Astron., Vol. 20, 357 - 361 (1972). In Polish.

The definition of the perturbation region in the case of the restricted three-body problem is generalized for the case of the four-body problem in Su Shu Huang's formulation. Based on this definition the perturbation region of the moon relative to the earth and the sun is given. The higher order effects in the generalized perturbation and interaction regions are also discussed.

042.071 **On a new form of coordinate expansion in elliptical non-perturbed motion.** É. A. Borisov.
Izv. vyssh. ucheb. zavedenij. Geod. i aehrofotos"emka, 1972, No. 1, p. 67 - 76. In Russian. − Abstr. in Referativ. Zhurn. 51. Astron., 12.51.105 (1972).

042.072 **Marginalien zum 3. Keplerschen Gesetz.**
R. Haase.
Kepler Festschrift 1971, (see 003.010), p. 159 - 165 (1971).

042.073 **On the motion of a satellite of a slowly rotating planet near the equatorial plane.**
Yu. I. Ivanov, V. K. Kajsin.
Trudy Astrofiz. Inst., *Alma-Ata,* Vol. 19, 115 - 119 (1972). In Russian.

The equations of motion for a satellite of a slowly rotating planet are simplified and integrated by means of a soluble problem of particle dynamics.

042.074 **A form of differential equations of perturbed satellite motion.** E. P. Aksenov, B. N. Noskov.
Astron. Zhurn. Akad. Nauk SSSR, Vol. 49, 1292 - 1299 (1972). In Russian. English translation in Soviet Astron. AJ, Vol. 16, No. 6.

Differential equations for the elements of an intermediate

orbit of a satellite are deduced. It is supposed that an intermediate orbit is constructed on the basis of the generalized problem of two fixed centers, and the perturbing forces have no force function.

042.075　Particular solutions of the motion of the three-spheroids problem having the same symmetry plane.
V. V. Vidyakin.
Astron. Zhurn. Akad. Nauk SSSR, Vol. 49, 1300 - 1310 (1972). In Russian. English translation in Soviet Astron. AJ, Vol. 16, No. 6.

The paper offers particular solutions of the problem concerning three homogeneous compressed spheroids close to the Lagrangian solutions of the three-body problem.

042.076　Acceleration of the numerical integration of equations of motion in celestial mechanics.
I. I. Karpov, A. K. Platonov.
Kosmich. Issled., Vol. 10, 811 - 826 (1972). In Russian.

042.077　Sur une théorie générale des éléments canoniques.
G. Ţopan.
Stud. Cerc. Astron., Vol. 17, 245 - 265 (1972).

Nous essayons de donner une théorie générale des éléments canoniques et nous montrons comment on peut obtenir des nouvelles variables canoniques. Deux exemples illustrent la théorie.

042.078　On the stability of symplectic mappings, I, II.
R. Nagai.
Univ. Tokyo, Tokyo Astron. Obs. Rep. (No. 60), Vol. 16, 1 - 3, 4 - 8 (1972). In Japanese.

042.079　Two forms of the equations for the problem of motions of finite bodies.　H. Kinoshita.
Univ. Tokyo, Tokyo Astron. Obs. Rep. (No. 60), Vol. 16, 222 - 229 (1972). In Japanese.

042.080　Integration of perturbed orbits by time series expansions.　E. Everhart.
Bull. American Astron. Soc., Vol. 4, 419 (1972). — Abstr. AAS.

042.081　The reduction of local truncation errors by time transformations.　P. E. Nacozy.
Bull. American Astron. Soc., Vol. 4, 421 (1972). — Abstr. AAS.

042.082　A new approach to the numerical integration of linear perturbed systems.　G. Scheifele.
Bull. American Astron. Soc., Vol. 4, 421 - 422 (1972). — Abstr. AAS.

042.083　Numerical solution of ordinary differential equations.　P. Sconzo.
Bull. American Astron. Soc., Vol. 4, 422 (1972). — Abstr. AAS.

042.084　Examples of transformations improving the numerical accuracy of the integration of differential equations.　E. L. Stiefel.
Bull. American Astron. Soc., Vol. 4, 422 (1972). — Abstr. AAS.

042.085　Determination of orbital elements through the known values of the velocity vector at three different moments of time.　V. S. Korolev.
Vestn. Leningr. un-ta, 1972, No. 13, p. 144 - 145. In Russian. — Abstr. in Referativ. Zhurn. 51. Astron., 1.51.167; 62. Issled. kosmich.prostranstva, 1.62.225 (1973).

042.086　Solutions particulières et invariants intégraux en mécanique céleste.　L. Losco.
Thesis, Sci. Math. Univ. Besançon. Centre Documentation C.N.R.S., (1972-02-29), 195 pp. (1972).

042.087　The invariance of Poincaré's generating function for canonical transformations.　A. Weinstein.
Inventiones Math., Vol. 16, 202 - 213 (1972).

042.088　Aplicación de las transformaciones de Lie a la eliminación de términos de corto período.
R. Cid, J. F. Lahulla.
Urania Barcelona, Año 56, No. 274, p. 177 - 184 (1971).

On the motions of artificial satellites moving in the gravitational field of the earth without drag, the elimination of the Hill's variable u (short-periodic perturbation) is obtained by application of Lie transforms. The formulary is carried out until the second order, following the iterative procedure proposed by Deprit.

042.089　Interstellar gravitational perturbations of cometary orbits.　M. B. Faintich.
Thesis, Univ. Illinois, Urbana—Champaign. [Available from Univ. Microfilms, Ann Arbor, Mich., USA. Order No. 72-12154], 126 pp. (1971).

Two approximate analytic methods and one numerical method, namely a restricted hyperbolic three-body problem, were developed in order to handle the encounter of a star (or interstellar cloud) with the sun-comet system. All three methods yield similar results, and the restrictions on a given hypothetical encounter determine the best method to be used.

042.090　Problems on the nature of relative equilibria in celestial mechanics.　S. Smale.
Lecture Notes in Math., [Springer-Verlag, Berlin], Vol. 197, 194 - 198 (1971).

042.091　Appendix to Smale's paper: "Diagonals and relative equilibria".　M. Shub.
Lecture Notes in Math., [Springer-Verlag, Berlin], Vol. 197, 199 - 201 (1971).

042.092　Stability of Kepler motion.　H. R. Schwarz.
Comput. Methods Applied Mechanics and Engineering, (*Netherlands*), Vol. 1, 279 - 299 (1972).

042.093　Differential coefficients in the variations of the velocity components when applying universal elements.　V. T. Khukhunaishvili.
Byull. Abastumansk. Astrofiz. Obs., No. 43, p. 247 - 252 (1972). In Russian.

042.094　Contributions to Lambertian mechanics.
T. Godal.
Inst. Mekanikk, Norges Tekn. Høgskole, Trondheim, Meddelelse No. 22, 2 + 55 pp. (1972). — Contents: Thesis survey paper; Conditions of compatibility of terminal positions and velocities (Repr. of a paper from 1960); Natural terminal velocity components; Variation of natural parameters (Repr. of a paper from 1963); Simple correctional maneuvres (Repr. of a paper from 1966); Method for determining the initial velocity corresponding to a given time of free flight transfer between given points in a simple gravitational field; The intrinsic equations of optimal two-impulse transfer; Undisturbed eccentric anomaly difference as the independent variable in the perturbation difference equations (Repr. of a paper from 1971).

042.095　Geometrical dynamics: A new approach to periodic orbits around L_4.　R. Rand, W. Podgorski.
Celestial Mechanics, Vol. 6, 416 - 420 (1972).

Geometrical dynamics is the study of the geometry of the orbits in configuration space of a dynamical system without reference to the system's motion in time. Generalized coordinates for the circular restricted problem of three bodies are taken as polar coordinates r, θ centered at the triangular libration point L_4. A time-independent nonlinear second order ordinary differential equation for r as a function of θ is derived. Approximations to periodic solutions are obtained by perturbations and Fourier series.

042.096 Orientation and resonance locks for satellites in the elliptic orbit. H.-S. Liu.
Celestial Mechanics, Vol. 6, 421 - 424 (1972).

In order to achieve the maximum strength of higher resonance locks for satellites in the elliptic orbit, the condition of satellite orientation during the process of deployment is established. It is shown that for maximum strength locks the axis of the minimum moment of inertia of satellites should point toward the attracting body at $\pm(5/8)\pi$ and 0 values of the true anomaly f. This condition of deployment is applicable to all cases of resonance rotation regardless of the value of lock number k and orbit eccentricity e.

042.097 Studies in the application of recurrence relations to special perturbation methods. I. Comparison with Runge-Kutta integration in the two-body problem.
A. E. Roy, P. E. Moran, W. Black.
Celestial Mechanics, Vol. 6, 468 - 482 (1972).

The integration by recurrent power series of certain differential equations occurring in celestial mechanics is shown to be very much more efficient and accurate than that produced by classical one step methods. It is shown that for any such system of differential equations the machine time taken to carry out an integration is a minimum for a certain choice of the number of terms taken in the recurrent power series. In the two-body orbits considered this number is about 15. For the same accuracy criterion the power series is faster than the Runge-Kutta method of the fourth order by a factor which varies between 6 and 15 depending on the eccentricity of the orbit.

042.098 A general solution of the equation of perturbations in rectangular coordinates. B. Popović.
Sci. Revuo, Beograd, Vol. 22, 157 - 172 (1971). In Esperanto.

042.099 Les séries nouvelles pour les coefficients f, g de Lagrange pour le mouvement des petites planètes ou comètes. B. Popović.
Mat. Vesnik, Vol. 9 (24), 173 - 177 (1972). — In Esperanto.

042.100 Zastosowanie rachunku krakowianowego do astronomii teoretycznej. J. Witkowski.
Zeszyt. Nauk. Akad. Gôrniczo-Hutniczej im. Stanisława Staszica No. 254 (Zeszyt Specjalny No. 18), p. 337 - 353 (1971).

Relativistic celestial mechanics.
See Abstr. 003.052.

Selected works of H. Poincaré. Vol. 2. New methods of celestial mechanics. Topology. Number theory.
See Abstr. 003.146.

An exercise in symbolic programming: Computation of general normalized inclination functions.
See Abstr. 021.006.

Use of Green's functions in the numerical solution of two-point boundary value problems.
See Abstr. 021.018.

About the motion of a heavy flexible string attached to a satellite in a central field of attraction.
See Abstr. 052.017.

Einige Konsequenzen der Mach-Einstein-Doktrin für Himmelsmechanik und Geophysik. See Abstr. 066.091.

Literal expressions for the co-ordinates of the moon. I. The first degree terms. See Abstr. 094.055.

Resonance rotation of celestial bodies and Cassini's laws. See Abstr. 094.137.

A study of commensurable motion in the asteroid belt. See Abstr. 098.019.

Resonances and encounters in the inner solar system. See Abstr. 098.033.

Orbit-orbit resonance capture in the solar system. See Abstr. 100.019.

The case against Planet X. See Abstr. 101.021.

The dependence on inclination of the planetary perturbations of the orbits of long-period comets. See Abstr. 102.002.

Comets and problems of numerical celestial mechanics. See Abstr. 102.042.

A physical model for HZ Her. See Abstr. 121.095.

Light curve of HZ Herculis in relation to Her X-1. See Abstr. 121.096.

Regularization in the N-body problem. See Abstr. 151.040.

L'aspect mécanique de l'expansion de l'Univers. See Abstr. 162.037.

Errata

042.901 Errata: 'Trigonometric linear second-order theory of the secular perturbations in the motion of the major planets' [Nablyud. iskusstv. nebesn. tel, No. 62, p. 42 ff. (1971)]. V. L. Brumberg, A. V. Egorova.
Byull. Inst. Teoret. Astron., *Leningrad*, Vol. 13, 336 (1972). In Russian.

043 Astronomical Constants

043.001 **On the motion of the equator and the ecliptic.**
W. Fricke.
IAU Symposium No. 48, (see 012.004), p. 196 (1972). – Abstract.

043.002 **Galactic rotation and the precession constant.**
S. V. M. Clube.
Monthly Notices Roy. Astron. Soc., Vol. 159, 289 - 314 (1972).
The effect of high proper motion stars on the determination of the precession constant is discussed, and an improved technique of determining the stellar velocity field in the solar neighbourhood is described. The technique is applied to stars in the FK4 and AGK3, and the results interpreted as implying an apparent rotation of the nearby stars in the opposite direction to galactic rotation. The investigation provides general support for the kind of correction to the precession constant proposed by Fricke and Vasilevskis and Klemola, and implies that observation of the motion of the equinox is confused by a stellar component.

043.003 **De massa van het stelsel aarde-maan.** W. de Rop.
Hemel en Dampkring, Vol. 70, 269 - 270 (1972).

043.004 **Die Astronomische Einheit.** F. Gondolatsch.
SuW, Vol. 11, 298 - 301 (1972).

Measurement of an optical frequency and the speed of light. See Abstr. 022.127.

Speed of light from direct frequency and wavelength measurements of the methane-stabilized laser.
See Abstr. 022.129.

A determination of the motion of the ecliptic.
See Abstr. 042.021.

The variability of the gravitational constant.
See Abstr. 066.045.

The excess secular change in the obliquity of the ecliptic and its relation to the internal motion of the earth.
See Abstr. 081.012.

Influence of the core on the nutations.
See Abstr. 081.066.

A system of planetary masses and related quantities.
See Abstr. 091.052.

Are the constants constant?
See Abstr. 162.035.

The cosmic numbers. See Abstr. 162.049.

044 Time, Rotation of the Earth

044.001 **Measures of time in astronomy.**
J. D. Mulholland.
Publ. Astron. Soc. Pacific, Vol. 84, 357 - 364 (1972).
Invited review paper.

044.002 **Past and future of research methods in problems of the earth's rotation.** P. Melchior.
IAU Symposium No. 48, (see 012.004), p. XI - XXII (1972). Presidential address.

044.003 **On the short period terms in the UT1 and those in the polar motion.** S. Iijima, S. Okazaki.
IAU Symposium No. 48, (see 012.004), p. 58 - 60 (1972).

044.004 **Seasonal effects observed in time determinations at Santiago.** F. Noël.
IAU Symposium No. 48, (see 012.004), p. 139 - 144 (1972).

044.005 **Historic variations in the rotation of the earth.**
R. R. Newton.
IAU Symposium No. 48, (see 012.004), p. 160 - 161 (1972).

044.006 **Rotational accelerations.** W. Markowitz.
IAU Symposium No. 48, (see 012.004), p. 162 - 164 (1972).

044.007 **Random variations in the earth rotation.**
G. E. O. Giacaglia.
IAU Symposium No. 48, (see 012.004), p. 165 - 171 (1972).

044.008 **Estimation of random changes in the earth's rotation.** B. D. Tapley, B. E. Schutz.
IAU Symposium No. 48, (see 012.004), p. 172 - 178 (1972).

044.009 **Effect of the change in the geomagnetic dipole moment on the rate of the earth's rotation.**
T. Yukutake.
IAU Symposium No. 48, (see 012.004), p. 229 - 230 (1972).

044.010 **On the relation between the rotation of the earth and solar activity.**
C. Sugawa, C. Kakuta, H. Matsukura.
IAU Symposium No. 48, (see 012.004), p. 231 - 233 (1972).

044.011 **On the effect of ocean tides on the secular retardation of the earth's rotation.**
N. N. Pariisky, M. V. Kuznetsov, L. V. Kuznetsova.
IAU Symposium No. 48, (see 012.004), p. 240 (1972). – Abstract.

044.012 **The role of a satellite swarm in the origin of the rotation of the earth.** E. L. Ruskol.
Astron. vestn., Vol. 6, 91 - 95 (1972). In Russsian. – Abstr. in Referativ. Zhurn. 51. Astron., 9.51.241 (1972).

044.013 **On the influence of the individual error on the determination of time from solar observations.**
V. V. Kirichuk.
Geod., kartogr. i aehrofotos"emka. Resp. mezhved. nauch.-

tekhn. sb., 1971, vyp. (No.) 14, p. 21 - 28. In Russian. – Abstr. in Referativ. Zhurn. 52. Geod. Aehros"emka, 9.52.119 (1972).

044.014   On the non-logarithmic calculation of corrections of a chronometer in azimuthal methods of time determination (method Struve, Struve-Pavlov, Döllen et al.). V. V. Kirichuk.
Geod., kartogr. i aehrofotos"emka. Resp. mezhved. nauch.-tekhn. sb., 1972, vyp. (No.) 15, p. 25 - 27. In Russian. – Abstr. in Referativ. Zhurn. 52. Geod. Aehros"emka, 9.52.120 (1972).

044.015   Élaboration du temps sidéral avec un diviseur de fréquence programmé. G. Jeansaume.
Ann. Françaises Chronométrie Micromécanique, Année 1972, 191 - 204.

044.016   Standard time. W. Palmer.
Journ. Navigation, *London,* Vol. 25, 535 - 536 (1972).

044.017   Aussendung und Empfang des Zeitmarken- und Normalfrequenzsenders DCF 77. G. Becker.
PTB Mitt., 82. Jahrgang, p. 224 - 229 (1972).
The German standard frequency and time signal transmitter DCF 77 on VLF 77.5 kHz transmits time signals according to the official Atomic Time Scale of the Physikalisch-Technische Bundesanstalt. A review is given about the transmitter programme, some fundamental problems of time scales, the performance of the emission, the precision and the reliability of the time signals and the standard frequency, the problems of reception of DCF 77 and the application possibilities of its emission.

044.018   On a relation between the variations of the rotational velocity of the earth and the anomalous behavior of the polar wobble around 1930. N. Sekiguchi.
Publ. Astron. Soc. Japan, Vol. 24, 545 - 547 = Tokyo Astron. Obs. Repr., No. 420 (1972). – Note.

044.019   About the choice of stars for time determination with a photoelectric transit instrument. K. A. Šteins, P. P. Rozenbergs.
Uch. zap. Latv. gos. un-ta imeni Petra Stuchki, Vol. 169, 40 - 52 (1972). In Russian.

044.020   A short-period irregularity of the earth's rotation and the motion of the earth's instantaneous pole. A. G. Fleer.
Astron. Zhurn. Akad. Nauk SSSR, Vol. 49, 1319 - 1321 (1972). In Russian. English translation in Soviet Astron. AJ, Vol. 16, No. 6.
On the basis of the reduction of materials of astronomical time determinations by time services of the USSR for 1955–1968 a formula connecting variations of the angular velocity of the earth's rotation with the motion of the earth's instantaneous pole is deduced.

044.021   Time and its measurement in Frederiction. J. E. Kennedy.
Journ. Roy. Astron. Soc. Canada, Vol. 66, 311 - 313 (1972).

044.022   Ein Einfluß des Sonnenwindes auf die Erdrotation. O. M. Burkard.
Gerlands Beiträge Geophys., Vol. 81, 277 - 280 (1972).
The statistical method of critical moment synchronization is applied to show that there are relationships between the interplanetary magnetic field frozen in the solar wind and the velocity of the earth's rotation. It is supposed that the deviations of the angular velocity of the earth are caused by radial shifts of the co-rotating parts of the magnetosphere.

044.023   The effects of earth tides on time and latitude observations. S. Iijima, Y. Niimi.
Univ. Tokyo, Tokyo Astron. Obs. Rep. (No. 60), Vol. 16, 190 - 202 (1972). In Japanese.

044.024   Ein neues Zeitsystem mit Schaltsekunden. G. Zimmermann.
SuW, Vol. 11, 342 - 343 (1972).

044.025   National Physical Research Laboratory, C.S.I.R. Time service notice. J. Hers.
Monthly Notes Astron. Soc. Southern Africa, Vol. 31, 86, 118, 138, 150, 156 (1972).

044.026   The general adoption of the atomic unit of time. C. Egidi.
Elettrotecnica, *(Italy),* Vol. 59, 430 - 433 (1972). In Italian.

044.027   Precision and accuracy of remote synchronization via network television broadcasts, Loran-C, and portable clocks. D. W. Allan, B. E. Blair, D. D. Davis, H. E. Machlan.
Metrologia, *(Germany),* Vol. 8, No. 2, p. 64 - 72 (1972).

044.028   Comité Consultatif pour la Définition de la Seconde. G. Becker.
PTB Mitt., 82. Jahrgang, p. 411 (1972).

044.029   Schaltsekunde am 1. 1. 1973 in der amtlichen Atomzeitskala der PTB. G. Becker.
PTB Mitt., 82. Jahrgang, p. 452 (1972).

044.030   Time and latitude service.
Polish Acad. Sci., Astron. Latitude Station, Borowiec, Circ. Nos. 121 - 123 (1972). – 1972 January – September.

044.031   Moderne Methoden des Zeitvergleichs. G. Hemmleb, F. Buckbesch.
Wiss. Zeitschr. Techn. Univ. Dresden, Vol. 21, No. 3, (see 012.021), 603 - 605 (1972).

044.032   Zur Bestimmung systematischer Fehler bei der Zeitbestimmung mit dem Passageinstrument. V. Milovanović.
Wiss. Zeitschr. Techn. Univ. Dresden, Vol. 21, No. 3, (see 012.021), 616 - 618 (1972).

044.033   Über thermisch bedingte instrumentelle Fehler bei Zeitbestimmungen aus Meridiandurchgangsbeobachtungen am Passageinstrument. J. Dittrich.
Vermessungstechnik, 20. Jahrgang, p. 270 - 272 = Mitt. Lohrmann-Obs. Techn. Univ. Dresden, No. 25a/I (1972).

044.034   The earth: a planet which does not rotate. K. Lambeck, F. Barlier.
Sci. Progr. Découverte, No. 3446, p. 40 - 47 (1972). In French.
Precession, nutation, displacement of the polar axes, the reduction in the rate of rotation. – Astronomers have not finished counting the irregularities in the movement of the earth. This paper discusses possible reasons for perturbations in the rate of rotation of the earth.

044.035   Corrections to Czechoslovak time signals. V. Ptáček.
Říše hvězd, Vol. 53, 182, 196, 222, 236 (1972). In Czech. 1972 June – September.

044.036   Astronomische Zeit- und Breitenbestimmungen. Empfangszeiten von Zeitsignalen.

Edited by Deutsches Hydrographisches Institut, Hamburg.
1972 January – September, 6 + 5 + 6 pp. (1972).

044.037 Astronomische Zeit- und Breitenbestimmungen. Empfangszeiten von Zeitsignalen.
Deutsche Akad. Wiss. Berlin, Zentralinst. Phys. Erde, Bereich II (Geod., Gravimetrie), Potsdam, Abt. Geod. Astron., Jahrgang 1971, Nos. 5 - 6 (1972). – 1971 September – December.

044.038 Time Service. C. Moranzino (Editor).
Oss. Astron. Torino (Pino Torinese), Bull. No. 2, 3 pp. (1972). – Results of the time determinations 1972 May – August.

044.039 Détermination astronomique de l'heure et heures demi-définitives de réception des signaux horaires.
L. Webrová, V. Ptáček.
Acad. Tchécoslov. Sci. Inst. Astron., Station de l'Heure, Prague, Sér. 5, No. 18; Sér. 6, Nos. 1 - 2, 10 + 12 + 14 pp. (1972). – 1971 November - 1972 April.

044.040 International Time and Latitude Service at the Tokyo Astronomical Observatory during 1972.
M. Huruhata.
Tokyo Astron. Obs., Time and Latitude Bull., Vol. 46, Nos. 1 - 8, p. I - IV, 1 - 53 (1972).
(1) Astronomical observation; (2) System of universal times; (3) System of UTC; (4) Time keeping; (5) Clock comparison; (6) Radio time signals emitted from Japan.

044.041 UTC time step on the 1st of July 1972. B. Guinot.
Bureau International de l'Heure, (B.I.H.), Paris, Circ. E 1, 2 pp. (1972). In English and French.

044.042 UTC time step on the 1st of January 1973. B. Guinot.
Bureau International de l'Heure, (B.I.H.), Paris, Circ. E 2, 2 pp. (1972). In English and French.

044.043 Universal time and coordinates of the pole; Emission time of time signals; Coordinated universal time;
Independent local atomic time scales AT (i).
Bureau International de l'Heure, (B.I.H.), Paris, Circ. D 69 - D 73 (1972). – 1972 June – September.

044.044 The reasons for the main errors in astronomical determinations of time in Beograd.
D. Dzhurovich, V. Radogostich.
Publ. 18th Astrometrical Conference 1969, (see 012.022), p. 214 - 221 (1972). In Russian.

044.045 Instituto Elettrotecnico Nazionale 'Galileo Ferraris'. Servizi di tempo e frequenza campione.
Boll. Geod. Sci. Affini, Anno 31, p. 405 - 412 (1972). – Circolare n. 18, 19.

044.046 Daily phase values.
U. S. Naval Obs., Washington, D. C., Time Service Publ., Ser. 4, Nos. 283 - 308 (1972). – 1972 July 5 – December 28.

044.047 Preliminary times and coordinates of the pole.
U. S. Naval Obs., Washington, D. C., Time Service Publ., Ser. 7, Nos. 236 - 261 (1972). – 1972 July 6 – December 28.

Automation of observations with a transit instrument. See Abstr. 032.055.

Results of observations made with the Danjon astrolabe at Santiago, 1967. See Abstr. 041.032.

Errors of star catalogues and the TU2 time system. See Abstr. 041.058.

The components of the tensor of the atmosphere's inertia and variations of the earth's rotation. See Abstr. 082.049.

The secular accelerations of the moon's orbital motion and the earth's rotation. See Abstr. 094.141.

045 Latitude Determination, Polar Motion

045.001 On the diurnal term in latitude variations.
A. M. Zulliev.
Astron. Zhurn. Akad. Nauk SSSR, Vol. 49, 886 - 889 (1972). In Russian. English translation in Soviet Astron. AJ, Vol. 16, No. 4.
A formula for an analytical expression of the diurnal term in latitude variations is derived. The results of the determination of its amplitude and initial phase from observations are given. A connection of the z-term and the inclination errors $\Delta \delta_a$ with the diurnal term is explained.

045.002 Revised values (1941–1961) of the coordinates of the pole referred to the CIO (*Conventional International Origin*). R. O. Vicente, S. Yumi.
IAU Symposium No. 48, (see 012.004), p. 10 - 11 (1972).

045.003 New determination of the polar motion from 1890 to 1969. E. P. Fedorov, A. A. Rorsun, S. P. Major, N. T. Panchenko, V. K. Tarady, Ya. A. Yatskiv.
IAU Symposium No. 48, (see 012.004), p. 12 - 13 (1972).

045.004 General considerations about the revision of all the calculations of the International Latitude Service.
P. Melchior, R. Dejaiffe, R. Verbeiren.
IAU Symposium No. 48, (see 012.004), p. 14 - 18 (1972).

045.005 Analysis of pole position from 1846 to 1970.
E. M. Gaposchkin.
IAU Symposium No. 48, (see 012.004), p. 19 - 32 (1972).

045.006 Spectral analyses of the Chandler wobble.
G. P. H. Pedersen, M. G. Rochester.
IAU Symposium No. 48, (see 012.004), p. 33 - 38 (1972).

045.007 The variation of latitude. H. Jeffreys.
IAU Symposium No. 48, (see 012.004), p. 39 - 42 (1972).

045.008　Analysis of the Chandler period of polar coordinates calculated by the Orlov method.
E. Proverbio, F. Carta, F. Mazzoleni.
IAU Symposium No. 48, (see 012.004), p. 43 - 45 (1972).

045.009　Comments on the changes in amplitude of the Chandlerian wobble.　B. Guinot.
IAU Symposium No. 48, (see 012.004), p. 46 - 48 (1972).

045.010　An interpretation of the ambiguity between annual terms obtained by time and latitude observations.
T. Okuda.
IAU Symposium No. 48, (see 012.004), p. 49 - 55 (1972).

045.011　Comparisons between results of polar coordinates derived from time data and those from latitude ones.　S. Okazaki, M. Nasaka.
IAU Symposium No. 48, (see 012.004), p. 56 - 57 (1972).

045.012　On the regularity of fluctuations in annual and secular polar motions.　H. J. Abraham.
IAU Symposium No. 48, (see 012.004), p. 61 - 67 (1972).

045.013　Separating the secular motion of the pole from continental drift. Where and what to observe?
I. I. Mueller, C. R. Schwarz.
IAU Symposium No. 48, (see 012.004), p. 68 - 77 (1972).

045.014　Non-periodic latitude variations and the secular motion of the earth's pole.
E. P. Fedorov, A. A. Korsun, N. T. Mironov.
IAU Symposium No. 48, (see 012.004), p. 78 - 85 (1972).

045.015　Secular and nonpolar variation of Washington latitude.　D. D. McCarthy.
IAU Symposium No. 48, (see 012.004), p. 86 - 96 (1972).

045.016　Secular and long-term variations of the polar motion.　E. Proverbio, F. Carta, F. Mazzoleni.
IAU Symposium No. 48, (see 012.004), p. 97 - 100 (1972).

045.017　Accuracy of Doppler determinations of station positions.　R. J. Anderle.
IAU Symposium No. 48, (see 012.004), p. 101 - 103 (1972).

045.018　Comparison of the coordinates of the pole as obtained by classical astrometry (IPMS, BIH) and as obtained by Doppler measurements on artificial satellites (Dahlgren Polar Monitoring Service).
M. Feissel, B. Guinot, N. Taton.
IAU Symposium No. 48, (see 012.004), p. 104 - 111 (1972).

045.019　A laser polar motion experiment.
D. E. Smith, P. J. Dunn, R. Kolenkiewicz.
IAU Symposium No. 48, (see 012.004), p. 121 - 122 (1972).

045.020　Polar motion from the tracking of close earth satellites.　K. Lambeck.
IAU Symposium No. 48, (see 012.004), p. 123 - 127 (1972).

045.021　Pole position studied with artificial earth satellites.
E. M. Gaposchkin.
IAU Symposium No. 48, (see 012.004), p. 128 - 130 (1972).

045.022　Old and new methods of observing polar motion.
R. O. Vicente.
IAU Symposium No. 48, (see 012.004), p. 133 - 135 (1972).

045.023　The use of the refractional pair observations.
G. Teleki.
IAU Symposium No. 48, (see 012.004), p. 145 - 146 (1972).

045.024　An explanation of the polar motion by a rigid core-mantle model.　Y. Kubo.
IAU Symposium No. 48, (see 012.004), p. 182 - 184 (1972).

045.025　Kimura's Z-term and the liquid core theory.
Y. Wako.
IAU Symposium No. 48, (see 012.004), p. 189 - 191 (1972).

045.026　Polar wandering and the earth's dynamical evolution cycle.　C. Pan.
IAU Symposium No. 48, (see 012.004), p. 206 - 211 (1972).

045.027　Polar wandering and mantle convection.
H. Takeuchi, N. Sugi.
IAU Symposium No. 48, (see 012.004), p. 212 - 214 (1972).

045.028　On the correlation between earthquake occurrence and disturbances in the path of the rotation pole.
M. A. Chinnery, F. J. Wells.
IAU Symposium No. 48, (see 012.004), p. 215 - 220 (1972).

045.029　On some natures of the excitation and damping of the polar motion.　N. Sekiguchi.
IAU Symposium No. 48, (see 012.004), p. 221 - 223 (1972).

045.030　Excitation of the Chandler wobble by large earthquakes.　K. Shimazaki, H. Takeuchi.
IAU Symposium No. 48, (see 012.004), p. 224 - 228 (1972).

045.031　Annual report of the International Polar Motion Service for the year 1970.　S. Yumi.
Published for the International Council of Scientific Unions by Central Bureau of the International Polar Motion Service, Mizusawa, 4 + 165 pp. (1972).
This volume is a continuation of earlier reports. It contains the results of latitude observations made in 1970 in collaboration of 54 stations and observatories all over the globe.

045.032　A method of the determination of earth's motions around its mass center from simultaneous laser and photographic observations of artificial earth satellites made by two stations.　B. Kołaczek.
Artificial Satellites, Polish Acad. Sci., Vol. 7, No. 2, p. 13 - 15 (1972).
The idea of determinations of the length and position of the chord joining two earth's stations from simultaneous laser and photographic observations of a satellite is presented. An application of these determinations for several choosen chords would enable investigations of the earth's wobbles with respect to the regarded geocentric coordinate system.

045.033　Polar motion from laser tracking of artificial satellites.　D. E. Smith, R. Kolenkiewicz, P. J. Dunn, H. H. Plotkin, T. S. Johnson.
Science, Vol. 178, 405 - 406 (1972).
Measurements of the range to the Beacon Explorer C spacecraft from a single laser tracking system at Goddard Space Flight Center have been used to determine the change in latitude of the station arising from polar motion. A precision of 0.03 arc second was obtained for the latitude during a 5-month period in 1970.

045.034　On processing of latitude observations by the least squares method.　V. V. Nesterov.
Soobshch. Gos. Astron. Inst. Shternberga, No. 178, p. 3 - 17 (1972).　In Russian.
A method of reducing latitude observations based on the principle of least squares is described. This method leads to large linear systems which are solved by the method of iterations. Latitude observations made at the Sternberg Astronomical Institute (1958 - 1963) were reduced by the described

method and good results were obtained. An ALGOL–60 program for the algorithm is presented.

045.035 **The periodic components of the polar motion.**
L. V. Rykhlova.
Soobshch. Gos. Astron. Inst. Shternberga, No. 179, p. 3 - 28 (1972). In Russian.

From the data of the instantaneous pole for 1846.0 - 1965.0 the secular, Chandler and seasonal components are separated. Following Iijimas' method the separation is made independently for the Chandler and for the seasonal component assuming a period of 1.2 years for the Chandler term.

045.036 **On systematic errors of the chain method.**
V. V. Nesterov.
Soobshch. Gos. Astron. Inst. Shternberga, No. 179, p. 29 - 34 (1972). In Russian.

045.037 **Simulation of latitude observations.**
V. V. Nesterov.
Soobshch. Gos. Astron. Inst. Shternberga, No. 179, p. 35 - 39 (1972). In Russian.

In order to analyse the absolute precision of the methods of reducing latitude observations, a 3-year series of fictitious observations with all parameters given beforehand has been constructed. There are described the principles of formation of this series and an example for reducing the series is presented.

045.038 **Earth tides and polar motions.** P. Melchior.
Tectonophysics, Vol. 13, 361 - 372 = Obs. Roy. Belgique, Commun., Sér. B, No. 76 = Sér. Géophys. No. 111 (1972).

Problems about the origin of polar motions and earth tides, and the accuracies reached in their determination, are discussed. The common reference system (C.I.O.) for the measurement of the instantaneous pole of rotation is defined. Hypotheses for the origin of the Chandler wobble are critically reviewed.

045.039 **The variation of the latitude of Tashkent owing to secular polar motion.** G. M. Kaganovskij.
Dokl. AN UzSSR, 1972, No. 7, p. 5 - 7. In Russian. – Abstr. in Referativ. Zhurn. 51. Astron., 12.51.135 (1972).

045.040 **On the spectrum of polar coordinates from 1846 to 1971.** Ya. S. Yatskiv, A. A. Korsun, L. V. Rykhlova.
Astron. Zhurn. Akad. Nauk SSSR, Vol. 49, 1311 - 1318 (1972). In Russian. English translation in Soviet Astron. AJ, Vol. 16, No. 6.

One-and two-component models for the Chandler motion of the earth's pole are discussed with the use of uniform data on polar coordinates. It is shown that a two-component model is not appropriate during 1846–1900.

045.041 **Results concerning the motion of the instantaneous pole of rotation of the earth using the data presented by R. Vicente and S. Yumi.** L. Rusu.
Stud. Cerc. Astron., Vol. 17, 189 - 195 (1972). In Romanian.

The author found some components of the periods: 1.14, 1.02, 0.78, 0.594, 0.456, 0.432, 1.26, 1.00. The amplitude of the last five components is very small. For the period 1910–1911 and for 1950–1952 a very great amplitude can be observed. All the reductions have been made for each coordinate x, y separately.

045.042 **On the polar motion of the earth and gravitational waves.** T. Mitani.
Mem. Japan Astron. Study Ass., No. 18, Vol. 5, 147 - 164 (1972). In Japanese.

A relation between polar motion of the earth and the gravitational waves from the nearest binary stars is suggested.

045.043 **Homogeneous polar motion 1900.0–1912.0 in the MD system.**
B. Buonocore, E. Fichera, A. Pugliano.
Separate print Ist. Univ. Navale, Astron. Generale e Sferica, Napoli. 40 pp. (1972). In Italian.

With the reduction to the MD system of the ILS observations since 1900.0 to 1912.0, and using only completed groups of all observations in the stations of Mizusawa, Carloforte, Gaithersburg, Ukiah, we obtain the values x and y of the polar motion. It is noted that the z term, calculated in the MD system, shows an oscillation with a period of about six years, a value which was already found in the 1935.0–1947.0 reductions.

045.044 **Astronomical determinations of latitude and longitude in 1961–1966.** E. Kääriäinen.
Publ. Finnish Geod. Inst., *Helsinki,* No. 71, 102 pp. (1971).

045.045 **Polar motion 1900-1912. Comparison between Albrecht-Wanach and MD system.**
B. Buonocore, A. Pugliano, R. Santamaria.
Separate print Ist. Univ. Navale, Astron. Generale e Sferica, Napoli. 16 pp. (1972). In Italian.

In this paper we have compared the polar motion on the period 1900–1912 drawn by Albrecht and by Wanach with those deduced in MD system. Using this comparison we have been able to analyse the mean values of latitudes respectively enclosed by Albrecht (1900.0–1906.0) and by Wanach (1906.0–1912.0).

045.046 **Monthly Notes of the International Polar Motion Service.**
IPMS Monthly Notes, International Latitude Obs. Mizusawa (Japan). 1972 Nos. 5–10, p. 37 - 86 (1972). – Announces the values of latitudes observed at the collaborating stations during 1972 May – October.

045.047 **Bestimmung der Polhöhenschwankungen des Geodätischen Observatoriums Pecný.** J. Rambousek.
Wiss. Zeitschr. Techn. Univ. Dresden, Vol. 21, No. 3, (see 012.021), 601 - 603 (1972).

045.048 **Latitude variation at Józefosław in the period of 1961–1970.**
B. Kołaczek, B. Chmielewska, J. Rogowski.
Wiss. Zeitschr. Techn. Univ. Dresden, Vol. 21, No. 3, (see 012.021), 605 - 607 (1972).

045.049 **Coordonnées du pôle instantané rapportées à l'origine conventionnelle internationale et corrections de longitude TU1–TU0, à 0h TU.**
Bureau International de l'Heure, (B.I.H.), Paris, Circ. B/C, Nos. 196 - 200 (1972). – Valeurs interpolées et extrapolées.

045.050 **Results of the determination of latitude in Józefosław.** L. Pieczyński.
Latitude Circ., Warsaw Techn. Univ., Astron.-Geod. Obs. Józefosław, Nos. 39 - 42 (1972). – 1971 October - 1972 September.

045.051 **Investigations of the effect of wind in astronomical observations at Pulkovo during IGY – IGC.**
G. S. Tyuterev.
Publ. 18th Astrometrical Conference 1969, (see 012.022), p. 270 - 274 (1972). In Russian.

045.052 **Breitenbestimmungen.**
Techn. Univ. Dresden, Lohrmann-Obs. Zirk. Nos. 56 - 60 (1972). – 1972 March – December.

Motion of the earth's pole from 1890.0 to 1969.0.
See Abstr. 003.069.

On the estimate of the frequency characteristic of a
smoothing operator.   See Abstr. 031.058.

Die Anwendung der Rechentechnik in der praktischen Astronomie.   See Abstr. 041.047.

Statistical evaluation of the results of latitude observations with a prismatic astrolabe at Pulkovo during
1963.2 – 1968.7.   See Abstr. 041.061.

On the declinations of the latitude program stars as
determined with the zenith telescope.   See Abstr. 041.071.

On the short period terms in the UT1 and those in
the polar motion.   See Abstr. 044.003.

On a relation between the variations of the rotational velocity of the earth and the anomalous behavior of the
polar wobble around 1930.   See Abstr. 044.018.

A short-period irregularity of the earth's rotation
and the motion of the earth's instantaneous pole.
See Abstr. 044.020.

The effects of earth tides on time and latitude
observations.   See Abstr. 044.023.

Time and latitude service. See Abstr. 044.030.

Astronomische Zeit- und Breitenbestimmungen,
Empfangszeiten von Zeitsignalen.   See Abstr. 044.036.

Astronomische Zeit- und Breitenbestimmungen.
Empfangszeiten von Zeitsignalen.   See Abstr. 044.037.

International Time and Latitude Service at the
Tokyo Astronomical Observatory during 1972.
See Abstr. 044.040.

Universal time and coordinates of the pole; Emission
time of time signals; Coordinated universal time; Independent
local atomic time scales AT (i).   See Abstr. 044.043.

Preliminary times and coordinates of the pole.
See Abstr. 044.047.

Simultanbestimmungen der Lotabweichungskomponenten ξ und η mit dem Prismenastrolabium. VIII.
See Abstr. 046.009.

Simultanbestimmungen der Lotabweichungskomponenten ξ und η mit dem Prismenastrolabium. IX.
See Abstr. 046.010.

Bestimmungen von astronomischen Längen,
Azimuten und Breiten in den Jahren 1965 bis 1970.
See Abstr. 046.036.

On the calculation of changes in the earth's inertia
tensor due to faulting.   See. Abstr. 081.007.

046   Geodetic Astronomy, Navigation

**046.001   On the variations of the data of vertical and azimu-
thal laying in astronomical determinations with the
position lines method.   A. Pericoli.**
Boll. Geod. Sci. Affini, Anno 31, p. 207 - 212 (1972).
In Italian.
    A nomogram is illustrated to obtain, for short intervals of
time, the variations of azimuth and of zenith distance in regard to the values of the programme given by the tables of the
Istituto Geografico Militare, and valid for the observation of
stars to the 60° altitude circle.

**046.002   The compiling of working ephemerides for latitude
determination on the Antarctic continent.**
F. D. Zablotskij.
Izv. vyssh. ucheb. zavedenij. Geod. i aehrofotos"emka, 1971,
No. 4, p. 47 - 53. In Russian. – Abstr. in Referativ. Zhurn. 51.
Astron., 8.51.142 (1972).

**046.003   Determination of the mutual position of points on
the earth's surface from synchronous laser observations of artificial earth satellites.   O. S. Razumov.**
Geod. kartogr. i aehrofotos"emka. Resp. mezhved. nauch.-
tekhn. sb., 1971, vyp. (No.) 14, p. 73 - 79. In Russian.
Abstr. in Referativ. Zhurn. 62. Issled. kosm. prostranstva,
8.62.302 (1972).

**046.004   Determination of a ship's position at sea by three
and more celestial bodies using the method of joint**

solution of equal altitude circle equations.
L. F. Cherniev, V. I. Danilov.
Astron. Tsirk., No. 702, p. 3 - 6 (1972). In Russian.

046.005   Ship navigation – The means and the end.
F. M. Foley.
Journ. Navigation, London, Vol. 25, 305 - 326 (1972).

**046.006   On compiling tables of the azimuths of Polaris for
the equatorial zone.   Tran Duy Thoan.**
Geod., kartogr. i aehrofotos"emka. Resp. mezhved. nauch.-
tekhn sb., 1972, vyp. (No.) 15, p. 91 - 96. In Russian. – Abstr.
in Referativ. Zhurn. 51. Astron., 9.51.151; 52. Geod. Aehro-
s"emka, 9.52.121 (1972).

**046.007   The application of the circumzenithal for latitude
and longitude determinations.   K.-G. Steinert.**
Izv. vyssh. ucheb. zavedenij. Geod. i aehrofotos"emka, 1971,
vyp. (No.) 5, p. 69 - 75. In Russian. – Abstr. in Referativ.
Zhurn. 51. Astron., 9.51.180; 52. Geod. Aehros"emka,
9.52.118 (1972).

**046.008   Bericht über die Arbeiten 1971 des Sonderforschungs-
bereiches 78 Satellitengeodäsie.**
Presented by M. Kneißl.
Veröff. Bayer. Kommission Internationale Erdmessung, Bayer.
Akad. Wiss., Astron.-Geod. Arbeiten, No. 29, 38 pp. (1972).

046.009 Simultanbestimmungen der Lotabweichungskomponenten ξ und η mit dem Prismenastrolabium. VIII.
Beobachtungen auf den Hauptdreieckspunkten Adenau, Dannenfels, Fürth, Gerolzhofen, Hilscheid, Langenbrand, Losheim, Maikammer, Mandern, Niederreifenberg, Nordenberg, Rotenfels, Spiegelberg, Untergrombach und Waldkatzenbach im Jahre 1966.   A. Rödde.
Deutsche Geod. Kommission Bayer. Akad. Wiss., Ser. B, No. 177, 28 + 71 pp. (1971).
The author carried out astronomical observations for simultaneous determination of longitude and latitude on 15 stations of the West-German section of the European network of primary triangulation. From these observations he computed the components ξ and η of the deflection of the vertical.

046.010 Simultanbestimmungen der Lotabweichungskomponenten ξ und η mit dem Prismenastrolabium. IX.
Beobachtungen auf den Hauptdreieckspunkten Auerbach, Bad Segeberg, Boostedt, Borlinghausen, Coburg, Hamburg-Blankenese, Haßwald, Havighorst, Heidelberg, Hoisdorf, Kaisersbach, Keidenzell, Kisdorf, Külsheim, Niederreifenberg, Niederstetten, Oberspeltach, Rohrbrunn, Rothmannsthal, Weiler, Weipoltshausen, Westensee, Zeegendorf und auf der Satellitenbeobachtungsstation Kloppenheim im Jahre 1967.   A. Rödde.
Deutsche Geod. Kommission Bayer. Akad. Wiss., Ser. B, No. 154, 27 + 119 pp. (1971).
The author carried out astronomical observations for the simultaneous determination of longitude and latitude on 23 stations of the West-German section of the European network of primary triangulation and on the satellite observing station Kloppenheim of the Institut für Angewandte Geodäsie, Frankfurt/Main. From these observations he computed the components ξ and η of the deflection of the vertical for the stations of the primary triangulation.

046.011 Über die Anwendung des sphärischen Bogenschnittes in der geodätischen Astronomie.   B. Kilar.
Deutsche Geod. Kommission Bayer. Akad. Wiss., Ser. C, No. 174, 98 pp. (1972). – Diss. Techn. Univ. München.

046.012 Zur geodätischen Nutzung der Laserortung zum Mond.   K. Arnold.
Gerlands Beiträge Geophys., Vol. 80, 401 - 405 (1971).
It is discussed how laser tracking to the moon can be used for the determination of spatial coordinates relative to the gravity centre of the earth.

046.013 Report of the DoD Geoceiver Test Program.
Defense Mapping Agency, Washington, D.C., DMA Report 0001, 18 + 184 pp. (1972).
In 1971, a test program was designed to evaluate the capabilities of an AN/PRR-14, a portable Doppler Geodetic Receiver (Geoceiver). Geoceiver data acquired jointly by each of the three Services were assembled and processed by each to determine the precise positions of the observing stations. The Geoceiver was shown to be an instrument of excellent geodetic capability.

046.014 Adjustment of large observation systems in networks of satellite triangulation.
W. Baran.
Artificial Satellites, Polish Acad. Sci., Vol. 7, No. 2, p. 17 - 22 (1972).

046.015 Analysis of the accuracy of determination of directions Poznań-Sofia and Riga-Sofia based on synchronous observations of the satellites Pageos and Echo 2 carried out during 1967–1969.
W. Baran.
Artificial Satellites, Polish Acad. Sci., Vol. 7, No. 2, p. 113 -

122 (1972).

046.016 Processing of results of observations of the satellite 65-011-04 carried out within the framework of INTEROBS program 1966.   W. Góral.
Artificial Satellites, Polish Acad. Sci., Vol. 7, No. 2, p. 123 - 134 (1972).

046.017 Une analyse numérique des projets du réseau mondial de triangulation satellitaire.   L. Minowska.
Geodezja i Kartografia, Vol. 21, 143 - 167 (1972). In Polish.

046.018 Transformation of rectangular coordinates x, y, z, into geodetic coordinates B, L, and ellipsoidal heights H.   A. Czarnecki.
Geodezja i Kartografia, Vol. 21, 195 - 198 (1972). In Polish.
Appropriate formulas are evolved for the solution of certain problems in satellite geodesy where coordinate transformation is required. The formulas are adapted for use of electronic digital computers.

046.019 Détermination de la différence d'altitude à l'aide de distances zénithales et d'une distance déterminée par le télémètre electromagnétique.   W. Dąbrowski.
Geodezja i Kartografia, Vol. 21, 199 - 206 (1972). In Polish.

046.020 Über das stochastische Verhalten von photographisch bestimmten Stern- und Satellitenkoordinaten.   G. Seeber.
Deutsche Geod. Kommission Bayer. Akad. Wiss., *München*, Ser. C, No. 178, 4 + 116 pp. (1972). – Diss. Landwirtschaftl. Fak., Rheinisch. Friedrich-Wilhelms-Univ. Bonn.

046.021 Application of laser ranging techniques to ground baselines.   J. C. Husson, M. Fournet, J. Gaignebet.
Bull. Groupe Recherches Géod. Spatiale, Obs. Meudon, No. 1, p. 170 - 176 (1971).

046.022 Application des méthodes de télémétrie laser spatiales à la mesure des distances terrestres.
J. C. Husson, M. Fournet, J. Gaignebet.
Bull. Groupe Recherches Géod. Spatiale, Obs. Meudon, No. 1, p. 177 - 183 (1971).

046.023 Precision geodesy via radio interferometry.
H. F. Hinteregger, I. I. Shapiro, D. S. Robertson, C. A. Knight, R. A. Ergas, A. R. Whitney, A. E. E. Rogers, J. M. Moran, T. A. Clark, B. F. Burke.
Science, Vol. 178, 396 - 398 (1972).
Very-long-baseline interferometry experiments, involving observations of extragalactic radio sources, were performed in 1969 to determine the vector separations between antenna sites in Massachusetts and West Virginia. The 845.130-kilometer baseline was estimated from two separate experiments. The experiments also yielded positions for nine extragalactic radio sources, most to within 1 arc second, and allowed the hydrogen maser clocks at the two sites to be synchronized a posteriori with an uncertainty of only a few nanoseconds.

046.024 Erfassung konstanter systematischer Fehlerkomponenten auf Grund parallel durchgeführter Beobachtungen.   J. Höpfner.
Vermessungstechnik, 19. Jahrgang, p. 379 - 381 = Mitt. Zentralinst. Physik der Erde, Potsdam, No. 167 (1971).

046.025 Zur geodätischen Nutzung der Entfernungs- und Radiointerferenzmessungen nach entfernten kosmischen Objekten.   K. Arnold.
Deutsche Akad. Wiss. Berlin. Forschungsbereich Kosmische Phys., Veröff. Zentralinst. Phys. Erde, No. 13, 62 pp. (1972).
Mathematical equations are derived which relate the ob-

served distances between a station on the earth and a laser reflector on the moon on the one hand and the geocentric station coordinates on the earth, the geocentric coordinates of the moon, the elements of the polar motion matrix, the light velocity and the sidereal time on the other hand. The error equations for the adjustment of such distance measurements to the moon are added, specified for different coordinate system. Likewise the observation of a quasar by a long base line interferometer yields the possibility to determine the difference between both the distances antenna—quasar for both the antennas.

046.026 **Very long baseline interferometry observations of radio emissions from geostationary satellites.**
R. D. Michelini, M. D. Grossi.
Space Research XII, (see 012.016), Vol. 1, 517 - 525 (1972).

046.027 **Reduction of refraction effects close to the horizon.**
J. Kakkuri.
Space Research XII, (see 012.016), Vol. 1, 575 - 579 (1972).

046.028 **Construction de grilles cartographiques à l'aide d'un ordinateur.** A. Delcroix.
Bull. Soc. Roy. Sci. Liège, Vol. 41, 319 - 330 = Univ. Mons, Dép. Astrophys., Commun. No. 30 (1972).
The purpose of the present paper is to explain how we have resolved the problem of geographic maps projections, by means of an electronic computer and a tracing table.

046.029 **Investigation of the accuracy of azimuth determination from the sun.**
N. A. Budenkov, N. F. Os'makov.
Sb. nauch. trudov po geodezii. Volgograd, 1971, p. 44 - 46. In Russian. — Abstr. in Referativ. Zhurn. 52. Geod. Aehros"-emka, 12.52.92 (1972).

046.030 **On a method of approximate azimuth determination from observations of the passage of the solar disc across the horizontal wire of a telescope.**
V. V. Kirichuk, A. S. Lavnikevich.
Geod. i kartografiya, 1972, No. 7, p. 11 - 15. In Russian. — Abstr. in Referativ. Zhurn. 52. Geod. Aehros"emka, 12.52.93 (1972).

046.031 **The relation of the European Datum to a geocentric reference system.**
J. G. Marsh, B. C. Douglas, S. M. Klosko.
Bull. Géod., Nouvelle Sér., Année 1972, No. 106, p. 407 - 424. — Presented at the International Union of Geodesy and Geophysics Meeting in Moscow, U.S.S.R., August 1971.

046.032 **Bemerkungen über das Geodätische Referenzsystem 1967.** V. K. Hristov.
Gerlands Beiträge Geophys., Vol. 81, 254 - 258 (1972).
The present paper is based on two official documents of the IUGG and IAG on the Geodetic Reference System 1967. The author is thoroughly convinced that on deducing the parameters of the reference ellipsoid and the parameters of the formula for the normal gravity there have to be established error equations containing all quantities derived by observation, i. e. geodetical, astronomical, gravimetrical and the values yielded by the satellite geodesy; here a joint balancing must be performed.

046.033 **Application in navigation of some considerable development into series.** E. Fichera.
Ist. Univ. Navale, Astron. Generale e Sferica, Napoli, Seminario No. 1, 11 pp. (1972). In Italian.

046.034 **La políode nelle applicazioni pratiche.**
E. Fichera.

Ist. Univ. Navale, Astron. Generale e Sferica, Napoli, Seminario No. 2, 9 pp. (1972).

046.035 **Plate reduction for the stellar triangulation.**
J. Kakkuri.
Publ. Finnish Geod. Inst., *Helsinki*, No. 72, 38 pp. (1972).
Three different reduction methods used in the Finnish Geodetic Institute and one used in the Tuorla Institute have been considered.

046.036 **Bestimmungen von astronomischen Längen, Azimuten und Breiten in den Jahren 1965 bis 1970.**
Prepared by H. Müller, edited by Schweiz. Geod. Kommission, Zürich. Astron.-geod. Arbeiten in der Schweiz, Vol. 29, 118 pp. (1972).

046.037 **Complements of modern navigation.** E. Fichera.
Ist. Univ. Navale, Astron. Generale e Sferica, Napoli, Seminario No. 3, 19 pp. (1972). In Italian.
In this paper we treat two complementary questions — Podaria and Volterra's integrals — necessities required in the construction of the modern nautical ephemerides.

046.038 **Beitrag zur Bestimmung geozentrischer Stations-koordinaten aus Satellitenbeobachtungen.**
H. Ludwig.
Deutsche Geod. Kommission, Bayer. Akad. Wiss., München, Ser. C, No. 177, 3 + 78 pp. (1972). — Diss. Techn. Univ. München.

046.039 **Zur Optimierung geodätischer Beobachtungen.**
H. Herzog.
Deutsche Geod. Kommission, Bayer. Akad. Wiss., München, Ser. C, No. 180, 4 + 63 pp. (1972).

046.040 **Descripción de un aparato para medidas de coordenadas.** R. Vieira, R. Ortiz.
Urania Barcelona, Año 56, No. 274, p. 227 - 240 = Univ. Madrid, Fac. Ci., Seminario Astron. Geod. Publ. No. 70 (1971).

046.041 **Systemgleichung und Modelltheorie.**
M. Schädlich.
Wiss. Zeitschr. Techn. Univ. Dresden, Vol. 21, No. 3, (see 012.021), 587 - 589 (1972).

046.042 **Erste Ergebnisse mit den transportablen Zenit-kameras der astronomisch-geodätischen Station Hannover.** J. Geßler, K. Pilowski.
Zeitschr. Vermessungswesen, 97. Jahrgang, p. 372 (1972).

046.043 **Beobachtungen nach dem Azimutstandlinienver-fahren mit dem geodätisch-astronomischen Universal-theodolit Theo 002, vom VEB Carl Zeiss Jena.** S. Knorke.
Wiss. Zeitschr. Techn. Univ. Dresden, Vol. 21, No. 3, (see 012.021), 595 - 597 (1972).

046.044 **Die Ausgleichung eines räumlichen terrestrischen Netzes nach der Methode der Satellitengeodäsie.**
J. Kabeláč.
Wiss. Zeitschr. Techn. Univ. Dresden, Vol. 21, No. 3, (see 012.021), 634 - 637 (1972).

Ein kleines Modell des Diazenitals Nušl—Frič und die Ortsbestimmung. See Abstr. 032.045.

Einsatz der Automatischen Kamera für Astrogeo-däsie aus Jena im ISAGEX-Programm. See Abstr. 032.049.

Astronomical determinations of latitude and longitude in 1961–1966. See Abstr. 045.044.

The ATS-F/Nimbus-E tracking experiment.
See Abstr. 054.002.

WESTA Net. Elaboration and preliminary analysis of observational assignment.   See Abstr. 055.004.

WESTA Net. Determination of absolute directions of lines connecting the points of the network.

See Abstr. 055.005.

Photographische Satellitenbeobachtungen mit der Automatischen Kamera für Astrogeodäsie.
See Abstr. 055.023.

Der Streit um die Figur der Erde. Zur Begründung der Geodäsie im 17. und 18. Jahrhundert.
See Abstr. 081.027.

047   Ephemerides, Almanacs, Calendars

047.001   Sterregids 1973.
Compiled by J. Meeus.
Hemel en Dampkring, Vol. 70, No. 7/8, 69 pp. (1972).

047.002   Astronomical Yearbook of the USSR for the year 1975.   V. K. Abalakin (Editor).
Institut Teoreticheskoj Astronomii Akademii Nauk SSSR.
Izdatel'stvo "Nauka", Leningradskoe Otdelenie, Leningrad. 718 pp. Price 7 Rbl. 72 Kop. (1972).   In Russian.

047.003   Supplement to the Astronomical Yearbook of the USSR for the year 1975.
Institut Teoreticheskoj Astronomii Akademii Nauk SSSR.
Izdatel'stvo "Nauka", Leningradskoe Otdelenie, Leningrad. 47 pp. (1972). In Russian.

047.004   Nautisches Jahrbuch für das Jahr 1973.
Edited by Deutsche Demokratische Republik, Seehydrographischer Dienst, Rostock. 23rd year. 29 + 365 pp. (1972).

047.005   Japanese Ephemeris 1973.
Compiled under the supervision of A. M. Sinzi, by T. Satō, A. Senda, T. Mori, Y. Harada, K. Inoue, K. Nagamori, Y. Suzuki, Y. Kubo.
Astronomical Division, Hydrographic Department, Tokyo, Japan. Pub. No. 684, 6 + 463 pp. (1972).

047.006   Nautisches Jahrbuch oder Ephemeriden und Tafeln für das Jahr 1973, zur Bestimmung der Zeit, Länge und Breite zur See nach astronomischen Beobachtungen.
Edited by "Deutsches Hydrographisches Institut", Hamburg. 122. Jahrgang, 3 + 43 + 366 + 30 pp. (1972).

047.007   The Star Almanac for Land Surveyors for the Year 1973.
Prepared by *H. M. Nautical Almanac Office,* published by Order of *The Science Research Council.* Her Majesty's Stationery Office, London. 16 + 76 pp. Price 45 p. (1972).

047.008   The Handbook of the British Astronomical Association 1973.
Prepared by the Computing Section of the Association under the supervision of C. Dinwoodie.
Office of the Association: Burlington House, Piccadilly, London. 87 pp. Price 15 s., $2.25 respectively (1972).

047.009   The American Ephemeris and Nautical Almanac for

the year 1972.
Issued by Nautical Almanac Office, United States Naval Observatory, Washington; Her Majesty's Nautical Almanac Office, Royal Greenwich Observatory, London. U.S. Government Printing Office, Washington. 8 + 568 pp. Price $ 6.00 (1970).

047.010   Efemérides Astronómicas para o ano de 1973.
Edited by Observatório Astronómico da Universidade de Coimbra.
Imprensa de Coimbra, Limitada, Coimbra. 13 + 240 pp. (1972)

047.011   Apparent Places of Fundamental Stars 1974, containing the 1535 stars in the Fourth Fundamental Catalogue (FK4).
Edited by Astronomisches Rechen-Institut, Heidelberg, under the supervision of T. Lederle. Published and produced by G. Braun GmbH, Karlsruhe. To be purchased from Verlag G. Braun, Karlsruhe. 44 + 510 pp. Price DM 42.00 (1971).

047.012   The Astronomical Ephemeris for the Year 1973.
Issued by Her Majesty's Nautical Almanac Office, London; Nautical Almanac Office, United States Naval Observatory, Washington. Her Majesty's Stationery Office, London. 8 + 584 pp. Price £ 3.50 net (1972).

047.013   The Nautical Almanac for the year 1974.
Issued by Her Majesty's Nautical Almanac Office, London; and Nautical Almanac Office United States Naval Observatory, Washington. Printed and published by Her Majesty's Stationery Office, London. A4 + 276 + 35 pp. Price £ 1.60 (1972).

047.014   Das Himmelsjahr. Sonne, Mond und Sterne im Jahr 1973.
Compiled by M. Gerstenberger.
Kosmos-Verlag, Franckh'sche Verlagshandlung, Stuttgart. 111 pp. Price DM 6.80 (1972).

047.015   The Indian Ephemeris and Nautical Almanac for the Year 1973.
Office of preparation: Nautical Almanac Unit, Regional Meteorological Centre, Alipore, Calcutta. Printed by the Manager, Government of India Press, Calcutta. 22 + 479 pp. Price Rs. 17.00, 39s, 8d., $ 6.12, respectively (1972).

047.016   The Air Almanac 1973, January – April.
Her Majesty's Stationery Office, London; United States Naval Observatory, Washington. 4 + 242 + A84 +

F4 pp. Price £ 2.00 (1972).

047.017 **1973 Nautical Almanac.** Pub. No. 681.
Published by Hydrographic Office of Japan, Tokyo.
4 + 463 pp. (1972).

047.018 **Efemérides Astronómicas 1973.**
Published by Instituto y Observatorio de Marina.
San Fernando (Cádiz). Printed in Spain by Imprenta del Observatorio de Marina, San Fernando. Vol. 182, 6 + 613 pp.
Price 200 pesetas (1972).

047.019 **Annuaire de l'Observatoire Royal de Belgique**
[Jaarboek van de Koninklijke Sterrenwacht van
België] 1973.
Imprimerie Hayez, Bruxelles. 140ᵉ année (jaargang), 233 pp.
(1972).

047.020 **Éphémérides astronomiques pour 1973.**
Publiées par la Société Astronomique de France,
with a preface by B. Morando.
L'Astronomie, 87ᵉ année, Suppl. pour Janvier 1973. 3 + 65 pp.
Price F 10.00 (1972).

047.021 **Astronomical calendar of the Observatory in Sofia**
for the year 1973. N. Bonev (Editor).
Izdatelstvo na B'lgarskata Akademiya na Naukite, Sofiya.
92 pp. Price 0.80 Lv. (1972). In Bulgarian.

047.022 **Rocznik Astronomiczny na Rok 1973.**
Prepared under the supervision of J. Radecki.
Instytut Geodezji i Kartografii, Państwowe przedsiębiorstwo
wydawnictw kartograficznych, Warszawa. Vol. 28, 132 pp.
Price zł 68.00 (1972).

047.023 **Himmelskalender 1973.**
Ein astronomisches Jahrbuch für Österreich.
H. Mucke, K. Mayrhofer (Editors).
Verlag, Astronomisches Büro, H. Mucke, Wien. 87 pp. Price
öS 25.00 (1972).

047.024 **The Observer's Handbook 1973.**
J. R. Percy (Editor).
Royal Astronomical Society of Canada, Toronto, Ontario,
Canada. 105 pp. Price $ 2.00 (1972).

047.025 **Hvězdářská ročenka 1973.**
Compiled by J. Ruprecht.
Ročník 49. Academia nakladatelství Československé akademie
věd, Praha. 231 pp. Price Kčs. 20.00 (1972).

047.026 **Anuarul Observatorului din Bucureşti – 1973.**
Editura Academiei Republicii Socialiste Românîa.
255 pp. Price Lei 25.00 (1972).

047.027 **Der Sternenhimmel 1973.**
Kleines astronomisches Jahrbuch für Sternenfreunde.
R. A. Naef (Editor).
Verlag Sauerländer, Aarau. 33. Jahrgang, 198 pp. Price Fr.
15.00 (1972).

047.028 **Astronomical Handbook for Southern Africa 1973.**
Published by the Astronomical Society of Southern
Africa, Cape Town. 2 + 46 pp. (1972).

047.029 **Anuário do Observatório de S. Paulo para 1973.**

Published by Instituto Astronômico e Geofísico,
Universidade de São Paulo, São Paulo, Brasil. 11 + 114 + 181 +
2 pp. (1972).

047.030 **Almanaque Nautico y Aeronautico para el año 1973.**
Observatorio Naval, Republica Argentina. Armada
Argentina, Servicio de Hidrografia Naval, Buenos Aires. 384 pp.
Price $ 15,00 (1972).

047.031 **Suplemento al Almanaque Nautico y Aeronautico**
para el año 1973. Sol, Planetas y Estrellas.
Observatorio Naval, Republica Argentina. Armada Argentina,
Servicio de Hidrografia Naval, Buenos Aires. 8 + 133 pp. Price
$ 4.00 (1972).

047.032 **Annuario 1973.**
Edited by Osservatorio Astronomico di Torino,
[printed by Tipografia Rubatto, Pino Torinese]. Pubbl. Varie
Fuori Ser., Oss. Astron. Torino, No. 51, 67 pp. (1972).

047.033 **Nautički Godišnjak 1973.**
Published by Hidrografski Institut Jugoslavenske
Ratne Mornarice, Split. H I–N–31, Godina 31, 10 + 213 +
68 pp (1972).

047.034 **Anuario del Observatorio Astronómico Nacional**
1972.
Published by Observatorio Astronómico Nacional, Univ. Nacional Colombia, Fac. Ciencias, Bogotá, Colombia, S.A.
112 pp. (1971).

047.035 **Astronomical ephemeris for the year 1973.**
R. Danić, I. Zoran.
Vasiona, Vol. 20, 82 - 96 (1972). In Serbo-Croatian.

047.036 **The Air Almanac 1973, May – August.**
Her Majesty's Stationery Office, London; United
States Naval Observatory, Washington. 2 + 248 + A 84 + F 4 pp.
Price £ 2.00 (1972).

047.037 **1973 Abridged Nautical Almanac.** Pub. No. 683.
Published by Hydrographic Office of Japan, Tokyo.
239 pp. (1972).

047.038 **Astronomical calendar. Yearbook. Variable part.**
1973. P. I. Bakulin (Editor).
Vsesoyuznoe astron.-geodezicheskoe obshchestvo, vyp. (No.)
76. Nauka, Moskva. 224 pp. Price 66 Kop. (1972). In Russian.

047.039 **Anuario del Observatorio Astronómico de Madrid**
para 1973.
Published by Instituto Geográfico y Catastral, Madrid. 500 pp.
Price 100 pesetas (1972).

047.040 **Connaissance des Temps ou des mouvements cé-**
lestes pour l'an 1974 à l'usage des astronomes et des
navigateurs.
Publiée par le Bureau des Longitudes. Gauthier-Villars Éditeur,
Paris. 64 + 493 + A 143 pp. Price F 220.00 (1972).

Annuaire 1973 du Bureau des Longitudes. Encyclopédie Physique et Spatiale. See Abstr. 003.043.

International Information Bureau on Astronomical
Ephemerides. See Abstr. 041.081.

Space Research

051 Extraterrestrial Research, Spaceflight Related to Astronomy

051.001 **Orbiting Astronomical Observatory: Review of scientific results.** A. D. Code, B. D. Savage.
Science, Vol. 77, 213 - 221 (1972).

051.002 **The strategy of space research.** H. Alfvén.
Priroda, No. 7.72, p. 11 - 14 (1972). In Russian.

051.003 **Sounding rockets in space astronomy.**
G. R. Carruthers.
Sky Telescope, Vol. 44, 218 - 221 (1972).

051.004 **Pioneer 10 observations of starlight and zodiacal light at large elongations: Preliminary results.**
J. L. Weinberg, M. S. Hanner, H. M. Mann, P. B. Hutchison.
Bull. American Astron. Soc., Vol. 4, 399 (1972). – Abstr. AAS.

051.005 **Projekt Helios.** H. G. Hasler.
Bild der Wiss., 9. Jahrgang, p. 592 - 599 (1972).
Within the scope of this German-American research project two space probes will be launched into an orbit around the sun in 1974/75. These space probes offer a favourable opportunity to explore the inner part of our planetary system and to study the interplanetary matter.

051.006 **Space physics and cooperation of scientists of socialist countries.** B. N. Petrov, M. G. Kroshkin.
Vestn. AN SSSR, 1972, No. 4, p. 76 - 84. In Russian.

051.007 **Strategy of space investigations.** H. Alfvén.
Zemlya i Vselennaya, 1972, No. 4, p. 13 - 17.
In Russian.

051.008 **The effects of drag on relativistic spaceflight.**
A. R. Martin.
Journ. British Interplanet. Soc., Vol. 25, 643 - 653 (1972).

051.009 **Introduction: Mission opportunities and modes.**
N. Sirri.
Comets. Proc. Tucson Conference 1970, (see 012.010), p. 1 - 3 (1972).

051.010 **1976 d'Arrest comet mission study.** J. A. Gardner.
Comets. Proc. Tucson Conference 1970, (see 012.010), p. 153 - 154 (1972).

051.011 **Trajectory requirements for comet rendezvous.**
A. L. Friedlander, J. C. Niehoff, J. I. Waters.
Comets. Proc. Tucson Conference 1970, (see 012.010), p. 155 (1972).

051.012 **Some scientific criteria for a cometary mission.**
A. H. Delsemme.
Comets. Proc. Tucson Conference 1970, (see 012.010), p. 156 - 161 (1972).

051.013 **Electronic balloon-payload cutdown timer.**
R. W. Fuchs, D. J. Hofmann, G. J. Erickson, J. S. Mentek, R. J. Archuleta.
Rev. Sci. Instruments, Vol. 41, 131 - 132 (1970). – Note.

051.014 **15 years of Soviet space investigations.**
M. P. Vasil'ev.
Zemlya i Vselennaya, 1972, No. 5, p. 2 - 6. In Russian.

051.015 **Interstellar flight and intelligence in the universe.**
C. Powell.
Spaceflight, Vol. 14, 442 - 447 (1972).

051.016 **Japan in space.** H. Shima.
Nature, Vol. 240, 215 - 217 (1972).
Japan's future plans for space research are outlined.

051.017 **The large space telescope program.** C. R. O'Dell.
Sky Telescope, Vol. 44, 369 - 371 (1972).

051.018 **A revised low-frequency cosmic noise spectrum.**
R. R. Weber.
Astron. Journ., Vol. 77, 707 - 710 (1972).
Cosmic radio noise spectra obtained from an Astrobee sounding rocket and from more detailed analysis of RAE-1 data are presented. The flux levels measured by the two spacecraft are consistent and fall off more rapidly below 1 MHz than previous data.

051.019 **Europäische Beteiligung am Apollo-Nachfolge-Programm.** Empfehlungen des Fachausschusses ,,Weltraumforschung und Weltraumtechnik''.
SuW, Vol. 11, 345 (1972).

051.020 **Kosmosa apgūšana.**
Zvaigžņōta debess, 1971. gada rudens, p. 22 - 29.
Cilvēks kosmosā (O. Gazenko), p. 22 - 25; Uz jaunu sasniegumu slieksņa, p. 25 - 29.

051.021 **Kosmosa apgūšana.**
Zvaigžņotā debess, 1972. gada pavasaris, p. 12 - 25.
Orbitālā stacija "Salūts", p. 12 - 14; Vimpelis ar PSRS gerboni uz Marsa, p. 15 - 17; Observatorijas pēta "sarkano" planētu, p. 17 - 21; Apollo-15 (J. Francmanis), p. 21 - 25.

051.022 **Kosmosa apgūšana. Celš uz Mēness kalniem.**
A. Guršteins.
Zvaigžņotā debess, 1972. gada vasara, p. 30 - 32.

051.023 **Astronautica.**
Coelum, Vol. 40, 154 - 156, 196 - 197, 234 - 236 (1972).

051.024 **Electrodynamic sailing: Beating into the solar wind.**
C. P. Sonett, U. Fahleson, H. Alfvén.
Science, Vol. 178, 1115 - 1119 (1972).

051.025 **The international ultraviolet explorer satellite.**
ESRO/ELDO Bull., No. 18, p. 13 - 17 (1972).
The general objectives for observations on this mission are summarized.

051.026 **Stellar X-ray source keeps rocket on course.**
R. D. Cooper, A. F. Janes.
Electron. Engineering, (GB), Vol. 44, No. 535, p. 58 - 60

(1972).

Discusses the use of stellar X-ray source to effect roll correction and thus stabilise rocket flight.

051.027 NASA's second decade in space. E. J. Manganiello. Proc. Instn. Mechanical Engineers, *(GB)*, Vol. 186, 527 - 545 (1972). – See Phys. Abstr., Vol. 76, No. 5618 (1973).

051.028 Space report.
Spaceflight, Vol. 14, 300 - 304, 343 - 344, 350, 380 - 383, 419 - 423, 432, 455 - 461 (1972).
Rock of lunar mountains, p. 300; India's first satellite, p. 300 - 301; Moon-strike recorded, p. 301; Has the moon a core?, p. 380; Space test of relativity, p. 380 - 381; The last Apollo, p. 420 - 421; International astronomy satellite, p. 421 - 422; Space check for atomic clocks, p. 422; Vast solar storm, p. 455; Life in space?, p. 455; Asteroid Gagarin, p. 457 - 458; Luna rock, p. 459; Metallurgists learn from moon, p. 459 - 460; Design of Venera 8, p. 460; Mars probes results, p. 460 - 461; Storing Mars data, p. 461.

Space research: Physical and technical principles.
See Abstr. 003.088.

Interplanetary flight and communication.
See Abstr. 003.119.

Space research 1969–70. See Abstr. 003.147.

Solar observations from Skylab.
See Abstr. 034.105.

Preliminary results of the third flight of the Soviet stratospheric solar observatory. See Abstr. 071.052.

The aftermath of Apollo: Science on the shelf?
See Abstr. 094.228

Astronomical observations with the Apollo 16 far-ultraviolet camera/spectrograph.
See Abstr. 114.056.

052 Astrodynamics and Navigation of Space Vehicles

052.001 Resonant attitude instabilities for a symmetric satellite in a circular orbit. D. L. Hitzl.
Celestial Mechanics, Vol. 5, 433 - 450 (1972).

The roll-yaw attitude motion of a spinning symmetric satellite in a circular orbit is investigated with particular emphasis on the behavior near resonance. Two sections of the resonance line $\omega_2 = 3\omega_1$ permitting the largest effects are determined and the equations of motion are integrated numerically as a check on the resonance theory. In particular, resonance-induced instabilities are confirmed.

052.002 Influence du champ magnétique terrestre sur le mouvement d'un satellite autour de son centre de gravité. I. Stellmacher.
Celestial Mechanics, Vol. 5, 470 - 482 (1972).

The effect of the earth's magnetic field on the motion of a satellite around its centre of mass is investigated. The satellite is assumed to be dynamically symmetric and to be magnetized in the same direction as that of a principal axis. The earth's magnetic field is assumed to be a dipole field whose poles coincide with the rotation poles of the earth. The satellite's orbit is circular and perturbations are neglected.

052.003 Stability criteria for a free dual-spin satellite. A. P. Wang.
Astrophys. Space Sci., Vol. 16, 413 - 420 (1972).

Stability criteria for a space-platform containing a spinning rotor are obtained in this paper. It is shown that the so-called energy sink condition is not necessary for stability.

052.004 On the tidal effects in the motion of artificial satellites. P. Musen, R. Estes.
Celestial Mechanics, Vol. 6, 4 - 21 (1972).

The general perturbations in the elliptic and vectorial elements of a satellite as caused by the tidal deformations of the non-spherical earth are developed into trigonometric series in the standard ecliptical arguments of Hill-Brown lunar theory and in the equatorial elements ω and Ω of the satellite. Graphs are presented of the tidal perturbations in the elliptic elements

of the BE-C satellite which illustrate long term periodic behavior. The tidal effects are clearly noticeable in the observations and their comparison with the theory permits improvement of the 'global' Love numbers for the earth.

052.005 Launch windows to the planets. H. Oja.
Spaceflight, Vol. 14, 386 - 389 (1972).

052.006 On the choice of measured parameters for determining the trajectory of a space vehicle.
I. K. Bazhinov, V. N. Pochukaev, A. I. Serdyukov.
Kosmich. Issled., Vol. 10, 467 - 476 (1972). In Russian.

052.007 Orbit determination from composite data.
Yu. S. Savrasov.
Kosmich. Issled., Vol. 10, 494 - 498 (1972). In Russian.

052.008 Accuracy estimate of the methods for construction of the local vertical, based on the application of the infrared radiation of the earth.
A. V. Pavlov, V. D. Permyakov.
Kosmich. Issled., Vol. 10, 499 - 503 (1972). In Russian.

052.009 Théorie analytique programmée de l'influence gravitationnelle de la lune et du soleil dans le mouvement des satellites artificiels. X. Berger, Y. Boudon.
Bull. Groupe Recherches Géod. Spatiale, Obs. Meudon, No. 5, p. 1 - 28 (1972).

The study reveals, besides the well known periods such as 27 days or 6 months, also resonance terms. The coupling between the earth, moon and sun has also been investigated and its importance and applications are demonstrated.

052.010 Expressions analytiquès des termes séculaires en J_2^3 et à longues periodes en J_2^2. X. Berger.
Bull. Groupe Géod. Spatiale, Obs. Meudon, No. 5, p. 30 - 58 (1972).

In the analytical theory of the motion of a satellite around an oblate spheroid, the secular terms of order J_2^3 and the long-period terms of order J_2^2 are the only non-negligible terms that have not been given at present, because their development requires a theory to the third order. This development and the results are given here and a comparison with numerical integration confirms their accuracy.

052.011 The influence of the Galilean satellites of Jupiter on the motion of the artificial satellite of Callisto.
N. B. Batueva.
Vestn. Leningr. un-ta, 1972, No. 7, p. 141 - 145. In Russian. Abstr. in Referativ. Zhurn. 51. Astron., 9.51.130 (1972).

052.012 Effects of gravity-gradient torque on the rotational motion of a triaxial satellite in a precessing elliptic orbit.   J. E. Cochran.
Celestial Mechanics, Vol. 6, 127 - 150 (1972).

A method of general perturbations, based on the use of Lie series to generate approximate canonical transformations, is applied to study the effects of gravity-gradient torque on the rotational motion of a triaxial, rigid satellite. The method is used to obtain the Hamiltonian for the nonresonant secular and long-period rotational motion of the satellite to second order in n/ω_0, where n is the orbital mean motion of the center of mass and ω_0 is a reference value of the magnitude of the satellite's rotational angular velocity. Geometrical aspects of the long-term rotational motion are discussed and a comparison of theoretical results with observations is made.

052.013 Earth satellites in resonance with the moon and the sun as objects of laser ranging. Analytical solution for their motion.   Yu. V. Batrakov, V. G. Sokolov.
Celestial Mechanics, Vol. 6, 247 - 251 (1972).

The analytical solution for the motion of a satellite in resonance both with the moon and the sun has been outlined, the periodic orbit of the planar restricted four-body problem being taken as an intermediary. The von Zeipel transformation gives the Hamiltonian not depending on the fast variables. The stationary solution for this Hamiltonian has been found. The solution of the variation equations has been obtained and its accuracy has been tested by numerical integration.

052.014 Evolution of a satellite orbit due to radiation pressure.   Yu. N. Isaev, A. L. Kunitsyn.
Mekh. tverd. tela. Resp. mezhved. sb., 1972, vyp. (No.) 4, p. 134 - 141. In Russian. – Abstr. in Referativ. Zhurn. 51. Astron., 10.51.140; 62 Issled. kosmich. prostranstva, 10.62.285 (1972).

052.015 On the solution of an extremal problem on flying to the planets.   V. P. Kuraev, A. A. Panchukov.
Kosm. Issled., Vol. 10, 679 - 683 (1972). In Russian.

052.016 On the stability of dual-spin bodies with unbalancing mass.   A. P.-I Wang.
Astrophys. Space Sci., Vol. 17, 459 - 466 (1972).

This paper contains a straight-forward spin stability analysis of a composite body composed of a platform, rotor and unbalancing mass.

052.017 About the motion of a heavy flexible string attached to a satellite in a central field of attraction.
R. B. Singh, V. G. Demin.
Celestial Mechanics, Vol. 6, 268 - 277 (1972).

The motion of a heavy inextensible flexible string attached to a satellite in a central gravitational field is discussed. It is supposed that the mass of the string is infinitesimally small compared to the mass of the satellite. Under the assumption that the satellite moves along a Keplerian elliptical orbit (in particular circular orbit), the relative motion of the string is investigated.

052.018 The general equilibria of a spinning satellite in a circular orbit.   H. J. Sperling.
Celestial Mechanics, Vol. 6, 278 - 293 (1972).

Let a rigid satellite move in a circular orbit about a spherically symmetric central body, taking into account only the main term of the gravitational torque. We shall investigate and find all solutions of the following problem: Let the satellite be permitted to spin about an axis that is fixed in the orbit frame; the satellite need not be symmetric, the spin not uniform, and the spin axis not a principal axis of inertia.

052.019 Application of the method of averaging to the study of the motion of artificial satellites.   M. Calvo.
Publ. Sem. mat., *Zaragoza*, Vol. 15, 68 pp. (1971). In Spanish.

The method of averaging is applied to describe the motion of an artificial satellite around an oblate planet. The problem is expressed in Hill's variables and, extending previous results given by Cid and Lahulla, the problem is completely integrated. The solution is free of small divisors and the results in Hill's variables are considerably shorter than in Delaunay's.

052.020 Analysis of satellite-orbit perturbations due to forces deriving from a potential.
Y. Genin, A. Thayse.
Philips Res. Rep., Vol. 24, 477 - 558 (1969).

A consistent second-order analysis of the orbital perturbation of a close earth satellite, due to the earth's gravitational irregularities and to the perturbating influence of external bodies, is carried out in closed form. The perturbating effects of the various, zonal as well as tesseral, earth gravitational harmonics and of the main terms in the development of the external-bodies gravitational field are successively computed. Significative generalizations of known results are obtained mainly by the systematic use of the osculating-orbit true anomaly as independent variable.

052.021 Analytical investigation of near-parabolic lunar trajectories between moon and cislunar libration point.   W. B. Blair.
AIAA Journ., Vol. 9, 2437 - 2442 (1971).

An approximate analytical theory is developed for analysis for near-parabolic trajectories between the moon and the cislunar-libration point (L_1). The analysis of the earth as a perturbative influence on the moon-referenced trajectories involves the local regularization of the unperturbed and perturbed two-body problem and development of general Encke-type perturbation theory in the regularized domain. The linearized perturbation equations are demonstrated to be analytically integrable in the regularized domain. The analysis investigates velocity requirements at L_1 and the moon for passage in either direction and passing either in front of or behind the moon. The approximate analytic theory gives close agreement to representative numerical results from a digital computer.

052.022 Sur l'optimisation de l'entrée des engins spatiaux dans l'atmosphère planétaire.
A. Marinescu.
Atti Accad. Nazionale Lincei, Ser. Ottava, Rend. Cl. Sci. fis., mat., nat., Vol. 51, 352 - 358 (1971).

052.023 Literal theory of the motion of an earth satellite with small orbital eccentricity under the influence of tesseral harmonics of the earth's potential.   L. L. Filenko.
Byull. Inst. Teoret. Astron., *Leningrad*, Vol. 13, 246 - 257 (1972). In Russian.

A method is developed which allows to calculate first order perturbations of the orbital elements of an earth satellite due to tesseral harmonics with arbitrary indices k, m for small orbital eccentricities. The coefficients of the trigonometric terms of these expressions have been given as polynomials in powers of sinus and cosinus of the orbital inclination and as a power series in the eccentricity with an accuracy up to e^6.

052.024   Linear perturbations of the coordinates of satellites in the standard gravitational field of the earth.
B. I. Lokhin.
Byull. Inst. Teoret. Astron., *Leningrad,* Vol. 13, 300 - 307 (1972). In Russian.

A method of determination of the perturbations of the coordinates of satellites in the gravitational field of two fixed centres was obtained in the form of a quadrature. The intermediate motion was taken in the form given by Kislik. The chief results were correct for the case of the description of the intermediate motion by the integrals of Vinti.

052.025   Application of Gauss' method to the determination of secular radiational perturbations of artificial earth satellites.   E. M. Polyakhova.
Byull. Inst. Teoret. Astron., *Leningrad,* Vol. 13, 308 - 317 (1972). In Russian.

A method for the determination of the secular perturbations of artificial earth satellites under the influence of solar radiation pressure forces is proposed, where the perturbing acceleration components are calculated by Gauss' method. For the expansion of the potential in series Subbotin's method is applied. An example of the perturbations determination is given.

052.026   Determination of a symmetrical intermediate orbit of an artificial earth satellite.
E. P. Aksenov, L. M. Domozhilova.
Trudy Gos. Astron. Inst. Shternberga, Vol. 43, 52 - 66 (1972). In Russian.

The symmetrical intermediate orbits of satellites based on the problem of the two fixed centres are discussed. A comparison is made between the results of the theory and numerical data.

052.027   Determination of an asymmetrical intermediate orbit of an artificial earth satellite.
E. P. Aksenov, L. M. Domozhilova.
Trudy Gos. Astron. Inst. Shternberga, Vol. 43, 67 - 78 (1972). In Russian.

The asymmetrical intermediate orbits of satellites based on the problem of the two fixed centres are discussed. A comparison is made between the results of the theory and numerical data.

052.028   The influence of the solar radiation pressure on the motion of some kinds of artificial earth satellites.
E. E. Mukin.
Trudy Gos. Astron. Inst. Shternberga, Vol. 43, 79 - 88 (1972). In Russian.

Investigation of the influence of radiation pressure on the motion of an artificial earth satellite in shape of a plate the normal of which is constantly directed to the centre of the earth. Each side of the surface may show a different type of reflection − either regular reflection or diffuse scattering.

052.029   Anomalous orbital accelerations of the Pageos spacecraft.   D. E. Smith, K. E. Kissell.
Space Research XII, (see 012.016), Vol. 2, 1523 - 1527 (1972).

052.030   To the problem of satellite's perturbed motion under the influence of solar radiation pressure.
Y. N. Isayev, A. L. Kunitsyn.
Celestial Mechanics, Vol. 6, 44 - 51 (1972).

A possibility of developing the analytical theory of perturbed motion for a balloon-satellite influenced by solar radiation pressure is analysed on the basis of the limit case modification of the two fixed centers problem whose force-field is a superposition of the Newtonian central field and a homogeneous one. The relations between canonical constants of the intermediate orbit and quasi-Keplerian elements coinciding in the absence of solar radiation pressure with Keplerian ones are derived. Numerical results and an illustration of the perturbations in the radius-vector of the intermediate orbit of a balloon-satellite of the Echo-I type are given.

052.031   Combined gravitational and solar radiation pressure effects on the semimajor axis of the earth's satellite.
P. Lála.
Bull. Astron. Inst. Czechoslovakia, Vol. 23, 342 - 345 (1972).

Perturbations of the semimajor axis due to the direct solar radiation pressure during no shadowing periods are investigated. Apart from the previously known short-periodic perturbations, long-periodic ones are found when the motion of the sun, of the perigee and of the ascending node are taken into account.

052.032   On some simplified representations of the normal gravitation potential used in the theory of motion of artificial earth satellites.   O. F. Malakhova.
Trudy Mosk. in-ta radiotekhn., ehlektron. i avtomatiki, 1972, vyp. (No.) 57, p. 83 - 90. In Russian. − Abstr. in Referativ. Zhurn. 51. Astron., 12.51.108; 52. Geod. Aehros"emka, 12.52.51 (1972).

052.033   General solution of the equations of relative motion of satellites in second approximation.
Yu. A. Ermilov.
Kosmich. Issled., Vol. 10, 850 - 858 (1972). In Russian.

052.034   On the stationary solutions for the problem of artificial satellites.   H. Kinoshita, G. Hori.
Univ. Tokyo, Tokyo Astron. Obs. Rep. (No. 60), Vol. 16, 230 - 240 (1972). In Japanese.

052.035   An analysis of error and efficiency of second-order generalized multistep methods and applications to orbit computation.   J. Dyer.
Bull. American Astron. Soc., Vol. 4, 419 (1972). − Abstr. AAS.

052.036   Multi-revolution methods for orbit computation.
O. Graf.
Bull. American Astron. Soc., Vol. 4, 420 (1972). − Abstr. AAS.

052.037   Comparison of numerical integration techniques for orbital applications.   W. H. Moore.
Bull. American Astron. Soc., Vol. 4, 421 (1972). − Abstr. AAS.

052.038   Numerical integration and interpolation considerations in space shuttle optimization with gradient-type methods.   J. Pesapane, W. F. Powers.
Bull. American Astron. Soc., Vol. 4, 421 (1972). − Abstr. AAS.

052.039   The Lahulla elimination of small divisors in the eccentricities in a second order artificial satellite theory.   J. Meffroy.
Bull. American Astron. Soc., Vol. 4, 425 (1972). − Abstr. AAS.

052.040   Librations of gravity-oriented satellites in elliptic orbits through atmosphere.
V. J. Modi, S. K. Shrivastava.
AIAA Journ., Vol. 9, 2208 - 2216 (1971).

052.041   A method of earth-pointing attitude control for elliptic orbits.   G. M. Connell.
AIAA Journ., Vol. 10, 258 - 263 (1972).

A method of attitude control utilizing active regulation of the satellite pitch moment of inertia for controlling a non spinning satellite in an elliptic orbit to an earth-pointing orientation is investigated.

052.042 Validez de la hipótesis de esfera de influencia. I. Comparación de las transferencias de Hohmann para radios nulos y finitos, de las esferas de influencia de los planetas terminales.   V. Camarena Badía.
Urania Barcelona, Año 56, No. 274, p. 140 - 158 (1971).

052.043 Orbit determinations from range measurements.   M. Caputo.
Mem. Soc. Astron. Italiana, Nuova Ser., Vol. 43, 425 - 445 (1972).

The problem of the determination of the orbit of satellites orbiting around planets far from an observer moving in space is treated in several cases. The special case of satellite-to-satellite tracking and also that of the determination of the orbits and relative motion of the two planets are treated. The observations are supposed to be range measurements.

052.044 The conjugate gradient method and its application to aerospace vehicle guidance and control – I. Basic results in the conjugate gradient method.
C. T. Leondes, C. A. Wu.
Astronaut. Acta, Vol. 17, 871 - 880 (1972).

052.045 The conjugate gradient method and its application to aerospace vehicle guidance and control – Part II. Mars entry guidance and control.   C. T. Leondes, C. A. Wu.
Astronaut. Acta, Vol. 17, 881 - 890 (1972).

The conjugate gradient method is shown to be an effective method for the guidance and control of aerospace vehicles. Mars re-entry is considered as an example.

052.046 Effects of the sun and the moon on a near-equatorial synchronous satellite.   C.-H. Zee.
Astronaut. Acta, Vol. 17, 891 - 906 (1972).

Trajectories of a geo-stationary synchronous satellite under the influence of the gravitational fields of an oblate earth, the sun and the moon are derived by the asymptotic method and the method of solving differential equations of monofrequency oscillations in nonlinear mechanics. Closed-form solutions for an improved first order approximation are obtained. Solutions in terms of osculating orbital elements are also derived.

052.047 Mouvement de translation rotation d'un satellite aimante dans le champ magnétique terrestre.
I. Stellmacher.
Bull. Groupe Recherches Géod. Spatiale, No. 6, p. 1 - 11 (1972).

052.048 A unified state model of orbital trajectory and attitude dynamics.   S. P. Altman.
Celestial Mechanics, Vol. 6, 425 - 446 (1972). − Presented at the Astrodynamics Specialists Conference 1971, Fort Lauderdale, Florida, August 17 - 19, 1971.

The trajectory and attitude dynamics of an orbital spacecraft are defined by a unified state model, which enables efficient and rapid machine computation for mission analysis, orbit determination and prediction, satellite geodesy and re-entry analysis. The analytic partials of position and velocity with the state and coordinate variables are presented, as well as representative perturbation functions such as air drag, gravitational potential harmonics, and propulsion thrust.

052.049 Analytic short period lunar and solar perturbations of artificial satellites.   D. Fisher.
Celestial Mechanics, Vol. 6, 447 - 467 (1972).

The short period luni-solar theory of Kozai is generalized for arbitrary obliquity of the ecliptic and inclination of the moon's orbit to the ecliptic. Analytic first order lunar perturbations to the elements are derived. The theory is illustrated by an application to the communication satellite Intelsat 3F3.

052.050 Calculation of the motion around an axisymmetric planet without secular terms.   B. Popović.
Sci. Revuo, Beograd, Vol. 23, 213 - 226 (1972). In Esperanto.

Outlines of the motion of cosmic bodies.
See Abstr. 003.041.

Zur vorläufigen Bahnbestimmung künstlicher Erdsatelliten.   See Abstr. 003.158.

Normality condition in the ideal resonance problem.
See Abstr. 042.034.

The difference equations for estimate biases in the filtering problem.   See Abstr. 042.066.

Aplicación de las transformaciones de Lie a la eliminación de términos de corto período.   See Abstr. 042.088.

A method for computing the gravitational attraction of three-dimensional bodies in a spherical or ellipsoidal earth.
See Abstr. 081.063.

Errata

052.901 Errata: 'Theory of an experiment in an orbiting space laboratory to determine the gravitational constant' [Celestial Mechanics, Vol. 5, 204 - 254 (1972)].
J. P. Vinti.
Celestial Mechanics, Vol. 5, 518 (1972).

052.902 Erratum: 'On the rotation of the orbital plane during ballistic re-entry' [Planet. Space Sci., Vol. 19, 1215 - 1224 (1971)].   R. Arho.
Planet. Space Sci., Vol. 20, 1797 (1972).

053 Lunar and Planetary Probes and Satellites

053.001 **Venus revisited.**   R. N. Watts, Jr.
Sky Telescope, Vol. 44, 151 (1972).

053.002 **Saturn-Jupiter rebound. A method of high-speed spacecraft ejection from the solar system.**
K. A. Ehricke.
Journ. British Interplanet. Soc., Vol. 25, 561 - 571 (1972).
High solar system ejection speed can be attained either by powered flight only or by a combination of powered flight and gravitational assist.

053.003 **Scientific objectives for atmospheric probe missions: Jupiter.**   J. S. Lewis.
Bull. American Astron. Soc., Vol. 4, 353 (1972). – Abstr. AAS.

053.004 **PAET, the planetary atmosphere experiments test project, mission and spacecraft description.**
D. E. Reese.
Bull. American Astron. Soc., Vol. 4, 353 (1972). – Abstr. AAS.

053.005 **Atmospheric composition – the mass spectrometer experiment.**   H. B. Niemann, N. W. Spencer.
Bull. American Astron. Soc., Vol. 4, 354 (1972). – Abstr. AAS.

053.006 **Viking ionospheric measurements.**   W. B. Hanson.
Bull. American Astron. Soc., Vol. 4, 355 (1972).
Abstr. AAS.

053.007 **Measurements of the structure of Mars' atmosphere during entry of the Viking lander.**   A. Seiff.
Bull. American Astron. Soc., Vol. 4, 355 (1972). – Abstr. AAS.

053.008 **The surviving Jupiter entry probe and experiments.**
T. N. Canning, M. E. Tauber, R. G. Nagler.
Bull. American Astron. Soc., Vol. 4, 355 (1972). – Abstr. AAS.

053.009 **The Planetary Explorer probe studies.**   G. M. Levin.
Bull. American Astron. Soc., Vol. 4, 355 (1972).
Abstr. AAS.

053.010 **The turbopause probe and experiments.**
G. M. Levin.
Bull. American Astron. Soc., Vol. 4, 356 (1972). – Abstr. AAS.

053.011 **Concepts for the exploration of Mars beyond Viking '75.**   W. T. Scofield.
Bull. American Astron. Soc., Vol. 4, 364 (1972). – Abstr. AAS.

053.012 **A study of missions to comets and asteroids, key to better understanding of the solar system.**
R. L. Newburn.
Bull. American Astron. Soc., Vol. 4, 365 - 366 (1972).
Abstr. AAS.

053.013 **Pioneer 10 – the mission to Jupiter.**
N. K. Kroschel.
Journ. Astron. Soc. Victoria, Vol. 25, 39 - 43 (1972).

053.014 **The "grand tour" quits the field for the "little" one.**
D. Yu. Gol'dovskij.
Zemlya i Vselennaya, 1972, No. 4, p. 29 - 31. In Russian.

053.015 **Flights to the sun.**   V. I. Levantovskij.
Zemlya i Vselennaya, 1972, No. 4, p. 41 - 45.
In Russian.

053.016 **Plans and objectives of the remaining Apollo mis-**sions.   L. R. Scherer.
IAU Symposium No. 47, (see 012.008), p. 94 - 103 (1972).

053.017 **Engineering potential for lunar missions after Apollo.**
J. D. Burke.
IAU Symposium No. 47, (see 012.008), p. 104 - 120 (1972).

053.018 **Progress report on project Viking.**   R. N. Watts, Jr.
Sky Telescope, Vol. 44, 302 - 303 (1972).

053.019 **Scientific photographic experiment of the "Zond" space probes.**   V. D. Bol'shakov, N. P. Lavrova.
Zemlya i Vselennaya, 1972, No. 5, p. 12 - 16. In Russian.

053.020 **Preliminary scientific results of the flight of Venera 8.**
Zemlya i Vselennaya, 1972, No. 5, p. 66 - 67.
In Russian.

053.021 **Methods of investigation of Lunokhod's mobility under terrestrial conditions.**
A. K. Alexandrov, G. B. Nikolayev, V. I. Grafov, O. G. Ivanov, V. P. Velikanov, V. G. Romov, P. S. Semenov, P. P. Artemyev, N. V. Perov, V. P. Puchkov.
Space Research XII, (see 012.016), Vol. 1, 65 - 72 (1972).

053.022 **Investigations of mobility of Lunokhod 1.**
A. K. Alexandrov, B. M. Borisov, I. S. Garin, V. I. Grafov, A. G. Ivanov, Yu. P. Kotlov, V. I. Komarov, A. F. Kuleshov, V. K. Mishkin, G. B. Nikolayev, L. N. Polenov, P. S. Semenov, F. P. Yakovlev.
Space Research XII, (see 012.016), Vol. 1, 73 - 82 (1972).

053.023 **In the atmosphere and on the surface of Venus.**
Aviatsiya i kosmonavtika, 1972, No. 9, p. 35, 40 - 41. In Russian.

053.024 **Cosmic observatories "Prognoz" on flight.**
V. S. Agalakov, V. N. Karachevskij.
Zemlya i Vselennaya, 1972, No. 6, p. 50 - 52. In Russian.

053.025 **Pioneer 10.**   G. Bodifee.
Hemel en Dampkring, Vol. 70, 349 (1972).

053.026 **Apollo lunar spacecraft–how it works.**
Spaceworld, (*USA*), Vol. 1, 4 - 100, p. 4 - 60 (1972).

053.027 **Apollo lunar spacecraft–how it works.**
Spaceworld, (*USA*), Vol. 1, 5-101, p. 4 - 59 (1972).

053.028 **Performance optimization technique for the 1975 Mars Viking lander.**
H. N. Zeiner, C. E. French, D. A. Howard.
Journ. Spacecraft and Rockets, (*USA*), Vol. 9, 364 - 369 (1972). – See Phys. Abstr., Vol. 75, No. 62422 (1972).

053.029 **A Venus planetology mission based on the planetary explorer spacecraft.**   R. G. Ziehm, J. R. Mellin.
Journ. Spacecraft and Rockets, (*USA*), Vol. 9, 280 - 283 (1972).

053.030 **Apollo lunar spacecraft–how it works.**
Spaceworld, (*USA*), Vol. 1, 6 - 102, p. 4 - 60 (1972).
Describes the spacecraft, the lunar module, crew equipment, environmental control, guidance and navigation main propulsion and reaction control, electric power, communications instrumentation, lighting, portable life support systems.

053.031 **Entry probe descent to the base of the Jovian clouds.**
H. B. Winkler, D. P. Fields, R. G. Gamache.
Journ. Spacecraft and Rockets, (*USA*), Vol. 9, 660 - 667 (1972).

A study was conducted to identify and describe feasible first-generation Jupiter entry probe missions that measure atmospheric phenomena below the cloud tops, and that tend to minimize engineering development.

053.032 **On the precision of determination of space probes coordinates at the Astrophysical Institute of the Kazakh Academy of Sciences.**
V. S. Matyagin, L. A. Usoltseva, A. A. Shipenshtein.
Astron. Tsirk., No. 727, p. 6 - 8 (1972). In Russian.

Satellites: Cosmic IMP.
Nature, Vol. 239, 186 - 187 (1972).

Mission to Taurus-Littrow. Apollo 17 landing site.
Spaceflight, Vol. 14, 342 - 343 (1972).

Station Venera 8 reached the goal.
Zemlya i Vselennaya, 1972, No. 5, p. 9. In Russian.

Les satellites artificiels de l'année 1971.
See Abstr. 054.007.

Astronautique 1971. See Abstr. 054.008.

PAET, the atmosphere structure experiment.
See Abstr. 082.053.

PAET, atmospheric composition by shock layer radiometry. See Abstr. 082.054.

PAET, effects of entry vehicle dynamics on the atmospheric data. See Abstr. 082.055.

Spacecraft techniques for lunar research.
See Abstr. 094.146.

Med månen som mål. V. – Surveyor 3.
See Abstr. 094.245.

Med månen som mål. VI. – Apollo 12.
See Abstr. 094.247.

Composition of the Martian upper atmosphere.
See Abstr. 097.021.

Missions to Jupiter – 1, 2. See Abstr. 099.005.

054 Artificial Earth Satellites

054.001 **An ecology satellite eyes the earth.**
R. N. Watts, Jr.
Sky Telescope, Vol. 44, 151 - 153 (1972).

054.002 **The ATS-F/Nimbus-E tracking experiment.**
F. O. von Bun.
IAU Symposium No. 48, (see 012.004), p. 112 - 120 (1972).

054.003 **HEOS 2 – der fünfte ESRO-Satellit.**
H.-J. Hoffmann.
SuW, Vol. 11, 275 - 278 (1972).

054.004 **The research satellite AZUR.** E. Keppler.
Zeitschr. Geophys., Vol. 36, 457 - 476 (1970).
In German.

A summary on the scientific aims to be reached with the satellite AZUR is given. The scientific instruments are described. The technical construction and the most important design parameters of the satellite are summarized. The observed flight performance is reported briefly.

054.005 **Kunstmanen.** J. Meeus.
Hemel en Dampkring, Vol. 70, 262 - 264 (1972).

054.006 **An astronomy satellite named Copernicus.**
R. N. Watts, Jr.
Sky Telescope, Vol. 44, 231 - 232, 235 (1972).

054.007 **Les satellites artificiels de l'année 1971.**
J. Thurnheer.
Orion, 30. Jahrgang, p. 157 - 162 (1972).

054.008 **Astronautique 1971.** J. Meeus.
Ciel et Terre, Vol. 88, 325 - 375 (1972).

054.009 **On permanent rotations of an equatorial satellite in a geomagnetic field.** A. A. Khentov.
Izv. vyssh. ucheb. zavedenij. Radiofizika, Vol. 15, 405 - 418 (1972). In Russian. – Abstr. in Referativ. Zhurn. 62. Issled. kosmich. prostranstva, 10.62.290; 51. Astron., 12.51.110 (1972).

054.010 **Tesseral resonance on IMP 4 orbit.**
B. E. Lowrey.
Journ. Geophys. Res., Vol. 77, 5820 - 5824 (1972). – Letter.

054.011 **The decisive step into space.** I. I. Yudin.
Zemlya i Vselennaya, 1972, No. 5, p. 10 - 11.
In Russian.

054.012 **On the choice of frequency of measurements for determining the orientation of an artificial satellite.**
B. L. Novak.
Kosmich. Issled., Vol. 10, 859 - 863 (1972). In Russian.

054.013 **Une nouvelle méthode pour le calcul de l'orbite préliminaire d'un satellite artificiel à l'aide d'observations simultanées.** A. Dinescu.
Stud. Cerc. Astron., Vol. 17, 183 - 188 (1972).

On propose une nouvelle méthode pour le calcul de l'orbite préliminaire d'un satellite artificiel à l'aide d'observations simultanées. On donne les formules pour le calcul des éléments orbitaux approximatifs et ensuite les formules pour leur compensation. La méthode a été appliquée pour le satellite

Echo I, observé simultanément pendant un intervalle de temps de 16^m, le 7 juin 1963.

054.014 Evolution of orbits and the lifetime of earth's artificial satellites. A. Drożyner.
Urania Kraków, Vol. 43, 258 - 262 (1972). In Polish.

054.015 Astronomische Satelliten-Observatorien.
Sternenbote, 15. Jahrgang, p. 163 - 168 (1972).

054.016 On the orbital evolution of a stationary artificial earth satellite. M. A. Vashkov'yak, M. L. Lidov.
In-t prikl. mat. AN SSSR. Preprint No. 45, Moskva. 42 pp. (1972). In Russian. − Abstr. in Referativ. Zhurn. 62. Issled. kosmich. prostranstva, 1.62.222 (1973).

054.017 Analysis of the orbit of Cosmos 316 (1969-108A).
D. G. King-Hele.
Proc. Roy. Soc. London, Ser. A, Vol. 330, 467 - 494 (1972).

054.018 Etude de la poursuite des satellites artificiels.
J.-L. Heudier.
Bull. Groupe Recherches Géod. Spatiale, No. 6, p. 12 - 38 (1972).

054.019 Satellite digest.
Compiled by G. Falworth.
Spaceflight, Vol. 14, 292 - 293, 345 - 349, 390 - 392, 425, 462 - 463 (1972). − A monthly listing of all known artificial satellites and spacecraft, 1972 February - July.

055 Observations of Earth Satellites, Lunar and Planetary Probes

055.001 Determination of n radii vectores of an earth satellite, whose elements are unknown, by a series n ($n \geqslant 3$) of altazimuth observations. A. Vassalo.
Boll. Geod. Sci. Affini, Anno 31, p. 129 - 154 (1972).
A method for determining $n \geqslant 3$ radii vectores of an earth satellite by means of not much spaced vertical−azimuthal observations is given. The magnitude of the radii vectores can be calculated with an error of about 2/parts per thousand for satellites within 1,000 km in altitude.

055.002 A method to improve ephemerides for observation of artificial earth satellites with the SBG camera.
G. G. Muzdrakov.
Izv. Gl. upr. geod. i kartograf., 1971, No. 3, p. 40 - 44. In Bulgarian.

055.003 Formulae for calculation of constants for observing a transit of an AES with an automatic camera.
G. Muzdrakov.
Geod., kartograf., zemleustr., 1971, No. 6, p. 7 - 9. In Bulgarian. − Abstr. in Referativ. Zhurn. 62. Issled. kosm. prostranstva, 8.62.301 (1972).

055.004 WESTA Net. Elaboration and preliminary analysis of observational assignment.
L. Minowska, K. Minowski.
Artificial Satellites, Polish Acad. Sci., Vol. 7, No. 2, p. 23 - 43 (1972).
The article indicates the first stage of general elaboration of the Experimental East European Satellite Triangulation Net (WESTA). The Net has been formed from simultaneous observations performed with NAFA - 3c/25 cameras on 13 stations during six observational campaigns starting from 20.04.1967.

055.005 WESTA Net. Determination of absolute directions of lines connecting the points of the network.
L. Minowska, K. Minowski.
Artificial Satellites, Polish Acad. Sci., Vol. 7, No. 2, p. 45 - 112 (1972).

055.006 Velocity aberration − application to laser measurements. J. C. Husson.
Bull. Groupe Recherches Géod. Spatiale, Obs. Meudon, No. 1,

p. 86 - 96 (1971). In French.

055.007 Interferometric observations of an artificial satellite.
R. A. Preston, R. Ergas, H. F. Hinteregger, C. A. Knight, D. S. Robertson, I. I. Shapiro, A. R. Whitney, A. E. E. Rogers, T. A. Clark.
Science, Vol. 178, 407 - 409 (1972).
Very-long-baseline interferometric observations of radio signals from the TACSAT synchronous satellite, even though extending over only 7 hours, have enabled an excellent orbit to be deduced. The results from this initial three-station experiment demonstrate the feasibility of using the method for accurate satellite tracking and for geodesy. Comparisons are made with other techniques.

055.008 Cosmos 71 − 1965 53 (rocket). Equatorial coordinates (January − October 1969).
Rezul'taty Nablyud. Sovet. Iskusstv. Sputnikov Zemli, No. 135, 71 pp. (1970). In Russian.

055.009 Cosmos 71 − 1965 53 F (rocket). Horizontal coordinates (January − November 1969).
Rezul'taty Nablyud. Sovet. Iskusstv. Sputnikov Zemli, No. 136, 67 pp. (1970). In Russian.

055.010 Cosmos 54 − 1965 11 D (rocket). Equatorial and horizontal coordinates (January − December 1969).
Rezul'taty Nablyud. Sovet. Iskusstv. Sputnikov Zemli, No. 137, 63 pp. (1970). In Russian.

055.011 Determination of the duration of visibility of artificial earth satellites.
V. I. Medvedev, G. I. Kazakov.
Trudy Mosk. vyssh. tekhn. uch-shcha im. N. Eh. Baumana, 1972, No. 150, p. 57 - 64. In Russian. − Abstr. in Referativ. Zhurn. 62. Issled. kosmich. prostranstva, 12.62.204 (1972).

055.012 Experimental station in Zvenigorod.
V. P. Osipenko, V. A. Yurevich.
Zemlya i Vselennaya, 1972, No. 6, p. 54 - 57. In Russian.

055.013 Observations photographiques du satellite Geos B, effectuées à l'astrographe double de 40 cm en 1970

et 1971.   H. Debehogne.
Bull. Astron. Obs. Roy. Belgique, Vol. 8, 41(1972).

055.014   Über die Satellitenbeobachtungen des deutschen
          Meßtrupps im Weltnetz des USC & GS.
W. Böhler.
Deutsche Geod. Kommission, Bayer. Akad. Wiss., München,
Ser. C, No. 182, 78 pp. (1972). – Diss. Techn. Univ. München.

055.015   Experience of tracking of the satellite Pageos with
          the AFU-75 camera in Ulan-Bator.
R. Radnaa, S. Sanzzav, V. T. Groschev.
Journ. Astronaut. Sci., Vol. 18, 397 - 398 (1971).

055.016   Bisherige Entwicklung der photographischen Satelli-
          tenbeobachtung und weitere Perspektiven.
L. Stange.
Wiss. Zeitschr. Techn. Univ. Dresden, Vol. 21, No. 3, (see
012.021), 619 - 622 (1972).

055.017   Zur Frage der Zeitzuordnung bei Satelliten-Kameras.
G. Karský.
Wiss. Zeitschr. Techn. Univ. Dresden, Vol. 21, No. 3, (see
012.021), 623 - 625 (1972). In Russian.

055.018   Ein Programm zur Plattenreduktion von Satelliten-
          beobachtungen mit dem SBG.   H. Pauscher.
Wiss. Zeitschr. Techn. Univ. Dresden, Vol. 21, No. 3, (see
012.021), 625 - 627 (1972).

055.019   Eine moderne Technologie zur photographischen
          Positionsbestimmung künstlicher Erdsatelliten.
K.-H. Marek.
Wiss. Zeitschr. Techn. Univ. Dresden, Vol. 21, No. 3, (see
012.021), 628 - 630 (1972).

055.020   Nachweiswahrscheinlichkeit bei Laserentfernungs-
          messung an bewegten Zielen.   M. Steinbach.
Wiss. Zeitschr. Techn. Univ. Dresden, Vol. 21, No. 3, (see
012.021), 631 - 634 (1972).

055.021   Die Vorhersage von Positionen künstlicher Satelliten
          mit einfachen Mitteln.   H. Pauscher.
Vermessungstechnik, 20. Jahrgang, p. 99 - 102 (1972).

055.022   Messung von Satellitenpositionen mit Laserimpulsen.
          M. Steinbach, R. Neubert.
Jenaer Rundschau, (Jena Rev.), 17. Jahrgang, p. 331 - 336
(1972).

055.023   Photographische Satellitenbeobachtungen mit der
          Automatischen Kamera für Astrogeodäsie.
K.-H. Marek.
Jenaer Rundschau, (Jena Rev.), 17. Jahrgang, p. 340 - 344
(1972).

055.024   Analysis of the accuracy of photographic observa-
          tions of a transit of an artificial earth satellite.
P. P. Pavlenko.
Visn. Kharkiv. Univ. No. 82, (Ser. Astron., No. 7), p. 78 - 83
(1972). In Ukrainian.

055.025   Samos 2 – 61011; Poljot 1 – 63431; Explorer 32 –
          66441; Explorer 19 – 63531. Visual observations.
Horizontal coordinates (July - November 1970).
Rezul'taty Nablyud. Iskusstv. Sputnikov Zemli, vyp. (No.) 1
(141), 61 pp. (1972). In Russian.

055.026   Samos 2 – 61011; Poljot 1 – 63431; Explorer 32 –
          66441; Explorer 19 – 63531. Equatorial coordi-
nates (June - October 1970).
Rezul'taty Nablyud. Iskusstv. Sputnikov Zemli, vyp. (No.) 2
(142), 58 pp. (1972). In Russian.

055.027   Explorer 19 – 1963-53-1. Visual observations.
          Equatorial coordinates (1950.0) January - June
1971.
Rezul'taty Nablyud. Iskusstv. Sputnikov Zemli, vyp. (No.) 3
(143), 60 pp. (1972). In Russian.

055.028   Explorer 32 – 1966-44-1; Explorer 19 – 1963-53-1.
          Visual observations. Equatorial coordinates
(1950.0) July 1971.
Rezul'taty Nablyud. Iskusstv. Sputnikov Zemli, vyp. (No.) 4
(144), 54 pp. (1972). In Russian.

055.029   Poljot 1 – 1963-431. Visual observations. Horizon-
          tal coordinates (March - August 1971).
Rezul'taty Nablyud. Iskusstv. Sputnikov Zemli, vyp. (No.) 5
(145), 74 pp. (1972). In Russian.

Laser goniometry on satellites. See Abstr. 034.087.

Über das stochastische Verhalten von photographisch bestimmten Stern- und Satellitenkoordinaten.
See Abstr. 041.042.

Processing of results of observations of the satellite
65-011-04 carried out within the framework of INTEROBS
program 1966.   See Abstr. 046.016.

Theoretical Astrophysics

061 General Theoretical Problems of Astrophysics, Gravitational Instability, Neutrino Astronomy, X Ray- and Gamma Ray-Astronomy, Frequency and Origin of Elements etc.

061.001 The threshold of disintegration of nuclei in a degenerate electron-neutron gas. Yu. L. Vartanian.
Astrofizika, Vol. 8, 117 - 121 (1972). In Russian. – English translation in Astrophysics, Vol. 8, No. 1.

The equation of state of superdense matter is considered. It is shown that the threshold of disintegration of nuclei in a degenerate electron-neutron gas strongly depends on the interaction between external nucleons.

061.002 On nuclear reactions in a degenerate electron-nuclear plasma. G. S. Sahakian, R. M. Avakian.
Astrofizika, Vol. 8, 123 - 138 (1972). In Russian. – English translation in Astrophysics, Vol. 8, No. 1.

It is known that a degenerate electron-nuclear plasma can exist in two kinds of thermodynamic equilibrium. It is shown that the state of absolute thermodynamic equilibrium can be achieved through the exchange of neutrons between nuclei. The probabilities of such reactions for liquid and solid phases are calculated.

061.003 The stability of a self-gravitating, nonrotating gas layer with stellar, magnetic, and cosmic-ray components. I. S. A. Kellman.
Astrophys. Journ., Vol. 175, 363 - 371 (1972).

A time-independent linear stability analysis is performed on a self-gravitating, plane-parallel, isothermal layer of nonrotating gas with magnetic and cosmic-ray components. The principal result to emerge from this study is that the magnetic field and cosmic-ray gas hinder gravitational instability, increasing the minimum length necessary to produce instability by the factor $(1 + \alpha + \beta)^{1/2}$, where α is the ratio of magnetic pressure to gas pressure and β is the ratio of cosmic-ray pressure to gas pressure. The possibility is discussed that gravitational instability of the gaseous component in the outer regions of galaxies excites density waves of the type described by Lin.

061.004 Xenon in irdischer und in extraterrestrischer Materie (Xenologie). H. Hintenberger.
Naturwissenschaften, 59. Jahrgang, p. 285 - 291 (1972).

Detailed studies of the isotopic abundances of the element xenon are of increasing importance about our knowledge of the formation of the heavy elements and the early history of the solar system. The new discipline which emerged from these studies is called "xenology".

061.005 The abundance of helium in the cosmos—I. M. Hack.
Sky Telescope, Vol. 44, 164 - 165 (1972).

061.006 Non-linear dynamo waves. M. Stix.
Astron. Astrophys., Vol. 20, 9 - 12 (1972).

The influence of the "cutoff α-effect" on the one-dimensional dynamo waves of Parker (1955) is studied. In particular oscillatory antisymmetric dynamos are computed and compared with the solar cycle.

061.007 The polarisation of synchro-Compton radiation. R. D. Blandford.
Astron. Astrophys., Vol. 20, 135 - 144 (1972).

Calculations are presented of the spectral and polarisation properties of synchro-Compton radiation. Both linear and circular polarisation are discussed. The results are applicable primarily to the case of a power law distribution of relativistic electrons radiating in a monochromatic, plane, strong wave of fixed elliptical polarisation. The relevance of the results to astrophysics is discussed.

061.008 Lagrangians, variational principles, and kinematic-dynamo equations. I. Lerche.
Astrophys. Journ., Vol. 176, 225 - 233 (1972).

The kinematic-dynamo equations are investigated from a Lagrangian point of view. We believe that the technique developed here – which is capable of handling complicated fluid flows and which does not have any numerical convergence problems – is a valuable addition to the methods used for discussing dynamo action as it occurs in astrophysical bodies.

061.009 The Nyquist criterion for kinematic-dynamo action. I. Lerche.
Astrophys. Journ., Vol. 176, 235 - 237 (1972).

Using the dispersion relation given elsewhere describing normal-mode frequencies of a large-scale magnetic field under incompressible, homogeneous, stationary velocity turbulence, we set up the general Nyquist criterion which guarantees the existence of growing modes. The criterion is valid for any assumed form of the correlation function of the velocity turbulence.

061.010 Die Schwerkraft der Sterne. A. Unsöld.
Phys. Blätter, 28. Jahrgang, p. 300 - 304 (1972).

061.011 Use of more universal, dimensionless formulae for spherical media. E. Woyk (Chvojková).
Bull. Astron. Inst. Czechoslovakia, Vol. 23, 268 - 278 (1972).

Physical problems are usually treated by proceeding from point to point. In spherical media where levels of equal physical characteristics are concentric spheres, the points should rather be replaced by spheres. This renders the method of computation much easier, the results much simpler. In addition a quite universal, dimensionless system of units has to be introduced, applicable to any cosmic medium regardless of its dimensions and other physical parameters. For comparison the quite general introductory formulae are applied to Kepler orbits, to the spiral motion of charged particles in extended magnetic and gravity fields and to radio paths in ionized media with varying electron distribution.

061.012 Trapping of condensed plasma loops and arcs in cosmic atmospheres. E. Woyk (Chvojková).
Bull. Astron. Inst. Czechoslovakia, Vol. 23, 278 - 293, with a correction, p. 378 (1972).

Big plasma masses, exploded from a stellar surface and

streaming along magnetic field lines can become trapped by higher parts of the field lines and turned into stationary arcs or clouds, hanging motionless above the stellar surface. This process can only occur when the mean particle velocity is very close to the escape velocity and when certain conditions are satisfied.

061.013 Highly excited electron levels of the H_2 molecule in astrophysics.
Ya. B. Zel'dovich, T. V. Ruzmajkina.
Pis'ma v ZhEhTF, Vol. 15, 283 - 285 (1972). In Russian.
Abstr. in Referativ. Zhurn. 51. Astron., 8.51.176 (1972).

061.014 On the origin of magnetic fields in astrophysics. (Turbulent mechanisms "dynamo").
S. I. Vajnshtejn, Ya. B. Zel'dovich.
Uspekhi fiz. nauk, Vol. 106, 431 - 457 (1972). In Russian.
Abstr. in Referativ. Zhurn. 51. Astron., 8.51.202 (1972).

061.015 The abundance of helium in the cosmos—II.
M. Hack.
Sky Telescope, Vol. 44, 233 - 235 (1972).

061.016 État actuel de la théorie de la nucléosynthèse. I. La formation des éléments jusqu'au groupe du fer.
M. Arnould.
Ciel et Terre, Vol. 88, 233 - 272 (1972).

061.017 The quantization of macroscopic rotators.
J. M. Barnothy.
Bull. American Astron. Soc., Vol. 4, 322 (1972). – Abstr. AAS.

061.018 Morphology of rapid cosmic processes. F. Zwicky.
Astronaut. Acta, Vol. 17, 307 - 314 (1972).
Presented at the international colloquium on gasdynamics of explosions, Marseille, September 12 - 17, 1971.

061.019 Nuclear surface energy and neutron-star matter.
D. G. Ravenhall, C. D. Bennett, C. J. Pethick.
Phys. Rev. Letters, Vol. 28, 978 - 981 (1972).
Direct calculations of the nuclear surface energy are made for a Hamiltonian containing the Skyrme nucleon-nucleon interaction. A plane surface separating nuclear matter and a neutron gas or a vacuum is considered in Hartree-Fock and Thomas-Fermi approximations. These surface energies are incorporated in the compressible liquid-drop model to obtain properties of neutron-star matter. The Hartree-Fock results lead to Z values for the nuclei roughly constant at around $Z \sim 36$–38.

061.020 Condensed π^- phase in neutron-star matter.
R. F. Sawyer.
Phys. Rev. Letters, Vol. 29, 382 - 385, with a correction, p. 823 (1972).
It is argued that at some density, estimated at about 1 baryon/F^3, superdense nuclear matter will make a transition to a phase of approximately equal numbers of protons, neutrons, and π^- particles, the latter condensed in one or two plane-wave states of momentum ≈ 170 MeV/c. These conclusions are based on the conventional theory of the pion-nucleon interaction.

061.021 π^- condensate in dense nuclear matter.
D. J. Scalapino.
Phys. Rev. Letters, Vol. 29, 386 - 388 (1972).
Sawyer has proposed a variational ground state for superdense nuclear matter which contains a coherent admixture of approximately equal numbers of protons, neutrons, and π^- particles. Here we investigate this problem by introducing a self-consistent mean π^- field.

061.022 An upper limit on the neutrino rest mass.
R. Cowsik, J. McClelland.
Phys. Rev. Letters, Vol. 29, 669 - 670 (1972).
In order that the effect of gravitation of the thermal background neutrinos on the expansion of the universe not be too severe, their mass should be less than 8 eV/c^2.

061.023 Theoretical curves-of-growth for truncated equivalent widths with a non-linear source function.
M. Ya. Orlov.
Astrometriya i Astrofiz., Kiev, No. 15, (see 003.001), p. 26 - 36 (1972). In Russian.

061.024 On superlight synchrotron radiation in vacuum.
V. Ya. Éjdman.
Izv. vyssh. uchebn. zavedenij. Radiofizika, Vol. 15, 634 - 635 (1972). In Russian. – Abstr. in Referativ. Zhurn. 51. Astron., 10.51.206 (1972).

061.025 Depolarization of synchrotron radiation in various sources. V. N. Sazonov.
Astron. Zhurn. Akad. Nauk SSSR, Vol. 49, 943 - 953 (1972). In Russian. English translation in Soviet Astron. AJ, Vol. 16, No. 5.
This paper presents the calculation of the dependence of the linear polarization on frequency for different models of sources. The results are compared with radio source (mainly extragalactic) polarization observations made in 1964—69. An interpretation of these observations is presented.

061.026 Astrophysical consequences of effects of electromagnetic emittance.
E. V. Antjukh, V. B. Braginsky, A. B. Manukin.
Astron. Zhurn. Akad. Nauk SSSR, Vol. 49, 1094 - 1097 (1972). In Russian. English translation in Soviet Astron. AJ, Vol. 16, No. 5.
The influence of effects of electromagnetic friction and rotational instability on the motion of astrophysical objects is considered. The first effect can shorten the life time of micrometeorites; the second effect, under certain conditions, can change considerably the angular velocity of the axial rotation of cosmic objects.

061.027 Systems with negative specific heat. W. Thirring.
Zeitschr. Physik, Vol. 235, 339 - 352 (1970).
Some systems for which the binding energy increases more rapidly than linearly with the number of particles, are shown to exhibit negative specific heat c for some energies. In thermal contact with larger systems, $c < 0$ creates an instability, and in the canonical ensemble one sees only a phase transition. It is argued that supernovae are, in essence, a phase transition of this origin.

061.028 The Kolar Gold Fields neutrino experiment. I. The interactions of cosmic ray neutrinos.
M. R. Krishnaswamy, M. G. K. Menon, V. S. Narasimham, K. Hinotani, N. Ito, S. Miyake, J. L. Osborne, A. J. Parsons, A. W. Wolfendale.
Proc. Roy. Soc. London, Ser. A, Vol. 323, 489 - 509 (1971).

061.029 The Kolar Gold Fields neutrino experiment. II. Atmospheric muons at a depth of 7000 hg cm^{-2} (Kolar).
M. R. Krishnaswamy, M. G. K. Menon, V. S. Narasimham, K. Hinotani, N. Ito, S. Miyake, J. L. Osborne, A. J. Parsons, A. W. Wolfendale.
Proc. Roy. Soc. London, Ser. A, Vol. 323, 511 - 522 (1971).

061.030 Relativistic hydrodynamics and gravitational instability. J. C. Jackson.
Proc. Roy. Soc. London, Ser. A, Vol. 328, 561 - 565 (1972).
An exact geometrical identity is used to derive an equa-

tion which is interpreted as that governing the propagation of finite amplitude sound in general relativity. It is used to obtain a relativistic version of Jeans's instability criterion.

061.031 Om solens kjemiske sammensetning og grunnstoffenes dannelse. O. Engvold.
Astron. Tidssk., Årg. 5, p. 121 - 124 (1972).

061.032 Hydrogen atom in intense magnetic field.
V. Canuto, D. C. Kelly.
Astrophys. Space Sci., Vol. 17, 277 - 291 (1972).

The structure of a hydrogen atom situated in an intense magnetic field is investigated. Energy eigenvalues are tabulated for field strengths of 2×10^{10} G and 2×10^{12} G.

061.033 Neutrino physics – Prospects.
C. L. Cowan, F. Reines.
G. Gamow Memorial Volume, (see 003.003), p. 150 - 168 (1972).

061.034 Reaction rates in the proton-proton chain.
R. W. Kavanagh.
G. Gamow Memorial Volume, (see 003.003), p. 169 - 185 (1972).

061.035 Analytic forms of the thermonuclear function.
C. L. Critchfield.
G. Gamow Memorial Volume, (see 003.003), p. 186 - 191 (1972).

061.036 A possible new mechanism for generating coherent emission of radio waves in astrophysical phenomena. M. Friedman.
Astrophys. Letters, Vol. 12, 129 - 133 (1972).

A new mechanism has been observed in recent laboratory experiments for generating coherent radiofrequency emission from relativistic electrons. Intense microwave emission has been detected when relativistic electrons interact with small-scale structures of a magnetic field. A qualitative theory suggests that the mechanism incorporates processes of generation and amplification of rf waves as well as a 'nonlinear' process that controls the spectral distribution. This mechanism may be important in astrophysical phenomena.

061.037 Neutron capture cross sections in F, Mg, Al, Si, P and S from 20 to 80 keV.
G. Nyström, B. Lundberg, I. Bergqvist.
Physica Scripta, (*Sweden*), Vol. 4, 95 - 99 (1971).

061.038 Thermal equilibrium states of a classical system with gravitation. E. B. Aronson, C. J. Hansen.
Astrophys. Journ., Vol. 177, 145 - 153 (1972).

We present here a set of calculations which describe the thermal equilibrium states of a classical hard-sphere gas contained in a spherical box where the coupling between local thermodynamics and gravitation is explicitly taken into account. We find that, under certain conditions, structural transformations may take place which give rise to core-halo configurations. Comparisons are made between this and other models. We comment on the possible implications for the statistical mechanics of gravitating systems.

061.039 On the equilibrium state and spectrum of pulsation frequencies of an electron cloud of spherical and cylindrical symmetry. R. S. Oganesian, M. G. Abramyan.
Izv. Akad. Nauk Armyan, SSR, Fizika, Vol. 7, 24 - 32 (1972). In Russian.

The problems of equilibrium distribution of density and radial oscillations of electron clouds around a charged sphere and an infinite cylinder are considered. The distribution law of electron cloud density in equilibrium state has been estab-

lished. The linearized equation is solved precisely and the total spectrum of the natural frequencies of the electron cloud are found.

061.040 The fluctuation–dissipation theorem and kinematic-dynamo activity. I. Lerche.
Astrophys. Journ., Vol. 177, 309 - 313 (1972).

Using the dispersion relation obtained elsewhere describing the regeneration of large-scale magnetic fields under isotropic velocity turbulence, we set up the equations describing the response of the system to an arbitrary externally applied current distribution. We demonstrate that the absorption coefficient (which characterizes the rate at which the impressed current supplies power) may be either positive or negative depending on the level of turbulence.

061.041 On the synthesis of neutron-rich iron-peak nuclei.
J. W. Truran.
Astrophys. Journ., Vol. 177, 453 - 458 (1972).

It is proposed that the isotopes ^{50}Ti, ^{54}Cr, ^{58}Fe, and ^{64}Ni might naturally be formed in neutron-rich equilibria ($\overline{Z}/\overline{N} \approx 0.8$) in matter subsequently ejected in supernova events. The character of these equilibria and astrophysical settings under which such conditions might be achieved are briefly explored.

061.042 Ultraviolet astronomy. R. C. Bless, A. D. Code.
Annual Rev. Astron. Astrophys., Vol. 10, (see 003. 005), 197 - 226 (1972). – Contents: (1) Introduction; (2) Solar system; (3) Stellar observations; (4) Interstellar extinction; (5) Interstellar gas; (6) Galaxies and globular clusters; (7) Conclusions.

061.043 Sull' equilibrio relativo di una massa di particelle elettrizzate (elettroni) soggetta alla propria gravitazione ed uniformemente rotante intorno ad un asse.
T. Zeuli.
Atti Accad. Nazionale Lincei, Ser. Ottava, Rend. Cl. Sci. fis., mat., nat., Vol. 51, 515 - 518 (1971).

In this paper we state the equations for the relative equilibrium of a mass of charged particles (electrons) subject to its own gravitation, uniformly rotating around an axis. Then we examine some particular cases.

061.044 Magnetic and gravitational energy release by resistive instabilities. M. A. Cross, G. Van Hoven.
Phys. Rev. A, General Phys., Vol. 4, 2347 - 2353 (1971).

The resistive magnetohydrodynamic "tearing" and "gravitational-interchange" instabilities are investigated in the linear incompressible limit in the absence of heating. A periodic model is used for the initial magnetic field. The requirement that the spatial Fourier series of the perturbation must converge uniquely determines the growth rate of the instability.

061.045 Problems of cosmic mineralogy. D. P. Grigor'ev.
Zap. Vses. mineral. o-va, Vol. 101, 264 - 280 (1972). In Russian. – Abstr. in Referativ. Zhurn. 62. Issled. kosmich. prostranstva, 11.62.162 (1972).

061.046 Anomalous motion of radiating particles in strong fields. J. Jaffe.
Phys. Rev. D, Particles and Fields, Vol. 5, 2909 - 2911 (1972).

The anomalous motion of radiatively damped particles in a uniform magnetic field is analyzed. A particularly simple method of solution is presented, and some possible astrophysical consequences of the motion are briefly discussed.

061.047 Gamma-ray production from proton-proton reactions in the galactic disk.
D. J. Levy, D. W. Goldsmith.
Astrophys. Journ., Vol. 177, 643 - 646 (1972).

We have used recent accelerator data to calculate the production rate of γ-rays from $p+p \to p+p+n\pi^0$ reactions, which produce γ-radiation by pion decay. The calculated γ-ray spectrum has a shape and peak value similar to the results found by Cavallo and Gould, but the spectrum is considerably broader than that calculated by Stecker. We discuss the reasons for the discrepancy with previous calculations.

061.048   Incoherent radiation from relativistic electrons with power energetic spectrum.
G. G. Getmantsev, Yu. V. Tokarev.
Astrophys. Space Sci., Vol. 18, 135 - 140 (1972).

It is shown that an incoherent high-frequency radiation from an ensemble of relativistic particles with the power energy distribution is described by a certain general expression which covers practically all the cases of particle radiation in random electromagnetic fields of cosmic radiation sources.

061.049   Direct simulation of collision processes.
R. J. Dodd, W. McD. Napier, A. A. Preece.
Astrophys. Space Sci., Vol. 18, 196 - 206 (1972).

A Monte Carlo technique for examining the mass distribution of mutually colliding bodies is described. The technique is simple to program and can handle a wide variety of physical circumstances. Some illustrative cases are given.

061.050   Plane-symmetric polytropic functions.
T. F. Tascione.
Astrophys. Journ., Suppl. Ser., No. 212, Vol. 24, 479 - 491 = Contr. Five College Obs., Univ. Mass., *Amherst,* No. 135 (1972).

The plane-symmetric polytropic functions are tabulated to five figures for 11 values of the polytropic index in the range $0.5 \leqslant n \leqslant 25$. The incomplete beta function $B_x(a, b)$ is also tabulated for $0 \leqslant x \leqslant 1$, $a=1/2, 2/3 \geqslant b \geqslant 1/26$.

061.051   Lorentz transformation properties of the Stokes parameters.   W. J. Cocke, D. A. Holm.
Nature, Phys. Sci., Vol. 240, 161 - 162 (1972).

The general Lorentz transformation characteristics of partially elliptically polarized radiation are investigated and the transformation properties of the Stokes parameters are studied. This problem has astrophysical interest, not only for the Crab nebula pulsar, but also for other cosmic sources emitting elliptically polarized radiation, such as the synchrotron sources studied by Pacholczyk and Swihart.

061.052   Gamma staru astronomijas problēmas.
A. Balklavs.
Zvaigžņotā debess, 1971./72. gada ziema, p. 1 - 11.

061.053   Structure of a cylindrical polytrope.
S. Srivastava.
Indian Journ. Pure Applied Math., Vol. 3, 443 - 450 (1972).

Structure of an infinite self-gravitating cylinder has been studied.

061.054   On the induced amplification of synchrotron radiation in cosmic sources.   V. N. Sazonov.
Astron. Zhurn. Akad. Nauk SSSR, Vol. 49, 1197 - 1204 (1972). In Russian. English translation in Soviet Astron. AJ, Vol. 16, No. 6.

The paper discusses the induced amplification of synchrotron radiation in the system of relativistic electrons in a magnetic field; the amplification is caused by anisotropy of the angular distribution of the electrons.

061.055   On the possibility of observation of circular polarization in the optical emission of some cosmic sources.
Yu. N. Gnedin, A. Z. Dolginov, N. A. Silant'ev.
Astron. Zhurn. Akad. Nauk SSSR, Vol. 49, 1205 - 1210

(1972). In Russian. English translation in Soviet Astron. AJ, Vol. 16, No. 6.

It is shown that the circular polarization and oscillations of the polarization plane (Faraday pulsations) may occur when the radiation passes through the anisotropic medium in case the angle between the direction of preferable oscillations of the electrical vector of the incident radiation and the plane in which the line of sight and the direction of orientation of scattering medium lays is not equal to 0 or $\pi/2$.

061.056   Schwingungsphänomene in der Astrophysik.
W. Deinzer.
Nachr. Akad. Wiss. Göttingen, II. Math.-Phys. Kl., Jahrgang 1972, (No. 8), p. 146 [20] - 148 [22] (1972).

061.057   Effects of erosion and fragmentation on the mass distribution of colliding particles.
L. W. Bandermann.
Monthly Notices Roy. Astron. Soc., Vol. 160, 321 - 338 (1972).

Effects of erosion and fragmentation on the distribution of the masses of an ensemble of colliding particles are investigated. An equation for the instantaneous rate of change in the number of particles per unit mass range is formulated and solved numerically. Results for a variety of physical conditions are presented and compared with results obtained by other authors.

061.058   r-process abundance effects caused by post-event processes.   J. B. Blake, D. N. Schramm.
Bull. American Astron. Soc., Vol. 4, 410 (1972). – Abstr. AAS.

061.059   An excited state in Clayton's nucleocosmochronology.   W. A. Fowler.
Bull. American Astron. Soc., Vol. 4, 412 (1972). – Abstr. AAS.

061.060   Transition radiation in astrophysics.
R. Ramaty, G. B. Yodh, X. Artru.
Bull. American Astron. Soc., Vol. 4, 414 (1972). – Abstr. AAS.

061.061   The astrophysical site of the s-process.   R. K. Ulrich.
Bull. American Astron. Soc., Vol. 4, 416 (1972). Abstr. AAS.

061.062   Freezing of matter from nuclear statistical equilibrium.   S. E. Woosley, W. D. Arnett.
Bull. American Astron. Soc., Vol. 4, 417 (1972). – Abstr. AAS.

061.063   Indirect determination of the photonuclear cross section above 20 GeV.
A. W. Wolfendale, E. C. M. Young, R. Davis, Jr.
Nature, Phys. Sci., Vol. 238, 130 - 131 (1972).

An analysis is made of the production of ^{37}Ar in tetrachloroethylene by cosmic ray muons at various underground depths, 25, 275, 408, 620, and 1080 hg cm^{-2}. The ^{37}Ar production is attributed to the $^{37}Cl(p, n)^{37}Ar$ reaction from protons resulting from nuclear disintegrations produced by fast muons via the virtual photon field accompanying the muon. The analysis shows the photon nuclear cross section is essentially constant from accelerator energies, 5 to 17 GeV, to the energies of 25 to 100 GeV used in this measurement. The production of ^{37}Ar in the Brookhaven solar neutrino detector by fast muons and cosmic ray neutrinos was evaluated.

061.064   Gerak photon dalam medan gravitasi jang kuat disekitar bintang jang mampat.   W. Sutantyo.
Proc. Bandung Inst. Techn., Vol. 6, No. 2, p. 57 - 63 = Publ. Bosscha Obs., *Lembang,* No. 5 (1971).

A review of the motion of photons in a strong gravitational field is given. Photons which are emitted from the neighborhood or from within a superdense star, may escape, falling or moving in a circular orbit, depending on the initial direction of the photons.

061.065 **High energy astrophysics research at the Max-Planck-Institute.** K. Pinkau.
Atomkernenergie, Vol. 19, No. 2, p. 133 - 138 (1972).

The following areas are studied: High energy cosmic rays, solar and atmospheric neutrons, and celestial gamma rays. Balloon and satellite experiments are described. Some results are presented and future plans are outlined.

061.066 **On non-geostrophic baroclinic stability. III. The momentum and heat transports.** P. H. Stone.
Journ. Atmosph. Sci., Vol. 29, 419 - 426 (1972).

The transports are calculated for both the conventional ('geostrophic') kind of baroclinic instability and for symmetric instability, without any restriction on the stratification, as measured by the Richardson number. The transports are calculated consistently to second order in the amplitude expansion of stability theory, so that the transports are the sum of the eddy transport term and a mean transport term.

061.067 **On stability of astrophysical toroidal magnetic fields.** V. Cadez.
Physics of ionized gases. 6th Yugoslav. Symposium, Split 1972, [Institute of Physics, Belgrade], p. 271 - 273 (1972). – See Phys. Abstr., Vol. 75, No. 62496 (1972).

061.068 **Identification of energetic heavy nuclei with solid dielectric track detectors: Applications to astrophysical and planetary studies.** P. B. Price, R. L. Fleischer.
Annual Rev. Nuclear Sci., Vol. 21, 295 - 334 (1971).

061.069 **On the termination of the r-process and the synthesis of superheavy elements from supernovae.**
R. Boleu, S. G. Nilsson, R. K. Sheline, K. Takahashi.
Phys. Letters B, (*Netherlands*), Vol. 40B, 517 - 521 (1972).

Predictions for spontaneous and neutron-induced fission, alpha and beta decay have been made on the basis of potential-energy surfaces for $Z \geqslant 90$ nuclei on the neutron-rich side of the beta stability line. In spite of their considerable quantitative uncertainties, these predictions appear to negate the possibility of superheavy element synthesis via a one– or two-step r-process from supernovae.

061.070 **Are the solar-neutrino experiments suggestive of the existence of a resonance in the $^3He + ^3He$ system?**
V. N. Fetisov, Yu. S. Kopysov.
Phys. Letters B, (*Netherlands*), Vol. 40B, 602 - 604 (1972).

061.071 **Astrophysical flow problems.** H. E. Johnson.
Thesis, Univ. California, San Diego. [Available from Univ. Microfilms, Ann Arbor, Mich., USA. Order No. 72-7593], 145 pp. (1971).

061.072 **Phenomenological causal model of nuclear decay, assuming interaction with neutrino sea.**
H. C. Dudley.
Nuovo Cimento Lettere, Ser. 2, Vol. 5, 231 - 232 (1972).

061.073 **The astrophysical applications of various high-energy electro-magnetic phenomena.**

G. R. Blumenthal.
Thesis, Univ. California, San Diego. [Available from Univ. Microfilms, Ann Arbor, Mich., USA. Order No. 71-18671], 154 pp. (1971).

The problems of the effects of various energy-loss mechanisms on the distribution of relativistic electrons in various astrophysical circumstances are considered.

061.074 **Gravitational instability of two streams.**
R. S. Sengar, H. C. Khare.
Ann. Soc. Sci. Bruxelles, (*Belgium*), Ser. 1, Vol. 86, 283 - 289 (1972).

061.075 **Metagalactic opacity to photons of energy larger than 10^{17} eV.** S. A. Bonometto, P. Marcolungo.
Nuovo Cimento Lettere, Ser. 2, Vol. 5, 595 - 603 (1972).

061.076 **Highly excited electronic levels of the H_2 molecule in astrophysics.**
Ya. B. Zel'dovich, T. V. Ruzmaikina.
JETP Letters, Vol. 15, 198 - 200 (1972).

061.077 **Convectively unstable layer in a periodical gravitational field. II.** L. N. Ivanov.
Vestn. Leningr. un-ta, 1972, No. 13, p. 126 - 133. In Russian. Abstr. in Referativ. Zhurn. 51. Astron., 2.51.229 (1973).

061.078 **Ten years X-ray astronomy.** S. Mandel'shtam.
Nauka i zhizn', 1972, No. 9, p. 25 - 32. In Russian.

The cosmic ν. See Abstr. 003.015..

Turbulence phenomena. See Abstr. 003.061.

Dimensional analysis and group theory in astrophysics. See Abstr. 003.089.

Astrophysics of high energies. See Abstr. 003.129.

K-shell photoionization cross-sections. See Abstr. 022.113.

Spectroscopic techniques in X-ray astronomy. See Abstr. 034.092.

On the cosmic abundance of helium. See Abstr. 065.062.

The generation of magnetic fields in astrophysical bodies. IX. A solar dynamo based on horizontal shear. See Abstr. 080.001.

Cosmic-ray production of deuterium, He^3, lithium, beryllium, and boron in the Galaxy. See Abstr. 143.020.

High-energy cosmic rays in the expanding universe. See Abstr. 143.032.

Proton-neutron concentration ratio in the expanding universe at the stages preceding the formation of the elements. See Abstr. 162.061.

062 Magneto-Hydrodynamics, Plasma

062.001 Time-dependent ionization equilibrium and line radiation under flarelike conditions.
M. C. Kafatos, W. H. Tucker.
Astrophys. Journ., Vol. 175, 837 - 841 (1972).
The results of calculations of time-dependent ionization equilibrium and line emission are presented and compared with the values obtained under the assumption that steady-state conditions prevail.

062.002 The polarization properties of the electromagnetic emission by relativistic electrons Compton-scattered on intensive turbulent plasma oscillations.
V. N. Tsytovich, S. A. Kaplan.
Astron. Zhurn. Akad. Nauk SSSR, Vol. 49, 890 - 892 (1972).
In Russian. English translation in Soviet Astron. AJ, Vol. 16, No. 4. – Short note.

062.003 On the population of the second level for the hydrogen atom in a plasma medium. E. B. Kleiman.
Astron. Zhurn. Akad. Nauk SSSR, Vol. 49, 892 - 895 (1972).
In Russian. English translation in Soviet Astron. AJ, Vol. 16, No. 4. – Short note.

062.004 Rayleigh-Taylor instability in a composite medium.
P. K. Bhatia.
Cosmic Electrodynamics, Vol. 3, 124 - 128 (1972).
The effects of collisions with neutral atoms have been studied on the dynamic stability of a composite medium in the presence of a uniform vertical magnetic field. The case of one-dimensional density gradient in the fluid medium has been studied. An explicit solution has been obtained by making use of a variational principle which is shown to characterize the problem. It is found that the stability criterion is the same as in the absence of collisions. The collisions, however, have a stabilizing influence.

062.005 A theoretical treatment of lunar particle shadows.
R. E. McGuire.
Cosmic Electrodynamics, Vol. 3, 208 - 239 (1972).
The interaction of particles having non-zero gyroradii with a spherical absorber such as the moon is theoretically treated. The moon is assumed to have no effect on ambient magnetic or electric fields, and collective interactions of the particles are ignored. Properties of particle shadows are investigated in a uniform magnetic field and also in the presence of a uniform particle drift normal to this field. Shadow properties are calculated for particles in specified narrow energy and pitch angle intervals. Numerous results are derived for the specific case of a detector in a low lunar orbit as well as some results on the structure of particle shadows as a general function of spatial position.

062.006 Nonlinear magnetosonic waves in a plasma with a finite conductivity. J.-I. Sakai.
Cosmic Electrodynamics, Vol. 3, 260 - 270 (1972).
Nonlinear magnetosonic waves in a turbulent plasma, characterized by an effective electrical conductivity $\sigma_{\rm eff}$ due to micro-instabilities, are governed by the Burgers type equation, which has a shock-like stationary solution. It is shown that initial magnetic pulses of gaussian and solitary wave forms first decay due to dissipation, and subsequently form a shock front whose steepness is determined by the opposing effects of nonlinearity and dissipation.

062.007 Nonlinear dissipation of Alfvén waves.
Y.-C. Chin, D. G. Wentzel.
Astrophys. Space Sci., Vol. 16, 465 - 477 (1972).

Alfvén waves are generated easily in many cosmic plasmas, but they possess no linear damping mechanism since they are not compressive. The most prominent nonlinear damping occurs when one Alfvén wave decays into another plus a slow magnetosonic wave, or two Alfvén waves combine into one fast magnetosonic wave; the resulting magnetosonic waves can then be dissipated.

062.008 Depression of the power spectrum of Compton radiation from relativistic particles in a rarefied plasma.
G. G. Getmantsev, Yu. V. Tokarev.
Astrophys. Letters, Vol. 12, 57 - 60 (1972).
An expression is obtained for the frequency spectrum of Compton radiation from an ensemble of relativistic charged particles with a given power spectrum. The refractive index of the medium in which the particles move is taken into account. The influence of the refractive index of the medium on the frequency spectrum in the case of synchrotron radiation is compared with that for Compton radiation.

062.009 Propagation of electromagnetic waves in a weakly ionized warm magnetoplasma.
S. P. Mishra, P. K. Shukla, K. D. Misra.
Australian Journ. Phys., Vol. 25, 265 - 273 (1972).
Modified wave equations for a weakly ionized warm magnetoplasma are obtained by linearizing Maxwell's equations with the help of the equation of state and the conservation equations for mass, momentum, and energy. The results are used to study wave propagation in the ionosphere and the magnetosphere.

062.010 Electron-plasma-wave shocks in a bounded plasma.
K. Saeki, H. Ikezi.
Phys. Rev. Letters, Vol. 29, 253 - 255 (1972).
Large-amplitude electron plasma waves in a bounded plasma are observed to steepen and to form shock waves with trailing wave trains. For larger-amplitude shocks, however, the trailing wave trains are found to disappear.

062.011 A modification and criticism of Petschek's mechanism. E. R. Priest.
Monthly Notices Roy. Astron. Soc., Vol. 159, 389 - 402 (1972).
Petschek's mechanism is shown to be internally inconsistent when variations in magnetic field strength along the current sheet are included. The inconsistency can be removed for both incompressible and compressible flow provided the magnetic field lines in the narrow region between a pair of shock waves are pulled out into long loops rather than being only slightly bowed. However, there are several difficulties which remain and, unless they can be overcome, Petschek's mechanism must be considered unworkable.

062.012 Diffusion of weak irregularities in a two-ion magnetoactive plasma. E. I. Ginzburg, V. F. Kim.
Geomagn. Aeronom., Vol. 12, 849 - 856 (1972). In Russian.

062.013 Excitation of metastable levels in low density nebular plasmas: [SII] and [ArIV].
S. J. Czyzak, T. K. Krueger, L. H. Aller.
Proc. National Acad. Sci., U.S.A., Vol. 66, 282 - 288 (1970).
Diagnostics of low density nebular plasma by means of its forbidden line spectrum require not only a knowledge of certain intensity ratios as a function of its density and temperature but also a knowledge of the occupation numbers of relevant ionic levels. We present the necessary data for important levels of ionized sulfur and triply ionized argon.

062.014   **Compton scattering in a moving gas.**
Yu. I. Morozov.
Astron. Zhurn. Akad. Nauk SSSR, Vol. 49, 954 - 964 (1972).
In Russian. English translation in Soviet Astron. AJ, Vol. 16,
No. 5.

In the case of a rarefied moving gas with high temperature which is in strong interaction with emitted radiation usually the system of equations of transfer and radiative hydrodynamics is used. In this paper the equations for the components of the energy-impulse tensor of radiation taking into account Compton scattering are deduced. For the case of Thomson's approximation precise formulae are deduced for the relativistic distribution of electrons. For the case of Compton scattering calculations were accomplished in non-relativistic approximation for both velocity distribution of electrons and macroscopical motion of the medium.

062.015   **Note about a revision of Inglis – Teller's formula.**
A. V. Mitrofanov.
Astron. Zhurn. Akad. Nauk SSSR, Vol. 49, 1063 - 1065 (1972). In Russian. English translation in Soviet Astron. AJ, Vol. 16, No. 5.

It is shown that the results of Kurochka (1967) and Inglis, Teller (1939) are all the same over a wide range of T_e and N_e. The reasons for this coincidence are discussed.

062.016   **Wave coupling at a collisionless plasma discontinuity.**   G. L. Kalra, R. Rajaram, J. N. Tandon.
Astrophys. Space Sci., Vol. 17, 87 - 116 (1972).

The coupling of magnetoacoustic waves at a plane interface that separates two semiinfinite collisionless fluids is studied. The fluids are characterized by different temperatures along and transverse to the ambient magnetic field. Continuum equations obtained by Chew et. al. (1956) are used and expressions for the reflection and transmission coefficients are derived. Extensive numerical computations are done to study the variation of the reflection coefficient, with the angle of incidence, for various temperature anisotropies of the media. Relevance of these investigations to the magnetosphere-solar wind boundary is discussed.

062.017   **Plasma emission processes in a magnetoactive plasma**
D. B. Melrose, W. N. Sy.
Australian Journ. Phys., Vol. 25, 387 - 402 (1972).

Plasma emission is treated taking into account the effects of the magnetic field on the electron plasma waves, on the conversion processes, and on the escaping radiation. The expected degrees of polarization of the fundamental and second harmonic are calculated in the weak field limit. The results are used to estimate the magnetic field strength B at the 80 MHz level from the observed polarization of type III bursts; the result $B < 0.04$ G is smaller than previous estimates. The possible importance of electron-cyclotron waves in an application to type I bursts is noted.

062.018   **Thermal arc plasmas as radiation sources.**
C. Goldbach, G. Nollez, R. Peyturaux.
Astron. Astrophys., Vol. 21, 299 - 302 (1972).

A description is given of recent work on the use of thermal arc plasmas as radiation sources, for the calibration of light intensities in astrophysics (ultra violet region) and for the metrology of high temperatures (determination of fixed points). A distinction is made between reference sources which must be calibrated using primary standards of lower temperature, and the primary standards themselves. The different possible approaches to the problem are considered.

062.019   **Stationary collisionless shock waves in an initial**
plasma with high ion temperature.   M. Kornherr.
Zeitschr. Physik, Vol. 233, 37 - 52 (1970).

062.020   **Dispersion relation for longitudinal waves in rela-**
tivistic plasmas.   R. M. Gupta.
Zeitschr. Physik, Vol. 235, 66 - 68 (1970).

Imre, Buti and others studied the longitudinal waves from the general dispersion relation. Here we have derived the dispersion relation for the longitudinal waves in relativistic plasmas, a result which cannot easily be brought in a physically more convenient form.

062.021   **Jump relations for shocks in an anisotropic magnet-**
ized plasma.   F. M. Neubauer.
Zeitschr. Physik, Vol. 237, 205 - 223 (1970).

The jump relations for shocks moving into a collision-free anisotropic magnetized plasma are investigated under the assumption of isotropy of the plasma behind the shock front. The plasma ahead of the shock is assumed to be stable against the fire-hose instability and the mirror instability.

062.022   **Modification of the Rankine-Hugoniot relations for**
shocks in space.   J. K. Chao, B. Goldstein.
Journ. Geophys. Res., Vol. 77, 5455 - 5466 (1972).

The Rankine-Hugoniot (R-H) equations for hydromagnetic shocks are extended to take into consideration the energy flux and momentum flux due to waves and/or turbulence and/or heat flow in the vicinity of shocks. Eighteen shocks observed in space were analyzed. It is the purpose of this paper to investigate how the fluctuations in the upstream and downstream states of shocks modify the Rankine-Hugoniot (R-H) relations.

062.023   **Measuring magnetic fields in plasmas by means of**
light scattering.   L. Kellerer.
Zeitschr. Physik, Vol. 239, 147 - 161 (1970).

The theory of light scattering in plasmas containing a magnetic field yields the special case of modulated scattering spectra. The modulation frequency is governed by the field in the plasma and is equal to the electron cyclotron frequency.

062.024   **La transformation réciproque des ondes magnéto-**
hydrodynamiques dans un plasma homogène situé
dans un champ dépendant d'une seule coordonnée.
D. Zoler, R. Maxim-Gazi.
Physica, Vol. 52, 246 - 252 (1971).

In this paper a study is made of the coupling of magnetohydrodynamic waves in a homogeneous plasma, in a field with a particular structure. The situation considered is characteristic of the upper layers of the atmosphere.

062.025   **Stabilization of magnetohydrodynamic instabilities**
by force-free magnetic fields. I. Plane plasma layer.
J. P. Goedbloed.
Physica, Vol. 53, 412 - 444 (1971).

A marginal-stability analysis is applied to the stability problem of a plane plasma layer under the influence of gravity. Complete stability criteria are derived from the marginal equation of motion and it is shown that this method is equivalent to the application of the energy principle.

062.026   **Stabilization of magnetohydrodynamic instabilities**
by force-free magnetic fields. II. Linear pinch.
J. P. Goedbloed.
Physica, Vol. 53, 501 - 534 (1971).

The marginal-stability analysis, developed in a previous paper, is applied to the stability problem of a linear pinch with a distributed current. Complete stability criteria are derived from the marginal equation of motion and it is shown that this method is equivalent to the application of the energy principle.

062.027   **Stabilization of magnetohydrodynamic instabilities**
by force-free magnetic fields. III. Shearless magnetic

fields.   J. P. Goedbloed.
Physica, Vol. 53, 535 - 570 (1971).

The marginal-stability analysis, as given in two previous papers, is further elaborated for magnetic fields of constant direction in the plane case and for magnetic fields of constant pitch in the cylindrical case.

062.028   Effects of collisions with neutrals on the dynamic stability of a finitely conducting hydromagnetic composite plasma in the presence of Hall currents.
P. K. Bhatia.
Publ. Astron. Soc. Japan, Vol. 24, 517 - 523 (1972).

The effects of collisions with neutrals are investigated on the dynamic stability of a composite plasma taking into account simultaneously the effects of Hall currents and finite electrical conductivity. Making use of the existence of a variational principle, an explicit solution has been obtained for a semi-infinite plasma of varying density.

062.029   Studies of a plasma wind tunnel for simulating the collisionless bow shock of the earth.
W. C. Condit, Jr.
Rev. Sci. Instruments, Vol. 41, 374 - 380 (1970).

Formulas are given for the plasma density and velocity required to simulate the flow of the solar wind over the magnetosphere of the earth. A description of a plasma source (coaxial arc-jet) intended to produce appropriate plasma flows is given. The operation of the device is described.

062.030   Hydromagnetic convection in a rapidly rotating fluid layer.   I. A. Eltayeb.
Proc. Roy. Soc. London, Ser. A, Vol. 326, 229 - 254 (1972).

The linear stability of a rotating, electrically conducting viscous layer, heated from below and cooled from above, and lying in a uniform magnetic field is examined, using the Boussinesq approximation. Several orientations of the magnetic field and rotation axes are considered under a variety of different surface conditions.

062.031   A simple stationary dynamo model.   D. Lortz.
Zeitschr. Naturforschung, Vol. 27a, 1350 - 1354 (1972).

Solutions of the stationary dynamo equations are derived such that outside a torus the magnetic field is the axisymmetric vacuum field of a circular loop, while inside the torus in the limit of large aspect ratio both the velocity and the magnetic fields have helical symmetry.

062.032   Accretion flows and their stability.   N. L. Balazs.
Monthly Notices Roy. Astron. Soc., Vol. 160, 79 - 87 (1972).

We show that in spherically symmetrical accretion flows the appearance of a maximum accretion rate is connected with the breakdown of the uniqueness of the solutions of the steady hydrodynamical equations. We analyze how this arises and what the properties of the resulting branches are. To understand which branch is realized in nature we analyze the stability of these flows under spherically symmetric perturbations for different polytropic indices γ. The results are summarized in a table.

062.033   Scattering of waves in a magnetoactive plasma.
D. B. Melrose, Wilson Sy.
Astrophys. Space Sci., Vol. 17, 343 - 356 (1972).

A general theory of scattering of waves in a magnetoactive plasma by particles of arbitrary energy is presented. The cross-section for the scattering of magnetoionic waves by thermal particles is derived and discussed. Conditions under which the effect of the spiralling motion of the scattering electron can be neglected in treating inverse Compton radiation are found.

062.034   Heating of charged particles by electric waves.
C. L. Longmire.
G. Gamow Memorial Volume, (see 003.003), p. 232 - 240 (1972).

062.035   Wavemechanical formulation of plasma dynamics in longitudinal electric fields.   H. E. Wilhelm.
Zeitschr. Physik, Vol. 241, 1 - 8 (1971).

The macroscopic dynamics of classical many-component plasmas, in which the particles interact by the self-consistent electric field, is formulated in terms of scalar, complex wave equations. The description through wave equations leads to a considerable mathematical simplification compared to the conventional many-fluid hydrodynamics. As an elementary illustration of the wave mechanical formalism, the dispersion of electrostatic waves in an electron plasma is treated.

062.036   Radiation losses of a theta pinch plasma in the wavelength range 10–200 Å.   W. Engelhardt.
Zeitschr. Physik, Vol. 244, 70 - 85 (1971).

The investigations were made over wide density and temperature ranges. It was found that the radiation losses are negligible compared with both the energy content of the electrons and the thermal conduction losses. This applies for the small degrees of impurity present in the deuterium plasma of about 0.1–0.5 % foreign atoms. For impurities of the order of 5 %, however, the radiation losses gain in influence and may exceed the thermal conduction losses in early phases of the discharge. Yet even then the electron temperature finally attained is not essentially affected.

062.037   Lack of partial thermal equilibrium in pinch discharges.
W. Engelhardt, W. Köppendörfer, J. Sommer.
Zeitschr. Physik, Vol. 246, 29 - 42 (1971).

Measurements of the deuterium Balmer line intensities in the early phase of a theta pinch discharge show strong deviations from steady state population of levels considerably above the collision limit, even if criteria for the relaxation times of these levels are well satisfied. Solutions of time dependent rate equations for a hydrogen plasma confirm these observations.

062.038   The effect of an electromagnetic wave-field pressure on the propagation of slow TM waves along the plasma layer.   V. V. Demtchenko, A. M. Hussein.
Zeitschr. Physik, Vol. 246, 216 - 224 (1971).

Slow electric-type waves are investigated in propagating along a plane plasma layer with account of the pressure of the high frequency field of wave. Expressions are obtained for the fields and the derived dispersion equations appreciably differ from those of linear theory. The high frequency pressure is shown to ensure electromagnetic field penetration into plasma, which is not brought about to a linear approximation (with no account of the high frequency pressure) due to skin-effect.

062.039   Damping of longitudinal plasma oscillations.
N. L. Varma, G. K. Khandpur.
Zeitschr. Physik, Vol. 246, 295 - 301 (1971).

Attenuation of electron oscillations in a fully ionized plasma is investigated by solving linearized kinetic equation without external fields. The general dispersion relation for longitudinal plasma oscillations is obtained using the BGK model. Damping due to electron ion collisions is obtained with a correction term. It is also observed that damping rate decreases as k increases, which is in agreement with McBride.

062.040   Stark broadening of singly ionized strontium and calcium lines.   J. Purić, M. Platiša, N. Konjević.
Zeitschr. Physik, Vol. 247, 216 - 222 (1971).

Profiles of five calcium II and six strontium II lines have

been measured in an argon plasma behind the reflected shock wave. The half halfwidths of the measured profiles of some Ca II and Sr II multiplets show large discrepancies with theoretical predictions.

062.041 **Total energy loss of a test charge in a collisional magnetoplasma.** P. K. Shukla, R. N. Singh.
Physica Scripta, (*Sweden*), Vol. 4, 281 - 289 (1971).
    A general theory of energy attenuation of a charged test particle in a cold, collisional plasma has been developed. The effect of collisions on the plasma properties have been investigated. The radiation characteristics of a spiralling test particle are discussed. Estimates of radiated and absorbed power by test charges of different energies have been made, and the role of collisions on the power spectrum has been studied in detail.

062.042 **Stabilization of hydromagnetic instabilities by magnetization currents?**
H.-R. Lehmann, R. Treumann.
Nature, Phys. Sci., Vol. 239, 146 - 147 (1972).

062.043 **Spectroscopy of laboratory plasmas.**
    D. D. Burgess.
Space Sci. Rev., Vol. 13, 493 - 527 (1972). − Invited paper IAU Colloquium No. 14 (see 012.012).

062.044 **Plasma polarization shift of the resonance lines of ionized helium.** S. Volonte.
Space Sci. Rev., Vol. 13, 528 - 530 (1972). − Conference paper IAU Colloquium No. 14 (see 012.012).

062.045 **Vacuum ultraviolet absorption of dense plasmas with resonance series of Be, B, C, N, Mg, Al and Si.**
G. Mehlman-Balloffet, J. M. Esteva.
Space Sci. Rev., Vol. 13, 531 (1972). − Abstract of conference paper IAU Colloquium No. 14 (see 012.012).

062.046 **Laboratory fundamental data.** W. R. S. Garton.
    Space Sci. Rev., Vol. 13, 532 - 552 (1972). − Invited paper IAU Colloquium No. 14 (see 012.012).

062.047 **Measurements of collisional rate coefficients in laboratory plasmas.** H.-J. Kunze.
Space Sci. Rev., Vol. 13, 565 - 583 (1972). − Invited paper IAU Colloquium No. 14 (see 012.012).

062.048 **Relation between laser flux, temperature and ionisation equilibrium in laser produced plasmas.**
M. H. Key, R. J. Hutcheon, D. A. Preston, T. P. Donaldson.
Space Sci. Rev., Vol. 13, 584 - 585 (1972). − Conference paper IAU Colloquium No. 14 (see 012.012).

062.049 **Low-frequency instabilities in magnetic pulses.**
N. A. Krall, P. C. Liewer.
Phys. Rev. A, General Phys., Vol. 4, 2094 - 2103 (1971).

062.050 **Plasma stream in the surroundings of a developing current layer.** S. I. Syrovatskij, B. V. Somov.
Trudy Mezhdunar. seminara po probl. "Uskorenie chastits v kosmich. prostranstve (okolozem. i mezhplanet. kosmich. prostranstve), Galaktike i metagalaktike". Moskva, 1972, p. 106 - 120. In Russian. − Abstr. in Referativ. Zhurn. 62. Issled. kosmich. prostranstva, 11.62.149 (1972).

062.051 **Heat flow and magnetic field diffusion in turbulent fluids.** F. Krause.
Astron. Nachr., Vol. 294, 83 - 87 (1972).
    The basic ideas of the mean-fields magnetohydrodynamics developed by M. Steenbeck, K.-H. Rädler and F. Krause are applied to heat conduction in turbulent fluids. As the main result it is shown that dynamo excitation caused by

homogeneous isotropic mirror-symmetric turbulence contradicts the second principle law of thermodynamics, whereas the non-mirrorsymmetric part of that turbulence does not influence the heat flow. Furthermore an anisotropic heat conductivity is calculated for the case that there is a strong magnetic field.

062.052 **Stability of a model current sheet with finite transverse field and finite flow velocity.**
D. F. Smith, M. A. Raadu.
Cosmic Electrodynamics, Vol. 3, 285 - 296 (1972).
    The stability of a model current sheet with finite transverse field and finite axial flow velocity which are taken as constant across the sheet was investigated for stability in the infinite conductivity MHD limit. The results may be useful in constructing models of parts of coronal streamers and the geomagnetic tail.

062.053 **On the formation of double layers in plasmas.**
    P. Carlqvist.
Cosmic Electrodynamics, Vol. 3, 377 - 388 (1972).
    A homogeneous plasma carrying a constant current is considered. It is demonstrated that the plasma is unstable if the current density exceeds a critical value which depends both on the plasma density and on the thermal energies of the ions and of the electrons in the plasma. A small one-dimensional disturbance in the form of a density dip will then give rise to a progressive local evacuation of the plasma. When the particle density in the dip has been reduced to a certain limit a double layer forms in the evacuated region. The stability of such a double layer is briefly discussed. The occurrence of double layers in the ionosphere and in the solar atmosphere is considered.

062.054 **Waves and resonances in magneto-active plasma.**
    V. L. Ginzburg, A. A. Ruhadze.
Encyclopedia of physics, Vol. 49/4, (see 003.009), 395 - 560 (1972).

062.055 **Gyro synchrotron radiation from thermal plasmas.**
    P. K. Shukla, K. P. Singh.
Physica Scripta, Vol. 5, 217 - 218 (1972).
    The gyro-synchrotron radiation from the solar coronal plasma is investigated with special regard to temperature effects. The influence of non-zero temperature is to enhance the radiated synchrotron power from the solar coronal plasma. We also observe that fast test charges radiate more power than slow charges.

062.056 **Continuum radiation from nonisothermal hydrogen plasmas.** H. F. Nelson, C. Y. Wang.
Journ. Quant. Spectrosc. Radiat. Transfer, Vol. 12, 1593 - 1608 (1972).
    In this account of an investigation of the continuum radiative flux from nonisothermal stagnation shock layers composed of a hydrogen plasma, the general equations for the composition are derived and the Rankine−Hugoniot equations are simplified and solved to give the thermodynamic conditions in the shock layers. The radiative flux is calculated by considering ground to free state radiative transitions in atomic hydrogen for conditions roughly representing those of a space probe entering Jupiter's atmosphere. The influence of Mach number, ambient density (altitude), spectral variation of the radiation absorption coefficient, shock layer thickness, excited state radiation, and temperature profiles are examined.

062.057 **Measurement of continuum radiation from an argon plasma.** D. N. Ghosh Roy, R. S. Tankin.
Journ. Quant. Spectrosc. Radiat. Transfer, Vol. 12, 1685 - 1699 (1972).
    The continuum radiation of atmospheric argon plasma

generated by a free burning arc has been investigated over the spectral range of 1000–9000 Å. Validity of Kramers–Unsöld model for the continuum radiation is investigated, and Biberman's ʃ functions are obtained and compared with theory.

062.058 Radiative modes of a weakly ionized, collision-dominated, turbulent plasma. I. Lerche.
Astrophys. Space Sci., Vol. 18, 94 - 103 (1972).

We obtain, and discuss, the roots of the dispersion relation describing normal mode propagation in a weakly ionized, collision dominated turbulent plasma with an isotropically distributed turbulent magnetic field. We demonstrate that, depending on the level of the turbulent field relative to the collision frequency, there may, or may not, be propagating, but decaying, modes present in the system. The structure and properties of the modes depend on both the precise level of the turbulent magnetic field and its spatial and temporal correlation.

062.059 Quantum theory of the dielectric constant of a magnetized plasma and astrophysical applications.
I. Theory.   V. Canuto, J. Ventura.
Astrophys. Space Sci., Vol. 18, 104 - 120 (1972).

A quantum mechanical treatment of an electron plasma in a constant and homogeneous magnetic field is considered, with the aim of (a) defining the range of validity of the magnetoionic theory (b) studying the deviations from this theory, in applications involving high densities, and intense magnetic field. While treating the magnetic field exactly, a perturbation approach in the photon field is used to derive general expressions for the dielectric tensor $\epsilon_{\alpha\beta}$. The properties of $\epsilon_{\alpha\beta}$ are explored in the various limits.

062.060 Stable electron density fluctuations in a plasma in the presence of a high-frequency electric field.
T. Hagfors, G. F. Gieraltowski.
Journ. Geophys. Res., Vol. 77, 6791 - 6803 (1972).

In this paper a theory is developed to describe the changes in the fluctuation spectrum that take place for excitation levels below those required for the instabilities to develop. We have developed theoretical expressions for the frequency spectra of both the ion line and the electron plasma oscillations under conditions when there is an electric field pump driving the charged particles into finite-amplitude oscillations.

062.061 Pair-producing electric fields and pulsars.
L. Parker, J. Tiomno.
Astrophys. Journ., Vol. 178, 809 - 817 (1972).

We consider a system in which a pair-producing electric field surrounding a central body gives rise to an oscillating coherent plasma of electron-positron pairs. A greatly simplified treatment of the model yields results indicating that there might possibly be some relation to pulsars. This crude model leads to expressions for the pulse half-width, lengthening of period, and power output, depending essentially on only one parameter proportional to the field amplitude, which is constant for periods from 10^{-2} to 10 s. Surprisingly, the above expressions lead to values which agree well with the corresponding observed quantities for pulsars.

062.062 On the transport properties of charged particles in one dimension in random electric fields.
I. Lerche.
Astrophys. Journ., Vol. 178, 819 - 835 (1972).

First-order smoothing theory is a commonly used approximation in obtaining estimates of the transport properties of charged particles in turbulent electric and magnetic fields. We discuss the problem of one-dimensional charged-particle evolution under a turbulent electric field. We set up and discuss the behavior of charged particles in a one-dimensional

turbulent electric field using a statistical pair of equations obtained by invoking the quasi-normality hypothesis. We also set up and discuss the behavior of charged particles in a one-dimensional turbulent electric field using a Gaussian distribution for the electric field. Comparison of the behavior of the three problems indicates several similarities and several differences.

062.063 Cosmic-ray evolution due to interactions with self-excited plasma waves.   M. A. Lee.
Astrophys. Journ., Vol. 178, 837 - 855 (1972).

We use kinetic equations describing a relativistic plasma embedded in a uniform ambient magnetic field to investigate the time evolution of the cosmic-ray momentum distribution due to interactions between the cosmic-ray particles and their self-excited waves. Within the cosmic-ray lifetime there is negligible change in the cosmic-ray energy spectrum, but for particles with energy less than about 20 GeV there occurs a marked reduction of the cosmic-ray streaming velocity to a value of the order of, but greater than, the Alfvén speed.

062.064 Behandlung ebener magnetohydrostatischer Gleich-gewichtsprobleme mittels komplexer Analysis.
R. Gorenflo.
Mitt. Inst. Theor. Geod. Univ. Bonn, No. 4, p. 2 - 18 (1972).

062.065 Relativistic corotating magnetosphere model.
F. C. Michel.
Bull. American Astron. Soc., Vol. 4, 413 (1972). – Abstr. AAS.

062.066 Energy transfer by fluctuations in a plasma.
B. R. Barkstrom, M. B. Lewis.
Bull. American Astron. Soc., Vol. 4, 424 (1972). – Abstr. AAS.

062.067 A photon rest mass and the dispersion of longitudinal electric waves in interstellar space.
R. Burman.
Journ. Phys. A, General Phys., Vol. 5, 162 - 163 (1972).

Treats the propagation of longitudinal electric waves in a cold plasma, which can occur if the photon has a nonzero rest mass.

062.068 Non-linear instabilities in streaming plasmas.
L. Stenflo, H. Wilhelmsson, K. Östberg.
Physica Scripta, (*Sweden*), Vol. 3, 231 - 232 (1971).

This paper outlines a theory, which is to be used for a numerical investigation of the non-linear interaction of three electrostatic plasma waves.

062.069 Propagation of radiative shock waves in an inhomo-geneous cosmic medium.   T. L. Chow.
Journ. Phys. A, General Phys., Vol. 5, 1409 - 1418 (1972).

The propagation of two strong radiative shocks moving opposite in an inhomogeneous cosmic medium is calculated by a 'modified' characteristic method.

062.070 Gravitational plasmas. II.   G. Severne.
Physica, Vol. 61, 307 - 313 (1972).

The non-Markovian time evolution of a weakly coupled homogeneous gravitational system was shown by Prigogine and Severne to present a steady creation of correlation energy superposed upon a weakly oscillatory relaxation process. The discussion of the relaxation process introduced an approximation, and is reconsidered. While some interesting modifications in detail appear, the overall behaviour of the system is unchanged.

062.071 Interpretation of spectral intensities from laboratory and astrophysical plasmas.
A. H. Gabriel, C. Jordan.
Case studies in atomic collision physics, Vol. 2, [North-Holland Publishing Company, Amsterdam], p. 209 - 291 (1972).

062.072 **Magnetic field and internal motions of a spherical fluid model of a celestial body with a convection zone. II.** E. Schmutzer.
Experim. Techn. Phys., Vol. 20, 301 - 319 (1972).
In German.

As a model the author considers the convection zone of the sun. By a convenient series expansion he succeeds in treating the nonlinear system of equations of electrodynamics and hydrodynamics. The result of the calculations is a simultaneous theory of the stationary magnetic field and the differential rotation of celestial bodies. The numerical scheme coincides surprisingly well with the empirical facts of the sun.

062.073 **Shock waves through self-gravitating gas spheres.** P. Chaturani.
Indian Journ. Pure Applied Math., Vol. 3, 379 - 383 (1972).

The author deals with the propagation of strong shock waves through the envelope of a generalized Roche model.

062.074 **Radio emission from cosmic plasma.** T. Takakura.
Butsuri, (*Japan*), Vol. 27, 369 - 374 (1972).
In Japanese.

062.075 **Heating of astrophysical plasma due to rotation.** R. Bondyopadhaya.
Czech. Journ. Phys. B, Vol. 22, 1199 - 1201 (1972).

It is shown that low-frequency transverse MHD waves may undergo damping inside a rotating stellar body. It is suggested that rotation may affect the heating of the stellar interior.

062.076 **On the realization of a force-free magnetic field configuration in the case of axial symmetric magnetohydrodynamic streams.** G. A. Rubo.
Problems of cosmic physics. Vyp. (No.) 7, (see 003.028), p. 16 - 22 (1972). In Russian.

The realization of a force-free magnetic field configuration in the case of axial symmetric steady vortex-free streams of fluid with infinite conduction has been investigated. The results are applied to the magnetic field of solar corpuscular streams.

062.077 **Oscillatory convection in a viscoelastic fluid layer in hydromagnetics.**
P. K. Bhatia, J. M. Steiner.
Australian Journ. Phys., Vol. 25, 695 - 702 (1972).

A study is made of the overstable mode of convection in an infinite horizontal layer of a viscoelastic fluid, heated from below, in the presence of a magnetic field. It is shown first that the problem is characterized by a variational principle. Proper solutions are then obtained for the case of two rigid boundaries using the variational method. It is found that the magnetic field has a stabilizing effect on the thermally induced overstability in a viscoelastic fluid, as in the case of an ordinary viscous fluid. The thermodynamic significance of the variational principle is also considered.

062.078 **On magnetic inhibition of thermal convection.**
R. Van der Borght, J. O. Murphy, E. A. Spiegel.
Australian Journ. Phys., Vol. 25, 703 - 718 (1972).

The effect of an imposed vertical magnetic field on convective transfer in a horizontal Boussinesq layer of fluid heated from below is studied in the mean field approximation. Solutions are found over a wide range of conditions, for free boundaries, by a combination of numerical and analytic techniques.

062.079 **Quantum statistical mechanics of dense partially ionized hydrogen.** H. E. Dewitt, F. J. Rogers.
Phys. Earth Planet. Interiors, Vol. 6, (see 012.023), 51 - 59 (1972).

062.080 **Radio noise generation by scattering of plasma waves.** R. A. Windsor.
Thesis Fac. Graduate School Univ. Minnesota. [Available from Univ. Microfilms, Ann Arbor, Mich.], 80 pp. (1972).

062.081 **On the instability of nonlinear longitudinal oscillations of magnetoactive plasma.**
V. V. Demchenko, I. A. El-Naggar.
Plasma Phys., Vol. 14, 1967 - 1970 (1972).

062.082 **Self-consistent electromagnetic waves in relativistic Vlasov plasmas.** B. B. Winkles, O. Eldridge.
Phys. Fluids, Vol. 15, 1790 - 1800 (1972).

062.083 **Generation of plasma oscillation by beam-plasma interaction.** T. P. Khan.
Phys. Fluids, Vol. 15, 1857 - 1859 (1972).

062.084 **On a resolution of Chandrasekhar's equations for a MHD turbulence.** S. Codreanu.
Stud. Univ. Babeş-Bolyai, Ser. Phys., Anul 17, Fasc. 2, p. 83 - 86 (1972). In Romanian.

The paper presents the possibility to solve the system of Chandrasekhar's equations for a magnetohydrodynamic turbulence, by using the method proposed by Smirnov and Shapiro in the case of ordinary hydrodynamic turbulence.

062.085 **Quantum theory of line formation in a magnetic field.**
E. Landi Degl'innocenti, M. Landi Degl'innocenti.
Solar Physics, Vol. 27, 319 - 329 (1972).

The transfer equations for the Stokes parameters in the presence of magnetic field and under the hypothesis of LTE are derived in an original way by the use of density matrix techniques. The results are substantially the same as those previously obtained by other authors. We finally compare our results to the previous ones in order to clarify some discrepancies still present in the literature.

Methods in nonlinear plasma theory.
See Abstr. 003.060.

Theory of fully ionized plasmas.
See Abstr. 003.067.

Plasma astrophysics. See Abstr. 003.083.

Absolute scale for Si I *gf*-values from wall-stabilized arc measurements. See Abstr. 022.045.

Nonlinear theory of the monochromatic circularly polarized VLF and ULF waves in the magnetosphere.
See Abstr. 084.209.

Errata

062.901 **Errata: "Root mean square fluctuation of a weak magnetic field in an infinite medium of homogeneous stationary turbulence"** [Astrophys. Journ., Vol. 173, 549 - 555 (1972)]. B.-C. Low.
Astrophys. Journ., Vol. 178, 277 (1972).

063 Radiative Transfer

063.001 On the theory of radiative heat exchange in polytropic atmospheres. V. P. Grinin.
Astrofizika, Vol. 8, 53 - 70 (1972). In Russian. — English translation in Astrophysics, Vol. 8, No. 1.

On the basis of a linearized heat conductivity equation the temperature conditions in polytropic atmospheres perturbed by a concentrated heat source are considered. The perturbation of the optical properties of the atmosphere as well as its optical nonhomogeneity is taken into account. Formulae for temperature perturbation produced by the point heat source are obtained. The heat-wave propagation velocity relations for the case of non-stationary perturbations are given. The main characteristics of the temperature perturbations in the solar atmosphere are given.

063.002 Noncoherent scattering. II. Anisotropic scattering. N. B. Yengibarian, A. G. Nicoghossian.
Astrofizika, Vol. 8, 71 - 90 (1972). In Russian. — English translation in Astrophysics, Vol. 8, No. 1.

The linear problem of noncoherent scattering in a plane-parallel layer of finite thickness has been discussed. The dependence of the redistribution function on scattering angle and nonsphericity of the phase function is taken into account. Some concrete calculations in the case of Doppler line broadening have been carried out.

063.003 Scattering of light in a sphere with arbitrary source distribution. N. B. Yengibarian.
Astrofizika, Vol. 8, 149 - 153 (1972). In Russian. — English translation in Astrophysics, Vol. 8, No. 1.

The problem of coherent and isotropic scattering of light in a homogeneous sphere with arbitrary distribution of sources has been discussed. The solution of this problem is reduced to integral equations with symmetrical kernels. One can solve these equations by application of the authors' method.

063.004 Circular polarization of light scattered from rough surfaces.
L. W. Bandermann, J. C. Kemp, R. D. Wolstencroft.
Monthly Notices Roy. Astron. Soc., Vol. 158, 291 - 304 (1972).

It is suggested that double reflection of light on rough absorbing surfaces is the principal cause of the circular polarization observed on Mars, Mercury and the moon as well as on rock samples studied in the laboratory by Pospergelis. Two simple models of rough surfaces are presented. The fractional circular polarization obtained with such models, using geometrical optics and Fresnel's laws, and its dependence on the scattering angles is compared with observations. Generally good agreement is obtained within the limitations of the models.

063.005 Résolution de l'équation de transfert en présence d'un champ de vitesses. E. Simonneau.
Comptes Rendus Acad. Sci. Paris, Sér. B, Vol. 275, 169 - 172 (1972).

On généralise au cas de la formation des raies spectrales dans une atmosphère avec un champ de vitesses, une méthode de résolution de l'équation de transfert, qui a été déjà exposée (1972). On se borne à prendre un profil Doppler pour le coefficient d'absorption.

063.006 Exploiting the linearity of radiative transfer.
A. C. Cogley, H. M. Domanus.
Journ. Quant. Spectrosc. Radiat. Transfer, Vol. 12, 1191 - 1204 (1972).

The present paper develops and uses the idea that the standard source function is an influence function for a given medium. The linearity of radiative transfer is then used to find certain general source functions in terms of the standard one. The usefulness of the above concept is demonstrated by four problems.

063.007 Probability distributions for photon exit: Photons formed with specified frequency. G. D. Finn.
Journ. Quant. Spectrosc. Radiat. Transfer, Vol. 12, 1217 - 1239 (1972).

A linear integral relation is formulated, for conditions of simple geometry, whose solution gives the distribution with frequency ν_e of the probability that a photon in a spectral line, which is formed at a specified depth below the surface of an atmosphere with a specified frequency, eventually escapes from the atmosphere with frequency ν_e. Numerical solutions are obtained and their physical meaning discussed under a variety of conditions.

063.008 On a semi-grey approximation to non-grey radiative transfer. D. H. Sampson, R. T. Marton.
Journ. Quant. Spectrosc. Radiat. Transfer, Vol. 12, 1389 - 1408 (1972).

A method suggested previously by one of us for treating non-grey radiative transfer by a semi-grey approximation is further developed and is extended to arbitrary geometry. The accuracy of the method is tested numerically for three idealized forms of the absorption coefficient.

063.009 Mathematical properties of the $K_n(\tau)$ functions. A. L. Crosbie, H. K. Khalil.
Journ. Quant. Spectrosc. Radiat. Transfer, Vol. 12, 1457 - 1464 (1972).

The $K_n(\tau)$ functions, which arise in the study of nongray radiative transfer, are investigated. Closed-form expressions for the triangular and exponential profiles are presented. The $K_4(\tau)$ and $K_5(\tau)$ functions for the rectangular, triangular, exponential, Doppler, and Lorenz profiles are tabulated.

063.010 On the validity of a generalized Kirchhoff's law for a nonisothermal scattering and absorptive medium.
J. L. Linsky, G. H. Mount.
Icarus, Vol. 17, 193 - 197 (1972).

The relationship of directional hemispherical reflectivity to emissivity is investigated for a nonisothermal medium with isotropic coherent scattering and absorption. Departures from a generalized Kirchhoff's law occur due to the long range nature of the scattering process. Such departures occur in lunar thermal emission at microwave but not at infrared frequencies.

063.011 Radiative transfer in a gray isothermal spherical layer. A. L. Crosbie, H. K. Khalil
Journ. Quant. Spectrosc. Radiat. Transfer, Vol. 12, 1465 - 1486 (1972).

The local radiative flux in an isothermal spherical layer is systematically investigated. The functions $h_1(\tau; \tau_1)/\tau^2$ and $h_2(\tau; \tau_1, \tau_2)/\tau^2$ are studied and tabulated. Limiting solutions are presented and evaluated. The radiative flux from an isothermal spherical layer is presented in graphical and tabular form. A comprehensive review of radiative transfer in a spherical geometry is performed.

063.012 Necessary conditions for steady state in radiation: Matter interaction and the role of entropy.
K. K. Sen.
Journ. Quant. Spectrosc. Radiat. Transfer, Vol. 12, 1487 - 1496 (1972).

A variational principle based on entropy considerations is developed for obtaining the necessary conditions for the existence of a stationary state in the case of time-dependent interactions between radiation and matter. The model is restricted to a gaseous system with two discrete levels and without continuum radiation. The role of the local potential is studied.

063.013 Approximate formulas for the intensity of radiation diffusely reflected by a semi-infinite atmosphere.
E. G. Yanovitsky.
Astrometriya i Astrofiz., *Kiev*, No. 15, (see 003.001), p. 63 - 75 (1972). In Russian.

Approximate formulas are found for the intensity of radiation diffusely reflected by a semi-infinite plane-parallel atmosphere. A comparison of exact and approximate calculations is given.

063.014 Internal dust in nebulae. III. Nonisotropic scattering.
J. S. Mathis.
Astrophys. Journ., Vol. 176, 651 - 658 (1972).

The equations of radiation transfer for spherical model nebulae, uniformly filled with dust except for a central hole, are solved numerically. The dust is assumed to have a forward-throwing Henyey-Greenstein phase function characterized by g, the average cosine of the angle between incident and scattered radiation. Values of g between 0.0 (isotropic scattering) and 0.75 are considered, and of albedo, $\tilde{\omega}$, between 0.4 and 1. The results are applied briefly to NGC 6514.

063.015 A probabilistic formulation of the noncoherent-scattering problem. R. G. Athay.
Astrophys. Journ., Vol. 176, 659 - 669 (1972).

The noncoherent-scattering problem is formulated in terms of the mean intensity J, averaged over the absorption profile and the mean number of scatterings required for photon escape. This leads to major simplifications in the resultant equations for J. The usual equations in three dimensions (optical depth, frequency, and direction angle) are replaced by a first-order differential equation in one dimension (mean number of scatterings). For certain idealized problems algebraic solutions are readily obtained. The method appears to offer major simplifications for problems dealing with multidimensional media.

063.016 On the calculation of the $H(a, v)$-function.
M. G. Gerbil'skij.
Solar Physics, Vol. 25, 274 - 276 (1972). – Research note.

063.017 Polarization of optical radiation in an absorbing medium in case of scattering.
Yu. N. Gnedin, A. Z. Dolginov, E. L. Potashnik, N. A. Silantjev.
Astron. Zhurn. Akad. Nauk SSSR, Vol. 49, 990 - 997 (1972). In Russian. English translation in Soviet Astron. AJ, Vol. 16, No. 5.

Stokes' parameters of the optical radiation, emerging from an absorbing semi-infinite medium, are determined. The degree of the linear polarization is calculated depending on the absorption value for an envelope in the form of a plane disk. An expected degree of linear polarization for radiation with wavelength of some microns is calculated for the nucleus of our Galaxy.

063.018 Transfer of resonance radiation and photon random walks. V. V. Ivanov, Sh. A. Sabashvili.
Astrophys. Space Sci., Vol. 17, 3 - 12, 13 - 22 (1972). In Russian and English.

Transfer of resonance radiation in an infinite medium is considered as a process of random walks of photons. Close relation is shown to exist between the problems of transfer of line radiation and the stable distributions of the probability

theory. This relation is used as a basis of a new method for the investigation of the asymptotic properties of the radiation field far from the sources.

063.019 Absorption and scattering in plane parallel turbid media. K. Klier.
Journ. Optical Soc. America, Vol. 62, 882 - 885 (1972).

The Kubelka-Munk hyperbolic equations for reflectance and transmittance of a turbid, isotropically scattering medium have been found formally identical with the solution for reflected and transmitted fluxes of Chandrasekhar's radiative-transfer equation for isotropically highly scattering media. The absorption and scattering coefficients of the two theories have been related through numerical coefficients.

063.020 Fourth moment of a wave propagating in a random medium. W. P. Brown, Jr.
Journ. Optical Soc. America, Vol. 62, 966 - 971 (1972).

063.021 Experiments on turbulence characteristics and multiwavelength scintillation phenomena.
J. R. Kerr.
Journ. Optical Soc. America, Vol. 62, 1040 - 1049 (1972).

063.022 Simplified equation for amplitude scintillations in a turbulent atmosphere.
J. W. Strohbehn, T. I. Wang.
Journ. Optical Soc. America, Vol. 62, 1061 - 1068 (1972).

063.023 Absorption profile of a planetary atmosphere: A proposal for a scattering independent determination.
A. L. Fymat, J. Lenoble.
Applied Optics, Vol. 11, 2249 - 2254 (1972).

The use of scattering theory to infer atmospheric optical parameters requires the separation of absorption and scattering. It is demonstrated that a gradient flux relation exists that would provide the absorption (altitude) profile independently of scattering and irrespective of the state of polarization of the light field. The relation is derived for an atmosphere of plane-parallel or spherical geometry and for broad (continuum) and narrow (spectral line) frequency bands.

063.024 A probabilistic model for time-dependent transfer problems in spherical shell media.
T. K. Leong, K. K. Sen.
Monthly Notices Roy. Astron. Soc., Vol. 160, 21 - 36 (1972).

A probabilistic model for solving time-dependent transfer problems in inhomogeneous, isotropic scattering, spherical shell media is proposed. Four probability functions are defined and an integro-differential equation for scattering function is deduced. Expressions for the emergent intensities in terms of scattering, transmission, back scattering and back transmission functions are derived.

063.025 General solution for polarized radiation in a homogeneous-slab atmosphere. R. Aronson.
Astrophys. Journ., Vol. 177, 411 - 421 (1972).

The transfer matrix method is extended to problems of radiative transfer in homogeneous-slab atmospheres in which the transmission and reflection operators are not the same for radiation incident from one side as from the other. This occurs for the azimuthally asymmetric components of the Stokes parameters U and V for polarized light. Physical interpretations of the various operators in the theory are given.

063.026 Analytically solvable problems in radiative transfer. III. C. van Trigt.
Phys. Rev. A, General Phys., Vol. 4, 1303 - 1316 (1971).

The density of excited atoms and the diffusely reemitted radiation is calculated for a homogeneous radiation field incident on a slab. The solutions of the Biberman-Holstein integral equation are used in the calculations. The optical thickness is assumed to be large. The line shapes of both the absorption line and the incident radiation are arbitrary.

063.027 Diffuse reflection of resonance radiation from a semi-infinite medium.
N. B. Engibaryan, A. G. Nikogosyan.
Dokl. AN ArmSSR, Vol. 54, No. 2, p. 91 - 95 (1972). In Russian. – Abstr. in Referativ. Zhurn. 51. Astron., 11.51.176 (1972).

063.028 Partial redistribution in radiation transport.
B. Kivel.
Journ. Quant. Spectrosc. Radiat. Transfer, Vol. 12, 1659 - 1672 (1972).

Radiation diffusion in a slab of purely scattering atoms with only one transition is calculated using a Monte Carlo method. Partial redistribution because of the Doppler effect is included. The scattering is assumed to be isotropic and incoherent. Results are obtained for (1) the amount of diffuse reflection and transmission as a function of optical depth, and (2) the angular distribution and spectrum of the reflected component.

063.029 Time-dependent radiation transfer and a possible explanation of the interpulse in CP 0950.
U. de Angelis, L. G. Taff.
Astrophys. Space Sci., Vol. 18, 21 - 33 (1972).

The time-dependent equation of radiative transfer is solved exactly and in the n-th Gaussian approximation. The atmosphere is plane-parallel and semi-infinite; isotropic scattering is assumed, but the boundary condition at $\tau = 0$ is arbitrary. The results are used to investigate a suggested mechanism for the origin of the secondary pulses in CP 0950; it is found that a binary system of neutron stars can indeed explain formation, time delay and intensity of the observed interpulse.

063.030 Solution of the transfer equation for interlocked multiplets by probabilistic method.
S. R. Das Gupta, S. Karanjai.
Astrophys. Space Sci., Vol. 18, 246 - 253 (1972).

Sobolev's probabilistic method – the method of quantum exit from the medium – has been applied to solve the transfer equation for the case of interlocking without redistribution. The solution contains the function $\phi(x)$ which is same as the H-function involved in the solution given by Busbridge and Stibbs by the method of principle of invariance.

063.031 Radiative transfer in a spherical shell.
N. B. Engibaryan.
Uch. zap. Erevan. un-ta Estestv. n., 1972, No. 1 (119), p. 26 - 37. In Russian. – Abstr. in Referativ. Zhurn. 51. Astron., 12.51.166 (1972).

063.032 Monte-Carlo treatment of Lyman-α radiation in an inhomogeneous spherical atmosphere. S. B. Modali.
Publ. Astron. Soc. Pacific, Vol. 84, 642 (1972). – Abstr. Astron. Soc. Pacific.

063.033 The Eddington factor for radiative transfer in spherical geometry.
S. J. Wilson, C. T. Tung, K. K. Sen.
Monthly Notices Roy. Astron. Soc., Vol. 160, 349 - 353 (1972).

The present paper shows that, contrary to the statement of Hummer & Rybicki, the half-range method (Wilson & Sen) can handle the outward peaking effect of the radiation field in extensive spherical media. The source function $J(r)$ and the Eddington factor $f(r) = [K(r)/J(r)]$ have been calculated by the half-range method with $k(r) = r^{-2}$ for $0 < r \leqslant R$, where $R = 3$

and $R = 10$. The values for $R = 10$ are compared with those of Hummer & Rybicki and are found to be in fairly good agreement.

063.034 Radiative transfer in spherically symmetric flows.
J. I. Castor.
Astrophys. Journ., Vol. 178, 779 - 792 (1972).

The transfer of radiation in a spherically symmetric moving medium is considered, including those corrections which are of the order of the flow velocity divided by the velocity of light. The formulation uses the radiation quantities which would be seen by an observer moving with the fluid. The fluid equations including the effects of matter-radiation coupling and the law of total energy conservation are discussed.

063.035 Backward Monte Carlo calculations of the polarization characteristics of the radiation emerging from spherical-shell atmospheres.
D. G. Collins, W. G. Blättner, M. B. Wells, H. G. Horak.
Applied Optics, Vol. 11, 2684 - 2696 (1972).

A Monte Carlo procedure, designated as FLASH, was developed for use in computing the intensity and polarization of the radiation emerging from spherical-shell atmospheres and is especially useful for investigating the sunlit sky at twilight time. The procedure is capable of computing the Stokes parameters for discrete directions at the receiver position. Both molecular and aerosol scattering are taken into account as well as ozone, aerosol, water vapor, and carbon dioxide absorption within the atmosphere. Some comparisons are made between FLASH calculations for a pure Rayleigh atmosphere, a combined Rayleigh and aerosol atmosphere, and calculations reported by other authors for plane-parallel atmospheres.

063.036 Information content of extinction and scattered-light measurements for the determination of the size distribution of scattering particles. D. Spänkuch.
Applied Optics, Vol. 11, 2844 - 2850 (1972).

The present study is, as a first step, concerned with the calculation of the optical properties of lognormally distributed particle collections as a function of the parameters that are typical for the distribution. The calculations were based upon the refractive index $m = 1.5$. The dependence of the optical parameters on the refractive index is discussed.

063.037 Degree and direction of polarization of multiple scattered light. 1: Homogeneous cloud layers.
G. W. Kattawar, G. N. Plass.
Applied Optics, Vol. 11, 2851 - 2865 (1972).

The degree of polarization as well as the direction of the polarization are calculated by a Monte Carlo method for homogeneous layers. Two solar zenith angles and a range of optical thicknesses up to 10 are considered. The results are compared with calculations for single scattered photons.

063.038 Efficient stream distributions in radiative-transfer theory. C. K. Whitney.
Journ. Optical Soc. America, Vol. 62, 1368 (1972). – Abstr. Optical Soc. America.

063.039 On the significance of a simplified theory of radiative transfer. R. W. Nelson.
Journ. Optical Soc. America, Vol. 62, 1368 (1972). – Abstr. Optical Soc. America,

063.040 A simplified radiative-dynamical model for the static stability of rotating atmospheres.
P. H. Stone.
Journ. Atmosph. Sci., Vol. 29, 405 - 418 (1972).

063.041 Absorption lines in a stratified Rayleigh atmosphere.
C. R. Molenkamp.

Thesis, Univ. Arizona, Tucson. [Available from Univ. Microfilms, Ann Arbor, Mich., USA. Order No. 72-11968], 189 pp. (1972).

In the present research a numerical model is developed that produces synthetic absorption lines in an inhomogeneous, finite, Rayleigh scattering model atmosphere. The transfer equation describing the interaction of solar radiation with the atmosphere including the effects of Lambert surface reflection is solved using the auxiliary equation method.

063.042   Line source functions with variable Doppler width and noncoherent scattering.

H. A. Beebe, D. R. Hollars.
Contr. Obs. New Mexico State Univ., Las Cruces, Vol. 1, 95 - 102 (1972).

The dependence of line source functions on Doppler widths variable with depth are discussed. We assume complete redistribution of photons in the line and apply the flux divergence method to the solution of the transfer equations. Results are given for strong lines formed in a solar-like atmosphere with several microturbulent velocity models.

063.043   Calculation of the functions characterizing the radiation transfer in a plane layer.

Sh. A. Sabashvili.
Byull. Abastumansk. Astrofiz. Obs., No. 43, p. 207 - 222 (1972). In Russian.

The isotropic monochromatic scattering of light in a plane layer of finite optical thickness τ_0 is considered. The tables of the mean number of photon-scatterings and of the resolvent function are given for several values of τ_0 and of the probability of photon survival. The functions were found by iterative solution of the integral equations.

Radiation transport in spectral lines.
See Abstr. 003.038.

Tabellen von Mie-Streufunktionen.
See Abstr. 003.159.

Transfer effects on X-ray lines in optically thick celestial sources.   See Abstr. 142.026.

064 Stellar Atmospheres, Stellar Envelopes

064.001 Convective zones and the depletion of Li in solar type stars. Eh. Ehrgma, A. A. Pamjatnih.
Nauchn. Informatsii, vyp. (No.) 20, p. 71 - 85 (1971).
In Russian.

Convective envelope models for main-sequence stars with masses M = 1.25, 1.0 and 0.85 $M_\odot$ are computed for various assumptions concerning the parameters of the mixing-length theory and mixing length itself. Absorption in autoionization lines is taken into account.

064.002 Model atmospheres for DA and DB white dwarfs.
D. T. Wickramasinghe.
Mem. Roy. Astron. Soc., Vol. 76, 129 - 179 (1972).

A grid of hydrogen line-blanketed model atmospheres for DA and DB white dwarfs is presented. Monochromatic fluxes at 185 wavelength points are tabulated for each of the models in order to enable a detailed comparison with observations.

064.003 Atmospheres of A-type supergiants. C. Aydin.
Astron. Astrophys., Vol. 19, 369 - 380 (1972).

Five A-type supergiant stars have been studied and compared with α Cygni. Microturbulence and radial velocity and their dependence upon electron pressure and time have been studied.

064.004 Chromospheric heating of very hot stars by radiation driven sound waves. A. G. Hearn.
Astron. Astrophys., Vol. 19, 417 - 426 (1972).

A mechanism for the chromospheric heating of very hot stars is suggested. A linearized theory shows that sound waves will be amplified by an interaction of the perturbations of the absorption coefficient caused by the sound wave with the radiative forces. It is suggested that this mechanism could explain some of the observed properties of Wolf-Rayet stars and possibly contributes to the formation of planetary nebulae and the ejection of matter from novae.

064.005 Blanketed model atmospheres for cool hydrogen-rich white dwarfs. R. Wehrse.
Astron. Astrophys., Vol. 19, 453 - 472 (1972).

Model atmospheres for cool hydrogen-rich white dwarfs have been constructed under the assumption of $T_{eff} = 5700°$K and T_{eff} = 5100°K, log g = 8, and solar composition. The blanketing effect of more than 23800 atomic and molecular lines has been taken into account in a statistical approach. Transition probabilities for all lines are derived from the Rowland Tables (Moore et al., 1966) by means of Holweger's empirical solar model (1967).

064.006 Studies of hydrodynamic events in stellar evolution. II. Dynamic instabilities in stellar envelopes.
W. M. Sparks, G. S. Kutter.
Astrophys. Journ., Vol. 175, 707 - 715 (1972).

For an evolved red giant of 1.1 $M_\odot$ (He and heavy-element core of 1.0 $M_\odot$, H-rich envelope of 0.1 $M_\odot$) we investigate the dynamic instabilities which result when the luminosity exceeds the limiting value L_0 for a static envelope. These instabilities may be responsible for the pulsation of long-period variables, the ejection of planetary nebulae, and the formation of certain infrared objects.

064.007 Model atmosphere analysis of the A 3Ia−O supergiant HD 33579 in the Large Magellanic Cloud.
B. Wolf.
Astron. Astrophys., Vol. 20, 275 - 285 = Veröff. Univ. Sternw. München, Vol. 7, No. 16 (1972).

A fine analysis of HD 33579, an A 3Ia−O supergiant in the Large Magellanic Cloud, has been done by evaluating 4 high dispersion spectra (12.3 Å/mm) taken at the European Southern Observatory. The observations are best fitted by a model atmosphere (radiative equilibrium, LTE, no line blanketing) with the parameters T_e = 8130° and log g = 0.7 and a microturbulence varying with the optical depth from 6 km/s at $\bar\tau$ = 2.0−26 km/s in the outermost layers. The computed continuum agrees reasonably with photoelectric measurements.

064.008 Non-LTE effects on continuum and hydrogen-line parameters in B and O stars. D. Mihalas.
Astrophys. Journ., Vol. 176, 139 - 152 (1972).

Using a homogeneous set of non-LTE models (which allow for bound-bound transitions) on the range $15,000° \leq T_{eff} \leq 55,000°$K at several gravities, the effects of departures from LTE upon the continuum and Balmer lines are evaluated. Several observational discriminants of non-LTE effects are discussed; in certain specific examples considered, the available data are well fitted by the non-LTE calculations.

064.009 A model for the chemical evolution of S and N star envelopes. R. K. Ulrich, J. M. Scalo.
Astrophys. Journ.,(Letters), Vol. 176, L37 - L42 (1972).

A model is presented for mixing in stars during the maximum convective phase of a nondegenerate He shell flash. This model requires stars of about 5 $M_\odot$ and can produce surface enrichments of ^{7}Li, ^{12}C, ^{13}C, and s-process elements.

064.010 La diffusion dans les atmosphères d'étoiles, ses possibilités et ses limites. G. Michaud.
Journ. Roy. Astron. Soc. Canada, Vol. 66, 217 - 218 (1972).
Abstr. Canadian Astron. Soc.

064.011 A search for density and pressure inversions in high-temperature, low-gravity model atmospheres.
E. F. Borra, L. Fortier.
Journ. Roy. Astron. Soc. Canada, Vol. 66, 219 - 220 (1972).
Abstr. Canadian Astron. Soc.

064.012 Transition from shallow to deep convection zones in stars. D. J. Mullan.
Astrophys. Letters, Vol. 12, 13 - 16 (1972).

It is generally believed that the existence of fast stellar rotation in main sequence stars with spectral types earlier than F4, and slow rotation in later types, can be explained by the onset of deep convection zones at spectral type F4. This letter points out that this belief is not valid when evolutionary effects are allowed for, and that the transition to deep convection zones must occur later than F4.

064.013 Relaxation oscillations in the envelopes of luminous red giants. R. L. Smith, W. K. Rose.
Astrophys. Journ., Vol. 176, 395 - 403 (1972).

Numerical hydrodynamic calculations are described which indicate that high radiation-pressure gradients at the base of the envelope, resulting from thermal instability in the core, are capable of driving envelope relaxation oscillations. The calculations suggest that planetary nebulae may result from the final phase of mass loss on the red-giant branch.

064.014 Transfer of resonance-line radiation in differentially expanding atmospheres. I. General considerations and Monte Carlo calculations.
L. J. Caroff, P. D. Noerdlinger, J. D. Scargle.
Astrophys. Journ., Vol. 176, 439 - 461 = Contr. Lick Obs.,

No. 348 (1972).

A Monte Carlo method is given for analyzing the transfer of resonance radiation in a rapidly and differentially expanding nebula. Results are presented in tabular form suitable for further use. An analytic solution is presented for the case of coherence in the fluid frame.

064.015 **Transfer of resonance-line radiation in differentially expanding atmospheres. II. Analytic solution for the case of coherence in the frame of the fluid.**
P. D. Noerdlinger, J. D. Scargle.
Astrophys. Journ., Vol. 176, 463 - 478 = Contr. Lick Obs., No. 356 (1972).

An exact solution is obtained for the problem of line transfer by resonant scattering (assumed coherent in the fluid frame) in a differentially expanding atmosphere. Generalization to finite destruction probability per scattering is immediate, allowing treatment of nonresonance lines such as Lβ. Some results of interest for the theory of expanding stellar atmospheres are presented graphically.

064.016 **On the calculation of the parameters of thermal waves.** I. A. Klimishin, V. I. Gromovik.
Visnik L'viv. un-tu. Ser. fiz., 1971, vyp. (No.) 6 (14), p. 10 - 15, 105. In Ukrainian. – Abstr. in Referativ. Zhurn. 51. Astron., 8.51.195 (1972).

064.017 **The cooling effect of CO in the atmospheres of red giants.** H. R. Johnson.
Bull. American Astron. Soc., Vol. 4, 324 (1972). – Abstr. AAS.

064.018 **Temporal variations of the flow parameters describing the envelopes about Be stars.**
T. H. Morgan, K.-Y. Chen.
Bull. American Astron. Soc., Vol. 4, 330 - 331 (1972). Abstr. AAS.

064.019 **A model for the chromosphere of Procyon.**
J. L. Linsky, T. R. Ayres.
Bull. American Astron. Soc., Vol. 4, 334 (1972). – Abstr. AAS.

064.020 **Fine structure of shock waves in an RR Lyrae model atmosphere.** S. J. Hill.
Bull. American Astron. Soc., Vol. 4, 337 - 338 (1972). Abstr. AAS.

064.021 **On the law of escape of a heat wave at a star surface.** I. A. Klimishin.
Astron. Tsirk., No. 708, p. 4 - 6 (1972). In Russian.

064.022 **An estimate of stellar wind mass loss during the red giant phase of evolution.**
J. N. Heasley, Jr., J. G. Mengel.
Observatory, Vol. 92, 93 - 96 (1972).

The purpose of this note is to estimate the amount of mass loss due to a stellar wind as the star evolves up the red giant branch.

064.023 **Model atmospheres for cool supergiant stars.**
D. R. Alexander, H. R. Johnson.
Astrophys. Journ., Vol. 176, 629 - 643 = Publ. Goethe Link Obs., Indiana Univ., *Bloomington*, No. 137 (1972).

Our purpose here is to present the results of an exploratory grid of model atmospheres for cool supergiant stars to illustrate the effect of varying the chemical composition of the atmosphere. In particular, we have studied the effects of composition changes (depletion of C and O, enrichment of N, and increase in the ratio C/O) which might be expected from processing of the original material of a star through the CNO cycle of nuclear burning. Our models also include, for the first time, the important CN opacity.

064.024 **A search for diffuse interstellar features in stars with circumstellar dust shells.**
T. P. Snow, Jr., G. Wallerstein.
Publ. Astron. Soc. Pacific, Vol. 84, 492 - 496 (1972).

We have searched for diffuse features at $\lambda\lambda$4430, 5780, 5796, and 6613 in the spectra of 17 stars which show infrared excesses or other grounds for believing that they have dusty envelopes. In no case could we identify diffuse bands that may be ascribed to the circumstellar envelope. We conclude that the grains in circumstellar envelopes are not identical with the grains in the general interstellar medium.

064.025 **Stellar winds and breezes.**
P. H. Roberts, A. M. Soward.
Proc. Roy. Soc. London, Ser. A, Vol. 328, 185 - 215 (1972).

064.026 **Polytropic subsonic stellar winds with magnetic fields.** B. Durney.
Astrophys. Space Sci., Vol. 17, 489 - 498 (1972).

In any complex magnetic field configuration it is to be expected that there will be not only regions of no flow and of supersonic flow, but also regions of subsonic flow. In the present paper the equations for the stellar wind are examined in the case of a polytropic relation between pressure and density and for small values of the parameter $\epsilon = \Omega^2 r_a^2 / u_a^2$. The radial distance ($r$) and the velocity at the Alfvén point ($4\pi\rho u^2 / B^2 = 1$) are denoted by r_a and u_a, and Ω is the angular velocity.

064.027 **Analyses of light-ion spectra in stellar atmospheres. I. Magnesium II in B and O stars.** D. Mihalas.
Astrophys. Journ., Vol. 177, 115 - 128 (1972).

A calculation of the spectrum of Mg II in B and O stars, using a rather complete model atom allowing for many explicitly calculated transitions in steady-state statistical equilibrium, has been performed. The results open again serious questions concerning the validity of LTE abundances for at least the O and early B stars, and suggest strongly that further analyses are needed to delineate the ions and spectral types for which the LTE assumption fails.

064.028 **A search for density and pressure inversions in high-temperature, low-gravity model atmospheres.**
E. F. Borra, L. Fortier.
Astrophys. Journ., Vol. 177, 129 - 135 (1972).

We have investigated the existence of pressure and density inversions in high-temperature, low-gravity model atmospheres. None of the hot unblanketed models we computed presented unstable layers. Our findings therefore seem to question a proposed mechanism for energization of a stellar corona and mass loss in hot stars. Rapidly rotating early-type stars are found to have cool unstable layers.

064.029 **The consequences of grains in the atmospheres of late-type stars. I. Intrinsic polarization, infrared excesses, and emission lines.** M. C. Jennings, H. M. Dyck.
Astrophys. Journ., Vol. 177, 427 - 440 (1972).

The suggestion of a relation between Ca II H and K emission and observational indicators of grain material has been investigated and confirmed. The high incidence of Balmer emission among polarized stars is discussed, but, contrary to previous suggestions, it is found not to be a necessary condition for polarization. It is argued that the disappearance of emission lines implies a weakening of the chromosphere usually associated with late-type stars and that grains are the primary agents responsible for the perturbation.

064.030 **Spectroscopic analysis of the weak-helium-line star α Sculptoris.** O. Vilhu.
Ann. Acad. Sci. Fennicae, Ser. A, VI. Phys., No. 394, 39 pp. = Contr. Obs. Astrophys. Lab. Univ. Helsinki (1972).

The purpose of the present study is to investigate more

closely the weak-helium line star, α Scl, on the basis of Mt. Wilson and Palomar spectrograms. Our method of analysis is conventional for the B stars. The atmospheric parameters, i.e. effective temperature and gravity, are derived from the continuum and the $H\gamma$-profile. Then the apparent abundances of individual elements are derived by model atmosphere techniques using the LTE-line formation. For comparison, on the basis of one spectrogram, an abundance analysis from a few lines is made also for the normal B7V star π Cet. Finally, the pole-on hypothesis is discussed for α Scl.

064.031 **On the stationary mass outflow from stars. I. The computational method and the results for** $1 M_\odot$ **star.**
A. Żytkow.
Acta Astron., Vol. 22, 103 - 139 (1972).

Solutions for spherically symmetric, time-independent (stationary) models of outflowing stellar envelopes with radiation pressure in continuum as the main driving force were investigated. The equations of conservation of mass, momentum and energy together with the radiative energy transport equation were integrated for the 1 solar mass star with luminosity at the surface $L < L_{crit} = 3.88 \times 10^4 L_\odot$. The calculations show that one formally can construct in the yellow supergiant region time-independent models leading to expansion velocities $(0 - 160$ km/s) covering well the observed range of velocities for planetary nebulae.

064.032 **Microturbulence in atmospheres of F, G. K type stars. I. Curve of growth analysis of G, K type subgiants.** R. Głębocki.
Acta Astron., Vol. 22, 141 - 154 (1972).

In order to obtain a complete picture of changes of microturbulence in the region of late type stars on the HR-diagram, available analyses have been supplemented by a study of 24 G, K type subgiants. They fill the "gap" in the HR-diagram for which very little was known about the microturbulence. The results indicate that the microturbulence in subgiants is markedly smaller than in giants.

064.033 **On the structure of envelopes of red supergiants.**
U. Uus.
Nauchn. Informatsii, vyp. (No.) 21, p. 46 - 50 (1972). In Russian.

Applying a method previously described, stellar models having degenerate carbon-oxygen cores and a thin double nuclear burning shell are computed (for stars with masses 1.5 and $5 M_\odot$, and with core masses 0.85 and 1.35 $M_\odot$ respectively.

064.034 **On the influence of nonlocal treatment of convection on the structure of the envelopes of red supergiants.** U. Uus.
Nauchn. Informatsii, vyp. (No.) 21, p. 51 - 57 (1972). In Russian.

The structure of the envelope of a $5 M_\odot$ star is calculated with a nonlocal treatment of convection for two values of luminosity: $L = 2 \times 10^5 L_\odot$ ($T_{eff} = 2640°$K) and $5 \times 10^5 L_\odot$ ($T_{eff} = 2510°$K). When the nonlocal features of convection are taken into account the boundary between the regions of efficient and nonefficient convection is smeared and density inversion is shifted to about 20% higher temperature regions.

064.035 **Molecular mysteries of red giants.** D. D. Clayton.
Comments Astrophys. Space Phys., Vol. 4, 137 - 140 (1972).

Little has been published of model atmospheres of red giants. In this comment I hope only to outline several related advances that have or promise to provide important constraints on their evolution.

064.036 **Element abundances in O- and early B-stars.**
M. Scholz.

Vistas in astronomy, Vol. 14, (see 003.008), 53 - 80 (1972).

064.037 **The splitting of lines in differentially rotating slabs.**
C. Magnan.
Astron. Astrophys., Vol. 21, 361 - 371 (1972).

Line profiles for rotating disks with continuous absorption are constructed. The source function is calculated by means of Monte-Carlo techniques. The effects of various parameters such as the dimension of the slab and the velocity are investigated. The splitting of the line into two or more components is a natural consequence of the rotation.

064.038 **Spectral line formation in extended stellar atmospheres.** I. P. Grant, A. Peraiah.
Monthly Notices Roy. Astron. Soc., Vol. 160, 239 - 248 (1972).

A method of numerical solution of the equation of radiative transfer for spectral line formation in extended spherically symmetric stellar atmospheres is described. A simple non-LTE two-level atom model is assumed. Results are presented for the cases $b/a \leqslant 2$ where a is the inner and b the outer radius of the atmosphere.

064.039 **Stellar winds and mass loss of a rotating star.**
J. P. de Grève, C. De Loore, C. de Jager.
Astrophys. Space Sci., Vol. 18, 128 - 134 (1972).

The mass loss to be expected from the corona of a rotating F2-star is calculated. The rotation is supposed to be rigid up to a certain distance s, as if it were maintained by a strong magnetic field. Dependent on the values of the rotational velocity the mass loss can increase to $26 - 40\%$ for v_{rot} up to 200 km s^{-1}.

064.040 **Stellar atmospheres with radiation incident at the surface.** P. F. Buerger.
Astrophys. Journ., Vol. 177, 657 - 663 (1972).

The usual surface boundary condition used in model stellar atmospheres is modified to allow for a plane-parallel beam of radiation incident at the surface. This involves a modification of the equation of radiative transfer and a consequent modification of the temperature-correction scheme. Models have been computed in order to show the effects of the incident radiation on the resultant emergent radiation. Relatively small amounts of incident radiation can significantly affect the formation of the hydrogen lines. Applications to the analysis of eclipsing binaries are discussed.

064.041 **Steady accretion in the presence of rotation and a magnetic field.** T. T. Chia, R. N. Henriksen.
Astrophys. Journ., Vol. 177, 699 - 711 (1972).

The solution for steady accretion with the stream lines antiparallel to the magnetic field lines onto a rotating, nontranslating, gravitating, magnetic monopole is given. The flow is everywhere "sub-Alfvénic," and the angular momentum flux can be outward despite the accretion. The solution is particularly simple in the limit that thermal pressure may be neglected (being free of critical points) and should allow ready estimates in various strong field situations. A numerical illustration of the general case is presented.

064.042 **The synthetic spectrum of the CN red system and its application to stellar spectra of moderate resolution.** I. Marenin, A. E. Greene.
Astrophys. Journ., Vol. 177, 841 - 846 (1972).

Assuming a Milne-Eddington model atmosphere, we have calculated synthetic spectra for the CN red system from $\lambda 5600$ to $\lambda 6800$. The spectra were calculated for a pure $^{12}C^{14}N$ spectrum and a CN spectrum with a $^{12}C/^{13}C = 4$. They were then compared to the spectra of several S, SC, and C stars.

064.043 **Intrinsic polarization in the atmospheres of super-giant stars.** G. V. Coyne.
Proc. Third Colloquium on astrophysics.Supergiant stars, Trieste 1971, p. 93 - 107 = Specola Vaticana, Comun. No. 53 (1972).

A statistical study is made of the polarization measured in early type supergiants. For a group of twenty such stars it is shown that their polarization is variable with amplitudes ranging from 0.2% to 0.5%. The changes in the polarization of Epsilon Aurigae are correlated with the phase of the eclipsing binary light curve.

064.044 **Free-free and Balmer line emission from optically thick stellar shells.** E. H. Avrett, J. R. Baldwin.
Publ. Astron. Soc. Pacific, Vol. 84, 633 - 634 (1972). — Presented at the Santa Cruz meeting of the Astronomical Society of the Pacific, 26–29 June 1972.

We consider a circumstellar shell with given inner and outer radii R_1 and R_2, illuminated by a central stellar black-body radiation field characterized by the temperature T_* and solve the radiative-transfer and statistical-equilibrium equations for a three-level hydrogen atom.

064.045 **Abundance anomalies as probes of envelopes in Am stars.** M. A. Smith.
Publ. Astron. Soc. Pacific, Vol. 84, 644 (1972). — Abstr. Astron. Soc. Pacific.

064.046 **Temperature reversal in a LTE atmosphere.** S. J. Hill.
Publ. Astron. Soc. Pacific, Vol. 84, 669 - 670 (1972).

It is shown that a temperature reversal in a LTE model atmosphere is possible.

064.047 **Kondensierte Materie im Kosmos. V.** J. Dorschner.
Sterne, 48. Jahrgang, p. 194 - 206 (1972). — Solid particles in stellar atmospheres; Crystalline stars.

064.048 **Electron scattering in spherically expanding envelopes.** L. H. Auer, D. Van Blerkom.
Astrophys. Journ., Vol. 178, 175 - 181 = Contr. Five College Obs., Univ. Mass., *Amherst*, No. 140 (1972).

Broad emission-line profiles from diverse astronomical objects have been interpreted as due to rapid radial outflows of matter. Estimates of characteristic electron densities lead to the conclusion that the effect of electron scattering on the line profiles may be substantial. A Monte Carlo method is employed to treat the transfer of radiation through an expanding envelope of comoving electrons and ions. Two models are considered. The results are applied to models of Wolf-Rayet stars and Seyfert galaxies.

064.049 **On the coupling of grains to the gas in circumstellar envelopes.** R. C. Gilman.
Astrophys. Journ., Vol. 178, 423 - 426 (1972).

Dust grains in circumstellar envelopes are not position coupled to the gas, but they are momentum coupled; therefore, radiation pressure on the grains can drive the mass loss.

064.050 **Hydrodynamic and radiative-transfer effects on an RR Lyrae atmosphere.** S. J. Hill.
Astrophys. Journ., Vol. 178, 793 - 808 (1972).

A nonlinear hydrodynamic model atmosphere was produced to understand the phenomenon of shock-wave development in Bailey type *a* RR Lyrae stars. Results of this model were used in a fine-structure steady-state model for the shock waves. These two models supplied the information for a successful comparison of light curves, absorption-line velocities, and emission strengths with X Ari. Significant shock-wave dis-sipation is noted, while there is an understandable lack of mass loss.

064.051 **Temperature scale for B-type supergiants.** R. Stalio.
Proc. Third Colloquium on astrophysics. Supergiant stars, Trieste 1971, (see 012.020), p. 28 - 37 (1972).

064.052 **Atmospheres of A-type supergiants.** C. Aydin.
Proc. Third Colloquium on astrophysics. Supergiant stars, Trieste 1971, (see 012.020), p. 48 - 49 (1972). Abstract.

064.053 **A fine analysis of two early type supergiants.** P. L. Dufton.
Proc. Third Colloquium on astrophysics. Supergiant stars, Trieste 1971, (see 012.020), p. 79 - 82 (1972).

064.054 **On possible separation of elements in the atmosphere of magnetic stars.** L. O. Lodén.
Proc. Third Colloquium on astrophysics. Supergiant stars, Trieste 1971, (see 012.020), p. 184 - 189 (1972).

064.055 **Atmospheres of very cool supergiants: Introductory report.** M. S. Vardya.
Proc. Third Colloquium on astrophysics. Supergiant stars, Trieste 1971, (see 012.020), p. 190 - 203 (1972).

064.056 **Changes of line intensities caused by variable velocity fields in atmospheres of A-type supergiants.**
Proc. Third Colloquium on astrophysics. Supergiant stars, Trieste 1971, (see 012.020), p. 204 - 206 (1972).

064.057 **Dissociation equilibrium and convection in cool supergiants.** M. S. Vardya.
Proc. Third Colloquium on astrophysics. Supergiant stars, Trieste 1971, (see 012.020), p. 207 - 210 (1972).

064.058 **Effect of atmospheric curvature on spectral line profiles formed in extended stellar atmospheres.** A. Peraiah, I. P. Grant.
Proc. Third Colloquium on astrophysics. Supergiant stars, Trieste 1971, (see 012.020), p. 211 - 221 (1972).

064.059 **Model atmosphere analysis of Alpha Persei.** S. B. Parsons.
Proc. Third Colloquium on astrophysics. Supergiant stars, Trieste 1971, (see 012.020), p. 222 - 230 (1972).

064.060 **Possible influence on the interstellar Lyman-α absorption measurements from hot star circumstellar envelopes.** F. Macchetto, N. Panagia.
Proc. Third Colloquium on astrophysics. Supergiant stars, Trieste 1971, (see 012.020), p. 238 - 249 (1972).

064.061 **Inhomogeneous photospheres.** G. Elste.
Bull. American Astron. Soc., Vol. 4, 424 - 425 (1972). — Abstr. AAS.

064.062 **Fluid dynamics of convective stellar envelopes.** N. S. Mel'nikova, I. M. Yavorskaya.
Fluid Mechanics Soviet Res., (*USA*), Vol. 1, 173 - 185 (1972).

Hydrodynamic methods, with allowance for the compressibility of the gas, rotation, gravitation and nonisotropic eddy viscosity, are used in the study of large-scale motions in convective stellar envelopes.

064.063 **Mixing between stellar envelope and core in advanced phases of evolution. IV. Effect of super-adiabaticity in convective envelope.** K. Nomoto, D. Sugimoto.
Progr. Theor. Phys., (*Japan*), Vol. 48, 46 - 65 (1972).

Model envelopes of red supergiants are integrated taking into account the effect of super-adiabatic temperature gradient by means of a standard mixing length theory.

064.064 A note on stellar winds and breezes.
N. C. Freeman, R. S. Johnson.
Proc. Roy. Soc. London, Ser. A, Vol. 329, 241 - 249 (1972).

The radial flow of an inviscid heat conducting gas with gravitational effects is considered. The gas is assumed non-monatomic, but the limit to the monatomic state is considered in great detail.

064.065 Optical and infrared observations of young stellar objects — an informal review.   S. E. Strom.
Publ. Astron. Soc. Pacific, Vol. 84, 745 - 756 (1972).

Invited symposium paper presented at the Santa Cruz meeting of the Astronomical Society of the Pacific, 26 - 29 June 1972.

064.066 Abundances in K giant stars.   D. M. Gottlieb.
Thesis, Univ. Maryland, College Park. [Available from Univ. Microfilms, Ann Arbor, Mich., USA. Order No. 72-12757], 130 pp. (1971).

The use of a computer program to calculate synthetic stellar spectra has made it possible to determine how physical atmospheric parameters will change narrow-band colors. The major result of this study is that there exists a class of stars with metal abundances considerably higher than those of the sun. This class of stars is also found to have space motions typical of moderately old Population I stars. The author finds the surface gravity of luminosity class III stars to decrease and the Doppler broadeningvelocity to increase as temperature decreases.

064.067 The physical consequences of grains in the atmospheres of late type irregular and semi-regular giants and supergiants.   M. C. Jennings.
Thesis, Univ. Arizona, Tucson. [Available from Univ. Microfilms, Ann Arbor, Mich., USA. Order No. 72-16151], 169 pp. (1972). — See Phys. Abstr., Vol. 75, No. 69084 (1972).

064.068 Atomic processes in astrophysical plasmas.
V. P. Myerscough, G. Peach.
Case studies in atomic collision physics, Vol. 2, [North-Holland Publishing Company, Amsterdam], p. 293 - 397 (1972).

The author begins by studying stellar atmospheres and deals with the general theory of radiative transfer, the construction of model atmospheres and line radiation. He next considers the continuous spectrum including the absorption by neutral atoms and positive ions, the photodetachment of negative ions and coherent scattering. This is followed by a study of the line spectrum, transition probabilities, line profiles,and pressure broadening.

064.069 A line blanketed model atmosphere for ε Virginis (HD 113226).   T. E. Morgan.
Thesis, Indiana Univ., Bloomington. [Available from Univ. Microfilms, Ann Arbor, Mich., USA. Order No. 71-17451], 119 pp. (1971).

064.070 Element abundances in A and F stars HR 5447, HR 6290, HR 6596, HR 6917, and HR 7061.
R. W. Avery.
Thesis, Univ. Pennsylvania, Philadelphia. [Available from Univ. Microfilms, Ann Arbor, Mich., USA. Order No. 72-6130], 136 pp. (1971).

Fine analyses of HR 5447, HR 6290, HR 6596, HR 6917, and HR 7061 are presented. Model atmospheres for the above stars plus HR 5986 are also given.

064.071 Metal abundances in the atmospheres of red giants in open clusters and dynamical groups.
A. E. Vasilevsky.
Byull. Abastumansk. Astrofiz. Obs., No. 43, p. 29 - 54 (1972). In Russian.

Relative abundances of iron in the atmospheres of red giants which are members of 15 stellar aggregates, were found on the basis of spectrograms, UBVRI-photometry, and cyanogen indices. The relations [Fe/H] — age, and [Fe/H] — distance from the galactic plane show, that the chemical composition of population I is nearly constant. The mean metal abundance of globular clusters is about 25 times less than that of open clusters and dynamical groups.

064.072 Numerical methods for computing stellar line-profiles and continuum fluxes.   J. B. Hutchings.
Publ. Dominion Astrophys. Obs., Victoria, Vol. 14, (No. 4), 59 - 96 (1972).

Procedures for numerical integration over spherical and distorted surfaces are fully described. It is shown how the method can be used to integrate the monochromatic flux from the surface of a star, or to compute the absorption profile of a line in the star's spectrum for various states of rotation or distortion of the star. The application of the method to the computation of line profiles in the spectrum of an extended, moving atmosphere is also described. Finally, the numerical methods for computing the effects of eclipses of one star by another are described, both for spherical and distorted stars. Examples are given of computer programs used for computations of this nature.

064.073 Concerning the atmosphere of magnetic neutron stars (pulsars).   V. L. Ginzburg, V. V. Usov.
JETP Letters, Vol. 15, 196 - 198 (1972).

064.074 Physical conditions in the atmospheres of M-giants.
N. S. Komarov, Yu. A. Medvedev, T. V. Mishenina.
Astron. Tsirk., No. 726, p. 4 - 6 (1972). In Russian.

064.075 Non-LTE model atmospheres for B and O stars.
D. Mihalas.
National Center Atmosph. Res., Boulder, Colorado, Techn. Notes. NCAR-TN/STR-76, 7 + 114 pp. (1972).

In this report, non-LTE model atmospheres for 15,000°K $\leqslant T_{eff} \leqslant$ 55,000°K, $3 \times 10^2 \leqslant g \leqslant 3 \times 10^4$ are tabulated, along with emergent energy distributions and Balmer-line profiles. Several refinements have been made over earlier non-LTE models, and this extensive homogeneous grid should be of considerable use in performing spectroscopic diagnostics of the observed spectrum of early-type stars, as well as in subsequent statistical-equilibrium calculations of ions of astrophysical interest.

064.076 Non-LTE central depths of the Mg II resonance lines near 2800 Å.   N. A. Sakhibullin.
Astron. Tsirk., No. 733, p. 5 - 7 (1972). In Russian.

On the theory of radiative heat exchange in polytropic atmospheres.   See Abstr. 063.001.

Penetration of a convective stellar envelope into the nuclear burning shell.   See Abstr. 065.005.

Calculation of stellar evolution with a convective envelope penetrating into a shell source of energy.   See Abstr. 065.006.

Stellar envelope models for computation of the evolution of stars of different masses.   See Abstr. 065.007.

On thermal waves in stars.   See Abstr. 065.037.

Uniformly rotating stars with hydrogen- and metallic-line blanketed model atmospheres.   See Abstr. 065.056.

Thermal instability of the hydrogen-burning shell in nondegenerate stars.   See Abstr. 065.065.

On mass loss from B stars.   See Abstr. 065.123.

Computation of self-consistent models of chromospheres of the sun and stars.   See Abstr. 073.101.

The effect of systematic gf-value errors on stellar curves of growth.   See Abstr. 114.006.

The helium-weak stars.   See Abstr. 114.009.

Anomalous abundance of lithium in the atmosphere of the star 105 Her.   See Abstr. 114.028.

On the abundances of noble gases in extreme population I matter and the sun.   See Abstr. 114.038.

Hα profiles for G-type dwarfs and subgiants. See Abstr. 114.048.

On the atmosphere of the star HD 103877. See Abstr. 114.070.

The spectrum of o^2 CMa, B3 Ia. II. Interpretation of the observations.   See Abstr. 114.078.

Computer programmes for differential curve-of-growth analysis.   See Abstr. 114.087.

A coarse analysis of HD 50896. See Abstr. 114.101.

Abundance anomalies in the ON type star HD 188209.   See Abstr. 114.110.

Effects of velocity and turbulence gradients on A-type supergiant spectra.   See Abstr. 114.112.

A model for the helium spectrum variable α Centauri.   See Abstr. 114.114.

Analysis of the spectrum of the metallic-line star 63 Tauri.   See Abstr. 114.115.

Line spectra of eight O stars from $\lambda 3059$ to $\lambda 6683$. See Abstr. 114.117.

Intensities of silicon lines in Ap-stars spectra. I. Calculated intensities.   See Abstr. 114.138.

Spectrum variations in A-type supergiants. II. A search for evidence for coupling between mass loss and turbulence in Alpha Cygni.   See Abstr. 114.144.

B-type stars with discrepant colors. See Abstr. 114.170.

On the study of the atmosphere of the bright component of β Lyr.   See Abstr. 121.025.

Evidence for the hydrodynamic character of microturbulence.   See Abstr. 122.077.

Model atmospheres of white dwarfs with convection.   See Abstr. 126.004.

Masses and radii of white dwarfs. See Abstr. 126.017.

Polarization of optical radiation of stars with non-spherical atmospheres and envelopes.   See Abstr. 131.013.

Polarization of light by circumstellar material. See Abstr. 131.092.

Circumstellar infrared emission. See Abstr. 131.093.

The dynamical effects of stellar mass loss on diffuse nebulae.   See Abstr. 132.005.

065 Stellar Structure, Stellar Evolution, Stellar Nucleosynthesis

065.001 **Stellar structure and the evolution of stars.**
A. G. Masevich.
Nauchn. Informatsii, vyp. (No.) 19, p. 3 - 31 (1971).
In Russian.

065.002 **Instationary processes in stars.** V. S. Imshennik.
Nauchn. Informatsii, vyp. (No.) 19, p. 41 - 44
(1971). In Russian.

065.003 **Some peculiarities of the evolution of stars with
large masses.** A. G. Masevich, A. V. Tutukov,
O. B. Dluzhnevskaya, V. I. Varshavskij, U. Kh. Uus, Eh. V.
Ehrgma, E. I. Popova, G. G. Rodionova.
Nauchn. Informatsii, vyp. (No.) 19, p. 45 - 77 (1971).
In Russian.

065.004 **Investigation of rapid processes in stars.**
G. Ruben.
Nauchn. Informatsii, vyp. (No.)20, p. 3 - 59 (1971).
In Russian.
The particle-in-cell-method is applied to study dynamical
processes in stars. Finite difference equations are derived and
discussed. Various examples of shock-wave propagation in an
initially polytropic model are computed. Effect of radiative
pressure, of changes of the polytropic index and specific heat
ratio on the propagation of shock waves and on mass loss is
discussed. Accuracy of the method is checked by changing
both time and space steps as well as the initial number of particles per cell.

065.005 **Penetration of a convective stellar envelope into
the nuclear burning shell.** U. Uus.
Nauchn. Informatsii, vyp. (No.), 20, p. 60 - 63 (1971).
In Russian.
It is shown that assuming sufficiently high conductivity
by convection, the convective envelope penetrates into the
thin hydrogen-burning shell in the stage of the growth of carbon-oxygen core in stars.

065.006 **Calculation of stellar evolution with a convective
envelope penetrating into a shell source of energy.**
U. Uus.
Nauchn. Informatsii, vyp. (No.) 20, p. 64 - 70 (1971).
In Russian.
The fitting method of constructing stellar models with
dense isothermal cores and thin nuclear burning shells, described previously, is adapted to calculate the evolution of
stars with a convective envelope penetrating into a nuclear
burning shell.

065.007 **Stellar envelope models for computation of the
evolution of stars of different masses.**
V. V. Musylev.
Nauchn. Informatsii, vyp. (No.) 20, p. 108 - 116 (1971).
In Russian.
A grid of stellar envelope models for stars evolving from
main sequence to the red giants region was computed for masses 64, 32, 24, 16, 8, 4, 2 and 1 $M_\odot$. Chemical composition
was assumed $X = 0.602$, $Z = 0.044$.

065.008 **Linear pulsations and stability of differentially rotating stellar models. I. Newtonian analysis.**
B. F. Schutz, Jr.
Astrophys. Journ., Suppl. Ser., No. 208, Vol. 24, 319 - 342
(1972).
A systematic method is presented for deriving the Lagrangian governing the evolution of small perturbations of
arbitrary flows of a self-gravitating perfect fluid. The method
is applied to a differentially rotating stellar model; the result
is a Lagrangian equivalent to that of Lynden-Bell and Ostriker.
A sufficient condition for stability of rotating stars is derived
from this Lagrangian.

065.009 **Linear pulsations and stability of differentially rotating stellar models. II. General-relativistic analysis.**
B. F. Schutz, Jr.
Astrophys. Journ., Suppl. Ser., No. 208, Vol. 24, 343 - 374
(1972).
The author has previously given an Eulerian velocity-po-
tential variational principle for relativistic perfect-fluid hydrodynamics. The second variation of the principle is here used
as the Lagrangian density governing the evolution of small
perturbations of fully relativistic, differentially rotating stellar
models. Noether's theorem is used to construct a globally conserved angular-momentum density, whose integral over a
spacelike hypersurface is the second-order correction to the
star's total angular momentum. From the Hamiltonian is constructed a globally conserved energy density, whose integral
is the second-order correction to the star's active gravitational
mass. By Liapunov's second theorem, positive-definiteness of
the energy density guarantees stability of the star.

065.010 **The structure and evolution of helium stars.**
A. S. Dinger.
Monthly Notices Roy. Astron. Soc., Vol. 158, 383 - 403
(1972).
Numerical models of helium stars have been constructed
in an effort to fit the models to the observed extreme-helium
stars, i.e. those which have equivalent spectral types of O7-B5,
absolute magnitudes brighter than −1.6, and weak or missing
hydrogen lines. Main-sequence models of 1, 2, 4 and 8 $M_\odot$
were computed for $Y = 0.98$, $Z = 0.02$. The effect of the inclusion of the $N^{14}(\alpha, \gamma)F^{18}(e^+\nu)O^{18}$ reaction was investigated.
One-solar-mass models with 2, 6, 14, 30 and 58 per cent addi-
tional carbon, and a four-solar-mass model with $Z = 0.60$
(0.02 solar metals and 0.58 additional carbon) was also computed.

065.011 **Studies of hydrodynamic events in stellar evolution.
I. Method of computation.**
G. S. Kutter, W. M. Sparks.
Astrophys. Journ., Vol. 175, 407 - 415 (1972).
We describe a numerical method of computation for hydrodynamic events in stellar evolution. The differential equations (spherical symmetry) of conservation of mass, momentum, and energy; energy transport by diffusion and convection; and the definition of velocity are replaced by finite-difference equations. We describe several test applications: Numerical stability, propagation of a shock wave, free-fall collapse, and nova outburst. The results are in satisfactory agreement with either analytical solutions or with numerical solutions obtained with an independent computer code.

065.012 **Analytic approximations to the mass-radius relation
and energy of zero-temperature stars.**
M. Nauenberg,
Astrophys. Journ., Vol. 175, 417 - 430 (1972).
The variational principle is applied to two approximate
forms of the energy integral of a star at zero temperature in
order to obtain analytic expressions for the mass-radius relation and for the energy. Several well-known equations of state
are considered, and the effects of rotation and of stored magnetic fields are included. Comparison of the analytic approximations with published numerical solutions of the exact equa-

tions show good agreement.

065.013 Fundamental data for massive stars compared with theoretical models. R. Stothers.
Astrophys. Journ., Vol. 175, 431 - 452 (1972).

New models for stars evolving on the upper main sequence have been constructed with a variety of masses, chemical compositions, and assumed modes of evolution. Reliable observational data for upper-main-sequence stars and supergiants have been compiled, including masses, luminosities, and spectral types — as determined from the members of double-line eclipsing binary systems, young clusters, and subgroups of associations. Comparison of the theoretical models with the observational data then permits the drawing of a number of conclusions regarding the initial chemical composition, axial rotation, loss of mass, mode of evolution, and age of a massive star.

065.014 The rate of star formation. J. Einasto.
Astrophys. Letters, Vol. 11, 195 - 199 (1972).

Formulae are found for the time derivative of the projected density and the total mass of stars in a galaxy, supposing for the local rate of star formation the relation $R_l = \gamma \rho_g{}^n$, where ρ_g is the density of interstellar gas, and γ and n are parameters.

065.015 Possible sites of star formation in Sgr B2.
A. H. M. Martin, D. Downes.
Astrophys. Letters, Vol. 11, 219 - 224 (1972).

Observations of Sgr B2 with the Cambridge One-Mile Telescope at 2.7 and 5 GHz reveal seven bright condensations of angular size $\lesssim 10$ arc sec immersed in an extended background of diameter 4 arc min. Each compact condensation requires for its excitation one to three O5 stars within a zone one light year across. These emission zones are discussed in the context of a six-component physical model for Sgr B2.

065.016 Structure et critères d'effondrement gravitationnel d'une sphère polytropique d'indice négatif.
Y.-P. Viala.
Comptes Rendus Acad. Sci. Paris, Sér. B, Vol. 275, 117 - 120 (1972).

On détermine la structure d'une sphère polytropique d'indice n inférieur à -1. On utilise les résultats obtenus pour calculer les masses critiques d'effondrement gravitationnel de telles sphères sous l'effet d'une pression extérieure.

065.017 Hadronic equilibrium-configurations.
C. Möllenhoff.
Astron. Astrophys., Vol. 19, 326 - 336 (1972).

Hagedorn's hadronic equation of state is used to construct relativistic spheres consisting of a pure hadron-gas with zero total baryon number. Their equilibrium configurations are uniquely determined by their central densities. The configurations made from a pure hadron gas are good approximations for the core of an extremely hot star in which the core structure is determined by hadrons, muons, photons, electrons and neutrinos.

065.018 Multiple solutions of the equations of stellar structure. D. Lauterborn.
Astron. Astrophys., Vol. 19, 473 - 481 (1972).

Three solutions are shown to exist, in a certain range of parameters, for a giant envelope for which four boundary conditions (two atmospheric boundary conditions, and radius and luminosity at the bottom of the envelope) are specified. Such triple solutions occur independent of whether or not the hydrogen burning region is included in the envelope.

065.019 Examples of multiple solutions for equilibrium stars with helium cores. M. L. Roth, A. Weigert.

Astron. Astrophys., Vol. 20, 13 - 18 (1972).

The different types of equilibrium configurations are discussed for stars with He cores and hydrogen-rich envelopes. As an example, a star of given mass and chemical composition is treated for which up to six solutions are shown to exist.

065.020 On the influence of the opacity values on static stellar models. I. Horizontal branch stars.
S. Refsdal, R. Stabell.
Astron. Astrophys., Vol. 20, 19 - 28 (1972).

A method is presented whereby the effect of opacity changes in different regions of static stellar models can be discussed. The method is applied to horizontal branch stars, and the effect on the luminosity and the effective temperature is discussed in detail.

065.021 The local Vogt-Russell theorem. H. Kähler.
Astron. Astrophys., Vol. 20, 105 - 110 (1972).

In the neighbourhood of any given equilibrium configuration fulfilling the basic stellar structure equations, the "local" uniqueness of this solution can be investigated by linearizing the stellar structure equations. The connection with the stellar stability problem is discussed with the result that a local Vogt-Russell theorem holds for each star which is stable.

065.022 Tidal perturbation of the non-radial oscillations of a star. J. Denis.
Astron. Astrophys., Vol. 20, 151 - 155 (1972).

A generalization of the perturbation method in which the distortion of the equilibrium configuration is taken into account is applied to the perturbation of a gaseous star by tidal action. Numerical results are obtained in the case of an homogeneous configuration. It is shown that the perturbation of the oscillation frequencies in the case of a synchronously rotating binary star can be obtained as a combination of the purely rotational and the purely tidal perturbation. The importance of this result for the interpretation of the beat phenomena in β Canis Majoris stars is discussed.

065.023 Hydrostatic oxygen burning in stars. II. Oxygen burning at balanced power.
S. E. Woosley, W. D. Arnett, D. D. Clayton.
Astrophys. Journ., Vol. 175, 731 - 749 (1972).

The nucleosynthesis that occurs during hydrostatic oxygen burning in a "typical" stellar zone is examined in detail under the assumption that such stellar zones are in an approximate state of balanced power from $^{16}O + {}^{16}O$ energy generation and from neutrino (or photon) energy loss. Special attention is paid to the amount of neutron enrichment that occurs and the implications for nucleosynthesis theory. Accuracy of a common approximation for $^{16}O + {}^{16}O$ energy generation is discussed also.

065.024 Stellar reaction rates for the $^{12}C + {}^{12}C$, $^{12}C + {}^{16}O$, and $^{16}O + {}^{16}O$ reactions. G. Michaud.
Astrophys. Journ., Vol. 175, 751 - 756 (1972).

Thermonuclear reaction rates have been calculated for the $^{12}C + {}^{12}C$, $^{12}C + {}^{16}O$, and $^{16}O + {}^{16}O$ reactions using optical models for the extrapolation from the recent measurements.

065.025 Shock wave propagation in a rotating star.
A. A. Rumjantsev.
Astron. Zhurn. Akad. Nauk SSSR, Vol. 49, 744 - 749 (1972). In Russian. English translation in Soviet Astron. AJ, Vol. 16, No. 4.

The output of a strong shock wave into external layers of a rotating star is considered. The influence of a regular magnetic field of a star on shell formation is discussed.

065.026 Slowly rotating relativistic stars. VI. Stability of the quasi-radial modes.

J. B. Hartle, K. S. Thorne, S. M. Chitre.
Astrophys. Journ., Vol. 176, 177 - 194 (1972).

Equations are given for calculating the effects of a slow and rigid rotation on the frequency of the radial modes of oscillation of a relativistic star. The rotation is treated to second order in the angular velocity, but no other approximations are made.

065.027 **Models of differentially rotating stars.**
J. R. Wilson.
Astrophys. Journ., Vol. 176, 195 - 204 (1972).

The binding energy is calculated for rotating stars with pressure equal to one-third the thermal energy density, by means of general relativity. For large central redshifts, the binding energy decreases rapidly as one proceeds from flat disk shapes to more nearly spherical stars. Binding-energy curves are also presented for the case of pressure equal to the thermal energy.

065.028 **Energy shifts of excited nucleons in neutron-star matter.** R. F. Sawyer.
Astrophys. Journ., Vol. 176, 205 - 211, with a correction Vol. 178, 279 (1972).

It is argued that baryon isobars embedded in superdense neutron-star matter will suffer large positive energy shifts, even in the absence of ordinary two-body interactions of the isobars with the other baryons of the matter. A numerical calculation is done for a zero-momentum $\Delta^-(1236$ MeV) immersed in a pure neutron medium, and energy shifts of from 145 to 250 MeV are found for the range of neutron densities which had been expected to prevail at the onset of the formation of Δ^-'s in superdense nuclear matter.

065.029 **The birth of stars.** B. J. Bok.
Sci. American, Vol. 227, No. 2, p. 48 - 61 (1972).

Since stars grow old and die, they are presumably born. In our Galaxy the most favorable conditions for their birth are found in the clouds of dust and gas along the central galactic plane.

065.030 **Nucleosynthesis and stellar evolution.**
G. Beaudet, M. Tassoul.
Journ. Roy. Astron. Soc. Canada, Vol. 66, 221 - 222 (1972).
Abstr. Canadian Astron. Soc.

065.031 **Secular stability in the presence of perturbations of chemical composition. I.**
M. L. Aizenman, J. Perdang.
Journ. Roy. Astron. Soc. Canada, Vol. 66, 222 (1972).
Abstr. Canadian Astron. Soc.

065.032 **Secular stability in the presence of perturbations of chemical composition. II.**
J. Perdang, M. Aizenman.
Journ. Roy. Astron. Soc. Canada, Vol. 66, 222 (1972).
Abstr. Canadian Astron. Soc.

065.033 **Extended horizontal branch loci.** V. Caloi.
Astron. Astrophys., Vol. 20, 357 - 360 (1972).

Models for extremely blue horizontal branch type stars are presented. An interpretation of B subdwarfs in the halo is attempted in terms of these models. The importance of reliable determinations of the luminosities of these objects for indications about the original He-content is stressed.

065.034 **Effects on semiconvection on the horizontal-branch.**
A. V. Sweigart, P. Demarque.
Astron. Astrophys., Vol. 20, 445 - 453 (1972).

The effects of semiconvection on the horizontal-branch evolution of a $0.60\,M_\odot$ population II star with the envelope composition parameters $X = 0.732$ and $Z = 0.001$ are investi-

gated. A comparison between sequences computed with and without a treatment of semiconvection indicates that the neglect of semiconvection will reduce the length of the horizontal-branch track by a factor of 1.8 and decrease the horizontal-branch lifetime by a factor of 1.9. A number of additional horizontal-branch sequences including a treatment of semiconvection are presented over the mass range from 0.51 – $0.65\,M_\odot$.

065.035 **On the conditions for homogeneous stellar models to have only real secular eigenvalues.**
M. Gabriel, A. Noels.
Astron. Astrophys., Vol. 20, 455 - 459 (1972).

The conditions to have only real eigenvalues in the secular spectrum of homogeneous stars are discussed numerically. Results highly suggest that they are identical to those for the existence of a nearly homologous solution as an eigensolution. When one deviates enough from these conditions, complex eigenvalues are generally present except when very special conditions on the opacity and energy generation laws are fulfilled.

065.036 **Perturbation of the radial and non-radial oscillations of a star by a magnetic field.** M. Goossens.
Astrophys. Space Sci., Vol. 16, 386 - 404 (1972).

A generalization of the perturbation method is applied to the problem of the radial and the non-radial oscillations of a gaseous star which is distorted by a magnetic field. An expression is derived for the perturbation of the oscillation frequencies due to the presence of a weak magnetic field when the equilibrium configuration is a spheroid. The main result is that the magnetic field has a large and almost stabilizing effect on unstable g-modes, particularly on higher order modes.

065.037 **On thermal waves in stars.** I. A. Klimishin.
Astrophys. Space Sci., Vol. 16, 432 - 436 (1972).

It is shown that by a certain combination of stellar parameters and energy which was released by an outburst, the transfer of that energy outwards is carried out by the thermal wave.

065.038 **Oscillations of a polytrope with a toroidal magnetic field.** N. K. Sood, S. K. Trehan.
Astrophys. Space Sci., Vol. 16, 451 - 464 (1972).

The radial and the non-radial ($l = 2$) modes of oscillation of a gaseous polytrope with a toroidal magnetic field are examined using a variational principle. It is found that the frequencies of oscillation of the radial mode and the Kelvin mode ($l = 2$) decrease due to the presence of the magnetic field. We study the effect of a small rotation and toroidal magnetic field on the structure of a polytrope. It is found that the resulting configuration is a prolate spheroid, a sphere or an oblate spheroid.

065.039 **Resonance effects in polytropes.**
N. R. Simon, V. K. Sastri.
Astron. Astrophys., Vol. 21, 39 - 44 (1972).

Second-order anharmonic pulsational amplitudes are investigated for a variety of polytropic models. An empirical asymptotic formula is given for the normal mode spectra of polytropes. The question of resonances is discussed for real stars, and a suggestion made for classifying cepheid-type pulsators on the basis of resonant properties. Invalid numerical results from an earlier work (Simon, 1971) are corrected in the appendix.

065.040 **Iterative nonlinear pulsations in massive stars. I. The iterative approach.** N. R. Simon.
Astron. Astrophys., Vol. 21, 45 - 49 (1972).

We consider "soft" pulsations in a massive main-sequence star, and attempt to investigate the regime in which the motion is nonlinear but not strongly so. A solution is sought in the form of a Fourier series over an interval which turns out not to deviate much from the fundamental period of pulsation. Arguments based on the size of successive terms are used to isolate a set of five equations, the iterative solution of which should provide a complete and consistent description of weakly nonlinear adiabatic pulsations to the order required.

065.041 **Iterative nonlinear pulsations in massive stars. II. Terms up to second order.** N. R. Simon.
Astron. Astrophys., Vol. 21, 51 - 55 (1972).

An iterative technique is used to calculate radial adiabatic pulsational amplitudes up to second order in a very massive main-sequence star. The second-order terms are displayed, and discussed in detail. A possible connection between the qualitative structure of these terms and the absence of observed massive main-sequence pulsators is suggested. Finally, we discuss the resonance properties of the present model and consider their effect on the iterative calculation.

065.042 **Ultraviolet stars and the interstellar gas.**
W. K. Rose, D. G. Wentzel.
Bull. American Astron. Soc., Vol. 4, 318 (1972). — Abstr. AAS.

065.043 **Pre-main sequence stellar evolution.** S. E. Strom.
Bull. American Astron. Soc., Vol. 4, 324 - 325 (1972). — Abstr. AAS.

065.044 **Low mass limit of the deuterium main sequence.**
H. C. Graboske, A. S. Grossman.
Bull. American Astron. Soc., Vol. 4, 326 (1972). — Abstr. AAS.

065.045 **On the "critical luminosity" in stellar interiors.**
P. C. Joss, E. E. Salpeter, J. P. Ostriker.
Bull. American Astron. Soc., Vol. 4, 327 (1972). — Abstr. AAS.

065.046 **On the helium-burning evolution of solar mass stars.**
D. R. Alexander.
Bull. American Astron. Soc., Vol. 4, 337 (1972). — Abstr. AAS.

065.047 **Determination of the s-process neutron capture time.** V. L. Peterson, D. A. Tripp.
Bull. American Astron. Soc., Vol. 4, 337 (1972). — Abstr. AAS.

065.048 **Effective temperatures of massive stars as a function of chemical composition and mass.**
F. D. A. Hartwick, D. A. Vanden Berg.
Astrophys. Journ., Vol. 176, 677 - 679 (1972).

Theoretical models for main-sequence stars of 50 and 100 $M_{\odot}$ have been constructed for a range in heavy-element abundance $10^{-4} < Z < 10^{-1}$. For a given mass, significant differences in the stellar effective temperature and size of H II region are found for moderate changes in Z. These differences are such that stars with lower Z have higher effective temperatures. The results suggest that the expected increase in excitation temperature in low-Z H II regions due to less efficient cooling will be enhanced.

065.049 **Advanced evolution of massive stars. I. Helium burning.** W. D. Arnett.
Astrophys. Journ., Vol. 176, 681 - 698 (1972).

This is the first paper in a series dealing with the thermonuclear evolution of massive stars, beginning at helium burning and proceeding through the final hydrodynamic stages. The paper contains a discussion of the theoretical framework within which the investigation will be conducted. The input physics and certain mathematical constraints used in constructing the evolutionary models, with emphasis on the first (helium burning) stage, are described and the results of evolutionary calculations of helium burning, with emphasis on comparison with previous work, and on aspects of importance for nucleosynthesis theory and for further evolution, are presented

065.050 **Advanced evolution of massive stars. II. Carbon burning.** W. D. Arnett.
Astrophys. Journ., Vol. 176, 699 - 710 (1972).

A simple approximate technique for calculating energy generation and nucleosynthesis during carbon burning is presented. It is found that (1) convection outside shell sources may have a different character during neutrino-dominated evolution, (2) under the influence of neutrino emission, carbon burning produces small post-carbon-burning cores; (3) very massive stars ($M_a \gtrsim 64\,M_{\odot}$ and $M \gtrsim 110\,M_{\odot}$ or so) encounter the electron-pair instability, and (4) neon burning will be more important for smaller-mass stars. Comparison is made with some other recent investigations of carbon burning in stars.

065.051 **Equation of state of neutron-star matter at subnuclear densities.**
Z. Barkat, J.-R. Buchler, L. Ingber.
Astrophys. Journ., Vol. 176, 723 - 738 (1972).

An extended nuclear Thomas-Fermi model is used in conjunction with a Wigner-Seitz model to determine the ground-state composition of cold catalyzed matter for densities ranging from 10^8 to $10^{14}\,\mathrm{g\,cm^{-3}}$, allowing for inhomogeneities on a nuclear scale (nuclei or "clusters"). As a function of the average density, the proton number Z of the clusters first increases from $Z \approx 29$ at $1.4 \times 10^8\,\mathrm{g\,cm^{-3}}$ to a maximum of $Z \approx 35$ at $3 \times 10^{12}\,\mathrm{g\,cm^{-3}}$, and then decreases until the clusters gradually evanesce just above $10^{14}\,\mathrm{g\,cm^{-3}}$. This transition appears to be smooth—in contrast to the "neutron drip" phase transition which occurs at $3.7 \times 10^{11}\,\mathrm{g\,cm^{-3}}$. A tabulation of the equation of state is presented.

065.052 **Untersuchungen stellarer Kernreaktionen im Beschleunigerlaboratorium.**
K. Ewen, B. Gonsior.
SuW, Vol. 11, 272 - 275 (1972).

065.053 **The structure of chemically homogeneous main-sequence stars.** L. Motz.
Publ. Astron. Soc. Pacific, Vol. 84, 465 - 488 (1972). — Invited review article. — The general differential equations for the structure of stars are developed. These are then applied to the analysis of the structure of chemically homogeneous main-sequence stars. Some general deductions are made about the dependence of the slope of the main sequence on the energy generating cycle. The variation in the position of the main sequence with chemical composition is also discussed.

065.054 **Evolution of a 16 $M_{\odot}$ star from the main sequence to a red supergiant.** V. I. Varshavsky.
Astron. Zhurn. Akad. Nauk SSSR, Vol. 49, 1055 - 1058 (1972). In Russian. English translation in Soviet Astron. AJ, Vol. 16, No. 5.

The evolutionary sequence of 16 $M_{\odot}$ star models has been computed up to helium exhaustion in the convective core. The mass of the growing carbon—oxygen core is of 3.0 $M_{\odot}$. At the main central helium burning phase instability occurs during the loop at the red supergiant region.

065.055 **On the models of core-helium-burning stars.**
V. Castellani, P. Giannone, A. Renzini.
Astrophys. Space Sci., Vol. 17, 80 - 86 (1972).

Some aspects concerning the core-mass increase and the appearance and development of an intermediate semiconvective zone in helium-burning stars are discussed. An iterative method of computation of core increase and semiconvection is also presented. Details of this procedure are given with regard to the horizontal-branch stars of globular clusters.

065.056 **Uniformly rotating stars with hydrogen- and metallic-line blanketed model atmospheres.**
A. Maeder, E. Peytremann.

Astron. Astrophys., Vol. 21, 279 - 284 (1972).

The energy distribution of uniformly rotating stars including continuous, hydrogen- and metallic-line opacities are computed for a 1.4 $M_\odot$ star. Colours and parameters are computed in the *UBV, uvby* and Geneva Observatory photometric systems. The effects of rotation predicted by these models are presented.

065.057 **Hydrodynamic model calculations for supermassive stars II. The collapse and explosion of a nonrotating** $5.2 \times 10^5 M_\odot$ **star.** I. Appenzeller, K. Fricke.
Astron. Astrophys., Vol. 21, 285 - 290 (1972).

The dynamical evolution of a $5.2 \times 10^5 M_\odot$ supermassive star is followed by solving numerically the general relativistic hydrodynamic equations. After initial quasistatic contraction general relativistic collapse below the static main-sequence radius takes place. The radiation pressure which is built up during the collapse phase by fast HCNO burning halts and reverses the collapse into a violent explosion which results in the complete dispersion of the star. The kinematics and energetics of the explosion suggest that a possible mechanism is found which may serve to explain the explosion phenomena in peculiar galaxies like M 82 and NGC 1275.

065.058 **Pulsational damping of neutron stars.**
H. Heintzmann, J. Nitsch.
Astron. Astrophys., Vol. 21, 291 - 298 (1972).

The damping of neutron star vibrations is reconsidered in the light of more recent, semirealistic neutron star models. It is assumed that the star possesses a strong magnetic field and that the interior is composed of essentially four components, namely an outer crust, a superfluid neutron and proton component, a normal neutron component and a quantum crystal. Possible mechanisms by which vibrations can be triggered over astronomical timescales are reviewed and a discussion of how other damping mechanisms previously discussed in the literature are affected through the more recently obtained results of neutron star matter is given. As a by-product of our calculations we derive the specific heat and thermal conductivity for normal neutron matter which might be important for the cooling of hot neutron stars.

065.059 **Meridional circulation in a rotating star.**
Y. Osaki.
Publ. Astron. Soc. Japan, Vol. 24, 509 - 516 (1972).

The time-dependent hydrodynamics of meridional circulation in the radiative zone of a rotating star has been studied. The Eddington-Sweet formulation of the circulation velocity is confirmed for a "slowly" rotating star. On the other hand, apparent singularities of the circulation velocity in a special case and near the stellar surface are found to be due to improper use of the Eddington-Sweet formulation.

065.060 **Note on general relativistic secular instability of supermassive stars.** Y. Osaki.
Publ. Astron. Soc. Japan, Vol. 24, 537 - 543 (1972).

A simple physical explanation of the secular instability of supermassive stars is given and it is demonstrated that this secular instability is not a new kind of instability but it is essentially one and the same with the dynamical instability. It is argued that dynamically stable supermassive stars are unlikely to become secularly unstable.

065.061 **Shock waves on neutron star cores.**
S. A. Colgate, Y.-H. Chen.
Astrophys. Space Sci., Vol. 17, 325 - 329 (1972).

A numerical model of the transient behavior of a radiation-dominated shock was calculated in order to demonstrate the relatively large initial escape of internal energy that takes place when the opacity law is a positive power of temperature, as for neutrinos, $K \sim T^2$. It is concluded that a shock formed

on the neutron star core of an imploding supernova may radiate its internal energy in electron neutrinos more effectively than had hithertofore been considered.

065.062 **On the cosmic abundance of helium.** P. G. Gross.
Bull. American Astron. Soc., Vol. 4, 341 (1972).
Abstr. AAS.

065.063 **Rotazione in oggetti stellari e quasi-stellari relativistici.** F. Occhionero.
Atti XV Riunione Soc. Astron. Italiana, Bologna 1971, (see 012.013), p. 105 - 136 (1972).

This is a review, with regard to the influence of rotation, of the models for stellar and quasi-stellar relativistic objects which are presently needed for an explanation of pulsars and quasars.

065.064 **On the contribution of autoionization lines to stellar opacities.** A. L. Merts, N. H. Magee, Jr.
Astrophys. Journ., Vol. 177, 137 - 143 (1972).

The opacity of a stellar mixture has been calculated including autoionization lines. The results of this calculation indicate that the Rosseland mean opacity is increased by less than 5 percent in the temperature density range of interest for stellar interiors.

065.065 **Thermal instability of the hydrogen-burning shell in nondegenerate stars.**
R. Stothers, C.-W. Chin.
Astrophys. Journ., Vol. 177, 155 - 160 (1972).

An investigation is made of thermal instability in the hydrogen-burning shell of stars of moderate to high mass evolving from the end of core hydrogen burning to the early stages of core helium burning, with the help of an approximate analytic criterion for thermal instability and full nonlinear numerical calculations of stellar evolution.

065.066 **Masses and magnetic fields of neutron stars.**
G. Greenstein.
Astrophys. Journ., Vol. 177, 251 - 253 = Contr. Five College Obs., *Amherst, Mass.*, No. 132 (1972).

Unless the mass of a pulsar is known beforehand, timing observations are able to determine its magnetic field only to within a range of 3–4 decades. Either this field is correlated with mass or both the field and the mass are constrained to lie within relatively narrow limits.

065.067 **The effect of Urca shells on the density of carbon ignition in degenerate stellar cores.** S. W. Bruenn.
Astrophys. Journ., Vol. 177, 459 - 471 (1972).

The Urca process is found to be ineffective in raising ρ_{ign}, the central density at which carbon ignites, in growing degenerate stellar cores unless the relative abundances of "Urca-active" nuclei have been increased far above their cosmic values at some prior evolutionary stage. Evolutionary calculations with increased abundances of "Urca-active" nuclei have been performed, and the effect on ρ_{ign} has been determined. Some implications of these results concerning the subsequent evolution of the core are discussed.

065.068 **Core helium burning in massive stars.**
J. W. Robertson.
Astrophys. Journ., Vol. 177, 473 - 488 (1972).

Nine evolutionary sequences have been computed from the initial main sequence well into the core-helium-burning stage, for stellar masses in the range 6–20 $M_\odot$ and for a variety of initial chemical compositions. The effects of semiconvection on the distribution of core-helium-burning stars in the H-R diagram have been studied for two cases: case A, in which the chemical composition gradient has been included, and case

B, in which it has been neglected, in the criterion for convection.

065.069 **Evolution of rotating stars.**
K. J. Fricke, R. Kippenhahn.
Annual Rev. Astron. Astrophys., Vol. 10, (see 003.005), 45 - 72 (1972). − Contents: (1) Introduction; (2) Mechanisms influencing stellar angular velocity distributions; (3) Stability of angular velocity distributions; (4) Evolution of rotating stars.

065.070 **Convection in stars. II. Special effects.**
E. A. Spiegel.
Annual Rev. Astron. Astrophys., Vol. 10, (see 003.005), 261 - 304 (1972). − Contents: (1) Therminology of stability theory; (2) Additional background; (3) Thermosolutal convection; (4) Thermodynamic effects; (5) Density variation; (6) Geometry; (7) Interaction with shear flow; (8) Rotation and magnetic fields; (9) Penetration and entrainment; (10) Time dependence; (11) Concluding remarks.

065.071 **On models for supermassive stars.** E. Anttila.
Ann. Acad. Sci. Fennicae, Ser. A, VI. Phys., No. 374, 44 pp. = Contr. Obs. Astrophys. Lab. Univ. Helsinki (1971).
It is shown that the solutions of the static field equations of general relativity can be divided into two main types. In the first type both pressure and density vanish at the surface. In the second type pressure vanishes but not the density. Thus, stars of this type have a halo surrounding them. Several supermassive star models have been calculated using different polytropic equations as equations of state in order to search for common features which are independent of the equation of state used.

065.072 **On stellar structure in non-uniform rotation. 1. Slow rotation with meridional circulation.**
I. V. Tuominen.
Ann. Acad. Sci. Fennicae, Ser. A, VI. Phys., No. 391, 49 pp. = Contr. Obs. Astrophys. Lab. Univ. Helsinki (1972).
The problem of the structure of differentially rotating non-magnetic stellar interiors is outlined generally from the point of view of finding a numerical solution by difference methods. When the gravitational potential is presented by means of Legendre polynomials and when barotropy holds at the equatorial plane, it is shown how the problem can be solved in the equatorial plane independently of the entire structure. This procedure greatly simplifies the general problem. The structural properties obtained compared with a nonrotating star of the same mass are given. As a numerical example the structure of a $2.5\,M_\odot$ star is studied using different angular velocities at the surface and different distributions of angular momentum within the star.

065.073 **On stellar structure in non-uniform rotation. 2. Properties of a fast rotating star averaged over spheres.**
I. V. Tuominen.
Ann. Acad. Sci. Fennicae, Ser. A, VI. Phys., No. 392, 27 pp. = Contr. Obs. Astrophys. Lab. Univ. Helsinki (1972).
The structure of a main sequence star in fast differential rotation, with different amounts and different distributions of angular momentum, is studied. A technique in which the variables are integrated over spheres, including first order perturbing terms only, is applied. The equilibrium equations reduce to a simple form, which is essentially the same as in the case of spherical symmetry. The results, however, give only the total luminosity and the mean radius of the star.

065.074 **On stellar structure in non-uniform rotation. 3. Nonbarotropic models in fast rotation.**
I. V. Tuominen.
Ann. Acad. Sci. Fennicae, Ser. A, VI. Phys., No. 393, 20 pp. =

Contr. Obs. Astrophys. Lab. Univ. Helsinki (1972).
Approximate two-dimensional solutions have been derived by numerical techniques for a $2.5\,M_\odot$ homogeneous nonmagnetic main-sequence star in differential rotation, including the rapid one. An increase of pressure and density, a decrease of temperature, and a strong decrease of luminosity and volume follow from an increase in the central angular velocity with respect to the surface velocity. Surface properties differ only slightly from those that can be obtained from the Roche-model and von Zeipel's relation. Distributions of angular velocity are nearly uniform in the surface regions, or else show weak polar acceleration.

065.075 **On the mixing of matter in the semiconvective zones of massive stars.** A. E. Dudorov, A. V. Tutukov.
Nauchn. Informatsii, vyp. (No.) 21, p. 3 - 10 (1972). In Russian.
The linearized equation of the motion of convective elements in a medium is obtained taking into account radiative heat exchange. A condition for instablility was found by means of Routh-Hurwitz criterium. The roots of the characteristic equation were calculated for several limiting cases. The analysis of mixing in the semiconvective zone of massive stars for $\nabla - \nabla\alpha << \nabla\mu$ is given. Possible mechanisms of excitation are discussed.

065.076 **Evolution of massive stars with differential rotation.**
V. V. Musylev.
Nauchn. Informatsii, vyp.(No.) 21, p. 11 - 24 (1972). In Russian.
The evolution of rotating massive stars is investigated. Spherical symmetry of models and local conservation of angular momentum throughout the star are assumed. Evolutionary models for a $32\,M_\odot$ star were calculated assuming two different laws of initial distribution of the angular velocity.

065.077 **Evolution of massive stars with semiconvection.**
V. I. Varshavsky.
Nauchn. Informatsii, vyp. (No.) 21, p. 25 - 45 (1972). In Russian.
The evolutionary sequence of $16\,M_\odot$ stellar models with initial chemical composition $X = 0.602$, $Y = 0.354$, $Z = 0.044$ ($X_{12} = 0,00619$, $X_{16} = 0,01847$) is computed from the main sequence up to the stage of complete helium exhaustion in the convective core in the red supergiant branch. In the course of calculations, particular attention was paid to the treatment of semiconvection. Simultaneous calculations were performed applying three different conditions for stability. The main results are presented in a table and in figures. Discussion of results and comparison with the results obtained by other authors is given.

065.078 **Construction of isochrones by interpolation. Age determination for some open clusters.**
O. B. Dluzhnevskaja, A. E. Piskunov.
Nauchn. Informatsii, vyp. (No.) 21, p. 58 - 67 (1972). In Russian.
A network of isochrones for $t = 4.73 \times 10^6 - 1.2 \times 10^9$ years is constructed by means of linear interpolation using the set of evolutionary tracks for stars with masses $0.8 \leqslant M/M_\odot \leqslant 15$ computed by Paczyński. The results are applied to HR-diagrams for 5 open clusters. Ages of these clusters are estimated.

065.079 **Ginzburg-Landau theory of anisotropic superfluid neutron-star matter.** R. W. Richardson.
Phys. Rev. D, Particles and Fields, Vol. 5, 1883 - 1896 (1972).
The Gor'kov procedure is used to obtain a generalized Ginzburg-Landau theory for anisotropic superfluid neutron-star matter. The resulting equations are solved for the case of superfluid flow past a plane boundary and the case of an iso-

lated vortex line. The structure of these solutions and their connection with rotating neutron stars is discussed.

065.080 Implications of tachyon-like matter for superdense stars. M. S. Bhatia, L. K. Pande.
Phys. Rev. D, Particles and Fields, Vol. 5, 2936 - 2938 (1972).

A new equation of state of superdense matter is proposed. It is derivable by treating superdense matter as a perfect, degenerate tachyon gas. Model calculations for superdense stars based on this equation of state are presented.

065.081 Destruction of the s process ($n\sigma$) correlations by electron capture after neutron irradiation.
M. Arnould.
Astron. Astrophys., Vol. 21, 401 - 412 (1972).

This paper studies the extent to which the isotopic or elemental ($n\sigma$) correlations for s-only nuclei can be affected by electron capture after neutron irradiation in media at various (even low) temperatures and densities. It is shown that the isotopic Te correlation can be destroyed to a relatively great extent in most part of the C–O and Ne–O regions of a great variety of stars which have evolved beyond central helium burning. This is also the case, but to a lesser extent, for the Sr and Os isotopic correlations.

065.082 On mass loss by sporadic ejection. J. A. Burke.
Monthly Notices Roy. Astron. Soc., Vol. 160, 233 - 237 (1972).

Mass loss by sporadic ejection of material can be effective in removing angular momentum from a star even in the absence of magnetic coupling. Under certain conditions a substantial transfer of angular momentum can occur with mass loss of only a few per cent.

065.083 Theoretical evolution of a hydrogen-helium star of $3\,M_\odot$ from the pre-main sequence to the core helium-exhaustion phase. D. Ezer.
Astrophys. Space Sci., Vol. 18, 226 - 245 (1972).

The present study was undertaken to follow the evolution of a first-generation $3\,M_\odot$ star from the threshold of stability through the final stages of its evolution. The results of the evolutionary study up to the core helium-exhaustion phase are presented. The subsequent evolution will be examined in another paper.

065.084 Dynamic stability of rotating configurations. D. M. Sedrakyan.
Uch. zap. Erevan. un-ta Estestv. n., 1972, No. 1 (119), p. 43 - 54. In Russian. – Abstr. in Referativ. Zhurn. 51. Astron., 12.51.393 (1972).

065.085 On a mechanism of stellar bursts. Yu. V. Vandakurov.
Trudy Mezhdunar. seminara po probl. "Uskorenie chastits v kosmich. prostranstve (okolozem. i mezhplanet. kosmich. prostranstve), Galaktike i Metagalaktike". Moskva, 1972, p. 133 - 145. In Russian. – Abstr. in Referativ. Zhurn. 51. Astron., 12.51.394 (1972).

065.086 Minimum mass of a neutron star. Y. C. Leung, C. G. Wang.
Nature, Phys. Sci., Vol. 240, 132 - 133 (1972).

The mass of a star is completely determined by its equation of state with appropriate boundary conditions. For neutron stars, because of their high density, the equation of state can usually be approximated by that of cold degenerate matter, because the stellar thermal energy is small compared with the Fermi kinetic energy of the matter. We start with solid iron which has the highest binding energy per nucleon; electric neutrality must be maintained for all densities. By our computation the stable minimum mass with a core of pure neu-

tron gas is $0.067\,M_\odot$, and with a core clustered with heavy nuclei it is $0.093\,M_\odot$. The real minimum mass is probably somewhere between these two numbers.

065.087 Accretion vortices and X-ray sources. R. N. Henriksen, T. T. Chia.
Nature, Phys. Sci., Vol. 240, 133 - 135 (1972).

We have studied accretion onto a rotating, magnetized "star", and have found some unexpected results; here we isolate their physical significance, and outline their particular implications for X-ray sources.

065.088 Metal-poor stars. IV. The evolution of red giants. R. T. Rood.
Astrophys. Journ., Vol. 177, 681 - 691 (1972).

Evolutionary tracks have been obtained for Population II red giants for the following sets of (M, X, Z): $(0.80\,M_\odot, 0.7, 10^{-4})$, $(0.80\,M_\odot, 0.7, 10^{-3})$, $(0.80\,M_\odot, 0.8, 10^{-3})$, $(0.60\,M_\odot, 0.7, 10^{-3})$, $(0.80\,M_\odot, 0.7, 10^{-2})$. The mass of the helium core at the flash is found to be $M_c/M_\odot = 0.475 + 0.23\,(X - 0.7) - 0.01\,(\log Z + 3) - 0.035\,(M/M_\odot - 0.8)$. If direct electron-neutrino interactions are neglected, the core masses are $0.033\,M_\odot$ larger. The lifetimes on the giant branch above a given luminosity are also determined. These are somewhat lower than in earlier work. Small bumps are predicted in red-giant luminosity functions.

065.089 On semiconvection. R. Mitalas.
Astrophys. Journ., Vol. 177, 693 - 697 (1972).

A condition for the occurrence of semiconvection is derived for a general opacity law. It is shown that a semiconvective zone is always initiated by the appearance of a convective zone in a region of varying mean molecular weight. The possibility of semiconvection during thermal helium flashes is suggested.

065.090 Ionization energies of hydrogen in magnetic white dwarfs.
A. K. Rajagopal, G. Chanmugam, R. F. O'Connell, G. L. Surmelian.
Astrophys. Journ., Vol. 177, 713 - 717 (1972).

The variational principle is used to determine the ionization energy of a hydrogen atom in its ground state, in the presence of magnetic fields of the order of those found in magnetic white dwarfs and some neutron stars. The results are shown to be better than those obtained by using perturbation theory for all fields and significantly better than those of Cohen et al., for fields less than $\sim 3 \times 10^{10}$ gauss. A general procedure for obtaining the energies of higher levels is outlined.

065.091 Polarized radiation from magnetic white dwarfs. II. Solution of Kemp's model at all temperatures.
G. Chanmugam, R. F. O'Connell, A. K. Rajagopal.
Astrophys. Journ., Vol. 177, 719 - 722 (1972).

Our exact solution of Kemp's model for polarized radiation from magnetic white dwarfs at low temperatures is extended to all temperatures. The implications of the results are discussed.

065.092 Formation of protostars by thermal instability.
R. F. Stein, R. McCray, J. Schwarz.
Astrophys. Journ., (Letters), Vol. 177, L125 - L128 (1972).

Spherically symmetric condensations driven by thermal instabilities in a cooling interstellar medium can produce gravitationally bound clouds, provided that the magnetic field has been previously removed from the gas. The resulting clouds have a very small ($\sim 10^{-4}$ pc), high-density ($\sim 10^{12}$ cm^{-3}), warm ($\sim 100°$K), stationary core, surrounded by a large (~ 1 pc), cool ($\sim 10°$K), infalling envelope. Their structure resembles that of an early stage in the protostar evolution calculated by Larson. Such objects may represent Bok globules.

065.093 Stellar evolution in elliptical galaxies.
B. M. Tinsley.
Publ. Astron. Soc. Pacific, Vol. 84, 645 (1972). – Abstr. Astron. Soc. Pacific.

065.094 Extrem junge Sterne. W. Wenzel.
Sterne, 48. Jahrgang, p. 213 - 223 (1972).

065.095 Thermal pulses in helium shell-burning stars.
D. J. Faulkner, P. R. Wood.
Astrophys. Journ., Vol. 178, 207 - 219 (1972).

The helium shell-burning evolution of a $0.8 M_\odot$ star has been calculated with a detailed treatment of the abundance profile in the shell source. The star undergoes a single thermal pulse before entering the white-dwarf region at the end of its evolution. The decay time of the thermal pulse is in good agreement with the observed rate of evolution for the nuclei of planetary nebulae, and it is suggested that planetary-nebula shells may be ejected by radiation pressure during such a pulse. The calculations show that further pulse events are possible even if the shell ejection removes all hydrogen-rich material. The present pulse was not sufficiently luminous, however, to eject any helium-rich material.

065.096 Stellar evolution in elliptical galaxies. B. M. Tinsley.
Astrophys. Journ., Vol. 178, 319 - 336 (1972).

Elliptical galaxies are described by models in which the evolution of stars is followed for 12 billion years. Stellar birthrates chosen are mainly power laws in stellar mass, cut off after 10^9 years; effects of these oversimplifications are discussed. Integrated magnitudes, broad-band colors, narrow-band scanner spectral energy distributions, and strengths of spectral features are derived as a function of time. Comparison of computed with empirical energy distributions sets limits to the galactic age for each initial mass function. Production of metals during the first 10^9 years is calculated, leading to an explanation of the observed increase in stellar metal abundance and redness toward the center.

065.097 Cepheids, presupernovae, and the $^{12}C(\alpha, \gamma)^{16}O$ reaction. I. Iben, Jr.
Astrophys. Journ., Vol. 178, 433 - 440 = Contr. Lick Obs., No. 376 (1972).

Models of mass 5, 7, and $9 M_\odot$ have been carried through core helium burning with several choices for the reduced width θ_a^2 of the 7.12-MeV level in ^{16}O. Comparison with classical cepheids in the Galaxy suggests that consistency is possible for $0.08 < \theta_a^2 < 0.25$. Models of mass $7 M_\odot$ have been carried further into the shell helium-burning and shell hydrogen-burning stage with the objective of determining whether or not carbon ignition and detonation can be initiated off-center. Also discussed are changes in envelope and surface abundances of the elements H, He, C, N, and O.

065.098 Comments on a *PLC* relationship for cepheids and on the comparison between pulsation and evolution masses for cepheids. I. Iben, Jr., R. S. Tuggle.
Astrophys. Journ., Vol. 178, 441 - 453 = Contr. Lick Obs., No. 375 (1972).

It is suggested that a period-luminosity-color (*PLC*) relationship currently in use for cepheids in our Galaxy is not a unique summary of the observational facts and may, in fact, be inconsistent with elementary pulsation theory. We suggest that either (1) the original estimates of luminosity (on which the normalization of the standard *PLC* relationship is based) be used to establish a mass-luminosity relationship for galactic cepheids or (2) the slope of the mass-luminosity relationship given by evolutionary calculations be accepted and be used to estimate relative luminosities. The effect on blue edges of

using spline interpolation in new Cox-Stewart opacities is shown for a single mass.

065.099 New cross-sections for s-process nucleosynthesis.
D. B. Stroud.
Astrophys. Journ., (*Letters*), Vol. 178, L93 - L94 (1972).

The 30-keV neutron-capture cross-sections of the separated isotopes ^{94}Mo, ^{96}Mo, ^{110}Cd, and ^{136}Ba have been measured.

065.100 Linear series of stellar models. I. Thermal stability of stars. B. Paczyński.

Similarity between a sequence of static stellar models and the linear series of Poincaré is emphasized. The points where the stellar mass (or the core mass) passes through an extremum correspond to the turning points of the linear series. Turning points separate the branches of a linear series occupied by thermally stable and thermally unstable stellar models. In the appendix the relation between the Schwarzschild and Henyey determinants is discussed.

065.101 Linear series of stellar models. II. Pure carbon stars.
B. Paczyński, M. Kozłowski.
Acta Astron., Vol. 22, 315 - 325 (1972).

Results of model computations are presented for pure carbon stars that are in thermal and hydrostatic equilibrium.

065.102 Evolution of single stars. VII. Evolution of massive stars. J. Ziółkowski.
Acta Astron., Vol. 22, 327 - 374 (1972).

The evolution of 15 and $30 M_\odot$ stars up to central carbon ignition and of a $60 M_\odot$ star up to central helium exhaustion is followed up using a density criterion for convective instability.

065.103 Advanced evolution of massive stars. IV. Secondary nucleosynthesis during helium burning.
R. G. Couch, W. D. Arnett.
Astrophys. Journ., Vol. 178, 771 - 777 (1972).

The thermonuclear processing of ^{14}N during core helium burning in massive stars is examined. A detailed discussion of the relevant reaction rates is given, including the evaluation of recent experimental data, and the presentation of analytic fits for $N_A \langle \sigma v \rangle$.

065.104 Zur Frage der Bestimmung der Oberfläche rotierender Polytropen als freies Randwertproblem.
K. H. Hauer.
Mitt. Inst. Theor. Geod. Univ. Bonn, No. 4, p. 19 - 24 (1972).

065.105 Structure and evolution of large mass supergiant stars. Review paper. N. Dallaporta.
Proc. Third Colloquium on astrophysics. Supergiant stars, Trieste 1971, (see 012.020), p. 250 - 278 (1972).

065.106 Late supergiant evolution.
I. W. Roxburgh, I. P. Williams.
Proc. Third Colloquium on astrophysics. Supergiant stars, Trieste 1971, (see 012.020), p. 279 - 287 (1972).

065.107 Some considerations concerning the ratio of blue to red supergiants.
G. Barbaro, C. Chiosi, L. Nobili.
Proc. Third Colloquium on astrophysics. Supergiant stars, Trieste 1971, (see 012.020), p. 313 - 322 (1972).

065.108 Surface enrichment from interior burning in supergiant stars. J. Sackmann.
Proc. Third Colloquium on astrophysics. Supergiant stars, Trieste 1971, (see 012.020), p. 323 - 333 (1972).

065.109 **Some features of the massive star evolution during the shell H-burning and core He-burning phases.**
G. Barbaro, C. Chiosi, L. Nobili.
Proc. Third Colloquium on astrophysics. Supergiant stars, Trieste 1971, (see 012.020), 334 - 354 (1972).

065.110 **Secular instabilities in supergiant stars.**
D. Lauterborn, S. Refsdal, M. L. Roth.
Proc. Third Colloquium on astrophysics. Supergiant stars, Trieste 1971, (see 012.020), p. 355 (1972).

065.111 **The effect of rotation on the evolution of red giants and supergiants.** E. Meyer-Hofmeister.
Proc. Third Colloquium on astrophysics. Supergiant stars, Trieste 1971, (see 012.020), p. 356 - 357 (1972).

065.112 **Models of helium and R Coronae Borealis stars.**
P. Biermann, R. Kippenhahn.
Proc. Third Colloquium on astrophysics. Supergiant stars, Trieste 1971, (see 012.020), p. 358 (1972). – Abstract.

065.113 **An interpretation of loop formation through homology relations.** N. Dallaporta.
Proc. Third Colloquium on astrophysics. Supergiant stars, Trieste 1971, (see 012.020), p. 359 - 367 (1972).

065.114 **Hot CNO-Ne cycle.** J. M. Audouze, G. R. Caughlan, W. A. Fowler, J. W. Truran, B. A. Zimmerman.
Bull. American Astron. Soc., Vol. 4, 410 (1972). – Abstr. AAS.

065.115 **Nuclear gamma rays from neutron stars.**
G. Borner, J. M. Cohen, R. Ramaty.
Bull. American Astron. Soc., Vol. 4, 410 (1972). – Abstr. AAS.

065.116 **Evolution of supermassive stars.**
K. J. Fricke, I. Appenzeller.
Bull. American Astron. Soc., Vol. 4, 412 (1972). – Abstr. AAS.

065.117 **Can neutron stars be formed with mass $\leqslant 0.2\,M_\odot$?**
Y. C. Leung, C. G. Wang.
Bull. American Astron. Soc., Vol. 4, 412 - 413 (1972). – Abstr. AAS.

065.118 **The production of discrete, quantized outflow velocities by radiation pressure in stars, Seyfert nuclei, and quasi-stellar objects.** J. D. Scargle.
Bull. American Astron. Soc., Vol. 4, 415 (1972). – Abstr. AAS.

065.119 **Aligned rotating magnetospheres.**
E. T. Scharlemann, R. V. Wagoner.
Bull. American Astron. Soc., Vol. 4, 415 (1972). – Abstr. AAS.

065.120 **The sensitivity of the Re-Os cosmochronology to recycling Re through stars.** R. J. Talbot, Jr.
Bull. American Astron. Soc., Vol. 4, 415 - 416 (1972). – Abstr. AAS.

065.121 **Convectively driven Urca neutrino losses and the carbon-detonation supernova.** J. C. Wheeler.
Bull. American Astron. Soc., Vol. 4, 416 (1972). – Abstr. AAS.

065.122 **Population II stars and the electron-neutrino weak interaction.** P. Demarque, J. G. Mengel.
Nature, Phys. Sci., Vol. 239, 55 - 56 (1972).
Current views on the internal structure of horizontal branch stars indicate that their luminosity is a function of the mass M_c in their helium cores. The value of M_c is determined by the onset of the helium flash in the cores of red giant stars. It is a sensitive function of the rate of neutrino losses from the electron-neutrino interaction near the center. A comparison with observed luminosities of horizontal branch stars seem to confirm the Feyman-Gell-Mann theory. It is pointed out, however, that this result may have to be revised if rotation plays a role in the evolution of the cores of red giants.

065.123 **On mass loss from B stars.**
S. P. Tarafdar, M. S. Vardya.
Observatory, Vol. 92, 238 - 239 (1972). – Letter.

065.124 **Überlagerung poloidaler magnetischer Multipolfelder in Neutronensternen.**
G. Dautcourt, K. Fritze.
Monatsber. Deutsch. Akad. Wiss. Berlin, Vol. 13, 749 - 753 (1971).

065.125 **The physical state of the interior of stars.**
J. Nemeth.
Fiz. Szemle, (*Hungary*), Vol. 22, No. 4, p. 97 - 104 (1972). In Hungarian.
The structure and evolution of stars can be described by four differential equations involving mass, energy, pressure and temperature together with equations of state. The evolution of stars is thought to be taking place through equilibrium states. Isothermal and adiabatic processes are examined in terms of radiative and convective energy transfer. Equations of state and boundary conditions are discussed.

065.126 **The structure of degenerate stars and their minimal properties.** J. P. Sharma.
Pure and Applied Geophys., (*Switzerland*), Vol. 96, 157 - 162 (1972).
The structure of massive fluid spheres (stars) which are composed of completely degenerate matters is investigated. Equations of equilibrium governing such configurations have been put in convenient forms. A few minimal properties of the stars, such as the maximum (or minimum) central energy density and temperature, etc., have been studied.

065.127 **Thermonuclear ignition of ^{12}C in dense stars.**
W. D. Arnett, R. Couch.
Bull. American Astron. Soc., Vol. 4, 409 - 410 (1972). Abstr. AAS.

065.128 **Evolution of a 0.5 $M_\odot$ population II star.** A. Noels.
Bull. Roy. Soc. Sci. Liège, (*Belgium*), Vol. 41, 50 - 65 (1972). In French.
The evolutionary sequence of a 0.5 $M_\odot$ star of chemical composition X = 0.9, Y = 0.099, Z = 0.001 was computed by L. G. Henyey's method, from the initial gravitational contraction phase to the phase of evolution towards the red giants. The results are presented for two values of mean free path, the greater of which results in the reappearance of a completely convective structure shortly following the start of nuclear reactions at the centre.

065.129 **Neutron stars.** H. Y. Chiu.
The physics of pulsars, (see 003.022), p. 111 - 118 (1972).

065.130 **Neutron stars, pulsar radiation and supernova remnants.** F. Pacini.
The physics of pulsars, (see 003.022), p. 119 - 134 (1972).

065.131 **Extreme states of matter.** D. Kirzhnits.
Soviet Sci. Rev., (*GB*), Vol. 3, 199 - 206 (1972).

065.132 **The physical state of the interior of stars. II.**
J. Nemeth.
Fiz. Szemle, (*Hungary*), Vol. 22, 135 - 140 (1972).
In Hungarian. – See Phys. Abstr., Vol. 75, No. 66083 (1972).

065.133 **Theories of star formation.** D. McNally.

Rep. Progr. Phys., Vol. 34, 71 - 108 (1971).

Four major groups of star formation theory are identified. These deal with formation by collapse under gravity, by random accretion, by condensation, and by processes associated with the activity of galactic nuclei.

065.134 A solution of the radiative blast wave in stellar interiors. S. N. Ojha.

Acta Phys. Acad. Sci. Hungaricae, Vol. 31, 375 - 383 (1972).

The author discusses the equations for the propagation of a symmetrically expanding spherical blast wave produced by a sudden explosion in a self-gravitating body, such as a star, with radiation effects taken into account. The medium is assumed to be a perfectly conducting plasma with radiative parameters independent of the magnetic field.

065.135 Stellar energy-loss rates in a convergent theory of weak and electromagnetic interactions.

D. A. Dicus.

Phys. Rev. D, Particles and Fields, Vol. 6, 941 - 949 (1972).

The stellar energy-loss rates due to the production of neutrino pairs are calculated in Weinberg's theory of electromagnetic and weak interactions.

065.136 Nuclear reactions in stars. II. D. Kisdi.

Fiz. Szemle, (Hungary), Vol. 22, 200 - 206 (1972). In Hungarian.

When the density of a red giant exceeds 10^5 gcm^{-3} at temperatures above 10^8 K, the burning of helium commences. Subsequent reactions to He burning are described, the order of reactions depending on the magnitude of the star.

065.137 Internal structure of neutron stars. J. Arponen.

Nuclear Phys. A, (Netherlands), Vol. A191, 257 - 282 (1972).

The equation of state and the structure and composition of neutron star matter are investigated in the density region $3.1 \times 10^{11} - 2 \times 10^{15}$ g/cm^3. It is shown that very small modifications in the details of the nuclear matter energy may lead to considerable differences in the resulting neutron star structure.

065.138 Stability of nonradial vibrational modes of relativistic neutron stars. P. Cazzola, L. Lucaroni.

Phys. Rev. D, Particles and Fields, Vol. 6, 950 - 956 (1972).

The authors establish an analytical connection between the amplitudes of gravitational waves scattered by a relativistic neutron star and its nonradial vibrational modes.

065.139 Detection of neutron stars. K. M. V. Apparao.

Nuovo Cimento Lettere, Ser. 2, Vol. 4, 809 - 810 (1972).

The author suggests that the X-radiation from the neutron stars in the nearest pulsars CP 0950, CP 1133, PSR 1929 and CP 0808 can be detected; in fact the flux from CP 0950 may already have been detected, if one identifies it with the X-ray source Leo XR-1.

065.140 Very-high-frequency gravitational radiation from neutron stars. D. Boccaletti.

Nuovo Cimento Lettere, Ser. 2, Vol. 4, 927 - 931 (1972).

The author considers the inner bremsstrahlung from the neutrons of a neutron star. The emission of soft gravitons in the scattering of nonrelativistic particles has been considered with the aim of evaluating the thermal gravitational radiation from the sun.

065.141 Thermodynamic critical field of superconducting neutron star matter. C. H. Yang, J. W. Clark.

Nuovo Cimento Lettere, Ser. 2, Vol. 4, 969 - 972 (1972).

The authors evaluate the zero-temperature thermodyna-

mic critical magnetic field of superconducting neutron star matter within the model of proton superfluidity.

065.142 Superfluid state in neutron star matter. III. Tensor coupling effect in $^3P_2$ energy gap. T. Takatsuka.

Progr. Theor. Phys., (Japan), Vol. 47, 1062 - 1064 (1972).

065.143 Nuclear reactions in stars. I. D. Kisdi.

Fiz. Szemle, (Hungary), Vol. 22, 171 - 182 (1972). In Hungarian.

The theory of thermonuclear reactions is presented which includes both resonant and non-resonant reactions. The rates of reactions are evaluated on the assumption of bare nuclei. The evolution of those reactions which play a major part in energy production in stars is outlined.

065.144 π^- condensation and neutron star cooling.

J. Kogut, J. T. Manassah.

Phys. Letters A, (Netherlands), Vol. 41A, 129 - 131 (1972).

065.145 Binding energy of neutron star matter.

D. Ellis, D. W. L. Sprung.

Canadian Journ. Phys., Vol. 50, 2277 - 2285 (1972).

The energy per particle of neutron star matter is calculated using reaction matrix elements deduced from Reid's potential. These are parameterized so that the calculation is presented in a simple form. A maximum concentration of nearly 9% protons is found at about twice normal nuclear matter density.

065.146 Stars and nuclei. II. O. Ames.

Phys. Teach., (USA), Vol. 10, 250 - 256 (1972). See Phys. Abstr., Vol. 75, No. 81860 (1972).

065.147 Nucleosynthesis and the formation of superheavy elements. A. G. W. Cameron.

Dynamic structure of nuclear states. Proc. MontTremblant Summer School 1971, [University of Toronto Press, Toronto, Canada], p. 365 - 394 (1972).

The author considers the processes of nucleosynthesis, stellar evolution, astrophysical conditions for heavy element nucleosynthesis, the chemical evolution of the Galaxy, the nuclear mass formula, the R-process and the existence of superheavy nuclei in nature.

065.148 Stellar evolution toward the main sequence. P. Bodenheimer.

Rep. Progr. Phys., Vol. 35, 1 - 54 (1972).

The review describes the evolution of stars during the period that starts just after formation in interstellar clouds and ends at the point of stabilization of the main sequence where nuclear reactions begin to provide the entire energy supply. The initial-boundary-value problem that must be solved to obtain the structure of a star as a function of time is discussed. The equations have been solved numerically for a wide range of stellar masses during both the hydrostatic and hydrodynamic phases, under the assumption of spherical symmetry with neglect of rotational or magnetic effects. The character of these solutions is described, with particular attention given to the effect upon the results of the assumed initial conditions. Remarks are also made on the effects of rotation in calculations of pre-main-sequence evolution.

065.149 Rotational energy, pulsars, and active nebulae. J. M. Cohen.

General Relativity and Gravitation, Vol. 3, 221 - 225 (1972).

A number of different expressions for the rotational energy of a slowly rotating star have appeared in the literature. The author shows that these expressions are equivalent. Application to pulsars and the Crab nebula are discussed.

065.150 **The evolution of stars and galaxies; condensation or expansion?** L. Mirzoyan.
Soviet Sci. Rev., (*GB*), Vol. 3, 300 - 308 (1972).

The two hypotheses concerning the origin of stars and galaxies, the idea of a contraction or an expansion, contradict one another in essential points, and have divided astronomers into two camps. But the new discoveries of quasars and central bodies in many galaxies seem to be gradually refuting the contraction hypotheses. The question is whether 'proto-stars' exist at the centres of galaxies.

065.151 **Hier entstehen Sterne.** W. Wenzel.
Jenaer Rundschau, (Jena Rev.), 17. Jahrgang, p. 319 - 321 (1972).

065.152 **Thermal waves in stars.** I. A. Klimishin.
Problems of cosmic physics. Vyp. (No.) 7, (see 003. 028), p. 108 - 114 (1972). In Russian.

The regularities of the motion of thermal waves in polytropic stellar models are discussed. The conditions of degeneration of a thermal wave into a shock wave at its excitation on the surface of the nonuniform medium are derived.

065.153 **Stellar evolution and galactic evolution. From THO theory to star-like problem.** M. Taketani.
Progr. Theor. Phys., Japan, Suppl. No. 50, p. 220 - 230 (1971). Reprint of paper from 1964.

065.154 **Cold neutron star model in Brans-Dicke theory of gravity.** M. Matsuda.
Progr. Theor. Phys., (*Japan*), Vol. 48, 341 - 343 (1972).

Various theories of gravity have been proposed after general relativity, but many of them have been disproved for their incompleteness or disagreement with observational data. The author investigates a cold spherical neutron star in the Brans-Dicke theory and compares the results with that of general relativity.

065.155 **On thermal X-ray emission of a neutron star.**
P. R. Amnuel, O. H. Gusejnov, F. K. Kasumov.
Astron. Tsirk., No. 731, p. 5 - 7 (1972). In Russian.

065.156 **On the convection in M-stars.** V. E. Panchuk.
Astron. Tsirk., No. 742, p. 6 (1972). In Russian.

Stellar evolution. See Abstr. 003.055.

Computer programs in astrophysics.
See Abstr. 021.028.

État actuel de la théorie de la nucléosynthèse. I. La formation des éléments jusqu'au groupe du fer.
See Abstr. 061.016.

Reaction rates in the proton-proton chain.
See Abstr. 061.034.

Analytic forms of the thermonuclear function.
See Abstr. 061.035.

Thermal equilibrium states of a classical system with gravitation. See Abstr. 061.038.

Heating of astrophysical plasma due to rotation.
See Abstr. 062.075.

Studies of hydrodynamic events in stellar evolution. II. Dynamic instabilities in stellar envelopes.
See Abstr. 064.006.

A model for the chemical evolution of S and N star envelopes. See Abstr. 064.009.

Transition from shallow to deep convection zones in stars. See Abstr. 064.012.

Relaxation oscillations in the envelopes of luminous red giants. See Abstr. 064.013.

An estimate of stellar wind mass loss during the red giant phase of evolution. See Abstr. 064.022.

On the structure of envelopes of red supergiants.
See Abstr. 064.033.

Kondensierte Materie im Kosmos. V.
See Abstr. 064.047.

Mixing between stellar envelope and core in advanced phases of evolution. IV. Effect of super-adiabaticity in convective envelope. See Abstr. 064.063.

Optical and infrared observations of young stellar objects – an informal review. See Abstr. 064.065.

On the stability of axisymmetric systems to axisymmetric perturbations in general relativity. I. The equations governing nonstationary, stationary, and perturbed systems.
See Abstr. 066.005.

Gravitational collapse of a rotating star. I. Formulation. See Abstr. 066.114.

The observed limits to gravitational collapse.
See Abstr. 066.153.

Red supergiants and neutrino emission. II.
See Abstr. 115.003.

The main sequence of the H-R diagram: Its significance and role in stellar evolution. See Abstr. 115.016.

The effect of the Coriolis force on the stability of rotating magnetic stars. See Abstr. 116.016.

Strongly magnetic, rapidly rotating main-sequence stars. See Abstr. 116.019.

Spectroscopic binaries and the evolution of supergiants. See Abstr. 119.009.

The eclipsing binary WW Cygni: An unlikely candidate for pre-main-sequence contraction.
See Abstr. 121.050.

Luminosity variation in the one-zone cepheid model.
See Abstr. 122.045.

On the location of pulsational blue edges and estimates of the luminosity and helium content of RR Lyrae stars. See Abstr. 122.100.

Some statistical properties and evolutionary considerations on the S-type stars. See Abstr. 122.119.

Quasi-radial pulsations of rotating white dwarfs and neutron stars in general relativity. See Abstr. 126.022.

Molekülwolken und Sternentstehung im interstellaren Raum. See Abstr. 131.064.

Electromagnetic radiation from the Crab nebula and

models of neutron stars.   See Abstr. 134.004.

Neutron star in Crab nebula – really?
See Abstr. 134.006.

Rotating neutron stars and the nature of pulsars.
See Abstr. 141.547.

Angular momentum and energy loss of a compact
star rotating in a thermal plasma.   See Abstr. 142.013.

GX 340+0 as a hot neutron star.
See Abstr. 142.044.

Shock waves in spiral arms and star formation.
See Abstr. 151.022.

Preliminary evidence for bursts of star formation in
giant elliptical galaxies.   See Abstr. 158.120.

New observations & old nucleocosmochronologies.
See Abstr. 162.036.

066 Relativistic Astrophysics (without Cosmology), Background Radiation, Gravitation Theory

066.001 The motion of massive bodies and light rays in the scalar-tensor gravitation theories.
A. M. Finkel'shtejn.
Nablyud. Iskusstv. Nebesn. Tel, No. 62, p. 138 - 152 (1971).
In Russian.

066.002 Floating orbits, superradiant scattering and the black-hole bomb.
W. H. Press, S. A. Teukolsky.
Nature, Vol. 238, 211 - 212 (1972).

066.003 Le equazioni del moto dei giroscopi in relatività generale. F. Occhionero.
Mem. Soc. Astron. Italiana, Nuova Ser., Vol. 43, 339 - 346 (1972).

066.004 The uniform transparent gravitational lens.
E. E. Clark.
Monthly Notices Roy. Astron. Soc., Vol. 158, 233 - 243 (1972).
An elementary discussion is given of the gravitational deflection of light due to radially and cylindrically symmetric masses. The effect of the deflection on apparent luminosity of distant sources is also considered. Detailed calculations are made for the simple case of the uniformly dense transparent sphere and comparisons are made with the opaque mass sphere.

066.005 On the stability of axisymmetric systems to axisymmetric perturbations in general relativity. I. The equations governing nonstationary, stationary, and perturbed systems. S. Chandrasekhar, J. L. Friedman.
Astrophys. Journ., Vol. 175, 379 - 405, with a correction Vol. 176. 768 (1972).
The field and the fluid equations that are appropriate to general nonstationary (but axisymmetric) systems are first derived. They are then specialized to yield the equations which govern stationary equilibrium. The equations which determine the evolution of small departures from equilibrium are also obtained. Related matters that are considered include the Landau-Lifshitz complex, the conserved quantities, and the constancy of the baryon number and the angular momentum (per baryon) of a fluid element as it moves.

066.006 Die Schwere des Lichtes und der vierte Einstein-Effekt. H.-J. Treder.
Naturwissenschaften, 59. Jahrgang, p. 311 (1972).

066.007 Around-the-world atomic clocks: Predicted relativistic time gains. J. C. Hafele, R. E. Keating.
Science, Vol. 177, 166 - 168 (1972).
During October 1971, four cesium beam atomic clocks were flown on regularly scheduled commercial jet flights around the world twice, once eastward and once westward, to test Einstein's theory of relativity with macroscopic clocks. From the actual flight paths of each trip, the theory predicts that the flying clocks, compared with reference clocks at the U.S. Naval Observatory, should have lost 40 ± 23 nanoseconds during the eastward trip, and should have gained 275 ± 21 nanoseconds during the westward trip.

066.008 Around-the-world atomic clocks: Observed relativistic time gains. J. C. Hafele, R. E. Keating.
Science, Vol. 177, 168 - 170 (1972).
Four cesium beam clocks flown around the world on commercial jet flights during October 1971, once eastward and once westward, recorded directionally dependent time differences which are in good agreement with predictions of conventional relativity theory. Relative to the atomic time scale of the U.S. Naval Observatory, the flying clocks lost 59 ± 10 (s.d.) nanoseconds during the eastward trip and gained 273 ± 7 (s.d.) nanoseconds during the westward trip.

066.009 The relativistic Roche problem.
L. G. Fishbone.
Astrophys. Journ., (*Letters*), Vol. 175, L155 - L159 (1972).
Bodies in orbit around Kerr black holes will experience tidal gravitational forces. The Roche limit for fluid bodies in stable circular orbits is qualitatively like that in Newtonian situations; for highly energetic circular orbits, the fluid density required for the existence of a body in equilibrium is magnified by the square of its energy-at-infinity per unit mass.

066.010 Quasi-radial pulsations of rotating relativistic models.
V. V. Papoian, D. M. Sedrakian, E. V. Chubarian.
Astron. Zhurn. Akad. Nauk SSSR, Vol. 49, 750 - 755 (1972).
In Russian. English translation in Soviet Astron. AJ, Vol. 16, No. 4.
The adiabatic radial pulsations of small amplitude with respect to an equilibrium state of slowly rotating relativistic models is considered. In case of homogeneous models it is shown that the relativistic rotational effects compensate the dynamical instability which arises due to the spherically symmetric distribution of matter.

066.011 Time varying gravitational satellite.
E. Groten, S. Thyssen-Bornemisza.
Geophys. Journ. Roy. Astron. Soc., Vol. 29, 237 - 239 (1972).
Research note.

066.012 Gravity waves: Are they real and what do they mean?
W. D. Metz.
Science, Vol. 177, 506 (1972).

066.013 Poincarés Relativität der Beschleunigung und die Mach-Einstein-Doktrin. (Eine historisch-kritische Studie). H.-J. Treder.
Monatsber. Deutsch. Akad. Wiss. Berlin, Vol. 13, 664 - 675 (1972).

066.014 Étude de processus d'émission de gravitons.
J.-L. Grossiord.
Comptes Rendus Acad. Sci. Paris, Sér. B, Vol. 275, 365 - 368 (1972).
Étude de la production de gravitons par annihilation d'une paire électron-positon.

066.015 Méthodes expérimentales pour la détection des masses négatives éventuelles. X. C. Hawecker.
Comptes Rendus Acad. Sci. Paris, Sér. B, Vol. 275, 373 - 374 (1972).
L'énergie émise par les quasars trouve une explication élégante grâce à l'hypothèse des masses négatives et à leurs propriétés physiques nouvelles.

066.016 Accretion onto black holes: The emergent radiation spectrum. S. L. Shapiro.
Bull. American Astron. Soc., Vol. 4, 321 (1972). – Abstr. AAS.

066.017 Conservation laws for a system of magnetically charged rotating bodies in general relativity.

A. P. Ryabushko.
Vestsi AN BSSR. Ser. fiz.-mat. n., Izv. AN BSSR. Ser. fiz.-mat. n., 1972, No. 2, p. 104 - 113. In Russian. – Abstr. in Referativ. Zhurn. 51. Astron., 9.51.831 (1972).

066.018 **Response of a gravitational-wave antenna to a polarized source.**   J. A. Tyson, D. H. Douglass.
Phys. Rev. Letters, Vol. 28, 991 - 994 (1972).

The response of a gravitational wave antenna to linear, mixed, and randomly polarized sources is studied as a function of sidereal time, source coordinates, and antenna location and orientation. We find that the gravitational signals reported by Weber cannot be highly polarized tensor radiation coming from a single source at the galactic nucleus.

066.019 **Interpretation of gravitational-wave observations.**   C. W. Misner.
Phys. Rev. Letters, Vol. 28, 994 - 997 (1972).

If Weber's gravitational-wave observations are interpreted in terms of a source at the galactic center, both the intensity and the frequency of the waves are more reasonable if the source is assumed to emit in a synchrotron mode (narrow angles, high harmonics). Although presently studied sources for such modes are astrophysically unsatisfactory – high-energy, nearly circular, scattering orbits – other possible sources are under study.

066.020 **Gravitational synchrotron radiation in the Schwarzschild geometry.**
C. W. Misner, R. A. Breuer, D. R. Brill, P. L. Chrzanowski, H. G. Hughes III, C. M. Pereira.
Phys. Rev. Letters, Vol. 28, 998 - 1001 (1972).

As described in an earlier paper, (see Abstr. 066.019), gravitational synchrotron radiation (GSR) is a crucial concept in searching for exotic astrophysical phenomena, which it might make visible through Weber's gravity telescopes. We here demonstrate that a gravitational synchrotron radiation mechanism exists as a consequence of Einstein's general relativity theory.

066.021 **Asymptotic limit for the speed of sound in a system of relativistically interacting particles.**
R. L. Bowers.
Phys. Rev. Letters, Vol. 29, 509 - 511 (1972).

A relativistic many-body theory is used to evaluate the equation of state to lowest order in the weak coupling constant for a dense system of electrons and neutrinos interacting through the universal Fermi interaction. The asymptotic limit for the speed of sound $v_s \leqslant c/\sqrt 3$ is obtained; it is conjectured that a large class of relativistic interactions leads to the same limit.

066.022 **Asymptotically flat space-times and elementary particles.**   P. J. McCarthy.
Phys. Rev. Letters, Vol. 29, 817 - 819 (1972).

It is suggested that elementary particles should be defined as irreducible unitary representations of the group of asymptotically flat space-times in general relativity. The implications of such a definition are described in physical terms.

066.023 **Limits on gravitational radiation from two gravitationally bound black holes.**
G. W. Gibbons, B. F. Schutz.
Monthly Notices Roy. Astron. Soc., Vol. 159, 41P - 45P (1972).

Upper bounds are computed for the energy emitted in gravitational radiation by two black holes falling into one another from rest at a range of separations. It is shown that before the ratio of total mass to proper separation exceeds 0.759 an event horizon must have formed around both.

066.024 **On the stability of axisymmetric systems to axisym-**

metric perturbations in general relativity. II. A criterion for the onset of instability in uniformly rotating configurations and the frequency of the fundamental mode in case of slow rotation.   S. Chandrasekhar, J. L. Friedman.
Astrophys. Journ., Vol. 176, 745 - 768 (1972).

The theory developed in paper I (Chandrasekhar and Friedman, 1972) is applied to solve two problems in general relativity: to obtain a criterion for the onset of instability in a uniformly rotating configuration via a neutral mode of axisymmetric oscillation; and to obtain an exact and an explicit formula for the square of the frequency of the fundamental axisymmetric mode of oscillation of a configuration rotating uniformly but slowly.

066.025 **Theoretical frameworks for testing relativistic gravity. IV. A compendium of metric theories of gravity and their post-Newtonian limits.**   W.-T. Ni.
Astrophys. Journ., Vol. 176, 769 - 796 (1972).

Metric theories of gravity are compiled and classified according to the types of gravitational fields they contain, and the modes of interaction among those fields. The gravitation theories considered are classified as (I) general relativity, (II) scalar-tensor theories, (III) conformally flat theories, and (IV) stratified theories with conformally flat space slices. The post-Newtonian limit of each theory is constructed and its Parametrized Post-Newtonian (PPN) values are obtained by comparing it with Will's version of the formalism.

066.026 **Studie zur Rotation eines fluiden Mediums mit gravitativen, elektromagnetischen und thermodynamischen Eigenschaften um eine feste Achse.**   E. Schmutzer.
Wiss. Zeitschr. Friedrich-Schiller-Univ. Jena, Math.-Nat. Ser., 21. Jahrgang, p. 85 - 104 (1972).

We treat the rotation of a fluid with gravitational, electromagnetic and thermodynamic properties around a fixed axis. Especially we stress the consequences which are caused by the electric acceleration current with respect to a magnetic field of the system. Some estimates of these effects are given.

066.027 **Innere kugelsymmetrische Lösung der Einsteinschen Feldgleichungen mit Wärmestrom (II).**   H. Strobel.
Wiss. Zeitschr. Friedrich-Schiller-Univ. Jena, Math.-Nat. Ser., 21. Jahrgang, p. 111 - 116 (1972).

In continuation of a paper with the same subject a nonstatic generalisation of Schwarzschild's exterior solution is considered, being interpreted as interior one. The static case (without thermal flow) possesses an interesting equation of state. By specialisation it leads back to Schwarzschild's exterior solution.

066.028 **Innere kugelsymmetrische Lösungen der Einsteinschen Feldgleichungen mit Energietransport durch Strahlung.**   H. Strobel.
Wiss. Zeitschr. Friedrich-Schiller-Univ. Jena, Math.-Nat. Ser., 21. Jahrgang, p. 117 - 123 (1972).

Interior solutions of field equations with transport of energy by radiation result under the assumption of a special spherical symmetrical, and shear-free line element. The simplest case gives a model with infinite central pressure and central density of radiation.

066.029 **The significance of the redshift rocket probe experiment to theories of gravitation.**   R. F. C. Vessot.
Smithsonian Astrophys. Obs., *Cambridge, Mass.*, Special Report, No. 343, 4 + 13 pp. (1972).

Advances in the development of atomic oscillators having great frequency stability and the progress in space technology now make possible direct measurements of the effects of gravitation on time. Using the earth's gravity these measurements can now be made to 20 parts per million. At present, there is a 1 % verification of the equivalence principle for clocks made over a 75-ft vertical distance by use of Mössbauer γ-ray emission and absorption. Measurements made to greater accuracy and spanning distances where appreciable curvature of the metric of spacetime will help verify the equivalence principle.

066.030 Discontinuities of the first derivatives of the space-time metric tensor.   V. I. Denisov.
Zhurn. ehksperim.i teor. fiz., Vol. 62, 1990 - 1997 (1972). In Russian. – Abstr. in Referativ. Zhurn. 51. Astron., 10.51.643 (1972).

066.031 Gravitational field equations.   P. Rastall.
Canadian Journ. Phys., Vol. 49, 678 - 684 (1971).
An earlier, scalar theory of gravitation is assumed to be valid for a class of static gravitational fields. The theory is written in tensor form, and generalized to the case of an arbitrary gravitational field. The interaction between the field and its sources is discussed, and the linearized form of the field equations is derived. Some possible alternative field equations are considered which are compatible with the linearized Einstein equations.

066.032 An action-at-a-distance theory of gravitation. II. Tensor interactions.   A. B. Volkov.
Canadian Journ. Phys., Vol. 49, 1697 - 1707 (1971).
A previous paper showed that the classic tests of the general theory of relativity can all be explained in terms of a special relativistic action-at-a-distance theory involving an appropriate mixture of a scalar and vector interaction. The theory has been generalized to tensor interactions of all orders. The demands of special relativity and the perihelion advance of Mercury lead to a unique specification of the scalar interaction strength, but since the limit for all other tensors is the same for the perihelion advance (one sixth the general relativity result) only the sum of the interaction strengths can be determined for these tensors. Some possible physical consequences are discussed.

066.033 Kinetic theory of small perturbations of spatio-uniformly expanding gravitating matter.
V. B. Magalinsky.
Astron. Zhurn. Akad. Nauk SSSR, Vol. 49, 1017 - 1025 (1972). In Russian. English translation in Soviet Astron. AJ, Vol. 16, No. 5.
Small perturbations of spatio-uniformly expanding matter are investigated by means of the kinetic equation with self-consistent Newton interaction (Vlasov's equation). An integral equation for density disturbances is derived, its kernel being defined by the unperturbed motion of matter and by unperturbed distribution function. Some general estimates are obtained and some particular solutions of this equation are investigated.

066.034 Conformal frames and field equations in a conformal theory of gravitation.   J. N. Islam.
Proc. Cambridge Philos. Soc., Vol. 67, 397 - 414 (1970).
It is the purpose of this paper to examine the use of conformal frames and to obtain a static solution for two particles in the conformal theory of gravitation put forward by Hoyle and Narlikar.

066.035 Quantum mechanics and the gravitational red shift.   H. F. Stoeckli.
Proc. Cambridge Philos. Soc., Vol. 69, 315 - 318 (1971).
It is shown that the formula for the gravitational red shift predicted by the theory of general relativity can also be derived by classical quantum mechanics combined with relativistic arguments.

066.036 Debye potentials for the gravitational field.
W. B. Campbell, T. Morgan.
Physica, Vol. 53, 264 - 288 (1971).
Debye potentials for the gravitational field of a bounded system are given. Their explicit dependence on the sources is given in terms of a multipole expansion and the general expression for the momentum loss of a bounded system due to gravitational radiation is calculated in terms of the multipole moments. The properties of the field in the near and far field zones are discussed.

066.037 Dynamics of extended bodies in general relativity. II. Moments of the charge-current vector.
W. G. Dixon.
Proc. Roy. Soc. London, Ser. A, Vol. 319, 509 - 547 (1970).
The problem is considered of defining multipole moments for a tensor field given on a curved spacetime, with the aim of applying this to the energy-momentum tensor and charge-current vector of an extended body.

066.038 Transfer of energy in general relativity.
T. Morgan, H. Bondi.
Proc. Roy. Soc. London, Ser. A, Vol. 320, 277 - 287 (1970).
This paper studies time sequences of axially symmetric static configurations which can be continuously deformed into each other. We consider only the regions of space surrounding a source where, for quasi-static systems, the exact, static, empty space solutions of Weyl and Levi-Civita are good approximations at all times, and exact whenever the motion stops. Our results are then valid for arbitrarily strong fields.

066.039 On the degree of sharpness in solutions of Einstein's field equations.   P. C. Waylen.
Proc. Roy. Soc. London, Ser. A, Vol. 321, 397 - 408 (1971).
Einstein's theory of gravitation predicts that small changes in the gravitational field will propagate both sharply along the light cone and diffusively through its interior. In this paper the relative importance of sharp and diffusive propagation is linked to an invariant which arises in Sciama, Waylen & Gilman's integral formulation of Einstein's field equations.

066.040 Radiation pressure on dust in general relativity.   H. Bondi.
Nature, Phys. Sci., Vol. 238, 58 - 59 (1972).
The problem of the possible equilibrium of absorbing dust in a shell surrounding a source of gravitation and light is examined via the field equations of general relativity.

066.041 A positive signature for the recognition of gravitational radiation.   W. L. Burke.
Nature, Phys. Sci., Vol. 239, 43 (1972).
This paper points out a peculiar feature of gravitational radiation which leaves a unique signature in a detector, namely, that the proper acceleration is not the second derivative of the proper strain.

066.042 Closed time as an explanation of the black body background radiation.   P. C. W. Davies.
Nature, Phys. Sci., Vol. 240, 3 - 5 (1972).
3 K isotropic black body radiation is explained in terms of the accumulated starlight from a subsequent time reversed cycle of the universe, thermalized in the adjoining dense state.

066.043 Cosmological vacuum solutions in Brans and Dicke's scalar-tensor theory.   H. Dehnen, O. Obregón.
Astrophys. Space Sci., Vol. 17, 338 - 342 (1972).
The exact cosmological vacuum solutions of Brans and Dicke's scalar-tensor theory are derived when a power law is valid between the gravitational constant and the radius of curvature of the universe.

066.044 Note concerning gravitation and electromagnetism.
C. C. Leiby, Jr.
Astrophys. Space Sci., Vol. 17, 368 - 377 (1972).

The purpose of this note is to demonstrate that it is possible to derive defining equations for gravitational fields which are identical to the electromagnetic Maxwell equations, which admit gravitational attraction, and which do not violate the conservation of energy. The possible relationship of such a linear vector theory of gravitation to that of electromagnetism is briefly discussed.

066.045 The variability of the gravitational constant.
P. A. M. Dirac.
G. Gamow Memorial Volume, (see 003.003), p. 56 - 59 (1972).

066.046 Cosmic effects of gravitational waves.
R. A. Isaacson, J. Winicour.
Nature, Vol. 239, 447 - 448 (1972).

We describe the gravitational radiation in terms of two parameters f_0 and n. The first parameter, f_0, gives the present fractional conversion rate of matter into gravitational waves, the second parameter, n, determines the profile of the mass loss rate as a function of time.

066.047 Operational determination of Schwarzschild coordinates. S. Banerji.
Nature, Phys. Sci., Vol. 239, 140 - 142 (1972).

066.048 Dependence of Doppler shift on photon rest mass.
A. R. Lee, J. Liesegang.
Nature, Phys. Sci., Vol. 240, 41 (1972).

Following renewed interest in the possible existence of a finite photon rest mass, Goldhaber and Nieto have presented an excellent review of the general methods which have been adopted in an attempt to establish an upper limit for this rest mass. We present here a novel method for pursuing this goal based on a departure from the conventional Doppler shift for photons.

066.049 The Kerr metric and its astrophysical consequences.
F. de Felice.
Atti XV Riunione Soc. Astron. Italiana, Bologna 1971, (see 012.013), p. 179 - 198 (1972).

A general review of the properties of the Kerr metric is made, with particular attention to their astrophysical implications. An original contribution is attempted to the problem of the Kerr metric source giving the family of all the possible perfect fluid boundaries, in the weak field approximation. The behaviour of the photon orbits in the equatorial plane is studied in more details with respect to the referred literature.

066.050 Investigation of a type of conformally reducible metrics. Yu. N. Kudrya, F. E. Khlystun.
Vestn. Kiev. Un-ta, Ser. Astron., No. 14, p. 89 - 93 (1972).
In Russian.

066.051 The increasing role of general relativity in astronomy.
S. Chandrasekhar.
Observatory, Vol. 92, 160 - 174 (1972). — Halley lecture for 1972, delivered in Oxford, May 2.

066.052 The short-wavelength spectrum of the microwave background. P. Thaddeus.
Annual Rev. Astron. Astrophys., Vol. 10, (see 003.005), 305 - 334 (1972). — Contents: (1) Introduction; (2) Interstellar molecules; (3) Radio observations of molecules; (4) Direct observations: Rocket and balloon work; (5) Ground-based searches for line radiation.

066.053 Gravitational-wave astronomy.
W. H. Press, K. S. Thorne.

Annual Rev. Astron. Astrophys., Vol. 10, (see 003.005), 335 - 374 (1972). — Contents: (1) Introduction; (2) Properties of gravitational waves; (3) Generation of gravitational waves; (4) Astrophysical sources of gravitational waves; (5) Gravitational-wave receivers; (6) The Weber experiment; (7) Conclusions.

066.054 Waves from nowhere or from somewhere else.
C. Vilain.
Sci. Progrès Découverte, 99ᵉ année, No. 3438, p. 4 - 10 (1971). In French.

The existence of gravitational waves, which now appears to be generally accepted, would be even more acceptable were it possible to explain their origin by a satisfactory theory. The author reviews the various mechanisms which have been proposed.

066.055 The two paradoxes of special relativity.
Ya. A. Smorodinskij, V. A. Ugarov.
Uspekhi fiz. nauk, Vol. 107, 141 - 152 (1972). In Russian.
Abstr. in Referativ. Zhurn. 51. Astron., 11.51.687 (1972).

066.056 Interaction of weak gravitational waves with gas.
A. G. Polnarev.
Zhurn. ehksperim. i teor. fiz., Vol. 62, 1598 - 1605 (1972). In Russian. — Abstr. in Referativ. Zhurn. 51. Astron., 11.51.690 (1972).

066.057 Measurement of general relativistic time delay with Mariners 6 and 7.
J. D. Anderson, P. B. Esposito, W. Martin, D. O. Muhleman.
Space Research XII, (see 012.016), Vol. 2, 1623 - 1630 (1972).

066.058 Scalar-tensor theory of gravitation. D. K. Ross.
Phys. Rev. D, Particles and Fields, Vol. 5, 284 - 290 (1972).

A scalar-tensor theory of gravitation is constructed using the Weyl formulation of Riemannian geometry. The field equations can be written down very simply in terms of a modified curvature tensor. The theory agrees with the usual Lagrangian formalism in its experimental predictions and offers a reformulation or reinterpretation of the transformation of units considered by Dicke.

066.059 Post-Newtonian gravitational radiation from point masses in a hyperbolic Kepler orbit. R. O. Hansen.
Phys. Rev. D, Particles and Fields, Vol. 5, 1021 - 1023 (1972).

The energy and the angular momentum radiated away in the form of gravitational waves from a system of two point particles with positive total energy are calculated in the lowest nonvanishing post-Newtonian approximation. By these radiations a particle arriving from infinity can be captured, and the cross section for such captures is determined as a function of the energy at infinity. The radiation from elliptic (bound) orbits can also be inferred, and is found to be in agreement with known results.

066.060 Conservation laws and symmetry properties of scalar-tensor gravitational theories. H. B. Hart.
Phys. Rev. D, Particles and Fields, Vol. 5, 1256 - 1262 (1972).

066.061 General-relativistic interior metric for a stable static charged matter fluid with large e/m.
M. Bailyn, D. Eimerl.
Phys. Rev. D, Particles and Fields, Vol. 5, 1897 - 1907 (1972).

The general-relativistic field equations are solved for a spherically symmetric static charged matter fluid in isotropic coordinates, and matched at the boundary R to the external Reissner-Nordström metric. It is found that the charged fluid can be in equilibrium even with a large e/m, such as for an electron considered as a fluid.

066.062 Solution of the scalar wave equation in a Kerr background by separation of variables. D. R. Brill, P. L. Chrzanowski, C. M. Pereira, E. D. Fackerell, J. R. Ipser.
Phys. Rev. D, Particles and Fields, Vol. 5, 1913 - 1915 (1972).

The effect of the Kerr gravitational field on wave phenomena is explored by examining the inhomogeneous wave equation for a scalar massive field in a Kerr background geometry. The equation is separated in Boyer-Lindquist coordinates. The angular functions are spheroidal harmonics, and the radial equation is reduced to a one-dimensional Schrödinger equation with an effective potential.

066.063 Nonexistence of baryon number for black holes. II. J. D. Bekenstein.
Phys. Rev. D, Particles and Fields, Vol. 5, 2403 - 2412 (1972).

In a previous paper we showed that a static (nonrotating) black hole cannot be endowed with exterior scalar-meson or massive vector-meson fields. Here we show that the same is true for massive spin-2 meson fields. We also extend the above results to the case of a rotating stationary black hole.

066.064 Relativistic disks. I. Background models. B. H. Voorhees.
Phys. Rev. D, Particles and Fields, Vol. 5, 2413 - 2418 (1972).

Relativistic kinetic theory is used in conjunction with the theory of relativistic surface layers in order to study relativistic disks of matter. After a brief general discussion, attention is restricted to the case of counter-rotating disks. The general surface stress-energy tensors of such disks are exhibited and a distribution function which generates these stress-energy tensors is deduced. This is followed by a discussion of stability, and a criteria for the stability of particle orbits is derived. Finally, the question of central red shift is considered.

066.065 Nonspherical perturbations of relativistic gravitational collapse. I. Scalar and gravitational perturbations. R. H. Price.
Phys. Rev. D, Particles and Fields, Vol. 5, 2419 - 2438 (1972).

When a nearly spherical star gravitationally collapses through its event horizon, it cannot leave behind a static gravitational field with nonspherical perturbations. The dynamics of these perturbations during collapse is studied with a scalar-field analog. An analysis is presented of the evolution of the exterior scalar field, based on a simple wave equation containing a space-time-curvature-induced potential barrier.

066.066 Nonspherical perturbations of relativistic gravitational collapse. II. Integer-spin, zero-rest-mass fields. R. H. Price.
Phys. Rev. D, Particles and Fields, Vol. 5, 2439 - 2454 (1972).

This paper treats fields of arbitrary integer spin and zero rest mass, using the Newman-Penrose tetrad formalism. It also treats gravitational perturbations in the same framework, and supplies some technical details missing in the gravitational-perturbation analysis of paper I.

066.067 Weak electromagnetic fields around a rotating black hole. E. D. Fackerell, J. R. Ipser.
Phys. Rev. D, Particles and Fields, Vol. 5, 2455 - 2458 (1972).

Foundations are laid for studying weak electromagnetic fields around a rotating black hole. In particular, Maxwell's equations in the Kerr geometry of the black hole are reduced to a single second-order partial differential equation. Although this differential equation is propably not separable, it is sufficiently simple to yield theorems about electromagnetic properties of black holes.

066.068 Hamiltonian formulation of spherically symmetric gravitational fields. B. K. Berger, D. M. Chitre, V. E. Moncrief, Y. Nutku.
Phys. Rev. D, Particles and Fields, Vol. 5, 2467 - 2470 (1972).

Hamiltonian methods are used to study spherically symmetric gravitational fields with electromagnetism and a massless scalar field as sources. The constraints reduce to a first-order linear ordinary differential equation which is solved to yield the dynamical Hamiltonian, or the complete solution when no dynamics is present in the problem.

066.069 Gravitational radiation from relativistic systems. P. C. Peters.
Phys. Rev. D, Particles and Fields, Vol. 5, 2476 - 2485 (1972).

Gravitational radiation is calculated for two selected systems, a pair of masses interacting through scalar or vector fields, to illustrate the behavior of gravitational radiation when the emitting bodies have arbitrarily large velocities. Explicit expressions are obtained for the radiation potentials and radiated power, which can be used for any orbit or trajectory. Numerical calculations are done for the cases of circular orbits and extreme hyperbolic trajectories, in order to show the transition from the nonrelativistic to extreme relativistic limits.

066.070 Backscattering caused by the expansion of the universe. L. Parker.
Phys. Rev. D, Particles and Fields, Vol. 5, 2905 - 2908 (1972).

We show that in a Robertson-Walker space-time, expansion-induced backscattering of electromagnetic waves is rigorously absent, contrary to earlier expectations.

066.071 Pulses of gravitational radiation of a particle falling radially into a Schwarzschild black hole. M. Davis, R. Ruffini, J. Tiomno.
Phys. Rev. D, Particles and Fields, Vol. 5, 2932 - 2935 (1972).

Using the Regge-Wheeler-Zerilli formalism of fully relativistic linear perturbations in the Schwarzschild metric, we analyze the radiation of a particle of mass $\dot m$ falling into a Schwarzschild black hole of mass $M \gg m$.

066.072 On lowering a rope into a black hole. G. W. Gibbons.
Nature, Phys. Sci., Vol. 240, 77 (1972).

Penrose and Bekenstein have considered the idea of slowly lowering a particle at the end of a rope into a black hole — or rather its ergosphere — and thereby letting the gravitational field do work on the apparatus at the upper end of the rope. Here I calculate the stresses in the rope for a general stationary spacetime. A simple formula is derived relating the tension at any point of the rope to the gravitational potential V.

066.073 Gravitation, strong interactions, and the creation of the universe. J. Sarfatt.
Nature, Phys. Sci., Vol. 240, 101 - 102 (1972).

The hypothesis that the strong interactions are gravitational in origin, and that unitary symmetry is a consequence of the equivalence principle of general relativity is considered. This hypothesis is a natural extension of the Misner-Wheeler "geometrodynamics" which gives a world view that describes all of physics in terms of space-time curvature and multiply connected space-like hypersurfaces or "wormholes".

066.074 A new way to observe gravitational waves. A. Ceapá.
Nature, Phys. Sci., Vol. 240, 102 - 103 (1972).

A new interpretation of the nature of gravitational waves of high frequencies and intensities and a new method of finding evidence for them is given.

066.075 Relativistic astrophysics. I. M. Demiański.
Postępy Astron., Vol. 20, 307 - 327 (1972). In Polish.

A description of the external gravitational field of relativistic astrophysical objects is given.

066.076 **Relativistic astrophysics. II.** M. Demiański.
Postępy Astron., Vol. 20, 329 - 350 (1972). In Polish.
Hydrodynamic equations are stated and discussed. Relativistic shock waves are also described.

066.077 **Gravitational field and metrics of collapsing objects.** I. D. Novikov.
Uspekhi fiz. nauk, Vol. 107, 521 (1972). In Russian. − Abstr. in Referativ. Zhurn. 51. Astron., 12.51.694 (1972).

066.078 **General covariant gage conditions for gravitational tetrad potentials.** O. S. Ivanitskaya.
Vestsi AN BSSR. Ser. fiz.-mat. n., Izv. AN BSSR. Ser. fiz.-mat. n., 1972, No. 4, p. 58 - 63. In Russian. − Abstr. in Referativ. Zhurn. 51. Astron., 12.51.710 (1972).

066.079 **Search for gravitational radiation of extraterrestrial origin.**
V. B. Braginskij, A. B. Manukin, E. I. Popov, V. N. Rudenko, A. A. Khorev.
Pis'ma v ZhEhTF, Vol. 16, 157 - 161 (1972). In Russian. Abstr. in Referativ. Zhurn. 51. Astron., 12.51.714 (1972).

066.080 **Einstein on the firing line.** C. M. Will.
Phys. Today, Vol. 25, No. 10, p. 23 - 29 (1972).

066.081 **General relativity in the equal proper time formalism.** J. L. Cook.
Australian Journ. Phys., Vol. 25, 469 - 477 (1972).
The general theory of relativity is discussed within the framework of the concept of surfaces of equal proper time as outlined in a previous paper. The three main tests of general relativity, namely the precession of the perihelion of Mercury, the gravitational shift of spectral lines, and the gravitational deflection of light rays by massive sources, are considered and it is shown that, though the modified equations are Lorentz-invariant with respect to distant observers, the deviations from conventional results are so minute as to be undetectable.

066.082 **On the stability of axisymmetric systems to axisymmetric perturbations in general relativity. III. Vacuum metrics and Carter's theorem.**
S. Chandrasekhar, J. L. Friedman.
Astrophys. Journ., Vol. 177, 745 - 756 (1972).
The analysis of Paper II is specialized to vacuum metrics appropriately for a discussion of their stability. And Carter's theorem, that asymptotically flat axisymmetric vacuum metrics, external to black holes, cannot allow nontrivial axisymmetric neutral deformations, is deduced.

066.083 **Conservation laws and preferred frames in relativistic gravity. I. Preferred-frame theories and an extended PPN formalism.** C. M. Will, K. Nordtvedt, Jr.
Astrophys. Journ., Vol. 177, 757 - 774 (1972).
We present and discuss a revised version of the Parametrized Post-Newtonian (PPN) formalism. By regrouping the old PPN parameters, we obtain a new set of parameters which are used to classify theories of gravity according to four attributes: curvature of space-geometry, nonlinearity of gravity, existence of a preferred universal rest-frame, and validity of conservation laws for momentum. We study a variety of preferred-frame theories of gravity.

066.084 **Conservation laws and preferred frames in relativistic gravity. II. Experimental evidence to rule out preferred-frame theories of gravity.**
K. Nordtvedt, Jr., C. M. Will.
Astrophys. Journ., Vol. 177, 775 - 792 (1972).
Some metric theories of gravity single out the mean rest-frame of the universe as a "preferred frame." These theories would satisfy the "three classical tests" of general relativity, if the solar system were at rest in the "preferred frame." But instead it probably moves with a speed of ~200 km s^{-1}. We use the new version of the Parametrized Post-Newtonian (PPN) formalism discussed in Paper I, to show that such motion may produce (1) an anomalous 12-hour sidereal tide of the solid earth; (2) an anomalous yearly variation in the earth's rotation rate; (3) an anomalous perihelion shift of the planets. We use gravimeter data, length-of-day data, and perihelion-shift data to put upper limits on the sizes of such anomalous effects.

066.085 **An experiment to detect related optical and radio pulses of astrophysical origin.** E. O'Mongain.
Publ. Astron. Soc. Pacific, Vol. 84, 642 - 643 (1972). − Abstr. Astron. Soc. Pacific.

066.086 **On the meaning of the Einstein- and the Lorentz-covariant derivation.** H.-J. Treder.
General Relativity Gravitation, Vol. 2, 313 - 319 (1971).

066.087 **Über eine Interpretation von Einsteins Hermite-symmetrischer Feldtheorie.** H.-J. Treder.
Tensor, New Ser., Vol. 23, 75 - 80 (1972).

066.088 **Rotating black holes: Locally nonrotating frames, energy extraction, and scalar synchrotron radiation.**
J. M. Bardeen, W. H. Press, S. A. Teukolsky.
Astrophys. Journ., Vol. 178, 347 - 369 (1972).
This paper outlines and applies a technique for analyzing physical processes around rotating black holes. The technique is based on the orthonormal frames of "locally nonrotating observers". The paper includes a number of useful formulae for particle orbits in the Kerr metric, many of which have not been published previously.

066.089 **On the problem of periods in the gravitational field of a rotating body.** Z. Kh. Kurmakaev.
Trudy Astrofiz. Inst., *Alma-Ata*, Vol. 19, 120 - 121 (1972). In Russian.

066.090 **Relativistic stress tensor of a solid body in general relativity.** T. S. Kozhanov.
Trudy Astrofiz. Inst., *Alma-Ata*, Vol. 19, 122 - 128 (1972). In Russian.
Analytical expressions of the relativistic stress tensor for three different motions of solid bodies: rectilinear, rotational and their combination, are obtained.

066.091 **Einige Konsequenzen der Mach-Einstein-Doktrin für Himmelsmechanik und Geophysik.**
H.-J. Treder.
Gerlands Beiträge Geophys., Vol. 81, 164 - 178 (1972).
In this paper secular consequences of the Mach-Einstein doctrine for palaeo-geophysics and celestial mechanics are examined. The research results in a secular decrease of the earth's flattening and a secular acceleration in the motion of the moon and of the planets. The numerical values of these secular effects are in very good agreement with the empirical facts.

066.092 **On the central singularity of a relativistic fluid sphere.** I.-M. Ganea.
Stud. Cerc. Astron., Vol. 17, 267 - 273 (1972). In Romanian.
This paper discusses the singularity which appears in the Schwarzschild interior metrics, when the radius of the sphere is $R = 9/8 \, R_s$, where R_s is Schwarzschild's radius.

066.093 **A new upper limit to the small scale spatial variations in the microwave cosmic background radiation.**
R. L. Carpenter, S. Gulkis, T. Sato, J. C. Pigg.
Bull. American Astron. Soc., Vol. 4, 411 (1972). − Abstr. AAS.

066.094 Relativistic precessions of macroscopic objects.
    D. P. Whitmire.
Nature, Vol. 239, 207 (1972).

The inherently relativistic rotations of macroscopic objects are considered from a microscopic point of view. It is concluded that whether or not a macroscopic object (for example a gyroscope) will undergo a relativistic rotation as a whole depends on its detailed microscopic structural properties, and that a metallic gyroscope (and possibly one made of any material) in orbit will not precess.

066.095 Black holes and temporal ordering.   F. D. Peat.
    Nature, Vol. 239, 387 (1972).

Directionality of time, or "Times Arrow" is a concept which has been accepted into General Relativity. It relies for its justification upon certain classical arguments, a feature of which is that a sink for energy may be provided for each system. It is suggested that such arguments must be reexamined in the case of black holes and the inevitability of a single temporal direction for collapse is questioned.

066.096 Relativistic runaway problem.   P. D. Noerdlinger.
    Nature, Phys. Sci., Vol. 239, 9 - 10 (1972).

If a test particle is initially in a circular orbit about a massive body that loses mass suddenly (e.g. by neutrino emission), it may escape. The required fraction of mass that must be lost by the central body varies smoothly from about one quarter (when the test particle is three Schwarzschild radii away) to the classical value $^1/_2$, as the initial orbit radius is taken larger and larger. Circumstances are also pointed out in which the "induction force" of Lindquist et al. could be large.

066.097 On the location of the source of Weber's gravitational events.   D. H. Douglass, J. A. Tyson.
Astrophys. Journ., Vol. 178, 341 - 346 (1972).

We find that the location of the source(s) of gravitational radiation reported by Weber depends strongly on assumptions concerning the nature of the polarization of the radiation.

066.098 Rotating black holes: Separable wave equations for gravitational and electromagnetic perturbations.
S. A. Teukolsky.
Phys. Rev. Letters, Vol. 29, 1114 - 1118 (1972).

Separable wave equations with source terms are presented for electromagnetic and gravitational perturbations of an uncharged, rotating black hole. These equations describe the radiative field completely, and also part of the nonradiative field. Future applications (stability of rotating black holes, "spin-down", superradiant scattering, floating orbits) are outlined.

066.099 New exact solution for the gravitational field of a spinning mass.   A. Tomimatsu, H. Sato.
Phys. Rev. Letters, Vol. 29, 1344 - 1345 (1972).

A solution for the gravitational field is presented, which reduces to one of the Weyl metrics in the limit of angular momentum $J = 0$ and reduces to the Kerr metric in the limit of $J = m^2$.

066.100 Gravitation im Universum.   J. B. Dance.
    Bild der Wiss., [Deutsche Verlags-Anstalt (Stuttgart)], 9. Jahrgang, p. 1172 - 1179 (1972). – Popular article.

066.101 Algol and relativity.   A. I. Kornilov.
    Sb. nauch. tr. Yaroslav. tekhnol. in-t, Vol. 32, 111 - 114 (1972). In Russian. – Abstr. in Referativ. Zhurn. 51. Astron., 1.51.762 (1973).

066.102 On reference systems, on the principle of non-holonomicity and the problem "Ptolemy–Copernicus".
R. F. Polishchuk.

Vestn. Mosk. un-ta. Fiz., astron., Vol. 13, 375 - 381 (1972). In Russian. – Abstr. in Referativ. Zhurn. 51. Astron., 1.51.786 (1973).

066.103 Static black holes in Brans-Dicke theory are Schwarzschild solutions.   M. Johnson.
Nuovo Cimento Lettere, Ser. 2, Vol. 4, 323 - 327 (1972).

The author extends Israel's theorem (1967) concerning the uniqueness of the exterior vacuum field of a static black hole of mass m to the Brans-Dicke theory.

066.104 Further evidence for an anomalous interaction between the electromagnetic and gravitational fields.
J. F. Woodward, W. Yourgrau.
Nuovo Cimento B, Ser. 11, Vol. 9B, 440 - 452 (1972).

066.105 Newtonian gravitational theory: Interaction with light.   S. L. Schwebel.
International Journ. Theor. Phys., (GB), Vol. 5, No. 1 – 3, p. 29 - 33 (1972).

066.106 A revised nonsymmetric unified field theory.
    A. H. Klotz.
Acta Phys. Polonica B, Vol. B3, 341 - 352 (1972).

A further development of the Einstein–Kaufman, nonsymmetric unified field theory is discussed. Static, spherically symmetric solutions of the field equations are considered.

066.107 Many-body forces and the effect of the matter distribution in the universe on the gravitational constant.   K. Hiida.
Nuovo Cimento Lettere, Ser. 2, Vol. 4, 398 - 400 (1972).

066.108 Gravitational collapse and black holes.
    N. K. Spyrou.
Techn. Chronika, No. 4, p. 259 - 267 (1972). In Greek.

Presents in brief what is known on the problem of gravitational collapse of a massive body and the formation of black holes. The author points out that the extremely high power of the energy emitted by quasars (100 times as high as the power of the energy emitted by a giant galaxy) is perhaps explained as a result of such a collapse.

066.109 The time symmetric initial value problem for black holes.   G. W. Gibbons.
Commun. Math. Phys. Vol. 27, 87 - 102 (1972). – See Phys. Abstr., Vol. 75, No. 58726 (1972).

066.110 Neutrinos in spherically symmetric gravitational fields.   S. Audretsch.
Nuovo Cimento Lettere, Ser. 2, Vol. 4, 339 - 343 (1972).

A microphysical description of the interaction between neutrinos and gravitational fields in curved space-time can be given by the general relativistic Weyl equation. In this paper the author studies the solutions of Ψ of the Weyl equation in the general spherically symmetric space-time without restrictions on the metric.

066.111 The oscillations of relativistic fluid masses.
    J. C. Jackson.
Nuovo Cimento Lettere, Ser. 2, Vol. 4, 494 - 496 (1972).

Chandrasekhar's study (see abstr. A15902 of 1964) of the small radial oscillations of strongly self-gravitating fluid spheres about their equilibrium, has shown that, according to general relativity, these bodies can be dynamically unstable under conditions which lead to stability according to Newtonian theory. The author presents here an exact relativistic hydrodynamical equation which should simplify the study of problems of this nature.

066.112 General relativity and Mach's principle.

P. C. Waylen.
Structure of matter. Rutherford Centennial Symposium, Christchurch 1971, [Univ. Canterbury, Christchurch, New Zealand], p. 330 - 336 (1972).

The author discusses a reformulation of the theory of gravitation in terms of a generally covariant form of Einstein's field equations. The connection between this theory and Mach's principle is outlined. The implications of these concepts to universe models are analyzed.

066.113 Gravitational waves come down to earth.
S. Mitton.
New Scient., (*GB*), No. 805, Vol. 55, 132 - 134 (1972).
See Phys. Abstr., Vol. 75, No. 62439 (1972).

066.114 Gravitational collapse of a rotating star. I. Formulation. K. Tomita.
Progr. Theor. Phys., (*Japan*), Vol. 48, 78 - 103 (1972).

A system of general relativistic hydrodynamical and gravitational field equations is generally formulated under Bondi-Sach's coordinate condition, and then it is applied to an approximate description of dynamical motions of a slowly rotating axisymmetric star.

066.115 Perihelion motion of planets and the new theory of gravitation. K. C. Kar, A. K. Bhattacharyya.
Indian Journ. Theor. Phys., Vol. 20, 1 - 8 (1972).

Einstein's formula for the advance of perihelion of planets has been deduced following his simple theory of relativity, without taking any help of the complicated so called generalised theory of relativity. In the new theory of gravitation after introducing relativistic correction in the law of area, Einstein's formula for perihelion advance has been deduced in a very simple manner. It is thus shown that Einstein's general theory is not necessary for solving this problem.

066.116 New theory of gravitation. H. Yilmaz.
Nuovo Cimento B, Ser. 11, Vol. 10B, 79 - 101 (1972).

066.117 Orbital and vortical motion in the Kerr metric.
F. de Felice, M. Calvani.
Nuovo Cimento B, Ser. 11, Vol. 10B, 447 - 458 (1972).

066.118 Generally covariant massive gravitation.
P. C. Aichelburg, R. Mansouri.
Nuovo Cimento B, Ser. 11, Vol. 10B, 483 - 497 (1972).

066.119 On the dependence of the gravitational effects of the heavy body mass distribution in Deser and Laurent's theory. H. Hafner.
Revue Rouman. Phys., Vol. 17, 433 - 441 (1972). – See Phys. Abstr., Vol. 75, No. 66423 (1972).

066.120 Some properties of a uniform fluid sphere in general relativity. A. Banerjee.
Journ. Phys. A, General Phys., Vol. 5, 1305 - 1311 (1972).

066.121 Relativity gyroscope experiment at arbitrary orbit inclinations. B. M. Barker, R. F. O'Connell.
Phys. Rev. D, Particles and Fields, Vol. 6, 956 - 961 (1972).

066.122 Best-fit estimate of relativistic effects in time-delay experiments. J.-P. Richard.
Phys. Rev. D, Particles and Fields, Vol. 6, 961 - 966 (1972).

066.123 Conservation of momentum in general relativity.
C. W. Price, Jr.
Thesis, Brown Univ., Providence, R.I. [Available from Univ. Microfilms, Ann Arbor, Mich., USA. Order No. 72-8163], 38 pp. (1971). – See Phys. Abstr., Vol. 75, No. 69002 (1972).

066.124 Colliding black holes. A. Cadez.
Thesis, Univ. North Carolina, Chapel Hill. [Available from Univ. Microfilms, Ann. Arbor, Mich., USA. Order No. 72-10696], 106 pp. (1971). – See Phys. Abstr., Vol. 75, No. 69006 (1972).

066.125 On the gravitational collapse in Brans-Dicke theory of gravity. T. Matsuda.
Progr. Theor. Phys., (*Japan*), Vol. 47, 738 - 740 (1972).

The author clarifies the discrepancies between the conjectures of Nariai and of Thorne and Dykla on phenomena for which the Brans-Dicke theory and the Einstein theory make qualitatively different predictions. He gives an example in which Thorne et al.'s discussion is not correct by analyzing the Brans type I solution.

066.126 Construction and operation of a Weber-type gravitational-wave detector and of a divided-bar prototype.
D. Bramanti, K. Maischberger.
Nuovo Cimento Lettere, Ser. 2, Vol. 4, 1007 - 1013 (1972).

The authors describe some relevant characteristics of a Weber-type detector and the observation of its thermal noise. For comparison they also report briefly on the operation of a divided-bar prototype.

066.127 Black holes and the second law. J. D. Bekenstein.
Nuovo Cimento Lettere, Ser. 2, Vol. 4, 737 - 740 (1972).

The author formulates the second law of thermodynamics in a form suitable for black-hole physics which resolves the transcendence problem. He also indicates why Geroch's gedanken experiment does not, in fact, violate the second law.

066.128 Polarization of gravitational synchrotron radiation. R. A. Breuer, J. Tiomno, C. V. Vishveshwara.
Nuovo Cimento Lettere, Ser. 2, Vol. 4, 857 - 860 (1972).

In this letter the authors propose a method of analysis of gravitational polarization in terms of Stokes parameters similar to those used for the electromagnetic case.

066.129 The dispersion of gravitational waves. J. Madore.
Commun. Math. Phys., Vol. 27, 291 - 302 (1972).

A definition is given of a plane gravitational wave in a curved background space-time manifold. For a particular background metric, a dispersion relation for the waves is derived analogous to that satisfied by plane electromagnetic waves in a dilute plasma.

066.130 The cosmological dependence of weak interactions. M. Novello, P. Rotelli.
Journ. Phys. A, General Phys., Vol. 5, 1488 - 1494 (1972).

A model for the cosmological time dependence of weak interactions is discussed and some experimental tests are suggested.

066.131 Angular momentum in general relativity and the Kerr metric. M. K. Moss, W. R. Davis.
Nuovo Cimento B, Ser. 11, Vol. 11B, 84 - 92 (1972).

066.132 An electromagnetic interpretation of the Kerr-Vaidya metric. P. A. Goodinson.
International Journ. Theor. Phys., (*GB*), Vol. 6, 47 - 51 (1972).

It is shown that at large distances from a rotating mass, the radiation may be associated with an Einstein-Maxwell null field with a non-zero null current.

066.133 A covariant scalar theory of gravitation and its implications for general relativity.
J. L. Pietenpol, D. Speiser.
Ann. Soc. Sci. Bruxelles, (*Belgium*), Ser. 1, Vol. 86, 220 - 226 (1972). In French.

066.134 **Advances in gravitational radiation detection.**
J. Weber.
General Relativity and Gravitation, Vol. 3, 59 - 62 (1972).
　　New experiments are reported which improve estimates of the bandwidth and energy flux of the observed radiation.

066.135 **An improved detector of gravitational radiation.**
P. S. Aplin.
General Relativity and Gravitation, Vol. 3, 111 - 113 (1972).
　　A detector of gravitational radiation is described, based on that of Weber but having enhanced time resolution.

066.136 **The weak-field approximation and the range of gravitational force.**　　D. Layzer, J. R. Burke.
General Relativity and Gravitation, Vol. 3, 121 (1972).
　　The authors consider the dynamics of a statistically homogeneous and isotropic universe with weak local gravitational fields.

066.137 **Present status of the theory of the relativity-gyroscope experiment.**　　R. F. O'Connell.
General Relativity and Gravitation, Vol. 3, 123 - 133 (1972).
　　The author considers the precession of a gyroscope in Einstein theory and presents analytical and numerical results. He then considers the precession of a gyroscope in an arbitrary theory of gravitation and concludes with a discussion of the results.

066.138 **Recent developments in the theory of gravitational radiation.**　　D. W. Sciama.
General Relativity and Gravitation, Vol. 3, 149 - 165 (1972).
　　The author reviews those papers which he feels to be most interesting or important. The material is devided into three sections, emission, propagation and absorption of gravitational radiation.

066.139 **On the new theory of gravitation.**　　H. Yilmaz.
　　Nuovo Cimento Lettere, Ser. 2, Vol. 5, 309 - 312 (1972).
　　The author presents a closed exponential form for the iteration solution of his theory.

066.140 **Electrodynamics in an expanding universe.**
G. E. Tauber.
General Relativity and Gravitation, (GB), Vol. 3, 17 - 27 (1972).

066.141 **On the gravitational field of simple spherical rotating layers in a linear approximation.**
P. Teyssandier.
Nuovo Cimento Lettere, Ser. 2, Vol. 5, 359 - 365 (1972).
In French.

066.142 **Solution of the Einstein-Maxwell equations for a static distribution of massive charged particles.**
K. Nordtvedt, Jr.
General Relativity and Gravitation, Vol. 3, 95 - 100 (1972).
　Exact solutions of the general relativistic field equations of Einstein and Maxwell have been found for a general static distribution of massive charged particles.

066.143 **Mach's principle in general relativity.**　　J. H. Higbie.
　　General Relativity and Gravitation, Vol. 3, 101 - 109 (1972).
　　

066.144 **Testing general relativity: Progress, problems, and prospects.**　　I. I. Shapiro.
General Relativity and Gravitation, Vol. 3, 135 - 148 (1972).
　　The author presents recent results from radar and radio gravity experiments for retardation of radar signals, deflection of radio waves, relativistic perihelion advance of Mercury and

time variation of gravitational constant.

066.145 **Schwarzschild interior solution in conformally flat co-ordinate system.**　　J. K. Rao, R. B. Patel.
Current Sci., (India), Vol. 41, 409 - 410 (1972).

066.146 **The sun and Brans-Dicke cosmology.**
S. N. Svolopoulos.
Technika Chronika, (Greece), No. 7, p. 601 - 604 (1972).
In Greek.
　　The Brans-Dicke theory is investigated by the exact study of the deviation of light in the solar gravitational field, the motion of Mercury's perihelion, the oblateness of the sun and the solar rotation.

066.147 **Energy and angular momentum flow into a black hole.**　　S. W. Hawking, J. B. Hartle.
Commun. Math. Phys., Vol. 27, 283 - 290 (1972).

066.148 **Gravitational collapse and related phenomena from an empirical point of view, or, black holes are where you find them.**　　P. J. E. Peebles.
General Relativity and Gravitation, Vol. 3, 63 - 82 (1972).
　　The author presents a review covering dead stars, pulsars, X-ray stars, neutron stars, close binary star systems, globular star clusters, M 31, elliptical galaxies, spectacular nuclei and the origin of spiral structure.

066.149 **Gravitational collapse, white holes and particle creation.**　　Ya. B. Zel'dovich, I. D. Novikov.
General Relativity and Gravitation, Vol. 3, 119 - 120 (1972).
　　The authors consider the outside picture of a collapsed body, the inescapability of singularity in collapse, the problem of identification of black holes in our Galaxy and outside, and white holes.

066.150 **Gravitational radiation from charged black holes.**
L. Basano, A. Morro.
Nuovo Cimento Lettere, Ser. 2, Vol. 5, 266 - 268 (1972).
　　The authors briefly summarize the results of Hawking. They show that the efficiency for the conversion of mass into gravitational radiation can be appreciably higher than that given by Hawking if the colliding black holes are charged.

066.151 **Astrophysical test for dilaton theory of non-Newtonian gravity.**　　D. Sugimoto.
Progr. Theor. Phys., (Japan), Vol. 48, 699 - 700 (1972).

066.152 **How to measure the earth's velocity with respect to absolute space.**　　S. Marinov.
Phys. Letters A, (Netherlands), Vol. 41A, 433 - 434 (1972).

066.153 **The observed limits to gravitational collapse.**
H. Andrillat.
Bull. Soc. Roy. Sci. Liège, (Belgium), Vol. 41, 292 - 301 (1972). In French.
　　The observational and elementary theoretical background to the stages of gravitational condensation of stellar matter, from nebula to neutron star, is reviewed and discussed.

066.154 **Gravitational radiation and neutrinos.**
J. B. Griffiths.
Commun. Math. Phys., Vol. 28, 295 - 299 (1972).

066.155 **Can gravitational waves change astronomy?**
M. Fujimoto.
Astron. Herald, (Japan), Vol. 65, 259 - 263 (1972).
In Japanese.
　　Reviews and considers Weber's observational results. The waves are considered in relation to the infrared 100 μ peak and the 3 kpc spiral arm. The paper also discusses the effect that

studies of gravitational radiation will make on developments in astronomy.

066.156 Model of gravitating sphere set in rotation by internal stress. P. A. Hogan, J. L. Synge.
General Relativity and Gravitation, (*GB*), Vol. 3, 269 - 280 (1972).

The systematic approximation technique of Synge is employed to construct a model of a spherical body at rest in the distant past and gradually attaining an angular velocity due to a modification of the internal stress.

066.157 On the apparent visual forms of relativistically moving objects. P. M. Mathews, M. Lakshmanan.
Nuovo Cimento B, Ser. 11, Vol. 12 B, 168 - 181 (1972).

066.158 A note on the gravitational field of a rotating radiating source. P. A. Goodinson.
Journ. Phys. A, General Phys., Vol. 5, L131 (1972).

The author presents here a generalization of an already known solution, given by Kramer (1972).

066.159 Gravitation theory: Empirical status from solar system experiments. K. L. Nordtvedt, Jr.
Science, Vol. 178, 1157 - 1164 (1972).

I have reviewed the historical and contemporary experiments that guide us in choosing a post-Newtonian, relativistic gravitational theory. A variety of experiments specify (or put limits on) the numerical values of the seven parameters in the post-Newtonian metric field, and other such experiments have been planned. The empirical results, to date, yield values of the parameters that are consistent with the predictions of Einstein's general relativity.

066.160 A classification of particle motions in the equatorial plane of a gravitational monopole-quadrupole field in Newtonian mechanics and general relativity.
A. Armenti, Jr.
Celestial Mechanics, Vol. 6, 383 - 415 (1972).

By an extension of the method used by Morton and Leavitt to obtain the Schwarzschild geodesics, exact solutions for the Newtonian orbits are obtained in terms of Jacobian elliptic functions, and a complete classification of the orbits is given. The motions are all quasi-Keplerian, except for a curious subclass. The lowest order solutions, which are essentially a superposition of relativistic monopole (Schwarzschild) and Newtonian quadrupole contributions, are shown to give accurate descriptions for motions in the solar system. The results are used to study the effect of a solar oblateness on the three classical tests of general relativity.

066.161 Gravitational radiation experiments. P. S. Aplin.
Contemporary Phys., (*GB*), Vol. 13, 283 - 293 (1972).

066.162 Gravitasjonskollaps. R. Stabell.
Naturen 1971, No. 1, p. 8 - 15 = Inst. Teor. Astrofys. Blindern — Oslo, Småtrykk No. 71 (1971).

066.163 Testing general relativity: Progress, problems, and prospects. I. I. Shapiro.
American Inst. Phys. Conference Proc. 1971, No. 2, p. 286 - 301.

066.164 Collision between an electromagnetic wave and a gravitational wave packet. Yu. G. Sbytov.
Zhurn. ehksperim. i teor. fiz., Vol. 63, 737 - 744 (1972). In Russian. — Abstr. in Referativ. Zhurn. 51. Astron., 2.51.786 (1973).

066.165 Gravitational collapse of homogeneous spheres.

R. M. Misra, D. C. Srivastava.
Nature, Phys. Sci., Vol. 238, 116 (1972).

The dynamics of perfect fluid spheres of uniform density has been discussed in literature under the assumption of spatial isotropy along with the usual regularity conditions at the centre. It is shown in this note that the assumption of spatial isotropy is redundant in the sense that for the uniform density models the requirement of regularity at the centre necessarily leads to it.

066.166 Computer analyses of gravitational radiation detector coincidences. J. Weber.
Nature, Vol. 240, 28 - 30 (1972).

Coincidence experiments at 1,661 Hz were carried out with two antennae, one situated at the University of Maryland and the other 1,000 km away at the Argonne National Laboratory. It is considered established beyond reasonable doubt that the gravitational radiation detectors at ends of a 1,000 km baseline are being excited by a common source as a result of interactions which are neither seismic, electromagnetic nor those of charged particles of cosmic rays.

066.167 Rigidity of a black hole. B. Carter.
Nature, Phys. Sci., Vol. 238, 71 - 72 (1972).

The author points out that the analogy between a black hole and an ordinarily rigidly rotating body can be pushed considerably further than previously done (1969).

066.168 Limit on the energy density in the submillimetre background radiation. R. Cowsik.
Nature, Phys. Sci., Vol. 239, 41 - 42 (1972).

The present analysis indicates that the energy density in the submillimetre quanta cannot exceed ~ 0.4 eV cm^{-3} averaged over the galactic dimensions, independent of the exact spectral distribution of this radiation. This corresponds to an upper limit of 3.4 K on the radiation temperature in the galactic neighbourhood.

066.169 Basis concepts of the kinematics of motor mechanics. P. Kustaanheimo.
Sci. Revuo, Beograd, Vol. 22, 173 - 188 (1971). In Esperanto.

066.170 On gravitational aberrations in stellar images. M. W. Cook.
Australian Journ. Phys., Vol. 25, 749 - 758 (1972).

On the basis of a cosmological model which is fundamentally of the Friedmann expanding type with a spherically symmetric inhomogeneity superimposed, a study is made of three gravitational aberrations of purely relativistic origin observed in the images of stellar objects: (1) the "gravitational lens" effect, (2) a dispersion effect whereby a point source would produce a diffuse image, and (3) an apparent systematic motion of all light sources towards (or away from) the inhomogeneity.

066.171 Black holes. S. E. Williams.
Journ. Astron. Soc. Western Australia, 1972 August, p. 4 - 5.

Einstein's theory of relativity.
See Abstr. 003.049.

Gravitation and cosmology: Principles and applications of the general theory of relativity. See Abstr. 003.124.

Gravitation. See Abstr. 003.164.

Possible effects of anisotropy of *G* on celestial orbits. See Abstr. 042.035.

Morphology of rapid cosmic processes.
See Abstr. 061.018.

Hadronic equilibrium-configurations.
See Abstr. 065.017.

Implications of tachyon-like matter for superdense stars.   See Abstr. 065.080.

Cold neutron star model in Brans-Dicke theory of gravity.   See Abstr. 065.154.

The perihelion of Mercury.
See Abstr. 092.009.

The luminosity of a collapsing star. Pt. I.
See Abstr. 115.028.

Variable stars and the photon rest mass.
See Abstr. 121.056.

Further evidence for a black hole in Beta Lyrae?
See Abstr. 121.068.

On the variability of X-ray radiation from black holes at disk accretion.   See Abstr. 142.106.

About Kerr's galactic expansion and Weber's gravitation signals.   See Abstr. 155.035.

Gravitational waves from the center of the Galaxy.
See Abstr. 155.072.

Cosmology and microwave astronomy.
See Abstr. 162.033.

Is the existence of a galaxy evidence for a black hole at its center?   See Abstr. 162.038.

Sun

071 Solar Photosphere, Spectrum

071.001 **The density dependent ionization balance of carbon, oxygen and neon in the solar atmosphere.**
H. P. Summers.
Monthly Notices Roy. Astron. Soc., Vol. 158, 255 - 275 (1972).

Density dependent ionization equilibria for carbon, oxygen and neon ions in the optically thin solar atmosphere are calculated. All relevant processes are included and the results are expected to be accurate to ~ 40 per cent. Excepting for the neutral and first ionized ions in each case, the results are applicable to the ionization equilibrium in any thermal plasma where the ionization is maintained by the free electrons. Fractional ion abundances on incorporating a chromosphere and corona model are also presented.

071.002 **The solar spectrum: Wavelengths and identifications from 60 to 385 Angstroms.**
W. E. Behring, L. Cohen, U. Feldman.
Astrophys. Journ., Vol. 175, 493 - 523 (1972).

The solar spectrum from 60 to 158 Å and 163 to 385 Å was photographed at a resolution of 0.04 Å or better on glass plates. The wavelengths of emission lines in these records were determined with a typical accuracy of 0.008 Å above 100 Å and 0.004 Å below 100 Å. The design considerations and limitations of this instrument are presented along with a description of the Aerobee 150 flight mission on 1969 May 16. The wavelengths for the 370 observed lines are listed together with intensity estimates. Of these, the 180 identified lines are due to ions of He, O, Ne, Mg, Si, S, Ar, Ca, Fe, and Ni. These identified lines are also tabulated separately for each isoelectronic sequence from Li I to K I.

071.003 **Calculation of solar CO vibration-rotation line profiles and equivalent widths.** R. A. Berger, P. Léna.
Astron. Astrophys., Vol. 20, 111 - 113 (1972).

Calculations of the profile and equivalent width of the P 36 line of the infrared fundamental band of CO have been undertaken assuming LTE. Results for the BCA and HSRA solar models are compared.

071.004 **Absolute measurement of the solar brightness in the spectral region between 100 and 500 microns.**
P. Stettler, F. K. Kneubühl, E. A. Müller.
Astron. Astrophys., Vol. 20, 309 - 312 (1972).

The solar brightness temperature at wavelengths between 110μ and 500μ was measured with a balloon-borne lamellar-grating interferometer. A high temperature black body served as absolute calibration source. The experimental results are compared with those of other research groups and with the empirical HSRA model.

071.005 **Solar seeing and the spatial properties of the five-minute oscillations.** J. H. Thomas.
Solar Physics, Vol. 24, 262 - 273 (1972).

A numerical simulation of observations of the spatial properties of the five-minute oscillations is carried out, assuming the oscillations are internal gravity waves excited by granular convection according to the theory of Thomas et al. (1971). The simulation includes the effects of seeing and finite aperture. The results show that the peak in the observed power spectrum of the oscillations can occur at a wavelength considerably longer than the true wavelength of the oscillations. In particular, the peak in Frazier's observed power spectra at wavelength $\lambda \approx 5000$ km is consistent with the considerably shorter true wavelength $\lambda \approx 1500$ km predicted by the gravity wave theory.

071.006 **Velocity oscillations in the solar atmosphere.**
J. C. Bhattacharyya.
Solar Physics, Vol. 24, 274 - 287 (1972).

From a series of long duration continuous Doppler records of selected spectral lines at various depths in the solar atmosphere, characteristics of solar velocity oscillations have been studied. Statistical distribution of the durations of the bursts of oscillations has been estimated. From the nature of distortion of the waveforms of the oscillation, the presence of disturbing impulses has been speculated. Constancy and homogeneity of the oscillations have been examined from detailed spectral density plots.

071.007 **Magnetic fields and helium-D_3 spectroheliograms.**
G. A. Chapman.
Solar Physics, Vol. 24, 288 - 300 (1972).

Spectroheliograms, having a resolution approaching 2″, have been obtained which show He-D_3 in absorption against the disk. The He-D_3 features are compared with the distribution of magnetic fields and with Hα structures.

071.008 **Characteristics of the Ca II K-line profiles in the quiet sun.** S.-Y. Liu, E. v. P. Smith.
Solar Physics, Vol. 24, 301 - 309 (1972).

Feature-to-feature identification is made on simultaneous Ca II K-line spectrograms (SG) and K_{2v} spectroheliograms (SHG). The line profiles in plages and in the network boundary nearly always have double-peaked reversal in the core, while those inside the cells present all possibilities: double-peaked, single-peaked on violet side, single-peaked on red side, and unreversed absorption. We call attention to the nontrivial contribution of these absorption profiles which are formed in 'dark regions' shown on SHG's. The physical conditions inferred from different kinds of profiles are briefly discussed.

071.009 **Raies nouvelles observées lors de l'éclipse du 7 mars 1970.** Z. Mouradian.
Solar Physics, Vol. 24, 368 - 369 (1972). – Research note.

071.010 **The solar abundance of calcium and collision broadening of Ca I- and Ca II-Fraunhofer lines by hydrogen.** H. Holweger.
Solar Physics, Vol. 25, 14 - 29 (1972).

An abundance analysis of the solar calcium spectrum is carried out using 46 lines with known f-values in the visible and near infrared spectral region. Resonance, forbidden and autoionizing lines are included. The solar abundance of calcium resulting from the 25 weaker, nearly damping-independent lines only is log $\epsilon_{Ca} = 6.36 \pm 0.07$, on the scale log $\epsilon_{H} = 12$. Together with the sodium abundance log $\epsilon_{Na} = 6.30$ determined earlier, the solar abundance ratio Ca/Na = 1.15 is obtained with an accuracy of $\approx 10\%$. Comparison with meteor-

ites (carbonaceous chondrites *I*) shows that solar and meteoritic ratio agree within these limits. The line broadening by
collisions with hydrogen atoms is determined empirically from
a comparison of weak and strong Fraunhofer lines of Ca I and
Ca II, thereby using the solar atmosphere as an absorption
tube of comparatively well-known physical state.

071.011   The solar abundance of silver.
      J. E. Ross, L. H. Aller.
Solar Physics, Vol. 25, 30 - 43 (1972).

Low noise, high resolution spectral scans have been ob-
tained for the resonance lines of silver $\lambda\,3280.7$ and $\lambda\,3382.9$,
observed at the centre of the solar disk. The data are analyzed
by the method of spectral synthesis wherein we employ a
model atmosphere resembling Elste's (1968) model and
checked by limb-darkening observations. The silver abundance
turns out to be $\langle Ag \rangle = \log[N(Ag)/N(H)] + 12 = 0.85$, a factor
of four under the value found from the type I carbonaceous
chondrites.

071.012   Spectral analyses of solar photospheric fluctuations.
      III. Bi-dimensional power, coherence and phase
spectra of deep-seated radial velocity and photometric fluctuations.   F. N. Edmonds, Jr., C. J. Webb.
Solar Physics, Vol. 25, 44 - 70 (1972).

Fluctuations measured from a time sequence of high-resolution, high-dispersion Sacramento Peak Observatory spectrograms and previously analyzed by computing one-dimensional
temporal and spatial spectra (Edmonds et al., 1965), are re-
analyzed using bi-dimensional (temporal and spatial) power,
coherence and phase spectra computed by fast-Fourier-trans-
form techniques. The fluctuations measured are radial velocity
for the Fe I 5049.83, Cr I 5051.91 and C I 5052.16 spectral
lines, continuum brightness, and equivalent width and central
intensity of the C I line. The bi-dimensional spectra, particularly those of coherence and phase, allow isolating different components of the fluctuations to a degree not possible in the
one-dimensional analyses. Six components of the fluctuations
have been isolated and are discussed in detail.

071.013   Micro-and macroturbulent motions and the velocity
      spectrum of the solar photosphere.   C. de Jager.
Solar Physics, Vol. 25, 71 - 80 (1972).

A given motion field in a stellar atmosphere is usually
observed through 'filters' defined by line shifts and -broaden-
ings and conventionally called macroturbulence and microturbulence. These 'filters' can be defined and computed
exactly, as a function of the wave number of the velocity
field. We apply the results to several cases of an assumed motion field spectrum, and to observations of broadenings and
displacements of solar Fraunhofer lines formed at a depth
$\tau_5 = 0.1$. The results show that a well-developed spectrum of
hydrodynamical turbulence extending over a large range of
wavelengths does not exist at that level of the photosphere.

071.014   The velocity fields at different levels in quiet solar
      regions.   S. I. Gopasyuk, T. T. Tsap.
Izv. Krymskoj Astrofiz. Obs., Vol. 44, 45 - 51 (1972). In Russian.

The velocity fields at different levels of quiet regions on
the sun are considered. The records of the radial velocities
were made in the lines H_α, K_3 Ca II, H_β, H_γ, H_δ, Mg I $\lambda5184$ Å,
Ca I $\lambda4227$ Å, Na I D_1, Ba II $\lambda4554$ Å, Ca I $\lambda6103$ Å, Fe I
$\lambda5250$ Å with the aid of a double-magnetograph.

071.015   On large-scale velocity fields in the solar photo-
      sphere.   S. I. Gopasyuk, B. Kalman.
Izv. Krymskoj Astrofiz. Obs., Vol. 44, 52 - 63 (1972). In Russian.

The velocity fields over the whole solar disk are consid-

ered on the basis of records of radial velocities made in the
Fe I $\lambda5250$ Å and Ca I $\lambda6103$ Å lines with a double-magnetograph. It is shown that the velocities in the solar photosphere
have isotropical distribution.

071.016   On the theory of radiative transfer in an inhomoge-
      neous magnetic field.   D. N. Rachkovsky.
Ivz. Krymskoj Astrofiz. Obs., Vol. 44, 64 - 69 (1972). In Russian.

The equations of radiative transfer in the presence of a
magnetic field derived by Kai (1968), Beckers (1969) and the
author (1970) are considered. It is shown that the equations
by Kai are not correct. The coefficient of anomalous dispersion
in Beckers' equations is wrong by the factor two. The transfer
equations are integrated numerically when the direction of a
pure transverse field is changing by 45° discontinuously. It is
found that in this case the line is split into two components
with opposite circular polarizations.

071.017   Uniform dodging of H-alpha filtergrams of the solar
      disk.   J. Goff, R. Hansen, L. Lacey.
AAS Photo-Bull., 1972, No. 2, p. 1, 3.

071.018   Thermal oscillations in the high solar photosphere.
      R. W. Noyes, D. N. B. Hall.
Astrophys. Journ., (*Letters*), Vol. 176, L89 - L92 (1972).

Prominent 5-minute oscillations have been detected in the
intensity of the fundamental vibration-rotation lines of CO at
4.67μ. These lines are formed near the temperature minimum
in the high photosphere. The intensity amplitude corresponds
to a temperature oscillation with peak-to-peak amplitude of
225° K.

071.019   The possible dependence of differential shifts of
      Fraunhofer telluric lines on the sun's zenith dis-
tance.   O. A. Melnikov, R. Kh. Salman-Zade, Iu. A. Solons-
kii, E. D. Khilov.
Dokl. Akad. Nauk SSSR, Ser. Mat. Fiz., Vol. 205, 1054 -
1056 (1972).   In Russian.

071.020   Line broadening by macroturbulence.
      J. C. Evans, L. W. Ramsey, D. F. Gray.
Bull. American Astron. Soc., Vol. 4, 333 (1972). – Abstr. AAS.

071.021   Improved wavelengths and identifications in the so-
      lar spectrum from 60–385 Å.
W. E. Behring, L. Cohen, U. Feldman.
Bull. American Astron. Soc., Vol. 4, 377 - 378 (1972).
Abstr. AAS.

071.022   Absolute intensity of the solar spectrum from
      1200 Å to 1790 Å derived from new rocket spectra.
G. E. Brueckner, K. Nicolas.
Bull. American Astron. Soc., Vol. 4, 378 (1972). – Abstr. AAS.

071.023   Observations on the relationship between the lati-
      tudinal variations of temperature and magnetic
field.   R. C. Canfield.
Bull. American Astron. Soc., Vol. 4, 378 - 379 (1972).
Abstr. AAS.

071.024   An investigation of the saturation of Fraunhofer
      lines.   C. R. Cowley, J. Toney.
Bull. American Astron. Soc., Vol. 4, 380 (1972). – Abstr. AAS.

071.025   A search for the photospheric origin of spicules.
      R. B. Dunn, J. B. Zirker.
Bull. American Astron. Soc., Vol. 4, 381 (1972). – Abstr. AAS.

071.026   Interferometric observations of small solar continu-
      um features.   J. W. Harvey.

Bull. American Astron. Soc., Vol. 4, 383 - 384 (1972).
Abstr. AAS.

071.027   **The solar iron abundance revisited: A determination from the weak Fe I line λ 5127.7.**
M. C. E. Huber, E. F. Tubbs.
Bull. American Astron. Soc., Vol. 4, 385 (1972). — Abstr. AAS.

071.028   **A mechanism for exploding solar granules.**
S. Musman.
Bull. American Astron. Soc., Vol. 4, 388 (1972). — Abstr. AAS.

071.029   **Spectrum synthesis and the solar abundance of gallium.**   J. P. Mutschlecner.
Bull. American Astron. Soc., Vol. 4, 388 (1972). — Abstr. AAS.

071.030   **Spectra of CO fundamental lines and the structure of the high photosphere.**
R. W. Noyes, D. N. B. Hall.
Bull. American Astron. Soc., Vol. 4, 389 (1972). — Abstr. AAS.

071.031   **Real-time analysis of flare-associated photospheric magnetic fields.**   D. M. Rust.
Bull. American Astron. Soc., Vol. 4, 390 (1972). — Abstr. AAS.

071.032   **The photospheric velocity field in and around sunspots.**   N. R. Sheeley, Jr.
Bull. American Astron. Soc., Vol. 4, 391 (1972). — Abstr. AAS.

071.033   **Spatial frequencies of the photosphere and low corona.**   S. M. Smith, G. C. J. Suffolk.
Bull. American Astron. Soc., Vol. 4, 392 (1972). — Abstr. AAS.

071.034   **TV registration of the solar spectrum. I.**
L. D. Parfinenko.
Solnechnye Dannye 1972 Byull., No. 6, p. 95 - 99 (1972).
In Russian.
    A method is described of obtaining simultaneous spectrophotometric sections across the solar spectrum using a TV system. Preliminary results are given.

071.035   **Fe I ionization and excitation equilibrium in the solar atmosphere.**   R. G. Athay, B. W. Lites.
Astrophys. Journ., Vol. 176, 809 - 831 (1972).
    An understanding of the ionization and excitation equilibrium of iron in the photosphere and low chromosphere is necessary for the correct interpretation of several diverse phenomena. The abundance of iron in the solar photosphere has recently been revised upward by a factor of 5 to 10 as a result of new determinations of Fe I f-values and the use of forbidden lines of Fe II. Revisions in the f-values have explained some, but not all, of the earlier results leading to low abundances. We will show that departures from LTE can help account for the remaining cases.

071.036   **Über die Bildung von Fraunhoferlinien bei Abweichungen vom lokalen thermodynamischen Gleichgewicht, dargestellt am Beispiel der infraroten O I-Linien im Sonnenspektrum.**   E. S. Sedlmayr.
Diss. Naturwiss. Gesamtfakultät Ruprecht-Karl-Univ., Heidelberg. 2 + 120 pp. (1972).
    The solar infrared triplets of neutral oxygen (λ 7773. and λ 8446.)—each reduced to one fictive line—are investigated with respect to deviations from LTE. It is shown that there is a marked non-LTE effect of line strengthening which can explain the discrepancies between LTE calculations and observations.

071.037   **The empirical determination of line source functions, β_L-values, and the microturbulent and convective velocity components as functions of depth in the photosphere-chromosphere transition region.**   C. de Jager, L. Neven.
Solar Physics, Vol. 25, 277 - 304 (1972).
    An empirical method for determining line source functions, previously applied by us to the cores of infrared lines, has now been extended to the whole line profile and was applied to centre-limb observations of sixteen lines of five infrared multiplets, mainly of high excitation potential. The present investigation was performed in two steps. In the first part of the paper approximate values are derived for the depth dependence of the four functions named in the title of this paper, where β_L is the ratio between the actual and the LTE population of the lower level of the transitions involved. In the second part of the paper we use these empirically derived functions to compute the line profiles. From the remaining differences between observed and computed profiles, corrections are derived to the four functions.

071.038   **Oscillatory motions in the solar photosphere and magnetic fields.**   S. I. Gopasjuk, T. T. Tsap.
Astron. Zhurn. Akad. Nauk SSSR, Vol. 49, 1066 - 1068 (1972). In Russian. English translation in Soviet Astron. AJ, Vol. 16, No. 5.
    From the data of measurements of radial velocities and longitudinal magnetic fields, carried out with the help of a magnetograph, the change of the amplitude of the oscillating velocity of the solar limb and the connection of the amplitude with the magnetic field are studied. It is found that there is a considerable horizontal velocity component of velocities of 5-minute oscillations.

071.039   **Effects of uncertainties in damping and microturbulence on theoretical deductions from solar equivalent widths.**   D. E. Blackwell, G. Calamai, R. B. Willis.
Monthly Notices Roy. Astron. Soc., Vol. 160, 121 - 127 (1972).
    The paper presents a quantitative discussion of the effect on the interpretation of the equivalent widths of solar lines, of present uncertainties in damping constants and in atmospheric microturbulence. Graphs are given showing the resultant uncertainty in the interpretation of equivalent widths as a function of excitation potential and wavelength.

071.040   **Identifications of emission lines in the EUV solar spectrum.**   C. Jordan.
Space Sci. Rev., Vol. 13, 595 - 605 (1972). — Invited paper IAU Colloquium No. 14 (see 012.012).

071.041   **Wavelengths of solar lines in the 50−380 Å region and their identifications.**
U. Feldman, W. Behring, L. Cohen.
Space Sci. Rev., Vol. 13, 608 - 609 (1972). — Conference paper IAU Colloquium No. 14 (see 012.012).

071.042   **High resolution solar spectra from 1780 to 1950 Å.**
H. C. McAllister, R. J. Wolff.
Space Sci. Rev., Vol. 13, 610 - 611 (1972). — Conference paper IAU Colloquium No. 14 (see 012.012).

071.043   **Observations of the profiles of solar UV emission lines and their analysis in terms of the heating and production of the corona.**
B. C. Boland, S. F. T. Engstrom, B. B. Jones, R. W. P. McWhirter, P. C. Thonemann, R. Wilson.
Space Sci. Rev., Vol. 13, 639 - 641 (1972). — Conference paper IAU Colloquium No. 14 (see 012.012).

071.044   **Calculations on the solar spectrum from 1 to 60 Å.**
R. Mewe.
Space Sci. Rev., Vol. 13, 666 - 667 (1972). — Conference paper IAU Colloquium No. 14 (see 012.012).

071.045 The interpretation of absorption-line shifts in the solar spectrum. R. I. Kostik, T. V. Orlova.
Solar Physics, Vol. 26, 42 - 51 (1972).

The shifts of Fraunhofer lines of different chemical elements in a homogeneous medium with plane monochromatic progressive adiabatic sound waves are derived. The results agree qualitatively and quantitatively with observations.

071.046 Some observational results on moustaches. A. Bruzek.
Solar Physics, Vol. 26, 94 - 107 = Mitt. Fraunhofer Inst., *Freiburg,* No. 114 (1972).

The results of new observations of moustaches in Hα filtergrams and in Hα spectra are presented and their relations to photospheric and chromospheric phenomena are studied.

071.047 The magnetic structure of arch filament systems. E. N. Frazier.
Solar Physics, Vol. 26, 130 - 141 (1972).

Photographic-type magnetograms are used in conjunction with Hα filtergrams to study the structure and evolution of magnetic fields associated with arch filament systems. Time lapse studies show the detailed process by which the flux tubes emerge through the surface.

071.048 Solar activity and the variations of the geomagnetic K_p-index. I: Photospheric activity.
J. T. Mariska, L. Oster.
Solar Physics, Vol. 26, 241 - 249 (1972).

A careful correlation analysis is made between various types of solar activity as observed at photospheric levels and the daily variations of the geomagnetic K_p-index which, in turn, is a measure of the solar wind speed. We find that in no case does a significant enough correlation exist to pin-point a physical relation between some aspect of photospheric activity and the solar wind speed. It is concluded that the physical processes that do determine the wind speed occur at coronal heights.

071.049 The damping of the Na D lines in the solar spectrum by atomic hydrogen.
D. E. Blackwell, J. H. Kirby, G. Smith.
Monthly Notices Roy. Astron. Soc., Vol. 160, 189 - 196 (1972).

The paper gives a brief discussion of the nature of the damping processes in the atmospheres of cooler stars, followed by a calculation of the absorption in the wings of the Na D lines in the solar spectrum using damping constants recently calculated by Lewis, McNamara and Michels. The agreement between theory and observation is discussed critically.

071.050 Structuur en dynamica van de zonnefotosfeer. C. de Jager.
Koninkl. Nederlandse Akad. Wetenschappen, Amsterdam. Afd. Natuurkunde, Vol. 79, 154 - 159 = Astrophys. Inst., Vrije Univ. Brussel, No. 65 (1970).

071.051 Theoretical rotational temperatures of molecules CH, NH, OH, C_2, CN and CO on the solar limb.
A. I. Khlystov.
Soobshch. Gos. Astron. Inst. Shternberga, No. 180, p. 20 - 22 (1972). In Russian.

It is shown that the center-to-limb variation of the rotational temperatures of molecules is equal nearly ±100°.

071.052 Preliminary results of the third flight of the Soviet stratospheric solar observatory.
V. A. Krat, V. N. Karpinsky, V. M. Sobolev.
Space Research XII, (see 012.016), Vol. 2, 1713 - 1717 (1972).

071.053 Isotopes of rubidium in the sun. Ö. Hauge.
Solar Physics, Vol. 26, 263 - 275 (1972).

The Rb I resonance lines at 7800 and 7947 Å in the photospheric spectrum of the sun have profiles which are influenced by the isotopic composition of rubidium. High resolution spectra obtained with the McMath Solar Telescope at Kitt Peak National Observatory have been studied. A solar isotopic composition $Rb_{87}/Rb = 0.27 ± 0.04$ was found using spectra of the Rb I line at 7800 Å obtained with the spectrograph slit in positions close to the solar limb. The other Rb I line was abandoned since it was seriously blended with a water vapour line and some additional faint unknown lines.

071.054 A search for the solar Sr 87 content and the solar Rb/Sr ratio. Ö. Hauge.
Solar Physics, Vol. 26, 276 - 282 (1972).

Some energy levels of Sr 87 show hyperfine splitting which broadens strontium lines in the solar spectrum. By analysis of two faint photospheric Sr I lines of Multiplet No. 3 an upper limit of the relative Sr 87 content (Sr 87/Sr) of $^1/_4$ has been found. The terrestrial value is 0.07−0.075. The solar abundance of strontium found from the two lines is log ϵ_{Sr} = 2.90 in the log ϵ_H = 12.00 scale. Using the solar rubidium abundance recently determined by the author (Hauge, 1972), one obtains $\epsilon_{Rb}/\epsilon_{Sr}$ = 0.5 ± 0.1. This value is larger than found even in chondrites showing high rubidium content.

071.055 Large-scale photospheric magnetic field: The diffusion of active region fields. K. H. Schatten, R. B. Leighton, R. Howard, J. M. Wilcox.
Solar Physics, Vol. 26, 283 - 289 (1972).

In the present investigation we compute the photospheric magnetic field using observed active regions as sources of the field. The random walk mechanism and the shearing effects of differential rotation are applied to the resulting magnetic flux. This computation is applied over ten consecutive solar rotations. The resulting magnetic field patterns are compared with the photospheric field observed at the same time with the Mount Wilson Observatory solar magnetograph.

071.056 A mechanism for the exploding granule phenomenon. S. Musman.
Solar Physics, Vol. 26, 290 - 298 (1972).

I suggest that the exploding granule phenomenon is a consequence of the observed internal granular motions and the conservation of angular momentum. When a granule rising from the convection zone penetrates into the overlying stable region it is stretched out horizontally. Conservation of angular momentum in the internal motions changes its form into a vortex ring. A time sequence of photographs showing an exploding solar granule is described. The proposed mechanism is illustrated by a laboratory simulation and a numerical calculation.

071.057 Time-averaged spectroheliograms. G. A. Chapman. Solar Physics, Vol. 26, 299 - 304 (1972).

The great improvement in signal-to-noise as a result of time-averaging a sequence of λ 6103-core spectroheliograms is shown. It is suggested that such a technique should greatly enhance the network seen on filtergrams made with the 3840 Å violet filter (Chapman, 1970). Finally, the evolution of a sunspot, observed with time-lapse spectroheliograms is discussed.

071.058 A possible new interpretation of power spectra of solar-granulation brightness fluctuations.
Y. Nakagawa, E. R. Priest.
Astrophys. Journ., Vol. 178, 251 - 255 (1972).

The brightness fluctuations of solar granulation are attributed to local temperature fluctuations through the photosphere. It is plausible to consider that the temperature fluctuations result from a passive response of temperature to turbulent fluid convections.

071.059 **Analysis of the extreme-ultraviolet quiet solar spectrum.** A. K. Dupree.
Astrophys. Journ., Vol. 178, 527 - 541 (1972).

The extreme-ultraviolet spectrum ($\lambda 304-1400$ Å) from a region 1' square at the quiet center of the solar disk is analyzed to obtain emission measures of the quiet chromosphere, transition region, and corona, and simultaneously to determine the relative abundances of the elements carbon, nitrogen, oxygen, neon, sodium, magnesium, aluminium, silicon, sulfur, and iron. Some aspects of line excitation and formation are also discussed.

071.060 **Effect of a progressive sound wave on the profiles of spectral lines. II. Asymmetry of faint Fraunhofer lines.** R. I. Kostik.
Astrometriya i Astrofiz., *Kiev*, Vyp. (No.) 17, (see 003.012), p. 50 - 54 (1972). In Russian.

The profile of the absorption coefficient is calculated for lines of different chemical elements in a medium with progressive sound waves. Degree and direction of the resulting asymmetry are discussed.

071.061 **Influence of deviation from local thermodynamic equilibrium on the Goldberg–Unno method.**
V. I. Troyan.
Astrometriya i Astrofiz., *Kiev*, Vyp. (No.) 17, (see 003.012), p. 54 - 59 (1972). In Russian.

071.062 **On the quasi-periodic (wave line) motions in the solar photosphere. I. Preliminary results.**
O. A. Melnikov, R. Kh. Salman-Zade, Y. A. Solonsky, E. D. Khilov.
Astron. Zhurn. Akad. Nauk SSSR, Vol. 49, 1275 - 1279 (1972). In Russian. English translation in Soviet Astron. AJ, Vol. 16, No. 6.

Direct measurements of the solar photosphere line displacements were carried out. Short-period variations of the line positions were detected. The mean period is ~8 min. A comparison of the observations for different instruments and moments of time showed that the phenomenon is probably not strictly periodic, but quasi-periodic or even only cyclic.

071.063 **The spectra of near-vertical structures on the solar disk.** O. R. White.
Solar Physics, Vol. 27, 27 - 33 (1972).

Bright emission arches in the spectra of Hα and the Ca II (H and K lines) are identified as the spectroscopic picture of the chromospheric network as it appears near the solar limb. Analysis of the geometrical properties of these spectroscopic arches indicates that the average network is a diverging sheet with a divergence angle of ~50°. This sheet extends to 2600 km and 2000 km as an opaque emission feature in Hα and the Ca II lines, respectively.

071.064 **Measurements of the solar spectrum between 30 and 128 Å.** J. E. Manson.
Solar Physics, Vol. 27, 107 - 129 (1972).

The results of two rocket flights of grazing incidence monochromators designed to measure solar line intensities in the wavelength region between 30 and 128 Å are compared. One of these flights sampled a very quiet sun, that of November 3, 1965, and has been reported previously. The other acquired data during a more active, but non-flaring, interval on August 8, 1967. The changes in line intensities observed in these two experiments follow a pattern which is in general qualitative agreement with theoretical ionization equilibrium calculations for the solar corona. An analysis of these and other observations of the C VI Lα line suggests that this line is particularly sensitive to local solar activity.

071.065 **Photoelectric study of absorption of the continuum by Fraunhofer lines. V. Monochromatic distribution of the blanketing effect with optical depth.**
P. P. Kozak, A. D. Kulchitsky.
Solnechnye Dannye 1972 Byull., No. 9, p. 73 - 78 (1972). In Russian.

The monochromatic absorption coefficients for different optical depths were calculated with and without blanketing effect. The functions of the absorption distribution, corrected for the lines, were obtained.

071.066 **TV registration of the solar spectrum. II.**
L. D. Parfinenko.
Solnechnye Dannye 1972 Byull., No. 9, p. 84 - 89 (1972). In Russian.

The possibility of using a TV system for solar spectroscopy is considered. The results of preliminary spectral observations and experimental errors are given.

071.067 **Iron in the sun and stars.** R. H. Garstang.
Structure of matter. Rutherford Centennial Symposium, Christchurch 1971, [Univ. Canterbury, Christchurch, New Zealand], p. 338 - 394 (1972).

071.068 **Theoretical explanation of the solar limb effect.**
C. Ferencz. G. Tarcsai.
Acta Techn. Acad. Sci. Hungaricae, Vol. 72, 171 - 181 (1972).

On the basis of the general theory of wave propagation in inhomogeneous moving media it is shown that the extreme red-shift values observed at the solar limb are produced by radial currents in the solar atmosphere due to an effect different from the familiar one which is responsible for the wavelength shifts observed at the inner parts of the solar disk.

071.069 **Solar photospheric abundances of problematical elements by spectrum synthesis.** R. P. Boyle.
Thesis, Georgetown Univ., Washington, D.C. [Available from Univ. Microfilms, Ann Arbor, Mich., USA. Order No. 72-16032], 183 pp. (1972).

New determinations of the solar photospheric abundances of seven heavier elements are presented: chromium, nickel, silver, erbium, ytterbium, lutetium, and thorium. Detailed line-profile calculation was used in matching the synthesized solar spectrum to the observed.

071.070 **Variations of solar granulation with wavelength (from λ 3900 Å to λ 6600 Å).**
V. N. Karpinsky, L. M. Pravdjuk.
Solnechnye Dannye 1972 Byull., No. 10, p. 79 - 92 (1972). In Russian.

Variations of the granulation with wavelength were investigated using the spectrograms taken during the third flight of the Soviet Stratospheric Solar Observatory on July 30, 1970 and direct photographs of the atmosphere, obtained in 1969 at Pulkovo simultaneously in two wavelengths (λ 4650 Å and λ 6000 Å) with the seeing monitor.

071.071 **Method and some results of determining the spatial spectrum of photometric tracings of solar granulation.** A. V. Andreiko, V. N. Karpinsky, L. M. Kotljar.
Solnechnye Dannye 1972 Byull., No. 10, p. 93 - 107 (1972). In Russian.

A method of determining the spatial spectrum of photometric tracings of solar granulation from a direct photograph of the solar photosphere, taken during the third flight of the Soviet Stratospheric Solar Observatory is described. The fast Fourier transform algorithm is used in data processing.

071.072 **The quiet sun emission at mm wavelengths.**
R. Barletti, P. Pampaloni.
Mem. Soc. Astron. Italiana, Nuova Ser., Vol. 43, 547 - 566

(1972).

The quiet sun emission in mm-λ region is described. The brightness temperature spectrum at the center of the disk and the centre-limb distribution are analysed. Both the most important observational results and interpreting models are emphasized. Possible observational efforts to improve our knowledge in the field are pointed out.

071.073 **The limb brightening of the sun at 8 mm.**
G. F. Eliseev.
Izv. Krymskoj Astrofiz. Obs., Vol. 45, 49 - 52 (1972).
In Russian.

The distribution of the radio brightness across the solar disk at 8 mm is considered. The ratio of the disk-averaged brightness temperature to the central brightness temperature at 8 mm is calculated to be 1.16 ± 0.03

071.074 **Isotopic composition of some metals in the sun.**
Ø. Hauge.
Inst. Theor. Astrophys. Blindern – Oslo, Rep. No. 35, 3 + 73 pp. (1972).

A discussion of the elements Rb, Sr, Sb and Eu are described. Studies of the solar isotopic composition of Cu from analysis of Cu I lines in photospheric spectra and CuH lines in sunspot spectra are described.

071.075 **Correction of solar intensity measurements for stray light.**   L. Staveland.
Inst. Theor. Astrophys. Blindern – Oslo, Rep. No. 36, 2 + 32 pp. (1972).

A computer program is described which corrects the observed intensity in the centre of an elliptical sunspot for stray light. The program also corrects the observed intensity profile of the solar limb. The analytical expression used for the relative solar limb darkening gives three to four correct digits for the Harvard–Smithsonian Reference Atmosphere model in the wavelength region 0.3 μm to 5.0 μm for $\cos \theta$ = 0.1 to $\cos \theta$ = 0.3.

071.076 **A comment on contribution functions.**
H. Sørli, O. Engvold.
Inst. Theor. Astrophys. Blindern – Oslo, Rep. No. 37, 8 pp. (1972).

The contribution function for the emergent radiation in spectral lines is the integrand of the equation of transfer $(\Delta I_\lambda (\tau))$. Numerical examples are used to compare this and various types of contribution functions that refer to the line contrast $(\Delta r_\lambda (\tau))$ (Gussmann, 1967, and Elste 1969).

071.077 **Further observations of the solar limb spectrum in the region 550–2000 Å.**
A. Ridgeley, W. M. Burton.
Solar Physics, Vol. 27, 280 - 285 (1972).

Further observations of the ultraviolet spectrum (550–2000 Å) of the solar limb and disc were obtained during a Skylark rocket flight on 5 August 1971. These observations have enabled several new spectral lines to be identified and classified.

071.078 **Solar isotopic composition and abundance of europium.**   Ø. Hauge.
Solar Physics, Vol. 27, 286 - 293 (1972).

High resolution spectra of six photospheric Eu II lines have been studied using the method of spectrum synthesizing. The isotope ratio is found to be Eu_{153}/Eu_{151} = (48 ± 6)/ (52 ∓ 6) and the solar abundance of europium equals $\log \epsilon_{Eu}$ = 0.7 ± 0.2 in the $\log \epsilon_H$ = 12.00 scale.

071.079 **The solar manganese abundance.**
T. E. Margrave, Jr.
Solar Physics, Vol. 27, 294 - 298 (1972).

A preliminary solar Mn abundance of $\log N$ (Mn) = 5.41 ($\log N$(H) = 12.00) is derived on the basis of fitting theoretical line profiles which include hyperfine structure (HFS) broadening to the profiles of the $\lambda\lambda$ 5394.7, 5432.6, and 5537.8 lines of Mn observed at the center of the solar disk with the double-pass spectrograph of the McMath solar telescope at Kitt Peak.

071.080 **Observed oddities in the lines H, K, b and Hβ.**
J. W. Evans, C. P. Catalano.
Solar Physics, Vol. 27, 299 - 302 (1972).

We compare microphotometer intensity traces perpendicular to dispersion in simultaneous spectrograms of good spatial resolution traced at various Δλ's in each of the lines. We have determined the coefficient of correlation, r, between each averaged tracing and all other averaged tracings from the same spectrogram.

071.081 **Suggested interpretation of the correlations in intensity fluctuations in the lines Ca II H and K, magnesium b, and hydrogen Hβ.**   R. N. Thomas.
Solar Physics, Vol. 27, 303 - 304 (1972). – Research note.

Radiative lifetimes for some resonance transitions of Fe I and Fe II in the region between 2300 Å and 3050 Å, and the application to iron abundance determinations in the sun and in the QSO PHL 938.   See Abstr. 022.009.

Ionization balance for ions of Na, Al, P, Cl, A, K, Ca, Cr, Mn, Fe and Ni.   See Abstr. 022.062.

Measurement of oscillator strengths in the singlet system of neutral magnesium.   See Abstr. 022.064.

Ionization equilibrium for ions of Na, Al, P, Cl, A, K, Ca, Cr and Mn.   See Abstr. 022.081.

The classification of Fe IX to XVI emission lines and isoelectronic lines in laboratory and solar spectra.   See Abstr. 022.084.

Further identifications in the Ar IX spectrum.   See Abstr. 022.125.

Hartree-Fock calculations of the Mg I spectrum in the extreme ultraviolet.   See Abstr. 022.151.

A heliostat for measuring the solar flux spectrum.   See Abstr. 034.111.

A subtractive double pass spectrograph for solar observations.   See Abstr. 034.160.

On the calculation of the $H(a, v)$-function.   See Abstr. 063.016.

Solar bright points in 3840 Å and Hα.   See Abstr. 072.037.

On the intensity ratio between sunspot umbrae and the photosphere in the 4000 - 8000 Å spectral region.   See Abstr. 072.067.

A method to calculate electric currents in quiescent prominences.   See Abstr. 073.010.

Flares and changing magnetic fields.   See Abstr. 073.020.

Velocity oscillations in solar plage regions.   See Abstr. 073.059.

Reply to 'The relations between chromospheric features and photospheric magnetic fields' by E. N. Frazier [Solar Physics, Vol. 24, 98 - 112 (1972)].   See Abstr. 073.082.

The solar active region from August 17–30, 1971 and associated events.   See Abstr. 073.102.

Missing solar ultraviolet opacity and diatomic molecules.   See Abstr. 076.027.

Dielectronic satellite spectra in the soft X-ray region.   See Abstr. 076.028.

Identifications of some highly-ionized iron and nickel lines in the 200–400 Å region of the solar spectrum. See Abstr. 076.029.

On the interpretation of the relative intensities of the solar XUV lines of lithium-like ions. See Abstr. 076.031.

Recent high resolution X-ray spectra of the sun. See Abstr. 076.032.

High angular resolution absolute intensity of the solar continuum from 1400 Å to 1790 Å. See Abstr. 076.036.

On the mean depth of line formation in a magnetic field.   See Abstr. 080.002.

Solar rotation: The photospheric height gradient. See Abstr. 080.024.

The five-minute oscillations as nonradial pulsations of the entire sun.   See Abstr. 080.036.

Identification of stratospheric NH. See Abstr. 082.211.

On the abundances of noble gases in extreme population I matter and the sun.   See Abstr. 114.038.

072 Sunspots, Faculae, Solar Activity

072.001 Graphical method of studying the distribution of the macrostructure of solar activity. P. Ambrož.
Bull. Astron. Inst. Czechoslovakia, Vol. 23, 232 - 237, 244a - 244d (1972).

The paper describes a photographic method which can be used for processing two-dimensional images in which it is necessary to suppress minor and sporadically occurring regions but, on the contrary, to stress extensive formations. A method of graphical surface integration is described, which makes use of optical transformation in processing the original image by an unfocussed camera. The method of processing the transformed image by means of the photographic isodensitometry is illustrated. The method is applied to the study of the macrostructural density of the distribution of Ca II flocculae on the solar surface.

072.002 The identification of the 1−0 and 2−1 bands of HCl in the infrared sunspot spectrum.
D. N. B. Hall, R. W. Noyes.
Astrophys. Journ., (*Letters*), Vol. 175, L95 - L97 (1972).

Observations of the infrared umbral spectrum between 2400 and 3000 cm^{-1} have permitted positive identification of 14 lines of the fundamental vibration-rotation bands of HCl. A preliminary solar ^{35}Cl abundance of $\log_{10}N(^{35}\text{Cl}) = 5.4 \pm 0.3$ [on a scale where $\log_{10}N(\text{H}) = 12$] has been obtained. Several weak features are consistent with the presence of ^{37}Cl with a terrestrial abundance ratio.

072.003 Eu, La and Sm in sunspot spectra. H. Molnar.
Astron. Astrophys., Vol. 20, 69 - 72 (1972).

Abundances of Eu, La and Sm were derived using high resolution sunspot spectra.

072.004 Photospheric faculae and the solar oblateness.
G. A. Chapman, A. P. Ingersoll.
Astrophys. Journ., Vol. 175, 819 - 829 (1972).

Photospheric faculae near the equatorial solar limb may provide the excess brightness which Ingersoll and Spiegel showed would explain Dicke and Goldenberg's oblateness measurement. Three lines of evidence support this statement.

072.005 Faculae and the solar oblateness. R. H. Dicke.
Astrophys. Journ., Vol. 175, 831 - 835 (1972).

Chapman and Ingersoll have suggested that the excess solar oblateness found by Dicke and Goldenberg is wholly or largely due to the presence of faculae near the solar limb. This contrasts with my earlier statement based on a statistical study that contributions from faculae are unimportant. A new statistical study based on Chapman and Ingersoll's own facular function supports my 1970 statement.

072.006 On the temperature distribution in an inhomogeneity with radiative equilibrium.
V. M. Dashevsky, V. N. Obridko.
Astron. Zhurn. Akad. Nauk SSSR, Vol. 49, 796 - 801 (1972). In Russian. English translation in Soviet Astron. AJ, Vol. 16, No. 4.

The problem of radiative equilibrium in a cylinder surrounded by a medium with any given temperature distribution is considered. Taking the absorption coefficient to depend linearly on the optical depth an analytical solution is obtained. The solution is applied to the problem of the temperature distribution in a photospheric tube inside a sunspot umbra fed from downside.

072.007 On the interpretation of the π-component splitting in sunspot spectra. V. N. Obridko, L. B. Demkina.

Solar Physics, Vol. 24, 336 - 341 (1972).

It is shown that in order to explain the observed splitting of the π-component in the sunspot umbra spectrum by the hypothesis of the coexistence in sunspots of weak- and strong-field regions with opposite polarities, one has to admit the additional assumption that in the weak-field regions the Doppler halfwidth and the ratio between line opacity and continuum opacity are both less than those in the strong-field regions.

072.008 On C$_2$ lines in sunspot spectra. H. Wöhl.
Solar Physics, Vol. 24, 342 - 353 (1972).

The questionable existence of C$_2$ absorption lines in spectra of sunspots was checked: In two umbral spectra of large single sunspots evidence for the presence of lines of the $(0,0)$ band and of the $(0,1)$ band was found.

072.009 C$_2$ in sunspots. J. W. Harvey.
Solar Physics, Vol. 24, 354 - 355 (1972).

This note presents results which show that C$_2$ lines are strengthened in the penumbra relative to the photosphere and are much weaker in the umbra in agreement with the predictions of molecular equilibrium calculations.

072.010 Observations of the horizontal velocity field surrounding sunspots. N. R. Sheeley, Jr.
Solar Physics, Vol. 25, 98 - 103 (1972).

During the summer and fall of 1971, Doppler spectroheliograms were obtained for several sunspots located near the solar limb. These observations confirm a previous result based on the study of only a few sunspots that in the plage-free photosphere surrounding sunspots the spatially-averaged, horizontal flow tends to be outward at 0.5−1.0 km s^{-1} for distances typically 10000−20000 km beyond the outer boundary of the penumbra. It is suggested that these material motions are the means by which small-scale fragments of magnetic flux are carried away from sunspots.

072.011 On practical representation of magnetic field.
Y. Nakagawa, M. A. Raadu.
Solar Physics, Vol. 25, 127 - 135 (1972).

Various manners of determination of a magnetic field are reviewed briefly from the standpoint of practicality and uniqueness. Then a practical representation of magnetic fields in terms of a class of force-free magnetic field is described. The applicability of the representation is demonstrated by examples and the limitations are discussed.

072.012 The vertical distribution of magnetic field strength in a bright region of sunspot penumbra. I.
M. J. Guseynov.
Izv. Krymskoj Astrofiz. Obs., Vol. 44, 70 - 76 (1972). In Russian.

The absolute magnetic field strength is derived from the study of 55 Fraunhofer lines originating at different photospheric depths. All spectral lines used for magnetic splitting measurements, except 5 lines, have normal Zeeman splitting. A change of magnetic field strength with depth is considered.

072.013 Determination of physical parameters of a sunspot.
E. A. Baranovsky.
Izv. Krymskoj Astrofiz. Obs., Vol. 44, 87 - 93 (1972). In Russian.

On four echelle grating spectra of two sunspots ($\lambda\lambda 5600-6300$ Å) the equivalent widths of 44 lines of Sc I, V I, Ti I, Cr I, Ca I, Na I, Fe I, Si I, Ni I, Sc II, Fe II, the intensity of the wings of 6 lines of Ca I and Fe I and the intensity of the continuous spectrum are measured and compared with the corresponding figures for the undisturbed photosphere. The temperature, the electron pressure and the gas pressure for sunspot umbrae are derived from the examination of the measured values.

072.014 **Observations of the helium lines D$_3$ and λ10830 Å in active regions.**   Nguen-Ngan.
Izv. Krymskoj Astrofiz. Obs., Vol. 44, 107 - 111 (1972).
In Russian.

Helium lines D$_3$ and λ10830 Å are considered on the spectrograms obtained with the horizontal solar telescope of the Crimean Astrophysical Observatory. The excitation temperature of the transition 2^3S-2^3P is determined on the basis of the equivalent widths of the D$_3$ line and of the weak component of λ10829 Å. Doppler widths determined for λ10829 and λ10830 lines are in good agreement with the assumption of a constant source function. It is shown that the turbulent velocity in helium regions is small.

072.015 **Crossover effect in the spectrum of sunspots.**
V. A. Golubjev, V. F. Tshistjakov.
Astron. Tsirk., No. 670, p. 1 - 3 (1972). In Russian.

072.016 **The region of darkening around solar faculae.**
I. F. Nikulin.
Astron. Tsirk., No. 670, p. 3 - 5 (1972). In Russian.

072.017 **On the geometry of facular granules.**
E. V. Kononovich.
Astron. Tsirk., No. 670, p. 5 - 7 (1972).

072.018 **On the Doppler displacements of magnetoactive lines in sunspots.**   A. V. Baranov.
Astron. Tsirk., No. 677, p. 1 - 3 (1972). In Russian.

072.019 **Theoretical interpretation of the Evershed flow on the basis of a rope model of a sunspot penumbra. I.**
A. A. Solovjev.
Solnechnye Dannye 1972 Byull., No. 4, p. 65 - 70 (1972).
In Russian.

Gas motions in a sunpot penumbra are investigated. Dark penumbra filaments are supposed to coincide with thin insulated magnetic ropes. This assumption permits to calculate the forces acting along the dark filaments. It is shown that Evershed motions can be explained both in terms of steady and unsteady flows. The velocities of unsteady gas motions are found to be close to sound velocity in the photosphere.

072.020 **On the behaviour of the magnetic field and radial velocities in a sunspot during its fragmentation.**
A. V. Baranov, G. F. Vjalshin, E. P. Surkov.
Solnechnye Dannye 1972 Byull., No. 4, p. 91 - 97 (1972).
In Russian.

The character of variations of the magnetic field intensity and radial velocity in the umbra of a breaking sunspot has been studied using spectrograms of a sunspot taken with a circular polarization analyser.

072.021 **On similarity of the 11-year solar cycles.**
Y. I. Vitinsky.
Solnechnye Dannye 1972 Byull., No. 5, p. 84 - 90 (1972).
In Russian.

The problem of possible detection of cycle-analogs according to some main parameters of the 11-year solar cycles is discussed. The method of analogs may appear not quite effective in prognoses of Wolf numbers for the next 11-year solar cycle.

072.022 **On the wings of the σ-components of magnetically sensitive lines in sunspots.**
L. B. Demkina, V. N. Obridko.

Solnechnye Dannye 1972 Byull., No. 5, p. 101 - 103 (1972).
In Russian.

The origin of the high intensity wings of the σ-components of magnetically sensitive lines in sunspots is discussed. This effect may be connected with the fine structure of the velocity or magnetic field.

072.023 **Theoretical interpretation of the Evershed flow on the basis of a rope model of a sunspot penumbra. II.**
A. A. Solovjev.
Solnechnye Dannye 1972 Byull., No. 5, p. 104 - 107 (1972).
In Russian.

The well known Evershed motions have been calculated as a steady gas flow in the dark penumbral filaments. According to the model suggested previously these dark filaments are identical with thin magnetic ropes. The data obtained are found to be in good agreement with those usually observed.

072.024 **Force-free magnetic-field structures and their role in solar activity.**   C. W. Barnes, P. A. Sturrock.
Bull. American Astron. Soc., Vol. 4, 377 (1972). – Abstr. AAS.

072.025 **Oscillatory phenomena in sunspots.**
J. M. Beckers, R. B. Schultz.
Bull. American Astron. Soc., Vol. 4, 377 (1972). – Abstr. AAS.

072.026 **Observations of sunspot umbral velocity oscillations.**
A. Bhatnagar, W. C. Livingston, J. W. Harvey.
Bull. American Astron. Soc., Vol. 4, 378 (1972). – Abstr. AAS.

072.027 **Photospheric faculae and the solar oblateness.**
G. A. Chapman, A. P. Ingersoll.
Bull. American Astron. Soc., Vol. 4, 379 (1972). – Abstr. AAS.

072.028 **Calculation of 5250.216 Å line profiles in sunspots.**
A. R. Dunn.
Bull. American Astron. Soc., Vol. 4, 381 (1972). – Abstr. AAS.

072.029 **Non-linear study of the dynamical behavior of a force-free magnetic field.**
M. J. Hagyard, Y. Nakagawa, S. T. Wu.
Bull. American Astron. Soc., Vol. 4, 383 (1972). – Abstr. AAS.

072.030 **An umbral model atmosphere derived from infrared observations.**   D. N. B. Hall, R. W. Noyes.
Bull. American Astron. Soc., Vol. 4, 383 (1972). – Abstr. AAS.

072.031 **Observations of moving magnetic features near sunspots.**   K. Harvey, J. Harvey.
Bull. American Astron. Soc., Vol. 4, 384 (1972). – Abstr. AAS.

072.032 **A new synoptic chart of solar activity.**
P. S. McIntosh.
Bull. American Astron. Soc., Vol. 4, 387 (1972). – Abstr. AAS.

072.033 **The dynamics of magnetic flux in a young active region.**   S. A. Schoolman.
Bull. American Astron. Soc., Vol. 4, 390 - 391 (1972).
Abstr. AAS.

072.034 **Bright running penumbral waves.**
A. Stein, H. Zirin.
Bull. American Astron. Soc., Vol. 4, 392 (1972). – Abstr. AAS.

072.035 **Methods for measurement of high fields.**
A. M. Title.
Bull. American Astron. Soc., Vol. 4, 394 (1972). – Abstr. AAS.

072.036 **Some effects of molecules in solar phenomena.**
E. B. Weston, G. W. Wares.
Bull. American Astron. Soc., Vol. 4, 395 (1972). – Abstr. AAS.

072.037 **Solar bright points in 3840 Å and Hα.**
J. Vorpahl, T. Pope.
Solar Physics, Vol. 25, 347 - 356 (1972).

Analysis of bright features in 3840 Å and Hα shows that for every Ellerman bomb (Hα−0.9 Å) there is a cospatial brightening in the 3840 Å network. We give properties of these bright points in both wavelengths as well as describe: (1) the appearance, and subsequent separation, of new elements in the 3840 Å network and (2) the direct transition from a 3840 Å bright point to a new sunspot.

072.038 **Evidence for an ultra-long cycle of solar activity.**
R. Henkel.
Solar Physics, Vol. 25, 498 - 499 (1972). − Research Note.

072.039 **Solar cycle forecast.** R. G. Zaikov.
Comptes Rendus Acad. Bulgar. Sci., (Dokl. Bolg. Akad. Nauk), Vol. 24, 1147 - 1149 (1971).

072.040 **Sunspots and planets.** K. D. Wood.
Nature, Vol. 240, 91 - 93 (1972).

Bigg has shown that the period of Mercury's orbit appears in the sunspot data, and that the influence of Mercury depends on the phases of Venus, Earth, and Jupiter. These four are the "tidal planets". Here the relationships between planetary tides on the sun and the number of sunspots are discussed. The effects attributed to Mercury have been neglected because of their very short period of about 3 months compared with the period of the sunspot cycle, which averages about 11.1 yr.

072.041 **Large sunspot groups.** R. S. Gnevysheva.
Solnechnye Dannye 1972 Byull., No. 7, p. 76 - 109 (1972). In Russian.

Catalogues of sunspot groups with a mean area of more than 500 and 1500 millionths of the visible solar hemisphere for the period from 1955 to 1969 are given; these catalogues serve as a continuation to the analogous Greenwich catalogues.

072.042 **On the correction for scattered light to sunspot observations.** F. G. Rozhavsky.
Solnechnye Dannye 1972 Byull., No. 7, p. 110 - 114 (1972). In Russian.

A simple method of the sunspot intensity correction for scattered light is described. Observations of aureole intensity at two points outside the solar limb are used in this method. The correlation of scattered light with atmospheric conditions appears to be weak. It indicates that scattered light is of instrumental origin.

072.043 **The magnetic strengthening of spectral lines in the spot observed on June 19, 1959.**
E. N. Zemanek, A. P. Stefanov.
Vestn. Kiev, Un-ta, Ser. Astron., No. 14, p. 23 - 27 (1972). In Russian.

The magnetic strengthening of 107 Fe I lines were obtained. The curve of growth was constructed by the equivalent widths corrected for magnetic strengthening. The physical parameters for the sunspot were derived.

072.044 **The structure of sunspots. I: Observational constraints; current sheet models.**
M. H. Gokhale, C. Zwaan.
Solar Physics, Vol. 26, 52 - 75 (1972).

In this paper we list observational features which seem both important and ascertained enough to be incorporated in, or predicted by, a satisfactory sunspot model. We then discuss the constraints which follow from the main observational features; these constraints are to be combined with theoretical constraints or assumptions to yield a model. We discuss some properties of such a model.

072.045 **Observations of the intensity of the penumbra of sunspots.** P. Maltby.
Solar Physics, Vol. 26, 76 - 82 (1972).

Observations of the penumbral intensity of sunspots in 13 wavelength regions are presented. In 4 wavelength regions 54 sunspots are measured. In the other wavelength regions the number of sunspots considered ranges from 3 to 19. The penumbra model of Kjeldseth Moe and Maltby (1969) with $\Delta\theta = 0.055$ is supported by the measurements.

072.046 **Calculation of 5250.216 Å line profiles in sunspots.**
A. R. Dunn.
Solar Physics, Vol. 26, 83 - 86 (1972).

The assumptions of pure absorption and local thermodynamic equilibrium are sometimes used to calculate approximate spectral line profiles in cases where a rigorous treatment is impractical or impossible. In certain conditions, the profile is not completely defined under these assumptions.

072.047 **On the minimum intensity of the Na D_2-5890 Å line in sunspot umbra.**
T. Fay, J. Remo, K. Czaja.
Solar Physics, Vol. 26, 87 - 89 = Publ. Goethe Link Obs., *Bloomington,* No. 139 (1972).

The line-to-continuum intensity ratio for the center of the Na D_2 absorption line in a spot of area 20×10^{-6} in the Rome Group 5847 was found to be 0.033 ± 0.05 on 6−7/7/70 by two methods.

072.048 **Water vapour in sunspots.** L. Staveland.
Solar Physics, Vol. 26, 90 - 93 (1972).

The line intensities are calculated at temperatures of 263 K and 3500 K for the H_2O band 201 at 0.94 μm. The possibility of detecting these lines in sunspots is discussed.

072.049 **Thallium in the solar atmosphere.**
D. L. Lambert, E. A. Mallia, G. Smith.
Solar Physics, Vol. 26, 250 - 256 (1972).

Umbral spectra are shown to contain an absorption feature attributable to the Tl I transition $6p\ ^2P^\bullet_{3/2}-7s^2S_{1/2}$ at 5350 Å. Analysis of the umbral spectrum suggests a solar abundance in the range of $0.72 < \log N(\text{Tl}) < 1.10$ on the standard scale.

072.050 **The motion of planets and solar activity.**
G. J. Vasilyeva, D. A. Kuznetsov, N. S. Petrova, A. A. Shpitalnaya.
Solnechnye Dannye 1972 Byull., No.8, p. 106 - 115 (1972). In Russian.

A preliminary attempt is made to compare the Wolf number range for solar cycles 18, 19, 20 and the electricity generated in a closed contour (sun−interplanetary space−planet) by the motion of the planets in the galactic magnetic field taking into account the motion of the solar system to the apex.

072.051 **On the sunspot structure.**
V. A. Krat, V. N. Karpinsky, L. M. Pravdjuk.
Solar Physics, Vol. 26, 305 - 317 (1972).

The fine structure of a sunspot is studied on a series of photographs obtained during the third flight of the Soviet Stratospheric Solar Station. The main results are as follows: (1) The micro-photometer tracings on the frames show extremely high Rayleigh resolution of small elements, the smallest distances being near to the theoretical limit. (2) The dimensions of the smallest dots are equal to the diffraction image of bright points. (3) The penumbra and umbra structure (dark and bright objects) is in good agreement with the picture of magnetic field splitting in a system of magnetic ropes giving rise to the magnetic arcs in the chromosphere and corona. Only in the umbra do we meet the large scale continuities.

072.052 **Correction of solar observations for stray light by numerical integration, with application to Mercury's drop.** R. Brahde.
Solar Physics, Vol. 26, 318 - 334 (1972).

A numerical method for correction of stray light in solar observations has been developed. In particular a regular sunspot, where the circular contours of penumbra and umbra are projected as ellipses, has been studied. By means of limb observations the stray light parameters may be improved, and finally a variation of the penumbra- and umbra intensities in the computation, enables a determination of these quantities by comparison with observations. The method is tested on observations of the transit of Mercury, May 9, 1970. Calculation of isophotes with Mercury close to the limb shows the black drop phenomenon; which thus may be explained as an effect of stray light only.

072.053 **Observations of running penumbral waves.** H. Zirin, A. Stein.
Astrophys. Journ., (*Letters*), Vol. 178, L85 - L87 (1972).

Quiet sunspots with well-developed penumbrae show running intensity waves with period running around 300 seconds. The waves appear connected with umbral flashes of exactly half the period. Waves are concentric, regular, with velocity constant around 10 km s^{-1}. They are probably sound waves and show intensity fluctuation in Hα centerline or wing of 10–20 percent. The energy is tiny compared to the heat deficit of the umbra.

072.054 **Oscillatory motions in sunspots.** J. M. Beckers, R. B. Schultz.
Solar Physics, Vol. 27, 61 - 70 (1972).

We observe vertical velocity oscillations in some sunspot umbrae with periods of about 180 s and peak to peak amplitudes up to 1 km s^{-1}. These oscillations are not visible in either the line depth, line width or the continuum intensity. In the spot penumbra there is an indication of a long period oscillation, the period increasing from about 300 s in the inner penumbra to nearly 1000 s at the penumbra-photosphere boundary. An attempt has been made to interpret these oscillations.

072.055 **Oscillations and waves in a sunspot.** R. G. Giovanelli.
Solar Physics, Vol. 27, 71 - 79 (1972).

Observations have been made in Hα of the vertical velocity distribution in a sunspot. Over the umbra the pattern consists of structures of scale-size 2–3''. The velocity distribution undergoes oscillations with a period of about 165 s and typical amplitude ±3 km s^{-1}. Transverse waves develop in the outer 0.1 of the umbral radius and propagate outwards with a velocity of about 20 km s^{-1}, becoming gradually invisible by or before the outer penumbral boundary; the amplitude is about ±1 km s^{-1} at the umbra-penumbra border. The penumbral waves are believed to be basically of the Alfvén type. The umbral oscillations presumably represent gravity waves.

072.056 **Observations of sunspot umbral velocity oscillations.** A. Bhatnagar, W. C. Livingston, J. W. Harvey.
Solar Physics, Vol. 27, 80 - 88 (1972).

Sunspot umbral molecular lines have been used to look for the oscillatory velocities in the umbra. Power spectrum analysis showed conspicuous power for periods in the range between 448 and 310 s. The maximum peak-to-peak amplitude of the umbral oscillatory velocity component is observed to be in the order of 0.5 km s^{-1}.

072.057 **Periodicities in the longitude distribution of sunspots.** W. Stanek.
Solar Physics, Vol. 27, 89 - 106 (1972).

The following investigation – based on observational data of rotations No. 1457–1568 (1962–1970) shows four main results: Northern and southern hemisphere behave independently; maximum spot occurrence is found in intervals of about 90° and 180°; second-order peaks can be found in intervals of 30° and multiples of it; areas of minimal spot occurrence can be traced over a long period of time. The importance of these 'streets of activity' is shown.

072.058 **Formation of B.M.R., appearance of sunspots, and Spörer's law.** M. H. Ibañez S.
Publ. Obs. Astron. Nacional, Univ. Nacional Colombia, Bogotá, No. 5, 19 pp. (1971). In Spanish.

On the basis of the dynamo model (Babcock) and the theoretical work of Kopecký, this paper attempts to explain certain solar phenomena, such as the formation of bipolar magnetic regions (B.M.R.), the appearance of sunspots, and Spörer's law from the general bipolar magnetic field whose intensity is of the order of 1 gauss.

072.059 **Keeping up with sunspots.** B. N. Parker.
Weather, (*GB*), Vol. 27, 247 - 251 (1972).

072.060 **Stability of a vertical magnetic rope.** A. A. Solovjev.
Solnechnye Dannye 1972 Byull., No. 9, p. 78 - 83 (1972). In Russian.

The stability of a vertical magnetic flux rope in connection with sunspots and faculae formation is considered. The condition of instability has been obtained on the basis of a variational principle. It is shown that the process of sunspots and faculae formation can be explained in terms of such an instability.

072.061 **A comment to the paper "Theoretical interpretation of the Evershed flow on the basis of a rope model of a sunspot penumbra. II" by A. A. Solovjev.** A. A. Solovjev.
Solnechnye Dannye 1972 Byull., No. 9, p. 114 (1972). In Russian.

072.062 **Dependence of some noise storms characteristics on the solar activity cycle.** O. S. Korolëv.
Sb. dokl. Sessii Nauch. Soveta po kompleks. probl. "Radioastronomiya", IZMIRAN, 1970. Moskva, 1972, p. 83 - 107. In Russian. – Abstr. in Referativ. Zhurn. 51. Astron., 1.51.442 (1973).

072.063 **On the correlation of solar activity with tidal forces.** A. Z. Dolginov, A. D. Kaminker, Yu. A. Shibanov.
Astron. vestn., Vol. 6, 195 - 199 (1972). In Russian. – Abstr. in Referativ. Zhurn. 51. Astron., 1.51.448 (1973).

072.064 **On the east-west asymmetry of white-light faculae.** D. H. Bruning.
Publ. Astron. Soc. Pacific, Vol. 84, 856 - 857 (1972).

We suggest that this east-west asymmetry is caused by a westward inclination of the facular surface normal to the solar radius. Appearing to be analogous with the inclination of sunspots as determined by Minnaert, the faculae inclination is approximately 0°.4.

072.065 **Solar activity prediction.** R. J. Slutz, T. B. Gray, M. L. West, F. G. Stewart, M. Leftin.
Rep. NASA-CR-1939, US Dep. Commerce, Boulder, Colorado. [Available from NTIS, Springfield, Va.], 115 pp. (1971). See Phys. Abstr., Vol. 75, No. 73141 (1972).

072.066 **On the correction of observed penumbral intensities for scattered light.** P. Maltby, L. Staveland.
Inst. Theor. Astrophys. Blindern – Oslo, Rep. No. 33, 13 pp.

(1972).

A simple method for correcting the observed penumbral intensities for light scattered in the atmosphere and the instrument is described. The correction is found to be proportional to the relative intensity at a fixed position outside the solar limb. The proportionality factor depends on the observing wavelength as well as the intensity gradient of the aureole.

072.067　**On the intensity ratio between sunspot umbrae and the photosphere in the 4000 - 8000 Å spectral region.**　F. G. Rozhavskij.
Astron. Tsirk., No. 726, p. 6 - 8 (1972). In Russian.

072.068　**New temperature models of sunspots.**
F. G. Rozhavskij.
Astron. Tsirk., No. 736, p. 1 - 3 (1972). In Russian.

072.069　**Determination of the dissipation function from sunspot observations.**
F. G. Rozhavskij, L. N. Reshetnyak.
Astron. Tsirk., No. 736, p. 3 - 5 (1972). In Russian.

072.070　**The cooling of a sunspot. I. A Carnot cycle and the hydromagnetic interactions.**　P. R. Wilson.
Solar Physics, Vol. 27, 354 - 362 (1972).

A mechanism is proposed to explain the cooling of a sunspot in terms of the detailed interactions between the magnetic field and the convective motions. The mechanism provides that an axially symmetric concentration of magnetic field deforms the normal supergranule cell pattern below the sunspot into a radial outflow of plasma over a region of diameter ~ 60 Mm. The flow occurs at depths where the magnetic and kinetic energy densities are approximately equal (≈ 5 Mm) and is described in terms of a Carnot refrigeration cycle. Observations of the outward drift of magnetic knots around sunspots and of supergranule-type surface motions extending radially outwards from the penumbra of a spot to the nearest faculae are discussed in relation to the mechanism.

072.071　**The cooling of a sunspot. II. Convection zone models and the magnetic power supply.**　P. R. Wilson.
Solar Physics, Vol. 27, 363 - 372 (1972).

In order to discuss the detailed interactions between the magnetic and velocity fields below a sunspot, several models of the convection zone are considered. The possibility that the surface supergranule motions are due to a counter-cell lying above a thermally driven supergranule cell are discussed and this concept is included in one of the models which may be typical of quiet regions of the convection zone. The magnetic power required by this cycle may be supplied by the upward drift of flux ropes expelled from and amplified by these elongated convection eddies.

Summertime observations of sunspots and auroras.
Sky Telescope, Vol. 44, 333 - 339 (1972).

The spectrum of FeH: Laboratory and solar identification.　See Abstr. 022.085.

High resolution solar infrared observations.
See ABstr. 031.046.

The systematic errors of the total vector H of magnetic field measurements with the Crimean Astrophysical Observatory magnetograph.　See Abstr. 034.007.

Non-linear dynamo waves.　See Abstr. 061.006.

Interferometric observations of small solar continuum features.　See Abstr. 071.026.

The photospheric velocity field in and around sunspots.　See Abstr. 071.032.

Time-averaged spectroheliograms.
See Abstr. 071.057.

Isotopic composition of some metals in the sun.
See Abstr. 071.074.

Correction of solar intensity measurements for stray light.　See Abstr. 071.075.

Photoelectric Ca II line profiles in solar plages and a sunspot and their preliminary interpretation.
See Abstr. 073.045.

What makes active regions grow?
See Abstr. 073.064.

Evidence for two maxima of activity in the 20th solar cycle.　See Abstr. 074.055.

Solar cycle variation and N–S asymmetry of $\lambda 5303$ coronal intensity.　See Abstr. 074.080.

Solar X-ray source unassociated with sunspots.
See Abstr. 076.030.

The polarization characteristics of two local sources at $\lambda = 9.0$ cm.　See Abstr. 077.033.

A search of a connection between the polarization of decam-type III bursts and magnetic fields in different heights of the solar atmosphere.　See Abstr. 077.037.

Thermal conductivity in solar magnetoplasmas.
See Abstr. 080.046.

Solar activity and the rotation of Jupiter.
See Abstr. 099.064.

073 Solar Chromosphere, Flares, Prominences

073.001 Periodic heating mechanism in solar flares.
W. M. Glencross, I. J. D. Craig.
Nature, Phys. Sci., Vol. 238, 50 - 52 (1972).
Observations of periodic X-ray flares are described. The flare origin is suggested to be energy transfer from oscillating layers low in the chromosphere.

073.002 A study of D_3 emission in a solar flare by use of narrow-band filtergrams.
R. W. Milkey, K. L. Harvey.
Publ. Astron. Soc. Pacific, Vol. 84, 400 - 405 (1972).
We report observations of helium D_3 emission in the solar flare of 11 February 1970 made at the Lockheed Solar Observatory. The morphological relationship between the D_3 and Hα emission is explored, and a photometric reduction technique is applied to the filtergrams to determine peak intensity of the D_3 emission relative to the local quiet sun continuum.

073.003 The solar chromosphere and its transition to the corona. H. Frisch.
Space Science Rev., Vol. 13, 455 - 483 (1972).
Our present knowledge on the average physical properties of the chromosphere and of the transition region between chromosphere and corona is reviewed. The chromosphere and the transition region are studied separately: for each region, the energy balance is considered and recent homogeneous models derived from ultra-violet, infrared and radio observations are discussed. Observational and theoretical evidence is given for the non-validity of the assumption of hydrostatic equilibrium which is commonly used in modeling the transition region. We conclude that a better understanding of the heating mechanism will come through a higher spatial resolution (less than 0.2″) and more accurate absolute measurements.

073.004 Composition and energy spectra of heavy nuclei with $0.5 < E < 40$ MeV per nucleon in the 1971 January 24 and September 1 solar flares.
H. J. Crawford, P. B. Price, J. D. Sullivan.
Astrophys. Journ., (Letters), Vol. 175, L149 - L153 (1972).
Energy spectra of O, Si, and Fe between 0.5 and 40 MeV per nucleon have been measured with stacks of plastic detectors exposed in a rocket during the 1971 January 24 flare. In both the January 24 flare and the September 1 flare, the abundances of elements $10 \leq Z \leq 28$ at ~ 15 MeV per nucleon are similar to high-energy cosmic-ray source abundances; S, Ne, and Ar are significantly lower than current solar values.

073.005 The energetic balance in a current layer of a solar flare and acceleration of cosmic rays by plasma waves. V. M. Tomozov.
Astron. Zhurn. Akad. Nauk SSSR, Vol. 49, 802 - 811 (1972).
In Russian. English translation in Soviet Astron. AJ, Vol. 16, No. 4.
The problems of the energetic balance and acceleration of the electrons by plasma waves in a current layer of a solar flare on the basis of Syrovatsky's model is considered.

073.006 Non-thermal X-ray radiation and electrical currents in solar flares. E. Ja. Vilkovisky, T. S. Kozhanov.
Astron. Zhurn. Akad. Nauk SSSR, Vol. 49, 812 - 816 (1972).
In Russian. English translation in Soviet Astron. AJ, Vol. 16, No. 4.
X-ray spectra have been calculated under the electrical current supposition for solar flares. Comparison with observational data shows that the model can be applied to the study of the first phase of flare phenomena.

073.007 Rocket observation of Ar XII–XVI, Ca XIV–XVIII, and Fe XIV, XV, XXIV in the extreme-ultraviolet spectrum of a solar flare. J. D. Purcell, K. G. Widing.
Astrophys. Journ., Vol. 176, 239 - 247 (1972).
The lithiumlike doublets of S XIV, Ar XVI, and Ca XVIII are identified in the 300–450 Å region by extrapolation of solar wavelengths and term intervals in the isoelectronic sequence. A possible observation of Fe XXIV in the flare is discussed. The solar blend at 417 Å involving S XIV and the Fe XV intersystem line is resolved. The singlet resonance lines of S XIII, Ar XV, and Ca XVII are also found in the flare spectrum by isoelectronic extrapolation. Ca XIV and Ca XV are identified on the basis of laboratory wavelengths. The problem of the identification of boron-like Ar XIV and Ca XVI is discussed.

073.008 Relativistic electrons associated with solar flares. K. Sakurai.
Planet. Space Sci., Vol. 20, 1229 - 1234 (1972).
Solar flares which produce relativistic electrons generally occur within sunspot groups which are active in the emission of meter type I noise storms. It is suggested that relativistic electrons in solar flares are accelerated from the keV-energy electrons responsible for the type I noise storms. The relationship between flare developments and the ejection of keV-electrons is briefly considered.

073.009 Ionized helium in prominences and in the chromosphere. T. Hirayama.
Solar Physics, Vol. 24, 310 - 323 (1972).
The purpose of this paper is to study whether the emission of ionized helium, in addition to that of neutral helium, is coming from the same cold region of ≤ 10000 K as the hydrogen and metallic emissions. We find that in quiescent prominences He II 4686 is emitted in a cold region of ≤ 10000 K, but in loop prominences it is not. In the case of the chromosphere we cannot reach a definitive answer because of the lack of observations.

073.010 A method to calculate electric currents in quiescent prominences. U. Anzer.
Solar Physics, Vol. 24, 324 - 335 (1972).
A 2-dimensional model of the magnetic field associated with quiescent prominences is presented. The coronal field is assumed to be current-free, currents are only allowed in the photosphere and inside the prominence. The prominence is taken to be infinitely thin. For this model a method is given to calculate the field configuration from the observed normal component of the field both in the photosphere and the prominence.

073.011 Observations and comments for the solar event of 24 October, 1969. A. E. Covington.
Solar Physics, Vol. 24, 405 - 410 (1972).
Observations of the flare of October 24, 1969 by Zirin et al. (Solar Physics, Vol. 19, 463 - 471 (1971)) are given with accompanying graphs of XUV emissions from satellites and four microwave radio emissions from Sagamore Hill Observatory. This event provides an excellent example of the development of two types of radio spectra within the same center and similar profiles for the hard X-ray burst and the 8800 MHz radio profile.

073.012 Thick-target processes and white-light flares.
H. S. Hudson.

Solar Physics, Vol. 24, 414 - 428 (1972).

Observations indicate that fast electrons in solar flares, which cause the hard X-ray burst and the impulsive microwave burst, lose energy predominantly by collisional processes. This requires a thick-target theory of the emission, for which the electron spectrum inferred from the X-ray spectrum becomes 1.5 powers steeper than in the usual thin-target theory.

073.013 New measurements of the polarization of X-ray solar flares.   I. P. Tindo, V. D. Ivanov,
S. L. Mandel'stam, A. I. Shuryghin.
Solar Physics, Vol. 24, 429 - 433 (1972).

Measurements of three X-ray flares in October/November 1970 aboard the Intercosmos 4 satellite confirm the existence of polarization in the initial phase of the X-ray bursts. The polarization can be observed for a few up to ten minutes, and an increase in polarization is observed during secondary maxima of the bursts as well.

073.014 Change of solar flare proton to alpha ratios during an energetic storm particle event.
M. Scholer, D. Hovestadt, B. Häusler.
Solar Physics, Vol. 24, 475 - 482 (1972).

Observations of the temporal behaviour of energetic storm protons and alpha particles are presented for the event associated with the storm sudden commencement observed on earth on March 8, 1970. The data are obtained on board the low altitude polar orbiting satellite GRS-A/AZUR by means of two particle telescopes. Large changes in the proton to alpha ratios for particles of equal energy and for particles of equal energy per nucleon are observed, whereas no significant change in the equal energy per charge ratio is observed. Electric fields, Fermi acceleration and cyclotron resonance are discussed as possible modulation mechanisms.

073.015 High resolution spectroscopy of the disk chromosphere. I. Observing procedures.
J. M. Beckers, H. A. Mauter, G. R. Mann, D. R. Brown.
Solar Physics, Vol. 25, 81 - 85 (1972).

This paper describes the details of an extensive observing program which is aimed at the precise photometric observation of chromospheric fine structures in the $\lambda 3933$, $\lambda 3968$, $\lambda 8498$, and $\lambda 8542$ lines of ionized calcium, the $\lambda 6563$ line of hydrogen, and the $\lambda 5890$ and $\lambda 5896$ lines of sodium.

073.016 High resolution spectroscopy of the disk chromosphere. II. Time-sequence observations of Ca II H and K emissions.
P. R. Wilson, D. E. Rees, J. M. Beckers, D. R. Brown.
Solar Physics, Vol. 25, 86 - 97 (1972).

Two independent sets of high resolution time series spectra of the Ca II H and K emission obtained at the solar tower and at the big dome of the Sacramento Peak Observatory on September 11th, 1971 are reported. The evolutionary behaviour of the emission is confirmed but the detail of the evolution is found to be more complex.

073.017 On the line intensity ratios $E(H\alpha)/E(D3)$ and $E(H\beta)/E(D3)$ in prominences.   G. Stellmacher.
Solar Physics, Vol. 25, 104 - 107 (1972).

The intensity ratios $E(H\alpha)/E(D3)$ and $E(H\beta)/E(D3)$ in prominences depend on the total optical thickness in $H\alpha$ of the layer. The emission of the He D3 line appears relatively enhanced in thin layers and in outer parts of the prominences.

073.018 Observations of prominences at 3.5 millimeter wavelength.   M. R. Kundu.
Solar Physics, Vol. 25, 108 - 115 (1972).

At 3.5 mm wavelength absorption features are observed in correspondence with $H\alpha$ dark filaments on the disk; beyond the limb the prominences correspond to emissive regions. The

absorption features are larger ($2' - 3'$ arc) than the corresponding $H\alpha$ dark filaments; the emissive regions at the limb have similar angular sizes. The emissive regions at the limb have electron temperatures of 5500 ± 500 K; the amount of absorption observed on the disk leads to mean electron densities of about 5×10^{10} per cm³.

073.019 Some aspects of flare properties versus magnetic boundary morphology.   S. W. Prata.
Solar Physics, Vol. 25, 136 - 140 (1972).

Flares often begin near lines of zero-longitudinal magnetic field. These boundary lines between opposite polarities can have the magnetic field either parallel to or transverse to the boundary. A study of 75 flares associated with one or the other type of boundary indicates that the difference in boundary morphology is not reflected in the frequency or properties of the associated flares.

073.020 Flares and changing magnetic fields.   D. M. Rust.
Solar Physics, Vol. 25, 141 - 157 (1972).

An observational study of maps of the longitudinal component of the photospheric fields in flaring active regions leads to eight conclusions, described in detail.

073.021 Quantitative analysis of the hydrogen lines in moustaches.   A. N. Koval.
Izv. Krymskoj Astrofiz. Obs., Vol. 44, 94 - 102 (1972). In Russian.

The profiles of the hydrogen emission lines $H_a - H_9$ in moustaches are considered. The characteristic feature of considered moustaches is the absence of absorption in the cores of H_8 and H_9 lines. It is shown that the Doppler broadening of emission is in good agreement with observations in the wings of moustaches.

073.022 On the character of the propagation of the excitation from flares.   N. N. Stepanyan.
Izv. Krymskoj Astrofiz. Obs., Vol. 44, 103 - 106 (1972). In Russian.

It is noted that in many flares we deal with two phenomena: a) the flare «itself», corresponding to the first maximum of the light curve, and b) the phenomena connected with the propagation of the excitation from the flare through the chromosphere (the decay part of the light curve). As a possible mechanism of the propagation of the excitation a strong shock wave is considered.

073.023 Current limitation in solar flares.
D. F. Smith, E. R. Priest.
Astrophys. Journ., Vol. 176, 487 - 495 (1972).

It is shown that the ion-sound current instability will be the relevant mechanism of current dissipation in most instances in the fully ionized part of the solar atmosphere (upper chromosphere and corona). The linear and nonlinear stages of the ion-sound current instability are examined and a turbulent resistivity is given for the saturated state which is in agreement with laboratory experiments. This turbulent resistivity is applied to the Alfvén-Carlqvist model of a cylindrical current filament and to the "hard" phase of Syrovatsky's dynamic dissipation model. The possibility that the ion-sound turbulence can systematically accelerate a select group of particles is investigated.

073.024 The D_3 helium line in the solar chromosphere from observations with the 53-cm Lyot coronograph.
V. I. Makarov.
Solnechnye Dannye 1972 Byull., No. 4, p. 71 - 77 (1972). In Russian.

An analysis of halfwidths, intensities and profiles of the D helium lines in spicules was made from observations with the 53-cm Lyot coronograph. The halfwidth of D_3 varies linearly with height from 0.38 Å at 1000 km to 0.68 Å at 5000 km. The maximum emission of helium might be at the height of 1200 km. The observed profiles of the D_3 line can be represented as sum of normal distribution curves with halfwidths from 0.17 Å to 0.22 Å.

073.025 On physical conditions in the large flare on July 12, 1961.   P. N. Polupan.
Solnechnye Dannye 1972 Byull., No. 4, p. 106 - 111 (1972). In Russian.
Spectra of the flare on July 12, 1961, obtained in the λλ 3400–7000 Å interval are discussed.

073.026 Comparison of electron density in quiet prominences at the places of emission of helium and metals.
N. N. Morozhenko.
Solnechnye Dannye 1972 Byull., No. 5, p. 94 - 101 (1972). In Russian.
Values of electron density n_e in 11 knots of quiet prominences for the places of emission of helium and metals were determined. The n_e values calculated according to the lines of metals were found to be higher by one order than those obtained according to helium lines.

073.027 The great solar flares of August, 1972.
D. M. Rust.
Sky Telescope, Vol. 44, 226 - 230 (1972).

073.028 Magnetic fields, electric currents and Lorentz forces in the chromosphere.   D. Dravins.
Bull. American Astron. Soc., Vol. 4, 309 (1972). – Abstr. AAS.

073.029 Reevaluation of relationships between solar flares and sporadic geomagnetic storms.
H. W. Dodson, E. R. Hedeman.
Bull. American Astron. Soc., Vol. 4, 309 - 310 (1972). Abstr. AAS.

073.030 The response of the helium triplet radiation in prominences to an increase in ultraviolet flux resulting from solar flares.   J. M. Beckers.
Bull. American Astron. Soc., Vol. 4, 377 (1972). – Abstr. AAS.

073.031 A short-lived chromosphere brightening observed in H-alpha and Ca II K.
G. L. Epstein, R. W. Hobbs, C. L. Hyder.
Bull. American Astron. Soc., Vol. 4, 382 (1972). – Abstr. AAS.

073.032 λ 10830 image tube filtergrams of plage activity and a small flare.   R. Fisher.
Bull. American Astron. Soc., Vol. 4, 382 (1972). – Abstr. AAS.

073.033 A current interruption, electrostatic discharge model of solar flares.   M. W. Haurwitz.
Bull. American Astron. Soc., Vol. 4, 384 (1972). – Abstr. AAS.

073.034 Hydrodynamics of the formation of quiescent prominences.   E. Hildner.
Bull. American Astron. Soc., Vol. 4, 384 (1972). – Abstr. AAS.

073.035 Infrared emission from solar flares.
H. S. Hudson, K. Ohki.
Bull. American Astron. Soc., Vol. 4, 385 (1972). – Abstr. AAS.

073.036 Flares, magnetic configurations, and magnetic energy release.   T. J. Janssens.
Bull. American Astron. Soc., Vol. 4, 385 (1972). – Abstr. AAS.

073.037 Acceleration of electrons in small solar flares.
S. R. Kane, R. P. Lin, K. A. Anderson.
Bull. American Astron. Soc., Vol. 4, 385 - 386 (1972). Abstr. AAS.

073.038 Phenomenological comparison between the solar flare and a laboratory discharge plasma.
T. N. Lie, R. C. Elton.
Bull. American Astron. Soc., Vol. 4, 386 (1972). – Abstr. AAS.

073.039 Action in the quiet solar chromosphere.   S.-Y. Liu.
Bull. American Astron. Soc., Vol. 4, 386 (1972). Abstr. AAS.

073.040 Force free magnetic field configurations in the chromosphere.   R. X. Meyer, E. B. Mayfield.
Bull. American Astron. Soc., Vol. 4, 387 (1972). – Abstr. AAS.

073.041 Overabundance of very heavy nuclei accelerated in solar flares.   A. Mogro-Campero, J. A. Simpson.
Bull. American Astron. Soc., Vol. 4, 388 (1972). – Abstr. AAS.

073.042 EUV spectra of prominences and filaments.
R. W. Noyes.
Bull. American Astron. Soc., Vol. 4, 388 - 389 (1972). Abstr. AAS.

073.043 Line profiles and microturbulence generated by acoustic waves in the solar chromosphere.
L. Oster, P. Ulmschneider.
Bull. American Astron. Soc., Vol. 4, 389 (1972). – Abstr. AAS.

073.044 An inconsistency in Petschek's mechanism.
E. R. Priest.
Bull. American Astron. Soc., Vol. 4, 389 (1972). – Abstr. AAS.

073.045 Photoelectric Ca II line profiles in solar plages and a sunspot and their preliminary interpretation.
R. A. Shine, J. L. Linsky.
Bull. American Astron. Soc., Vol. 4, 391 (1972). – Abstr. AAS.

073.046 Correlation between the intensity fields of the chromospheric and coronal networks.
G. W. Simon, D. K. Lynch.
Bull. American Astron. Soc., Vol. 4, 391 (1972). – Abstr. AAS.

073.047 On the quantitative description of the fluctuating solar atmosphere. I. Regression analysis and calibration of multi-channel observations.
A. Skumanich, C. Smythe, E. N. Frazier.
Bull. American Astron. Soc., Vol. 4, 391 - 392 (1972). Abstr. AAS.

073.048 Current limitation in solar flares.
D. F. Smith, E. R. Priest.
Bull. American Astron. Soc., Vol. 4, 392 (1972). – Abstr. AAS.

073.049 Videomagnetograph studies of magnetic field diffusion.   R. C. Smithson.
Bull. American Astron. Soc., Vol. 4, 392 (1972). – Abstr. AAS.

073.050 Chromospheric oscillation in the Hα plage.
K. Tanaka.
Bull. American Astron. Soc., Vol. 4, 393 (1972). – Abstr. AAS.

073.051 Simultaneous X-ray, EUV, and H-alpha spectroheliograms of solar flares observed by OSO-7.
R. J. Thomas.
Bull. American Astron. Soc., Vol. 4, 393 (1972). – Abstr. AAS.

073.052 Multiple components in the impulsive phase of solar flares.   J. Vorpahl.

Bull. American Astron. Soc., Vol. 4, 394 - 395 (1972).
Abstr. AAS.

073.053 **On the production of solar flares.** M. L. White.
Bull. American Astron. Soc., Vol. 4, 395 (1972).
Abstr. AAS.

073.054 **H$_\alpha$-flares: The response of the chromosphere to a downward shock wave.**
S. T. Wu, S. M. Han, Y. Nakagawa.
Bull. American Astron. Soc., Vol. 4, 396 (1972). – Abstr. AAS.

073.055 **Some parameters of flocculi characteristics for flare activity.**
I. Y. Brailovskaya, A. N. Koval, M. B. Ogir, N. N. Stepanjan.
Solnechnye Dannye 1972 Byull., No. 6, p. 88 - 94 (1972).
In Russian.
There have been found the flocculi characteristics required in forecasting the total flare activity of a flocculus during its transit over the disk.

073.056 **Comparison of solar-flare energy estimates made by analytical and numerical techniques.** M. Dryer.
Journ. Geophys. Res., Vol. 77, 4851 - 4854 (1972). – Brief report.

073.057 **Temperature structure and conductive flux in the chromosphere-corona transition region.**
G. Elwert, P. K. Raju.
Solar Physics, Vol. 25, 319 - 328 (1972).
Observations of UV-line intensities referring to the whole undisturbed sun are used to investigate the chromosphere-corona transition region. For the evaluation of the integral representing the theoretical line intensities it appears to be an improvement to consider not the temperature gradient but the conductive flux to be nearly constant in the line-forming region. The results are used for determining the temperature structure of the transition region.

073.058 **Physical properties of solar chromospheric plages. I. Line profiles of the Ca II H, K, and infrared triplet lines.** R. A. Shine, J. L. Linsky.
Solar Physics, Vol. 25, 357 - 379 (1972).
Double pass photoelectric observations are presented of five Ca II lines (H, K, 8498 Å, 8542 Å, and 8662 Å) in a number of solar plages of different degrees of activity, quiet regions, and a sunspot. The data are compared with previous work. The question of source function equality is considered and the differences and similarities among plage profiles and between plage and quiet profiles are shown qualitatively and quantitatively.

073.059 **Velocity oscillations in solar plage regions.**
C.-J. Chen, P. S. Lykoudis.
Solar Physics, Vol. 25, 380 - 401 (1972).
The properties of wave propagation in a perfectly electrically conducting, plane-stratified, inviscid, compressible atmosphere premeated by a horizontal magnetic field which varies with height are investigated. It is shown that a diagnostic diagram can be constructed through a generalization of the propagation equation to account for the presence of a magnetic field. The effect of the magnetic field on the oscillations in solar plages around the temperature minimum is studied and compared with the non-magnetic case based on the Bilderberg Continuum Atmosphere.

073.060 **The loop prominence of May 13, 1971 and its associated effects.**
M. E. Machado, H. Grossi Gallegos, A. F. Silva.
Solar Physics, Vol. 25, 402 - 412 (1972).
A study is presented of the formation of a loop promi-

nence system on May 13, 1971. The development of the phenomenon is found to be consistent with the model of Jefferies and Orrall.

073.061 **The acceleration of ions in the current layer of a solar flare.** V. M. Tomozov.
Astron. Zhurn. Akad. Nauk SSSR, Vol. 49, 1069 - 1072 (1972). In Russian. English translation in Soviet Astron. AJ, Vol. 16, No. 5.
The problem of the acceleration of heavy multiply charged ions in the current layer of a solar flare by plasma oscillations is considered. It is shown that ion-sonic oscillations are able to inject ions into the region of acceleration by Langmuir plasmons.

073.062 **Solar flare photographed at Boyden Observatory on the 11th August 1972, at 14h 44m SAST.**
H. Bacik, J. P. Eksteen.
Monthly Notes Astron. Soc. Southern Africa, Vol. 31, 120 (1972).

073.063 **Further iron-line observations during solar flares.**
G. A. Doschek, J. F. Meekins, R. D. Cowan.
Astrophys. Journ., Vol. 177, 261 - 269 (1972).
Transitions in Fe XXIV of the type $1s^2 2l\ ^2L - 1s^2\ 4l'\ ^2L'$, and the Fe XXIII transition, $1s^2 2s^2\ ^1S_0 - 1s^2\ 2s4p\ ^1P_1$, are identified in soft X-ray spectra of solar flares. The relative line strengths of Fe XXIV are compared with theory, and the temporal behavior of the Fe XXIV lines and the Fe XXIII line is discussed, particularly with reference to the continuum emission near 8 Å.

073.064 **What makes active regions grow?** S. Weart.
Astrophys. Journ., Vol. 177, 271 - 276 (1972).
We have studied the growth, or failure to grow, of well over 100 active regions. Most growth is connected with the emergence of a large batch of flux in the shape of a new Arch Filament System (AFS). We present evidence for the hypotheses that (1) a twist in the flux tubes of new AFSs is a key factor in determining which new AFSs will grow, and (2) this twist is related to the well-known asymmetry of sunspot groups.

073.065 **The abundances of solar accelerated nuclei from carbon to iron.** A. Mogro-Campero, J. A. Simpson.
Astrophys. Journ., (*Letters*), Vol. 177, L37 - L41 (1972).
From revised observation periods and new data we again find that the overabundance of solar-flare nuclei with respect to solar photospheric or coronal abundances increases with increasing atomic number.

073.066 **The preferential acceleration of heavy nuclei in solar flares.** B. G. Cartwright, A. Mogro-Campero.
Astrophys. Journ., (*Letters*), Vol. 177, L43 - L47 (1972).
Several recently reported measurements of the relative abundances of heavy nuclei ($Z \gtrsim 8$) accelerated in solar flares have shown a significant enhancement of these nuclei compared with solar photospheric and coronal abundances. We propose a mechanism for the preferential acceleration of heavy nuclei.

073.067 **Models of active regions in the transition zone from UV observations.**
A. M. Cantú, G. Poletto, G. L. Tagliaferri.
Space Sci. Rev., Vol. 13, 638 (1972). – Conference paper
IAU Colloquium No. 14 (see 012.012).

073.068 **Experiment to determine the temperature structure in the solar chromosphere and corona.**
C. R. Negus.
Space Sci. Rev., Vol. 13, 668 - 669 (1972). – Conference pa-

per IAU Colloquium No. 14 (see 012.012).

073.069 Temperature structure of the chromosphere—corona transition region.   G. Elwert, P. K. Raju.
Space Sci. Rev., Vol. 13, 670 - 671 (1972). — Conference paper IAU Colloquium No. 14 (see 012.012).

073.070 Laboratory-produced radiation related to the solar flare emission.   R. C. Elton, T. N. Lie.
Space Sci. Rev., Vol. 13, 747 - 760 (1972). — Invited paper IAU Colloquium No. 14 (see 012.012).

073.071 The classification of Fe XVIII to XXIV emission lines in solar flare spectra.   B. C. Fawcett.
Space Sci. Rev., Vol. 13, 763 - 764 (1972). — Conference paper IAU Colloquium No. 14 (see 012.012).

073.072 The solar flare plasma: Observation and interpretation.   G. A. Doschek.
Space Sci. Rev., Vol. 13, 765 - 821 (1972). — Invited paper IAU Colloquium No. 14 (see 012.012).

073.073 Production of different non-thermal electron groups in small solar flares.   S. R. Kane.
Space Sci. Rev., Vol. 13, 822 - 823 (1972). — Conference paper IAU Colloquium No. 14 (see 012.012).

073.074 The expected behaviour of the hydrogen Lyman lines in solar flares.   Z. Švestka, L. D. De Feiter.
Space Sci. Rev., Vol. 13, 824 (1972). — Abstract of conference paper IAU Colloquium No. 14 (see 012.012).

073.075 The transient highly excited solar flare plasma.   L. D. De Feiter.
Space Sci. Rev., Vol. 13, 827 - 842 (1972). — Invited paper IAU Colloquium No. 14 (see 012.012).

073.076 The ionized helium radiation in prominences and flares.   N. A. Yakovkin, M. Yu. Zeldina.
Vestn. Kiev. Un-ta, Ser. Astron., No. 14, p. 3 - 16 (1972). In Russian.
In quiescent prominences the faint emission of ionized helium in the 4686 Å line is due to second ionization by far-ultraviolet radiation. In flare-like events the emission is caused by the process of charge-exchange collisions.

073.077 Spectrophotometry of eruptive prominences.   A. N. Sergeeva.
Vestn. Kiev. Un-ta, Ser. Astron., No. 14, p. 17 - 22 (1972). In Russian.
Photometric results of two eruptive prominences spectra in $H_a - H_{20}$, 5876, 4471, 3889 Å, He and Mg, Na, Fe, Ca^+, Ti^+, Fe^+, Sr^+ lines are reduced. The tables contain central intensities in units of the continuous spectrum of the disk centre, complete and equivalent widths and number of atoms at the upper level on line of sight.

073.078 Analysis of some aspects of 25 chromospheric events. I: Reduction of the optical data.
R. Falciani, M. Rigutti, C. J. Macris.
Solar Physics, Vol. 26, 108 - 113 (1972).
Some twenty five solar events photographed at the National Observatory, Athens, have been used for the production of isophotal contours. Curves showing the development with time of the energy emitted by the events at several intensity levels have been obtained. Examples are shown and general remarks about measurements of flares are made.

073.079 Analysis of some aspects of 25 chromospheric events. II. Discussion on the optical data.
R. Falciani, M. Rigutti.
Solar Physics, Vol. 26, 114 - 129 (1972).
The discussion of the development curves of several solar flares observed in the Hα-line leads to some new information about the photometric structure and behaviour of flares of importance from 1 to 2. Analytical representations for both the pulsation periods and the exponential decreases are given. A discussion on the present day state of solar patrol and some suggestions to improve it to be useful for solar physics is also made in the light of the present analysis.

073.080 Comments on the paper 'Fine structure of solar magnetic fields' by H. Zirin [Solar Physics, Vol. 22, 34 - 48 (1972)].   E. N. Frazier.
Solar Physics, Vol. 26, 142 - 144 (1972). — Research note.

073.081 Response to Dr Frazier's comments [Solar Physics, Vol. 26, 142 - 144 (1972)].   H. Zirin.
Solar Physics, Vol. 26, 145 - 147 (1972). — Research note.

073.082 Reply to 'The relations between chromospheric features and photospheric magnetic fields' by
E. N. Frazier [Solar Physics, Vol. 24, 98 - 112 (1972)].
P. Foukal, H. Zirin.
Solar Physics, Vol. 26, 148 - 150 (1972). — Research note.

073.083 Spectra of solar flares.   Z. Švestka.
Annual Rev. Astron. Astrophys., Vol. 10, (see 003.005), 1 - 24 (1972). — Contents: (1) General features of the spectrum; (2) Simplifying assumptions; (3) The broadening of Balmer lines; (4) Electron density; (5) The optical thickness; (6) Deviations from LTE and electron temperature; (7) Hydrogen density; (8) The filamentary structure of flares; (9) The helium lines; (10) H and K (Ca II) lines; (11) Metallic lines; (12) Line asymmetry; (13) Continuous emission.

073.084 Solar spicules.   J. M. Beckers.
Annual Rev. Astron. Astrophys., Vol. 10, (see 003.005), 73 - 100 (1972). — Contents: (1) Introduction; (2) Morphology of spicules; (3) Spectroscopic properties of spicules; (4) Physical conditions in spicules; (5) The relation of spicules to their surroundings; (6) Spicules as seen against the solar disk; (7) Mechanical models of spicules; (8) Concluding remarks.

073.085 Solar flare spectrum in the 1.85−1.87 Å region.   B. N. Vasil'ev, Yu. I. Grineva, I. A. Zhitnik, V. I. Karev, V. V. Korneev, V. V. Krutov, S. L. Mandel'shtam.
Kratkie soobshch. po fiz., 1972, No. 3, p. 29 - 34. In Russian. Abstr. in Referativ. Zhurn. 51. Astron., 11.51.347 (1972).

073.086 Excitation of the solar flare spectrum in the 1.8 Å region.
L. A. Vajnshtejn, I. A. Zhitnik, V. V. Korneev, S. L. Mandel'shtam.
Kratkie soobshch. po fiz., 1972, No. 3, p. 35 - 39. In Russian. Abstr. in Referativ. Zhurn. 51. Astron., 11.51.348 (1972).

073.087 Neutron and gamma-ray emission from white-light flares.   L. D. de Feiter, Z. Švestka.
Space Research XII, (see 012.016), Vol. 2, 1547 - 1551 (1972).

073.088 Funnel prominences.   J. Kleczek.
Bull. Astron. Inst. Czechoslovakia, Vol. 23, 315 - 318 (1972).
Properties of funnel prominences are deduced from prominence surveys and prominence films. Association with active regions is discussed for 284 funnel prominences observed in H-alpha filter from January 1956 to December 1961. No conspicuous preference for specific spotgroup type, neither in Zürich, nor in Mt. Wilson magnetic classification, has been found.

073.089 **Über die große Sonneneruption Anfang August 1972.** R. Born, P. Brandt, W. Mattig, H. Wöhl.
SuW, Vol. 11, 339 - 342 (1972).

073.090 **Acceleration of 10–100 keV electrons in solar flares.** R. P. Lin.
Trudy Mezhdunar. seminara po probl. "Uskorenie chastits v kosmich. prostranstve (okolozem. i mezhplanet. kosmich. prostranstve), Galaktike i Metagalaktike". Moskva, 1972, p. 9 - 40. – Abstr. in Referativ. Zhurn. 51. Astron., 12.51.357 (1972).

073.091 **Spectrum of a solar flare in the region from 1.85– 1.87 Å.**
B. N. Vasil'ev, Yu. I. Grineva, I. A. Zhitnik, V. I. Karev, V. V. Korneev, V. V. Krutov, S. L. Mandel'shtam.
Kratkie soobshch. po fiz., 1972, No. 3, p. 29 - 34. In Russian. Abstr. in Referativ. Zhurn. 62. Issled. kosmich. prostranstva, 12.62.114 (1972).

073.092 **Excitation of the spectrum of a solar flare in the region of 1.8 Å.**
L. A. Vajnshtejn, I. A. Zhitnik, V. V. Korneev, S. L. Mandel'shtam.
Kratkie soobshch. po fiz., 1972, No. 3, p. 35 - 39. In Russian. Abstr. in Referativ. Zhurn. 62. Issled. kosmich. prostranstva, 12.62.115 (1972).

073.093 **Equator-pole differences in the solar chromosphere from Lyman-continuum data.**
J. E. Vernazza, R. W. Noyes.
Solar Physics, Vol. 26, 335 - 342 (1972).
In the present analysis, we study pole-equator differences in the region of Lyman-continuum formation in the middle chromosphere. In previous papers a model of the solar chromosphere was derived on the basis of the Harvard OSO-4 data on the Lyman-continuum. Here we extend the analysis to explain the observed differences between equator and poles, and derive comparative models of the two regions.

073.094 **The absence of flares in λ3835 and the heating of the chromosphere.** H. Zirin.
Solar Physics, Vol. 26, 393 - 396 (1972).
Simultaneous observations of flares in Hα and a band 15 Å wide centered on 3835 Å show no change whatever in 3835 Å at the time of several flares, although the chromospheric network is easily visible. Flares are therefore transparent in this wavelength.

073.095 **X-radiation ($E > 10$ keV), Hα and microwave emission during the impulsive phase of solar flares.**
J. A. Vorpahl.
Solar Physics, Vol. 26, 397 - 413 (1972).
The present work shows that the impulsive phase of solar flares has an optical component in addition to the already well-known X-ray and microwave impulsive bursts. Properties of these intense Hα brightenings, as well as additional characteristics of the impulsive X-ray and microwave bursts, are given. Finally all three types of radiation are used to construct a flare picture of the impulsive phase.

073.096 **Search for weak white-light flares by time-wise photographic cancellation.**
Y. Uchida, H. Hudson.
Solar Physics, Vol. 26, 414 - 417 (1972).
This study proposes as a working hypothesis that small white-light flares accompany all major (proton) flare events and suggests a new method for systematically finding these 'patches' of white-light emission. The new technique consists of the time-wise application of the photographic cancellation method to detect small time-varying features around the time

of the impulsive phase of a flare.

073.097 **The magnetic configuration of the November 18, 1968 loop prominence system.** J.-R. Roy.
Solar Physics, Vol. 26, 418 - 430 (1972).
Computed current-free magnetic fields are compared to the loop prominence associated with the west limb proton flare of 18 November 1968. Successive sets of fitting fieldlines closely resemble the loop prominence system throughout its growth and lifetime. The successive position bases of the fieldlines reproduce the drift rate of spreading two-ribbon flares.

073.098 **Solar flares in the EUV observed from OSO-5.**
P. T. Kelly, W. A. Rense.
Solar Physics, Vol. 26, 431 - 440 (1972).
Solar flares in three broad EUV spectral bands have been observed from OSO-5 with a grating spectrophotometer. Results are given for three large flares. Absolute EUV intensities for the flares are estimated, and a comparison made with the 2800 mc s^{-1} radio emission. A flare model is proposed to account for the EUV time variations during a large flare.

073.099 **Extreme-ultraviolet emission from solar prominences.** R. W. Noyes, A. K. Dupree, M. C. E. Huber, W. H. Parkinson, E. M. Reeves, G. L. Withbroe.
Astrophys. Journ., Vol. 178, 515 - 525 (1972).
Spectra and spectroheliograms of prominences have been obtained at wavelengths 300 Å $< \lambda <$ 1400 Å from instruments aboard the OSO 4 and OSO 6 spacecraft. This introductory report describes our observations and discusses some conclusions that can be readily drawn from the data.

073.100 **Spectrophotometry of prominences in the phase preceding decay.** A. S. Rakhubovsky.
Astrometriya i Astrofiz., *Kiev*, Vyp. (No.) 17, (see 003.012), p. 64 - 95 (1972). In Russian.
Results are presented of spectrophotometric processing of prominence spectra in both their quiet and decay phases. A catalogue of equivalent widths, central intensities, Doppler half-widths and half-widths of emission lines is compiled.

073.101 **Computation of self-consistent models of chromospheres of the sun and stars.**
S. A. Kaplan, L. A. Ostrovskii, N. S. Petrukhin, V. E. Friedman.
Astron. Zhurn. Akad. Nauk SSSR, Vol. 49, 1267 - 1274 (1972). In Russian. English translation in Soviet Astron. AJ, Vol. 16, No. 6.
A model of a self-consistent solution of the problem of propagation of sound waves in an inhomogeneous medium, and of the development of discontinuities and dissipation of energy at these discontinuities is given.

073.102 **The solar active region from August 17–30, 1971 and associated events.** E. Țifrea, A. Dimitriu.
Stud. Cerc. Astron., Vol. 17, 197 - 202 (1972). In Romanian.
The paper gives a morphological analysis of the McMath 11482 active region. There are some evident changes in the photospheric structure, but not a very high activity in the chromosphere. The surges and bright points are main events of this region. The data of 3 flares observed at the Bucharest Observatory are also given.

073.103 **Ca II K emission arches.**
H. A. Beebe, H. R. Johnson.
Solar Physics, Vol. 27, 34 - 38 = Publ. Goethe Link Obs., Indiana Univ., *Bloomington*, No. 144 (1972).
Arch-like features are often seen in spectrograms of very strong lines near the solar limb when the slit crosses the chromospheric network. We show how earlier kinetic-equilibrium (non-LTE) calculations for Ca II can be used to predict such features for the K line with two-component atmospheric

models.

073.104 On the size of the structure elements in the solar chromosphere. V. A. Krat.
Solar Physics, Vol. 27, 39 - 43 (1972).

Photometric reductions of the spectrograms obtained during the third flight of the Soviet Stratospheric Solar Station are discussed. A comparison of photometric scans in Hα and its far wings near to the continuum leads to the conclusion that chromospheric 'mottles' are at least several times broader than photospheric granules. The optimum size of mottles is about $0\rlap{.}''8 - 1\rlap{.}''1$. The Hα profiles of mottles are practically the same as those obtained from the ground observations. The broadening of mottles is considered as an effect of expansion of magnetic arcs growing up to chromospheric levels.

073.105 Flares, magnetic configurations, and magnetic energy release. T. J. Janssens.
Solar Physics, Vol. 27, 149 - 163 (1972).

Using a newly developed Aerospace digital videomagnetograph, three solar active regions are studied as to their magnetic configurations and their flare productivity. These three regions have very different types of magnetic configurations and different types of flare productivity. We review previous theoretical and experimental research on flares and magnetic energy storage, and discuss various ways to observe magnetic energy release due to flares. Results for six subflares are presented.

073.106 Polarization structure of a solar flare region at 9.5 mm wavelength. M. R. Kundu, T. P. McCullough.
Solar Physics, Vol. 27, 182 - 191 (1972).

Polarization structure of an active region that produced a minor flare around 1900 UT on September 28, 1971 was measured at 9.5 mm wavelength using the 85-ft telescope of the Naval Research Laboratory Maryland Point Observatory. The flare region underwent changes both in the degree of polarization as well as in its polarization structure before and after the start of the flare. These changes in the degree of polarization correspond to a decrease of longitudinal magnetic field of about 200 G at the chromospheric levels where the 9.5 mm radiation originates. Observations on the polarization structure of active regions for several days before and after September, 1971 are also presented.

073.107 Investigation of the He I 3888.65 Å and H_8 3889.05 Å lines in the spectrum of the solar chromosphere from observations with a large coronograph.
V. V. Makarova, V. I. Makarov.
Solnechnye Dannye 1972 Byull., No. 9, p. 99 - 103 (1972). In Russian.

073.108 Major solar eruption observed.
R. Tousey, G. E. Brueckner, M. J. Koomen, D. J. Michels.
Naval Res. Rev., (USA), Vol. 25, No. 1, p. 8 - 13 (1972).

A large region of the sun's outer atmosphere erupted on 13 Dec. 1971 and energetic debris from this massive disturbance collided with the earth three and a half days later. A review of this observation is given.

073.109 A theoretical study of formation of prominences.
E. G. Hildner III.
Thesis, Univ. Colorado, Boulder. [Available from Univ. Microfilms, Ann Arbor, Mich., USA. Order No. 72-3662], 209 pp. (1971).

Numerical computations are presented for the formation of quiescent solar prominences by condensation from the solar corona. The governing equations for one and two dimensional geometries are integrated to obtain solutions which describe the condensation of model coronal material to a cooler,

denser state.

073.110 A model of the chromosphere and of the transition zone from chromosphere to solar corona.
E. E. Dubov.
Izv. Krymskoj Astrofiz. Obs., Vol. 45, 20 - 39 (1972). In Russian.

A model of the chromosphere and of the transition zone from chromosphere to corona for quiet regions of the sun is developed. The model corresponds to observed solar emission in optical and radio regions of the spectrum. Besides, this model accounts for the brightness distribution on spectroheliograms in Hα, K_{232} Ca II, and several UV lines. The solar emission flux calculated using this model for the O VI λ 1032 Å line corresponds to the flux obtained with spacecraft OSO 4 for different sites of the chromospheric network.

073.111 Hot regions in solar flares. K. Oki.
Astron. Herald, (Japan), Vol. 65, 231 - 235 (1972). In Japanese.

Describes observations of soft X-rays in comparison with radiowave measurements, and considers the relation between hot clouds and optical flares, and that between the non-thermal particles and the origin of the hot clouds.

073.112 On the measurement of prominence light polarization in the CaI λ 4227 Å emission line.
Ts. S. Khetsuriani.
Byull. Abastumansk. Astrofiz. Obs., No. 43, p. 171 - 174 (1972). In Russian.

The polarization degree of prominence light is measured. In the CaI λ 4227 Å emission line the polarization degree turned out to be about 4 % and the continuous emission near this line gives 10 %. The temperature and electron concentration are computed.

073.113 Flares and shock waves in interplanetary space.
A. T. Nesmyanovich, E. I. Nesmyanovich.
Problems of cosmic physics. Vyp. (No.) 7, (see 003.028), p. 3 - 15 (1972). In Russian.

The comparison of 45 cases of disturbances in interplanetary space, registered by the satellite "Vela" and by the ground techniques as SC, with flares has not allowed to establish a definite connection between these two events. The results of comparisons of 190 cases SC with flares (according to the catalogue of flare activity for 1957–1967 prepared by the authors) are presented.

073.114 Magnetic fields in solar prominences – a review.
D. M. Rust.
U. S. Air Force Cambridge Res. Labs. Air Force Systems Command, Sacramento Peak Obs. Air Force Surveys Geophys. No. 237. AFCRL-72-0048, 26 pp. (1972).

073.115 Energy balance in the chromosphere-corona transition region. R. A. Kopp.
Solar Physics, Vol. 27, 373 - 393 (1972).

It is shown that the downward heat flux between the chromosphere and corona cannot be nearly as large as the value 6×10^5 erg cm^{-2} s^{-1} derived in previous studies by assuming a planar atmosphere, and in fact is insufficient to balance transition-region radiative losses. An alternative picture is developed, consisting of a 'transition region network' covering only a small fraction of the solar disk.

073.116 A model for the polar transition layer and corona for November 1967.
G. L. Withbroe, Y.-M. Wang.
Solar Physics, Vol. 27, 394 - 401 (1972).

A model for the chromospheric-coronal transition layer and lower corona has been constructed for the south polar region. EUV observations acquired by the Harvard OSO-4 experiment in the fall of 1967 were used in the analysis. The observations can be explained with a simple model consisting of two types of regions.

073.117 Lifetime of solar flare particles in coronal storage regions. K. A. Anderson.
Solar Physics, Vol. 27, 442 - 445 (1972).

Most discussions of lifetime of flare particles in the solar corona have assumed that collision loss is the dominant means of slowing and stopping these particles. It is quite possible that the solar cosmic rays are not imbedded in 10^6 K coronal material but rather all particles in the storage region are energetic. Collision times are sufficiently short so that the energy spectrum may approach a maxwellian distribution with kT on the order of 30 keV. If this is the case, the rate of collision loss will be greatly reduced. Bremsstrahlung and magneto-bremsstrahlung then will be the important energy losses.

Dielectronic satellite spectra for highly-charged helium-like ion lines.   See Abstr. 022.061.

High resolution solar observations from space.   See Abstr. 032.032.

Time-dependent ionization equilibrium and line radiation under flarelike conditions.   See Abstr. 062.001.

On the formation of double layers in plasmas.   See Abstr. 062.053.

Calculation of solar CO vibration-rotation line profiles and equivalent widths.   See Abstr. 071.003.

Velocity oscillations in the solar atmosphere.   See Abstr. 071.006.

Magnetic fields and helium-D_3 spectroheliograms.   See Abstr. 071.007.

Characteristics of the Ca II K-line profiles in the quiet sun.   See Abstr. 071.008.

The velocity fields at different levels in quiet solar regions.   See Abstr. 071.014.

A search for the photospheric origin of spicules.   See Abstr. 071.025.

Fe I ionization and excitation equilibrium in the solar atmosphere.   See Abstr. 071.035.

The empirical determination of line source functions, β_L-values, and the microturbulent and convective velocity components as functions of depth in the photosphere-chromosphere transition region.   See Abstr. 071.037.

Some observational results on moustaches.   See Abstr. 071.046.

The magnetic structure of arch filament systems.   See Abstr. 071.047.

A mechanism for the exploding granule phenomenon.   See Abstr. 071.056.

The spectra of near-vertical structures on the solar disk.   See Abstr. 071.063.

Some effects of molecules in solar phenomena.   See Abstr. 072.036.

The derivation of temperature gradient and electron density maps from EUV spectroheliograms.   See Abstr. 074.013.

The magnetic field configuration of the solar corona after a proton flare.   See Abstr. 074.016.

The coronal origin of a solar flare.   See Abstr. 074.028.

Émissions 'froides' dans la couronne solaire.   See Abstr. 074.052.

The solar EUV-emitting plasma.   See Abstr. 074.068.

Coronal abundance of elements and a model of the quiet sun from radio observations.   See Abstr. 074.082.

Temperature-density structure in coronal helmets: The quiescent prominence and coronal cavity.   See Abstr. 074.088.

On the structure of the solar transition zone in active regions.   See Abstr. 076.003.

Solar soft X-rays and solar activity. II. Observational assessment of the role of the type III acceleration mechanism in establishment of the soft X-ray source volume.   See Abstr. 076.008.

Accurate calculations of wavelengths of certain helium-like lines in the X-ray region.   See Abstr. 076.010.

A comparison of thermal and non-thermal solar flare X-ray emission observed on OSO-7.   See Abstr. 076.014.

The soft X-ray flare of 12 August 1970.   See Abstr. 076.015.

Hard X-rays and limits on gamma line emission from a solar flare on 22 November 1971.   See Abstr. 076.018.

EUV emissions of solar flares: A comparison of OSO 6 observations and SFD's.   See Abstr. 076.023.

EUV emissions of solar flares: OSO 4 and OSO 6 observations.   See Abstr. 076.024.

Polarization of hard X-rays from solar flares.   See Abstr. 076.026.

The directivity and polarisation of thick target X-ray bremsstrahlung from solar flares.   See Abstr. 076.038.

Preliminary interpretation of the polarization measurements performed on 'Intercosmos-4' during three X-ray solar flares.   See Abstr. 076.049.

A simulation of the directivity effect to be expected in hard X-ray flares.   See Abstr. 076.050.

The time behaviour of the continua during the initial stage of type IV bursts.   See Abstr. 077.006.

The quiet sun brightness distributions at millimeter wavelengths and chromospheric inhomogeneities. See Abstr. 077.016.

Brightness distribution of the sun at 8.6 mm wavelength.   See Abstr. 077.039.

Free-free absorption of gyrosynchrotron radiation in solar microwave bursts.   See Abstr. 077.047.

A 350 MHz radio event associated with the solar flare photographed at the Boyden Observatory on the 11th August, 1972, at 14h 44m SAST.   See Abstr. 077.050.

Centimeter radiation associated with the solar limb prominence of 8 February 1972.   See Abstr. 077.051.

Some studies on solar microwave bursts of different types in relation to optical flares and other allied events. See Abstr. 077.066.

A search for neutrons of solar origin using balloon borne detectors 1967–69.   See Abstr. 078.003.

Azimuthal propagation of low-energy solar-flare protons as observed from spacecraft very widely separated in solar azimuth.   See Abstr. 078.004.

Several observations of low-energy solar-proton spectra and possible interpretations.   See Abstr. 078.005.

Polar-cap measurements of solar-flare protons with energies down to 12.4 kev.   See Abstr. 078.019.

Evidence for a two-component injection of cosmic rays from the solar flare of 1969, March 30. See Abstr. 078.033.

Free oscillations of the sun and their possible stimulation by solar flares.   See Abstr. 080.021.

On the small-scale structure of solar magnetic fields.   See Abstr. 080.062.

A model of the quiet solar atmosphere. See Abstr. 080.063.

August solar activity and its geophysical effects. See Abstr. 085.003.

074 Solar Corona, Solar Wind

074.001 List of observations of the yellow coronal line made at Lomnický Štít. M. Rybanský.
Bull. Astron. Inst. Czechoslovakia, Vol. 23, 238 (1972).

The paper gives a list of all observations of the yellow coronal line made at Lomnický Štít between 1966 and 1970.

074.002 Reduction of the intensities of the 5303 Å and 6374 Å lines observed at Lomnický Štít to a new scale. M. Rybanský.
Bull. Astron. Inst. Czechoslovakia, Vol. 23, 239 - 241 (1972).

The paper describes the method of reducing the intensities of the 5303 Å and 6374 Å coronal lines, observed during the period 1965 - 1970, to a new scale used at Lomnický Štít since 1 January 1971.

074.003 Velocity and flux dependence of the solar-wind helium abundance.
J. Hirshberg, J. R. Asbridge, D. E. Robbins.
Journ. Geophys. Res., Vol. 77, 3583 - 3588 (1972).

This report presents data demonstrating that the relative helium abundance varies with the solar-wind velocity and also data confirming that the percentage of helium does not show the decrease with decreasing proton flux expected from the calculations performed by *Geiss et al.* [1970].

074.004 Detection of solar-wind electron plasma frequency fluctuations in an oblique nonlinear magnetohydrodynamic wave.
R. W. Fredricks, F. L. Scarf, C. T. Russell, M. Neugebauer.
Journ. Geophys. Res., Vol. 77, 3598 - 3601 (1972). – Letter.

074.005 Temperature and emission measure deduced by coronal visible lines. M. Landini, B. C. Fossi.
Astron. Astrophys., Vol. 20, 157 - 158 (1972).

Coronal electron temperature versus emission measure is obtained from visible lines, by means of recent computations of ionization balance for coronal ions. Comparison is made with similar results obtained by Pottasch for EUV lines.

074.006 Density of the solar corona from occultations of NP 0532. C. C. Counselman III, J. M. Rankin.
Astrophys. Journ., Vol. 175, 843 - 856 (1972).

We report the first determination of the mean free-electron density of the solar corona based on pulsar observations. We observed pulsar NP 0532 during the solar near-occultations in mid-June of both 1969 and 1970. From measurements of the radiofrequency dispersion over a wide range of solar elongation we have been able to determine solar corona densities in the interval from 5 to $20 \, r_\odot$.

074.007 The inner solar corona polarization at the total eclipse on March 7, 1970. A. A. Sazanov.
Astron. Zhurn. Akad. Nauk SSSR, Vol. 49, 827 - 832 (1972). In Russian. English translation in Soviet Astron. AJ, Vol. 16, No. 4.

Successive pictures of the inner solar corona were taken through a polarizer at three position angles during the total eclipse on March 7, 1970 in Miahuatlan (Mexico). From the isophotes the degree of polarization was calculated for the heights $0.1 - 0.5 \, R_\odot$.

074.008 Solar wind heating.
A. Barnes, J. C. Brandt, R. E. Hartle, C. L. Wolff.
Cosmic Electrodynamics, Vol. 3, 254 - 259 (1972).

A number of workers have developed detailed models aimed at understanding energy transport in the solar wind and we analyze, in this note, the differences in objectives, physical

assumptions, and results among the most recent of these models: Barnes, Hartle, and Bredekamp (1971); Cuperman and Harten (1971); Wolff, Brandt, and Southwick (1971).

074.009 Z-dependence of the level intervals in $2s^2 \, 2p^2$, $2s^2 \, 2p^3$ and $2s^2 \, 2p^4$. B. Edlén.
Solar Physics, Vol. 24, 356 - 367 (1972).

Values of the level intervals in the ground configurations have been critically compiled from laboratory observations and from observations of nebular and coronal forbidden transitions. The data are represented within experimental errors by means of semi-empirical extrapolation formulae which contain from 3 to 5 adjusted parameters. The results provide means for checking laboratory and astrophysical identifications and measurements. Tables of 'best' level values are given for the configurations concerned.

074.010 Atlas of magnetic fields in the solar corona.
G. Newkirk, Jr., D. E. Trotter, M. D. Altschuler, R. Howard.
Solar Physics, Vol. 24, 370 - 372 (1972).

An atlas providing maps of the coronal (current-free) magnetic field from August 1959 to June 1970 is now available. The atlas contains some 11000 maps depicting magnetic field lines from $1.0 \, R_\odot$ to $3.0 \, R_\odot$ from the solar center, at $10°$ intervals of longitude. There are two presentations for each longitudinal position. The maps are preceded by an explanation of their construction and limitations and a legend to facilitate their use. To be obtained from D. E. Trotter, High Altitude Observatory, Boulder, Co., U.S.A. Price $ 50.00.

074.011 A hydrodynamical study of the large NE coronal streamer observed during the 7th March eclipse.
S. Koutchmy.
Solar Physics, Vol. 24, 373 - 384 (1972).   In French.

The large coronal streamer located on the N-E limb of the sun is located in space. The density distribution in the core of the streamer and for $r > 1.5 \, R_\odot$ is given as well as the half-widths of the density distribution perpendicular to the axis of the streamer. The variation of the expansion velocity has been derived from the hydrodynamical continuity equation. The values of the corpuscular flux of the streamer are calculated and agree with the value of the flux corresponding to the density enhancements of the solar 'wind', measured in situ.

074.012 Coronagraphic observations of an enhanced coronal region. II. Temperature and density structure through the enhanced region. R. R. Fisher.
Solar Physics, Vol. 24, 385 - 394 (1972).

Ratios of emission line intensities are used to calculate the variation of temperature and the variation of electron density as a function of ion class for differing paths through a coronal enhancement. The data indicate (a) a peak mean electron density of $2.3 \times 10^9 \, cm^{-3}$, (b) a temperature maximum greater than $2.3 \times 10^6 \, K$, and (c) the non-coincidence of the peak temperature and peak mean electron density. The abundance of Ni was found to be equal to 0.045 that of Fe from the line ratio $I(\lambda 6702)/I(\lambda 7059)$ and a density model based on the variation of the ratio $I(\lambda 8024)/I(\lambda 6702)$.

074.013 The derivation of temperature gradient and electron density maps from EUV spectroheliograms.
G. L. Withbroe.
Solar Physics, Vol. 25, 116 - 126 (1972).

We discuss spatial variations in electron density at the base of the corona and in the temperature gradient in the chromospheric-coronal transition layer as determined from analysis of maps constructed from Mg X and O VI spectroheliograms. Both the mapping techniques and results of analyzing EUV spectra from OSO 6 observations are presented. Comparisons of these maps with photospheric magnetograms and spectroheliograms made in chromospheric EUV lines and continua indicate that the electron density and temperature gradient in the transition layer tend to be enhanced in areas where the photospheric magnetic field and chromospheric EUV emission are enhanced.

074.014 The heating of the solar plasma due to microwave phenomena correlated with type II meter bursts.
E. Fürst.
Solar Physics, Vol. 25, 178 - 187 (1972).

The heating of the solar plasma of those layers is considered where the microwave bursts are emitted. Bursts of this kind are excited by shock-waves initiated near the optical flare region. These shock-waves spread out into the higher corona, and if the shock strength is sufficiently high, the microwave region is heated to 10^7 K. A plasma of this temperature with an electron density about 5×10^9 cm^{-3} and a magnetic induction of 300 G is optically thick even at frequencies about 10 GHz, because the gyromagnetic absorption is very high.

074.015 A possibly direct measurement of coronal magnetic field strengths. H. Rosenberg.
Solar Physics, Vol. 25, 188 - 196 (1972).

During a unique solar radio event on March 2, 1970, among very diverse features, a very regular pattern was observed in the solar radio spectrographic record between 220 and 320 MHz. The proposed explanation is emission by slightly relativistic electrons at the sum frequency of the plasma frequency (≈ 160 MHz) and the lower harmonics of the local gyrofrequency (n from 3 to 10), rather than emission at pure harmonics of the gyrofrequency (n from 15 to 20). The derived magnetic field strengths range from 3 to 8 G.

074.016 The magnetic field configuration of the solar corona after a proton flare. J.-R. Roy.
Journ. Roy. Astron. Soc. Canada, Vol. 66, 220 (1972).
Abstr. Canadian Astron. Soc.

074.017 Collisionless solar wind protons: A comparison of kinetic and hydrodynamic descriptions.
E. Leer, T. E. Holzer.
Journ. Geophys. Res., Vol. 77, 4035 - 4041 (1972).

In this report, we compare equivalent kinetic and hydrodynamic models of a collisionless solar wind proton gas, in an attempt to discover the degree of validity of the hydrodynamic description. Viscosity and thermal conduction are neglected in the hydrodynamic treatment, but a proton thermal anisotropy is included.

074.018 Solar-wind properties at the earth as predicted by one-fluid models. B. R. Durney.
Journ. Geophys. Res., Vol. 77, 4042 - 4051 (1972).

The spiraling magnetic field of the sun reduces the electron conductivity κ by the factor $\cos^2 \theta$, where θ is the spiral field angle. For a variety of values of the density and temperature at the base of the corona, we compute one-fluid solar-wind models for thermal conductivities equal to κ and $\kappa \cos^2 \theta$ For both cases, the values of the computed solar-wind parameters at the earth are compared with observed properties.

074.019 Helium abundance variations. K. W. Ogilvie.
Journ. Geophys. Res., Vol. 77, 4227 - 4232 (1972).

In this note, we discuss and present observations of helium ions in the solar wind made by experiments aboard the satellites Explorer 34 and Explorer 43. Besides presenting new results, we shall compare these results with those recently published by Hirshberg et al. (1972), obtained from the ana-

lysis of data from Vela 3A and 3B.

074.020 Current sheets in coronal streamers.
E. R. Priest, D. F. Smith.
Astrophys. Letters, Vol. 12, 25 - 29 (1972).

The current sheet in a coronal streamer has been analyzed by Pneuman. An improvement to his model, in which the equations are solved in a more consistent manner, is presented here. As a result, the radial magnetic field drops off more rapidly with distance, the velocity at the base of the current sheet is larger, and the transverse magnetic field is significantly smaller than in Pneuman's model.

074.021 The F and K components of the solar corona.
R. Calbert, D. B. Beard.
Astrophys. Journ., Vol. 176, 497 - 509 (1972).

By taking advantage of the different spatial dependencies of electrons and dust in the solar system, we have developed a reliable method of separating the K component from the F component in measurements of the total solar coronal light as a function of elongation angle. By assuming a sum of inverse power of distance terms for electron density and using the same observational data, we obtain dust and electron density separation in good agreement with others who have used polarization and spectroscopic methods for the separation.

074.022 Properties of a coronal "hole" derived from extreme-ultraviolet observations.
R. H. Munro, G. L. Withbroe.
Astrophys. Journ., Vol. 176, 511 - 520 (1972).

The present paper describes results of an analysis of EUV observations of a large coronal "hole" observed by the Harvard College Observatory experiment on OSO-4 (Goldberg et al. 1968; Reeves and Parkinson 1970) in 1967 November.

074.023 The influence of velocity of electrons on polarization in the solar corona theory. M. M. Molodensky.
Solnechnye Dannye 1972 Byull., No. 4, p. 78 - 84 (1972).
In Russian.

An attempt is made to explain the fact that in some regions the level of the solar corona polarization is higher than the maximum one assumed by Thomson scattering and when the direction of E-oscillations does not coincide with the tangential one. The velocity of the scattering electrons may be responsible for this fact. The relativistic formula of polarization is given.

074.024 On photographing the form of the solar corona taken in the 5302.8 Å line at the High-Altitude Station (near Kislovodsk).
M. N. Gnevyshev, S. B. Ioffe, V. I. Makarov, T. A. Smirnova.
Solnechnye Dannye 1972 Byull., No. 4, p. 105 (1972).
In Russian.

074.025 Variations of solar wind parameters, magnetic activity and electrons of the magnetosphere's tail and of the outer radiation zone. K. G. Ivanov, N. V. Mikerina.
Geomagn. Aeronom., Vol. 12, 688 - 692 (1972). In Russian.

074.026 On determining the electron density distribution of the solar corona from K-coronameter data.
M. D. Altschuler, R. M. Perry.
Bull. American Astron. Soc., Vol. 4, 377 (1972). – Abstr. AAS.

074.027 A general program for numerical integration of Thomson scattering from coronal streamer models.
J. D. Bohlin, L. M. Garrison.
Bull. American Astron. Soc., Vol. 4, 378 (1972). – Abstr. AAS.

074.028 The coronal origin of a solar flare.
G. E. Brueckner.

Bull. American Astron. Soc., Vol. 4, 378 (1972). – Abstr. AAS.

074.029 **Coronal abundances and a model of the quiet sun from radio observations.**
C. Chiuderi, F. Chiuderi Drago, G. Noci.
Bull. American Astron. Soc., Vol. 4, 379 (1972). – Abstr. AAS.

074.030 **Interplanetary solar wind electrons to 1 MeV.**
T. L. Cline.
Bull. American Astron. Soc., Vol. 4, 379 - 380 (1972).
Abstr. AAS.

074.031 **Eclipse measurements of the coronal iron abundance and electron temperature.**   J. A. Eddy.
Bull. American Astron. Soc., Vol. 4, 382 (1972). – Abstr. AAS.

074.032 **Evolution of coronal helmets during the ascending phase of solar cycle 20.**
S. F. Hansen, R. T. Hansen, C. Garcia.
Bull. American Astron. Soc., Vol. 4, 383 (1972). – Abstr. AAS.

074.033 **Observations of coronal forms: 7 March - 7 June 1970.**   R. T. Hansen, S. F. Hansen, G. A. Newkirk, R. M. MacQueen, J. T. Gosling, A. I. Poland.
Bull. American Astron. Soc., Vol. 4, 383 (1972). – Abstr. AAS.

074.034 **Observations of coronal forms: 31 July - 13 September 1971.**
R. M. MacQueen, C. L. Ross, R. T. Hansen, A. Dollfus, Z. Mouradian, A. Worden.
Bull. American Astron. Soc., Vol. 4, 387 (1972). – Abstr. AAS.

074.035 **Properties of a coronal "hole" from EUV observations.**   R. H. Munro, G. L. Withbroe.
Bull. American Astron. Soc., Vol. 4, 388 (1972). – Abstr. AAS.

074.036 **Observations of the solar corona using extreme ultraviolet and X-ray spectroheliographs on the OSO-7 satellite.**   W. M. Neupert, J. H. Underwood, R. J. Thomas.
Bull. American Astron. Soc., Vol. 4, 388 (1972). – Abstr. AAS.

074.037 **Temperature – density structure of the solar corona.**   G. W. Pneuman.
Bull. American Astron. Soc., Vol. 4, 390 (1972). – Abstr. AAS.

074.038 **Equatorial coronal arches.**
C. Sawyer, S. F. Hansen.
Bull. American Astron. Soc., Vol. 4, 390 (1972). – Abstr. AAS.

074.039 **A search for compressional waves in the inner white light corona.**
H. U. Schmidt, W. J. Wagner, G. Newkirk, Jr.
Bull. American Astron. Soc., Vol. 4, 390 (1972). – Abstr. AAS.

074.040 **The coronal transient of 1970 March 21.**
K. Sheridan, C. Garcia, R. Hansen.
Bull. American Astron. Soc., Vol. 4, 391 (1972). – Abstr. AAS.

074.041 **Rotation of active regions in the corona.**
G. W. Simon.
Bull. American Astron. Soc., Vol. 4, 391 (1972). – Abstr. AAS.

074.042 **Movement of a bright source in the white-light corona.**   R. Tousey, M. Koomen.
Bull. American Astron. Soc., Vol. 4, 394 (1972). – Abstr. AAS.

074.043 **Photometric and polarimetric analysis of the coronal streamers observed at the March 7, 1970 Mexican eclipse.**   K. Saito.
Ann. Tokyo Astron. Obs., Second Ser., Vol. 13, 93 - 148 (1972).

In order to investigate the three-dimensional structure and physical conditions of streamers in the solar corona, three regular and four polarized photographs of the white-light corona were obtained at the 1970 Mexican eclipse. These photographs covered the corona out to 5 solar radii. A horizontal camera of $f = 5$ m equipped with a radially-graded neutral density filter and a rotatable polarizing filter at the focus was used.

074.044 **Polarigraphic observations of the solar corona at the total eclipse on March 7, 1970 in Mexico.**
S. Hata, A. Tojo.
Ann. Tokyo Astron. Obs., Second Ser., Vol. 13, 149 - 167 (1972).

At the total eclipse on March 7, 1970 in Puerto Escondido the white light corona was observed photographically with two sets of quadruple-lens cameras ($f = 228$ cm and $f = 30$ cm, respectively). The following data were obtained: (1) brightness data up to 21 solar radii and at 36 position angles every ten degrees apart; (2) polarization data up to 6 solar radii and at the above-noted position angles; (3) the magnetic vector of the polarization in the white light corona; (4) the sky brightness during totality, its polarization value and orientation.

074.045 **Shock waves in the solar system.**   J. R. Spreiter.
Astronaut. Acta, Vol. 17, 321 - 338 (1972).
Presented at the international colloquium on gasdynamics of explosions, Marseille, September 12 - 17, 1971.

074.046 **On physical conditions in the corona from spectral observations of the solar eclipse on March 7, 1970.**
E. A. Gurtovenko, K. V. Alikayeva.
Astrometriya i Astrofiz., *Kiev*, No. 15, (see 003.001), p. 55 - 63 (1972). In Russian.

The emission lines λ 5303 Fe XIV, λ 4231 Ni XII, λ 4567 Cr IX and λ 3987 Fe XI of the solar corona were investigated. The halfwidths and total numbers of particles in the line of sight are determined from the line profiles. Emission lines of various classes are considered arising in the various parts of the corona. Physical conditions in the region of the corona within the heights of 48000 to 100000 km and position angles 36°E to 42°E are studied by treating the λ 5303 Å and λ 4231 Å lines. The distributions of n_e and T_e in the corona are found from excitation and ionization equilibrium.

074.047 **Analysis of three-station interplanetary scintillation.**
J. W. Armstrong, W. A. Coles.
Journ. Geophys. Res., Vol. 77, 4602 - 4610 (1972).

Measurements of the scintillation of small-diameter radio sources caused by the solar wind can be used to derive information about the source structure and to measure some of the parameters of the solar wind. The purpose of this paper is to discuss the methods of analysis of three-station interplanetary scintillation that can be applied to the latter problem.

074.048 **Solar-wind and interplanetary electron measurements on the Apollo 15 subsatellite.**
K. A. Anderson, L. M. Chase, R. P. Lin, J. E. McCoy, R. E. McGuire.
Journ. Geophys. Res., Vol. 77, 4611 - 4626 (1972).

Measurements of high-energy solar-wind electrons have been made from a low orbit around the moon. We describe briefly the main features of the spacecraft and the particle detectors and give the results obtained to date from the plasma and energetic-particle detectors concerning the interaction of the solar wind with the moon.

074.049 **Upper limit of the torque of the solar wind on the earth.**   J. Hirshberg.
Journ. Geophys. Res., Vol. 77, 4855 - 4857 (1972). – Letter.

074.050 **A note on large velocity discontinuities in the solar wind.** L. F. Burlaga.
Geomagn. Aeronom., Vol. 12, 797 - 799 (1972). In Russian.

074.051 **The results of coronal investigation at the September 22, 1968 solar eclipse.**
Ts. S. Khetsuriani, E. I. Tetruashvili.
Solar Physics, Vol. 25, 343 - 346 (1972).

The 1968 Abastumani Astrophysical Observatory solar eclipse expedition obtained photographic records of the polarization and intensity of the solar corona on September 22. A photometric study of the corona was carried out. Polarization has been computed both in a total corona and in some of its streamers. The coronal intensity I_K and I_F components are separated. Electron concentrations and temperatures are computed.

074.052 **Émissions 'froides' dans la couronne solaire.**
J.-L. Leroy.
Solar Physics, Vol. 25, 413 - 417 (1972).

Prominences have been photographed through a coronagraph and an Hα Lyot filter with long exposure times. Faint Hα emissions are often detected down to the threshold 2×10^{-6} times of the sun's brightness; they show definite structures but their relations to the low-level ordinary prominences are not very clear. Estimates are given for the density and thickness of such cool regions.

074.053 **Particle motions in coronal streamers and type III radio bursts.** D. F. Smith, G. W. Pneuman.
Solar Physics, Vol. 25, 461 - 477 (1972).

Both individual and collective motions of electron and proton streams in the current sheet which is thought to exist near the center of a coronal streamer are considered. Unlike previous analyses, closed field lines which must exist when finite conductivity is taken into account as well as a B_ϕ field due to solar rotation are present. The possibility that the stream could collectively drag the closed field lines out with itself is considered.

074.054 **Magnetic and electric waves in space.**
C. T. Russell.
Astrophys. Space Sci. Library, Vol. 32, (see 012.009), 39 - 50 (1972).

074.055 **Evidence for two maxima of activity in the 20th solar cycle.** S. Cuperman, A. Sternlieb.
Solar Physics, Vol. 25, 493 - 497 (1972).

Analysis of the 5303 Å coronal line intensity and of the sunspot activity during the period 1962–1970 confirms the existence of two distinct maxima of solar activity, in accordance with the previous findings of Gnevyshev for the period 1954–1960.

074.056 **Electron density distribution in a coronal condensation.** V. P. Vasilyev.
Astron. Zhurn. Akad. Nauk SSSR, Vol. 49, 1073 - 1077 (1972). In Russian. English translation in Soviet Astron. AJ, Vol. 16, No. 5.

On the basis of solar corona eclipse photometry made on September 22, 1968, an empirical law of the distribution of the surface brightness of a coronal condensation has been obtained which points to spherical symmetry of the condensation clots. According to the method worked out for the corona on the whole, an analytical expression for electron concentration as function of the distance from the condensation clot core has been obtained.

074.057 **Temperature distribution in the solar corona.**
E. Ja. Vilkovisky.
Astron. Zhurn. Akad. Nauk SSSR, Vol. 49, 1125 - 1127 (1972). In Russian. English translation in Soviet Astron. AJ, Vol. 16, No. 5.

Theoretical and observed temperature distributions are compared both for open and closed structure elements. The coincidence certifies the accuracy of the heat balance analysis.

074.058 **Contribution à l'étude de la couronne solaire en expansion.** S. Koutchmy.
Thesis, Sci. Phys. Univ. Paris. Centre Documentation C.N.R.S. (1972-05-10), 180 pp. (1972).

074.059 **Interaction of the solar wind with the neutral component of the interstellar gas.** T. E. Holzer.
Journ. Geophys. Res., Vol. 77, 5407 - 5431 (1972).

A model is constructed to represent the interaction between the solar wind and the neutral component of the interstellar gas. It is found that the neutral gas has several important effects on the solar-wind expansion beyond the orbit of the earth and that it should be possible to infer the presence of the neutral gas from observations of the solar wind made by a space probe traveling into the outer solar system.

074.060 **Energy and mass content of high-speed solar-wind streams.**
M. D. Montgomery, S. J. Bame, A. J. Hundhausen.
average energy and mass fluxes for dominant high-speed streams are found to be 9×10^{25} erg/sec and 8×10^9 g/sec, respectively. The solar-wind energy and mass flux density averaged over a solar rotation remained remarkably constant.

074.061 **Compressions and rarefactions in the solar wind: Vela 3.**
J. T. Gosling, A. J. Hundhausen, V. Pizzo, J. R. Asbridge.
Journ. Geophys. Res., Vol. 77, 5442 - 5454 (1972).

Large non-shock-associated proton density enhancements in the solar wind observed by the Vela 3 satellites have been studied by a superposed epoch analysis. We present the results of an investigation of the origin of density variations in the solar wind, using 3-hour averages of Vela 3 data. We find that the interaction of interplanetary streams of different speeds is one of two major sources of the high densities observed at 1 AU. (Flare-associated shock waves are the other major source). Average time-space profiles of interacting streams have been obtained; the profiles compare favorably with the nonlinear models of *Matsuda* and *Sakurai* [1972], *Goldstein* [1971], and *Hundhausen* [1971].

074.062 **Solar-wind velocity from IPS observations.**
W. A. Coles, S. Maagoe.
Journ. Geophys. Res., Vol. 77, 5622 - 5624 (1972).

The solar-wind velocity has been measured daily from March to June 1972 by using interplanetary scintillation (IPS) of the radio sources 3C48 and 3C144, which have ecliptic latitudes of +21° and −1°, respectively. These preliminary results show that the velocity is higher and more variable in the direction of the high-latitude source but that there is no significant deviation from radial outflow.

074.063 **The recurrent solar wind streams observed by interplanetary scintillation of 3C 48.**
T. Watanabe, T. Kakinuma.
Publ. Astron. Soc. Japan, Vol. 24, 459 - 467 (1972).

The interplanetary scintillation of 3C 48 was observed by two spaced receivers (69.3 MHz) during February and March 1971. The recurrent property of the observed velocity increases if the solar wind is clearly seen, and their recurrent period is 24 to 25 days. A comparison with the data of the wind velocity obtained by space probes shows that the observed enhancements are associated with two high velocity streams corotating around the sun.

074.064 Solar wind observations on the lunar surface with the Apollo-12 ALSEP.
M. Neugebauer, C. W. Snyder, D. R. Clay, B. E. Goldstein.
Planet. Space Sci., Vol. 20, 1577 - 1591 (1972).

The Apollo-12 ALSEP solar wind spectrometer obtained data from the lunar surface starting November 20, 1969. In this report, we present initial results on solar wind observations, sunrise and sunset observations, and an anomalous presunrise plasma flux.

074.065 Long-lived sectors of enhanced density irregularities in the solar wind.   Z. Houminer, A. Hewish.
Planet. Space Sci., Vol. 20, 1703 - 1716 (1972).

Observations of interplanetary scintillation on 32 radio sources over a period of 8 months indicate the presence of enhanced scintillation sectors which usually persist for one solar rotation or longer. The structure of these sectors, both in and out of the plane of the ecliptic is described. A strong correlation is found between these sectors and the velocity structure of the solar wind derived from spacecraft observations. This suggests an origin of enhanced scintillation in terms of a fast-slow stream interaction.

074.066 The sun and its corona.   R. Michard.
Sci. Progrès Découverte, 98e année, No. 3428, p. 33 - 41 (1970). In French.

This is a general review of our knowledge of the solar corona. In particular, the author reviews the various types of emissions for which the corona is the source, both electromagnetic (light) and radioelectric. He then examines the various hypotheses on the origin of the very high temperatures which are observed in this part of the solar atmosphere.

074.067 Interplanetary gas. XVII. An astrometric determination of solar-wind velocities from orientations of ionic comet tails.
J. C. Brandt, R. G. Roosen, R. S. Harrington.
Astrophys. Journ., Vol. 177, 277 - 284 (1972).

A statistical analysis of ionic comet-tail position angles on the sky using no weights and no assumptions of coplanarity yields a model of the solar wind with mean properties of a radial bulk velocity $w_r = 415$ km s^{-1}, azimuthal bulk velocity $w_\phi = 6$ km s^{-1}, polar bulk velocity $w_\theta = 0$ km s^{-1}, and an isotropic, peculiar velocity of about 40 km s^{-1}. Thus, the solar wind slows the solar rotation with an e-folding time comparable to the age of the sun.

074.068 The solar EUV-emitting plasma.
R. W. Noyes, G. L. Withbroe.
Space Sci. Rev., Vol. 13, 612 - 637 (1972). – Invited paper IAU Colloquium No. 14 (see 012.012).

074.069 The coronal X-spectrum: Problems and prospects.
A. B. C. Walker, Jr.
Space Sci. Rev., Vol. 13, 672 - 730 (1972). – Invited paper IAU Colloquium No. 14 (see 012.012).

074.070 Mapping the solar corona in X-ray lines of O VII and Ne IX.
R. C. Catura, L. W. Acton, A. J. Meyerott, J. L. Culhane.
Space Sci. Rev., Vol. 13, 742 - 743 (1972). – Conference paper IAU Colloquium No. 14 (see 012.012).

074.071 Radio evidence of twisted bi-polar magnetic fields in the solar corona.
D. J. McLean, K. V. Sheridan.
Solar Physics, Vol. 26, 176 - 182 (1972).

Observations of bi-polar type I storm centers with the Culgoora radioheliograph operating at 80 MHz show that in many cases they are not oriented as we should expect for emission in the ordinary mode and for the simplest magnetic

field geometry. We interpret this as evidence for a twist in the magnetic field.

074.072 Coronal survey in X-rays of O VII and Ne IX.
L. W. Acton, R. C. Catura, A. J. Meyerott, C. J. Wolfson, J. L. Culhane.
Solar Physics, Vol. 26, 183 - 201 (1972).

We report some results of a rocket experiment flown on 29 April, 1971. A survey of the solar corona was carried out with a pair of collimated Bragg spectrometers to study the resonance, intersystem and forbidden line emission from the helium-like ions O VII (22 Å) and Ne IX (13 Å).

074.073 Evolution of coronal helmets during the ascending phase of solar cycle 20.
S. F. Hansen, R. T. Hansen, C. J. Garcia.
Solar Physics, Vol. 26, 202 - 224 (1972).

The principal polar-crown coronal helmet structures were selected from nearly three years (May, 1965 – January, 1968) of K-coronameter observations made at Haleakala and Mauna Loa, Hawaii. Six isolated and long-lived helmet systems were found at latitudes of 45° and above. Their developments are compared with underlying chromospheric and photospheric activity and a simple phenomenological model is presented showing that a coronal system is formed over an active region. By comparison of these coronal helmets with observations of the outer corona (to circa 4 $R_\odot$), it appears that ground-based K-coronameter measurements to a distance of 1.5–2.0 $R_\odot$ are sufficient to detect the coronal streamers.

074.074 Note on solar plasma irregularities and plasma instabilities.   S. K. Alurkar.
Solar Physics, Vol. 26, 225 - 228 (1972).

Observations of interplanetary scintillation of radio sources are used to estimate the size of plasma irregularities down to a distance of about 6 $R_\odot$ from the sun. This is compared with the values of the ion gyro-radius estimated for a range of distance from 1 AU to about 6 $R_\odot$ from the sun. The results of the calculations are discussed in the context of the hypothesis of plasma instability.

074.075 Characteristics of the quiet solar wind beyond the earth's orbit. S. Cuperman, A. Harten, M. Dryer.
Astrophys. Journ., Vol. 177, 555 - 566 (1972).

Solutions of one-fluid model equations for the quiet solar wind between 1 and 12 a.u. are presented. The solar wind is treated as a steady spherically expanding flow, and effects such as viscosity, magnetic fields, solar rotation, interaction with the interstellar medium, and fluctuations are neglected. The results of the integration in the range 1 a.u. $\leqslant r \leqslant$ 12 a.u. are presented in both graphical and numerical form. Analytical expressions describing the space dependence of the solutions obtained are also discussed. Finally, values of the gross features of the solar wind at distances larger than 12 a.u. are presented. They are obtained by the extrapolation of those which resulted from the direct integration up to 12 a.u. noted above.

074.076 Near sun observations of the solar wind.
P. S. Callahan, P. F. MacDoran, A. I. Zygielbaum.
Space Research XII, (see 012.016), Vol. 2, 1529 - 1533 (1972).

074.077 Azimuthal structure in the solar wind.
I. H. Urch.
Cosmic Electrodynamics, Vol. 3, 316 - 329 (1972).

The magneto-hydrodynamic equations which describe an azimuthally dependent solar wind near the equatorial plane are obtained in a simplified form. Some solutions for the solar wind flow resulting from several simple coronal configurations are presented. In particular the mean azimuthal velocity of the solar wind has been calculated and, in the cases studied, the interaction between fast and slow streams does not increase the mean azimuthal velocity as has been found by Sakurai (1971).

074.078 **X-ray emission of coronal condensations during the eclipse on 20 May 1966 and its connection with optical and radio observations.** V. Letfus, M. A. Livshits.
Bull. Astron. Inst. Czechoslovakia, Vol. 23, 307 - 314 (1972).

The electron temperature and the emission measure of three coronal condensations over active regions have been derived from X-ray observations made on the SOLRAD 8 satellite during the 20 May 1966 eclipse. For the analysis it was necessary to distinguish the "quiet" level values from those affected eventually by any activity phenomena. The X-ray data are compared with the optical and radio data, characterizing the investigated coronal condensations. Good correlation between the 8 - 12 Å X-ray fluxes and green line intensities of all three coronal condensations signifies that both types of emission are controlled by the corresponding emission measures.

074.079 **Electrostatic ion cyclotron waves in an anisotropic plasma.** K. L. Vithal, J. N. Tandon.
Astrophys. Space Sci., Vol. 18, 49 - 58 (1972).

In the solar wind, electrostatic ion cyclotron waves can be excited by electrons or ions when the flow velocity becomes supersonic. The instability of these waves is investigated for a situation in which ions are streaming in opposite directions along the interplanetary magnetic field in a uniform background of relatively stationary electrons. Many modes become unstable under the existing conditions. It is conjectured that the excitation of this instability may lead to a steady state electrostatic turbulence in the solar wind.

074.080 **Solar cycle variation and N−S asymmetry of λ5303 coronal intensity.** P. N. Pathak.
Solar Physics, Vol. 25, 489 - 492 (1972).

It is shown that during the present solar cycle (No. 20), the λ5303 coronal intensity at heliographic latitudes between 15°−40° in both hemispheres had two maxima. The N−S asymmetry of λ5303 intensity for the period 1957−1970 is studied and its implications to solar-terrestrial relationships are discussed.

074.081 **Remarks on the heating of solar-wind protons.** J.-K. Chao, N. D'Angelo.
Journ. Geophys. Res., Vol. 77, 6226 - 6229 (1972).
Brief report.

074.082 **Coronal abundance of elements and a model of the quiet sun from radio observations.**
C. Chiuderi, F. Chiuderi Drago, G. Noci.
Solar Physics, Vol. 26, 343 - 353 (1972).

It is shown that the combined use of radio observations of the quiet sun and UV line intensities allows to compute the absolute coronal abundance of the elements. The abundances found by this method agree very well with the most recent determinations. A model of the transition region and corona in hydrostatic equilibrium is also presented. Similarities and differences with models based on UV observations are discussed.

074.083 **Coronal holes.** M. D. Altschuler, D. E. Trotter, F. Q. Orrall.
Solar Physics, Vol. 26, 354 - 365 (1972).

Coronal holes are extensive regions of extremely low density in the solar corona within 60° of latitude from the equator. We have superposed maps of the calculated current-free (potential) coronal magnetic field with maps of the coronal electron density for the period of November 1966, and find that coronal holes are generally characterized by weak and diverging magnetic field lines. The chromosphere underlying the holes is extremely quiet, being free of weak plages and filaments. The existence of coronal holes clearly has important implications for the energy balance in the transition region and the solar wind.

074.084 **On emission lines of hydrogen, helium and ionized calcium seen on a coronal spectrogram of the March 7, 1970 eclipse.**
M. K. V. Bappu, J. C. Bhattacharyya, K. R. Sivaraman.
Solar Physics, Vol. 26, 366 - 369 (1972).

Emission lines of the Balmer series, D_3 and H and K are reported present on a coronal spectrogram obtained at the March 7, 1970 eclipse. Arguments are presented to show that these could not have originated from scattering in the earth's atmosphere and hence possibly have a coronal origin.

074.085 **Equatorial coronal arches and geomagnetic disturbance.** C. Sawyer, S. F. Hansen.
Solar Physics, Vol. 26, 370 - 377 (1972).

Central-meridian passage of transequatorial arches detected in the K-corona tends to be followed after about 7 days by a slight but probably significant enhancement of geomagnetic activity. The arches show no preferred location relative to the equatorial dipole. Metric noise-storm sources tend to cluster at the arches, or just west of them. Together with analyses of earlier and later periods, this suggests that the relation between coronal enhancements and sector boundaries may evolve during the solar cycle.

074.086 **The fine structure of the solar corona on March 7, 1970.** V. I. Ivanchuk.
Astron. Tsirk., No. 716, p. 1 - 4 (1972). In Russian.

074.087 **Fine streamers, coronal background and solar wind.** V. I. Ivanchuk.
Astron. Tsirk., No. 716, p. 4 - 7 (1972). In Russian.

074.088 **Temperature-density structure in coronal helmets: The quiescent prominence and coronal cavity.**
G. W. Pneuman.
Astrophys. Journ., Vol. 177, 793 - 805 (1972).

Physical processes which are most important in determining temperature and density distributions in closed coronal loop systems are studied within the context of how this type of magnetic geometry influences the local balance of mechanical energy flux, conductive flux, and radiative losses. It is demonstrated that the quiescent prominences normally observed at the base of coronal helmets as well as the low-density cavity surrounding these structures are the natural consequences of the adjustment of density and temperature structure in the helmet required to satisfy energy and mass balance conditions as modulated by the magnetic field.

074.089 **Mesures directionnelles du vent solaire par la sonde HEOS I, S 58-73, de l'ESRO.**
R. Coutrez, A. Joukoff, W. Scholiers.
Ciel et Terre, Vol. 88, 426 - 457 (1972).

The twofold experiment S 58-73 aboard the european satellite HEOS 1 measures the energy spectrum (S 73) and the angular distribution (S 58) of the protonic component of the plasma emitted by the sun. Results achieved during the period from Dec. 11, 1968 to Oct. 1970 have made it possible to set up a map of the flux in the magnetosheath and to obtain an overall view of the interplanetary plasma during the solar maximum and the decreasing phase of solar activity.

074.090 **A solar-wind model including proton thermal anisotropy.** Y. C. Whang.
Astrophys. Journ., Vol. 178, 221 - 239 (1972).

The present model divides the interplanetary space into two regions: near 1 a.u. the solar-wind plasma is observed to be two-fluid in nature: the electrons and protons have two different temperatures; near the sun, there is sufficient interaction between them so that in the region $r < 0.1$ a.u. the solar wind is one-fluid in nature: the electrons and protons have the same isotropic temperature. The exchange of energy weakens at distances far away from the sun. Values of the solar-wind velocity, the electron and proton temperature, etc. are given. They are in good agreement with observations.

074.091 Classification, peculiarities of orientation and some examples of rotation discontinuities in the solar wind. K. G. Ivanov.
Geomagn. Aeronom., Vol. 12, 984 - 988 (1972). In Russian.

074.092 Luminosity conditions of Cr IX and Fe XI in the corona. K. V. Alikayeva, Z. M. Bikchantayeva.
Astrometriya i Astrofiz., *Kiev*, Vyp. (No.) 17, (see 003.012), p. 59 - 64 (1972). In Russian.
The coronal lines λ 4566 Å Cr IX and λ 3896 Å Fe XI were investigated. The electron temperature in the corona was found to be 10^6 °K and the electron density equal to ~0.8×10^9 cm^{-3}. The calculations were carried out for Cr IX ion level populations.

074.093 Threadlike coronal streamers. M. Waldmeier.
Solar Physics, Vol. 27, 143 - 148 (1972).
Threadlike streamers have small cross sections that are almost constant over their whole length. A typical value of their diameter is 30000 km. Threadlike streamers may reach a length of several hundred thousand kilometers or even more than a solar radius. A photometric analysis of a threadlike streamer, observed at the eclipse of 1963 July 20, yields an electron density 8 times larger than that of the undisturbed corona. This ratio undergoes only small variations with the distance from the sun.

074.094 High-sensitivity measurement of the emission spectrum of quiescent active regions. R. S. Wolff.
Bull. American Astron. Soc., Vol. 4, 416 (1972). – Abstr. AAS.

074.095 On the theory of magnetohydrodynamic turbulence of the solar wind. L. I. Dorman, M. E. Kats.
Trudy Mezhdunar. seminara po probl. "Uskorenie chastits v kosmich. prostranstve (okolozem. i mezhplanet. kosmich. prostranstve), Galaktike i Metagalaktike". Moskva, 1972, p. 96 - 105. In Russian. – Abstr. in Referativ. Zhurn. 51. Astron., 1.51.401 (1973).

074.096 Latitude distribution of the coronal activity. N. P. Tsimakhovich.
Trudy Radioastrophys. Obs., *Riga,* Vol. 14, (see 003.019), 105 - 109 (1972). In Russian.
From observations of the green coronal line the presence of two antipodal longitudinal intervals of increased activity is established.

074.097 Rotational discontinuities in the solar wind. J. M. Turner.
Rep. NASA-CR-119172, Mass. Inst. Technology, Cambridge, Mass. [Available from NTIS, Springfield, Va.], 32 pp. (1971).
Solar wind plasma and magnetic field data from Mariner 5 are used to identify rotational discontinuities. Of the 40 discontinuities found, 35 were clustered in three distinct three to six day intervals and were characterized by high solar wind bulk velocities, high magnetic field magnitudes, low densities, and high correlation between velocity and magnetic field changes.

074.098 Tangential discontinuities in the solar wind. J. M. Turner.
Rep. NASA-CR-119173, Mass. Inst. Technology, Cambridge, Mass. [Available from NTIS, Springfield, Va.], 26 pp. (1971).
Solar wind plasma and magnetic field data were used to identify tangential discontinuities during the first 40 days of the flight of Mariner 5.

074.099 Observations of coronal magnetic field strengths and flux tubes and their stability. H. Rosenberg.
Cosmic Plasma Physics, [Plenum Publ. Corporation, New York], p. 191 - 193 = Utrechtse Sterrekundige Overdrukken, No. 189 (1972).

074.100 Mathematical analysis and interpretation of the solar corona. R. Calbert.
Thesis, Univ. Kansas, Lawrence. [Available from Univ. Microfilms, Ann Arbor, Mich., USA. Order No. 72-11736], 157 pp. (1971). – See Phys. Abstr., Vol. 75, No. 73153 (1972).

074.101 Helium abundance variations. K. W. Ogilvie.
Rep. NASA-TM-X-65822, NASA, Greenbelt, Md. [Available from NTIS, Springfield, Va.], 18 pp. (1972).

074.102 Solar-wind flow past the planets earth, Mars, and Venus. A. W. Rizzi.
Thesis, Stanford Univ., Stanford, Calif. [Available from Univ. Microfilms, Ann Arbor, Mich., USA. Order No. 72-5982], 211 pp. (1971).
The hydromagnetic theory of solar-wind flow past the earth has been extended and modified so as to be applicable to non-magnetic planets, such as Venus and Mars, that have a sufficient ionosphere to deflect the solar plasma around the planet and its atmosphere.

074.103 Latitude effects in the solar wind. C. R. Winge, Jr.
Thesis, Univ. California, Los Angeles. [Available from Univ. Microfilms, Ann Arbor, Mich., USA. Order No. 72-5873], 113 pp. (1971).
The Weber-Davis model of the solar wind is generalized to include the effects of latitude. The principal assumptions of perfect electrical conductivity, rotational symmetry, a polytropic relation between pressure and density, and a flow aligned magnetic field in a system rotating with the sun, are retained.

074.104 Review of the heliosphere. I. T. Obayashi.
Astron. Herald, (*Japan*), Vol. 65, 115 - 120 (1972). In Japanese.
Reviews the theory of the solar wind, the properties of the solar wind, the interplanetary magnetic field, the tidal wave of solar flares and observations of the solar wind from the earth.

074.105 Review of the heliosphere. II. T. Obayashi.
Astron. Herald, (*Japan*), Vol. 65, 147 - 152 (1972). In Japanese.
Reviews the studies of H II regions, the influencing sphere of the solar wind and the boundary region and neutral particle groups, the interplanetary glow and the interstellar wind, and considers a model for the solar system.

074.106 On the polarization of solar coronal radiation in λ 5303 Å. E. I. Tetruashvili.
Byull. Abastumansk. Astrofiz. Obs., No. 43, p. 175 - 178 (1972). In Russian.
Polarization observations in the red and green coronal lines were performed with the aid of the Lyot-type coronograph of the Abastumani Astrophysical Observatory. The polarization in the green line λ 5303 Å is shown to be small, except for rare cases, when it amounts to 10 % and even more in some active knots. In the red coronal line the polarization is close to zero within the errors of measurements.

074.107   **The determination of electron densities and temperatures in the corona of September 22, 1968.**
Ts. S. Khetsuriani, E. I. Tetruashvili, R. I. Kiladze.
Byull. Abastumansk. Astrofiz. Obs., No. 43, p. 179 - 190 (1972). In Russian.

The F and K components of the solar corona have been separated on the basis of photometric and polarization data obtained by the treatment of September 22, 1968 total solar eclipse observations. The variation of electron concentration from the center of the sun with distance has been computed by Baumbach's (1937) method as well as by taking into account Thomson scattering anisotropy. The temperatures at various distances have been computed.

074.108   **An investigation of the properties of some streamers in the corona of September 22, 1968.**
Ts. S. Khetsuriani, E. I. Tetruashvili.
Byull. Abastumansk. Astrofiz. Obs., No. 43, p. 191 - 202 (1972). In Russian.

A detailed study of two streamers has been made on the basis of September 22, 1968 total solar eclipse observations. In particular the polarization, intensity, electron density and temperature have been studied.

074.109   **The solar corona from observations of an expedition of the Kharkov Astronomical Observatory**
during the solar eclipse on September 22, 1968.
V. I. Bistrits'kij, V. P. Vasil'ehv.
Visn. Kharkiv. Univ. No. 82, (Ser. Astron., No. 7), p. 60 - 66 (1972). In Ukrainian.

074.110   **Preliminary report on the coronal line 6374 and 5303 Å interferograms obtained during the total**
eclipse of July 11, 1972.   A. B. Delone, E. A. Makarova.
Astron. Tsirk., No. 727, p. 1 - 2 (1972). In Russian.

074.111   **On the energy balance in the solar corona.**
E. E. Dubov.
Astron. Tsirk., No. 727, p. 2 - 3 (1972). In Russian.

074.112   **Interferometric investigations of emission lines of the solar corona during the eclipse of 1972 July 10.**
I. S. Kim.
Astron. Tsirk., No. 731, p. 7 - 8 (1972). In Russian.

074.113   **Core electron densities of coronal polar plumes.**
C. J. Psujek, R. G. Teske.
Solar Physics, Vol. 27, 420 - 425 (1972).

The electron density in the cores of coronal polar plumes that is determined from observations will depend upon the assumed electron density distribution through the plume in a direction normal to its axis. Core electron densities obtained by Saito (1965) and by Newkirk and Harvey (1968) were derived using different assumed electron density profiles, and are not in agreement. We have re-discussed Saito's data using Newkirk and Harvey's electron density profile and find that the disagreement persists.

074.114   **The asymptotic behavior of the supersonic solutions of the two-fluid solar wind equations.**
I. W. Roxburgh.
Solar Physics, Vol. 27, 478 - 480 (1972). – Research note.

Coronal expansion and solar wind.
See Abstr. 003.077.

Interpolation formulae for the electron impact excitation of ions in the H-, He-, Li-, and Ne-sequences.
See Abstr. 022.007.

Ionization balance for ions of Na, Al, P, Cl, A, K,
Ca, Cr, Mn, Fe and Ni.   See Abstr. 022.062.

Investigation of the photographic congruency in a system of equidensities.   See Abstr. 031.088.

Adaption de la caméra électronique au coronographe.   See Abstr. 034.076.

Wave coupling at a collisionless plasma discontinuity.   See Abstr. 062.016.

Stabilization of hydromagnetic instabilities by magnetization currents?   See Abstr. 062.042.

Stability of a model current sheet with finite transverse field and finite flow velocity.   See Abstr. 062.052.

Gyro synchrotron radiation from thermal plasmas.
See Abstr. 062.055.

The density dependent ionization balance of carbon, oxygen and neon in the solar atmosphere.
See Abstr. 071.001.

Spatial frequencies of the photosphere and low corona.   See Abstr. 071.033.

Observations of the profiles of solar UV emission lines and their analysis in terms of the heating and production of the corona.   See Abstr. 071.043.

Calculations on the solar spectrum from 1 to 60 Å.
See Abstr. 071.044.

Solar activity and the variations of the geomagnetic K_p-index. I: Photospheric activity.   See Abstr. 071.048.

The solar chromosphere and its transition to the corona.   See Abstr. 073.003.

Correlation between the intensity fields of the chromospheric and coronal networks.   See Abstr. 073.046.

Temperature structure and conductive flux in the chromosphere-corona transition region.   See Abstr. 073.057.

Models of active regions in the transition zone from UV observations.   See Abstr. 073.067.

Experiment to determine the temperature structure in the solar chromosphere and corona.   See Abstr. 073.068.

Temperature structure of the chromosphere—corona transition region.   See Abstr. 073.069.

A model of the chromosphere and of the transition zone from chromosphere to solar corona.   See Abstr. 073.110.

Energy balance in the chromosphere-corona transition region.   See Abstr. 073.115.

A model for the polar transition layer and corona for November 1967.   See Abstr. 073.116.

Lifetime of solar flare particles in coronal storage regions.   See Abstr. 073.117.

Large scale coronal X-ray structures.
See Abstr. 076.019.

The structure of the X-ray corona surrounding

quiescent filaments.   See Abstr. 076.020.

Coronal active region models and relative abundances based on soft X-ray emission line and continuum fluxes. See Abstr. 076.022.

Exciter of type III bursts and coronal temperature. See Abstr. 077.001.

Radio bursts from the solar corona. See Abstr. 077.041.

On the long-term behaviour of the circular polarization from coronal condensation radio emission at 4.3 cm wavelength.   See Abstr. 077.053.

A model of the quiet solar atmosphere. See Abstr. 080.063.

Heat conduction in a turbulent magnetic field, with application to solar-wind electrons.   See Abstr. 084.202.

Explorer 33 and 35 plasma observations of magnetosheath flow.   See Abstr. 084.203.

Radial penetration of a hot plasma associated with a large-scale electric field in the magnetosphere, and some related problems.   See Abstr. 084.210.

Plasma sheet at lunar distance: Structure and solar-wind dependence.   See Abstr. 084.219.

Simultaneous solar-wind plasma and magnetic-field measurements in the expected region of the extended geomagnetic tail.   See Abstr. 084.244.

Consistency of fields and particle motion in the 'Speiser' model of the current sheet.   See Abstr. 084.294.

Geomagnetic field variations caused by changes in the quiet-time solar wind pressure.   See Abstr. 084.299.

Unipolar interaction of Mercury with the solar wind: The steady state bow shock problem.   See Abstr. 092.010.

From lunar samples to solar evolution. See Abstr. 094.140.

Upper limits on the lunar atmosphere determined from solar-wind measurements.   See Abstr. 094.183.

Type I tails — solar wind interactions. See Abstr. 102.062.

Correlation between brightness of comets and fluctuations of the solar wind.   See Abstr. 102.071.

Lyman-alpha radiation of comet Bennett 1969 i and determination of the solar wind flux.   See Abstr. 103.100.

An analysis of Pioneer 9 low-frequency wave observations near interplanetary discontinuities. See Abstr. 106.001.

Pioneer 7 observations of the August 29, 1966, interplanetary shock-wave ensemble.   See Abstr. 106.004.

Magnetic field structure in flare-associated solar-wind disturbances.   See Abstr. 106.016.

Influence of interstellar hydrogen on the location of the heliospheric shock front.   See Abstr. 131.077.

Off-ecliptic control of cosmic ray modulation. See Abstr. 143.046.

Errata

074.901 **Errata: 'New method of photometry of coronal lines 5303 Å and 6374 Å at Lomnický Štít'**
[Bull. Astron. Inst. Czechoslovakia, Vol. 22, 122 (1971].
M. Rybanský.
Bull. Astron. Inst. Czechoslovakia, Vol. 23, 243 (1972).

075   Solar Patrol

075.001 **Catalogue of solar activity for the year 1969.** R. S. Gnevysheva.
Trudy Glav. Astron. Obs. Pulkovo, 156 pp. Price 1 Rbl. 53 Kop. (1972). In Russian.

075.002 **Reporte de actividad solar durante los meses de febrero y marzo del 72.**   F. J. Mandujano O.
El Universo, No. 99, Vol. 26, 53 - 54 (1972).

075.003 **Photoheliographic results 1962, 1963 and 1964** prepared under the direction of the Astronomer Royal.
Roy. Obs. Ann., [Herstmonceux: Royal Greenwich Obs.], No. 6, 138 pp. Price £ 2.20 (1971).
For each year the following data are given: Positions and areas of sunspots for each day; General catalogue of sunspots; Total areas of sunspots and faculae; Mean areas of sunspots and faculae; Mean heliographic latitude of sunspots; Summary of solar activity.

075.004 **Activité solaire en 1970.**
G. Evrard, C. Gonze, A. Koeckelenbergh.
Ciel et Terre, Vol. 88, 273 - 291 (1972).

075.005 **Solar activity during 1971.**   H. Hill.
Journ. British Astron. Ass., Vol. 82, 449 - 454, with a correction in Vol. 83, 60 (1972). — Report of the Solar Section of the British Astron. Ass.

075.006 **Heliographic maps of the photosphere for the year 1971.**   M. Waldmeier.
Publ. Sternw. Zürich, Vol. 14, (No. 1), 1 - 31 (1972).
The present publication gives heliographic maps of the photosphere and evolution tables of sunspot-groups for the year 1971. Maps and tables are based on daily drawings of spots and faculae using a projected solar image with a diameter of 25 cm. Such drawings are carried out at the Swiss Federal Observatory Zürich and at its two branch stations, the

Astrophysical Observatory Arosa and the Specola Solare Locarno-Monti.

075.007 Sunspots (sunspot relative numbers and sunspot-areas); Synoptic charts of solar magnetic fields (Mount Wilson Observatory); **Eruptions chromosphériques brillantes; Intensité de la couronne solaire; Solar radio emission.** M. Waldmeier, R. Howard, R. Michard, G. Olivieri, M. Bernot. Quarterly Bull. Solar Activity (published by Eidgen. Sternw. Zürich), Nos. 175 - 176, p. 73 - 147 (1972). – Observations of the co-operating observatories for 1971 July - December are given.

075.008 Sunspot numbers. Sky Telescope, Vol. 44, 129, 203, 272, 339, 411 (1972). – 1972 June – October.

075.009 Summary of solar observations for the first half of 1969. M. K. V. Bappu. Kodaikanal Obs. Bull., Ser. B, No. 206, B141 - B168 (1971).

075.010 Summary of magnetic and ionospheric observations January – June, 1967. M. K. V. Bappu. Kodaikanal Obs. Bull., Ser. C, No. 198, 304 pp. (1970/71).

075.011 Solar observations made at Catania Astrophysical Observatory during 1971. G. Godoli, V. Sciuto, M. L. Sturiale, R. A. Zappalà, E. Catinoto, G. Domina, G. Celeani, G. Sapienza. Oss. Astrofis. Catania, Pubbl. No. 148, 118 pp. (1972). This bulletin includes all the data deduced from the solar observations made during 1971 at Catania Astrophysical Observatory: Sunspots; $H\alpha$ and K faculae; $H\alpha$ flares; $H\alpha$ quiescent prominences; K quiescent prominences; $H\alpha$ active prominences on disc and at limb; $H\alpha$ disk and limb patrol hours.

075.012 Nota sobre actividad solar en 1971. J. A. Gutierrez. Obs. Astron. Municipal Rosario, (*Argentina*), Bol. No. 2, p. 14 - 22 (1972).

075.013 Fenomeni solari. F. Mazzucconi, S. Delli Santi, M. L. Sturiale, A. Abrami, V. Sciuto. Coelum, Vol. 40, 157 - 163, 203 - 209, 242 - 246 (1972). 1972 March - August.

075.014 Osservatorio Magnetico de l'Aquila. Bollettino magnetico. Coelum, Vol. 40, 169, 212, 248 (1972). – 1972 February - July.

075.015 Centro Universitario Fenomeni fluttuanti – Firenze. Test P. Coelum, Vol. 40, 170, 211, 247 (1972). – 1972 March – August.

075.016 L'activité solaire. M.-J. Martres. L'Astronomie, 86e année, p. 376 - 377, 441 - 443, 498 - 499, 551 - 553 (1972). – Rotations 1582 - 1588.

075.017 Geomagnetic and solar data. J. V. Lincoln (Editor). Journ. Geophys. Res., Vol. 77, 3630, 4280, 4903, 5639, 6301, 6930 (1972). – 1972 March – August.

075.018 What new on the sun? W. Szymański. Urania Kraków, Vol. 43, 338 - 342 (1972). In Polish.

075.019 Solar and solar system activity.

R. J. J. Langton, J. R. Smith, K. F. Tapping. Journ. British Astron. Ass., Vol. 82, 379 - 381, 470 - 473; Vol. 83, 54 - 57 (1972). – 1972 March – August.

075.020 Report on sunspots for 1971. P. S. Laurie. Quarterly Journ. Roy. Astron. Soc., Vol. 13, 524 (1972).

075.021 Solar prominences in 1971. M. K. V. Bappu. Quarterly Journ. Roy. Astron. Soc., Vol. 13, 525 (1972).

075.022 Cartes synoptiques de la chromosphère solaire et catalogues des filaments et des centres d'activité. M. J. Martres. Obs. Paris, Section d'Astrophys. Meudon, Vol. 5, Fasc. 3, années 1970 - 1971, 86 pp. (1972). – Rotations Nos. 1556 - 1582, 1969 Décembre 24 – 1971 Décembre 31.

075.023 Datos relativos a la actividad solar y geomagnética en 1970. Urania Barcelona, Año 56, No. 274, p. 176 (1971).

075.024 Solare Beobachtungsergebnisse (Solar Data). C.-U. Wagner, A. Böhme, F. Fürstenberg, D. Scholz, S. Böhm. Zentralinst. für Solar-Terrestrische Physik (Heinrich-Hertz-Inst.), Deutsche Akad. Wiss. Berlin, HHI Solar Data, Vol. 23, March – August (1972). – Solar radio emission.

075.025 Curve of the solar radio radiation from observations of the Observatory of the Department of Astronomy at the Kiev University in Lesnikakh. Kometn. Tsirk., *Kiev*, No. 136 (1972). In Russian.

075.026 Daily maps of the sun and geophysical graphs. Solnechnye Dannye 1972 Byull., No. 4, p. 1 - 58; No. 5, p. 1 - 78; No. 6, p. 1 - 82; No. 7, p. 1 - 75; No. 8, p. 1 - 99; No. 9, p. 1 - 66; No. 10, p. 1 - 78 (1972). In Russian.

075.027 Magnetic fields of sunspots. Prilozhenie k Byulletenyu "Solnechnye Dannye", 1972, Nos. 4–10. In Russian.

075.028 Map of the sun. Edited by Fraunhofer Institut, Freiburg. 1972 July 1 – December 31.

075.029 Grafikoni izlaza i zalaza sunca i mjeseca 1973. Edited by Hidrografski Institut Jugoslavenske Ratne Mornarice, Split. HI-N-32, 29 pp. (1972).

075.030 Observations solaires. C. Popovici, E. Ţifrea, V. Dinulescu, A. Dimitriu, S. Nicolescu, G. Mariş, I. Niţă. Obs. Bucarest, Secteur Solaire, Acad. République Socialiste Roumanie. 46 pp. (1972). – Rotations 1570 - 1582 (11 janvier - 31 decembre 1971).

075.031 Solar phenomena. M. Cimino, M. Torelli, A. Cacciani, V. Croce, R. Flamini, U. Bartolini. Oss. Astron. Roma, Monthly Bull. Nos. 169 - 175 (1972). 1972 May – November: Daily total areas of sunspot-groups; Heliographic position, classification and area of sunspot-groups; Longitudinal sunspot magnetic fields; Hours of K-line cinematographic patrol; Hours of $H\alpha$ cinematographic patrol; S.C.N.A. and S.E.A.; Explanation.

075.032 Daily $H\alpha$ chromosphere pictures, daily K_{232} chromosphere pictures, daily white light photosphere pictures. M. Cimino (Editor).

Photographic Journ. of the Sun, Oss. Astron. Roma, Nos. 53 - 56, 62 - 65 (1971/72). − 1971 November 7 - 1972 February 24, 1972 July 9 - 1972 October 26. − Rotations 1581 - 1584, 1590 - 1593.

075.033 Solar photospheric observations.
F. Bruin, H. Hourani, N. G. Bustati.
Lee Obs., American Univ. Beirut, Monthly Bull. Astron. Section, 1972 April - September (1972).

Sunspot relative numbers; Heliographic mean position and classification of the sunspot groups; Number of facular zones.

075.034 Catalogo general de grupos de manchas solares 1971.
J. A. Gutiérrez, V. Capolongo, R. Barbarroja, C. F. Sosa.
Obs. Astron. Municipal, Rosario − Argentina, Dep. Fis. Solar, Contr. Ser. 1, No. 4, 18 pp. (1972).

075.035 Actividad solar enero - junio 1972.
R. Barbarroja, V. Capolongo, J. A. Gutiérrez, O. F. Liesche, L. A. Mansilla.

Obs. Astron. Municipal, Rosario − Argentina, Dep. Fis. Solar, Contr. Ser. 1, No. 5, 44 pp. (1972).

075.036 Indices of geomagnetic activity.
Journ. Atmosph. Terr. Phys., Vol. 34, 1433, 1561, 1953 - 1954 (1972). − 1972 April - July.

075.037 Actividad solar en 1970. J. Pensado.
Bol. Astron. Obs. Madrid, Vol. 8, No. 1, p. 3 - 142 (1972).
I. Números relativos de Wolf; II. Estadistica de manchas y superficie de las mismas; III. Fáculas cromosféricas brillantes; IV. Filamentos de hidrógeno; V. Protuberancias.

Sonnenbeobachtung und -forschung mit Raketen.
See Abstr. 003.160.

Observing programs in solar physics during the 1973 ATM (*Apollo Telescope Mount*) **Skylab program.**
See Abstr. 080.060.

076 Solar UV, X Rays, Gamma Radiation

076.001 Signal-to-energy conversion function in the photometry of solar soft X-radiation with broad-band detectors. V. Letfus.
Bull. Astron. Inst. Czechoslovakia, Vol. 23, 223 - 231 (1972).

The paper describes the method of determining the function which defines the conversion of a signal, observed by a broad-band X-ray detector, to the incident energy flux. As an example, the conversion function with two ionization chambers has been investigated in the 0−8 Å and 8−12 Å ranges; the chambers were installed in the Solrad 8 satellite. Theoretical spectra of the coronal plasma for a model of a homogeneous isothermic source were used.

076.002 Two X-ray bursts (1 August 1967 and 30 January 1968) and some associated VLF disturbances.
R. Barletti, G. L. Tagliaferri.
Mem. Soc. Astron. Italiana, Nuova Ser., Vol. 43, 301 - 308 (1972).

Time evolutions of two solar X-ray bursts in 0.5−3 Å and 1−8 Å bands as observed by the SOLRAD 9 satellite are the starting point for deriving the solar spectrum responsible of the D-region extra ionization. The ion production rate, the electron density and the expected VLF phase anomaly on a given radio link are obtained. The computed anomalies are compared with the observed ones.

076.003 On the structure of the solar transition zone in active regions.
A. M. Cantù, G. Poletto, G. L. Tagliaferri.
Mem. Soc. Astron. Italiana, Nuova Ser., Vol. 43, 371 - 374 (1972).

Two solar active regions are investigated on 4 spectroheliograms obtained by OSO IV on Nov. 2, 1967. These spectroheliograms refer to spectral lines formed in the transition zone of the solar atmosphere.

076.004 Solar XUV fluxes of SOLRAD 10 satellite from July 1971 to December 1971.
M. Landini, B. C. Fossi, G. Poletto, G. L. Tagliaferri.

Mem. Soc. Astron. Italiana, Nuova Ser., Vol. 43, 383 - 388 (1972). − Letter.

076.005 Correlation studies of solar X-ray and radio bursts.
D. L. McKenzie.
Astrophys. Journ., Vol. 175, 481 - 492 (1972).

Solar X-ray bursts observed by the UCSD OSO-III instrument between 1967 March 9 and 1968 January 31 are studied using a sample of more than 300 bursts having associated microwave emission. Spectral observations suggest that X-rays and associated microwave emission early in flare events arise from nonthermal electron distributions.

076.006 The solar X-ray spectrum deduced from a proportional counter experiment and the resultant production of ionization in the mesosphere.
D. Lepine, J. E. Hall.
Journ. Atmosph. Terr. Phys., Vol. 34, 1507 - 1523 (1972).

This paper describes the measurement of the 1−6 keV (2−12 Å) solar X-ray spectrum made by Petrel rocket on 19 June 1969. The incident spectrum was found from data obtained above 130 km, and data collected at lower heights were used to verify laboratory values of absorption coefficient. The production of ionization by X-rays was calculated for the height range 65−90 km and a comparison with other published results is presented.

076.007 Photographs of the sun in the XUV-region.
M. Burger, J. H. Dijkstra.
Solar Physics, Vol. 24, 395 - 404 (1972).

X-ray photographs obtained with a zone plate camera on October 3, 1967 in the wavelength band 49.5−52.5 Å have been investigated photometrically. The most intense X-ray emission corresponds with active regions in Hα and Ca II. About one quarter of the total solar flux is emitted by the three brightest X-ray sources (A, E and J). X-ray emission from quiet regions is also observed. Limb brightening is found, also at the poles, which indicates a higher electron density at the poles than during solar minimum.

076.008 **Solar soft X-rays and solar activity. II. Observational assessment of the role of the type III acceleration mechanism in establishment of the soft X-ray source volume.**
R. G. Teske, R. J. Thomas.
Solar Physics, Vol. 24, 434 - 443 (1972).

Peak fluxes of flare-associated 8–12 Å X-ray bursts occur at or near the time of the maximum energy content of the soft X-ray source volume. The amplitudes of flare-associated bursts may thus be used as a measure of the energy deposited in the source volume by non-thermal electrons and other processes. In the mean, the soft X-ray burst amplitude is apparently independent of the occurrence of a type III event. This is interpreted to indicate that electrons accelerated by the type III process do not directly participate in establishing the soft X-ray source volume.

076.009 **The decay characteristics of models of solar hard X-ray bursts.** J. C. Brown.
Solar Physics, Vol. 25, 158 - 177 (1972).

Models of solar hard X-ray bursts are considered in which non-thermal electrons are impulsively injected into a coronal magnetic trap. It is shown that the X-ray spectra will initially soften with time, due to collisions, when this non-uniformity is strong enough. This removes a well-known discrepancy in models with uniform density. It is shown also that non-uniformity steepens the electron spectrum required to produce a given observed X-ray spectrum.

076.010 **Accurate calculations of wavelengths of certain helium-like lines in the X-ray region.**
A. M. Ermolaev, M. Jones, K. J. H. Phillips.
Astrophys. Letters, Vol. 12, 53 - 56 (1972).

Accurate calculations have been made of the wavelengths of certain lines in the helium-like ions Si XIII, Ca XIX and Fe XXV, which have been observed in solar flares. The calculations include relativistic corrections and an estimate of the Lamb shift. The accuracy of the calculations is such that it is possible to indicate misidentifications of the line $2^3P_1 - 1^1S_0$ in Ca XIX and Fe XXV by Neupert.

076.011 **Pulsed intensity increase of X-ray radiation of the sun on December 10, 1970.**
G. E. Kocharov, Yu. E. Charikov, A. A. Kharchenko, G. V. Gusev, A. V. Baskakov.
Pis'ma v ZhEhTF, Vol. 15, 153 - 156 (1972). In Russian. Abstr. in Referativ. Zhurn. 62. Issled. kosm. prostranstva, 8.62.169 (1972).

076.012 **Analysis of X-ray line emission from individual solar active regions.**
R. C. Catura, L. W. Acton, C. J. Wolfson, J. L. Culhane.
Bull. American Astron. Soc., Vol. 4, 379 (1972). – Abstr. AAS.

076.013 **Further classifications in the XUV spectrum of Fe XV.** R. D. Cowan, K. G. Widing.
Bull. American Astron. Soc., Vol. 4, 380 (1972). – Abstr. AAS.

076.014 **A comparison of thermal and non-thermal solar flare X-ray emission observed on OSO-7.**
D. Datlowe, D. McKenzie.
Bull. American Astron. Soc., Vol. 4, 380 (1972). – Abstr. AAS.

076.015 **The soft X-ray flare of 12 August 1970.**
G. A. Doschek, J. F. Meekins.
Bull. American Astron. Soc., Vol. 4, 381 (1972). – Abstr. AAS.

076.016 **Analysis of the EUV quiet solar spectrum.**
A. K. Dupree.
Bull. American Astron. Soc., Vol. 4, 381 (1972). – Abstr. AAS.

076.017 **Comparison of solar X-ray polarization measure-**ments of Intercosmos 4 with the simultaneous high energy X-ray measurements of OSO-5.
K. J. Frost, B. R. Dennis.
Bull. American Astron. Soc., Vol. 4, 382 - 383 (1972). Abstr. AAS.

076.018 **Hard X-rays and limits on gamma line emission from a solar flare on 22 November 1971.**
D. D. Guo, S. V. Damle, J. C. Kish, J. Lezniak, W. R. Webber.
Bull. American Astron. Soc., Vol. 4, 383 (1972). – Abstr. AAS.

076.019 **Large scale coronal X-ray structures.**
A. Krieger, T. Barrett, A. F. Timothy, G. S. Vaiana, L. van Speybroeck.
Bull. American Astron. Soc., Vol. 4, 386 (1972). – Abstr. AAS.

076.020 **The structure of the X-ray corona surrounding quiescent filaments.**
A. F. Timothy, T. Barrett, A. Krieger, G. S. Vaiana, L. van Speybroeck.
Bull. American Astron. Soc., Vol. 4, 393 - 394 (1972). Abstr. AAS.

076.021 **On the difference between the poles and equator in the region of formation of the Lyman continuum.**
J. E. Vernazza, R. W. Noyes.
Bull. American Astron. Soc., Vol. 4, 394 (1972). – Abstr. AAS.

076.022 **Coronal active region models and relative abundances based on soft X-ray emission line and continuum fluxes.** A. B. C. Walker, Jr., H. R. Rugge.
Bull. American Astron. Soc., Vol. 4, 395 (1972). – Abstr. AAS.

076.023 **EUV emissions of solar flares: A comparison of OSO 6 observations and SFD's.**
A. T. Wood, Jr., R. F. Donnelly.
Bull. American Astron. Soc., Vol. 4, 396 (1972). – Abstr. AAS.

076.024 **EUV emissions of solar flares: OSO 4 and OSO 6 observations.** A. T. Wood, Jr.
Bull. American Astron. Soc., Vol. 4, 396 (1972). – Abstr. AAS.

076.025 **The impulsive X-ray burst of October 10, 1970.**
S. R. Kane, S. W. Kahler, J. D. Kurfess.
Solar Physics, Vol. 25, 418 - 424 (1972).

An impulsive burst of 100–400 keV solar X-rays associated with a small solar flare was observed on October 10, 1970 with a large area scintillator aboard a balloon floating at an altitude of 4.2 g cm^{-2} above the earth's surface. The X-ray burst was also observed simultaneously in 10–80 keV range by the OGO-5 satellite and in 8–20 Å range by the SOLRAD-9 satellite. The spectral characteristics of this event are examined in the light of the earlier X-ray observations of small solar flares.

076.026 **Polarization of hard X-rays from solar flares.**
E. Haug.
Solar Physics, Vol. 25, 425 - 434 (1972).

The polarization of hard solar X-radiation (> 10 keV) is calculated on the assumption that electrons get a non-isotropic velocity distribution in the initial phase of a flare.

076.027 **Missing solar ultraviolet opacity and diatomic molecules.** S. P. Tarafdar, M. S. Vardya.
Space Sci. Rev., Vol. 13, 651 (1972). – Abstract of conference paper IAU Colloquium No. 14 (see 012.012).

076.028 **Dielectronic satellite spectra in the soft X-ray region.** A. H. Gabriel.
Space Sci. Rev., Vol. 13, 655 - 664 (1972). – Invited paper IAU Colloquium No. 14 (see 012.012).

076.029 Identifications of some highly-ionized iron and nickel lines in the 200—400 Å region of the solar spectrum.   K. G. Widing, G. Sandlin, R. Cowan.
Space Sci. Rev., Vol. 13, 665 (1972). – Abstract of conference paper IAU Colloquium No. 14 (see 012.012).

076.030 Solar X-ray source unassociated with sunspots.   G. A. Gurzadyan, K. V. Vartanian.
Space Sci. Rev., Vol. 13, 731 - 737 (1972). – Conference paper IAU Colloquium No. 14 (see 012.012).

076.031 On the interpretation of the relative intensities of the solar XUV lines of lithium-like ions.
D. R. Flower.
Space Sci. Rev., Vol. 13, 738 - 739 (1972). – Conference paper IAU Colloquium No. 14 (see 012.012).

076.032 Recent high resolution X-ray spectra of the sun.   J. H. Parkinson, K. Evans, K. A. Pounds.
Space Sci. Rev., Vol. 13, 740 - 741 (1972). – Conference paper IAU Colloquium No. 14 (see 012.012).

076.033 On special features of non-thermal solar X radiation above 10 keV.   G. Elwert, E. Haug.
Space Sci. Rev., Vol. 13, 761 - 762 (1972). – Conference paper IAU Colloquium No. 14 (see 012.012).

076.034 Thermal and non-thermal soft X-ray bursts.   M. Landini, B. C. Fossi, R. Pallavicini.
Space Sci. Rev., Vol. 13, 825 - 826 (1972). – Conference paper IAU Colloquium No. 14 (see 012.012).

076.035 Observations of solar X-ray emission from the satellite Intercosmos 4 and the rocket Vertical 1.
Yu. I. Grineva, V. I. Karev, V. V. Korneyev, V. V. Krutov, S. L. Mandel'stam, L. A. Vainstein, B. N. Vasilyev, I. A. Žitnik.
Space Research XII, (see 012.016), Vol. 2, 1553 - 1557 (1972).

076.036 High angular resolution absolute intensity of the solar continuum from 1400 Å to 1790 Å.
G. E. Brueckner, O. K. Moe.
Space Research XII, (see 012.016), Vol. 2, 1595 - 1602 (1972).

076.037 High angular resolution observations from rockets: Solar XUV observations.   R. Tousey.
Space Research XII, (see 012.016), Vol. 2, 1719 - 1737 (1972).

076.038 The directivity and polarisation of thick target X-ray bremsstrahlung from solar flares.
J. C. Brown.
Solar Physics, Vol. 26, 441 - 459 (1972).
In this paper detailed calculations of the directivity and polarisation of a continuous injection model are made on the assumption of a purely vertical field. Non-uniform plasma density and a thick-target scattering model are, however, incorporated in the calculations and the consequences of a non-vertical field are discussed.

076.039 On ultraviolet stellar fluxes. III. Importance of H_2 Lyman-band absorption in the sun and other stars.
S. P. Tarafdar, M. S. Vardya.
Astrophys. Journ., Vol. 178, 509 - 513 (1972).
Lyman-band absorption of H_2, when compared with the combined opacity of H, C I, Mg I, and Si I, has been found to contribute significantly to the total opacity in the range 912–1900 Å in the solar atmosphere. The Lyman-band absorption may also be important in stars as hot as A0.

076.040 Solar X-rays in July 1964 and November–December 1965 from data of Electron 4 and Venera 2.
V. Letfus, I. P. Tindo.
Kosmich. Issled., Vol. 10, 920 - 924 (1972). In Russian.

076.041 Thermal and non-thermal soft X-ray bursts.   M. Landini, B. C. Fossi, R. Pallavicini.
Solar Physics, Vol. 27, 164 - 173 (1972).
X-ray bursts observed for energies lower than 25 keV are usually interpreted as being produced by a thermal plasma with several million degrees of temperature. A small number of events recorded at Arcetri by real time telemetry of SOLRAD 9 satellite agrees with a thermal interpretation and gives temperatures ranging between 10×10^6 and 30×10^6 K and emission measures between 10^{47} and $10^{48} cm^{-3}$. An impulsive event recorded on January 7, 1969 shows an anomalous behaviour. A tentative interpretation of the event is suggested and the way to produce non-relativistic electrons with a power law energy distribution is investigated.

076.042 Evidence for a common origin of the electrons responsible for the impulsive X-ray and type III radio bursts.   S. R. Kane.
Solar Physics, Vol. 27, 174 - 181 (1972).
Observations of impulsive solar flare X-rays $\gtrsim 10$ keV made with the OGO-5 satellite are compared with ground based measurements of type III solar radio bursts in 10–580 MHz range. It is shown that the times of maxima of these two emissions, when detectable, agree within ~ 18 s. This maximum time difference is comparable to that between the maxima of the impulsive X-ray and impulsive microwave bursts. The observations indicate that the non-thermal electron groups responsible for the impulsive X-ray, impulsive microwave, and type III radio bursts are accelerated simultaneously in essentially the same region of the solar atmosphere.

076.043 The importance of iron opacities in the solar balloon ultraviolet.   A. N. Cox, N. H. Magee, A. L. Merts, J. N. Dragon, J. P. Mutschlecner.
Bull. American Astron. Soc., Vol. 4, 425 (1972). – Abstr. AAS.

076.044 The SOLRAD satellite.
Spaceworld, (USA), Vol. 1, 1-97, p. 22 - 25 (1972).
See Phys. Abstr., Vol. 75, No. 51844 (1972).

076.045 Height distribution of soft X-ray emission in the solar atmosphere.   C. P. Catalano.
Thesis, Univ. Iowa, Iowa City. [Available from Univ. Micro-films, Ann Arbor, Mich., USA. Order No. 72-8225], 209 pp. (1971). – See Phys. Abstr., Vol. 75, No. 69184 (1972).

076.046 Satellite measurements of solar X-ray flux and their use for interpretation of sudden ionospheric disturbances.   J. Lastovicka, J. Smilauer.
Pure and Applied Geophys., (Switzerland), Vol. 98, 178 - 183 (1972).
The apparatus for the reception and evaluation of the real-time telemetry data on the X-ray flux, measured by the satellite Solrad-9, is briefly described.

076.047 Pulsed increase of intensity of solar X-ray on 10 December 1970.   G. E. Kocharov, Yu. E. Charikov, A. A. Kharchenko, G. V. Gusev, A. V. Baskakov.
JETP Letters, Vol. 15, 105 - 107 (1972).

076.048 Ultraviolet ion chamber measurements of the solar minimum brightness temperature.
J. H. Carver, B. H. Horton, G. W. A. Lockey, B. Rofe.
Solar Physics, Vol. 27, 347 - 353 (1972).
The solar ultraviolet flux in the wavelength bands 1580–1640 Å and 1430–1470 Å (FWHM) has been measured using photon ion chambers carried on the satellite WRESAT I (1967-118A). These observations of the integrated ultraviolet

flux from the entire disk indicate a value of (4570 ± 50) K for the solar temperature minimum. The results are compared with other estimates of the minimum value of the solar brightness temperature.

076.049 Preliminary interpretation of the polarization measurements performed on 'Intercosmos-4' during three X-ray solar flares.
I. P. Tindo, V. D. Ivanov, B. Valníček, M. A. Livshits.
Solar Physics, Vol. 27, 426 - 435 (1972).

Analysis of the X-ray polarization data at $\lambda \simeq 0.8$ Å for three major chromospheric flares shows that during the 'hard' phase of the flare the X-rays are polarized in the plane, the projection of which on the solar disc is going approximately from the flare region to the center of the disc. Simultaneously performed measurements of the spectral energy distribution have proved that observed X-rays are produced by the bremsstrahlung of the accelerated electrons with the energies in the range 10−100 keV.

076.050 A simulation of the directivity effect to be expected in hard X-ray flares. M. L. Shaw.
Solar Physics, Vol. 27, 436 - 441 (1972).

A Monte Carlo technique has been used to predict the relative visibility of solar hard X-ray flares as a function of solar longitude assuming the model of Takakura and Kai to be realistic. Comparison is made with previous statistical studies of observations. A discernable longitudinal variation in the relative visibility of flares is shown to be expected but the probability of flares being visible towards the limb is shown to be higher than had previously been evident.

New lines of neon ions in the range 50−200 Å. See Abstr. 022.044.

The classification of Fe IX to XVI emission lines and isoelectronic lines in laboratory and solar spectra. See Abstr. 022.084.

Identifications of emission lines in the EUV solar spectrum. See Abstr. 071.040.

Wavelengths of solar lines in the 50−380 Å region and their identifications. See Abstr. 071.041.

High resolution solar spectra from 1780 to 1950 Å. See Abstr. 071.042.

Observations of the profiles of solar UV emission lines and their analysis in terms of the heating and production of the corona. See Abstr. 071.043.

Calculations on the solar spectrum from 1 to 60 Å. See Abstr. 071.044.

Analysis of the extreme-ultraviolet quiet solar spectrum. See Abstr. 071.059.

Measurements of the solar spectrum between 30 and 128 Å. See Abstr. 071.064.

Non-thermal X-ray radiation and electrical currents in solar flares. See Abstr. 073.006.

Observations and comments for the solar event of 24 October, 1969. See Abstr. 073.011.

Thick-target processes and white-light flares. See Abstr. 073.012.

New measurements of the polarization of X-ray solar flares. See Abstr. 073.013.

The classification of Fe XVIII to XXIV emission lines in solar flare spectra. See Abstr. 073.071.

The expected behaviour of the hydrogen Lyman lines in solar flares. See Abstr. 073.074.

Solar flares in the EUV observed from OSO-5. See Abstr. 073.098.

Properties of a coronal "hole" derived from extreme-ultraviolet observations. See Abstr. 074.022.

The solar EUV-emitting plasma. See Abstr. 074.068.

The coronal X-spectrum: Problems and prospects. See Abstr. 074.069.

X-ray emission of coronal condensations during the eclipse on 20 May 1966 and its connection with optical and radio observations. See Abstr. 074.078.

High-sensitivity measurement of the emission spectrum of quiescent active regions. See Abstr. 074.094.

The role of energetic electrons in the correlation of meter and decimeter type III bursts with 4 keV X-ray emission. See Abstr. 077.036.

The self-absorption of gyro-synchrotron emission in a magnetic dipole field: Microwave impulsive burst and hard X-ray burst. See Abstr. 077.040.

On the S- and B-components of solar radio and X-emission and their relationships to energetic solar events. See Abstr. 077.055.

A search for neutrons of solar origin using balloon borne detectors 1967−69. See Abstr. 078.003.

Solar rotation as measured in EUV chromospheric and coronal lines. See Abstr. 080.032.

X-ray emission and D-region "sluggishness". See Abstr. 083.058.

Backscatter of solar resonance radiation − II. See Abstr. 131.061.

OGO 5 determination of the local interstellar wind parameters. See Abstr. 131.075.

New interpretations of extraterrestrial Lyman-alpha observations. See Abstr. 131.076.

077 Solar Radio Radiation

077.001 Exciter of type III bursts and coronal temperature.
M. Aubier, A. Boischot.
Astron. Astrophys., Vol. 19, 343 - 353 (1972).

We show that, under certain hypothesis, the time profiles of type III solar radio bursts give independently the duration of the exciter and the decay time constant of the burst. New observations of type III bursts in the decametre range are presented. The analysis of the time profiles leads to average total durations of the exciter function of 3.9, 6.2 and 7.5 s and average decay time constants of 1.2, 1.7 and 2.0 s at 60, 36.9 and 29.3 MHz respectively. A relation is found between the duration and the decay time constant. This relation brings some doubt on the significance of the coronal temperature derived from the decay time constant with the hypothesis of a collisional damping of the plasma waves. Other explanations are proposed.

077.002 The type IIIb burst: a precursor of decametre type III radio-burst. J. de la Noë, A. Boischot.
Astron. Astrophys., Vol. 20, 55 - 62 (1972).

The main characteristics of type IIIb bursts are studied. It is shown in particular that type IIIb bursts are often precursors of a normal type III burst. Some conclusions about the nature of the exciter of type IIIb and type III bursts are presented.

077.003 On the origin of chains of type I bursts.
V. V. Zaitsev, V. V. Fomichev.
Astron. Zhurn. Akad. Nauk SSSR, Vol. 49, 817 - 826 (1972). In Russian. English translation in Soviet Astron. AJ, Vol. 16, No. 4.

A mechanism of emission of chains of type I radio bursts with weak collisionless shock waves in a plasma with a strong magnetic field is discussed. From the data on chains the distribution of magnetic fields and parameters of shock waves in the interval of heights from 250×10^3 to 1.5×10^6 km above the photosphere is calculated. Estimates of dimensions of inhomogeneities in the solar corona are obtained.

077.004 Measurement of the electron temperature of small 3-cm radio bursts. P. Foukal.
Solar Physics, Vol. 24, 411 - 413 (1972). – Research note.

077.005 A dynamic theory of type III solar radio bursts.
V. V. Zaitsev, N. A. Mityakov, V. O. Rapoport.
Solar Physics, Vol. 24, 444 - 456 (1972).

In the present paper we solve the relativistic quasilinear equations under the initial conditions of a local explosion type. In the deep coronal layers where the meter electromagnetic waves are generated, the condition $\nu_{eff}^{-1} \ll \Delta x/V_s$ (ν_{eff}^{-1} the characteristic time, Δx and V_s the extent of the stream in the corona and its mean speed respectively) is satisfied. In this case the solution reduces to the results obtained by Zheleznyakov and Zaitsev (1970). The other situation arises in the decameter and longer wave ranges. In this case the collisions play an insignificant role ($\nu_{eff}^{-1} \gg \Delta x/V_s$); the plasma wave generation becomes non-stationary and dependent on the dynamics of the electron stream; it is described by simple formulae of similar motion and permits a detailed comparison of the theory with experimental data of the low-frequency type III bursts.

077.006 The time behaviour of the continua during the initial stage of type IV bursts. A. Böhme.
Solar Physics, Vol. 24, 457 - 474 (1972).

The existence of a group of broad-band continua during the initial stage of some type IV bursts can be shown which are polarized in the extraordinary sense. It can be shown that the time behaviour of the broad-band continua is clearly different from that of the moving type IV bursts recorded directly by interferometers. Therefore, it must be concluded that both types of continua originate from different physical processes at the sun, though both are caused by synchrotron radiation.

077.007 Type III solar noise observed below 100 kHz on OGO 3. I. Description of events.
N. Dunckel, R. A. Helliwell, J. Vesecky.
Solar Physics, Vol. 25, 197 - 209 (1972).

Type III noise bursts observed with OGO 3 in the frequency range 25−100 kHz have the following characteristics: (1) Above about 50 kHz their spectrum is smooth and slowly changing. (2) Their drift and decay is comparable to that extrapolated from high-frequency bursts. (3) They are closely associated with west limb flares, ground-based type III burst observations, and in some cases, with energetic interplanetary particle events. (4) They tend to occur in association with type III bursts at 2−4 MHz which are related to flares in the western-hemisphere.

077.008 Peculiar absorption and emission microstructures in the type IV solar radio outburst of March 2, 1970.
Solar Physics, Vol. 25, 210 - 231 (1972).

The high resolution dynamic spectrogram between 320 and 160 MHz of the Type IV event which started at 13:35 UT on the 2nd of March 1970 shows a remarkable richness of absorption-emission microstructures. These are morphologically analyzed into structure elements and patterns. The elements are normal and reversed intermediate drift bursts, which we call fiber bursts, medium band shortlived absorptions, broadband shortlived absorptions, broadband wedge shaped absorptions and a type which we call tadpole.

077.009 Quasi-periodic solar radio pulsations at decimetric wavelengths. B. L. Gotwols.
Solar Physics, Vol. 25, 232 - 236 (1972).

Observations are reported of radio-pulsations at dm-wavelengths. The pulsations are quasi-periodic with a period of 0.5 s, they have a bandwidth > 300 MHz and show up to a 50 % enhancement of the underlying type IV continuum.

077.010 On the geometry of local sources on the sun from observations of the eclipse on September 22, 1968.
N. N. Eruchev, L. I. Tsvetkov.
Izv. Krymskoj Astrofiz. Obs., Vol. 44, 112 - 121 (1972). In Russian.

The results regarding the positions of local sources from observations of the partial solar eclipse of September 22, 1968 are given. Close connection of the radio sources with sunspot groups and local magnetic fields at the photospheric level are found. The heights of the local sources as well as of the radio limb above the photosphere are determined. The position of the most intense part of the source in the total emission is coinciding with the region of polarized emission.

077.011 The distribution of radio emission over the disk of the sun at the wavelengths 2, 4, 6 and 8 mm.
V. A. Efanov, A. G. Kislyakov, G. V. Lebskij, I. G. Moiseev, A. J. Naumov.
Izv. Krymskoj Astrofiz. Obs., Vol. 44, 137 - 173 (1972). In Russian.

Solar radio images (radio isophotes) at the wavelengths 2, 4, 6 and 8 mm are presented. They were obtained as results of observations made with the 22-m radiotelescope of Crimean

Astrophysical Observatory. The wavelength dependences of the brightness temperature, of the flux density, and of the angular sizes of observed features on the disk have been considered.

077.012 Observations radioélectriques solaires faites sur 600 MHz en 1971 au Laboratoire de Radioastronomie de Humain-Rochefort. C. Gonze, R. Gonze.
Bull. Astron. Obs. Roy. Belgique, Vol. 8, 43 - 57 (1972).

077.013 An attempt to study solar oscillations at 31.4 GHz.
W. L. H. Shuter, W. H. McCutcheon.
Journ. Roy. Astron. Soc. Canada, Vol. 66, 221 (1972).
Abstr. Canadian Astron. Soc.

077.014 Solar radio emission at 10.7 cm, 1947–1972.
A. E. Covington.
Journ. Roy. Astron. Soc. Canada, Vol. 66, 221 (1972).
Abstr. Canadian Astron. Soc.

077.015 Local sources on the sun from observations of the eclipse on Sept. 22, 1968 at $\lambda = 2$ cm.
I. F. Belov, E. I. Lebedev, N. A. Prokofjeva, B. V. Timofeev, V. M. Fridman.
Solnechnye Dannye 1972 Byull., No. 5, p. 79 - 83 (1972).
In Russian.
   The data on the solar radio radius and brightness temperature of the "quiet" sun obtained from observations at $\lambda = 2$ cm are given. They have been compared with the results of observations at the wavelength of 3 cm and 50 cm.

077.016 The quiet sun brightness distributions at millimeter wavelengths and chromospheric inhomogeneities.
P. Lantos, M. R. Kundu.
Astron. Astrophys., Vol. 21, 119 - 124 (1972).
   The quiet sun brightness distributions at 9 and 3.5 mm wavelengths are presented. These distributions are interpreted in terms of a two-component chromospheric model.

077.017 Radio observations of the sun at 350 MHz 1972 April 25 to May 26. A. N. Kelly.
Monthly Notes Astron. Soc. Southern Africa, Vol. 31, 99 - 103 (1972).

077.018 Free-free absorption of gyrosynchrotron radiation in solar microwave bursts.
R. Ramaty, V. Petrosian.
Bull. American Astron. Soc.,Vol. 4, 310 (1972). – Abstr. AAS.

077.019 High resolution measurements of the sun at 3.71 and 11.1 cm wavelength.
R. W. Hobbs, S. D. Jordan, W. J. Webster, Jr.
Bull. American Astron. Soc., Vol. 4, 310 (1972). – Abstr. AAS.

077.020 High resolution maps of the sun and moon at 1 millimeter wavelength.
J. D. G. Rather, P. A. R. Ade, P. E. Clegg.
Bull. American Astron. Soc., Vol. 4, 322 (1972). – Abstr. AAS.

077.021 Spectral characteristics of solar radio bursts associated with the emission of energetic electrons from the sun. S. Basu.
Bull. American Astron. Soc., Vol. 4, 377 (1972). – Abstr. AAS.

077.022 Simultaneous observations at 2800 MHz and 13,500 MHz of the circular polarization of solar active regions. M. B. Bell.
Bull. American Astron. Soc., Vol. 4, 378 (1972). – Abstr. AAS.

077.023 The distribution of peak flux-density spectra of solar radio bursts. J. P. Castelli, D. A. Guidice.
Bull. American Astron. Soc., Vol. 4, 379 (1972). – Abstr. AAS.

077.024 The 2800 MHz microwave solar index.
A. E. Covington.
Bull. American Astron. Soc., Vol. 4, 380 (1972). – Abstr. AAS.

077.025 Position observations of simultaneous continuum and type III bursts at decametric wavelengths.
R. J. Fitzenreiter, J. Fainberg.
Bull. American Astron. Soc., Vol. 4, 382 (1972). – Abstr. AAS.

077.026 Low frequency radiation from relativistic electrons in a cold magnetoplasma – implications for U shaped type IV bursts. G. Kalman, S. Yukon.
Bull. American Astron. Soc., Vol. 4, 385 (1972). – Abstr. AAS.

077.027 Evidence for electron excitation of type III radio bursts. R. P. Lin, S. R. Kane, F. T. Haddock.
Bull. American Astron. Soc., Vol. 4, 386 (1972). – Abstr. AAS.

077.028 Rayleigh and Raman scattering of large amplitude plasma waves and solar radio burst. B. Prasad.
Bull. American Astron. Soc., Vol. 4, 389 (1972). – Abstr. AAS.

077.029 Some characteristics of microwave solar type IV radio bursts and generation of solar cosmic rays.
K. Sakurai.
Bull. American Astron. Soc., Vol. 4, 390 (1972). – Abstr. AAS.

077.030 On the source of the slowly varying component at centimeter and millimeter wavelengths.
F. I. Shimabukuro, G. A. Chapman, S. Edelson, E. B. Mayfield.
Bull. American Astron. Soc., Vol. 4, 391 (1972). – Abstr. AAS.

077.031 Polarization scans of active regions at 3.8 cm.
R. M. Straka, D. W. Richards, K. K. Arora.
Bull. American Astron. Soc., Vol. 4, 392 - 393 (1972).
Abstr. AAS.

077.032 Temperature depressions at $\lambda = 3.3$ mm.
K. P. White III.
Bull. American Astron. Soc., Vol. 4, 395 (1972). – Abstr. AAS.

077.033 The polarization characteristics of two local sources at $\lambda = 9.0$ cm. Sh. B. Akhmedov, A. V. Temirova.
Solnechnye Dannye 1972 Byull., No. 6, p. 110 - 116 (1972).
In Russian.
   The radio characteristics of two local sources connected with developed unipolar sunspot groups in polarized and nonpolarized light are investigated.

077.034 Some characteristics of microwave type IV radio bursts and the acceleration of solar cosmic rays.
K. Sakurai.
Publ. Astron. Soc. Pacific, Vol. 84, 531 - 533 (1972).
   This paper discusses the relationships between some characteristics of microwave type IV radio bursts and solar cosmic ray protons of MeV energy. It is shown that the peak flux intensity of those bursts is almost linearly correlated with the MeV proton peak flux observed by satellites near the earth.

077.035 On quasi-periodic components with periods from 30 to 60 min of amplitude fluctuations of X-band solar radio emission. M. M. Kobrin, A. I. Korshunov.
Solar Physics, Vol. 25, 339 - 342 (1972).
   Special experiments have been performed to investigate the fluctuations of the intensity difference of the solar radio emission at two close frequencies. The autocorrelation functions and their spectra are obtained. The latter shows the presence of quasi-periodical components with periods of about 50 min in the solar radio emission. The possibility of explaining the observed quasi-periodical components by the supergranula-

tion oscillations and the solar self oscillations is considered.

077.036 The role of energetic electrons in the correlation of meter and decimeter type III bursts with 4 keV X-ray emission. S. W. Kahler.
Solar Physics, Vol. 25, 435 - 451 (1972).

The correlation of type III burst-groups with 4 keV solar X-ray emission is examined. A total of 151 burst-groups reported by the Fort Davis Observatory were compared with X-ray emission observed by the Naval Research Laboratory experiment on the OGO-5 satellite. A higher X-ray correlation is found for type III burst-groups when: (1) the bursts are observed on the decimeter band and (2) the bursts are more intense.

077.037 A search of a connection between the polarization of decam-type III bursts and magnetic fields in different heights of the solar atmosphere.
I. M. Chertok, V. V. Fomichev, A. Krüger, W. Willimczik.
Solar Physics, Vol. 25, 452 - 460 (1972).

Polarization measurements of type III bursts at 23.5 and 29.5 MHz have been compared for several years with indicators of magnetic fields in different height levels such as sunspot data, S-component characteristics, and noise storm data. By applying the Mount-Wilson and Brunner types of the related spot groups there results a positive relationship between the average degree of type III burst polarization and the magnitude or complexity of photospheric magnetic fields. For other parameters (leading spot area, peak intensity of the S-component at 9.1 cm wavelength) such a clear monotonic relation has not been found.

077.038 Spectral behaviour and proton effects of the type IV broad-band continua. A. Böhme.
Solar Physics, Vol. 25, 478 - 488 (1972).

The spectral behaviour of a group of broad-band continua at metre and decametre waves is discussed. These broad-band continua are polarized in the extraordinary mode and occur during the explosive phase of some strong flares.

077.039 Brightness distribution of the sun at 8.6 mm wavelength. K.-a. Kawabata, Y. Sofue.
Publ. Astron. Soc. Japan, Vol. 24, 469 - 481 (1972).

The brightness distribution of the quiet sun at wavelength λ=8.6 mm is synthesized from off-meridian observations using a four-element east-west interferometer with equal spacing of 273 λ. The observed brightness distribution at 8.6 mm is essentially flat from the disk center to 0.95 $R_\odot$. In addition to the nearly uniform component over the optical disk there exists an excess component just outside the optical limb. The component outside the limb is in agreement with the coronal emissions at this wavelength. The physical condition of the spicules is discussed by comparing the present observational data with the plane-parallel models of the chromosphere by Gingerich and de Jager (1968; BCA) and by Vernazza and Noyes (1972).

077.040 The self-absorption of gyro-synchrotron emission in a magnetic dipole field: Microwave impulsive burst and hard X-ray burst. T. Takakura.
Solar Physics, Vol. 26, 151 - 175 (1972).

The aim of the present paper is to investigate the effect of self-absorption on the radio spectrum emitted from a source with a non-uniform magnetic field. The radio spectrum and its variation due to the change of parameters of the radio source are shown in order to interpret the time variation of the spectrum of the microwave impulsive burst. It is shown that the present model may almost solve the discrepancy between the number of non-thermal electrons estimated from the microwave impulsive burst and that estimated from the hard X-ray burst.

077.041 Radio bursts from the solar corona.
J. P. Wild, S. F. Smerd.
Annual Rev. Astron. Astrophys., Vol. 10, (see 003.005), 159 - 196 (1972). − Contents: (1) Type III bursts; (2) Type II bursts; (3) Moving type IV bursts; (4) Coronal pulsations− a clue to particle acceleration? (5) Appendix.

077.042 Regions with low radio brightness on the sun from observations at 8 mm.
G. P. Apushkinsky, A. N. Tsyganov.
Solnechnye Dannye 1972 Byull., No. 8, p. 100 - 106 (1972). In Russian.

The results of observations of the solar region with low radio brightness during August−September 1970 are given. The development of a radio filament during 21−28 September 1970 has been studied in detail. Regions with low radio brightness which have no optical analogues have been detected.

077.043 Fast polarized pulses in decameter-wave radiation from the sun. C. H. Barrow, H. Saunders.
Astrophys. Letters, Vol. 12, 211 - 214 (1972).

Solar radio bursts have been observed at fixed frequencies close to 18 MHz (left- and right-hand polarization components with an adjacent frequency total-power channel) and 22 MHz (total-power). Fast narrow-band pulses may appear superimposed on some type III bursts particularly at the beginning and the end of the main burst. Typical durations are in the range 5 to 40 msec with bandwidths less than 150 kHz. Many of the pulses are circularly polarized; consecutive pulses within a sequence may show rapid polarization changes and reversals. The pulses appear to be similar, in some respects, to those which occur in the decametric emission from Jupiter.

077.044 Catalogue of 260 MHz solar radio noise storms (Ondřejov 1962 - 1971).
J. Olmr, J. Šebl, A. Tlamicha.
Bull. Astron. Inst. Czechoslovakia, Vol. 23, 323 - 327 (1972).

A complete list is given of solar radio noise storms on the 115 cm wave-length recorded at the Astronomical Institute Observatory Ondřejov, over the years 1962 - 1971. The list is a continuation of the catalogue of 130 cm solar radio noise storms, recorded at the Ondřejov Observatory during the years 1959 - 1961.

077.045 Some studies on the solar microwave bursts in relation to the slowly varying component.
M. K. Das Gupta, S. K. Sarkar.
Solar Physics, Vol. 26, 378 - 385 (1972).

The purpose of the present investigation is to examine the 27-day periodicity, if any, in the number of occurrences and the average energy excesses of the microwave bursts at 2695, 4995 and 8800 MHz and to compare the results thus obtained with that of the average S-component over a period of several years covering the maximum phase of the current solar cycle. Average spectrum of the microwave bursts having a spectral distribution of the inverted U-type with a peak at 4995 MHz has also been examined in relation to the average spectrum of the S-component. The results obtained from the statistical investigation are presented here.

077.046 Evidence for electron excitation of type III radio burst emission.
H. Alvarez, F. Haddock, R. P. Lin.
Solar Physics, Vol. 26, 468 - 473 (1972).

Type III radio bursts observed at kilometric wavelengths ($\lesssim 0.35$ MHz) by the OGO-5 spacecraft are compared with >45 keV solar electron events observed near 1 AU by the IMP-5 and Explorer 35 spacecraft for the period March 1968 − November 1969. We have presented evidence that essentially a one-to-one correspondence exists between kilometric wavelengths type III bursts above a threshold of approximately $10^{-13}\,\mathrm{Wm^{-2}Hz^{-1}}$ and >45 keV electrons observed at 1 AU. We conclude that streams of $\sim 10-100$ keV electrons are the exciting agent for type III bursts, and that $\gtrsim 5 \times 10^{32}$ electrons with energy >45 keV are emitted in a strong type III burst.

077.047 Free-free absorption of gyrosynchrotron radiation in solar microwave bursts.
R. Ramaty, V. Petrosian.
Astrophys. Journ., Vol. 178, 241 - 249 (1972).

A model for solar microwave bursts is considered in which, over a broad frequency band, the flux density is a slowly varying function of frequency. It is proposed that such an essentially flat spectrum could result from free-free absorption of gyrosynchrotron emission of nonthermal electrons accelerated in solar flares. The theory of gyrosynchrotron radiation is reviewed. The thermal electron density, temperature, emission measure, magnetic fields, and number of nonthermal electrons in the emitting region are evaluated for a flare with a flat microwave spectrum.

077.048 About time dependence of cosmic ray intensity on the anisotropic stage of solar bursts.
I. N. Toptygin.
Geomagn. Aeronom., Vol. 12, 989 - 995 (1972). In Russian.

077.049 On the time dependence of the degree of polarization of type III solar radio bursts. I. M. Chertok.
Astron. Zhurn. Akad. Nauk SSSR, Vol. 49, 1280 - 1286 (1972). In Russian. English translation in Soviet Astron. AJ, Vol. 16, No. 6.

The examination of the time dependence of the degree of polarization of type III radio bursts carried out previously (1968) is generalized for the case, when the stream exciting the burst is inhomogeneous. The influence of the group delay of ordinary and extraordinary waves on the dependence is analysed.

077.050 A 350 MHz radio event associated with the solar flare photographed at the Boyden Observatory on the 11th August,1972, at 14h 44m SAST. A. N. Kelly.
Monthly Notes Astron. Soc. Southern Africa, Vol. 31, 152 (1972). – Letter.

077.051 Centimeter radiation associated with the solar limb prominence of 8 February 1972. M. B. Bell.
Solar Physics, Vol. 27, 137 - 142 (1972).

Detailed maps of the sun have been made on 8 and 9 February 1972 with a 2.2' pencil beam. Superimposed upon a disc that shows only slight limb brightening, there are a number of radio emissive regions associated with centers of activity. Relatively intense radiation from the position of a limb prominence is apparent on 8 February but is absent on the 9th. Between observations the prominence became active leading to the ejection of a cloud of plasma. When the emission from the prominence at 2.2 cm is combined with observations at 10.7 cm, the spectral index obtained is much flatter than that usually associated with a center of activity.

077.052 Meter-wavelength observations of the solar radio burst storm of August 17 - 22, 1968.
R. T. Stewart, N. R. Labrum.
Solar Physics, Vol. 27, 192 - 202 (1972).

The data comprise dynamic spectra and high-resolution brightness distributions from the 80 MHz radioheliograph. It is found that the storm consisted essentially of type III bursts at the lower frequencies and type I at the higher frequencies. The type I source was located over an active region associated with a large sunspot group. The type III position was displaced about $0.5\,R_\odot$ transversely from the type I, in a region of low

magnetic field. A model is proposed to explain the close association between the two types of emission. A comparison of these results with the hectometer-wavelength satellite observations of the 1968 August event makes possible a qualitative estimate of the outward path of the type III exciters through the corona.

077.053 On the long-term behaviour of the circular polarization from coronal condensation radio emission at 4.3 cm wavelength. M. H. Paes de Barros, P. Kaufmann.
Solar Physics, Vol. 27, 203 - 207 (1972).

The circular polarization from coronal condensations at $\lambda = 4.3$ cm correspond to the extraordinary mode of propagation, due to the contribution of preceding spots' polarities, being usually left-handed. The fewer cases of right-handed polarization are normally associated to an excess of sunspot plages in the southern hemisphere, thus making it difficult to give evidence for magnetoionic coupling phenomena as a general rule.

077.054 The time-latitude distribution of solar flares accompanied by type IV radio bursts during the period 1956 to 1969.
M. D. Papagiannis, C. S. Zerefos, C. C. Repapis.
Solar Physics, Vol. 27, 208 - 216 (1972).

A list of nearly 350 flares accompanied by type IV radio bursts by Krüger et al. (1971), which covers a period of 14 yr (1956 - 1969), was expanded to include all PCA and solar cosmic ray events during this entire period. This list, which includes practically all of the most energetic events during the maxima of two consecutive solar cycles, was used to investigate the latitudinal distribution of the above-mentioned flares, as well as of all PCA events, solar cosmic ray events and plage regions associated with them.

077.055 On the S- and B-components of solar radio and X-emission and their relationships to energetic solar events. A. Krüger.
Solar Physics, Vol. 27, 217 - 226 (1972).

The slowly varying solar radio and X-ray emissions are considered theoretically and statistically concerning their significance as potential indicators of energy storage processes in active regions leading to large flare-burst events. A correlation analysis has been carried out in order to test different global emission parameters accessible by daily routine measurements regarding their connection with type IV bursts and proton flares.

077.056 On some outlooks for solar investigations based on the study of quasi-periodic components of radio emission fluctuations. M. M. Kobrin.
Sb. dokl. Sessii Nauch. Soveta po kompleks. probl. "Radioastronomiya", IZMIRAN, 1970. Moskva, 1972, p. 27 - 39. In Russian. – Abstr. in Referativ. Zhurn. 51. Astron., 1.51.377 (1973).

077.057 Measurements of solar radio emission fluctuations at 3 cm wavelength.
A. A. Bezotosnyj, O. G. Gontarëv, E. F. Rizov.
Trudy Sektora ionosfery. AN KazSSR, Vol. 3, 132 - 136 (1972). In Russian. – Abstr. in Referativ. Zhurn. 51. Astron., 1.51.395 (1973).

077.058 Sporadic solar radio emission. State of theory and problems. V. V. Zajtsev.
Sb. dokl. Sessii Nauch. Soveta po kompleks. probl. "Radioastronomiya", IZMIRAN, 1970. Moskva, 1972, p. 6 - 26. In Russian. – Abstr. in Referativ. Zhurn. 51. Astron., 1.51.435 (1973).

077.059 On the polarization of type III solar radio bursts.

V. V. Fomichev, I. M. Chertok.
Sb. dokl. Sessii Nauch. Soveta po kompleks. probl. "Radioastronomiya", IZMIRAN, 1970. Moskva, 1972, p. 56 - 64. In Russian. – Abstr. in Referativ. Zhurn. 51. Astron., 1.51.437 (1973).

077.060 On some peculiarities of the groups of type III solar radio bursts.
A. K. Markeev, V. A. Styazhkin, I. M. Chertok.
Sb. dokl. Sessii Nauch. Soveta po kompleks. probl. "Radioastronomiya", IZMIRAN, 1970. Moskva, 1972, p. 65 - 76. In Russian. – Abstr. in Referativ. Zhurn. 51. Astron., 1.51.438 (1973).

077.061 The main groups of type IV solar radio bursts.
S. T. Akin'yan.
Sb. dokl. Sessii Nauch. Soveta po kompleks. probl. "Radioastronomiya", IZMIRAN, 1970. Moskva, 1972, p. 77 - 82. In Russian. – Abstr. in Referativ. Zhurn. 51. Astron., 1.51.440 (1973).

077.062 Two phases in the local source development and a model with an autonomous magnetic field.
A. S. Grebinskij, O. V. Korobchuk.
Sb. dokl. Sessii Nauch. Soveta po kompleks. probl. "Radioastronomiya", IZMIRAN, 1970. Moskva, 1972, p. 125 - 146. In Russian. – Abstr. in Referativ. Zhurn. 51. Astron., 1.51.441 (1973).

077.063 Multidimensional classification of solar radio bursts.
N. P. Tsimakhovich.
Trudy Radioastrophys. Obs., *Riga,* Vol. 14, (see 003.019), 95 - 103 (1972). In Russian.
The multidimensional classification of radio bursts of the sun is described, which are given in a table of coded symbols of the basic characteristics. This classification is convenient for telegraphic transmissions of the characteristics of radio bursts.

077.064 Data on routine radio observations of the sun in eight observatories of the USSR in 1964 - 1968.
N. P. Tsimakhovich.
Trudy Radioastrophys. Obs., *Riga,* Vol. 14, (see 003.019), 111 - 117 (1972). In Russian.

077.065 Scattering of Langmuir waves produced by a beam with finite transverse dimensions.
G. Berthomieu.
Journ. Plasma Phys., Vol. 7, 523 - 543 (1972).
It has been proposed that type II radio bursts may be produced by streams of protons with radius of the order of 7 km, which is of the same order as the characteristic lengths of the nonlinear processes which are supposed to take place in the dynamics of these bursts. In this paper, a method is considered for studying systems with finite transverse dimensions and apply it to a simple model: the scattering of a beam of plasma waves by acoustic turbulence and by the particles of the plasma.

077.066 Some studies on solar microwave bursts of different types in relation to optical flares and other allied events. M. K. Das Gupta, S. K. Sarkar.
Indian Journ. Pure Applied Phys., Vol. 10, No. 2, p. 153 - 160 (1972).
Occurrences of some common types of solar microwave bursts at 4995 and 8800 MHz in relation to flares of different areas and intensities have been examined for the period November 1966 to September 1969.

077.067 The structure of the noise storm source according to observations of the solar eclipse on the 22nd of September 1968 at 1.37 m. L. I. Yurovskaya.
Izv. Krymskoj Astrofiz. Obs., Vol. 45, 40 - 48 (1972). In Russian.
From the examination of the observations of the solar eclipse on the 22nd of September 1968 at 1.37 m wavelength the noise storm source is found to consist of two regions separately observed on the solar disk: one is the region of the enhanced background continuum, the other is that of the burst.

077.068 Solar mapping at the Haystack Observatory on the wavelength of 3.8 cm. D. W. Richards.
U. S. Aire Force Cambridge Res. Labs. Air Force Systems Command Ionospher. Phys. Lab., Phys. Sci. Res. Papers No. 479. AFCRL-72-0090, 23 pp. (1972).
Details of observations and presentation of results, includes maps of brightness temperature and polarization for October 1970 and February 1971. – *DMCL*

Plasma emission processes in a magnetoactive plasma.
See Abstr. 062.017.

Relativistic electrons associated with solar flares.
See Abstr. 073.008.

Observations and comments for the solar event of 24 October, 1969. See Abstr. 073.011.

Observations of prominences at 3.5 millimeter wavelength. See Abstr. 073.018.

The heating of the solar plasma due to microwave phenomena correlated with type II meter bursts.
See Abstr. 074.014.

A possibly direct measurement of coronal magnetic field strengths. See Abstr. 074.015.

Particle motions in coronal streamers and type III radio bursts. See Abstr. 074.053.

Correlation studies of solar X-ray and radio bursts.
See Abstr. 076.005.

Evidence for a common origin of the electrons responsible for the impulsive X-ray and type III radio bursts.
See Abstr. 076.042.

Evidence for a two-component injection of cosmic rays from the solar flare of 1969, March 30.
See Abstr. 078.033.

Direct observations of low-energy solar electrons associated with a type III solar radio burst.
See Abstr. 078.046.

Solar eclipse observed at cm- and dm-wavelengths.
See Abstr. 079.104.

Errata

077.901 Addendum: 'Peculiar absorption and emission microstructures in the type IV solar radio outburst of March 2, 1970' [Solar Physics, Vol. 25, 210 - 231 (1972)].
C. Slottje.
Solar Physics, Vol. 26, 259 (1972).

078 Solar Cosmic Radiation

078.001 Entry of high-energy solar protons into the distant geomagnetic tail.
A. C. Durney, G. E. Morfill, J. J. Quenby.
Journ. Geophys. Res., Vol. 77, 3345 - 3360 (1972).
During the solar proton events of November 18, 1968, and February 25, 1969, a detector on board the low-altitude polar-orbiting Esro 2 spacecraft observed marked structure in the enhanced intensities of 100- to 300-Mev protons at high latitudes. While concentrating mainly on the results of the northern hemisphere for November 18, this structure has been investigated in an attempt to discover the access mechanisms for these protons.

078.002 A comparison of measurements of the charge spectrum of solar cosmic rays from nuclear emulsions and the Explorer 35 solid-state detector.
T. P. Armstrong, S. M. Krimigis, D. V. Reames, C. E. Fichtel.
Journ. Geophys. Res., Vol. 77, 3607 - 3612 (1972). − Letter.

078.003 A search for neutrons of solar origin using balloon borne detectors 1967−69.
C. J. Eyles, A. D. Linney, G. K. Rochester.
Solar Physics, Vol. 24, 483 - 497 (1972).
A series of telescopes having approximately a 30° half opening angle and responding to neutrons in the energy range 50 MeV to 350 MeV has been flown to the top of the atmosphere on balloons released from an equatorial launching site at Kampala, Uganda, between 1967 and 1969. The aim of the experiment was to attempt to detect solar neutrons during periods of enhanced solar activity. No neutrons of solar origin were detected, but an upper limit of the order of 30 neutrons $m^{-2} s^{-1}$ at the earth has been placed on the continuous solar neutron flux in the above energy range, and a limit of four photons $m^{-2} s^{-1}$ has also been placed on the corresponding γ-ray flux above 80 MeV. Limits have likewise been placed on the total emission from various flares.

078.004 Azimuthal propagation of low-energy solar-flare protons as observed from spacecraft very widely separated in solar azimuth. R. B. McKibben.
Journ. Geophys. Res., Vol. 77, 3957 - 3984 (1972).

078.005 Several observations of low-energy solar-proton spectra and possible interpretations.
P. Verzariu, S. M. Krimigis.
Journ. Geophys. Res., Vol. 77, 3985 - 3998 (1972).
It is the purpose of this paper to show that, for a large class of solar-particle events, distortions in the proton spectrum not attributable to velocity dispersion do not occur and to show by a qualitative analysis that either continuous emission at the sun coupled with adiabatic deceleration or storage in the vicinity of the sun could explain the flattening in the proton spectrum at the low energies. The differential energy spectrum of solar protons in the range 0.3−25 Mev obtained with the Injun 5 polar-orbiting satellite is examined for several solar-particle events. Of these, the events of February 25, 1969, April 26, 1969, and March 6, 1970, are discussed in detail.

078.006 Coordinate system for use with high-latitude energetic-particle phenomena. G. Morfill.
Journ. Geophys. Res., Vol. 77, 4010 - 4020 (1972).
By using trajectory integration in a model geomagnetic field a new coordinate system is developed that is not based on internal geomagnetic-field components alone but also takes into account the currents flowing in the magnetopause and the neutral sheet.

In this paper, the concentration will be on observations, which consist of simultaneous measurements of fluxes and energy spectra of ~10−30 Mev protons made with University of Chicago cosmic-ray telescopes on board the deep-space probes Pioneer 6 and Pioneer 7, in orbit about the sun, and the earth satellite IMP 4 in the period December 1967 through August 1968.

078.007 Statistical analysis of Forbush-decreases and foregoing increases of cosmic ray intensity.
A. E. Kuzmicheva, L. I. Dorman, N. S. Kaminer.
Geomagn. Aeronom., Vol. 12, 593 - 597 (1972). In Russian.

078.008 Rocket measurements of a particle stream during the burst of solar cosmic rays in April 1969.
V. V. Tulyakov.
Kosmich. Issled., Vol. 10, 629 - 630 (1972). In Russian.
Brief information.

078.009 International cooperative survey of energetic solar particle events. − A progress report.
H. W. Dodson, E. R. Hedeman.
Bull. American Astron. Soc., Vol. 4, 380 - 381 (1972).
Abstr. AAS.

078.010 Solar cosmic ray composition measured with nuclear emulsions flown on sounding rockets during 1971.
C. E. Fichtel, D. L. Bertsch, C. J. Pellerin, D. V. Reames.
Bull. American Astron. Soc., Vol. 4, 382 (1972). − Abstr. AAS.

078.011 The onset of the November 18, 1968 solar event.
J. A. Lezniak, W. R. Webber.
Bull. American Astron. Soc., Vol. 4, 386 (1972). − Abstr. AAS.

078.012 Pitch angle distribution of energetic solar particles.
R. H. Maurer, S. P. Duggal, M. A. Pomerantz.
Bull. American Astron. Soc., Vol. 4, 387 (1972). − Abstr. AAS.

078.013 A study of solar cosmic ray microevents.
F. B. McDonald, M. van Hollebeke, J. Wang.
Bull. American Astron. Soc., Vol. 4, 387 (1972). − Abstr. AAS.

078.014 Observations of solar proton events made from very widely separated spacecraft. R. B. McKibben.
Bull. American Astron. Soc., Vol. 4, 387 (1972). − Abstr. AAS.

078.015 The relative abundance of helium in solar cosmic rays.
T. T. von Rosenvinge, B. J. Teegarden, F. B. McDonald.
Bull. American Astron. Soc., Vol. 4, 390 (1972). − Abstr. AAS.

078.016 Composition and energy spectra of heavy nuclei with $1.0 \leq E \leq 40.0$ MeV/nuc in the January 24, 1971 solar flare. J. D. Sullivan, P. B. Price, H. J. Crawford.
Bull. American Astron. Soc., Vol. 4, 393 (1972). − Abstr. AAS.

078.017 Satellite measurements of the composition of flare accelerated particles in the charge range $6 \leq Z \leq 26$.
B. J. Teegarden, T. T. von Rosenvinge, F. B. McDonald.
Bull. American Astron. Soc., Vol. 4, 393 (1972). − Abstr. AAS.

078.018 The coronal transport of the scatter-free electrons.
J. R. Wang.
Bull. American Astron. Soc., Vol. 4, 395 (1972). − Abstr. AAS.

078.019 Polar-cap measurements of solar-flare protons with energies down to 12.4 kev.

P. F. Mizera, J. F. Fennell, J. B. Blake.
Journ. Geophys. Res., Vol. 77, 4845 - 4850 (1972). — Brief report.

078.020 **The influence of ionization losses on the conditions of cosmic ray generation on the sun.** P. Velinov.
Geomagn. Aeronom., Vol. 12, 806 - 813 (1972). In Russian.

078.021 **About the possibility of detection of weak streams of solar cosmic rays by ground-based radio technical methods.** V. M. Driatzky, A. V. Shirochkov.
Geomagn. Aeronom., Vol. 12, 823 - 829 (1972). In Russian.

078.022 **Solar particle injection at medium energies $(25 < E < 250\,\text{MeV})$.** J. Engelmann.
Astrophys. Space Sci. Library, Vol. 32, (see 012.009), 95 - 100 (1972).

078.023 **Entry of energetic solar protons into the tail.** A. C. Durney, G. E. Morfill.
Astrophys. Space Sci. Library, Vol. 32, (see 012.009), 101 - 106 (1972).

078.024 **Sudden increase of solar cosmic radiation on July 7, 1966 and its measurement aboard Proton 3.**
N. N. Volodichev, Yu. F. Galaktionova, M. A. Zel'dovich, O. M. Kovrizhnykh, M. O. Madeev, O. Yu. Nechaev, I. A. Savenko, Yu. T. Slyusarev.
Kosm. Issled., Vol. 10, 737 - 745 (1972). In Russian.

078.025 **Characteristics of quiet as well as enhanced diurnal anisotropy of cosmic radiation.**
U. R. Rao, A. G. Ananth, S. P. Agrawal.
Planet. Space Sci., Vol. 20, 1799 - 1816 (1972).
It is shown that the model which has been successful in explaining the anisotropy of low energy cosmic radiation of solar origin in terms of simple convection and field aligned diffusion can also be extended to explain both the quiet and enhanced diurnal variation of cosmic radiation at higher energies. The enhanced diurnal variation which shows a maximum around ~ 2000 hr is shown to be caused by the superposition of normal convection and enhanced field aligned diffusion due to an enhanced positive density gradient of approximately ~ 10 per cent/AU.

078.026 **North/south asymmetric entry of solar proton during the November 18, 1968 event.**
K. Aarsnes, R. Amundsen.
Planet. Space Sci., Vol. 20, 1835 - 1841 (1972).
Solar proton observations by the ESRO IA satellite are presented for the November 18, 1968 event. The time history of proton influx over the polar regions, showing a clear north/south asymmetry during the onset phase of the event, is presented.

078.027 **Observation of solar particle fluxes over extended solar longitudes.**
R. P. Bukata, U. R. Rao, K. G. McCracken, E. P. Keath.
Solar Physics, Vol. 26, 229 - 240 (1972).
Detailed particle observations from various Pioneer spacecrafts located at different helio-longitudes during the complex solar flare events of March 30—April 10, 1969 have been utilised to investigate the energy dependence of azimuthal gradients of cosmic ray particles and its effect on the decay of the flare intensity.

078.028 **Flux and energy spectra of solar protons observed aboard the ESRO 2 satellite in 1968—1969.**
J. Engelmann.
Space Research XII, (see 012.016), Vol. 2, 1479 - 1485 (1972).

078.029 **Measurements of the isotopic composition of particle fluxes carried out on spacecrafts Soyuz, Zond 8 and Luna 16.** B. S. Boltenkov, V. N. Gartmanov, G. E. Kocharov, B. A. Mamyrin, V. O. Naidenov.
Space Research XII, (see 012.016), Vol. 2, 1487 - 1491 (1972).

078.030 **Solar cosmic ray bursts in November—December 1970 according to data from Venus 7 space probe and Lunokhod 1 station.** S. N. Vernov, N. N. Kontor, G. P. Lyubimov, N. V. Pereslegina, E. A. Chuchkov.
Space Research XII, (see 012.016), Vol. 2, 1535 - 1544 (1972).

078.031 **Peculiarities of particles accelerated on the sun.** L. I. Miroshnichenko.
Trudy Mezhdunar. seminara po probl. "Uskorenie chastits v kosmich. prostranstve (okolozem. i mezhplanet. kosmich. prostranstve), Galaktike i Metagalaktike". Moskva, 1972, p. 41 - 54. In Russian. — Abstr. in Referativ. Zhurn. 62. Issled. kosmich. prostranstva, 12.62.158 (1972).

078.032 **Energy changes of solar cosmic rays.** R. C. Englade.
Journ. Geophys. Res., Vol. 77, 6266 - 6270 (1972).— Letter.

078.033 **Evidence for a two-component injection of cosmic rays from the solar flare of 1969, March 30.**
I. D. Palmer, S. F. Smerd.
Solar Physics, Vol. 26, 460 - 467 = Radiophys. Publ., *Sydney*, RPP 1590 (1972).
The solar flare of 1969 March 30, occurring ≈20° behind the west limb, produced very extensive 80 MHz radio emission at the sun, and gave rise to the deployment of cosmic radiation over 360° long. in interplanetary space. The wide spread of this event may reflect a similar spread of coronal magnetic fields from the flare site. We interpret the solar proton data recorded by spacecraft at two separate points both at ≲ 1 AU, in terms of a two-component injection of particles at the sun. The radio spectral and positional data provide evidence of shock waves which propagated far and wide from the flare.

078.034 **Evidence for the existence of adiabatic energy loss in interplanetary space from observations of the decay of the February 25—March 2, 1969 series of solar cosmic ray events.** J. A. Lezniak, W. R. Webber.
Solar Physics, Vol. 26, 474 - 483 (1972).
The integral flux of low energy protons (> 10 MeV) observed by the University of New Hampshire cosmic ray detector aboard the Pioneer 9 spaceprobe has been compared with similar measurements of the near-earth spacecraft Explorer 34 during the decay phase of the February 25—March 2, 1969 series of solar cosmic ray events. A comparison of the observed and theoretically calculated ratios suggests that the adiabatic energy loss process is operative.

078.035 **Radio tracking of solar energetic particles through interplanetary space.**
J. Fainberg, L. G. Evans, R. G. Stone.
Science, Vol. 178, 743 - 745 (1972).
Energetic particles ejected from the sun generate radio waves as they travel out through the interplanetary medium. Satellite observations of this emission at long radio wavelengths provide a means of investigating properties of the interplanetary medium, including the gross magnetic field configuration over distances of 1 astronomical unit. Results of such observations are illustrated.

078.036 **Anisotropy of low-energy solar protons at the boundary of the magnetotail.**
W. A. Cooper, G. P. Haskell.
Journ. Geophys. Res., Vol. 77, 6849 - 6853 (1972). — Brief report.

078.037 Solar electrons and alpha particles during polar-cap absorption events. T. A. Potemra, A. J. Zmuda.
Journ. Geophys. Res., Vol. 77, 6916 - 6921 (1972).

This letter reviews some characteristics of solar electrons and α particles and assesses the importance of these particles for D region enhancements at times when solar protons are also present. Calculations for some ionospheric effects are also made with the D region model developed by Potemra et al. (1970).

078.038 Sectorial anisotropy of solar cosmic rays. S. P. Duggal, M. A. Pomerantz.
Solar Physics, Vol. 27, 227 - 241 (1972).

A hitherto unobserved sectorial pattern of anisotropy that was limited to a narrow and stable region was displayed by the ground level event (GLE) of January 24, 1971. For the entire $1\,^1/_2$ h interval following onset before isotropy set in, the anisotropy was limited to a $10°$ cone centered about $60°$ from the spiral magnetic field line. It is also the first solar particle event for which it is possible, by analytical procedures based upon a theoretical propagation model, to distinguish between two rival candidates for the parent flare.

078.039 Features of spectra of particles accelerated on the sun. L. I. Miroshnichenko.
Trudy Mezhdunar. seminara po probl. "Uskorenie chastits v kosmich. prostranstve (okolozem. i mezhplanet. kosmich. prostranstve), Galaktike i Metagalaktike". Moskva, 1972, p. 41 - 54. In Russian. — Abstr. in Referativ. Zhurn. 51. Astron., 1.51.408 (1973).

078.040 Energy spectra of proton generation on the sun. A. N. Charakhch'yan, T. N. Charakhch'yan.
Trudy Mezhdunar. seminara po probl. "Uskorenie chastits v kosmich. prostranstve (okolozem. i mezhplanet. kosmich. prostranstve), Galaktike i Metagalaktike". Moskva, 1972, p. 85 - 95. In Russian. — Abstr. in Referativ. Zhurn. 51. Astron., 1.51.409 (1973).

078.041 Cosmic ray and solar activity variations associated with the rotation of the planets around the sun. E. V. Kolomeets, Yu. A. Shakhova.
Trudy Mezhdunar. seminara po probl. "Uskorenie chastits v kosmich. prostranstve (okolozem. i mezhplanet. kosmich. prostranstve), Galaktike i Metagalaktike". Moskva, 1972, p. 253 - 258. In Russian. — Abstr. in Referativ. Zhurn. 51. Astron., 1.51.449 (1973).

078.042 Particle acceleration by magnetohydrodynamic waves of small amplitude. I. N. Toptygin.
Trudy Mezhdunar. seminara po probl. "Uskorenie chastits v kosmich. prostranstve (okolozem. i mezhplanet. kosmich. prostranstve), Galaktike i Metagalaktike". Moskva, 1972, p. 55 - 84. In Russian. — Abstr. in Referativ. Zhurn. 62. Issled. kosmich. prostranstva, 1.62.177 (1973).

078.043 Composition of energetic solar particles. S. Biswas.
Indian Journ. Radio and Space Phys., Vol. 1, No. 1, p. 19 - 24 (1972).

078.044 The absorption length for solar particles in the earth's atmosphere. Solar proton event, November 18, 1968. J. Ilencik.
Fyz. Cas Bratislava, Vol. 22, 190 - 192 (1972).

078.045 Observational evidence of the existence of quasi-stationary solar corpuscular streams. M. S. Bobrov.
Astron. Tsirk., No. 730, p. 6 - 8 (1972). In Russian.

078.046 Direct observations of low-energy solar electrons associated with a type III solar radio burst. L. A. Frank, D. A. Gurnett.
Solar Physics, Vol. 27, 446 - 465 (1972).

Direct observations of the impulsive ejection of a highly anisotropic, low-energy solar electron packet which was accompanied by a type III solar radio noise burst detected simultaneously with instrumentation on the satellite IMP-6 have been examined here. These events were associated with a solar flare commencing at 0935 UT, 6 April 1971, on the western limb of the sun.

078.047 Shock wave effects in solar cosmic ray events. I. D. Palmer.
Solar Physics, Vol. 27, 466 - 477 (1972).

Two low-energy ($\lesssim 1$ MeV) solar proton events which display a gradual intensity increase to a maximum near the time of an SSC, followed by an abrupt, large decrease, are interpreted in terms of a population of cosmic rays which are 'swept' ahead of an interplanetary shock wave. A model which describes the variation with time of intensity and anisotropy at the earth is developed using a Monte Carlo technique which traces the histories of particles released impulsively at the sun.

078.048 On the reality of influence of solar corpuscular streams upon the lower layers of the earth's atmosphere. E. R. Mustel.
Nauchn. Informatsii, vyp. (No.) 24, p. 5 - 55 (1972). In Russian.

On the realization of a force-free magnetic field configuration in the case of axial symmetric magnetohydrodynamic streams. See Abstr. 062.076.

Composition and energy spectra of heavy nuclei with $0.5 < E < 40$ MeV per nucleon in the 1971 January 24 and September 1 solar flares. See Abstr. 073.004.

Change of solar flare proton to alpha ratios during an energetic storm particle event. See Abstr. 073.014.

The abundances of solar accelerated nuclei from carbon to iron. See Abstr. 073.065.

The preferential acceleration of heavy nuclei in solar flares. See Abstr. 073.066.

Collisionless solar wind protons: A comparison of kinetic and hydrodynamic descriptions. See Abstr. 074.017.

Some characteristics of microwave type IV radio bursts and the acceleration of solar cosmic rays. See Abstr. 077.034.

Evidence for electron excitation of type III radio burst emission. See Abstr. 077.046.

About time dependence of cosmic ray intensity on the anisotropic stage of solar bursts. See Abstr. 077.048.

A mixed-up sun and solar neutrinos. See Abstr. 080.040.

Effects of sudden mixing in the solar core on solar neutrinos and ice ages. See Abstr. 080.041.

A measurement of the atmospheric neutron flux in the energy range $50 < E < 350$ MeV. See Abstr. 082.107.

Bombardment of the polar-cap ionosphere by solar cosmic rays. See Abstr. 083.057.

Model for the uneven illumination of polar caps by solar protons. See Abstr. 084.037.

Reconnection of the geomagnetic tail deduced from solar-particle observations. See Abstr. 084.218.

Solar proton intensity structures in the magnetosphere during interplanetary anisotropies. See Abstr. 084.309.

Penetration of solar protons into the geomagnetic tail. See Abstr. 084.338.

Decametre-wave radiation from Jupiter and solar activity. See Abstr. 099.067.

Detection of interplanetary electrons from 18 keV to 1.8 MeV during solar quiet times. See Abstr. 106.014.

Origin of 200-keV interplanetary electrons. See Abstr. 106.015.

Association between interplanetary shock waves and delayed solar particle events. See Abstr. 106.019.

The future of balloons in cosmic-ray research. See Abstr. 143.004.

Observations of the radial gradient of galactic cosmic radiation over a solar cycle. See Abstr. 143.005.

Peculiarities of cosmic ray variations near the plane of the solar equator. See Abstr. 143.050.

079 Solar Eclipses

079.001 **Remarques sur le retour des éclipses de soleil et de lune.** J. Dommanget.
Ciel et Terre, Vol. 88, 298 - 300 (1972).

079.002 **Some contributions of Canadian astronomers to solar eclipse expeditions.** J. E. Kennedy.
Journ. Roy. Astron. Soc. Canada, Vol. 66, 261 - 274 (1972).

079.003 **Total solar eclipses of great duration.** E. S. Light.
Journ. Roy. Astron. Soc. Canada, Vol. 66, 295 - 302 (1972).
The conditions necessary for a solar eclipse to have the maximum possible duration of totality are examined for the present as well as a remote epoch. All total solar eclipses occurring between the years −3000 and +5000 were examined and all those having maximum durations of totality exceeding seven minutes are tabulated.

079.004 **Solformørkelser.** O. P. Sveen.
Naturen 1971, No. 5, p. 279 - 302 = Inst. Teor.
Astrofys. Blindern − Oslo, Småtrykk No. 75 (1971).

The most ancient eclipse.
Southern Stars, Vol. 24, 123 - 124 (1972).

079.100 **Solar eclipse, 1971 February 25**

Eclipse parcial de sol de 25 de febrero de 1971.
R. Carrasco, M. M. Lorón, J. Pensado, M. de Pascual.
Bol. Astron. Obs. Madrid, Vol. 8, No. 1, p. 143 - 144 (1972).

Ancora sui tempi dei contatti nell'eclisse parziale di sole del 25 febbraio 1971.
M. Boggio, M. G. Fracastoro, G. Francese.
Mem. Soc. Astron. Italiana, Nuova Ser., Vol. 43, 381 - 382 (1972). − Letter.

The correlation between X rays and ultraviolet ionizing radiation in the E region by data of the solar eclipse of February 25, 1971. See Abstr. 083.027.

079.101 **Solar eclipse, 1972 July 10**

Observations of the solar eclipse on July 10, 1972.
G. N. Salukvadze.
Astron. Tsirk., No. 725, p. 1 - 2 (1972). In Russian.

Observations of the solar eclipse on Chukotka on July 10, 1972. S. K. Vsekhsvyatskij, A. T. Nesmyanovich, N. I. Dzyubenko.
Astron. Tsirk., No. 725, p. 2 - 3 (1972). In Russian.

Eclipse total de sol en Canadá: Un viaje inolvidable.
F. Diego Q.
El Universo, No. 101, Vol. 26, 131 - 148 (1972).

De totale zonsverduistering van 10 juli 1972.
P. J. Bruijn.
Hemel en Dampkring, Vol. 70, 327 - 335 (1972).

CRESS solar eclipse expedition. R. W. Nicholls.
Journ. Roy. Astron. Soc. Canada, Vol. 66, 278 - 279 (1972).

The Toronto Centre Eclipse Expedition 8 to 11 July, 1972. B. R. Chou.
Journ. Roy. Astron. Soc. Canada, Vol. 66, L26 - L28 (1972).

Solar eclipse July 10 from Prince Edward Island.
B. F. Shinn.
Journ. Roy. Astron. Soc. Canada, Vol. 66, L29 - L30 (1972).

Solar corona at the eclipse of July 10, 1972.
Mercury, (Journ. Astron. Soc. Pacific), Vol. 1, No. 6, p. 17 (1972).

The eclipse across North America.
Sky Telescope, Vol. 44, 142 - 149, 173 - 176 (1972).

The Olympia's voyage to darkness.
E. M. Brooks, G. S. Mumford, L. J. Robinson.
Sky Telescope, Vol. 44, 154 - 157 (1972).

A widely observed partial solar eclipse.

Sky Telescope, Vol. 44, 222 - 224 (1972).

Interferometric investigations of emission lines of the solar corona during the eclipse of 1972 July 10. See Abstr. 074.112.

079.102   Solar eclipse, 1970 March 7

Equidensities of the circumsolar region at the 1970 March 7 eclipse.   N. I. Dzyubenko, A. T. Nesmyanovich, Yu. A. Chomenko.
Astron. Tsirk., No. 727, p. 3 - 6 (1972). In Russian.

Y se hizo la oscuridad... J. G. Hernández.
El Universo, No. 98, Vol. 26, 19 - 24 (1972).

Observation of the solar eclipse on March 7, 1970 at 1.37 m wavelength.
L. I. Yurovskaya, Yu. F. Yurovskij.
Sb. dokl. Sessii Nauch. Soveta po kompleks. probl. "Radioastronomiya", IZMIRAN, 1970. Moskva, 1972, p. 147 - 152. In Russian. − Abstr. in Referativ. Zhurn. 51. Astron., 1.51.396 (1973).

Results of an observation of the solar eclipse on March 7, 1970 at 10 cm wavelength.   Yu. F. Yurovskij.
Sb. dokl. Sessii Nauch. Soveta po kompleks. probl. "Radioastronomiya", IZMIRAN, 1970. Moskva, 1972, p. 153 - 163. In Russian. − Abstr. in Referativ. Zhurn. 51. Astron., 1.51.397 (1973).

Observations of the solar eclipse on March 7, 1970 at 3.07 cm wavelength in Cuba.
M. M. Kobrin, E. I. Lebedev, B. V. Timofeev, V. M. Fridman.
Sb. dokl. Sessii Nauch. Soveta po kompleks. probl. "Radioastronomiya", IZMIRAN, 1970. Moskva, 1972, p. 164 - 179. In Russian. − Abstr. in Referativ. Zhurn. 51. Astron., 1.51.398 (1973).

Raies nouvelles observées lors de l'éclipse du 7 mars 1970.   See Abstr. 071.009.

The inner solar corona polarization at the total eclipse on March 7, 1970.   See Abstr. 074.007.

A hydrodynamical study of the large NE coronal streamer observed during the 7th March eclipse. See Abstr. 074.011.

Photometric and polarimetric analysis of the coronal streamers observed at the March 7, 1970 Mexican eclipse. See Abstr. 074.043.

Polarigraphic observations of the solar corona at the total eclipse on March 7, 1970 in Mexico. See Abstr. 074.044.

On physical conditions in the corona from spectral observations of the solar eclipse on March 7, 1970. See Abstr. 074.046.

On emission lines of hydrogen, helium and ionized calcium seen on a coronal spectrogram of the March 7, 1970 eclipse.   See Abstr. 074.084.

Sodium emission from the atmosphere during a solar eclipse.   See Abstr. 082.017.

079.103   Solar eclipse, 1965 May 30

Skylight intensity, polarization and airglow measurements during the total solar eclipse of 30 May 1965. See Abstr. 082.021.

079.104   Solar eclipse, 1968 September 22

Solar eclipse observed at cm- and dm-wavelengths.
M. S. Durasova, T. S. Podstrigach, A. K. Chandaev.
Solnechnye Dannye 1972 Byull., No. 4, p. 59 - 65 (1972). In Russian.
The results of observations of the solar eclipse on Sept. 22, 1968 at seven frequencies in the wavelength range from 3 cm to 3 m are given. The radio radii of the sun have been determined. Local radio sources connected with sunspot groups, flocculi and coronal condensations were detected. An attempt is made to explain the configuration of these radio sources.

Observations of the partial solar eclipse on September 22, 1968 at 75 cm wavelength.
M. G. Kamenskij, N. P. Tsimakhovich.
Trudy Radioastrophys. Obs., *Riga,* Vol. 14, (see 003.019), 119 - 126 (1972). In Russian.

The results of coronal investigation at the September 22, 1968 solar eclipse.   See Abstr. 074.051.

On the geometry of local sources on the sun from observations of the eclipse on September 22, 1968. See Abstr. 077.010.

Local sources on the sun from observations of the eclipse on Sept. 22, 1968 at λ = 2 cm.   See Abstr. 077.015.

079.105   Solar eclipse, 1966 November 12

Eclipse measurements of the coronal iron abundance and electron temperature.   See Abstr. 074.031.

079.106   Solar eclipse, 1973 June 30

Århundradets solförmörkelse den 30 juni 1973.
P. Å. Björklund.
Astron. Tidssk., Årg. 5, p. 138 - 140 (1972).

L'éclipse de soleil du 30 juin 1973.   J. Meeus.
Ciel et Terre, Vol. 88, 458 - 466 (1972).

De zonsverduistering van 30 juni 1973.
J. Meeus.
Hemel en Dampkring, Vol. 70, 336 - 341 (1972).

Slovak solar eclipse expedition − Africa 1973.
J. Sýkora.
Kozmos, Vol. 3, 100 - 102 (1972).   In Slovak.

Die totale Sonnenfinsternis vom 30. Juni 1973.
R. A. Naef.
Orion Schaffhausen, 30. Jahrgang, p. 184 - 185 (1972).

The total solar eclipse of June 30, 1973 − one of the best.   C. H. Smiley.
Sky Telescope, Vol. 44, 282 - 283 (1972).

Weather prospects for next year's total eclipse.
E. M. Brooks.
Sky Telescope, Vol. 44, 284 - 285 (1972).

Notes on the eclipse of the sun in June.
F. Liems, A. Rigby, R. W. Tuthill.
Sky Telescope, Vol. 44, 285 - 287 (1972).

Atmospheric gravity waves to be expected from the solar eclipse of June 30, 1973.　　See Abstr. 082.104.

079.107 **Solar eclipse, 1973 January 4**

Eclipse anular de sol del 4 de enero de 1973.
G. M. Iannini, A. Niell.
Edited by Obs. Astron., Inst. Mat., Astron., Fís., Univ. Nacional Córdoba, Comisión Nacional Estud. Geo-Heliofís., Córdoba. 20 pp. (1972).

079.108 **Solar eclipse, 1974 June 20**

Total eclipse of the sun, 1974 June 20th.
Journ. Astron. Soc. Western Australia, 1972 December, p. 3 - 5.

080 Solar Figure, Internal Constitution, Rotation, Miscellanea

080.001 The generation of magnetic fields in astrophysical bodies. IX. A solar dynamo based on horizontal shear. I. Lerche, E. N. Parker.
Astrophys. Journ., Vol. 176, 213 - 223 (1972).

In view of the uncertainty in the angular velocity in the solar interior, we have worked out the general behavior of the solar dynamo for horizontal shear for comparison with the earlier models based on vertical shear. The horizontal shear excites the even and odd modes equally, whereas vertical shear excites the odd mode before the even mode.

080.002 On the mean depth of line formation in a magnetic field. J. Staude.
Solar Physics, Vol. 24, 255 - 261 (1972).

Probability interpretation of radiative transfer is used to calculate the contribution of different layers of the solar atmosphere to the emergent intensity. Generally the mean depths of line formation increase with increasing intensity; this is valid also for arbitrarily polarized constituents of a line formed in a magnetic field.

080.003 Polar magnetic fields of the sun: 1960—1971. R. Howard.
Solar Physics, Vol. 25, 5 - 13 (1972).

Observations of the magnetic fields in the polar regions of the sun are presented for the period 1960–1971. At the start of this interval the fields at the two poles were consistently of opposite sign and averaged around 1 G. Early in 1961 the field in the south decreased suddenly and the field in the north decreased in strength slowly over the next few years. A comparison of field strengths in the east and west quadrants in the north suggests that even at the extreme polar latitudes the following polarity fields are inclined slightly toward the rotation and the preceding polarity field lines are inclined slightly to trail the rotation.

080.004 A non imaging approach to solar oblateness measurements. T. J. Janssens.
Solar Physics, Vol. 25, 237 - 241 (1972).

A design is presented for an instrument to measure solar oblateness without forming a solar image and having two identical prisms as the only optical elements. Feasibility calculations indicate that this might be sensitive and quite free from instrumental induced errors.

080.005 Solar magnetic fields derived from hydrogen alpha filtergrams. P. S. McIntosh.
Rev. Geophys. Space Phys., Vol. 10, 837 - 846 (1972).

The distribution and polarity of solar magnetic fields can be inferred from photographs taken with narrow band-pass filters centered on the Hα spectral line. Structures observed in Hα, including filaments, filament channels, fibrils, arch-filament systems, and plage corridors are used to infer positions of neutral lines in the radial component of the solar magnetic fields.

080.006 Energía solar. N. Fragoso.
El Universo, No. 98, Vol. 26, 29 - 30 (1972).

080.007 Horizontal propagation of solar atmospheric oscillations.
J. H. Thomas, P. A. Clark, A. Clark, Jr.
Astrophys. Letters, Vol. 12, 31 - 34 (1972).

The three basic types of resonant wave modes that have been proposed as the mechanism for solar five-minute oscillations have inherent differences in their horizontal propagation. Observations of horizontal propagation may help to identify the true mode of oscillation.

080.008 Reduction of the solar neutrino flux by primordial ^{3}He in the sun. R. Mitalas.
Astrophys. Letters, Vol. 12, 35 - 36 (1972).

The presence of primordial ^{3}He in the sun would decrease the hydrogen burning age of the sun and would reduce the solar neutrino flux significantly.

080.009 A new solar fluctuation. R. H. Dicke.
Astrophys. Journ., Vol. 176, 479 - 486 (1972).

The fluctuation in the solar-oblateness signal during 1966 has been found to be strongly autocorrelated with a 25.3-day lag. The autocovariance functions show that the disturbed regions of the sun occur in the vicinity of ±45° latitude for which the synodic rotation period is ~29.5 days. It is concluded that the 25-day correlation may be due to density fluctuations occurring near the top of the radiative zone.

080.010 Time variations of the magnetic field in the undisturbed solar atmosphere.
S. I. Gopasjuk, T. T. Tsap.
Solnechnye Dannye 1972 Byull., No. 4, p. 84 - 90 (1972).
In Russian.

Time variations of the magnetic field in the undisturbed regions of the sun have been investigated. It is shown that the maximum intensity of the magnetic field at a hill increases with area. The distribution of the magnetic field inside hills with different sizes is similar.

080.011 Neutrino and the sun.
S. Gershtejn, V. Folomeshkin.
Nauka i zhizn', 1972, No. 4, p. 49 - 55. In Russian.

080.012 On possible measurements of weak solar magnetic fields using radio astronomical methods.
G. B. Gelfrejkh.
Astron. Tsirk., No. 699, p. 3 - 5 (1972). In Russian.

080.013 The minimum temperature in the solar atmosphere. R. C. Altrock, C. J. Cannon.
Bull. American Astron. Soc., Vol. 4, 310, with a correction, p. 426 (1972). – Abstr. AAS.

080.014 Solar rotation as measured by EUV spectrohelio-grams. W. Henze, Jr., A. K. Dupree.
Bull. American Astron. Soc., Vol. 4, 384 (1972). – Abstr. AAS.

080.015 Solar rotation: The height gradient. W. C. Livingston.
Bull. American Astron. Soc., Vol. 4, 387 (1972). – Abstr. AAS.

080.016 Energy spectral analyses of small scale solar magnetic fields. Y. Nakagawa.
Bull. American Astron. Soc., Vol. 4, 388 (1972). – Abstr. AAS.

080.017 Comparisons of the mean solar magnetic field and the interplanetary field observed during 1969.
P. H. Scherrer, J. M. Wilcox, A. B. Severny.
Bull. American Astron. Soc., Vol. 4, 390 (1972). – Abstr. AAS.

080.018 Some comments on the cross section of ^{37}Cl for solar neutrino absorption.
W. A. Lanford, B. H. Wildenthal.
Phys. Rev. Letters, Vol. 29, 606 - 608 (1972).

Nuclear wave functions from recent shell-model calculations are used to evaluate the log (ft) values relevant to the neutrino absorption cross section of ^{37}Cl.

080.019 On origination of magnetic force tubes.
S. I. Gopasjuk.
Solnechnye Dannye 1972 Byull., No. 6, p. 83 - 87 (1972).
In Russian.
The possibility of origination of magnetic force tubes by streams of gas moving in the solar atmosphere is considered.

080.020 Some peculiarities of the fine structure of the differential rotation of the sun.
Y. I. Vitinsky, R. N. Ikhsanov.
Solnechnye Dannye 1972 Byull., No. 6, p. 99 - 109 (1972).
In Russian.
The fine structure of the differential rotation of the sun inside the 11-year cycle is investigated on the basis of the Greenwich data on the sunspot groups for the 18th solar cycle.

080.021 Free oscillations of the sun and their possible stimulation by solar flares. C. L. Wolff.
Astrophys. Journ., Vol. 176, 833 - 842 (1972).
A large solar flare can raise the temperature of the underlying photosphere by 10 percent. The resulting thermal expansion exerts upon the solar interior a mechanical impulse whose kinetic energy is $\sim 10^{28}$ ergs. This will stimulate free modes of oscillation of the entire sun, and the initial amplitudes of the more easily observable modes are calculated. Preliminary estimates of the effects of radiation damping and turbulent viscosity indicate that damping times will typically be longer than 1 day for the modes of most interest.

080.022 Recent solar research. R. Howard.
Science, Vol. 177, 1157 - 1163 (1972). – Review paper. – Results from new methods of observation prove the sun to be a beautifully complex astronomical body.

080.023 The $(B-V)$ and $(U-B)$ color indices of the sun.
S. K. Croft, D. H. McNamara, K. A. Feltz, Jr.
Publ. Astron. Soc. Pacific, Vol. 84, 515 - 518 (1972).
Two photometric (khg) indices have been measured for 42 F, G, and K luminosity class V stars and Uranus. The indices correlate well with the $(B-V)$ and $(U-B)$ values of the stars. The (khg) indices of Uranus are used to derive $(B-V)$ and $(U-B)$ values for the sun. The results are $(B-V)_\odot = 0^m631 \pm 0^m003$ (p.e.) and $(U-B)_\odot = 0^m14 \pm 0^m007$ (p.e.).

080.024 Solar rotation: The photospheric height gradient.
W. Livingston, R. Milkey.
Solar Physics, Vol. 25, 267 - 273 (1972).
For selected pairs of Fraunhofer lines the height of formation has been calculated corresponding to that portion of the profile intercepted by the magnetograph exit slits. A photospheric height discrimination of 150–300 km is realized. In 1971 simultaneous measurements of equatorial angular velocity from spectroscopic displacements of these line pairs indicate no height gradient in excess of 1%. The disturbing influence of telluric line blends is analyzed.

080.025 On the adjustment of outer solar layer models.
I. A. Krinberg, R. B. Teplitskaja.
Solar Physics, Vol. 25, 305 - 318 (1972).
Using the electron density n_e as an independent variable agreement between the models of the convective zone, photosphere, chromosphere, corona and solar wind is obtained. As a base the known data about the mean models of the individual layers of the quiet sun are taken. A plot of the gas density ρ versus n_e permits to get a clear representation about the rate of change of the degree of ionization x and to evaluate quickly the numerical values of x.

080.026 Tunnel-effect and propagation of 5-min oscillations in the solar atmosphere. Y. D. Zhugzhda.
Solar Physics, Vol. 25, 329 - 338 (1972).
Tunneling of surface waves in an isothermal atmosphere is considered. Tunneling of 5-min oscillations in the solar atmosphere is discussed.

080.027 Precision measurement of the sun's gravitational field by means of the twin probe method.
B. Bertotti, G. Colombo.
Astrophys. Space Sci., Vol. 17, 223 - 237 (1972).
An accurate measurement of the gravitational field of the sun, needed for the verification of the theories of gravitation, requires the use of a geodesic test body. To eliminate the effect of non-gravitational forces (mainly the solar radiation pressure) we propose to use two twin space probes, whose surface has identical geometrical and optical properties, but with different mass. Their differential motion leads to the determination of the motion of an ideal geodesic point.

080.028 Solar rotation: Phenomenology and superficial models.
G. Belvedere, G. Godoli, S. Motta, L. Paternò.
Atti XV Riunione Soc. Astron. Italiana, Bologna 1971, (see 012.013), p. 45 - 70 (1972).
After a critical review of the observations of the solar rotation angular velocity versus heliographic latitude, depth, and time, the authors deal with the possibility that a superficial circulation may be responsible for the angular momentum transport.

080.029 The sun's rotation and the Brans-Dicke cosmology.
V. de Sabbata.
Atti XV Riunione Soc. Astron. Italiana, Bologna 1971, (see 012.013), p. 71 - 90 (1972).
We examine the implications that the Brans-Dicke scalar-tensor theory has on the physics of solar interior. Questions of lithium depletion and rotation in solar-type stars are considered.

080.030 Alfvénic motions in the solar atmosphere.
J. V. Hollweg.
Astrophys. Journ., Vol. 177, 255 - 259 (1972).
The amplitude of bulk velocities associated with upward-propagating Alfvén waves in the lower solar atmosphere is discussed. We show that for a given wave energy flux, the bulk velocities can be appreciably lower in cases when the wavelength is much larger than the scale height, than in situations where the wavelength is smaller than the scale height. In the chromosphere and lower corona, the former case pertains to waves with dominant timescales of 1–2 hours, as is found for Alfvén waves observed in the solar wind at 1 a.u.

080.031 On the sun's differential rotation and pole-equator temperature difference. B. Durney.
Solar Physics, Vol. 26, 3 - 7 (1972).
The sun's differential rotation can be understood in terms of preferential stabilization of convection (by rotation) in the polar regions of the lower part of the convection zone (where the Taylor number is large).

080.032 Solar rotation as measured in EUV chromospheric and coronal lines. G. W. Simon, R. W. Noyes.
Solar Physics, Vol. 26, 8 - 14 (1972).
Active regions were followed across the disk on OSO 4 spectroheliograms in the Lyman continuum and in Mg X λ625. These observations indicate differential rotation with latitude, but not with height in the atmosphere.

080.033 Differential rotation in the solar atmosphere inferred from optical, radio and interplanetary data.
M. El-Raey, P. H. Scherrer.

Solar Physics, Vol. 26, 15 - 20 (1972).

Autocorrelation analysis of sunspot number, solar radio flux, and interplanetary field in the period 1967 to 1970 yields new information concerning solar atmospheric rotation.

080.034 The formation of Mg I 4571 Å in the solar atmosphere. I: A model analysis of a one-dimensional static atmosphere.   R. C. Altrock, C. J. Cannon.
Solar Physics, Vol. 26, 21 - 29 (1972).

A one-dimensional analysis of the 4571 Å line of neutral magnesium is presented. The Harvard-Smithsonian Reference Atmosphere and the Bilderberg Continuum Atmosphere are used to compute the emergent line profiles at various positions on the solar disc. The resultant profiles are compared to the observations.

080.035 Mean values in inhomogeneous atmospheres.   P. R. Wilson, N. V. Williams.
Solar Physics, Vol. 26, 30 - 41 (1972).

A brief summary of observations of inhomogeneities in the solar atmosphere and progress in the theoretical analysis of two-dimensional model atmospheres is given. In particular, it is asserted that reliable reference models of the mean temperature, pressure, etc. may be derived only by averaging over the horizontal coordinates of a two- or three-dimensional model.

080.036 The five-minute oscillations as nonradial pulsations of the entire sun.   C. L. Wolff.
Astrophys. Journ., (*Letters*), Vol. 177, L87 - L91 (1972).

Calculations of the stability coefficient show that the sun should be pulsating as a unit in nonradial modes of high order. The pulsations are driven by the superadiabatic gradient of the low photosphere and by the same sensitive changes in opacity that are known to be important in variable stars. The existence of solar pulsations can explain many of the large scale features observed in the well known 5-minute oscillations of the solar atmosphere.

080.037 Solar neutrinos.   J. N. Bahcall, R. L. Sears.
Annual Rev. Astron. Astrophys., Vol. 10, (see 003. 005), 25 - 44 (1972). − Contents: (1) Introduction; (2) Methods of detection; (3) Cross sections for neutrino capture; (4) Solar models; (5) Implications of the observations.

080.038 Why does the sun sometimes look like a magnetic monopole?   J. M. Wilcox.
Comments Astrophys. Space Phys., Vol. 4, 141 - 147 (1972).

A recent model of magnetic fields in the sunspot cycle, based on interplanetary field polarities from geomagnetic observations during an interval of forty-five years, suggests that an interval of several months at sunspot minimum during which the solar field is very predominantly directed out of the sun may be a characteristic of each sunspot cycle. In the present note we will attempt to describe and develop these interesting circumstances. Observations obtained with different solar magnetographs are reported and discussed.

080.039 The solar spoon.   F. W. W. Dilke, D. O. Gough.
Nature, Vol. 240, 262 - 264, 293 - 294 (1972).

Overstability causes the sun's core to mix every few hundred million years. This induces geological ice ages and temporarily depresses the solar neutrino flux.

080.040 A mixed-up sun and solar neutrinos.   R. T. Rood.
Nature, Phys. Sci., Vol. 240, 178 - 180 (1972).

Fowler has suggested that the negative result of the Davis et al. effort to detect neutrinos from the sun might be due to the sun being in a transient phase after having been mixed some millions of years ago. Here I examine that hypo-

thesis in greater detail with stellar evolutionary calculations.

080.041 Effects of sudden mixing in the solar core on solar neutrinos and ice ages.   D. Ezer, A. G. W. Cameron.
Nature, Phys. Sci., Vol. 240, 180 - 182 (1972).

A continuous mixing throughout much of the solar interior might maintain a large abundance of hydrogen at the centre of the sun, thus lowering the temperature relative to an unmixed model, and lowering the ^{8}B neutrino flux by a factor 4 or so. Although this suggestion is unconventional, and some aspects of it are not very appealing in terms of present stellar evolution theory, once again the solar neutrino problem is facing a crisis. We report on some numerical experiments with a solar model. These experiments involve a variant of our original mixing hypothesis.

080.042 Waves in the solar atmosphere. II. Large-amplitude acoustic pulse propagation.   R. F. Stein, R. A. Schwartz.
Astrophys. Journ., Vol. 177, 807 - 828 (1972).

Numerical experiments are performed with vertically propagating acoustic pulses by solving the nonlinear equations of fluid motion using a finite-difference technique. The pulse energy, dissipation, wake, and atmospheric heating are investigated, and the results compared with weak-shock theory.

080.043 Interaction contributions to the solar proton-proton reaction.   M. Gari, A. H. Huffman.
Astrophys. Journ., Vol. 178, 543 - 549 (1972).

Interaction contributions (meson-exchange effects) to the solar p-p reaction are evaluated using the low-energy theorem results. A correction to the cross-section S_{11} of approximately 9 percent is found.

080.044 On the dynamo action of the global convection in the solar convection zone.   H. Yoshimura.
Astrophys. Journ., Vol. 178, 863 - 886 (1972).

The very large-scale nonaxisymmetric convection zone, called here the global convection, has been conceived to explain the solar equatorial acceleration, the clustering in longitudes of magnetic activities, and the proper motions of sunspots. The magnetohydrodynamics of the convection is examined in the framework of a slowly rotating spherical geometry, and the dynamo action is shown to be operated by the global convection when modified by rotation in such a way as to regenerate and reverse the poloidal general magnetic field in the middle layer in the solar convection zone.

080.045 Nonlinear Boussinesq convective model for large scale solar circulations.   P. A. Gilman.
Solar Physics, Vol. 27, 3 - 26 (1972).

We present extensive numerical calculations for a model of thermal convection of a Boussinesq fluid in an equatorial annulus of a rotating spherical shell. The convection induces and maintains differential rotation and meridian circulation. Our model represents a useful compromise between the Busse (1970) and Durney (1970, 1971) models, and those of Davies-Jones and Gilman (1970, 1971) in that it takes into account in an approximate way the Coriolis forces as they would act on convection in a spherical shell, while retaining the much simpler Cartesian geometry. This is done by restricting consideration to an annular region of fluid, bounded by two latitude circles, which symmetrically straddles the equator.

080.046 Thermal conductivity in solar magnetoplasmas.   H. S. Yun, A. A. Wyller.
Solar Physics, Vol. 27, 44 - 60 (1972).

The work of Ulmschneider (1970) on the effective thermal conductivity of partially ionized plasmas has been generalized to include the effects of magnetic fields. The contributions of electron, ion and neutral heat flows have been evaluated within the framework of the binary collision theory of Burgers (1960). It is shown that the neutral component dominates heat transport in a solar magnetoplasma even at temperatures as high as 15000 K and that the magnetic modulations are of a secondary character. The application of the present theories to the sunspot models of Yun (1972) shows pronounced maxima in both electrical and thermal conductivities at the interface between umbra and penumbra.

080.047 On the radio optical depth of the layer where the temperature equals the brightness temperature.
F. Chiuderi-Drago.
Solar Physics, Vol. 27, 132 - 136 (1972).

It is usually believed that the radio optical depth of the layer h^* where the temperature equals the brightness temperature is independent of the frequency. This assumption is criticized from the theoretical point of view and the behaviour of $\tau(h^*)$ as a function of the frequency ν is computed for two different solar models.

080.048 More on solar neutrinos.
R. T. Rood, W. A. Fowler, F. Hoyle.
Bull. American Astron. Soc., Vol. 4, 415 (1972). – Abstr. AAS.

080.049 The centre–limb variations of turbulent velocities on the solar disk.   N. N. Kondrashova.
Solnechnye Dannye 1972 Byull., No. 9, p. 104 - 109 (1972).
In Russian.

The depth dependencies of the turbulent velocity were derived from comparing the observed profiles with the computed ones for five positions on the solar disk. The turbulent velocity was found to be increased with depth. The tangential component of the anisotropic turbulent velocity was larger than the radial one.

080.050 Unstable solutions to Laplace's tidal equation with negative equivalent depth.   W. L. Jones.
Journ. Atmosph. Sci., Vol. 29, 457 - 462 (1972).

The complex eigenvalues of frequency are computed for Hough functions of negative equivalent depth, using the method of Longuet-Higgins. The positive kinetic wave energy and negative wave potential energy of unstable modes cancel. These unstable modes may be important in the solar convective zone as a means of sustaining solar differential rotation.

080.051 On the acceleration time of particles in the solar atmosphere.   P. I. Velinov.
Comptes Rendus Acad. Bulg. Sci., Vol. 25, 495 - 498 (1972).

The conditions necessary for an acceleration of particles in the solar atmosphere have been examined and the effect of the ionization losses acting jointly with the acceleration mechanism was studied in detail.

080.052 On the stability of magnetic baroclinic flow of an unbounded fluid with a linear velocity profile.
S. D. Gedzelman.
Journ. Atmosph. Sci., Vol. 29, 971 - 976 (1972).

The stability analysis of baroclinic magnetic flow for an unbounded fluid with a linear velocity profile is presented. This work has application to the differential rotation of the sun.

080.053 Solar neutrino and dilaton theory of non-Newtonian gravity.   M. Fujimoto, D. Sugimoto.
Progr. Theor. Phys., (Japan), Vol. 48, 705 - 707 (1972).

Recently the apparent discrepancy between observation and theory of solar neutrino flux has become serious. The authors discuss the effect of a change in the gravitational constant upon the solar model and the solar neutrino.

080.054 Effect of a neutrino-photon interaction on the solar-neutrino flux.   W. A. Perkins.
Nuovo Cimento Lettere, Ser. 2, Vol. 5, 672 - 674 (1972).

The results of the solar-neutrino experiments of Davis et. al. (1968, 1971) could be explained if there were a severe attenuation in the high-energy solar-neutrino flux in its traversal of the sun. The neutrino-photon cross-section required to reduce the solar neutrino flux by a factor of 10 is calculated.

080.055 The large-scale velocity fields, magnetic fields and brightness in the solar atmosphere.
S. I. Gopasyuk, T. T. Tsap.
Izv. Krymskoj Astrofiz. Obs., Vol. 45, 3 - 13 (1972).
In Russian.

A correlation between the radial velocities, magnetic fields and brightness in active and undisturbed regions on the sun is studied. The records of these quantities were made in the λ 5250 Å Fe I, λ 6103 Å Ca I, and Hα lines with the aid of a double magnetograph.

080.056 The velocity field in solar active regions.
S. I. Gopasyuk, T. T. Tsap.
Izv. Krymskoj Astrofiz. Obs., Vol. 45, 14 - 19 (1972).
In Russian.

The velocity fields at different levels of active regions are studied. The records of the radial velocities were made with the double magnetograph in the lines Hα, Hβ, Hγ, Hδ, λ 5184 Å Mg I, λ 4227 Å Ca I, D$_1$ Na I, λ 4554 Å Ba II, λ 6103 Å Ca I, and λ 5250 Å Fe I. It is found that there is a good correlation between the velocities at the adjacent levels.

080.057 Neutrino archaelogy. The simulation of double beta-decay by solar neutrinos.   G. Bozoki, K. Lande.
Nuovo Cimento B, Ser. 11, Vol. 12B, 65 - 71 (1972).

The authors suggest the possibility that neutrinos from solar thermo-nuclear reactions can induce apparent double beta-decay. The rate $A^z \rightarrow A^{z+2}$ is computed for a group of likely target isotopes. Apparent double positron decay induced by antineutrinos emitted by the radioactive component of the earth is also evaluated.

080.058 Spin down of radiation-penetrated, opaque, compressible fluid in a circular cylinder.
T. Sakurai.
Phys. Fluids, Vol. 15, 555 - 564 (1972).

080.059 Solar neutrinos.   J. N. Bahcall.
American Inst. Phys. Conference Proc. 1971, No. 2, p. 243 - 246.

080.060 Observing programs in solar physics during the 1973 ATM (*Apollo Telescope Mount*) Skylab program.
E. M. Reeves, R. W. Noyes, G. L. Withbroe.
Solar Physics, Vol. 27, 251 - 270 (1972).

In this paper we shall describe the ATM and its capabilities, the principal scientific goals and the planned observing program, and methods of coordinating the observations with ground-based and other types of solar observations.

080.061 Solar rotation as determined from OSO-4 EUV spectroheliograms.   A. K. Dupree, W. Henze, Jr.
Solar Physics, Vol. 27, 271 - 279 (1972).

Spectroheliograms obtained in extreme ultraviolet (EUV) lines and the Lyman continuum are used to determine the rotation rate of the solar chromosphere, transition region, and corona. A cross-correlation analysis of the observations indicates the presence of differential rotation through the chromosphere and transition region. The rotation rate does not vary with height. The average sidereal rotation rate is given by ω (deg day^{-1}) = 13.46 − 2.99 sin$^2 B$ where B is the solar latitude. The corona does not clearly show differential rotation.

080.062   **On the small-scale structure of solar magnetic fields.**
          E. N. Frazier, J. O. Stenflo.
Solar Physics, Vol. 27, 330 - 346 (1972).

The small-scale structure of solar magnetic fields has been studied using simultaneous recordings in the spectral lines Fe I 5250 Å and Fe I 5233 Å, obtained with the Kitt Peak multi-channel magnetograph. We find that more than 90 % of the magnetic flux in active regions (excluding the sunspots), observed with a 2.4 by 2.4″ aperture, is channelled through narrow filaments. This percentage is even higher in quiet areas. The field lines in a magnetic filament diverge rapidly with height, and part of the flux returns back to the neighbouring photosphere. Therefore the strong fields within a magnetic filament are surrounded by weak fields of the order of a few gauss of the opposite polarity.

080.063   **A model of the quiet solar atmosphere.**
          J. H. Piddington.
Solar Physics, Vol. 27, 402 - 419 (1972).

The velocity field in the SBRs (supergranule boundary regions) is discussed. There are continuous gas streaming motions up and down between the photosphere and the corona; spicules may be mainly downward moving gas. A unifying model is developed of these various components, as well as the heating mechanism of the whole quiet atmosphere. The model atmosphere has a chromosphere-corona transition layer which bulges upwards above the SBRs and so conforms with

EUV data. The energy and mass balances in this solar atmosphere are considered, and it is also shown to be consistent with the radio data.

Electron impact excitation rates for helium.
See Abstr. 022.152.

Instrumental polarization concerning magnetographic measurements.   See Abstr. 031.091.

Quantum theory of line formation in a magnetic field.   See Abstr. 062.085.

On the theory of radiative heat exchange in polytropic atmospheres.   See Abstr. 063.001.

The density dependent ionization balance of carbon, oxygen and neon in the solar atmosphere.
See Abstr. 071.001.

Photospheric faculae and the solar oblateness.
See Abstr. 072.004.

Faculae and the solar oblateness.
See Abstr. 072.005.

Photospheric faculae and the solar oblateness.
See Abstr. 072.027.

Earth

081 Figure, Composition, and Gravity of the Earth

081.001 **The solution of the inverse problem of gravimetry based on harmonic moments of the gravitational field.** A. V. Kudria.
Dokl. Akad. Nauk SSSR, Ser. Mat. Fiz., Vol. 205, 574 - 577 (1972). In Russian.

081.002 **Self-consistent statistical models for the gravity anomaly, vertical deflections, and undulation of the geoid.** S. K. Jordan.
Journ. Geophys. Res., Vol. 77, 3660 - 3670 (1972).

Mathematical relationships exist between the anomaly, deflections, and undulation that constrain the models for the autocorrelation and cross-correlation functions of these processes. In this paper, new models are proposed that take these constraints into account. One of these new models, the 'third-order Markov undulation model' is suggested for the analysis of inertial navigation system errors caused by uncertainties in the gravity field.

081.003 **A geopotential model (APL 5.0-1967) determined from satellite Doppler data at seven inclinations.**
S. M. Yionoulis, F. T. Heuring, W. H. Guier.
Journ. Geophys. Res., Vol. 77, 3671 - 3677 (1972).

Results of a determination of the nonzonal harmonic coefficients of the earth's gravitational potential are presented. This model is based on an analysis of Doppler data obtained from tracking satellites at seven different inclinations. These coefficients are combined with a set of zonal harmonic coefficients determined by R. J. Anderle and S. J. Smith of the Naval Weapons Laboratory to produce a geoidal-heights contour map.

081.004 **Composition and state of matter in the deep interior of the earth.** L. V. Al'tshuler.
Phys. Earth Planet. Interiors, Vol. 5, 295 - 300 (1972).

The origin of the central zone of the earth is explained by the more rapid growth of iron-nickel bodies and iron-enriched planetary protocores in the protoplanetary cloud. The primary division of the circumsolar matter into iron and stone bodies naturally explains the present chemical composition of the moon, which was formed from the circumterrestrial swarm.

081.005 **Archaean greenstone belts may include terrestrial equivalents of lunar maria?** D. H. Green.
Earth Planet. Sci. Letters, Vol. 15, 263 - 270 (1972).

I have used the distinctive characters of the ultramafic and mafic intrusives of some Archaean greenstone belts to argue for a particularly catastrophic type of magmatism adding the hypothesis that this can best be interpreted in terms of major impact. I have attempted to support this hypothesis with the argument that events recorded in some Archaean greenstone belts are nearly synchronous with very large impacts forming the maria basins on the lunar surface.

081.006 **Equations for 15th-order geopotential coefficients from the orbit of Transit 1B.**
H. Hiller, D. G. King-Hele.
Planet. Space Sci., Vol. 20, 1213 - 1228 (1972).

The main object of the study is to investigate the region of 15th-order resonance revealed by a steep decrease in inclina-tion of about 0.04° between March and November 1962. The analysis leads to equations for the 15th-order geopotential coefficients and to an improved value for the average rotation rate Λ of the upper atmosphere at a mean height of 380 km between 1963 and 1967.

081.007 **On the calculation of changes in the earth's inertia tensor due to faulting.**
J. R. Rice, M. A. Chinnery.
Geophys. Journ. Roy. Astron. Soc., Vol. 29, 79 - 90 (1972).

The paper is concerned with calculating the changes in the inertia tensor of an elastic sphere due to mass displace-ments accompanying slip over an interior surface. The calcula-tion is of interest to the theory of the earth's rotation, in that sudden changes in the earth's inertia accompanying earth-quake faulting will alter the position and, more importantly, the subsequent motion of the rotation pole relative to an earth-bound observer.

081.008 **A simple earth model.** C.-Y. Wang.
Journ. Geophys. Res., Vol. 77, 4318 - 4329 (1972).

081.009 **Creep in the earth and planets.** H. Jeffreys.
IAU Symposium No. 48, (see 012.004), p. 1 - 9 (1972). – Invited lecture.

081.010 **Possible changes in the core-mantle and inner-outer core boundaries.** J. A. Jacobs.
IAU Symposium No. 48, (see 012.004), p. 179 - 181 (1972).

081.011 **Motion of the core, and its influence on the earth's axis.** J. Mateo.
IAU Symposium No. 48, (see 012.004), p. 185 - 188 (1972).

081.012 **The excess secular change in the obliquity of the ecliptic and its relation to the internal motion of the earth.** C. Kakuta, S. Aoki.
IAU Symposium No. 48, (see 012.004), p. 192 - 195 (1972).

081.013 **Nearly diurnal nutation derived from the observa-tions of time and latitude.** S. Débarbat.
IAU Symposium No. 48, (see 012.004), p. 197 - 199 (1972).

081.014 **On the comparison of diurnal nutation derived from separate series of latitude and time observations.**
Ya. S. Yatskiv.
IAU Symposium No. 48, (see 012.004), p. 200 - 205 (1972).

081.015 **The determination of Love's number K from tidal variations of rotation of a compressible earth.**
N. N. Pariisky, B. P. Pertsev.
IAU Symposium No. 48, (see 012.004), p. 234 (1972). – Ab-stract.

081.016 **On the torques due to tidal friction of the oceans and adjacent seas.** P. Brosche, J. Sündermann.
IAU Symposium No. 48, (see 012.004), p. 235 - 239 (1972).

081.017 **Field investigations of the reflective and polarime-tric capacity of some volcanic depositions.**

V. V. Novikov.
Astron. Tsirk., No. 692, p. 4 - 7 (1972). In Russian.

Deep-sea drilling in the SW Pacific has revealed a vast regional unconformity of Oligocene age that resulted from deep sea erosion induced by palaeocirculation changes related to Australian continental drift and Antarctic glacial episodes. These palaeocirculation changes played a major role in the distribution and evolution of Cainozoic planktonic assemblages.

081.018 Australian–Antarctic continental drift, palaeocirculation changes and Oligocene deep-sea erosion.
J. P. Kennett, R. E. Burns, J. E. Andrews, M. Churkin, Jr., T. A. Davies, P. Dumitrica, A. R. Edwards, J. S. Galehouse, G. H. Packham, G. J. van der Lingen.
Nature, Phys. Sci., Vol. 239, 51 - 55 (1972).

081.019 Het ontstaan van de maan. J. van Diggelen.
Hemel en Dampkring, Vol. 70, 209 - 215 (1972).

081.020 On the problem of cosmogenic diamonds in terrestrial sedimentary rocks.
E. A. Vitrichenko, Yu. A. Polkanov.
Astron. Tsirk., No. 698, p. 2 - 5 (1972). In Russian.

081.021 Comments on a paper by K. Lambeck: 'Comparison of surface gravity data with satellite data' [Bull. Géod., Nouvelle Sér., Année 1971, No. 100, p. 203 - 219].
R. H. Rapp.
Bull. Géod., Nouvelle Sér., Année 1972, No. 105, p. 343 - 349, with further comments by K. Lambeck, p. 351 - 358.

081.022 Transformation of Stokes' formula. V. V. Brovar.
Geod., kartograf. i aehrofotos"emka. Resp. mezhved. nauch.-tekhn. sb., 1971, vyp. (No.) 14, p. 3 - 7. In Russian. Abstr. in Referativ. Zhurn. 52. Geod. Aehros"emka, 9.52.87 (1972).

081.023 Differential formulae of altitude systems.
N. K. Migal', R. R. Il'kiv.
Geod., kartogr. i aehrofotos"emka. Resp. mezhved. nauch.-tekhn. sb., 1972, vyp. (No.) 15, p. 50 - 52. In Russian. – Abstr. in Referativ. Zhurn. 52. Geod. Aehros"emka, 9.52.88 (1972).

081.024 On the displacement of a level surface (of the geoid) under the influence of topographic reduction.
Eh. M. Shatalova.
Geod., kartogr. i aehrofotos"emka. Resp. mezhved. nauch.-tekhn. sb., 1971, vyp. (No.) 14, p. 80 - 85. In Russian. – Abstr. in Referativ. Zhurn. 52. Geod. Aehros"emka, 9.52.89 (1972).

081.025 On the problem of the deformation of the geoid.
Eh. M. Shatalova.
Geod., kartogr. i aehrofotos"emka. Resp. mezhved. nauch.-tekhn. sb., 1972, vyp. (No.) 15, p. 109 - 113. In Russian. Abstr. in Referativ. Zhurn. 52. Geod. Aehros"emka, 9.52.90 (1972).

081.026 Development of Schmidt's theory. V. S. Safronov.
Zemlya i Vselennaya, 1972, No. 4, p. 18 - 23. In Russian.

081.027 Der Streit um die Figur der Erde. Zur Begründung der Geodäsie im 17. und 18. Jahrhundert.
V. Bialas.
Deutsche Geod. Kommission Bayer. Akad. Wiss., Ser. E, No. 14, 4 + 39 pp. (1972).
In the present paper the development of geodesy, from 1670 until nearly 1750, is expounded on the basis of the main publications.

081.028 An earth model consistent with free oscillation and surface wave data. H. Mizutani, K. Abe.
Phys. Earth Planet. Interiors, Vol. 5, 345 - 356 (1972).

081.029 Electrical conduction in physical and chemical mixtures. Application to planetary mantles.
H. G. Tolland, R. G. J. Strens.
Phys. Earth Planet. Interiors, Vol. 5, 380 - 386 (1972).
The mantle of the earth is believed to consist of a physical mixture of ferromagnesian silicates, which are themselves chemical mixtures (solid solutions) of iron and magnesium end-members. The interpretation of the electrical properties of the mantle therefore requires a knowledge of the composition-dependence of resistivity in both physical mixtures of phases of differing resistivity, and in solid solutions between end-members of different resistivity. We have investigated the behaviour of binary mixtures of both types in an attempt to assess the factors determining the properties of planetary mantles.

081.030 Le sondage vertical du champ de gravité de la terre.
J. B. Zieliński.
Artificial Satellites, Polish Acad. Sci., Vol. 7, No. 2, p. 3 - 12 (1972).
On propose d'étudier le champ de gravité de la terre par observation de la chute libre d'un corps depuis une altitude de quelques centaines de kilomètres. On donne la précision exigée de la mesure et on la confronte avec les possibilités techniques existantes.

081.031 The dynamical properties and internal structures of the earth, the moon and the planets. A. H. Cook.
Proc. Roy. Soc. London, Ser. A, Vol. 328, 301 - 336 (1972). Review lecture.
The aim of this review is to bring together and relate recent progress in three subjects – the internal structure of the earth, the behaviour of materials at very high pressures and the dynamical properties of the planets.

081.032 Paleomagnetic data analysis and continental drift.
D. P. Zidarov.
Comptes Rendus Acad. Bulgar. Sci., (Dokl. Bolg. Akad. Nauk), Vol. 24, 1637 - 1640 (1971).

081.033 Evaluation of gravity anomalies directly from satellite observations. G. Obenson.
Geophys. Journ. Roy. Astron. Soc., Vol. 30, 69 - 83 (1972).
In order to introduce suitable equations for evaluating gravity anomalies directly from satellite observations, the equations of motion of an artificial earth satellite are written out with terms involving gravity anomalies. Differential equations are then derived which express the variation of the satellite position and gravity anomalies. Results of simulation computations done to test the equations derived show that it is possible to determine gravity anomalies quite accurately directly from satellite observations.

081.034 What weight tells us.
K. Lambeck, F. Barlier.
Sci. Progrès Découverte, 99e année, No. 3438, p. 24 - 33 (1971). In French.

081.035 Form and dimensions of the earth from observations with artificial earth satellites. N. Georgiev.
Priroda (NRB), Vol. 21, No. 2, p. 8 - 11 (1972). In Bulgarian.

081.036 On some simplified representations of the normal gravitational potential used in the theory of motion of artificial earth satellites. O. F. Malakhova.
Trudy Mosk. in-ta radiotekhn., ehlektron. i avtomatiki, 1972, vyp. (No.) 57, p. 83 - 90. In Russian. – Abstr. in Referativ.

Zhurn. 62. Issled. kosmich. prostranstva, 11.62.211 (1972).

081.037 Estimates of the coefficients of spherical harmonic series for the geopotential. M. S. Petrovskaya.
Byull. Inst. Teoret. Astron., *Leningrad,* Vol. 13, 225 - 230 (1972). In Russian.
Estimates have been obtained for the values of harmonic coefficients in the geopotential expansion. The method is based on the use of the data characterizing the deviation of the earth's potential from the ellipsoidal one.

081.038 Les recherches gravimétriques dans la région des stations antarctiques soviétiques Lasarev et Novola-
sarevskaja. A. L. Kogan, P. A. Stroev.
Trudy Gos. Astron. Inst. Shternberga, Vol. 43, 3 - 7 (1972).
In Russian.

081.039 An approximate method of determining the spectral density of geophysical fields. V. L. Panteleev.
Trudy Gos. Astron. Inst. Shternberga, Vol. 43, 8 - 17 (1972).
In Russian.

081.040 Le développement des anomalies de la force d'attraction en séries ellipsoïdales de Lamé.
L. A. Savrov.
Trudy Gos. Astron. Inst. Shternberga, Vol. 43, 18 - 29 (1972).
In Russian.

081.041 Tectonique des plaques, expansion de la Terre et origine des continents. J. Lagrula.
Comptes Rendus Acad. Sci. Paris, Sér. B, Vol. 275, 881 - 885 (1972).
Le principal argument en faveur de l'hypothèse de la Terre en expansion est qu'elle fournit à la tectonique des plaques un moteur puissant. Mais pour expliquer l'existence des dorsales il semble efficace de faire intervenir une autre hypothèse, par exemple celle de l'origine polaire des continents, associée à une rotation d'ensemble des couches superficielles autour de la ligne des pôles et à des mouvements dirigés vers l'équateur, déclenchés par l'effet Eötvös et amplifiés par une expansion dont l'importance sera mesurée grâce aux progrès des études concernant la croissance des coraux et éventuellement d'autres phénomènes.

081.042 Density differentiation of the earth's matter and processes at the core-mantle interface.
E. V. Artyushkov.
Journ. Geophys. Res., Vol. 77, 6454 - 6458 (1972).
The formation of the earth's core was due to the density differentiation of the primitive matter of the earth. This differentiation should be in progress at the core-mantle interface in the present epoch. Redistribution of substances of various densities gives rise to large-scale convective motions in the core and mantle. These motions create the geomagnetic field, and they may be a cause of disturbances at the core-mantle interface. It is shown that the outer core cannot be composed of a mixture or an alloy of two or several substances.

081.043 The solution to Professor Runcorn's problem.
R. A. Lyttleton.
Nature, Vol. 240, 459 - 460 (1972).

081.044 Truncation error formulas for the geoidal height and the deflection of the vertical. Y. Hagiwara.
Bull. Géod., Nouvelle Sér., Année 1972, No. 106, p. 453 - 466.

081.045 Determination of Mohorovičič discontinuity from deep seismic soundings and the characteristics of
the outer earth's gravity field. M. Pick, I. Jakubcová.
Bull. Géod., Nouvelle Sér., Année 1972, No. 106, p. 477 - 488.

081.046 Method of integral equations for the geodetic boundary value problem. K. R. Koch.
Mitt. Inst. Theor. Geod. Univ. Bonn, No. 4, p. 38 - 49 (1972).

081.047 On the stochastic approach to the determination of the gravity potential. S. L. Lauritzen.
Mitt. Inst. Theor. Geod. Univ. Bonn, No. 4, p. 50 - 59 (1972).

081.048 Die freie geodätische Randwertaufgabe und das Problem der Integrationsfläche innerhalb der Inte-
gralgleichungsmethode. E. Grafarend.
Mitt. Inst. Theor. Geod. Univ. Bonn, No. 4, p. 60 - 85 (1972).

081.049 A solution for the Lambert problem in the gravitational field of the oblate earth. M. P. Batra.
Bull. American Astron. Soc., Vol. 4, 417 (1972). – Abstr. AAS.

081.050 Potsdam correction from the satellite determined geopotential. M. A. Khan.
Nature, Phys. Sci., Vol. 239, 43 - 45 (1972).
A new method is suggested to determine the scale error in the Potsdam Gravity System by making use of the satellite determined gravity field and the theoretical gravity values on the appropriate reference surfaces. This approach is entirely different from the experimental techniques hitherto employed which involve actual measurements of absolute gravity. With the input of appropriate parameters, the value of Potsdam correction thus determined is 13.67 milligals.

081.051 Analysis of accuracy of calculating the harmonic coefficients of gravity. I. G. Vovk.
Geod. i kartografiya, 1972, No. 8, p. 13 - 18. In Russian.
Abstr. in Referativ. Zhurn. 52. Geod. Aehros"emka, 1.52.70 (1973).

081.052 Die schrittweise Spektralausgleichung, eine Methode zur Erdgezeitenanalyse. B.-S. Schulz.
Deutsche Geod. Kommission, Bayer. Akad. Wiss., München, Ser. C, No. 184, 121 pp. (1972). – Diss. Rheinisch. Friedrich-Wilhelms-Univ. Bonn.

081.053 The estimation of plate motions by astronomical methods. R. W. Tanner.
Canadian Journ. Earth Sci., Vol. 9, 1052 - 1054 = Contr. Earth Phys. Branch, Ottawa, No. 403 (1972).
Some of the problems involved in the detection of plate motions by astronomical methods are discussed with particular reference to a Herstmonceux—Calgary cooperative program using photographic zenith tubes. Analysis of the observations obtained in 1969 and 1970 shows that several decades of continued observation may suffice with present techniques to demonstrate and measure the relative motion of the North American—Eurasian plates.

081.054 Creep in the earth and planets. H. Jeffreys.
Tectonophysics, (*Netherlands*), Vol. 13, 569 - 582 (1972).
Proposes a law of creep for the earth and the planets. This law can account for the right amount of damping (or internal friction) in various phenomena such as earth's free nutation, elastic waves passing through the earth, and figure of the moon.

081.055 Determinación de armónicos zonales impares del potencial terrestre. C. Simó Torres.
Urania Barcelona, Año 56, No. 274, p. 167 - 175 (1971).

081.056 Dua luno de Tero. Y. Väisälä.
Sci. Revuo, Beograd, Vol. 22, 141 - 144 = Astron.-Optika Inst. Univ. Turku, Informo No. 37/II (1971).

081.057 **Marées terrestres.**   P. Melchior (Editor).
Bull. d'Informations, (Obs. Roy. Belgique, Bruxelles),
No. 63, p. 3239 - 3332 (1972).

081.058 **The origin and chemical composition of the earth's core.**   V. R. Murthy, H. T. Hall.
Phys. Earth Planet. Interiors, Vol. 6, (see 012.023), 123 - 130 (1972).

081.059 **Partitioning of potassium between silicates and sulphide melts: Experiments relevant to the earth's core.**   K. A. Goettel.
Phys. Earth Planet. Interiors, Vol. 6, (see 012.023), 161 - 166 (1972).

081.060 **Melting of iron by significant structure theory.**   D. A. Leppaluoto.
Phys. Earth Planet. Interiors, Vol. 6, (see 012.023), 175 - 181 (1972).

081.061 **Viscous stratification of the earth and convection.**   W. M. Elsasser.
Phys. Earth Planet. Interiors, Vol. 6, (see 012.023), 198 - 204 (1972).

081.062 **Viscosity of the earth's core.**   R. Hide.
Phys. Earth Planet. Interiors, Vol. 6, (see 012.023), 205 (1972). – Abstract.

081.063 **A method for cumputing the gravitational attraction of three-dimensional bodies in a spherical or ellipsoidal earth.**   L. R. Johnson, J. J. Litehiser.
Journ. Geophys. Res., Vol. 77, 6999 - 7009 (1972).
A method is presented for calculating the gravitational attraction of three-dimensional bodies of arbitrary shape within a spherical or ellipsoidal earth. The method is well suited for evaluation on a digital computer, and calculations are presented for a couple of examples that demonstrate in a quantitative manner the relative effects of sphericity and ellipticity.

081.064 **Zu Stellung, Aufgaben und Problemen der geodätischen Forschung.**   E. Buschmann.
Vermessungstechnik, 20. Jahrgang, p. 281 - 284 = Mitt. Zentralinst. Phys. Erde, No. 262 (1972).

081.065 **A model for continental drift.**   W. V. R. Malkus.
Mantle and core in planetary physics, Varenna 1970, (see 012.024), p. 1 - 16 (1971).

081.066 **Influence of the core on the nutations.**   R. O. Vicente.
Mantle and core in planetary physics, Varenna 1970, (see 012.024), p. 17 - 26 (1971).

Earth sciences.   See Abstr. 003.118.

Origin and history of the earth. (Collection of articles).   See Abstr. 003.128.

Sampling functions for geophysics.
See Abstr. 021.029.

Polar wandering and the earth's dynamical evolution cycle.   See Abstr. 045.026.

The elastic energy and character of quakes in solid stars and planets.   See Abstr. 091.054.

Compressibility and planetary interiors.
See Abstr. 091.055.

On the interaction between tectonic processes of the earth and the moon.   See Abstr. 094.074.

On the possible differences in the bulk chemical composition of the earth and the moon forming in the circumterrestrial swarm.   See Abstr. 094.091.

Global tectonics of the moon and earth.
See Abstr. 094.099.

The history of the earth.   See Abstr. 107.008.

Geological evidence relating to the origin and secular rotation of the solar system.   See Abstr. 107.017.

082 The Earth's Atmosphere including Refraction, Scintillation, Extinction, Airglow, Site Testing

082.001 Stardust.
C. L. Hemenway, D. S. Hallgren, D. C. Schmalberger.
Nature, Vol. 238, 256 - 260 (1972).
Heavy metal particles have been detected in noctilucent clouds. Possible sources for these particles are considered and a solar origin is suggested.

082.002 Tests and experiments on the model of the island of Lampione.
F. Bavagnoli, C. Bistagnino, G. Cignolo.
Mem. Soc. Astron. Italiana, Nuova Ser., Vol. 43, 233 - 246 (1972).
Results obtained in the wind tunnel on a model of the island of Lampione are presented. The island has been under test as a possible 'JOSO-site'. A program of new tests to be performed on the model is also presented.

082.003 On the dependence of quietness and sharpness of solar image on local and large scale atmospheric circulation.
G. Belvedere, L. Paternò, V. Sciuto.
Mem. Soc. Astron. Italiana, Nuova Ser., Vol. 43, 375 - 380 (1972). − Letter.

082.004 Seeing astronomico e fluttuazioni microtermiche dell'atmosfera.
G. Ceppatelli, A. Righini.
Mem. Soc. Astron. Italiana, Nuova Ser., Vol. 43, 389 - 402 (1972).
We describe the calculations necessary to obtain the modulation transfer function of a telescope in a thermally turbulent atmosphere, in order to evaluate the astronomical seeing. Some practical examples and suggestions how to measure the thermal turbulence of the atmosphere are given.

082.005 Observations of the airglow continuum.
J. R. Sternberg, M. F. Ingham.
Monthly Notices Roy. Astron. Soc., Vol. 159, 1 - 20 (1972).
A programme of observations of the night sky continuum over the range 4100−8200 Å is described. As well as confirming the existence of an airglow continuum, the results show evidence of discrete structure at certain wavelengths and a large increase in brightness in the near infra-red. A preliminary study of the temporal variations of the continuum was also made.

082.006 Photochemistry of the airglow continuum.
J. R. Sternberg.
Monthly Notices Roy. Astron. Soc., Vol. 159, 21 - 29 (1972).
Recent observations of the airglow continuum and related phenomena are discussed in an effort to identify the photochemical processes responsible for the emission.

082.007 Thermospheric molecular oxygen from solar extreme-ultraviolet occultation measurements.
R. G. Roble, R. B. Norton.
Journ. Geophys. Res., Vol. 77, 3524 - 3533 (1972).
The molecular-oxygen number density distribution in the 95- to 120-km region of the atmosphere has been determined from solar-occultation measurements made by the Solrad 8 satellite. The solar intensity in the region 1040−1350 Å was monitored by a photometer aboard the satellite during occultation of the sun by the earth. The results show a seasonal variation of molecular oxygen in the lower thermosphere.

082.008 Geomagnetic effect on the neutral temperature of the F region during the magnetic storm of September 1969.
J. E. Blamont, J. M. Luton.
Journ. Geophys. Res., Vol. 77, 3534 - 3556 (1972).
A spherical Fabry-Perot interferometer on OGO 6 gives the profile of the 6300-Å oxygen line in the dayglow at different altitudes between 200 and 320 km. The temperature of the neutral components of the atmosphere is deduced from these measurements with an accuracy of ±65°K.

082.009 Analysis of OGO 6 observations of the O I 5577-Å tropical nightglow.
R. J. Thomas, T. M. Donahue.
Journ. Geophys. Res., Vol. 77, 3557 - 3565 (1972).
Atomic oxygen green line data from the horizon scanning photometer on OGO 6 have been examined. Unfolding the satellite data from the tropical F region yields altitude and latitude variations of the $O(^1S)$ emissions. The spatial variations of the tropical F-region electron density are then calculated by assuming dissociative recombination and using a model atmosphere.

082.010 Absorption of the 4- to 6-millimeter wavelength band in the atmosphere. E. E. Reber.
Journ. Geophys. Res., Vol. 77, 3831 - 3845 (1972).

082.011 Large scale circulation patterns of the stratosphere.
B. Kriester.
Space Science Rev., Vol. 13, 258 - 273 (1972). − Conference paper (see 012.002).

082.012 Air motions in the tropical stratosphere deduced from satellite tracking of horizontally floating balloons. J. K. Angell.
Space Science Rev., Vol. 13, 274 - 289 (1972). − Conference paper (see 012.002).

082.013 Aeronomical balloon experiments. M. Ackerman.
Space Science Rev., Vol. 13, 290 - 294 (1972). Conference paper (see 012.002).

082.014 An interferometric seeing test on Mt. Wilson.
I. S. Glass, J. L. Elliot.
Astron. Journ., Vol. 77, 523 (1972).
Using our quantitative detector of interference fringes, we have found that the steadier seeing conditions at Mt. Wilson make it a better site for stellar interferometry than the Agassiz Station at Harvard, Massachusetts.

082.015 Earth and Mars: Evolution of atmospheres and surface temperatures. C. Sagan, G. Mullen.
Science, Vol. 177, 52 - 56 (1972).
Solar evolution implies, for contemporary albedos and atmospheric composition, global mean temperatures below the freezing point of seawater less than 2.3 aeons ago, contrary to geologic and paleontological evidence. Ammonia mixing ratios of the order of a few parts per million in the middle Precambrian atmosphere resolve this and other problems. Possible temperature evolutionary tracks for earth and Mars are described.

082.016 The variation of air density at 240 and 280 km from April to November 1967. B. R. May.
Planet. Space Sci., Vol. 20, 1077 - 1084 (1972).
Changes in the orbital periods of two satellites, 1962-β7ₐ6 (Injun 3 rocket) and 1965-11D (Cosmos 54 rocket), have been used to deduce the air density at heights of 240 and 280 km during April−November 1967. The ratio of the maximum

(October) to minimum (July) density was about 1.8 at 240 km and 2.2 at 280 km. A diurnal variation of density was also detected with a maximum density at 14 hr and a maximum to minimum ratio of 1.7 at 280 km.

082.017 Sodium emission from the atmosphere during a solar eclipse. M. Gadsden.
Journ. Atmosph. Terr. Phys., Vol. 34, 1309 - 1320 (1972).
Observations of the *D*-lines of sodium in the spectrum of the zenith sky during the solar eclipse of 7 March 1970 are discussed and examined for the presence of atmospheric emission, both continuum (the 'Ring effect') and line (dayglow). It is concluded that observational data indicate dayglow of an intensity approximately to be expected from reflection and scattering of penumbral light.

082.018 Atmospheric pressure, density and scale height calculated from H Lyman-α absorption allowing for the variation in cross-section with wavelength. J. E. Hall.
Journ. Atmosph. Terr. Phys., Vol. 34, 1337 - 1348 (1972).
Observations of the extinction in the earth's upper atmosphere of hydrogen Lyman-α radiation from the sun has been used to calculate atmospheric pressure, density and scale height. Atmospheric parameters are presented for the height range 67–95 km determined from Lyman-α observations made aboard a rocket launched from South Uist, Scotland, on 19 June 1969.

082.019 Airglow fluctuations at 2.2 μ.
A. W. Peterson, L. M. Kieffaber.
Journ. Atmosph. Terr. Phys., Vol. 34, 1357 - 1364 (1972).
The OH airglow has been observed in the infrared, from high altitude sites, with a bandpass filter from 2.0 to 2.5 μ. An average spatial fluctuation of ±4 per cent and an average time fluctuation of ±3 per cent is found about the mean nightly level. The evening fluctuations are more pronounced than the morning ones.

082.020 Altitude profile of $O_2(^1\Delta_g)$ at night.
R. H. Bishop, K. D. Baker, R. Y. Han.
Journ. Atmosph. Terr. Phys., Vol. 34, 1477 - 1482 (1972).

082.021 Skylight intensity, polarization and airglow measurements during the total solar eclipse of 30 May 1965. R. E. Miller, W. G. Fastie.
Journ. Atmosph. Terr. Phys., Vol. 34, 1541 - 1546 (1972).
The absolute skylight intensity and polarization were measured at 5577, 5780, 6100 and 6300 Å using an interference filter photometer mounted in a jet aircraft flying at an altitude of 12.2 km during the total solar eclipse of 30 May 1965. The results are compared with other recent measurements made during total solar eclipses.

082.022 A comparison of the observed twilight with the vertical scatter distribution. S. L. Seaton.
Journ. Atmosph. Terr. Phys., Vol. 34, 1553 - 1555 (1972). Short paper.

082.023 Atmospheric CH_4, CO, and CO_2.
S. C. Wofsy, J. C. McConnell, M. B. McElroy.
Journ. Geophys. Res., Vol. 77, 4477 - 4493 (1972).
The chemistry of atmospheric CH_4, CO, and CO_2 is treated with a one-dimensional model incorporating the effects of eddy diffusion in the altitude region of 0–120 km.

082.024 Diurnal and seasonal variation of the atmospheric temperature at the 90-kilometer altitude.
F. Verniani, E. Reggiani Viani.
Journ. Geophys. Res., Vol. 77, 4581 - 4585 (1972). – Brief report.

082.025 A calculated hydrogen distribution in the exosphere.
A. Vidal-Madjar, J. L. Bertaux.
Planet. Space Sci., Vol. 20, 1147 - 1162 (1972).
With the use of Liouville's theorem, the exospheric density may be calculated with a triple integral. The results of a numerical evaluation of this integral are presented for three different exobase models for the case of atomic hydrogen in the earth's exosphere.

082.026 Airglow observations with a Hadamard photometer.
Y. P. Neo, G. G. Shepherd.
Planet. Space Sci., Vol. 20, 1351 - 1355 (1972). – Research note.

082.027 Atmospheric phenomena and zodiacal light.
M. V. Gavin.
Journ. British Astron. Ass., Vol. 82, 353 - 356 (1972).

082.028 Modulation transfer function for solar telescopes and atmospheric turbulence. A. Righini.
Solar Physics, Vol. 25, 242 - 251 (1972).
Results on the structure coefficient of the temperature field present in the low atmosphere are presented. Measurements have been performed during the national Italian expedition for solar site testing in Isola delle Correnti (southern Sicily). Calculations have been carried out to show the effect of the observed thermal properties of the low atmosphere on telescope performances, with various assumptions as the structure at greater heights.

082.029 Gravity waves in the atmosphere. C. O. Hines.
Nature, Vol. 239, 73 - 78 (1972).
Atmospheric gravity waves similar to waves that occur on the surface of a body of water have been known since the nineteenth century, but their significance for the study of atmospheric dynamics has been realized only in the past ten years.

082.030 Atmospheric transmission in the 10–12μm window.
J. T. Houghton, A. C. L. Lee.
Nature, Phys. Sci., Vol. 238, 117 - 118 (1972).

082.031 Superrotation of the upper atmosphere.
H. Rishbeth.
Rev. Geophys. Space Phys., Vol. 10, 799 - 819 (1972).
This article outlines several theories that have been advanced to explain the so-called 'superrotation'.

082.032 Preliminary results of the photoelectric method of site testing at Mt. Maidanak. S. B. Novikov.
Astron. Tsirk., No. 672, p. 1 - 4 (1972). In Russian.

082.033 Sky conditions on isolated mountains.
G. V. Novikova.
Astron. Tsirk., No. 672, p. 5 - 7 (1972). In Russian.

082.034 Photométrie simultanée à deux niveaux de 21 et 6 mb du ciel crépusculaire. M. Fehrenbach,
D. Frimout, F. Link, C. Lippens, G. Weill.
Comptes Rendus Acad. Sci. Paris, Sér. B, Vol. 275, 223 - 226 (1972).

082.035 Contribution à l'étude de l'aérosol atmosphérique dans le domaine submicronique à l'aide de méthodes photoélectriques. J. Bricard, P. Cazes, P. Reiss, P.-Y. Turpin.
Comptes Rendus Acad. Sci. Paris, Sér. B, Vol. 275, 263 - 266 (1972).

082.036 Tunable infrared laser spectroscopy of atmospheric water vapor. F. A. Blum, K. W. Nill, P. L. Kelley,
A. R. Calawa, T. C. Harman.
Science, Vol. 177, 694 - 695 (1972).

Absorption lines in the ν_2 band of water vapor at 6.3 micrometers have been fully resolved by using a tunable semiconductor laser. Three atmospheric water vapor lines near 5.32 micrometers were studied in detail and found to have line widths two to four times narrower than the width calculated by Benedict and Kaplan.

082.037 **Azimuth angle dependence of equatorial ultraviolet airglow.**
J. A. Quessette, F. Millier, J. M. Ajello.
Nature, Vol. 239, 157 (1972).

082.038 **On the production of N_2O from the reaction of $O(^1D)$ with N_2.**
R. Simonaitis, E. Lissi, J. Heicklen.
Journ. Geophys. Res., Vol. 77, 4248 - 4250 (1972).
Brief report.

082.039 **Extinction on Mt. Maidanak.**
I. S. Isakov, V. I. Kardopolov, V. S. Shevchenko.
Astron. Tsirk., No. 685, p. 1 - 4 (1972). In Russian.

082.040 **Some questions of economical efficiency constructing new astronomical observatories.**
B. D. Kokarev, V. S. Shevchenko.
Astron. Tsirk., No. 685, p. 4 - 7 (1972). In Russian.

082.041 **On the determination of anomalous refraction out of astrometrical measurements in the zenith zone.**
G. Teleki, B. Ševarlić.
IAU Symposium No. 48, (see 012.004), p. 147 - 149 (1972).

082.042 **Apollo 16 far-ultraviolet camera/spectrograph: Earth observations.**
G. R. Carruthers, T. Page.
Science, Vol. 177, 788 - 791 (1972).
A far-ultraviolet camera/spectrograph experiment was operated on the lunar surface during the Apollo 16 mission. Among the data obtained were images and spectra of the terrestrial atmosphere and geocorona in the wavelength range below 1600 angstroms. These gave the spatial distributions and relative intensities of emissions due to atomic hydrogen, atomic oxygen, molecular nitrogen, and other species — some observed spectrographically for the first time.

082.043 **Charged particles in the lower atmosphere.**
Yu. A. Bragin.
Priroda, No. 8.72, p. 42 - 49 (1972). In Russian.

082.044 **Measurements of the atmospheric background emission between 10μ and 100μ.**
A. F. M. Moorwood.
Astrophys. Space Sci. Library, Vol. 30, (see 012.005), 60 - 62 (1972).

082.045 **The light night sky.** A. H. Jarrett.
Monthly Notes Astron. Soc. Southern Africa,
Vol. 31, 112 - 118 (1972). — Presidential address.

082.046 **Energy releases in the upper atmosphere during geomagnetic disturbances.** V. I. Krassovsky.
Planet. Space Sci., Vol. 20, (see 012.006), 1363 - 1367 (1972).
Invited paper.

082.047 **Sur un phénomène d'optique atmosphérique.**
P. Renson.
Ciel et Terre, Vol. 88, 395 - 396 (1972).

082.048 **On the formation mechanism of telluric O_2 lines.**
A. I. Khlystov.
Astron. Tsirk., No. 698, p. 5 - 7 (1972). In Russian.

082.049 **The components of the tensor of the atmosphere's inertia and variations of the earth's rotation.**
N. S. Sidorenkov.
Astron. Tsirk., No. 701, p. 3 - 5 (1972). In Russian.

082.050 **The second positive system of nitrogen bands in the daylight luminescence of the atmosphere from data of Cosmos 224.** V. A. Krasnopolsky.
Geomagn. Aeronom., Vol. 12, 603 - 607 (1972). In Russian.

082.051 **Daytime luminescence at λ 3914 Å N_2^+ from data of Cosmos 224.** V. A. Krasnopolsky.
Geomagn. Aeronom., Vol. 12, 608 - 612 (1972). In Russian.

082.052 **On some seeing features at Baikal Lake.**
P. B. Kovadlo, V. I. Ivanov, Sh. P. Darchiya.
Astron. Tsirk., No. 706, p. 3 - 6 (1972). In Russian.

082.053 **PAET, the atmosphere structure experiment.**
S. C. Sommer.
Bull. American Astron. Soc., Vol. 4, 353 - 354 (1972).
Abstr. AAS.

082.054 **PAET, atmospheric composition by shock layer radiometry.** E. E. Whiting.
Bull. American Astron. Soc., Vol. 4, 354 (1972). – Abstr. AAS.

082.055 **PAET, effects of entry vehicle dynamics on the atmospheric data.** D. B. Kirk.
Bull. American Astron. Soc., Vol. 4, 354 (1972). – Abstr. AAS.

082.056 **Automated solar seeing measurements.**
S. A. Colgate, E. P. Moore.
Bull. American Astron. Soc., Vol. 4, 380 (1972). – Abstr. AAS.

082.057 **The stratospheric emission spectrum at millimeter wavelengths.**
I. G. Nolt, J. V. Radostitz, R. J. Donnelly.
Bull. American Astron. Soc., Vol. 4, 399 (1972). – Abstr. AAS.

082.058 **Contributions to the theory of atmospheric refraction.** J. Saastamoinen.
Bull. Géod., Nouvelle Sér., Année 1972, No. 105, p. 279 - 298.
A derivation of the general formula for astronomical refraction is given and an application is made for three different atmospheric models.

082.059 **The development of a theoretical model of the atmosphere and the ionosphere.**
S. Chandra, P. Stubbe.
Journ. Atmosph. Terr. Phys., Vol. 34, 1627 - 1633 (1972).
Conference paper (see 012.007).

082.060 **Distribution of hydrogen and helium in the upper atmosphere.** G. Kockarts.
Journ. Atmosph. Terr. Phys., Vol. 34, 1729 - 1743 (1972).
Conference paper (see 012.007).

082.061 **A three-dimensional model of thermosphere dynamics—I. Heat input and eigenfunctions.**
H. Volland, H. G. Mayr.
Journ. Atmosph. Terr. Phys., Vol. 34, 1745 - 1768 (1972).
Conference paper (see 012.007).

082.062 **A three-dimensional model of thermosphere dynamics—II. Tidal waves.**
H. Volland, H. G. Mayr.
Journ. Atmosph. Terr. Phys., Vol. 34, 1769 - 1799 (1972).
Conference paper (see 012.007).

082.063 **A three-dimensional model of thermosphere dy-**

namics–III. Planetary waves.
H. Volland, H. G. Mayr.
Journ. Atmosph. Terr. Phys., Vol. 34, 1797 - 1816 (1972).
Conference paper (see 012.007).

082.064 **A thermospheric model from satellite orbital decay densities and incoherent scatter temperatures.**
W. E. Swartz, J. L. Rohrbaugh, J. S. Nisbet.
Journ. Atmosph. Terr. Phys., Vol. 34, 1817 - 1826 (1972).
Conference paper (see 012.007).

082.065 **The spectral coefficient of the atmospheric transparency in the region of the Astrophysical Institute [of the Kazakh Soviet Republic] in 1970–1971.**
V. M. Tereshchenko.
Vestn. AN KazSSR, 1972, No. 3, p. 73 - 76. In Russian.
Abstr. in Referativ. Zhurn. 51. Astron., 9.51.122 (1972).

082.066 **Theoretical calculations of the F-region tropical ultraviolet airglow intensity.** D. N. Anderson.
Journ. Geophys. Res., Vol. 77, 4782 - 4789 (1972).

082.067 **Electron deposition in water vapor, with atmospheric applications.**
J. J. Olivero, R. W. Stagat, A. E. S. Green.
Journ. Geophys. Res., Vol. 77, 4797 - 4811 (1972).
  We examine the consequences of electron impact on water vapor in terms of the microscopic details of excitation, dissociation, and ionization and combinations of these processes. We consider several applications of electron and water-vapor interactions in the atmospheric sciences, in particular, H_2O comets, aurora and airglow, and lightning.

082.068 **Neutral composition in the thermosphere.**
D. R. Taeusch, G. R. Carignan.
Journ. Geophys. Res., Vol. 77, 4870 - 4876 (1972). – Letter.

082.069 **Altitude distribution of the $O_2(^1\Delta)$ nightglow emission.**
W. F. J. Evans, E. J. Llewellyn, A. V. Jones.
Journ. Geophys. Res., Vol. 77, 4899 - 4901 (1972). – Letter.

082.070 **Laser probing of the atmosphere.** V. E. Zuev.
Priroda, No. 10.72, p. 86 - 93 (1972). In Russian.

082.071 **Vertical-ray structure (horizontal inhomogeneity) of the earth's upper atmosphere radiation according to observations from Soyuz 3.**
G. T. Beregovoi, A. A. Buznikov, K. J. Kondratiev, A. I. Lazarev, O. I. Smoktii.
Dokl. Akad. Nauk SSSR, Ser. Mat. Fiz., Vol. 206, 601 - 604 (1972). In Russian.

082.072 **Measurements of polarization at artificial barium clouds.** G. F. Möller, L. Haser.
Zeitschr. Geophys., Vol. 36, 451 - 456 (1970). In German.

082.073 **About a peculiarity of determination of astronomical refraction from solar observations.** V. V. Kirichuk.
Geod., kartogr. i aehrofotos"emka. Resp. mezhved. nauch.-tekhn. sb., 1972, vyp. (No.) 15, p. 20 - 24. In Russian.
Abstr. in Referativ. Zhurn. 51. Astron., 10.51.160 (1972).

082.074 **Die Komponenten des Himmelslichts vom Zenit im Verlaufe der Dämmerung.** G. Dietze.
Gerlands Beiträge Geophys., Vol. 79, 483 - 488 (1970).

082.075 **Mesosphärische Zirkulation und Leuchtende Nachtwolken im Frühjahr 1967.** W. Schröder.
Gerlands Beiträge Geophys., Vol. 79, 489 - 492 (1970).

082.076 **Über die Berechnung der Größenverteilung von absorbierenden Teilchen aus dem Extinktionskoeffizienten.** A. Y. Perelman, V. A. Punina, L. Foitzik.
Gerlands Beiträge Geophys., Vol. 80, 345 - 356 (1971).

082.077 **Die spektrale Extinktion für Modelle mit kombinierten logarithmischen Gauß-Verteilungen.**
L. Foitzik, D. Spänkuch, E. Unger.
Gerlands Beiträge Geophys., Vol. 80, 361 - 370 (1971).

082.078 **Zur Vorgeschichte der Erforschung der Leuchtenden Nachtwolken.** W. Schröder.
Gerlands Beiträge Geophys., Vol. 80, 371 - 374 (1971).

082.079 **Darstellung der Mieschen Streuwerte $i_1(\varphi)$ und $i_2(\varphi)$ nach einer modifizierten van de Hulstschen Beziehung.** L. Foitzik, E. Unger.
Gerlands Beiträge Geophys., Vol. 80, 509 - 514 (1971).

082.080 **Twilight airglow measurements of the OH and O_2 bands by means of balloon-borne instruments.**
D. R. Pick, E. J. Llewellyn, A. V. Jones.
Canadian Journ. Phys., Vol. 49, 897 - 905 (1971).
  This paper describes an experiment to obtain detailed information on the 1.2 to 2.0 μ airglow spectrum by means of a grating spectrometer and several specialized photometers. In particular the 1.27 μ (0,0) O_2 band, the 5,3 and 8,6 OH bands were measured with photometers while the spectrum was scanned continuously at a resolution of 200 Å by a grating spectrometer. In this paper the experimental data are presented.

082.081 **Study of the (4−1) and (5−2) hydroxyl bands in the night airglow.**
A. W. Harrison, W. F. J. Evans, E. J. Llewellyn.
Canadian Journ. Phys., Vol. 49, 2509 - 2517 (1971).
  It is shown that the time-averaged total band intensity fluctuations during a single night are quite large, sometimes a factor of 2, and are not definitely correlated with the rotational temperature during the same period.

082.082 **Scintillation in an earth-to-space propagation path.**
P. O. Minott.
Journ. Optical Soc. America, Vol. 62, 885 - 888 (1972).
  Results of an experiment to measure the scintillation at a satellite from a ground-based laser transmitter are presented. A detector aboard the satellite measured the incident light and telemetered the data to recording equipment on the ground. Log-amplitude variance, probability distributions, and scintillation frequency distributions are derived from the data. The probability distribution is shown to be log-normal. Log-amplitude variance and normalized power spectral density are shown to be within the limits measured for stellar scintillation.

082.083 **Beam spreading in a turbulent medium.**
J. L. Poirier, D. Korff.
Journ. Optical Soc. America, Vol. 62, 893 - 898 (1972).
  The aim of this paper is to calculate the broadening of a gaussian beam propagating in a turbulent medium using a modified von Karman spectrum to characterize the turbulent medium.

082.084 **Comments on "Irradiance fluctuations in optical transmission through the atmosphere".** J. R. Kerr.
Journ. Optical Soc. America, Vol. 62, 916 (1972). – Concerning a recent letter by D. J. Torrieri, L. S. Taylor, Journ. Optical Soc. America, Vol. 62, 145 - 147 (1972). – See Abstr., 07.082.105.

082.085 **Measurements of turbulence profiles in the troposphere.**
J. L. Bufton, P. O. Minott, M. W. Fitzmaurice, P. J. Titterton.

Journ. Optical Soc. America, Vol. 62, 1068 - 1070 (1972).

Temperature structure coefficients were measured with balloon-borne temperature sensors. Data converted to refractive-index-structure coefficients are reported. These extend knowledge of this coefficient to the upper troposphere. The results are discussed with reference to possible meteorological origins for turbulence.

082.086 **Tilting-filter measurements in dayglow rocket photometry.** R. C. Schaeffer, W. G. Fastie.
Applied Optics, Vol. 11, 2289 - 2293 (1972).

A rocket-borne photometer containing two tilting-filter channels for the measurement of the [OI] $\lambda\lambda 6300$ Å and 5577 Å emission lines in the day airglow is described. The results of one flight substantiate the employment of tilting filters to determine accurate corrections for background continuum and provide reliable height profiles of emission intensity down to approximately 90 km. Discussions on the calibration of the instrument and its baffling against sunlight are also presented.

082.087 **Aerosol extinction contribution to atmospheric attenuation in infrared wavelengths.** J. A. Hodges.
Applied Optics, Vol. 11, 2304 - 2310 (1972).

The investigation deals with the estimation of atmospheric attenuation over moderately long, approximately horizontal slant paths near sea level for instruments operating in the $3-5$ μm and $8-14$ μm transmission windows.

082.088 **Meteorological observations on La Silla in 1970.**
B. E. Westerlund.
European Southern Obs., Bull. No. 9, 18 pp. (1972).

The observations concern cloudiness, wind velocity, wind direction, temperature, and humidity as in previous years.

082.089 **The diurnal variations of hydrogen and oxygen constituents in the mesosphere and lower thermosphere.** L. Thomas, M. R.Bowman.
Journ. Atmosph. Terr. Phys., Vol. 34, 1843 - 1858 (1972).

082.090 **Photometric weather at Toronto.** J. D. Fernie.
Journ. Roy. Astron. Soc. Canada, Vol. 66, 249 - 253 = Commun. David Dunlap Obs., Univ. Toronto, *Richmond Hill*, No. 338 (1972).

082.091 **Remote sounding of atmospheric temperature from satellites. I. Introduction.**
J. T. Houghton, S. D. Smith.
Proc. Roy. Soc. London, Ser. A, Vol. 320, 23 - 33 (1970).

Atmospheric temperature can be sounded remotely by observing the emission from carbon dioxide in its 15 μm band. The principles of design of suitable instrumentation for such observation are reviewed.

082.092 **Remote sounding of atmospheric temperature from satellites. II. The selective chopper radiometer for Nimbus D.** P. G. Abel, P. J. Ellis, J. T. Houghton, G. Peckham, C. D. Rodgers, S. D. Smith, E. J. Williamson.
Proc. Roy. Soc. London, Ser. A, Vol. 320, 35 - 55 (1970).

The design of a six-channel radiometer for remote temperature sounding to be mounted on the Nimbus D satellite is described.

082.093 **Remote sounding of atmospheric temperature from satellites. III. Measurements up to 35 km altitude with a balloon-borne selective chopper radiometer.**
P. G. Abel, J. T. Houghton, J. B. Matley, E. J. Williamson.
Proc. Roy. Soc. London, Ser. A, Vol. 320, 57 - 69 (1970).

082.094 **Atmospheric oscillations – IV.** M. N. Jones.
Planet. Space Sci., Vol. 20, 1627 - 1650 (1972).

082.095 **Aeronomic chemistry of the stratosphere.**
M. Nicolet.
Planet. Space Sci., Vol. 20, 1671 - 1702 (1972).

082.096 **A technique for recovering the vertical number density profile of atmospheric gases from planetary occultation data.** R. G. Roble, P. B. Hays.
Planet. Space Sci., Vol. 20, 1727 - 1744 (1972).

The occultation technique of determining the properties of the atmosphere using absorption spectroscopy is examined. The intensity of a star, in certain atmospheric absorption bands, is monitored by a satellite tracking the star during occultation by the earth's atmosphere. We describe a technique for retrieving the number density profile of the absorbing species from occultation intensity data. The data-reduction procedure is similar to the method of computing the radial emission distribution of a cylindrical plasma source, as treated by the plasma physicists. The similarity exists because both techniques reduce to the problem of solving the Abel integral equation.

082.097 **Annual and sub-annual effects of EUV heating – I. Harmonic analysis.** B. K. Ching, Y. T. Chiu.
Planet. Space Sci., Vol. 20, 1745 - 1759 (1972).

In this paper we compute the rate of solar EUV heating in the upper atmosphere by photo-dissociation and photo-ionization, taking care to include properly the effects of oblique incidence of solar flux, sphericity of the atmosphere and ellipticity of the earth's orbit.

082.098 **Annual and sub-annual effects of EUV heating – II. Comparison with density variations.**
B. K. Ching, Y. T. Chiu.
Planet. Space Sci., Vol. 20, 1761 - 1771 (1972).

By a detailed comparison of annual and sub-annual components of EUV absorption heat input with those of the Jacchia density models, we consider the importance of EUV heating in the annual and sub-annual variations of the upper atmosphere.

082.099 **Atomic oxygen concentration from the [OI] 5577 Å line emission at the auroral zone latitude.**
B. S. Dandekar.
Planet. Space Sci., Vol. 20, 1781 - 1784 (1972). – Research note.

082.100 **Study of aerosols in the atmosphere by twilight scattering.** G. M. Shah.
Tellus, Vol. 22, 82 - 93 (1970).

With the idea of detecting the presence and levels of dust layers and their day-to-day changes, by the method of twilight scattering, twilight intensity measurements were started at Mt. Abu in India from 1957–59. A twilight photometer in conjunction with a telescope of small aperture and a colour filter transmitting in red (centred at 6000 Å) was used.

082.101 **Apparent solar constant variations and their relation to the variability of atmospheric transmission.**
A. Ångström.
Tellus, Vol. 22, 205 - 218 (1970).

The considerations contained in the following article are all based on the large number of observations of sun radiation and atmospheric transmission collected by the Astrophysical Observatory of the Smithsonian Institution. The author has shown that there is a close correlation between the variability of transmission and the apparent variability of the solar constant. The relationship is explained on the basis of theoretical deductions presented by Lundblad (1922) and by the present author (1921) at an early date. Further results, concerning the connection between the spectral distribution of intensity of the apparent solar constant and the variations of transmission, are based on the named theory.

082.102 Circular polarization of twilight.
J. R. P. Angel, R. Illing, P. G. Martin.
Nature, Vol. 238, 389 - 390 (1972).

Observations of twilight sky in an infrared band (7800–8800 Å), where Mie scattering is predominant, show an ellipticity V/I of up to 2×10^{-3}; in an ultraviolet band (3500–4000 Å), where Rayleigh scattering dominates, no significant polarization is found (V/I < 0.016%). In the infrared, the polarization increases with increasing angle of the sun below the horizon. This is expected, since both the geometry of multiple scattering becomes more favorable for the production of circular polarization and single scattering is reduced.

082.103 Absence of rapid fluctuations in the ground level
γ-ray background. A. N. James.
Nature, Phys. Sci., Vol. 238, 91 - 92 (1972).

The radiation signature of antimatter micrometeorite annihilation is considered. In Nature, Vol. 230, 180 - 182 (1971) Ashby et. al. have reported events lasting about one second in which the background γ-ray rate increased several orders of magnitude. A similar experiment lasting 230 days detected no events corresponding to a 99.75% significant absence of events. NaI(Tl) crystals flown at 15 km would constitute sensitive antimatter dust grain detectors.

082.104 Atmospheric gravity waves to be expected from the
solar eclipse of June 30, 1973. T. Beer, A. N. May.
Nature, Vol. 240, 30 - 32 (1972).

082.105 Et infrarødt observatorium i Norden?
J.-E. Solheim.
Astron. Tidssk., Årg. 5, p. 101 - 106 (1972).

082.106 Enhanced N_2 vibrational temperatures in the thermo-
sphere. W. S. Varnum.
Planet. Space Sci., Vol. 20, 1865 - 1873 (1972).

082.107 A measurement of the atmospheric neutron flux in
the energy range $50 < E < 350$ MeV.
C. J. Eyles, A. D. Linney, G. K. Rochester.
Planet. Space Sci., Vol. 20, 1915 - 1922 (1972).

082.108 Low energy atmospheric gamma rays near geomag-
netic equator.
K. Kasturirangan, U. R. Rao, P. D. Bhavsar.
Planet. Space Sci., Vol. 20, 1961 - 1977 (1972).

Based on the results from three balloon flights, made at Hyderabad using omnidirectional gamma ray spectrometers, the different aspects of the low energy atmospheric gamma rays at equatorial latitudes in the energy interval 100 keV to 1 MeV are investigated and detailed discussion is presented. The contribution of the cosmic gamma rays to the observed count rates at 6 g cm^{-2} is shown to be negligible in the case of the omnidirectional spectrometers of the type used in the present observations even for low latitude stations.

082.109 Further comments on a parallel study of 6300 Å
airglow emission and ionospheric scintillation.
H. Mullaney, M. D. Papagiannis, J. F. Noxon.
Planet. Space Sci., Vol. 20, 1982 - 1984 (1972). — Research note.

082.110 Estimation of atomic nitrogen and nitric oxide in
the night-time atmosphere in the altitude region 80–
110 kms and their contribution to the nightairglow continuum.
P. P. Saxena.
Ann. Géophys., Vol. 26, 771 - 776 (1970).

Production and loss mechanisms of atomic nitrogen and nitric oxide are investigated in the altitude region 80–110 kms of the night–time atmosphere. Using night–time NO-densities, the chemiluminescent reaction: $NO + O \rightarrow NO_2 + h\nu$, is discussed for the production of the yellow-green spectral region of the nightairglow continuum.

082.111 Post-twilight decay of 6300 Å emission.
S. R. Pal.
Ann. Géophys., Vol. 26, 791 - 793 (1970).

The post-twilight decay rates of 6300 Å nightglow emission observed in different seasons have been examined using appropriate parameters in Chamberlain's relation for post-twilight decay. In the normal cases, the F-layer ionization at the time of layer sunset is found to control the post-twilight decay rate.

082.112 Study of the variation of intensity with solar activity
of the 1 PG bands of nitrogen molecule excited by
photoelectron impacts in day-glow.
V. N. Sharma, R. Singh, S. N. Singh, S. K. Tolpadi.
Ann. Géophys., Vol. 27, 45 - 48 (1971).

The equilibrium photoelectron fluxes have been obtained in the F-region of the ionosphere using CIRA models 2, 5 and 8, 12 hours, $x = 0$ corresponding to low, medium and high solar activity. Using the absolute excitation cross-sections for the emission of 1 PG bands of nitrogen molecule under electron impact, the height profiles for the volume emission rates have been obtained by including cascade contribution from the higher levels. The integrated overhead intensity height profiles have also been calculated. Results are compared with those of the earlier workers and the intensity variation with the solar activity has been discussed.

082.113 On noctilucent clouds. W. Schröder.
Ann. Géophys., Vol. 27, 57 - 59 (1971).

The major results of study of noctilucent clouds (NLC) are described. The connection between mesospheric circulation and frequency of NLC are discussed in relation with new concepts of the nature and origin of NLC.

082.114 Transmission functions for twilight.
D. A. Graham, T. Ichikawa, J. S. Kim.
Ann. Géophys., Vol. 27, 223 - 228 (1971).

New atmospheric transmission functions have been calculated which are suitable for the study of the twilight airglow of the alkalis sodium, lithium, and potassium. The calculated results are presented primarily to demonstrate the improved transmission results which can be obtained from the application of the new methods described; the methods are easily applicable to a number of problems in planetary atmospheric optics.

082.115 Étude en ballon des neutrons de 1 à 10 MeV et des
rayons gamma de 0.7 à latitude équatoriale.
F. Albernhe, I. M. Martin, R. Talon, G. Vedrenne.
Ann. Géophys., Vol. 27, 339 - 343 (1971).

The results of a balloon .flight at an equatorial latitude of 10°N in Guiana are given in this paper. The variations of the flux and spectrum versus the altitude for 1 – 10 MeV neutrons and 0.7 to 4.5 MeV gamma rays are analysed.

082.116 Méthode pour le calcul de l'évolution de l'altitude
d'émission de la raie interdite $\lambda = 6300$ Å de l'oxy-
gène atomique, au cours du renforcement matinal.
M.-L. Duboin.
Ann. Géophys., Vol. 27, 391 - 399 (1971).

A computation method for altitude and intensity variations of OI 6300 Å airglow is proposed which uses data from a rotating photometer. It is based on a correlation between conjugate solar depression and the red emission during its predawn enhancement. The results are compared with those which would be obtained if excitation mechanism were radiative recombination. This hypothesis does not seem to be confirmed by the first data from the Observatoire de Haute-Provence which have been studied.

082.117 **Observations of sodium, lithium and potassium twilightglow at Moscow, Idaho, U. S. A.**
D. A. Graham, T. Ichikawa, J. S. Kim.
Ann. Géophys., Vol. 27, 483 - 491 (1971).
The results of an investigation of the twilight airglow of the alkalies are presented. The data were taken at Moscow, Idaho during 1967 and 1968 and subsequently analysed in some detail to determine the vertical abundance characteristics.

082.118 **Ultraviolett-Halo der Erde erstmals photographiert.**
Umschau, 72. Jahrgang, p. 698 - 699 (1972).
A far-ultraviolet camera/spectrograph experiment was operating on the lunar surface during the Apollo 16 mission. Among the data obtained were images and spectra of the terrestrial atmosphere and geocorona in the wavelength range below 1600 Å, giving the spatial distributions and relative intensities of emissions due to atomic hydrogen, atomic oxygen, molecular nitrogen, and other species, some observed spectrographically for the first time.

082.119 **On the determination of the transfer function of the atmosphere using spectrophotometry of a** planet's surface from cosmos.
K. J. Kondratiev, O. I. Smoktii.
Dokl. Akad. Nauk SSSR, Ser. Mat. Fiz., Vol. 206, 1102 - 1105 (1972). In Russian.

082.120 **Energetic calibration of instruments used for measurements of scattered light in the upper atmosphere.**
F. Rössler.
Optik, Vol. 35, 445 - 458 (1972). In German.
Because of the calibrating technique used, measurements of scattered light made in the upper atmosphere provide results that are primarily recorded as photometric units. In the present paper, these units are converted into energetic units and/or photon numbers. For this, the spectral distribution of the standard lamp used for calibration as well as the spectral sensitivity of the photomultiplier must be known. The measurements performed are described. The Rayleigh scattering is computed by means of these energetic units. Finally, the special case of a receiver with the sensitivity of the eye is investigated.

082.121 **Some results of measurements of short-wave and long-wave radiation fluxes aboard Cosmos 320.**
G. P. Faraponova.
Izv. AN SSSR. Fiz. atmosf. i okeana, Vol. 8, 626 - 633 (1972). In Russian. – Abstr. in Referativ. Zhurn. 62. Issled. kosmich. prostranstva, 11.62.177 (1972).

082.122 **Zeldzame optische verschijnselen.**
P.-P. H. Verschure.
Hemel en Dampkring, Vol. 70, 307 - 309 (1972).

082.123 **Profiles and half-widths of telluric lines.**
N. I. Kozhevnikov, G. F. Sitnik, A. I. Khlystov.
Soobshch. Gos. Astron. Inst. Shternberga, No. 180, p. 3 - 19 (1972). In Russian.
The dependence of the half-width and depth of a line on observational conditions is considered. An absorption line contour is calculated for an isothermal atmosphere with density changing according to the barometric law. Formulae for the dependence of the half-width and depth of a line on the height of the observational site and the value of the air mass in the moment of observation are given.

082.124 **About the connection of meteorological conditions**

and of the relief with the astroclimate of mountain areas. V. S. Shevchenko, V. E. Slutsky, V. I. Kardopolov, V. G. Khetselius.
Soobshch. Gos. Astron. Inst. Shternberga, No. 180, p. 23 - 39 (1972). In Russian.
The contribution of different components of the turbulent atmosphere to the forming of mountain astroclimate and in particular the climatic and meteorological factors are considered. The authors point out the advantages of separate peaks for astronomical stations building.

082.125 **Changes in thermospheric molecular oxygen abundance inferred from twilight 6300 Å airglow.**
J. F. Noxon, A. E. Johanson.
Planet. Space Sci., Vol. 20, 2125 - 2151 (1972).

082.126 **Analysis of the orbit of Cosmos 268 rocket (1969-20B).** D. G. King-Hele, A. N. Winterbottom.
Planet. Space Sci., Vol. 20, 2153 - 2163 (1972).
This report describes the orbit determinations, and analyses the resulting values of inclination to determine atmospheric rotation rates at heights near 240 km.

082.127 **Air density at heights near 200 km from the orbit of 1969-20B.** D. M. C. Walker.
Planet. Space Sci., Vol. 20, 2165 - 2173 (1972).
In the present paper the computed perigee heights of Cosmos 268 rocket were used, together with the orbital decay rates calculated from USAF Spacetrack elements: 103 values of air density at heights between 185 km and 253 km were obtained at dates between 17 July 1969 and 8 February 1970.

082.128 **Correcting the OH contribution in emission line measurements in the night airglow filter photometry.** V. R. Rao, P. V. Kulkarni.
Planet. Space Sci., Vol. 20, 2198 - 2201 (1972). – Research note.

082.129 **Daytime laser radar measurements of the atmospheric sodium layer.**
A. J. Gibson, M. C. W. Sandford.
Nature, Vol. 239, 509 - 511 (1972).

082.130 **Cosmic dust in the mesosphere.**
N. H. Farlow, G. V. Ferry.
Space Research XII, (see 012.016), Vol. 1, 369 - 380 (1972).

082.131 **Upper atmospheric dust concentrations in polar regions.** G. V. Ferry, N. H. Farlow.
Space Research XII, (see 012.016), Vol. 1, 381 - 390 (1972).

082.132 **Combined dust collection and detection experiment during a noctilucent cloud display above Kiruna,** Sweden. P. Rauser, H. Fechtig.
Space Research XII, (see 012.016), Vol. 1, 391 - 402 (1972).

082.133 **Aerosol layers in the atmosphere.** F. Rössler.
Space Research XII, (see 012.016), Vol. 1, 423 - 431 (1972).

082.134 **The influence of cosmic dust on twilight phenomena.**
F. Link, R. Robley.
Space Research XII, (see 012.016), Vol. 1, 433 - 435 (1972).

082.135 **A global model of atmospheric temperature, chemical composition and density (25 – 1000 km altitude).**
D. K. Weidner, J. L. Chambers, G. Y. Lou.
Space Research XII, (see 012.016), Vol. 1, 565 - 574 (1972).

082.136 **A comparison between total ozone as measured by Nimbus 3 and that computed from a numerical**

model.   L. Berkofsky, S. Gyoeri.
Space Research XII, (see 012.016), Vol. 1, 645 - 650 (1972).

082.137 O_2 densities from solar hydrogen Lyman α absorption measurements by Intercosmos 4 and Vertical 1.
D. Felske, L. Martini, B. Stark, J. Taubenheim.
Space Research XII, (see 012.016), Vol. 1, 651 - 656 (1972).

082.138 Mass-spectrometric investigations of upper atmosphere neutral composition at equatorial, middle and polar latitudes.   A. A. Pokhunkov.
Space Research XII, (see 012.016), Vol. 1, 657 - 663 (1972).

082.139 Experimental measurements of emission (OI) $\lambda = 5577$ Å and scattered day-time sky radiation.
V. I. Konkov, G. I. Kuznetsov, G. F. Sitnik, A. H. Hrgian, A. F. Chizhov, O. V. Shtyrkov.
Space Research XII, (see 012.016), Vol. 1, 685 - 690 (1972).

082.140 Atmospheric density and the accuracy of determination of the drag coefficient of a satellite.
M. Ya. Marov, A. A. Pyarnpuu, G. I. Zmievskaya.
Space Research XII, (see 012.016), Vol. 1, 721 - 726 (1972).

082.141 Upper atmosphere density determination from the Cosmos satellite deceleration results.
P. E. Elyasberg, B. V. Kugaenko, V. M. Synitsyn, M. I. Voiskovsky.
Space Research XII, (see 012.016), Vol. 1, 727 - 731 (1972).

082.142 Semi-annual density variations of the atmosphere at heights of 200–300 km.
M. Ya. Marov, A. M. Alpherov.
Space Research XII, (see 012.016), Vol. 1, 803 - 808 (1972).

082.143 Direct measurement of the semi-annual variation during 1968.   J. P. McIsaac, K. S. W. Champion.
Space Research XII, (see 012.016), Vol. 1, 809 - 813 (1972).

082.144 The latitudinal dependence of the semi-annual effect.   C. Wulf-Mathies.
Space Research XII, (see 012.016), Vol. 1, 815 - 819 (1972).

082.145 Structure and motion of the thermosphere deduced from satellite drag.
M. Ya. Marov, A. M. Alpherov.
Space Research XII, (see 012.016), Vol. 2, 823 - 840 (1972).

082.146 Density variations and atmospheric rotation below 200 km from the drag on the satellite OV1-15.
B. K. Ching.
Space Research XII, (see 012.016), Vol. 2, 841 - 846 (1972).

082.147 Measurements of upper-atmosphere rotational speed from changes in satellite orbits.   D. G. King-Hele.
Space Research XII, (see 012.016), Vol. 2, 847 - 855 (1972).

082.148 Upper atmosphere heating at high latitudes.
R. R. Allan.
Space Research XII, (see 012.016), Vol. 2, 857 - 866 (1972).

082.149 Structure and motion of the thermosphere shown by density data from the Low-G Accelerometer Calibration System (LOGACS).   L. L. DeVries.
Space Research XII, (see 012.016), Vol. 2, 867 - 879 (1972).

082.150 Measurement of thermospheric composition.
A. O. Nier.
Space Research XII, (see 012.016), Vol. 2, 881 - 889 (1972).

082.151 Diurnal variations of atmospheric neutral composi-

tion at altitudes of 130–200 km.   A. D. Danilov.
Space Research XII, (see 012.016), Vol. 2, 891 - 898 (1972).

082.152 The atmospheric extinction at Cerro Tololo, 1967 - 1969.   A. Gutiérrez-Moreno, H. Moreno.
Publ. Dep. Astron., Univ. Chile, Obs. Astron. Nacional, Cerro Calán, Santiago de Chile, Vol. 2,(No. 1), 22 - 26 (1970).
The values of the monochromatic extinction for the wavelength interval 3100 Å to 5800 Å are presented.

082.153 Oxidation of SO_2 and other atmospheric gases by ozone in aqueous solution.   S. A. Penkett.
Nature, Phys. Sci., Vol. 240, 105 - 106 (1972).
It is concluded that reactions of ozone in solution are a novel feature of atmospheric chemistry. With the findings of an efficient oxidation of SO_2 by the ozone-olefin reaction in the gas phase, the general importance of ozone as an atmospheric oxidant is emphasized.

082.154 Secular variation of the stratospheric ozone layer over Middle Europe during the solar cycles from 1951 to 1972.   H. K. Paetzold, F. Piscalar, H. Zschörner.
Nature, Phys. Sci., Vol. 240, 106 - 107 (1972).
We have suggested a typical secular variation of the ozone in the terrestrial stratosphere with the solar activity (1961). We report here observations with optical ozone radio sonde which give further evidence.

082.155 Atmospheric helium and geomagnetic field reversals.   W. R. Sheldon, J. W. Kern.
Journ. Geophys. Res., Vol. 77, 6194 - 6201 (1972).
The problem of the earth's helium budget is examined in the light of recent work on the interaction of the solar wind with nonmagnetic planets. It is proposed that the dominant mode of helium (^{4}He) loss is ion pumping by the solar wind during geomagnetic field reversals, when the earth's magnetic field is very small.

082.156 Excitation of the Herzberg bands of O_2 in laboratory afterglow and night airglow.   V. Degen.
Journ. Geophys. Res., Vol. 77, 6213 - 6218 (1972).

082.157 Possibility of O III λ304-A emissions in the extreme ultraviolet airglow.   R. W. Carlson.
Journ. Geophys. Res., Vol. 77, 6282 - 6283 (1972). – Letter.

082.158 On the atomic oxygen measurements by rocket-borne mass spectrometers.   D. Offermann.
Journ. Geophys. Res., Vol. 77, 6284 - 6286 (1972). – Letter.

082.159 On optical methods for sounding of the upper atmosphere and of the circumterrestrial dust cloud.
V. G. Fesenkov.
Byull. Abastumansk. Astrofiz. Obs., No. 41, (see 012.018), p. 7 - 21 (1972). In Russian.

082.160 On the variation of the scattering coefficient in the stratosphere with height from measurements of Soyuz 3.   A. B. Sandomirskij, G. V. Rozenberg.
Byull. Abastumansk. Astrofiz. Obs., No. 41, (see 012.018), p. 23 - 26 (1972). In Russian.

082.161 Qualitative proof for the existence of layers of dust particles of cosmic origin in the lower thermosphere by means of the twilight method.   G. Kohl.
Byull. Abastumansk. Astrofiz. Obs., No. 41, (see 012.018), p. 27 - 48 (1972). In German.

082.162 Results of spectral investigations of the twilight aureole of the earth's atmosphere from Soyuz 5.
K. Ya. Kondrat'ev, B. V. Volynov, A. P. Gal'tsev, O. I.

Smoktij, E. V. Khrunov.
Byull. Abastumansk. Astrofiz. Obs., No. 41, (see 012.018),
p. 49 - 51 (1972). In Russian.

082.163 On variations of scattered light during twilight in
the spectral region between 5500 and 6600 Å from
spectral observations in 1962–1968 in Abastumani.
T. G. Megrelishvili.
Byull. Abastumansk. Astrofiz. Obs., No. 41, (see 012.018),
p. 53 - 75 (1972). In Russian.

082.164 Analysis of the variation of polarization during
twilight with regard to influences from the upper
atmosphere. G. Dietze.
Byull. Abastumansk. Astrofiz. Obs., No. 41, (see 012.018),
p. 77 - 86 (1972). In German.

082.165 On the determination of the atmospheric dust concentration and its scattering indicatrix from brightness measurements of primary twilight. N. B. Divari.
Byull. Abastumansk. Astrofiz. Obs., No. 41, (see 012.018),
p. 87 - 98 (1972). In Russian.

082.166 Polarization measurements of bright sky light and
comparison with theoretical data. J. Lukáč.
Byull. Abastumansk. Astrofiz. Obs., No. 41, (see 012.018),
p. 99 - 104 (1972). In Russian.

082.167 On the effective screening height applying the twilight method for studying the earth's atmosphere.
T. I. Toroshelidze.
Byull. Abastumansk. Astrofiz. Obs., No. 41, (see 012.018),
p. 105 - 113 (1972). In Russian.

082.168 On atmospheric emission of atomic oxygen (5577 Å)
during night and its connection with sporadic appearing micrometeorites. B. A. Mirtov.
Byull. Abastumansk. Astrofiz. Obs., No. 41, (see 012.018),
p. 117 - 124 (1972). In Russian.

082.169 On some results of brightness measurements of the
twilight sky made for separation of primary twilight. Yu. I. Zaginajlo.
Byull. Abastumansk. Astrofiz. Obs., No. 41, (see 012.018),
p. 125 - 135 (1972). In Russian.

082.170 Some statistical properties of the hydroxyl emission.
N. N. Shefov.
Byull. Abastumansk. Astrofiz. Obs., No. 42, (see 012.019),
p. 9 - 24 (1972). In Russian.

082.171 Variations of the rotational temperature of hydrox-
yl during geomagnetic disturbed periods above
Yakutsk.
V. M. Ignatiev, L. D. Sivtseva, V. A. Yugov, K. V. Atlasov.
Byull. Abastumansk. Astrofiz. Obs., No. 42, (see 012.019),
p. 25 - 28 (1972). In Russian.

082.172 Diurnal variations of the night airglow Hα emission.
L. M. Fishkova, P. V. Shcheglov.
Byull. Abastumansk. Astrofiz. Obs., No. 42, (see 012.019),
p. 29 - 38 (1972). In Russian.

082.173 Spatial distribution of the upper atmosphere's Hα
emission, its variations during the solar cycle and
dependence on geomagnetic disturbances.
N. M. Martsvaladze.
Byull. Abastumansk. Astrofiz. Obs., No. 42, (see 012.019),
p. 39 - 46 (1972). In Russian.

082.174 Results of investigations of the λ 5577 Å emission in
Ashkhabad during 1957–1968.
M. P. Korobeynikova, G. A. Nasirov.
Byull. Abastumansk. Astrofiz. Obs., No. 42, (see 012.019),
p. 47 - 58 (1972). In Russian.

082.175 The predawn enhancement of the airglow λ 6300 Å
[OI] emission according to observations at
Abastumani. L. M. Fishkova, Yu. D. Mateshvili.
Byull. Abastumansk. Astrofiz. Obs., No. 42, (see 012.019),
p. 59 - 76 (1972). In Russian.

082.176 Spectral investigation of λ 6300 Å [OI] emission in
twilight at Abastumani. T. G. Megrelishvili.
Byull. Abastumansk. Astrofiz. Obs., No. 42, (see 012.019),
p. 77 - 90 (1972). In Russian.

082.177 Anisotropy and energetic spectra of fresh photoelectrons.
Yu. I. Galperin, T. M. Mulyarchik, F. K. Shujskaya.
Byull. Abastumansk. Astrofiz. Obs., No. 42, (see 012.019),
p. 97 - 102 (1972). In Russian.

082.178 Observations of the upper atmosphere's Hα emission
at Abastumani in 1962–1969.
N. M. Martsvaladze, L. M. Fishkova.
Byull. Abastumansk. Astrofiz. Obs., No. 42, p. 119 - 130
(1972). In Russian.

082.179 Investigation of the hydrogen of the upper atmosphere and geocorona by observations of the Hα
emission line in the airglow spectrum. L. M. Fishkova.
Byull. Abastumansk. Astrofiz. Obs., No. 42, p. 131 - 181
(1972). In Russian.

082.180 The nocturnal variation and the decay coefficient
of the 6300 Å emission in nightglow. S. R. Pal.
Ann. Géophys., Vol. 28, 51 - 56 (1972).
The different seasonal features in the diurnal variations
of the 6300 Å emission in nightglow are discussed from the
observations over a period from 1965 to 1967 at Mont Abu
(India). An exponential relation is found to represent the
post-twilight decay of 6300 Å emission.

082.181 Laboratory measurements of reactions related to
ozone photochemistry. H. I. Schiff.
Ann. Géophys., Vol. 28, 67 - 77 (1972). – Paper presented at
I.A.G.A. symposium on 'Aurora and airglow', Moscow, August 1971.

082.182 Molecular absorption cross-sections.
M. Ackerman.
Ann. Géophys., Vol. 28, 79 - 83 (1972). – Paper presented at
I.A.G.A. symposium on 'Aurora and airglow', Moscow, Au-
gust 1971.

082.183 Hydroxyl emission. N. N. Shefov.
Ann. Géophys., Vol. 28, 137 - 143 (1972). – Paper
presented at I.A.G.A. symposium on 'Aurora and airglow',
Moscow, August 1971.

082.184 Mise en évidence expérimentale d'une structure in-
homogène à petite échelle dans la couche émissive
de l'oxygène atomique à 5577 Å.
J. Barat, J.-E. Blamont, M. Petitdidier, C. Sidi, H. Teitelbaum.
Ann. Géophys., Vol. 28, 145 - 148 (1972). – Paper presented
at I.A.G.A. symposium on 'Aurora and airglow', Moscow, August 1971.

082.185 Short time-interval spectrometric hydroxyl emission
studies. K. A. Dick.
Ann. Géophys., Vol. 28, 149 - 153 (1972). – Paper presented

at I.A.G.A. symposium on 'Aurora and airglow', Moscow, August 1971.

082.186 OI and NI allowed transitions in the airglow and aurora. B. A. Tinsley.
Ann. Géophys., Vol. 28, 155 - 168 (1972). – Paper presented at I.A.G.A. symposium on 'Aurora and airglow', Moscow, August 1971.

082.187 Predawn airglow enhancement project: Perspective. H. C. Carlson.
Ann. Géophys., Vol. 28, 179 - 186 (1972). – Paper presented at I.A.G.A. symposium on 'Aurora and airglow', Moscow, August 1971.

082.188 The 6300 Å predawn enhancement: Excitation by photoelectrons from the magnetic conjugate point. V. B. Wickwar.
Ann. Géophys., Vol. 28, 187 - 192 (1972). – Paper presented at I.A.G.A. symposium on 'Aurora and airglow', Moscow, August 1971.

082.189 Manifestations optiques des aérosols météoriques. I. – Orionides 1970. M. Fehrenbach, D. Frimout, F. Link, C. Lippens.
Ann. Géophys., Vol. 28, 363 - 375 (1972).
During the period of Orionids between 19th and 26th October 1970 photometrical measurements of the twilight sky has been performed at the height of 30 km with aid of balloons launched at Aire-sur-l'Adour (Landes, France). The results of 6 flights show clearly the influence of Orionids on the optical properties of the upper atmosphere in the green region (5100 Å) of the spectrum.

082.190 Review of laboratory measurements of aeronomic ion-neutral reactions. E. E. Ferguson.
Ann. Géophys., Vol. 28, 389 - 395 (1972). – Paper presented at I.A.G.A. symposium on 'Aurora and airglow', Moscow, August 1971.

082.191 6300 Å night airglow emission over the magnetic equator. P. V. Kulkarni, V. R. Rao.
Ann. Géophys., Vol. 28, 475 - 481 (1972). – Paper presented at I.A.G.A. symposium on 'Aurora and airglow', Moscow, August 1971.

082.192 Research of the emission at 5577 Å in the period of 1958–1967 in Ashkhabad. M. P. Korobeynikova, G. A. Nasirov.
Ann. Géophys., Vol. 28, 483 - 487 (1972). – Paper presented at I.A.G.A. symposium on 'Aurora and airglow', Moscow, August 1971.

082.193 Rocket measurements of low energy electrons and optical emissions in the dayglow and aurora. P. D. Feldman.
Ann. Géophys., Vol. 28, 489 - 495 (1972). – Paper presented at I.A.G.A. symposium on 'Aurora and airglow', Moscow, August 1971.

082.194 The predawn enhancement of the airglow λ 6300 Å (OI) emission according to observations in Abastumani. L. M. Fishkova, Yu. D. Mateshvili.
Ann. Géophys., Vol. 28, 497 - 501 (1972). – Paper presented at I.A.G.A. symposium on 'Aurora and airglow', Moscow, August 1971.

082.195 MgI emission in the night sky spectrum. T. R. Hicks, B. H. May, N. K. Reay.
Nature, Vol. 240, 401 - 402 (1972).
During September, October 1971 and April 1972 we observed a narrow region of the night sky spectrum in the vicinity of the 5183.62 Å MgI wavelength. Careful examination of the 350 spectra revealed the presence of a small, variable, but undoubtedly real emission core at the centre of the zodiacal light Fraunhofer line.

082.196 Normal incidence radiation trends on Mauna Loa, Hawaii. R. F. Pueschel, L. Machta, G. F. Cotton, E. C. Flowers, J. T. Peterson.
Nature, Vol. 240, 545 - 547 (1972). – Letter.

082.197 The influence of atmospheric haze on the colour of the underlaying surface observed from a manned spaceship. K. J. Kondratiev, O. I. Smoktii.
Dokl. Akad. Nauk SSSR, Ser. Mat. Fiz., Vol. 207, 86 - 89 (1972). In Russian.

082.198 On the depth of Fraunhofer lines in day sky glow. V. E. Pavlov, J. A. Teifel, V. P. Golovachev.
Dokl. Akad. Nauk SSSR, Ser. Mat. Fiz., Vol. 207, 566 - 568 (1972). In Russian.

082.199 Circular polarization of the nightsky radiation. R. D. Wolstencroft, J. C. Kemp.
Astrophys. Journ., (*Letters*), Vol. 177, L137 - L140 (1972).
The fractional circular polarization, q, of the night sky has been measured in 13 directions. Definite values (at least 3σ) were found in four regions of the zodiacal light and two areas of the Milky Way. The values of q, when corrected for foreground radiation, range from -0.6 to $+0.5$ percent and agree well with results reported recently by Staude, Wolf, and Schmidt. Scattering of light by nonspherical particles that are partially aligned can explain these observations.

082.200 Vela 4 Lyman-α observations: Evidence for an aspherical hydrogen geocorona at 18 R_E. P. E. Fehlau, W. H. Chambers, W. E. Kunz, J. C. Fuller.
Journ. Geophys. Res., Vol. 77, 6665 - 6670 (1972).
Observations of scattered solar hydrogen Lyman-α radiation made with detectors carried by satellites Vela 4A and 4B show a repetitive and marked asymmetry in the Lyman-α flux. Detailed analysis of data for many orbits indicates that the most precisely located feature of the data, the leading edge of the broad maximum, moves in right ascension at about the same rate and in the same direction as the motion of the earth projected on the celestial sphere.

082.201 Theoretical model for the latitude dependence of the thermospheric annual and semiannual variations. H. G. Mayr, H. Volland.
Journ. Geophys. Res., Vol. 77, 6774 - 6790 (1972).
A three-dimensional model for the annual and semiannual variations of the thermosphere is presented in which the energy and diffusive mass transport associated with global circulation are considered in a self-consistent form. It is shown that these processes play a major role in thermosphere dynamics and thus account for a number of temperature and composition phenomena.

082.202 Dye-laser observations of the nighttime atomic sodium layer. R. D. Hake, Jr., D. E. Arnold, D. W. Jackson, W. E. Evans, B. P. Ficklin, R. A. Long.
Journ. Geophys. Res., Vol. 77, 6839 - 6848 (1972).
Two new interesting features are shown: a) a sharp decrease in density that terminates the layer on the bottom side at a variable altitude near the mesopause; b) a fourfold increase in sodium-layer content during a 4-hour period surrounding the transit of the radiant of the Geminids meteor shower on the night of December 13–14, 1971, when the shower was at its peak.

082.203 **Measurements of atmospheric absorption of millimeter waves.**
G. G. Gimmestad, H. A. Gebbie, R. A. Bohlander, E. E. Mendoza V.
Astrophys. Journ., Vol. 178, 267 - 270 (1972).

Spectral measurements of atmospheric absorption in a slant path from the sun show maxima at 7.3 and 9.5 wavenumbers. The level of absorption is greater than predictions for water vapor alone and the excess too great to be attributed to known minor atmospheric constituents.

082.204 **About temperature variations of the thermosphere with day time and solar activity.**
M. N. Izakov, I. A. Yashchenko.
Geomagn. Aeronom., Vol. 12, 1037 - 1041 (1972). In Russian.

082.205 **Determination of astronomical refraction near the horizon in different seasons of the year.**
N. A. Vasilenko.
Astrometriya i Astrofiz., *Kiev*, Vyp. (No.) 17, (see 003.012), p. 96 - 108 (1972). In Russian.

Results are presented of measuring astronomical refraction for zenith distances 80−90° carried out by the author in different seasons by means of a 2″ universal instrument.

082.206 **Scintillation spectra of stars and planets and dependence of their characteristics on meteorological conditions.** Yu. K. Filippov.
Astrometriya i Astrofiz., *Kiev*, Vyp. (No.) 17, (see 003.012), p. 108 - 115 (1972). In Russian.

839 scintillation spectra of stars and 102 scintillation spectra of planets were obtained at Golosejevo. The dependences of spectral density of scintillation, amplitude and frequency characteristics on zenith distance and aerological parameters were studied. Differences in amplitude characteristics of scintillation spectra for «white» and «red» stars were found.

082.207 **On the distribution law of star image motions.**
M. L. Divinsky, I. G. Kolchinsky.
Astrometriya i Astrofiz., *Kiev*, Vyp. (No.) 17, (see 003.012), p. 115 - 122 (1972). In Russian.

The distribution of the deviations from the mean directions of star trails was investigated using 105 star trails. It was shown that about 93 % of the trails give a normal law distribution. About 4 % of star trails are in agreement with a Charlier distribution.

082.208 **Introduction to practical computation of astronomical refraction. Part II.** J. Saastamoinen.
Bull. Géod., Nouvelle Sér., Année 1972, No. 106, p. 383 - 397.

082.209 **Approximate analytical solutions for the diffusion equation of static models of Jacchia's upper atmosphere.** M. A. Degtyarev.
Kosmich. Issled., Vol. 10, 901 - 904 (1972). In Russian.

082.210 **Light flare excited in the upper atmosphere by a pulse source of X-rays.**
A. V. Zhemerev, Yu. A. Medvedev, B. M. Stepanov.
Kosmich. Issled., Vol. 10, 916 - 919 (1972). In Russian.

082.211 **Identification of stratospheric NH.**
R. W. Nicholls.
Nature, Phys. Sci., Vol. 240, 142 - 143 (1972).

082.212 **Über den Kontrast und die Sichtbarkeit Leuchtender Nachtwolken am Dämmerungshimmel.**
G. Dietze.
Gerlands Beiträge Geophys., Vol. 81, 414 - 422 (1972).

For a noctilucent cloud (NLC) with its typical features the brightness contrast was calculated and plotted for all points in the sky and for different twilight phases. With the same conditions the quality of visibility for a normal observer is calculated and represented. The results, which agree fairly well with observations, are discussed. The figures enable a critical examination of problems of the spatial and temporal distribution of the perceptibility of NLC by means of the eye as well as by other optical methods.

082.213 **Otto Jesse und die Erforschung der Leuchtenden Nachtwolken.** W. Schröder.
Gerlands Beiträge Geophys., Vol. 81, 423 - 432 (1972).

Extracts from the papers of O. Jesse relating to noctilucent cloud research are presented here so as to give an idea of his contribution to this subject made during the years 1884 - 1901.

082.214 **Spectrophotometric investigations of mean atmospheric extinction at Piwnice.** W. Wegner.
Stud. Soc. Sci. Torunensis, Sectio F (Astron.), Vol. 5, 9 - 20 = Bull. Astron. Obs. Toruń No. 49/II (1972).

Extinction coefficients at different weather conditions for 12 wavelength values are derived.

082.215 **Sunshine duration in Thessaloniki−Greece.**
G. C. Livadas, A. A. Flocas.
Sci. Ann. Fac. Math. Phys. Aristotelian Univ. Thessaloniki, Vol. 12, 109 - 146 = Publ. Meteorol. Inst. Univ. Thessaloniki No. 17 (1972).

082.216 **On the origin of the bad seeing at the Cape.**
J. B. Alexander.
Monthly Notes Astron. Soc. Southern Africa, Vol. 31, 159 - 161 (1972). − Note.

082.217 **Calculated Mie scattering properties in the visible and infrared of measured Los Angeles aerosol size distributions.** F. S. Harris, Jr.
Applied Optics, Vol. 11, 2697 - 2705 (1972).

Aerosol size distributions of varying types selected from those measured in clear air, smog, and fog in the Los Angeles Basin have been used with Lorenz-Mie scattering theory to predict radiation scattering by aerosols. Eleven different indices of refraction were assumed for wavelength from 0.488 μm to 8.4 μm for aerosol materials with varying humidity, and for water and quartz. The effect on the scattering by the type of size distribution and the complex index of refraction is shown as a function of the polarization parameters of polarization, polarization ratio, ellipticity, and the inclination angle of the polarization ellipse.

082.218 **Verification of an approximation method for calculating multiple scattering of sky radiation.**
E. de Bary.
Applied Optics, Vol. 11, 2717 - 2718 (1972). − Letter.

082.219 **Degree and direction of polarization of multiple scattered light. 2: Earth's atmosphere with aerosols.**
G. N. Plass, G. W. Kattawar.
Applied Optics, Vol. 11, 2866 - 2879 (1972).

The degree of polarization as well as the direction of the polarization are calculated by a Monte Carlo method for the reflected and transmitted photons from the earth's atmosphere. The solar photons are followed through multiple collisions with the aerosols and the Rayleigh scattering centers in the atmosphere. Three different aerosol number densities are used to study the effects of aerosol variations. Results are given for a solar zenith angle of 81.37° and various surface albedos.

082.220 **Airborne laser-beam scintillation measurements at high altitudes.** G. J. Morris.
Journ. Optical Soc. America, Vol. 62, 1339 (1972). – Abstr. Optical Soc. America.

082.221 **Slant-path scintillation in the planetary boundary layer.** W. F. Dabberdt.
Journ. Optical Soc. America, Vol. 62, 1340 (1972). – Abstr. Optical Soc. America.

082.222 **Correlation of vertical profile turbulence data with stellar observations.** J. L. Bufton.
Journ. Optical Soc. America, Vol. 62, 1353 (1972). – Abstr. Optical Soc. America.

082.223 **General case of temporal spectra of phase difference fluctuations.** G. W. Reinhardt, S. A. Collins, Jr.
Journ. Optical Soc. America, Vol. 62, 1354 (1972). – Abstr. Optical Soc. America.

082.224 **Calculated effects of aerosols on sunlight.** L. M. Shotkin, H. Ludewig, J. F. Thompson, Jr.
Journ. Optical Soc. America, Vol. 62, 1367 (1972). – Abstr. Optical Soc. America,

082.225 **Effect of aerosols on the radiance and the color of the light in the twilight sky.**
G. N. Plass, G. W. Kattawar, C. N. Adams.
Journ. Optical Soc. America, Vol. 62, 1368 (1972). – Abstr. Optical Soc. America.

082.226 **Notas sobre algunos aspectos de la radiación solar difusa y su variación diurna.** M. Puigcerver.
Urania Barcelona, Año 56, No. 274, p. 197 - 214 (1971).
Two short series of clear-sky scattered radiation measurements, made at different places and times and using different measuring techniques, are analyzed. A slack dependence on solar height and a strong influence of even traces of cloudiness are both found.

082.227 **Considerazioni sulla classificazione visuale delle tracce di Walker.** E. Moroder, A. Righini.
Mem. Soc. Astron. Italiana, Nuova Ser., Vol. 43, 535 - 537 (1972).
A comparison is presented between the visual classification of some polar star trails and the measure of the seeing figure based on the wave structure of the trail.

082.228 **Extra-atmospheric observations of the night-sky brightness at the sputniks Cosmos 51 and Cosmos 213. I. Method and calibration of the measurements.**
N. A. Dimov, A. B. Severny.
Izv. Krymskoj Astrofiz. Obs., Vol. 45, 53 - 66 (1972). In Russian.
Descriptions of the photomultipliers and of the electronic scheme are given. The limiting sensitivity for the photoelectric photometry is about ± (20 − 30) stars of $m_V = 10^m$ per square degree.

082.229 **Extra-atmospheric observations of night-sky brightness with the sputniks Cosmos 51 and Cosmos 213. II. Results and discussion.**
N. A. Dimov, A. M. Zvereva, A. B. Severny.
Izv. Krymskoj Astrofiz. Obs., Vol. 45, 67 - 89 (1972). In Russian.
With the aid of the photometers installed on board Cosmos 51 and Cosmos 213 the brightness of night sky was measured in the visual and ultraviolet regions. The data of telemetry permit us to estimate the observed flux ratios in the ultraviolet and the visual ranges $(U/V)_0$ as well as the brightness of night sky V_0. These observed values are compared with the expected ratio $(U/V)_c$ and the values V_c calculated on the basis of data on the population of stars of different spectral types and magnitudes, and on the expected ratio of fluxes $(U/V)_i$, for different spectral types according to the models of stellar atmospheres. The influence of the zodiacal light and of the interstellar absorption is taken into account.

082.230 **Astronomical seeing quality at Blue Mesa Observatory.** J. Cuffey.
Contr. Obs. New Mexico State Univ., Las Cruces, Vol. 1, 73 - 76 (1972).

082.231 **On the correction due to aerosol in Abastumani measurements of the total amount of atmospheric ozone.** V. M. Iskandarova.
Byull. Abastumansk. Astrofiz. Obs., No. 43, p. 203 - 204 (1972). In Russian.

082.232 **Resultados de mediciones con globos estratosfericos. I. (Balloon flights data. I.)**
Inst. Astron. Fís. Espacio, Buenos Aires, Ser. Publ. Registros No. 1, 18 + 44 + 56 + 17 + 17 pp. (1972).
In this Data Report we present the results of two series of measurements carried out with balloon borne equipment. The balloon launchings were made at different places of the Argentine Republic, and a group of flights was launched in the Atlantic Ocean. Data presented in this report result from measurements made since the end of 1963 up to about the middle of 1969. The first series of flights corresponds to measurements of the total charged ionizing component throughout and near the top of the atmosphere; the second series corresponds to measurements of the secondary cosmic ray photon component.

082.233 **A study of cloudiness in Mexico through meteorological satellites.**
E. E. Mendoza V., J. Luna, T. Gómez.
Bol. Obs. Tonantzintla y Tacubaya, Vol. 6, (No. 38), 215 - 227 (1972).
The Mexican sites with the highest number of cloudless days are located in the northwest of Mexico, which probably also contains the best place on the whole planet.

082.234 **Dichteschwankungen der Hochatmosphäre während geomagnetischer Störungen im Höhenbereich 250 bis 800 km.** M. Römer.
Veröff. Astron. Inst. Bonn, No. 85, 74 pp. (1972).
More than 200 density fluctuations following geomagnetic disturbances, derived from drag data of the Explorer 1, Explorer 9, Injun 3, Explorer 17, Explorer 19, and Explorer 24 satellites were used to deduce the reaction time of the upper atmosphere and the amplitude of the density variation. The altitude range from 250 to 800 km is covered by the data in the time interval 1961 through 1966. The results on the time lag are compared in detail with results by Jacchia et al. (1967). In addition to the standard extrema procedure an integral method is used to determine the delay time giving the same result for the average time lag.

082.235 **Condiciones meteorologicas de Lima con relación a la contaminación del aire.** M. Casaverde.
Bol. As. Peruana Astron., Vol. 6, (No. 148), 536 - 539 (1971).

082.236 **A note on Eulerian-Lagrangian time scale transformation for large-scale atmospheric turbulence.**
S. K. Kao.
Journ. Geophys. Res., Vol. 77, 7188 - 7189 (1972).
An expression for the transformation parameter for the Eulerian-Lagrangian time scale transformation is derived for the large-scale atmospheric turbulence through integral time scale approach.

082.237 **Astroclimatic conditions on Crimea deduced from observations at different telescopes at the Crimean Astrophysical Observatory and Southern Station of the Sternberg Institute.** R. E. Gershberg, V. I. Pronik, E. A. Dibaj. Astron. Tsirk., No. 729, p. 1 - 5 (1972). In Russian.

082.238 **Observations météorologiques de Thessaloniki, 1968–1969.** Published by G. C. Livadas. Univ. Thessaloniki, Annuaire Inst. Météorologique et Climatologique, Nos. 36 - 37, 31 + 31 pp. (1972).

082.239 **Meteorological observations of the German weather station in Thessaloniki 1941 - 1944.** G. C. Livadas, A. Arseni-Papadimitriou. Publ. Meteorol.-Climatol. Inst. Univ. Thessaloniki, No. 1, 71 pp. (1972). Tables in French.

082.240 **Programs for the processing of meteorological data by computers.** T. J. Makroyannis, A. Arseni-Papadimitriou. Publ. Meteorol.-Climatol. Inst. Univ. Thessaloniki, No. 2, 15 pp. (1972). In Greek.

082.241 **Results of determination of sky conditions at the Astronomical Observatory of the MPR Academy of Sciences at Khureltogot.** D. Gangbator, N. Tugzhsurehn, D. Khaltar. Astron. Tsirk., No. 740, p. 4 - 6 (1972). In Russian.

082.242 **Variations of the barometric pressure field in the troposphere and in the lower stratosphere after geomagnetic disturbances.** R. F. Usmanov. Nauchn. Informatsii, vyp. (No.) 24, p. 56 - 63 (1972). In Russian.

082.243 **Statistical analysis of the relationship between geomagnetic activity and variations of atmospheric pressure.** V. E. Chertoprud, N. B. Mulyukova. Nauchn. Informatsii, vyp. (No.) 24, p. 64 - 98 (1972). In Russian.

Atmospheric Optics. Vol. 2. See Abstr. 003.064.

The earth: Our physical environment. See Abstr. 003.065.

Electronic excitation of N_2 and dissociative excitation of O_2 by proton impact. See Abstr. 022.050.

Effect of temperature on atomic hydrogen lines excited in the afterglow of water vapour. See Abstr. 022.055.

Measurement of the photodetachment cross section for O_4^- at high pressure. See Abstr. 022.108.

Contribution à la recherche d'un site d'observatoire astrométrique. See Abstr. 041.020.

The mean refraction and periodic atmospherical oscillations. See Abstr. 041.077.

On the determination of the influence of anomalous refraction. See Abstr. 041.078.

Monte-Carlo treatment of Lyman-α radiation in an inhomogeneous spherical atmosphere. See Abstr. 063.032.

The far-infrared and submillimeter background. See Abstr. 155.053.

083   Ionosphere

083.001   Seasonal, diurnal and magnetic dependence of iono-
spheric scintillation at 64° invariant latitude.
J. Aarons, J. P. Mullen, H. E. Whitney, F. Steenstrup.
Planet. Space Sci., Vol. 20, 957 - 964 (1972).

083.002   Electron temperature and density determination
from RF impedance probe measurements in the
lower ionosphere.   R. H. Bishop, K. D. Baker.
Planet. Space Sci., Vol. 20, 997 - 1013 (1972).

083.003 D-region electron densities and collision frequencies
from Faraday rotation and differential absorption
measurements.
F. D. G. Bennett, J. E. Hall, P. H. G. Dickinson.
Journ. Atmosph. Terr. Phys., Vol. 34, 1321 - 1335 (1972).
    A rocket technique for measuring electron density and
collision frequency in the lower ionosphere is described. Absolute values of electron density were determined by a radio
propagation experiment, and detailed height variations were
measured with a Langmuir probe.

083.004   A describing function of the diurnal variation of
$N_m(E)$ for solar zenith angles from 0 to 90°.
L. M. Muggleton.
Journ. Atmosph. Terr. Phys., Vol. 34, 1379 - 1384 (1972).

083.005   Regression-line studies of E-region seasonal anomaly.
L. M. Muggleton.
Journ. Atmosph. Terr. Phys., Vol. 34, 1385 - 1391 (1972).

083.006   Results of a comparison between radar meteor wind
measurements and simultaneous lower ionosphere
drift measurements in the same area.
I. A. Lysenko, Yu. I. Portnyagin, K. Sprenger, K. M. Greisiger,
R. Schminder.
Journ. Atmosph. Terr. Phys., Vol. 34, 1435 - 1444 (1972).

083.007   Electron and positive ion density altitude distribu-
tions in the equatorial D-region.
A. C. Aikin, R. A. Goldberg, Y. V. Somayajulu, M. B.
Avadhanulu.
Journ. Atmosph. Terr. Phys., Vol. 34, 1483 - 1494 (1972).

083.008   Magnetically symmetric detection of the mid-latitude
electron density trough by Ariel 3 satellite.
Y. (Kabasakal) Tulunay.
Journ. Atmosph. Terr. Phys., Vol. 34, 1547 - 1551 (1972).
Short paper.

083.009   Electron energy transfer rates in the ionosphere.
P. Stubbe, W. S. Varnum.
Planet. Space Sci., Vol. 20, 1121 - 1126 (1972).

083.010   Molecular ions in the $F2$ layer.
H. Rishbeth, P. Bauer, W. B. Hanson.
Planet. Space Sci., Vol. 20, 1287 - 1297 (1972).

083.011   Diurnal variation of the H^+ flux between the iono-
sphere and the plasmasphere.
A. F. Nagy, P. M. Banks.
Journ. Geophys. Res., Vol. 77, 4277 - 4279 (1972). – Letter.

083.012   A method for determining the electron density
distribution about the F_2 peak of the ionosphere.
P. L. Dyson.
Australian Journ. Phys., Vol. 25, 293 - 297 (1972).
    A method is given for extending the analysis of topside
ionograms to yield ionospheric electron density profiles down
to and below the peak of the F_2 layer, by analysis of the
ground echoes.

083.013   Ionospheric substorms.
V. M. Driatsky, O. I. Shumilov.
Planet. Space Sci., Vol. 20, (see 012.006), 1375 - 1389
(1972). – Invited paper.

083.014   About global irregularities of the ionosphere.
I. S. Vsekhsvyatskaya, Y. K. Kalinin, N. P. Sergeen-
ko, L. A. Yudovitch.
Geomagn. Aeronom., Vol. 12, 622 - 624 (1972). In Russian.

083.015   N(h)-profiles and the horizontal irregularity of the
ionosphere.   N. P. Danilkin, O. A. Maltzeva.
Geomagn. Aeronom., Vol. 12, 625 - 630 (1972). In Russian.

083.016   Dependence of absorption on frequency. II. Numerical experiment.   G. K. Kalinovskaya.
Geomagn. Aeronom., Vol. 12, 645 - 650 (1972). In Russian.

083.017   Variations of planetary values of the F2-layer thick-
ness and parameters of the neutral atmosphere.
B. S. Shapiro, T. A. Anufrieva.
Geomagn. Aeronom., Vol. 12, 752 - 754 (1972). In Russian.
Brief information.

083.018   Theoretical model of the D-region.
C. F. Sechrist, Jr.
Journ. Atmosph. Terr. Phys., Vol. 34, 1565 - 1589 (1972).
Conference paper (see 012.007).

083.019 E-region model parameters.   W. Swider.
Journ. Atmosph. Terr. Phys., Vol. 34, 1615 - 1626
(1972). – Conference paper (see 012.007).

083.020   A theoretical model of the ionosphere dynamics
with interhemispheric coupling.
H. G. Mayr, E. G. Fontheim, L. H. Brace, H. C. Brinton, H. A.
Taylor, Jr.
Journ. Atmosph. Terr. Phys., Vol. 34, 1659 - 1680 (1972).
Conference paper (see 012.007).

083.021   Auroral ionosphere models.   J. K. Walker.
Journ. Atmosph. Terr. Phys., Vol. 34, 1681 - 1689
(1972). – Conference paper (see 012.007).

083.022   Photoionization and photoabsorption cross sections
for ionospheric calculations.
R. S. Stolarski, N. P. Johnson.
Journ. Atmosph. Terr. Phys., Vol. 34, 1691 - 1701 (1972).
Conference paper (see 012.007).

083.023   Sunrise effects on the latitudinal variations of top-
side ionospheric densities and scale heights.
H. Soicher.
Nature, Phys. Sci., Vol. 239, 93 - 95 (1972).

083.024   Observations of simultaneous auroral D and E layers
with incoherent scatter radar.
R. D. Hunsucker, H. F. Bates, A. E. Belon.
Nature, Phys. Sci., Vol. 239, 102 - 104 (1972).
    We present here electron density profiles ($N_e(h)$) for the
aurorally-disturbed ionosphere in the height range 80 - 150 km
obtained by the incoherent-scatter technique. Simultaneous
data obtained with an approximately co-located vertical iono-

sonde, magnetometer, 30 MHz riometer, and an all-sky camera are used to interpret the $N_e(h)$ data.

083.025 Spatial and temporal variations of the thermal plasma between 3000 and 5700 kilometers at $L = 2$ to 4.
A. B. Bewersdorff, R. C. Sagalyn.
Journ. Geophys. Res., Vol. 77, 4734 - 4745 (1972).

083.026 Power-law wavenumber spectrum deduced from ionospheric scintillation observations.
C. L. Rufenach.
Journ. Geophys. Res., Vol. 77, 4761 - 4772 (1972).

Ionospheric scintillation observations near Boulder, Colorado, from the radio source Cygnus A were spectral analyzed at 26 MHz. The scintillation spectral features are used to calculate the diffraction effects and hence deduce the F-region irregularity wavenumber spectrum for sizes from about 4 to 0.6 km.

083.027 The correlation between X rays and ultraviolet ionizing radiation in the E region by data of the solar eclipse of February 25, 1971.
Tz. Gogosheva, K. Kazakov, K. Serafimov.
Geomagn. Aeronom., Vol. 12, 836 - 842 (1972). In Russian.

083.028 On the theory of sounding measurements in the lower ionosphere. A. A. Jastrebov.
Geomagn. Aeronom., Vol. 12, 864 - 870 (1972). In Russian.

083.029 About a possibility of observing disintegrated interaction of plasma waves in experiments on upper sounding. V. V. Vaskov, G. A. Gusev.
Geomagn. Aeronom., Vol. 12, 924 (1972). In Russian.
Brief information.

083.030 Altitude profiles of the effective number of collisions in the E- and F-regions of the ionosphere.
Sh. G. Shlionsky, T. N. Soboleva, T. Yu. Letshinskaya.
Geomagn. Aeronom., Vol. 12, 925 - 927 (1972). In Russian.
Brief information.

083.031 Space structure of the E_s layer. O. Ovezgeldiev.
Geomagn. Aeronom., Vol. 12, 927 - 928 (1972).
In Russian. – Brief information.

083.032 Distribution of the electron concentration with altitude in the northern hemisphere at the geomagnetic pole (from data of ionosphere sounding from above and from below).
O. P. Kolomiitzev, G. N. Pushkova, L. A. Yudovitch.
Geomagn. Aeronom., Vol. 12, 928 - 931 (1972). In Russian.
Brief information.

083.033 Solar flare effects in the ionosphere observed at Lindau from October 27 to November 2, 1968.
H. Schwentek.
Zeitschr. Geophys., Vol. 36, 125 - 134 (1970). In German.

083.034 Interactions between the ionosphere and the magnetosphere for solar regular daily geomagnetic variations. S. Matsushita.
Gerlands Beiträge Geophys., Vol. 80, 91 - 110 (1971).
Review paper.

083.035 Plasma drifts in the auroral ionosphere derived from barium releases. G. Haerendel.
Astrophys. Space Sci. Library, Vol. 32, (see 012.009), 246 - 257 (1972).

083.036 VLF phenomena. T. R. Kaiser.
Astrophys. Space Sci. Library, Vol. 32, (see

012.009), 340 - 350 (1972).

083.037 Changes of lower ionosphere electron concentrations with solar activity.
E. A. Mechtly, S. A. Bowhill, L. G. Smith.
Journ. Atmosph. Terr. Phys., Vol. 34, 1899 - 1907 (1972).

083.038 Radioastronomical measurements of ionospheric electron content. J. R. Baker.
Journ. Atmosph. Terr. Phys., Vol. 34, 1923 - 1933 (1972).

Observations are presented of the variation with local time of the position angle of celestial polarization at the north pole. The observations show that for radioastronomical studies, if frequent polarization measurements at the pole are incorporated into an observing schedule, then the ionospheric rotation of position angle, and thus the extraionospheric position angles of polarization of celestial sources, may be conveniently determined.

083.039 Ionospheric confirmation of the modulation processes in the plasma around the earth.
G. T. Nestorov.
Comptes Rendus Acad. Bulgar. Sci., (Dokl. Bolg. Akad. Nauk), Vol. 24, 1317 - 1320 (1971).

083.040 Measurement of the ion concentration in the earth's ionosphere at heights from 200 to 6000 km.
O. I. Babkov, E. A. Bashkin, S. I. Borisov, I. A. Dubov, V. D. Nikolaev, E. G. Ul'yanov.
Kosm. Issled., Vol. 10, 746 - 750 (1972). In Russian.

083.041 Latitude-time variations of the total electron number and its gradients in the ionosphere of high latitudes. G. K. Solodovnikov, V. M. Migunov, A. R. Yagovkin, Yu. G. Ivanov.
Kosm. Issled., Vol. 10, 796 - 799 (1972). In Russian. – Brief information.

083.042 High altitude releases of barium vapour using a rubis rocket.
E. Rieger, H. Neuss, R. Lüst, R. Meyer, L. Haser, H. Loidl, J. Stöcker, G. Haerendel.
Ann. Géophys., Vol. 26, 845 - 852 (1970).

Two barium cloud experiments at an altitude of about 2000 km over the Algerian Sahara are described. The neutral barium atoms expanded like spherical shells with velocities of 0.93 and 1.2 km/sec, respectively. They became ionized by the solar UV radiation with a time scale of 16 ± 3 sec. The ion clouds could be observed to their maximum extension of nearly 2000 km and be used to derive inclination and declination of the magnetic field. The sedimentation of the ions at altitudes below 350 km is discussed. The expected variation of the drift velocity with height was observed.

083.043 VLF wave propagation and its interaction with the magnetoplasma. P. K. Shukla, R. N. Singh.
Physica Scripta, (Sweden), Vol. 5, 81 - 89 (1972).

Propagation of low frequency electromagnetic waves in the magnetoplasma is discussed. Dispersion relations for cold, collisional plasmas as well as for a cold beam-plasma system are derived. A mechanism is proposed in terms of wave-particle interactions. It is shown that the VLF signal is amplified by a few orders of magnitude due to resonance interactions. We outline a possible theory for the Cerenkov radiation from a beam of charged particles. The detectibility of VLF waves is also discussed.

083.044 Equatorial scintillation. J. R. Koster.
Planet. Space Sci., Vol. 20, 1999 - 2014 (1972).

In this paper it is proposed to summarize briefly the occurrence characteristics of scintillation effects at Legon, Ghana, and to review what has been experimentally determined concerning the ionospheric irregularities which give rise to the scintillation phenomenon.

083.045 **Observations of ionospheric electron content at medium latitude geomagnetically conjugate stations.**
K. C. Yeh.
Planet. Space Sci., Vol. 20, 2045 - 2050 (1972).

083.046 **Brief review of scintillation studies.**
G. K. Hartmann.
Space Research XII, (see 012.016), Vol. 2, 1221 - 1228 (1972).

083.047 **E region electron density profiles.** K. Maeda.
Space Research XII, (see 012.016), Vol. 2, 1229 - 1240 (1972).

083.048 **Critical survey of electron and ion temperatures measured with probes.** W. Pfister.
Space Research XII, (see 012.016), Vol. 2, 1261 - 1274 (1972).

083.049 **Observed solar geomagnetic control of the ionosphere: Implications for reference ionospheres.**
H. A. Taylor, Jr.
Space Research XII, (see 012.016), Vol. 2, 1275 - 1290 (1972).

083.050 **Flare time models of ionization profiles in the D region.** A. P. Mitra, S. D. Deshpande.
Space Research XII, (see 012.016), Vol. 2, 1291 - 1298 (1972).

083.051 **Ion composition and photochemistry of the E region.** A. D. Danilov.
Space Research XII, (see 012.016), Vol. 2, 1299 - 1304 (1972).

083.052 **Simultaneous rocket measurements of corpuscular and solar electromagnetic ionizing radiation at altitudes up to 150–160 km.** V. F. Tulinov, V. M. Feigin.
Space Research XII, (see 012.016), Vol. 2, 1305 - 1309 (1972).

083.053 **H_2O^+ ions and H_2O molecules in the lower thermosphere and ionosphere.** G. M. Martynkevich.
Space Research XII, (see 012.016), Vol. 2, 1311 - 1314 (1972).

083.054 **Solar activity variation of $[NO^+]/[O_2^+]$ and $[NO^+]/[O^+]$ in E and F regions.**
A. P. Mitra, P. Banerjee.
Space Research XII, (see 012.016), Vol. 2, 1315 - 1320 (1972).

083.055 **Shallow-solar-zenith-angle control of topside ionospheric parameters.** H. Soicher.
Nature, Phys. Sci., Vol. 240, 107 - 109 (1972).

Strong latitudinal variations in topside ionospheric electron densities and scale heights occurred during the overall sunrise period October 28, 1963, to November 21, 1963. Here I report variations in topside electron densities, scale heights, and related parameters at fixed latitudes (45°N, 55°N, 60°N geomagnetic latitude) with solar zenith angle at local sunrise during the time period mentioned.

083.056 **Potential double layers in the ionosphere.**
L. P. Block.
Cosmic Electrodynamics, Vol. 3, 349 - 376 (1972).

In this paper Langmuir's old theory of double layers is reviewed and extended to ionospheric conditions, including effects of gravity and expansion in diverging geomagnetic flux tubes. Bohm's self-consistency condition is refined by including the temperature of the beam particles.

083.057 **Bombardment of the polar-cap ionosphere by solar cosmic rays.** A. J. Zmuda, T. A. Potemra.
Rev. Geophys. Space Phys., Vol. 10, 981 - 991 (1972).

Solar flares often produce bursts of energetic particles (mainly protons but also heavier nuclei and electrons). Comparison of ion and electron concentration profiles in the lower ionosphere (specifically in the D region) with inferences drawn from both ground-based radio data and satellite particle data lead to the conclusion that the ion chemistry at such times is much simpler than during undisturbed times. The diverse methods seem to be compatible in defining major characteristics and reactions of the D region during periods when it is bombarded by solar-flare protons. Thus these events provide the framework for a detailed investigation of D region chemical processes.

083.058 **X-ray emission and D-region "sluggishness".**
B. Valníček, P. Ranzinger.
Bull. Astron. Inst. Czechoslovakia, Vol. 23, 318 - 322 (1972).

By comparing the direct flow of solar X-rays in the 1 - 6 Å range, recorded by the Intercosmos 1 satellite, with the records of atmospherics and other ionospheric effects the value of the sluggishness Δt of the ionospheric D-region was found to vary. This value clearly depends on the spectral composition of the X-rays, so that Δt is small with phenomena displaying larger energies, and Δt increases for lower emission energies and may reach values in excess of 10 minutes.

083.059 **D region lunar variations deduced from long-path 10.2-kilohertz phase measurements.**
V. R. Noonkester.
Journ. Geophys. Res., Vol. 77, 6592 - 6598 (1972).
Brief report.

083.060 **Isis 1 observations of the high-latitude ionosphere during a geomagnetic storm.** J. H. Whitteker,
L. H. Brace, J. R. Burrows, T. R. Hartz, W. J. Heikkila, R. C. Sagalyn, D. M. Thomas.
Journ. Geophys. Res., Vol. 77, 6121 - 6128 (1972).

083.061 **Monte Carlo simulation of D region sampling.**
R. L. Long, Jr., F. W. Vogenitz.
Journ. Geophys. Res., Vol. 77, 6181 - 6193 (1972).

This paper reports the results of calculations made by the Monte Carlo direct simulation technique for rocket payloads designed to measure the neutral composition of the D region with a mass spectrometer.

083.062 **Measurements of $O(^1D)$ quenching rates in the F region.** D. P. Sipler, M. A. Biondi.
Journ. Geophys. Res., Vol. 77, 6202 - 6212 (1972).

Determinations of the quenching of $O(^1D)$ atoms by N_2 and O_2 molecules in the F region have been made from observations of 6300–A night glow intensity enhancements produced by the Platteville 1.6–Mw transmitter.

083.063 **Revised calculations of F region ambient electron heating by photoelectrons.**
W. E. Swartz, J. S. Nisbet.
Journ. Geophys. Res., Vol. 77, 6259 - 6261 (1972).
Brief report.

083.064 **Energetic metastable molecular oxygen as a source of ionization in the D region.**
R. B. Norton, G. C. Reid.
Journ. Geophys. Res., Vol. 77, 6287 - 6290 (1972). – Letter.

083.065 **The light-ion trough, the main trough, and the plasmapause.** H. A. Taylor, Jr., W. J. Walsh.
Journ. Geophys. Res., Vol. 77, 6716 - 6723 (1972).

Extensive observations of midlatitude depletions in electron and total ion density by both direct and indirect techniques have prompted numerous studies of the possible association between these troughs, observed both in the F region and in the topside ionosphere, and the plasmapause. Within this work, one basic problem arises in that, although the plasmapause has been detected as a global phenomenon by both VLF and ion composition measurements, the electron and ion density troughs have been identified primarily as nightside features.

083.066 **Dayglow [OI] λλ 6300 and 5577 Å lines in the early morning ionosphere.**
R. C. Schaeffer, P. D. Feldman, E. C. Zipf.
Journ. Geophys. Res., Vol. 77, 6828 - 6838 (1972).

083.067 **Ion composition and photochemistry of the E region.** A. D. Danilov.
Geomagn. Aeronom., Vol. 12, 1004 - 1008 (1972).
In Russian.

083.068 **Space-time distribution of E_s formations connected with visual forms of aurorae.**
A. S. Besprozvannaya, T. I. Shchuka.
Geomagn. Aeronom., Vol. 12, 1009 - 1012 (1972).
In Russian.

083.069 **Some questions on the method of determining the parameters of dispersing irregularities.**
V. D. Gusev, Nguen Bik Lan.
Geomagn. Aeronom., Vol. 12, 1013 - 1019 (1972).
In Russian.

083.070 **Fluctuations of the level of ion density in the earth's ionosphere at heights between 200 and 1300 km.**
S. I. Borisov, V. D. Nikolaev.
Kosmich. Issled., Vol. 10, 892 - 896 (1972). In Russian.

083.071 **On the possibility of radar observations of non-linear interaction between waves in the ionospheric plasma.**
G. A. Gusev, Yu. V. Kushnerevskij.
Kosmich. Issled., Vol. 10, 950 (1972). In Russian. – Brief information.

083.072 **Solar cycle control of the ionospheric E-region.**
M. Bossolasco, A. Elena.
Gerlands Beiträge Geophys., Vol. 81, 403 - 406 (1972).
The index of the daily and annual variations of f_0E at mid-latitudes behaves somewhat irregular during the last sunspot minimum. The winter anomaly of the E-region becomes maximum at sunspot minimum. These irregularities are studied in relation to the solar ionizing radiation and aeronomical parameters.

083.073 **On the vertical motion of the lower ionosphere at a solar eclipse.** B. A. Turkeeva.
Trudy Sektora ionosfery. AN KazSSR, Vol. 3, 74 - 77 (1972).
In Russian. – Abstr. in Referativ. Zhurn. 51. Astron., 1.51.470 (1973).

083.074 **Ionospheric D region, a sensitive detector of hard X-rays of solar subflares.** G. T. Nestorov.
Comptes Rendus Acad. Bulg. Sci., Vol. 25, 317 - 320 (1972).

Royal Astronomical Society meeting on the D and E regions of the ionosphere. See Abstr. 011.006.

Waves and resonances in magneto-active plasma. See Abstr. 062.054.

Stable electron density fluctuations in a plasma in the presence of a high-frequency electric field. See Abstr. 062.060.

Two X-ray bursts (1 August 1967 and 30 January 1968) and some associated VLF disturbances. See Abstr. 076.002.

The solar X-ray spectrum deduced from a proportional counter experiment and the resultant production of ionization in the mesosphere. See Abstr. 076.006.

Analysis of OGO 6 observations of the O I 5577-Å tropical nightglow. See Abstr. 082.009.

The development of a theoretical model of the atmosphere and the ionosphere. See Abstr. 082.059.

Further comments on a parallel study of 6300 Å airglow emission and ionospheric scintillation. See Abstr. 082.109.

A theory on the latitude and local time distribution of precipitating electrons during a sudden commencement. See Abstr. 084.281.

Recent satellite measurements of the morphology and dynamics of the plasmasphere. See Abstr. 084.320.

Possibility of continuous monitoring of celestial X-ray sources through their ionization effects in the nocturnal D-region ionosphere. See Abstr. 142.089.

084 Aurorae, Geomagnetic Field, Radiation Belts

Aurorae

084.001 Fermi acceleration of auroral particles.
J. R. Sharber, W. J. Heikkila.
Journ. Geophys. Res., Vol. 77, 3397 - 3410 (1972).

084.002 ELF noise bands associated with auroral electron precipitation. D. A. Gurnett, L. A. Frank.
Journ. Geophys. Res., Vol. 77, 3411 - 3417 (1972).

084.003 Observations of the auroral oval by the Alaskan meridian chain of stations.
A. L. Snyder, S.-I. Akasofu.
Journ. Geophys. Res., Vol. 77, 3419 - 3430 (1972).
   All-sky photographs from the chain of stations between the corrected geomagnetic latitudes of 60° and 85°N have provided and clarified several important features of auroral morphology, in particular the dynamics of the night-sector auroral oval.

084.004 Satellite and ground-based observations of a red arc.
A. F. Nagy, W. B. Hanson, R. J. Hoch, T. L. Aggson.
Journ. Geophys. Res., Vol. 77, 3613 - 3617 (1972). – Letter.

084.005 The $H\beta/N_2^+$ 4709-Å intensity ratio and vibrational enhancement of N_2^+ first negative bands in proton-excited auroras. V. Degen, A. E. Belon, G. J. Romick.
Journ. Geophys. Res., Vol. 77, 3618 - 3620 (1972). – Letter.

084.006 The brilliant aurora on June 17 - 18. R. L. Berry.
Sky Telescope, Vol. 44, 126 - 129 (1972).

084.007 Formation of auroral patches in the midday sector during a substorm.
A. L. Snyder, J. Buchau, S.-I. Akasofu.
Planet. Space Sci., Vol. 20, 1116 - 1119 (1972). – Research note.

084.008 Spectroscopic studies of the arc of March 8–9, 1970. G. Hernandez.
Planet. Space Sci., Vol. 20, 1309 - 1321 (1972).

084.009 Auroral photometric observations at geomagnetically conjugate points. A. L. Spitz, M. D. Watson.
Planet. Space Sci., Vol. 20, 1337 - 1343 (1972).

084.010 Photometric observations of the March aurora.
B. S. Dandekar, D. J. Davis, Jr.
Sky Telescope, Vol. 44, 203 (1972).

084.011 Dispersive auroral hiss. J. C. Siren.
Nature, Phys. Sci., Vol. 238, 118 - 119 (1972).

084.012 Auroral emissions and particle precipitation in the noon sector. W. J. Heikkila, J. D. Winningham, R. H. Eather, S.-I. Akasofu.
Journ. Geophys. Res., Vol. 77, 4100 - 4115 (1972).

084.013 Local-time survey of plasma at low altitudes over the auroral zones. L. A. Frank, K. L. Ackerson.
Journ. Geophys. Res., Vol. 77, 4116 - 4127 (1972).

084.014 An evaluation of the intensity of Cerenkov radiation from auroral electrons with energies down to 100 ev.
T. L. Lim, T. Laaspere.
Journ. Geophys. Res., Vol. 77, 4145 - 4157 (1972).

084.015 Alignment of auroral arcs. S.-I. Akasofu, D. S. Kimball, J. Buchau, R. W. Gowell.
Journ. Geophys. Res., Vol. 77, 4233 - 4236 (1972).
Brief report.

084.016 Auroral morphology. T. N. Davis.
Planet. Space Sci., Vol. 20, (see 012.006), 1369 - 1373 (1972). – Invited paper.

084.017 Ray structures of aurorae and their connection with the instability of a drift stream in a plasma cluster.
I. N. Menshutina.
Geomagn. Aeronom., Vol. 12, 613 - 617 (1972). In Russian.

084.018 About the nature of pulsations of aurorae intensity luminescence connected with geomagnetic pulsations of Pi2 type.
V. K. Koshelevsky, O. M. Raspopov, V. K. Roldugin.
Geomagn. Aeronom., Vol. 12, 618 - 621 (1972). In Russian.

084.019 Mean daily variation of aurorae energy radiation in the region of the Tiksi bay. N. I. Dzubenko.
Geomagn. Aeronom., Vol. 12, 746 - 748 (1972). In Russian.
Brief information.

084.020 Longitudinal currents and auroral electrojet.
L. L. Van'yan, A. S. Debabov, I. L. Osipova.
Kosmich. Issled., Vol. 10, 628 - 629 (1972). In Russian.
Brief information.

084.021 VHF power spectra of the radar aurora.
B. B. Balsley, W. L. Ecklund.
Journ. Geophys. Res., Vol. 77, 4746 - 4760 (1972).

084.022 Extreme ultraviolet emissions from an aurora.
F. Paresce, M. Lampton, J. Holberg.
Journ. Geophys. Res., Vol. 77, 4773 - 4781 (1972).

084.023 Simultaneity of appearance of field and aurora pulsations.
B. N. Kazak, V. K. Roldugin, S. A. Chernous.
Geomagn. Aeronom., Vol. 12, 941 - 944 (1972). In Russian.
Brief information.

084.024 Dependence of aurora-ray length on the level of aurora activity. N. N. Blisnyuk, N. I. Dzyubenko.
Geomagn. Aeronom., Vol. 12, 944 - 945 (1972). In Russian.
Brief information.

084.025 Raketenexperiment zur Untersuchung von Nordlichtern. Meßergebnisse des Protonendetektors EI 101.
E. Kirsch.
Zeitschr. Geophys., Vol. 36, 165 - 173 (1970).

084.026 Proton measurements in the morning sector of the auroral zone during slowly varying cosmic noise absorption. H. Raethjen.
Zeitschr. Geophys., Vol. 37, 195 - 210 (1971). In German.

084.027 Comment on ion-acoustic waves in the auroral plasma. R. S. Unwin, F. B. Knox.
Canadian Journ. Phys., Vol. 49, 848 (1971).
   Correction is made to a statement concerning a deduction about the cause of radio aurora. A possible way is suggested of reconciling the conflict of evidence for and against the occurrence of the ion-acoustic instability in the evening.

084.028 **Power spectral analyses of auroral light and X-ray pulsations.** M. W. J. Scourfield, N. R. Parsons.
Canadian Journ. Phys., Vol. 49, 2195 - 2201 (1971).

084.029 **New results on particle arrival at the polar caps.**
D. E. Page, V. Domingo.
Astrophys. Space Sci. Library, Vol. 32, (see 012.009), 107 - 114 (1972).

084.030 **Auroral particle precipitation patterns.**
W. Riedler.
Astrophys. Space Sci. Library, Vol. 32, (see 012.009), 133 - 140 (1972).

084.031 **Electron intensities over auroral arcs.**
D. A. Bryant, G. M. Courtier, G. Bennett.
Astrophys. Space Sci. Library, Vol. 32, (see 012.009), 141 - 146 (1972).

084.032 **The pre-midnight asymmetry in the 40 keV electron flux profiles and its relation to magnetospheric substorms.** L. Rossberg.
Astrophys. Space Sci. Library, Vol. 32, (see 012.009), 147 - 152 (1972).

084.033 **Angular distributions of precipitating electrons.**
G. Paschmann.
Astrophys. Space Sci. Library, Vol. 32, (see 012.009), 168 - 174 (1972).

084.034 **Photometric auroral particle measurements.**
S. B. Mende, R. H. Eather.
Astrophys. Space Sci. Library, Vol. 32, (see 012.009), 179 - 186 (1972).

084.035 **Acceleration of auroral particles by electric double layers.** L. P. Block.
Astrophys. Space Sci. Library, Vol. 32, (see 012.009), 258 - 267 (1972).

084.036 **X-ray observations and interpretations.**
G. R. Pilkington.
Astrophys. Space Sci. Library, Vol. 32, (see 012.009), 391 - 399 (1972).

084.037 **Model for the uneven illumination of polar caps by solar protons.**
R. Gall, S. Bravo, A. Orozco.
Journ. Geophys. Res., Vol. 77, 5360 - 5373 (1972).
A model for the uneven illumination of polar caps by solar proton fluxes of 5–300 Mev is proposed. It is based on observations and on the assumption that cosmic rays move with the direct mode in the geomagnetic field. The model applies to the anisotropic phase of the proton event.

084.038 **Rocket measurements of electron influx during a major magnetic storm with type A aurora.**
D. J. McEwen, G. G. Sivjee.
Journ. Geophys. Res., Vol. 77, 5523 - 5529 (1972).

084.039 **A possible method for estimating any indirect process in the production of the O(^{1}S) atoms in aurora.**
A. Brekke, H. Pettersen.
Planet. Space Sci., Vol. 20, 1569 - 1576 (1972).

084.040 **Spatial coherency in pulsating aurora.**
M. W. J. Scourfield, W. F. Innes, N. R. Parsons.
Planet. Space Sci., Vol. 20, 1843 - 1848 (1972).

084.041 **A dual wavelength ground-based auroral scanner.**
W. Sawchuk, C. D. Anger.

Planet. Space Sci., Vol. 20, 1935 - 1940 (1972).
A scanner capable of observing in two separate wavelengths the auroral emissions of the complete night sky hemisphere is described. This instrument can detect auroras from subvisual to IBC IV. Data are displayed in pictorial form, recorded on magnetic tape and later processed by computer.

084.042 **Local time dependence of auroral zone electron spectra.** G. A. Kuck.
Ann. Géophys., Vol. 26, 689 - 695 (1970).
The energy spectra from 10 to 100 KeV of quasi-trapped and precipitating electrons have been measured by a polar orbiting satellite. Spectral parameters were obtained for 15 electron events. These events indicate that the electron spectra are softer before local midnight than after local midnight. More events were seen at higher K_p than lower K_p, indicating that the intensity of the events depended upon K_p. The satellite results were in qualitative agreement with balloon measurements of X-rays caused by precipitating particles.

084.043 **Metastable oxygen ions distribution and related optical emission in the aurora.** J.-C. Gérard.
Ann. Géophys., Vol. 26, 777 - 781 (1970).
The importance of metastable species $O_2^+ : a^4\Pi_u$, $O^{+2}D°$ and $^2P°$ and related line intensities are discussed. The results are illustrated for a typical IBC II aurora and the 7319–7330 Å ($^2D° - {}^2P°$) and 3726–3729 Å ($^4S° - {}^2D°$) emission rates compared. The efficiency of the charge transfer from $O^+({}^2D°)$ to O (^{3}P) is evaluated in normal auroral conditions and its contribution for the excitation of the λ 6300 line estimated at less than 1 per cent between ionization peak and 300 kilometers.

084.044 **Observations d'une aurore de basse-latitude.**
Nguyen-Huu-Doan.
Ann. Géophys., Vol. 26, 783 - 789 (1970).
A low latitude aurora has been observed at the Haute-Provence Observatory (latitude 43°56') during the night from the 8th to the 9th of March 1970. Its spectrum is represented by very intense bands and line of N_2^+ [OI], [OII] and [NI]. The hydrogen lines of the Balmer series seem to be absent while the violet helium line is present. The non intensified lines are the yellow line of NaI and the green line of [OI].

084.045 **Enregistrement spectrophotographique d'émissions ultraviolettes aurorales durant un vol de fusée-sonde.**
J.-M. Vreux.
Ann. Géophys., Vol. 27, 493 - 498 (1971).
Spectrophotographic recording of ultra-violet auroral emissions has been taken during a rocket flight by means of an off-plane inverse Wadsworth mounting. It has been shown that recorded intensities in the first negative systeme of N_2^+ and second positive system of N_2 agree with an excitation by electron impact on molecular nitrogen in the zero vibrational level of the fundamental state.

084.046 **Energy spectra in relativistic electron precipitation events.**
T. J. Rosenberg, L. J. Lanzerotti, D. K. Bailey, J. D. Pierson.
Journ. Atmosph. Terr. Phys., Vol. 34, 1977 - 1990 (1972).

084.047 **Electron excitation and auroral emission parameters.**
R. J. R. Judge.
Planet. Space Sci., Vol. 20, 2081 - 2092 (1972).

084.048 **The morphology of auroral particle precipitation.**
W. J. Heikkila.
Space Research XII, (see 012.016), Vol. 2, 1343 - 1355 (1972).

084.049 **Electron density and temperature profiles during a moderate auroral event.**

J. J. Berthelier, R. Godard.
Space Research XII, (see 012.016), Vol. 2, 1357 - 1367 (1972).

084.050 **Rocket measurements of low energy electrons during auroral events.** D. J. McEwen.
Space Research XII, (see 012.016), Vol. 2, 1385 - 1390 (1972).

084.051 **Evidence of a systematic variation of the ratio between the intensities of the auroral emissions at 5577 Å and 3914 Å.** A. Monfils, J. M. Vreux.
Space Research XII, (see 012.016), Vol. 2, 1391 - 1395 (1972).

084.052 **Comparison of 4278 Å N_2^+ emission and low energy electrons observed from the satellite Aurorae.**
G. Gustafsson, A. Egeland, C. S. Deehr.
Space Research XII, (see 012.016), Vol. 2, 1405 - 1411 (1972).

084.053 **Proton energy spectra from recent rocket measurements in the night and morning time auroral zone.**
H. Raethjen.
Space Research XII, (see 012.016), Vol. 2, 1449 - 1457 (1972).

084.054 **Angular distributions of auroral electrons in the energy range 0.8 to 16 kev.** G. Paschmann,
R. G. Johnson, R. D. Sharp, E. G. Shelley.
Journ. Geophys. Res., Vol. 77, 6111 - 6120 (1972).

084.055 **Auroral X-ray and conjugate ionospheric absorption observations of an electron precipitation event accompanying a sudden impulse in the geomagnetic field.**
J. R. Barcus, R. R. Brown, R. H. Karas, T. J. Rosenberg,
H. Trefall, K. Brønstad, M. Kodama.
Journ. Geophys. Res., Vol. 77, 6294 - 6297 (1972). – Letter.

084.056 **Interferometric measurements of Doppler temperature from λ 6300 Å width in aurorae.**
V. M. Ignatiev, V. A. Yugov, V. N. Alekseev, K. V. Atlasov.
Byull. Abastumansk. Astrofiz. Obs., No. 42, (see 012.019),
p. 91 - 96 (1972). In Russian.

084.057 **Low intensity auroral red arcs.**
A. V. Jones, R. L. Gattinger.
Ann. Géophys., Vol. 28, 85 - 89 (1972). – Paper presented at
I.A.G.A. symposium on 'Aurora and airglow', Moscow, August 1971.

084.058 **The intensity ratios of auroral emission features.**
R. L. Gattinger, A. V. Jones.
Ann. Géophys., Vol. 28, 91 - 97 (1972). – Paper presented at
I.A.G.A. symposium on 'Aurora and airglow', Moscow, August 1971.

084.059 **Auroral spectroscopy and excitation.**
G. G. Shepherd.
Ann. Géophys., Vol. 28, 99 - 107 (1972). – Paper presented
at I.A.G.A. symposium on 'Aurora and airglow', Moscow, August 1971.

084.060 **Auroral spectra recorded between 2000 and 3000 Å with a fast scanning spectrometer.**
R. Duysinx, A. Monfils.
Ann. Géophys., Vol. 28, 109 - 110 (1972). – Paper presented
at I.A.G.A. symposium on 'Aurora and airglow', Moscow, August 1971.

084.061 **The radio aurora.** R. S. Unwin, W. J. Baggaley.
Ann. Géophys., Vol. 28, 111 - 127 (1972). – Paper
presented at I.A.G.A. symposium on 'Aurora and airglow',
Moscow, August 1971.

084.062 **Oxygen emission at 6300 Å.** Yu. L. Truttse.

Ann. Géophys., Vol. 28, 169 - 177 (1972). – Paper
presented at I.A.G.A. symposium on 'Aurora and airglow',
Moscow, August 1971.

084.063 **The excitation of O_2 in auroras.**
D. C. Cartwright, S. Trajmar, W. Williams.
Ann. Géophys., Vol. 28, 397 - 401 (1972). – Paper presented
at I.A.G.A. symposium on 'Aurora and airglow', Moscow, August 1971.

084.064 **The IR emission spectrum of N_2 excited under auroral conditions.**
D. C. Cartwright, W. Williams, S. Trajmar.
Ann. Géophys., Vol. 28, 403 - 407 (1972). – Paper presented
at I.A.G.A. symposium on 'Aurora and airglow', Moscow, August 1971.

084.065 **The infrared spectrum of nitrogen excited by fast electrons.**
N. P. Danilevskii, L. I. Popova, A. G. Koval, N. I. Fedorova,
V. T. Koppe, Ya. M. Fogel.
Ann. Géophys., Vol. 28, 409 - 414 (1972). – Paper presented
at I.A.G.A. symposium on 'Aurora and airglow', Moscow, August 1971.

084.066 **Auroral morphology.** C. S. Deehr, A. Egeland.
Ann. Géophys., Vol. 28, 415 - 425 (1972). – Paper
presented at I.A.G.A. symposium on 'Aurora and airglow',
Moscow, August 1971.

084.067 **Variations of the auroral electron energy spectra during substorms.** V. E. Ivanov, G. V. Starkov.
Ann. Géophys., Vol. 28, 427 - 431 (1972). – Paper presented
at I.A.G.A. symposium on 'Aurora and airglow', Moscow, August 1971.

084.068 **Periodically varying radio aurora.**
D. R. McDiarmid, A. G. McNamara.
Ann. Géophys., Vol. 28, 433 - 441 (1972). – Paper presented
at I.A.G.A. symposium on 'Aurora and airglow', Moscow, August 1971.

084.069 **Precipitation patterns in the arctic ionosphere determined from airborne observations.**
J. Buchau, G. J. Gassmann, C. P. Pike, R. A. Wagner, J. A. Whalen.
Ann. Géophys., Vol. 28, 443 - 453 (1972). – Paper presented
at I.A.G.A. symposium on 'Aurora and airglow', Moscow, August 1971.

084.070 **L'aurore du 8 mars 1970 à Kerguelen et Sogra.**
M. Fehrenbach, V. K. Roldugin.
Ann. Géophys., Vol. 28, 473 (1972). – Paper presented at
I.A.G.A. symposium on 'Aurora and airglow', Moscow, August 1971.

084.071 **Auroral helium precipitation.**
W. I. Axford, F. Bühler, H. J. A. Chivers, P. Eberhardt, J. Geiss.
Journ. Geophys. Res., Vol. 77, 6724 - 6730 (1972).
   We have adapted the metal foil sampling technique to the problem of measuring the fluxes of helium and neon in the auroral primary radiation. We have only detected ^{4}He, but with a more sensitive mass spectrometer it should be possible to detect ^{3}He.

084.072 **On the distinction between the auroral electrojet and partial ring current systems.**
N. U. Crooker, R. L. McPherron.
Journ. Geophys. Res., Vol. 77, 6886 - 6889 (1972).
   This analysis allows for interpretation of some systematic

behavior of disturbance parameters presented by Crooker (1972). Furthermore, it is shown that observations of auroral positive and low-latitude negative bays are consistent with two separate current systems.

084.073 **Behavior of aurora in Germany, 1880–1964.**
   W. Schröder.
Journ. Geophys. Res., Vol. 77, 6890 - 6892 (1972). — Letter.

084.074 **Comments on paper by J. R. Sharber and W. J. Heikkila, 'Fermi acceleration of auroral particles'**
[Journ. Geophys. Res., Vol. 77, 3397 - 3410 (1972)].
G. P. Haskell, D. J. Southwood, with a reply by J. R. Sharber, W. J. Heikkila.
Journ. Geophys. Res., Vol. 77, 6926 - 6927, 6928 - 6929 (1972).

084.075 **Modulation of auroral electron streams and geomagnetic pulsations during the storm of March 8, 1970.** J. Gasset, I. A. Zhulin, F. Chambou, Kh. D. Kanonidi, N. G. Kleimenova, O. M. Raspopov, A. Saint-Marc, J.-P. Treilhou.
Geomagn. Aeronom., Vol. 12, 1059 - 1066 (1972).
In Russian.

Summertime observations of sunspots and auroras.
Sky Telescope, Vol. 44, 333 - 339 (1972).

Aurorae and processes in the magnetosphere.
See Abstr. 003.140.

Solar particle injection at medium energies
$(25 < E < 250 \text{ MeV})$.   See Abstr. 078.022.

OI and NI allowed transitions in the airglow and aurora.   See Abstr. 082.186.

Rocket measurements of low energy electrons and optical emissions in the dayglow and aurora.
See Abstr. 082.193.

Geomagnetic Field

084.201 On the state of the geomagnetic field and its reversals. E. H. Levy.
Astrophys. Journ., Vol. 175, 573 - 581 (1972).

Stationary solutions of the kinematic-dynamo equations are discussed when the fluid velocity consists of a spherical shell of nonuniform rotation and two or three pairs of rings of cyclonic convective cells. In addition the reverse toroidal magnetic flux produced by these dynamo fields is discussed in terms of its relevance to the phenomenon of geomagnetic reversal. We estimate that some stationary dipole modes are unstable to a change in sign by fluctuations of 20 or 30 percent in the distribution of cells of cyclonic convection.

084.202 Heat conduction in a turbulent magnetic field, with application to solar-wind electrons.
J. V. Hollweg, J. R. Jokipii.
Journ. Geophys. Res., Vol. 77, 3311 - 3316 (1972).

We consider random, long-wavelength fluctuations in a turbulent magnetic field and show that they can appreciably decrease the heat conductivity of a plasma along the magnetic field. Application to solar-wind electrons indicates that this reduction in heat conductivity due to observed fluctuations in the interplanetary magnetic field may be of the order of a factor of 2.

084.203 Explorer 33 and 35 plasma observations of magnetosheath flow.
H. C. Howe, Jr., J. H. Binsack.
Journ. Geophys. Res., Vol. 77, 3334 - 3344 (1972).

Simultaneous plasma data are used to map the magnetosheath flow pattern in the region $-20 \, R_E \geq X_{SE} \geq -60 \, R_E$. The effects of temporal variations in the solar wind are removed by normalizing each magnetosheath measurement to a simultaneous solar-wind measurement.

084.204 Changes in magnetospheric configuration during the substorm growth phase.
F. V. Coroniti, C. F. Kennel.
Journ. Geophys. Res., Vol. 77, 3361 - 3370 (1972).

084.205 Anisotropy of high-latitude electron fluxes during substorms and structure of the magnetotail.
I. B. McDiarmid, A. Hruška.
Journ. Geophys. Res., Vol. 77, 3377 - 3383 (1972).

084.206 Conjugate features of magnetospheric electron dynamics observed at balloon altitudes. J. R. Barcus.
Space Science Rev., Vol. 13, 295 - 312 (1972). – Conference paper (see 012.002).

084.207 Satellite measurements of high latitude convection electric fields. D. P. Cauffman, D. A. Gurnett.
Space Science Rev., Vol. 13, 369 - 410 (1972).

This paper reviews the first results of satellite experiments to measure magnetospheric convection electric fields using the double-probe technique. The earliest successful measurements were made with Injun-5 spacecraft. The Injun-5 results are compared with the initial findings of the electric field experiment on OGO-6 satellite. The implications of the electric field measurements for magnetospheric and auroral structure are summarized, and a list of specific recommendations for improving future experiments is presented.

084.208 Fluctuating magnetic fields in the magnetosphere. II. ULF waves.
R. L. McPherron, C. T. Russell, P. J. Coleman, Jr.
Space Science Rev., Vol. 13, 411 - 454 (1972).

We consider separately initial observations, magneto-spheric wave phenomena occurring during quiet times, during magnetospheric substorms, and during magnetic storms, and discuss the measurement of the transfer function of a field line. Finally, the techniques used in the analysis of ULF waves are described.

084.209 Nonlinear theory of the monochromatic circularly polarized VLF and ULF waves in the magnetosphere. N. I. Bud'ko, V. I. Karpman, O. A. Pokhotelov.
Cosmic Electrodynamics, Vol. 3, 147 - 164, 165 - 183 (1972). In English and Russian.

The interaction of resonant particles with a monochromatic circularly polarized wave propagating along the magnetic field in a homogeneous plasma is investigated.

084.210 Radial penetration of a hot plasma associated with a large-scale electric field in the magnetosphere, and some related problems. T. Tamao.
Planet. Space Sci., Vol. 20, 973 - 996 (1972).

The inward penetration of a hot plasma as an energy accumulation process for magnetospheric substorms, and some related topics are discussed: I. Induced electric field in the magnetosphere due to solar wind disturbances; II. Earthward penetration of hot plasma associated with the enhancement of the large-scale electric field in the magnetosphere; III. Plasma instabilities, stochastic penetration, and acceleration in the inhomogeneous magnetosphere; IV. Runaway electrons in the topside ionosphere as a possible source of the magnetospheric particles; V. Critical discussions and further studies for magnetospheric substorms.

084.211 Persistent particle anisotropies and magnetospheric models. M. Scholer, G. Morfill.
Planet. Space Sci., Vol. 20, 1051 - 1059 (1972).

084.212 Signatures for substorm development of the growth phase and expansion phase.
T. Iijima, T. Nagata.
Planet. Space Sci., Vol. 20, 1095 - 1112 (1972).

084.213 Lunar magnetic variations at Trelew (Argentina). H. R. Affolter, O. Schneider.
Journ. Atmosph. Terr. Phys., Vol. 34, 1349 - 1356 (1972).

084.214 Sudden impulses in the geomagnetotail and the vicinity. V. L. Patel.
Planet. Space Sci., Vol. 20, 1127 - 1136 (1972).

084.215 On the physical mechanism of the magnetospheric substorm development. A. P. Kropotkin.
Planet. Space Sci., Vol. 20, 1245 - 1257 (1972).

084.216 Kinematic dynamos and geomagnetism. D. Gubbins.
Nature, Phys. Sci., Vol. 238, 119 - 122 (1972).

084.217 Erdmagnetische Variationen. H. Volland.
SuW, Vol. 11, 228 - 232 (1972).

084.218 Reconnection of the geomagnetic tail deduced from solar-particle observations.
G. Morfill, M. Scholer.
Journ. Geophys. Res., Vol. 77, 4021 - 4026 (1972).

The paper is concerned with the structure of the geomagnetic tail as deduced from observations of energetic solar particles inside the magnetosphere.

084.219 Plasma sheet at lunar distance: Structure and solar-wind dependence. A. Nishida, E. F. Lyon.
Journ. Geophys. Res., Vol. 77, 4086 - 4099 (1972).

Explorer 35 observation of low-energy (0.1–3 kev) electrons in the distant geomagnetic tail at $60 R_E$ is analyzed and compared with the solar-wind conditions monitored simultaneously by Explorer 33. The flux is correlated with the solar-wind dynamic pressure, reflecting the dynamic balance between the solar wind and the geomagnetic tail.

084.220 Spectrum of the geomagnetic activity index Ap.
A. C. Fraser-Smith.
Journ. Geophys. Res., Vol. 77, 4209 - 4220 (1972).

084.221 Index of the activity of PcI geomagnetic pulsations ("pearls"). E. T. Matveyeva.
Solnechnye Dannye 1972 Byull., No. 5, p. 108 - 112 (1972). In Russian.
The P-index of the activity of PcI pulsations is discussed. This index is proportional to the logarithm of hydromagnetic waves energy produced in the magnetosphere in the PcI range. Tables of the P-index are given for several years.

084.222 Plasma sheet variations during substorms.
E. W. Hones, Jr.
Planet. Space Sci., Vol. 20, (see 012.006), 1409 - 1431 (1972). – Invited paper.

084.223 The behavior of low-energy particles during substorms. R. D. Sharp, R. G. Johnson.
Planet. Space Sci., Vol. 20, (see 012.006), 1433 - 1442 (1972). – Invited paper.

084.224 Polar magnetic substorms. N. Fukushima.
Planet. Space Sci., Vol. 20, (see 012.006), 1443 - 1454 (1972). – Invited paper.

084.225 Magnetic field fluctuations during substorms.
D. H. Fairfield.
Planet. Space Sci., Vol. 20, (see 012.006), 1455 - 1474 (1972). – Invited paper.

084.226 Electric field variations during substorms: OGO-6 measurements. J. P. Heppner.
Planet. Space Sci., Vol. 20, (see 012.006), 1475 - 1498 (1972). – Invited paper.

084.227 Micropulsations and VLF emissions during substorms. V. A. Troitskaya, N. G. Kleimenova.
Planet. Space Sci., Vol. 20, (see 012.006), 1499 - 1519 (1972). – Invited paper.

084.228 Substorm related changes in the geomagnetic tail: The growth phase. R. L. McPherron.
Planet. Space Sci., Vol. 20, (see 012.006), 1521 - 1539 (1972). – Invited paper.

084.229 Noise in the geomagnetic tail. C. T. Russell.
Planet. Space Sci., Vol. 20, (see 012.006), 1541 - 1553 (1972). – Invited paper.

084.230 Observations au Portugal de phénomènes lumineux se rapportant à une expérience de lacher de barium dans la magnétosphère. E. Van Hemelrijck, H. Debehogne.
Ciel et Terre, Vol. 88, 292 - 297 (1972).

084.231 Evidence of a geomagnetic excursion 30,000 yr BP.
M. Barbetti, M. McElhinny.
Nature, Vol. 239, 327 - 330 (1972).

084.232 Secular variations of the geomagnetic field by archeomagnetic and paleomagnetic data.
S. P. Burlatzkaya.
Geomagn. Aeronom., Vol. 12, 662 - 675 (1972). In Russian.

084.233 The description of the main geomagnetic field.
L. O. Turmina.
Geomagn. Aeronom., Vol. 12, 676 - 687 (1972). In Russian.

084.234 About current instability in the neutral layer of the tail of the earth's magnetosphere.
O. A. Pokhotelov.
Geomagn. Aeronom., Vol. 12, 693 - 698 (1972). In Russian.

084.235 The families of geomagnetic storms, direction of the interplanetary magnetic field and solar activity.
V. I. Afanasieva.
Geomagn. Aeronom., Vol. 12, 712 - 719 (1972). In Russian.

084.236 About the influence of the interplanetary magnetic field on geomagnetic activity. R. N. Kulieva.
Geomagn. Aeronom., Vol. 12, 770 - 771 (1972). In Russian.
Brief information.

084.237 The influence of the interplanetary magnetic field on the characteristics of Pc2-4 type pulsations.
T. A. Plyasova-Bakunina.
Geomagn. Aeronom., Vol. 12, 772 - 774 (1972). In Russian.
Brief information.

084.238 On the influence of the sector structure of the interplanetary magnetic field upon the field component normal to the ecliptic plane. M. S. Bobrov.
Astron. Tsirk., No. 705, p. 1 - 3 (1972). In Russian.

084.239 Solar activity and the variation of the geomagnetic k_p-index.
L. Oster, J. T. Mariska, M. D. Altschuler, D. E. Trotter.
Bull. American Astron. Soc., Vol. 4, 389 (1972). – Abstr. AAS.

084.240 Plasma in Labor und Weltraum. R. Sagdejew.
Bild der Wiss., 9. Jahrgang, p. 818 - 827 (1972).
The author gives an outlook on cosmic experiments carried out in a laboratory which try to imitate the earth's magnetic field.

084.241 Parametric excitation of electromagnetic waves.
D. W. Forslund, J. M. Kindel, E. L. Lindman.
Phys. Rev. Letters, Vol. 29, 249 - 252 (1972).
Electromagnetic waves propagating along a dc magnetic field are shown to excite parametrically decay, purely growing, modulational and beat wave instabilities. Particular attention is given to whistler parametric instabilities, including a discussion of their nonlinear development and saturation.

084.242 Daily variation of electron and proton geomagnetic cutoffs calculated for Fort Churchill, Canada.
D. F. Smart, M. A. Shea.
Journ. Geophys. Res., Vol. 77, 4595 - 4601 (1972).

084.243 Magnetohydrodynamic theory for the interaction of an interplanetary double-shock ensemble with the earth's bow shock. W.-W. Shen, M. Dryer.
Journ. Geophys. Res., Vol. 77, 4627 - 4644 (1972).
The interactions between interplanetary shocks and the earth's bow shock are formulated first on a gas-dynamic basis. The problem is repeated on a magnetohydrodynamic basis for the special case of perpendicular shocks. Interaction dynamics are applied, in the case of the geomagnetic storm's sudden commencement, to the characteristic time and energy of the instantaneous compression. Several magnetic-storm sudden-commencement events (including the March 8, 1970, event) are discussed on the basis of spacecraft plasma observations and magnetogram records on the earth.

084.244 **Simultaneous solar-wind plasma and magnetic-field measurements in the expected region of the extended geomagnetic tail.**
D. S. Intriligator, J. H. Wolfe, D. D. McKibbin.
Journ. Geophys. Res., Vol. 77, 4645 - 4649 (1972).

A comparison is made of simultaneous plasma and magnetic-field measurements obtained by Pioneer 7 in the expected region of the geomagnetic tail at $\sim 1000\,R_E$. The simultaneous plasma and magnetic-field data clearly indicate that the characteristics of the geomagnetic tail can be very different in the extended regions of the tail ($\sim 1000\,R_E$) from those in the near-earth regions ($< 80\,R_E$). Unlike the near-earth tail data, the plasma and magnetic-field data have no unique signature.

084.245 **Spatial distribution of energetic plasma sheet electrons.** R. J. Walker, T. A. Farley.
Journ. Geophys. Res., Vol. 77, 4650 - 4660 (1972).

084.246 **On the diamagnetic effect of the plasma sheet near $60\,R_E$.** C.-I. Meng, J. D. Mihalov.
Journ. Geophys. Res., Vol. 77, 4661 - 4669 (1972).

084.247 **Onset of magnetospheric substorms.**
B. Tsurutani, F. Bogott.
Journ. Geophys. Res., Vol. 77, 4677 - 4681 (1972).

084.248 **Coordinated observations of the magnetosphere: The development of a substorm.**
S. B. Mende, R. D. Sharp, E. G. Shelley, G. Haerendel, E. W. Hones.
Journ. Geophys. Res., Vol. 77, 4682 - 4699 (1972).

084.249 **Self-consistent description of the magnetotail current system.** M. K. Bird, D. B. Beard.
Journ. Geophys. Res., Vol. 77, 4864 - 4866 (1972). – Letter.

084.250 **About the excitation of natural oscillations of the earth's magnetic field.**
A. I. Ershkovitch, A. A. Nusinov.
Geomagn. Aeronom., Vol. 12, 877 - 881 (1972). In Russian.

084.251 **About the structure of pulsations of the earth's electromagnetic field as a random function of time.**
V. V. Boretz.
Geomagn. Aeronom., Vol. 12, 882 - 885 (1972). In Russian.

084.252 **The connection of parameters of geomagnetic pulsations Pi2 with processes in the auroral zone.**
V. K. Koshelevsky, O. M. Raspopov, G. V. Starkov.
Geomagn. Aeronom., Vol. 12, 886 - 891 (1972). In Russian.

084.253 **About the propagation velocity of shock waves causing geomagnetic storms and Forbush-decreases.**
A. E. Kuzmicheva, L. I. Dorman, N. S. Kaminer.
Geomagn. Aeronom., Vol. 12, 918 - 920 (1972). In Russian. Brief information.

084.254 **About the influence of the magnetic field irregularity on oscillations of the plasma layer of the magnetosphere's tail.** A. I. Ershkovitch, V. V. Kuzmin.
Geomagn. Aeronom., Vol. 12, 945 - 947 (1972). In Russian. Brief information.

084.255 **About the dependence of the magnetopause location on the orientation of the interplanetary magnetic field.** M. S. Kovner.
Geomagn. Aeronom., Vol. 12, 947 - 949 (1972). In Russian. Brief information.

084.256 **About a possibility of experimental division of a variable geomagnetic field into a poloidal and a** toroidal part. M. N. Berdichevsky, E. F. Fainberg.
Geomagn. Aeronom., Vol. 12, 950 - 954 (1972). In Russian. Brief information.

084.257 **Model of the geomagnetic field of the epoch 1955.**
V. N. Sipko.
Geomagn. Aeronom., Vol. 12, 957 - 959 (1972). In Russian. Brief information.

084.258 **A secular oscillation of the earth's magnetic quadrupole field causing a change of the earth's rotation.**
H. Wilhelm.
Zeitschr. Geophys., Vol. 36, 697 - 723 (1970). In German.

084.259 **Motions within the earth's core.** F. H. Busse.
Zeitschr. Geophys., Vol. 37, 153 - 177 (1971).
In German. – Review paper.

084.260 **Investigation of the angular distribution of particles from data of Cosmos 219.**
L. V. Zubareva, O. I. Savun, P. I. Shavrin, V. S. Paplin.
Vestn. Mosk. un-ta. Fiz., astron., Vol. 13, 366 - 367 (1972).
In Russian. – Abstr. in Referativ. Zhurn. 62. Issled. kosmich. prostranstva, 10.62.246 (1972).

084.261 **The calculation of current systems of daily geomagnetic variations.** D. J. Stone.
Gerlands Beiträge Geophys., Vol. 80, 117 - 128 (1971).

084.262 **Morphology of the lunar daily geomagnetic variation and its relation to S.** S. R. C. Malin.
Gerlands Beiträge Geophys., Vol. 80, 151 - 154 (1971).

084.263 **An analysis of the geomagnetic diurnal variation during the International Geophysical Year.**
W. D. Parkinson.
Gerlands Beiträge Geophys., Vol. 80, 199 - 232 (1971).

084.264 **Magnetospheric structure.** L. D. Kavanagh, Jr.
Astrophys. Space Sci. Library, Vol. 32, (see 012.009), 3 - 15 (1972).

084.265 **Magnetospheric processes.** C.-G. Fälthammar.
Astrophys. Space Sci. Library, Vol. 32, (see 012.009), 16 - 28 (1972).

084.266 **The interrelationship of magnetospheric processes.**
V. M. Vasyliunas.
Astrophys. Space Sci. Library, Vol. 32, (see 012.009), 29 - 38 (1972).

084.267 **Magnetospheric particle populations.**
J. I. Vette.
Astrophys. Space Sci. Library, Vol. 32, (see 012.009), 53 - 67 (1972).

084.268 **Characteristics of magnetosheath plasma observed at low altitudes in the dayside magnetospheric cusps.** J. D. Winningham.
Astrophys. Space Sci. Library, Vol. 32, (see 012.009), 68 - 80 (1972).

084.269 **Mechanisms for the injection of protons into the magnetosphere.** G. P. Haskell, R. J. Hynds.
Astrophys. Space Sci. Library, Vol. 32, (see 012.009), 81 - 94 (1972).

084.270 **ESRO IA/B observations at high latitudes of trapped and precipitating protons with energies above 100 keV.** F. Søraas.
Astrophys. Space Sci. Library, Vol. 32, (see 012.009), 120 -

132 (1972).

084.271 Review of magnetic field observations.
N. F. Ness.
Astrophys. Space Sci. Library. Vol. 32, (see 012.009), 189 - 199 (1972).

084.272 A self-consistent theory of the tail of the magnetosphere. K. Schindler.
Astrophys. Space Sci. Library, Vol. 32, (see 012.009), 200 - 209 (1972).

084.273 Theory of neutral sheets. J. W. Dungey.
Astrophys. Space Sci. Library, Vol. 32, (see 012.009), 210 - 220 (1972).

084.274 Critical review of electric field measurements.
U. V. Fahleson.
Astrophys. Space Sci. Library, Vol. 32, (see 012.009), 223 - 232 (1972).

084.275 INJUN 5 observations of magnetospheric electric fields and plasma convection. D. A. Gurnett.
Astrophys. Space Sci. Library. Vol. 32, (see 012.009), 233 - 245 (1972).

084.276 Plasma convection in the vicinity of the geosynchronous orbit. C. E. McIlwain.
Astrophys. Space Sci. Library, Vol. 32, (see 012.009), 268 - 279 (1972).

084.277 Thermal ions in the magnetosphere.
C. R. Chappell.
Astrophys. Space Sci. Library, Vol. 32, (see 012.009), 280 - 290 (1972).

084.278 Magnetic field variations at micropulsation frequencies. D. J. Southwood.
Astrophys. Space Sci. Library, Vol. 32, (see 012.009), 302 - 310 (1972).

084.279 Changes in the distribution function of magnetospheric particles associated with gyroresonant interactions. R. Gendrin.
Astrophys. Space Sci. Library, Vol. 32, (see 012.009), 311 - 328 (1972).

084.280 Electrostatic waves in the magnetosphere.
F. L. Scarf, R. W. Fredricks.
Astrophys. Space Sci. Library, Vol. 32, (see 012.009), 329 - 339 (1972).

084.281 A theory on the latitude and local time distribution of precipitating electrons during a sudden commencement. G. E. Perona.
Astrophys. Space Sci. Library, Vol. 32, (see 012.009), 351 - 354 (1972).

084.282 A short review of magnetospheric substorms.
M. P. Aubry.
Astrophys. Space Sci. Library, Vol. 32, (see 012.009), 357 - 364 (1972).

084.283 Substorm behavior of plasma sheet particles.
E. W. Hones, Jr.
Astrophys. Space Sci. Library, Vol. 32, (see 012.009), 365 - 378 (1972).

084.284 Interpretation of magnetic field variations during substorms. G. Rostoker.
Astrophys. Space Sci. Library, Vol. 32, (see 012.009), 379 -

390 (1972).

084.285 Excitation of polar substorms by northward interplanetary magnetic field. A. Nishida.
Astrophys. Space Sci. Library, Vol. 32, (see 012.009), 400 - 405 (1972).

084.286 Binary index for assessing local bow shock obliquity.
E. W. Greenstadt.
Journ. Geophys. Res., Vol. 77, 5467 - 5479 (1972).

084.287 Evidence of a diffuse magnetopause boundary.
D. S. Intriligator, J. H. Wolfe.
Journ. Geophys. Res., Vol. 77, 5480 - 5486 (1972).

084.288 Outer magnetosphere near midnight at quiet and disturbed times. M. P. Aubry, M. G. Kivelson, R. L. McPherron, C. T. Russell, D. S. Colburn.
Journ. Geophys. Res., Vol. 77, 5487 - 5502 (1972).

084.289 Measurements of magnetotail plasma flow made with Vela 4B. E. W. Hones, Jr., J. R. Asbridge, S. J. Bame, M. D. Montgomery, S. Singer, S.-I. Akasofu.
Journ. Geophys. Res., Vol. 77, 5503 - 5522 (1972).

084.290 Electron polar cap and the boundary of open geomagnetic field lines. L. C. Evans, E. C. Stone.
Journ. Geophys. Res., Vol. 77, 5580 - 5584 (1972). – Brief report.

084.291 Preconditions for the triggering of polar magnetic substorms by storm sudden commencements.
J. L. Burch.
Journ. Geophys. Res., Vol. 77, 5629 - 5632 (1972). – Letter.

084.292 Detection of earthward flow of kev protons in the geomagnetic tail at lunar distances.
A. Prakash.
Journ. Geophys. Res., Vol. 77, 5633 - 5637 (1972). – Letter.

084.293 Relationship between the various indices of geomagnetic activity and the interplanetary plasma parameters. R. P. Kane.
Journ. Atmosph. Terr. Phys., Vol. 34, 1941 - 1943 (1972). – Short paper.

084.294 Consistency of fields and particle motion in the 'Speiser' model of the current sheet.
J. W. Eastwood.
Planet. Space Sci., Vol. 20, 1555 - 1568 (1972).

084.295 Spectral analysis of geomagnetic variations to study the tidal and the storm modulation effects.
J. C. Gupta.
Planet. Space Sci., Vol. 20, 1613 - 1625 (1972).
The upper atmospheric disturbances, caused by the impact of the solar wind emanating from the sun's atmosphere, lead to geomagnetic storms. In the present paper a discussion on the 27 day modulation of the geomagnetic field by the storms and by the tides has been made.

084.296 On the influence of inclination of geomagnetic force lines on the field of a magnetospheric-ionospheric current system. L. A. Abramov, A. S. Debabov.
Kosm. Issled., Vol. 10, 711 - 720 (1972). In Russian.

084.297 Shadowing of electron azimuthal-drift motions near the noon magnetopause.
H. I. West, Jr., R. M. Buck, J. R. Walton.
Nature, Phys. Sci., Vol. 240, 6 - 7 (1972).

084.298 **Asymmetric eigenmodes in a simple model plasma-sphere.** A. H. Craven, J. A. Lawrie.
Planet. Space Sci., Vol. 20, 1875 - 1882 (1972).

084.299 **Geomagnetic field variations caused by changes in the quiet-time solar wind pressure.**
P. Verzariu, M. Sugiura, I. B. Strong.
Planet. Space Sci., Vol. 20, 1909 - 1914 (1972).

Solar wind data from the Vela 2A and 2B satellites are compared with hourly D_{st} values for the period July 1964—July 1965. During periods of positive D_{st} on magnetically quiet days the square root of the solar wind pressure is found to be statistically linearly correlated to the hourly D_{st} values. This relation is in accordance with the theoretical expectation from the pressure balance at the magnetopause between the solar wind plasma and the earth's magnetic field.

084.300 **Model development of supersonic trough wind with shocks.** J. M. Grebowsky.
Planet. Space Sci., Vol. 20, 1923 - 1934 (1972).

084.301 **Magnetospheric field fluctuations and the penetration of solar protons to low geomagnetic latitude.**
J. J. Quenby.
Planet. Space Sci., Vol. 20, 1979 - 1982 (1972). – Research note.

084.302 **Development of magnetic storms and the state of the magnetosphere according to the data of ground-based observations.**
M. I. Pudovkin, S. I. Isaev, S. A. Zaitzeva.
Ann. Géophys., Vol. 26, 761 - 770 (1970).

In this report we try to consider the following problems: 1) morphology and origin of the beginning of auroral substorms; 2) development of the auroral substorm and dynamics of the auroral plasma in the earth's magnetosphere; 3) intensity and direction of the electric field in the magnetosphere and its origin; 4) formation and dynamics of the DR-current, its location in the magnetosphere; 5) influence of the DR-currents on the location and structure of the auroral zone.

084.303 **Evidence for strong pitch-angle diffusion within the polar cusps.** H.-R. Lehmann, R. Treumann.
Nature, Phys. Sci., Vol. 239, 143 - 146 (1972).

084.304 **Seasonal variations in the solar and lunar daily geomagnetic variations.**
J. C. Gupta, S. R. C. Malin.
Geophys. Journ. Roy. Astron. Soc., Vol. 30, 11 - 18 (1972).

Seasonal variations of the solar, S, and lunar, L, daily geomagnetic variations are discussed in terms of the ratio of seasonal range to annual mean range, for data based on hourly mean values from 100 observatories for the interval 1957.5 to 1960.0. The conflicting views of previous workers on the difference between seasonal variations of L and S are discussed in the light of the present results and those from longer series of data from a few observatories.

084.305 **Magnetosphäre.** M. Siebert.
Phys. Blätter, 28. Jahrgang, p. 399 - 409 (1972).

084.306 **A quantitative model of the geomagnetic tail.**
D. M. Willis, R. J. Pratt.
Journ. Atmosph. Terr. Phys., Vol. 34, 1955 - 1976 (1972).

A quantitative model of the geomagnetic tail is presented that is realistic to distances beyond the orbit of the moon. This model magnetosphere is examined in detail and it is shown that it provides a good representation of the geomagnetic field under quiet conditions.

084.307 **Étude analytique de l'évolution de la normale à**

l'onde lors du trajet magnétosphérique des PC 1.
J.-L. Lacoume.
Comptes Rendus Acad. Sci. Paris, Sér. B, Vol. 275, 809 - 812 (1972).

084.308 **The self-consistent geomagnetic tail under static conditions.** M. K. Bird, D. B. Beard.
Planet. Space Sci., Vol. 20, 2057 - 2072 (1972).

084.309 **Solar proton intensity structures in the magnetosphere during interplanetary anisotropies.**
G. Morfill, M. Scholer.
Planet. Space Sci., Vol. 20, 2113 - 2123 (1972).

Detailed comparisons of measured solar particle structure in the magnetosphere during interplanetary anisotropies are made with the predictions of open tail models and diffusion calculations. Using observed and predicted intensity structures and their time variations, it is found that the nonadiabatic reconnection model shows the closest agreement with observations in all cases so far examined.

084.310 **Characteristics of the geomagnetic activity effect in the thermosphere.** M. Roemer, G. Lay.
Space Research XII, (see 012.016), Vol. 1, 797 - 802 (1972).

084.311 **Corpuscular radiation as an upper atmospheric energy source.** W. P. Olson.
Space Research XII, (see 012.016), Vol. 2, 1007 - 1013 (1972).

084.312 **Magnetospheric processes and the behavior of the neutral atmosphere.** P. M. Banks.
Space Research XII, (see 012.016), Vol. 2, 1051 - 1067 (1972).

084.313 **The international magnetospheric study 1975—1977: Scientific fundaments and objectives.**
J. G. Roederer.
Space Research XII, (see 012.016), Vol. 2, 1419 - 1436 (1972).

084.314 **Plasmasphere dynamics inferred from OGO 5 observations.**
C. R. Chappell, K. K. Harris, G. W. Sharp.
Space Research XII, (see 012.016), Vol. 2, 1513 - 1521 (1972).

084.315 **Étude de l'effet de la composante Est-Ouest du champ magnétique interplanétaire sur les courants équivalents aux perturbations magnétiques des régions de haute latitude.** A. Berthelier.
Comptes Rendus Acad. Sci. Paris, Sér. B, Vol. 275, 841 - 844 (1972).

Nous mettons en évidence l'existence de perturbations de la composante horizontale du champ magnétique terrestre dans les stations de haute latitude liées à une composante Est-Ouest du champ magnétique interplanétaire. Nous tentons d'interpréter les courants équivalents à ces perturbations par des modifications des champs électriques de convection.

084.316 **On the equilibrium configuration of the geomagnetic tail.** G. H. A. Cole, K. Schindler.
Cosmic Electrodynamics, Vol. 3, 275 - 284 (1972).

Conditions of momentum balance in the geomagnetic tail are investigated on the basis of a simple steady two-dimensional model. It is concluded that available observations on the tail structure can be explained within this frame-work during times when the tail is sufficiently quiet.

084.317 **The earth's magnetosphere.** H. Poeverlein.
Encyclopedia of physics, Vol. 49/4, (see 003.009), 7 - 113 (1972).

084.318 **Variations rapides du champ magnétique terrestre.**
E. Selzer.

Encyclopedia of physics, Vol. 49/4, (see 003.009), 231 - 394 (1972).

084.319 **Geomagnetic indices.** G. Rostoker.
Rev. Geophys. Space Phys., Vol. 10, 935 - 950 (1972).

The purpose of this review is to outline the definition and method of derivation of the indices of geomagnetic activity that are most commonly used at present. The first part of the review presents the methods of derivation of Kp, ap, AE, and Dst; the second part involves a discussion of the strengths and weaknesses of each index.

084.320 **Recent satellite measurements of the morphology and dynamics of the plasmasphere.** C. R. Chappell.
Rev. Geophys. Space Phys., Vol. 10, 951 - 979 (1972).

Plasmasphere morphology and dynamics can be understood in terms of a time-varying convection electric-field model of the magnetosphere that includes the bulge region as part of the main circulation pattern of the plasmasphere.

084.321 **Structure of the earth's bow shock.**
D. Biskamp, H. Welter.
Journ. Geophys. Res., Vol. 77, 6052 - 6059 (1972).

084.322 **Reinterpretation of the Pioneer 6 bow shock crossing.** F. Mariani, N. F. Ness, J. K. Chao.
Journ. Geophys. Res., Vol. 77, 6060 - 6070 (1972).

This paper re-examines and reinterprets the interesting bow shock crossing of Pioneer 6 on December 16, 1965, by combining the high-resolution data for the magnetic field measurements and all the available plasma data.

084.323 **Magnetopause motions at lunar distance determined from the Explorer 35 plasma experiment.**
H. C. Howe, G. L. Siscoe.
Journ. Geophys. Res., Vol. 77, 6071 - 6086 (1972).

From solar-wind plasma observations on the lunar orbiting Explorer 35 satellite, multiple boundary crossings have been analyzed to give statistical information on the tail boundary motions at lunar distance.

084.324 **Equatorial current sheet in the magnetosphere.**
M. Sugiura.
Journ. Geophys. Res., Vol. 77, 6093 - 6103 (1972).

The purpose of this paper is to show that the quiet-time ring current is a quasi-permanent structure of the magnetosphere and that what has been referred to as the quiet-time ring current is actually an equatorial sheet current that is essentially an extension to a geocentric distance of at least 2.5 R_E of the neutral sheet current in the magnetospheric tail.

084.325 **On the plasma sheet contribution to the force balance requirements in the geomagnetic tail.**
G. L. Siscoe.
Journ. Geophys. Res., Vol. 77, 6230 - 6234 (1972).
Brief report.

084.326 **Comparison of magnetosphere models with storm observations from Explorer 26.**
B. Parady, L. J. Cahill, Jr.
Journ. Geophys. Res., Vol. 77, 6235 - 6242 (1972).
Brief report.

084.327 **Nonuniqueness of magnetic field line motion.**
V. M. Vasyliunas.
Journ. Geophys. Res., Vol. 77, 6271 - 6274 (1972). – Letter.

084.328 **Comments on the growth phase of magnetospheric substorms.** S.-I. Akasofu, A. L. Snyder.
Journ. Geophys. Res., Vol. 77, 6275 - 6277 (1972). – Letter.

084.329 **The distance of the magnetospheric boundary to the subsolar point for various indices of geomagnetic activity.** N. M. Rudneva, Ya. I. Feldstein.
Byull. Abastumansk. Astrofiz. Obs., No. 42, (see 012.019), p. 103 - 117 (1972). In Russian.

084.330 **Some peculiarities of the development of the magnetic storm on March, 5–10 1970.**
M. I. Pudovkin, O. M. Raspopov, S. V. Leontyev, V. A. Troitskaya.
Ann. Géophys., Vol. 28, 455 - 463 (1972). – Paper presented at I.A.G.A. symposium on 'Aurora and airglow', Moscow, August 1971.

084.331 **Precambrian geomagnetic reversal stratigraphy.**
H. Spall.
Nature, Vol. 240, 402 - 403 (1972).

084.332 **Outflow of plasma from the magnetotail into the magnetosheath.**
E. W. Hones, Jr., S.-I. Akasofu, S. J. Bame, S. Singer.
Journ. Geophys. Res., Vol. 77, 6688 - 6695 (1972).

It is inferred from measurements of proton fluxes with electrostatic analyzers on the Vela satellites that plasma leakage from the magnetosphere into the magnetosheath is rather commonplace and may be strongly enhanced during magnetospheric substorms. It is suggested that this leakage accounts for the loss of at least some of the plasma that leaves the thinning plasma sheet early in a substorm. Simultaneous measurements by two Vela satellites, one in the magnetosheath and one in the solar wind, show conclusively that these protons do not originate in the solar wind. Spatial variations of their intensity across the magnetosheath suggest further that they come not from the bow shock but from the magnetosphere.

084.333 **Morphology and interpretation of magnetospheric plasma waves at conjugate points during December solstice.** L. J. Lanzerotti, A. Hasegawa, N. A. Tartaglia.
Journ. Geophys. Res., Vol. 77, 6731 - 6745 (1972).

084.334 **Two substorm studies of relations between westward electric fields in the outer plasmasphere, auroral activity, and geomagnetic perturbations.**
D. L. Carpenter, S.-I. Akasofu.
Journ. Geophys. Res., Vol. 77, 6854 - 6863 (1972). – Brief report.

084.335 **The aa indices: A 100-year series characterizing the magnetic activity.** P.-N. Mayaud.
Journ. Geophys. Res., Vol. 77, 6870 - 6874 (1972).

In a previous note (Mayaud, 1971) we described the possibility of characterizing the magnetic activity at the earth's surface by using the K indices of two antipodal observatories. Here we present some results from a 100-year series. One result permits us to forecast the intensity of the magnetic activity; the other result is a clear confirmation of the annual variation.

084.336 **A note on the geomagnetic spectrum.**
R. G. Currie.
Journ. Geophys. Res., Vol. 77, 6893 - 6895 (1972). – Letter.

084.337 **Nonlinear waves in geomagnetic wake.**
A. I. Ershkovich, A. A. Nusinov, A. A. Chernikov.
Journ. Geophys. Res., Vol. 77, 6907 - 6910 (1972). – Letter.

084.338 **Penetration of solar protons into the geomagnetic tail.**
I. I. Alexeev, A. P. Kropotkin, V. P. Shabansky.
Geomagn. Aeronom., Vol. 12, 974 - 978 (1972). In Russian.

084.339 **Identification of discontinuities at the boundary**

of the magnetosphere. K. G. Ivanov.
Geomagn. Aeronom., Vol. 12, 979 - 983 (1972). In Russian.

084.340 The magnetic storm of March 8 - 10, 1970 from observations aboard Cosmos 321 and on the surface of the earth. I. Morphology of the disturbance.
Sh. Sh. Dolginov, L. N. Zhigalov, L. V. Strunnikova, Ya. I. Feldstein, T. N. Cherevko, V. A. Sharova.
Geomagn. Aeronom., Vol. 12, 1046 - 1058 (1972). In Russian.

084.341 Some statistic characteristics of spectra of polar magnetic substorms.
Yu. G. Turbin, Yu. I. Vakulin, V. D. Urbanovich.
Geomagn. Aeronom., Vol. 12, 1070 - 1073 (1972). In Russian.

084.342 Origin and decay of a narrow belt of energetic electrons in the earth's magnetosphere.
E. V. Gorchakov, M. V. Ternovskaya, I. V. Getselev.
Geomagn. Aeronom., Vol. 12, 1124 - 1125 (1972). In Russian. – Brief information.

084.343 Physical conditions in the magnetosphere and in the interplanetary space under excitation of geomagnetic pulsations of Pc1 type.
E. T. Matveeva, A. L. Kalisher, B. V. Dovbnya.
Geomagn. Aeronom., Vol. 12, 1125 - 1127 (1972). In Russian. – Brief information.

084.344 On the position of the magnetopause on the day side.
M. S. Kovner, N. M. Rudneva, Ya. I. Fel'dshtejn.
Kosmich. Issled., Vol. 10, 864 - 870 (1972). In Russian.

084.345 On the demarcation layer in the magnetosphere.
M. S. Bobrov.
Kosmich. Issled., Vol. 10, 879 - 887 (1972). In Russian.

084.346 On the theory of daily variations of the earth's magnetic tail. A. I. Ershkovich, S. V. Sobolev.
Kosmich. Issled., Vol. 10, 888 - 891 (1972). In Russian.

084.347 Über die zeitliche Veränderlichkeit des geomagnetischen Hauptfeldes (auf Grund von sphärisch-harmonischen Analysen geomagnetischer Weltkarten für den Zeitraum von 1880 bis 1960). H. Kautzleben.
Gerlands Beiträge Geophys., Vol. 81, 233 - 239 (1972).

084.348 On the relationship between the solar wind velocity and some indices of the geomagnetic activity.
G. I. Ol.
Solnechnye Dannye 1972 Byull., No. 9, p. 110 - 113 (1972). In Russian.
The relationship between the solar wind velocity and indices of geomagnetic activity for polar magnetic disturbances and ring current intensity has been obtained. This relationship is derived for the descending branch of the 11-year cycle.

084.349 Polar magnetic substorms 03–06, U.T. December 5, 1968. E. I. Loomer, G. J. van Beek.
Publ. Earth Phys. Branch, Dep. of Energy, Mines and Resources, Ottawa, Canada, Vol. 42, (No. 4), 157 - 165 (1972).
Magnetic effects associated with a westward travelling surge are illustrated by analysis of simple bays which occurred during three moderately weak substorms, 03–06 U.T. on December 5, 1968.

084.350 Record of observations at Meanook Magnetic Observatory 1969. A. B. Cook, S. J. Sprysak.
Publ. Earth Phys. Branch, Dep. of Energy, Mines and Re-

sources, Ottawa, Canada, Vol. 42, (No. 9), 239 - 290 (1972).
(1) Hourly values of horizontal intensity, declination and vertical intensity; hourly, daily and monthly means; (2) Mean hourly values H, D and Z, for month and year; all days, international quiet days and disturbed days; (3) Three-hour range indices in H, D and Z and K-indices.

084.351 Record of observations at Alert Magnetic Observatory 1969. G. J. van Beek, H. R. Reny.
Publ. Earth Phys. Branch, Dep. of Energy, Mines and Resources, Ottawa, Canada, Vol. 43, (No. 1), 1 - 74 (1972).

084.352 Record of observations at Victoria Magnetic Observatory 1970. D. R. Auld, B. D. Lowe.
Publ. Earth Phys. Branch, Dep. of Energy, Mines and Resources, Ottawa, Canada, Vol. 43, (No. 7), 447 - 499 (1972).

084.353 A simple activity index for short-period geomagnetic fluctuations. D. R. Auld, B. Caner.
Journ. Geomagn. Geoelectr., Vol. 23, 369 - 389 = Contr. Earth Phys. Branch, Ottawa, No. 360 (1971).
An index for the quantitative expression of short-period geomagnetic activity is being proposed. The index is based on the peak-to-peak range in 2 (or 2.5)min intervals, and is derived from amplitude-linear (fluxgate) instrumentation.

084.354 Geomagnetic observations at the Kanozan and Mizusawa Geodetic Observatories (1970).
Bull. Geograph. Survey Inst., Tokyo, (*Japan*), Vol. 17, Part 2, 51 pp. (1972).

084.355 Solar particle access to the magnetosphere – how?
L. J. Lanzerotti, F. C. Michel.
Comments Astrophys. Space Phys., Vol. 4, 161 - 165 (1972).
Two magnetospheric models are considered: the "open" and the "closed" magnetosphere. The electrons tend to support the open model while the protons tend to support the closed model! Naturally there have been attempts to reconcile (or discount) the failures of each model, but we will here be concerned with the prospects of understanding the problem from further observations.

084.356 Catalogue of the geomagnetic disturbance index for the period 1841 - 1864, 1870. I. D. Zosimovich.
Problems of cosmic physics. Vyp. (No.) 7, (see 003.028), p. 61 - 73 (1972). In Russian.

084.357 Evidence for short geomagnetic polarity intervals in the early Cenozoic. R. J. Blakely, A. Cox.
Journ. Geophys. Res., Vol. 77, 7065 - 7072 (1972).
To test the validity of the theory that geomagnetic reversals are produced by a stochastic process in the earth's core, a search was made for heretofore undetected short polarity intervals in marine magnetic anomalies over segments of ocean floor with ages between 55 and 63 m.y. Six new short events were determined by modeling the profiles with a single crustal layer. All the new events were found to be <60,000 years.

084.358 Knowledge of the earth's core from geomagnetism.
F. J. Lowes.
Mantle and core in planetary physics, Varenna 1970, (see 012.024), p. 27 - 37 (1971).

084.359 Motions in the fluid core. W. V. R. Malkus.
Mantle and core in planetary physics, Varenna 1970, (see 012.024), p. 38 - 63 (1971).

Geomagnetic daily variations: Sources, current and induction. See Abstr. 011.007.

A theoretical treatment of lunar particle shadows.

See Abstr. 062.005.

La transformation réciproque des ondes magnéto-hydrodynamiques dans un plasma homogène situé dans un champ dépendant d'une seule coordonnée. See Abstr. 062.024.

Reevaluation of relationships between solar flares and sporadic geomagnetic storms. See Abstr. 073.029.

Variations of solar wind parameters, magnetic activity and electrons of the magnetosphere's tail and of the outer radiation zone. See Abstr. 074.025.

Shock waves in the solar system. See Abstr. 074.045.

Magnetic and electric waves in space. See Abstr. 074.054.

Equatorial coronal arches and geomagnetic disturbance. See Abstr. 074.085.

Entry of high-energy solar protons into the distant geomagnetic tail. See Abstr. 078.001.

Coordinate system for use with high-latitude energetic-particle phenomena. See Abstr. 078.006.

Entry of energetic solar protons into the tail. See Abstr. 078.023.

Anisotropy of low-energy solar protons at the boundary of the magnetotail. See Abstr. 078.036.

Atmospheric helium and geomagnetic field reversals. See Abstr. 082.155.

Statistical analysis of the relationship between geomagnetic activity and variations of atmospheric pressure. See Abstr. 082.243.

Spatial and temporal variations of the thermal plasma between 3000 and 5700 kilometers at L = 2 to 4. See Abstr. 083.025.

Interactions between the ionosphere and the magnetosphere for solar regular daily geomagnetic variations. See Abstr. 083.034.

VLF wave propagation and its interaction with the magnetoplasma. See Abstr. 083.043.

Characteristics of the lunar photoelectron layer in the geomagnetic tail. See Abstr. 094.203.

Non-dipole terms in the magnetic fields of Jupiter and the earth. See Abstr. 099.065.

Critical component of the interplanetary magnetic field responsible for large geomagnetic effects in the polar cap. See Abstr. 106.002.

Relationship of interplanetary magnetic field structure with development of substorm and storm main phase. See Abstr. 106.005.

Interplanetary magnetic-sector structure, 1926 - 1971. See Abstr. 106.006.

Variations of the interplanetary magnetic field and the substorm of June 17–18, 1965. See Abstr. 106.018.

On the possibility of sounding of the interplanetary shock waves and forecasting of the geomagnetic storms on the basis of the ground observations of cosmic rays. See Abstr. 106.024.

Inferring the interplanetary magnetic field by observing the polar geomagnetic field. See Abstr. 106.028.

Use of an electron beam for low-temperature plasma measurement in the magnetosphere and interplanetary space. See Abstr. 106.031.

Interaction of the interplanetary medium with the geomagnetosphere. See Abstr. 106.039.

Geomagnetic cutoffs for cosmic-ray protons for seven energy intervals between 1.2 and 39 Mev. See Abstr. 143.007.

Cosmic-ray scintillations. 1. Inside the magnetosphere. See Abstr. 143.047.

Radiation Belts

084.401 **Geomagnetically trapped alpha particles. 1. Off-equator particles in the outer zone.**
J. B. Blake, G. A. Paulikas.
Journ. Geophys. Res., Vol. 77, 3431 - 3440 (1972).

084.402 **Inner-zone energetic-electron repopulation by radial diffusion.**
A. D. Tomassian, T. A. Farley, A. L. Vampola.
Journ. Geophys. Res., Vol. 77, 3441 - 3454 (1972).

084.403 **Pitch-angle diffusion of radiation belt electrons within the plasmasphere.**
L. R. Lyons, R. M. Thorne, C. F. Kennel.
Journ. Geophys. Res., Vol. 77, 3455 - 3474 (1972).

084.404 **Observation of very-low-frequency whistler-mode waves in the region of the radiation-belt slot.**
H. C. Koons, D. A. McPherson.
Journ. Geophys. Res., Vol. 77, 3475 - 3482 (1972).

084.405 **Motion and diffusion of energetic particles in the outer zone.** J. C. Kosik.
Planet. Space Sci., Vol. 20, 1025 - 1031 (1972).

084.406 **Studies of outer belt and slot region protons at low altitudes.** W. L. Imhof, J. B. Reagan.
Journ. Geophys. Res., Vol. 77, 4128 - 4144 (1972).

084.407 **Behavior of outer radiation zone and a new model of magnetospheric substorm.**

G. K. Parks, G. Laval, R. Pellat.
Planet. Space Sci., Vol. 20, (see 012.006), 1391 - 1408
(1972). — Invited paper.

084.408   Dynamics of the outer radiation belt during the
          International Year of the Quiet Sun.
S. N. Vernov, S. N. Kuznetsov, Yu. I. Logachev, G. B. Lopa-
tina, V. G. Stolpovskij.
Kosmich. Issled., Vol. 10, 545 - 554 (1972). In Russian.

084.409   Spatial intensity distribution of the excess radiation
          at small heights.
V. S. Tsaplin, P. I. Shavrin, L. V. Zubareva, O. I. Savun.
Kosmich. Issled., Vol. 10, 555 - 560 (1972). In Russian.

084.410   Earth albedo neutrons from 10 to 100 MeV.
          A. M. Preszler, G. M. Simnett, R. S. White.
Phys. Rev. Letters, Vol. 28, 982 - 985 (1972).
    We report the measurement of the energy and angular
distributions of earth albedo neutrons from 10 to 100 MeV at
40° N geomagnetic latitude from a balloon at 120000 ft, below 4.65 g/cm². The absolute neutron energy distribution is of
the correct strength and shape for the albedo neutrons to be
the source of the protons trapped in earth's inner radiation
belt.

084.411   Resonance acceleration of low-energy protons in
          the earth's radiation belts.
S. N. Vernov, M. I. Panasyuk, E. N. Sosnovetz, L. V. Tvers-
kaya, O. V. Khorosheva.
Geomagn. Aeronom., Vol. 12, 785 - 789 (1972). In Russian.

084.412   Electron acceleration in the outer radiation belt.
          S. N. Vernov, S. N. Emeljanenko, S. N. Kuznetzov,
V. G. Stolpovsky.
Geomagn. Aeronom., Vol. 12, 790 - 796 (1972). In Russian.

084.413   Measurements of radiation belt protons in the
          energy range 0.25 to 1.65 MeV on board the satel-
lite "Azur".   J. Moritz.
Zeitschr. Geophys., Vol. 37, 179 - 194 (1971). In German.
    Directional proton intensities in the energy range 0.25
to 1.65 MeV are measured with an experiment on board the
satellite Azur (1969–97 A). Within the inner radiation belt
(L< 1.9) the count rates have to be corrected for a background
due to higher energy protons and electrons whose magnitude
can be derived from the count rate of one of the detectors.
The count rate correction leads to low proton intensities within the inner radiation belt and a rising energy spectrum in the
energy range 0.25 to 1.65 MeV.

084.414   New observations of the proton population of the
          radiation belt between 1.5 and 104 MeV.
D. Hovestadt, E. Achtermann, B. Ebel, B. Häusler, G. Paschmann.
Astrophys. Space Sci. Library, Vol. 32, (see 012.009), 115 -
119 (1972).

084.415   Pitch angles and spectra of particles in the outer
          zone near noon.
J. R. Burrows, I. B. McDiarmid, M. D. Wilson.
Astrophys. Space Sci. Library, Vol. 32, (see 012.009), 153 -
167 (1972).

084.416   Some parameters affecting the poleward boundary
          of trapped electrons.   D. E. Page, M. L. Shaw.
Astrophys. Space Sci. Library, Vol. 32, (see 012.009), 175 -
178 (1972).

084.417   High energy proton model for the inner radiation
          belt.   M. Walt, T. A. Farley.
Astrophys. Space Sci. Library, Vol. 32, (see 012.009), 293 -
301 (1972).

084.418   Energy density of excess radiation at mean latitudes.
          V. S. Tsaplin, P. I. Shavrin, V. M. Gribkov.
Kosm. Issled., Vol. 10, 732 - 736 (1972). In Russian.

084.419   Proton radiation belt variations in July—August
          1970.
S. N. Vernov, I. Ya. Kovalskaya, M. I. Panasyuk, A. I. Rubin-
stein, E. N. Sosnovets, L. V. Tverskaya, O. V. Khorosheva.
Space Research XII, (see 012.016), Vol. 2, 1493 - 1497 (1972).

084.420   A study of self-consistent ring current models.
          N. Sckopke.
Cosmic Electrodynamics, Vol. 3, 330 - 348 (1972).
    A number of stationary, axially symmetric, ring-current
models have been computed, to study the influence of certain
parameters on the self-consistent solutions. These parameters
were the particles' pitch-angle distribution and the vacuum
field. The pitch-angle distribution has been varied for particle
belts with identical equatorial energy-density profiles, and for
belts with equal energy content. Most of the models have been
computed for two different vacuum fields B_0: for an unbounded dipole field, and for a dipole confined by a spherical magnetopause. The virial theorem has been rewritten for this latter
field model.

084.421   The earth's radiation belt.   W. N. Hess.
          Encyclopedia of physics, Vol. 49/4, (see 003.009),
115 - 230 (1972).

084.422   Effects of the secular magnetic variation on the
          distribution function of inner-zone protons.
T. A. Farley, M. G. Kivelson, M. Walt.
Journ. Geophys. Res., Vol. 77, 6087 - 6092 (1972).
    The equation governing the distribution function of high-
energy protons in the earth's radiation belt has been modified
to include the effects of the secular change of the earth's magnetic dipole moment in addition to the effects of albedo neutron decay, collisional energy loss, and radial diffusion, which
have been studied in the past.

084.423   Series expansion of the magnetic vector potential
          for a ring current.   V. A. Blednov.
Geomagn. Aeronom., Vol. 12, 1106 - 1111 (1972).
In Russian.

084.424   Fast injection of energetic particles into the gap
          between the inner and the outer radiation belts.
S. N. Kuznetsov, L. V. Tverskaya, O. V. Khorosheva.
Geomagn. Aeronom., Vol. 12, 1113 - 1115 (1972).
In Russian. — Brief information.

Errata

084.901   Correction:'Multiple crossings of the earth's bow
          shock at large geocentric distances' [Journ. Geophys.
Res., Vol. 76, 5970 - 5977 (1971)].
B. Bavassano, F. Mariani, U. Villante, N. F. Ness.
Journ. Geophys. Res., Vol. 77, 4902 (1972).

085 Solar-Terrestrial Relations

085.001 **Effects of solar activity and zonal wind in the strato-**
sphere and lower mesosphere.
E. S. Kazimirovsky, V. F. Loginov.
Astron. Zhurn. Akad. Nauk SSSR, Vol. 49, 860 - 865 (1972).
In Russian. English translation in Soviet Astron. AJ, Vol. 16,
No. 4.

On the basis of the data of the American rocket net,
models W–E of the wind component for high and low solar
activity are constructed. The radio emission flux at 10.7 cm
and the planetary geomagnetic index were taken as indices of
solar activity. An interpretation of regularities obtained is
given and methods of the choice of reference data for the
analysis of the atmospheric circulation are proposed. Some
considerations concerning the mechanism of solar-atmospheric
connections are stated.

085.002 **Anomalous increase in the total X-ray background**
at balloon altitude.   R. K. Manchanda.
Journ. Geophys. Res., Vol. 77, 4254 - 4258 (1972). – Letter.

085.003 **August solar activity and its geophysical effects.**
P. S. McIntosh.
Sky Telescope, Vol. 44, 214 - 217 (1972).

085.004 **The sunspot cycle influence on the solar and lunar**
daily geomagnetic variations.
S. Chapman, J. C. Gupta, S. R. C. Malin.
Proc. Roy. Soc. London, Ser. A, Vol. 324, 1 - 15 (1971).

The relative importance of the sunspot cycle influence on
solar and lunar daily geomagnetic variations has been discussed
for over a century, but hitherto inconclusively. The problem is
re-examined using results from a number of new analyses of
data obtained during the IGY/C and the preceding sunspot
minimum, as well as results from analyses of longer series of
data. Particular attention is paid to the significance of the results.

085.005 **On the relationship of the tropospheric circulation**
indices with changes of solar wind velocity.
R. V. Smirnov.
Astron. Tsirk., No. 719, p. 3 - 5 (1972). In Russian.

085.006 **12.5-minute periodicity in solar proton fluxes at**
balloon altitude and in magnetic micropulsations.
J. G. Greenhill, K. B. Fenton, A. G. Fenton, K. S. White.
Journ. Geophys. Res., Vol. 77, 6656 - 6664 (1972).

This paper discusses the association between 12.5-min-
ute pulsations in the geomagnetic tail field and the intensity
of solar cosmic radiation observed at balloon altitudes over
Wilkes, Antarctica, during the large PCA event of September 3,
1966.

085.007 **Mechanismen geophysikalischer Periodizitäten.**
M. Siebert.
Nachr. Akad. Wiss. Göttingen, II. Math.-Phys. Kl., Jahrgang
1972, (No. 8), p. 149 [23] - 155 [29] (1972).

085.008 **Korrelation und Kausalität bei lunaren Periodizi-**
tätserscheinungen in Biologie und Geophysik.
H.-J. Lang.
Nachr. Akad. Wiss. Göttingen, II. Math.-Phys. Kl., Jahrgang
1972, (No. 8), p. 156 [30] - 160 [34] (1972).

085.009 **Dependences between cycles of solar activity and**
processes in sun-earth space.   P. I. Velinov.
Comptes Rendus Acad. Bulg. Sci., Vol. 25, 321 - 324 (1972).

085.010 **Effect of solar activity delays on the processes in**
solar-terrestrial space.   P. I. Velinov.
Comptes Rendus Acad. Bulg. Sci., Vol. 25, 1045 - 1048
(1972).

The effect of accelerations of solar activity on the
courses of the processes in solar-terrestrial space are studied.

Solar-terrestrial physics. An account of the wave
and particle radiations from the quiet and the active sun, and
of the consequent terrestrial phenomena.
See Abstr. 003.034.

The sun and the ionosphere.
See Abstr. 003.079.

Keeping up with sunspots.   See Abstr. 072.059.

Planetary System

091 Physics of the Planetary System (Planetary Atmospheres, Figure, Interior, Magnetic Fields, Rotation, etc.)

091.001 **Minor constituents in planetary atmospheres: Ultraviolet spectroscopy from the Orbiting Astronomical Observatory.** T. Owen, C. Sagan.
Icarus, Vol. 16, 557 - 568 (1972).

Orbiting Astronomical Observatory data between 2000 and 3600 Å, obtained by the Wisconsin Experimental Package objective grating scanning spectrometer, are used to set upper limits to the abundances of many minor constituents in the atmospheres of Mars, Jupiter, Saturn, and Venus.

091.002 *P-, T*-invariance of electromagnetic interaction and circular polarization of the radiation of planets.
V. N. Sazonov.
Astron. Zhurn. Akad. Nauk SSSR, Vol. 49, 833 - 836 (1972).
In Russian. English translation in Soviet Astron. AJ, Vol. 16, No. 4.

The interpretation of the common properties of the circular polarization observed in the optical radiation of the moon, Jupiter and other planets different by their physical characteristics is given on the basis of the general principles of *P-, T*-invariance of electromagnetic interaction.

091.003 **The spherical albedo of a planetary atmosphere.** E. G. Yanovitsky.
Astron. Zhurn. Akad. Nauk SSSR, Vol. 49, 844 - 849 (1972).
In Russian. English translation in Soviet Astron. AJ, Vol. 16, No. 4.

The problem is considered of the determination of the spherical albedo of a semi-infinite planetary atmosphere with arbitrary scattering indicatrix. The four terms in the expansion of the spherical albedo in powers of $(1-\lambda)^{1/2}$ are obtained, λ being the single scattering albedo of the particles. Using this expansion the spectral values of λ of the Venus atmosphere are obtained. It is shown that the atmosphere of Venus is practically purely scattering in the continuum spectral region $0.6 - 1 \mu$.

091.004 **Rayleigh and Raman scattering by H_2 in a planetary atmosphere.** L. Wallace.
Astrophys. Journ., Vol. 249 - 257 (1972).

Approximate calculations of the reflected intensities of the various frequency-shifted Raman components are presented for a planetary atmosphere of pure H_2 in which there is no absorption of photons.

091.005 **Determination of planetary masses from the motions of comets.** W. J. Klepczynski.
IAU Symposium No. 45, (see 012.003), p. 209 - 226 (1972).

091.006 **On the determination of planetary masses.**
P. Herget.
IAU Symposium No. 45, (see 012.003), p. 244 - 245 (1972).

091.007 **On the temperature distribution in a planetary atmosphere.** L. D. G. Young, A. T. Young.
Astrophys. Journ., Vol. 176, 533 - 554 (1972).

We present a sensitive method of estimating the range of temperature in the part of the atmosphere where spectral lines are formed, from observations made in reflected sunlight. This method is applied to the best available spectroscopic observations of Venus made in the 7820 Å CO_2 band.

091.008 **Water on the planets.**
V. Derpgol'ts, G. Katterfel'd.
Nauka i zhizn', 1972, No. 4, p. 33 - 34. In Russian. – Abstr. in Referativ. Zhurn. 51. Astron.,8.51.58 (1972).

091.009 **The equation of the figure of planets in Clairaut's problem.** V. P. Trubitsyn.
Izv. AN SSSR. Fiz. Zemli, 1972, No. 4, p. 10 - 15. In Russian. Abstr. in Referativ. Zhurn. 51. Astron., 8.51.128 (1972).

091.010 **Frequencies of occultations of stars by planets, satellites and asteroids.** B. O'Leary.
Bull. American Astron. Soc., Vol. 4, 362 (1972). – Abstr. AAS.

091.011 **Reflectivities of ammonium hydrosulfides: Application to the interpretation of reflection spectra of outer solar system bodies.** L. A. Lebofsky.
Bull. American Astron. Soc., Vol. 4, 362 (1972). – Abstr. AAS.

091.012 **Two dimensional digital vidicon photometry of the planets.**
T. B. McCord, J. A. Westphal, J. Kunin.
Bull. American Astron. Soc., Vol. 4, 362 (1972). – Abstr. AAS.

091.013 **Submillimeter radiometry of the planets.**
T. Z. Martin, W. M. Sinton, C. W. Wood.
Bull. American Astron. Soc., Vol. 4, 362 (1972). – Abstr. AAS.

091.014 **Ionization threshold spectrometer – an instrument for characterizing outer planet atmospheres.**
H. Myers.
Bull. American Astron. Soc., Vol. 4, 364 (1972). – Abstr. AAS.

091.015 **The probability distribution of path lengths traveled by photons diffusely reflected from a semi-infinite atmosphere.** J. F. Appleby, W. M. Irvine.
Bull. American Astron. Soc., Vol. 4, 365 (1972). – Abstr. AAS.

091.016 **Results of matrix-operator theory for Rayleigh and maritime haze phase functions.**
G. N. Plass, G. W. Kattawar, F. E. Catchings.
Bull. American Astron. Soc., Vol. 4, 365 (1972). – Abstr. AAS.

091.017 **A practical matrix-operator theory for realistic radiative transfer problems.**
G. W. Kattawar, G. N. Plass.
Bull. American Astron. Soc., Vol. 4, 365 (1972). – Abstr. AAS.

091.018 **The natural satellites: A review of physical properties.** D. P. Cruikshank.
Bull. American Astron. Soc., Vol. 4, 366 (1972). – Abstr. AAS.

091.019 **Satellites of the outer planets: A review of spectrophotometric observations.** T. V. Johnson.

Bull. American Astron. Soc., Vol. 4, 366 (1972). – Abstr. AAS.

091.020 Physico-chemical models for the satellites of the outer planets. J. S. Lewis.
Bull. American Astron. Soc., Vol. 4, 366 (1972). – Abstr. AAS.

091.021 Eclipses of outer planet satellites: Calculations. R. T. Brinkmann.
Bull. American Astron. Soc., Vol. 4, 368 (1972). – Abstr. AAS.

091.022 Satellites and asteroids: Topography on small planetary bodies. T. V. Johnson, T. R. McGetchen.
Bull. American Astron. Soc., Vol. 4, 368 (1972). – Abstr. AAS.

091.023 Radiative transfer in planetary atmospheres with three-term scattering indicatrix.
A. S. Anikonov.
Vestn. Leningr. un-ta, 1972, No. 7, p. 133 - 140. In Russian. Abstr. in Referativ. Zhurn. 51. Astron., 9.51.192 (1972).

091.024 Phase equilibria in gas mixtures at high pressures: Implications for planetary structures.
W. B. Streett, A. L. Erickson.
Phys. Earth Planet. Interiors, Vol. 5, 357 - 366 (1972).
An understanding of phase equilibria in gas mixtures under pressure is essential to the study of the deep-atmosphere and interior structures of the outer planets. The results of experiments on phase equilibria in mixtures of He with Ar and N_2, at pressures up to 10000 atm, are presented, and serve as the basis for a discussion of the characteristics of phase diagrams for two-component gas mixtures at high pressures. The application of these results to problems of planetary interiors is briefly discussed. A brief description of the experimental method is included.

091.025 Coulomb interaction in the solar system?
N. N. Malov.
Mat. i fizika (NRB), Vol. 15, No. 2, p. 2 (1972). In Bulgarian. Abstr. in Referativ. Zhurn. 51. Astron., 10.51.143 (1972).

091.026 The Laplacean plane and the masses of the planets.
S. V. Serova.
Vestn. Leningr. un-ta, 1972, No. 7, p. 150 - 153. In Russian. Abstr. in Referativ. Zhurn. 51. Astron., 10.51.144; 62. Issled. kosmich. prostranstva, 10.62.281 (1972).

091.027 Determination of the refraction angle of light rays in the atmosphere of a planet. I. F. Kushtin.
Geod. i kartografiya, 1972, No. 3, p. 21 - 31. In Russian. Abstr. in Referativ. Zhurn. 62. Issled. kosmich. prostranstva, 10.62.139 (1972).

091.028 Steady neutral planetary boundary layer forced by a horizontally non-uniform flow. M. K. Mak.
Journ. Atmosph. Sci., Vol. 29, 707 - 717 (1972).

091.029 On the determination of the transfer function for the spectral albedo of the system planetary surface–atmosphere. K. J. Kondratiev, O. I. Smoktii.
Dokl. Akad. Nauk SSSR, Ser. Mat. Fiz., Vol. 206, 1349 - 1352 (1972). In Russian.

091.030 High resolution planetary observations.
W. A. Baum.
Space Research XII, (see 012.016), Vol. 2, 1671 - 1682 (1972).

091.031 High resolution planetary observation.
G. P. Kuiper.
Space Research XII, (see 012.016), Vol. 2, 1683 - 1687 (1972).

091.032 Bedekkingen van sterren door planeten.

J. Meeus.
Hemel en Dampkring, Vol. 70, 235 (1972).

091.033 Observations of planets, nebulae, and galaxies at 350 microns. D. A. Harper, Jr., F. J. Low, G. H. Rieke, K. R. Armstrong.
Astrophys. Journ., (*Letters*), Vol. 177, L21 - L25 (1972).
Ground-based measurements have been made of the $350\text{-}\mu$ flux from Venus, Mars, Jupiter, Saturn, Uranus, and the Orion nebula. Upper limits were obtained for the flux from NGC 2024, NGC 1068, NGC 4151, and the galactic center.

091.034 Far-infrared brightness temperatures of the planets.
K. R. Armstrong, D. A. Harper, Jr., F. J. Low.
Astrophys. Journ., (*Letters*), Vol. 178, L89 - L92 (1972).
Brightness temperatures in several bands from 30 to 300μ are determined for Venus, Mars, Jupiter, and Saturn.

091.035 Satellites of the outer planets. I. Halliday.
Journ. Roy. Astron. Soc. Canada, Vol. 66, 314 - 316 (1972).

091.036 Radiation transport in a single-layer spherical atmosphere of a planet. L. G. Titarchuk.
Kosmich. Issled., Vol. 10, 905 - 915 (1972). In Russian.

091.037 On the variation of the Laplacian spheres of activity of planets on using a new intermediate orbit.
S. V. Al'shevskij.
Kosmich. Issled., Vol. 10, 948 - 949 (1972). In Russian. – Brief information.

091.038 On the calculation of light scattering in planetary atmospheres taking into account refraction.
I. N. Minin.
Izv. AN SSSR. Fiz. atmosf. i okeana, Vol. 8, 985 - 987 (1972). In Russian. – Abstr. in Referativ. Zhurn. 51. Astron., 1.51.210 (1973).

091.039 The propagation of electromagnetic optical waves in the earth's and planetary atmospheres.
V. E. Zuev.
Vestn. AN SSSR, 1972, No. 8, p. 18 - 24. In Russian. – Abstr. in Referativ. Zhurn. 51. Astron., 1.51.296 (1973).

091.040 The early history of the Titius-Bode law.
S. T. Jaki.
American Journ. Phys., Vol. 40, 1014 - 1023 (1972).

091.041 A new method for the determination of the mixing ratio hydrogen to helium in the giant planets.
D. Gautier, K. Grossman.
Journ. Atmosph. Sci., Vol. 29, 788 - 792 (1972).
By using a numerical iterative method, it is demonstrated that the mixing ratio H_2/He on the giant planets can be inferred from spectral measurements of the intensity emitted by these plants in the far infrared range. The method is sucessfully applied to synthetic spectra of Saturn computed from atmospheric thermal models. The effect of random and systematic measurement errors on the determination of the mixing ratio is also studied.

091.042 Planetary structures in general relativity.
J. P. Sharma.
Pure and Applied Geophys., (*Switzerland*), Vol. 97, No. 5, p. 14 - 24 (1972). – See Phys. Abstr., Vol. 75, No. 69143 (1972).

091.043 When the planets were only dust.
J. Lequeux, H. Reeves.
Sci. Progr. Découverte, No. 3445, p. 12 - 20 (1972). In French.

091.044 Planetary resonances, bi-stable oscillation modes and solar activity cycles. H. P. Sleeper, Jr.
Rep. CR-2035, Northrop Services Inc., Huntsville, Ala. [Available from NTIS, Springfield, Va.], 56 pp. (1972).

091.045 The influence of line shape and band structure on temperatures in planetary atmospheres.
A. Arking, K. Grossman.
Journ. Atmosph. Sci., Vol. 29, 937 - 949 (1972).

Numerical experiments are performed to examine the effects of line shape and band structure on the radiative equilibrium temperature profile in planetary atmospheres. In order to accurately determine these effects, a method for calculating radiative terms is developed which avoids the usual approximations.

091.046 Temperature sounding experiments for the Jovian planets. F. W. Taylor.
Journ. Atmosph. Sci., Vol. 29, 950 - 958 (1972).

The possibilities for vertical temperature sounding experiments by medium-resolution measurements of outgoing radiance are examined for non-scattering models of Jupiter, Saturn, Uranus and Neptune.

091.047 Das Titius-Bodesche Gesetz im Licht der Originaltexte. S. L. Jaki.
Nachr. Olbers-Ges. Bremen, No. 86, p. 1 - 8 (1972).

091.048 Bode's law and the missing planet.
M. W. Ovenden.
Nature, Vol. 239, 508 - 509 (1972).

The author has shown that the point-mass perturbations (without dissipative forces) are a sufficient explanation of the existing distribution of satellite and planetary orbits, affording a reason for the present planetary distances to an accuracy ~1%.

091.049 Hydrogen planets and high-pressure physics.
R. Wildt.
Phys. Earth Planet. Interiors, Vol. 6, (see 012.023), 1 - 4 (1972).

091.050 Planets and satellites – a survey of fundamental facts. Z. Kopal.
Phys. Earth Planet. Interiors, Vol. 6, (see 012.023), 5 - 9 (1972).

091.051 A survey of dynamical data for the major planets and satellites. P. J. Message.
Phys. Earth Planet. Interiors, Vol. 6, (see 012.023), 17 - 20 (1972).

091.052 A system of planetary masses and related quantities. J. Kovalevsky.
Phys. Earth Planet. Interiors, Vol. 6, (see 012.023), 29 - 35 (1972).

091.053 Evidence on the deeper planetary interiors.
S. K. Runcorn.
Phys. Earth Planet. Interiors, Vol. 6, (see 012.023), 100 - 102 (1972).

091.054 The elastic energy and character of quakes in solid stars and planets. D. Pines, J. Shaham.
Phys. Earth Planet. Interiors, Vol. 6, (see 012.023), 103 - 115 (1972).

091.055 Compressibility and planetary interiors.
K. E. Bullen.
Phys. Earth Planet. Interiors, Vol. 6, (see 012.023), 131 - 135 (1972).

091.056 Shockwave determination of the shear velocity at very high pressures. O. L. Anderson.
Phys. Earth Planet. Interiors, Vol. 6, (see 012.023), 136 - 140 (1972).

091.057 The present thermal state of the terrestrial planets.
D. C. Tozer.
Phys. Earth Planet. Interiors, Vol. 6, (see 012.023), 182 - 197 (1972).

091.058 The method of "conic intersections" in determining the geometric figure of a planet by photographs made from space.
Yu. G. Karpushin, S. V. Lebedev, Yu. M. Nejman.
Izv. vyssh. ucheb. zavedenij. Geod. i aehrofotos"emka, 1972, No. 3, p. 53 - 60. In Russian. – Abstr. in Referativ. Zhurn. 62. Issled. kosmich. prostranstva, 2.62.163 (1973).

Physics of the solar system.
See Abstr. 003.130.

Light scattering in planetary atmospheres.
See Abstr. 003.138.

Physics of the moon and planets. International symposium. Kiev, 15–22 October, 1968.
See Abstr. 003.144.

Total absorption cross sections of several gases of aeronomic interest at 584 Å. See Abstr. 022.031.

Electron impact excitation cross sections and energy degradation in CO. See Abstr. 022.033.

Structure of the pressure-induced infrared spectrum of hydrogen in the first overtone region.
See Abstr. 022.035

Isentropic compression of fused quartz and liquid hydrogen to several Mbar. See Abstr. 022.141.

Correlation of theory and experiment for high-pressure hydrogen. See Abstr. 022.142.

Thermodynamics of hydrogen–helium mixtures at high pressure and finite temperature. See Abstr. 022.143.

Phase equilibria in fluid mixtures at high pressures: the He–CH$_4$ system. See Abstr. 022.144.

Lattice model calculation of elastic and thermodynamic properties at high pressure and temperature.
See Abstr. 022.148.

Observations of Jupiter, Saturn, Uranus and Neptune made with the zonal astrograph at the Nikolayev Observatory during 1964–1967. See Abstr. 041.002.

Determination of right ascensions of major planets with the meridian circle of the Kharkov Astronomical Observatory in 1968–1970. See Abstr. 041.050.

Secular perturbations of the major planets.
See Abstr. 042.006.

Quantum statistical mechanics of dense partially ionized hydrogen. See Abstr. 062.079.

Circular polarization of light scattered from rough surfaces. See Abstr. 063.004.

Absorption profile of a planetary atmosphere: A proposal for a scattering independent determination. See Abstr. 063.023.

Sunspots and planets.   See Abstr. 072.040.

The motion of planets and solar activity. See Abstr. 072.050.

The dynamical properties and internal structures of the earth, the moon and the planets.   See Abstr. 081.031.

Creep in the earth and planets. See Abstr. 081.054.

A calculated hydrogen distribution in the exosphere. See Abstr. 082.025.

Origin and evolution of the earth-moon system. See Abstr. 094.029.

Color information from lunar and planetary space missions.   See Abstr. 094.212.

Orbit-orbit resonance capture in the solar system. See Abstr. 100.019.

Metal/silicate fractionation in the solar system. See Abstr. 107.003.

092 Mercury

092.001 **Report on the planet Mercury. 1970 February to 1972 February.**   J. H. Robinson.
Journ. British Astron. Ass., Vol. 82, 373 - 376 (1972).
Report of the Mercury and Venus Section of the British Astron. Ass.

092.002 **Determination of parameters related to the interior of Mercury.**   S. J. Peale.
Icarus, Vol. 17, 168 - 173 (1972).
Bounds on $(C-A)/C$ for Mercury as a function of the uncertainty in the value of the obliquity are determined. The high precision of $1'$ of arc which is required for reasonable bounds on $(C-A)/C$ cannot be obtained by either earth-based observations or the television imagery of the Mariner 73 flyby.

092.003 **Über die maximale Helligkeit von Merkur.**   G. P. Können, J. Meeus.
Orion, 30. Jahrgang, p. 137 - 138 (1972).

092.004 **Mercury: Surface composition from the reflection spectrum.**   T. B. McCord, J. B. Adams.
Bull. American Astron. Soc., Vol. 4, 370 (1972). – Abstr. AAS.

092.005 **Thermal elastic deformations of the planet Mercury.**   H.-S. Liu.
Journ. Geophys. Res., Vol. 77, 6482 - 6485 (1972).
The variation in solar heating due to the resonance rotation of Mercury produces periodic elastic deformations on the surface of the planet. The thermal stress and strain fields under Mercury's surface are calculated after certain simplifications. It is found that deformations penetrate to a greater depth than the variation of solar heating, and that the thermal strain on the surface of the planet pulsates with an amplitude

of 4×10^{-3} and a period of 176 days.

092.006 **Mercury: Surface composition from the reflection spectrum.**   T. B. McCord, J. B. Adams.
Science, Vol. 178, 745 - 747 (1972).
The reflection spectrum for the integral disk of the planet Mercury was measured and was found to have a constant positive slope from 0.32 to 1.05 micrometers, except for absorption features in the infrared. The reflectivity curve matches closely the curve for the lunar upland and mare regions. Thus, the surface of Mercury is probably covered with a lunar-like soil rich in dark glasses of high iron and titanium content. Pyroxene is probably the dominant mafic mineral.

092.007 **Cartography of the surface markings of Mercury.**   J. B. Murray, A. Dollfus, B. Smith.
Icarus, Vol. 17, 576 - 584 (1972).
Accurate determination of the rotation period, the orientation axis, the coordinates of major surface features and the cartography of the surface markings lead to an improved knowledge of Mercury, and provide basic data for space missions scheduled to fly by the planet. They are feasible from ground-based observations.

092.008 **Mercury: Interpretation of optical observations.**   T. B. McCord, J. B. Adams.
Icarus, Vol. 17, 585 - 588 = Contr. MIT Planet. Astron. Lab., No. 59 (1972).
The spectral reflectivity of Mercury has constant positive slope from 0.32 to 1.05 μm, except for the possibility of an absorption feature in the infrared. The reflectivity curve matches closely the curve for lunar upland and mare regions. Thus, the surface of Mercury is probably covered with lunar-

like soil rich in dark glasses of high iron and titanium content. If the absorption band is real, pyroxene is the dominant mafic mineral.

092.009 The perihelion of Mercury.
R. E. Moore, D. Greenspan.
Bull. American Astron. Soc., Vol. 4, 421 (1972). – Abstr. AAS.

092.010 Unipolar interaction of Mercury with the solar wind: The steady state bow shock problem.
D. S. Colburn, C. P. Sonett, K. Schwartz.
Rep. NASA-TM-X-62028, NASA Moffett Field, California. [Available from NTIS, Springfield, Va.], 40 pp. (1971).
The steady state electromagnetic interaction of the solar wind with the planet Mercury is computed for a spectrum of electrical conductivity functions using the assumption that no atmosphere or magnetic field damps the direct interaction. The form of the induction is described by the unipolar effect and corresponds to the zero frequency limit of a transverse magnetic mode. Calculations are included to determine the effective surface temperature of the planet.

092.011 Transit of Mercury across the sun on November 10, 1973. I. Zoran.
Vasiona, Vol. 20, 59 - 60 (1972). In Serbo-Croatian.

092.012 Merkurpassasjer. T. Hansen.
Naturen 1971, No. 2, p. 99 - 109 = Inst. Teor. Astrofys. Blindern – Oslo, Småtrykk No. 72 (1971).

092.013 Transit of Mercury across the solar disk on May 9, 1970. V. Kh. Pluzhnikov.
Visn. Kharkiv. Univ. No. 82, (Ser. Astron., No. 7), p. 84 (1972). In Ukrainian.

092.014 Paso de Mercurio por el disco del sol en dia 9 de mayo de 1970.
M. M. Lorón, J. Pensado, J. Claver.
Bol. Astron. Obs. Madrid, Vol. 8, No. 1, p. 143, 144 (1972).

Some questions on the treatment of observations of major planets.   See Abstr. 042.065.

Correction of solar observations for stray light by numerical integration, with application to Mercury's drop. See Abstr. 072.052.

Review of surface and atmosphere studies of Venus and Mercury.   See Abstr. 093.030.

Orbital eccentricity of Mercury and the origin of the moon.   See Abstr. 094.180.

093 Venus

093.001 **Spectral observations of Venus in the frequency interval 18.5−24.0 GHz; 1964 and 1967−68.**
D. E. Jones, D. M. Wrathall, B. L. Meredith.
Publ. Astron. Soc. Pacific, Vol. 84, 435 - 442 (1972).

During July 1964, Venus was observed with a tunable radiometer at eight frequencies in the interval 20.6−24.0 GHz and again at six frequencies in the interval 18.5−24.0 GHz during an eight-month period beginning approximately one month after the inferior conjunction of 1967. Microwave spectra for these frequency intervals were computed using the results from Mariner V and Veneras 4−7 and compared to these data.

093.002 **Venus: a perspective at the beginning of planetary exploration.** M. Ya. Marov.
Icarus, Vol. 16, 415 - 461 (1972).

The results of measurements of the Venus atmosphere, made by the entry probes Venera 4, 5, 6 and 7, and data from the Mariner 5 fly-by, are presented. The mechanism of radiative and convective heat transfer in the Venus lower atmosphere and the greenhouse and deep circulation models are discussed. A model of the Venus atmosphere up to several hundred kilometers above the surface, based on results of Venera and Mariner 5 measurements and data from groundbased optical and radio observations, is developed.

093.003 **How to measure surface and atmospheric conditions on Venus by microwave interferometry.**
W. A. Gale, A. C. E. Sinclair.
Astrophys. Journ., Vol. 175, 535 - 554 (1972).

Predictions of the polarized and unpolarized interferometric visibilities of Venus are made for wavelengths from 3 cm to 100 cm. The basis for the predictions is a multiparameter planetary model of brightness temperature. We reviewed earth-based and spacecraft observations of Venus to establish particular values of all parameters in the general model, thus yielding a reference model. We also used experimental interferometric results at 11.1 cm to help select the parameter values of the reference model.

093.004 **Precision interferometric observations of Venus at 11.1-centimeter wavelength.**
A. C. E. Sinclair, J. P. Basart, D. Buhl, W. A. Gale.
Astrophys. Journ., Vol. 175, 555 - 572 (1972).

Precision measurements were made of Venus's interferometric visibility for both the total and the polarized radiation at 11.1-cm wavelength. The observations showed that at this wavelength the brightness of the planet's disk has substantially circular symmetry. By analysis of the data according to a simple planetary model we deduced values of two planetary parameters, obtaining 3.90 ± 0.15 for the dielectric constant of the surface and 0.65 percent ± 0.12 percent for the water content of Venus's atmosphere, where the quoted uncertainties reflect only experimental error. We discuss the effects of errors in assumptions for the model parameters.

093.005 **The phase anomaly of Venus. Does refraction play a part ?** F. V. Davies.
Journ. British Astron. Ass., Vol. 82, 341 - 352 (1972).

093.006 **On mercury clouds in the atmosphere of Venus.**
J. F. Potter.
Icarus, Vol. 17, 79 - 87 (1972).

In most of the models of the atmosphere of Venus the upper layer consists of highly reflecting clouds of Hg_2Cl_2 and $HCl−H_2O$ solution above a highly absorbing cloud of liquid mercury droplets. We have calculated the reflectivity of these cloud systems to determine the limits placed upon such models by the requirement that they reproduce the observed high albedo of Venus. In this way an upper limit is placed on the mercury droplet cloud mass. Relations are given which allow one to determined whether a particular model (i. e. a set of cloud masses and particle sizes) satisfies the constraints imposed by the requirement that it produce the high albedo of Venus.

093.007 **New optical measurements of planetary diameters. Part II: Planet Venus.** A. Dollfus.
Icarus, Vol. 17, 104 - 115 (1972).

The survey of planetary diameter measurements with the double-image birefringent micrometer and large diameter refractors provided, between 1953 and 1964, a total of 82 independent measurements of the planet Venus, all selected under excellent seeing conditions. Their reduction gives the semidiameter 6115 ± 13 km corresponding to the level in the atmosphere at the upper boundary of the opaque clouds. From comparisons with values given by other techniques, a vertical model of the atmosphere is deduced and summarized by a figure.

093.008 **A model for the lower atmosphere of Venus based on high resolution interferometric observations at radio wavelengths.**
G. S. Orton, D. O. Muhleman, G. L. Berge.
Bull. American Astron. Soc., Vol. 4, 321 (1972). − Abstr. AAS.

093.009 **On the determination of characteristics of scattering particles in the Venus atmosphere based on photometric measurements.**
Yu. L. Biryukov, L. G. Titarchuk.
Kosmich. Issled., Vol. 10, 576 - 579 (1972). In Russian.

093.010 **Prediction of atmospheric parameters on Venus from high resolution spectroscopy of a CO_2 line.**
N. P. Carleton, W. A. Traub.
Bull. American Astron. Soc., Vol. 4, 362 (1972). − Abstr. AAS.

093.011 **Detecting temperature gradients in a planetary atmosphere spectroscopically.** A. T. Young.
Bull. American Astron. Soc., Vol. 4, 362 - 363 (1972). Abstr. AAS.

093.012 **Spectroscopic observations of weather on Venus.** L. D. G. Young.
Bull. American Astron. Soc., Vol. 4, 363 (1972). − Abstr. AAS.

093.013 **Variability and lateral inhomogeneity in the Venus clouds.** M. J. S. Belton.
Bull. American Astron. Soc., Vol. 4, 363 (1972). − Abstr. AAS.

093.014 **Spectroscopic evidence against pure ice clouds on Venus.**
L. P. Giver, R. W. Boese, J. H. Miller, J. L. Regas.
Bull. American Astron. Soc., Vol. 4, 363 (1972). − Abstr. AAS.

093.015 **Spectroscopic observations of Venus at superior conjunction.** R. A. J. Schorn, G. E. Hunt.
Bull. American Astron. Soc., Vol. 4, 363 (1972). − Abstr. AAS.

093.016 **The far ultraviolet emission spectra of Venus.** G. J. Rottman, H. W. Moos.
Bull. American Astron. Soc., Vol. 4, 363 - 364 (1972). Abstr. AAS.

093.017 "Atomic oxygen on Venus". D. J. Strickland.
Bull. American Astron. Soc., Vol. 4, 364 (1972).
Abstr. AAS.

093.018 Determination of the turbulence characteristics of
the Venusian atmosphere from Mariner V S-band
occultation data. R. Woo, A. Ishimaru.
Bull. American Astron. Soc., Vol. 4, 364 (1972). — Abstr. AAS.

093.019 What is today known about Venus.
M. Ya. Marov.
Priroda, No. 10.72, p. 9 - 20 (1972). In Russian.

093.020 De donkere wolken van Venus. G. A. T. Heillegger.
Hemel en Dampkring, Vol. 70, 266 - 267 (1972).

093.021 News from Venus. R. N. Watts, Jr.
Sky Telescope, Vol. 44, 303 (1972).

093.022 Thermal equilibrium calculations of the lower Venus
atmosphere. E. Roeckner, P. Fabian.
Contr. Atmosph. Phys. (Beiträge Phys. Atmosphäre), Vol. 45,
230 - 243 (1972).
Global average temperature profiles of the Venus atmosphere were calculated in a time dependent model on the basis
of solar and thermal radiation fluxes and eddy heat fluxes in
a 100% CO_2-atmosphere with additional water vapor up to a
content of 0.5%. A comparison of calculations with in situ
measurements of temperature by space probes indicates that
the Venus atmosphere is likely to contain water vapor.

093.023 On possibilities of determining the temperature
profile in the Venus atmosphere from the thermal
radio radiation of the planet.
Yu. M. Timofeev, M. A. Gruzdeva, O. M. Pokrovskij.
Kosm. Issled., Vol. 10, 751 - 759 (1972). In Russian.

093.024 Über die Phasen der Venus. E. Wischnewski.
SuW, Vol. 11, 324 (1972).

093.025 On the choice of the optimal distribution of radar
and optical observations of Venus. V. A. Izvekov.
Byull. Inst. Teoret. Astron., Leningrad, Vol. 13, 201 - 209
(1972). In Russian.
The distribution of the observations of Venus allowing
to define its ephemeris in a short time is chosen experimentally
The periods favourable for making observations of Venus are
shown. Estimates of the expected precision of the determination and interdependences of the values sought for are given.

093.026 Results of the Venus atmosphere measurements
made by the landing station Venera 7.
M. Ya. Marov, V. S. Avduevsky, M. K. Rozhdestvensky, V. V.
Kerzhanovich, N. F. Borodin, O. L. Ryabov.
Space Research XII, (see 012.016), Vol. 1, 281 - 292 (1972).

093.027 Diurnal variation of the exospheric temperatures on
Venus and Mars. M. Shimizu.
Space Research XII, (see 012.016), Vol. 1, 293 - 298 (1972).

093.028 Report on the elongation of Venus, 1972 April.
J. H. Robinson.
Journ. British Astron. Ass., Vol. 83, 38 - 43 (1972). — Report
of the Mercury and Venus Section of the British Astron. Ass.

093.029 The thermal structure within the stratospheres of
Venus and Mars. R. D. Cess.
Icarus, Vol. 17, 561 - 569 (1972).
A relatively simple radiative-convective model has been
formulated to predict the vertical temperature profile within
the stratospheres of Venus and Mars. For Venus this yields a

tropopause temperature which is essentially independent of
latitude with a value comparable to the Mariner 5 occultation
experiment for a tropopause altitude of 58 km. With respect
to Mars, the calculated profiles are in good agreement with
the Mariner 6 and 7 entry profiles below 30 km, producing
subadiabatic lapse rates throughout most of this region due to
relatively low tropopause altitudes. In the upper stratosphere,
however, the calculated profiles do not predict carbon dioxide
condensation as implied by the Mariner 6 and 7 experiments.

093.030 Review of surface and atmosphere studies of Venus
and Mercury. R. M. Goldstein.
Icarus, Vol. 17, 571 - 575 (1972).
This review paper covers material published since the
last symposium on 'planetary atmospheres and surfaces',
which was held in August, 1969. Only ground-based observations are reviewed, and emphasis is kept close to the data.

093.031 Venus: Atmospheric rotation
A. H. Scott, E. J. Reese.
Icarus, Vol. 17, 589 - 601 (1972).
There is an obvious difference between the mean rotation period of about 4.5 days derived from short-interval displacements and the 4.067-day period based on Boyer and
Guerin's study of features which were reported to endure for
many years. It was this conflict of results which prompted a
new study of ultraviolet plates taken at this station.

093.032 Ultraviolet clouds on Venus: Observational bias.
R. F. Beebe.
Icarus, Vol. 17, 602 - 607 (1972).
A sidereal model for calculating the observational bias
imposed on an assumed distribution of periods of rotation
for the ultraviolet cloud markings on Venus has been investigated. The results are compared to the observed distribution.
Observational constraints are varied in the model in order to
determine the influence of each parameter. The observational bias is investigated for single stations and pairs of stations
located 6 and 12 hr apart.

093.033 Retrograde rotation of the upper atmosphere of
Venus. J. Caldwell.
Icarus, Vol. 17, 608 - 616 (1972).
From June through September 1970, ultraviolet photographs of Venus, many of very high quality, were obtained
by the International Planetary Patrol Program. A study of
these photographs, together with some from other observatories, confirms previous reports that there is a retrograde circulation of the upper atmosphere of Venus.

093.034 Water vapor in the atmosphere of Venus.
U. Fink, H. P. Larson, G. P. Kuiper, R. F. Poppen.
Icarus, Vol. 17, 617 - 631 (1972).
Infrared spectra of Venus produced by a Fourier spectrometer flown aboard the NASA CV 990 jet aircraft were analyzed for water-vapor content by comparison with calculated
model spectra. The reflecting layer model gave an abundance
of $1.6 \pm 0.4 \, \mu$ of precipitable water for the two-way transmission of the Venus atmosphere. The scattering model resulted in a value of $0.25 \pm 0.10 \, \mu$ of water per scattering mean
free path.

093.035 High resolution spectra of Venus—a review.
L. D. G. Young.
Icarus, Vol. 17, 632 - 658 (1972).
The existing high resolution spectra of Venus show considerable variations in the abundance of CO_2. As this is the
major constituent of the atmosphere, these day-to-day variations indicate the effective reflecting cloud layer is undergo-
ing substantial vertical displacements over a relatively short
time scale. The published observations indicate that CO_2 is

uniformly mixed with the CO_2, but that the lines of HF may be formed deeper in the atmosphere and HCl lines considerably deeper.

093.036 Data on dynamics of the subcloud Venus atmosphere from Venera spaceprobe measurements.
V. V. Kerzhanovich, M. Ya. Marov, M. K. Rozhdestvensky.
Icarus, Vol. 17, 659 - 674 (1972).
This paper presents the principal results of wind velocity and turbulence measurements in the Venus atmosphere during the Venera flights.

093.037 High resolution interferometric observations of Venus at three radio wavelengths.
G. L. Berge, D. O. Muhleman, G. S. Orton.
Icarus, Vol. 17, 675 - 681 (1972).
We present new 6.0 and 21.1 cm interferometric observations of Venus. When combined with our previous 3.12 cm work they provide a self-consistent set of high-resolution observations at three wavelengths covering a range in which the opacity of the Venus atmosphere varies by a factor of 50. Model calculations indicate that a model atmosphere of CO_2 in adiabatic equilibrium containing uniformly mixed gaseous absorbers surrounding a dielectric sphere cannot simultaneously and adequately predict the radio interferometric measurements at all wavelengths together with the radar and radio occultation measurements.

093.038 Venus : Measurements of brightness temperatures in the 7−15 cm wavelength range and theoretical radio and radar spectra for a two-layer subsurface model.
W. W. Warnock, J. R. Dickel.
Icarus, Vol. 17, 682 - 691 (1972).
The objectives of the present research were to define more accurately the radio spectrum of Venus in the wavelength range of 7−15 cm and to interpret the resulting radio spectrum and the radar spectrum in terms of atmospheric and subsurface parameters.

093.039 Some characteristics of the Venus surface.
N. N. Kroupenio.
Icarus, Vol. 17, 692 - 698 (1972).
From data on the atmospheric temperature, pressure and chemical composition, direct measurements on which have been carried out by the Soviet interplanetary automatic stations of the 'Venera' series, radiowave attenuation and refraction in the Venus atmosphere are calculated. This allows a correction to Venus radar measurements from the earth and determines the characteristics of radiowave scattering by the planetary surface. From these data a dielectric constant and density of the surface layer were calculated as well as root-mean-square angles of surface inclination from 40 cm up to 7.0 m.

093.040 A radar image of Venus.
R. M. Goldstein, H. C. Rumsey.
Icarus, Vol. 17, 699 - 703 (1972).
Radar scans of Venus, performed at the Jet Propulsion Laboratory's Goldstone Tracking Station, have yielded a brightness map of a large portion of the surface.

093.041 On the observations of transits of Venus over the sun with particular emphasis on the December 9, 1874 event observed in Japan − Part I. (Collective review).
K. Saito, S. Shinozawa.
Univ. Tokyo, Tokyo Astron. Obs. Rep. (No. 60), Vol. 16, 72 - 162 (1972). In Japanese.

093.042 Intensities and spectroscopic constants for some weak CO_2 bands. L. D. G. Young.
Journ. Optical Soc. America, Vol. 62, 1382 (1972). − Abstr.

Optical Soc. America.

093.043 No oceans of oil on Venus. O. Borisov.
Spaceworld, (USA), Vol. 1, 1-97, p. 41 (1972).
See Phys. Abstr., Vol. 75, No. 51832 (1972).

093.044 Stability of thin non-rotating Hadley circulations.
J. E. Hart.
Journ. Atmosph. Sci., Vol. 29, 687 - 697 (1972).

093.045 Venus: Measurements of microwave brightness temperatures and interpretations of the radio and radar spectra. W. W. Warnock.
Thesis, Univ. Illinois, Urbana–Champaign. [Available from Univ. Microfilms, Ann Arbor, Mich., USA. Order No. 72-12425], 70 pp. (1971).
Measurements of the brightness temperature and radar cross section of Venus at radio frequencies reveal a wealth of information concerning its atmospheric and subsurface properties.

093.046 A possibility of multiple exospheric temperatures for Venus and Mars. J. R. Herman.
Rep. NASA-TM-X-65690, NASA Goddard Space Flight Center, Greenbelt, Md. [Available from NTIS, Springfield, Va.] 17 pp. (1971). − See Phys. Abstr.,Vol. 75, No. 73090 (1972).

093.047 Venus atmosphere: structure and stability of the ClOO radical. R. G. Prinn.
Journ. Atmosph. Sci., Vol. 29, 1004 - 1007 (1972).
The ClOO radical has been suggested as an important intermediate in a Cl atom catalyzed combination of CO and O_2 enabling conservation of the observed low mixing ratios for CO and O_2 in the predominantly CO_2 Venusian atmosphere. The author reviews evidence concerning the existence, structure and stability of this radical.

093.048 Radar interferometric observations of Venus.
D. B. Campbell.
Thesis, Cornell Univ., Ithaca, New York. [Available from Univ. Microfilms, Ann Arbor, Mich., USA. Order No. 72-7547], 131 pp. (1971).
During the 1969 inferior conjunction of Venus both single-station and interferometric range-Doppler radar mapping of the planet were carried out at a wavelength of 70 cm at the Arecibo Observatory.

093.049 A method of improvement of the Venus ephemeris by optical and radar observations. V. A. Izvekov.
Publ. 18th Astrometrical Conference 1969, (see 012.022), p. 37 - 43 (1972). In Russian.

093.050 Venus Section report: The eastern (evening) apparition of 1966−1967. D. P. Cruikshank.
Strolling Astronomer, Vol. 24, 1 - 14 (1972).
The evening apparition of Venus of 1966−1967 includes the period from November 9, 1966 (superior conjunction) to August 30, 1967 (inferior conjunction). The planet appeared unusually active in that the number of persistent dark markings was large, the extended cusps and "halo" presented an interesting show, and the Ashen Light was recorded by several observers.

Soviet science: Venus-8 data.
Nature, Vol. 239, 125 - 126 (1972).

Some observations of Venus.
Sky Telescope, Vol. 44, 197 - 199 (1972).

Branching ratios in photoelectron spectroscopy.
See Abstr. 022.037.

The Planetary Explorer probe studies.
See Abstr. 053.009.

Electrical conduction in physical and chemical mixtures. Application to planetary mantles.
See Abstr. 081.029.

The spherical albedo of a planetary atmosphere.
See Abstr. 091.003.

On the temperature distribution in a planetary atmosphere.   See Abstr. 091.007.

Compressibility and planetary interiors.
See Abstr. 091.055.

The specific effective scattering area of the surfaces of the moon, Mars and Venus in the radio region.
See Abstr. 094.046.

Observations of O_2 on Mars and Venus.
See Abstr. 097.046.

094 Moon

094.001 Moon model – an offset core.
G. Ransford, W. Sjogren.
Nature, Vol. 238, 260 - 262 (1972).

A lunar model having an asymmetric core explains the offset centre of gravity, moment of inertia, mascons and maria. A possible evolution theory is also presented.

094.002 Fragments of terra rock in the Apollo 12 soil samples and a structural model of the moon.
J. A. Wood.
Icarus, Vol. 16, 462 - 501 (1972).

Samples of lunar soil collected by Apollo 12 in Oceanus Procellarum contain not only basalt fragments, obviously locally derived, but also a component of light-colored rock fragments that probably derive from the lunar terrae. Most of these are noritic in composition, not anorthositic as were the light-colored soil particles of Apollo 11. Texturally, they are recrystallized breccias. An evolutionary model is developed that is consistent with the properties of this and lunar rock types, and with the geophysical properties of the moon.

094.003 Possible production mechanisms of lunar magnetic fields. F. F. Cap.
Journ. Geophys. Res., Vol. 77, 3328 - 3333 (1972).

The impossibility of the production of local surface magnetic fields on the moon by conduction currents in the lunar soil and in local lunar atmospheres by volcanic eruption is shown. However, it is suggested that convection currents produced by the ionization of volcanic-ash-particle flows may produce the local magnetic fields of about 1000 γ.

094.004 Lunar ash flows: Isothermal approximation.
S. I. Pai, T. Hsieh, J. A. O'Keefe.
Journ. Geophys. Res., Vol. 77, 3631 - 3649 (1972).

A general description of the lunar ash flow is given, and a simple mathematical model of the isothermal lunar ash flow is worked out with numerical examples to show the differences between the lunar and the terrestrial ash flow. It appears that the lunar surface layer in the maria is not a residual mantle rock (regolith) but a series of ash flows due, at least in part, to great meteorite impacts.

094.005 Apollo 16 geochemical X-ray fluorescence experiment: Preliminary report. I. Adler, J. Trombka,
J. Gerard, P. Lowman, R. Schmadebeck, H. Blodget, E. Eller, L. Yin, R. Lamothe, G. Osswald, P. Gorenstein, P. Bjorkholm, H. Gursky, B. Harris.
Science, Vol. 177, 256 - 259 (1972).

The lunar surface was mapped with respect to magnesium, aluminum, and silicon as aluminum/silicon and magnesium/silicon intensity ratios along the projected ground tracks swept out by the orbiting Apollo 16 spacecraft. The results confirm the observations made during the Apollo 15 flight and provide new data for a number of features not covered before.

094.006 The concentric craters of the moon.
A. V. Bugaevskij.
Astron. Zhurn. Akad. Nauk SSSR, Vol. 49, 850 - 852 (1972). In Russian. English translation in Soviet Astron. AJ, Vol. 16, No. 4.

Some statistic regularities for small (to 500 m diameter) craters with sharply traced concentric walls are considered.

094.007 Highly aluminous glasses in lunar soils and the nature of the lunar highlands.
A. M. Reid, W. I. Ridley, R. S. Harmon, J. Warner, R. Brett, P. Jakeš, R. W. Brown.

Geochim. Cosmochim. Acta, Vol. 36, 903 - 912 (1972).

Approximately 25 per cent of the glasses in two Apollo 14 soil samples and in the soils at two levels in the Luna 16 core have compositions equivalent to anorthositic gabbro. Anorthositic gabbro glasses have the same major element composition at all four sites, and resemble the Surveyor 7 analysis from a 'highland' site. Thus, strong presumptive evidence exists that material with this specific composition is abundant in the lunar highlands.

094.008 The lunar temperature profile.
A. Duba, H. C. Heard, R. N. Schock.
Earth Planet. Sci. Letters, Vol. 15, 301 - 304 (1972).

Data on the electrical conductivity of a natural olivine having little or no Fe^{3+} indicate that the temperature of the lunar interior can be near the solidus of some lunar models.

094.009 The average $^{130}Ba(n, \gamma)$ cross section and the origin of ^{131}Xe on the moon.
W. A. Kaiser, B. L. Berman.
Earth Planet. Sci. Letters, Vol. 15, 320 - 324 (1972).

The average $^{130}Ba(n, \gamma)$ cross section has been measured for a neutron spectrum similar to the one at the lunar surface. The large cross section obtained probably explains the anomalously high concentration of ^{131}Xe found in the lunar rocks.

094.010 Rates of solidification of Apollo 11 basalt and Hawaiian tholeiite.
A. Provost, Y. Bottinga.
Earth Planet. Sci. Letters, Vol. 15, 325 - 337 (1972).

Cooling histories of Apollo 11 basalt flows have been evaluated numerically. It is concluded that the Apollo 11 basalts are samples from less than 70 cm depth in a flow, and that the sample 10020 comes from a depth less than 10 cm. To check the computations, the cooling of a Hawaiian tholeiitic lava lake was calculated; the results are in good agreement with the observed data on the cooling of the Kilauea Iki and Alea lava lakes.

094.011 Fission xenon from extinct ^{244}Pu in 14301.
R. Drozd, C. M. Hohenberg, D. Ragan.
Earth Planet. Sci. Letters, Vol. 15, 338 - 346 (1972).

Xenon extracted in step-wise heating of lunar breccia 14301 contains a fission-like component in excess of that attributable to uranium decay during the age of the solar system. It appears that ^{244}Pu was extant at the time lunar crustal material cooled sufficiently to arrest the thermal diffusion of xenon. Subsequent history has apparently maintained the isotopic integrity of plutonium fission xenon.

094.012 Recent developments in the theory of the motion of the moon. J. S. Griffith.
Journ. British Interplanet. Soc., Vol. 25, 553 - 556 (1972).

The following questions are considered: Why is a new theory being produced that is known to be inadequate? How do we know that discrepancies do exist? What high precision purposes require a theory that appears to be almost impossible to produce?

094.013 Mission to Descartes – 2. D. Baker.
Spaceflight, Vol. 14, 287 - 291 (1972).

094.014 Classes of lunar transient phenomena. P. Moore.
Journ. British Astron. Ass., Vol. 82, 372 (1972).

094.015 The prospects for mineral analysis by remote infrared spectroscopy. J. R. Aronson, A. G. Emslie.

The Moon, Vol. 5, 3 - 15 (1972). — Commun. presented at the Conference on Lunar Geophysics 1971, Houston, Texas.

094.016 Infrared observations of the moon and their interpretation. M. J. Pugh, J. A. Bastin.
The Moon, Vol. 5, 16 - 30 (1972). — Commun. presented at the Conference on Lunar Geophysics 1971, Houston, Texas.

The lunar spectrum, resulting from both the directly scattered solar radiation and the moon's intrinsic thermal radiation, is described. The variations of the thermal component with latitude and phase, and during eclipse conditions, are described and compared with a plane homogeneous model of temperature-independent thermal constants.

094.017 The effect of a thermal and ultrahigh vacuum environment on the strength of precompressed granular materials. E. A. Nowatzki.
The Moon, Vol. 5, 31 - 40 (1972). — Commun. presented at the Conference on Lunar Geophysics 1971, Houston, Texas.

094.018 Lunar thermal history revisited.
R. K. McConnell, Jr., P. W. Gast.
The Moon, Vol. 5, 41 - 51 (1972). — Commun. presented at the Conference on Lunar Geophysics 1971, Houston, Texas.

The internal temperatures, heat fluxes, and rates of evolution of volcanic liquids for lunar models with initial radioactivities and temperatures that decrease going downward in the moon are calculated.

094.019 Spectrophotometry (0.3 to 1.1 μ) of visited and proposed Apollo lunar landing sites.
T. B. McCord, M. P. Charette, T. V. Johnson, L. A. Lebofsky, C. Pieters.
The Moon, Vol. 5, 52 - 89 (1972). — Commun. presented at the Conference on Lunar Geophysics 1971, Houston, Texas.

094.020 The moon's thermal state and an interpretation of the lunar electrical conductivity distribution.
D. C. Tozer.
The Moon, Vol. 5, 90 - 105 (1972). — Commun. presented at the Conference on Lunar Geophysics 1971, Houston, Texas.

094.021 On the applicability of lunar breccias for paleomagnetic interpretations.
W. A. Gose, G. W. Pearce, D. W. Strangway, E. E. Larson.
The Moon, Vol. 5, 106 - 120 (1972). — Commun. presented at the Conference on Lunar Geophysics 1971, Houston, Texas.

094.022 The chemical composition and structure of the moon. P. W. Gast.
The Moon, Vol. 5, 121 - 148 (1972). — Commun. presented at the Conference on Lunar Geophysics 1971, Houston, Texas.

The main objective of this paper will be to determine constraints on the composition of the lunar interior that derive from the chemical composition of the lunar igneous rocks.

094.023 Comments on the figure of the moon from Apollo landmark tracking.
W. R. Wollenhaupt, R. K. Osburn, G. A. Ransford.
The Moon, Vol. 5, 149 - 157 (1972). — Commun. presented at the Conference on Lunar Geophysics 1971, Houston, Texas.

094.024 The significance of the magnetism observed in lunar rocks. C. E. Helsley.
The Moon, Vol. 5, 158 - 160 = Geosci. Division, Univ. Texas, Dallas, Contr. No. 201 (1972). — Commun. presented at the Conference on Lunar Geophysics 1971, Houston, Texas.

094.025 The sunlit lunar surface. I. Albedo studies and full moon temperature distribution.
J. M. Saari, R. W. Shorthill.

The Moon, Vol. 5, 161 - 178 (1972).

Certain studies were made of the lunar surface based on measurements at visible and far infrared wavelengths obtained from a scan over the full moon. The simultaneous measurements at the two wavelengths provided a factor for converting full moon photometric brightness to bolometric albedo. The disk-averaged value of this parameter is consistent with previous photometric measurements. The brightness temperature statistics over the disk led to a replacement of the traditional $\cos^{1/6} \theta$ law of temperature variation over the full moon by an expression which is linear in $\cos \theta$.

094.026 The sunlit lunar surface. II. A study of far infrared brightness temperatures.
J. M. Saari, R. W. Shorthill, D. F. Winter.
The Moon, Vol. 5, 179 - 199 (1972).

Directional infrared emission from the sunlit lunar surface is determined for the thermal meridian and as a function of observer elevation and azimuth angles at three sun elevation angles. A study of selected mare sites at full moon suggests that brightness temperatures are relatively insensitive to changes in certain surface parameters, such as the photometric function, emissivity, and thermophysical properties of the soil.

094.027 Cosmic influences on the early history of the lunar surface. Z. Kopal.
The Moon, Vol. 5, 200 - 205 (1972).

It is pointed out that the observed random distribution of low-angle impact craters over the lunar surface rules out the possibility that particles initially responsible for the origin of such craters had, prior to impact, been in heliocentric orbits.

094.028 Origin of the moon by tidal capture and some geophysical consequences. S. F. Singer.
The Moon, Vol. 5, 206 - 209 (1972).

Of the many proposed modes of origin of the moon, perhaps the least improbable is capture of the moon as it passed near the earth in a direct (prograde) orbit, shortly after the formation of moon and earth, about 4.5 billion years ago. The effects of capture on the earth would have been cataclysmic, leading to intensive heating of its interior, to volcanism, and to the immediate formation of an atmosphere and hydrosphere. Thus capture of a moon may have given rise to the unique properties of the earth and to the early evolution of life, about 3.5 billion years ago.

094.029 Origin and evolution of the earth-moon system.
H. Alfvén, G. Arrhenius.
The Moon, Vol. 5, 210 - 230 (1972).

The origin and evolution of the earth-moon system is studied by comparing it to the satellite systems of other planets. The earth should have a 'normal' satellite system that consists of about half a dozen satellites each with a mass of a fraction of a percent of the lunar mass. Hence, the moon is not likely to have been generated in the environment of the earth by a normal accretion process. There is little doubt that the moon is a captured satellite. The earth and the moon are likely to have been formed from planetesimals accreting in particle swarms in Kepler orbits (jet streams). This process leads to the formation of a cool lunar interior with an outer layer accreted at increasingly higher temperatures. The primeval earth should similarly have formed.

094.030 Sternwarte Passau photographierte rätselhafte Lichtfontäne im Aristarch-Gebiet.
G. Küveler, R. Klemm.
SuW, Vol. 11, 238 - 239 (1972).

Neues vom Mondprogramm der Sternwarte Gummersbach; Neue Intensitätsskalen für das sekundäre Mondlicht.

094.031 Superficie lunar. N. Fragoso.

El Universo, No. 99, Vol. 26, 60 - 62 (1972).

094.032 Reflection spectra of lunar dust grains with amorphous coatings. B. Hapke.
Science, Vol. 177, 535 - 536 (1972).

094.033 Occurrence of chromian, hercynitic spinel ("pleonaste") in Apollo-14 samples and its petrologic implications. E. Roedder, P. W. Weiblen.
Earth Planet. Sci. Letters, Vol. 15, 376 - 402 (1972).

Spinels are a minor but important component of most lunar samples. Their importance lies in the wide range of compositions possible (Mg and Fe; Al, Cr, and Ti), that reflect both the stage of differentation and the state of oxidation of the host rock. We describe here the composition and occurrence of a group of spinels that are beyond the compositional range shown by Haggerty.

094.034 Uranium distribution in basalt fragments of five lunar samples. K. Thiel, W. Herr, J. Becker.
Earth Planet. Sci. Letters, Vol. 16, 31 - 44 (1972).

The U-distribution in selected basaltic fragments from lunar fines 10084, rocks 12021 and 12053 and breccias 14305 and 14321 was studied by means of fission tracks in mica detectors. U-enriched mineral phases were localized with an accuracy of better than 1 μm by applying a photographic mapping technique.

094.035 Lunar reflectivity at 0.86-centimeter wavelength. J. J. G. McCue, E. A. Crocker.
Journ. Geophys. Res., Vol. 77, 4069 - 4078 (1972).

Backscattering of 0.86-cm circularly polarized radiation by 29 regions on the moon has been measured with better signal-to-noise ratio than was available heretofore. Both circular polarizations were received. Our measurements do show that the highlands and the seas depart systematically from the smoothed reflectivity curve. We find a more rapid decrease in reflectivity with increasing angle of incidence ϕ.

094.036 Orbital search for lunar volcanism. R. R. Hodges, Jr., J. H. Hoffman, T. T. J. Yeh, G. K. Chang.
Journ. Geophys. Res., Vol. 77, 4079 - 4085 (1972).

The total rate of volcanic release of gases into the lunar atmosphere is estimated to be less than 60 g/sec. One of the implications of this degassing is that, if it occurs as sporadic releases of large quantities of gas, these events can be detected by an orbiting mass spectrometer. The nature of a volcanic perturbation of the lunar atmosphere is discussed, and a lower bound is derived for the expected time between detected events.

094.037 Aluminum 26 and manganese 53 produced by solar-flare particles in lunar rock and cosmic dust.
S. Tanaka, K. Sakamoto, K. Komura.
Journ. Geophys. Res., Vol. 77, 4281 - 4288 (1972).

In this report, by a somewhat different approach and by using different input data the production rates of ^{26}Al ($T_{1/2}$ = 7.4 $\times$ 10^5 years) and ^{53}Mn ($T_{1/2}$ = 3.7 $\times$ 10^6 years) in lunar rocks and cosmic dust have been recalculated, and the results have been compared with the observed depth variations of ^{26}Al and ^{53}Mn in lunar rocks and with the observed limit of ^{26}Al in marine sediments.

094.038 A multiple-scattering model of the diffuse component of lunar radar echoes.
J. B. Pollack, L. Whitehill.
Journ. Geophys. Res., Vol. 77, 4289 - 4303 (1972).

We propose that the average diffuse component of lunar radar echoes results from the cumulative effect of multiple scattering within the ejecta blanket of fresh young craters. We compare our model with a variety of observations of the diffuse component and find a general agreement between the two.

094.039 Craters formed in mineral dust by hypervelocity microparticles. J. F. Vedder.
Journ. Geophys. Res., Vol. 77, 4304 - 4309 (1972).

As a simulation of erosion processes on the lunar surface, impact craters were formed in dust targets by 2- to 5-μm-diameter polystyrene spheres with velocities between 2.5 and 12 km/sec. The results are useful in studies of mixing of the fine fraction of the lunar regolith and shielding of dust-coated lunar rock surfaces under bombardment by micrometeoroids.

094.040 Maps of lunar hemispheres. Giving the views of the lunar globe from six cardinal directions in space.
A. Rükl, with a foreword by Z. Kopal.
D. Reidel Publishing Company, Dordrecht – Holland.
Astrophys. Space Sci. Library, Vol. 33, 5 + 24 pp. + 6 charts.
Price hfl. 70.00 (1972). – Contents: Mapping of the moon; Maps of lunar hemispheres: 1. Coordinates of places on the lunar surface; 2. Orientation of the six maps of the lunar hemisphere and their representation; 3. Construction of the maps; 4. Lunar nomenclature; Index of named formations.

094.041 The ground of lunar continents. L. S. Tarasov.
Priroda, No. 7.72, p. 102 (1972). In Russian.

094.042 Preliminary data on lunar ground returned by the Luna 20 automatic space probe. A. P. Vinogradov.
Priroda, No. 8.72, p. 2 - 10 (1972). In Russian.

094.043 Some differentials in Brown's lunar theory. J. S. Griffith.
Celestial Mechanics, Vol. 6, 111 - 116 (1972).

Differences between expressions derived by Brown and Griffith are shown to be a consequence of the different methods of derivation. Other differentials used in Brown's method are derived.

094.044 The lunar conductivity profile and the nonuniqueness of electromagnetic data inversion.
R. J. Phillips.
Icarus, Vol. 17, 88 - 103 (1972).

A review of the theory for the electromagnetic functional used to date to determine the lunar conductivity profile from spectral analyses of lunar magnetometer data is presented. The use of the spectral data in conjunction with the functional to find a least squares conductivity profile is examined from the point of view of the nonuniqueness of nonlinear estimation. An unsuccessful search was carried out to find models that were both geophysically reasonable and consistent with the data. It is asserted that the failure to find these models is due to the inadequacy of the functional.

094.045 The origin of the moon. D. L. Anderson.
Nature, Vol. 239, 263 - 265 (1972).

Explanation of the moon's composition in a model for the condensation of the planets from a cloud of solar composition suggests that the moon accreted at higher temperatures and lower pressures than the terrestrial planets. This may be explained if the lunar orbit initially had a high inclination.

094.046 The specific effective scattering area of the surfaces of the moon, Mars and Venus in the radio region.
N. N. Krupenio.
Kosmich. Issled., Vol. 10, 569 - 575 (1972). In Russian.

094.047 Bistatic radar lunar surface observations with Apollos 14 and 15. H. T. Howard, G. L. Tyler.
Bull. American Astron. Soc., Vol. 4, 369 (1972). – Abstr. AAS.

094.048 Description of an anomalous region in the lunar highlands from ground based and orbital measurements. A. F. H. Goetz, J. W. Head III.
Bull. American Astron. Soc., Vol. 4, 370 (1972). – Abstr. AAS.

094.049 High resolution scans showing intensity and color in selected lunar areas. L. A. Riley, J. S. Hall.
Bull. American Astron. Soc., Vol. 4, 370 (1972). – Abstr. AAS.

094.050 Lunar samples: Progress in interpreting telescopic spectral reflectivity measurements.
J. B. Adams, T. B. McCord.
Bull. American Astron. Soc., Vol. 4, 370 (1972). – Abstr. AAS.

094.051 Lunar spectral types explained.
T. B. McCord, J. B. Adams, C. Pieters.
Bull. American Astron. Soc., Vol. 4, 370 (1972). – Abstr. AAS.

094.052 Geologic and photoballistic studies at Mt. Etna and Stromboli. T. R. McGetchin, M. Settle, B. Chouet.
Bull. American Astron. Soc., Vol. 4, 373 (1972). – Abstr. AAS.

094.053 The accuracy in the determination of contact moments of the earth's shadow with lunar formations based on the observations of 18 eclipses of 1910–1959. S. V. Drozdov.
Astron. vestn., Vol. 6, 133 - 134 (1972). In Russian. – Abstr. in Referativ. Zhurn. 51. Astron., 9.51.174 (1972).

094.054 Störungen des Gitterbaus in Mondmineralen.
H. Jagodzinski, M. Korekawa.
Umschau, 72. Jahrgang, p. 663 (1972).
The crystal lattice of some lunar minerals which have been returned by Apollo 15, shows a disorder greater than in terrestrial minerals. This requires further examinations.

094.055 Literal expressions for the co-ordinates of the moon. I. The first degree terms. S. R. Bourne.
Celestial Mechanics, Vol. 6, 167 - 186 (1972).
This paper describes a method for finding literal expressions for the first order terms in the moon's co-ordinates. The method is based on the use of rectangular co-ordinates and was originally proposed by Euler. The variation curve and the terms dependent on the first power of the lunar eccentricity have been obtained. These results are compared with those of Hill and a number of errors in Hill's results have been found.

094.056 Numerical isolation of flaws in the lunar theory.
J. D. Mulholland.
Celestial Mechanics, Vol. 6, 242 - 246 (1972).
The general concept of a numerical study of the lunar motion is examined and a particular case is discussed in greater detail than was possible in an earlier paper. Particular emphasis is placed upon the distinction to be made between model inadequacies and numerical artifacts.

094.057 Dynamics of the moon. H. Jeffreys.
IAU Symposium No. 47, (see 012.008), p. 11 - 12 (1972).

094.058 On the inclination of the lunar axis.
M. Moutsoulas.
IAU Symposium No. 47, (see 012.008), p. 13 - 21 (1972).

094.059 The shape of the moon.
S. K. Runcorn, S. Hofmann.
IAU Symposium No. 47, (see 012.008), p. 22 - 31 (1972).

094.060 Some differences between geometrical and dynamical figures of the moon. I. V. Gavrilov.
IAU Symposium No. 47, (see 012.008), p. 32 - 34 (1972).

094.061 Large disks as representations for the lunar mascons with implications regarding theories of formation.
P. M. Muller, W. L. Sjogren.
IAU Symposium No. 47, (see 012.008), p. 35 - 40 (1972).

094.062 The geomorphic evolution of the lunar surface.
L. B. Ronca.
IAU Symposium No. 47, (see 012.008), p. 43 - 54 (1972).

094.063 Erosion, transportation and the nature of the maria.
T. Gold.
IAU Symposium No. 47, (see 012.008), p. 55 - 67 (1972).

094.064 Radar mapping of the moon at 162 MHz.
J. E. B. Ponsonby, I. Morrison, A. R. Birks, J. K. Landon.
IAU Symposium No. 47, (see 012.008), p. 68 - 71 (1972).

094.065 Introductory remarks: The Apollo 14 mission.
H. E. Newell.
IAU Symposium No. 47, (see 012.008), p. 75 - 80 (1972).

094.066 The Apollo 14 mission and preliminary results.
L. R. Scherer.
IAU Symposium No. 47, (see 012.008), p. 81 - 93 (1972).

094.067 Experimental petrology and petrogenesis of Apollo 12 basalts. D. H. Green.
IAU Symposium No. 47, (see 012.008), p. 123 (1972). Abstract.

094.068 Electronic microscopic studies of some lunar minerals. P. E. Champness, G. W. Lorimer.
IAU Symposium No. 47, (see 012.008), p. 124 - 128 (1972).

094.069 Maria lavas, mascons, layered complexes, achondrites and the lunar mantle.
G. M. Biggar, M. J. O'Hara, D. J. Humphries, A. Peckett.
IAU Symposium No. 47, (see 012.008), p. 129 - 164 (1972).

094.070 Petrochemistry and chemical features of lunar glassy spherules. R. Trigila.
IAU Symposium No. 47, (see 012.008), p. 165 - 179 (1972).

094.071 Formation of lunar glassy spherules: a dynamical model.
A. Carusi, A. Coradini, M. Fulchignoni, G. Magni.
IAU Symposium No. 47, (see 012.008), p. 180 - 184 (1972).

094.072 Lunar mare ridges, rings and volcanic ring complexes. R. G. Strom.
IAU Symposium No. 47, (see 012.008), p. 187 - 215 (1972).

094.073 A possible mechanism of the generating of the unusually long lunar seismic oscillations.
I. P. Passechnik, D. D. Sultanov.
IAU Symposium No. 47, (see 012.008), p. 216 - 219 (1972).

094.074 On the interaction between tectonic processes of the earth and the moon. N. A. Kozyrev.
IAU Symposium No. 47, (see 012.008), p. 220 - 225 (1972).

094.075 On the origin of central peaks in the crater formations filled with melt after impacts.
L. Křivský.
IAU Symposium No. 47, (see 012.008), p. 226 - 230 (1972).

094.076 Geologic interpretation of the study of lunar rocks.
L. Kopecký.
IAU Symposium No. 47, (see 012.008), p. 231 - 245 (1972).

094.077 Interferometric studies on Apollo 11 and Apollo 12 lunar glass objects. S. Tolansky.
IAU Symposium No. 47, (see 012.008), p. 249 - 263 (1972).

094.078 The valence states of 3*d*: Transition elements in Apollo 11 and 12 rocks. A. J. Cohen.
IAU Symposium No. 47, (see 012.008), p. 264 - 278 (1972).

094.079 Luminescence excitation by protons and electrons, applied to Apollo lunar samples.
J. E. Geake, G. Walker, A. A. Mills.
IAU Symposium No. 47, (see 012.008), p. 279 - 297 (1972).

094.080 The solar irradiation record in lunar dust grains.
J. Borg, B. Vassent.
IAU Symposium No. 47, (see 012.008), p. 298 - 308 (1972).

094.081 Low energy solar nuclear particle irradiation of lunar and meteoritic breccias.
J. C. Dran, J. P. Duraud, M. Maurette.
IAU Symposium No. 47, (see 012.008), p. 309 - 323 (1972).

094.082 The thermoluminescence of lunar samples.
G. F. J. Garlick, I. Robinson.
IAU Symposium No. 47, (see 012.008), p. 324 - 329 (1972).

094.083 The particle track record of the lunar surface.
G. M. Comstock.
IAU Symposium No. 47, (see 012.008), p. 330 - 352 (1972).

094.084 Lunar magnetic field measurements, electrical conductivity calculations and thermal profile inferences.
D. S. Colburn.
IAU Symposium No. 47, (see 012.008), p. 355 - 371 (1972).

094.085 Thermal gradients in the outer lunar layers.
J. A. Bastin, S. J. Pandya, D. A. Upson.
IAU Symposium No. 47, (see 012.008), p. 372 - 376 (1972).

094.086 Convection in the moon. S. K. Runcorn.
IAU Symposium No. 47, (see 012.008), p. 377 - 383 (1972).

094.087 Possible thermal history of the moon.
P. E. Fricker, R. T. Reynolds, A. L. Summers.
IAU Symposium No. 47, (see 012.008), p. 384 - 391 (1972).

094.088 The role of occultations in the improvement of the lunar ephemeris. L. V. Morrison.
IAU Symposium No. 47, (see 012.008), p. 395 - 401 (1972).

094.089 On the initial distance of the moon forming in the circumterrestrial swarm. E. L. Ruskol.
IAU Symposium No. 47, (see 012.008), p. 402 - 404 (1972).

094.090 Fluidization on the moon and planets. A. A. Mills.
IAU Symposium No. 47, (see 012.008), p. 407 - 425 (1972).

094.091 On the possible differences in the bulk chemical composition of the earth and the moon forming in the circumterrestrial swarm. E. L. Ruskol.
IAU Symposium No. 47, (see 012.008), p. 426 - 428 (1972).

094.092 The origin of the moon and solar system.
H. C. Urey.
IAU Symposium No. 47, (see 012.008), p. 429 - 440 (1972).

094.093 Evolution of the moon: Recent modification of previous ideas. B. J. Levin.
IAU Symposium No. 47, (see 012.008), p. 441 - 449 (1972).

094.094 Lunar tidal phenomena and the lunar rille system.
B. M. Middlehurst.
IAU Symposium No. 47, (see 012.008), p. 450 - 457 (1972).

094.095 Dating of mechanical events by deformation-induced erasure of particle tracks.
R. L. Fleischer, G. M. Comstock, H. R. Hart, Jr.
Journ. Geophys. Res., Vol. 77, 5050 - 5053 (1972).
Natural fine-scale plastic deformation of lunar pyroxenes has been observed to fragment pre-existing charged-particle tracks. In one sample from the Apollo 12 mission the deformation was estimated to have occurred 20–25 m.y. ago.

094.096 The banded face of Alhazen α. L. E. Fitton.
Journ. British Astron. Ass., Vol. 82, 434 - 435 (1972).

094.097 High resolution measures of polarization and color of selected lunar areas. L. A. Riley, J. S. Hall.
Lowell Obs. Bull., *Flagstaff, Arizona*, No. 159, Vol. 7, (No. 22) 255 - 271 (1972).
High resolution observations of intensity, color (UBV) and polarization have been obtained with scanning techniques for a number of lunar areas of special interest. The general form of the relation between intensity and polarization was established at a number of different phase angles.

094.098 Origin, evolution and present thermal state of the moon. T. C. Hanks, D. L. Anderson.
Phys. Earth Planet. Interiors, Vol. 5, 409 - 425 (1972).
The evidence for extensive volcanic and igneous activity in the outer reaches of the moon and its relative absence thereafter, thermal history models, the early thermal history, an origin of the moon, and the present thermal state of the moon are discussed.

094.099 Global tectonics of the moon and earth.
V. V. Kozlov, Yu. Ya. Kuznetsov, E. D. Sulidi-Kondratiev.
Priroda, No. 10.72, p. 21 - 28 (1972). In Russian.

094.100 On the estimate of electric conductivity of the lunar interior. L. L. Van'yan, M. N. Berdichevskij, M. B. Gokhberg, G. G. Obukhov.
Izv. AN SSSR. Fiz. Zemli, 1972, No. 4, p. 78 - 79. In Russian.
Abstr. in Referativ. Zhurn. 51. Astron., 10.51.276 (1972).

094.101 Die Tumortheorie als Schlüssel zur Erklärung der Urkrustenbildung und Erstarrung von Mond und Erde.
H. Borchert.
Gerlands Beiträge Geophys., Vol. 79, 425 - 464 (1970).
It is to be presumed that planets and their moons have been formed in a "hot" manner by progressive cooling and condensation of solar matter. The development of the moon's crust by processes of fractional crystallization of basaltoid silicate melts is shown on the basis of petrological and volcanological considerations.

094.102 The structure of lunar maria from data of their albedo. N. N. Evsjukov.
Astron. Zhurn. Akad. Nauk SSSR, Vol. 49, 1088 - 1093 (1972). In Russian. English translation in Soviet Astron. AJ, Vol. 16, No. 5.
The albedo distribution in lunar maria for the red region of the spectrum is given. Two types of distributions – chaotic and with concentric structure – are picked out. Possible reasons of distinction of albedo distributions in maria, as well as the influence on the albedo of continental ejections and that of age of the surface are discussed.

094.103 Paleocratering of the moon: Review of post-Apollo

data. W. K. Hartmann.
Astrophys. Space Sci., Vol. 17, 48 - 64 (1972). – Paper given at Philadelphia meeting of American Association for Advancement of Science, December, 1971.

We will briefly comment on early attempts to categorize surface features as endogenetic, exogenetic, etc. Rather we will attempt to show that present knowledge can be used to determine the moon's past bombardment history, which has direct implications upon the theory of the origin of the solar system and of the earth–moon system.

094.104 **Lunar notes.** J. E. Westfall, H. D. Jamieson.
Strolling Astronomer, Vol. 23, 208 - 215 (1972).
Concerning: Additions to the A.L.P.O. Lunar Photograph Library: Amateur and Apollo-14 photographs; The lunar dome survey: A progress report.

094.105 **Mineralogy, petrology and chemistry of lunar rock 12039.** T. E. Bunch, K. Keil, M. Prinz.
Meteoritics, Vol. 7, 245 - 255 (1972).
A more detailed description of rock 12039 is given and its relationship to other Apollo 12 rocks is discussed.

094.106 **Metastable growth patterns in some terrestrial and lunar rocks.** H. I. Drever, R. Johnston.
Meteoritics, Vol. 7, 327 - 340 (1972).
With reference to new information on the textures in some terrestrial mafic and ultramafic igneous rocks of Tertiary age, textural patterns attributable to metastable crystallization are described, illustrated and reviewed. Compared with these textural patterns are the skeletal crystal growth and intergrowth in igneous lunar rock samples obtained, mainly by the Apollo 12 mission.

094.107 **Solid solution, subsolidus reduction and compositional characteristics of spinels in some Apollo 15 basalts.** S. E. Haggerty.
Meteoritics, Vol. 7, 353 - 370 (1972).
A reflection microscopy and electron microprobe study has been carried out on spinels in 3 large basalts and on spinels in 48 basalt fragments in 4 soil samples from the lunar Hadley-Appenine site. The results are reported in detail.

094.108 **Major element composition of glasses in three Apollo 15 soils.**
A. M. Reid, J. Warner, W. I. Ridley, R. W. Brown.
Meteoritics, Vol. 7, 395 - 415 (1972).
Approximately 180 glasses from three lunar soils were analyzed by electron microprobe techniques. All the data are from a random survey of glasses in the 75-125 micron size range. The glasses were analyzed for Si, Ti, Al, Cr, Fe, Mg, Ca, Na, and K and a complete listing of the analyses will be made available as a NASA Technical Memorandum.

094.109 **Ultraviolet photometry of the moon with the Celescope experiment on the OAO-II.**
I. A. Ahmad, W. A. Deutschman.
Astron. Journ., Vol. 77, 692 - 694, 703 (1972).
Thirteen television frames of the moon taken in the wavelength regions of 1350 to 2150 Å by filter U_3 and 1050 to 2150 Å by filter U_4 of the Smithsonian Astrophysical Observatory's Celescope experiment are discussed. The data include phase angles from $-81°$ to $-26°$. The brightness dependence on phase angle differs slightly from that of visual observations. The brightness dependence on selenographical longitude is compared with the theoretical photometric function derived by Hapke.

094.110 **The lithification and metamorphism of lunar breccias.** R. J. Williams.
Earth Planet. Sci. Letters, Vol. 16, 250 - 256 (1972).

Experimental data have been used to establish the temperature coordinates for Warner's scale of lunar metamorphic grades. A model of an ejecta blanket with a hot base layer overlain by a cold top layer can produce the observed features of the Apollo 14 breccias.

094.111 **Trace element profiles, notably Hg, from a preliminary study of the Apollo 15 deep-drill core.**
S. Jovanovic, G. W. Reed, Jr.
Earth Planet. Sci. Letters, Vol. 16, 257 - 262 (1972).
Previously reported results on Hg in Apollo 11, 12 and 14 and Luna 16 core and trench samples are interpreted in terms of the diurnal thermal pulse and layering processes. Concentrations of Os, Ru and U in the deep-drill core samples are also reported.

094.112 **The composition of the lunar highlands: Evidence from modal and normative plagioclase contents in anorthositic lithic fragments and glasses.** G. J. Taylor.
Earth Planet. Sci. Letters, Vol. 16, 263 - 268 (1972).
Modal analyses of 33 crystalline anorthositic fragments from Apollo 11 soil sample 10085, 24 are reported.

094.113 **A depth profile of ^{14}C in the lunar rock 12002.** R. S. Boeckl.
Earth Planet. Sci. Letters, Vol. 16, 269 - 272 (1972).
A depth profile of ^{14}C in lunar rock 12002 has been measured using six specimens with well-defined depth below the exposed surface of the rock. The results obtained show an excess surface activity attributed to solar particle flux.

094.114 **X-ray study and Mössbauer spectroscopy on lunar ilmenites (Apollo 11).**
G. Bayer, J. Felsche, H. Schulz, P. Rüegsegger.
Earth Planet. Sci. Letters, Vol. 16, 273 - 274 (1972).
A comparison of lunar ilmenites (Apollo 11, 10047, 13) with terrestrial ilmenites by means of electron microprobe analysis, X-ray and Mössbauer spectrometry showed that the lunar samples contained no Fe^{3+} but excess Ti^{3+}.

094.115 **Electrical properties of lunar soil dependence on frequency, temperature and moisture.**
D. W. Strangway, W. B. Chapman, G. R. Olhoeft, J. Carnes.
Earth Planet. Sci. Letters, Vol. 16, 275 - 281 (1972).
We have examined the dielectric constant and loss tangent of a lunar soil sample in the frequency range from 100 Hz to 1 MHz. These results suggest that there is very little dispersion in the dielectric properties and that the loss tangent values are nearly a factor of 10 less than those measured by earlier studies. The d.c. conductivity is very low at room temperature and is strongly temperature-dependent with an activation energy in the range of $0.4 - 0.9$ e.V. The dielectric constant and loss tangent increase at frequencies below 10 kHz due to the presence of the moisture.

094.116 **Stability relations of ilmenite and ulvöspinel in the Fe–Ti–O system and application of these data to lunar mineral assemblages.**
L. A. Taylor, R. J. Williams, R. H. McCallister.
Earth Planet. Sci. Letters, Vol. 16, 282 - 288 (1972).
The f_{O_2} stability relations of ilmenite and ulvöspinel were determined using C–O–H–N gas-flow apparatus with f_{O_2} measured by a solid ceramic oxygen electrolyte cell. The occurrences of fayalite reduction to SiO_2 + Fe in lunar rock 14053, as well as a new finding of this assemblage in 14072, are evidence for extreme subsolidus reduction, whereas ulvöspinel breakdown alone occurs under less reducing conditions.

094.117 **The Rb–Sr age of a crystalline rock from Apollo 16.**

D. A. Papanastassiou, G. J. Wasserburg.
Earth Planet. Sci. Letters, Vol. 16, 289 - 298 (1972).

A precise internal isochron was determined for 68415, a plagioclase rich basalt or anorthosite returned by the Apollo 16 mission. The age and initial $^{87}Sr/^{86}Sr$ are T = 3.84 ± 0.01 AE, and I = 0.69920 ∓ 3.

094.118 Density of the lunar interior.
P. W. Gast, R. T. Giuli.
Earth Planet. Sci. Letters, Vol. 16, 299 - 305 (1972).

We attempt to derive the constraints that can be placed on the density of the lunar interior from: (1) the mean density, (2) the moment of inertia, and (3) the mass and density of the lunar crust that have been inferred from the seismic refraction data recorded by the passive seismometer.

094.119 The carbon chemistry of the moon.
G. Eglinton, J. R. Maxwell, C. T. Pillinger.
Sci. American, Vol. 227, No. 4, p. 80 - 90 (1972).

Exhaustive analysis of the Apollo samples reveals various simple organic compounds. These substances did not originate with life, but they add to the store of information on how life originated.

094.120 Magnetic dynamo in the moon: A comparison with the earth. E. H. Levy.
Science, Vol. 178, 52 - 53 (1972).

The assumption that the moon had an internal magnetic field produced in the same way as the geomagnetic field requires that the moon rotated faster than the angular velocity at which it would break up. This suggests that a lunar dynamo is not a tenable explanation for the magnetic remanence observed on the moon.

094.121 The remanent magnetization of lunar soils.
M. Fuller.
Science, Vol. 178, 154 - 156 (1972).

A mechanism whereby hard natural remanent magnetization my be acquired by material buried in the regolith is proposed.

094.122 Soil mechanical properties at the Apollo 14 site.
J. K. Mitchell, L. G. Bromwell, W. D. Carrier III, N. C. Costes, R. F. Scott.
Journ. Geophys. Res., Vol. 77, 5641 - 5664 (1972).

Studies of the soil (regolith) at the Apollo 14 site have been made (1) to obtain data on the compositional, textural, and mechanical properties of lunar soils and the variations of these properties with depth and location at and among Apollo landing sites; (2) to use these data to formulate, verify, or modify theories of lunar history and lunar processes; (3) to develop information that may aid in the interpretation of data obtained from other surface activities or experiments.

094.123 Magnetic properties of Apollo 12 lunar samples.
S. K. Runcorn, D. W. Collinson, W. O'Reilly, A. Stephenson, M. H. Battey, A. J. Manson, P. W. Readman.
Proc. Roy. Soc. London, Ser. A, Vol. 325, 157 - 174 (1971).

094.124 Comparative analysis of the magnetic properties of lunar rocks and meteorites.
Eh. S. Gorshkov, E. G. Gus'kova, V. I. Pochtarev.
Kosm. Issled., Vol. 10, 760 - 765 (1972). In Russian.

094.125 Characteristics of scattering of radio waves by the lunar surface at the landing sites of the automatic stations Luna 16 and Luna 17.
N. N. Krupenio, V. V. Cherkasov.
Kosm. Issled., Vol. 10, 794 - 796 (1972). In Russian. – Brief information.

094.126 Radon-222 in the lunar atmosphere.
R. L. Brodzinski.
Nature, Phys. Sci., Vol. 238, 107 - 109 (1972).

The equilibrium concentration of ^{222}Rn above the lunar surface and of ^{210}Po of atmospheric origin on the lunar surface at Oceanus Procellarum have been calculated to be 0.082 ± 0.012 and 0.0238 ± 0.0035 disintegrations $cm^{-2} sec^{-1}$ respectively from a determination of the ^{210}Po concentration in exposed Solar Wind Composition experiment foils and in reflective coatings of lunar orbiting spacecraft.

094.127 Lunar electrical conductivity.
A. C. Reisz, D. L. Paul, T. R. Madden.
Nature, Vol. 238, 144 - 145 (1972).

Boundary condition asymmetries inherent in the solar wind flow past the moon have been included in a cylindrical model of the interplanetary magnetic field–moon interaction. Numerical examination of the sunward frequency response of this asymmetric model lead us to conclude that the experimentally determined lunar response is characteristic of the electrical conductivity of the bulk crust and is insensitive to a possible core of high conductivity. These conclusions are in disagreement with a previous inversion of the experimental response based on a spatially uniform, symmetric model of the interaction.

094.128 Lunar electrical conductivity – reply.
C. P. Sonett, B. F. Smith, D. S. Colburn, G. Schubert, K. Schwartz.
Nature, Vol. 238, 145 - 147 (1972).

Day-night asymmetry in lunar magnetic induction is reported to insignificantly affect the estimated deep temperature of the moon based on inversion of the day side lunar response using a spherically symmetric confining current theory. The influence of higher order magnetic multipole induction (Mie scattering) i.e. wavelengths comparable to and smaller than the lunar radius, is reviewed and shown to affect the estimated near surface conductivity profile, whereas the deep temperature is determined by induction in the Rayleigh limit (long wavelength).

094.129 Mascons and isostasy. G. Hulme.
Nature, Vol. 238, 448 - 450 (1972).

It is shown that the existence of mascons does not necessarily imply that the moon has been rigid to great depths for a long period. A model of a circular mare is described which gives rise to a gravity anomaly characteristic of mascons and yet could have existed for a large part of the moon's history on a moon unable to support non-hydrostatic stress below a depth of about 100 km. The model requires a moon with a low density surface layer. A thick layer of high density lava on top of this depresses it until isostatic equilibrium achieved but a positive gravity anomaly remains.

094.130 Crater Copernicus. S. V. Landau.
Zemlya i Vselennaya, 1972, No. 5, p. 40 - 41.
In Russian.

094.131 Ultrabasic lunar samples.
I. M. Steele, J. V. Smith.
Nature, Phys. Sci., Vol. 240, 5 - 6 (1972).

Four rock fragments from the Apollo 14 samples have ultrabasic affinities. The principal distinguishing features are olivine and pyroxene with an Fe/(Mg + Fe) atomic ratio below 0.2 by contrast with Fe-rich minerals of recognized lunar rock types including norite, mare basalt, gabbroic anorthosite, and anorthosite.

094.132 Implications of lunar aseismicity. L. Thomsen.
Nature, Vol. 240, 94 - 95 (1972).

The aseismicity of the moon has been interpreted as an

indication that the moon is relatively cold in the interior. Here I present an alternative argument which avoids any assumptions or conclusions on strength properties or stress levels, and which suggests the reverse conclusion, that the moon is too warm to be seismic.

094.133 Volatilization losses from lunar lava 14310.
M. J. O'Hara.
Nature, Vol. 240, 95 - 96 (1972).

The petrography of sample 14310 provides the strongest evidence for the real operation of a process which was always a common sense probability, namely the significant alteration of magma compositions by the selective volatilization of water, alkalis and oxygen from silicate liquids erupted at 1,100–1,300° C into a vacuum of harder than 10^{-10} torr on the lunar surface.

094.134 Refrigeration of lunar samples destined for thermoluminescence studies. S. A. Durrani.
Nature, Vol. 240, 96 - 97 (1972).

Quantitative data are provided supporting the need for refrigeration of lunar samples, retrieved by both manned and unmanned missions, which are meant for thermoluminescence investigations.

094.135 Response of lunar atmosphere to volcanic gas releases. R. R. Hodges, Jr.
Planet. Space Sci., Vol. 20, 1849 - 1864 (1972).

If lunar volcanism is presently extant, there is a good possibility of its detection via atmospheric sensors left on the moon in the Apollo program. A theory of transport of gases emanating from a source on the lunar surface is developed, in which the distribution of a gas is given by the convolution of the time and space description of its source with a Green's function.

094.136 Possible sidereal period for the seismic lunar activity.
D. Sadeh.
Nature, Vol. 240, 139 - 140 (1972).

Maximum lunar seismic activity corresponds to the rise of the pulsar CP 1133 at the site. Other agreement with earth-based experiments suggests that they have a common origin.

094.137 Resonance rotation of celestial bodies and Cassini's laws. V. V. Beletskii.
Celestial Mechanics, Vol. 6, 356 - 378 (1972).

Celestial body rotation about its center of mass, taking into account the body orbit evolution, is considered. Nonlinear evolution equations of motion are constructed. Empirical Cassini's laws describing the moon's motion result from these equations as their stationary points. Bifurcation conditions of steady motions are written out and conditions of their stability are investigated. The hypothesis of Mercury's resonance motion analogous to the 'motion by Cassini' is discussed. Consequences of this hypothesis are considered.

094.138 Aperture synthesis polarimetry of the moon at 21 cm. P. H. Moffat.
Monthly Notices Roy. Astron. Soc., Vol. 160, 139 - 154 (1972).

We present aperture synthesis maps of the moon's polarized thermal radio emission at 1420 MHz which have a resolution of 47″arc in right ascension by 105″arc in declination. The method by which the motion of the moon was taken into account is outlined. A simple model of the emission gives a good description of the emission.

094.139 Osservazioni preliminari sulle rocce stratificate degli appennini di Hadley fotografate dagli astronauti dell' Apollo 15. P. Leonardi.
Atti Accad. Nazionale Lincei, Ser. Ottava, Rend. Cl. Sci. fis., mat., nat., Vol. 51, 519 - 524 (1971).

The Apollo 15 photographs bear evidence that at least some of these mountains are formed by stratified rocks, which because of various considerations – being forced to exclude the existence of marine or alluvial sediments – one is inclined to consider pyroclastic.

094.140 From lunar samples to solar evolution.
M. Maurette.
Sci. Progrès Découverte, 99ᵉ année, No. 3430, p. 3 - 13 (1971). In French.

094.141 The secular accelerations of the moon's orbital motion and the earth's rotation.
L. V. Morrison.
The Moon, Vol. 5, 253 - 264 (1972).

The results and methods of determining the secular accelerations of the moon's orbital motion and the earth's rotation from astronomical observations are critically reviewed. In particular, the effect on these results is considered should Spencer Jones' value for the secular acceleration of the moon be revised. General relationships are deduced between these accelerations, the rate of dissipation of energy in the earth and the fractional change in the rate of rotation of the earth.

094.142 Photometry of the lunar surface. F. Link.
The Moon, Vol. 5, 265 - 285 (1972).

The photometry of the moon gives us some information about the properties of the lunar surface. The photometric uniformity of the lunar surface as a scattering screen is determined by the shadow phenomena on small irregularities due to the dust layer covering the whole surface. A small component of light (<10%) exhibits the features of the luminescence excited by solar radiations.

094.143 Radar images of the moon at 75 and 185 cm wavelengths.
J. E. B. Ponsonby, I. Morison, A. R. Birks, J. K. Landon.
The Moon, Vol. 5, 286 - 293 (1972).

The librations of the moon allow it to be mapped using a continuous wave radar by an aperture synthesis method particularly suited to long wavelengths. Maps, as seen in the depolarised return at wavelengths of 75 and 185 cm, are presented. Both are broadly similar and show that most of the depolarised return comes from the highland regions with no significant return from the maria.

094.144 Rotation of the moon and lunar coordinate systems.
M. Moutsoulas.
The Moon, Vol. 5, 302 - 331 (1972).

The need for precise definition of lunar reference systems is stressed and the principles on which systems of lunar coordinates could be based are established. Differences between coordinate systems defined by the dynamical properties of the lunar configuration and the rotational motion of the lunar globe about its centre of gravity are outlined, and rigorous mathematical formulae relating those systems have been developed.

094.145 Spectroscopic remote sensing of lunar surface composition. J. W. Salisbury.
The Moon, Vol. 5, 332 - 347 (1972).

The various regions of the electromagnetic spectrum currently being used to determine the composition of the lunar surface remotely are reviewed. It is concluded that it would be most useful to apply all of these techniques to the problem of remotely exploring the moon because of the complementary nature of the data obtained.

094.146 Spacecraft techniques for lunar research.
L. D. Jaffe, R. Choate, R. B. Coryell.

The Moon, Vol. 5, 348 - 367 (1972).

The most significant findings about the moon obtained by spacecraft so far, have resulted from measurements of gravity, electromagnetic properties, seismicity, mechanical properties, geologic features, composition, ages, and the lunar environment. The questions discussed include: What can these measurements tell us that is significant? To what kinds of missions are they appropriate?

094.147 **A photometric and polarimetric study of the moon's surface.** N. Sekiguchi.
The Moon, Vol. 5, 368 - 389 (1972).

The results of the photometric and polarimetric observation of the moon's surface with high resolution from October 1969 to March 1971 are discussed. A formula is found for the observed brightness valid for a wide part of the moon's illuminated surface. This formula can be interpreted by assuming the luminescence on the lunar surface stimulated by the solar activity.

094.148 **An empirically derived lunar gravity field.** A. J. Ferrari.
The Moon, Vol. 5, 390 - 410 (1972).

A new method for determining the spherical harmonic coefficients of the lunar gravitational potential is applied to Doppler tracking data from the Lunar Orbiter satellites. This selenodesy scheme consists of two separate data reduction and estimation processes. First Doppler data are reduced and estimates obtained for the long-period Kepler elements and element rates of the orbit. These rates are used as input to a second processor which utilizes the long-period Lagrange perturbation equations to determine a finite set of lunar gravity coefficients.

094.149 **Comparison of the analytical results from the Surveyor, Apollo, and Luna missions.**
A. L. Turkevich.
The Moon, Vol. 5, 411 - 421 (1972).

The principal chemical element composition and inferred mineralogy of the powdered lunar surface material at seven mare and one terra sites on the moon are compared. Analytical information for the principal chemical elements is available at present from eight sites on the moon. The location of these sites, their characteristics, and the nature of the analytical technique used, are summarized in a table.

094.150 **The mineralogy, petrology and geochemistry of lunar samples – A review.** J. Zussman.
The Moon, Vol. 5, 422 - 435 (1972).

The principal chemical and mineralogical features of the Apollo 11, 12 and 14 basaltic crystalline rocks are described, and an account is given of other rock types and minerals which are represented among the coarser particles in the lunar soils. A comparison is made between the chemical compositions (major, minor and trace element concentrations) of rocks and soils.

094.151 **Surface features on glass spherules from the Luna 16 sample.** J. B. Hartung, F. Hörz,
D. S. McKay, F. L. Baiamonte.
The Moon, Vol. 5, 436 - 446 (1972).

A variety of small-scale surface features have been described from lunar fines returned by Apollo missions 11, 12, and 14 (Carter and MacGregor, 1970; McKay, 1970; Cloud et al., 1970; Frondel et al., 1970; McKay et al., 1970; Devaney and Evans, 1970; Neukum et al., 1970; Bloch et al., 1971). Results of similar studies of lunar material returned from Mare Fecunditatis by Luna 16 are reported here.

094.152 **On surface photometry of the moon.** K. Lumme.
The Moon, Vol. 5, 447 - 456 (1972).

The surface photometric observations of the moon made by various authors are compared with the author's theoretical scattering law (Lumme, 1971), and a good agreement is obtained. The integrated brightness of the moon has also been calculated from these different sets of data and then compared with the observed brightness. Some differences between the different observations were found.

094.153 **Apollo 17 explorations and experiments.** R. Hillenbrand.
Sky Telescope, Vol. 44, 372 - 375 (1972).

094.154 **Lunar photographs from Oklahoma.** E. K. Owen.
Sky Telescope, Vol. 44, 406 - 408 (1972).

094.155 **Comparative petrology of Apollo 16 sample 68415 and Apollo 14 samples 14276 and 14310.**
A. J. Gancarz, A. L. Albee, A. A. Chodos.
Earth Planet. Sci. Letters, Vol. 16, 307 - 330 (1972).

Petrographic and electron microprobe studies of Apollo 16 igneous rock 68415 and Apollo 14 rocks 14276 and 14310 show that all three samples differ from the mare basalts and are characterized by plagioclase as the first liquidus phase and by the abundance of plagioclase which is in part cumulate in origin. Major and minor element abundances and isotopic data prohibit the derivation of rocks like any of these samples from one another by magmatic fractionation during their crystallization.

094.156 **The lunar neutron flux revisited.**
R. E. Lingenfelter, E. H. Canfield, V. E. Hampel.
Earth Planet. Sci. Letters, Vol. 16, 355 - 369 (1972).

Cosmic ray-produced neutron equilibrium spectra are calculated for a variety of lunar surface compositions. From these spectra production rates of ^{80}Kr, ^{82}Kr, ^{114}Cd, ^{131}Xe, ^{150}Sm, ^{152}Sm, ^{152}Gd, ^{156}Gd, ^{158}Gd and ^{187}Re due to neutron capture on ^{79}Br, ^{81}Br, ^{113}Cd, ^{130}Ba, ^{149}Sm, ^{151}Eu, ^{155}Gd, ^{157}Gd and ^{186}W are determined and compared with measurements.

094.157 **A point of phase equilibria interpretation in connection with lavas from the Apollo 12 site.**
M. J. O'Hara, G. M. Biggar.
Earth Planet. Sci. Letters, Vol. 16, 388 - 390 (1972).

The patterns actually observed in Apollo 12 lava samples belong to liquids related by accumulation of olivine plus pyroxene into a cotectic liquid and result principally from local accumulation of a small proportion of olivine, pyroxene and spinel phenocrysts present in liquids which were also close to saturation with plagioclase at low pressure.

094.158 **Lunar "dunite", "pyroxenite" and "anorthosite".**
H. G. Wilshire, E. D. Jackson.
Earth Planet. Sci. Letters, Vol. 16, 396 - 400 (1972).

Monomineralic aggregates of olivine, clinopyroxene, orthopyroxene and plagioclase with granoblastic textures are widespread minor constituents of Apollo 14 breccias. Our examination of 200 thin sections of Apollo 14 rocks suggests that these monomineralic aggregates are actually recrystallized single-mineral grains and that their source rocks are more likely to be moderately coarse-grained gabbroic rather than ultramafic rocks.

094.159 **Temperature-time relationships from lunar two phase metallic particles (14310, 14163, 14003).**
H. J. Axon, J. I. Goldstein.
Earth Planet. Sci. Letters, Vol. 16, 439 - 447 (1972).

094.160 **Investigation of the moon with the Lunokhod 1 space vehicle.** G. I. Petrov.
Space Research XII, (see 012.016), Vol. 1, 1 - 12 (1972).

094.161 **Investigation of the chemical composition of lunar surface along the route of Lunokhod 1.**
G. E. Kocharov, S. V. Victorov, O. M. Voropayev, A. Yu. Dzevanovskaya, G. V. Kirian, V. V. Petrov, V. A. Sakulsky.
Space Research XII, (see 012.016), Vol. 1, 13 - 22 (1972).

094.162 **Rare gases in the regolith from the Sea of Fertility.**
A. P. Vinogradov, I. K. Zadorozhny.
Space Research XII, (see 012.016), Vol. 1, 23 - 31 (1972).

094.163 **Investigation of the composition and radioactivity of the lunar rock from the Sea of Fertility.**
Yu. A. Surkov, A. S. Stan, F. F. Kirnozov, I. N. Ivanov, L. P. Moskaleva.
Space Research XII, (see 012.016), Vol. 1, 33 - 38 (1972).

094.164 **Age determinations on samples from the Apollo 14 landing site.** G. J. Wasserburg, J. C. Huneke, D. A. Papanastassiou, F. A. Podosek, F. Tera, G. Turner.
Space Research XII, (see 012.016), Vol. 1, 39 - 41 (1972).

094.165 **Results of investigations of the physical and mechanical properties of the lunar sample from Luna 16.** V. V. Gromov, A. K. Leonovich, V. A. Lozhkin, A. V. Rybakov, P. S. Pavlov, A. D. Dmitryev, V. V. Shvarev.
Space Research XII, (see 012.016), Vol. 1, 43 - 52 (1972).

094.166 **Investigations of the mechanical properties of the lunar soil along the path of Lunokhod 1.**
A. K. Leonovich, V. V. Gromov, A. V. Rybakov, V. N. Petrov, P. S. Pavlov, I. I. Cherkasov, V. V. Shvarev.
Space Research XII, (see 012.016), Vol. 1, 53 - 64 (1972).

094.167 **Magnetic properties of lunar specimens returned by ALS Luna 16.**
E. S. Gorshkov, E. G. Gus'kova, V. I. Pochtarev.
Space Research XII, (see 012.016), Vol. 1, 83 - 85 (1972).

094.168 **The infrared reflection, emission and absorption spectra of regolith from the Sea of Fertility and its scattering coefficient.**
M. V. Akhmanova, B. V. Dementyev, A. V. Karyakin, M. N. Markov, V. S. Petrov, A. M. Prokhorov, M. M. Sushchinsky.
Space Research XII, (see 012.016), Vol. 1, 87 - 93 (1972).

094.169 **The optical parameters of Mare Foecunditatis regolith.**
I. I. Antipova-Karatayeva, Yu. I. Stakheyev, K. P. Florensky.
Space Research XII, (see 012.016), Vol. 1, 95 - 98 (1972).

094.170 **Observations of lunar atmosphere.**
F. S. Johnson, D. E. Evans, J. M. Carroll.
Space Research XII, (see 012.016), Vol. 1, 99 - 105 (1972).

094.171 **Geomorphological analysis of the area of Mare Imbrium explored by the automatic roving vehicle Lunokhod 1.**
K. P. Florensky, A. T. Basilevsky, A. A. Gurshtein, V. V. Zasetsky, R. B. Zezin, A. A. Pronin, Z. V. Popova.
Space Research XII, (see 012.016), Vol. 1, 107 - 121 (1972).

094.172 **The morphology, types and distribution of sizes of regolith particles in the Sea of Fertility.**
K. P. Florensky, A. V. Ivanov, Yu. I. Stakheyev, L. S. Tarasov.
Space Research XII, (see 012.016), Vol. 1, 123 - 136 (1972).

094.173 **Mössbauer spectroscopy of regolith from the Sea of Fertility.** T. V. Malysheva, V. V. Kurash.
Space Research XII, (see 012.016), Vol. 1, 137 - 140 (1972).

094.174 **Petrology and mineralogy of lunar rocks from the Sea of Fertility.**
A. S. Pavlenko, L. S. Tarasov, I. D. Shevaleyevsky, A. V. Ivanov.
Space Research XII, (see 012.016), Vol. 1, 141 - 153 (1972).

094.175 **The composition and crystalline structure of the minerals of regolith from the Sea of Fertility.**
E. S. Makarov, N. P. Ilyin, V. I. Ivanov.
Space Research XII, (see 012.016), Vol. 1, 155 - 162 (1972).

094.176 **Lunar gravity field as determined by orbiters.** A. S. Liu, P. A. Laing.
Space Research XII, (see 012.016), Vol. 1, 163 - 175 (1972).

094.177 **Meteoroid activity on the lunar surface from the Surveyor 3 sample examination.**
B. G. Cour-Palais, H. A. Zook, R. E. Flaherty.
Space Research XII, (see 012.016), Vol. 1, 319 - 331 (1972).

094.178 **Erosion phenomena on the lunar surface and meteorites.** J. A. M. McDonnell, D. G. Ashworth.
Space Research XII, (see 012.016), Vol. 1, 333 - 347 (1972).

094.179 **The Apollo 17 landing site.** B. K. Lucchitta.
Nature, Vol. 240, 259 - 260 (1972).
The geology of the Apollo 17 landing site in the Taurus–Littrow region of the moon is described.

094.180 **Orbital eccentricity of Mercury and the origin of the moon.** A. G. W. Cameron.
Nature, Vol. 240, 299 - 300 (1972).
If Anderson is correct in his conclusion that the bulk composition of the moon resembles that of the Allende inclusions, then the natural place for the formation of the moon in the solar system is inside the orbit of Mercury, through planetary accumulation from the condensed material to be found there. This explains the anomalous large eccentricity (0.206) of the orbit of Mercury.

094.181 **Comparison of linear polarization and the albedo of igneous rocks in the spectral range $0.75-3.0\,\mu$ with those values for the moon.**
V. V. Novikov, A. P. Popov.
Astron. vestn., Vol. 6, 160 - 164 (1972). In Russian. − Abstr. in Referativ. Zhurn. 51. Astron., 12.51.238 (1972).

094.182 **Lunar winds.** D. E. Rehfuss.
Journ. Geophys. Res., Vol. 77, 6303 - 6315 (1972).
The purpose of this study is to estimate the characteristics of impact-generated gas flows as a function of the mass, the velocity, and the composition of the impacting body and as correlated with crater size. A thermodynamic model is used to find the amount of irreversible heat available for the production of vapor, which is considered to expand adiabatically.

094.183 **Upper limits on the lunar atmosphere determined from solar-wind measurements.**
G. L. Siscoe, N. R. Mukherjee.
Journ. Geophys. Res., Vol. 77, 6042 - 6051 (1972).
Ionization of the lunar atmosphere by solar photons and solar-wind particles will create perturbations in the solar wind. Observations by lunar-orbiting Explorer 35 show that any such perturbations are generally <10% in the magnetic field (the practical threshold of detectability) but that they may occasionally be >10%.

094.184 **Interplanet variations in scale of crater morphology.− Earth, Mars, moon.** W. K. Hartmann.
Icarus, Vol. 17, 707 - 713 (1972).
Crater on the earth, Mars, and the moon show a spectrum of morphologies with diameter increasing from simple, bowl-

shaped craters through craters with increasingly complex central peaks, to craters with 'peak rings' and basins with multiple concentric scarps. In each category there is a range of diameters, centered around a characteristic diameter, D_c. It is found that D_c decreases as the size of the planet increases. Several possible explanations are considered.

094.185 Antipodes on the moon.
    Yu. N. Lipskij, J. F. Rodionova.
Astron. Tsirk., No. 712, p. 3 - 7 (1972). In Russian.

094.186 Are the lunar seismic signals compatible with a deep layer of fine powder?   B. W. Jones.
Nature, Vol. 240, 458 - 459 (1972).
    The seismic wave velocity may vary with depth in the moon in the way derived from the lunar seismic signals by Toksöz et al. It is customary to seek sudden changes in velocity and interpret these as evidence for sudden changes in mineral type. Here it is shown that one such sudden change at 25 km and the more gradual increase in velocity from the surface to it can be interpreted in terms of a single type of material, namely the fine rock powder that is so abundant on the lunar surface.

094.187 "Lietuvas ezers" Mēness otrajā pusē.   V. Straižys.
    Zvaigžņotā debess, 1971. gada rudens, p. 54 - 55.

094.188 Kosmosa apgūšana. "Apollo-14" Mēness ekspedīcija.
    I. Daube.
Zvaigžņotā debess, 1971./72. gada ziema, p. 20 - 25.

094.189 Lunar topography: Global determination by radar.
    I. I. Shapiro, S. H. Zisk, A. E. E. Rogers, M. A.
Slade, T. W. Thompson.
Science, Vol. 178, 939 - 948 (1972).
    Our main purpose in this article is to describe new radar methods devised to obtain surface-height variations with higher accuracy and better global fidelity than has heretofore been possible. We describe previous methods for mapping the moon with radar, and then the new techniques for adding altitude information to the radar map. Finally we present sample results for altitude contours on the moon obtained with the Haystack and Westford radar systems of the Massachusetts Institute of Technology.

094.190 Lunar topography: First radar-interferometer measurements of the Alphonsus-Ptolemaeus-Arzachel region.   S. H. Zisk.
Science, Vol. 178, 977 - 980 (1972).
    Radar interferometry is a new technique for accurately measuring the topography of the lunar surface from the earth. Measurements have been made with this technique of an area including the craters Ptolemaeus, Alphonsus, and Arzachel and a portion of Mare Nubium. There is evidence for a late episode of volcanism that partially filled two of the craters through a crustal fault of Imbrian origin. Several other features of the topography, particularly those coinciding with local gravitational anomalies, can be correlated with flow events.

094.191 The geologic setting of the Luna 20 site.
    G. Heiken, M. C. McEwen.
Earth Planet. Sci. Letters, Vol. 17, 3 - 6 (1972).
    Preliminary analysis of samples and photography indicates some similarities between the Luna 20 and Apollo 16 landing site. The abundance of anorthosite and feldspar-rich breccias in the soil and the implied structure of the hummocky terrain at the Luna 20 site are also characteristic of the Apollo 16 site. The implication is that the highlands at the Luna 20 site consist of heavily cratered, subdued hills consisting of complex cataclastic anorthosites and feldspar-rich breccias.

094.192 Major element composition of Luna 20 glasses.
    J. Warner, A. M. Reid, W. I. Ridley, R. W. Brown.
Earth Planet. Sci. Letters, Vol. 17, 7 - 12 (1972).
    Ten percent of the 50−150 micron size fraction of Luna 20 soil is glass. A random suite of 270 of these glasses has been analyzed by electron microprobe techniques. The major glass type forms a strong cluster around a mean value corresponding to highland basalt (anorthositic gabbro) with 70% normative feldspar. Minor glass groups have the compositions of mare basalts and of low-K Fra Mauro type basalts. The glass data indicate that highland basalt is the major rock type in the highlands north of Mare Fecunditatis.

094.193 Luna 20 and Apollo 16 core fines: Large-ion lithophile trace-element abundances.
J. A. Philpotts, S. Schuhmann, A. L. Bickel, R. K. L. Lum.
Earth Planet. Sci. Letters, Vol. 17, 13 - 18 (1972).
    Li, K, Rb and Sr abundances have been determined by mass-spectrometric stable isotope dilution analysis for a sample of Luna 20 fines and for eight fines from the Apollo 16 deep drill core. Ba and rare-earth abundances have also been determined for the Luna 20 sample and two of the Apollo 16 samples. Luna 20 and Apollo 16 were the first missions to interior highlands sites. The trace-element similarity of the fines samples may indicate that the lunar highlands are fairly homogeneous.

094.194 Teneurs en K, Rb, Sr, Ba et terres rares des échantillons ramenés par la sonde soviétique Luna 20 de la région du cratère Apollonius (montagnes lunaires).
M. Loubet, J.-L. Birck, C. J. Allegre.
Earth Planet. Sci. Letters, Vol. 17, 19 - 23 (1972).
    Rare earth and K, Rb, Sr, Ba concentrations are given for Luna 20 samples. Considering these samples to comprise a major part of the lunar crust, we propose a model to explain the chemical evolution of the moon.

094.195 "Ages" ^{87}Rb−^{87}Sr du sol et des fragments rocheux ramenés des montagnes lunaires par la sonde automatique Luna 20, (région du cratère Apollonius).
J.-L. Birck, C. J. Allegre.
Earth Planet. Sci. Letters, Vol. 17, 24 - 28 (1972).
    ^{87}Rb−^{87}Sr results are reported for soil and fragments from Luna 20 mission. The results show a low radiogenic character but demonstrate clearly the presence of the high K components. A model is presented for lunar evolution.

094.196 The chemical composition of soil from the Apollo 16 and Luna 20 sites.   B. M. Bansal, S. E. Church,
P. W. Gast, N. J. Hubbard, J. M. Rhodes, H. Wiesmann.
Earth Planet. Sci. Letters, Vol. 17, 29 - 35 (1972).
    The concentrations of the rare earth elements K, Rb, Sr, Ba, U, Zr and Cr for the Luna 20 soil and four different Apollo 16 soils are reported. These trace element abundances imply: (1) that the lunar highlands consist of a mixture of rocks rich in large ion lithophile (LIL) elements and LIL-element impoverished anorthosites; or (2) that the bulk of the aluminum-rich crust did not originate by upward segregation of plagioclase in a primitive liquid shell. The Luna 20 soil is distinguished from the Apollo 16 soil by lower aluminum and LIL element abundances.

094.197 U−Th−Pb systematics in lunar highland samples from the Luna 20 and Apollo 16 missions.
F. Tera, G. J. Wasserburg.
Earth Planet. Sci. Letters, Vol. 17, 36 - 51 (1972).
    In this study we report analyses of the concentration of Pb, U and Th and the isotopic composition of Pb in soil samples from two highland sites sampled by the Apollo 16 and Luna 20 missions and an anorthosite from Apollo 16. We also

report a further improvement in the analytical techniques used for Pb.

094.198 Rb–Sr systematics of Luna 20 and Apollo 16 samples. D. A. Papanastassiou, G. J. Wasserburg.
Earth Planet. Sci. Letters, Vol. 17, 52 - 63 (1972).

Lunar soils from all missions indicate that an early lunar differentiation process occurred in the interval 4.3–4.6 AE and resulted in the formation of a crust rich in K, Rb, U and Th. An attempt has been made to identify rocks which were formed during this early differentiation event.

094.199 Extralunar materials in Apollo 16 soils and the decay rate of the extralunar flux 4.0 Gy ago.
P. A. Baedecker, C.-L. Chou, L. L. Sundberg, J. T. Wasson.
Earth Planet. Sci. Letters, Vol. 17, 79 - 83 (1972).

The concentration of extralunar materials in the Apollo 16 regolith is about 3.5%, about 50% higher than that observed at the Apollo 14 site, and three times higher than values at mare landing sites. The integrated flux of extralunar materials is 2.5 times higher at the Apollo 16 than at the Apollo 14 site. These data support the hypothesis that the flux of extralunar materials decreased rapidly between 4.0 and 3.7 Gy ago, and demonstrate the presence of appreciable unaccreted materials near 1 AU 600 my after the formation of the solar system.

094.200 The origin and stability of lunar goethite, hematite and magnetite. R. J. Williams, E. K. Gibson.
Earth Planet. Sci. Letters, Vol. 17, 84 - 88 (1972).

Extra-lunar contamination, fumarolic activity, and exposure to oxidizing gases from comet or carbonaceous meteorite impacts have been previously proposed as the causes of magnetite, hematite, and goethite in lunar materials. However, these minerals can occur in the stable low temperature gas-solid equilibrium assemblages of lunar rocks.

094.201 The magnetic properties and morphology of metallic iron produced by subsolidus reduction of synthetic Apollo 11 composition glasses.
G. W. Pearce, R. J. Williams, D. S. McKay.
Earth Planet. Sci. Letters, Vol. 17, 95 - 104 (1972).

This study was undertaken to test the hypothesis that the excess iron in lunar soils and breccias relative to the igneous rocks could have been generated by reduction of glass at subsolidus temperatures sufficiently high to impart a thermoremanent magnetization to the materials and to test whether variations of temperature, degree of reduction, and length of annealing might produce iron with grain sizes similar to those found in lunar soils and breccias.

094.202 An experimental investigation of the significance of zirconium partitioning in lunar ilmenite and ulvöspinel. L. A. Taylor, R. H. McCallister.
Earth Planet. Sci. Letters, Vol. 17, 105 - 109 (1972).

The present study was conducted in an attempt to explain the significance of the zirconium partitioning between coexisting oxides as it relates to the crystallization and evolution of the parent magma, the cooling history of the rocks and to the crystal chemistry of zirconium within the oxide phases.

094.203 Characteristics of the lunar photoelectron layer in the geomagnetic tail.
D. L. Reasoner, W. J. Burke.
Journ. Geophys. Res., Vol. 77, 6671 - 6687 (1972).

In this paper, we report on observations of stable photoelectron fluxes with energies between 40 and 200 eV by the Apollo 14 charged-particle lunar environment experiment. Numerically calculated density and potential distributions, when compared with our measured values, help us estimate the photoelectron yield function of the dust layer covering the moon.

094.204 Petrología de la luna. F. J. Escandón.
El Universo, Vol. 26, 92 - 96 (1972).

094.205 Distant train of the moon. A. M. Moskalenko.
Kosmich. Issled., Vol. 10, 871 - 878 (1972).
In Russian.

094.206 Investigation of the gamma-radiation of the lunar soil returned by the automatic station Luna 16.
Yu. A. Surkov, G. A. Fedoseev, O. P. Sobornov.
Kosmich. Issled., Vol. 10, 938 - 942 (1972). In Russian.

094.207 Neutron-activation analysis of the lunar ground returned by Luna 16 from Mare Foecunditatis.
Yu. A. Surkov, F. F. Kirnozov, I. N. Ivanov, G. M. Kolesov, B. N. Ryvkin, A. P. Shpanov.
Kosmich. Issled., Vol. 10, 943 - 947 (1972). In Russian.

094.208 Uranium–lead systematics in lunar basalts. N. H. Gale.
Earth Planet. Sci. Letters, Vol. 17, 65 - 78 (1972).

A discussion is given of the extremely limited value, and misleading nature, of the concept of 'model age' as ordinarily used. The possibility of deriving an independent formal 'age of the moon' from lunar basalt U–Pb data is discussed, and compared with similar attempts to derive an independent 'age of the earth' from terrestrial oceanic basalts. In both cases it is demonstrated that the validity of the 'age of the planetary body' derived is dependent on a strictly two-stage episodic U–Pb evolution. No firm conclusions can yet be drawn from the Apollo 14 U–Pb data about the 'age of the moon'.

094.209 Method of computations for relative positions on the lunar disk. K. Hurukawa.
Univ. Tokyo, Tokyo Astron. Obs. Rep. (No. 60), Vol. 16, 40 - 48 (1972). In Japanese.

094.210 Trace element geochemistry of Apollo 16 soil 68501. S. R. Taylor, M. P. Gorton, P. Muir, W. Nance, R. Rudowski, N. Ware.
Nature, Vol. 239, 205 - 207 (1972).

Data are given for Cs, Rb, Ba, Pb, Sr, REE, Y, Th, U, Zr, Hf, Nb, W, Cr, V, Sc, Ni, Co and Cu. The soil composition can be represented mainly as a two component mixture of gabbroic anorthosite (highland basalt) and low K Fra Mauro basalt, as represented by the components of Apollo 15 black and white breccia 15455. The Cayley Formation is thus considered to be not volcanic but derived by mixing during intense highland cratering. The highlands are postulated to form by melting and fractionation of the outer few hundred km of the moon.

094.211 The structure and geological-morphological features of the landing region of the Luna 20 automatic probe.
K. P. Florenskii, A. A. Gurshtein, A. T. Bazilevskii, V. V. Zasetskii.
Dokl. Akad. Nauk SSSR, Ser. Mat. Fiz., Vol. 207, 1078 - 1081 (1972). In Russian.

094.212 Color information from lunar and planetary space missions. J. J. Rennilson.
Journ. Optical Soc. America, Vol. 62, 1365 (1972). – Abstr. Optical Soc. America.

094.213 About a method to determine the position of the lunar mass centre by ground observations.
N. L. Makarenko.

Astron. vestn., Vol. 6, 153 - 159 (1972). In Russian. — Abstr. in Referativ. Zhurn. 51. Astron., 1.51.178 (1973).

094.214 **Single photon detection and timing in the Lunar Laser Ranging Experiment.** S. K. Poultney.
IEEE Trans. Nuclear Sci., Vol. NS-19, No. 3, p. 12 - 17 (1972).

094.215 **Subsurface temperature of the moon.**
W. W. Salisbury, D. L. Fernald.
Journ. Astronaut. Sci., Vol. 18, 236 - 243 (1971).
Measurements of thermal radiation from the lunar interior were made in the meter-wavelength range with the Arecibo radio telescope during the period between July 24 and August 13, 1969.

094.216 **Numerical calculation of the lunar wake in a magnetohydrodynamic model.** B. L. Beers.
Phys. Fluids, Vol. 15, 1450 - 1456 (1972).
A model is developed which has the solar wind flowing directly into the lunar surface and the magnetic field passing unimpeded through the moon. The collapse of the plasma void behind the moon is calculated numerically using the equations of magnetohydrodynamics.

094.217 **Where on the moon? An Apollo systems engineering problem.** J. O. Cappellari, Jr.
Bell. System Techn. Journ., (USA), Vol. 51, 961 - 1126 (1972). — See Phys. Abstr., Vol. 75, No. 62644 (1972).

094.218 **A new look at the origin of lunar surface breccias.**
G. J. H. McCall.
Astron. and Space, (GB), Vol. 2, No. 1, p. 5 - 17 (1972).

094.219 **Phase transformations and exsolution in lunar and terrestrial calcic plagioclases.**
A. H. Heuer, S. V. Radcliffe, J. S. Lally, J. Christie.
Phil. Mag., Ser. 8, Vol. 26, 465 - 482 (1972).
A high voltage transmission electron microscopic study of lunar plagioclase from Apollo 11, 12 and 14 basalts, an Apollo 14 breccia and an Apollo 15 anorthosite has disclosed the presence of two types of crystallographic domain structure, both caused by subsolidus phase transformations. The analysis of these features has clarified the sequence of structural evolution following crystallization in this complex mineral.

094.220 **A lunar crater dedicated to Luis Enrique Erro.**
F. O. Tovar.
Acta Politecn. Mexico, Vol. 12, 173 - 176 (1972). In Spanish.
Brief biographical information is given about L. E. Erro (1897), whose name has been given to a crater on the far side of the moon.

094.221 **Winding and meandering furrows on the lunar surface.** P. Leonardi.
Modern Geol., (GB), Vol. 3, 151 - 156 (1972).

094.222 **Magnetic properties of lunar glasses, terrestrial glasses and tektites.** S. Sullivan.
Thesis, Howard Univ., Washington, D.C. [Available from Univ. Microfilms, Ann Arbor, Mich., USA. Order No. 72-14048], 178 pp. (1971).

094.223 **Lunar tridymite and cristobalite.** B. Mason.
American Mineral., Vol. 57, 1530 - 1535 (1972).

094.224 **Spectral reflectance and emittance of Apollo 11 and 12 lunar material.** R. C. Birkebak.
AIAA Journ., Vol. 10, 1064 - 1067 (1972).
The thermal radiation properties of Apollo 11 and 12 fines (soils) are reviewed and presented as a function of wavelength, angle of illumination and bulk density.

094.225 **Determination of lunar orbital elements by the method of equal altitudes.** J. Vondrák.
Wiss. Zeitschr. Techn. Univ. Dresden, Vol. 21, No. 3, (see 012.021), 608 - 609 (1972).

094.226 **Lunar rock.** A. Vinogradov.
Trans. American Geophys. Union, Vol. 53, 820 - 822 (1972).
The author describes the results of the investigation of the rock delivered to the earth by the Luna 20 probe.

094.227 **Is the moon hot or cold?** D. L. Anderson, T. C. Hanks.
Science, Vol. 178, 1245 - 1249 (1972).
We find that an iron-deficient, highly resistive, hot lunar interior, capped by a cool, rigid lunar lithosphere with a thickness of several hundred kilometers, can explain the relevant observations and is a reasonable model of the moon today.

094.228 **The aftermath of Apollo: Science on the shelf?**
R. Gillette.
Science, Vol. 178, 1265 - 1268 (1972).

094.229 **Compositional zoning in pyroxenes from lunar rock 12021, Oceanus Procellarum.**
F. R. Boyd, D. Smith.
Journ. Petrol., (GB), Vol. 12, 439 - 464 (1971).

094.230 **A search for amino acids in Apollo 11 and 12 lunar fines.** C. W. Gehrke, R. W. Zumwalt, D. L. Stalling, D. Roach, W. A. Ave, C. Ponnamperuma, K. A. Kvenvolden.
Journ. Chromatogr., (Netherlands), Vol. 59, 305 - 319 (1971).

094.231 **Sedimentology of Apollo 11 and 12 lunar soils.**
J. F. Lindsay.
Journ. Sedimentary Petrology, (USA), Vol. 41, 780 - 797 (1971).

094.232 **Composition and origin of lithic fragments and glasses in Apollo 11 samples.**
M. Prinz, T. E. Bunch, K. Keil.
Beiträge Mineral. Petrograph., (Germany), Vol. 32, 211 - 230 (1971).

094.233 **Magmatic differentiation of the moon.**
B. J. Levin.
Chemie der Erde, (Germany), Vol. 30, 251 - 257 (1971).

094.234 **The geologic evolution of the moon.**
P. D. Lowman, Jr.
Journ. Geol., (USA), Vol. 80, 125 - 166 (1972).

094.235 **Le lien gravitationnel terre-lune à travers les âges.**
J.-M. Chevallier, A. Cailleux.
Canadian Journ. Earth Sci., Vol. 9, 479 - 485 (1972).

094.236 **Distribution and significance of carbon compounds on the moon.** S. Chang, K. A. Kvenvolden.
Frontiers of Biology, (Netherlands), Vol. 23, 400 - 430 (1972).

094.237 **Geologic setting and petrology of Apollo 15 anorthosite (15415).** H. G. Wilshire, G. G. Schaber, L. T. Silver, W. C. Phinney, E. D. Jackson.
Bull. American Geol. Soc., Vol. 83, 1083 - 1092 (1972).

094.238 **Isotopic analysis of chromium in lunar materials by mass spectrometry of the trifluoroacetylacetonate.**
N. M. Frew, J. J. Leary, T. L. Isenhour.

Analytical Chemistry, (*USA*), Vol. 44, 665 - 671 (1972).

094.239 Glass in the bottom of small lunar craters: An observation from Apollo 15.
G. G. Schaber, D. R. Scott, J. B. Irwin.
Bull. American Geol. Soc., Vol. 83, 1573 - 1578 (1972).

094.240 Apollo mission 15 lunar photography index maps.
Prepared under the direction of the Department of Defense by the Aeronautical Chart and Information Center, United States Air Force for the National Aeronautics and Space Administration, Greenbelt, Maryland. 10 sheets (1972).

094.241 Apollo 15 lunar photography.
Prepared by W. S. Cameron, M. A. Niksch.
Data User's Note, NSSDC 72-07, National Space Science Data Center, National Aeronautics and Space Administration, Goddard Space Flight Center, Greenbelt, Md. 6 + 33 + A6 + B20 pp. (1972).
The purposes of this Data Users' Note are to announce the availability of Apollo 15 pictorial data and to aid an investigator in the selection of Apollo 15 photographs for study. As background information, the Note includes brief descriptions of the Apollo 15 mission objectives, photographic equipment, and photographic coverage and quality.

094.242 Physik und Geochemie des Mondes. Seminar über neuere Fragen der Physik, Universität Heidelberg Sommer–Semester 1972.
Conducted by T. Kirsten.
Edited by Max-Planck-Institut für Kernphysik, Heidelberg. 4 + 119 pp. (1972).

094.243 On the influence of the internal structure of the moon on its rotation.   K. S. Shakirov.
Astron. Tsirk., No. 730, p. 3 - 5 (1972). In Russian.

094.244 Tranquillity base map.   W. H. Arant.
Photogrammetric Engineering, (*USA*), Vol. 37, 1131 - 1137 (1971).

094.245 Med månen som mål. V. – Surveyor 3.
T. Ringnes.
Naturen 1971, No. 3, p. 131 - 160 = Inst. Teor. Astrofys. Blindern – Oslo, Småtrykk No. 73 (1971).

094.246 Nordmenn på månen.   T. Ringnes.
Naturen 1971, No.4, p. 215 - 253 = Inst. Teor. Astrofys. Blindern – Oslo, Småtrykk No. 74 (1971).

094.247 Med månen som mål. VI. – Apollo 12.
T. Ringnes.
Naturen 1972, No. 1, p. 15 - 60 = Inst. Teor. Astrofys. Blindern – Oslo, Småtrykk No. 78 (1972).

094.248 Das Nördlinger Ries-Modell für die Bildung der Mondkrater und der Gesteine der Mondoberfläche.
D. Stöffler.
Zeiss-Inform., No. 79, Vol. 19, 54 - 57 (1972).

094.249 Microscopic investigation of the Luna 16 sample.
A.Cimbálková, A. Maštalka.
Vesmír, Vol. 51, 269 - 271 (1972).   In Czech.
Preliminary results of studying the Luna 16 sample at various institutes of the Czechoslovak Academy of Sciences.

094.250 Remarkable chains of craters on the moon.
M. M. Shemiakin.
Říše hvězd, Vol. 53, 162 - 168 (1972).   In Czech.

094.251 Experimental evidence against the role of selective volatilization on the lunar surface.
D. H. Green, N. G. Ware, W. O. Hibberson.
Nature, Vol. 238, 450 - 452 (1972).
The matching of the natural crystallization sequence in 14310, like that of mare basalts previously studied, with those observed experimentally, demonstrates that there has been no quantitative loss of alkalis by vacuum volatilization during eruption of magmas at the lunar surface.

094.252 Lunar ultramafic glasses, chondrules and rocks.
T. E. Bunch, W. Quaide, M. Prinz, K. Keil, E. Dowty.
Nature, Phys. Sci., Vol. 239, 57 - 59 (1972).

094.253 Catalogue of the normal albedo and of the brightness gradient of lunar surface sections.
M. P. Barabashov, V. O. Ehzers'ka, V. J. Ehzers'kij, I. I. Latinina.
Visn. Kharkiv. Univ. No. 82, (Ser. Astron., No. 7), p. 12 - 35 (1972). In Ukrainian.

094.254 Comparison of photometric catalogues of sections of the lunar surface.
M. P. Barabashov, V. O. Ehzers'ka, V. J. Ehzers'kij, N. P. Stadnikova.
Visn. Kharkiv. Univ. No. 82, (Ser. Astron., No. 7), p. 36 - 48 (1972). In Ukrainian.

094.255 Some results of lunar observations with the heliometer of the Engelhardt Astronomical Observatory.
A. A. Nefed'ev.
Publ. 18th Astrometrical Conference 1969, (see 012.022), p. 281 - 283 (1972). In Russian.

094.256 On the values of the parameters g' and f of the moon from tracking artificial lunar satellites.
Sh. T. Habibullin, Yu. A. Chikanov.
Publ. 18th Astrometrical Conference 1969, (see 012.022), p. 284 - 287 (1972). In Russian.

094.257 On the position of the mass center in the moon.
I. V. Gavrilov.
Publ. 18th Astrometrical Conference 1969, (see 012.022), p. 287 - 292 (1972). In Russian.

094.258 Comparison of the absolute heights of the equipotential and physical lunar surfaces.
I. V. Gavrilov, G. T. Yanovitskaya.
Publ. 18th Astrometrical Conference 1969, (see 012.022), p. 292 - 298 (1972). In Russian.

094.259 An analysis of the precision of selenodetic control systems.   S. G. Valeev.
Publ. 18th Astrometrical Conference 1969, (see 012.022), p. 298 - 301 (1972). In Russian.

094.260 On compiling a catalogue of lunar objects in the system of reference stars.   S. G. Valeev.
Publ. 18th Astrometrical Conference 1969, (see 012.022), p. 301 - 304 (1972). In Russian.

094.261 Interposition of the centers of reference circles on the charts of the lunar marginal zones.
S. G. Valeev, A. A. Nefed'ev.
Publ. 18th Astrometrical Conference 1969, (see 012.022), p. 304 - 307 (1972). In Russian.

094.262 The geometrical figure of the marginal zone of the moon.   N. L. Makarenko.
Publ. 18th Astrometrical Conference 1969, (see 012.022), p. 307 - 314 (1972). In Russian.

094.263 **On the precision of lunar observations reduced with Watts' marginal zone charts.**
D. P. Duma, L. N. Kizyun.
Publ. 18th Astrometrical Conference 1969, (see 012.022), p. 315 - 321 (1972). In Russian.

094.264 **Positional observations of the moon with a horizontal telescope at the Engelhardt Observatory.**
N. G. Rizvanov.
Publ. 18th Astrometrical Conference 1969, (see 012.022), p. 321 - 324 (1972). In Russian.

094.265 **Positional photographic observations of the moon with the (f = 8 m) horizontal telescope of the Engelhardt Observatory.** N. F. Bystrov, N. G. Rizvanov.
Publ. 18th Astrometrical Conference 1969, (see 012.022), p. 325 - 327 (1972). In Russian.

094.266 **Lunar notes.**
W. S. Cameron, K. J. Delano, J. E. Westfall.
Strolling Astronomer, Vol. 24, 18 - 22 (1972).
LTP program status; Dark-haloed craters; A.L.P.O. Lunar Orbiter reference library; Luna Incognita for 1973.

094.267 **The lunar interior.** D. L. Anderson, R. L. Kovach.
Phys. Earth Planet. Interiors, Vol. 6, (see 012.023), 116 - 122 = Contr. Division Geol. Planet. Sci., California Inst. Techn., Pasadena, No. 2122 (1972).

094.268 **Preliminary mapping of the lunar magnetic field.**
P. J. Coleman, Jr., C. T. Russell, L. R. Sharp, G. Schubert.
Phys. Earth Planet. Interiors, Vol. 6, (see 012.023), 167 - 174 (1972).

094.269 **Thermal structure of the moon.**
D. L. Turcotte, A. T. Hsui, K. E. Torrance, E. R. Oxburgh.
Journ. Geophys. Res., Vol. 77, 6931 - 6939 (1972).
Numerical calculations for the structure of convection cells within a self-gravitating, fluid sphere are used to determine the temperature distribution within the moon. The distribution of surface heat flux is also given. The results are compared with the temperatures deduced from magnetic induction within the moon and with the surface heat flow measurement carried out on Apollo 15.

094.270 **Mascons: Progress toward a unique solution for mass distribution.** R. J. Phillips, J. E. Conel, E. A. Abbott, W. L. Sjogren, J. B. Morton.
Journ. Geophys. Res., Vol. 77, 7106 - 7114 (1972).
Through a series of analyses with high-altitude Lunar Orbiter and low-altitude Apollo 15 Doppler gravity data, it is shown that the Serenity mascon is a thin body whose horizontal dimensions are well determined and show a strong correlation with circular wrinkle ridge structure. Analysis to date has not uniquely determined the depth of the anomalous mass. However, geologic evidence strongly suggests that the mass excess is near the surface, because (1) the surface solution has a geometry highly suggestive of the partial filling of a ringed circular basin, and (2) the boundaries of the anomalous mass separate regions of shallow and deep mare flooding.

Physics of the moon and planets. International symposium. Kiev, 15–22 October, 1968.
See Abstr. 003.144.

Modern ideas on the moon.
See Abstr. 003.145.

Atlas & gazetteer of the near side of the moon.
See Abstr. 003.150.

Analysis of X-ray fluorescence without dispersion for investigations of cosmic and terrestrial objects.
See Abstr. 022.123.

On the possibility of utilizing a narrow-angle television camera for determination of elements of the moon's rotation. See Abstr. 031.006.

On automatic angle measurements and a proposition of their application into zenith distance measurements on the surface of the moon. See Abstr. 031.083.

A device for photographic positional observations of the moon. See Abstr. 031.090.

A rugged, stable, differential platinum resistance thermometer. See Abstr. 034.083.

A preliminary system of lunar laser ranging.
See Abstr. 034.097.

Le réflecteur laser de Lunokhod.
See Abstr. 034.104.

The effects of viscous friction on the precession and nutation of celestial bodies. See Abstr. 042.022.

Plans and objectives of the remaining Apollo missions. See Abstr. 053.016.

Methods of investigation of Lunokhod's mobility under terrestrial conditions. See Abstr. 053.021.

A theoretical treatment of lunar particle shadows.
See Abstr. 062.005.

On the validity of a generalized Kirchhoff's law for a nonisothermal scattering and absorptive medium.
See Abstr. 063.010.

Solar-wind and interplanetary electron measurements on the Apollo 15 subsatellite. See Abstr. 074.048.

Solar wind observations on the lunar surface with the Apollo-12 ALSEP. See Abstr. 074.064.

High resolution maps of the sun and moon at 1 millimeter wavelength. See Abstr. 077.020.

Composition and state of matter in the deep interior of the earth. See Abstr. 081.004.

Archaean greenstone belts may include terrestrial equivalents of lunar maria? See Abstr. 081.005.

Creep in the earth and planets.
See Abstr. 081.009.

The dynamical properties and internal structures of the earth, the moon and the planets. See Abstr. 081.031.

Two dimensional digital vidicon photometry of the planets. See Abstr. 091.012.

Martian, lunar, & terrestrial crusts: A three-dimensional exercise in comparative geophysics.
See Abstr. 097.059.

Magnetism of meteorites and lunar rocks.
See Abstr. 105.023.

Search for extinct natural radioactivity of ^{205}Pb via thallium-isotope anomalies in chondrites and lunar soil.
See Abstr. 105.047.

Ar^{40} in meteorites, fines and breccias from the moon. See Abstr. 105.108.

Transition element distribution in stony meteorites and in terrestrial and lunar rocks. See Abstr. 105.113.

Comment on paper by Y. C. Whang and N. F. Ness, 'Magnetic field anomalies in the lunar wake'.
See Abstr. 106.032.

Hard rock cosmic ray archaeology.

See Abstr. 143.069.

Errata

094.901 **Addendum and erratum: 'Thermal evolution of the moon' [The Moon, Vol. 4, 190 - 213 (1972)].**
M. N. Toksöz, S. C. Solomon, J. W. Minear, D. H. Johnston.
The Moon, Vol. 5, 231 - 232 (1972).

094.902 **Erratum: "The hypsometric features of the visible side of the moon" [Astron. Zhurn. Akad. Nauk SSSR, Vol. 49, 621 - 623 (1972). In Russian].**
J. F. Rodionova.
Astron. Zhurn. Akad. Nauk SSSR, Vol. 49, 909 (1972).
In Russian.

095   Lunar Eclipses

095.001 **Reporte de la observación del eclipse total de luna del 30 de enero de 1972.** F. Diego Quintana.
El Universo, No. 99, Vol. 26, 50 - 52 (1972).

095.002 **Eclipse calculations of lunar features.** J. Meeus.
Vistas in astronomy, Vol. 14, (see 003.008), 1 - 11 (1972).

095.003 **Enlargement of the earth's shadow during the lunar eclipse of January 30, 1972.**
V. A. Golubjev, V. A. Lukashenko.

Astron. Tsirk., No. 710, p. 7 - 8 (1972). In Russian.

095.004 **Observation of the lunar eclipse of 6 August 1971.**
V. Novotný.
Říše hvězd, Vol. 53, 218 (1972). In Czech.

Notes from lunar eclipse observers.
Sky Telescope, Vol. 44, 264 - 268 (1972). – 1972 July 25/26.

Remarques sur le retour des éclipses de soleil et de lune. See Abstr. 079.001.

096 Lunar Occultations

096.001 **Observación de ocultaciones. I, II.**
F. Diego Quintana.
El Universo, No. 98 - 99, Vol. 26, 15 - 18, 55 - 57 (1972).

096.002 **Ocultaciones de estrellas durante el eclipse total de luna del 30 de enero de 1972.**
F. Diego Quintana.
El Universo, No. 98, Vol. 26, 40 - 41 (1972).

096.003 **Grazing occultation, 1971 August 6 – ZC 3091 – Bloemfontein.** G. N. Walker.
Monthly Notes Astron. Soc. Southern Africa, Vol. 31, 93 - 98 (1972).

096.004 **Occultation des Pléiades par la lune du 19 mars 1972.** A. Hamon.
L'Astronomie, 86ᵉ année, p. 518 - 520 (1972).

096.005 **Een bedekking van de Pleiaden.** G. W. E. Beekman.
Hemel en Dampkring, Vol. 70, 238 - 239 (1972).

096.006 **Occultation series of five stars.**
G. P. Können, J. Meeus.
Journ. British Astron. Ass., Vol. 82, 431 - 433 (1972).

096.007 **Occultations rasantes, janvier–juin 1973.**
J. Meeus.
L'Astronomie, 86ᵉ année, p. 462 - 464 (1972).

096.008 **Månen i Plejaderne 19. marts 1972.** P. Darnell.
Astron. Tidssk., Årg. 5, p. 140 (1972).

096.009 **Den norske seksjon for okkultasjoner 1971.**
H. Brubak.
Astron. Tidssk., Årg. 5, p. 141 (1972).

096.010 **Rakende sterbedekkingen, januari - juli 1973.**
J. Meeus.
Hemel en Dampkring, Vol. 70, 302 - 304 (1972).

096.011 **Catalogue of theoretical lunar-occultation diffraction patterns for single and double stars occulted by the lunar limb.** C. L. Morbey.
Publ. Dominion Astrophys. Obs., Victoria, Vol. 14, (No. 3), 45 - 58 (1972).
Theoretical diffraction curves are computed for occultations of double stars of which each component has an angular radius of $0.''001$ for a range of angular separations of the components from $0.''001$ to $0.''020$. Curves are also computed for single stars of angular radius $0.''001$ that are occulted by simple step irregularities on the lunar limb. Both groups have been computed for a monochromatic source and a source emitting light with a 'Gaussian' bandpass 1000 Å wide; for the double star group the centre wavelength was chosen at 4500 Å; for the single star group curves were computed for centre wavelengths at 4500 Å and 7000 Å.

096.012 **Occultations observed at Sydney Observatory during 1970.** K. P. Sims.
Journ. Proc. Roy. Soc. New South Wales, Vol. 103, 91 - 92 (1970) = Sydney Obs. Papers No. 64 (1971).

096.013 **Two grazing occultations on 1972 May 19. ZC 1441 – Kingswood, SAO 98794 – Wolmaransstad.** J. Hers.
Monthly Notes Astron. Soc. Southern Africa, Vol. 31, 145 - 149 (1972).

096.014 **Occultations of stars by the moon observed at the Cracow Astronomical Observatory in the year 1971.**
T. Z. Dworak.
Acta Astron., Vol. 22, 309 - 313 (1972).
This paper contains 178 reduced observations of occultations of stars by the moon made at the old Cracow Observatory and the new observatory "Fort Skała".

096.015 **Report of occultation observations in 1969.**
S. Kimura.
Mem. Japan Astron. Study Ass., No. 17, Vol. 5, 5 - 6 (1971).
In Japanese.

096.016 **Grazing occultation of SAO 97918, 1972 June 14.**
J. Hers.
Monthly Notes Astron. Soc. Southern Africa, Vol. 31, 153 - 155 (1972).

096.017 **Observations of eclipses in the Pleiades cluster stars on March 19, 1972.** L. Zajdler.
Urania Kraków, Vol. 43, 214 - 219 (1972). In Polish.

096.018 **Observations of eclipses of stars by the moon.**
L. Zajdler.
Urania Kraków, Vol. 43, 279 - 281 (1972). In Polish.

096.019 **Occultations d'étoiles par la lune observées à l'équatorial de 45 cm, en 1969, 1970 et 1971.**
J. Dommanget, E. Van Dessel.
Bull. Astron. Obs. Roy. Belgique, Vol. 8, 42 (1972).

096.020 **Ocultaciones, tercera parte: (Ocultaciones totales).**
F. Diego Q.
El Universo, No. 101, Vol. 26, 154 - 160 (1972).

096.021 **Ocultación rasante de Antares del 28 de mayo de 1972.** F. Diego Q.
El Universo, No. 101, Vol. 26, 166 (1972).

096.022 **Ocultación parcial de Venus: Una interesante visita.**
F. Diego Q.
El Universo, No. 101, Vol. 26, 167 - 169 (1972).

096.023 **Sternbedeckungen durch den Mond 1972.**
Techn. Univ. Dresden, Lohrmann-Obs. Zirk. No. 60, p. 5 - 6 (1972).

096.024 **Lunar occultations.**
Contr. Obs. Valongo, Univ. Federal Rio de Janeiro, Sér. III, No. 22 (1972). – 1972 April – June.

The role of occultations in the improvement of the lunar ephemeris. See Abstr. 094.088.

Rocznik Astronomiczny Obserwatorium Krakowskiego 1973. International Supplement No. 44.
See Abstr. 120.010.

A study of the lunar occultations of eleven radio sources. See Abstr. 141.005.

097 Mars

097.001 **Bistatic radar measurements of the surface of Mars with Mariner 1969.** G. Fjeldbo, A. Kliore, B. Seidel.
Icarus, Vol. 16, 502 - 508 (1972).
The detection of echoes produced by oblique reflection of the RF (2300 MHz) spacecraft carrier from the Martian surface as Mariner 6 and 7 flew behind Mars in 1969, is described. Changes in echo center frequency and bandwidth are utilized to study the radius and roughness of the surface along a quasi-specular radar track that led from an optically dark and densely cratered region of Meridiani Sinus over into a smoother and brighter looking area of Thymiamata.

097.002 **Diurnal and seasonal behavior of discrete white clouds on Mars.** S. A. Smith, B. A. Smith.
Icarus, Vol. 16, 509 - 521 (1972).
Two distinct types of discrete white "clouds" on Mars are found to show periodic activity that is highly correlated with the Martian seasons. In addition to the seasonal cycle, type I clouds exhibit a repetitive diurnal variation, forming in the late morning or early afternoon and continuing to brighten for several hours thereafter.

097.003 **An orthographic photomap of the south pole of Mars from Mariner 7.** A. R. Gillespie, J. M. Soha.
Icarus, Vol. 16, 522 - 527 (1972).
Television pictures of the south polar regions of Mars obtained by the Mariner 6 and 7 spacecraft in 1969 are rectified to a standard mapping projection using computer image processing techniques. Mosaicking of these pictures produces the first photomap of the entire south polar cap.

097.004 **The location of the Mountains of Mitchel and evidence for their nature in Mariner 7 pictures.**
J. A. Cutts, J. Veverka, J. D. Goguen.
Icarus, Vol. 16, 528 - 534 (1972).
A critical literature search indicates that of the two mutually exclusive locations usually quoted for the Mountains of Mitchel: ($267° -293°$W, $\sim73°$S) and ($\sim320°$W, $\sim75°$S), the first is definitely spurious and must be rejected. In Mariner 7 far-encounter photographs, the second location appears as an elongated bright marking which in higher resolution photographs is seen to correspond with relatively rugged terrain.

097.005 **Mariner 7 ultraviolet spectrometer experiment: Topographic slopes of Mars' polar region.**
K. Pang.
Icarus, Vol. 16, 535 - 542 (1972).
The Mariner 7 ultraviolet spectrometer observed the south polar cap of Mars. The near ultraviolet polar spectrum showed that light reflected from the surface predominates over light scattered from the atmosphere. A method has been formulated to determine the average slope of an area from intensity ratio measurements made at any two different sets of illumination and observation angles.

097.006 **The composition of the Martian atmosphere: Minor constituents.** D. Horn, J. M. McAfee, A. M. Winer, K. C. Herr, G. C. Pimentel.
Icarus, Vol. 16, 543 - 556 (1972).
The Mariner 6 and 7 infrared spectrometers provided data which, in principle, determine upper limits on the possible atmospheric abundance of every gaseous substance that was undetected but which has recognized absorptions in the accessible spectral region, 1.9 to 14.4μ. Through supporting laboratory determinations of curves of growth under pressure broadening conditions appropriate to Mars, upper limits can be spec-

ified for the following gases: NO_2, NH_3, C_3O_2, SO_2, OCS, NO, O_3, CH_4, N_2O, HCl, HBr, H_2S. In addition, considerations of band contours, moments of inertia, and experimental absorption coefficients permit us to place useful upper limits on twenty-seven additional substances that were not detected and for which curves of growth have not been measured.

097.007 **The new Mariner 9 map of Mars.**
W. Hartmann, F. E. Bristow.
Sky Telescope, Vol. 44, 77 - 82, with a folding topographic chart (1972).

097.008 **La planète Mars en 1971.** J. Dragesco.
L'Astronomie, 86ᵉ année, p. 337 - 353 (1972).
Report of the Commission des Surfaces Planétaires de la Société Astronomique de France.

097.009 **Mariner 6 and 7 ultraviolet spectrometer experiment: Analysis of the O I 1304- and 1356-Å emissions.** D. J. Strickland, G. E. Thomas, P. R. Sparks.
Journ. Geophys. Res., Vol. 77, 4052 - 4068 (1972).
An analysis of the O I 1304- and 1356-Å data from Mariner 6 and 7 is presented, and the atomic oxygen concentration of the Martian atmosphere is estimated. At an exospheric temperature of 350°K the atomic oxygen concentration at 135 km is estimated to be between 0.5 and 1%. The likely source of the 1304-Å intensity is resonance scattering of solar 1304-Å photons.

097.010 **Some properties of the Martian atmosphere during dust storms.** A. V. Morozhenko.
Astron. Tsirk., No. 683, p. 1 - 3 (1972). In Russian.

097.011 **Monochromatic albedo of the central zone of the Martian disk in the range of 3325 - 7475Å.**
S. M. Omarov, V. V. Avramchuk, A. R. Gajduk, M. S. Gadzhiev.
Astron. Tsirk., No. 687, p. 1 - 3 (1972). In Russian.

097.012 **Observations of Mars in 1971.**
N. P. Barabashov, N. N. Kiselev, D. F. Lupishko.
Astron. Tsirk., No. 687, p. 3 - 6 (1972). In Russian.

097.013 **Der Mars.**
Naturwissenschaften, 59. Jahrgang, p. 394 - 395 (1972). – Some photos of surface structures on Mars are reproduced by permission of NASA.

097.014 **Coordinates of features on the Mariner 6 and 7 pictures of Mars.** M. E. Davies.
Icarus, Vol. 17, 116 - 167 (1972).
A control net of Mars has been computed from measurements of 115 control points identified on the Mariner 6 and 7 pictures. Most of these points are located with respect to topographic features on the surface of Mars, and their areographic coordinates were computed by photogrammetric techniques.

097.015 **Mars, wie ihn Mariner IX gesehen hat.** H. Müller.
Orion, 30. Jahrgang, p. 127 - 129 (1972).

097.016 **Height of dust clouds on Mars during the dust storm in 1971 from groundbased observations.**
V. I. Moroz, O. G. Taranova.
Astron. Tsirk., No. 697, p. 1 - 4 (1972). In Russian.

097.017 **Oxidized basalt is a possible component of the Martian soil.** V. V. Novikov.

Astron. Tsirk., No. 701, p. 5 - 8 (1972). In Russian.

097.018 **The variation in contrast of Syrtis Major in 1971.**
D. T. Thompson.
Bull. American Astron. Soc., Vol. 4, 313 (1972). – Abstr. AAS.

097.019 **Variable features on Mars: Preliminary Mariner 9 television results.** C. Sagan, J. Veverka, P. Fox,
R. Dubisch, J. Lederberg, E. Levinthal, L. Quam, R. Tucker,
J. Pollack, B. Smith.
Bull. American Astron. Soc., Vol. 4, 313 (1972). – Abstr. AAS.

097.020 **Infrared brightness temperature of Mars in the 8–13 micron region at the time of a dust storm according** to earthbased observations.
A. A. Liberman, V. I. Moroz, G. S. Khromov.
Astron. Tsirk., No. 705, p. 3 - 5 (1972). In Russian.

097.021 **Composition of the Martian upper atmosphere.**
N. W. Spencer.
Bull. American Astron. Soc., Vol. 4, 354 - 355 (1972).
Abstr. AAS.

097.022 **Mariner 9 mission.** R. H. Steinbacher.
Bull. American Astron. Soc., Vol. 4, 356 (1972).
Abstr. AAS.

097.023 **Mariner 9 Mars television experiment.**
H. Masursky, R. M. Batson, M. H. Carr, J. F.
McCauley, D. J. Milton, L. A. Soderblom, R. L. Wildey, D. E.
Wilhelms, J. Lederberg, E. Levinthal, G. de Vaucouleurs, G. A.
Briggs, A. T. Young, B. A. Smith, J. A. Cutts, R. B. Leighton,
B. C. Murray, R. P. Sharp, W. K. Hartmann, C. B. Leovy, M. E.
Davies, C. Sagan, J. Veverka, E. N. Shipley, J. B. Pollack.
Bull. American Astron. Soc., Vol. 4, 356 (1972). – Abstr. AAS.

097.024 **Ultraviolet spectroscopy of Mars.** C. A. Barth.
Bull. American Astron. Soc., Vol. 4, 356 (1972).
Abstr. AAS.

097.025 **Structure of upper atmosphere and its variations.**
A. I. Stewart.
Bull. American Astron. Soc., Vol. 4, 356 (1972). – Abstr. AAS.

097.026 **Ultraviolet reflectance of Mars' south polar cap and opacity of the 1971 dust cloud.** K. D. Pang.
Bull. American Astron. Soc., Vol. 4, 357 (1972). – Abstr. AAS.

097.027 **Observations of ozone in the north polar region.**
A. L. Lane.
Bull. American Astron. Soc., Vol. 4, 357 (1972). – Abstr. AAS.

097.028 **Overview of results from the infrared spectroscopy experiment on Mariner 9.**
R. A. Hanel, B. J. Conrath, W. A. Hovis, V. G. Kunde, P. D.
Lowman, J. C. Pearl, C. Prabhakara, B. Schlachman, G. Levin,
T. E. Burke.
Bull. American Astron. Soc., Vol. 4, 357 (1972). – Abstr. AAS.

097.029 **Thermal structure of the Martian atmosphere obtained from the Mariner 9 infrared spectroscopy experiment (IRIS).**
B. J. Conrath, R. A. Hanel, V. G. Kunde, J. C. Pearl.
Bull. American Astron. Soc., Vol. 4, 357 (1972). – Abstr. AAS.

097.030 **Mariner IRIS Martian surface pressure and topographic mapping.**
J. C. Pearl, B. J. Conrath, R. A. Hanel, V. G. Kunde.
Bull. American Astron. Soc., Vol. 4, 357 (1972). – Abstr. AAS.

097.031 **Preliminary results from Mariner 9 IRIS on water**
vapor and other minor constituents in the atmosphere of Mars. V. G. Kunde, B. J. Conrath, R. A. Hanel,
W. C. Maguire, J. C. Pearl, T. E. Burke.
Bull. American Astron. Soc., Vol. 4, 357 (1972). – Abstr. AAS.

097.032 **Mariner 9 infrared radiometer: Observations of general properties of Mars.** H. Kieffer.
Bull. American Astron. Soc., Vol. 4, 357 (1972). – Abstr. AAS.

097.033 **Mariner 9 infrared radiometer: Observations of specific Martian features.** E. D. Miner.
Bull. American Astron. Soc., Vol. 4, 358 (1972). – Abstr. AAS.

097.034 **Preliminary results on the atmosphere and topography of Mars from Mariner 9 radio occultation** measurements. A. Kliore, B. L. Seidel, M. J. Sykes.
Bull. American Astron. Soc., Vol. 4, 358 (1972). – Abstr. AAS.

097.035 **The shape of Mars.** D. L. Cain.
Bull. American Astron. Soc., Vol. 4, 358 (1972).
Abstr. AAS.

097.036 **Mariner 9 radio occultation measurements of the upper atmosphere of Mars.**
G. Fjeldbo, A. Kliore, D. Cain, B. Seidel.
Bull. American Astron. Soc., Vol. 4, 358 (1972). – Abstr. AAS.

097.037 **Gravity field of Mars from Mariner 9 tracking data.**
J. Lorell, G. H. Born, E. J. Christensen, P. B.
Esposito, J. F. Jordan, P. A. Laing, W. L. Martin, W. L. Sjo-
gren, S. K. Wong, I. I. Shapiro, R. D. Reasonberg, G. L. Slater.
Bull. American Astron. Soc., Vol. 4, 358 (1972). – Abstr. AAS.

097.038 **Laboratory studies of carbon suboxide polymer applications to Mars.**
B. Khare, M. Khare, C. Sagan.
Bull. American Astron. Soc., Vol. 4, 362 (1972). – Abstr. AAS.

097.039 **Approaches for exploration of Phobos and Deimos.**
E. F. Harrison, W. T. Scofield.
Bull. American Astron. Soc., Vol. 4, 364 - 365 (1972).
Abstr. AAS.

097.040 **Determination of the orbits of Phobos and Deimos from Mariner IX.** G. Born, T. Duxbury.
Bull. American Astron. Soc., Vol. 4, 368 (1972). – Abstr. AAS.

097.041 **Martian satellite debris.** S. Soter.
Bull. American Astron. Soc., Vol. 4, 368 (1972).
Abstr. AAS.

097.042 **IR spectrophotometric mapping of Mars.**
A. B. Binder, J. C. Jones.
Bull. American Astron. Soc., Vol. 4, 370 - 371 (1972).
Abstr. AAS.

097.043 **Photometry of Martian surface features in 1971.**
P. B. Boyce.
Bull. American Astron. Soc., Vol. 4, 371 (1972). – Abstr. AAS.

097.044 **Why Mars is red.** R. L. Huguenin, T. B. McCord.
Bull. American Astron. Soc., Vol. 4, 371 (1972).
Abstr. AAS.

097.045 **Infrared spectra of Mars: Effects of experimental conditions.**
G. R. Hunt, L. M. Logan, J. W. Salisbury.
Bull. American Astron. Soc., Vol. 4, 371 (1972). – Abstr. AAS.

097.046 **Observations of O_2 on Mars and Venus.**
W. A. Traub, N. P. Carleton.

Bull. American Astron. Soc., Vol. 4, 371 (1972). – Abstr. AAS.

097.047 Detection of O$_2$ in the Martian atmosphere with the echelle-coudé scanner of the 107-inch telescope.
E. S. Barker.
Bull. American Astron. Soc., Vol. 4, 371 - 372 (1972). Abstr. AAS.

097.048 An upper limit for O$_2$ on Mars. T. D. Parkinson.
Bull. American Astron. Soc., Vol. 4, 372 (1972). Abstr. AAS.

097.049 Ground-based photoelectric measures of H$_2$O on Mars during the Mariner 9 encounter.
R. G. Tull, E. S. Barker.
Bull. American Astron. Soc., Vol. 4, 372 (1972). – Abstr. AAS.

097.050 Studies of Martian water vapor in the 8200 Å band with a Varo image tube. W. M. Sinton, R. Carson.
Bull. American Astron. Soc., Vol. 4, 372 (1972). – Abstr. AAS.

097.051 High altitude aircraft infrared spectrometric observations of water of hydration on Mars.
J. Houck, J. B. Pollack, C. Sagan, D. Schaack, J. Dekker.
Bull. American Astron. Soc., Vol. 4, 372 (1972). – Abstr. AAS.

097.052 Circular polarization spectrum of Mars.
R. D. Wolstencroft, J. C. Kemp, J. B. Swedlund.
Bull. American Astron. Soc., Vol. 4, 372 (1972). – Abstr. AAS.

097.053 Martian topography: Final contour map from all data sources through 1969. R. A. Wells.
Bull. American Astron. Soc., Vol. 4, 372 - 373 (1972). Abstr. AAS.

097.054 CO$_2$ distribution on Mars for the 1971 opposition.
T. D. Parkinson, D. M. Hunten.
Bull. American Astron. Soc., Vol. 4, 373 (1972). – Abstr. AAS.

097.055 Evidence from Mariner '69 pictures concerning eolian processes as a crater degradation mechanism on the Martian surface. M. Gaffey, L. Lebofsky, T. McCord.
Bull. American Astron. Soc., Vol. 4, 373 (1972). – Abstr. AAS.

097.056 Possible volcanic landforms in Mariner 6 and 7 photographs and implications.
T. R. McGetchin, M. Settle, D. Francis, W. S. Baldridge.
Bull. American Astron. Soc., Vol. 4, 373 (1972). – Abstr. AAS.

097.057 The Martian lineament systems as defined by the Mariners 4, 6, and 7 imagery.
A. B. Binder, D. W. McCarthy.
Bull. American Astron. Soc., Vol. 4, 373 - 374 (1972). Abstr. AAS.

097.058 Martian regional contrasts in 1971.
D. T. Thompson.
Bull. American Astron. Soc., Vol. 4, 374 (1972). – Abstr. AAS.

097.059 Martian, lunar, & terrestrial crusts: A three-dimensional exercise in comparative geophysics.
R. A. Wells.
Bull. American Astron. Soc., Vol. 4, 374 (1972). – Abstr. AAS.

097.060 Martian dust storm: Its depth on November 25, 1971. T. D. Parkinson, D. M. Hunten.
Bull. American Astron. Soc., Vol. 4, 374 (1972). – Abstr. AAS.

097.061 Survey of Martian yellow storms.
C. F. Capen, L. J. Martin.
Bull. American Astron. Soc., Vol. 4, 374 (1972). – Abstr. AAS.

097.062 Where will the Martian dust be when Viking arrives?
W. A. Baum.
Bull. American Astron. Soc., Vol. 4, 374 - 375 (1972). Abstr. AAS.

097.063 Determination of the Mars spin axis direction from Mariner IX.
G. Born, E. Christensen, T. Duxbury, J. Jordan, S. Mohan.
Bull. American Astron. Soc., Vol. 4, 375 (1972). – Abstr. AAS.

097.064 Bericht über erste Ergebnisse der Mars-Sonden 1971.
H. Lambrecht, S. Marx.
Sterne, 48. Jahrgang, p. 129 - 138 (1972).

097.065 Raumsonden photographieren die Marsmonde.
E. Krug.
Sterne, 48. Jahrgang, p. 138 - 141 (1972).

097.066 Stability of the martian atmosphere.
M. B. McElroy, T. M. Donahue.
Science, Vol. 177, 986 - 988 (1972).
A detailed chemical dynamic model is presented for a moist martian atmosphere. Recombination of carbon dioxide is catalyzed by trace amounts of water. The abundances of carbon monoxide and molecular oxygen should vary in response to changes in atmospheric water and atmospheric mixing.

097.067 Detection of molecular oxygen on Mars.
N. P. Carleton, W. A. Traub.
Science, Vol. 177, 988 - 992 (1972).
Molecular oxygen was detected in martian spectra near 7635 Å and its abundance measured both during and after the 1971 dust storm. Its column abundance in the clear martian atmosphere is about 10.4 ± 1.0 centimeters amagat, giving a mixing ratio of molecular oxygen to carbon dioxide of 1.3×10^{-3}.

097.068 Photographic observations of Mars at the Main Astronomical Observatory of the Ukrainian Academy of Sciences in 1963–1967. E. M. Sereda.
Astrometriya i Astrofiz., *Kiev*, No. 15, (see 003.001), p. 76 - 90 (1972). In Russian.
130 positions of Mars in the system of reference stars of the Yale Catalogue are given. The equatorial coordinates of Mars were compared with the ephemeris positions from «Astronomical Almanac».

097.069 Photocatalytic production of organic compounds from CO and H$_2$O in a simulated Martian atmosphere.
J. S. Hubbard, J. P. Hardy, N. H. Horowitz.
Proc. National Acad. Sci., *U.S.A.*, Vol. 68, 574 - 578 (1971).
We have performed organic synthesis experiments with mixtures of CO$_2$, CO, and H$_2$O exposed to UV in the presence of soil or powdered vycor glass. The purpose of these tests was to uncover possible sources of error in an experiment, planned for the first Mars lander, designed to detect biosynthesis of organic matter in Martian soil. The findings, presented here, show that UV radiation of wavelengths above 2000 Å produces organic compounds in the substratum.

097.070 Peculiarities of some solar halos on Mars in the marginal zone and near the terminator.
V. D. Davydov.
Astron. Zhurn. Akad. Nauk SSSR, Vol. 49, 1078 - 1087 (1972). In Russian. English translation in Soviet Astron. AJ, Vol. 16, No. 5.
Conditions for visibility of solar halos on the planet from cosmic distances are considered with the aim of searching for criteria that allow in some cases to distinguish solar halos on Mars from bright spots of another origin. Examples are given

of halo-like spots on Martian disc margins obtained from ground-based observations and from pictures made by Mariner 6 and Mariner 7.

097.071 Aeolian deposition in Martian craters.
H. T. U. Smith.
Nature, Phys. Sci., Vol. 238, 72 - 74 (1972).

Mariner 6 and 7 photos of the Meridiani Sinus region show conspicuously asymmetric albedo in many craters. Light areas appear to contrast with their surroundings also in slope and configuration, and to be topographically unconformable on normal crater form. Partial filling by wind-blown particulate material provides the best explanation. Confirmation is found in the analogy of certain sand-modified craters of West Africa.

097.072 Mars: Are observed white clouds composed of H_2O?
R. A. Wells.
Nature, Vol. 238, 324 - 326 (1972).

Results of seasonal Martian cloud frequencies from the author's revised list of white cloud occurrences are presented. A comparison with the observed seasonal variation of Martian H_2O abundances, compiled over the past decade, gives a rank correlation coefficient of 0.74. The calculated z deviate 3.23 falls very close to the value 3.29 at the 0.1 % level of chance occurrence. A comparison between these cloud frequencies and H_2O abundances distributed by latitude for the season corresponding to the northern Martian hemisphere summer and southern winter yields a correlation of 0.94 with the calculated z deviate 2.10 falling between the 5 % and 2 % levels of chance occurrence. These two correlations imply that Martian white clouds are most probably composed of water ice crystals.

097.073 Resultater fra Marssonden Mariner 9. K. Messell.
Astron. Tidssk., Årg. 5, p. 113 - 116 (1972).

097.074 Preliminary results of investigations on the Martian atmosphere by means of the Mars 2 satellite.
M. A. Kolosov, O. I. Iakovlev, Iu. M. Kruglov, B. P. Trusov, A. I. Efimov, V. V. Kerzhanovich.
Dokl. Akad. Nauk SSSR, Ser. Mat. Fiz., Vol. 206, 1071 - 1073 (1972). In Russian.

097.075 New facts on Martian geology. P. N. Kropotkin.
Priroda, No. 11.72, p. 88 - 89 (1972). In Russian.

097.076 On the integral radio emission of Mars during the opposition of 1971.
V. D. Krotikov, O. B. Shchuko.
Izv. vyssh. ucheb. zavedenij. Radiofizika, Vol. 15, 487 - 489 (1972). In Russian. — Abstr. in Referativ. Zhurn. 51. Astron., 11.51.223 (1972).

097.077 An amateur's map of Mars for 1971. B. Salmon.
Sky Telescope, Vol. 44, 410 - 411 (1972).

097.078 The planet Mars. N. K. Kroschel.
Journ. Astron. Soc. Victoria, Vol. 25, 74 - 75 (1972).

097.079 Brief history of the Martian 'violet haze' problem.
D. T. Thompson.
Rev. Geophys. Space Phys., Vol. 10, 919 - 933 (1972).

A brief but thorough survey of the literature on the Martian 'violet haze' problem is presented. It is evident that both the normal lack of contrast of the surface features in violet light and their occasional appearance are phenomena intrinsic to Mars. Models involving simple uniform layers of scattering or absorbing materials are inadequate to account for the observations. We suggest that the role of haze has historically been misinterpreted.

097.080 The dust-storms of Mars. P. Moore.
Journ. British Astron. Ass., Vol. 83, 31 - 32 (1972).

097.081 Meteorological observations of Mars during the 1971 opposition. S. Miyamoto.
Contr. Kwasan and Hida Obs., Univ. Kyoto, Nos. 206/207, p. 1 - 71 (1972).

This is the eighth report of our series about the meteorological observation of Mars during the 1971 apparition. In this apparition, we secured records of meteorological phenomena covering the Martian season from southern winter to autumn.

097.082 Life cycle of Martian polar cap, sand storm and general circulation. S. Miyamoto.
Contr. Kwasan and Hida Obs., Univ. Kyoto, Nos. 206/207, p. 73 - 80 (1972).

The early developing stage of the southern polar cap was described. The cap is formed not in midwinter, but just before the vernal equinox by the arrival of moisture to the pole. This fact supports the ordinary snow or frost hypothesis of the cap. The pattern of general circulation in Mars is the exchange of air masses between summer and winter hemispheres in solstitial season and between the equator and poles in equinoxial season. This gives the theoretical background for W. H. Pickering's migration theory of water vapour from pole to pole with the advance of season.

097.083 Preliminary Mariner 9 report on the geology of Mars.
J. F. McCauley, M. H. Carr, J. A. Cutts, W. K. Hartmann, H. Masursky, D. J. Milton, R. P. Sharp, D. E. Wilhelms.
Icarus, Vol. 17, 289 - 327 (1972). — Presented at the 'Symposium on planetary atmospheres and surfaces' (see 012.017).

Mariner 9 pictures indicate that the surface of Mars has been shaped by impact, volcanic, tectonic, erosional and depositional activity. The moonlike cratered terrain, identified as the dominant surface unit from the Mariner 6 and 7 flyby data, has proven to be less typical of Mars than previously believed, although extensive in the mid- and high-latitude regions of the southern hemisphere. Martian craters are highly modified but their size-frequency distribution and morphology suggest that most were formed by impact. Circular basins encompassed by rugged terrain and filled with smooth plains material are recognized. Mariner 9 has thus revealed that Mars is a complex planet with its own distinctive geologic history and that it is less primitive than the moon.

097.084 Geological framework of the south polar region of Mars.
B. C. Murray, L. A. Soderblom, J. A. Cutts, R. P. Sharp, D. J. Milton, R. B. Leighton.
Icarus, Vol. 17, 328 - 345 (1972). — Presented at the 'Symposium on planetary atmospheres and surfaces' (see 012.017).

The purpose of this paper is to present an initial formulation of the geology of the south polar region based on Mariner 9 pictures. A preliminary comparison with the north polar region is also presented.

097.085 Variable features on Mars: Preliminary Mariner 9 television results.
C. Sagan, J. Veverka, P. Fox, R. Dubisch, J. Lederberg, E. Levinthal, L. Quam, R. Tucker, J. B. Pollack, B. A. Smith.
Icarus, Vol. 17, 346 - 372 (1972). — Presented at the 'Symposium on planetary atmospheres and surfaces' (see 012.017).

Systematic Mariner 9 photography of a range of Martian surface features, observed with all three photometric angles approximately invariant, reveals three general categories of albedo variations: (1) an essentially uniform contrast enhancement due to the dissipation of the dust storm; (2) the appearance of splotches, irregular dark markings at least partially related to topography; and (3) the development of both bright and dark linear streaks, generally emanating from craters.

097.086 The Martian atmosphere: Mariner 9 television experiment progress report.

C. B. Leovy, G. A. Briggs, A. T. Young, B. A. Smith, J. B. Pollack, E. N. Shipley, R. L. Wildey.

Icarus, Vol. 17, 373 - 393 (1972). – Presented at the 'Symposium on planetary atmospheres and surfaces' (see 012.017).

Mariner 9 television pictures revealed a wide variety of atmospheric phenomena, particularly after the clearing of the global dust storm, which had obscured much of the atmosphere, as well as the surface of Mars, at Mariner 9 arrival (Masursky et al., 1972). Television pictures illustrating these phenomena will be presented in this report.

097.087 Mariner 9 television observations of Phobos and Deimos.

J. B. Pollack, J. Veverka, M. Noland, C. Sagan, W. K. Hartmann, T. C. Duxbury, G. H. Born, D. J. Milton, B. A. Smith.

Icarus, Vol. 17, 394 - 407 (1972). – Presented at the 'Symposium on planetary atmospheres and surfaces' (see 012.017).

Mariner 9 photographs of Phobos and Deimos have yielded new information about the orbits, rotation periods, sizes, shapes, and surface characteristics of the satellites. Both satellites appear to be in synchronous rotation. They are irregular, heavily cratered bodies whose shapes appear to have been determined largely by impact fragmentation and spalling. The surfaces of both satellites have crater densities close to saturation and nearly identical, very low albedos.

097.088 Preliminary results of astrophysical observations of Mars from Mars 3.

V. I. Moroz, L. V. Ksanfomaliti.

Icarus, Vol. 17, 408 - 422 (1972). – Presented at the 'Symposium on planetary atmospheres and surfaces' (see 012.017).

On board the Automatic Interplanetary Stations (AIS) Mars 2 and 3 four experiments intended for studying the surface and the lower atmosphere of the planet with optical techniques were installed. Preliminary results of measurements with these experiments on Mars 3 in seven periares passages between December and March are presented.

097.089 Investigation of the Martian environment by infrared spectroscopy on Mariner 9.

R. Hanel, B. Conrath, W. Hovis, V. Kunde, P. Lowman, W. Maguire, J. Pearl, J. Pirraglia, C. Prabhakara, B. Schlachman, G. Levin, P. Straat, T. Burke.

Icarus, Vol. 17, 423 - 442 (1972). – Presented at the 'Symposium on planetary atmospheres and surfaces' (see 012.017).

The infrared spectroscopy experiment on Mariner 9 provides extensive information on the Martian environment, including spatial, diurnal, and secular dependences of atmospheric and surface parameters. Measurements obtained during and after the planet-wide dust storm indicate that large diurnal variations in atmospheric temperature existed up to at least 30 km; winds inferred from the temperature fields show a strong tidal component and significant ageostrophic behavior. Preliminary analysis has thus far revealed spectral features due to CO_2, water vapor, and silicate dust suspended in the atmosphere.

097.090 Mariner 9 ultraviolet spectrometer experiment: Photometry and topography of Mars.

C. W. Hord, C. A. Barth, A. I. Stewart, A. L. Lane.

Icarus, Vol. 17, 443 - 456 (1972). – Presented at the 'Symposium on planetary atmospheres and surfaces' (see 012.017).

Reflectance properties of Mars are measured in a 100-Å band centered at 3050 Å by the ultraviolet spectrometer. The transition from dusty conditions, which prevailed at the time of arrival of Mariner 9 on 14 November 1971, began on 1 January 1972, and relatively clear conditions existed after 23 January 1972. As the atmosphere became clearer, the scattering properties began to show a morning enhancement in both terminator and illuminated disk reflectance. A topographic map of Mars based on the scattering of ultraviolet light from the Mars atmosphere is shown.

097.091 Mariner 9 ultraviolet spectrometer experiment: Mars airglow spectroscopy and variations in Lyman alpha.

C. A. Barth, A. I. Stewart, C. W. Hord, A. L. Lane.

Icarus, Vol. 17, 457 - 468 (1972). – Presented at the 'Symposium on planetary atmospheres and surfaces' (see 012.017).

Starting on 14 November 1971, the Mariner 9 ultraviolet spectrometer measured the upper atmosphere airglow of Mars over a time period of 120 days. The large number of Mariner 9 observations has produced not only spectroscopic data of greater precision, but, in addition, measurements of variations that occur in the Mars airglow over an extended time period. Some of the results of these observations will be reported here.

097.092 Mariner 9 ultraviolet spectrometer experiment: Structure of Mars' upper atmosphere.

A. I. Stewart, C. A. Barth, C. W. Hord, A. L. Lane.

Icarus, Vol. 17, 469 - 474 (1972). – Presented at the 'Symposium on planetary atmospheres and surfaces' (see 012.017).

The present paper is concerned with measurements of the Cameron bands and of the CO_2^+ $(\tilde{B}-\tilde{X})$ doublet, and their relationship to the solar extreme ultraviolet (EUV) flux and to the temperature in the Martian upper atmosphere.

097.093 Preliminary results of measurements of UV emissions scattered in the Martian upper atmosphere.

N. N. Dementyeva, V. G. Kurt, A. S. Smirnov, L. G. Titarchuk, S. D. Chuvahin.

Icarus, Vol. 17, 475 - 483 (1972). – Presented at the 'Symposium on planetary atmospheres and surfaces' (see 012.017).

The results of measurements of UV emissions in the Martian upper atmosphere in the wavelength ranges of $\lambda\lambda$ 1050–1340 Å and 1225–1340 Å, which were obtained by means of equipment mounted on the Mars 2 and Mars 3 artificial satellites, are presented.

097.094 The atmosphere of Mars from Mariner 9 radio occultation measurements.

A. J. Kliore, D. L. Cain, G. Fjeldbo, B. L. Seidel, M. J. Sykes, S. I. Rasool.

Icarus, Vol. 17, 484 - 516 (1972). – Presented at the 'Symposium on planetary atmospheres and surfaces' (see 012.017).

The Mariner 9 spacecraft was used to perform 160 radio occultation measurements in orbit about Mars during November and December of 1971. At that time, Mars was experiencing a severely obscuring global dust storm. The effect of dust in the atmosphere was reflected in the reduced temperature gradients that were measured in the daytime near-equatorial atmosphere, indicating heating of the atmosphere by solar radiation being absorbed by dust and a simultaneous cooling of the surface. The surface pressures in the near equatorial regions ranged from a high of 8.9 mbar in Hellas to a low of 2.8 mbar in the Claritas and Tharsis areas, with a mean pressure of 4.95 mbar. A daytime ionosphere having a peak density of about $1.5-1.7 \times 10^5$ el/cm^3 at an altitude of 140–134 km over a range of solar zenith angles of 56–47° was measured.

097.095 The shape of Mars from the Mariner 9 occultations.

D. L. Cain, A. J. Kliore, B. L. Seidel, M. J. Sykes.

Icarus, Vol. 17, 517 - 524 (1972). – Presented at the 'Symposium on planetary atmospheres and surfaces' (see 012.017).

The extinction time of the radio signal, as the Mariner 9 spacecraft was occulted by Mars, together with an accurate ephemeris of the spacecraft were used to determine radii from the mass center to the occulting feature. Similarly estimations were made of the radius to a point where the pressure reached a certain fixed value. Several simple models were proposed to fit both sets of radii data.

097.096 **New optical measurements of planetary diameters – Part IV: Planet Mars.** A. Dollfus.
Icarus, Vol. 17, 525 - 539 (1972).
Optical measurements of the diameter of Mars were made using a double-image micrometer with large refractors from 1952 to 1971. Discussion of the 90 independent series of measurements gives nine determinations of radius with an accuracy of ±7–8 km for different latitudes from pole to equator. The Mariner 4, 6 and 7 occultation results and the radar results available in 1970 added seven further determinations of comparable accuracy. All these values, within the accuracy of measurement, fit an ellipsoid with R_{eq} = 3398 ± 3 km and R_{pol} = 3371 ± 4 km. The mean density of Mars is thus $3.940 \pm 0.012 \mathrm{g\,cm^{-3}}$.

097.097 **Microwave radiometry of Mars from the Mars 2 and 3 orbiters. (Preliminary results).**
A. E. Basharinov, I. B. Drozdowskaya, S. T. Egorov, V. N. Galaktionov, M. A. Kolosov, V. D. Krotikov, N. N. Kroupenio, A. D. Kuzmin, V. A. Lodygin, L. I. Malafeev, E. I. Omelchenko, O. B. Schuko, N. Y. Shapirovskaya, A. M. Shutko, V. S. Troitskiy, Yu. N. Vetukhnovskaya.
Icarus, Vol. 17, 540 - 542 (1972). – Presented at the 'Symposium on planetary atmospheres and surfaces' (see 012.017).
A 3.4-cm wavelength radiometer aboard the Mars 2 and 3 orbiters observed the brightness temperature due to planetary thermal emission in two orthogonal polarizations as a function of position on Mars. Preliminary results for two orbits of Mars 3 show a correlation between subsurface temperature and dielectric constant, interpreted as an effect of porosity.

097.098 **Interferometric observations of Mars at 21-cm wavelength.** F. H. Briggs, F. D. Drake.
Icarus, Vol. 17, 543 - 547 (1972).
Continuum observations of Mars at 21 cm with the NRAO interferometer during the 1971 opposition give an integrated brightness temperature of 179 ± 11 K, a value for the surface dielectric constant in agreement with those determined by radar, and lead to a flat microwave spectrum.

097.099 **Photographic observations of Mars in the period of the great opposition of 1971.**
V. V. Avramchuk, A. R. Gajduk, Yu. D. Davudov, N. B. Ibragimov, I. K. Koval', V. D. Krugov.
Astron. vestn., Vol. 6, 165 - 167 (1972). In Russian. – Abstr. in Referativ. Zhurn. 51. Astron., 12.51.210 (1972).

097.100 **Origin of the Martian chaotic terrains.**
A. Woronow.
Science, Vol. 178, 649 - 650 (1972).
If one accepts McElroy's (Science, Vol. 175, 443 (1972)) data, certain restrictions can be imposed on the theory that the Martian "chaotic terrains" (rough, uncratered terrains, apparently caused by vertical subsidence) were produced by withdrawal of permafrost.

097.101 **Formation and evolution of the Martian dust storm of 1971.** A. N. Abramenko, M. N. Naugolnaja.
Astron. Tsirk., No. 717, p. 3 - 5 (1972). In Russian.

097.102 **On the decrease of the Martian dust storm of 1971.**
V. V. Prokofieva, V. A. Fenchak.
Astron. Tsirk., No. 717, p. 5 - 7 (1972). In Russian.

097.103 **Algunas observaciones de Marte por el Mariner 9.**
A. D. Lara.
El Universo, Vol. 26, 107 - 110 (1972).

097.104 **Dust storm on Mars from photometric observations aboard the automatic interplanetary station Mars 3.**
V. I. Moroz, L. V. Ksanfomaliti, A. M. Kasatkin, A. Eh. Nadzhip.
Kosmich. Issled., Vol. 10, 925 - 929 (1972). In Russian.

097.105 **Mars invaded by humans.** S. R. Brzostkiewicz.
Urania Kraków, Vol. 43, 238 - 245 (1972).
In Polish.

097.106 **Preliminary results of the Mariner 9 mission.**
Z. Paprotny.
Urania Kraków, Vol. 43, 301 - 305 (1972). In Polish.

097.107 **The microwave spectrum and nature of the subsurface of Mars.** J. N. Cuzzi, D. O. Muhleman.
Icarus, Vol. 17, 548 - 560 (1972).
Expected microwave spectra of Mars are computed using an improved thermal model and accurate aspect geometry. It is found that when seasonal polar cap effects are included in the calculations, the observable spectrum of Mars is flat from 0.1–21 cm to within the accuracy of present data. The spectra obtained from this model are consistent with all the data and are obtainable from a relatively simple model (homogeneous, dry, smooth dielectric sphere). This result differs from that predicted by the analytical theory in common use which is in apparent conflict with the observed spectra.

097.108 **Der rote Planet und Mariner 9.** G. de Vaucouleurs.
Bild der Wiss., [Deutsche Verlags-Anstalt (Stuttgart)], 9. Jahrgang, p. 1196 - 1209 (1972).
Observations of Mars, made with a powerful telescope in Arizona, and the first results of the experiments with space probes are described.

097.109 **The magnetic field in the nearest environment of Mars according to the data from Mars 2 and Mars 3 satellites.**
Sh. Sh. Dolginov, E. G. Eroshenko, L. N. Zhuzgov.
Dokl. Akad. Nauk SSSR, Ser. Mat. Fiz., Vol. 207, 1296 - 1299 (1972). In Russian.

097.110 **Electromagnetic wave propagation in the Martian ionosphere.** D. C. Agarwal.
Journ. Instn. Telecommun. Engineers, (India), Vol. 18, No. 1, p. 26 - 28 (1972).

097.111 **Mars: an active planet?** P. Moore.
Astron. and Space, (GB), Vol. 2, No. 1, p. 18 - 26 (1972).
Discusses the doubt cast upon the activity of Mars by Mariner 4 data.

097.112 **Mariner 9 photographs of Mars.** D. Verguese.
Sci. Progr. Découverte, No. 3444, p. 22 - 27 (1972).
In French.

097.113 **Martian doublet craters.**
V. R. Oberbeck, M. Aoyagi.
Rep. NASA-TM-X-62027, NASA Moffett Field, California. [Available from NTIS, Springfield, Va.], 40 pp. (1971).

097.114 **Mars seen by the Soviet space probes.**
V. Kurt, V. Moroz.
New Scient., (GB), Vol. 55, 441 - 443 (1972).

097.115 **The annual heat balance of the north polar cap of**

Mars.   V. I. Aleshin.
Rep. NASA-TT-F-13974, Scientific Translation Service, Santa Barbara, California. [Available from NTIS, Springfield, Va.], 7 pp. (1971). In Russian. – See Phys. Abstr., Vol. 75, No. 73084 (1972).

097.116   **Infrared spectroscopy experiment on Mariner 9:**
Preliminary results.   R. A. Hanel, B. J. Conrath, W. A. Hovis, V. G. Kunde, P. D. Lowman, J. C. Pearl, C. Prabhakara, B. Schlachman, G. Levin.
Rep. NASA-TM-X-65826, NASA, Greenbelt, Md. [Available from NTIS, Springfield, Va.], 12 pp. (1972).

The Mariner 9 infrared spectroscopy experiment has provided good quality spectra of many areas of Mars, predominantly in the southern hemisphere. Large portions of the thermal emission spectra are significantly affected by dust with a silicon oxide content approximately corresponding to that of an intermediate igneous rock, implying that Mars has undergone a substantial geochemical differentiation.

097.117   **Mars: The effects of topography on baroclinic**
instability.   S. L. Blumsack, P. J. Gierasch.
Journ. Atmosph. Sci., Vol. 29, 1081 - 1089 (1972).

The effects of a sloping lower boundary on the quasi-geostrophic baroclinic instability model of Eady are considered.

097.118   **Circularity of Martian craters.**
V. R. Oberbeck, M. Aoyagi, J. B. Murray.
Modern Geol., (GB), Vol. 3, 195 - 199 (1972).

A modified method of calculating crater circularity based on the average deviation of crater radius from mean radius is presented and used to calculate circularity of Martian craters. Calculation of the circularity indices of over 200 Martian craters yields values similar to circularity indices of terrestrial meteorite craters and different from indices of terrestrial calderas.

097.119   **Aeolian processes on Mars: Erosive velocities, sett-**
ling velocities, and yellow clouds.
R. E. Arvidson.
Bull. American Geol. Soc., Vol. 83, 1503 - 1508 (1972).

097.120   **Mariner 9, segunda parte.**   F. D. Lara.
El Universo, No. 101, Vol. 26, 149 - 151 (1972).

097.121   **Detection of molecular oxygen in the Martian atmo-**
sphere.   E. S. Barker.
Nature, Vol. 238, 447 - 448 (1972).

Molecular oxygen (O_2) has been detected spectroscopically in the atmosphere of Mars, using lines in the 7620 Å A-band. Weak lines in the wings of the strong terrestrial atmospheric $^PQ(7)$, $^PP(7)$, $^PQ(9)$, and $^PP(9)$ oxygen transitions show equivalent widths of 2 to 4 mÅ. Positive identification of the Martian atmosphere as the source of these lines is based on several facts pointed out.

097.122   **Variation of the optical density of the Martian**
gaseous atmosphere with height.
O. M. Starodubtseva.
Visn. Kharkiv. Univ. No. 82, (Ser. Astron., No. 7), p. 48 - 53 (1972). In Ukrainian.

097.123   **On the nature of the "blue clearing" on Mars.**
V. V. Prokofieva.
Astron. Tsirk., No. 720, p. 4 - 6 (1972). In Russian.

097.124   **Four months on the Martian orbit.**
V. I. Moroz, L. V. Ksanfomaliti.
Vestn. AN SSSR, 1972, No. 9, p. 10 - 25. In Russian. – Abstr. in Referativ. Zhurn. 62. Issled. kosmich. prostranstva, 2.62. 169 (1973).

Mariner 9 photography ends.
Sky Telescope, Vol. 44, 360 (1972).

Vistas of Mars.
Spaceflight, Vol. 14, 384 - 385 (1972).

Electron energy deposition in CO_2.
See Abstr. 022.032.

Branching ratios in photoelectron spectroscopy.
See Abstr. 022.037.

Temperature and pressure dependence of CO_2 ex-
tinction coefficients.   See Abstr. 022.109.

An electronic 'Brownie' in Martian orbit.
See Abstr. 034.130.

Measurements of the structure of Mars' atmosphere
during entry of the Viking lander.   See Abstr. 053.007.

Progress report on project Viking.
See Abstr. 053.018.

Electrical conduction in physical and chemical mix-
tures. Application to planetary mantles.
See Abstr. 081.029.

Earth and Mars: Evolution of atmospheres and
surface temperatures.   See Abstr. 082.015.

Diurnal variation of the exospheric temperatures on
Venus and Mars.   See Abstr. 093.027.

The thermal structure within the stratospheres of
Venus and Mars.   See Abstr. 093.029.

A possibility of multiple exospheric temperatures
for Venus and Mars.   See Abstr. 093.046.

The specific effective scattering area of the surfaces
of the moon, Mars and Venus in the radio region.
See Abstr. 094.046.

Interplanet variations in scale of crater morpholo-
gy. – Earth, Mars, moon.   See Abstr. 094.184.

Errata

097.901   **Correction: 'Mariner 6 and 7 ultraviolet spectro-**
meter experiment: Analysis of hydrogen Lyman-
alpha data' [Journ. Geophys. Res., Vol. 76, 6666 - 6673 (1971)]. D. E. Anderson, Jr., C. W. Hord.
Journ. Geophys. Res., Vol. 77, 5638 (1972).

098 Minor Planets

098.001 First order absolute perturbations of the minor planet Juno. M. P. Boris.
Nablyud. Iskusstv. Nebesn. Tel, No. 62, p. 31 - 41 (1971). In Russian.

098.002 Minor planets (1969).
N. S. Samoilova-Yakhontova.
Byull. Inst. Teoret. Astron., *Leningrad*, Vol. 13, 137 - 144 (1972). In Russian.

098.003 Improvement of orbits of 161 minor planets.
F. B. Khanina.
Byull. Inst. Teoret. Astron., *Leningrad*, Vol. 13, 160 - 185 (1972). In Russian.

098.004 Observations of minor planets made at the Observatory of Hradec Králové. J. Židů.
Byull. Inst. Teoret. Astron., *Leningrad*, Vol. 13, 187 (1972). In Russian.

098.005 Observations of minor planets made at the Crimean Astrophysical Observatory (18th report).
L. I. Chernykh.
Byull. Inst. Teoret. Astron., *Leningrad*, Vol. 13, 188 - 197 (1972). In Russian.

098.006 Posizioni di pianetini nel 1970.
M. A. Vogliotti, V. Zappalà.
Mem. Soc. Astron. Italiana, Nuova Ser., Vol. 43, 365 - 370 (1972).
181 positions for 119 asteroids are given, as deduced from plates taken at the Observatory of Turin (Pino Torinese) with the Zeiss Astrograph during 1970.

098.007 The orbit of Polyhymnia and the mass of Jupiter.
P. M. Janiczek.
Astron. Papers, U.S. Naval Obs., Washington, D.C., Vol. 21, (Part 1), 1 - 47 (1971). [For sale by the Superintendent of Documents, U.S. Government Printing Office, Washington, D.C., Price 55 cents].
In this study, observations of the minor planet, extending from 1854 to 1969, are discussed and compared with numerically integrated ephemerides. Solution of 940 conditional equations yields corrections to the orbital elements of Polyhymnia and to the mass of Jupiter.

098.008 Neue sonnennahe Planetoiden.
H. C. Courten.
Umschau, 72. Jahrgang, p. 562 (1972).
Photographic observations made during the solar eclipses of 1963, 1966, 1968 and 1970 indicate the presence of previously unidentified celestial objects in the near-angular vicinity of the sun. Observational data and equipment are described. Possible origins of the "objects" are enumerated, as well as suggested future experiments to confirm the results.

098.009 Investigation of the orbital stability of minor planets with cometary eccentricities.
G. A. Chebotarev, N. A. Belyaev, R. P. Eremenko.
IAU Symposium No. 45, (see 012.003), p. 431 - 436 (1972).

098.010 Evolution of the orbits of selected minor planets during an interval of 1000 years. M. A. Dirikis.
IAU Symposium No. 45, (see 012.003), p. 437 - 439 (1972).

098.011 Secular perturbations on the minor bodies of the solar system. I. V. Galibina.
IAU Symposium No. 45, (see 012.003), p. 440 (1972).

098.012 On the possible common origin of minor planets, comets, and meteors. S. Gaska.
IAU Symposium No. 45, (see 012.003), p. 515 (1972).

098.013 Beobachtungen der Annäherung des Planetoiden 1685 Toro an die Erde im August 1972. R. A. Naef.
Orion, 30. Jahrgang, p. 155 (1972).

098.014 The rotation of asteroids.
J. A. Burns, D. C. McAdoo.
Bull. American Astron. Soc., Vol. 4, 321 - 322 (1972). Abstr. AAS.

098.015 Asteroids: Albedo and size from photometric models. D. L. Matson.
Bull. American Astron. Soc., Vol. 4, 368 (1972). – Abstr. AAS.

098.016 Asteroids: Discussion of the infrared, thermal emission data for Vesta, Iris and Bamberga.
D. L. Matson.
Bull. American Astron. Soc., Vol. 4, 368 - 369 (1972). Abstr. AAS.

098.017 Asteroid surface compositions.
C. R. Chapman, T. B. McCord.
Bull. American Astron. Soc., Vol. 4, 369 (1972). – Abstr. AAS.

098.018 Positions of bright minor planets.
R. S. Harrington, B. F. Mintz.
Publ. United States Naval Obs., *Washington,* Second Ser., Vol. 22, Part 3, 48 pp. (1972). – Observing list: 1 Ceres, 2 Pallas, 3 Juno, 4 Vesta, 6 Hebe, 7 Iris, 10 Hygiea, 14 Irene, 18 Melpomene, 24 Themis, 31 Euphrosyne, 33 Polyhymnia, 39 Laetitia, 40 Harmonia, 48 Doris, 49 Pales, 52 Europa, 57 Mnemosyne, 62 Erato, 65 Cybele, 76 Freia, 86 Semele, 87 Sylvia, 90 Antiope, 91 Aegina, 92 Undina, 104 Klymene, 106 Dione, 108 Hecuba, 196 Philomela, 197 Arete.

098.019 A study of commensurable motion in the asteroid belt. R. B. Giffen.
Diss. Naturwiss. Gesamtfakultät Ruprecht-Karl-Univ., Heidelberg. 3 + 101 pp. (1972).
Schubart's averaging method for studying the secular effects of commensurable motion is applied to the 2/1 Hecuba gap and the 3/2 Hilda group in the asteroid belt. The development of this method, which is based on the elliptic restricted three-body problem, and the assumptions used to arrive at this model are discussed in detail. A study of periodic solutions to the averaged elliptic restricted problem indicates significant differences in the behavior of the motion for both cases.

098.020 Object Huchra = 1036 Ganymed.
Kometn. Tsirk., *Kiev*, No. 135 (1972). In Russian.

098.021 New object Klemola.
Kometn. Tsirk., *Kiev*, No. 136 (1972). In Russian.

098.022 Analytical theory of the first order motion of the minor planet (3) Juno. M. P. Boris.
Moskovskij Gosudarstvennyj Universitet, Moskva. 62 pp. (1972). In Russian. – Review in Referativ. Zhurn. 51. Astron., 10.51.145 (1972).

098.023 **Structure and evolution of the asteroid belt.**
G. A. Chebotarev, M. J. Shmakova.
Astron. Zhurn. Akad. Nauk SSSR, Vol. 49, 1107 - 1111
(1972). In Russian. English translation in Soviet Astron. AJ,
Vol. 16, No. 5.

Based on investigations of the structure of the asteroid
belt it is possible to distinguish four main stages in its evolution.

098.024 **De kleine planeet Toro.**   J. Meeus.
Hemel en Dampkring, Vol. 70, 260 - 262 (1972).

098.025 **Stability of the solar system: Evidence from the
asteroids.**   S. F. Dermott, A. P. Lenham.
The Moon, Vol. 5, 294 - 301 (1972).

An analysis of the distribution of the orbital periods of
the asteroids has shown that there is a preference for these
periods to be near-commensurate with that of Mars. We suggest that this preference is associated with a formation process
and implies that the orbital period of Mars has not changed
greatly since the time of asteroid formation. We deduce from
this that the solar system is highly stable and long-period gravitational perturbations have probably had little influence on the
gross evolution of the solar system.

098.026 **Ephemerides of minor planets for 1973.**
Editor: Institut Teoreticheskoj Astronomii Akade-
mii Nauk SSSR, under the editorship of G. A. Chebotarev.
Izdatel'stvo "Nauka", Leningradskoe Otdelenie, Leningrad.
188 pp. Price 2 Rbl. 41 Kop. (1972). In Russian and English.
Contents: Introduction, p. 3 - 8; Information on new elements,
p. 9 - 11; Elements, p. 12 - 44; Opposition dates, p. 45 - 55;
Ephemerides, p. 56 - 171; Ephemerides of bright planets, p.
172 - 183; Ephemerides of some unusual planets, p. 184 - 185;
Critical list, p. 186.

098.027 **Precise observations of minor planets at Sydney
Observatory during 1969 and 1970.**
W. H. Robertson.
Journ. Proc. Roy. Soc. New South Wales, Vol. 104, 5 - 10
(1971) = Sydney Obs. Papers No. 65 (1972).

Positions of 1 Ceres, 3 Juno, 11 Parthenope, 18 Melpomene, 39 Laetitia, 40 Harmonia and 433 Eros obtained
with the 23 cm camera are given.

098.028 **Observations of minor planets made at the Crimean
Astrophysical Observatory (19th report).**
L. I. Chernykh.
Byull. Inst. Teoret. Astron., *Leningrad,* Vol. 13, 258 - 270
(1972). In Russian.

098.029 **Improved elements of 52 minor planets.**
M. A. Dirikis.
Byull. Inst. Teoret. Astron., *Leningrad,* Vol. 13, 277 - 293
(1972). In Russian.

098.030 **Photographic observations of minor planets at the
Nikolayev Observatory during 1969.**
V. I. Voronenko, F. F. Kalihevich.
Byull. Inst. Teoret. Astron., *Leningrad,* Vol. 13, 318 - 321
(1972). In Russian.

39 positions of minor planets 2, 39, 40 are given.

098.031 **Observations of minor planets made at the Crimean
Astrophysical Observatory (20th report).**
L. I. Chernykh.
Byull. Inst. Teoret. Astron., *Leningrad,* Vol. 13, 322 - 335
(1972). In Russian.

098.032 **Minor planets positions.**
H. Wroblewski, L. Panaiotov, S. Vásquez.

Publ. Dep. Astron., Univ. Chile, Obs. Astron. Nacional, Cerro
Calán, Santiago de Chile, Vol. 2,(No. 2), 43 - 58 (1971).

Precise positions of ten minor planets (1) Ceres, (2)
Pallas, (3) Juno, (4) Vesta, (6) Hebe, (7) Iris, (11) Parthenope,
(39) Laetitia, (40) Harmonia, (51) Nemausa, are given. The
dependences and data of the comparison stars are also included.

098.033 **Resonances and encounters in the inner solar system.**
P. M. Janiczek, P. K. Seidelmann, R. L. Duncombe.
Astron. Journ., Vol. 77, 764 - 773 (1972).

Elements of the numbered minor planets were scanned
for the possibility of resonance with, and close approaches to,
the earth and Venus. The interesting cases were integrated
over a 600-year period. The results were examined for the
existence of librations involving multiples of the mean longitudes of the major and minor planets. Osculating mean motions and their perturbations were also examined. The results
for Toro (1685), Alinda (887), and Amor (1221) are presented in this paper.

098.034 **The origin of asteroids.**   V. A. Bronshtehn.
Priroda (NRB), Vol. 21, No. 3, p. 6 - 12 (1972).
In Bulgarian.

098.035 **Pazudušās mazās planētas atkal atrodas.**
M. Dīriķis.
Zvaigžņotā debess, 1972. gada pavasaris, p. 1 - 5.

098.036 **Mazo planētu vārdi.**   Ā. Alksne.
Zvaigžņotā debess, 1972. gada pavasaris, p. 52 - 56.

098.037 **Calculs d'éphémérides de petites planètes à l'aide
d'un ordinateur IBM 1130.**   F. Puel.
Ann. Obs. Besançon, Univ. Besançon, Vol. 8,(Fasc. 5), 154 -
194 (1971).

Le but de ce travail est l'élaboration d'un programme de
calcul d'éphémérides, de petites planètes principalement, d'un
accès aussi simple et rapide que possible, c'est-à-dire réduisant
au maximum les perforations.

098.038 **Heldere kleine planeten in 1973.**   J. Meeus.
Hemel en Dampkring, Vol. 70, 349 - 351 (1972).

098.039 **New minor planet discovered.**   A. R. Klemola.
Mercury, (Journ. Astron. Soc. Pacific), Vol. 1, No.
6, p. 16 (1972).

098.040 **New object Klemola (1972RA).**
Kometn. Tsirk., *Kiev,* No. 138 (1972).   In Russian.

098.041 **The light-curve for the minor planet (4) Vesta.**
C. Cristescu.
Stud. Cerc. Astron., Vol. 17, 177 - 181 (1972).

098.042 **Index of names of minor planets.**
S. Kanda, T. Urata.
Mem. Japan Astron. Study Ass., No. 17, Vol. 5, 42 - 52
(1971).   In Japanese.

098.043 **On the relation between some orbital elements of
minor planets and meteors.**   S. Gąska.
Stud. Soc. Sci. Torunensis, Sectio F (Astron.), Vol. 5, 29 - 32 =
Bull. Astron. Obs. Toruń No. 49/IV (1972).

The relations $\sigma_i(a)$, ($\sigma_i(a)$ is the dispersion of inclination
angle i calculated for arbitrary intervals of semimajor axis a),
calculated for minor planets, fireballs, photographic meteors
and radio-meteors are compared and the results are reported.

098.044 **Observations photographiques de petites planètes,
effectuées à l'astrographe double de 40 cm au cours**

de l'année 1971.   H. Debehogne.
Bull. Astron. Obs. Roy. Belgique, Vol. 8, 2 - 8 (1972).

098.045   **Observations photographiques de petites planètes, effectuées en 1971 en Argentine à l'équatorial de la Carte du Ciel de l'Observatoire National de Cordoba.**
H. Debehogne.
Bull. Astron. Obs. Roy. Belgique, Vol. 8, 9 - 39 (1972).

098.046   **Observations photographiques de petites planètes, effectuées en 1971 en Argentine au grand télescope de Bosque Alegre (filiale de l'Observatoire de Cordoba).**
H. Debehogne.
Bull. Astron. Obs. Roy. Belgique, Vol. 8, 39 (1972).

098.047   **Fast-moving object Huchra.**
J. P. Huchra, R. Green.
IAU Circ., No. 2423 (1972).

098.048   **1685 Toro.**   E. Roemer, A. H. Ferguson, R. A. McCallister, B. Milet.
IAU Circ., No. 2424 (1972).

098.049   **Object Huchra.**   R. E. McCrosky, C. Y. Shao.
IAU Circ., No. 2425 (1972).

098.050   **Object Huchra = 1036 Ganymed.**   M. Antal, R. E. McCrosky, M. Mattei, B. G. Marsden.
IAU Circ., No. 2426 (1972).

098.051   **1036 Ganymed.**
J. Huchra, R. Green, T. Seki, N. Kojima, M. Antal.
IAU Circ., No. 2428 (1972).

098.052   **433 Eros.**   B. Milet.
IAU Circ., No. 2428 (1972).

098.053   **1971 UA.**   J. Gibson, U. T. Gibson.
IAU Circ., No. 2438 (1972).

098.054   **1685 Toro.**
B. Milet, A. Mrkos, R. Petrovičová.
IAU Circ., No. 2438 (1972).

098.055   **433 Eros.**   B. Milet.
IAU Circ., No. 2439 (1972).

098.056   **Fast-moving object Klemola.**
A. R. Klemola, B. G. Marsden.
IAU Circ., No. 2442 (1972).

098.057   **1971 UA.**
E. Roemer, R. E. McCrosky, C. Y. Shao.
IAU Circ., No. 2443 (1972).

098.058   **1556 Wingolfia.**   B. G. Marsden.
IAU Circ., No. 2443 (1972).

098.059   **1972 RA (object Klemola).**
A. R. Klemola, E. A. Harlan.
IAU Circ., No. 2445 (1972).

098.060   **1972 RA.**
R. E. McCrosky, C. Y. Shao, H. L. Giclas.
IAU Circ., No. 2447 (1972).

098.061   **Fast-moving object Gehrels.**   T. Gehrels, C. Vesely, R. Sather, R. Capen, B. G. Marsden.
IAU Circ., No. 2448 (1972).

098.062   **1972 RA.**   A. R. Klemola, K. Rao, B. McNa-

mara, T. Seki, J. Gibson, U. Gibson.
IAU Circ., No. 2448 (1972).

098.063   **1953 EA.**   B. G. Marsden.
IAU Circ., No. 2448 (1972).

098.064   **1972 RB (object Gehrels).**   C. Y. Shao.
IAU Circ., No. 2450 (1972).

098.065   **1972 RA.**
J. Gibson, U. Gibson, H. L. Giclas, E. A. Harlan.
IAU Circ., No. 2450 (1972).

098.066   **433 Eros.**   C. Cristescu, V. Ionescu.
IAU Circ., No. 2450 (1972).

098.067   **1972 RB.**
C. Y. Shao, E. Roemer, B. G. Marsden.
IAU Circ., No. 2452 (1972).

098.068   **1971 UA.**   C. Y. Shao, R. E. McCrosky.
IAU Circ., No. 2453 (1972).

098.069   **1971 FA.**   B. G. Marsden.
IAU Circ., No. 2454 (1972).

098.070   **1972 RA.**
J. Gibson, U. Gibson, A. R. Klemola, B. G. Marsden.
IAU Circ., No. 2455 (1972).

098.071   **1972 RB.**   C. Y. Shao.
IAU Circ., No. 2458 (1972).

098.072   **1972 RB.**
C. Vesely, R. Sather, T. Gehrels.
IAU Circ., No. 2463 (1972).

098.073   **1971 FA.**   E. Roemer, R. McCallister.
IAU Circ., No. 2465 (1972).

098.074   **Fast-moving object Wild.**   P. Wild.
IAU Circ., No. 2467 (1972).

098.075   **Object Wild.**   R. E. McCrosky, B. G. Marsden.
IAU Circ., No. 2468 (1972).

098.076   **1972 RB.**   T. Gehrels, F. Dossin.
IAU Circ., No. 2468 (1972).

098.077   **1953 EA.**   B. G. Marsden.
IAU Circ., No. 2468 (1972).

098.078   **Object Wild.**   P. Wild, C. Y. Shao, B. Milet, R. E. McCrosky, T. Gehrels.
IAU Circ., No. 2471 (1972).

098.079   **1972 XA (object Wild).**   T. Seki, P. Wild, H. Rickman, Tidblad, C. Y. Shao, B. G. Marsden.
IAU Circ., No. 2474 (1972).

098.080   **Posizioni de pianetini australi nel 1968–1969.**
M. A. Vogliotti, V. Zappalà.
Mem. Soc. Astron. Italiana, Nuova Ser., Vol. 43, 423 - 424 (1972).
   26 positions of southern asteroids are given, as deduced from plates taken at the Observatory of Turin (Pino Torinese) with the Zeiss astrograph during 1968–1969.

098.081   **The possibility of asteroid heating by short-lived radioactive isotopes.**   B. Yu. Levin, S. V. Maeva.
Meteoritika, vyp. (No.) 31, p. 18 - 23 (1972). In Russian.

098.082 **Photographic position observations of minor planet 1620 Geographos.** A. Sh. Khatisov.
Byull. Abastumansk. Astrofiz. Obs., No. 43, p. 205 - 206 (1972). In Russian.

098.083 **Objects Gehrels.** T. Gehrels.
Yamamoto Circ., No. 1751, p. 1 (1972). In Japanese.

098.084 **Elliptic orbits of minor planets.** T. Urata.
Japan Astron. Study Ass., Circ. 276 (1972). In Japanese.

098.085 **Minor planet 1948 EA.**
Japan Astron. Study Ass., Circ. 277 (1972). In Japanese.

098.086 **Elliptic orbits of minor planets.** T. Urata.
Japan Astron. Study Ass., Circ. 278 - 279 (1972). In Japanese.

098.087 **Minor Planet Circulars, (MPC), Nos. 3353 - 3406 (1972).**
Edited by Cincinnati Observatory, under the supervision of P. Herget.
A repository of nearly all new data for numbered and unnumbered minor planets: Observations, elements and ephemerides, identifications, newly assigned numbers and names, occultations.

An approximate analytic theory of Trojan motion. See Abstr. 042.002.

A numerical method of integration by means of Taylor-Steffensen series and its possible use in the study of the motions of comets and minor planets. See Abstr. 042.023.

Computation of Schwarzschild's periodic solutions in the restricted three-body problem. See Abstr. 042.061.

Direct simulation of collision processes. See Abstr. 061.049.

Satellites and asteroids: Topography on small planetary bodies. See Abstr. 091.022.

Determination of the mass of Jupiter from observations of 10 Hygiea during 1932–1969. See Abstr. 099.009.

The motion of Hidalgo and the mass of Saturn. See Abstr. 100.005.

The influence of minor planets on the motions of comets. See Abstr. 102.026.

The origin and evolution of the comets and other small bodies in the solar system. See Abstr. 102.047.

On the dividing line between cometary and asteroidal orbits. See Abstr. 102.050.

Observations of comets and asteroids at the Kleť Observatory in the year 1970. See Abstr. 103.007.

Interplanetary objects in review: Statistics of their masses and dynamics. See Abstr. 106.009.

Numerical simulation of jetstreams. I: The three-dimensional case. See Abstr. 107.010.

099 Jupiter

099.001 A revision of Jupiter brightness temperatures in the frequency interval 18.5−24.0 GHz (1968).
D. E. Jones.
Publ. Astron. Soc. Pacific, Vol. 84, 434 (1972).

099.002 Sheath effects and related charged-particle acceleration by Jupiter's satellite Io. D. A. Gurnett.
Astrophys. Journ., Vol. 175, 525 - 533 (1972).
This paper considers the effects of the plasma sheath around Io on the interaction of Io with the Jovian magnetosphere. It is found that under some conditions the plasma sheath around Io can effectively insulate the magnetospheric plasma from the motional electric field generated within Io, thus preventing the plasma from being frozen to the motion of Io. A simplified model illustrating the basic effects is discussed.

099.003 Properties of the Red Spot of Jupiter in 1971.
C. J. Banos, D. G. Dialetis, C. E. Alissandrakis.
Astron. Astrophys., Vol. 19, 381 - 383 (1972).
Isodensity tracings of Jupiter and its Red Spot during the year 1971 as well as photometric profiles are given. The variation of the relative intensity of the Red Spot as the Spot moves from the one edge of the planetary disc to the other for 1971 is studied.

099.004 Electrodynamic effects of Jupiter's satellite Io.
J. H. Piddington.
Cosmic Electrodynamics, Vol. 3, 240 - 253 (1972).
The motion of Jupiter's satellite Io in the magnetosphere creates a disturbance which provides the decametric radio emission. Recent measurements tend to confirm a disturbance model in which a magnetic force tube is partially frozen into Io and so moves through the ionosphere almost as a solid in a fluid. A number of mechanisms have been proposed to explain this emission, but none appears to have achieved general acceptance. It is the purpose of this paper to discuss critically and modify the model of Piddington and Drake (1968), which was extended by Goldreich and Lynden-Bell (1969).

099.005 Missions to Jupiter − 1, 2. A. R. Martin.
Spaceflight, Vol. 14, 294 - 299, 325 - 332 (1972).
Concerning research program of Pioneer 10 launched 1972 March 3 which will take pictures of Jupiter during flyby on 1973 December 3.

099.006 Limits to energetic proton fluxes trapped in Jupiter's magnetosphere. J. D. Mihalov.
Planet. Space Sci., Vol. 20, 1345 - 1347 (1972). − Research note.

099.007 Sternbedeckungen durch Jupiter und seine Monde.
W. Sandner.
SuW, Vol. 11, 237 (1972).

099.008 The determination of Jupiter's mass from large perturbations on cometary orbits in Jupiter's sphere of action. E. I. Kazimirchak-Polonskaya.
IAU Symposium No. 45, (see 012.003), p. 227 - 232 (1972).

099.009 Determination of the mass of Jupiter from observations of 10 Hygiea during 1932−1969.
N. S. Chernykh.
IAU Symposium No. 45, (see 012.003), p. 233 - 238 (1972).

099.010 Spectral dependence of single scattering albedo of the Jovian cloud layer. V. F. Kartashov.
Astron. Tsirk., No. 694, p. 1 - 3 (1972). In Russian.

099.011 Change of limb darkening of Jupiter. 1.1966/67.
V. F. Kartashov.
Astron. Tsirk., No. 694, p. 3 - 6 (1972). In Russian.

099.012 Determination of scale heights of Jupiter's atmosphere from spectrograms with high dispersion.
A. N. Aksenov, L. S. Galkin, I. K. Koval.
Astron. Tsirk., No. 695, p. 1 - 3 (1972). In Russian.

099.013 New set of elements of Jupiter VI. L. E. Bykova.
Astron. Tsirk., No. 695, p. 5 - 7 (1972). In Russian.

099.014 Statistical mechanics of light elements at high pressure. II. Hydrogen and helium alloys.
W. B. Hubbard.
Astrophys. Journ., Vol. 176, 525 - 531 (1972).
A Monte Carlo method described in a previous paper is used to evaluate various thermodynamic quantities for hydrogen-helium mixtures at temperatures and pressures characteristic of the Jovian interior.

099.015 A tentative identification of $^{13}CH_4$ and an estimate of $^{12}C/^{13}C$ in the atmosphere of Jupiter.
K. Fox, T. Owen, A. W. Mantz, K. N. Rao.
Astrophys. Journ., (Letters), Vol. 176, L81 - L84 (1972).
$^{13}CH_4$ has been identified in the atmosphere of the planet Jupiter by comparison of several J-multiplets in new laboratory and Jovian spectra of the $3\nu_3$ band at 1.1μ. From the equivalent width of $R(2)$, measured on the two best spectrograms, a Jovian $^{12}C/^{13}C$ isotopic abundance ratio of 110 ± 35 has been derived.

099.016 Jupiter: New evidence of long-term variations of its decimeter flux density.
M. J. Klein, S. Gulkis, C. T. Stelzried.
Astrophys. Journ., (Letters), Vol. 176, L85 - L88 (1972).
Jupiter's flux density at 12.6 cm was measured at weekly intervals from 1971 May through October. When compared with previous decimetric measurements, these data indicate that Jupiter's total flux density has decreased approximately 20 percent since 1964. No short-term variations greater than a few percent were observed.

099.017 The physical properties of the Jovian atmosphere inferred from eclipses of the Galilean satellites. II. 1971 apparition.
M. J. Price, J. S. Hall, P. B. Boyce, R. Albrecht.
Icarus, Vol. 17, 49 - 56 (1972).
Further simultaneous two-color photoelectric photometry of Ganymede was carried out at blue (λ 4500 Å) and yellow (λ 5790 Å) wavelengths during its eclipse by Jupiter on March 10, 1971. The observations have been interpreted in terms of the local optical transmission properties of the Jovian atmosphere. No unique scattering model exists for the Jovian atmosphere.

099.018 Jupiter: Its Red Spot and disturbances in 1970−1971. E. J. Reese.
Icarus, Vol. 17, 57 - 72 (1972).
Photographic observations of Jupiter and its Red Spot between 11 December 1970 and 12 October 1971 are reported. The Red Spot had a mean rotation period of $9^h55^m38.5$, the shortest to be observed since the apparition of 1931−32. The outstanding event of the apparition was an outburst of activity in the South Equatorial Belt which was similar in many ways

to the great disturbance of 1928.

099.019 Structure of Jupiter's decametric radio sources:
Two-dimensional probability and flux studies,
1957–1970.   H. R. Miller, A. G. Smith, R. J. Leacock.
Icarus, Vol. 17, 73 - 78 (1972).

Decameter-wavelength observations of Jupiter extending from 1957 through 1970 are combined to present a two-dimensional picture of the radio source structure at frequencies of 15, 18, 22.2 and 27.6 MHz. An analysis based solely on the probability of detecting emission is compared and contrasted with a second analysis reflecting the intensities of the received signals.

099.020 The occultation of β Sco by Jupiter.
K. C. Freeman, N. R. Stokes.
Icarus, Vol. 17, 198 - 201 (1972).

The occultation of the bright star β Sco by Jupiter on May 13, 1971 was observed photoelectrically. The scale height for Jupiter's upper atmosphere appears to be about 8 km, in agreement with an earlier estimate. The bright flashes of starlight during the emersion are discussed briefly.

099.021 The determination of the diameter of Io from its
occultation of β Scorpii C on May 14, 1971.
G. E. Taylor.
Icarus, Vol. 17, 202 - 208 (1972).

All the available observations of the times of occultation of β Scorpii C by Jupiter satellite I (Io) on May 14, 1971 are analyzed to determine the diameter. Additional events noted on the photoelectric recordings are analyzed to indicate that the star may be double, the fainter component being in position angle 312°, separation 0″.10.

099.022 Io's triaxial figure.
B. O'Leary, T. C. van Flandern.
Icarus, Vol. 17, 209 - 215 (1972).

If Io were a homogeneous fluid body in hydrostatic equilibrium and in synchronous rotation, which we consider to be reasonable assumptions, it would show a bulge along a line to Jupiter about 20 km in radius greater than the polar radius.

099.023 Observation of the occultation of β Sco C by Io.
F. W. Fallon, E. J. Devinney, Jr.
Icarus, Vol. 17, 216 - 217 (1972).

099.024 Upper limits for an atmosphere on Io.
B. A. Smith, S. A. Smith.
Icarus, Vol. 17, 218 - 222 (1972).

The 14 May 1971 occultation of β Scorpii C by Io was successfully observed in ultraviolet light near Kingston, Jamaica. Upper limits for the surface pressure for N_2, CH_4, and H_2 atmospheres and the corresponding number densities have been derived.

099.025 The electrical conductivity of Io.   S. F. Dermott.
Icarus, Vol. 17, 223 - 224 (1972).

If Io has a thin crust of ice then the electrical resistance of the satellite is determined by an outer layer of thickness ~8 km and is higher by a factor ~10^{15} than that needed to account for the modulation of Jupiter's decametric radio emission in the unipolar inductor model of Goldreich and Lynden-Bell. The modulation, however, could possibly be accounted for if the surface composition of Io is chondritic or if it has an ionosphere.

099.026 Du rôle des centres éruptifs de l'atmosphère de
Jupiter dans la détermination de vitesse de rotation
du noyau.   C. Botton.
Orion, 30. Jahrgang, p. 130 - 135 (1972).

099.027 On some parameters of similarity characterizing
the dynamics of the Jovian atmosphere.
L. P. Sorokina.
Astron. Tsirk., No. 696, p. 2 - 4 (1972). In Russian.

099.028 Photoelectric observations of the occultation of $β_1$
Sco by Jupiter.   G. V. Zajtseva, V. M. Lyutyj,
A. M. Cherepashchuk, V. F. Esipov, O. G. Taranova.
Astron. Tsirk., No. 696, p. 4 - 6 (1972). In Russian.

099.029 New set of elements of Jupiter VII.
T. V. Bordovitsyna.
Astron. Tsirk., No. 696, p. 6 - 8 (1972). In Russian.

099.030 Jovian decametric pulse distribution.
H. R. Miller, A. G. Smith.
Bull. American Astron. Soc., Vol. 4, 312 (1972). – Abstr. AAS.

099.031 Photometry of Jupiter VI and Phoebe (Saturn IX).
L. Andersson.
Bull. American Astron. Soc., Vol. 4, 313 (1972). – Abstr. AAS.

099.032 Change of Jupiter limb darkening. 2.1967/68.
V. F. Kartashov.
Astron. Tsirk., No. 704, p. 5 - 8 (1972). In Russian.

099.033 Observations of Jupiter in the spectral region λ 5 μ.
I. A. Maslov.
Astron. Tsirk., No. 705, p. 5 - 7 (1972). In Russian.

099.034 Quadrupole H_2 absorption in the spectra of Jupiter
and Saturn.   L. Trafton.
Bull. American Astron. Soc., Vol. 4, 359 (1972). – Abstr. AAS.

099.035 Observations of the beaming of Jupiter's emission at
12.6 cm wavelength.
M. J. Klein, S. Gulkis, C. T. Stelzried.
Bull. American Astron. Soc., Vol. 4, 359 (1972). – Abstr. AAS.

099.036 On the level of formation of the hydrogen quadru-
pole lines on Jupiter.   J. S. Margolis, G. E. Hunt.
Bull. American Astron. Soc., Vol. 4, 359 - 360 (1972).
Abstr. AAS.

099.037 Spectroscopic evidence for the structure of the visi-
ble Jovian clouds from observations of methane and
hydrogen quadrupole lines.   G. E. Hunt.
Bull. American Astron. Soc., Vol. 4, 360 (1972). – Abstr. AAS.

099.038 Spectrophotometry of the 1.5 μm window of Jupi-
ter: NH_3 variations over the disk and limb darken-
ing.   A. B. Binder.
Bull. American Astron. Soc., Vol. 4, 360 (1972). – Abstr. AAS.

099.039 The Jovian atmosphere aerosol distribution inferred
from Jovian satellite eclipse observations.
T. F. Greene, R. W. Shorthill.
Bull. American Astron. Soc., Vol. 4, 360 (1972). – Abstr. AAS.

099.040 Solar occultation experiments with the Jovian at-
mosphere.
M. J. Price, J. S. Hall, P. B. Boyce, R. Albrecht.
Bull. American Astron. Soc., Vol. 4, 360 (1972). – Abstr. AAS.

099.041 The far ultraviolet emission spectrum of Jupiter.
H. W. Moos, G. J. Rottman.
Bull. American Astron. Soc., Vol. 4, 360 (1972). – Abstr. AAS.

099.042 Spectroscopy of Jupiter: 3200 Å to 11200 Å.
C. B. Pilcher, R. G. Prinn, T. B. McCord.
Bull. American Astron. Soc., Vol. 4, 361 (1972). – Abstr. AAS.

099.043 **Observations of isolated five micron sources on Jupiter.** J. A. Westphal.
Bull. American Astron. Soc., Vol. 4, 361 (1972). – Abstr. AAS.

099.044 **Thermal radio emission from Jupiter and Saturn.** R. Poynter, S. Gulkis.
Bull. American Astron. Soc., Vol. 4, 361 (1972). – Abstr. AAS.

099.045 **A tentative identification of $C^{13}H_4$ and an estimate of C^{12}/C^{13} in the atmosphere of Jupiter.**
K. Fox, A. W. Mantz, T. Owen, K. N. Rao.
Bull. American Astron. Soc., Vol. 4, 361 (1972). – Abstr. AAS.

099.046 **The infrared spectrum of Jupiter and radiative properties of the clouds.** F. W. Taylor, G. E. Hunt.
Bull. American Astron. Soc., Vol. 4, 361 (1972). – Abstr. AAS.

099.047 **The abundance of CH_3D and the deuterium-hydrogen ratio in Jupiter.** R. Beer, F. W. Taylor.
Bull. American Astron. Soc., Vol. 4, 361 (1972). – Abstr. AAS.

099.048 **An observational test of absorption line formation models in the Jovian atmosphere.**
J. T. Bergstralh.
Bull. American Astron. Soc., Vol. 4, 366 (1972). – Abstr. AAS.

099.049 **Io's triaxial figure.** B. O'Leary.
Bull. American Astron. Soc., Vol. 4, 366 (1972). Abstr. AAS.

099.050 **Infrared observations of the Galilean satellites.**
O. L. Hansen.
Bull. American Astron. Soc., Vol. 4, 367 (1972). – Abstr. AAS.

099.051 **Thermal properties of Io and Ganymede from 20-μm eclipse radiometry in 1971.**
D. Morrison, D. P. Cruikshank, R. E. Murphy.
Bull. American Astron. Soc., Vol. 4, 367 (1972). – Abstr. AAS.

099.052 **Temperatures of Titan and the Galilean satellites at 20 microns.**
D. Morrison, D. P. Cruikshank, R. E. Murphy.
Bull. American Astron. Soc., Vol. 4, 367 (1972). – Abstr. AAS.

099.053 **Relation between Jupiter's decametric radio emission and its Great Red Spot.**
D. Basu, J. L. Pires.
Astrophys. Letters, Vol. 12, 99 - 100 (1972).
Good correlation has been found between the visibility of Jupiter's Great Red Spot and the Jovian decametre radiation at 18.0, 22.2, and 27.6 MHz.

099.054 **Optical properties and the structure of Jupiter's atmosphere. V. A possible structure of the ammonia aerosol layer.**
L. P. Sorokina, V. G. Tejfel', L. A. Usol'tseva.
Astron. vestn., Vol. 6, 77 - 84 (1972). In Russian. – Abstr. in Referativ. Zhurn. 51. Astron., 9.51.278 (1972).

099.055 **Photometric characteristics of Jupiter and Saturn in the region of 0.48 −0.33 μ.**
V. D. Krugov.
Astron. vestn., Vol. 6, 85 - 90 (1972). In Russian. – Abstr. in Referativ. Zhurn. 51. Astron., 9.51.287 (1972).

099.056 **The activity of Jupiter during 1965−1970.**
S. K. Vsekhsvyatskij, A. V. Karpenko.
Astron. vestn., Vol. 6, 112 - 119 (1972). In Russian. – Abstr. in Referativ. Zhurn. 51. Astron., 9.51.281 (1972).

099.057 **Zum Stand der Jupiter-Forschung.** O. Röhrig.
SuW, Vol. 11, 266 - 271 (1972).

099.058 **The 1970 apparition of Jupiter.** P. K. Mackal.
Strolling Astronomer, Vol. 23, 189 - 205 (1972).

099.059 **Synoptic observations of Jupiter.** V. A. Zinov'ev.
Astron. vestn., Vol. 6, 119 - 127 (1972). In Russian.
Abstr. in Referativ. Zhurn. 51. Astron., 9.51.280 (1972).

099.060 **Infrared maps of Jupiter.**
C. S. L. Keay, F. J. Low, G. H. Rieke.
Sky Telescope, Vol. 44, 296 - 297 (1972).

099.061 **Occultation of Beta Scorpii by Jupiter on May 13, 1971.** J. C. Bhattacharyya.
Nature, Phys. Sci., Vol. 238, 55 - 56 (1972).
Using a fast recording device, photoelectric light curve of the emersion phase of the occultation of β Sco by Jupiter has been obtained. The emersion light curve shows a steep slope from which a scale height of 3 km in the refracting layers of the Jovian atmosphere has been estimated. The light curve also indicates stratifications in these layers.

099.062 **The D/H ratio in Jupiter's atmosphere.**
H. Reeves, Y. Bottinga.
Nature, Vol. 238, 326 - 327 (1972).
The recent determination of CH_3D abundance in Jupiter is discussed in terms of molecular exchange reaction, in relation with the problem of the deuterium abundance in the protosolar nebula.

099.063 **Jupiter og Io okkulterer Beta Scorpii.** P. Darnell.
Astron. Tidssk., Årg. 5, p. 117 - 120 (1972).

099.064 **Solar activity and the rotation of Jupiter.**
L. Křivský, Z. Pokorný.
Astrophys. Letters, Vol. 12, 173 - 175 (1972).
The relation between solar activity and the rotational periods of surface features on Jupiter were investigated over the period 1880−1968. Chree's superposition analysis shows that the mean rotational periods of the features display a double-maximum relationship in the course of the 11-yr solar cycle.

099.065 **Non-dipole terms in the magnetic fields of Jupiter and the earth.** R. G. Conway, D. Stannard.
Nature, Phys. Sci., Vol. 239, 142 - 143 (1972).

099.066 **The 8 to 13 μm spectrum of Jupiter.**
D. K. Aitken, B. Jones.
Nature, Vol. 240, 230 - 232 (1972).
We present observations with resolution $\Delta\lambda/\lambda = 0.007$ in which details of the v_2 band of NH_3 are seen in absorption, resembling the model atmosphere of Encrenaz with monotonically decreasing temperature.

099.067 **Decametre-wave radiation from Jupiter and solar activity.** C. H. Barrow.
Planet. Space Sci., Vol. 20, 2051 - 2056 (1972).
The problem of a possible short term correlation between Jupiter's decametre-wave radio emission and solar activity is re-examined and the complicating factors are discussed. The results are found to be consistent with the idea that at least some of the Jupiter radiation may be stimulated by solar particles having radial velocities in the range 350−700 km/sec. travelling outward as far as the orbit of Jupiter.

099.068 **Jupiter occultation of Beta Scorpii: Are the flashes time-symmetric?** J. Veverka, J. Elliot, C. Sagan, L. Wasserman, W. Liller,
Nature, Vol. 240, 344 - 345 (1972).

099.069 The Jovian decametric rotation period.
M. L. Kaiser, J. K. Alexander.
Astrophys. Letters, Vol. 12, 215 - 217 (1972).

A new power-spectrum analysis method is used to determine the Jovian decametric rotation period from nearly 17 yr of observations. The value deduced is $9^h 55^m 29.70^s \pm 0.02^s$, in agreement with most recent determinations.

099.070 Farbwahrnehmungen auf Jupiter.   R. Sopper.
SuW, Vol. 11, 350 - 351 (1972).

099.071 Jupiter in 1966–1968.
V. D. Bezuglova, N. G. Bondar', Z. L. Novak.
Astron. vestn., Vol. 6, 200 - 207 (1972). In Russian. – Abstr. in Referativ. Zhurn. 51. Astron., 12.51.222 (1972).

099.072 Time variations in ultraviolet absorption in the continuous spectrum of Jupiter and Saturn.
V. D. Krugov.
Astron. vestn., Vol. 6, 168 - 171 (1972). In Russian. – Abstr. in Referativ. Zhurn. 51. Astron., 12.51.225 (1972).

099.073 On radio emission of Callisto.
A. D. Kuz'min, B. Ya. Losovskij.
Astron. vestn., Vol. 6, 177 - 179 (1972). In Russian. – Abstr. in Referativ. Zhurn. 51. Astron., 12.51.226 (1972).

099.074 An earlier generation of long-enduring south temperate ovals on Jupiter.   E. J. Reese.
Icarus, Vol. 17, 704 - 706 (1972).

A study of three early, long-enduring , bright ovals (from 1914 to 1935) suggests that they may be precursors of the present south temperate ovals: FA, BC, and DE.

099.075 Discrepancies in measurements of the Jupiter atmospheric scale height.
D. S. Evans, W. B. Hubbard.
Nature, Phys. Sci., Vol. 240, 162 (1972).

The scale heights given by different authors vary from 3 km to 31 km. We wish to point out several factors and independent lines of evidence which indicate that some results for the mean Jovian scale height are far more reliable than others.

099.076 Deuterium–hydrogen ratio in Jupiter.
R. Beer, F. W. Taylor.
Nature, Vol. 240, 465 (1972). – Letter.

099.077 The geometry and dynamic spectra of Io-modulated Jovian decametric radio emissions.
R. A. Smith, C. S. Wu, J. S. Žmuidzinas.
Astrophys. Journ., (*Letters*), Vol. 177, L131 - L136 (1972).

An explanation of the asymmetry of Io's ephemeris positions for the early and main source emissions is given. The asymmetry is attributed to a time lag caused by the small group velocity which controls the propagation of the radiation inside the Jovian magnetosphere. As observed, the radiation has propagated through a rotating medium. The overall propagation characteristics determine the sense of frequency drift of the individual sources.

099.078 Photographic observations of the occultation of Beta Scorpii by Jupiter.   S. M. Larson.
Contr. Bosscha Obs., *Lembang*, No. 45, 11 pp. (1972).

The occultation of the multiple star Beta Scorpii by Jupiter was observed visually and photographically from the Bosscha Observatory in Lembang (Java), Indonesia, on May 13, 1971. The photographs recorded the dimming of the stars as the light was differentially refracted by the Jovian atmosphere, and gave support to a scale height greater than 8 km. Measurement of the position of the brightest component of Beta Scorpii during ingress shows refraction of approximately

$1.''4$ before becoming unobservable.

099.079 Jovian rotational profiles for 1970 and 1971.
J. L. Inge.
Publ. Astron. Soc. Pacific, Vol. 84, 641 (1972). – Abstr. Astron. Soc. Pacific.

099.080 Osservazioni di pianeti.   P. Campaner.
Coelum, Vol. 40, 236 - 238 (1972).

099.081 Galilean satellites: Identification of water frost.
C. B. Pilcher, S. T. Ridgway, T. B. McCord.
Science, Vol. 178, 1087 - 1089 (1972).

Water frost absorptions have been detected in the infrared reflectivities of Jupiter's Galilean satellites J II (Europa) and J III (Ganymede). We have determined the percentage of frost-covered surface area to be 50 to 100 percent for J II, 20 to 65 percent for J III, and possibly 5 to 25 percent for J IV (Callisto). The reflectivity of the material underlying the frost on J II, J III, and J IV resembles that of silicates. The surface of J I (Io) may be covered by frost particles much smaller than those on J II and J III.

099.082 On the Great Red Spot of Jupiter and the gravitational waves.   T. Mitani.
Mem. Japan Astron. Study Ass., No. 17, Vol. 5, 53 - 64 (1971).   In Japanese.

099.083 Very long baseline interferometry of Jupiter at 18 MHz.   T. D. Carr, M. A. Lynch, M. P. Paul,
G. W. Brown, J. May, N. F. Six, V. M. Robinson, W. F. Block.
Radio Sci., Vol. 5, 1223 - 1226 = Univ. Chile, Dep. Astron. *Santiago,* Separata 8 (1970).

099.084 Inhomogeneous models of the atmosphere of Jupiter.   H. Axel.
Thesis, Princeton Univ., Princeton, N.J. [Available from Univ. Microfilms, Ann Arbor, Mich., USA. Order No. 72-2683], 66 pp. (1971).

099.085 Jupiter: a physico-mathematical study.   O Borisov.
Spaceworld, (*USA*), Vol. 1, 9 - 105, p. 42 - 43 (1972).

099.086 Summary of Jovian latitude and rotation period observations from 1898 to 1970.   E. J. Reese.
Contr. Obs. New Mexico State Univ., Las Cruces, Vol. 1, 83 - 94 (1972).

A ten-year summary of recent measurements of the latitudes of Jupiter's belts and the rotation periods of the various atmospheric currents is presented. Comparison of these measured values with earlier visual observations indicates that the latter are sufficiently accurate to provide a reliable and continuous record of latitudes since 1908 and rotation periods since 1898. A summary of the earlier observations is also given.

099.087 On the structure of the Jovian clouds.
M. F. Khodyachikh.
Visn. Kharkiv. Univ. No. 82, (Ser. Astron., No. 7), p. 69 - 75 (1972). In Ukrainian.

099.088 On the optical properties of the atmosphere of Jupiter from photometric observations.
Yu. D. Davudov, V. D. Krugov.
Astron. Tsirk., No. 724, p. 1 - 3 (1972). In Russian.

099.089 About the single-scattering albedo of Jupiter's cloud layer.   V. F. Kartashov.
Astron. Tsirk., No. 724, p. 3 - 5 (1972). In Russian.

099.090 About the contrasts on the disk of Jupiter.

V. F. Kartashov.
Astron. Tsirk., No. 724, p. 5 - 7 (1972). In Russian.

099.091 **Nonthermal radio observations of the major planets.**
A. G. Smith, R. J. Leacock, T. D. Carr, G. R. Lebo, C. N. Olsson.
Phys. Earth Planet. Interiors, Vol. 6, (see 012.023), 10 - 16 (1972).

099.092 **Jupiter's decametric rotation period and the source-A emission beam.** T. D. Carr.
Phys. Earth Planet. Interiors, Vol. 6, (see 012.023), 21 - 28 (1972).

099.093 **Thermal radio emission from Jupiter and Saturn.**
S. Gulkis, R. Poynter.
Phys. Earth Planet. Interiors, Vol. 6, (see 012.023), 36 - 43 (1972).

099.094 **Electrical conductivity of condensed molecular hydrogen in the giant planets.** R. Smoluchowski.
Phys. Earth Planet. Interiors, Vol. 6, (see 012.023), 48 - 50 (1972).

099.095 **An evolutionary calculation of Jupiter.**
A. S. Grossman, H. Graboske, J. Pollack, R. Reynolds, A. Summers.
Phys. Earth Planet. Interiors, Vol. 6, (see 012.023), 91 - 98 (1972).

099.096 **Jupiter's Great Red Spot revisited.** R. Hide.
Phys. Earth Planet. Interiors, Vol. 6, (see 012.023), 99 (1972). − Abstract.

099.097 **On the density of the upper cloud layer of Jupiter.**
V. G. Tejfel'.
Astron. Tsirk., No. 735, p. 6 - 8 (1972). In Russian.

099.098 **The brightness coefficients of the Jovian cloud cover in 1964 - 1965.**
Yu. A. Egorov, L. P. Sorokina.
Astron. Tsirk., No. 737, p. 4 - 7 (1972). In Russian.

099.099 **Photometry of the Great Red Spot on Jupiter.**
V. F. Kartashov.
Astron. Tsirk., No. 742, p. 1 - 3 (1972). In Russian.

099.100 **White oval spots in STB on Jupiter.**
V. F. Kartashov.
Astron. Tsirk., No. 742, p. 4 - 6 (1972). In Russian.

099.101 **Satellites of Jupiter.**
Contr. Obs. Valongo, Univ. Federal Rio de Janeiro, Sér II, No. 14 (1972). − 1972 April − June.

Shapes and widths of ammonia lines collision-broadened by hydrogen. See Abstr. 022.017.

A modern version of the Ole Roemer experiment. See Abstr. 022.057.

Fabry-Perot spectrometer adjustment for the compensation of Doppler shift from rapidly rotating and rapidly flowing sources. See Abstr. 034.060.

Observed positions of Jupiter III (Ganymede) and corrections to the tabular place of the star SAO 186800. See Abstr. 041.041.

The surviving Jupiter entry probe and experiments. See Abstr. 053.008.

The turbopause probe and experiments. See Abstr. 053.010.

Pioneer 10 − the mission to Jupiter. See Abstr. 053.013.

The orbit of Polyhymnia and the mass of Jupiter. See Abstr. 098.007.

The transit of the Jupiter system across the multiple star β Scorpii. See Abstr. 117.009.

100 Saturn

100.001 Un'ipotesi sull'origine degli anelli di Saturno.
V. Banfi.
Mem. Soc. Astron. Italiana, Nuova Ser., Vol. 43, 247 - 262 (1972).
A hypothesis on Saturn's rings formation is presented. Instead of assuming a mechanical basic process, the formation is explained as a consequence of an electromagnetic phenomenon.

100.002 **Titan and its atmosphere.**
D. P. Cruikshank, D. Morrison.
Sky Telescope, Vol. 44, 83 - 85 (1972).

100.003 **Polarimetry of giant planets. II. Phase dependence of polarization for selected regions of the disk of Saturn.** O. I. Bugaenko, L. S. Galkin.
Astron. Zhurn. Akad. Nauk SSSR, Vol. 49, 837 - 843 (1972). In Russian. English translation in Soviet Astron. AJ, Vol. 16, No. 4.
Phase dependence of polarization is obtained in ten spectral regions between 360 and 750 mμ for four parts of Saturn's disk: centre, south pole and east and west limbs in the equatorial belt of the planet. Large fluctuations of the polarization vector are found in the ultraviolet.

100.004 **Some properties of the cloud cover on Saturn.**
V. G. Teifel, G. A. Kharitonova.
Astron. Tsirk., No. 683, p. 4 - 7 (1972). In Russian.

100.005 **The motion of Hidalgo and the mass of Saturn.**
B. G. Marsden.
IAU Symposium No. 45, (see 012.003), p. 239 - 243 (1972).

100.006 **UBV photometry of Iapetus.** R. L. Millis.
Bull. American Astron. Soc., Vol. 4, 313 (1972). Abstr. AAS.

100.007 **Limb darkening of Saturn and thermal properties of the rings from 10 and 20 micron radiometry.**
R. E. Murphy, D. P. Cruikshank, D. Morrison.
Bull. American Astron. Soc., Vol. 4, 358 - 359 (1972). Abstr. AAS.

100.008 **Circular polarization of Saturn.**
J. B. Swedlund, J. C. Kemp, R. D. Wolstencroft.
Bull. American Astron. Soc., Vol. 4, 359 (1972). – Abstr. AAS.

100.009 **Jupiter turbopause probe applicability to investigations of Saturn.** R. S. Wiltshire.
Bull. American Astron. Soc., Vol. 4, 365 (1972). – Abstr. AAS.

100.010 **Albedos, radii, and temperature of Iapetus and Rhea.**
R. E. Murphy, D. P. Cruikshank, D. Morrison.
Bull. American Astron. Soc., Vol. 4, 367 (1972). – Abstr. AAS.

100.011 **Newly discovered absorptions in Titan's infrared spectrum.** L. Trafton.
Bull. American Astron. Soc., Vol. 4, 367 (1972). – Abstr. AAS.

100.012 **Titan: Polarimetric evidence for an optically thick atmosphere.** J. Veverka.
Bull. American Astron. Soc., Vol. 4, 367 - 368 (1972). Abstr. AAS.

100.013 **An elementary greenhouse argument for H_2 in the atmosphere of Titan.** C. Sagan, G. Mullen.
Bull. American Astron. Soc., Vol. 4, 368 (1972). – Abstr. AAS.

100.014 **The 1970–71 apparition of Saturn.**
J. L. Benton, Jr.
Strolling Astronomer, Vol. 23, 215 - 222 (1972).

100.015 **The mass and figure of Saturn by photographic astrometry of its satellites.** H. A. Garcia.
Astron. Journ., Vol. 77, 684 - 691 (1972).
Photographic observations of the Saturnian satellites collected by the Yale University, southern station at Johannesburg and the Allegheny Observatory, University at Pittsburgh are used to improve the astronomical constants of the satellite orbits: Tethys, Dione, Rhea, and Titan. The revised orbits of these satellites are used to compute new values for Saturn's mass and the second and fourth zonal harmonic coefficients of Saturn's figure.

100.016 **On the rings of Saturn.** J. Trulsen.
Astrophys. Space Sci., Vol. 17, 330 - 337 (1972).
Results from numerical simulations of jetstreams are used to discuss certain aspects of the dynamics of the rings of Saturn.

100.017 **The atmosphere of Titan.** D. M. Hunten.
Comments Astrophys. Space Phys., Vol. 4, 149 - 154 (1972).
Twenty-eight years ago, Kuiper detected methane in the atmosphere of Titan, and a few years later he published a laboratory comparison, indicating an amount of 200 m-atm. From recent papers it is now clear that Titan has a much more massive atmosphere than Mars. The available basic data on Titan are shown in a table and discussed in detail.

100.018 **On the origin of the commensurabilities amongst the satellites of Saturn.** A. T. Sinclair.
Monthly Notices Roy. Astron. Soc., Vol. 160, 169 - 187 (1972).
The hypothesis that the commensurabilities amongst the satellites of Saturn are due to the action of tidal forces is examined. It is shown that this hypothesis provides a satisfactory explanation of the origin of the commensurabilities between Mimas and Tethys, and between Enceladus and Dione. The origin of the commensurability between Titan and Hyperion cannot be explained in this way, but it is possibly the result of a close approach between these satellites.

100.019 **Orbit-orbit resonance capture in the solar system.**
R. J. Greenberg, C. C. Counselman, III, I. I. Shapiro.
Science, Vol. 178, 747 - 749 (1972).
A realistic model involving mutual gravitation and tidal dissipation for the first time provides a detailed explanation for satellite orbit-orbit resonance capture. Although applying directly only to Saturn's satellites Titan and Hyperion, the model reveals general principles of resonance capture, evolution, and stability which seem applicable to other orbit-orbit resonances in the solar system.

100.020 **Circular polarization of Saturn.**
J. B. Swedlund, J. C. Kemp, R. D. Wolstencroft.
Astrophys. Journ., Vol. 178, 257 - 265 (1972).
Circular polarization in the scattered light from Saturn has been discovered and measured before and after the opposition of 1971 November. Spectra of the circular polarization at two phases were taken with interference filters. Compared to Jupiter, the values are opposite in sign and the variation with phase angle is unexpectedly complex. The rings show little or no circular polarization.

100.021 **Radii, albedos, and 20-micron brightness temperatures of Iapetus and Rhea.**
R. E. Murphy, D. P. Cruikshank, D. Morrison.
Astrophys. Journ., (*Letters*), Vol. 177, L93 - L95 (1972).

From infrared flux measurements at a wavelength of 20 μ we deduce that the radius of Iapetus is 850 ± 100 km and that the remarkable light variations are caused by an albedo difference between the leading and trailing hemispheres. The visual albedo is 0.04 ± 0.01 on the leading side and 0.28 ± 0.05 on the trailing side. For Rhea we find a radius of 725 ± 100 km and an albedo of 0.57 ± 0.07.

100.022 **The 1969–70 apparition of Saturn.** J. L. Benton.
Strolling Astronomer, Vol. 24, 27 - 35 (1972).

This report covers the period from June 15, 1969 to April 5, 1970, during which the value of B (the axial tilt of Saturn) varied between −16°.5 and −18°.7. Opposition occurred on October 29, 1969, when Saturn was a zero magnitude object in southern Aries. At this time, the equatorial diameter of the planet was 20″, while the major axis of the ring system extended to 46″.

100.023 **The height of the cloud cover and equatorial acceleration in Saturn's atmosphere.**
V. G. Tejfel', G. A. Kharitonova.
Astron. Tsirk., No. 735, p. 4 - 6 (1972). In Russian.

100.024 **Phase curves of Saturn's rings in the short-wave region of the spectrum.** V. D. Krugov.
Astron. Tsirk., No. 738, p. 1 - 3 (1972). In Russian.

On the determination of planetary masses.
See Abstr. 091.006.

Statistical mechanics of light elements at high pressure. II. Hydrogen and helium alloys. See Abstr. 099.014.

Photometry of Jupiter VI and Phoebe (Saturn IX).
See Abstr. 099.031.

Quadrupole H_2 absorption in the spectra of Jupiter and Saturn. See Abstr. 099.034.

Thermal radio emission from Jupiter and Saturn.
See Abstr. 099.044.

Temperatures of Titan and the Galilean satellites at 20 microns. See Abstr. 099.052.

Photometric characteristics of Jupiter and Saturn in the region of $0.48 - 0.33\mu$. See Abstr. 099.055.

Time variations in ultraviolet absorption in the continuous spectrum of Jupiter and Saturn.
See Abstr. 099.072.

Nonthermal radio observations of the major planets.
See Abstr. 099.091.

Thermal radio emission from Jupiter and Saturn.
See Abstr. 099.093.

Electrical conductivity of condensed molecular hydrogen in the giant planets. See Abstr. 099.094.

101 Uranus, Neptune, Pluto, Transplutonian Planet

101.001 **Search for transplutonian planets by means of periodic comets.** G. A. Chebotarev.
Byull. Inst. Teoret. Astron., *Leningrad*, Vol. 13, 145 - 147 (1972). In Russian.

As a result of the investigation of orbital elements of some periodic comets with large aphelion distances a suggestion is made about the existence of two yet undiscovered major planets, whose mean distances from the sun would have to be approximately 53.7 and 100 a.u. The method of orbit determination of these planets is described.

101.002 **On a criterion for the prediction of Neptune.**
C. J. Brookes.
Monthly Notices Roy. Astron. Soc., Vol. 158, 79 - 83 (1972).

The present paper indicates that the derivation of the predicted position via the simplified method, which neglected terms in e and e', is unfortunately not valid and that it is essential that these terms be included in any analysis of the problem.

101.003 **Comments on the interior of Pluto.** J. D. Fix.
Icarus, Vol. 16, 569 - 570 (1972).

Models of the interior of Pluto have been constructed for a variety of combinations of mass and radius. The equations of state used in these models were those of terrestrial core and mantle material.

101.004 **Accurate positions of the planet Pluto in the years 1969–1970.**
C. Barbieri, M. Capaccioli, R. Ganz, G. Pinto.
Astron. Journ., Vol. 77, 521 - 522 (1972).

Sixteen positions of the planet Pluto in the years 1969–1970 are given referred to AGK 3 stars. The positions, which have been measured on plates taken with the 67-92-cm Asiago Schmidt telescope, have a standard error of 0″.34 in R. A. and 0″.26 in Dec.

101.005 **Neptune and ν Scorpii 1972.** E. G. Moore.
Journ. British Astron. Ass., Vol. 82, 367 - 368 (1972).

101.006 **New observations on the Kuiper bands of Uranus.**
B. L. Lutz, D. A. Ramsay.
Astrophys. Journ., Vol. 176, 521 - 524 (1972).

The 7500 Å Kuiper bands of Uranus have been rephotographed at 10 Å mm^{-1} dispersion utilizing an electrostatic image intensifier. Comparison of these bands with new long-path laboratory spectra of methane (up to 8.45 km-atm) confirms that it is responsible for the bands. Twenty new Uranian bands are reported, of which at least eight, and possibly all, are due to methane. Equivalent widths of some of the planetary features are presented for comparison with future laboratory spectra.

101.007 Search for a trans-Plutonian planet.
A. P. O. Foss, J. S. Shawe-Taylor, D. P. D. Whitworth.
Nature, Vol. 239, 266 (1972).

101.008 Spectrophotometry of Pluto, 1972.
L. J. Kelsey, J. D. Fix, J. S. Neff.
Bull. American Astron. Soc., Vol. 4, 321 (1972). – Abstr. AAS.

101.009 Linear polarization measurements of Pluto.
L. A. Kelsey, J. D. Fix.
Bull. American Astron. Soc., Vol. 4, 321 (1972). – Abstr. AAS.

101.010 Photometric observations of pressure-induced absorptions in the spectrum of Uranus.   L. Trafton.
Bull. American Astron. Soc., Vol. 4, 361 (1972). – Abstr. AAS.

101.011 The atmosphere of Uranus.
R. G. Prinn. J. S. Lewis.
Bull. American Astron. Soc., Vol. 4, 361 (1972). – Abstr. AAS.

101.012 The masses, densities and moments of inertia of Uranus and Neptune.   A. H. Cook.
Observatory, Vol. 92, 84 - 85 (1972).
Recent observations have given better values of the masses, mean densities and moments of inertia than those available so far and it may be useful to collect the relevant data in this note.

101.013 The planet Uranus.   J. L. Perdrix.
Journ. Astron. Soc. Victoria, Vol. 25, 56 - 57 (1972).

101.014 Limb and polar brightening of Uranus at 8870 Å.
W. M. Sinton.
Astrophys. Journ., (Letters), Vol. 176, L131 - L133 (1972).
Photographs of Uranus in the 8870 Å band of methane exhibit limb brightening and brightening at the south pole. The polar brightening is explained by an upper atmospheric haze in addition to Rayleigh scattering by the upper atmosphere. An estimate of the geometric albedo at this wavelength yields a value near 0.01.

101.015 A near-infrared view of the Uranus system.
W. M. Sinton.
Sky Telescope, Vol. 44, 304 - 305 (1972).

101.016 Planets beyond Pluto?   A. T. Lawton.
Spaceflight, Vol. 14, 454 - 455 (1972).

101.017 On the construction of models of Neptune.
A. B. Makalkin.
Astron. vestn., Vol. 6, 172 - 176 (1972). In Russian. – Abstr. in Referativ. Zhurn. 51. Astron., 1.51.320 (1973).

101.018 Is there a tenth planet in the solar system?
D. Rawlins, M. Hammerton.
Nature, Vol. 240, 457 (1972).
We have been analysing the longitudinal residuals of Neptune (N), from sheer interest aroused by the discordance between the pre-discovery position of N and its currently accepted orbit. We attempted to establish the limits of the possible position of a tenth planet.

101.019 Temperatures of Uranus and Neptune at 24 microns.
D. Morrison, D. P. Cruikshank.
Publ. Astron. Soc. Pacific, Vol. 84, 642 (1972). – Abstr. Astron. Soc. Pacific.

101.020 Search for Brady's hypothetical trans-Plutonian planet.   A. R. Klemola, E. A. Harlan.
Publ. Astron. Soc. Pacific, Vol. 84, 736 = Lick Obs. Bull., No. 631 (1972).
No object brighter than magnitude 17 - 18 has been found within about 3° of the ephemeris position for the hypothetical trans-Plutonian planet.

101.021 The case against Planet X.
P. Goldreich, W. R. Ward.
Publ. Astron. Soc. Pacific, Vol. 84, 737 - 742 (1972).
In this paper, we show that the present configuration of the solar system is incompatible with the existence of Planet X. Our demonstration consists of two parts. First, we show that Planet X would force the plane of the known solar system planets to precess in a few times 10^7 years about an axis which lies close to this plane. Second, the coplanar configuration of the outer planets would be destroyed on a time scale of order 10^6 years by the perturbations due to Planet X.

101.022 High-resolution imagery of Uranus obtained by Stratoscope II.
R. E. Danielson, M. G. Tomasko, B. D. Savage.
Astrophys. Journ., Vol. 178, 887 - 900 (1972).
From 17 photographs of Uranus obtained by Stratoscope II, a composite image has been produced having a Gaussian point spread function with a half-maximum intensity diameter of 0".2. No certain surface markings are visible. The measured limb darkening does not agree with either a deep Rayleigh atmosphere or with clouds high in the atmosphere; a cloud deck under a finite Rayleigh atmosphere seems to be indicated. The equatorial diameter of Uranus is measured to be 51,800 ± 600 km, and the ellipticity is estimated to be 0.01 ± 0.01.

101.023 Note on Brady's hypothetical trans-Plutonian planet.
P. K. Seidelmann, B. G. Marsden, H. L. Giclas.
Publ. Astron. Soc. Pacific, Vol. 84, 858 - 864 (1972).
It is shown that a planet having the mass and orbital elements derived by Brady (1972) from the residuals in the times of perihelion passage of Halley's comet will affect the motions of the five planets Jupiter-Pluto to a degree that is completely incompatible with the observations. Furthermore, careful reexamination of plates obtained during the Lowell Proper-Motion Survey does not reveal a planetary object anywhere near the position predicted by Brady and brighter than magnitude $16^m.5$. It is concluded that the anomalous motion of Halley's comet has some other cause.

101.024 The planet Uranus.   H. Menzel.
Journ. Astron. Soc. Western Australia, 1972 June, p. 2 - 4.

Astrometric ephemeris of Pluto 1970–1990.
See Abstr. 041.046.

A method of integration of the equations of planetary motion in rectangular coordinates. The Pluto perturbations from Neptune.   See Abstr. 042.001.

Nonthermal radio observations of the major planets.
See Abstr. 099.091.

The effect of a trans-Plutonian planet on Halley's comet.   See Abstr. 103.104.

Über den Einfluß eines transplutonischen Planeten auf den Halley'schen Kometen.   See Abstr. 103.104.

102 Comets

102.001 **On the Kelvin–Helmholtz instability of type I comet tails.**
A. I. Ershkovich, A. A. Nusinov, A. A. Chernikov.
Astron. Zhurn. Akad. Nauk SSSR, Vol. 49, 866 - 871 (1972).
In Russian. English translation in Soviet Astron. AJ, Vol. 16,
No. 4.

The magnetohydrodynamic instability of Kelvin–Helmholtz type in comet tails is considered. The tail is supposed to be a plasma cylinder immersed into the interplanetary plasma. Both the criterion of instability and the growth rate of magnetohydrodynamic waves are obtained in the comet tail. Using the suggested model, it is possible to interpret, at least qualitatively, a number of observed effects.

102.002 **The dependence on inclination of the planetary perturbations of the orbits of long-period comets.**
S. Yabushita.
Astron. Astrophys., Vol. 20, 205 - 214 (1972).

Planetary perturbations of the orbital elements of long-period comets are calculated on the assumption that the orbits are parabolas. The perturbation of the binding energy of Halley's comet calculated by the parabolic approximation is compared with Cowell and Crommelin's calculation. It is shown that the agreement between the two calculations is such that the parabolic approximation can be used at least qualitatively for comets with periods as short as 77 years. Applying the Lyttleton-Hammersley random-walk theory of cometary energy, direct comets are more rapidly expelled from the solar system than retrograde ones. This theoretical deduction is compared with the distribution of inclinations of observed long-period comets. The *deficiency* of observed comets near i (inclination) ~130° excess might be explicable by the weakness of the planetary perturbation near $i = 120°$.

102.003 **Oscillations of type-1 comet tails.**
A. I. Ershkovich, A. A. Nusinov, A. A. Chernikov.
Planet. Space Sci., Vol. 20, 1235 - 1243 (1972).

Wave motions in type-1 comet tails are studied. It is shown that quasiperiodical variations of type-1 comet tail direction with a period of some days, first described by Bessel in 1836, may be the eigenmodes, excited by solar wind parameter fluctuations.

102.004 **Un nouveau catalogue d'orbites cométaires.**
J. Meeus.
L'Astronomie, 86ᵉ année, p. 363 - 365 (1972).

102.005 **Motion of comets.** G. Sitarski.
Postępy Astron., Vol. 20, 193 - 204 (1972).
In Polish.

Recent results of investigations on the motion of comets with near-parabolic orbits and of short-period comets are presented. The hypothesis of origin of comets by Oort and by Witkowski, results of investigations by Everhart and studies by Marsden and Sekanina on the nongravitational anomalies in the motion of short-period comets are discussed.

102.006 **Wave motions of type I comet tails.**
A. I. Ershkovich, A. A. Nusinov, A. A. Chernikov.
Astron. Tsirk., No. 671, p. 6 - 8 (1972). In Russian.

102.007 **Evolution of cometary orbits on a cosmogonic time scale.** G. A. Chebotarev.
IAU Symposium No. 45, (see 012.003), p. 1 - 5 (1972).

102.008 **Cometary observations and variations in cometary brightness.** S. K. Vsekhsvyatskij.

IAU Symposium No. 45, (see 012.003), p. 9 - 15 (1972).

102.009 **Cometary brightness variations and conditions in interplanetary space.**
D. A. Andrienko, A. A. Demenko, I. M. Demenko, I. D. Zosimovich.
IAU Symposium No. 45, (see 012.003), p. 16 - 21 (1972).

102.010 **General remarks on orbit and ephemeris computation.** B. G. Marsden.
IAU Symposium No. 45, (see 012.003), p. 36 - 38 (1972).

102.011 **A series-solution method for cometary orbits.**
P. E. Nacozy.
IAU Symposium No. 45, (see 012.003), p. 43 - 51 (1972).

102.012 **On the application of Hansen's method of partial anomalies to the calculation of perturbations in cometary motions.** V. I. Skripnichenko.
IAU Symposium No. 45, (see 012.003), p. 52 - 54 (1972).

102.013 **Orbital characteristics of comets passing through the 1:1 commensurability with Jupiter.** E. Rabe.
IAU Symposium No. 45, (see 012.003), p. 55 - 60 (1972).

102.014 **The motions of bodies close to commensurabilities with Jupiter.** A. T. Sinclair.
IAU Symposium No. 45, (see 012.003), p. 61 (1972).

102.015 **On the motion of short-period comets in the neighbourhood of Jupiter.** V. M. Chepurova.
IAU Symposium No. 45, (see 012.003), p. 62 - 65 (1972).

102.016 **Secular perturbations on periodic comets.**
G. E. O. Giacaglia.
IAU Symposium No. 45, (see 012.003), p. 66 - 80 (1972).

102.017 **The solution of problems of cometary astronomy on electronic computers.** N. A. Belyaev.
IAU Symposium No. 45, (see 012.003), p. 90 - 94 (1972).

102.018 **A numerical interpretation of the homogenization of observational material for one-apparition comets.**
G. Sitarski.
IAU Symposium No. 45, (see 012.003), p. 107 - 111 (1972).

102.019 **The problem of elaboration and classification of observational material for one-apparition comets.**
M. Bielicki.
IAU Symposium No. 45, (see 012.003), p. 112 - 117 (1972).

102.020 **The influence of properties of a set of observations on the weights of determination of the orbital elements of a one-apparition comet.** M. Bielicki.
IAU Symposium No. 45, (see 012.003), p. 118 - 122 (1972).

102.021 **On the differential correction of nearly parabolic orbits.** P. Herget.
IAU Symposium No. 45, (see 012.003), p. 123 (1972).

102.022 **Standardization of the calculation of nearly parabolic cometary orbits.**
L. E. Nikonova, N. A. Bokhan.
IAU Symposium No. 45, (see 012.003), p. 124 - 126 (1972).

102.023 **Détermination d'orbites paraboliques à partir de N observations au moyen de l'ordinateur électronique.**

H. Debehogne.
IAU Symposium No. 45, (see 012.003), p. 127 - 129 (1972).

102.024 **Nongravitational effects on comets: The current status.** B. G. Marsden.
IAU Symposium No. 45, (see 012.003), p. 135 - 143 (1972).

102.025 **On the determination of nongravitational forces acting on comets.** P. E. Zadunaisky.
IAU Symposium No. 45, (see 012.003), p. 144 - 151 (1972).

102.026 **The influence of minor planets on the motions of comets.** K. A. Shtejns, I. E. Zal'kalne.
IAU Symposium No. 45, (see 012.003), p. 246 - 250 (1972).

102.027 **Physical processes in cometary atmospheres.** A. Z. Dolginov.
IAU Symposium No. 45, (see 012.003), p. 253 - 259 (1972).

102.028 **Some remarks on the liberation of gases from cometary nuclei.** B. Yu. Levin.
IAU Symposium No. 45, (see 012.003), p. 260 - 264 (1972).

102.029 **The chemical composition of cometary nuclei.** L. M. Shul'man.
IAU Symposium No. 45, (see 012.003), p. 265 - 270 (1972).

102.030 **The evolution of cometary nuclei.** L. M. Shul'man.
IAU Symposium No. 45, (see 012.003), p. 271 - 276 (1972).

102.031 **On the sizes of cometary nuclei.** V. P. Konopleva, L. M. Shul'man.
IAU Symposium No. 45, (see 012.003), p. 277 - 282 (1972).

102.032 **Splitting and sudden outbursts of comets as indicators of nongravitational effects.** E. M. Pittich.
IAU Symposium No. 45, (see 012.003), p. 283 - 286 (1972).

102.033 **On nongravitational effects in two classes of models for cometary nuclei.**
O. V. Dobrovol'skij, M. Z. Markovich.
IAU Symposium No. 45, (see 012.003), p. 287 - 293 (1972).

102.034 **Rotation effects in the nongravitational parameters of comets.** Z. Sekanina.
IAU Symposium No. 45, (see 012.003), p. 294 - 300 (1972).

102.035 **Laboratory simulation of icy cometary nuclei.** E. A. Kajmakov, V. I. Sharkov.
IAU Symposium No. 45, (see 012.003), p. 308 - 315 (1972).

102.036 **A nongravitational effect in the simulation of cometary phenomena.**
E. A. Kajmakov, V. I. Sharkov, S. S. Zhuravlev.
IAU Symposium No. 45, (see 012.003), p. 316 - 323 (1972).

102.037 **On the stability of the Oort cloud.** E. M. Nezhinskij.
IAU Symposium No. 45, (see 012.003), p. 335 - 340 (1972).

102.038 **Determination of the form of the Oort cometary cloud as the Hill surface in the galactic field.**
V. A. Antonov, I. N. Latyshev.
IAU Symposium No. 45, (see 012.003), p. 341 - 345 (1972).

102.039 **On 'new' comets and the size of the cometary cloud.** G. T. Yanovitskaya.
IAU Symposium No. 45, (see 012.003), p. 346 (1972).
Abstract.

102.040 **Diffusion of comets from parabolic into nearly parabolic orbits.** K. A. Shtejns.
IAU Symposium No. 45, (see 012.003), p. 347 - 351 (1972).

102.041 **New estimates of cometary disintegration times and the implications for diffusion theory.**
O. V. Dobrovol'skij.
IAU Symposium No. 45, (see 012.003), p. 352 - 355 (1972).

102.042 **Comets and problems of numerical celestial mechanics.** S. K. Vsekhsvyatskij.
IAU Symposium No. 45, (see 012.003), p. 356 - 359 (1972).

102.043 **The effect of the ellipticity of Jupiter's orbit on the capture of comets to short-period orbits.**
E. Everhart.
IAU Symposium No. 45, (see 012.003), p. 360 - 363 (1972).

102.044 **Evolution of short-period cometary orbits due to close approaches to Jupiter.** O. Havnes.
IAU Symposium No. 45, (see 012.003), p. 364 - 369 (1972).

102.045 **A new orbital classification for periodic comets.** M. Bielicki.
IAU Symposium No. 45, (see 012.003), p. 370 - 372 (1972).

102.046 **The major planets as powerful transformers of cometary orbits.** E. I. Kazimirchak-Polonskaya.
IAU Symposium No. 45, (see 012.003), p. 373 - 397 (1972).

102.047 **The origin and evolution of the comets and other small bodies in the solar system.**
S. K. Vsekhsvyatskij.
IAU Symposium No. 45, (see 012.003), p. 413 - 418 (1972).

102.048 **On the problem of the origin of comets.** J. M. Witkowski.
IAU Symposium No. 45, (see 012.003), p. 419 - 425 (1972).

102.049 **Formation of comets in meteor streams.** J. Trulsen.
IAU Symposium No. 45, (see 012.003), p. 487 - 490 (1972).

102.050 **On the dividing line between cometary and asteroidal orbits.** L. Kresák.
IAU Symposium No. 45, (see 012.003), p. 503 - 514 (1972).

102.051 **Once more about the formation of comets due to outbursts on the surface of Jupiter's satellite Io.**
M. A. Mamedov.
Izv. AN AzSSR. Ser. fiz.-tekhn. i mat. n., 1971, No. 4, p. 114 - 121. In Russian. — Abstr. in Referativ. Zhurn. 51. Astron., 8.51.302 (1972).

102.052 **The destruction of a superficial dust matrix of a cometary nucleus.**
O. V. Dobrovol'skij, Kh. Ibadinov.
Dokl. AN TadzhSSR, Vol. 14, No. 12, p. 16 - 19 (1971). In Russian. — Abstr. in Referativ. Zhurn. 51. Astron., 8.51.304 (1972).

102.053 **Comet formation.** L. W. Fullerton, W. F. Huebner.
Bull. American Astron. Soc., Vol. 4, 369 (1972).
Abstr. AAS.

102.054 **Statistical connections between observed characteristics of comets.** R. S. Osherov.
Astrometriya i Astrofiz., *Kiev*, No. 15, (see 003.001), p. 91 - 95 (1972). In Russian.
Statistical connections are established between observed

characteristics of comets on the basis of numerous observations made by M. Beyer.

102.055 The size of dust particles of cometary atmospheres and heliocentric distance.
O. V. Dobrovol'skij, Kh. Ibadinov.
Dokl. AN TadzhSSR, Vol. 15, No. 2, p. 15 - 18 (1972). In Russsian. – Abstr. in Referativ. Zhurn. 51. Astron., 10.51.300 (1972).

102.056 Cometary nuclei - models. F. L. Whipple.
Comets. Proc. Tucson Conference 1970, (see 012.010), p. 4 - 15 (1972).

102.057 Reality of comet nucleus. R. A. Lyttleton.
Comets. Proc. Tucson Conference 1970, (see 012.010), p. 16 - 19 (1972).

102.058 Nature and origin of cometary heads.
A. H. Delsemme.
Comets. Proc. Tucson Conference 1970, (see 012.010), p. 32 - 47 (1972).

102.059 Photometry of comets. F. Miller.
Comets. Proc. Tucson Conference 1970, (see 012.010), p. 48 - 50 (1972).

102.060 Comments on photochemistry. B. D. Donn.
Comets. Proc. Tucson Conference 1970, (see 012.010), p. 55 - 56 (1972).

102.061 Further comments on photochemistry.
W. M. Jackson.
Comets. Proc. Tucson Conference 1970, (see 012.010), p. 57 - 58 (1972).

102.062 Type I tails – solar wind interactions. L. Biermann.
Comets. Proc. Tucson Conference 1970, (see 012.010), p. 59 - 64 (1972).

102.063 Type I tails – further comments. J. C. Brandt.
Comets. Proc. Tucson Conference 1970, (see 012.010), p. 65 - 68 (1972).

102.064 Comet tails of type II. R. F. Probstein.
Comets. Proc. Tucson Conference 1970, (see 012.010), p. 69 - 83 (1972).

102.065 Comet spectra. C. Arpigny.
Comets. Proc. Tucson Conference 1970, (see 012.010), p. 84 - 111 (1972).

102.066 Spectroscopic observations of comets. T. C. Owen.
Comets. Proc. Tucson Conference 1970, (see 012.010), p. 112 - 121 (1972).

102.067 Comet orbits: Prediction, nongravitational effects.
B. G. Marsden.
Comets. Proc. Tucson Conference 1970, (see 012.010), p. 123 - 141 (1972).

102.068 Shape and orientation of nucleus. T. Gehrels.
Comets. Proc. Tucson Conference 1970, (see 012.010), p. 142 - 144 (1972).

102.069 Present understanding of comets. A. H. Delsemme.
Comets. Proc. Tucson Conference 1970, (see 012.010), p. 174 - 182 (1972).

102.070 Cometary hydrogen and hydroxyl comas.
M. K. Wallis, A. H. Delsemme.

Science, Vol. 178, 78 (1972).

102.071 Correlation between brightness of comets and fluctuations of the solar wind.
O. V. Dobrovol'skij, Yu. N. Gnedin, G. G. Novikov.
Kosm. Issled., Vol. 10, 791 - 792 (1972). In Russian. – Brief information.

102.072 Bright comets −86 to +1950. Ephemerides and concise descriptions. H. Mucke.
Edited by Astronomisches Büro, Wien. 98 pp. Price öS 95.00 (1972). In German.

102.073 Cometary astronomy in the USSR.
O. V. Dobrovol'skij.
Astron. vestn., Vol. 6, 137 - 152 (1972). In Russian. – Abstr. in Referativ. Zhurn. 51. Astron., 1.51.356 (1973).

102.074 On the disintegration of periodic comets.
S. K. Vsekhsvyatskij.
Problems of cosmic physics. Vyp. (No.) 7, (see 003.028), p. 74 - 78 (1972). In Russian.

102.075 Periodic comets. J. Bouška.
Říše hvězd, Vol. 53, 191 - 196 (1972). In Czech.

Dynamics of cometary atmospheres. Neutral gas.
See Abstr. 003.134.

Radiative lifetimes of the $A^2\Pi$ state of CO^+.
See Abstr. 022.040.

A comet photometer for the amateur.
See Abstr. 034.158.

A numerical method of integration by means of Taylor-Steffensen series and its possible use in the study of the motions of comets and minor planets.
See Abstr. 042.023.

Interstellar gravitational perturbations of cometary orbits. See Abstr. 042.089.

Some scientific criteria for a cometary mission.
See Abstr. 051.012.

Electron deposition in water vapor, with atmospheric applications. See Abstr. 082.067.

Determination of planetary masses from the motions of comets. See Abstr. 091.005.

On the possible common origin of minor planets, comets, and meteors. See Abstr. 098.012.

The determination of Jupiter's mass from large perturbations on cometary orbits in Jupiter's sphere of action.
See Abstr. 099.008.

The case against Planet X. See Abstr. 101.021.

Theoretical cometary radiants and the structure of meteor streams. See Abstr. 104.009.

Statistics of the orbits of meteor streams and comets. See Abstr. 104.010.

Laser radar studies of upper atmosphere dust layers and the relation of temporary increases in dust to cometary micrometeoroid streams. See Abstr. 105.054.

On the rate of ejection of dust by long-period comets. See Abstr. 106.008.

Interplanetary objects in review: Statistics of their masses and dynamics. See Abstr. 106.009.

Ejection of bodies from the solar system in the course of the accumulation of the giant planets and the formation of the cometary cloud. See Abstr. 107.004.

The origin of comets. See Abstr. 107.005.

On the origin of comets and their importance for the cosmogony of the solar system. See Abstr. 107.006.

On the relation between comets and meteoroids. See Abstr. 107.007.

103   Comets: Listed Objects

103.001 Observations of comets at the Crimean Astrophysical Observatory.   N. S. Chernykh.
IAU Symposium No. 45, (see 012.003), p. 22 - 24 (1972).

103.002 L'observation des comètes à l'astrographe de l'Observatoire de Nice.   B. Milet.
IAU Symposium No. 45, (see 012.003), p. 25 - 26 (1972).

103.003 On establishing an international service for cometary observations and ephemerides.
M. P. Candy.
IAU Symposium No. 45, (see 012.003), p. 35 (1972).

103.004 Comets in 1971.   B. G. Marsden.
Quarterly Journ. Roy. Astron. Soc., Vol. 13, 415 - 435 (1972).

103.005 Telegram on the discovery of a new comet.
Zhukov.
Kometn. Tsirk., *Kiev*, No. 135 (1972). In Russian.

103.006 Comet notes.
Journ. Astron. Soc. Victoria, Vol. 25, 65 - 69 (1972).

103.007 Observations of comets and asteroids at the Klet' Observatory in the year 1970.   A. Mrkos.
Acta Univ. Carolinae Math. Phys., Vol. 12, 67 - 78 = Astron. Inst. Charles Univ., Praha, Publ. No. 66 (1971).
The routine programme of comet observations at the Klet̂ Observatory is continued. A list of precise observations of comets and two asteroids obtained at the Observatory during the year 1970 is presented.

103.008 Discovery of a comet.
Astron. Tsirk., No. 716, p. 1 (1972). In Russian.
Zhukov reported the possible discovery of a new comet.

103.009 Comet notes.   E. Roemer.
Mercury, (Journ. Astron. Soc. Pacific), Vol. 1, No. 4, p. 16 - 18; No. 5, p. 17 - 18; No. 6, p. 18 - 19 (1972).

103.010 Definitive designations of comets of 1970.
Kometn. Tsirk., *Kiev,* No. 138 (1972). In Russian.

103.011 Possible comet Edwards.   L. R. Edwards.
IAU Circ., No. 2432 (1972).

103.012 Possible comet Zhukov.   Zhukov.
IAU Circ., No. 2454 (1972).

103.013 Possible comet Torres.   C. Torres.
IAU Circ., No. 2473 (1972).

103.014 Possible comet Torres.   C. Torres.
IAU Circ., No. 2475 (1972).

103.015 Comets in the year 1969.   M. Antal.
Kozmos, Vol. 3, 110 - 111, 115 (1972).   In Slovak.

103.016 Comet observations.   R. L. Waterfield, R. H. South, I. M. Purcell, A. Griffiths.
British Astron. Ass., Circ. No. 543 (1972). — Concerning the comets 1972a, 1972d, 1972h.

103.100 Comet 1970 II Bennett

The scattered-light continuum of comet Bennett 1969i.   G. M. Stokes.
Astrophys. Journ., Vol. 177, 829 - 834 (1972).
The scattered-light continuum of comet Bennett has been observed in the region between 4500 and 8000 Å. This continuum is found to be distinctly redder than sunlight, and analysis leads to a lower limit on mean particle size of 0.1 μ. The contribution to the reddening due to the C_2 Phillips bands and blackbody thermal radiation is considered and found to be negligible.

Comets as interplanetary probes — comet Bennett, March, April 1970.   F. D. Miller.
Bull. American Astron. Soc., Vol. 4, 425 (1972). — Abstr. AAS.

Spectrophotometry of comet Bennett.
G. S. D. Babu, P. P. Saxena.
Bull. Astron. Inst. Czechoslovakia, Vol. 23, 346 - 349 (1972).
Photoelectric spectrum scans of the head of comet Bennett (1969i) covering the range 352 ÷ 612 mμ are presented and the emission features of CN, CH, C_2 and Na have been identified. The scattered light becomes redder as the phase angle decreases and matches with the solar light scattered according to the λ^2 law at a particular phase angle, indicating that the size of the scattering particles is of the order of 0.5 μ.

Observations of comet Bennett (1969i) at the Observatory of the Sternberg Institute. B. S. Vozdvizhensky.
Byull. Inst. Teoret. Astron., *Leningrad*, Vol. 13, 186 (1972). In Russian.

Lα photometry of comet Bennett. M. Dubin.
Comets. Proc. Tucson Conference 1970, (see 012.010), p. 51 - 54 (1972).

Photographic observations of comet Bennett, 1970 II.
S. M. Larson, R. B. Minton.
Comets. Proc. Tucson Conference 1970, (see 012.010), p. 183 - 208 (1972).

Spectral investigations of comet Bennett, 1970 II (1969 i). O. V. Dobrovol'skij, O. Mamadov.
Dokl. AN TadzhSSR, Vol. 15, No. 4, p. 14 - 17 (1972). In Russian. − Abstr. in Referativ. Zhurn. 51. Astron., 12.51.269 (1972).

Lyman-alpha radiation of comet Bennett 1969i and determination of the solar wind flux.
A. Z. Dolginov, Yu. N. Gnedin, G. G. Novikov.
Kosm. Issled., Vol. 10, 793 (1972). In Russian. − Brief information.

The tail orientation of comet Bennett (1969i).
D. R. L. Jones.
Observatory, Vol. 92, 181 - 183 (1972). − Note.

A 1.1 micron spectrogram of the central coma of comet Bennett (1969i). D. D. Meisel, J. L. Deutsch.
Publ. Astron. Soc. Pacific, Vol. 84, 732 - 735 (1972).
Image-tube slit spectra of the moon and the central coma of comet Bennett (1969i) and transverse tracings of the comet spectrum are presented. While the diatomic species CN can be positively identified, a number of faint band-like features originally assigned to OH and atomic lines remain unidentified. It is suggested that these may be due to vibrational overtones of polyatomic species containing CH or OH bonds, but the necessary laboratory spectra are not yet available to investigate this in any detail.

Photographic observations of comet Bennett.
M. P. Bondarevs'kij, K. N. Kuz'menko, G. A. Orlov.
Visn. Kharkiv. Univ. No. 82, (Ser. Astron., No. 7), p. 84 (1972). In Ukrainian.

Infrared measures of comets. See Abstr. 103.121.

Evidence from polarization. See Abstr. 103.122.

Infrared observations of comets Ikeya-Seki (1965 f) and Bennett (1969i). See Abstr. 103.122.

103.101 Comet 1969 IV Churyumov-Gerasimenko

Physical observations of the short-period comet 1969 IV. K. I. Churyumov, S. I. Gerasimenko.
IAU Symposium No. 45, (see 012.003), p. 27 - 34 (1972).

103.102 Comet 1967 XII Wolf 1

A method of integrating the equations of motion in special coordinates and the elimination of a discontinuity in the theory of the motion of periodic comet Wolf.
E. I. Kazimirchak-Polonskaya.

IAU Symposium No. 45, (see 012.003), p. 95 - 102 (1972).

103.103 Comet 1971 II Encke

Periodic comet Encke (1970l).
E. Roemer, G. McCorkle.
IAU Circ., No. 2435 (1972).

Periodic comet Encke. E. Roemer, G. McCorkle, M. Gonzales, R. E. McCrosky, C. Y. Shao.
IAU Circ., No. 2446 (1972).

A search for Encke's comet in ancient Chinese records: A progress report. F. L. Whipple, S. E. Hamid.
IAU Symposium No. 45, (see 012.003), p. 152 - 154 (1972).

A model for the nucleus of Encke's comet.
Z. Sekanina.
IAU Symposium No. 45, (see 012.003), p. 301 - 307 (1972).

Comet Encke near aphelion.
Kometn. Tsirk., *Kiev,* No. 138 (1972). In Russian.

103.104 Comet 1910 II Halley

The motion of Halley's comet from 837 to 1910.
J. L. Brady.
IAU Symposium No. 45, (see 012.003), p. 155 (1972). Abstract.

The effect of a trans-Plutonian planet on Halley's comet.
Journ. Astron. Soc. Victoria, Vol. 25, 55 - 56 (1972).

Über den Einfluß eines transplutonischen Planeten auf den Halley'schen Kometen. E. Wiedemann.
Orion, 30. Jahrgang, p. 136 - 137 (1972). − Extract of a paper by J. L. Brady, Publ. Astron. Soc. Pacific, Vol. 84, 314 - 322 (1972).

Note on Brady's hypothetical trans-Plutonian planet.
See Abstr. 101.023.

103.105 Comet 1960 VI Brooks 2

A numerical analysis of the motion of periodic comet Brooks 2. P. Stumpff.
IAU Symposium No. 45, (see 012.003), p. 156 - 166 (1972).

103.106 Comet 1969 VI Faye

Linkage of seven apparitions of periodic comet Faye 1925−1970 and investigation of the orbital evolution during 1660−2060. N. A. Belyaev, F. B. Khanina.
IAU Symposium No. 45, (see 012.003), p. 167 - 172 (1972).

Periodic comet Faye, 1843 III = 1969a.
J. E. Bortle.
Strolling Astronomer, Vol. 24, 22 - 25 (1972).

103.107   Comet 1972d Giacobini-Zinner

P/Giacobini-Zinner 1972d.
R. L. Waterfield, R. W. Panther, P. Doherty.
British Astron. Ass., Circ. No. 543 (1972).

P/Giacobini-Zinner (1972d).
R. H. South, D. Griffiths, R. H. Rutter, R. L. Waterfield.
British Astron. Ass., Circ. No. 544 (1972).

Periodic comet Giacobini-Zinner (1972d).
T. Seki, M. Takeishi, R. L. Waterfield, I. M. Purcell, R. H. South, B. Milet.
IAU Circ., No. 2423 (1972).

Periodic comet Giacobini-Zinner (1972d).
A. Mrkos, B. Milet.
IAU Circ., No. 2426 (1972).

Periodic comet Giacobini-Zinner (1972d).
M. Antal, N. Kojima, T. Seki.
IAU Circ., No. 2429 (1972).

Periodic comet Giacobini-Zinner (1972d).
T. Seki.
IAU Circ., No. 2433 (1972).

Periodic comet Giacobini-Zinner (1972d).
B. Milet, R. H. S. South, D. Griffiths, R. L. Waterfield, D. Sykes, A. Mrkos, T. Seki, T. Urata.
IAU Circ., No. 2437 (1972).

Periodic comet Giacobini-Zinner (1972d).
R. E. McCrosky, C. Y. Shao, N. Kojima, T. Seki, J. Bortle.
IAU Circ., No. 2445 (1972).

Periodic comet Giacobini-Zinner (1972d).
M. Antal, G. H. Rutter, R. L. Waterfield, T. Urata.
IAU Circ., No. 2449 (1972).

Periodic comet Giacobini-Zinner (1972d).
T. Kleine, O. Guthier, V. Kasten.
IAU Circ., No. 2451 (1972).

Periodic comet Giacobini-Zinner (1972d).
M. Antal, T. Seki, N. Kojima.
IAU Circ., No. 2458 (1972).

Periodic comet Giacobini-Zinner (1972d).
T. Seki, T. Furuta, T. Urata.
IAU Circ., No. 2464 (1972).

Periodic comet Giacobini-Zinner (1972d).
B. Milet.
IAU Circ., No. 2467 (1972).

Investigation of the motion of periodic comet Giacobini-Zinner and the origin of the Draconid meteor showers of 1926, 1933 and 1946.   Yu. V. Evdokimov.
IAU Symposium No. 45, (see 012.003), p. 173 - 180 (1972).

Nongravitational forces and periodic comet Giacobini-Zinner.   D. K. Yeomans.
IAU Symposium No. 45, (see 012.003), p. 181 - 186 (1972).

Comet P/Giacobini-Zinner (1972d).
Japan Astron. Study Ass., Cir. 277 (1972). In Japanese.

Observations of comet 1972 d at the Alpine Station of the Sternberg Astronomical Institute.
Kometn. Tsirk., *Kiev*, No. 135 (1972). In Russian.

Periodic comet Giacobini-Zinner, 1972 d.
Kometn. Tsirk., *Kiev*, No. 135 (1972). In Russian.

Short-period comet Giacobini-Zinner, 1972 d.
Kometn. Tsirk., *Kiev*, No. 136 (1972). In Russian.

Periodic comet Giacobini-Zinner, 1972d.
Kometn. Tsirk., *Kiev*, No. 137 (1972).   In Russian.– Observations at the Kleť Observatory, *A. Mrkos;* Observations by the Alpine expedition of the State Astronomical Sternberg Institute, *L. Markova, V. Kovalenko;* Observations at the Skalnaté Pleso Observatory, *M. Antal.*

Periodic comet Giacobini-Zinner, 1972d.
Kometn. Tsirk., *Kiev*, No. 139 (1972).   In Russian.

103.108   Comet 1960 V Borrelly

A non-Newtonian orbit for periodic comet Borrelly.
D. K. Yeomans.
IAU Symposium No. 45, (see 012.003), p. 187 - 189 (1972).

An investigation of the motion of periodic comet Borrelly from 1904 to 1967.   L. M. Belous.
IAU Symposium No. 45, (see 012.003), p. 190 - 194 (1972).

103.109   Comet 1954 VII Pons-Brooks

The motion of periodic comet Pons-Brooks, 1812–1954.   P. Herget, H. J. Carr.
IAU Symposium No. 45, (see 012.003), p. 195 - 199 (1972).

103.110   Comet 1866 I Tempel-Tuttle

Periodic comet Tempel-Tuttle and the Leonid meteor shower.   E. D. Kondrat'eva.
IAU Symposium No. 45, (see 012.003), p. 200 - 202 (1972).

103.111   Comet 1942 IX Stephan-Oterma

Investigation of the motion of periodic comet Stephan-Oterma.   M. Ya. Shmakova.
IAU Symposium No. 45, (see 012.003), p. 203 - 205 (1972).

103.112   Comet 1972f Bradfield

Comet Bradfield 1972f.
Astron. Tsirk., No. 690, p. 1 (1972).   In Russian.

Comet Bradfield (1972f).   J. A. Bruwer.
IAU Circ., No. 2429 (1972).

Comet Bradfield (1972f).
Japan Astron. Study Ass., Circ. 277 (1972). In Japanese.

Observations of comet Bradfield 1972f.
J. C. Bennett.
Monthly Notes Astron. Soc. Southern Africa, Vol. 31, 74 (1972).

Comet Bradfield (1972f).
Yamamoto Circ., No. 1752, p. 1 - 2 (1972). In Japanese.

103.113 Comet 1937 V Finsler

Definitive orbit of comet Finsler (1937 V).
V. N. Klevetskij.
Astron. vestn., Vol. 6, 96 - 98 (1972). In Russian. − Abstr. in
Referativ. Zhurn. 51. Astron., 8.51.131 (1972).

103.114 Comet 1971 I Gehrels

Comet Gehrels 1972e.
Astron. Tsirk., No. 690, p. 1 (1972). In Russian.

Comet Gehrels (1972e).
E. Roemer, L. M. Vaughn, A. Ferguson.
IAU Circ., No. 2438 (1972).

Comet Gehrels (1972e).   B. G. Marsden.
IAU Circ., No. 2447 (1972).

Comet Gehrels (1972e).   C. T. Kowal.
IAU Circ., No. 2469 (1972).

Comet Gehrels (1972e).
Japan Astron. Study Ass., Circ. 277 (1972). In Japanese.

Comet Gehrels, 1972e.
Kometn. Tsirk., *Kiev*, No. 138 (1972).   In Russian.

Comet Gehrels (1972e).
Yamamoto Circ., No. 1751, p. 1 (1972). In Japanese.

Comet Gehrels (1972e).
Yamamoto Circ., No. 1752, p. 2 (1972). In Japanese.

103.115 Comet 1962 III Seki-Lines

The dust tail of comet Seki-Lines.   B. J. Jambor.
Bull. American Astron. Soc., Vol. 4, 322 (1927). − Abstr. AAS.

103.116 Comet 1972h Sandage

Discovery of a comet.
Astron. Tsirk., No. 710, p. 1 (1972). In Russian. − Concerning comet Sandage 1972h.

New comet Sandage 1972h.
A. R. Sandage, B. G. Marsden, D. R. L. Jones.
British Astron. Ass., Circ. No. 543 (1972).

Comet Sandage 1972h.   B. G. Marsden.
British Astron. Ass., Circ. No. 544 (1972).

Comet Sandage (1972h).
R. L. Waterfield, G. H. Rutter, I. M. Purcell.
British Astron. Ass., Circ. No. 544 (1972).

Comet Sandage (1972h).
T. Seki, N. Kojima, C. Cristescu, B. Milet.
IAU Circ., No. 2422 (1972).

Comet Sandage (1972h).
A. Mrkos, R. L. Waterfield, I. M. Purcell, T. Seki, C. Cristescu.
IAU Circ., No. 2425 (1972).

Comet Sandage (1972h).   M. Antal, N. Kojima,
T. Seki, A. Griffiths, R. L. Waterfield.
IAU Circ., No. 2428 (1972).

Comet Sandage (1972h).
T. Seki, N. Kojima, H. Dürbeck.
IAU Circ., No. 2434 (1972).

Comet Sandage (1972h).   C. Cristescu, B. Milet,
A. Mrkos, N. Kojima, T. Seki, B. G. Marsden.
IAU Circ., No. 2436 (1972).

Comet Sandage (1972h).
R. E. McCrosky, C. Y. Shao.
IAU Circ., No. 2438 (1972).

Comet Sandage (1972h).   V. Kravchenko.
IAU Circ., Nos. 2441, 2443 (1972).

Comet Sandage (1972h).
R. L. Waterfield, G. H. Rutter.
IAU Circ., No. 2443 (1972).

Comet Sandage (1972h).
H. L. Giclas, R. L. Waterfield.
IAU Circ., No. 2452 (1972).

Comet Sandage (1972h).   A. Mrkos.
IAU Circ., No. 2454 (1972).

Comet Sandage (1972h).   S. K. Vsekhsvyatskij.
IAU Circ., No. 2467 (1972).

Comet Sandage (1972h).   B. G. Marsden.
IAU Circ., No. 2472 (1972).

Comet Sandage (1972h).
Japan Astron. Study Ass., Circ. 278 - 279 (1972). In Japanese.

Comet Sandage, 1972 h.
Kometn. Tsirk., *Kiev*, No. 135 (1972). In Russian.

Comet Sandage, 1972 h.
Kometn. Tsirk., *Kiev*, No. 136 (1972). In Russian.

Comet Sandage, 1972h.
Kometn. Tsirk., *Kiev*, No. 137 (1972). In Russian. − Observations at the Kleť Observatory, *A. Mrkos;* Observations by the Alpine expedition of the State Astronomical Sternberg Institute, *G. Zhukov, L. Markova, V. Kovalenko;* Observations at the Crimean Astrophysical Observatory.

103.117 Comet 1972a Tempel 1

Periodic comet Tempel 1 (1972a).
R. L. Waterfield, I. M. Purcell, R. H. South, A. Mrkos.
IAU Circ., No. 2427 (1972).

Periodic comet Tempel 1 (1972a).
K. Tomita, T. Seki, M. Antal.
IAU Circ., No. 2430 (1972).

Comet P/Tempel 1 (1966 VII = 1972a).
Japan Astron. Study Ass., Circ. 276 (1972). In Japanese.

Comet P/Tempel 1 (1972a).
Japan Astron. Study Ass., Circ. 277 (1972). In Japanese.

Comet Tempel 1, 1972 a.
Kometn. Tsirk., *Kiev*, No. 135 (1972). In Russian.

103.118 Comet 1972i Reinmuth 1

Improved orbit and ephemeris of the periodic comet Reinmuth 1 for its apparition in 1972–73. G. Sitarski. Acta Astron., Vol. 22, 155 - 162 (1972).

Three apparitions of the comet were linked using 38 observations made in 1949, 1957–58 and 1965. The equations of motion of the comet were integrated by Cowell's method in the period 1949–1973 taking into account the perturbations caused by all the planets from Mercury to Pluto. The ephemeris of the comet for its next return in 1972/73 is computed; the conditions for observations are presented in a graph.

Comet P/Reinmuth 1 1972i. E. Roemer, M. R. Gonzales. British Astron. Ass., Circ. No. 544 (1972).

Periodic comet Reinmuth 1 (1972i). E. Roemer, M. R. Gonzales. IAU Circ., No. 2444 (1972).

Comet P/Reinmuth 1 (1972i). Japan Astron. Study Ass., Circ. 278 - 279 (1972). In Japanese.

Ephemeris of periodic comet Reinmuth 1. Kometn. Tsirk., *Kiev*, No. 135 (1972). In Russian.

Rediscovery of comet Reinmuth 1, 1972i. Kometn. Tsirk., *Kiev*, No. 137 (1972). In Russian.

103.119 Comet 1954 VIII Vozárová

On the decay of the nucleus of comet 1954 VIII. V. A. Golubev. Kometn. Tsirk., *Kiev*, No. 136 (1972). In Russian.

103.120 Comet 1971c Kearns-Kwee

Periodic comet Kearns-Kwee (1971c). N. Kojima, T. Seki. IAU Circ., No. 2453 (1972).

Periodic comet Kearns-Kwee (1971c). S. K. Vsekhsvyatskij, L. V. Zhuravleva, G. R. Kastel', T. Seki, T. Furuta, T. Urata, N. Kojima. IAU Circ., No. 2465 (1972).

Periodic comet Kearns-Kwee (1971c). R. H. S. South, R. L. Waterfield, T. Seki, A. Mrkos, R. Petrovičová, T. Urata, B. Milet. IAU Circ., No. 2474 (1972).

Comet P/Kearns-Kwee (1971c). Japan Astron. Study Ass., Circ. 278 - 279 (1972). In Japanese.

Ephemeris of comet Kearns-Kwee 1963 VII. Kometn. Tsirk., *Kiev*, No. 136 (1972). In Russian.

Comet Kearns-Kwee, 1971c. Kometn. Tsirk., *Kiev*, No. 138 (1972). In Russian.

Comet Kearns-Kwee, 1971c. Kometn. Tsirk., *Kiev*, No. 139 (1972). In Russian.

103.121 Comet 1969 IX Tago-Sato-Kosaka

The kinematical behaviour of the plasma tail of comet Tago-Sato-Kosaka 1969 IX. K. Jockers, Rh. Lüst, T. Nowak. Astron. Astrophys., Vol. 21, 199 - 207 (1972).

20 pictures of comet Tago-Sato-Kosaka, taken on 4 days between January 5 and 12, 1970 with the Hamburg Schmidt-telescope of Boyden Observatory, and 4 exposures made with the Curtis Schmidt telescope at Cerro Tololo were used for an analysis of the kinematical behaviour of the tail material. The aberration angles between the tail and the comet-sun-line were brought into connection with the solar wind parameters taken from satellite observations and geophysical data.

Infrared measures of comets. T. Lee. Comets. Proc. Tucson Conference 1970, (see 012.010), p. 20 - 22 (1972).

Spectrophotometry of comet 1969g (Tago-Sato-Kosaka). C. R. O'Dell. Comets. Proc. Tucson Conference 1970, (see 012.010), p. 122 (1972). – Abstract.

Spectrophotometry of comet Tago-Sato-Kosaka, 1969 IX. O. V. Dobrovol'skij, O. Mamadov, G. P. Chernova. Dokl. AN TadzhSSR, Vol. 15, No. 16, p. 18 - 21 (1972). In Russian. – Abstr. in Referativ. Zhurn. 51. Astron., 1.51.358 (1973).

Evidence from polarization. See Abstr. 103.122.

103.122 Comet 1965 VIII Ikeya-Seki

Infrared observations of comets Ikeya-Seki (1965f) and Bennett (1969i). J. A. Westphal. Comets. Proc. Tucson Conference 1970, (see 012.010), p. 23 - 31 (1972).

Evidence from polarization. T. Gehrels. Comets. Proc. Tucson Conference 1970, (see 012.010), p. 152 (1972).

A spectacular comet: Ikeya-Seki in 1965. Mercury, (Journ. Astron. Soc. Pacific), Vol. 1, No. 4, p. 19 (1972).

103.123 Comet 1970 III Kohoutek

A note on the spectrum of comet Kohoutek 1970 III. J. Bouška, A. Mrkos. Acta Univ. Carolinae Math. Phys., Vol. 12, 65 = Astron. Inst. Charles Univ., Praha, Publ. No. 65 (1971).

103.124 Comet 1972j Kojima

New comet Kojima 1972j. N. Kojima, B. G. Marsden, D. R. L. Jones. British Astron. Ass., Circ. No. 545 (1972).

Comet Kojima. N. Kojima. IAU Circ., No. 2457 (1972).

Comet Kojima (1972j).   N. Kojima, H. Kosai.
T. Seki, K. Tomita, K. Hurukawa.
IAU Circ., No. 2459 (1972).

Comet Kojima (1972j).
H. L. Giclas, M. L. Kantz, G. Araya, B. M. Blanco.
IAU Circ., No. 2461 (1972).

Comet Kojima (1972j).   N. Kojima, T. Seki,
T. Furuta, T. Urata, K. Tomita, H. L. Giclas, M. L. Kantz,
K. Suzuki, K. Hurukawa, B. G. Marsden.
IAU Circ., No. 2462 (1972).

Comet Kojima (1972j).
N. Kojima, T. Seki, T. Urata, K. Hurukawa.
IAU Circ., No. 2464 (1972).

Comet Kojima (1972j).
B. Milet, T. Seki, A. F. Jones.
IAU Circ., No. 2466 (1972).

Comet Kojima (1972j).   K. Suzuki, T. Urata,
B. Milet, T. Seki, J. Gibson, U. Gibson.
IAU Circ., No. 2472 (1972).

Comet Kojima (1972j).
T. Seki, B. Milet, H. Hatanaka, T. Urata.
IAU Circ., No. 2475 (1972).

Comet Kojima (1972j).
Japan Astron. Study Ass., Circ. 278 - 279 (1972). In Japanese.

New comet Kojima (1972j).
Kometn. Tsirk., *Kiev*, No. 138 (1972).   In Russian.

Comet Kojima, 1972j.
Kometn. Tsirk., *Kiev*, No. 139 (1972).   In Russian.

**Observations by the Alpine expedition of the State
Astronomical Sternberg Institute.**
V. M. Kovalenko, I. N. Potapov, O. N. Kovalenko.
Kometn. Tsirk., *Kiev*, No. 139 (1972). In Russian.

103.125   Comet 1964 IX Everhart

Definitive orbit of comet Everhart (1964 IX).
L. A. Markelov.
Kometn. Tsirk., *Kiev*, No. 138 (1972).   In Russian.

103.126   Comet 1972k Gehrels

Comet Gehrels (1972k).
T. Gehrels, R. Adams, C. Vesely, R. Sather, B. G. Marsden.
IAU Circ., No. 2460 (1972).

Periodic comet Gehrels (1972k).
C. Vesely, R. Sather, T. Gehrels, B. G. Marsden.
IAU Circ., No. 2463 (1972).

Periodic comet Gehrels (1972k).
R. E. McCrosky, C. Y. Shao.
IAU Circ., No. 2471 (1972).

New comet Gehrels, 1972k.
Kometn. Tsirk., *Kiev*, No. 139 (1972).   In Russian.

103.127   Comet 1972*l* Araya

Comet Araya (1972*l*).   G. Araya, W. E. Kunkel.
IAU Circ., No. 2469 (1972).

Comet Araya (1972*l*).
B. M. Blanco, G. Araya, W. Kunkel.
IAU Circ., No. 2470 (1972).

Comet Araya (1972*l*).
B. M. Blanco, G. Araya, B. G. Marsden.
IAU Circ., No. 2473 (1972).

New comet Araya, 1972*l* on the southern sky.
Kometn. Tsirk., *Kiev*, No. 139 (1972).   In Russian.

103.128   Comet 1962 V Tuttle-Giacobini-Kresák

Periodic comet Tuttle-Giacobini-Kresák.
B. G. Marsden.
IAU Circ., No. 2458 (1972).

Periodic comet Tuttle-Giacobini-Kresák.
Kometn. Tsirk., *Kiev*, No. 139 (1972).   In Russian.

103.129   Comet 1971 V Toba

**Observations photographiques de la comète Toba
(1971a), effectuées en 1971 en Argentine à l'équatorial de la
Carte du Ciel de l'Observatoire National de Cordoba (CO) et
au grand télescope de Bosque Alegre (Observatoire National de
Cordoba) (BA).**   H. Debehogne.
Bull. Astron. Obs. Roy. Belgique, Vol. 8, 40 (1972).

103.130   Comet 1957 IV Schwassmann-Wachmann 1

Periodic comet Schwassmann-Wachmann 1.
Z. M. Pereyra.
IAU Circ., No. 2424 (1972).

**Possibility of observing outbursts of P/Schwass-
mann-Wachmann 1.**   M. Dryer.
IAU Circ., No. 2432 (1972).

Periodic comet Schwassmann-Wachmann 1.
E. Roemer, E. Weber, R. Barnes.
IAU Circ., No. 2439 (1972).

Periodic comet Schwassmann-Wachmann 1.
Z. M. Pereyra.
IAU Circ., No. 2440 (1972).

Periodic comet Schwassmann-Wachmann 1.
F. Dossin, J. P. Swings.
IAU Circ., No. 2464 (1972).

103.131   Comet 1972c Tempel 2

Periodic comet Tempel 2 (1972c).
IAU Circ., No. 2424 (1972).

Periodic comet Tempel 2 (1972c).
R. E. McCrosky, C. Y. Shao
IAU Circ., No. 2439 (1972).

Comet P/Tempel 2 (1972c).
Japan Astron. Study Ass., Circ. 277 (1972). In Japanese.

103.132 **Comet 1971e Shajn-Schaldach**

Periodic comet Shajn-Schaldach (1971e).
A. C. Danks.
IAU Circ., No. 2438 (1972).

103.133 **Comet 1971b Holmes**

Periodic comet Holmes (1971b).   E. Roemer.
IAU Circ., No. 2446 (1972).

103.134 **Comet 1968 III Wild**

Periodic comet Wild (1960 I).   B. G. Marsden.
IAU Circ., No. 2456 (1972).

103.135 **Comet 1879 I Brorsen**

Periodic comet Brorsen.   B. G. Marsden.
IAU Circ., No. 2470 (1972).

103.136 **Comet 1965 VII de Vico-Swift**

Periodic comet De Vico-Swift.   B. G. Marsden.
IAU Circ., No. 2474 (1972).

103.137 **Comet 1972g Neujmin 3**

Comet P/Neujmin (1972g).
Japan. Astron. Study Ass., Circ. 277 (1972). In Japanese.

103.138 **Comet 1971f Tsuchinshan 1**

Comet P/Tsuchinshan 1 (1971f).
Japan Astron. Study Ass., Circ. 276 (1972). In Japanese.

103.139 **Comet 1972b Grigg-Skjellerup**

Comet P/Grigg-Skjellerup (1972b).
Japan. Astron. Study Ass., Circ. 276 (1972). In Japanese.

104 Meteors, Meteor Streams

104.001 The effect of meteoric ion processes on radio studies of meteoroids. W. J. Baggaley.
Monthly Notices Roy. Astron. Soc., Vol. 159, 203 - 217 (1972).
   The processes governing the loss of ionization are examined for the case where meteoric ions remain in the atomic form. Numerical solutions are presented demonstrating the variations expected in electron and negative ion concentrations during meteoric diffusion.

104.002 Les Giacobinides du 9 octobre 1972. P. Muller.
L'Astronomie, 86ᵉ année, p. 354 (1972).

104.003 Giacobinid meteor spectra. P. M. Millman.
Journ. Roy. Astron. Soc. Canada, Vol. 66, 201 - 211 (1972).
   Measures of ten Giacobinid spectra photographed in 1946 suggest that these meteoroids have a relative distribution by weight of the four elements iron, magnesium, calcium and sodium that is very similar to the carbonaceous chondrites and to the olivine-bronzite chondrites.

104.004 Giacobinid meteor spectra. P. M. Millman.
Journ. Roy. Astron. Soc. Canada, Vol. 66, 221 (1972). - Abstr. Canadian Astron. Soc.

104.005 The use of the Halphen-Goryachev method in the study of the evolution of the orbits of the Quadrantid and δ Aquarid meteor streams. A. F. Zausaev.
IAU Symposium No. 45, (see 012.003), p. 441 (1972).

104.006 Evolution séculaire des orbites de particules météoriques. J. Delcourt.
IAU Symposium No. 45, (see 012.003), p. 447 - 453 (1972).

104.007 Deformation of a meteor stream caused by an approach to Jupiter.
B. Yu. Levin, A. N. Simonenko, L. M. Sherbaum.
IAU Symposium No. 45, (see 012.003), p. 454 - 461 (1972).

104.008 Orbital evolution of the α Virginid and α Capricornid meteor streams.
E. I. Kazimirchak-Polonskaya, N. A. Belyaev, A. K. Terent'eva.
IAU Symposium No. 45, (see 012.003), p. 462 - 471 (1972).

104.009 Theoretical cometary radiants and the structure of meteor streams. E. N. Kramer.
IAU Symposium No. 45, (see 012.003), p. 472 - 481 (1972).

104.010 Statistics of the orbits of meteor streams and comets. V. N. Lebedinets, V. N. Korpusov, A. K. Sosnova.
IAU Symposium No. 45, (see 012.003), p. 491 - 497 (1972).

104.011 On the production of meteor streams by cometary nuclei. L. A. Katasev, N. V. Kulikova.
IAU Symposium No. 45, (see 012.003), p. 498 - 502 (1972).

104.012 Observations of meteors by means of TV technique. M. Begkhanov, H. Gulmedov, M. Polyakov, S. Mukhammed-Nasarov.
Astron. Tsirk., No. 690, p. 6 - 7 (1972). In Russian.

104.013 On the utilization of TV-equipment for meteor observations. M. Begkhanov, Kh. Gul'medov.
Izv. AN TurkmSSR. Ser. fiz.-tekhn., khim. i geol. n., 1972, No. 2, p. 125 - 128. In Russian. – Abstr. in Referativ. Zhurn. 51. Astron., 8.51.310 (1972).

104.014 Results of radar measurements of meteor train drifts in Turkmenistan in 1969 - 1970.
Kh. D. Gul'medov, G. G. Zaevskij, S. S. Takhaudinova, A. A. Khamzin.
Izv. AN TurkmSSR. Ser. fiz.-tekhn., khim. i geol. n., 1972, No. 1, p. 124 - 125. In Russian. – Abstr. in Referativ. Zhurn. 51. Astron., 8.51.312 (1972).

104.015 Perseid radiants in 1969. O. P. Batylova, N. I. Bondar', N. K. Kremneva, V. V. Martynenko, V. V. Frolov.
Astron. vestn., Vol. 6, 63 - 69 (1972). In Russian. – Abstr. in Referativ. Zhurn. 51. Astron., 8.51.315 (1972).

104.016 What gebeurt er's nachts aan de hemel tussen 20 en 30 april? B. Apeldoorn.
Hemel en Dampkring, Vol. 70, 236 - 238 (1972).

104.017 On the age of meteor streams. I. S. Astapovich.
Astron. Tsirk., No. 706, p. 6 - 8 (1972). In Russian.

104.018 Results of high-speed modelling for the interpretation of data on direct investigations of meteor matter. T. N. Nazarova, A. K. Rybakov, V. F. Agarkov, R. V. Lebedeva, L. G. Lukashov, A. A. Samodurov.
Kosmich. Issled., Vol. 10, 596 - 599 (1972). In Russian.

104.019 On the calculation of the angle of encounter of a meteor particle with clamp and piezoelectric counters for estimating the spatial density of meteor matter.
T. N. Nazarova.
Kosmich. Issled., Vol. 10, 600 - 603 (1972). In Russian.

104.020 The origin of short-period meteors. L. Kresak.
Astron. vestn., Vol. 6, 73 - 76 (1972). In Russian.
Abstr. in Referativ. Zhurn. 51. Astron., 9.51.341 (1972).

104.021 Mean-square error determination of the velocity of a radio meteor.
A. A. D'yakov, B. L. Kashcheev.
Astron. vestn., Vol. 6, 99 - 103 (1972). In Russsian. – Abstr. in Referativ. Zhurn. 51. Astron., 9.51.342 (1972).

104.022 α-Lyrid meteor shower in 1969. N. I. Bondar', N. M. Kremneva, V. V. Martynenko, V. V. Frolov.
Astron. vestn., Vol. 6, 127 - 130 (1972). In Russian. – Abstr. in Referativ. Zhurn. 51. Astron., 9.51.345 (1972).

104.023 An international centre for meteor observations. K. B. Hindley.
Journ. British Astron. Ass., Vol. 82, 459 - 463 (1972). – Report of the Meteor Section of the British Astron. Ass.

104.024 Pre-atmospheric dimensions of meteoric bodies. A. K. Lavrukhina, T. A. Ibraev.
Astron. vestn., Vol. 6, 104 - 111 (1972). In Russian. – Abstr. in Referativ. Zhurn. 51. Astron., 10.51.319 (1972).

104.025 Evidence from stream meteoroids. R. McCrosky.
Comets. Proc. Tucson Conference 1970, (see 012.010), p. 145 - 151 (1972).

104.026 A bright meteor in Sagittarius. D. Nicell.

Monthly Notes Astron. Soc. Southern Africa, Vol. 31, 134 (1972). – Letter.

104.027 The meteoroid influx and the maintenance of the solar system dust cloud. D. W. Hughes.
Planet. Space Sci., Vol. 20, 1949 - 1959 (1972).

The total mass influx onto the earth's surface has been calculated over the mass range $10^{-13}-10^{10}$ g by using cumulative influx rate data, Interpolation formulae have been fitted to this data and the resulting influx is found to vary between 3×10^8 and 5×10^{13} g (earth surface)$^{-1}$ yr^{-1} depending on which data are used to obtain the formulae. By considering the latest measurements of comet masses and estimates for the rate of cometary decay, the mass influx to the solar system dust cloud has been calculated.

104.028 Photometry of meteors by the equidensity method. V. V. Benyuch, A. N. Shaido.
Vestn. Kiev. Un-ta, Ser. Astron., No. 14, p. 28 - 32 (1972). In Russian.

The equidensity method of photometry of meteors is proposed. The results of the photometry of two meteor flares are given.

104.029 On the question of interpretation of spectra of faint meteors. V. V. Kalenichenko.
Vestn. Kiev. Un-ta, Ser. Astron., No. 14, p. 33 - 38 (1972). In Russian.

Cases of direct excitation and re-charge in excited states under collision of atoms and ions of coma are considered. The expressions for population of excited levels and intensities of meteor spectral lines have been obtained.

104.030 Increase of some meteor spectral lines by excitation transfer. V. V. Kalenichenko.
Vestn. Kiev. Un-ta, Ser. Astron., No. 14, p. 39 - 42 (1972). In Russian.

It is shown that the excitation transfer from metastable states of OI and NI can notably increase the lines MgI λ 2852 Å, SiI λ 3020 Å, NaI λ 3302 Å, CaII λλ 3968 Å, 3933 Å (H and K) in meteor spectra.

104.031 On the application of the heat conductivity equation to the calculation of densities of meteor bodies. V. G. Kruchinenko, S. S. Tryashin.
Vestn. Kiev. Un-ta, Ser. Astron., No. 14, p. 43 - 47 (1972). In Russian.

The article deals with the determination of meteor body densities on the basis of the solution of the heat conductivity equation with taking into account the energy liberated by evaporation.

104.032 The beginning of intensive vaporization of meteor bodies. V. G. Kruchinenko.
Vestn. Kiev. Un-ta, Ser. Astron., No. 14, p. 48 - 51 (1972). In Russian.

The article deals with the determination of the heights of beginning intensive vaporization of meteor bodies. It is shown that the calculated heights conform to the results of observations for small meteor bodies which heat up throughout before the beginning of vaporization. There is no such conformity for bright meteors: calculated heights as a rule are less then observed ones.

104.033 Evolution of meteor streams with different inclinations of their orbits. L. M. Sherbaum.
Vestn. Kiev. Un-ta, Ser. Astron., No. 14, p. 52 - 58 (1972). In Russian.

The evolution of two hypothetical filament-like meteor streams which differ in inclination ($i = 2°$, $i = 30°$) is considered. 6 meteor particles being chosen on orbit of each

meteor stream, the differential equations of their motions were integrated by Cowell's method taking into account the perturbations of Jupiter and Saturn.

104.034 On the nature of polarization phenomena at scattering of radio waves on overdense meteor trails. I. V. Bajrachenko.
Vestn. Kiev. Un-ta, Ser. Astron., No. 14, p. 59 - 63 (1972). In Russian.

An analysis of approximate values of reflection coefficients for overdense meteor trails has shown that the polarization phenomena do correspond neither to plasma resonance of the type of underdense trails nor to the metallic model. The possible mechanism of appearance of resonance is an interaction of the radio wave with the surface plasma waves arising between the "metallic region" of the trail and the boundary "plasma-free space". A resonance of similar kind was observed in laboratory for scattering of radio waves by gas discharge tubes.

104.035 On the influence of intermediate meteor trails upon the results of measurements of initial radii. R. I. Moisya.
Vestn. Kiev. Un-ta, Ser. Astron., No. 14, p. 71 - 75 (1972). In Russian.

An analysis of the influence of intermediate meteor trails upon the results of initial trails radii measurements is carried out. It is shown, that if the radio echo from underdense meteor trails were not selected correctly, the mean initial radius would be 30% less than the real value.

104.036 Investigation of the radial electron distribution in a meteor trail. G. I. Kolomiets, E. I. Fialko, R. I. Moisya.
Vestn. Kiev. Un-ta, Ser. Astron., No. 14, p. 76 - 78 (1972). In Russian.

The method of experimental determination of the radial electron distribution law of an ionized meteor trail is described. The experimental data show that the electron distribution in meteor trails is close to Gaussian.

104.037 Distribution of kinetic energies of meteor bodies taking into account the physical factor. E. I. Fialko, V. F. Romanyuk.
Vestn. Kiev. Un-ta, Ser. Astron., No. 14, p. 79 - 82 (1972). In Russian.

Distribution of kinetic energies of meteor bodies with taking account dependence of minimum mass velocity is examined in this work.

104.038 Light distribution in photographic meteors by means of multicolour photometry. M. Hajduková.
Bull. Astron. Inst. Czechoslovakia, Vol. 23, 350 - 356 (1972).

The radiation of meteors is studied on the basis of their photographic magnitudes determined through different colour filters at the Skalnaté Pleso Observatory. The colour indices of meteors defined by the comparison of meteor radiation in different spectral regions are discussed. The results confirm Ceplecha's suggestion of a general reddening of meteors towards fainter magnitudes.

104.039 Shock waves and flares by meteors. J. Rajchl.
Bull. Astron. Inst. Czechoslovakia, Vol. 23, 357 - 366 (1972).

A new relation for the shock wave interaction onset by meteors is deduced. Comparison with observations leads to the principal result that the onset of this type of interaction corresponds to the flare occurrence by photographic meteors. Possible mechanisms leading to the observed sudden increase of the CaII emission in flares are shortly discussed.

104.040 On the source of the 3840 Å persistent emission by meteors. J. Rajchl.
Bull. Astron. Inst. Czechoslovakia, Vol. 23, 366 - 368 (1972).
From the fundamental properties of the observed long-term emission of about 3840 Å and others, and from laboratory experiments, it is concluded that the Herzberg metastable system of O_2 is the more probable source of these emissions than the Vegard–Kaplan metastable bands of N_2.

104.041 Die Perseiden 1972. J. Hellwig.
SuW, Vol. 11, 348 - 349 (1972).

104.042 Weitere Beobachtungen der Perseiden.
W. Kramer.
SuW, Vol. 11, 349 (1972).

104.043 Circumterrestrial meteor dust.
T. N. Nazarova.
Byull. Abastumansk. Astrofiz. Obs., No. 41, (see 012.018), p. 115 - 116 (1972). In Russian.

104.044 Meteoros. J. Rubi Garza.
El Universo, Vol. 26, 121 - 123 (1972).

104.045 De Boötiden 1973; iets voor binoculairbezitters.
B. Apeldoorn.
Hemel en Dampkring, Vol. 70, 352 - 353 (1972).

104.046 The density of meteor bodies. V. V. Benyukh.
Kometn. Tsirk., *Kiev*, No. 138 (1972). In Russian.

104.047 Observations of the Draconid meteor stream at the Ussuriisk Solar Station.
A. V. Mazhuga, V. S. Mazhuga, V. A. Golubev.
Kometn. Tsirk., *Kiev*, No. 138 (1972). In Russian.

104.048 Report of meteor observations in 1969.
H. Kobayashi.
Mem. Japan Astron. Study Ass., No. 17, Vol. 5, 7 - 8 (1971). In Japanese.

104.049 Statistical investigation of orbital elements of meteors. Part II.
L. Adamiak, S. Gaska, K. Wiercioch.
Stud. Soc. Sci. Torunensis, Sectio F (Astron.), Vol. 5, 21 - 28 = Bull. Astron. Obs. Toruń No. 49/III (1972).
The results of statistical investigation of orbital elements of 1530 radio-meteors are given. These are compared with those for photographic meteors.

104.050 Draconid meteors 1972.
Y. Kozai, K. Simmons.
IAU Circ., No. 2451 (1972).

104.051 Draconid meteors 1972.
P. D. Maley, I. Saulietis.
IAU Circ., No. 2452 (1972).

104.052 The influence of selection on the observed distribution of orbital parameters of radio meteors.
O. I. Bel'kovich, M. F. Lagutin, D. M. Smagin.
Astron. vestn., Vol. 6, 180 - 185 (1972). In Russian. – Abstr. in Referativ. Zhurn. 51. Astron., 1.51.367 (1973).

104.053 On the ablation coefficient and maximum brightness of a meteor. I. S. Shestaka.
Astron. vestn., Vol. 6, 186 - 194 (1972). In Russian. – Abstr. in Referativ. Zhurn. 51. Astron., 1.51.368 (1973).

104.054 Visual observations of meteors in January – September.
Y. Yabu, T. Mōri, H. Yamamoto, M. Hayakawa.
Provisional Rep. Nippon Meteor Soc., Nos. 81 - 83, 29 + 39 + 54 pp. (1971).

104.055 Conditions of an encounter of the Leonid meteor stream with the earth in apparitions 1898 - 2000.
I. S. Astapovich, A. K. Terenteva.
Problems of cosmic physics. Vyp. (No.) 7, (see 003.028), p. 100 - 107 (1972). In Russian.
On the basis of results of a numerical integration of the differential equations of motion for the 11 meteor groups of the Leonid stream the conditions of an encounter of these groups with the earth over the apparitions 1898 - 1900, 1930 - 1934, 1963 - 1967, 1996 - 2000 were studied. A forecast has been made for the times of maximum activity of the Leonid shower in the years 1997 - 2000.

104.056 Formulation of the problem of formation and decay of meteor trains in terms of the kinetic theory.
Yu. I. Portnyagin.
Trudy In-t ehksperim. meteorol. Gl. upr. gidrometeorol. sluzhby pri Sov. Min. SSSR, 1972, vyp. (No.) 1 (34), p. 64 - 70. In Russian. – Abstr. in Referativ. Zhurn. 51. Astron., 2.51.405 (1973).

104.057 On the heights of appearance and disappearance of meteors. Yu. I. Portnyagin.
Trudy In-t ehksperim. meteorol. Gl. upr. gidrometeorol. sluzhby pri Sov. Min. SSSR, 1972, vyp. (No.) 1 (34), p. 70 - 76. In Russian. – Abstr. in Referativ. Zhurn. 51. Astron., 2.51.406 (1973).

104.058 Limiting resolving power of meteor radars.
D. M. Smagin, A. Kh. Khanberdiev.
Izv. AN TurkmSSR. Ser. fiz.-tekhn. khim. i geol. n., 1972, No. 5, p. 53 - 56. In Russian. – Abstr. in Referativ. Zhurn. 51. Astron., 2.51.407 (1973).

104.059 Reception recording radar device for measuring meteor train drifts.
A. D. Orlyanskij, B. I. Petrov.
Trudy In-t ehksperim. meteorol. Gl. upr. gidrometeorol. sluzhby pri Sov. Min. SSSR, 1972, vyp. (No.) 1 (34), p. 80 - 88. In Russian. – Abstr. in Referativ. Zhurn. 51. Astron., 2.51.408 (1973).

104.060 Streams of radio meteors.
V. N. Lebedinets, V. N. Korpusov, A. K. Sosnova.
Trudy In-t ehksperim. meteorol. Gl. upr. gidrometeorol. sluzhby pri Sov. Min. SSSR, 1972, vyp. (No.) 1 (34), p. 88 - 171. In Russian. – Abstr. in Referativ. Zhurn. 51. Astron., 2.51.409 (1973).

104.061 Determination of the attachment coefficient of electrons to neutral particles of meteor matter.
R. Sh. Bibarsov, P. B. Babadzhanov.
Dokl. AN TadzhSSR, Vol. 15, No. 6, p. 29 - 31 (1972). In Russian. – Abstr. in Referativ. Zhurn. 51. Astron., 2.51.410 (1973).

104.062 Some data on meteors of the Perseid shower.
P. B. Babadzhanov, V. S. Getman.
Dokl. AN TadzhSSR, Vol. 15, No. 7, p. 22 - 24 (1972). In Russian. – Abstr. in Referativ. Zhurn. 51. Astron., 2.51.413 (1973).

104.063 On the dependence of the height of appearance of sporadic meteors on their geocentric velocity.
E. N. Kramer.
Astron. Tsirk., No. 730, p. 1 - 3 (1972). In Russian.

104.064   Meteor stream of Draconids in 1972.
          A. K. Terent'eva.
Astron. Tsirk., No. 742, p. 1 (1972). In Russian.

104.065   Apparent short-duration meteor shower.
          M. Buhagiar.
Journ. Astron. Soc. Western Australia, 1972 November, p. 7 - 8.

          A great daylight fireball over the northwest.
Sky Telescope, Vol. 44, 269 - 272 (1972).

          The Perseids make a good showing.
Sky Telescope, Vol. 44, 340 - 343 (1972).

          The influence of cosmic dust on twilight phenomena.
See Abstr. 082.134.

          Manifestations optiques des aérosols météoriques.
I. – Orionides 1970.   See Abstr. 082.189.

          Secular perturbations on the minor bodies of the solar system.   See Abstr. 098.011.

          On the possible common origin of minor planets, comets, and meteors.   See Abstr. 098.012.

          On the relation between some orbital elements of minor planets and meteors.   See Abstr. 098.043.

          Formation of comets in meteor streams.
See Abstr. 102.049.

          On the dividing line between cometary and asteroidal orbits.   See Abstr. 102.050.

          Investigation of the motion of periodic comet Giacobini-Zinner and the origin of the Draconid meteor showers of 1926, 1933 and 1946.   See Abstr. 103.107.

          Periodic comet Tempel-Tuttle and the Leonid meteor shower.   See Abstr. 103.110.

          Gas dynamics of the flight and explosion of meteorites.   See Abstr. 105.025.

          Results of a 1970 Geminid dust particle rocket experiment and analysis of OGO 3 dust particle velocity measurements.   See Abstr. 105.053.

          On the relation between comets and meteoroids.
See Abstr. 107.007.

          Numerical simulation of jetstreams. I: The three-dimensional case.   See Abstr. 107.010.

Errata

104.901   Erratum: "On the correlation of hourly rates of radio meteors recorded at different sensitivity levels" [Astron. Tsirk., No. 664, p. 5 - 7 (1971). In Russian].
          V. N. Donij, E. I. Fialko.
Astron. Tsirk., No. 680, p. 7 - 8 (1972). In Russian.

105   Meteorites, Meteorite Craters

105.001   Radiation effects in ^{129}I–^{129}Xe dating of Bjurböle chondrules by neutron irradiation.
B. Srinivasan, E. C. Alexander, Jr., O. K. Manuel.
Icarus, Vol. 16, 571 - 576 (1972).

105.002   Argon 37/argon 39 activity ratios in meteorites and the spatial constancy of the cosmic radiation.
F. Begemann.
Journ. Geophys. Res., Vol. 77, 3650 - 3659 (1972).

Argon 37 and argon 39 have been measured in a bulk sample and a metal-rich fraction of the Lost City stone meteorite. The activity ratio of 0.75 ± 0.10 is considerably higher than the previously published values. Consequently the radial heliocentric gradient of the galactic cosmic radiation is found to be much smaller. The data available for the activity ratios in the metal of different meteorites are discussed.

105.003   An October influx of submicron particles into the lower stratosphere.
E. K. Bigg, Z. Kviz, W. J. Thompson.
Journ. Geophys. Res., Vol. 77, 3916 - 3923 (1972).

The particles have modal diameters of about 0.05 μm but are often joined together to form chains. One of the possible sources of the particles is the infall of cometary or meteoritic debris.

105.004   Bismuth in stony meteorites and standard rocks.
P. M. Santoliquido, W. D. Ehmann.
Geochim. Cosmochim. Acta, Vol. 36, 897 - 902 (1972).

Bismuth has been determined by alpha counting of the ^{210}Po daughter activity of the ^{210}Bi formed by thermal neutron activation. In the present work, we have attempted to complete the study of the Bi abundance pattern in meteorites by analyzing previously omitted chondrite classes as well as a number of achondrites and separated meteoritic phases.

105.005   Iron transport in chondrites: Evidence from the Warrenton meteorite.   J. F. Kerridge.
Geochim. Cosmochim. Acta, Vol. 36, 913 - 916 (1972).

Veins of iron enrichment in an olivine chondrule in Warrenton are attributed to migration of iron prior to final consolidation. Aqueous transport along cracks is suggested as the most plausible mechanism responsible for such migration, with possible implications for 'metamorphic' theories of iron distribution in chondrites.

105.006   Niobium in meteorites.
A. L. Graham, B. Mason.
Geochim. Cosmochim. Acta, Vol. 36, 917 - 922 (1972).

Spark source mass spectrographic analyses of six chondrites and six achondrites for niobium are reported. The geochemical behaviour of niobium shows a relative coherence between the Ta, Ti, Zr contents of meteorites and lunar samples and a lack of correlation with Ca and Al.

105.007   Activity–composition relations in the fayalite-forsterite solid solution between 900° and 1300° C at low pressures.   R. J. Williams.
Earth Planet. Sci. Letters, Vol. 15, 296 - 300 (1972).

The purpose of this work is to combine the available experimental data on the olivine solid solutions into an equation from which the activity of fayalite in olivine solid solution can be calculated over a range of geologically interesting conditions. The minimum oxygen fugacities in equilibrium with the olivines from basaltic achondrites, type G chondrites and lunar materials have been calculated using this equation and the formation constant of fayalite.

105.008   Ucera meteorite: Determination of differential atmospheric heating using its natural thermoluminescence.   J. E. Vaz.
Meteoritics, Vol. 7, 77 - 86 (1972).

The purpose of this study was to analyze the variations of the natural thermoluminescence as a function of depth from the fusion crust of a stony meteorite, to determine if the results of the thermoluminescence measurements could be related to the atmospheric history of the meteorite.

105.009   The Seoni chondrite.
T. E. Bunch, A. P. Mall, C. F. Lewis.
Meteoritics, Vol. 7, 87 - 95 (1972).

The Seoni (India) chondrite is an H6 group ordinary chondrite that contains olivine, orthopyroxene, clinopyroxene, plagioclase, together with chromite, troilite, kamacite, taenite, chlorapatite, and whitlockite. It fell about mid-day on January 16, 1966.

105.010   Savonoski crater, Alaska: A possible meteorite impact structure.
B. M. French, E. H. Muller, P. L. Ward.
Meteoritics, Vol. 7, 97 - 107 (1972).

Savonoski crater, a small basin about 1700 feet in diameter, is located on a ridge of gently-dipping jurassic sandstone in Katmai National Monument, southwestern Alaska. It was formed either by volcanic processes or meteorite impact prior to the end of the latest glaciation in the region.

105.011   Mineralogical and chemical researches on L-chondrites: Girgenti.
G. R. Levi-Donati, E. Jarosewich.
Meteoritics, Vol. 7, 109 - 125 (1972).

Girgenti is a severally metamorphosed stone, whose total iron content (23.5 %) is somewhat higher than the average for hypersthene chondrites.

105.012   On investigating the shock history of ataxites.
R. Knox, Jr.
Meteoritics, Vol. 7, 127 - 129 (1972).

It is suggested that transmission electron microscopy studies of ataxites and shock-loaded artificial nickel-iron alloys could develop criteria whereby the shock history of this class of meteorites might be established.

105.013   Elemental abundances in stone meteorites.
R. A. Schmitt, G. G. Goles, R. H. Smith, T. W. Osborn.
Meteoritics, Vol. 7, 131 - 213 (1972).

Abundances of Na, Al, Sc, Cr, Mn, Fe, Co and Cu have been measured by instrumental neutron activation analyses of 103 chondrites and 17 achondrites. The results are given in many tables.

105.014   The Meteoritical Bulletin, No. 51.
R. S. Clarke, Jr. (Editor).
Meteoritics, Vol. 7, 215 - 232 (1972).

New falls and discoveries of meteorites are given in this facsimile copy of the Bulletin printed by the U.S.S.R. Academy of Sciences.

105.015   Formation of ordinary chondrites.   J. T. Wasson.
Rev. Geophys. Space Phys., Vol. 10, 711 - 759 (1972).

In this paper I review the evidence for four independent types of fractionation that are recorded in the properties of the ordinary chondrites, as well as evidence on other properties. It appears that most of the evidence can best be understood in terms of processes that occurred in the primitive solar nebula.

105.016 **Meteoritenfälle in Deutschland.** Zeitliche Verteilung der Fälle und Rückschlüsse daraus auf ihre Herkunft – Hypothesen über die Entstehung der Meteorite sowie weitere Fälle und Funde aus Deutschland.
H. Eisenlohr.
SuW, Vol. 11, 216 - 219 (1972).

105.017 **Double moldavites in southern Bohemia.**
V. Bouška, R. Rost.
Science, Vol. 177, 519 - 520 (1972).

105.018 **Cosmogenic radionuclides in stones and meteorite orbits.** A. K. Lavrukhina, G. K. Ustinova.
Earth Planet. Sci. Letters, Vol. 15, 347 - 360 (1972).

Depth distributions of 14 cosmic-ray produced radionuclides in chondrites and achondrites of different sizes have been calculated. The dependence of depth distributions of different radionuclides on compositions and pre-atmospheric sizes of meteorites, on temporal and spatial variations of cosmic rays has been investigated. The proposed method for calculating the sizes of meteorite orbits combined with the method of visible radiants makes it possible to obtain all orbit parameters (a, e, i, q, q') for the meteorites with known atmospheric trajectories.

105.019 **Complex irradiation history of the Weston chondrite.**
L. Schultz, P. Signer, J. C. Lorin, P. Pellas.
Earth Planet. Sci. Letters, Vol. 15, 403 - 410 (1972).

Helium, neon and argon were analysed in seven different light inclusions of the gas-rich chondrite Weston. We investigate the history of gas rich meteoritic breccias by a study of the lithic fragments therein imbedded. This goal is approached by a combination of noble gas analysis and track counting. With the application of the two corroborative methods to the study of noble-gas-rich meteorites we aim particularly at the pre-compaction history of the meteoritic material.

105.020 **Crystalline inclusions in a Muong Nong-type indochinite.** B. P. Glass.
Earth Planet. Sci. Letters, Vol. 16, 23 - 26 (1972).

Zircon, corundum, rutile, monazite and quartz have been recovered from a Muong Nong-type indochinite. X-ray asterism studies indicate that all of the minerals are shocked. The presence of these minerals in Muong Nong-type tektites seems to strengthen the case for a terrestrial origin for tektites.

105.021 **Isotopic compositions of rare gases in the carbonaceous chondrites Mokoia and Allende.**
O. K. Manuel, R. J. Wright, D. K. Miller, P. K. Kuroda.
Geochim. Cosmochim. Acta, Vol. 36, 961 - 983 (1972).

The isotopic compositions have been measured mass spectrometrically for neon, argon, krypton and xenon released from the carbonaceous chondrites Mokoia and Allende in stepwise heating experiments. A marked enrichment of Xe^{129} due to the decay of extinct nuclide I^{129} was observed in both meteorites. The two different gases being considered here are most likely the so-called solar and planetary rare gases, whose isotopic compositions are quite different from each other.

105.022 **Cosmic iron on earth?** M. I. Kalganov.
Priroda, No. 7.72, p. 64 - 67 (1972). In Russian.

105.023 **Magnetism of meteorites and lunar rocks.**
E. G. Gus'kova, I. D. Vetoshkin, É. S. Gorshkov, M. A. Grabovskij, O. N. Zherdenko, V. I. Pochtarev.
Izv. AN SSSR, Fiz. Zemli, 1972, No. 4, p. 3 - 9. In Russian.
Abstr. in Referativ. Zhurn. 51. Astron., 8.51.262 (1972).

105.024 **The chemical classification of iron meteorites. VI. A reinvestigation of irons with Ge concentrations lower than 1 ppm.**
R. Schaudy, J. T. Wasson, V. F. Buchwald.
Icarus, Vol. 17, 174 - 192 (1972).

We report Ni, Ga, Ge, and Ir concentrations for 65 irons with Ge concentrations less than 1 ppm, including 36 members of chemical group IVA and 10 members of group IVB.

105.025 **Gas dynamics of the flight and explosion of meteorites.**
V. P. Korobeinikov, P. I. Chushkin, L. V. Shurshalov.
Astronaut. Acta, Vol. 17, 339 - 348 (1972). – Presented at the international colloquium on the gasdynamics of explosions, Marseille, September 12 - 17, 1971.

105.026 **Meteoritic craters.** I. Kostov.
Priroda (NRB), Vol. 21, No. 2, p. 12 - 16 (1972).
In Bulgarian.

105.027 **Non-destructive neutron activation analysis of troilite of the Sikhote-Alin meteorite.**
A. Ljulj.
Latv. PSR Zinātņu Akad. vestis. Fiz. un tehn. zinātņu sěr., Izv. AN LatvSSR. Ser. fiz. i tekhn. n., 1972, No. 2, p. 14 - 17. In Russian. – Abstr. in Referativ. Zhurn. 51. Astron., 9.51.353 (1972).

105.028 **Cosmogenic radionuclides in the Allende and Murchison carbonaceous chondrites.** P. J. Cressy, Jr.
Journ. Geophys. Res., Vol. 77, 4905 - 4911 (1972).

Measurements were made of ^{22}Na, ^{26}Al, ^{54}Mn, and ^{60}Co produced by cosmic rays in five samples of the Allende C3 meteorite and in one specimen of Murchison (C2); ^{46}Sc, ^{48}V, ^{51}Cr and ^{57}Co were also measured in several of these samples.

105.029 **Analogy of nucleic acid in meteorites.**
A. P. Vinogradov, G. P. Vdovykin.
Dokl. Akad. Nauk SSSR, Ser. Mat. Fiz., Vol. 206, 563 - 565 (1972). In Russian.

105.030 **Chemical fractionation in iron meteorites and its interpretation.** E. R. D. Scott.
Geochim. Cosmochim. Acta, Vol. 36, 1205 - 1236 (1972).

105.031 **New data on selected Ivory Coast tektites.**
F. Cuttitta, M. K. Carron, C. S. Annell.
Geochim. Cosmochim. Acta, Vol. 36, 1297 - 1309 (1972).

The chemical composition of fourteen Ivory Coast (IVC) tektites is analyzed. The data show that compositional similarities between the IVC tektites and green or black Bosumtwi Crater glasses strongly support the hypothesis of a common impact origin–i.e. the Bosumtwi Crater site. Comparison of the IVC tektite composition with those of returned lunar materials do not support a lunar origin for the Ivory Coast tektites.

105.032 **Determination of Ni, Ga, and Ge in iron meteorites by X-ray fluorescence analysis.** S. J. B. Reed.
Meteoritics, Vol. 7, 257 - 262 (1972).

It is shown that Ga, Ge, and Ni can be determined by X-ray fluorescence on metallographic polished mounts sufficiently accurately for classification according to Wasson's chemical groups. Results are given for 45 irons, including some not previously classified.

105.033 **Magnetism of meteorites: A review of Russian studies.**
J. M. Herndon, M. W. Rowe, E. E. Larson, D. E. Watson.
Meteoritics, Vol. 7, 263 - 284 (1972).

105.034 **The Mundrabilla meteorite shower.**
J. R. DeLaeter.
Meteortitics, Vol. 7, 285 - 294 (1972).
A detailed examination of the Loongana Station, Mundrabilla and Premier Downs siderites has led to the conclusion that they are all members of the one meteorite shower.

105.035 **Rare earth and other abundances in the Murchison carbonaceous meteorite.**
D. L. Showalter, H. Wakita, R. A. Schmitt.
Meteoritics, Vol. 7, 295 - 301 (1972).
The abundances of 27 elements are reported for the Murchison meteorite. Nine of these elements (Al, Ca, Fe, Mn, Na, K, Cr, Co, and Sc) have been determined previously for different Murchison specimens. Abundances for 18 elements (In, Cd, V, Y, and REE) are new data.

105.036 **Laguna Guatavita: Not meteoritic, probable salt collapse crater.**   R. S. Dietz, J. F. McHone.
Meteoritics, Vol. 7, 303 - 307 (1972).

105.037 **The mineralogy of meteorites.**   B. Mason.
Meteoritics, Vol. 7, 309 - 326 (1972).
Review article.

105.038 **Evidence for vapor fractionation in the origin of chondrules.**   L. S. Walter, R. T. Dodd.
Meteoritics, Vol. 7, 341 - 352 (1972).

105.039 **Quartz and feldspar glasses produced by natural and experimental shock.**
D. Stöffler, U. Hornemann.
Meteoritics, Vol. 7, 371 - 394 (1972).

105.040 **A possible impact crater associated with Darwin glass.**   R. J. Ford.
Earth Planet. Sci. Letters, Vol. 16, 228 - 230 (1972).

105.041 **Electron microscope photographs of extraterrestrial particles.**   E. K. Bigg, Z. Kviz, W. J. Thompson.
Tellus, Vol. 23, 247 - 260 (1971).
With suitable collection methods it is possible to distinguish particles resident in the stratosphere from contaminants and to identify those of extraterrestrial origin. A catalogue of physical appearances under an electron microscope is presented which shows the most common types of such particles encountered in about 40 collections made from balloons in the course of three years.

105.042 **Cosmic abundance of iron and nature of primitive material in meteorites.**   J. Kerridge.
Nature, Vol. 239, 44 - 45 (1972).
It is suggested that type I carbonaceous meteorites consist of a primitive component, containing the condensable elements in essentially their cosmic proportions, with which has been mixed some secondary, fractionated material. The primitive component is identified with ill-defined, sub-microscopic phyllosilicate grains which comprise about 65 wt% of such a meteorite. This leads to a value for the primordial Fe/Si ratio of 0.53 ± 0.6.

105.043 **New meteorites.**   I. T. Zotkin.
Zemlya i Vselennaya, 1972, No. 5, p. 28 - 31.
In Russian.

105.044 **Uranium-lead chronology of chondritic meteorites.**
N. H. Gale, J. Arden, R. Hutchison.
Nature, Phys. Sci., Vol. 240, 56 - 57 (1972).
We present the lead and uranium concentrations for three L6 chondrites, measured on the identical sample on which the lead isotopic abundance measurements were made. This shows unequivocally for the first time that there is indeed a real problem in the uranium/lead evolution in meteorites, in that in each of these meteorites there is now insufficient uranium to support the lead isotopic composition.

105.045 **The Angra dos Reis (stone) mineral assemblage and the genesis of stony meteorites.**   R. Hutchison.
Nature, Phys. Sci., Vol. 240, 58 - 59 (1972).
This unique olivine-clinopyroxenite meteorite (angrite) has been intensively studied first because it contains evidence of the existence of extinct radionuclides and second because it is the meteorite richest in Ti and Ca and is therefore most similar to lunar rocks. Surprisingly, in spite of this interest, no analyses of Angra dos Reis (stone) minerals have yet been published, the data presented here remedy this deficiency.

105.046 **Cosmogenic rare gas production rates in chondritic meteorites.**   D. E. Fisher.
Earth Planet. Sci. Letters, Vol. 16, 391 - 395 (1972).
Comparison of chondrite radiation ages directly measured by either ^{3}He/^{3}H or ^{38}Ar/^{39}Ar with those calculated on the basis of the most recent estimation of ^{3}He and ^{21}Ne constant production rates leads to the conclusion that these constant rates are too high by a factor ~ 1.75.

105.047 **Search for extinct natural radioactivity of ^{205}Pb via thallium-isotope anomalies in chondrites and lunar soil.**   J. M. Huey, T. P. Kohman.
Earth Planet. Sci. Letters, Vol. 16, 401 - 412 (1972).
Thallium and ^{204}Pb contents were determined by stable-isotope—dilution analysis in 16 chondrites, one achondrite, and Apollo 11 and 12 lunar fines. The ^{205}Tl/^{203}Tl ratio was determined in most of the samples. The chondritic isochron slope for ^{205}Pb (13.8-my half-life) is ≤ 0.00009 (99% confidence level), corresponding to an interval of at least 60 my and possibly exceeding 120 my between the termination of s-process nucleosynthesis and the lead-thallium fractionations.

105.048 **The study of cosmic dust.**
M. Shima, H. Yabuki, A. Okada, S. Yabuki.
Space Research XII, (see 012.016), Vol. 1, 301 - 307 (1972).

105.049 **Meteor dust motion in the upper atmosphere and in the vicinity of the earth's orbit.**
V. N. Lebedinets, A. V. Manochina, V. B. Shushkova.
Space Research XII, (see 012.016), Vol. 1, 309 - 312 (1972).

105.050 **Craters in Surveyor 3 glass surfaces.**
P. W. Hodge, D. E. Brownlee, W. Bucher.
Space Research XII, (see 012.016), Vol. 1, 313 - 317 (1972).

105.051 **Lunar Explorer 35: 1970 dust particle data and shower related picogram ejecta orbits.**
W. M. Alexander, C. W. Arthur, J. L. Bohn, J. H. Johnson, B. J. Farmer.
Space Research XII, (see 012.016), Vol. 1, 349 - 355 (1972).

105.052 **Some peculiarities of cosmic dust distribution.**
L. V. Leontyev, A. V. Tarasov, I. A. Teryeshkin.
Space Research XII, (see 012.016), Vol. 1, 357 - 360 (1972).

105.053 **Results of a 1970 Geminid dust particle rocket experiment and analysis of OGO 3 dust particle velocity measurements.**
C. W. Arthur, W. M. Alexander, J. L. Bohn, J. H. Johnson, B. J. Farmer.
Space Research XII, (see 012.016), Vol. 1, 361 - 367 (1972).

105.054 **Laser radar studies of upper atmosphere dust layers and the relation of temporary increases in dust to cometary micrometeoroid streams.**   S. K. Poultney.

Space Research XII, (see 012.016), Vol. 1, 403 - 421 (1972).

105.055 Xenon in carbonaceous chondrites.
O. K. Manuel, E. W. Hennecke, D. D. Sabu.
Nature, Phys. Sci., Vol. 240, 99 - 101 (1972).
Carbonaceous chondrites contain two isotopically distinct components of trapped xenon which cannot be explained by the occurrence of nuclear or fractionation processes within these meteorites.

105.056 Shock-metamorphosed rocks and impactites of the Popigai meteorite crater.
V. L. Masajtis, T. V. Selivanovskaya.
Zap. Vses. mineralog. o-va, Vol. 101, 385 - 393 (1972). In Russian. – Abstr. in Referativ. Zhurn. 51. Astron., 12.51.293 (1972).

105.057 Chlorine, bromine, iodine, and uranium in tektites, obsidians, and impact glasses.
V. J. Becker, O. K. Manuel.
Journ. Geophys. Res., Vol. 77, 6353 - 6359 (1972).
This study on the concentrations of halogens in tektites and natural terrestrial glasses was undertaken to obtain reliable data on these relatively volatile elements in order to examine the current framework of ideas on possible source material and on the thermal history of tektites.

105.058 Relation between solar and planetary neon in carbonaceous chondrites. G. F. Herzog.
Journ. Geophys. Res., Vol. 77, 6219 - 6225 (1972).
We have attempted to calculate production ratios for the nuclear reactions best able to effect the transformation of solar Ne to planetary Ne. Formation of the light planetary gases in the context of the other theories will also be discussed briefly.

105.059 Natural remanent magnetizations of carbonaceous chondrites and the magnetic field in the early solar system. S. K. Banerjee, R. B. Hargraves.
Earth Planet. Sci. Letters, Vol. 17, 110 - 119 (1972).
The natural remanent magnetizations (NRM) of three carbonaceous chondrites, Allende (type III), Murchison (type II) and Orgueil (type I), have been studied in an attempt to explain the origin of the stable component of their NRM which is greatest in Allende and least in Orgueil. Thermal demagnetization of NRM and remagnetization in a known field were observed.

105.060 Natural remanent magnetization and thermomagnetic properties of the Allende meteorite.
R. F. Butler.
Earth Planet. Sci. Letters, Vol. 17, 120 - 128 (1972).
Paleomagnetic examination of a 1-kg specimen of the Allende meteorite has revealed the existence of natural remanent magnetization of constant direction and moderate intensity. Thermomagnetic analysis indicates that the ferromagnetic minerals in Allende consist of 95 wt% taenite containing 67% Ni plus 5 wt% taenite with 36% Ni. These iron–nickel minerals amount to only 0.46 wt% of the whole rock.

105.061 Impact glasses from the suevite of the Nördlinger Ries. V. Stähle.
Earth Planet. Sci. Letters, Vol. 17, 275 - 293 (1972).
The main aim of this study was to learn more about the condition of formation, internal structure, melting and reaction processes of the inclusions and chemical composition of the Ries glasses.

105.062 Fundamental optical properties of the Bruderheim meteorite. W. G. Egan, T. Hilgeman.
Bull. American Astron. Soc., Vol. 4, 425 (1972). – Abstr. AAS.

105.063 New data on the cosmic history of the Sikhote-Alin meteorite.
E. M. Kolesnikov, A. K. Lavrukhina, L. K. Levskii, A. V. Fisenko.
Dokl. Akad. Nauk SSSR, Ser. Mat. Fiz., Vol. 207, 1300 - 1302 (1972). In Russian.

105.064 Magnetic properties of meteorites. Meteorites in laboratory. E. G. Gus'kova.
Nauka, Leningrad. 108 pp. Price 95 Kop. (1972). In Russian. Review in Referativ. Zhurn. 51. Astron., 1.51.374 (1973).

105.065 Evidence for an impact origin for Lake Wanapitei, Ontario. M. R. Dence, J. Popelar.
Geol. Ass. Canada, Special Paper No. 10, p. 117 - 124 = Contr. Earth Phys. Branch, Ottawa, No. 398 (1972).

105.066 Meteorite impact craters and the structure of the Sudbury Basin. M. R. Dence.
Geol. Ass. Canada, Special Paper No. 10, p. 7 - 18 = Contr. Earth Phys. Branch, Ottawa, No. 402 (1972).
The structure of hypervelocity impact craters such as Clearwater Lake, Charlevoix and Vredefort when compared with that of the Sudbury Basin shows important similarities and some differences. Outcrops of Huronian rocks northwest of the Basin are interpreted as remnants of a downdropped rim. A model is proposed in which the Sudbury structure is attributed to hypervelocity impact of an asteroid into a target with high thermal gradient.

105.067 Achievements and actual problems of meteoritics.
V. G. Fesenkov.
Meteoritika, vyp. (No.) 31, p. 3 - 17 (1972). In Russian.

105.068 Form of the air wave of the Tunguska meteorite.
I. T. Zotkin.
Meteoritika, vyp. (No.) 31, p. 35 - 41 (1972). In Russian.

105.069 Cosmic matter influx onto the earth.
A. N. Simonenko, B. Yu. Levin.
Meteoritika, vyp. (No.) 31, p. 45 - 56 (1972). In Russian.

105.070 Spherical microparticles in the Antarctic ice cap.
V. D. Vilenskij.
Meteoritika, vyp. (No.) 31, p. 57 - 61 (1972). In Russian.

105.071 Four years of new investigations on fall and collection of particles of the Sikhote-Alin meteorite shower. E. L. Krinov.
Meteoritika, vyp. (No.) 31, p. 62 - 67 (1972). In Russian.

105.072 Shock meteorite craters. A. O. Aaloe.
Meteoritika, vyp. (No.) 31, p. 68 - 73 (1972). In Russian.

105.073 The Popigai meteorite crater in North Siberia.
V. L. Masajtis, M. V. Mikhajlov, T. V. Selivanovskaya.
Meteoritika, vyp. (No.) 31, p. 74 - 78 (1972). In Russian.

105.074 Selective erosion as main factor for formation of the structure of Konder.
V. I. Tsvetkov, A. A. Pronin.
Meteoritika, vyp. (No.) 31, p. 79 - 82 (1972). In Russian.

105.075 Mineralographic investigation of the meteorite Pilistfer. I. A. Yudin.
Meteoritika, vyp. (No.) 31, p. 83 - 89 (1972). In Russian.

105.076 Microscopic and roentgenographic investigation of the chondrite Tennasilm.
I. A. Yudin, G. V. Pal'gueva.

Meteoritika, vyp. (No.) 31, p. 90 - 95 (1972). In Russian.

105.077 **Features of similarity in the mineral composition of calcined sulfid nickel ore and meteorites.**
V. M. Grigor'eva, I. E. Gorbunova, A. A. Yasinskaya.
Meteoritika, vyp. (No.) 31, p. 96 - 100 (1972). In Russian.

105.078 **Investigation of pyroxenes in meteorites by means of Mössbauer spectroscopy.** T. V. Malysheva,
V. V. Kurash, A. K. Lavrukhina, L. D. Akol'zina.
Meteoritika, vyp. (No.) 31, p. 101 - 103 (1972). In Russian.

105.079 **The iron meteorite Sejmchan.**
O. A. Kirova, M. I. D'yakonova.
Meteoritika, vyp. (No.) 31, p. 104 - 108 (1972). In Russian.

105.080 **The eucrite of Pomozdino.**
L. G. Kvasha, M. I. D'yakonova.
Meteoritika, vyp. (No.) 31, p. 109 - 115 (1972). In Russian.

105.081 **New data on the chemical composition of the stone meteorites Timokhina and Zhigajlovka from the** collection of the Committee on Meteorites of the USSR Academy of Sciences. V. Ya. Kharitonova.
Meteoritika, vyp. (No.) 31, p. 116 - 118 (1972). In Russian.

105.082 **The chemical composition of the chondrites Vengerovo, Grossliebenthal, Doroninsk, Zaboritsa, Nerft.**
M. I. D'yakonova.
Meteoritika, vyp. (No.) 31, p. 119 - 121 (1972). In Russian.

105.083 **Mercury in meteorites.** L. G. Kvasha, N. A. Ozerova, N. Kh. Ajdin'yan, N. D. Shikina.
Meteoritika, vyp. (No.) 31, p. 122 - 136 (1972). In Russian.

105.084 **The distribution of uranium in various minerals of the meteorites Sikhote-Alin, Gressk and Aarus.**
L. I. Genaeva, L. L. Kashkarov, A. K. Lavrukhina, L. V. Yukina.
Meteoritika, vyp. (No.) 31, p. 137 - 140 (1972). In Russian.

105.085 **Chemical groups of iron meteorites and their peculiarities.** A. A. Yavnel'.
Meteoritika, vyp. (No.) 31, p. 141 - 148 (1972). In Russian.

105.086 **New data on the isotopic content of rare gases of stony meteorites.** L. K. Levskij.
Meteoritika, vyp. (No.) 31, p. 149 - 150 (1972). In Russian.

105.087 **Cosmogenic isotopes in the Sikhote-Alin meteorite.**
A. K. Lavrukhina, T. A. Ibraev, A. P. Zajtseva,
L. V. Yukina, T. V. Malysheva, V. I. Mednikov, N. G. Mednikova, I. E. Dubinin.
Meteoritika, vyp. (No.) 31, p. 151 - 156 (1972). In Russian.

105.088 **Investigation of the nature of the tracks in the olivine meteorite Ilimaez.**
L. L. Kashkarov, A. K. Lavrukhina, L. I. Genaeva.
Meteoritika, vyp. (No.) 31, p. 157 - 161 (1972). In Russian.

105.089 **Two supposed East Midlands meteorites. I. The Yaddlethorpe (Lincolnshire) stone.**
R. D. Morton, W. A. S. Sarjeant.
Mercian Geologist, (GB), Vol. 4, No. 1, p. 37 - 40 (1971).

105.090 **Twinning and intergrowth of olivine crystals in chondritic meteorites.**
R. T. Dodd, C. Calef.
Mineral. Mag., (GB), Vol. 38, 324 - 327 (1971).

105.091 **The Sitathali meteorite.**
T. V. Viswanathan, N. R. Sen Gupta, D. R. Das

Gupta, S. Benerjee.
Mineral. Mag., (GB), Vol. 38, 335 - 343 (1971).

105.092 **Les acides aminés dans les météorites.**
P. Arpino.
Recherche, (France), 1972, No. 19, p. 74 - 75.

105.093 **The Bagnone meteorite.**
S. Bonatti, M. Franzini, L. Schiaffino.
Atti Soc. Tosc. Sci. Nat. Pisa, Ser. A, Vol. 77, 123 - 133 (1970).

105.094 **An examination of the structure and response to heat treatment of the iron meteorite La Primitiva** (B. M. 1927, 77). H. J. Axon, J. Boustead.
Chemie der Erde, (Germany), Vol. 30, 1 - 11 (1971).

105.095 **Description and origin of large tektite from Thailand.** V. E. Barnes.
Chemie der Erde, (Germany), Vol. 30, 13 - 19 (1971).

105.096 **Chondrule groundmass in the Parnallee meteorite.**
R. A. Binns.
Chemie der Erde, (Germany), Vol. 30, 21 - 31 (1971).

105.097 **Tritium loss resulting from cosmic annealing, compared with the microstructure and microhardness of six iron meteorites.** V. F. Buchwald.
Chemie der Erde, (Germany), Vol. 30, 33 - 57 (1971).

105.098 **Dark-zoned chrondrules.**
R. T. Dodd, W. R. Van Schmus.
Chemie der Erde, (Germany), Vol. 30, 59 - 69 (1971).

105.099 **Der Kayakent-Meteorit.**
G. Dörfler, W. Kiesl.
Chemie der Erde, (Germany), Vol. 30, 71 - 75 (1971).

105.100 **Djerfisherite composition in Bishopville, Peña Blanca Springs, St. Marks and Toluca meteorites.**
A. El Goresy, N. Grögler, J. Ottemann.
Chemie der Erde, (Germany), Vol. 30, 77 - 82 (1971).

105.101 **Fresh meteorites in 1970 and the cosmic-ray gradient.** E. L. Fireman, G. Spannagel.
Chemie der Erde, (Germany), Vol. 30, 83 - 101 (1971).

105.102 **Mineralogy, petrology, and chemistry of the Burdett, Kansas, chondrite.**
R. V. Fodor, K. Keil, E. Jarosewich, G. I. Huss.
Chemie der Erde, (Germany), Vol. 30, 103 - 113 (1971).

105.103 **Total nitrogen content of ordinary chondrites.**
E. K. Gibson, Jr., C. B. Moore.
Chemie der Erde, (Germany), Vol. 30, 115 - 131 (1971).

105.104 **A review of meteoritic abundances of sodium, potassium, rubidium and caesium.** G. G. Goles.
Chemie der Erde, (Germany), Vol. 30, 133 - 137 (1971).

105.105 **Über Schemata meteoritischer Strukturen.**
D. M. Grigor'ev.
Chemie der Erde, (Germany), Vol. 30, 139 - 144 (1971).

105.106 **Kosmochemische Meteoritenuntersuchungen mittels Neutronenaktivierungsanalyse.**
F. Hecht, W. Kiesl.
Chemie der Erde, (Germany), Vol. 30, 145 - 155 (1971).

105.107 **Die Anwendung glaschemischer Methoden bei der Untersuchung von Tektiten.**

K. Heide, H.-P. Bruckner.
Chemie der Erde, (Germany), Vol. 30, 157 - 174 (1971).

105.108 Ar⁴⁰ in meteorites, fines and breccias from the
moon. D. Heymann, A. Yaniv.
Chemie der Erde, (Germany), Vol. 30, 175 - 189 (1971).

105.109 Zur Erfassung der Meteorite. G. Hoppe.
Chemie der Erde, (Germany), Vol. 30, 191 - 197
(1971).

105.110 Shock history of iron meteorites and their parent
bodies: A review, 1967 - 1971.
A. V. Jain, M. E. Lipschutz.
Chemie der Erde, (Germany), Vol. 30, 199 - 215 (1971).

105.111 Sekundärstrukturen opaker Mineralien in Stein-
meteoriten. I. A. Yudin.
Chemie der Erde, (Germany), Vol. 30, 217 - 233 (1971).

105.112 Die chemische Zusammensetzung von Gläsern und
Chondrenmatrizes im Chondriten von Tieschitz.
G. Kurat.
Chemie der Erde, (Germany), Vol. 30, 235 - 249 (1971).

105.113 Transition element distribution in stony meteorites
and in terrestrial and lunar rocks.
B. Mason, E. Jarosewich, J. Nelen.
Chemie der Erde, (Germany), Vol. 30, 259 - 268 (1971).

105.114 Einiges über den Meteoriten von Mundrabilla in
Westaustralien.
P. Ramdohr, A. El Goresy.
Chemie der Erde, (Germany), Vol. 30, 269 - 285 (1971).

105.115 The iron meteorites of Northern Chile.
R. Schaudy, J. T. Wasson.
Chemie der Erde, (Germany), Vol. 30, 287 - 296 (1971).

105.116 On the radiogenic argon in iron meteorites.
L. Schultz, H. Funk, P. Signer.
Chemie der Erde, (Germany), Vol. 30, 297 - 304 (1971).

105.117 Shape analysis of moldavites and their impact
origin. J. Konta.
Mineral. Mag., (GB), Vol. 38, 408 - 417 (1971).

105.118 Isotopic ratio of lithium in chondrite measured by
an ion probe mass spectrometer.
H. Nishimura, J. Okano.
Japanese Journ. Applied Phys., Vol. 10, 1613 - 1622 (1971).

105.119 Correction de l'âge apparent ⁴He de la whitlockite
de Saint-Séverin. Y. Cantelaube, P. Pellas.
Comptes Rendus Acad. Sci. Paris, Sér. D, Vol. 274, 1125 -
1127 (1972).

105.120 Occurrence of wollastonite, rhönite, and andradite
in the Allende meteorite. L. H. Fuchs.
American Mineralogist, Vol. 56, 2053 - 2068 (1971).

105.121 Noble metal determination in meteorites.
G. E. Gillum, W. D. Ehmann.
Radiochim. Acta, (Germany), Vol. 16, 123 - 128 (1971).

105.122 Extraterrestrial abiogenic organization of organic
matter: The hollow spheres of the Orgueil meteorite. M. Rossignol-Strick, E. S. Barghoorn.
Space Life Sci., (Netherlands), Vol. 3, 89 - 107 (1971).

105.123 Techniques for the study of fossil tracks in extra-

terrestrial and terrestrial samples. I. Methods of
high contrast and high resolution study.
D. J. MacDougall, D. Lal, L. L. Wilkening, S. G. Bhat,
S. S. Liang, G. Arrhenius, A. S. Tamhane.
Geochem. Journ., (Japan), Vol. 5, 95 - 112 (1971).

105.124 The Oktibbeha County iron meteorite.
S. J. B. Reed.
Mineralog. Mag., (GB), Vol. 38, 623 - 626 (1972).

105.125 The metal phase of the Bustee enstatite achondrite. C. M.Wai, C. R. Knowles.
Mineralog. Mag., (GB), Vol. 38, 627 - 629 (1972).

105.126 Platinum and gold in chondritic meteorites.
W. D. Ehmann, D. E. Gillum.
Chem. Geol., (Netherlands), Vol. 9, No. 1, p. 1 - 11 (1972).

105.127 Distribution of uranium and thorium among com-
ponents of some chondrites.
H. Matsuda, M. Shima, M. Honda.
Geochem. Journ., (Japan), Vol. 6, 37 - 42 (1972).

105.128 Olivine content of chondrites measured by X-ray
diffraction. S. S. Pollack, R. D. Chi.
American Mineralogist, Vol. 57, 584 - 591 (1972).

105.129 "Organics" in meteorites. I. S. Astapovich.
Problems of cosmic physics. Vyp. (No.) 7, (see 003.
028), p. 79 - 99 (1972). In Russian.
A short review of modern state in the study of "organized
elements" and of the "organic substances" in meteorites
(chiefly carbonacious ones) is given, accompanied by some
considerations of the author.

105.130 Evidence for association between Ir and Al in L
chondrites. T. W. Osborn.
Nature, Phys. Sci., Vol. 239, 10 - 11 (1972).

105.131 Hardness of kamacite and shock histories of 119
meteorites.
A. V. Jain, R. B. Gordon, M. E. Lipschutz.
Journ. Geophys. Res., Vol. 77, 6940 - 6954 (1972).
We have used metallographic and X-ray diffraction techniques to study the shock histories of 119 iron and stony-iron
meteorites and have measured the hardness of kamacite in
these specimens and in artificially shocked unannealed and an-
nealed meteorite specimens.These results, together with those
obtained previously, indicate that the plurality, if not the majority, of all iron and stony-iron meteorites sampled by the
earth were shocked to pressures of ⩾ 130 kb during preterres-
trial collisions between asteroidal-sized objects.

105.132 Bottle green microtektites. B. P. Glass.
Journ. Geophys. Res., Vol. 77, 7057 - 64 (1972).
Transparent bottle green microtektites with SiO₂ contents of as low as 48% and MgO contents of as high as 24%
have been found in deep-sea sediments in association with
both the normal Australasian and the Ivory Coast microtektites. Thus the presence of these unusual glasses in both the
Australasian and the Ivory Coast strewnfield places an important constraint on theories concerning the origin of tektites.

105.133 Anomalous optical phenomena connected with the
fall of the Tunguska meteorite.
N. V. Vasil'ev, N. P. Fast.
Gerlands Beiträge Geophys., Vol. 81, 433 - 438 (1972).
In Russian.
It is shown that the "bright nights" observed in summer
of 1908 markedly differ from optical anomalies caused by

volcanic events. Neither are they able to be interpreted by the seasonal maximum of the noctilucent clouds, which appeared over the northern hemisphere at the end of June. Most probable is the supposition that they are connected with the fall of the Tunguska meteorite. Synchronous with this meteorite a cloud of highly disperse cosmic matter appears to have penetrated the earth's atmosphere; after having scattered in the upper atmospheric layers, it may have given rise to the optical anomalies.

The Allende, Mexico, meteorite shower. See Abstr. 003.057.

Astrophysical consequences of effects of electromagnetic emittance. See Abstr. 061.026.

Cosmic dust in the mesosphere. See Abstr. 082.130.

On atmospheric emission of atomic oxygen (5577 Å) during night and its connection with sporadic appearing micrometeorites. See Abstr. 082.168.

Low energy solar nuclear particle irradiation of lunar and meteoritic breccias. See Abstr. 094.081.

Comparative analysis of the magnetic properties of lunar rocks and meteorites. See Abstr. 094.124.

Meteoroid activity on the lunar surface from the Surveyor 3 sample examination. See Abstr. 094.177.

Erosion phenomena on the lunar surface and meteorites. See Abstr. 094.178.

Interplanet variations in scale of crater morphology. – Earth, Mars, moon. See Abstr. 094.184.

Interplanetary objects in review: Statistics of their masses and dynamics. See Abstr. 106.009.

Hard rock cosmic ray archaeology. See Abstr. 143.069.

106 Interplanetary Matter, Interplanetary Magnetic Field, Zodiacal Light

106.001 An analysis of Pioneer 9 low-frequency wave observations near interplanetary discontinuities.
F. L. Scarf, E. W. Greenstadt, J. H. Wolfe, D. S. Colburn.
Journ. Geophys. Res., Vol. 77, 3317 - 3327 (1972).

106.002 Critical component of the interplanetary magnetic field responsible for large geomagnetic effects in the polar cap. E. Friis-Christensen, K. Lassen, J. Wilhjelm, J. M. Wilcox, W. Gonzalez, D. S. Colburn.
Journ. Geophys. Res., Vol. 77, 3371 - 3376 (1972).

106.003 Circular polarisation measurements of the zodiacal light. J. Staude, T. Schmidt.
Astron. Astrophys., Vol. 20, 163 - 164 (1972).
The circular polarisation component of the zodiacal light is shown to be zero within the accuracy of $\pm 1^o/_{oo}$.

106.004 Pioneer 7 observations of the August 29, 1966, interplanetary shock-wave ensemble.
M. Dryer, Z. K. Smith, G. H. Endrud, J. H. Wolfe.
Cosmic Electrodynamics, Vol. 3, 184 - 207 (1972).
It is the purpose of this paper to present the detailed plasma (except for electrons) and magnetic field data from Pioneer 7 for a proposed double-shock ensemble. These data are then compared with MHD theory to show that the discontinuities themselves satisfy the magnetic Hugoniot equations. In addition, the data are compared with the theoretical predictions for the double-shock structure.

106.005 Relationship of interplanetary magnetic field structure with development of substorm and storm main phase. S. Kokubun.
Planet. Space Sci., Vol. 20, 1033 - 1049 (1972).
In the present paper the relationship between the development of a polar magnetic substorm and magnetic field change in interplanetary space is first discussed in greater detail to confirm the result obtained in the previous study (Kokubun, 1971). The second purpose is to examine the development of

a magnetic storm and its relationship to the kinetic properties of the solar wind and structures of the interplanetary magnetic field behind the shock wave.

106.006 Interplanetary magnetic-sector structure, 1926 - 1971. L. Svalgaard.
Journ. Geophys. Res., Vol. 77, 4027 - 4034 (1972).
The influence of the direction of the interplanetary magnetic field on the geomagnetic field at high latitudes is used to study the long-term behavior of the sector structure during nearly four solar cycles. It is found that the rotation period of the sector structure varies from about 28.5 days in the beginning of a solar cycle to 27.0 days in the end.

106.007 Cosmic-ray diffusion coefficient in interplanetary space. L. J. Gleeson, I. H. Urch.
Journ. Geophys. Res., Vol. 77, 4259 - 4263, with a reply by J. J. Burger, P. N. Swanenburg, p. 4264 - 4267 (1972). Letter.

106.008 On the rate of ejection of dust by long-period comets. V. N. Lebedinets.
IAU Symposium No. 45, (see 012.003), p. 442 - 446 (1972).

106.009 Interplanetary objects in review: Statistics of their masses and dynamics. J. S. Dohnanyi.
Icarus, Vol. 17, 1 - 48 (1972). – An *Icarus* invited review paper.
Present knowledge on the spatial distribution of the masses of interplanetary bodies and their current evolution is reviewed. The following list of problems and the progress on their solution are discussed in this review. First, from the most reliable data, a best estimate of the distribution of interplanetary objects, including meteoroids, asteroids, and comets, is derived and some of the uncertainties are discussed. Following this the influence of collisions on the population of interplanetary objects is discussed. The influence of gravitational perturbations on the motion of interplanetary objects is then reviewed. Finally, the influence of radiation forces on micrometeoroids

is discussed.

106.010 Observations of the interplanetary medium and of the structure of radio sources using higher moments of interplanetary scintillations. G. Bourgois, C. Cheynet.
Astron. Astrophys., Vol. 21, 25 - 31 (1972).

We present a new method for studying interplanetary scintillations of radio sources based upon the observation of moments of order 3 and 4 of the fluctuations of the received signal.

106.011 A theoretical study of the higher moments of the interplanetary scintillations. G. Bourgois.
Astron. Astrophys., Vol. 21, 33 - 38 (1972).

Informations on the interplanetary medium can be derived from the study of the skewness coefficient of the probability density function of the received intensity: γ_1. Generalizing the work of Mercier (1962), we derive an expression of γ_1 which is valid for all conditions of observation of interplanetary scintillations at sufficiently large distances from the sun.

106.012 Interplanetary sector structure during 4 solar cycles. L. Svalgaard.
Bull. American Astron. Soc., Vol. 4, 393 (1972). – Abstr. AAS.

106.013 An annual and a solar-magnetic-cycle variation in the inferred interplanetary magnetic field, 1926–1971. J. M. Wilcox, P. H. Scherrer.
Bull. American Astron. Soc., Vol. 4, 396 (1972). – Abstr. AAS.

106.014 Detection of interplanetary electrons from 18 keV to 1.8 MeV during solar quiet times.
R. P. Lin, K. A. Anderson, T. L. Cline.
Phys. Rev. Letters, Vol. 29, 1035 - 1038 (1972).

We have observed a quiet-time component of interplanetary electrons having energies above solar-wind energies and below those characterized as cosmic radiation. Its energy spectrum generally falls with energy from 18 keV to 1.8 MeV, but shows a feature in the 100–300-keV range. The observed temporal variations of the intensity suggest that the 18–100-keV portion is solar and the 0.3–1.8-MeV portion is galactic in origin.

106.015 Origin of 200-keV interplanetary electrons.
R. Ramaty, T. L. Cline, L. A. Fisk.
Phys. Rev. Letters, Vol. 29, 1039 - 1042 (1972).

A spectral feature at ~200 keV in the recently observed intensity of interplanetary electrons is examined as to its possible solar, interplanetary, or galactic origin.

106.016 Magnetic field structure in flare-associated solar-wind disturbances.
K. H. Schatten, J. E. Schatten.
Journ. Geophys. Res., Vol. 77, 4858 - 4863 (1972).

It is the purpose of this study to determine what information and conclusions can be drawn about the proposed solar-wind disturbance models by examining the interplanetary magnetic field.

106.017 About the spectrum of small-scale irregularities of the interplanetary plasma.
N. A. Lotova, I. V. Chashei.
Geomagn. Aeronom., Vol. 12, 800 - 805 (1972). In Russian.

106.018 Variations of the interplanetary magnetic field and the substorm of June 17–18, 1965.
Ya. I. Feldshtein, P. V. Sumaruk, N. F. Shevnina.
Geomagn. Aeronom., Vol. 12, 962 - 963 (1972). In Russian. Brief information.

106.019 Association between interplanetary shock waves and delayed solar particle events.
D. Datlowe.
Journ. Geophys. Res., Vol. 77, 5374 - 5384 (1972).

This paper points out cosmic-ray phenomena for which the steady-state picture of the solar wind is not a valid approximation and shows that large-scale temporary changes in the solar-wind flow due to flares may have large effects on the cosmic rays observed at the earth. In these events, abrupt changes in the configuration of the interplanetary medium after the passage of a shock will be reflected by a large increase in the cosmic-ray flux.

106.020 Annual and solar-magnetic-cycle variations in the interplanetary magnetic field, 1926–1971.
J. M. Wilcox, P. H. Scherrer.
Journ. Geophys. Res., Vol. 77, 5385 - 5388 (1972).

The present analysis of 45 years of inferred field polarity clearly shows an annual variation and also a variation of about 20 years, which we associate with the solar-magnetic cycle. On the average the phase of the annual variation of the interplanetary field changes about $2\,^2/_3$ years after sunspot maximum.

106.021 Annual variation of the interplanetary He⁺ velocity distribution at 1 AU.
W. C. Feldman, D. Goldstein, F. Scherb.
Journ. Geophys. Res., Vol. 77, 5389 - 5398 (1972).

The interplanetary velocity distribution of He^+ ions of interstellar origin is calculated for various ecliptic longitudes at 1 AU and for various temperatures T of the local interstellar medium. We find that these He^+ fluxes should be observable at 1 AU with present observation techniques.

106.022 Electromagnetic instabilities produced by neutral-particle ionization in interplanetary space.
C. S. Wu, R. C. Davidson.
Journ. Geophys. Res., Vol. 77, 5399 - 5406 (1972).

The interaction between solar wind and interstellar particles is a topic of considerable interest, as evidenced by recent observations that confirm the existence of interstellar gas in the inner solar system. We consider a different but highly relevant aspect of the subject, namely, the collective plasma effects that occur immediately after the interstellar atoms become ionized. It is shown that newly ionized particles can result in the local generation of electromagnetic waves in interplanetary space.

106.023 Depth of an interplanetary shock wave front.
K. G. Ivanov.
Kosm. Issled., Vol. 10, 788 - 789 (1972). In Russian. – Brief information.

106.024 On the possibility of sounding of the interplanetary shock waves and forecasting of the geomagnetic storms on the basis of the ground observations of cosmic rays.
L. I. Dorman, N. S. Kaminer.
Ann. Géophys., Vol. 26, 697 - 701 (1970).

A small increase in the cosmic ray intensity due to acceleration of the galactic cosmic rays, reflected from shock wave propagating from the sun, is shown to arise before the commencement of geomagnetic storms. Analysis of the cosmic ray ground observation data, for more than twenty storms, permitted the main regularities of the increase effect to be revealed. Good agreement between the obtained and expected regularities shows that simultaneous precise ground observations of cosmic rays, at many stations, make it possible to obtain continuous information on powerful shock wave in the interplanetary space, and to give short-term forecasts of geomagnetic storms.

106.025 Intensité et polarisation de la lumière solaire diffusée par un volume isolé de matière inter-

planétaire. R. Dumont.
Comptes Rendus Acad. Sci., Paris, Sér. B, Vol. 275, 765 - 768 (1972).

On montre que les gradients photométrique et polarimétrique de la lumière zodiacale le long de l'écliptique et à 90° du Soleil permettent de remonter à l'intensité et au degré de polarisation de la lumière diffusée à angle droit par 1 km³ de matière interplanétaire situé à 1 U. A. du Soleil.

106.026 Zodiacal light and interplanetary particle number densities. R. H. Giese.
Space Research XII, (see 012.016), Vol. 1, 437 - 443 (1972).

106.027 The zodiacal light as seen from the Pioneer F/G and Helios probes. M. S. Hanner, C. Leinert.
Space Research XII, (see 012.016), Vol. 1, 445 - 455 (1972).

106.028 Inferring the interplanetary magnetic field by observing the polar geomagnetic field. J. M. Wilcox.
Rev. Geophys. Space Phys., Vol. 10, 1003 - 1014 (1972).

L. Svalgaard and S. M. Mansurov have shown that it is possible to infer the polarity of the interplanetary magnetic field quite reliably from observations of the diurnal variation of polar geomagnetic fields. The fact that observations of the polar geomagnetic field have existed without interruption since 1926 at the Danish Meteorological Institute station at Godhavn, Greenland, means that in effect the inferred solar magnetic field during five sunspot cycles is available for analysis. The specific nature of the relation between the interplanetary field and the polar geomagnetic field should aid toward physical understanding of the interaction between the two fields.

106.029 On the energy spectrum of small-scale irregularities in the interplanetary plasma. N. A. Lotova.
Izv. vyssh. ucheb. zavedenij. Radiofizika, Vol. 15, 826 - 831 (1972). In Russian. – Abstr. in Referativ. Zhurn. 51. Astron., 12.51.523 (1972).

106.030 Precipitation of low-energy electrons at high latitudes: Effects of interplanetary magnetic field and dipole tilt angle. J. L. Burch.
Journ. Geophys. Res., Vol. 77, 6696 - 6707 (1972).

The purpose of this study is to examine the movement in latitude of low-energy electron precipitation on the dayside in response to substorm activity and a southward-turning interplanetary magnetic field. With these effects eliminated, the tilt-angle dependence of the location of the dayside soft zone, or polar cusp, is re-evaluated and is found to be somewhat weaker than was found by Maehlum (1968) and inferred from Feldstein and Starkov (1970).

106.031 Use of an electron beam for low-temperature plasma measurement in the magnetosphere and interplanetary space. N. Kawashima.
Journ. Geophys. Res., Vol. 77, 6896 - 6899 (1972). – Letter.

106.032 Comment on paper by Y. C. Whang and N. F. Ness, 'Magnetic field anomalies in the lunar wake'.[Journ. Geophys. Res., Vol. 77, 1109 - 1115 (1972)].
C. P. Sonett, J. D. Mihalov, with a reply by N. F. Ness, Y. C. Whang.
Journ. Geophys. Res., Vol. 77, 6922 - 6923, 6924 - 6925 (1972).

106.033 Luz zodiacal y luz antisolar. F. J. Mandujano O.
El Universo, Vol. 26, 105 - 106 (1972).

106.034 The evolution of dust particles near the sun.
O. V. Dobrovolsky, P. Jegibekov.
Astron. Zhurn. Akad. Nauk SSSR, Vol. 49, 1287 - 1291 (1972). In Russian. English translation in Soviet Astron. AJ, Vol. 16, No. 6.

The influence of the particle size variation during its evaporization and sputtering on the evolution of the orbit under Poynting–Robertson and corpuscular drag is investigated.

106.035 A photometric model of the zodiacal light.
F. E. Roach.
Astron. Journ., Vol. 77, 887 - 891 (1972).

An intercomparison of published photometric investigations of the zodiacal light and gegenschein is made, leading to a compilation of brightnesses over the entire sky in a reseau with 5-deg intervals. The results are given in absolute units. Several graphical representations illustrate the compilation.

106.036 The effective modulating layer in interplanetary space.
A. K. Lavrukhina, G. K. Ustinova, A. N. Simonenko.
Meteoritika, vyp. (No.) 31, p. 24 - 34 (1972). In Russian.

106.037 On the mass of the circumterrestrial cosmic cloud.
V. G. Fesenkov.
Meteoritika, vyp. (No.) 31, p. 42 - 44 (1972). In Russian.

106.038 Interplanetary debris. K. Saito.
Astron Herald, (*Japan*), Vol. 65, 179 - 185 (1972). In Japanese.

Reviews the current state of investigations. Debris is scattered by perturbation forces, by radiation pressure and by collisions. The confirmation of the size distribution by systematic observations of meteor showers using telescopes is considered to be useful.

106.039 Interaction of the interplanetary medium with the geomagnetosphere. J. V. Kovalevsky.
Problems of cosmic physics. Vyp. (No.) 7, (see 003.028), p. 23 - 60 (1972). In Russian.

Some theoretical and experimental aspects of the interaction of the interplanetary medium with the geomagnetic field are discussed: the formation of a collisionless bow shock of the transition region (or magnetosheath), of the magnetopause and of the earth's magnetic tail.

106.040 Comparison of two methods of determination of zodiacal light intensity. S. N. Krylova.
Astron. Tsirk., No. 726, p. 1 - 3 (1972). In Russian.

106.041 Sectorial structure and disturbances in interplanetary space.
A. T. Nesmyanovich, E. I. Nesmyanovich.
Astron. Tsirk., No. 737, p. 1 - 4 (1972). In Russian.

The Pioneer 8 cosmic dust experiment.
See Abstr. 034.081.

Pioneer 10 observations of starlight and zodiacal light at large elongations: Preliminary results.
See Abstr. 051.004.

Effects of erosion and fragmentation on the mass distribution of colliding particles. See Abstr. 061.047.

Modification of the Rankine-Hugoniot relations for shocks in space. See Abstr. 062.022.

Waves and resonances in magneto-active plasma.
See Abstr. 062.054.

Flares and shock waves in interplanetary space.
See Abstr. 073.113.

Analysis of three-station interplanetary scintillation. See Abstr. 074.047.

Magnetic and electric waves in space. See Abstr. 074.054.

The recurrent solar wind streams observed by interplanetary scintillation of 3C 48.   See Abstr. 074.063.

Interplanetary gas. XVII. An astrometric determination of solar-wind velocities from orientations of ionic comet tails.   See Abstr. 074.067.

Note on solar plasma irregularities and plasma instabilities.   See Abstr. 074.074.

Mesures directionnelles du vent solaire par la sonde HEOS I, S 58-73, de l'ESRO.   See Abstr. 074.089.

Direct observations of low-energy solar electrons associated with a type III solar radio burst. See Abstr. 078.046.

Comparisons of the mean solar magnetic field and the interplanetary field observed during 1969. See Abstr. 080.017.

Why does the sun sometimes look like a magnetic monopole?   See Abstr. 080.038.

Atmospheric phenomena and zodiacal light. See Abstr. 082.027.

On optical methods for sounding of the upper atmosphere and of the circumterrestrial dust cloud. See Abstr. 082.159.

MgI emission in the night sky spectrum. See Abstr. 082.195.

Circular polarization of the nightsky radiation. See Abstr. 082.199.

New results on particle arrival at the polar caps. See Abstr. 084.029.

The families of geomagnetic storms, direction of the interplanetary magnetic field and solar activity. See Abstr. 084.235.

About the influence of the interplanetary magnetic field on geomagnetic activity.   See Abstr. 084.236.

The influence of the interplanetary magnetic field on the characteristics of Pc2-4 type pulsations. See Abstr. 084.237.

On the influence of the sector structure of the interplanetary magnetic field upon the field component normal to the ecliptic plane.   See Abstr. 084.238.

Magnetohydrodynamic theory for the interaction of an interplanetary double-shock ensemble with the earth's bow shock.   See Abstr. 084.243.

About the dependence of the magnetopause location on the orientation of the interplanetary magnetic field. See Abstr. 084.255.

Relationship between the various indices of geomagnetic activity and the interplanetary plasma parameters. See Abstr. 084.293.

Aluminum 26 and manganese 53 produced by solar-flare particles in lunar rock and cosmic dust. See Abstr. 094.037.

Cometary brightness variations and conditions in interplanetary space.   See Abstr. 102.009.

The meteoroid influx and the maintenance of the solar system dust cloud.   See Abstr. 104.027.

Off-ecliptic control of cosmic ray modulation. See Abstr. 143.046.

Modulation of cosmic-ray electrons. See Abstr. 143.051.

107 Cosmogony of the Planetary System

107.001 On the formation of planets from a solar nebula.
R. A. Lyttleton.
Monthly Notices Roy. Astron. Soc., Vol. 158, 463 - 483
(1972).

The development of planets within a solar nebula of both gas and dust is shown to occur in two main stages. First, in a time comparable with the orbital period at any distance the dust comes to move with Keplerian motion in a thin plane disc. Second, the range-of-influence of a planet (of mass m moving at distance R) out to which its gravitation dominates over that of the sun is $(m/3M)^{1/3} R$. At any stage material within this distance can be captured. It seems probable that many small bodies would begin to form separately, and then themselves combine by collisions to yield larger bodies fewer in number.

107.002 Accretionary processes in the early solar system: An experimental approach.
J. F. Kerridge, J. F. Vedder.
Science, Vol. 177, 161 - 163 (1972).

Micrometer-size silicate flakes do not accrete during impacts in the velocity range 1.5 to 9.5 kilometers per second. Conventional accretionary theories for silicate bodies are applicable only to particles whose orbits are similar. Metal-silicate fractionation in the solar system may have been affected by differences in the accretionary behavior of the metal and silicate particles.

107.003 Metal/silicate fractionation in the solar system.
J. S. Lewis.
Earth Planet. Sci. Letters, Vol. 15, 286 - 290 (1972).

Fractionation between the metal and silicate components of objects in the inner solar system has long been recognized as a necessity in order to explain the observed density variations of the terrestrial planets and the H-group, L-group dichotomy of the ordinary chondrites. This paper discusses the densities of the terrestrial planets in light of current physical and chemical models of processes in the solar nebula.

107.004 Ejection of bodies from the solar system in the course of the accumulation of the giant planets and the formation of the cometary cloud. V. S. Safronov.
IAU Symposium No. 45, (see 012.003), p. 329 - 334 (1972).

107.005 The origin of comets. F. L. Whipple.
IAU Symposium No. 45, (see 012.003), p. 401 - 408 (1972).

107.006 On the origin of comets and their importance for the cosmogony of the solar system.
V. G. Fesenkov.
IAU Symposium No. 45, (see 012.003), p. 409 - 412 (1972).

107.007 On the relation between comets and meteoroids.
H. Alfvén.
IAU Symposium No. 45, (see 012.003), p. 485 - 486 (1972).

107.008 The history of the earth. F. Hoyle.
Quarterly Journ. Roy. Astron. Soc., Vol. 13, 328 - 345 (1972). – Presidential address delivered at the anniversary meeting of 1972 February 11.

107.009 Physical conditions in the inner solar nebula.
E. Anders, J. C. Laul, R. Ganapathy, J. W. Morgan.
Bull. American Astron. Soc., Vol. 4, 369 (1972). – Abstr. AAS.

107.010 Numerical simulation of jetstreams. I. The three-
dimensional case. J. Trulsen.
Astrophys. Space Sci., Vol. 17, 241 - 262 (1972).

In two articles we shall present the results from numerical simulations of the dynamical evolution of small body populations with a preferential revolution around a central gravitating body, the small bodies interacting through hard, partially inelastic collisions. It will be assumed that the total mass of the small bodies is much less than that of the central body so that self-gravitational effects can be ignored and the small bodies considered to move in a pure central gravitational force field.

107.011 Numerical simulation of jetstreams. II. The two-dimensional case. J. Trulsen.
Astrophys. Space Sci., Vol. 18, 3 - 20 (1972).

The dynamics of two dimensional jetstreams have been studied by following the evolution of simulation particle populations for different collision models. Collisions, independent of details of the collision model, rapidly lead to the establishment of a distribution of perihelion vectors. The characteristic time for this process being of the order of magnitude equal to the mean free collision time. Under appropriate conditions a radial focusing takes place. In terms of the varians of semimajor axis a focusing exceeding a factor 2 has been achieved. Necessary conditions for the existence of this radial focusing are discussed.

107.012 The floccule theory for planetary formation.
I. P. Williams.
Astrophys. Space Sci., Vol. 18, 223 - 225 (1972).

The accumulation of floccules into protoplanets is discussed, and it is pointed out that the simplifications which have been introduced into recent numerical models may result in the incorrect conclusion being reached.

107.013 On the tides and the Roche limit. T. Kwast.
Urania Kraków, Vol. 43, 290 - 294 (1972).
In Polish.

107.014 Elastic and inelastic scattering in orbital clustering.
D. C. Baxter, W. B. Thompson.
Trans. American Geophys. Union, Vol. 53, 433 (1972).

Loss of orbital kinetic energy through inelastic collisions can cause clustering of particles into similar Kepler orbits, a process which may be relevant in the formation and stability of planets, satellites, comets, or asteroid streams. A Fokker-Planck equation is found which predicts either radial clustering or diffusion.

107.015 Production of light elements in the solar system.
C. Koike.
Progr. Theor. Phys., (Japan), Vol. 48, 66 - 77 (1972).

107.016 Volcanism and the history of the planets.
S. Vsekhsvyatskii.
Soviet Sci. Rev., (GB), Vol. 3, 239 - 245 (1972).

Recent research and discoveries are casting more and more doubt on the hypothesis that solar systems have formed as condensation products of intergalactic gases. The idea that even our own solar system could have originated as a result of a disintegration of stellar material may seem astonishing. This is, however, supported by the high volcanic activity of the planets, the indications of which are only now being discovered.

107.017 Geological evidence relating to the origin and secular rotation of the solar system. G. E. Williams.

Modern Geol., *(GB)*, Vol. 3, 165 - 181 (1972).

107.018 Jet streams and the development of the solar system.
M. L. White.
Nature, Phys. Sci., Vol. 238, 104 - 105 (1972).

Hydrodynamic considerations show that jet streams can develop in a rotating gaseous disk only at certain discrete orbital distances, given by a geometric progression (a^n where a is a constant and n is integer). This progression is a good representation for the distance, from their parent body, of the planets (asteroid belt included) as well as the satellites of low inclination of Jupiter, Saturn and Uranus. The theory also yields streamlines which are logarithmic spirals and, as shown in a more detailed report, may be justifiably applied to the form and structure as well as the dynamics of hurricanes and spiral galaxies.

107.019 The internal structure of planetary bodies.
G. H. A. Cole.
Contemporary Phys., *(GB)*, Vol. 13, 585 - 600 (1972).

Each planet in the solar system is represented as a mechanically spherical cold body, possibly in rotation. It is seen that, on this basis, the properties of the planets can be understood within the already established principles of the physics of matter under high pressures. Internal structures of the planets can be deduced which are consistent with observed data, although no unique detailed structural model can be isolated at this stage.

Xenon in irdischer und in extraterrestrischer Materie (Xenologie).   See Abstr. 061.004.

The astrophysical state of the s-process.
See Abstr. 061.061.

Destruction of the *s* process (*nσ*) correlations by electron capture after neutron irradiation.
See Abstr. 065.081.

Review of the heliosphere. II.   See Abstr. 074.105.

Composition and state of matter in the deep interior of the earth.   See Abstr. 081.004.

Development of Schmidt's theory.
See Abstr. 081.026.

Planets and satellites — a survey of fundamental facts.   See Abstr. 091.050.

Fluidization on the moon and planets.
See Abstr. 094.090.

The origin of the moon and solar system.
See Abstr. 094.092.

Die Tumortheorie als Schlüssel zur Erklärung der Urkrustenbildung und Erstarrung von Mond und Erde.
See Abstr. 094.101.

Paleocratering of the moon: Review of post-Apollo data.   See Abstr. 094.103.

Stability of the solar system: Evidence from the asteroids.   See Abstr. 098.025.

An evolutionary calculation of Jupiter.
See Abstr. 099.095.

Orbit-orbit resonance capture in the solar system.
See Abstr. 100.019.

Diffusion of comets from parabolic into nearly parabolic orbits.   See Abstr. 102.040.

The origin and evolution of the comets and other small bodies in the solar system.   See Abstr. 102.047.

Stars

111 Stellar Parallaxes

111.001 **Experimental trigonometric parallaxes in the Hyades region using a Schmidt telescope of moderate size.**   A. N. Argue, C. M. Kenworthy.
Monthly Notices Roy. Astron. Soc., Vol. 159, 31 - 49 (1972).

As an experiment, parallaxes and proper motions were measured for 51 stars on 64 plates with a single field centre in the Hyades region, using the Cambridge 43/61 cm Schmidt telescope. The internal standard errors for the parallaxes were ±0.017 arc sec at brightness 11^m increasing to ±0.030 at 16^m, and for the proper motions ±0.015 and ±0.025 arc sec yr^{-1} respectively. The mean parallax for 12 of Van Altena's probable main sequence members with apparent magnitude between 10^m3 and 14^m5 was 0.032 ± 0.005 arc sec. Five of his probable or possible subdwarf members were measured and shown to form a group spatially and dynamically separate from the main sequence members.

111.002 **Secular parallaxes of stars and solar motion determined from absolute proper motions with reference to galaxies of 14600 stars.**   N. V. Fatchikhin.
Astron. Tsirk., No. 668, p. 5 - 7 (1972). In Russian.

111.003 **Trigonometric parallaxes of 6 stars.**
I. I. Kanaev, V. A. Sokolova.
Astron. Tsirk., No. 705, p. 7 - 8 (1972). In Russian.

111.004 **Pioneer stellar parallax determinations. On the priority of a certain discovery.**
Z. K. Sokolovskaya.
Vestn. AN SSSR, 1972, No. 3, p. 132 - 136. In Russian.
Abstr. in Referativ. Zhurn. 51. Astron., 9.51.12 (1972).

111.005 **Parallaxes and proper motions. VIII.**
J. C. Titter, W. S. Mesrobian, A. R. Upgren.

Astron. Journ., Vol. 77, 875 - 877 (1972).

Relative parallaxes and proper motions are given for 25 stars, 11 of which have no previous trigonometric parallax determination. The precision determined in the first three papers in this series is found to be maintained for the following ones including this present list.

111.006 **The first measurements of stellar parallax.**
N. S. Hetherington.
Ann. Sci., *(GB)*, Vol. 28, 319 - 325 (1972).

The author discusses the work, mainly carried out by Herschel, which preceded the first successful measurements of stellar parallax. The successful determinations of the parallax of 61 Cygni by Bessel in 1838 and of α Lyrae by Struve in 1840 are discussed. The work of Henderson on the measurement of the parallax of α Centauri is also mentioned.

111.007 **Die Hyaden als Basis der kosmischen Entfernungsskala.**   J. Wempe.
Wiss. Zeitschr. Techn. Univ. Dresden, Vol. 21, No. 3, (see 012.021), 609 - 610 (1972).

Correlation of parallax errors and intrinsic dispersion of the K3—M2 main sequence.
See Abstr. 113.015.

Catalogue of stellar dimensions.
See Abstr. 115.009.

Parallax of Ross 986 and G 87-25.
See Abstr. 119.002.

The parallax and proper motion of Scorpius X-1.
See Abstr. 142.110.

112 Proper Motions, Radial Velocities, Space Motions

112.001 **Radial velocities of some stars in NGC 1893.**
F. S. Jones.
Publ. Astron. Soc. Pacific, Vol. 84, 459 - 460 (1972).
Radial velocity measures of eight stars in the field of NGC 1893 are reported.

112.002 **Radial velocities of southern B stars determined at the Radcliffe Observatory — VI. Stars in H II regions.** D. Crampton.
Monthly Notices Roy. Astron. Soc., Vol. 158, 85 - 98 (1972).
Radial velocities are given for 72 stars which are probably associated with H II regions. Approximately one-half of the stars are suspected of being variable in velocity. One of the stars appears to be an extreme example of a 'runaway star'.

112.003 **Proper motions of field M stars in the region of the Pleiades.** B. F. Jones.
Monthly Notices Roy. Astron. Soc., Vol. 159, 3P - 5P (1972).
In a recent paper, Murray and Sanduleak found a large space density of dwarf M stars with a low velocity dispersion in the solar neighbourhood. Here we present results for a similar investigation in the region of the Pleiades. Good agreement is found with the results of Murray and Sanduleak, but due to several sources of uncertainty, the results should be viewed with some caution.

112.004 **Radial velocities measurement with the coudé spectrograph of the 152-cm telescope (Observatoire de Haute Provence).** C. Fehrenbach.
Astron. Astrophys., Vol. 19, 427 - 433 (1972). In French.
We describe the method of analysis used at the Haute Provence Observatory for plates taken with the coudé spectrograph of the 152-cm telescope. We publish wavelength tables for F5 to M (III to V), F Ia to G Ia stars and a special table for A 8 to F 5 stars. We compare our radial velocity measures for 25 IAU standard stars. These tables have been used for RV determinations of 13 Large Magellanic Cloud supergiants observed at the European Observatory in Chile (ESO) (8 stars of classes B 9 to A 8 — 5 stars F to G).

112.005 **Richtungsabhängige Genauigkeitsunterschiede bei Eigenbewegungen.** K. Ferrari d'Occhieppo.
Anzeiger Österreich. Akad. Wiss. Math.-nat. Kl., 106. Jahrgang, p. 220 - 223 (1970).

112.006 **Reduced-proper-motion diagrams. II. Luyten's white-dwarf catalog.** E. M. Jones.
Astrophys. Journ., Vol. 177, 245 - 249 (1972).
Spectroscopic and photometric surveys by Eggen, Greenstein, and others have yielded almost no classical degenerates with Luyten color class redder than class g. Reduced-proper-motion diagrams made from Luyten's catalog and from a large number of photometrically observed objects show that Luyten's selection criterion will naturally yield a large proportion of subdwarfs and "Eggenites" among the red and yellow candidates.

112.007 **Radial velocities from objective prism plates.**
J. Stock, W. Osborn.
The role of Schmidt telescopes in astronomy. Conference Hamburg 1972, (see 012.014), p. 63 - 64 (1972).

112.008 **Radial velocity observations of southern stars.**
C. Contreras, J. Stock.
Publ. Dep. Astron., Univ. Chile, Obs. Astron. Nacional, Cerro Calán, Santiago de Chile, Vol. 2, (No 2), 40 - 42 (1971).

Radial velocities for twenty-three stars selected from an objective prism survey are communicated. The data indicate that the peculiar G- and K-stars included in the program constitute a high velocity group.

112.009 **Proper Motion Survey with the forty-eight inch Schmidt telescope. XXXI. Proper motions for 2520 faint stars.** W. J. Luyten.
Separate print Univ. Minnesota, Minneapolis, Minnesota. 24 pp. (1972).
In continuation of the earlier list of proper motions for 1,357 stars, the present publication gives similar data for another 2,520 such stars. All of these data were obtained with the automated-computerized plate scanner and measuring machine. Included is a list of errata to No. XXX of this series.

112.010 **A catalogue of stars with proper motions exceeding 0.″2 annually, between +70 and +90.** W. J. Luyten.
Separate print Univ. Minnesota, Minneapolis, Minnesota. 48 pp. (1972).
The catalogue contains 2120 entries.

112.011 **Proper Motion Survey with the forty-eight inch Schmidt telescope. XXXII. Faint proper motion stars near the South Galactic Pole.**
W. J. Luyten, A. E. La Bonte.
Separate print Univ. Minnesota, Minneapolis, Minnesota. 8 pp. (1972).
In continuation of the list published in No. XXVII of this series we are herewith giving data for 744 further proper motion stars near the South Galactic Pole.

112.012 **Three stars of large proper motion.**
W. J. Luyten, A. E. La Bonte.
IAU Circ., No. 2438 (1972).

112.013 **The establishment of 21 new ninth magnitude IAU standard radial velocity stars.**
J. F. Heard, C. Fehrenbach.
Publ. David Dunlap Obs., Univ. Toronto, *Richmond Hill*, Vol. 3, 113 - 123 (1972).
Twenty-four stars of photographic magnitude 8.26 to 9.64 and of spectral types F, G and K, in the declination zone +25° to +30° have been investigated for suitability as an extension of the IAU lists of standard velocity stars. The stars were observed with dispersions of 12, 20 and 15 A/mm at the David Dunlap, Haute Provence and Dominion Astrophysical Observatories respectively, and the spectrograms were measured with systems tested against the IAU system. Three of the 24 stars are believed to have small-range velocity variations. The remaining 21 are presented as new IAU standard velocity stars.

112.014 **On the apparent connection between space velocity and rotational velocity in early type stars.**
O. Havnes.
Inst. Theor. Astrophys. Blindern—Oslo, Rep. No. 32, 18 + 14 pp. (1971).
An apparent increase in space velocity with increasing rotational velocity for single stars earlier than B4 was reported by Havnes (1968). Here we consider a larger sample of stars and find the same tendency for field stars earlier than B6, while the stars from B6 to B9 do not show the same tendency. Some increase of average space velocity with increasing rotational velocity is also found in double stars, cluster members etc. The possibility is considered that systematic errors in the determination of spectral class (absolute magnitude) may be present due to the rotation of the star.

112.015　The results of determination of absolute proper motions of stars with respect to galaxies.
N. V. Fatchikhin.
Publ. 18th Astrometrical Conference 1969, (see 012.022), p. 53 - 60 (1972). In Russian.

A polarimetric method of measuring radial velocities. See Abstr. 034.110.

Catalogue of proper motions of 12590 faint stars in the +25° to −20° declination zone.　See Abstr. 041.001.

Fundamental systems of positions and proper motions.　See Abstr. 041.022.

Comparison between MD and GC catalogues and analysis of the proper motion errors.　See Abstr. 041.044.

Parallaxes and proper motions. VIII. See Abstr. 111.005.

A peculiar B star in the old disk population: 38 Draconis.　See Abstr. 114.073.

A-type supergiants: a list of line intensities and radial velocity measurements.　See Abstr. 114.085.

Effects of velocity and turbulence gradients on A-type supergiant spectra.　See Abstr. 114.112.

The C-classification of the spectra of carbon stars. See Abstr. 114.130.

On differential shifts of lines in the spectrum of the supergiant β Ori.　See Abstr. 114.147.

Luminosities and motions of the F-type stars. II. Metal-deficient stars.　See Abstr. 115.004.

Luminosities and motions of A0 to A2 stars. See Abstr. 115.025.

Spectroscopic-binary frequency among extreme high-velocity dwarfs.　See Abstr. 119.008.

Proper motions of 122 eclipsing variable stars. See Abstr. 121.042.

Proper motion of UX Monocerotis. See Abstr. 122.011.

The kinematics of semi-regular red variables in the solar neighbourhood.　See Abstr. 122.016.

The parallax and proper motion of Scorpius X-1. See Abstr. 142.110.

On the nature of X Persei; evidence from the 1957 outburst.　See Abstr. 142.144.

Membership in the extremely young open cluster NGC 6530 (M8).　See Abstr. 153.010.

Note sur les vitesses radiales des étoiles d'un amas galactique en direction du Grand Nuage de Magellan. See Abstr. 153.015.

113 Stellar Magnitudes, Colors, Photometry

113.001 Photometric and polarimetric study of IR-stars in the optical and infrared parts of the spectrum.
V. A. Dombrovsky, G. V. Khozov.
Astrofizika, Vol. 8, 5 - 16 (1972). In Russian. – English translation in Astrophysics, Vol. 8, No. 1.

The paper presents the results of photometric and polarimetric observations of four IR-stars (NML Cyg, NML Tau, IRC + 10216, VY CMa) in the optical and infrared parts of the spectrum. The interpretation of the results is based on the hypothesis of cool stars having circumstellar dust envelopes.

113.002 Spectral classification through seven-colour photometry. M. Golay.
Vistas in astronomy, Vol. 14, (see 003.008), 13 - 51 (1972).

113.003 Four-colour and Hβ photometry of some bright southern stars. N. R. Stokes.
Monthly Notices Roy. Astron. Soc., Vol. 159, 165 - 177 (1972).

Observations of 370 bright southern A, F and B stars in the four-colour (y, b, v, u) and Hβ systems are reported.

113.004 Infrared photometry of HBV 475 and MHα 328 – 116. R. F. Knacke.
Astrophys. Letters, Vol. 11, 201 - 202 (1972).

HBV 475 and MHα 328–116 have been observed at 3.5, 11.7, and 22 μm. The strong infrared emission from both objects is similar to that observed from planetary nebulae and symbiotic stars.

113.005 A photometric study of the peculiar A star 21 Comae. J. R. Percy.
Journ. Roy. Astron. Soc. Canada, Vol. 66, 218 (1972). Abstr. Canadian Astron. Soc.

113.006 Infrared photometry of northern Wolf–Rayet stars. D. A. Allen, J. P. Swings, P. M. Harvey.
Astron. Astrophys., Vol. 20, 333 - 336 (1972).

Observations of Wolf–Rayet stars have been made at 1.6 and 2.2 μ. The [2.2 μ]–[1.6 μ] colors are found to be systematically greater in WC than in WN stars.

113.007 UBV photometry of some southern stars.
A. W. J. Cousins.
Monthly Notes Astron. Soc. Southern Africa, Vol. 31, 75 - 80 (1972).

113.008 Photometry of the supergiant HR 5171.
G. M. Harvey.
Monthly Notes Astron. Soc. Southern Africa, Vol. 31, 81 - 85 (1972).

113.009 Individual reddening laws from O type stars. I. Computation method, first results. G. Goy.
Astron. Astrophys., Vol. 21, 11 - 23 (1972). In French.

We obtain the individual interstellar extinction laws by means of the Geneva 7-colour photometry. In view of insuring good precision in the determination of the parameters, we restrict our sample to O type stars. We obtain the reddening law by simulation.

113.010 Strömgren photometry of weak-G-band stars.
R. B. Herr, D. J. MacConnell.
Bull. American Astron. Soc., Vol. 4, 310 - 311 (1972). Abstr. AAS.

113.011 Far ultraviolet photometry of the blue Ap stars from OAO-2. D. S. Leckrone.
Bull. American Astron. Soc., Vol. 4, 311 (1972). – Abstr. AAS.

113.012 The relation between red and [3.5] – [11.0] micron colors for carbon stars.
T. D. Faÿ, R. K. Honeycutt, W. H. Warren, Jr.
Bull. American Astron. Soc., Vol. 4, 323 (1972). – Abstr. AAS.

113.013 Correlation coefficients between [3.5] – [11] micron color and the 1.87 and 2.3 micron band intensities for 17 K and M stars. T. D. Faÿ.
Bull. American Astron. Soc., Vol. 4, 323 (1972). – Abstr. AAS.

113.014 Intrinsic ultraviolet colors from OAO-II Celescope observations.
K. Haramundanis, C. Payne-Gaposchkin.
Bull. American Astron. Soc., Vol. 4, 331 (1972). – Abstr. AAS.

113.015 Correlation of parallax errors and intrinsic dispersion of the K3–M2 main sequence. A. R. Upgren.
Bull. American Astron. Soc., Vol. 4, 399 - 400 (1972). Abstr. AAS.

113.016 The application of synthetic spectra to colour system transformations–I. The RGU and UBV systems.
R. A. Bell.
Monthly Notices Roy. Astron. Soc., Vol. 159, 349 - 355 (1972).

Synthetic spectra, computed for F and G dwarf stars, have been used to compute theoretical UBV and RGU colours. These colours have been used to obtain the transformation equations between the two systems. These equations have been checked against UBV and RGU colours computed for hotter stellar models. The effects of metal abundance and gravity on the theoretical colours of the cooler stars have been examined. The effect of reddening on the colours is examined.

113.017 The application of synthetic spectra to colour system transformations–II. The photographic and photoelectric UBV systems. R. A. Bell.
Monthly Notices Roy. Astron. Soc., Vol. 159, 357 - 360 (1972).

Synthetic spectra and the photoelectric and various photographic UBV sensitivity functions are used to compute theoretical colours for F and G dwarfs. Similar calculations are made for hotter stars.

113.018 Detection of far-infrared astronomical sources.
I. Furniss, R. E. Jennings, A. F. M. Moorwood.
Astrophys. Journ., (*Letters*), Vol. 176, L105 - L108 (1972).

Four sources not previously observed in the far-infrared have been detected in the range 40–350 μ from balloon altitude. Flux measurements were also obtained for the infrared objects in the Orion nebula and NGC 2024.

113.019 Photometry of symbiotic and VV Cephei stars in the near infrared (with a note on MWC 56).
J. P. Swings, D. A. Allen.
Publ. Astron. Soc. Pacific, Vol. 84, 523 - 527 (1972).

The present near-infrared survey of symbiotic stars, VV Cephei stars, and "related" peculiar objects indicates that only very strong thermal radiation from dust shells can be detected by infrared photometry at wavelengths shorter than 3.5 μ. Two such objects were found: RX Pup and V1016 Cyg.

113.020 Miscellaneous UBV observations of symbiotic stars.

J. D. Fernie.
Publ. Astron. Soc. Pacific, Vol. 84, 528 = Commun. David Dunlap Obs., Univ. Toronto, *Richmond Hill*, No. 331 (1972).

Occasional photometric observations of nine symbiotic stars during the years 1967−69 are reported.

113.021 Multicolor photometry of the M dwarf Proxima Centauri.
J. A. Frogel, D. E. Kleinmann, W. Kunkel, E. P. Ney, D. W. Strecker.
Publ. Astron. Soc. Pacific, Vol. 84, 581 - 582 (1972).

Photometric observations of Proxima Centauri between $0.3\,\mu$ and $5\,\mu$ show that this star has an effective temperature of $2700°K$, a bolometric luminosity of 6.7×10^{30} ergs sec^{-1}, and a radius of 1.3×10^{10} cm, agreeing with the calibration of the faint end of the main sequence made by Greenstein, Neugebauer, and Becklin (1970).

113.022 A registration of utmostly faint stars with photographic materials of various types.
I. I. Breido, O. M. Mikhailova.
Astron. Zhurn. Akad. Nauk SSSR, Vol. 49, 1098 - 1101 (1972). In Russian. English translation in Soviet Astron. AJ, Vol. 16, No. 5.

On the basis of laboratory models of stars and the background of the sky it is shown that fine-grain, not sensitive photomaterials with large contrast allow at the given background brightness to register considerably fainter stars, than coarse-grain, high sensitive materials.

113.023 Zero point stars in the Harvard E and F regions for UBV photometry. A. W. J. Cousins.
Monthly Notes Astron. Soc. Southern Africa, Vol. 31, 127 - 133 (1972).

113.024 Two *UBV* photoelectric sequences in Cygnus.
B. Stenholm, S. Söderhjelm.
Astron. Astrophys., Suppl. Ser., Vol. 7, 385 - 391 (1972).

Magnitudes and colours in the *UBV* system are given for altogether 73 stars. The magnitude range is approximately $V = 7$ to $V = 13$ and the stars are situated near $\alpha = 21^h20^m$, $\delta = +50°$.

113.025 The application of objective gratings to faint star photometry.
G. A. Harding, R. S. Harbour, K. P. Tritton.
Royal Obs. Bull., [Herstmonceux: Royal Greenwich Obs.], No. 172, p. 25 - 60 (1971).

The use of an objective grating on a 74-inch telescope for the extension of magnitude sequences beyond those available photoelectrically is investigated. The existing two-colour photometry in ω Centauri and NGC 6397 is extended to $V = 19^m.5$ in ω Cen and $V = 19^m0$ in NGC 6397. In both cases the turn-off point from the main sequence is indicated.

113.026 20-micron fluxes of bright stellar standards.
T. Simon, D. Morrison, D. P. Cruikshank.
Astrophys. Journ., *(Letters)*, Vol. 177, L17 - L20 (1972).

Observations made at Mauna Kea Observatory between 1971 July and 1972 May establish relative fluxes in the 17−25-μ band. Based on an absolute flux for α Boo at $20\,\mu$ of 1.54×10^{-16} W cm$^{-2}\,\mu^{-1}$ ($m_{20} = -3.32$) derived from shorter-wavelength observations and from model atmospheres, we obtain 20-μ magnitudes for nonvariable standards as follows: α Tau, -3.21; α Ori, -5.67; β Gem, -1.24; and β Peg, -2.71. Preliminary data on eight other stars are also presented.

113.027 Photographic R magnitudes of 228 stars in Orion.
M. T. Brück.
Publ. Roy. Obs. Edinburgh, Vol. 7, (No. 7), 85 - 91 (1972).

Photographic R magnitudes down to $R = 13^m5$ have been measured for stars in the Orion I association. Combined with V magnitudes of the same stars taken from Walker's observations they allow the lower end of the colour magnitude diagram (V against (V−R)) to be drawn.

113.028 The use of Schmidt telescopes in the southern hemisphere. P. A. Wayman.
The role of Schmidt telescopes in astronomy. Conference Hamburg 1972, (see 012.014), p. 101 - 105 (1972).

113.029 A blue-infrared survey of the sky south of $-36°$. H. Haffner.
The role of Schmidt telescopes in astronomy. Conference Hamburg 1972, (see 012.014), p. 107 - 108 (1972).

113.030 Identification of selected IRC objects.
W. Wiemer.
The role of Schmidt telescopes in astronomy. Conference Hamburg 1972, (see 012.014), p. 111 - 112 (1972).

113.031 Four-colour and Hβ photometry of some bright southern stars−II. N. R. Stokes.
Monthly Notices Roy. Astron. Soc., Vol. 160, 155 - 168 (1972).

Observations of 511 bright southern B, A, F and G stars in the four-colour (y, b, v, u) and Hβ systems are reported.

113.032 Blaue Objekte in der Umgebung von M 31. I. Entdeckungswahrscheinlichkeit neu aufgefundener Blauer Objekte. G. A. Richter, I. Meinunger.
Astron. Nachr., Vol. 294, 39 - 45 (1972).

This paper presents the first part of an extensive investigation of blue objects in the 5C3 area (centre near $0^h\,40^m + 39°$) in order to find out the probability of discovery, the total of quasi-stellar objects, the abundance in correlation with the apparent brightness, as well as the analysis of a relation to radio sources of small diameter in the 5C3 catalogue. Up to the limiting magnitude ($U = 20.1$) 205 objects with $0.00 \geqq U-B \geqq -0.39$ and 11 objects with $U-B \leqq -0.40$ were discovered. Furthermore the relation of the cumulative numbers of stars to the U magnitude has been deduced.

113.033 Photographic photometry of T Tauri stars and related objects in Orion. M. T. Brück.
Publ. Roy. Obs. Edinburgh, Vol. 7, (No.8), 93 - 104 (1972).

The paper completes determinations of photographic V and R magnitudes of 130 T Tauri and Hα emission stars in Orion I down to $R = 13.8$; photographic B and U magnitudes have also been observed in some cases. By comparison with normal stars in the association, 9 of the stars are shown to have colour anomalies. Variability is found to be greater in V than in R, and interstellar extinction is also heavier for these stars than for the normal field. Where luminosities can be computed, results suggest that the stars are more luminous and younger than the normal stars in the field.

113.034 *UBVRI* photometry of North Galactic Pole K giants. II. C. R. Sturch, H. L. Helfer.
Astron. Journ., Vol. 77, 726 - 729 (1972).

UBVRI photometry for 70 more K giants in the North Galactic Pole is presented. A total of 200 giants have now been observed. Both the slight decrease in [Fe/H] with height above the plane and the lack of Upgren giants with extreme metal deficiency reported earlier are confirmed. New space densities are calculated for these stars.

113.035 *UBVRI* photometry of North Galactic Pole K giants. III. Search for halo stars.
C. R. Sturch, H. L. Helfer.
Astron. Journ., Vol. 77, 730 - 732 (1972).

Previous *UBVRI* photometry of K giants in Upgren's area

near the North Galactic Pole has revealed no extremely metal-deficient stars. Since Upgren's spectral survey may have discriminated against such objects, direct visual-red Schmidt plates covering 34 sq deg of the area were searched for red stars not included in the Upgren catalogue. Subsequent $UBVRI$ photometry of the most promising candidates yielded only four stars with large ultraviolet excess. We conclude that the frequency of extremely metal-deficient halo giants among the normal giants with $B < 13$ mag is ~2% at the N. G. P.

113.036 Photometric standards for the southern hemisphere.
II. B. J. Bok, P. F. Bok, E. W. Miller.
Astron. Journ., Vol. 77, 733 - 744, 775 - 796 (1972).

Three years ago Bok and Bok (1969) published Paper I of the present series. Since then, additional photo-electric observations by Bok and Bok in 1971 and by Miller in 1970 and 1972 have resulted in 12 new standard sequences for the Southern Milky Way from Vela to Norma. New standard stars have been added to the previously published sequences. The present paper also includes a minor adjustment to the $U-B$ colors for some of the standard sequence stars that are listed in Paper I. We find that the adjustment is necessary to bring the earlier $U-B$ colors into better agreement with our more recently derived values. Identification charts for 20 sequence fields are included.

113.037 Photometric standard stars. A. W. J. Cousins.
Roy. Obs. Ann., Greenwich — Cape Obs., No. 7, 86 pp. (1971).

(1) Introduction; (2) Mean magnitudes and colours of bright stars south of +10° declination; (3) Mean magnitudes and colours of fainter stars in the equatorial zone; (4) $U - B$ for stars in the E regions; (5) Errata to the Washington photoelectric catalogue (Blanco et al.).

113.038 Infrared excesses in early-type stars: Free-free
emission. H. M. Dyck, R. W. Milkey.
Publ. Astron. Soc. Pacific, Vol. 84, 597 - 612 (1972). — Invited symposium paper presented at the Santa Cruz meeting of the Astronomical Society of the Pacific, 26–29 June 1972.

We demonstrate that the infrared color excesses of many hot stars may be explained as low-temperature ($T \lesssim 3000°$K) free-free emission. We present a model for the circumstellar region in which metal atoms are radiatively ionized by the stellar flux and infrared emission occurs from H-minus free-free transitions.

113.039 Infrared excesses in supergiant stars: Evidence for
silicates. E. P. Ney.
Publ. Astron. Soc. Pacific, Vol. 84, 613 - 618 (1972).

Invited symposium paper presented at the Santa Cruz meeting of the Astronomical Society of the Pacific, 26–29 June 1972.

113.040 Photometric variability of the Wolf-Rayet star CV
Serpentis. N. D. Morrison, S. C. Wolff.
Publ. Astron. Soc. Pacific, Vol. 84, 635 - 637 (1972). — Presented at the Santa Cruz meeting of the Astronomical Society of the Pacific, 26–29 June 1972.

Four-color ($uvby$) photometry in 1970 and 1971 of CV Ser indicates that the only major light variation (amplitude ~ $0^{\mathrm{m}}16$) occurs in b, and is apparently caused by variability in the blend of C III lines at $\lambda 4652$. Continuum measurements in u, v, and y show a shallow minimum (~ $0^{\mathrm{m}}03$) at the same phase as minimum light in b. There is no evidence of a secondary minimum in any of the wavelength regions observed.

113.041 Reddening and infrared excesses in T Tauri stars.
C. L. Imhoff.
Publ. Astron. Soc. Pacific, Vol. 84, 641 (1972). — Abstr. Astron. Soc. Pacific.

113.042 Far-infrared and $uvby$ photometry of V1057 Cygni.
T. Simon, N. D. Morrison, S. C. Wolff, D. Morrison.
Publ. Astron. Soc. Pacific, Vol. 84, 644 (1972). — Abstr. Astron. Soc. Pacific.

113.043 Infrared photometry of the Pleiades lower main
sequence. R. R. Zappala.
Publ. Astron. Soc. Pacific, Vol. 84, 647 (1972). — Abstr. Astron. Soc. Pacific.

113.044 A direct photometric comparison of M 67 and
Hyades stars. C. Sturch.
Publ. Astron. Soc. Pacific, Vol. 84, 666 - 668 (1972).

Photoelectric photometry of stars in M 67 and the Hyades when transformed to the UBV system implies an observational inconsistency of $-0^{\mathrm{m}}05$ in $(U - B)$ in published values for the two clusters. If one accepts this correction, the $(U - B)$, $(B - V)$ diagrams for the two clusters are nearly indistinguishable. An interpretation of the new color indices leads to negligible reddening for M 67 and a mild metal dificiency.

113.045 Photometry of the early-type stars from the Tonant-
zintla lists, near the south galactic pole.
A. G. D. Philip.
Publ. Astron. Soc. Pacific, Vol. 84, 677 - 679 (1972).

A spectroscopic survey of a 230-square-degree region surrounding the south galactic pole (Sanduleak and Philip 1968) enabled spectral classifications to be made for the 16 early-type stars in common with the Tonantzintla lists. These stars have now been measured photoelectrically in the Strömgren four-color system. Most of the stars on the list are normal, early-type Population I stars. Four of the stars have δc_1 indices > 0.15, implying higher luminosity.

113.046 Ultraviolet photometry from the Orbiting Astro-
nomical Observatory. V. The helium-weak stars.
P. L. Bernacca, M. R. Molnar.
Astrophys. Journ., Vol. 178, 189 - 201 (1972).

Ultraviolet filter photometry with the Wisconsin Experiment Package aboard OAO-2 has been carried out for the helium-weak stars 3 Cen A, 3 Sco. HD 144334, HD 21699, HD 144844, α Scl, HR 8535, HR 8770, and HD 144661. The flux distribution is compared with that of B3 to B8 main-sequence and giant stars in terms of color-color diagrams. After taking into consideration line blocking shortward of 2800 Å due to lines of P II, Ga II, Si II, Si III, Ti II, and Sr II, we find that most of the helium-weak stars have normal fluxes in good agreement with their ground-based colors.

113.047 UBV photometry of some selected stars in the
Hyades. P. Pesch.
Astrophys. Journ., Vol. 178, 203 - 206 (1972).

Consideration of both new and existing photoelectric UBV photometry shows that systematic errors in earlier color indices are responsible for most of the points which, in a color-magnitude diagram of the Hyades cluster, fall between the main sequence and the white-dwarf sequence. The light curve of a flare of a Hyades member not previously known to be a flare star is presented.

113.048 Interstellar absorptions and colour-excesses in the
Vilnius photometric system. G. Kavaliauskaitė.
Bull. Vilnius Astron. Obs., No. 33, p. 3 - 14 (1972). In Russian.

The total interstellar absorptions and colour-excesses of the Vilnius photometric system are computed and plotted in figures against $(B-V)_0$ for 213 stars. The variations of A and E of the Vilnius system show considerably smaller range than in the system UBV due to narrowness of the response bands. However, they cannot be neglected when exact investigation of interstellar absorption effects is necessary.

113.049 **Photographic photometry of stars in the region of the open cluster NGC 6871 in the Vilnius photometric system. Part I.** A. Bogdanovičius, V. Straižys.
Bull. Vilnius Astron. Obs., No. 33, p. 15 - 27 (1972).
In Russian.

The seven-color photometric system UPXYZVS is applied photographically to classify the stars in spectral types and luminosities and to investigate the interstellar absorption in the $1.^\circ 1 \times 1.^\circ 3$ area around the open cluster NGC 6871 in Cygnus.

113.050 **Photoelectric photometry of stars in the system UPXYZVTS. VI.** K. Zdanavičius, V. B. Nikonov, J. Sūdžius, V. Straižys, Z. Sviderskienė, R. Kalytis, E. Jodinskienė, E. Meištas, G. Kavaliauskaitė, V. Jasevičius, G. Kakaras, A. Bartkevičius, A. Gurklytė, R. Bartkus, A. Ažusienis, J. Sperauskas, A. Kazlauskas, V. Žitkevičius.
Bull. Vilnius Astron. Obs., No. 34, p. 3 - 29 (1972).
In Russian.

The previous papers of the present series contained the catalogues of the stars intended to use for calibration of the Vilnius photometric system in spectral classes, absolute magnitudes, chemical composition and for determination of ZAMS. With the present paper we start to publish the results of multicolor observations of BS stars having no published two-dimensional classification. The results of photometry and two-dimensional classification of 530 stars are given in a table.

113.051 **Photometric effects of rapid stellar axial rotation.** V. Žitkevičius, V. Straižys.
Bull. Vilnius Astron. Obs., No. 34, p. 30 - 48 (1972).
In Russian.

The effects of stellar axial rotation with break-up velocity on color indices of the photometric system UBV and the Vilnius photometric system are calculated.

113.052 **Ultraviolet photometry from the Orbiting Astronomical Observatory. IV. Photometry of late-type stars.** L. R. Doherty.
Astrophys. Journ., Vol. 178, 727 - 742 (1972).

Broad-band interference-filter photometry obtained with OAO-2 at four ultraviolet wavelengths is presented for 57 normal, bright stars of spectral type A2 to M. Relative digital counting rates are given for selected filter-photometer combinations with effective wavelengths of 3320, 2980, 2460, and 1910 Å. Color-color diagrams show that there is remarkably little intrinsic variation in ultraviolet properties among stars with the same visual color and luminosity class, if class I supergiants are excepted. OAO and ground observations of ι UMa (A7 V), 24 UMa (G4 IV), and α Ari (K2 III) from 1910 Å to 1μ are compared with blanketed models from the Smithsonian grid.

113.053 **Four-color and Hβ photometry for open clusters. VIII. IC 4665.** D. L. Crawford, J. V. Barnes.
Astron. Journ., Vol. 77, 862 - 868 (1972).

Photoelectric $uvby\beta$ photometry of 45 B-, A-, and F-type stars contained in or near the cluster IC 4665 were obtained at the Kitt Peak National Observatory between 1964 and 1972. Thirty-two of these stars appear to be cluster members.

113.054 **UBV and Hβ photometry of the galactic cluster NGC 2343.** J. J. Clariá.
Astron. Journ., Vol. 77, 868 - 874, 893 (1972).

The results are presented of a photoelectric study of the galactic cluster NGC 2343. UBV observations of 55 stars in the vicinity of the cluster confirm a minimum membership of 32 stars. Photoelectric measurements of the Hβ line intensity of 16 early-type stars are given.

113.055 **Photométrie photoélectrique U.B.V. de 160 étoiles supergéantes O–B membres probables du Grand Nuage de Magellan.**
J. H. Bigay, A. Bernard, G. Paturel, S. Roux.
Proc. Third Colloquium on astrophysics. Supergiant stars, Trieste 1971, (see 012.020), p. 108 - 118 (1972).

113.056 **Photométrie en six couleurs de 12 supergéantes F et G du Grand Nuage de Magellan.**
P. Mianes, J. Rousseau.
Proc. Third Colloquium on astrophysics. Supergiant stars, Trieste 1971, (see 012.020), p. 147 (1972). – Abstract.

113.057 **A–F supergiants in the Geneva Observatory photometric system.** B. Hauck, C. Nicollier.
Proc. Third Colloquium on astrophysics. Supergiant stars, Trieste 1971, (see 012.020), p. 153 - 159 (1972).

113.058 **Eight color narrow band infrared photometry of M-type supergiants.** N. M. White.
Proc. Third Colloquium on astrophysics. Supergiant stars, Trieste 1971, (see 012.020), p. 160 - 167 (1972).

113.059 **Ultraviolet photometry of B-type stars.** R. L. Bottemiller.
Bull. American Astron. Soc., Vol. 4, 423 (1972). – Abstr. AAS.

113.060 **UBV photoelectric observations: I. Stars within 25 parsecs of the sun; II. Stars in quasar and galaxy fields; III. Stars in Kapteyn Selected Areas; IV. Miscellaneous stars.** E. A. Epps.
Royal Obs. Bull., [Herstmonceux: Royal Greenwich Obs.], No. 176, p. 127 - 145 (1972).

UBV photometry is presented for 115 nearby stars (within 25 pc of the sun), for bright secondary standards in ten quasar fields and ten galaxy fields, and for stars in eleven Kapteyn Selected Areas. In addition UBV data are given for several miscellaneous stars including R CrB, δ Cep C, six M supergiants, five red giants in M 67 and four horizontal branch stars.

113.061 **On the ratio of total to selective absorption for carbon stars.** R. K. Honeycutt.
Publ. Astron. Soc. Pacific, Vol. 84, 823 - 826 = Publ. Goethe Link Obs., Indiana Univ., *Bloomington,* No. 145 (1972).

The ratio of total to selective absorption for two carbon stars is evaluated by numerical integration of photoelectric spectral scans over the B and V bandpasses of the UBV system. It is found that $R = 3.8$ for carbon stars with very little dependence of R on color excess.

113.062 **Comparative UBVR photometry of Orion flare stars and Hα emission-line stars.** A. D. Andrews.
Bol. Obs. Tonantzintla y Tacubaya, Vol. 6, (No. 38), 161 - 178 (1972).

Multi-colour photographic photometry for 279 flare stars and emission stars in the Orion aggregate is presented. These objects depart entirely from the standard UBVR colour relations for main sequence stars. A comparative colorimetric study indicates that the presence of eHα is a useful criterion on a statistical basis for separating out the more anomalous objects in the colour-magnitude and two-colour arrays. The blueward trend of stars of lower luminosity towards the main sequence in the B–V/V diagram, already evident at $V \sim 15^m$, takes the form of a band crossing the main sequence entirely at $V \sim 16^m$, $B-V \sim 1.^m3$. Non-emission flare stars generally appear at constant brightness at photographic accuracies in the UBVR bands. Evolutionary interpretations in terms of pre-main sequence tracks for stars of discrete mass are suggested.

113.063 **Photometric observations of AG Pegasi.** E. E. Mendoza V.

Bol. Obs. Tonantzintla y Tacubaya, Vol. 6, (No. 38), 211 - 214 (1972).

We have made *UBVJHKL* broad-band photometry, and Hβ and Hδ narrow-band photometry of the symbiotic star AG Pegasi. The β and δ-indices for this object are the strongest among nearly 200 Be stars (which include P Cygni) and standard stars. The *UBV* photometry of AG Peg is similar to a reddened early type star. Its infrared photometry corresponds very closely to an M3 III star. This confirms the results of several spectroscopic studies which conclude that this nova-like star is a binary, comprising a hot WN6 component and a cool M3 III object embedded in a gas cloud.

113.064 Die blauen Objekte. N. Richter.
Jenaer Rundschau, (Jena Rev.), 17. Jahrgang, p. 322 - 327 (1972).

113.065 UBV-system for a 12 inch telescope.
S. Sivertsen.
Inst. Theor. Astrophys. Blindern — Oslo, Rep. No. 34, 16 pp. (1972).

In this report the UBV-system, established at the Oslo Solar Observatory, and the transformation and extinction results obtained are shortly described.

113.066 The corrections for B and V magnitudes of α Ori and μ Gem due to absorption in bands and lines.
A. V. Dragunova.
Astron. Tsirk., No. 734, p. 1 - 2 (1972). In Russian.

113.067 On the relation between colour and morphology of a photographic star image. S. B. Vladimirov.
Astron. Tsirk., No. 734, p. 2 - 6 (1972). In Russian.

Digitization of photoelectric photometry at the Okayama Astrophysical Observatory. See Abstr. 031.059.

Digital recording system for photoelectric photometry at the Dodaira Station. See Abstr. 031.060.

Reduction program of digital data in photoelectric photometry. See Abstr. 031.061.

Uniformly rotating stars with hydrogen- and metallic-line blanketed model atmospheres. See Abstr. 065.056.

The $(B-V)$ and $(U-B)$ color indices of the sun. See Abstr. 080.023.

Reduced-proper-motion diagrams. II. Luyten's white-dwarf catalog. See Abstr. 112.006.

On the apparent connection between space velocity and rotational velocity in early type stars. See Abstr. 112.014.

Identification of IRC-objects. See Abstr. 114.024.

Stellar compositions from narrow-band photometry — III. Iron abundances in a further 120 evolved stars. See Abstr. 114.062.

Long wavelength spectrometry and photometry of M, S and C-stars. See Abstr. 114.079.

Spectra of Upgren's unclassified stars. See Abstr. 114.081.

Study of a sample of faint M stars in the direction of the North Galactic Pole. See Abstr. 114.098.

Stellar compositions from narrow-band photometry— IV. Calcium abundances in G and K giants. See Abstr. 114.103.

Additional observations of some stars with strong helium lines. See Abstr. 114.166.

Five-colour photometry of the eclipsing binary HO Telescopii. See Abstr. 121.026.

Infrared excesses in eclipsing binaries of the RS Canum Venaticorum type. See Abstr. 121.075.

Near-infrared photometry of Mira variables. See Abstr. 122.015.

Variable stars of small amplitude. I. Supergiants and OB stars. See Abstr. 122.031.

Intrinsic colours of classical cepheids and yellow supergiants. See Abstr. 122.037.

***uvby* photometry of EH Librae.** See Abstr. 122.096.

HR 6684: A new Beta Cephei-type variable star. See Abstr. 122.097.

Photoelectric monitoring of BL Lacertae. See Abstr. 122.098.

***UBV* observations of the RR Lyrae variable HD 176387 (MT Telescopii).** See Abstr. 122.124.

Photometric systems for quantitative studies of white dwarfs. See Abstr. 126.019.

Interstellar reddening near the north galactic pole. See Abstr. 131.052.

***UBV* observations of 3 C 273.V.** See Abstr. 141.123.

The pulse shape of the Crab nebula pulsar NP 0532 as a function of color. See Abstr. 141.532.

The *UBV* colors of Sco X-1 at the time of an X-ray observation. See Abstr. 142.053.

On the color-magnitude locus of Scorpius X-1. See Abstr. 142.100.

A search for short-period light variability among stars in h and χ Persei. See Abstr. 153.002.

Photographic photometry of NGC 6866. See Abstr. 153.027.

***UBV* and Hβ photometry of OB stars in a Milky Way region in Norma.** See Abstr. 155.004.

Distribution of O to A0 stars in a region in Vela $(l = 268°.5, b = -0°.3)$. See Abstr. 155.059.

Light variations of high luminosity O and B stars in the Large Magellanic Cloud. See Abstr. 159.008.

Photometric standards in the Magellanic Clouds. See Abstr. 159.010.

114 Stellar Spectra, Temperatures, Spectroscopy

114.001 On the spectrum of the Be star MWC 939.
V. P. Arhipova, O. D. Dokuchaeva.
Peremennye Zvezdy, Vol. 18, 289 - 292 (1972). In Russian.

On objective-prism spectrograms of the Bep-star MWC 939, taken in 1960, 1968 and 1970, the identification of the emission lines in Hα−Hε region is made. The absolute intensities of 10 emissions were measured. The appearance of the spectrum as a whole is similar to that of η Car. It shows considerable changes as compared with that observed by R. Minkowski. The observed spectrophotometric temperature of the star is obtained (T = 10400°K).

114.002 Discoveries on southern objective-prism plates, III.
Three new hydrogen-deficient stars and a bright
B-type subdwarf.
D. J. MacConnell, R. L. Frye, W. P. Bidelman.
Publ. Astron. Soc. Pacific, Vol. 84, 388 - 391 (1972).

Four relatively bright peculiar stars are presented; three show no hydrogen lines, and the fourth is a subdwarf B star. *UBV* photometry is given for three of the stars.

114.003 Is HR 5491 really a pulsating Am star?
M. Breger, H. M. Maitzen, A. P. Cowley.
Publ. Astron. Soc. Pacific, Vol. 84, 443 - 445 (1972).

The Am star at the south pole, HR 5491, reported by Bessell and Eggen (1972) to be pulsating has been reinvestigated. The metallic-line nature is confirmed, but careful photometric observations on 10 November 1971 showed no evidence of light variability.

114.004 Fourteen new peculiar stars. H. E. Bond.
Publ. Astron. Soc. Pacific, Vol. 84, 446 - 447 =
Contr. Louisiana State Univ. Obs., *Baton Rouge,* No. 62 (1972).

A list of stars with peculiar spectra noted on objective-prism plates is given. The list includes seven peculiar A-type stars, six metallic-line stars, and a peculiar B-type star.

114.005 On the Wolf-Rayet stars HD 90657 and HD 117688.
V. S. Niemela.
Publ. Astron. Soc. Pacific, Vol. 84, 450 - 453 (1972).

Recent spectrographic observations of HD 90657 and HD 117688 do not confirm their earlier classification as W-R objects that are intermediate to those that belong to the WN and WC sequences. HD 90657 is very probably a spectroscopic binary with a period of a few days.

114.006 The effect of systematic gf-value errors on stellar
curves of growth.
J. C. Evans, L. W. Schroeder.
Publ. Astron. Soc. Pacific, Vol. 84, 454 - 458 (1972).

From an analysis of 101 Fe I lines in the spectrum of the star θ UMa, a value has been derived for the magnitude of the systematic increase, noted by several laboratory studies, in the Corliss and Tech (1968) gf-values with increasing energy of the upper level. The effect of this systematic error on the curve of growth is to spread the observational data along the direction of the abscissa.

114.007 The application of Michelson interferometers to
high resolution astronomical spectrometry.
J. Ring, C. L. Stephens.
Montly Notices Roy. Astron. Soc., Vol. 158, 5P - 9P (1972).

Griffin has shown that the instrumental profiles of existing high resolution spectrographs are such as to introduce errors in the measurement of absorption line parameters. It is shown here that a Michelson interferometer is theoretically

much better suited to the observation of stellar line profiles at high resolution.

114.008 The analysis of the Small Magellanic Cloud super-
giant HD 7583. A. Przybylski.
Monthly Notices Roy. Astron. Soc., Vol. 159, 155 - 163 (1972).

In a coarse analysis the Small Magellanic Cloud star HD 7583 has been compared with α Cygni and HD 67456. It has been found that: 1. Helium is normal or nearly normal in HD 7583; 2. Metals are deficient by a factor of 10.

114.009 The helium-weak stars. M. R. Molnar.
Astrophys. Journ., Vol. 175, 453 - 464 (1972).

In an extensive investigation of spectral classification, the class of helium-weak stars is described. Their spectra are found to resemble the Bp-type stars with the distinction that helium-weak group does not have the profuse and strongly enhanced metal lines characteristic of the Bp stars. In a quantitative analysis using model atmospheres, several helium-weak stars are found to have apparent helium deficiencies. On the other hand, HD 37129, HD 36629, and HD 37807, previously classified as "helium-weak," appear to be normal early B-type stars.

114.010 Remarks on the proposed identification of prome-
thium in HR 465.
S. C. Wolff, N. D. Morrison.
Astrophys. Journ., Vol. 175, 473 - 475 (1972).

Spectrograms of HR 465 obtained in 1966−1967 show that Pm II was not present at that time. A rediscussion of the observations made in 1960−1961 indicates that there is doubt that Pm was ever present.

114.011 Comments on the identification of promethium in
HR 465. C. R. Cowley, M. F. Aller.
Astrophys. Journ., Vol. 175, 477 - 480 (1972).

The general principles of line identification are discussed in connection with the identification of promethium in HR 465.

114.012 On the N III λλ4640, 4097 lines in Of stars.
D. Mihalas, D. G. Hummer, P. S. Conti.
Astrophys. Journ., (*Letters*), Vol. 175, L99 - L104 (1972).

Detailed calculations based on non−LTE plane-parallel model atmospheres show that the N III emission lines at λλ4634, 4640, 4641 observed in Of stars are produced primarily by dielectronic recombination to $3d\ ^3D$ followed by the $3d$-$3p$ transition in a compact atmosphere.

114.013 Dwarf K and M stars in the southern hemisphere.
A. R. Upgren, R. Grossenbacher, W. S. Penhallow,
D. J. MacConnell, R. L. Frye.
Astron. Journ., Vol. 77, 486 - 499 (1972).

In the course of the near-complete Michigan Spectral Survey of the Southern Sky, 624 dwarf stars of type K 2 V and later have been identified. Many of these stars are previously unrecognized as nearby late dwarfs.

114.014 Two new He I lines in the spectra of B-type super-
giants. R. van Helden.
Astron. Astrophys., Vol. 19, 388 - 392 (1972).

Seven supergiants are found with He I 8582.65 in their spectra.

114.015 Abundances in late-type giants.
D. M. Gottlieb, R. A. Bell.

Astron. Astrophys., Vol. 19, 434 - 452 (1972).

The narrow-band photometry of Spinrad and Taylor (1969) is analyzed by the synthetic spectra technique. Surface gravities, Doppler broadening velocities, effective temperatures and metal abundances are determined for 131 K giants. Evolutionary tracks and gravities determined from the photometry are used to give space positions, space velocities, and absolute magnitudes of the 131 stars.

114.016 **A search for He-weak stars in very young clusters.**
P. L. Bernacca, F. Ciatti.
Astron. Astrophys., Vol. 19, 482 - 487 (1972).

Spectral classification has been carried out for B-type stars in Ori OBl Ib, NGC 2362, IC 5146 and NGC 2264. Six stars with weak helium lines for their UBV colors were found in Ori OBl Ib, one in NGC 2362 which might be helium variable, none in IC 5146.

114.017 *UBV* **photometry of** o **Andromedae.** E. H. Olsen.
Astron. Astrophys., Vol. 20, 167 - 168 (1972).

253 *UBV* observations of the shell star o And have been made. The data do not show the eclipses with a period around $1\overset{d}{.}6$, which has been indicated by earlier observations. A possible period of about $1\overset{d}{.}018$ has been found. The evidence for the binary nature of o And is still very weak.

114.018 **On the chemical composition of** ε **Pegasi.**
J. van Paradijs, H. de Ruiter.
Astron. Astrophys., Vol. 20, 169 - 171 (1972).

Spectra of the star ε Peg (K 2 Ib) have been analyzed, using simple model atmospheres, in order to determine the effective temperature, gravity and the chemical composition.

114.019 **Ten-micron spectroscopy of circumstellar shells.**
R. H. Gammon, J. E. Gaustad, R. R. Treffers.
Astrophys. Journ., Vol. 175, 687 - 691 (1972).

The infrared spectra of several stars with 10-μ "silicate" excesses have been observed at high resolution. A broad emission feature extending from 850 to 1100 cm^{-1} was found, but no fine structure was seen. The excess is consistent with radiation from dust composed of basic silicates but not with that from acidic minerals such as quartz.

114.020 **The measurement of polarized 10-micron radiation from cool stars with circumstellar shells.**
R. W. Capps, H. M. Dyck.
Astrophys. Journ., Vol. 175, 693 - 697 (1972).

We report observations of the linear polarization of 8-14-μ radiation from 10 cool stars. We find that the level of measurable polarization is small, about 1 percent, but that a possibly significant detection has been achieved for several stars. The implications of the data are discussed briefly.

114.021 **Observations of the He II** λ10124 **line in O and Of stars.** D. Mihalas, G. W. Lockwood.
Astrophys. Journ., Vol. 175, 757 - 764 (1972).

Observed equivalent widths of He II λ10124 in absorption-line O stars are found to be in agreement with non-LTE computations assuming no overlap of H and He II lines, and in disagreement with the predictions of LTE and of non-LTE calculations assuming exact coincidence of even-even He II lines and the corresponding hydrogen lines. It is therefore inferred that pumping of He II transitions by H lines does not actually occur in the O stars.

114.022 **A method for absolute calibration of objective-prism plates suitable for computer reduction.**
A. Ardeberg, B. Virdefors.
Astron. Astrophys., Vol. 20, 177 - 188 (1972).

A method for photometric calibration of objective-prism plates is presented. Use is made of available information on absolute and relative continuum energy distribution of standard stars. All dependence of the calibration on wave-length is avoided. Atmospheric extinction does not influence the results. Thus, both monochromatic magnitudes and absolute gradients in international systems can be directly obtained from the objective-prism plates. A computer programme is discussed and a programme outline is presented.

114.023 **Transuranium elements in HD 25354.**
M. Jaschek, E. Brandi.
Astron. Astrophys., Vol. 20, 233 - 235 (1972).

Arguments are presented for the presence of U, Th and W in the Ap star HD 25354. Evidence is also shown for the presence of some transuranium elements, like Am and Cm. The presence of several elements not usually found in Ap stars, with Z about 40 might lend support to the occurrence of nuclear fission in this star.

114.024 **Identification of IRC-objects.**
G. V. Schultz, W. Wiemer.
Astron. Astrophys., Vol. 20, 317 - 319 (1972).

Six early type stars selected from the Caltech Infrared Catalog were identified as late type stars by infrared photometry and spectroscopy.

114.025 **Observations of the stellar Mg II resonance doublet at 2795 and 2802 Å.**
Y. Kondo, R. T. Giuli, J. L. Modisette, A. E. Rydgren.
Astrophys. Journ., Vol. 176, 153 - 164 (1972).

Spectrophotometry of the Mg II resonance doublet at 2795.5 and 2802.7 Å has been performed for five stars with a balloon-borne telescope and spectrometer. Spectral resolution attained was between 0.25 and 0.5 Å. A description of the payload and flight operations is given. The stars observed were β Lyr, γ Lyr, β Cas, α CMi, and α Ori.

114.026 **The unusual hot star BD−7°3007.**
A. P. Cowley, D. J. MacConnell.
Astrophys. Journ.,(*Letters*), Vol. 176, L27 - L30 (1972).

The almost continuous spectrum of BD−7°3007 is shown to have very broad hydrogen lines with weak and variable central emission. In addition, both Ca II and He I are definitely present. The spectroscopic and photometric properties are compared with white dwarfs, old novae, and other subluminous objects.

114.027 **High-frequency stellar oscillations. IX. The peculiar blue variable, BD−7°3007.**
J. E. Hesser, B. M. Lasker, P. S. Osmer.
Astrophys. Journ.,(*Letters*), Vol. 176, L31 - L36 (1972).

The peculiar star BD−7°3007 varies rapidly with amplitudes less than 0.10 mag. The power spectra show no harmonic features; however, significant low-amplitude transient activity exists in the 100−1000 s range; and there are persistent broad-banded peaks at about 21 and 10.3 minutes. Scanner observations imply that the star is degenerate and deficient in hydrogen. The available data admit the presence of two degenerate stars with circumstellar material.

114.028 **Anomalous abundance of lithium in the atmosphere of the star 105 Her.** M. E. Boyarchuk.
Izv. Krymskoj Astrofiz. Obs., Vol. 44, 18 - 32 (1972). In Russian.

Five spectrograms of the K4 II star 105 Her in the region 6870−5800 Å have been investigated by the curve-of-growth method. The equivalent widths of 580 absorption lines were measured. The physical parameters of the stellar atmosphere have been obtained. Lithium is about 6 times overabundant in comparison with the sun and $\sim 2 \times 10^2$ times overabundant, when compared with the average Li abundance in K4 stars.

114.029 Early data from the ultraviolet sky-scan telescope in the TD1 satellite. R. Wilson, S. Gardier, C. Jamar, J. P. Macau, D. Malaise, A. Monfils, H. E. Butler, C. M. Humphries, K. Nandy, G. I. Thompson, P. J. Barker, H. Wroe, L. Houziaux, A. Boksenberg.
Nature, Phys. Sci., Vol. 238, 34 - 36 (1972).

An initial examination of data from the TD1 satellite telescope shows it to be functioning satisfactorily. The spectra observed are in good agreement with information obtained from rocket-borne telescopes.

114.030 X-ray flux from HR 465.
R. Mitalas, J. M. Marlborough.
Journ. Roy. Astron. Soc. Canada, Vol. 66, 217 (1972).
Abstr. Canadian Astron. Soc.

114.031 The line profiles of an oblique rotator.
A. E. Falk, W. H. Wehlau.
Journ. Roy. Astron. Soc. Canada, Vol. 66, 218 (1972).
Abstr. Canadian Astron. Soc.

114.032 New variable carbon stars. III.
Z. Alksne, A. Alksnis.
Astron. Tsirk., No. 670, p. 7 - 8 (1972). In Russian.

114.033 On the accuracy of the spectrophotometric standard of α Lyr. N. S. Komarov, V. A. Pozigun.
Astron. Tsirk., No. 671, p. 5 - 6 (1972). In Russian.

114.034 Étude des étoiles à raies métalliques par une méthode photométrique à bandes passantes étroites. M. Gerbaldi.
Comptes Rendus Acad. Sci. Paris, Sér. B, Vol. 275, 295 - 298 (1972).

On examine le comportement de la raie K du Ca II et les variations d'abondance des métaux pour les étoiles à raies métalliques par rapport aux étoiles de type A normales.

114.035 Ultraviolet absorption lines in the spectrum of Vega.
G. A. Gurzadyan, J. B. Ohanesyan.
Astron. Astrophys., Vol. 20, 321 - 324 (1972).

Ultraviolet spectrograms of Vega, covering the wavelength 2000–3800 Å, obtained with the help of the orbital astrophysical observatory "Orion", installed in the space station "Salyut". More than ten ultraviolet absorption lines and bands have been detected and identified. The strongest line is the resonance doublet of ionized magnesium. All the other absorption lines or bands (blended lines) are produced by metals, mainly neutral and ionized iron and nickel.

114.036 Quantitative analysis of γ Capricorni.
P. L. Selvelli.
Astron. Astrophys., Vol. 20, 325 - 332 (1972).

Our purpose is to determine the chemical composition of the atmosphere of γ Cap. We have used the method of the differential curve of growth, comparing γ Cap with the standard F0 V star 30 L Mi. A comparison has been made also with the Am star 63 Tau.

114.037 Equivalent width of interstellar molecular lines. II. CO, SiO, CS, H_2 and HD in the interstellar spectrum of ξ Persei. K. S. K. Swamy, S. P. Tarafdar.
Astron. Astrophys., Vol. 20, 341 - 345 (1972).

It appears that the density of H_2 is larger than that of H in the direction of ξ Per. Calculation of the equivalent widths of the lines of the bands $(C-X)$ of H_2 and $(B-X)$ and $(C-X)$ of CO and HD reveal that they may be observable shortward of 1150 Å.

114.038 On the abundances of noble gases in extreme population I matter and the sun. M. Scholz.
Astron. Astrophys., Vol. 20, 465 - 467 (1972).

Recent abundance determinations in extreme population I matter on the one hand and the sun on the other indicate that helium has about the same abundance but neon and argon are enhanced in young objects.

114.039 Highly excited hydrogen lines in stellar spectra. II.
S. Barcza.
Astrophys. Space Sci., Vol. 16, 372 - 385 (1972).

In the first part of the paper the shifts of the last visible Balmer lines as a function of the gas density were predicted theoretically. Here spectroscopic research for these shifts is reported: 10 Å/mm^{-1} spectra of six stars were examined and published material of other authors was also used.

114.040 The spectrum of the supergiant ϵ Orionis (B0 Ia). I. Identifications, equivalent-widths, line profiles.
H. J. Lamers.
Astron. Astrophys., Suppl. Ser., Vol. 7, 113 - 132 (1972).

The visual spectrum of the supergiant star ϵ Ori (B0 Ia, HD 37128) is described. A list of observed lines, their identification and blend-free equivalent-width is presented. The two unidentified emission lines at $\lambda 4486$ and $\lambda 4504$ are found to be present. The profiles of a number of important spectral lines are given.

114.041 On the accuracy of energy distribution in spectra of stars recommended as "primary standards".
V. M. Tereshchenko, L. D. Frishberg, A. V. Kharitonov.
Astron. Tsirk., No. 686, p. 1 - 3 (1972). In Russian.

114.042 Infrared spectroelectrophotometry of zirconium stars. R. I. Chuprina.
Astron. Tsirk., No. 686, p. 3 - 5 (1972). In Russian.

114.043 Interstellar lines in the ultraviolet spectrum of Zeta Ophiuchi. A. M. Smith.
Astrophys. Journ., Vol. 176, 405 - 423 (1972).

Interstellar lines arising in C^0, C^+, O^0, Si^+, and S^+ observed in the ultraviolet spectrum of ζ Oph by rocket spectrographic techniques are analyzed. The main results are that within a factor of 10 the abundances of C^+, O^0, and Si^+ outside the H II region surrounding ζ Oph relative to the hydrogen abundance are equal to solar values. The lines in C^0 and S^+ imply that the interstellar matter is distributed among several clouds as indicated by high-resolution visible spectra.

114.044 Review of results in infrared space astronomy.
W. F. Hoffmann.
Astrophys. Space Sci. Library, Vol. 30, (see 012.005), 5 - 25 (1972).

114.045 A second CH star in Omega Centauri and the C^{12}/C^{13} ratio. R. J. Dickens.
Monthly Notices Roy. Astron. Soc., Vol. 159, 7P - 10P (1972).

A red giant in the globular cluster ω Cen, star no. 70 in the Herstmonceux catalogue, shows carbon features and strong CN and CH bands. Both for this star and the previously known CH star, RGO no. 55, the absorption features around $\lambda 4737$ Å probably indicate a relatively high C^{13} content.

114.046 Molecules in Eta Aquilae. M. C. Pande, G. C. Joshi, B. M. Tripathi, V. P. Gaur.
Bull. Astron. Inst. Czechoslovakia, Vol. 23, 301 - 305 (1972).

The total numbers of a given molecule in a column extending between $\tau = 0$ to $\tau = 3.00$ have been calculated for various phases of light variations in η Aql for the molecules CO, CN, C_2, OH, NH and CH.

114.047 The influence of ultraviolet line blanketing on the neutral helium triplet lines in B-type stars.

R. W. Simpson, B. J. O'Mara, A. G. Hearn.
Astron. Astrophys., Vol. 21, 57 - 60 (1972).

Calculations show that the inclusion of line blanketing of the ultraviolet radiation field between 913 Å and 1500 Å in the calculation of the photoionization rates significantly improves the agreement between observed and calculated triplet line profiles in B0-type stars.

114.048   Hα profiles for G-type dwarfs and subgiants.
J. B. Hearnshaw, E. G. Schmidt.
Astron. Astrophys., Vol. 21, 111 - 117 (1972).

The Hα wing profiles have been computed from model atmospheres for G-type dwarfs and subgiants of various compositions. The profiles are insensitive to gravity and to helium abundance, but quite sensitive to metal abundance in addition to effective temperature. A study of the observed profiles in eighteen stars whose metal abundances are also measured, shows that the observed sensitivity of the Hα wings to metal abundance is somewhat greater than that predicted by the models.

114.049   Spectrum variations in A-type supergiants. II. A search for evidence for coupling between mass loss and turbulence in Alpha Cygni. J. D. Rosendhal.
Bull. American Astron. Soc., Vol. 4, 311 (1972). – Abstr. AAS.

114.050   On the identification of trace elements in Ap stars with application to HR 465.
M. R. Hartoog, C. R. Cowley.
Bull. American Astron. Soc. Vol. 4, 311 - 312 (1972). Abstr. AAS.

114.051   A search technique for faint OB stars.
J. A. Graham, E. W. Miller.
Bull. American Astron. Soc., Vol. 4, 312 (1972). – Abstr. AAS.

114.052   CPD –31°1701; a newly discovered, helium-rich subluminous O-type star.
R. F. Garrison, W. A. Hiltner.
Bull. American Astron. Soc., Vol. 4, 312 (1972). – Abstr. AAS.

114.053   Observed and computed spectra of three carbon stars 6700–6712 Å.
H. R. Johnson, W. L. Kelch.
Bull. American Astron. Soc., Vol. 4, 323 (1972). – Abstr. AAS.

114.054   Carbon stars in the globular cluster Omega Centauri.
J. Stock, R. F. Wing.
Bull. American Astron. Soc., Vol. 4, 324 (1972). – Abstr. AAS

114.055   Ca II interstellar lines in the southern sky.
J. J. Rickard.
Bull. American Astron. Soc., Vol. 4, 327 (1972). – Abstr. AAS.

114.056   Astronomical observations with the Apollo 16 far-ultraviolet camera/spectrograph.
G. R. Carruthers, T. L. Page.
Bull. American Astron. Soc., Vol. 4, 331 (1972). – Abstr. AAS.

114.057   Recent progress in the interpretation of band O star spectra. D. Mihalas.
Bull. American Astron. Soc., Vol. 4, 333 (1972). – Abstr. AAS.

114.058   Tests of the Minnaert formula for spectral analysis.
J. G. Collins, J. P. Mutschlecner.
Bull. American Astron. Soc., Vol. 4, 334 (1972) – Abstr. AAS.

114.059   The Ca II–FeI λ3969 resonance in T-Tauri stars.
L. A. Willson.
Bull. American Astron. Soc., Vol. 4, 334 (1972). – Abstr. AAS.

114.060   Detection of gravity darkening and axial inclination of rapidly rotating stars from photographic line profiles. T. R. Stoeckley.
Bull. American Astron. Soc., Vol. 4, 334 - 335 (1972). Abstr. AAS.

114.061   An inexpensive data system for a rapid scanner.
J. S. Neff, G. L. Clements.
Bull. American Astron. Soc., Vol. 4, 335 (1972). – Abstr. AAS.

114.062   Stellar compositions from narrow-band photometry – III. Iron abundances in a further 120 evolved stars. P. M. Williams.
Monthly Notices Roy. Astron. Soc., Vol. 158, 361 - 373 (1972).

Using narrow-band indices observed with the Cambridge spectrophotometer and analyzed using synthetic spectra computed from model atmospheres, the iron abundances of a further 120 evolved stars have been determined. The results compare well with those from high-dispersion spectroscopic analyses. A number of super-metal-rich stars have been isolated, including three high-velocity stars lying below the M67 subgiant branch in the H–R diagram.

114.063   Computer analyses of low-resolution image dissector spectrophotometric data.
R. K. Oines, L. W. Schroeder, J. C. Evans.
Bull. American Astron. Soc., Vol. 4, 397 (1972). – Abstr. AAS.

114.064   OH infrared stars. M. M. Litvak, D. F. Dickinson.
Astrophys. Letters, Vol. 12, 113 - 117 (1972).

The strong 1612-MHz maser emission from infrared stars is calculated to be the result of excitation mainly by near-infrared radiation and partly by collisions.

114.065   OH in the Hoffman infrared sources.
E. J. Chaisson, D. F. Dickinson.
Astrophys. Letters, Vol. 12, 119 - 122 (1972).

We have observed OH in the direction of two of the unidentified objects in the Hoffman 100-μm survey.

114.066   Classification of stars by not broadened low-dispersion spectra. III. Technique and criteria of classification. V. I. Voroshilov, N. B. Kalandadze, V. I. Kuznetsov.
Astrometriya i Astrofiz., Kiev, No. 15, (see 003.001), p. 15 - 22 (1972). In Russian.

Technique and system of criteria are worked out for classification of spectra obtained with a 4°-prism attached to the Abastumani 70-cm Maksutov telescope. The classification is carried out by means of microphotometer tracings of not broadened spectra on the basis of central line depth ratios. Accuracy of the classification is about three subclasses. The technique makes it possible to carry out the MK classification up to B-magnitude 15^m.

114.067   Classification of stars by not broadened low-dispersion spectra. IV. Catalogue of spectra of faint stars around NGC 2129. V. I. Kuznetsov.
Astrometriya i Astrofiz., Kiev, No. 15, (see 003.001), p. 22 - 26 (1972). In Russian.

Spectral classification of faint stars in the region of a square degree around NGC 2129 is extended up to 15^m0. Not broadened low-dispersion spectra were used to obtain 337 spectra.

114.068   Determination of the vibrational temperature of the stars δ^2Lyr and α Her A from spectral bands of the TiO α-system. A. V. Shavrina.
Astrometriya i Astrofiz., Kiev, No. 15, (see 003.001), p. 37 - 39 (1972). In Russian.

Vibrational temperatures for δ^2Lyr and α Her A were determined by means of band intensities of the TiO molecule α-system. These values of the temperature are compared with those obtained by band intensity measurements of the TiO γ-system.

114.069 The barium and R type stars.   O. J. Eggen.
Monthly Notices Roy. Astron. Soc., Vol. 159, 403 - 427 (1972).

(*UBVRI*) photometry of R and barium stars (including CH stars) is discussed in connection with the red giant branch of the old disk population (Wolf 630 Group and M67 cluster). Luminosity calibrations are obtained from membership in moving groups, clusters or wide binaries and from Olin Wilson's observations of H and K emission widths.

114.070 On the atmosphere of the star HD 103877.
D. J. Stickland.
Monthly Notices Roy. Astron. Soc., Vol. 159, 29P - 33P (1972).

A study of the spectrum of HD 103877 suggests that it is an evolved early F type star with apparent surface abundance anomalies typical of the classical Am stars. This agrees with the predictions of the element separation mechanism and recent model envelope calculations.

114.071 On circumstellar gas emission among pre-main-sequence stars in Ic Orionis and NGC 2264.
M. A. Smith.
Astrophys. Journ., Vol. 176, 617 - 622 = Contr. Lick Obs., No. 363 (1972).

In this communication the probable presence of Balmer emission is considered in over 40 mainly A-type stars observed in the Ic Orionis (Outer Sword) association. This observation will be related to a similar one in NGC 2264 by Strom, Strom, and Yost (1971). The procurement and reduction procedure of our Kitt Peak *ubvy* photometry and classification spectrograms for these stars is discussed in Smith (1972).

114.072 Infrared surveys of the southern Milky Way. II. Suspected supergiant M stars.   H. Albers.
Astrophys. Journ., Vol. 176, 623 - 628 (1972).

An infrared objective-prism survey has been carried out along the southern Milky Way. Suspected M-type supergiants have been selected from the presence of a CN absorption feature at λ7900. A catalog of 170 M stars showing this feature is given.

114.073 A peculiar B star in the old disk population: 38 Draconis.   S. J. Adelman, W. L. W. Sargent.
Astrophys. Journ., Vol. 176, 671 - 676 (1972).

Eggen has found that the sixth-magnitude late B-type star 38 Draconis has a space motion typical of the old disk population. A study of coudé spectrograms has shown that this star has a peculiar composition which is similar, but not identical, to the Population I manganese-mercury stars.

114.074 Studies of heavy-element synthesis in the Galaxy. I. Separation of *r*-and *s*-process abundances.
H. R. Butcher.
Astrophys. Journ., Vol. 176, 711 - 722 (1972).

The data are obtained with an échelle grating system and image-tube in the coudé of the Mount Stromlo 74-inch reflector. Initial results of the survey suggest that only very small, if any, relative abundance variations are to be found among dwarf stars of the galactic disk, but small relative excesses of the *r*-process element Eu have been discovered in two of the three metal-poor, intermediate-population stars so far examined.

114.075 A measurement of the interstellar $^{12}C/^{13}C$ ratio.
P. A. Vanden Bout.
Astrophys. Journ., (*Letters*), Vol. 176, L127 - L129 (1972).

The λ4232 line of $^{13}CH^+$ has been detected photoelectrically in the interstellar line spectrum of ζ Oph. The equivalent width of this line is 0.35 ± 0.10 mÅ; comparison with the strength of 22.3 ± 0.4 mÅ of the $^{12}CH^+$ line yields $a \equiv {}^{12}C/^{13}C$ = 75 (+25, −15).

114.076 The metal deficiency of two members of M67.
M. S. Bessell.
Publ. Astron. Soc. Pacific, Vol. 84, 489 - 491 (1972).

Hydrogen line profiles and K-line strengths for F 131 and F 90 probable members of M 67 indicate a metal deficiency of a factor of 1.7 relative to the sun and a reddening for these two stars of about $E(b-y) = 0.05$.

114.077 On emission-line stars in the ζ Sculptoris cluster.
H. E. Bond.
Publ. Astron. Soc. Pacific, Vol. 84, 583 = Contr. Louisiana State Univ. Obs., *Baton Rouge*, No. 65 (1972).

New objective-prism plates fail to confirm the reported presence of Hα-emission stars in the region of the ζ Scl cluster.

114.078 The spectrum of o^2 CMa, B3 Ia. II. Interpretation of the observations.   R. van Helden.
Astron. Astrophys., Vol. 21, 209 - 222 (1972).

It is one aim of this paper to present a fine analysis using LTE model atmosphere techniques and to establish, among other things, the chemical composition, the effective temperature and the surface gravity. For this purpose Klinglesmith's (1971) models, which are in hydrostatic, as well as in radiative equilibrium, are used. In order to investigate whether also the outermost layers of the atmosphere of o^2 CMa are in hydrostatic equilibrium the radial velocities of the hydrogen lines are discussed.

114.079 Long wavelength spectrometry and photometry of M, S and C-stars.   J. A. Hackwell.
Astron. Astrophys., Vol. 21, 239 - 248 (1972).

Spectral measurements of 11 stars corrected for atmospheric absorption from $8-13$ μ and independent photometric measurements in 7 bands between 2 μ and 20 μ of 45 selected M, S and C-stars are presented.

114.080 Observations of the peculiar emission object HBV 475.   N. Richter.
Inform. Bull. Variable Stars (I.A.U. Commission 27), Konkoly Obs., Budapest, No. 708 (1972).

114.081 Spectra of Upgren's unclassified stars.
C. R. Sturch, S. Sharpless.
Astron. Journ., Vol. 77, 669 - 671 (1972).

Moderate dispersion spectra of Upgren's G5–G7 unclassified stars indicate that most are evolved for $B \leqslant 10.2$, but that a majority may be dwarfs in the next half-magnitude interval. When combined with densities of the other Upgren stars, these data suggest a doubling of G-dwarf space density beyond $z \sim 80$ pc.

114.082 The abundances of the elements in the oldest disk stars.   J. B. Hearnshaw.
Mem. Roy. Astron. Soc., Vol. 77, 55 - 108 (1972).

Nineteen old disk G-stars in the solar neighbourhood have been selected which, according to their *UBV* photometry and their trigonometric parallaxes, have evolved above the zero-age main sequence so as to be apparently older than, or nearly as old as, the old galactic cluster NGC 188. Effective temperatures for these stars have been obtained from (a) $(R-I)$ colours, (b) photoelectric spectral scans, and (c) Hα wing profiles, the latter two being calibrated by means of flux constant LTE model atmospheres, computed for this project. A coarse differential curve of growth analysis has been carried out with respect to the sun for 20 elements in each of the stars from measurements of 376 lines on high dispersion plates of the blue and red spectral regions. The element-to-iron ratios for several elements are discussed. A model atmosphere analysis of neutral and ionized iron lines has been carried out, and a calibration set up between gravity and electron pressure; hence , spectroscopic gravities have been obtained. The ages of the stars have been derived by comparing their positions in the (M_{bol}, log T_e) diagram with isochrones interpolated from evolutionary tracks, using the premiss that their helium contents are normal. The rates of formation of different elements during the old disk epoch are discussed.

114.083   **Statistics on the solar spectrum suitable for the study of the blanketing effect in stars of spectral types F, G and K.**   A. Natta, M. Ranieri.
Astrophys. Space Sci., Vol. 17, 390 - 408 (1972).

A complete set of statistics useful to the study of the blanketing effect in stars of spectral type F, G and K have been obtained on the basis of the solar spectrum. In the spectral range 2250 Å − 6400 Å the following distributions have been derived: distribution of lines with chemical elements, percentage of lines originating from ionized atoms, distribution of lines with the excitation potential and with the oscillator strength. Also the distribution with the equivalent width has been obtained in the range 3060 Å − 6400 Å. With these distributions, samples of lines have been derived for intervals of about 200 Å in the range 2250 Å − 6400 Å. Evaluations for the spectral density of lines and the percentage of non-overlapping lines are also included.

114.084   **The spectrum of o^2 CMa, B3 Ia. I. Equivalent widths and line profiles.**   R. van Helden.
Astron. Astrophys., Suppl. Ser., Vol. 7, 311 - 329 (1972).

Eleven spectrograms with a dispersion of 2.2 Å mm^{-1} in the blue and 17.7 Å mm^{-1} in the infrared spectral region were used to perform an analysis of the spectrum of o^2 CMa, B3 Ia. Equivalent widths and identifications of 285 spectral features in the wavelength interval 3500−8870 Å are given as well as the profiles of 60 absorption lines.

114.085   **A-type supergiants: a list of line intensities and radial velocity measurements.**   C. Aydin.
Astron. Astrophys., Suppl. Ser., Vol. 7, 331 - 354 (1972).

The equivalent widths of the spectral lines used for the construction of the curve of growth for five A-type supergiants and the radial velocity of the Balmer and metallic lines for the same stars are given. The scientific discussion of these data is published in the Main Journal.

114.086   **Discovery of infrared emission from the radio source near Cygnus X-3.**
E. E. Becklin, J. Kristian, G. Neugebauer, C. G. Wynn-Williams.
Nature, Phys. Sci., Vol. 239, 130 - 131 (1972).

Four weeks after the recent radio outburst near Cygnus X-3, a search was made for infrared emission at the radio position. On the night of October 2–3, 1972, an infrared source was discovered at 2.2 and 1.6 μm. The measured flux densities are given in a table.

114.087   **Computer programmes for differential curve-of-growth analysis.**   A. L. T. Powell.
Royal Obs. Bull., [Herstmonceux: Royal Greenwich Obs.], No. 171, p. 3 - 21 (1971).

The paper describes computer programmes which undertake a complete curve-of-growth analysis, using measurements of the dimensions of stellar lines as input data. The programmes perform the analysis in a number of distinct stages and the final results are presented in graphic and tabular form. This procedure permits a very flexible analysis by exercising options at each stage.

114.088   **Rapid changes in the new shell star HR 6000.**
M. S. Bessell, O. J. Eggen.
Astrophys. Journ., Vol. 177, 209 - 217 (1972).

Large quasi-periodic variations in color and brightness have been found for HR 6000 which is also a shell star with bright Hα. Analysis of the light and polarization changes suggests that the variations are caused by obscuration due to circumstellar dust clouds. The relative position in the color-magnitude diagram of HR 6000 and its companion HR 5999 indicates that HR 6000 may have evolved prematurely.

114.089   **Rocket-ultraviolet spectra of eight stars in Ophiuchus and Scorpius.**
D. C. Morton, E. B. Jenkins, T. A. Matilsky, D. G. York.
Astrophys. Journ., Vol. 177, 219 - 234 (1972).

Rocket ultraviolet spectra longward of 1100 Å have been obtained for the O and B stars ζ Oph and β^1, δ, ν, π, σ, τ, and ω^1 Sco. Detailed discussions are given for ζ Oph (O9.5 V) and δ Sco (B0 V), for which resolutions of 0.5 and 0.6 Å, respectively, were achieved. The interstellar Lα absorption line is the strongest feature in all the spectra except τ Sco which does not extend to 1216 Å. Several other interstellar absorption lines are present in the high-resolution spectra including C II, O I, Si II, and possibly C I and N I in δ Sco, and C II and O I in ζ Oph. It seemed desirable to obtain additional ultraviolet spectra of the Scorpius stars with improved resolution to study both the stellar and interstellar lines.

114.090   **Rocket-ultraviolet spectra of six stars in Perseus.**
D. C. Morton, E. B. Jenkins, W. W. Macy.
Astrophys. Journ., Vol. 177, 235 - 244 (1972).

Spectra of the O and B stars ε, ζ, ξ, o, and 40 Per, and HD 24640 have been photographed between 1130 and 2300 Å with a pair of rocket-borne objective spectrographs. Photospheric lines of He II, C II, C III, C IV, N III, Ni IV, Si III, and Si IV and also possibly N V, O IV, Al II, and Al III have been identified in one or more of the stars. The interstellar Lα line was measured in ξ, o, and 40 Per, giving an average volume density of 1.7 ± 0.5 hydrogen atoms cm^{-3} in this direction.

114.091   **Observations of carbon monoxide in cool stars at 4.7 microns.**
T. R. Geballe, E. R. Wollman, D. M. Rank.
Astrophys. Journ., (Letters), Vol. 177, L27 - L32 (1972).

Spectra of six stars with high infrared luminosities have been obtained, with spectral resolution of 0.18 cm^{-1}, near the fundamental vibration-rotation band origin of $^{12}C^{16}O$ at 2140 cm^{-1}. They have been analyzed for rotational and vibrational temperatures, isotopic abundances, and column densities, the latter being in all cases substantially lower than theoretical predictions.

114.092   **Spectrograms of α Lyr and β Cen in the region of 2000–3800 Å.**
G. A. Gurzadyan, J. B. Ohanesyan.
Space Sci. Rev., Vol. 13, 647 - 650 (1972). − Conference paper IAU Colloquium No. 14 (see 012.012).

114.093   **An abundance analysis of α Centauri.**
V. A. French, A. L. T. Powell.
Royal Obs. Bull., [Herstmonceux: Royal Greenwich Obs.], No. 173, p. 63 - 103 (1971).

The chemical composition of the components A and B of α Centauri has been examined. Abundances of twenty-two elements have been determined from 921 and 1029 spectral lines respectively and both stars are found to have small overabundances of metals compared with the sun.

114.094   **Objective-prism spectral work.**   W. P. Bidelman.
The role of Schmidt telescopes in astronomy. Con-

ference Hamburg 1972, (see 012.014), p. 53 - 60 (1972).

114.095 An Hα atlas of the southern Milky Way.
G. Lyngå.
The role of Schmidt telescopes in astronomy. Conference Hamburg 1972, (see 012.014), p. 109 (1972).

114.096 Five Schmidt telescope programmes in progress and in planning. W. C. Seitter.
The role of Schmidt telescopes in astronomy. Conference Hamburg 1972, (see 012.014), p. 113 - 116 (1972).

114.097 The effect of the surface H/He ratio on spectra of carbon stars. R. I. Thompson.
Astrophys. Journ., Vol. 177, 509 - 513 (1972).

It is shown that observations of CH in N irregular and semiregular carbon stars are consistent with hydrogen depletion by a factor of 10 as predicted by a CNO bi-cycle model. Chemical abundances for carbon stars at constant temperature and pressures have been calculated for hydrogen mass fractions from 0.739 to 10^{-4}.

114.098 Study of a sample of faint M stars in the direction of the North Galactic Pole. P. Pesch.
Astrophys. Journ., Vol. 177, 519 - 522 (1972).

Slit spectra and UBV photometry of 27 M stars (ranging in V from 9.2 to 15.1) found in Sanduleak's low-dispersion objective-prism survey of the north galactic polar region show that most are nearby dwarfs; only six are halo giants. Since these 27 stars were selected on the basis of not showing appreciable proper motion, it appears that numerous nearby low-velocity M dwarf stars exist.

114.099 Infrared stars with strong 1665/1667-MHz OH microwave emission. W. J. Wilson, P. R. Schwartz, G. Neugebauer, P. M. Harvey, E. E. Becklin.
Astrophys. Journ., Vol. 177, 523 - 540 (1972).

Microwave and infrared observations are presented of seven stars which have strong infrared and 1665- and 1667-MHz main-line OH emission. The sources were selected by the fact that their 1665/1667-MHz OH emission is stronger than any 1612-MHz OH emission. Five of these sources have associated H_2O microwave emission. The sources are associated with M-type long-period variable stars and generally show only a small excess infrared radiation from 3 to 20μ above that expected from the photosphere.

114.100 Vibration-rotation bands of NH in the spectrum of Alpha Orionis. D. L. Lambert, R. Beer.
Astrophys. Journ., Vol. 177, 541 - 545 (1972).

Several lines of the 1−0 and 2−1 fundamental vibration-rotation bands of the NH radical are identified on a high-resolution spectrum of the supergiant α Orionis.

114.101 A coarse analysis of HD 50896.
D. Van Blerkom, G. (Yanchak) Patton.
Astrophys. Journ., Vol. 177, 547 - 554 = Contr. Five College Obs., Univ. Mass., Amherst, No. 136 (1972).

A coarse analysis of the He II spectrum of the WN5 star HD 50896 is presented. The model obtained is in agreement with current ideas of the state of excitation, dimensions, and densities characteristic of Wolf-Rayet stars. Intensities of the ultraviolet lines in the Fowler ($n \to 3$) series, observed by OAO-2, are correctly predicted by the theory. The good agreement between theory and observation argues favorably for the expanding envelope model of Wolf-Rayet stars.

114.102 Recent changes in the nature of V1057 Cygni.
R. D. Schwartz, T. P. Snow, Jr.
Astrophys. Journ., (*Letters*), Vol. 177, L85 - L86 (1972).

Photometric and spectroscopic data show that V1057 Cyg,

which had a spectrum of an A-type giant or supergiant after its flare-up in late 1969, now has an early F-type spectrum of similar luminosity class. This is consistent with the history of FU Ori which underwent a similar flare-up in 1936, except that V1057 Cyg now shows no evidence of shell features in its spectrum.

114.103 Stellar compositions from narrow-band photometry– IV. Calcium abundances in G and K giants.
M. E. Rego, P. M. Williams, D. W. Peat.
Monthly Notices Roy. Astron. Soc., Vol. 160, 129 - 137 (1972).

Narrow-band calcium photometric indices have been measured with the Cambridge spectrometer for 147 G- and K-type giant stars. Calcium abundances have been derived from the indices using model atmospheres. The variation of calcium abundance from the most metal-poor to the metal-rich stars is less that of sodium or iron.

114.104 Recent results of the Goddard rocket program for observing stars. A. B. Underhill.
Space Research XII, (see 012.016), Vol. 2, 1589 - 1594 (1972).

114.105 Large-field study in the ultraviolet (2650 Å) from a sounding rocket. J. P. Sivan, M. Viton.
Space Research XII, (see 012.016), Vol. 2, 1603 - 1608 (1972).

114.106 Über die Veränderlichkeit der Sterne des Spektraltyps A. G. Jackisch.
Astron. Nachr., Vol. 294, 1 - 8 (1972).

215 stars with spectral types A0 to F5, among them 65 members of the open star clusters NGC 2548, Praesepe and Coma, were repeatedly measured photoelectrically, in order to determine the amount of objects with light variations of small amplitude. The method of measuring and the analysis of the observations by means of the F-distribution are explained.

114.107 Spectrophotometric parameters of early-type stars. III. Additional data.
A. Gutiérrez-Moreno, H. Moreno, J. Stock.
Publ. Dep. Astron., Univ. Chile, Obs. Astron. Nacional, Cerro Calán, Santiago de Chile, Vol. 2, (No. 2), 27 - 39 (1971).

Equivalent widths of Hβ, Hγ and Hδ and relative spectral energy distributions are given for 59 stars of spectral type earlier than F3.

114.108 A southern objective prism survey.
J. Stock, H. Wroblewski.
Publ. Dep. Astron., Univ. Chile, Obs. Astron. Nacional, Cerro Calán, Santiago de Chile, Vol. 2, (No. 3), 59 - 129 (1972).

An objective prism survey intended to cover the entire southern sky within the magnitude range from the 10th to the 13th photographic magnitude is described. The survey includes luminous stars, planetary nebulae, Me-stars, carbon stars, and other objects of interest. The first five sections of the survey, covering an area of approximately 4000 square degrees have been completed. The spectral types, magnitudes, and positions for the objects of interest are given in five separate catalogues.

114.109 Spectral classification of the bright B8 stars.
A. Cowley.
Astron. Journ., Vol. 77, 750 - 755 (1972).

MK spectral types have been assigned to all of the stars originally classified as B8 which are in the Bright Star Catalogue and north of −20°. This spectral region is of special interest because of the high percentage of mercury-manganese stars within it.

114.110 Abundance anomalies in the ON type star HD 188209. B. Baschek, K. Kodaira, M. Scholz.

Astrophys. Letters, Vol. 12, 227 - 229 (1972).

Comparison of the line spectrum of the ON type star HD 188209 with ζ Ori A shows that its atmosphere has a peculiar chemical composition. The anomalous abundances can be explained by an admixture of about 20 to 40 per cent of matter processed by the CNO cycle.

114.111 **Optical observations of three new infrared sources.** D. A. Allen.

Astrophys. Letters, Vol. 12, 231 - 234 (1972).

Low-dispersion slit spectra and direct photographs have been secured of the objects associated with the recently discovered infrared sources in NGC 2264 and M 1 - 82. The former is interpreted as a heavily reddened star. The sources in M 1 - 82 are probably young Be stars with circumstellar envelopes.

114.112 **Effects of velocity and turbulence gradients on A-type supergiant spectra.** H. G. Groth.

Astron. Astrophys., Vol. 21, 337 - 353 = Veröff. Univ.-Sternw. München, Vol. 7, No. 15 (1972).

The observed variations of radial velocities of some supergiants are compiled. Measurements of differential line displacements in spectra of α Cyg and φ Cas are presented. These observations can be interpreted, if one assumes a depth dependent radial velocity, which is varying with time. The microturbulence is also depending on optical depth. A series of models for an A2-type supergiant have been computed, assuming different velocity fields and microturbulence fields. Using these models the equivalent widths and the differential displacements of lines of some ions of several elements were calculated.

114.113 **A spectroscopic study of the peculiar stars in the open cluster NGC 2516.** J. Dachs.

Astron. Astrophys., Vol. 21, 373 - 383 (1972).

Coudé spectrograms with a dispersion of 20 Å/mm were taken for 22 stars in the open cluster NGC 2516 which is known to contain three red giants and several stars with peculiar spectra. Three new peculiar B– and A-type stars of the Si and the Ti–Si groups as well as a second B-emission line star were detected in the cluster. The peculiar stars in NGC 2516 are located on the main sequence with absolute visual magnitudes between $-1^m8 < M_v < +1^m5$.

114.114 **A model for the helium spectrum variable α Centauri.** J. Norris, B. Baschek.

Astron. Astrophys., Vol. 21, 385 - 392 (1972).

We propose an empirical model to explain the observed variations of the helium spectrum variable α Centauri. The model consists of a rotating star, the surface of which consists of a small helium rich region (spot), and a large helium deficient region (non-spot). The non-spot resembles a weak-helium-line star. The variations of metallic lines and radial velocities can then be explained with this model. In particular, we can simply obtain the observed constancy of Si II lines and the variability of the Si III lines.

114.115 **Analysis of the spectrum of the metallic-line star 63 Tauri.** E. Hundt.

Astron. Astrophys., Vol. 21, 413 - 430 (1972).

A fine analysis of 63 Tau based on high dispersion spectra was carried out. The derived model atmosphere parameters are: effective temperature T_{eff}= 7750° ± 200°, surface gravity log g = 4.3 ± 0.3, microturbulent velocity ξ = 5.0 ± 0.5 km/s (depth independent). The atmospheric structure does not deviate from that of a normal main sequence star, except for a lower surface temperature caused by stronger line blanketing which is derived empirically. The influence of the secondary component of the binary system on the analysis is investigated, and found to be negligible.

114.116 **UV interstellar absorption lines in the spectrum of ζ Pup.** K. S. de Boer, R. Hoekstra, K. A. van der Hucht, T. M. Kamperman, H. J. Lamers, S. R. Pottasch.

Astron. Astrophys., Vol. 21, 447 - 448 (1972).

Interstellar absorption lines have been detected in the spectrum of the O5 star ζ Pup. They are due to the resonance lines of Mg II λ 2803, 2796, Mg I λ 2852, Fe II λ 2586 and Mn II λ 2576, and possibly Fe I λ 2522.

114.117 **Line spectra of eight O stars from λ3059 to λ6683.** M. Scholz.

Astron. Astrophys., Suppl. Ser., Vol. 7, 469 - 486 (1972).

Measurements from high-dispersion photographic spectrograms of ζ Ori A (O 9.5 Ib), HD 188209 (O 9.5 III), HD 34078 (O 9.5 V), HD 57682 (O 9 V), λ Ori A (O 8), 15 Mon A (O 7), HD 54662 (O 6) and 9 Sgr (O 5) are presented. Identifications, equivalent widths and profiles of absorption and emission lines in the wavelength range λ 3059 to λ 6683 are given. HD 188209 seems to be linked to Walborn's ON stars.

114.118 **Image tube spectra of shell stars.** G. V. Coyne.

Ric. Astron.,Specola Vaticana, *Castel Gandolfo*, Vol. 8,(No. 16), 353 - 358 (1972).

A study of image-tube spectra at a reciprocal dispersion of 117 Å/mm of the stars HD 12882 and HD 13590 show that both are shell stars with hydrogen lines in emission, a shell spectrum of about class A2, and an underlying star probably later than B2. When comparison is made with previous published classification of these stars, there is evidence that the shells are changing.

114.119 **The He I λ5876 line in O-star spectra.** C. M. Anderson.

Astrophys. Journ., (*Letters*), Vol. 177, L121 - L124 (1972).

New data are presented which resolve one of the more serious discrepancies between the theory and observations of O-star spectra.

114.120 **Spectrum variability in B and A type supergiants.** J. B. Hutchings.

Proc. Third Colloquium on astrophysics. Supergiant stars, Trieste 1971, p. 38 - 47 = Contr. Dominion Astrophys. Obs., Victoria, No. 180 (1972).

Photoelectric scans of line profiles in five B type supergiants of various luminosities and mass-loss rates show them all to be varying in time. Spectrographic observations of six late B and early A type supergiants indicate mass-loss at a much lower rate, and atmospheric velocity fields which vary in times of several days. The transition between B and A type supergiants is discussed briefly.

114.121 **High-resolution stellar and circumstellar spectra of SiO and CO at 4μ and 5μ.** T. R. Geballe, E. R. Wollman, D. M. Rank.

Publ. Astron. Soc. Pacific, Vol. 84, 640 (1972). – Abstr. Astron. Soc. Pacific.

114.122 **Carbon stars with peculiar red spectra.** P. M. Rybski.

Publ. Astron. Soc. Pacific, Vol. 84, 643 (1972). – Abstr. Astron. Soc. Pacific.

114.123 **Twenty-micron fluxes of bright stellar standards.** T. Simon, D. Morrison, D. P. Cruikshank.

Publ. Astron. Soc. Pacific, Vol. 84, 643 - 644 (1972). – Abstr. Astron. Soc. Pacific.

114.124 **Low-dispersion spectroscopic classification of the unidentified sources in the "two-micron sky survey".** S. S. Vogt.

Publ. Astron. Soc. Pacific, Vol. 84, 645 (1972). – Abstr. Astron. Soc. Pacific.

114.125 Two heavily reddened M supergiants near the galactic center. R. F. Wing, J. W. Warner.
Publ. Astron. Soc. Pacific, Vol. 84, 646 (1972). – Abstr. Astron. Soc. Pacific.

114.126 Spectroscopic observations of EZ Pegasi.
N. J. Irvine.
Publ. Astron. Soc. Pacific, Vol. 84, 671 - 672 (1972).

This G5 V variable exhibits short-term fluctuations in the intensity of Hα emission. The star is a possible bright U Geminorum-like system.

114.127 An unusual emission object and a new S star.
W. P. Bidelman, L. E. Krumenaker.
Publ. Astron. Soc. Pacific, Vol. 84, 685 (1972).

114.128 New ultraviolet line identifications for early-type stars. W. R. Luebke, Jr.
Publ. Astron. Soc. Pacific, Vol. 84, 697 - 717 (1972).

New data on the energy level systems of C III, N III, and N IV were searched for dipole-allowed transitions coinciding with previously unidentified features in Princeton rocket ultraviolet spectra of early-type stars. Possible identifications are tabulated and discussed.

114.129 Uranium in the spectrum of HR 465.
C. R. Cowley, M. R. Hartoog.
Astrophys. Journ. (*Letters*), Vol. 178, L9 - L10 (1972).

Uranium has been identified in the 1960–1961 spectrum of HR 465 at a confidence level of 3.8 σ (one chance in 10^4 of fortuitous occurrence). The uranium abundance is about a million times the solar-system value. This should be compared to the abundances of the lanthanides, for which M. Aller finds excesses $\leqslant 10^5$.

114.130 The C-classification of the spectra of carbon stars.
Y. Yamashita.
Ann. Tokyo Astron. Obs., Second Ser., Vol. 13, (No. 3), 169 - 217 (1972).

Spectra of about 180 carbon stars were observed and were classified on the system of the C-classification set up by Keenan and Morgan. The temperature sequence was set up by means of the absolute intensities of D lines of Na I and Ca I $\lambda 4227$. Intensities of Hβ and other spectral features were also taken into account. The carbon-abundance class was defined by the intensity of the Swan bands of C_2. Three new members were added to the group of CH-stars. In addition, several CH-like stars were also found. They show some of characteristic spectral features of CH-stars, however, their proper motions and radial velocities are quite small. It was pointed out that the temperature sequence of carbon stars also forms a sequence with respect to abundance of various elements. The discrepancy between the effective temperature obtained by Mendoza and Johnson, and the C-classification was carefully studied. There is a possibility that the effective temperature of Mendoza and Johnson is an indicator not only of a temperature but also of a CN-intensity. Preliminary values of radial velocities are given to a few stars for which no radial velocity could be found in literature.

114.131 The Balmer progression during Pleione's shell episode, 1938–1954.
J. M. Marlborough, P. R. Gredley.
Astrophys. Journ., Vol. 178, 477 - 480 (1972).

The hypothesis that the Balmer progression during Pleione's shell phase is due to different effects of reemission in the Balmer lines is investigated numerically. It is shown that inclusion of reemission in the Balmer lines for the specific

model due to Limber of Pleione's circumstellar shell does not account for the Balmer progression as it was observed.

114.132 Titanium isotopes in Omicron Ceti.
S. Wyckoff, P. Wehinger.
Astrophys. Journ., Vol. 178, 481 - 489 (1972).

Profiles of the TiO γ_3 0–0 band (7054 Å) have been computed for a variety of assumed fractional abundances of the five stable titanium isotopes. From a comparison of the computed spectra and an observed spectrum of the long-period variable o Ceti (M6e) at maximum light, limits for the isotopic abundance ratios were determined. Evidence for the presence of ^{18}O could not definitely be established.

114.133 A newly discovered variable: the Ap star HR 5153.
E. W. Burke, Jr., J. T. Howard.
Astrophys. Journ., Vol. 178, 491 - 493 (1972).

The Ap star HR 5153 was discovered to be a periodic variable. The mean amplitude (maximum to minimum) is 0.056 mag in B and 0.052 mag in U, but there appears to be no variation in V. The ephemeris for minimum light in B is JD (hel.) = $2440753.713 + 1\overset{d}{.}706E \pm 0\overset{d}{.}001$.

114.134 Ultraviolet photometry from the Orbiting Astronomical Observatory. VI. Magnesium II 2800 Å emission in cool stars. L. R. Doherty.
Astrophys. Journ., Vol. 178, 495 - 501 (1972).

Spectral scans of the 2800 Å region made with 25 Å resolution are presented for 19 giants and supergiants of spectral types G, K, and M. Mg II emission fluxes are given for eight of these stars and compared with the emission in Ca II K.

114.135 On ultraviolet stellar fluxes. II. Importance of CO band absorption and bound-free absorption of Al I in A, F, G, and K stars. S. P. Tarafdar, M. S. Vardya.
Astrophys. Journ., Vol. 178, 503 - 507 (1972).

The discrepancy between observed and calculated ultraviolet fluxes of A to G stars can be largely alleviated between $\lambda\lambda 1376$–2400 Å, if absorption coefficients of the $A-X$ band of CO and bound-free transitions of Al I are incorporated in the model calculations. However, the discrepancy around $\lambda 2800$ needs further investigation.

114.136 Zonal spectrophotometric standards. Distribution of energy in the spectra of 109 stars in absolute units. V. M. Tereshchenko, A. V. Kharitonov.
Trudy Astrofiz. Inst., *Alma-Ata*, Vol. 19, 98 - 103 (1972). In Russian.

114.137 Variations of Hα in the Be star ψ Per.
L. M. Sapargalieva.
Trudy Astrofiz. Inst., *Alma-Ata*, Vol. 19, 104 - 109 (1972). In Russian.

Changes of form and intensity of the Hβ profiles are considered. Profiles computed in rest intensities are given. Equivalent widths of the photosphere and of the shell components were measured. All the components of the line are characterized by a fairly rapid variation during one night and from night to night. A correlation between the total intensity of the emission line and the narrow shell absorption is detected.

114.138 Intensities of silicon lines in Ap-stars spectra. I. Calculated intensities.
V. L. Khokhlova, A. A. Krivosheina.
Astron. Zhurn. Akad. Nauk SSSR, Vol. 49, 1168 - 1174 (1972). In Russian. English translation in Soviet Astron. AJ, Vol. 16, No. 6.

Calculations of theoretical intensities and effective optical depths of line formation have been made for six lines of neutral, ionized, and doubly ionized silicon in the atmospheres of main-sequence stars in the range of T_{eff} from 11000 to 20000°K, log g = 4.0 and microturbulent velocities 3, 6 and 10 km per sec.

114.139 Photospheric and circumstellar Hα profiles in the spectra of M-supergiants.
M. Ya. Orlov, M. H. Rodriguez.
Astron. Zhurn. Akad. Nauk SSSR, Vol. 49, 1184 - 1187 (1972). In Russian. English translation in Soviet Astron. AJ, Vol. 16, No. 6.

Hα profiles for 6 supergiants M1—M5 have been determined using high-dispersion spectrograms. Equivalent widths, central depths and half-widths have been measured. Separation of circumstellar cores of Hα lines permitted to obtain equivalent widths of CS-lines and their radial velocities relative to the photosphere.

114.140 Catalogue of stars in the region of λ Orionis.
R. Bartkus.
Bull. Vilnius Astron. Obs., No. 33, p. 28 - 49 (1972). In Russian.

Spectral types for 2322 stars in the region $\alpha_{1950} = 5^h16^m - 5^h48^m$, $\delta_{1950} = +6° - +13.°5$ are presented in a catalogue or on charts. The stars have been classified from objective prism spectra up to the limiting magnitude about 12^m. For the early-type stars magnitudes V and indices B—V are also given.

114.141 Two new probable symbiotic stars with variable spectra. D. J. MacConnell.
Inform. Bull. Variable Stars, (IAU Commission 27), Konkoly Obs., Budapest, No. 734, 3 pp. (1972).

114.142 Spectral changes in V 1057 Cygni.
G. F. Gahm, G. Welin.
Inform. Bull. Variable Stars, (IAU Commission 27), Konkoly Obs., Budapest, No. 741, 5 pp. (1972).

114.143 MHα 73 — 59. E. Splittgerber.
Inform. Bull. Variable Stars, (IAU Commission 27), Konkoly Obs., Budapest, No. 747 (1972).

114.144 Spectrum variations in A-type supergiants. II. A search for evidence for coupling between mass loss and turbulence in Alpha Cygni. J. D. Rosendhal.
Astrophys. Journ., Vol. 178, 707 - 714 (1972).

An extensive time sequence of Hα spectrograms of α Cygni has been analyzed for changes in line strengths and radial velocities in order to look for evidence for coupling between macroturbulence, microturbulence, and mass loss.

114.145 A spectroscopic study of the strong helium-line star HD 37017. J. B. Lester.
Astrophys. Journ., Vol. 178, 743 - 761 (1972).

The peculiar early-type star HD 37017 has been studied using model stellar-atmosphere techniques in an attempt to determine the reason for the enhanced strength of its helium absorption lines. It is found that the helium abundance is approximately twice as great as that found for the nearby control star HD 37016. The mass and radius of HD 37017 are found to be inconsistent with its effective temperature when it is compared to the control star. The possibility of binary mass transfer is discussed.

114.146 Four southern A-type supergiants.
W. Buscombe.
Proc. Third Colloquium on astrophysics. Supergiant stars, Trieste 1971, (see 012.020), p. 50 (1972). – Abstract.

114.147 On differential shifts of lines in the spectrum of the supergiant β Ori.

E. L. Chentsov, L. I. Snezhko.
Proc. Third Colloquium on astrophysics. Supergiant stars, Trieste 1971, (see 012.020), p. 51 - 57 (1972).

114.148 Variations rapides du spectre de HD 198 478 (55 Cyg). P. Granes, R. Herman.
Proc. Third Colloquium on astrophysics. Supergiant stars, Trieste 1971, (see 012.020), p. 58 - 67 (1972).

114.149 Spectral line variations in F0—K5 supergiants.
J. Smoliński.
Proc. Third Colloquium on astrophysics. Supergiant stars, Trieste 1971, (see 012.020), p. 68 - 78 (1972).

114.150 The visual spectrum of the supergiant ε Orionis (B 0 Ia). H. J. Lamers.
Proc. Third Colloquium on astrophysics. Supergiant stars, Trieste 1971, (see 012.020), p. 83 - 84 (1972).

114.151 Preliminary study on early-type supergiants in h and χ Per. R. Stalio.
Proc. Third Colloquium on astrophysics. Supergiant stars, Trieste 1971, (see 012.020), p. 85 - 90 (1972).

114.152 Discovery of two new He I lines in the infrared spectra of B-type supergiants. R. van Helden.
Proc. Third Colloquium on astrophysics. Supergiant stars, Trieste 1971, (see 012.020), p. 91 - 92 (1972).

114.153 Hα-characteristics of luminous late-type stars.
G. F. Gahm, L. Hultquist.
Proc. Third Colloquium on astrophysics. Supergiant stars, Trieste 1971, (see 012.020), p. 148 - 152 (1972).

114.154 On the spectral analysis of the superluminous emission line stars. R. Viotti, O. Ricciardi.
Proc. Third Colloquium on astrophysics. Supergiant stars, Trieste 1971, (see 012.020), p. 233 - 237 (1972).

114.155 Helium spectra in large magnetic fields.
R. H. Garstang, S. B. Kemic.
Structure of matter. Rutherford Centennial Symposium, Christchurch 1971, [Univ. Canterbury, Christchurch, New Zealand], p. 396 - 404 (1972).

The discovery of circular polarization in the continuum of the DC white dwarf Grw + 70°8247 has been reported by Kemp et al. (1970). This polarization is taken as indicating the presence of a magnetic field of the order of 10^7 gauss. The authors examine the effects of a field of up to 10^7 gauss on the spectrum of helium.

114.156 ε Pegasi. R. J. Wood,
IAU Circ., No. 2450 (1972).

114.157 MWC 349. D. A. Allen.
IAU Circ., No. 2454 (1972).

114.158 MWC 349. G. H. Herbig.
IAU Circ., No. 2457 (1972).

114.159 Absolute flux measurements in astronomy.
J. B. Oke.
Journ. Optical Soc. America, Vol. 62, 1342 - 1343 (1972). Abstr. Optical Soc. America.

114.160 The identification problem in astronomical spectra.
W. P. Bidelman.
Journ. Optical Soc. America, Vol. 62, 1377 - 1378 (1972). Abstr. Optical Soc. America.

114.161 Spectrometric standards for observations of planets

and comets and some questions of stellar spectro-photometry. A. V. Kharitonov, E. A. Glushkova, L. N. Knyazeva, N. N. Morozova, V. T. Rebristyj, T. V. Solodovnikova, V. M. Tereshchenko, L. D. Frishberg. Izdatel'stvo "Nauka" Kazakhskoj SSR. Trudy Astrofiz. Inst., *Alma-Ata,* Vol. 22, 123 pp. Price 58 Kop. (1972). In Russian.

114.162 **Mass loss and infrared excesses in hot stars.** S. Kleinmann, L. V. Kuhi.
Publ. Astron. Soc. Pacific, Vol. 84, 766 - 767 (1972). − Paper presented at the Santa Cruz meeting of the Astronomical Society of the Pacific, 26 - 29 June 1972.

114.163 **The infrared spectrum of Alpha Herculis from 4000 to 4800 cm⁻¹.** H. L. Johnson, R. I. Thompson, F. F. Forbes, D. L. Steinmetz. Publ. Astron. Soc. Pacific, Vol. 84, 775 - 778 (1972).
We present the portion of the α Her spectrum from 4000 to 4800 cm⁻¹. A number of atomic and molecular absorption features have been identified. Most prominent are CO $\Delta V = 2$ bands of $^{12}C^{16}O$; a feature which may be identified with $^{13}C^{16}O$ has been pointed out.

114.164 **The infrared spectrum of Alpha Herculis from 5700 to 6700 cm⁻¹.** R. I. Thompson, H. L. Johnson, F. F. Forbes, D. L. Steinmetz. Publ. Astron. Soc. Pacific, Vol. 84, 779 - 783 (1972).

114.165 **Peculiar southern emission-line objects with strong [O III] λ 4363.** N. Sanduleak, C. B. Stephenson. Publ. Astron. Soc. Pacific, Vol. 84, 816 - 817 (1972).
Five stellar-like peculiar emission-line objects, having exceptionally strong [O III] λ 4363 emission, were detected in a recently conducted objective-prism survey of the southern Milky Way. This feature and other characteristics suggest that these may be similar in nature to IC 4997 and several other strong λ 4363 objects suspected of being protoplanetary nebulae.

114.166 **Additional observations of some stars with strong helium lines.** P. Lee, P. Daigle. Publ. Astron. Soc. Pacific, Vol. 84, 842 - 843 = Contr. Louisiana State Univ. Obs., *Baton Rouge,* No. 70 (1972).

114.167 **A new symbiotic-like object behind the Coalsack.** W. B. Weaver. Publ. Astron. Soc. Pacific, Vol. 84, 854 - 855 (1972).
A new object with a combination spectrum has been discovered. It probably lies about 2.3 kpc behind the Coalsack.

114.168 **A study of B6 stars.** A. B. Underhill. Rep. NASA-TM-X-65823, NASA, Greenbelt, Md. [Available from NTIS, Springfield, Va.], 59 pp. (1972).

114.169 **A study of SC stars.** A. E. Greene. Thesis, Ohio State Univ., Columbus. [Available from Univ. Microfilms, Ann Arbor, Mich., USA. Order No. 72-4505], 278 pp. (1971).
Examines several aspects of the atomic and molecular spectra of stars that appear to be intermediate between the spectral classes S and C. Image-tube spectrograms covering the region from 3840 Å to 7200 Å at a dispersion of 66 Å/mm were obtained for 51 stars including S, SC and carbon stars.

114.170 **B-type stars with discrepant colors.** M. R. Molnar. Thesis, Univ. Wisconsin, Madison. [Available from Univ. Microfilms, Ann Arbor, Mich., USA. Order No. 71-14156], 92 pp. (1971).
The color-spectrum discrepant stars, usually referred to as 'weak helium-line' stars, have intrinsic colors too blue for the MK type by at least two-tenths of a spectral type. The

physical characteristics of this anomalous group were investigated by observing the stars energy distributions line profiles, and equivalent widths of prominent spectral lines. These data were used in a model atmosphere analysis and abundance determination. Another objective of this work was to better define the class and its members.

114.171 **Variations of the hydrogen Balmer lines in the spectrum of the silicon Ap star CU Vir.** T. A. Ryabchikova. Izv. Krymskoj Astrofiz. Obs., Vol. 45, 146 - 151 (1972). In Russian.
The phase variations of the Balmer lines $H_\gamma - H_{13}$ were examined in the spectrum variable silicon Ap star CU Vir (HD 124224).

114.172 **Automatic classification of G5-K5 stars by means of 166 Å/mm objective prism spectra (4000−4550 Å).** R. M. West. Byull. Abastumansk. Astrofiz. Obs., No. 43, p. 109 - 170 (1972).
An exercise in automatic classification utilizing optimized quantitative spectral criteria in the 4000−4550 Å region of 142 G5-K5 stars (mainly giants) has been carried out by means of 239 objective prism spectra of dispersion 166 Å/mm at Hγ.

114.173 **A finding list of stars of spectral type A 7 and earlier in regions at high galactic latitudes. VI.** A. G. D. Philip, J. Stock. Bol. Obs. Tonantzintla y Tacubaya, Vol. 6, (No. 38), 201 - 209 (1972).
An objective prism survey has been made in a 434 square degree region extending from right ascension 22 hours to 4 hours at a declination of −28°. This trip extends from $l = 22°$, $b = −54°$, to $l = 225°$, $b = −50°$ and goes through the south galactic pole. A finding list containing positions, magnitudes and spectral types for 539 stars is presented.

114.174 **Fünf Jahre Spektroskopie am Karl-Schwarzschild-Observatorium Tautenburg.** E. Bartl. Jenaer Rundschau, (Jena Rev.), 17. Jahrgang, p. 327 - 330 (1972).

114.175 **Spectrophotometry of 12 bright stars.** V. G. Karetnikov. Problems of cosmic physics. Vyp. (No.) 7, (see 003.028), p. 115 - 122 (1972). In Russian.
From spectrograms with a dispersion of 166 Å/mm the energy distribution in the spectra of 12 stars has been investigated. The values of the spectrophotometric gradients for three regions of the spectrum, the value of the Balmer discontinuity, as well as electron densities of stellar atmospheres have been determined by the methods of Inglis-Teller and Unsöld. All the star characteristics have been compared with the data of Code, Kopylov, Karyagina and Kharitonov, Chalonge and Divan.

114.176 **Observational, theoretical and predicted data on the infrared and microwave spectra of early-type stars.** J. C. Pecker. JILA Rep., No. 109, 83 pp. (1971).

114.177 **On the selective mechanism of excitation of the λ 5696 CIII line in spectra of some stars.** A. A. Nikitin, T. Kh. Feklistova. Vestn. Leningr. un-ta, 1972, No. 13, p. 134 - 139. In Russian. Abstr. in Referativ. Zhurn. 51. Astron., 2.51.208 (1973).

114.178 **Observed energy distribution of α Lyra and β Cen at 2000−3800 Å.**

G. A. Gurzadyan, J. B. Ohanesyan.
Nature, Vol. 239, 90 (1972).

**114.179 On the accuracy of the spectrophotometric stand-
ard of η UMa.** N. S. Komarov, V. A. Pozigun.
Astron. Tsirk., No. 726, p. 3 - 4 (1972). In Russian.

**114.180 On fast changes of the Hβ profile in the spectrum
of RW Aurigae.** Z. A. Ismailov.
Astron. Tsirk., No. 734, p. 6 - 8 (1972). In Russian.

**114.181 The mean relative energy distribution in the spectra
of G8III-and K0III-stars.**
T. V. Solodovnikova, V. M. Tereshchenko, A. V. Kharitonov.
Astron. Tsirk., No. 738, p. 3 - 4 (1972). In Russian.

114.182 New carbon stars. Z. Alksne, V. Ozolinya.
Astron. Tsirk., No. 738, p. 7 - 8 (1972). In Russian.

114.183 On the variability of Ap-stars. L. I. Snezhko.
Astron. Tsirk., No. 741, p. 3 - 5 (1972). In Russian.

A helium-rich star in Puppis.
Sky Telescope, Vol. 44, 305 (1972).

**Zonal spectrophotometric standards. A study of the
energy distribution in absolute units in the spectra of 109
stars.** See Abstr. 003.121.

**A search for diffuse interstellar features in stars with
circumstellar dust shells.** See Abstr. 064.024.

**Analyses of light-ion spectra in stellar atmospheres.
I. Magnesium II in B and O stars.** See Abstr. 064.027.

**Spectroscopic analysis of the weak-helium-line star
α Sculptoris.** See Abstr. 064.030.

Element abundances in O- and early B-stars.
See Abstr. 064.036.

**The synthetic spectrum of the CN red system and
its application to stellar spectra of moderate resolution.**
See Abstr. 064.042.

**Free-free and Balmer line emission from optically
thick stellar shells.** See Abstr. 064.044.

Temperature scale for B-type supergiants.
See Abstr. 064.051.

Atmospheres of A-type supergiants.
See Abstr. 064.052.

Abundances in K giant stars. See Abstr. 064.066.

Ultraviolet stars and the interstellar gas.
See Abstr. 065.042.

Extrem junge Sterne. See Abstr. 065.094.

Iron in the sun and stars. See Abstr. 071.067.

**On ultraviolet stellar fluxes. III. Importance of H$_2$
Lyman-band absorption in the sun and other stars.**
See Abstr. 076.039.

**Photometric and polarimetric study of IR-stars in
the optical and infrared parts of the spectrum.**
See Abstr. 113.001.

**The application of synthetic spectra to colour sys-
tem transformations—I. The RGU and UBV systems.**
See Abstr. 113.016.

**The application of synthetic spectra to colour sys-
tem transformations—II. The photographic and photoelectric
UBV systems.** See Abstr. 113.017.

**Photographic photometry of T Tauri stars and re-
lated objects in Orion.** See Abstr. 113.033.

**Infrared excesses in early-type stars: Free-free
emission.** See Abstr. 113.038.

Far-infrared and $uvby$ photometry of V1057 Cygni.
See Abstr. 113.042.

**Photometry of the early-type stars from the Tonant-
zintla lists, near the south galactic pole.**
See Abstr. 113.045.

**Ultraviolet photometry from the Orbiting Astro-
nomical Observatory. V. The helium-weak stars.**
See Abstr. 113.046.

Luminosity classification of stars earlier than O9.
See Abstr. 115.013.

**Supergéantes. Rapport introductif – Resultats des
observations.** See Abstr. 115.018.

The TiO absorption and the magnitudes of dM stars.
See Abstr. 115.027

**The unique magnetic and spectrum variations of
HD 24712.** See Abstr. 116.001.

**Magnetic-field variations in 78 Virginis, Beta Coro-
nae Borealis, and 73 Draconis.** See Abstr. 116.005.

**The periodic variability of the peculiar A star
HD 111133.** See Abstr. 116.006.

**Rotational velocities and spectral types of some
A-type stars.** See Abstr. 116.011.

The orbit of 53 Tauri (B9p-Mn).
See Abstr. 119. 012.

**Variable stars of small amplitude. I. Supergiants and
OB stars.** See Abstr. 122.031.

**The influence of metal lines on the colors of ce-
pheids in the Small Magellanic Cloud.** See Abstr. 122.034.

**The classification of intrinsic variable stars. II. The
red variables of S and related types.** See Abstr. 122.080.

**Infrared radiation from RV Tauri stars. I. An infrared
survey of RV Tauri stars and related objects.**
See Abstr. 122.114.

Microwave celestial water-vapor sources.
See Abstr. 131.008.

**Millimeter wavelength molecular line observations of
infrared sources and dust clouds.** See Abstr. 131.037.

New OH sources associated with IR-late-type stars.
See Abstr. 131.058.

Infra-red sources in the H II region W3.
See Abstr. 131.062.

Observations of the $^2\Pi_{3/2}$, $J = 5/2$ state of interstellar OH. See Abstr. 131.065.

Very-low-excitation compact nebulae.
See Abstr. 132.026.

On the problem of connection of diffuse nebulae with stars of early spectral classes. See Abstr. 132.036.

Radio observations of early-type stars.
See Abstr. 141.025.

Infrared observations of variable radio objects.
See Abstr. 141.032.

Digicon spectrophotometry of the quasi-stellar object PHL 957. See Abstr. 141.101.

The distribution of optical spectral indices for QSOs from U, B, V data. See Abstr. 141.104.

Spectroscopic changes in the suspected X-ray source X Persei. See Abstr. 142.145.

On the reality of a group of carbon stars in Auriga.
See Abstr. 152.009.

Some characteristics of the moderately old open cluster NGC 752. See Abstr. 153.026.

Distribution of O to A0 stars in a region in Vela ($l = 268°.5$, $b = -0°.3$). See Abstr. 155.059.

Light variations of high luminosity O and B stars in the Large Magellanic Cloud. See Abstr. 159.008.

Errata

114.901 **Erratum: 'The nature of the Herbig Ae- and Be-type stars associated with nebulosity'** [Astrophys. Journ., Vol. 173, 353 - 366 (1972)]. S. E. Strom, K. M. Strom, J. Yost, L. Carrasco, G. Grasdalen.
Astrophys. Journ., Vol. 176, 845 (1972).

114.902 **Errata: "Infrared spectroelectrophotometry of zirconium stars"** [Astron. Tsirk., No. 686, p. 3 - 5 (1972)]. R. I. Chuprina.
Astron. Tsirk., No. 717, p. 8 (1972). In Russian.

114.903 **Erratum: 'On the Wolf-Rayet stars HD 90657 and HD 117688'** [Publ. Astron. Soc. Pacific, Vol. 84, 450 - 453 (1972)]. V. S. Niemelä.
Publ. Astron. Soc. Pacific, Vol. 84, 888 (1972). – See Abstr. 08.114.005.

115 Stellar Luminosities, Masses, Diameters, HR-Diagrams and Others

115.001 **A recalibration of the absolute magnitudes of supergiants.** R. Stothers.
Publ. Astron. Soc. Pacific, Vol. 84, 373 - 378 (1972).
Because of the usefulness of supergiants in studies of galactic structure, an improved calibration of the visual absolute magnitudes for luminosity class I stars is derived. Significant differences from the results of Blaauw and Schmidt-Kaler are found at some MK classes.

115.002 **Stellar angular diameters.** B. Warner.
Monthly Notices Roy. Astron. Soc., Vol. 158, 1P - 3P (1972).
It is shown that the equation $\log d'' = 0.730(B-V) - 0.2V - 2.512$ gives excellent estimates of stellar angular diameters (d), independent of luminosity and reddening, for stars with $B-V > -0.1$.

115.003 **Red supergiants and neutrino emission. II.** R. Stothers.
Astrophys. Journ., Vol. 175, 717 - 730 (1972).
The variation with stellar mass of the ratio of the numbers of blue and red supergiants is investigated. Statistical data for supergiants in young open clusters and subgroups of associations are collected to supplement a more restricted list in Paper I. Improved methods are used to identify hydrogen-burning supergiants, as well as faint supergiant remnants of binary mass exchange, and to arrange the bright evolved supergiants in order of their masses. Relevant published work on stellar evolution, rotation, mass loss, and duplicity is used to predict upper and lower limits on the blue-to-red ratio.

115.004 **Luminosities and motions of the F-type stars. II. Metal-deficient stars.** O. J. Eggen.
Astrophys. Journ., Vol. 175, 787 - 807 (1972).
An improved method of computing luminosities of F- and early G-type metal-deficient stars from intermediate-band photometry is applied to the available photometry for some 250 objects with $\Delta [m_1] > + 0.025$, including all of those brighter than visual magnitude 6.5 for which accurate apparent motions are available. Twenty halo stars, for which $\Delta [m_1] > + 0.080$, are included. The distribution of the resulting space motions is in excellent agreement with that found from previous studies, based on UBV photometry, and used in discussing the dynamics of a collapsing galaxy.

115.005 **The absolute flux of Canopus.** S. B. Parsons.
Bull. American Astron. Soc., Vol. 4, 335 (1972).
Abstr. AAS.

115.006 **Stars HD 28052, HD 31109 and HD 205767 with**

anomalous high luminosity.
O. H. Gusejnov, H. I. Novruzova.
Astron. Tsirk., No. 709, p. 6 - 7 (1972). In Russian.

115.007 **The diameter of an occulting body from a single observation.** R. E. Nather.
Monthly Notes Astron. Soc. Southern Africa, Vol. 31, 135 - 138 (1972).

115.008 **Luminosity and mass functions for low main-sequence stars.** I. Mazzitelli.
Astrophys. Space Sci., Vol. 17, 378 - 389 (1972).

The luminosity function derived for main sequence stars in the neighbourhood of the sun shows evidence for a flattening in the interval $5 \lesssim M_v \lesssim 8$. An interpretation of this feature by means of theoretical models enables us to deduce the mass function for stars belonging to the main sequence.

115.009 **Catalogue of stellar dimensions.**
A. J. Wesselink, K. Paranya, K. DeVorkin.
Astron. Astrophys., Suppl. Ser., Vol. 7, 257 - 289 (1972).

The catalogue presents angular diameters of 2392 stars, linear diameters, absolute magnitudes and spectroscopic parallaxes for 2301 stars. Previous attempts of computing stellar diameters, using the colour and the surface brightness, are briefly discussed. For the diameters in the present catalogue, the colour index is again used; the method, based on a calibrated surface brightness, was used before by the first author (Wesselink 1969).

115.010 **Construction of theoretical diagrams for stellar groups.**
O. B. Dluzhnevskaja, V. V. Musylev, G. G. Rodionova.
Nauchn. Informatsii, vyp. (No.) 21, p. 68 - 79 (1972). In Russian.

This paper is the first part of an investigation concerning construction of theoretical HR-diagrams for different stellar groups. Different possibilities of comparison between theoretical evolutionary calculations and observational data are discussed. Some papers dealing with the same problem are briefly reviewed. A new method for the calculation of isochrones using theoretical evolutionary sequences is suggested.

115.011 **The influence of space velocity on the M_v, $R-I$ main sequences of K3–M2 dwarfs.**
A. R. Upgren.
Astron. Journ., Vol. 77, 745 - 750 (1972).

New data based on the McCormick and Michigan catalogues of late dwarf stars have substantially increased the total number for which space motions can be found. For 205 dwarfs of spectral types K3–M2 with $UBVRI$ photometry, space motions have been determined which show that the M_v, $R-I$ main sequence is linear and varies significantly with the velocity perpendicular to the galactic plane, but only marginally with stellar motion in the galactic plane.

115.012 **The angular diameter of X Cancri.**
P. Bartholdi, D. S. Evans, R. I. Mitchell, E. C. Silverberg, D. C. Wells, J. R. Wiant.
Astron. Journ., Vol. 77, 756 - 759 (1972).

An occultation reappearance of the semiregular variable X Cancri was observed with the 107-inch telescope of McDonald Observatory on 9 November 1971 by the laser group using slightly modified laser equipment. The trace has been analyzed by the occultation group and yields a timing at $08^h47^m31\overset{s}{.}045$ (three sec after predicted time), a lunar slope of $-3\overset{\circ}{.}1$ and an angular diameter of 7.9 arc msec for a uniform disk and 9.0 for a fully darkened disk. The statistical accuracy of fit is discussed. The problem of the effective temperature of this carbon star is discussed, and estimates near 2500 K are derived both from the occultation data and from an integra-

tion based on the multicolor photometry reported by Mendoza.

115.013 **Luminosity classification of stars earlier than O9.**
G. V. Coyne.
Ric. Astron.,Specola Vaticana, *Castel Gandolfo,* Vol. 8,(No. 15), 343 - 351 (1972).

Spectra of O stars ranging from dwarfs to supergiants have been obtained at a dispersion of 117 Å/mm with an image-tube spectrograph at the Cassegrain focus of the Vatican Observatory 60 cm reflector. Luminosity criteria established at lower reciprocal linear dispersions are applicable to this material to the extent that supergiant stars can be readily detected. The evolution of the spectra from late to early spectral types and towards higher luminosity seems to indicate empirically that the Of stars fit naturally into the O star sequence at high luminosities.

115.014 **On the luminosity functions of giant stars in globular clusters.** J. N. Bahcall, A. Yahil.
Astrophys. Journ., Vol. 177, 647 - 651 (1972).

The observed luminosity function for stars on the giant branch of M15, as determined by Sandage, Katem, and Kristian, is compared with simulated luminosity functions generated from a smooth curve. Apparent features are seen in both the observed and simulated luminosity functions. A crude fit to the observed differential luminosity function is given by $N(L) \propto L^{-2.4}$.

115.015 **The luminosities of late-type stars of differing metal abundance.** D. H. P. Jones, M. E. Dixon.
Astrophys. Journ., Vol. 177, 665 - 680 (1972).

A technique is described which determines approximate luminosities of G- and K-type stars. It is based on narrowband photometric observations of the Mg b-lines and the blue CN band, and upon broad-band observations of ultraviolet excess, $\delta(0.6)$, and the blanketing-free index, $(B-V)_c$. The technique may be applied to stars of normal as well as of low metal abundance, but in this paper special attention is given to stars of low metal abundance. The use of the technique in kinematic studies is illustrated by reference to stars in Sandage's list of newly discovered subdwarfs.

115.016 **The main sequence of the H-R diagram: Its significance and role in stellar evolution.** E. B. Weston.
Publ. Astron. Soc. Pacific, Vol. 84, 645 - 646 (1972). – Abstr. Astron. Soc. Pacific.

115.017 **The luminosity function and density distribution of disk population stars.** D. Weistrop.
Astron. Journ., Vol. 77, 849 - 862 (1972).

The density distribution and luminosity function of the disk population are investigated, using star counts as a function of V and $B-V$ for several thousand stars near the North Galactic Pole. Having estimated the contributions to the counts due to population II stars and disk giants and subgiants, the remaining stars are assumed to constitute a pure disk-dwarf population. Density distributions for stars in successive $(B-V)_0$ intervals are calculated and combined to form a composite disk density distribution. The agreement with Oort's K-giant distribution is satisfactory.

115.018 **Supergéantes. Rapport introductif – Resultats des observations.** C. Fehrenbach.
Proc. Third Colloquium on astrophysics. Supergiant stars, Trieste 1971, (see 012.020), p. 7 - 21 (1972).

115.019 **The absolute magnitudes of A and F supergiants.**
G. D. Bouw, S. B. Parsons.
Proc. Third Colloquium on astrophysics. Supergiant stars, Trieste 1971, (see 012.020), p. 22 - 27 (1972).

**115.020 The M_{bol} − log T_e diagram of the brightest super-
giants in the Large Magellanic Cloud.**
J. P. Brunet, L. Prévot.
Proc. Third Colloquium on astrophysics. Supergiant stars,
Trieste 1971, (see 012.020), p. 119 - 124 (1972).

**115.021 The surface gravities of F and G supergiants and
classical cepheids. S. B. Parsons.**
Proc. Third Colloquium on astrophysics. Supergiant stars,
Trieste 1971, (see 012.020), p. 231 - 232 (1972). − Abstract.

115.022 The absolute magnitude of Gamma Velorum.
R. Rajamohan.
Observatory, Vol. 92, 232 - 233 (1972). − Note.

**115.023 The mass-luminosity relationship of the Hyades and
other stars. J. B. Alexander.**
Royal Obs. Bull., [Herstmonceux: Royal Greenwich Obs.], No.
175, p. 117 - 123 (1972).
    The mass-luminosity relationship of the Hyades cluster is
rediscussed and the conclusions of a previous paper are confirmed. Although the mass-luminosity relationship of the
cluster is poorly determined at present, it is probable that the
unevolved stars in the Hyades are somewhat undermassive
when compared with typical unevolved field stars of the
corresponding luminosity. For stars in the general field, there
is no evidence for two discrete mass-luminosity relationships.

115.024 The Hertzsprung-Russell diagram. K. Barlai.
    Fiz. Szemle, (*Hungary*), Vol. 22, No. 3, p. 65 - 75
(1972). In Hungarian.

115.025 Luminosities and motions of A0 to A2 stars.
O. J. Eggen.
Publ. Astron. Soc. Pacific, Vol. 84, 757 - 765 (1972).
    The space motions of 182 A0 to A2 stars (Strömgren's 'in-
termediate group') are derived from luminosities determined by
intermediate-band photometry. The accurate luminosities
(mean error of ±0^{m}2) confirm a previous conclusion that the
young disk population stars near the sun occur in only some
half-dozen groups. A few blue stragglers of the older disk
population are also included.

115.026 High velocity stars population.
V. Castellani, A. Martini.
Mem. Soc. Astron. Italiana, Nuova Ser., Vol. 43, 447 - 453
(1972).
    Stars with radial velocity $V_R \geqslant 65$ km/sec show in the
H R diagram a behaviour which can be attributed to a well defined evolutionary stage, perhaps peculiar with respect to the
globular and old-disk clusters. This behaviour can be related to
similar characteristics of some moving groups found by Eggen.
The variable stars belonging to this population as well as some

general consequences are shortly discussed.

115.027 The TiO absorption and the magnitudes of dM stars.
P. F. Chugainov.
Izv. Krymskoj Astrofiz. Obs., Vol. 45, 130 - 134 (1972).
In Russian.
    The assumption is made that the magnitudes R_J, I_J
(Johnson's system) of the red dwarf stars depend not only on
the bolometric luminosities and the effective temperatures but
also on the strength of the TiO absorption bands. The colour
temperature scale for the dwarf stars obtained by Eggen (1967,
1968) should be revised.

115.028 The luminosity of a collapsing star. Pt. I.
H. Dwivedi, R. Kantowski.
Lecture Notes Phys., Vol. 14, 126 - 136 (1972).

**Microturbulence in atmospheres of F, G, K type
stars. I. Curve of growth analysis of G, K type subgiants.**
See Abstr. 064.032.

Effects of semiconvection on the horizontal-branch.
See Abstr. 065.034.

**The structure of chemically homogeneous main-se-
quence stars. See Abstr. 065.053.**

**Multicolor photometry of the M dwarf Proxima
Centauri. See Abstr. 113.021.**

Twenty-micron fluxes of bright stellar standards.
See Abstr. 114.123.

Spectral line variations in F0−K5 supergiants.
See Abstr. 114.149.

**On the spectral analysis of the superluminous emis-
sion line stars. See Abstr. 114.154.**

Hidden mass in the solar neighborhood.
See Abstr. 117.019.

**Variable stars of small amplitude. I. Supergiants and
OB stars. See Abstr. 122.031.**

**Determination of the mean absolute magnitude of
RR Lyrae stars. See Abstr. 122.061.**

The red giants in the Hyades group.
See Abstr. 153.001.

**Des étoiles supergéantes à très fortes raies d'hydro-
gène dans le Grand Nuage de Magellan. See Abstr. 159.011.**

116   Stellar Magnetic Field, Figure, Rotation

116.001   The unique magnetic and spectrum variations of HD 24712.   G. W. Preston.
Astrophys. Journ., Vol. 175, 465 - 472 (1972).

A series of coudé spectrograms is used to show that the effective magnetic field of HD 24712 varies in a period of 12.45 days with no polarity reversal. Strong antiphase variations between the lines of Mg and Eu also occur in this period. Finally, it is shown that marked spectrum variations in the absence of a magnetic polarity reversal do not pose a difficulty for the oblique-rotator model.

116.002   On the period of γ Equ.   P. Renson.
Astron. Astrophys., Vol. 20, 173 - 174 (1972).
In French.

The respective worths of the values 314 d and 1786 d for the period of γ Equ are weighed. The dispersion around a mean curve is smaller for the 1786 d period, but the photometric data seem rather to support $P = 314$ d.

116.003   Rotational velocities of Ap stars.
H. A. Abt, F. H. Chaffee, G. Suffolk.
Astrophys. Journ., Vol. 175, 779 - 785 (1972).

Projected rotational velocities were measured on coudé spectra for essentially all the northern bright Ap stars. Differences in these rotational velocities between subgroups are found to be marginal, except that Guthrie's discovery of a deficiency of very narrow-lined Si stars is confirmed. A comparison of projected rotational velocities with equatorial rotational velocities computed from periods of variations in light, magnetic field, or spectrum leads to evidence for random orientations of rotational axes.

116.004   Photoelectric measurements of stellar magnetic fields.
E. F. Borra, J. D. Landstreet, A. H. Vaughan, Jr.
Journ. Roy. Astron. Soc. Canada, Vol. 66, 217 (1972).
Abstr. Canadian Astron. Soc.

116.005   Magnetic-field variations in 78 Virginis, Beta Coronae Borealis, and 73 Draconis.
S. C. Wolff, W. K. Bonsack.
Astrophys. Journ., Vol. 176, 425 - 432 (1972).

Photographic observations of the Zeeman effect in stellar spectra have been made with new equipment at the Mauna Kea Observatory. Measurements of the effective longitudinal field H_e have been obtained for 78 Vir, β CrB, and 73 Dra. Comparison of the values obtained for 78 Vir and β CrB with the measurements made by Preston at Lick Observatory establishes the absence of any significant systematic difference between the two observatories. The results do not support the suggestion of long-term systematic changes in the cycles of magnetic variation of β CrB and 73 Dra, which were marginally indicated by previous investigations.

116.006   The periodic variability of the peculiar A star HD 111133.   S. C. Wolff, R. J. Wolff.
Astrophys. Journ., Vol. 176, 433 - 438 (1972).

The period of the peculiar A star HD 111133 is 16.31 days. Maximum light coincides in phase with minimum longitudinal magnetic field. The strengths of the lines of the iron-peak elements vary and are greatest at maximum light. This star belongs to that subgroup of Ap stars in which the magnetic and rotation axes are nearly aligned.

116.007   On magnetic stars.   G. S. D. Babu, S. D. Sinvhal.
Bull. Astron. Inst. Czechoslovakia, Vol. 23, 297 - 301 (1972).

A statistical study of magnetic stars reveals that their mean rotational velocities do not depend on their magnetic field strengths. Stars with low magnetic field strengths are mostly Mn stars and those with high field strengths are mostly Cr-Sr stars.

116.008   Lorentz broadening of the hydrogen lines and possible astrophysical applications.
A. E. Sonnanstine, C. R. Cowley.
Bull. American Astron. Soc., Vol. 4, 333 - 334 (1972).
Abstr. AAS.

116.009   Magnetic star dynamos.   K. L. McDonald.
Bull. American Astron. Soc., Vol. 4, 338 (1972).
Abstr. AAS.

116.010   Remarks on magnetic intensification and starspots in the periodic magnetic variable HD 188041.
L. C. Green, C. D. Littleton.
Bull. American Astron. Soc., Vol. 4, 338 (1972). – Abstr. AAS.

116.011   Rotational velocities and spectral types of some A-type stars.   O. H. Levato.
Publ. Astron. Soc. Pacific, Vol. 84, 584 - 588 (1972).

The rotational velocities and MK spectral types of more than 100 southern A-type stars are described.

116.012   Photoelectric observations of magnetic stars. IV. 41 Tau and 21 Com.   C. Blanco, F. A. Catalano.
Astron. Journ., Vol. 77, 666 - 668 (1972).

Photoelectric observations in three colors of the magnetic stars 41 Tau and 21 Com are reported. For both stars improved periods are obtained. The oblique-rotator model seems to be the most reliable for 41 Tau.

116.013   Stellar rotation.   M. Hack.
Atti XV Riunione Soc. Astron. Italiana, Bologna 1971, (see 012.013), p. 91 - 103 (1972).

A general review of the phenomenology of stellar rotation is given. The dependence of rotation upon spectral type and luminosity class and the importance of large-scale motions like macroturbulence are discussed. The effect of the rotational velocity and of the position of the rotation axis on the observable spectroscopic characteristics is examined.

116.014   Statistical studies in stellar rotation. II. A method of analyzing rotational coupling in double stars and an introduction to its applications.   P. L. Bernacca.
Astrophys. Journ., Vol. 177, 161 - 175 (1972).

The coupling between the rotational velocities v_1 and v_2 of the components of double stars has bearing on their origin, on the problem of synchronism, and on the distribution of angular momentum in dust and gas clouds. Since the observations give the apparent velocity $v_k \sin i_k$ ($k = 1, 2$), the only large-scale approach is statistical and it requires the knowledge of the probability density $\psi(i_1, i_2)$. A formula of the probability density was derived and application was made for a sample of visual and closely spaced binaries.

116.015   Circular polarimetry of fifteen interesting objects.
J. C. Kemp, R. D. Wolstencroft, J. B. Swedlund.
Astrophys. Journ., Vol. 177, 177 - 189 (1972).

A search is described for circular polarization of visible light in 15 objects: two eclipsing binaries (β Lyr and BM Ori); six magnetic Ap stars; three planetary nebulae; NGC 2261 (Hubble's nebula); M87; Sirius; and the Orion A region. The results are given in a table. The paper also contains a complete description of the photoelastic polarimeter, with special attention to the incidental linear-circular conversion.

116.016 The effect of the Coriolis force on the stability of rotating magnetic stars.   K. Sakurai.
Astrophys. Journ., Vol. 177, 423 - 425 (1972).
The effect of the Coriolis force on the stability of rotating magnetic stars in hydrostatic equilibrium is investigated by using the method of the energy principle. It is shown that this effect is to inhibit the onset of instability.

116.017 Magnetic stars.   S. B. Pikel'ner, V. L. Khokhlova.
Uspekhi fiz. nauk, Vol. 107, 389 - 404 (1972).
In Russian.

116.018 On the spectral distribution of the known magnetic stars.
C. Blanco, F. A. Catalano, G. Godoli, S. Vaccari.
Mem. Soc. Astron. Italiana, Nuova Ser., Vol. 43, 545 - 546 (1972). − Letter.

116.019 Strongly magnetic, rapidly rotating main-sequence stars.   K. W. Robson.
Bull. Australian Math. Soc., Vol. 7, 313 - 314 (1972).
The author studies the effects of uniform rotation and strong dipolar magnetic fields on the structure and properties of main-sequence stars using a perturbation method.

On possible separation of elements in the atmosphere of magnetic stars.   See Abstr. 064.054.

Perturbation of the radial and non-radial oscillations of a star by a magnetic field.   See Abstr. 065.036.

Oscillations of a polytrope with a toroidal magnetic field.   See Abstr. 065.038.

Accretion vortices and X-ray sources.   See Abstr. 065.087.

On the apparent connection between space velocity and rotational velocity in early type stars.   See Abstr. 112.014.

The rotation of stars in binary systems.   See Abstr. 117.027.

New orbit and mass of the visual binary and magnetic Ap star ADS 7334 = HR 3724.   See Abstr. 118.005.

On the polarization of magnetic stars radiation.   See Abstr. 131.054.

117 Binary and Multiple Stars, Theory

117.001 Evolution of close binaries with mass loss from the system. I.   A. V. Tutukov, L. R. Yungelson.
Nauchn. Informatsii, vyp. (No.) 20, p. 86 - 93 (1971).
In Russian.
Effects of mass and angular momentum loss from close binary systems on the evolution of components are studied. It is shown that under certain assumptions it is possible to obtain a close binary composed of two white dwarf stars.

117.002 Evolution of close binaries with mass loss from the system. II.   L. R. Yungelson.
Nauchn. Informatsii, vyp. (No.) 20, p. 94 - 100 (1971).
In Russian.
The evolution of the originally more massive star of a close binary system was computed through the phase of mass exchange. Two cases were considered: 1) 25 per cent of mass, lost by the star, leaves the system, 2) mass of system is conserved.

117.003 Mass accretion by a neutron star in a double system. II.   P. R. Amnuel, O. H. Guseinov.
Astrofizika, Vol. 8, 107 - 115 (1972). In Russian. − English translation in Astrophysics, Vol. 8, No. 1.
The accretion of plasma flow by a magnetic field of a neutron star (white dwarf) in a double system is analyzed. We analyze the flows arising from the main component under different angles to the line of centres, and cases of different inclinations of the magnetic axis of the second component.

117.004 VY Canis Majoris as a multiple star.
C. E. Worley.
Astrophys. Journ., (*Letters*), Vol. 175, L93 - L94 (1972).
The nature of VY Canis Majoris as a multiple star is discussed, and new observations are reported.

117.005 Mass flow and period changes of contact binaries.   F. van't Veer.
Astron. Astrophys., Vol. 19, 337 - 342 (1972).
Period changes of eclipsing contact binaries are interpreted as the consequence of mass flows from the first and second Lagrangian points L_1 and L_2. The possibilities of material flow are studied. Formulae are given for the rate of increase or decrease of the orbital radius and the period as a function of the rate of mass transport and the mass ratio of the components in a certain number of idealized cases. A short discussion of some evolutionary aspects is also given.

117.006 Eruptive binaries. Part II.   J. Smak.
Postępy Astron., Vol. 20, 205 - 218 (1972).
In Polish.
This article contains a review of problems connected with the physical properties of the components and the circumstellar matter, and with the mechanisms of outbursts.

117.007 Physical pairs among stars with large proper motions.   I. N. Latyshev.
Astron. Tsirk., No. 672, p. 7 (1972). In Russian.

117.008 The function $\alpha(k, p)$ for a sphere-ellipsoid model.   M. I. Lavrov.
Astron. Tsirk., No. 677, p. 3 - 5 (1972). In Russian.

117.009 The transit of the Jupiter system across the multiple star β Scorpii. Z. Klimek.
Acta Astron., Vol. 22, 49 - 54 (1972).

The paper contains the results of observations of occultations of the stars β_1 Sco and β_2 Sco by Jupiter and of the close approach of Io (Jupiter I) to β_2 Sco. The photoelectric, visual and photographic observations indicate that no occultation of β_2 Sco by Io occurred at Cracow.

117.010 Wide moving pairs among nearby K and M dwarfs.
P. K. Lü, A. R. Upgren.
Bull. American Astron. Soc., Vol. 4, 329 (1972). – Abstr. AAS.

117.011 Observations of binaries and trapezium systems using a focal plane modulation technique.
F. Rosenberg, L. Fredrick, F. Villamediana.
Bull. American Astron. Soc., Vol. 4, 400 (1972). – Abstr. AAS.

117.012 On the pointlessness of stellar masses: (i) AM CVn (= HZ 29), a semi-detached double white dwarf binary system, (ii) the general relativistic Roche model.
J. Faulkner, B. P. Flannery.
Bull. American Astron. Soc., Vol. 4, 401 (1972). – This abstract was presented at the AAS meeting in Seattle, Washington, instead of the abstract published earlier in Bull. American Astron. Soc., Vol. 4, 222 (1972). – See 07.066.051.

117.013 Enge Doppelsterne. R. Henkel.
BAV Rundbrief, 21. Jahrgang, Sonder-Rundbrief, (see 012.011), p. 51 - 53 (1972).

117.014 The peculiar O6f star HD 148937 and the symmetrically surrounding nebulae. H. M. Johnson.
Astrophys. Journ., Vol. 176, 645 - 649 (1972).

Peculiarities of the system of nebular shells around HD 148937, of which NGC 6164–5 are the inner-most, are discussed with reference to optical and radiofrequency data. A standard extrapolation from the optical flux density of NGC 6164–5 predicts a marginally detectable radio source, and it does not appear in the relevant surveys. The Of star and its two companions are photometered around 4640, 4686, and 4861 Å. The ultraviolet continuum of the star is observed and the spectrum is compared with some models.

117.015 The disc model of gaseous accretion on a relativistic star in a close binary system. N. I. Shakura.
Astron. Zhurn. Akad. Nauk SSSR, Vol. 49, 921 - 929 (1972). In Russian. English translation in Soviet Astron. AJ, Vol. 16, No. 5.

A close binary system model with a relativistic neutron or collapsed star is considered. The formation of the disc surrounding the relativistic star and the radiated energy flux are calculated. It is shown that this model is capable to explain the properties of galactic X-ray sources with thermal emission spectrum.

117.016 Evolution of a close binary with mass loss from the system. L. R. Yungel'son.
Astron. Zhurn. Akad. Nauk SSSR, Vol. 49, 1059 - 1062 (1972). In Russian. English translation in Soviet Astron. AJ, Vol. 16, No. 5.

The evolution of the originally more massive component (primary) of a close binary system has been computed. Two distinct cases have been discussed: 1) total mass and angular momentum of the system are conserved; 2) twenty-five per cent of mass lost by the primary leaves the system. When mass loss from the whole system is taken into account, the fitting of theoretical models to observational data is improved.

117.017 On the mass-luminosity relation.
G. E. McCluskey, Jr., Y. Kondo.

Astrophys. Space Sci., Vol. 17, 134 - 149 (1972).

The results of a least-squares study of the mass-luminosity relation for eclipsing and visual binary stars consisting of main sequence components are presented. Two methods are discussed. The results and a comparison of the two methods are discussed. It is found that the following mass-luminosity relation represents the observational data satisfactorily:
$\log M = 0.504 - 0.103 M_{BOL}$, $-8 \leqslant M_{BOL} \leqslant +10.5$. A discussion of the data and of the possibility that separate mass-luminosity relations may exist for visual and eclipsing binaries is given. The possibility that more than one mass-luminosity relation is required in the range $-8 \leqslant M_{BOL} \leqslant +13$ is also discussed.

117.018 Tidal evolution in close binary systems.
Z. Kopal.
Astrophys. Space Sci., Vol. 17, 161 - 185 = Lunar Sci. Inst., Houston, Texas, Contr. No. 90 (1972).

The aim of the present paper will be to give a mathematical outline of the theory of tidal evolution in close binary systems of secularly constant total momentum – an evolution activated by viscous friction of dynamical tides raised by the two components on each other. The first section contains a general outline of the problem; and in Section 2 we shall establish the basic expressions for the energy and momenta of close binaries consisting of components of arbitrary internal structure. In Section 3 we shall investigate the maximum and minimum values of the energy (kinetic and potential) which such systems can attain for given amount of total momentum; while in Section 4 we shall compare these results with the actual facts encountered in binaries with components whose internal structure (and, therefore, rotational momenta) are known to us from evidence furnished by the observed rates of apsidal advance.

117.019 Hidden mass in the solar neighborhood.
S. S. Kumar.
Astrophys. Space Sci., Vol. 17, 219 - 222 (1972).

It is suggested that the minimum mass of a star at the time of its formation is approximately 0.01 $M_\odot$. Making use of this fact and the stellar mass function $F(M) \propto M^{-\alpha}$, it is found that the hidden mass (or the missing mass) in the solar neighborhood may be explained by the presence of a large number of invisible stars of very low mass (0.01 $M_\odot \lesssim M < 0.07\ M_\odot$).

117.020 The dynamical evolution of triple star systems: A numerical study. E. M. Standish, Jr.
Astron. Astrophys., Vol. 21, 185 - 191 (1972).

The initial conditions for 800 triple star systems are created and the evolution is followed, usually until the escape of one member. It is found that the angular momentum of the system is the quantity most influential in determining the following distributions: the time of disintegration, the velocity and mass of the escaping particle, the semi-major axis, eccentricity and angular momentum of the remaining binary. These quantities are displayed graphically and discussed.

117.021 Evolution of close binaries.
V. S. Bychkova, S. B. Pikel'ner.
Zemlya i Vselennaya, 1972, No. 5, p. 22 - 26. In Russian.

117.022 The location and size of the hot spot in cataclysmic variable stars. B. Warner, W. L. Peters.
Monthly Notices Roy. Astron. Soc., Vol. 160, 15 - 20 (1972).

Single particle trajectories have been calculated for material leaving the inner Lagrangian point of close binary systems at thermal velocities. The dimensions and location of the stream as it reaches the ring around the primary were calculated for a range of mass ratios. Applications to VV Pup and U Gem are given.

117.023 Contact binaries: Opacity and rotation.

J. A. J. Whelan.
Monthly Notices Roy. Astron. Soc., Vol. 160, 63 - 77 (1972).

The main purpose of this paper is to investigate whether, in the framework of equal adiabatic constants, consideration of opacity changes or the inclusion of rotation (both uniform and non-uniform) will alleviate the age zero problem. A secondary purpose of the paper is to consider the recent objections to the Lucy model made by Mauder (1972) and Robinson (1972) and further, to point out the reasons why the Biermann–Thomas models have incorrect light curves.

117.024 Relative positions of stars of trapezium-type multiple systems. G. N. Salukvadze.
Byull. Abastumansk. Astrofiz. Obs., No. 43, p. 95 - 108 (1972). In Russian.

The present paper contains relative positions of the following stars of trapezium-type multiple systems: ADS 1869, 1920, 2159, 2843, 3940, 4728, 5322, 5682, 5685, 6216, 6366, 11169, 13374, 13376, 13626, 14526, 16953.

117.025 Stability criteria for triple stars.
R. S. Harrington.
Celestial Mechanics, Vol. 6, 322 - 327 (1972).

A wide variety of equal-mass stellar triple systems has been numerically integrated in order to establish factors pertinent to stability. The significant parameters appear to be whether the relative revolution is direct or retrograde, and the ratio of the periastron distance in the outer orbit to the semimajor axis of the inner orbit. For stability, this ratio must be at least 3.5 for direct orbits and at least 2.75 for retrograde orbits.

117.026 Evolution of close binaries. VIII. Mass exchange on the dynamical time scale.
B. Paczyński, R. Sienkiewicz.
Acta Astron., Vol. 22, 73 - 91 (1972).

The process of mass exchange in a close binary system is studied in the case where the more massive component has a deep convective envelope while filling up its Roche lobe. It is shown that a considerable fraction of mass is transferred on a dynamical time scale, provided the surface boundary condition for the mass losing component (i.e. the mixing length in its convective envelope) is not affected by the mass loss. Following Jędrzejec (1969) the relation is derived between the rate of mass transfer in a close binary and the extent to which the mass losing component overflows its Roche lobe. This relation was used in the evolutionary computations presented in this paper.

117.027 The rotation of stars in binary systems.
S. L. Piotrowski, S. M. Ruciński.
Postępy Astron., Vol. 20, 351 - 355 (1972). In Polish.

Rediscussion of the paper by Nariai (Publ. Astron. Soc. Japan, Vol. 23, 529 - 538 (1971)) shows that the apparent dependence of the asynchronism parameter V/V_{syn} on the orbital period for B0V-A3V binaries can be most easily explained by almost constant rotational velocities V for each of the four arbitrarily introduced spectral subgroups.

117.028 On the formation of double stars. S. S. Kumar.
Astrophys. Space Sci., Vol. 17, 453 - 458 (1972).

From a calculation of Roche densities in visual binaries, it is concluded that the stars in these systems would be tidally unstable at the time of formation if they were formed with their observed separations (semi-major axes). To avoid this difficulty, it is proposed that the visual binaries are formed by the disintegration of small clusters. Binaries formed this way are shown to have some interesting upper limits to their mass ratios. Comments are also made on the differences between the processes of star and planet formation.

117.029 Parallax and orbital motion of the unresolved astrometric binary BS +43°4305.
P. van de Kamp, M. D. Worth.
Astron. Journ., Vol. 77, 762 - 763 (1972).

The astrometric binary BD +43°4305 is found to have a period of 28.9 years, and a semi-axis major $0''039 \pm 0''002$ (p.e.) for the photocentric orbit. The absolute parallax is $+0''200 \pm 0''002$. The mass of the unseen companion is between 0.01 and 0.03 $M_\odot$.

117.030 Mass transfer in close binaries. III. Gaseous rings in Algol-like binaries. S. Kříž.
Bull. Astron. Inst. Czechoslovakia, Vol. 23, 328 - 331 (1972).

In the present paper an attempt is made at applying the theoretical results from the first two papers to models of Algol-like binaries, at determining when gaseous rings are generated in these systems and at comparing the results with observations.

117.031 Structure of close binaries. III. Interior models of uniformly rotating binaries.
M. D. T. Naylor, S. P. S. Anand.
Astrophys. Space Sci., Vol. 18, 59 - 84 (1972).

Internal models have been obtained for uniformly rotating synchronous close binary systems using a modified double approximation scheme. We have considered primaries of $10\,M_\odot$, $5\,M_\odot$, and $2\,M_\odot$ with mass ratios of 0.0 to 1.0 in steps of 0.1, and some results are given for a $1\,M_\odot$ primary with a mass ratio of 1.0. A maximum luminosity reduction of 2.3 % was found for a $10\,M_\odot$ primary with a mass ratio of 1.0 and 7.7 % for a mass ratio of 0.0. The corresponding values for $5\,M_\odot$ are 2.0 % and 7.0 %, and for $2\,M_\odot$ they are 1.6 % and 5.3 %, respectively. These values were not found to be sensitive to small changes in composition. The effect of gravity darkening on the apparent position of the primary in the theoretical H-R diagram was investigated. Two models of the secondaries are given and their dimensions are compared with their critical Roche lobes.

117.032 Structure of close binaries. IV. Non-synchronous polytropes. M. D. T. Naylor.
Astrophys. Space Sci., Vol. 18, 85 - 88 (1972).

Models of uniformly rotating non-synchronous binary polytropes have been calculated and compared with the results obtained from the Roche model. It is found that the relative dimensions of the non-synchronous systems are adequately portrayed by a Roche model but the degree of central condensation affects the minimum separation by an amount which is probably not insignificant.

117.033 Photometry and intrinsic period of HZ 29 (= AM CVn). W. Krzemiński.
Acta Astron., Vol. 22, 387 - 403 (1972).

Photoelectric observations of the ultrashort-period variable HZ 29 obtained in 1967 show a double-humped light curve of low amplitude and reveal antiphase colour variations. The period variations of this close pair observed in a time span of nine years are interpreted as a result of the orbital motion of HZ 29 about another star. A model of HZ 29 as a triple system is advanced.

117.034 New double systems. V. A. Sokolova.
Astron. Zhurn. Akad. Nauk SSSR, Vol. 49, 1326 - 1327 (1972). In Russian. English translation in Soviet Astron. AJ, Vol. 16, No. 6.

Three star pairs which are with high probability physical double systems have been found on the basis of common proper motions.

117.035 Gasströme in engen Doppelsternsystemen.
K. Walter.

BAV Rundbrief, 21. Jahrgang, p. 41 - 46 (1972).

117.036 CI Cyg as an eclipsing variable. A. Pučinskas.
Bull. Vilnius Astron. Obs., No. 33, p. 50 - 55 (1972).
In Russian.
The light curve of the symbiotic star CI Cyg is very similar to the curves of eclipsing variables. However, long duration of "eclipses" (about 0.3 P) makes such an interpretation difficult. If the system is composed of a red giant and a very hot dwarf, the light variations can be understood as eclipses of the hot component mainly by gas streams or clouds.

117.037 On the Napier method for the photometric reflection effect in close binary stars. V. Ureche.
Stud. Cerc. Astron., Vol. 17, 213 - 220 (1972).
In this paper some modifications of Napier's method (1968) are brought, in order to make it more easy for applications. An application to the close binary system AI Draconis is made. The results obtained with Napier's method and with Kopal's method (1959) are compared.

117.038 Formation of binaries from triple systems. V. Szebehely.
Proc. National Acad. Sci., USA, Vol. 69, 1077 - 1080 (1972).
The dynamical behavior of three masses moving under their mutual gravitational attraction in a plane is investigated by a systematic series of numerical experiments. It is shown that in 73% of the cases, a triple system disintegrates in less than 150 time units (corresponding to about 150 crossing times), and a binary is formed with the third star that escapes at hyperbolic velocity. The average time for disintegration is of the order of 10^9 years for triple stellar systems, as well as for triple galaxies.

117.039 The aberration of components of double stars. L. Janossy.
Acta Phys. Acad. Sci. Hungaricae, Vol. 31, 353 - 359 (1972).
So as to clear a primitive misconception about the nature of aberration of light a short derivation of the facts is given. In particular the effects are discussed which appear when the observer or alternatively the source are set to move.

117.040 Results of observations of the star Lalande 21185 made with the 26″ refractor at Pulkovo.
N. A. Shakht.
Astron. Tsirk., No. 736, p. 5 - 7 (1972). In Russian.

117.041 On the possible mass of the invisible component of δ Gem. G. V. Akhundova, O. H. Gusejnov.
Astron. Tsirk., No. 728, p. 1 - 2 (1972). In Russian.

117.042 61 ω Eri – a star with a possible relativistic component. G. V. Akhundova, O. H. Gusejnov.
Astron. Tsirk., No. 728, p. 2 - 4 (1972). In Russian.

The effects of viscous friction on the precession and nutation of celestial bodies. See Abstr. 042.022.

Tidal perturbation of the non-radial oscillations of a star. See Abstr. 065.022.

The occultation of β Sco by Jupiter. See Abstr. 099.020.

The determination of the diameter of Io from its occultation of β Scorpii C on May 14, 1971. See Abstr. 099.021.

Observation of the occultation of β Sco C by Io. See Abstr. 099.023.

Photoelectric observations of the occultation of β_1 Sco by Jupiter. See Abstr. 099.028.

Occultation of Beta Scorpii by Jupiter on May 13, 1971. See Abstr. 099.061.

Jupiter og Io okkulterer Beta Scorpii. See Abstr. 099.063.

Jupiter occultation of Beta Scorpii: Are the flashes time-symmetric? See Abstr. 099.068.

Photographic observations of the occultation of Beta Scorpii by Jupiter. See Abstr. 099.078.

Statistical studies in stellar rotation. II. A method of analyzing rotational coupling in double stars and an introduction to its applications. See Abstr. 116.014.

Binary stars as X-ray sources. See Abstr. 142.043.

The dynamical evolution of a stellar cluster with initial subclustering. See Abstr. 151.014.

The stability of certain model binary stellar systems in galactic gravitational fields. See Abstr. 151.025.

Zur Struktur der näheren Sonnenumgebung. Untersuchungen über Sterntrupps, Sternfamilien und Sternströme. See Abstr. 155.042.

118 Visual Binaries

118.001 **The orbit of the visual binary B 1909.**
G. A. Starikova.
Astron. Tsirk., No. 684, p. 3 - 6 (1972). In Russian.

118.002 **Orbit of the visual double star ADS 7896 =
JDS 2052 = A 2768.** J. Dommanget.
Astron. Astrophys., Suppl. Ser., Vol. 6, 415 - 418 (1972).
In French.
The orbital elements of the visual binary ADS 7896 =
JDS 2052 = A 2768 are computed by using the Thiele-Innes
method. The individual observations, the residuals as well as
an ephemeris concerning the relative positions of the compo-
nents and their relative radial velocities are given.

118.003 **New double stars (8th series) discovered at Nice
with the 50-cm refractor.** P. Couteau.
Astron. Astrophys., Suppl. Ser., Vol. 6, 419 - 428 (1972).
In French.
We give a list of 100 double stars discovered at the 50-cm
refractor.

118.004 **Visual observations of double stars.** J. Bem.
Acta Astron., Vol. 22, 41 - 47 (1972).
The results of measurements of position angles and
distances of 168 double stars made at the Wrocław Observatory
are given.

118.005 **New orbit and mass of the visual binary and magnetic Ap star ADS 7334 = HR 3724.** E. L. van Dessel.
Astron. Astrophys., Vol. 21, 155 - 157 (1972). – Research
note.

118.006 **Energy distribution in spectra of some binaries in a
wide spectral region ($\lambda\lambda$ 3300–7300 Å and 0.88–
1.53 mkm).** I. N. Glushneva.
Astron. Zhurn. Akad. Nauk SSSR, Vol. 49, 1037 - 1045
(1972). In Russian. English translation in Soviet Astron. AJ,
Vol. 16, No. 5.
6 visual and spectroscopic binaries are investigated for
the purpose of obtaining the energy distribution in their spectra in absolute units.

118.007 **Orbit, mass ratio, and parallax of the visual binary
Ross 614.** S. L. Lippincott, J. L. Hershey.
Astron. Journ., Vol. 77, 679 - 683 (1972).
Photographs taken from 1938 to 1972 with the Sproul
24-inch refractor yield a definitive photocentric orbit for Ross
614 AB. The absolute parallax for the system, the sum of the
masses and the mass ratio have been determined.

118.008 **Orbites nouvelles.** P. Muller.
Circ. Inform. (U.A.I. Commission des Étoiles
Doubles), Obs. Meudon, No. 58 (1972).

118.009 **Étoiles doubles nouvelles.**
R. L. Walker, P. Couteau, P. Muller.
Circ. Inform. (U.A.I. Commission des Étoiles Doubles), Obs.
Meudon, No. 58 (1972).

118.010 **Parallax, proper motion, and orbital motion of the
visual binary Σ 1321.** K. Chang.
Astron. Journ., Vol. 77, 759 - 761 (1972).
The relative positions of Σ 1321 are represented by an orbit with a period of 975 years, and a semi-axis major of
$16\overset{''}{.}725$. The 489 Sproul plates covering an interval of 52 years
yield a relative parallax of $+0\overset{''}{.}160\pm0\overset{''}{.}002$ (p.e.), and the component masses $M_A = 0.41\pm0.03\,M_\odot$ and $M_B = 0.73\pm0.05\,M_\odot$.

118.011 **The atmospheres of the visual binary HR 1886 and
HR 1887.** G. J. Peters.
Publ. Astron. Soc. Pacific, Vol. 84, 643 (1972). – Abstr.
Astron. Soc. Pacific.

118.012 **Micrometer measures of 401 double stars.**
C. E. Worley.
Astron. Journ., Vol. 77, 878 - 887 (1972).
Micrometer measures of 401 double stars, made with the
24-, 36-, and 60-inch reflectors of the Cerro Tololo Inter-
American Observatory, are presented.

118.013 **L'orbite parabolique d'une étoile double visuelle.**
S. Arend, R. R. de Freitas Mourão.
Bull. Astron. Obs. Roy. Belgique, Vol. 8, 58 - 62 (1972).

118.014 **Éléments d'une perturbation dans le mouvement
relatif des composantes du couple visuel ADS 2111 =
BDS 1420 = β 83.** J. Dommanget.
Bull. Astron. Obs. Roy. Belgique, Vol. 8, 63 - 66 (1972).

118.015 **β 1163 = ADS 1123.** E. B. Weston.
IAU Circ., No. 2443 (1972).

118.016 **β 1163 = ADS 1123.** J. M. Fletcher.
IAU Circ., Nos. 2447, 2451 (1972).

118.017 **Micrometer measures of 1,056 double stars.**
C. E. Worley.
Publ. United States Naval Obs., *Washington,* Second Ser., Vol.
22, Part 4, 111 pp. (1972).
This paper lists 3,947 measures of 1,056 double stars
made principally with the Naval Observatory's 12-inch and 26-
inch refractors in Washington, D.C., but including a few measures made with the 40-inch and 61-inch reflectors in Flagstaff,
Arizona. These measures represent a continuation of the two
series presented earlier (Worley 1967, 1971).

118.018 **A comparison of photoelectric and visual measurements of plates with the double star ADS 7251.**
N. A. Shakht.
Publ. 18th Astrometrical Conference 1969, (see 012.022), p.
243 - 246 (1972). In Russian.

Les mesures par double image.
See Abstr. 031.034.

De nærmeste stjerner. See Abstr. 041.040.

Rapid changes in the new shell star HR 6000.
See Abstr. 114.088.

An abundance analysis of α Centauri.
See Abstr. 114.093.

MWC 349. See Abstr. 114.157, 114.158.

**Statistical studies in stellar rotation. II. A method
of analyzing rotational coupling in double stars and an introduction to its applications.** See Abstr. 116.014.

On the mass-luminosity relation.
See Abstr. 117.017.

On the formation of double stars.
See Abstr. 117.028.

Orbital elements of the spectroscopic binary in ADS 14893. See Abstr. 119.004.

The complicated giant binary, SX Cassiopeiae. See Abstr. 121.028.

The light variation and orbital elements of TZ Lyrae. See Abstr. 121.053.

Observations of rapid blue variables—VIII. The companion to Mira. See Abstr. 122.022.

119   Spectroscopic Binaries

119.001   Is the shell star 88 Herculis a binary?
P. Harmanec, P. Koubský, J. Krpata.
Bull. Astron. Inst. Czechoslovakia, Vol. 23, 218 - 223 (1972).

A periodical variability of the radial velocities, determined from the hydrogen lines of the shell, was found on 31 high dispersion spectrograms covering the period from January to December 1971. The elements of the hypothetical spectroscopic binary are given.

119.002   Parallax of Ross 986 and G 87-25. L. T. Appelbaum
Astron. Journ., Vol. 77, 518 - 520 (1972).

Measurement and reduction of 80 plates of the Ross 986 region taken with the McCormick refractor during the period 1940–1970 yield a relative parallax of $0.''169 \pm 0.''007$ (m.e.) for Ross 986 and $0.''018 \pm 0.''005$ (m.e.) for G 87-25. Ten and 12 reference stars, respectively, were used in the parallax determinations of Ross 986 and G 87-25. The Ross 986 remainders suggest the possibility of orbital motion.

119.003   HD 152667 and Sco X-2.   E. N. Walker.
Monthly Notices Roy. Astron. Soc., Vol. 159, 253 - 259 (1972).

The spectroscopic binary P Cygni star HD 152667 lies close in the sky to the X-ray source Sco X-2. We have analysed the existing photometry of this star and find that two solutions are possible. One is a conventional one and predicts that the unseen secondary is a B3 V star. The alternative solution suggests that all photometric variations are due only to one, gravity darkened star filling its Roche lobe and accompanied by what could be a dense, massive companion. HD 152667 has many similarities to HDE 226868 which has been identified with the X-ray source Cyg X-1.

119.004   Orbital elements of the spectroscopic binary in ADS 14893.   F. R. West.
Bull. American Astron. Soc., Vol. 4, 330 (1972). – Abstr. AAS.

119.005   A possible astrometric spectroscopic binary.
R. M. Catchpole.
Observatory, Vol. 92, 125 - 127 (1972).

An astrometric orbit is proposed to explain the discrepancy between the trigonometric and spectroscopic parallax of HD 1273.

119.006   Déplacement vers le violet de la raie d'absorption Hα de l'étoile HD 30353 dépourvue d'hydrogène.
K. Nariai.
Publ. Astron. Soc. Japan, Vol. 24, 495 - 501 = Tokyo Astron. Obs. Repr., No. 418 (1972).

The hydrogen-deficient single-line binary HD 30353 shows a P Cygni type profile of Hα when the primary is farthest from us, while Hα appears only in emission at other phases.

119.007   Light variations of three spectroscopic binaries.
N. K. Rao.
Publ. Astron. Soc. Pacific, Vol. 84, 563 - 565 = Lick Obs. Bull., No. 628 (1972).

Three spectroscopic binaries HD 208095, 14 Cep, and HD 217312 have been observed to see if they show any of the periodic light variations of a few hours in length that are characteristic of β Cephei stars. HD 208095 and 14 Cep do not show any such variations with amplitude greater than $0^{m}01$, but 14 Cep and HD 217312 seem to show variations of eclipsing binary type.

119.008   Spectroscopic-binary frequency among extreme high-velocity dwarfs.
D. Crampton, F. D. A. Hartwick.
Astron. Journ., Vol. 77, 590 - 594 (1972).

The conclusion of Abt and Levy that short period spectroscopic binaries occur relatively less frequently among Population II stars is confirmed by considering an enlarged sample of extremely metal-deficient subdwarfs.

119.009   Spectroscopic binaries and the evolution of supergiants.   T. Lloyd Evans.
Proc. Third Colloquium on Astrophysics. Supergiant Stars, Trieste 1971, p. 288 - 291 = Radcliffe Obs. Repr., No. 107 (1972).

119.010   Apsidal motion in the binary Delta Orionis.
V. Natarajan, R. Rajamohan.
Kodaikanal Obs. Bull., Ser. A, No. 208, A219 - A225 (1971).

A velocity curve of the spectroscopic binary 'Delta Orionis' has been obtained from the spectrograms taken at Kodaikanal during the years 1968–70. The orbital elements derived are in good agreement with earlier values except for the longitude of periastron. The period of rotation of the line of apsides is found to be 208 years.

119.011   On the stability of the spectroscopic binary α Vir.
T. S. Galkina.
Nauchn. Informatsii, vyp. (No.) 21, p. 80 - 88 (1972). In Russian.

Several spectrograms of α Vir with dispersion 15 Å/mm and 37 Å/mm were obtained by a diffraction spectrograph attached to the 50" telescope of the Crimean Astrophysical Observatory. On the basis of the spectral analysis, the spectral classes, luminosities, radii and masses of both components were evaluated. A schematical model of the system is constructed.

119.012   The orbit of 53 Tauri (B9p-Mn).

M. M. Dworetsky.
Publ. Astron. Soc. Pacific, Vol. 84, 652 - 655 (1972).

The orbit of the Mn star 53 Tau has been determined from measures of 30 Mount Wilson, Lick, and Palomar coudé spectrograms obtained by several observers between 1946 and 1972. 53 Tau has the shortest orbital period yet found for a Mn star, $4^d.45$.

119.013 **The structure and apsidal constant of Spica.**
J. S. Mathis, A. P. Odell.
Bull. American Astron. Soc., Vol. 4, 423 - 424 (1972). — Abstr. AAS.

119.014 **The spectroscopic binary system H.D. 11860.**
A. H. Batten, B. Szeidl.
Publ. Dominion Astrophys. Obs., Victoria, Vol. 14, (No. 5), 97 - 105 (1972).

The spectroscopic binary system H.D. 11860 has been observed with the 48-inch and 72-inch telescopes of the Dominion Astrophysical Observatory. Twenty-six spectrograms were obtained between 1968 and 1971. From these spectrograms the orbital elements have been derived. The difference in magnitude between the two components is $0^m.04$ in the normal photographic region of the spectrum.

Photometric variability of the Wolf-Rayet star CV Serpentis.   See Abstr. 113.040.

Photometric observations of AG Pegasi.
See Abstr. 113.063.

On the Wolf-Rayet stars HD 90657 and HD 117688.
See Abstr. 114.005.

The absolute magnitude of Gamma Velorum.
See Abstr. 115.022.

Energy distribution in spectra of some binaries in a wide spectral region ($\lambda\lambda$ 3300–7300 Å and 0.88–1.53 mkm).
See Abstr. 118.006.

A nitrogen B supergiant in a peculiar eclipsing system.   See Abstr. 121.049.

Eclipsing binary HR 6611.   See Abstr. 121.055.

A spectroscopic study of AS Camelopardalis.
See Abstr. 121.061.

A photometric and spectroscopic investigation of SV Centauri.   See Abstr. 121.076.

A non-LTE analysis of the O-type subdwarf, HD 49798.   See Abstr. 126.001.

Radio behavior of β Persei.   See Abstr. 141.111.

B and V photometry of Cygnus X-1.
See Abstr. 142.048.

Dimensions of the binary system HDE 226868 = Cygnus X-1.   See Abstr. 142.088.

Identification of 2U0525–06 with ϑ^2 Orionis.
See Abstr. 142.092.

Errata

119.901 **Erratum: 'Orbit of the double-lined binary mercury star Chi Lupi'** [Publ. Astron. Soc. Pacific, Vol. 84, 254 - 259 (1972)].   M. M. Dworetsky.
Publ. Astron. Soc. Pacific, Vol. 84, 464 (1972).

120 Variable Stars: Catalogues, Ephemerides, Miscellanea

120.001 A propos d'étoiles variables. A. Brun.
L'Astronomie, 86ᵉ année, p. 361 - 362 (1972).

120.002 New identifications of variable stars with stars from the "Two-Micron Sky Survey – a Preliminary Catalog" by G. Neugebauer et al. (IRC). B. V. Kukarkin.
Astron. Tsirk., No. 674, p. 4 - 7 (1972). In Russian.

120.003 List of the stars from the "Two-Micron Sky Survey – a Preliminary Catalog" by G. N. Neugebauer et al. (IRC), the identification of which is not correct.
B. V. Kukarkin.
Astron. Tsirk., No. 674, p. 7 - 8 (1972). In Russian.

120.004 Amateur observations of variable stars.
T. B. Tregaskis.
Journ. Astron. Soc. Victoria, Vol. 25, 46 - 55 (1972).
Presidential address delivered at the General Meeting on 1972 June 15.

120.005 Auswertungsverfahren und -Probleme bei Kurzperiodischen. J. Hübscher.
BAV Rundbrief, 21. Jahrgang, Sonder-Rundbrief, (see 012. 011), p. 16 - 20 (1972).

120.006 Periodenableitung bei Kurzperiodischen unter Beachtung des Scheinperiodenproblems. W. Braune.
BAV Rundbrief, 21. Jahrgang, Sonder-Rundbrief, (see 012. 011), p. 21 - 28 (1972).

120.007 Zur Praxis der photographischen Veränderlichen-Beobachtung. I. Irisblenden-Mikrophotometer;
II. Photographische Überwachung von Veränderlichen Sternen.
P. Frank.
BAV Rundbrief, 21. Jahrgang, Sonder-Rundbrief, (see 012. 011), p. 29 - 35 (1972).

120.008 58ᵗʰ name-list of variable stars. B. V. Kukarkin, P. N. Kholopov, N. P. Kukarkina, N. B. Perova.
Inform. Bull. Variable Stars (I.A.U. Commission 27), Konkoly Obs., Budapest, No. 717, 36 pp. (1972).
The present 58ᵗʰ name-list of variable stars has been composed in accordance with the rules established in the 56ᵗʰ list. It contains all necessary identifications for 1836 new variable stars designated in 1972.

120.009 Photographische Überwachung von Veränderlichen.
P. Frank.
SuW, Vol. 11, 316, 318 (1972).

120.010 Rocznik Astronomiczny Obserwatorium Krakowskiego 1973. International Supplement No. 44.
Prepared under the supervision of K. Kozieł.
Komitet Astronomii, Polskiej Akademii Nauk, Kraków, 5 + 135 pp. Price zł 72.00 (1972). – Contents: Eclipsing binaries (K. Kordylewski); Informations concerning 240 eclipsing variables (R. Szafraniec); Ephemerides of eclipsing binaries among cataclysmic variables 1973 (J. M. Kreiner); RR-Lyrae-type variables (W. Zessewitsch, J. M. Kreiner); Auxiliary tables; Occultation of stars by the moon 1973 (L. Orkisz); Geocentric ephemeris of the libration points L_4 and L_5 in the earth-moon system for the year 1973 (A. Szczepanowska).

120.011 The place of the amateur in variable star observing.
F. M. Bateson.
Journ. Astron. Soc. Victoria, Vol. 25, 62 - 64 (1972).

120.012 Die Dr. Remeis-Sternwarte Bamberg und die „ Veränderlichen Sterne" als die Objekte ihrer Forschung.
W. Strohmeier, R. Knigge.
Separate print Remeis-Sternw. Bamberg, 32 pp. (1972).

120.013 List of preliminary designations of variable stars in the SVS system.
Astron. Tsirk., No. 711, p. 8; No. 713, p. 7 - 8 (1972). In Russian.

120.014 Gemeinschafts-Lichtkurven für Veränderliche geringer Amplitude? G. Pfeiffer.
BAV Rundbrief, 21. Jahrgang, p. 46 - 49 (1972).

120.015 Über die photographische Beobachtung der Veränderlichen Sterne. H. Muthsam.
BAV Rundbrief, 21. Jahrgang, p. 50 - 53 (1972).

120.016 B–R bei Kurzperiodischen. W. Braune.
BAV Rundbrief, 21. Jahrgang, p. 60 - 61 (1972).

120.017 Programme of cooperative flare star observations for 1973. P. F. Chugainov.
Inform. Bull. Variable Stars, (IAU Commission 27), Konkoly Obs., Budapest, No. 744 (1972).

120.018 Observer's guide: Visual observations of variable stars. P. Flin.
Urania Kraków, Vol. 43, 204 - 214 (1972). In Polish.

120.019 The Delta Scuti stars. An annotated catalogue and bibliography. M. A. Seeds, G. A. Yanchak.
Published by Bartol Research Foundation of the Franklin Institute, Swarthmore, Pennsylvania, 47 pp. (1972).
This catalogue contains data on 155 stars which are either Delta Scuti variables or suspects. The 58 stars in list 1 are known Delta Scuti stars, meaning that their variability and the justification of the Delta Scuti appellation have been confirmed. The 97 stars included in list 2 are termed suspects, meaning: (1) the star is suspected of being a variable; or (2) the star is a variable which is suspected of being a Delta Scuti variable; or (3) the star has been termed a Delta Scuti star but may not actually belong to that class. Also included in list 2 are 14 apparently non-variable stars falling in the Delta Scuti instability regions.

120.020 Sequences for southern variables.
F. M. Bateson, P. J. Gordon, B. Menzies.
Roy. Astron. Soc. New Zealand, Variable Star Section, Circ. No. 187, 3 pp. (1972).

120.021 The calculation of heliocentric corrections.
A. U. Landolt, K. L. Blondeau.
Publ. Astron. Soc. Pacific, Vol. 84, 784 - 809 = Contr. Louisiana State Univ. Obs., Baton Rouge, No. 71 (1972).
Tables which will permit the determination of the light time correction to be applied to the time of observation are presented. Examples are given.

120.022 Preliminary designations of SVS variables discovered in the USSR.
Astron. Tsirk., No. 725, p. 7 - 8 (1972). In Russian.

An astronomical interface for a small computer.
See Abstr. 031.077.

121 Eclipsing Variables

121.001 CW Sagittae. R. K. Kanishcheva, G. A. Lange.
Peremennye Zvezdy, Prilozhenie, Vol. 1, 139 - 146 (1971). In Russian.
1150 visual observations of the β Lyr type variable CW Sge were obtained in 1960 - 1965 by G. A. Lange. Moments of minima, O−C diagram, mean light curves are given. New elements of light variations were determined.

121.002 TV observations of the eclipse of GR Cassiopeiae. V. N. Ivchenko.
Peremennye Zvezdy, Prilozhenie, Vol. 1, 197 - 200 (1971). In Russian.

121.003 Eclipsing variable CSV 4205 = Ross 304. V. P. Tsesevich.
Peremennye Zvezdy, Prilozhenie, Vol. 1, 249 - 251 (1971). In Russian.

121.004 Variations of the periods of U Cephei and RZ Cassiopeiae.
M. A. Svechnikov, L. P. Surkova, V. M. Danilov.
Peremennye Zvezdy, Vol. 18, 237 - 260 (1972). In Russian.
The variations of the periods of the eclipsing binaries U Cep and RZ Cas were investigated. New light elements for U Cep are given; along with rapid irregular variations there are also secular increases of the period of the order of $1\overset{d}{.}25 \times 10^{-8}$ for a period. The period of RZ Cas was found to change spontaneously. The ejected mass causing the observed variations was estimated to be of the order of $10^{-6} M_\odot$ per year. To reveal possible latent sine waves in the period variations the autocorrelation functions were calculated. The autocorrelograms show the existence of sine waves with periods of the order of some years and amplitudes of the order of 10^{-6} of a day.

121.005 Variations of the periods of ST Persei and Y Leonis.
M. A. Svechnikov, L. P. Surkova.
Peremennye Zvezdy, Vol. 18, 261 - 268 (1972). In Russian.
The period variations of ST Per and Y Leo were investigated. Changes of the periods of these stars occur spontaneously. The ejected mass causing the observed variations was estimated to be of the order of $10^{-6} M_\odot$ per year.

121.006 Photometric investigation of the eclipsing binary HS Herculis. D. Ja. Martynov, M. I. Lavrov.
Peremennye Zvezdy, Vol. 18, 269 - 287 (1972). In Russian.
The eclipsing binary HS Herculis was observed in the three colours (U, B, V) at the Crimean Station of the Sternberg Institute in the years 1969−1970. A detailed analysis of these observations and the best photometric elements are given. Moreover, the photometric elements are determined from the observations made at the Kitt Peak Observatory. The distortions of the light curves resulting from objective sources are analyzed. An apsidal rotation with a period of 110−130 years is suggested. The absolute dimensions of the system are determined.

121.007 A spectroscopic study of AG Virginis. G. Hill, J. V. Barnes.
Publ. Astron. Soc. Pacific, Vol. 84, 382 - 387 (1972).
Radial velocities determined at the Kitt Peak National Observatory have been analyzed to yield revised orbital elements for the eclipsing binary AG Virginis. The possible evolutionary state of the system is also discussed.

121.008 UBV observations of the eclipsing binary LY Auri- gae. A. U. Landolt, K. L. Blondeau.
Publ. Astron. Soc. Pacific, Vol. 84, 394 - 399 = Contr. Louisiana State Univ. Obs., *Baton Rouge*, No. 63 (1972).

121.009 The 1971 eclipse of 32 Cygni: *UBV* **and spectrophotometric observations.**
D. W. Griffiths, R. E. Stencel.
Publ. Astron. Soc. Pacific, Vol. 84, 427 - 429 (1972).
Standard *UBV* and 20 Å spectrophotometry at four blue wavelengths obtained during the November 1971 eclipse of 32 Cyg is reported.

121.010 A spectroscopic study of AT Pegasi. G. Hill, J. V. Barnes.
Publ. Astron. Soc. Pacific, Vol. 84, 430 - 433 (1972).
Radial velocities determined at the Kitt Peak National Observatory and the Cerro Tololo Inter-American Observatory have been analyzed to yield new orbital elements for the eclipsing binary AT Peg. A revised photometric ephemeris, based on published data, has also been computed.

121.011 A time of minimum for AH Virginis. A. U. Landolt.
Publ. Astron. Soc. Pacific, Vol. 84, 448 - 449 = Contr. Louisiana State Univ. Obs., *Baton Rouge,* No. 64 (1972).
A time of minimum light at secondary eclipse is given for the eclipsing binary AH Virginis.

121.012 Photoelectric light-curves of TW Cas. B. Cester, M. Pucillo.
Mem. Soc. Astron. Italiana, Nuova Ser., Vol. 43, 291 - 300 (1972).
New photoelectric observations of TW Cas have been carried out at Trieste in three colours mostly in 1970. New epochs of minimum have been observed in order to improve the period. This is variable and parabolic light elements are proposed. The best set of elements indicates that TW Cas should be a semi-detached system composed of a late B main sequence star and a late F larger and less massive subgiant companion.

121.013 A statistical method for the determination of the rate of changes of period for eclipsing binaries.
F. van 't Veer.
Astron. Astrophys., Vol. 20, 131 - 134 (1972).
A statistical treatment for the quantitative analysis of period changes ΔP for eclipsing binaries is given. The method can be applied to a series of measurements of minima times. It allows one to determine the rate of period change $\Delta P/\Delta t$ and also gives a test of the reality of sudden period changes of contact binaries. Examples are given for SW Lac and 44i Boo.

121.014 Secondary fluctuations in the light curve of ϵ Aur. H. Stub.
Astron. Astrophys., Vol. 20, 161 - 162 (1972).
Colour dependent fluctuations in the light curve of ϵ Aur was found outside the eclipse. The maximum value of the amplitude equal to $0\overset{m}{.}2$ occurred in the blue and ultraviolet.

121.015 Photoelectric observations of the 1971-eclipse of 32 Cyg. U. K. Gehlich, J. Prölss, R. Wehmeyer.
Astron. Astrophys., Vol. 20, 165 - 166 (1972).
From photoelectric observations of the 1971 eclipse of 32 Cyg the time of mid-eclipse, the maximum duration of totality and the spectral types of the components have been derived.

121.016 **Preliminary analysis of the atmospheric eclipse of 32 Cygni.** A. Galatola.
Astrophys. Journ., Vol. 175, 809 - 818 (1972).

The blue and ultraviolet light curves of 32 Cygni have been analyzed by using as a model an opaque core surrounded by a semitransparent envelope. The effects of the atmospheric eclipse were removed from the light curve to obtain a light curve of the body eclipse. In order to get a solution the body eclipse was assumed to be total. An expression for $k\rho$, the opacity-density relation, could be derived.

121.017 **The primary spectrum of the eclipsing binary LR Centauri.** M. S. Bessell.
Astrophys. Journ., (*Letters*), Vol. 175, L133 - L136 (1972).

Coudé spectra and photoelectric scans have been obtained during maximum light for LR Cen, an eclipsing binary of period $2^d.095595$ variously suspected and discounted as being the pulsating X-ray source Cen X-3. If one assumes synchronous rotation and orbital revolution, a radius of $12\,R_\odot$ is determined for the primary, in agreement with the X-ray results of Cen X-3. The star is filling its Roche lobe and possibly losing mass.

121.018 **The Wolf—Rayet eclipsing binary CQ Cep in monochromatic light of emission lines and of the continuum.** Kh. F. Khaliullin.
Astron. Zhurn. Akad. Nauk SSSR, Vol. 49, 777 - 785 (1972). In Russian. English translation in Soviet Astron. AJ, Vol. 16, No. 4.

Narrow-band mean light curves of CQ Cep at $\lambda4795$ (continuum), $\lambda4686$ (emission He II), $\lambda6320$ (continuum), $\lambda6563$ (emissions He II, Hα) are given. A qualitative discussion is carried out of some observational data from V 444 Cyg on the basis of the development of a proposed model of CQ Cep for cases of separated Wolf—Rayet eclipsing binary systems.

121.019 **Photoelectric observations of Y Cygni (1959—1961).** T. Herczeg.
Astron. Astrophys., Vol. 20, 201 - 204 (1972).

A set of 184 photoelectric observations of Y Cygni is presented, obtained in 1959—1961; the measurements define 4 epochs of minimum. An ensuing short discussion of the period changes indicates that Dugan's formula still represents the apsidal rotation very well and that, further, no light-time effect of any significant amplitude can be found. Contrary to previous suggestions based on radial velocity data, the existence of a third body in the system seems rather improbable.

121.020 **The effect of the time shift of the min II in eclipsing binaries.** T. Z. Dworak.
Postępy Astron., Vol. 20, 251 - 256 (1972). In Polish.

121.021 **Observations of circumstellar matter in the eclipsing binary U Cephei.** B. W. Baldwin.
Journ. Roy. Astron. Soc. Canada, Vol. 66, 218 - 219 (1972). Abstr. Canadian Astron. Soc.

121.022 **A model for U Cephei.** A. H. Batten.
Journ. Roy. Astron. Soc. Canada, Vol. 66, 219 (1972). — Abstr. Canadian Astron. Soc.

121.023 **Spectral variations of Algol.** C. T. Bolton.
Journ. Roy. Astron. Soc. Canada, Vol. 66, 219 (1972). — Abstr. Canadian Astron. Soc.

121.024 **Radio flare on Beta Persei.** V. A. Hughes, A. Woodsworth.
Journ. Roy. Astron. Soc. Canada, Vol. 66, 220 (1972). Abstr. Canadian Astron. Soc.

121.025 **On the study of the atmosphere of the bright component of β Lyr.** M. Yu. Skulskij.
Astron. Tsirk., No. 668, p. 2 - 5 (1972). In Russian.

121.026 **Five-colour photometry of the eclipsing binary HO Telescopii.** T. A. T. Spoelstra, C. J. van Houten.
Astron. Astrophys., Suppl. Ser., Vol. 7, 83 - 102 (1972).

528 five-colour photoelectric measurements are presented of the eclipsing binary HO Telescopii, observed during 13 nights in August and September 1965 at the Leiden Southern Station. Four epochs of the minimum were used to derive a new period of 1.6131409 days. The light-curve is quite normal; primary and secondary minimum are about equally deep in all five colours.

121.027 **Photoelectric narrow-band five-colour ($\lambda\lambda 4260-7500$ Å) continuum observations of the Wolf-Rayet eclipsing binary V444 Cyg.**
A. M. Cherepashchuk, Kh. Khaliullin.
Astron. Tsirk., No. 680, p. 1 - 4 (1972). In Russian.

121.028 **The complicated giant binary, SX Cassiopeiae.** R. H. Koch.

The origin of the 1965—1967 cooperative observing campaign of SX Cas is based on obscurities or paradoxes in compiling older independent photometric and spectrographic studies. The journals of new or newly published astrometric, photoelectric, and spectrographic observations are given. It is shown that the light curves, both within and outside the eclipses, are complicated by gas stream and ring effects and the hypothesis is explored that Rayleigh scattering of the A-star radiation pervades the system. Finally, an appreciation of the evolutionary condition of the binary is described and it is shown that differential evolution and mass exchange are not greatly advanced for this system.

121.029 **Variable star S8315.** P. Kalv, L. Leis.
Astron. Tsirk., No. 686, p. 8 (1972). In Russian.

121.030 **Analysis of the period variations of U Cephei.** N. E. Kurochkin.
Astron. Tsirk., No. 688, p. 4 - 7 (1972). In Russian.

121.031 **Colour-radius relation for W UMa systems and the possibility of estimating their absolute elements.**
L. F. Istomin, M. A. Svechnikov.
Astron. Tsirk., No. 693, p. 3 - 6 (1972). In Russian.

121.032 **The monochromatic phase effect of Algol-type binaries.** V. P. Merezhin.
Astron. Tsirk., No. 695, p. 3 - 5 (1972). In Russian.

121.033 **Observations of rapid blue variables—XI. DQ Herculis.** B. Warner, W. L. Peters, W. B. Hubbard, R. E. Nather.
Monthly Notices Roy. Astron. Soc., Vol. 159, 321 - 335 (1972).

A study has been made of the phase, amplitude and shape of the 71-s periodic light variations around several binary cycles of DQ Her. It is shown that, relative to the usually adopted period of 71.06550 s, there is a 360° phase shift through eclipse. The observed phase variations are difficult to explain except in terms of a grazing eclipse of a white dwarf pulsating in a non-radial quadrupole mode.

121.034 **Elements of three eclipsing stars.** V. P. Tsesevich.
Astron. Tsirk., No. 698, p. 7 - 8 (1972). In Russian.

121.035 **F-, G- and K-type components of close binaries.** R. H. Koch.
Bull. American Astron. Soc., Vol. 4, 329 - 330 (1972). Abstr. AAS.

121.036 TX Cancri. J. Whelan, S. P. Worden, S. W. Mochnacki.
Bull. American Astron. Soc., Vol. 4, 330 (1972). − Abstr. AAS.

121.037 Ultraviolet photometry of LY Aur from OAO-2.
S. R. Heap.
Bull. American Astron. Soc., Vol. 4, 330 (1972). − Abstr. AAS.

121.038 On a model of the system β Lyr.
V. G. Gorbatskij.
Astron. Tsirk., No. 707, p. 1 - 3 (1972). In Russian.

121.039 β Lyr is a triple system. Observations during spring 1972. M. Yu. Skulskij.
Astron. Tsirk., No. 707, p. 3 - 5 (1972). In Russian.

121.040 Radio flares from Algol.
R. M. Hjellming, C. M. Wade, E. Webster.
Bull. American Astron. Soc., Vol. 4, 384 - 385 (1972).
Abstr. AAS.

121.041 A model for the contact binary W UMa.
S. W. Mochnacki.
Bull. American Astron. Soc., Vol. 4, 399 (1972). − Abstr. AAS.

121.042 Proper motions of 122 eclipsing variable stars.
M. Yu. Volyanskaya.
Astrometriya i Astrofiz., *Kiev*, No. 15, (see 003.001), p. 96 - 100 (1972). In Russian.
On the basis of star positions from the author's catalogue and numerous other meridian and photographic catalogues proper motions were determined for 102 eclipsing variable stars and improved for 20 stars. The components $\mu_\alpha \cos \delta$ and μ_δ with probable errors are given in the GC system.

121.043 Observations of rapid blue variables−XII. UX Ursae Majoris. B. Warner, R. E. Nather.
Monthly Notices Roy. Astron. Soc., Vol. 159, 429 - 444 (1972).
Photoelectric photometry of the nova-like variable UX UMa with time resolution of 1-5 s is reported. Times of mideclipse, compared with a previous ephemeris, indicate that the mean period of UX UMa may be shorter by $\sim 7 \times 10^{-8}$ than previously believed. Two observing runs made in January 1971 show a very clear periodicity in the light curve of 29.44 and 28.92 s respectively. Several observing runs made two months later show no sign of the 29-s periodicity. A discussion is given of known periodicities in other white dwarfs.

121.044 Der Bedeckungsveränderliche BM Cassiopeiae.
R. Diethelm.
BAV Rundbrief, 21. Jahrgang, Sonder-Rundbrief, (see 012.011), p. 36 - 38 (1972).

121.045 Der Veränderliche Z Camelopardalis. H. Feijth.
BAV Rundbrief, 21. Jahrgang, Sonder-Rundbrief, (see 012.011), p. 39 - 41 (1972).

121.046 Umkehr der Periodenänderung bei TW Draconis.
K. Wälke.
BAV Rundbrief, 21. Jahrgang, Sonder-Rundbrief, (see 012.011), p. 42 - 46 (1972).

121.047 Verlauf der B−R von X Trianguli. K. Wälke.
BAV Rundbrief, 21. Jahrgang, Sonder-Rundbrief, (see 012.011), p. 47 - 50 (1972).

121.048 Der Bedeckungsveränderliche U Cephei.
H. Zipprich.
BAV Rundbrief, 21. Jahrgang, Sonder-Rundbrief, (see 012.011), p. 54 - 56 (1972).

121.049 A nitrogen B supergiant in a peculiar eclipsing system. N. R. Walborn.
Astrophys. Journ., (*Letters*), Vol. 176, L119 - L121 (1972).
The absorption-line spectrum of the interesting southern spectroscopic binary HD 163181 has been found to be nitrogen-enhanced and extremely deficient in carbon and oxygen. Since there is evidence for mass loss or mass transfer in the system, the possibility exists that these spectroscopic anomalies are due to material processed through the CNO cycle in one of the massive components.

121.050 The eclipsing binary WW Cygni: An unlikely candidate for pre-main-sequence contraction.
D. S. Hall, A. S. Wawrukiewicz.
Publ. Astron. Soc. Pacific, Vol. 84, 541 - 551 (1972).
The eclipsing binary WW Cygni has become particularly interesting ever since Field (1969) suggested it may be a binary in pre-main-sequence contraction. In this paper we present a new *UBV* photoelectric light curve and its solution, and examine critically the likelihood of Field's suggestion.

121.051 *UBV* photometry of the ring in SW Cygni.
D. S. Hall, L. M. Garrison, Jr.
Publ. Astron. Soc. Pacific, Vol. 84, 552 - 562 (1972).
We present new *UBV* photometry of SW Cyg obtained in 1967, 1968, and 1969. The bottom of primary eclipse exhibits several complications: asymmetries, nonconstant light between second and third contact, and an apparent ultraviolet excess in the cool subgiant of $\delta(U-B) = 0^m3$. The most satisfactory explanation for this excess is contamination by light from an envelope surrounding the hot star that is not completely eclipsed at midprimary minimum.

121.052 Photoelectric light curves of AK Herculis.
B. B. Bookmyer.
Publ. Astron. Soc. Pacific, Vol. 84, 566 - 580 (1972).
AK Herculis is a W Ursae Majoris-type eclipsing binary system which undergoes complete eclipses; secondary eclipse is total. Approximately 2000 photoelectric observations of this system were obtained with B and V filters within a two-week interval in 1966. Three analyses of the observations for the system's orbital elements are presented and discussed.

121.053 The light variation and orbital elements of TZ Lyrae.
L. Binnendijk.
Astron. Journ., Vol. 77, 595 - 602 (1972).
A total of 581 observations in yellow light and a total of 581 observations in blue light of TZ Lyrae are presented. A consistent set of orbital elements is presented which shows that secondary minimum is caused by a total eclipse.

121.054 The light variation of AU Serpentis.
L. Binnendijk.
Astron. Journ., Vol. 77, 603 - 609 (1972).
A total of 833 observations in yellow light and a total of 828 observations in blue light of AU Ser are presented. The light curve changes its shape in a rather short time interval. The components are too distorted to give reliable orbital elements according to the Russell model.

121.055 Eclipsing binary HR 6611. R. Zissell.
Astron. Journ., Vol. 77, 610 - 616 (1972).
One thousand photoelectric V observations were made of the eclipsing binary HR 6611. The period was found to be 3.894977 days. The amplitude of primary eclipse was 0.18 mag and that of secondary 0.17 mag. The spectroscopic orbit was recomputed and the results combined with those from the newly determined eclipsing orbit to give masses and radii of the components.

121.056 Variable stars and the photon rest mass.

R. Breinhorst, M. Reinhardt.
Ann. Phys., 7. Ser., Vol. 27, 122 - 124 (1971).

121.057 V 745 Centauri. M. J. M. Saraber.
Astron. Astrophys., Vol. 21, 311 - 312 (1972).

From an investigation on Franklin Adams plates, the star V 745 Centauri (HD 126344) has been found to be an eclipsing binary with a period of $3^d025101$. The secondary minimum of the light-curve appears to be wider than the primary one.

121.058 Note on the period of OO Aquilae.
T. Herczeg.
Inform. Bull. Variable Stars (I.A.U. Commission 27), Konkoly Obs., Budapest, No. 699, 5 pp. (1972).

121.059 New light elements for V Crateris.
M. Parthasarathy, N. B. Sanwal.
Inform. Bull. Variable Stars (I.A.U. Commission 27), Konkoly Obs., Budapest, No. 719 (1972).

121.060 A U, B, V photoelectric investigation of the eclipsing–binary system V505 Sagittarii.
C. R. Chambliss.
Astron. Journ., Vol. 77, 672 - 679 (1972).

The Algol-type eclipsing-binary system V505 Sgr was observed at Cerro Tololo during July and August of 1969. Approximately 470 observations each were obtained in U, B, and V. It is found that a consistent set of geometrical orbital elements will satisfy all light curves. The orbital elements determined in this investigation are in good agreement with those determined in previous studies of this system. Using the available spectroscopic data for V505 Sgr, the masses and radii of the components are estimated.

121.061 A spectroscopic study of AS Camelopardalis.
R. W. Hilditch.
Publ. Astron. Soc. Pacific, Vol. 84, 519 - 522 (1972).

Spectrographic observations of this eclipsing binary system are presented for the first time. It is shown that both components are main-sequence stars which conform to the theoretical calculations of Iben (1967).

121.062 On the chromospheric structure of the K-type component of 32 Cygni. M. Saitō, H. Sato.
Publ. Astron. Soc. Japan, Vol. 24, 503 - 508 = Tokyo Astron. Obs. Repr., No. 419 (1972).

From a discussion of the UBV light curves of 32 Cygni obtained during the 1968 and 1971 eclipses the number density of atoms at the bottom of the chromosphere of the K-type component is deduced. The light curve reveals that the eclipse is grazing and even at mideclipse about half the disk of the B-type component is eclipsed by the photospheric body of the K-type component. The mechanism of a brightness excess appearing at a phase 40–60 days after mideclipse is considered.

121.063 YY Geminorum – Et problematisk system.
B. R. Pettersen.
Astron. Tidssk., Årg. 5, p. 125 - 128 (1972).

121.064 Photometry and differential corrections analysis of Algol.
R. E. Wilson, M. R. de Luccia, K. Johnston, S. A. Mango.
Astrophys. Journ., Vol. 177, 191 - 208 (1972).

Two narrow-band light curves of Algol, at $\lambda\lambda 5500$ and 4350, and one broad-band ($\lambda 5500$) light curve are presented, along with their transformations to the B, V system. Two times of primary minima were determined accurately, while five others were estimated from segments of primary eclipse. The bolometric albedo of the secondary component was found to be 0.52 ± 0.02 p.e., in good agreement with Rucinski's theoretical estimate of 0.4 to 0.5. However, it seems that the presently available photometry does not permit a meaningful determination of the gravity-darkening parameter for the secondary.

121.065 Six ultrashort-period binary stars. B. Warner.
Sky Telescope, Vol. 44, 358 - 360 (1972).

121.066 UBV photometry of 32 Cygni during the 1971 eclipse. M. Saitō, H. Sato, N. Sato.
Tokyo Astron. Bull., Second Ser., No. 219, p. 2557 - 2561 (1972).

The eclipse of the long-period binary system 32 Cygni occurred in October and November of 1971. This was the eighth eclipse after the eclipse nature was first recognized in 1949 (McLaughlin, 1950). This system consists of the supergiant primary component of spectral type K5 and the main-sequence secondary of type B3 and revolves with the orbital period of about 1147 days. During the 1971 eclipse the UBV photoelectric observation was made with the 91 cm reflector at the Dodaira Station of Tokyo Astronomical Observatory, the 30 cm reflector at the Okayama Astrophysical Observatory and the 25 cm reflector at Akita University.

121.067 Photoelectric observations of the close binary system SZ Camelopardalis.
M. Kitamura, A. Yamasaki.
Tokyo Astron. Bull., Second Ser., No. 220, p. 2563 - 2575 (1972).

The variable star SZ Cam, the northern component of the visual binary $\Sigma 485 = $ ADS 2984, is an eclipsing binary system of early spectral type with the period 2.698 days, and it is conjectured to be the brightest member of the open cluster NGC 1502. This system shows total and annular eclipses alternatively, and is known as an important example that the limb-darkening coefficients of the early type components may be accurately determined. The purpose of the present work is to obtain UBV photoelectric light curves of SZ Cam.

121.068 Further evidence for a black hole in Beta Lyrae?
Y. Kondo, G. E. McCluskey, Jr., T. E. Houck.
Nature, Phys. Sci., Vol. 240, 119 - 120 (1972).

The secondary component of β Lyr is apparently a massive object ($M > 10\,M_\odot$), probably more massive than the primary component, yet its spectrum is not observable, except possibly in mysterious emission lines. It may involve a star collapsing toward its Schwarzschild radius (a black hole).

121.069 The binary systems of W Ursae Majoris type (W UMa). I. S. Ruciński.
Postępy Astron., Vol. 20, 275 - 296 (1972). In Polish.

The article briefly discusses the observational facts concerning W UMa stars and contains a review of recent attempts to give a theoretical model (zero age or evolutionary) of these systems.

121.070 Gravity-darkening of the secondary component of RW Monocerotis. M. Parthasarathy.
Astrophys. Space Sci., Vol. 18, 190 - 195 (1972).

The two near-infrared light curves of the secondary minimum of RW Monocerotis are analyzed for the gravity-darkening coefficient of the distorted subgiant secondary component of later spectral type. It is found that the monochromatic gravity-darkening coefficient of the secondary component of RW Monocerotis at wavelengths of 7800 Å and 7000 Å is about twice that obtained from von Zeipel theory.

121.071 Preliminary observations of variable polarization in Epsilon Aurigae. G. V. Coyne.
Ric. Astron., Specola Vaticana, *Castel Gandolfo*, Vol. 8, (No. 13), 311 - 318 (1972).

Measures of the polarization of ϵ Aur over a four year period show an increase of about 0.5% polarization beginning at a phase of about 0.4 and the polarization remains at this increased value through the non-eclipse conjunction. It is expected that the polarization will again decrease on the other side of this conjunction. Further observations should be made and may help to establish whether the opacity at eclipse is due to electron scattering or to scattering from molecules and grains.

121.072 **Photometric orbit and apsidal motion of DR Vulpeculae.** D. J. K. O'Connell.
Ric. Astron.,Specola Vaticana, *Castel Gandolfo,* Vol. 8,(No. 14), 319 - 342 (1972).

A preliminary orbit of DR Vulpeculae is based on 1334 photoelectric observations made by the writer with the 60 cm Zeiss Cassegrain reflector of the Vatican Observatory. From the times of minima a value of 0.10 is derived for the orbital eccentricity, and a period of about 37.8 years for the rotation of the line of apsides. There is a small, but definite, variation in the depths of both minima, which is certainly not due to the apsidal motion. The analysis of the light curves shows that a third star contributes to the total light of the system.

121.073 **Computer solution of eclipsing-binary light curves by the method of differential corrections.**
D. D. Proctor, A. P. Linnell.
Astrophys. Journ., Suppl. Ser., No. 211, Vol. 24, 449 - 477 (1972).

A group of computer programs rectifies individual observations defining an eclipsing-binary light curve according to the Russell model and produces a final set of parameters by differential corrections.The present program applies to systems which can be represented to first order in their oblateness. A sample solution of MR Cygni permits a comparison with other published solutions and demonstrates the ability of the present programs to fit a light curve to within the accuracy of its observational definition.

121.074 **Variations of the period of XZ And.**
N. E. Kurochkin.
Astron. Tsirk., No. 713, p. 1 - 4 (1972). In Russian.

121.075 **Infrared excesses in eclipsing binaries of the RS Canum Venaticorum type.**
H. L. Atkins, D. S. Hall.
Publ. Astron. Soc. Pacific, Vol. 84, 638 (1972). − Abstr. Astron. Soc. Pacific.

121.076 **A photometric and spectroscopic investigation of SV Centauri.** J. B. Irwin, A. U. Landolt.
Publ. Astron. Soc. Pacific, Vol. 84, 686 - 696 = Contr. Louisiana State Univ. Obs., *Baton Rouge,* No. 68 (1972).

The purposes of this study were to obtain both a *UBV* light curve and radial velocity data for this remarkable system as a follow-up to earlier photoelectric work by Irwin (1966) and the spectroscopic observation by Popper (1966) that the lines were possibly double, a fact independently confirmed by Bond (1970) from Tololo Schmidt plates.

121.077 **Het duivelsoog Algol.** J. W. Wijbenga.
Hemel en Dampkring, Vol. 70, 320 - 323 (1972).

121.078 **The period and light curve of HZ Herculis.**
J. N. Bahcall, N. A. Bahcall.
Astrophys. Journ., (*Letters*), Vol. 178, L1 - L4 (1972).

It is shown that HZ Herculis has an eclipse period and phase equal to the eclipse period and phase of the pulsating X-ray source Hercules X-1. Data for HZ Herculis are also given for the light curve, colors, proper motion, and the influence of the 35-day X-ray cycle on the optical light. A model for the binary system is described and the optical light curve is calculated.

121.079 **Spectroscopic observations of HZ Herculis.**
B. W. Bopp, G. Grupsmith, P. Vanden Bout.
Astrophys. Journ., (*Letters*), Vol. 178, L5 - L8 (1972).

Spectra have been obtained of HZ Her, the optical counterpart of the pulsating X-ray source Her X-1 (= 2U 1705 + 34), which show variable Ca II K absorption, $\lambda\lambda 4640-4650$ and $\lambda 4686$ emission, and He I absorption lines. The emission lines vary on a timescale as short as 30^m. The Ca II and the He I features indicate a composite spectrum combining A- and B-star characteristics.

121.080 **The period variations of VW Cephei.**
I. Todoran, V. Pop.
Acta Astron., Vol. 22, 267 - 271 (1972).

It is shown that the orbital period variations of VW Cephei are better represented by periodic elements than by parabolic ones. A lower limit for the long period is found to be P = 57 years.

121.081 **The light-curves of 11 eclipsing variables.**
R. Szafraniec.
Acta Astron., Vol. 22, 273 - 303 (1972).

The paper presents the results of visual estimates of brightness made in Cracow for 11 eclipsing variables: RT Leo, T LMi, RV, TZ, UZ, EW Lyr, BO, FS Mon, SW, V449, V501 Oph. These are fainter stars, for which no photoelectric observations are available. Their mean light-curves are determined and, for some of them, these are the first light–curves ever published.

121.082 **Photoelectric observations of β Lyrae.**
T. Z. Dworak, Z. Klimek.
Acta Astron., Vol. 22, 305 - 308 (1972).

The paper contains b, v differential magnitudes of β Lyr obtained in Cracow in the frame of the second international programme for the observations of this star.

121.083 **Intrinsic polarization of the eclipsing binary W Serpentis.** A. Kruszewski.
Acta Astron., Vol. 22, 405 - 410 (1972).

Six-colour polarimetric and photometric observations of the peculiar eclipsing binary W Ser are presented.

121.084 **Three eclipsing binaries with eccentric orbits.**
H. Bossen, P. Klawitter.
Acta Astron., Vol. 22, 411 - 418 (1972).

During a reexamination of plates taken by Wachmann, the eclipsing variables V 501 Mon, V 1136 Cyg, and EQ Vul were found to have eccentric orbits. Light elements could be found and the resulting light curves are given in figures.

121.085 **Investigation of RZ Scuti from low dispersion spectrograms.** V. G. Karetnikov.
Astron. Zhurn. Akad. Nauk SSSR, Vol. 49, 1188 - 1196 (1972). In Russian. English translation in Soviet Astron. AJ, Vol. 16, No. 6.

A spectrophotometric investigation has been carried out based on 83 spectrograms of RZ Scuti. The energy distribution in the spectral range $\lambda\lambda$ 3460 − 6800 Å, spectrophotometric gradients in three regions of the spectrum, and Balmer discontinuities have been determined. Monochromatic curves of light change of RZ Scuti are plotted and their particularities are found. Equivalent spectral line widths of hydrogen and other elements have been measured, and electron densities have been determined by the methods of Inglis − Teller and Unsöld. The structure of the star is discussed.

121.086 **Das System IQ Persei.** W. Bischof.

BAV Rundbrief, 21. Jahrgang, p. 53 - 56 (1972).

121.087 The co-ordinated program for observing the Zeta Aurigae stars, 1971–72. K. O. Wright.
Journ. Roy. Astron. Soc. Canada, Vol. 66, 289 - 294 = Contr. Dominion Astrophys. Obs., Victoria, No. 194 (1972). – Concerning 32 Cygni, ζ Aurigae and 31 Cygni.

121.088 Photoelectric observations of the eclipsing variable U Coronae Borealis.
S. N. Svolopoulos, S. Kapranidis.
Inform. Bull. Variable Stars, (IAU Commission 27), Konkoly Obs., Budapest, No. 731, 3pp. (1972).

121.089 Minima of eclipsing variables. P. Flin.
Inform. Bull. Variable Stars, (IAU Commission 27), Konkoly Obs., Budapest, No. 740 (1972).

121.090 Epochs of photoelectric minima of Y Cygni.
H. Ogata, T. Hayasaka, N. Sato, M. Koga, M. Kitamura.
Inform. Bull. Variable Stars, (IAU Commission 27), Konkoly Obs., Budapest, No. 746 (1972).

121.091 Optical observations of HZ Herculis.
E. J. Groth, M. R. Nelson.
Astrophys. Journ., (*Letters*), Vol. 178, L111 - L114 (1972).
Optical observations of HZ Herculis show no evidence of pulsation (to better than 0.1 percent) and no evidence of nonperiodic modulation.

121.092 Remarks on the period of TW Draconis.
I. Todoran.
Stud. Cerc. Astron., Vol. 17, 203 - 211 (1972).
The diagram of the differences O–C is examined. The following hypotheses are made concerning the cause of the variation of the orbital period: presence of a third component, apsidal motion and exchange of matter, but none of these may be accepted without reservation.

121.093 The determination of the apsidal motion coefficient from stellar models and its comparison with observations. R. Dinescu.
Stud. Cerc. Astron., Vol. 17, 221 - 226 (1972). In Romanian.
A comparison is made between the theoretical coefficient of the apsidal motion and same coefficients obtained from eclipsing binary observations. Using new stellar models for $1M_\odot \leqslant M \leqslant 4M_\odot$ one gets better fits with the observations.

121.094 Differential UBV photometry of Zeta Aurigae in the 1971–72 eclipse. M. Kiyokawa, M. Kitamura, M. Saito, H. Sato, N. Sato, H. Ogata.
Tokyo Astron. Bull., Second Ser., No. 221, p. 2577 - 2588 (1972).
Between September of 1971 and January of 1972, more than three hundred differential UBV photoelectric observations were made at Dodaira Station, Okayama Astrophysical Station, Akita Univ., and at Kanagawa. These cooperative observations were undertaken with the purpose of covering as many phases as possible during the eclipse.

121.095 A physical model of HZ Her.
A. Davidsen, J. P. Henry.
Bull. American Astron. Soc., Vol. 4, 411 (1972). – Abstr. AAS.

121.096 Light curve of HZ Herculis in relation to Her X-1.
V. H. Regener.
Bull. American Astron. Soc., Vol. 4, 414 (1972). – Abstr. AAS.

121.097 Hα and Hβ photoelectric photometry of β Lyrae.
E. F. Guinan, G. P. McCook, E. J. O'Donnell.

Bull. American Astron. Soc., Vol. 4, 423 (1972). – Abstr. AAS.

121.098 31 Cygni. K. O. Wright.
IAU Circ., No. 2421 (1972).

121.099 Remarks on the Wolf-Rayet binary CV Serpentis.
A. Cowley.
Publ. Astron. Soc. Pacific, Vol. 84, 772 - 774 (1972).
Recent photometric observations by Cherepashchuk have been published, making it possible to intercompare the spectroscopic and photometric behavior of this peculiar system.

121.100 Minima for the eclipsing binary RT Persei.
J. S. Drilling, A. U. Landolt.
Publ. Astron. Soc. Pacific, Vol. 84, 810 - 812 = Contr. Louisiana State Univ. Obs., *Baton Rouge*, No. 72 (1972).

121.101 On the metal deficiency of the secondary component of S Velorum. H. E. Bond.
Publ. Astron. Soc. Pacific, Vol. 84, 839 - 841 = Contr. Louisiana State Univ. Obs., *Baton Rouge*, No. 73 (1972).
An ultraviolet spectrogram of the Algol-type binary S Vel was obtained during total eclipse. The spectrum shows no inconsistency with the hypothesis that the ultraviolet excess of the secondary star is due to a moderate metal deficiency.

121.102 Photometric researches on 3 eclipsing variables.
S. Taffara.
Mem. Soc. Astron. Italiana, Nuova Ser., Vol. 43, 481 - 486 (1972).
The photometric elements of three eclipsing variables are given: V406, V407 Aql are already known; the third is probably new.

121.103 Photoelectric light-curves and elements of U Sge.
B. Cester, M. Pucillo.
Mem. Soc. Astron. Italiana, Nuova Ser., Vol. 43, 501 - 521 (1972).
On the basis of *UBV* photoelectric observations made at Trieste from 1969 to 1971, the elements have been derived. U Sge can be understood as a semi-detached system consisting of a brighter, smaller and more massive B 8.5 IV-V star surrounded perhaps by external matter and a G 2-5 III-IV companion filling its Roche lobe.

121.104 Photoelectric investigations of the faint eclipsing variable stars TY Ursae Majoris and AZ Virginis and the determination of the spectral classification for the eclipsing variable stars WY Cancri, AZ Virginis and TZ Bootis.
H. E. Durgin.
Thesis, Georgetown Univ., Washington, D.C. [Available from Univ. Microfilms, Ann Arbor, Mich., USA. Order No. 72-4218], 206 pp. (1971).

121.105 A numerical analysis of the variations in the light curves of the close binary systems W Ursae Majoris and U Pegasi. P. V. Rigterink.
Thesis, Univ. Pennsylvania, Philadelphia. [Available from Univ. Microfilms, Ann Arbor, Mich., USA. Order No. 72-6218], 173 pp. (1971).
Photoelectric observations of yellow and blue magnitudes of W Ursae Majoris and U Pegasi are presented. The light curves are compared with previous observations, and the changes in the light curves are noted. New orbital elements are derived for both systems using the mean binary light curve.

121.106 Quantitative analysis of β Lyrae spectra. I. Variations of some hydrogen and helium lines.
M. Yu. Skulsky.
Izv. Krymskoj Astrofiz. Obs., Vol. 45, 135 - 145 (1972).

In Russian.

Using spectrograms with dispersions 14 and 34 Å/mm in the regions $\lambda\lambda$ 3600 − 4900 Å and $\lambda\lambda$ 5400 − 6800 Å respectively, contours of lines λ 3888 He I, H_δ, H_γ, λ 4471 He I, H_β, λ 5875 He I, and H_α have been obtained in absolute units for different phases of β Lyrae.

121.107 **β Persei: Radio star and probable X-ray star.**
R. M. Hjellming.
Nature, Phys. Sci., Vol. 238, 52 - 55 (1972).

It is pointed out that the simplest interpretation of virtually all of the β Persei data is in terms of a variable thermal bremsstrahlung source with such high temperatures during peak levels of the recent flaring that it must have become a variable X-ray source.

121.108 **Photometric elements of V444 Cyg (WN5 +O6) − a Wolf-Rayet eclipsing binary.**
A. M. Cherepashchuk.
Astron. Tsirk., No. 739, p. 1 - 4 (1972). In Russian.

121.109 **Wolf-Rayet eclipsing binary V444 Cyg (WN5 +O6).**
Photometric structure of the extended WR photosphere in the continuum λ 4244, 4789, 6320, 7512 Å.
A. M. Cherepashchuk, Kh. F. Khaliullin.
Astron. Tsirk., No. 739, p. 5 - 8 (1972). In Russian.

121.110 **On the lines of the secondary component in the spectrum of β Lyr.** M. Yu. Skulskij.
Astron. Tsirk., No. 741, p. 1 - 3 (1972). In Russian.

121.111 **Lists of minima of eclipsing binaries.**
R. Diethelm, R. German, M. Giger, K. Locher, H. Peter, F. Schäpper.
BBSAG Bull. No. 5, p. 1 - 3; No. 6, p. 1 - 4 (1972).

121.112 **Probable period change of RW Tauri in 1970.**
R. Diethelm.
BBSAG Bull. No. 5, p. 4 (1972).

121.113 **New light elements for the eclipsing binary U Sagittae.** R. Diethelm.
BBSAG Bull. No. 5, p. 5 (1972).

121.114 **Improved results on BV 1481 Ceti.**
K. Locher.
BBSAG Bull. No. 5, p. 5 - 6 (1972).

121.115 **A new interpretation of VY Ceti.** K. Locher.
BBSAG Bull. No. 6, p. 6 (1972).

Stellar atmospheres with radiation incident at the surface. See Abstr. 064.040.

Intrinsic polarization in the atmospheres of supergiant stars. See Abstr. 064.043.

Numerical methods for computing stellar line-profiles and continuum fluxes. See Abstr. 064.072.

Photometry of symbiotic and VV Cephei stars in the near infrared (with a note on MWC 56). See Abstr. 113.019.

Mass transfer in close binaries. III. Gaseous rings in Algol-like binaries. See Abstr. 117.030.

Observations of rapid blue variables−VII. EX Hydrae. See Abstr. 122.020.

Observations of rapid blue variables−IX. AM CVn (HZ 29). See Abstr. 122.023.

Autocorrelation analysis of EX Hya brightness. See Abstr. 122.040.

Über das Sonneberger Programm zur Bestimmung von Spektraltypen veränderlicher Sterne auf Platten mit geringfügig verbreiterten Spektrogrammen. See Abstr. 122.126.

Ultrashort-period binaries. II. HZ 29 (= AM CVn): A double-white-dwarf semidetached postcataclysmic nova? See Abstr. 126.002.

Polarization of light by circumstellar material. See Abstr. 131.092.

Radio behavior of β Persei. See Abstr. 141.111.

Some observational distinctions among models of pulsing X-ray binaries. See Abstr. 142.005.

On the optical search for Centaurus X-3. See Abstr. 142.010.

The characteristics of the X-ray source Cen X-3 obtained on the supposition of its identity with the Algol-type variable LR Cen. See Abstr. 142.022.

Accretion disc models for compact X-ray sources. See Abstr. 142.023.

A new X-ray binary associated with an O-type star. See Abstr. 142.028.

X-ray spectra of binary sources. See Abstr. 142.029.

Centaurus X-3, possible reactivation of an old neutron star by mass exchange in a close binary. See Abstr. 142.036.

On the nature of the optical variations of HZ Her = Her X1. See Abstr. 142.042.

Her X-1: A precessing binary pulsar? See Abstr. 142.047.

Identification of the X-ray pulsar in Hercules: a new optical pulsar. See Abstr. 142.096.

Optical studies of UHURU sources. III. Optical variations of the X-ray eclipsing system HZ Herculis. See Abstr. 142.097.

Measurement of the position and spectrum of Hercules X-1 from the OSO-7 satellite. See Abstr. 142.098.

Hard X-ray observations of Hercules X-1 by OSO-7. See Abstr. 142.103.

Spectroscopic observations of HZ Herculis and a model for Hercules X-1. See Abstr. 142.104.

Optical variability of Her X-1 (HZ Her). See Abstr. 142.107.

Identification of the X-ray pulsar in Hercules: A new optical pulsar. See Abstr. 142.114.

BD +34°3815. See Abstr. 142.121.

HZ Her as a possible optical pulsar. See Abstr. 142.122.

Possible identifications of X-ray sources. See Abstr. 142.123.

HZ Herculis. See Abstr. 142.124.

HZ Herculis. See Abstr. 142.125.

HZ Herculis. See Abstr. 142.126.

HZ Herculis. See Abstr. 142.127.

HZ Herculis. See Abstr. 142.128.

HZ Herculis. See Abstr. 142.129.

HZ Herculis. See Abstr. 142.132.

HZ Herculis. See Abstr. 142.136.

Röntgen-Quelle Her X-1 = HZ Her. See Abstr. 142.148.

122 Physical Variables, Flare Stars, Pulsation Theory

122.001 Visual observations of RR Lyrae type variables in 1934 - 1935. R. K. Kanishcheva, G. A. Lange.
Peremennye Zvezdy, Prilozhenie, Vol. 1, 107 - 138 (1971). In Russian.

122.002 Improved elements of AX Aquilae. N. N. Samus.
Peremennye Zvezdy, Prilozhenie, Vol. 1, 147 - 150 (1971). In Russian.

122.003 MM Aquilae. B. L. Shaganyan.
Peremennye Zvezdy, Prilozhenie, Vol. 1, 169 - 171 (1971). In Russian.

122.004 Observations of three RR Lyrae type variables in Delphinus.
V. F. Karamish, V. G. Karetnikov, N. S. Komarov.
Peremennye Zvezdy, Prilozhenie, Vol. 1, 175 - 189 (1971). In Russian.

122.005 On two cepheids in Sagitta.
B. A. Dragomiretskaya, V. P. Tsesevich.
Peremennye Zvezdy, Prilozhenie, Vol. 1, 201 - 221 (1971). In Russian.

Two cepheids SVS 1673 and SVS 1674 had been investigated. Photographic and photovisual light curves were obtained.

122.006 On two RR Lyrae type variables. V. P. Tsesevich.
Peremennye Zvezdy, Prilozhenie, Vol. 1, 227 - 234 (1971). In Russian.

It is shown that CE Vul and SVS 1675 Sgr belong to RR Lyr type variables. The earlier classifications of these stars were erroneous.

122.007 RR Lyrae type variables in the globular cluster NGC 5466. T. I. Gryzunova.
Peremennye Zvezdy, Prilozhenie, Vol. 1, 253 - 285 (1972). In Russian.

The period changes of 20 RR Lyr type variables in the globular cluster NGC 5466 have been investigated. It is shown that the period variations may be represented by a Poisson distribution.

122.008 On five variable stars in the vicinity of the globular cluster NGC 5466. T. I. Gryzunova.
Peremennye Zvezdy, Prilozhenie, Vol. 1, 287 - 297 (1972). In Russian.

122.009 Zonal character of activity and axial rotation of RW Aurigae. I. M. Ishchenko.
Peremennye Zvezdy, Vol. 18, 293 - 302 (1972). In Russian.

A statistical method is proposed for the determination of the period of the axial rotation of irregular variable stars from a long series of brightness estimates. The efficiency of the method is proved by using it for the independent determination of the period of the solar axial rotation. The method was used for the treatment of the six longest and most dense series of observations of the brightness of RW Aur. We obtained a period of axial rotation of 100^d.

122.010 Light curves of RV Tauri stars with constant mean brightness. G. E. Erleksova.
Peremennye Zvezdy, Vol. 18, 303 - 313 (1972). In Russian.
The light curves of 19 RV Tauri stars were examined. Photoelectric, photographic and visual observations were used for constructing the light curves. The mean square deviations of the stellar magnitudes at light extrema from corresponding

mean values were calculated and analyzed.

122.011 Proper motion of UX Monocerotis.
N. M. Artiukhina.
Peremennye Zvezdy, Vol. 18, 315 - 316 (1972). In Russian.

122.012 On the masses of β CMa (β Cep) stars.
V. I. Varshavskij, A. V. Tutukov.
Nauchn. Informatsii, vyp. (No.) 20, p. 101 - 107 (1971). In Russian.

Using four independent methods we estimated the masses of β CMa stars. They were found to be in the range of $10 \lesssim M/M_\odot \lesssim 25$. These results satisfy Chandrasekhar-Lebovitz's theory for the beat phenomenon of pulsations in β CMa stars.

122.013 Influence of shock waves upon profiles of Hγ spectral lines and light curves for stars of RR Lyrae and W Virginis types. V. I. Golinko.
Astrofizika, Vol. 8, 91 - 105 (1972). In Russian. – English translation in Astrophysics, Vol. 8, No. 1.

A formula is derived for calculating the emission due to the shock-wave front at the frequencies of spectral lines and at the frequencies of the continuous spectrum. Taking into consideration the radiation due to the shock-wave front the profiles of a Hγ spectral line are calculated with and without taking into account the absorption in the above-lying layers. The shock wave is shown to be responsible for the ultraviolet excess. Humps of the light curves are the result of heating the stellar atmosphere with the compression wave following the shock wave. A conclusion is made that in the atmospheres of the population II type cepheids, shock waves originate which have the strength of several units.

122.014 Minimum-light spectra of nine M-type variable stars. S. Wyckoff, P. Wehinger.
Publ. Astron. Soc. Pacific, Vol. 84, 424 - 426 (1972).

Spectral types are given for eight M-type long-period variables and one SRa variable all of which were observed at or near minimum light. Several explanations of the apparently anomalous spectral type (M8) found for R Vir are discussed.

122.015 Near-infrared photometry of Mira variables.
G. W. Lockwood.
Astrophys. Journ., Suppl. Ser., No. 209, Vol. 24, 375 - 419 (1972).

A five-color narrow-band photometric system is described which has been used for determining magnitudes at 1.04 μ, near-infrared colors, and molecular band-strength indices for M stars. The system gives accurate spectral types from M4 to M10 and useful, though less accurate, spectral types from M0 to M4. Mean 1.04-μ magnitudes, near-infrared colors, and spectral-type indices are given for photometric and spectroscopic standard stars, and 1795 individual sets of five-color measurements are given for 292 M- and S-type Mira variables.

122.016 The kinematics of semi-regular red variables in the solar neighbourhood.
M. W. Feast, R. Woolley, N. Yilmaz.
Monthly Notices Roy. Astron. Soc., Vol. 158, 23 - 46 (1972).

Radial velocities for 67 northern M type SR variables observed at the Kottamia Observatory and for 53 similar southern stars observed at the Radcliffe Observatory are given. The kinematics of the SR variables are investigated on the basis of these and previously published radial velocities.

122.017 Photometry of RR Lyrae variables in the globular cluster NGC 6981. R. J. Dickens, R. Flinn.
Monthly Notices Roy. Astron. Soc., Vol. 158, 99 - 123 (1972).

Two-colour (B, V) photometry for 21 RR Lyrae variables in the globular cluster NGC 6981 is presented. The light and colour curves are derived and correlations between some of their parameters are discussed and compared with those of other globular clusters. A mean mass of $\sim 0.4\,M_\odot$ and a helium abundance of $Y \gtrsim 30$ per cent are derived by comparison of the observations with pulsation theory.

122.018 A spectroscopic and photometric study of the pulsating R Coronae Borealis type variable, RY Sagittarii. J. B. Alexander, P. J. Andrews, R. M. Catchpole, M. W. Feast, T. Lloyd Evans, J. W. Menzies, P. N. J. Wisse, M. Wisse.
Monthly Notices Roy. Astron. Soc., Vol. 158, 305 - 360 (1972).

Results are given of an intensive study of RY Sgr during the period 1967–70. The rapid decline in brightness at the beginning of this period and the gradual rise to maximum light are covered.

122.019 The R Coronae Borealis variables in the Large Magellanic Cloud. M. W. Feast.
Monthly Notices Roy. Astron. Soc., Vol. 158, 11P - 13P (1972).

Spectra were obtained of three RCB stars in the LMC. W Men ($M_v = -5.1$) and HV 12842 ($M_v \sim -4$) have weak C_2 absorption bands. HV 5637 ($M_v = -4.0$) has very strong C_2 bands.

122.020 Observations of rapid blue variables – VII. EX Hydrae. B. Warner.
Monthly Notices Roy. Astron. Soc., Vol. 158, 425 - 430 (1972).

Photometric observations of EX Hya with time resolutions of 2 and 5 s are reported. EX Hya is very active on time scales of 10–100 s. A comparison with WZ Sge is made. It is suggested that the X-ray source 2ASE 1253-28 may be identified with EX Hya.

122.021 V553 Centauri: a type II cepheid with an overabundance of carbon.
T. Lloyd Evans, P. N. J. Wisse, M. Wisse.
Monthly Notices Roy. Astron. Soc., Vol. 159, 67 - 78 (1972).

The cepheid V553 Centauri ($P = 2^{d}06$) has strong C_2, CH and CN bands in its spectrum. Their presence implies a considerable overabundance of carbon in the star's atmosphere. The light and colour curves show the characteristic features of a globular cluster cepheid of this period. RT TrA ($P = 1^{d}495$), for which fragmentary data are available, is very similar. The possibility that V553 and RT TrA are related to the cepheids found in dwarf spheroidal galaxies, which are confined to short periods, is discussed.

122.022 Observations of rapid blue variables – VIII. The companion to Mira. B. Warner.
Monthly Notices Roy. Astron. Soc., Vol. 159, 95 - 100 (1972).

Ultra-violet photoelectric photometry with 2 s time resolution of Mira near minimum is reported. The light variations observed are attributed to o Cet B.

122.023 Observations of rapid blue variables – IX. AM CVn (HZ 29). B. Warner, E. L. Robinson.
Monthly Notices Roy. Astron. Soc., Vol. 159, 101 - 111 (1972).

Photometric observations with 1- and 3-s time resolution demonstrate the existence of rapid flickering in AM CVn. Mean light curves confirm the existence of a double-humped light curve with a period near 18 min. Periodogram analyses show the existence of long-lived coherent periodicities in the range 115–120 s. A binary star model is suggested for AM CVn, with an orbital period of 18 min. It is proposed that AM CVn represents a late stage of evolution of a cataclysmic variable star, in which all hydrogen has been removed from the system.

122.024 Far-infrared and $uvby$ photometry of V 1057 Cygni. T. Simon, N. D. Morrison, S. C. Wolff, D. Morrison.
Astron. Astrophys., Vol. 20, 99 - 104 (1972).

Four-color ($uvby$) photometry and broad-band observations at $5\,\mu m$, $11\,\mu m$, and $20\,\mu m$ are reported for V 1057 Cyg (LkHα 190).

122.025 Non-thermal bremsstrahlung of fast electrons and flare of stars. G. A. Gurzadyan.
Astron. Astrophys., Vol. 20, 145 - 149 (1972).

The author compares the amount of inverse Compton radiation with that of non-thermal bremsstrahlung also produced by the fast electrons. It is concluded that, again if it can be shown that fast electrons are present, the optical flares in stars can be brought about by the inverse Compton effect alone owing to these electrons.

122.026 The mass of TU Cassiopeiae. E. G. Schmidt.
Astrophys. Journ., Vol. 176, 165 - 168 (1972).

By use of a measured temperature and the value of Q_0 implied by the period ratio, the mass of the beat cepheid, TU Cas, is determined. The value found ($4.0\,M_\odot$) compares well with values for evolutionary models (4.0–$4.7\,M_\odot$). On the other hand, the mass calculated from pulsation theory using the period, luminosity, and temperature is smaller than either of these two estimates ($M_Q = 3.3\,M_\odot$).

122.027 Spectrocolorimetric observations of flares of EV Lac. P. F. Chugainov.
Izv. Krymskoj Astrofiz. Obs., Vol. 44, 3 - 10 (1972). In Russian.

Absolute measurements of continuous radiation of three flares of EV Lac have been obtained. The device used, a photoelectric spectrocolorimeter, gave the possibility to record simultaneously the stellar brightnesses in the following spectral regions: $\lambda\lambda 3350$–3650, 4155–4280 and 5120–5320 Å.

122.028 Balmer decrement in flare spectra of UV Cet-type stars. R. E. Gershberg, S. A. Kaplan.
Izv. Krymskoj Astrofiz. Obs., Vol. 44, 11 - 17 (1972). In Russian.

The results of a spectrophotometric investigation of the EV Lac flare on 12.VII.1966, relative intensities and equivalent widths of emission lines, are given. For preliminary interpretation of the observed variety of Balmer decrements a model of a stationary radiating medium having a velocity gradient and large optical thicknesses in the lines of several first spectral series has been considered. A more complicated theory of Balmer decrement (Boyarchuk, 1966) allows to interpret both Balmer line intensity ratios and Hβ equivalent widths observed in the flare.

122.029 The composite period-age relation for cepheids of the Magellanic Clouds, M 31, and the Galaxy.
Yu. N. Efremov.
Astron. Tsirk., No. 671, p. 1 - 3 (1972). In Russian.

122.030 Heterogeneity of the complex of δ Sct stars. M. S. Frolov.
Astron. Tsirk., No. 671, p. 3 - 4 (1972). In Russian.

122.031 Variable stars of small amplitude. I. Supergiants and OB stars. A. Maeder, F. Rufener.
Astron. Astrophys., Vol. 20, 437 - 443 (1972).

The aim of this work is to detect new kinds of stars which show light variation of small amplitude, similar to those observed for the Be, β C Ma, δ Scuti and Ap stars. A basic data to test for stellar light variability, we use the whole set of colour indices measurements now available in the Geneva Observatory photometric system. In this paper, we essentially discuss the case of supergiants and OB stars.

122.032   On the interpretation of the photometry of irregular variable stars observed intermittently.
P. Fellgett.
Astrophys. Space Sci., Vol. 16, 437 - 444 (1972).

This note is concerned with the inherent possibilities and limitations in the characterisation of irregular variables when, as is almost inevitably the case, the observations are not effectively continuous. The historical tendency to interpret almost everything in terms of strict periodicity is criticised.

122.033   On the pulsation amplitude of cepheid variables.
N. Nikolov, Ts. Tsvetkov.
Astrophys. Space Sci., Vol. 16, 445 - 450 (1972).

On the basis of more abundant data a relation $\log P \, \Delta V/ \log \Delta R$ for cepheid variables (Fernie, 1965) is constructed. A linear relation between $\log P \, \Delta V$ and $\log \Delta R$ for classical cepheids is found, which perhaps has a break at $\Delta R = 10 \, R_\odot$.

122.034   The influence of metal lines on the colors of cepheids in the Small Magellanic Cloud.
R. A. Bell, S. B. Parsons.
Astrophys. Letters, Vol. 12, 5 - 8 (1972).

Synthetic spectra are used to confirm that the blueness of cepheids in the SMC relative to cepheids in the Galaxy and the LMC is entirely explained by reduced line blocking, corresponding to a factor of four in metal deficiency. Strömgren photometry should provide a test of this composition difference.

122.035   Photometric studies of UV Ceti.
A. H. Jarrett, J. P. Eksteen.
Astron. Astrophys., Suppl. Ser., Vol. 7, 103 - 112 (1972).

Further flare observations of UV Ceti have confirmed the profiles obtained from earlier observations at Boyden Observatory.

122.036   Observations of BL Lac in 1969–1971.
N. E. Kurochkin.
Astron. Tsirk., No. 678, p. 6 - 7 (1972). In Russian.

122.037   Intrinsic colours of classical cepheids and yellow supergiants.   N. N. Yakimova, Yu. N. Shvetzov.
Astron. Tsirk., No. 679, p. 5 - 7 (1972). In Russian.

122.038   On the two "strange" long-period variable stars CL Aquilae and SVS 1739 Scuti.
V. P. Tsesevich.
Astron. Tsirk., No. 685, p. 7 - 8 (1972). In Russian.

122.039   On the RR Lyrae-type variable star SVS 567 = DL Com.   B. D. Pochinok.
Astron. Tsirk., No. 690, p. 7 - 8 (1972).   In Russian.

122.040   Autocorrelation analysis of EX Hya brightness.
A. F. Pugach.
Astron. Tsirk., No. 693, p. 6 - 8 (1972).   In Russian.

122.041   New elements of three short-period cepheids.
G. A. Lange, P. P. Gusev.
Astron. Tsirk., No. 695, p. 7 - 8 (1972). In Russian.

122.042   Pulsating variables in the Pleiades cluster.
M. Breger.

Astrophys. Journ., Vol. 176, 367 - 371 (1972).

Thirty-one B and A stars in the Pleiades cluster were tested for short-period variability. Five δ Scuti stars and one Be shell variable were detected. The new variables extend the hot border of the instability strip to spectral type A2 on the zero-age main sequence.

122.043   Main-sequence pulsation in open clusters.
M. Breger.
Astrophys. Journ., Vol. 176, 373 - 380 (1972).

Studies of short-period pulsation in open clusters show that pulsation occurs in clusters of all ages provided stars are present in the instability strip (A2 V–F0 V). About 30 percent of the main-sequence cluster stars inside the instability strip show detectable variability. Statistically extremely significant correlations between slow rotation, metallicity, and pulsational stability are established for different clusters. The observed relations can be explained by a diffusion hypothesis.

122.044   Photometry and spectroscopy of red variables in Omega Centauri.
R. J. Dickens, M. W. Feast, T. Lloyd Evans.
Monthly Notices Roy. Astron. Soc., Vol. 159, 337 - 348 (1972).

Spectroscopy and infra-red photographic photometry have been carried out for a number of red variables in the globular cluster ω Centauri (NGC 5139). V2 and V129 are nonmembers. V53, V152, V162, V138, V148 and V161 lie near the red giant tip of the $I_K/(V-I_K)$ diagram. V164, a radial velocity member with a similar spectrum, lies below the giant branch. V6, V17 and V42, all radial velocity members, have large $(V-I_K)$ colours and M-type spectra. RGO 4789 (membership uncertain) is also a red variable lying in the same region of the HR diagram.

122.045   Luminosity variation in the one-zone cepheid model.
R. F. Stellingwerf.
Astron. Astrophys., Vol. 21, 91 - 96 (1972).

The one-zone stellar pulsation model is proposed as a tool to investigate the factors affecting luminosity variations of pulsating stars. Linear and non-linear analyses of the resulting equations are described. The results are in very good agreement with observations.

122.046   Variations of the Hα-emission profile in three irregular variable stars.
G. V. Zajtseva, E. A. Kolotilov.
Astron. Tsirk., No. 699, p. 1 - 3 (1972). In Russian.

122.047   Variations of the period of X Tri.
N. E. Kurochkin.
Astron. Tsirk., No. 699, p. 5 - 8 (1972). In Russian.

122.048   On the U Gem type variable in the NGC 3147 region.   P. N. Kholopov.
Astron. Tsirk., No. 700, p. 1 (1972). In Russian.

122.049   Photographic observations of AP Lib.
N. E. Kurochkin.
Astron. Tsirk., No. 702, p. 6 - 8 (1972). In Russian.

122.050   The period and near-infrared variations of IK Tauri.
G. W. Lockwood, R. F. Wing.
Bull. American Astron. Soc., Vol. 4, 322 - 323 (1972).
Abstr. AAS.

122.051   Possible constraints on the evolution of semiregular variable N-stars.   B. M. Schlesinger.
Bull. American Astron. Soc., Vol. 4, 323 - 324 (1972).
Abstr. AAS.

122.052 **A photoelectric investigation of the double cepheid CE Cas.** O. G. Franz.
Bull. American Astron. Soc., Vol. 4, 329 (1972). – Abstr. AAS.

122.053 **A new approach to periodogram analyses.** D. F. Gray, K. Desikachary.
Bull. American Astron. Soc., Vol. 4, 337 (1972). – Abstr. AAS.

122.054 **A search for ultrashort-periodic light variations of stars in the region of Cyg X-1.**
A. N. Abramenko, O. P. Gollandsky, V. V. Prokofieva.
Astron. Tsirk., No. 708, p. 6 - 8 (1972). In Russian.

122.055 **A comment on the use of the displacement potential in pulsational analysis.** G. J. Vaughan.
Monthly Notices Roy. Astron. Soc., Vol. 159, 375 - 377 (1972).

A mathematical error in Rosseland's 'Pulsational theory of variable stars' (Rosseland) is pointed out and a cautionary word on the use of the displacement potential technique is given.

122.056 **Gründe von visuellen Beobachtungsdifferenzen bei Langperiodischen.** E. Heiser.
BAV Rundbrief, 21. Jahrgang, Sonder-Rundbrief, (see 012. 011), p. 2 - 12 (1972).

122.057 **U Boötis, 1940–69.** J. E. Isles.
Journ. British Astron. Ass., Vol. 82, 464 - 469, with a correction in Vol. 83, 63 (1972). – Report of the Variable Star Section of the British Astron. Ass.

122.058 **HD 90386: A new short-period variable star.** M. Jerzykiewicz.
Publ. Astron. Soc. Pacific, Vol. 84, 529 - 530 (1972).

HD 90386 has been found to vary in yellow light with the period of about 0^d0799. The light range is close to 0^m01 and appears to be slightly variable. The star is probably a member of the δ Scuti group of variable stars.

122.059 **The photometric behaviour of RU Cam during 1966–1971 and an evolutional interpretation of the star's characteristics.**
G. V. Zaitseva, V. M. Lyutyj, Yu. N. Efremov.
Astron. Zhurn. Akad. Nauk SSSR, Vol. 49, 1049 - 1054 (1972). In Russian. English translation in Soviet Astron. AJ, Vol. 16, No. 5.

The composite light curve of RU Cam for 1966–1971 shows that periods of small nonperiodic fluctuations have a tendency to become more prolongated. Probably the star will be a cepheid nevermore.

122.060 **Sur les déplacements radiaux de différents éléments dans α^2 CVn.** P. Renson.
Astrophys. Space Sci., Vol. 17, 69 - 79 (1972).

On compare les courbes de déplacements radiaux déduites des mesures de Struve et Swings, celles déduites des mesures de Babcock et Burd et celle d'Eu déduite des mesures de Pyper (en tenant compte de tous les résultats photométriques, on a un $P = 5,4693 \pm 0,0001$ j). Les différences sont surtout attribuables, dans le cadre du modèle binaire, à l'effet Ovenden. Celui-ci ainsi que les courants circumstellaires peuvent aussi être responsables des trop grands déplacements constatés et d'autres particularités.

122.061 **Determination of the mean absolute magnitude of RR Lyrae stars.** A. Heck.
Astron. Astrophys., Vol. 21, 231 - 238 (1972). In French.

The method used here is that based on the principle of maximum likelihood developed by Jung (1970) with some improvements. The bias and the precision of the estimates were calculated by numerical experiments. We have applied the method to a sample of 102 RR Lyrae stars, after elimination of high velocity stars and of stars with large errors on the proper motions, in order to determine its mean absolute photographic magnitude and the corresponding dispersion. We also applied it to the sample defined by Clube and Jones (1971).

122.062 **Evidence for a suspected period change of SX Phoenicis.**
J. Stock, W. E. Kunkel, J. E. Hesser, B. M. Lasker.
Astron. Astrophys., Vol. 21, 249 - 253 (1972).

New photoelectric observations in the UBV- and H_β-systems of the short-period variable star SX Phoenicis (HD 223065) are presented. When combined with earlier data, the UBV observations indicate the existence of a small variation of the basic periods.

122.063 **Pleione, October 1971–April 1972.**
A. S. Sharov, V. M. Lyutyj.
Inform. Bull. Variable Stars (I.A.U. Commission 27), Konkoly Obs., Budapest, No. 698 (1972).

122.064 **Spectral class and metal-abundance index determination for 20 RR Lyrae stars at light maximum.**
I. F. Alania.
Inform. Bull. Variable Stars (I.A.U. Commission 27), Konkoly Obs., Budapest, No. 702, 3 pp. (1972).

The spectral classes of 20 RR Lyrae type stars determined according to hydrogen as well as K(Ca II) lines are given in a table. The observations were performed with the 70 cm meniscus telescope (dispersion 166 A/mm near Hγ) of the Abastumani Observatory.

122.065 **32 Virginis: a pulsating Am star.**
C. Bartolini, F. Grilli, G. Parmeggiani.
Inform. Bull. Variable Stars (I.A.U. Commission 27), Konkoly Obs., Budapest, No. 704 (1972).

122.066 **The observation of a stellar flare in the dM5 star Ross 128.** T. A. Lee, D. T. Hoxie.
Inform. Bull. Variable Stars (I.A.U. Commission 27), Konkoly Obs., Budapest, No. 707, 4 pp. (1972).

122.067 **Photometry of V1216 Sagittarii.**
A. H. Jarrett, J. P. Eksteen.
Inform. Bull. Variable Stars (I.A.U. Commission 27), Konkoly Obs., Budapest, No. 711 (1972).

122.068 **Photoelectric observations of the flare stars BD+13°2618 and BD+16°2708.**
G. Asteriadis, L. N. Mavridis.
Inform. Bull. Variable Stars (I.A.U. Commission 27), Konkoly Obs., Budapest, No. 712, 3 pp. (1972).

122.069 **The R-I color of V 1057 Cygni before the outburst, and its brightness and spectral changes.** G. Haro.
Inform. Bull. Variable Stars (I.A.U. Commission 27), Konkoly Obs., Budapest, No. 714, 4 pp. (1972).

122.070 **New flare stars in the Pleiades region (a re-examination of the Tonantzintla photographic material: 1963–1970).** G. Haro, G. González.
Inform. Bull. Variable Stars (I.A.U. Commission 27), Konkoly Obs., Budapest, No. 715 (1972).

122.071 **New flare stars in the Pleiades region (1971–1972).**
G. Haro, E. Chavira.
Inform. Bull. Variable Stars (I.A.U. Commission 27), Konkoly Obs., Budapest, No. 716 (1972).

122.072 **Photoelectric observations of V 1216 Sgr during the 1972, July 3 - 17 international patrol.**

S. Cristaldi, M. Rodonò.
Inform. Bull. Variable Stars (I.A.U. Commission 27), Konkoly Obs., Budapest, No. 721 (1972).

122.073 Continuous photoelectric monitoring of EV Lac during the international patrol, September 1 – 15, 1972. B. N. Andersen, B. R. Pettersen.
Inform. Bull. Variable Stars (I.A.U. Commission 27), Konkoly Obs., Budapest, No. 723 (1972).

122.074 Photoelectric UBV observations of Cepheid variable BE Monocerotis.
N. Buchancowa, P. Kunchev, W. Wisniewski.
Inform. Bull. Variable Stars (I.A.U. Commission 27), Konkoly Obs., Budapest, No. 727, 3 pp. (1972).

122.075 De veranderlijke van de maand:T Herculis.
H. Feijth.
Hemel en Dampkring, Vol. 70, 256 - 258 (1972).

122.076 UV Cet-type flare stars.
R. E. Gershberg, S. B. Pikel'ner.
Comments Astrophys. Space Phys., Vol. 4, 113 - 120 (1972).
The observational data of UV Cet-type stars are described. The nature of their flares are explained by the so called nebular or chromospheric model, and an analogy is drawn with solar chromospheric flares. Therefore it is an important evidence for the electro-magnetic nature of the flares. Some aspects of the evolution of flare stars are given.

122.077 Evidence for the hydrodynamic character of micro-turbulence. J. van Paradijs.
Nature, Phys. Sci., Vol. 238, 37 - 38 (1972).
From curve of growth analyses of classical cepheids, as given in the astronomical literature, a correlation is found between the variation of the microturbulence and the pulsation velocity, but not with the electron pressure. This correlation suggests that microturbulence is, at least in part, a real hydrodynamic effect.

122.078 Optical and radio flares in BL Lacerta (VRO 42.22.01).
R. L. Hackney, K. R. Hackney, A. G. Smith, G. H. Folsom, R. J. Leacock, R. L. Scott, E. E. Epstein.
Astrophys. Letters, Vol. 12, 147 - 152 (1972).
In late 1971 the peculiar source BL Lac underwent an isolated optical flare of a type previously observed in OVV quasars. Although this event was followed in 55 days by a similar radio flare, examination of all available data indicates that the observations cannot be correlated by assuming a constant radio-optical delay.

122.079 Importance of high time resolution in flare star observations. T. J. Moffett.
Nature, Phys. Sci., Vol. 240, 41 - 43 (1972).
Flare star observations have been made here of about 75 events on 7 UV Ceti type stars, using time resolutions as short as 50 ms. Striking examples of the advantages of a high speed system are shown in figures obtained on the stars Wolf 359 and EQ Peg.

122.080 The classification of intrinsic variable stars. II. The red variables of S and related types. O. J. Eggen.
Astrophys. Journ., Vol. 177, 489 - 507 (1972).
Observations in the (*UBVRI*) system, mainly obtained with the 40-inch (102-cm) reflector at Siding Spring Mountain, are presented for 34 stars of S or related type. Twenty-four of these stars are named variables (first light curves were obtained for two of these, π^1 Gru and Henize 120), suspected variation in three stars (HD 58881 = HV 100, HD 75021, and HD 168227 = HV 170) is confirmed, and five new variables

(HD 34738, HD 49368, HD 52432, HD 216672, and Henize 244) were found. Membership in the Wolf 630 group (R And and HR 1556), σ Puppis group (HR 363) and Hyades group (π^1 Gru) is used to demonstrate that the S stars occur on the giant sequence of the old disk population.

122.081 Pulsation constants for the population I cepheids. M. Takeuti, Y. Shibata.
Sci. Rep. Tôhoku Univ., First Ser., Vol. 54, 207 - 212 (1971).
Pulsation constants are calculated for ten model envelopes of population I cepheids to examine the period-mass-radius relation proposed by Cogan. The relation is hardly affected by the difference in chemical compositions and procedures of model construction.

122.082 Deveranderlijke van de maand: RT Cygni.
H. Feijth.
Hemel en Dampkring, Vol. 70, 297 - 300 (1972).

122.083 Red supergiant variables of large amplitude in the SMC. T. Lloyd Evans.
Proc. Third Colloquium on Astrophysics. Supergiant Stars, Trieste 1971, p. 125 - 127 = Radcliffe Obs. Repr., No. 108 (1972).
New observations with the 1.88 m reflector of the Radcliffe Observatory, Pretoria, have shown that some of the variable red supergiants (Shapley and Nail, 1955) in the SMC have amplitudes of 4^m or more (Evans, 1971). The Gaposchkins (1966) determined periods and magnitudes at maximum for 23 of them.

122.084 Photometry of cepheids in the Magellanic Clouds. C. J. Butler.
Proc. Third Colloquium on Astrophysics. Supergiant Stars, Trieste 1971, p. 128 - 143 = Dunsink Obs. Repr., No. 63 (1972).
A new photographic survey in V and B has been made of cepheids in the Magellanic Clouds. Period-luminosity relations are derived which have very similar slopes for the cepheids in both Clouds. A comparison is made with the results of Gascoigne and it is concluded that differences in the period-luminosity relations derived by the two observers for the same Cloud, are due, not to photometric differences, but to change selection effects amongst the brighter variables.

122.085 Herschel's red star No. 10 (HD 20234).
B. F. Marino, W. S. G. Walker.
Southern Stars, Vol. 24, 124 - 126 = Auckland Obs. Observational Note No. 3 (1972).

122.086 Lichtwechsel und Extinktion des Veränderlichen WW Vulpeculae. S. Rössiger, W. Wenzel.
Astron. Nachr., Vol. 294, 29 - 38 (1972).
Photoelectric UBV observations of the A star WW Vulpeculae show three components of its brightness variations: permanent short-term fluctuations, slow variations of the "normal light" and unperiodic minima. The hypothesis that these minima are produced by clouds of circumstellar condensation products leads to an R value of 5. From photometry and spectral-type determination of a number of surrounding stars we conclude that outside of the minima (during "normal light") the extinction of WW Vulpeculae has no circumstellar component.

122.087 I Zw 1727 +50: a new lacertid?
A. M. Le Squéren, F. Biraud, R. Lauqué.
Nature, Phys. Sci., Vol. 240, 75 - 76 (1972).
BL Lac (= VRO 422201), AP Lib (= PKS 1514−24), W Coma (= ON 231), OJ 287 and B2 1215+30 probably belong to a new class of objects, which might be called "lacertids", from BL Lacertae, the first discovered and most extensively

studied. Their main characteristics are: a featureless optical spectrum; a rather flat or even "inverted" radio spectrum; and very fast and irregular variations, both at radio and optical wavelengths. The overall energy distribution of I Zw 1727 + 50 closely resembles these spectra.

122.088 Further observations of HV 13055.
E. M. Lindsay, P. A. Wayman.
Irish Astron. Journ., Vol. 10, 141 - 148 (1971).

122.089 Observations of linear polarization of optical radiation of BL Lac-type objects.
Yu. S. Efimov, N. M. Shakhovskoy.
Astron. Tsirk., No. 710, p. 2 - 5 (1972). In Russian.

122.090 Hα emission in the spectra of some rapid irregular variable stars. G. V. Zaitseva, V. F. Esipov.
Astron. Tsirk., No. 712, p. 7 - 8 (1972). In Russian.

122.091 The heating of the interstellar medium by Mira stars.
L. Carrasco, G. Grasdalen.
Publ. Astron. Soc. Pacific, Vol. 84, 639 (1972). − Abstr. Astron. Soc. Pacific.

122.092 Infrared photometry of R Coronae Borealis during its recent decline.
R. F. Wing, J. H. Baumert, S. E. Strom, K. M. Strom.
Publ. Astron. Soc. Pacific, Vol. 84, 646 - 647 (1972). − Abstr. Astron. Soc. Pacific.

122.093 Titanium isotopes in o Ceti.
S. Wyckoff, P. Wehinger.
Publ. Astron. Soc. Pacific, Vol. 84, 647 (1972). − Abstr. Astron. Soc. Pacific.

122.094 Motions in the outer layers of the 27-day cepheid T Monocerotis. G. Wallerstein.
Publ. Astron. Soc. Pacific, Vol. 84, 656 - 663 (1972).

Radial velocities and Hα profiles are presented for the entire cycle of T Mon. Several interpretations in terms of rising or falling material are presented, but no simple physical model can be derived and fully justified. The metallic absorption lines show the same depth effects as seen for SV Vul. The sodium D lines are resolved near minimum light into stellar and interstellar components.

122.095 Light elements for DK Velorum − a corrigendum.
D. H. P. Jones.
Publ. Astron. Soc. Pacific, Vol. 84, 644 - 665 (1972).

122.096 uvby photometry of EH Librae.
W. J. Boardman, A. M. Heiser.
Publ. Astron. Soc. Pacific, Vol. 84, 680 - 684 (1972).

The short-period variable EH Lib has been observed photoelectrically in the uvby system. The times of maximum brightness were found to agree, within observational errors, with the elements obtained by Oosterhoff and Walraven (1966). No regular variation was detected in m_1, but an analysis of the $(b-y)$ and c_1 variations indicate an expansion of the star as maximum brightness is reached.

122.097 HR 6684: A new Beta Cephei-type variable star.
M. Jerzykiewicz.
Publ. Astron. Soc. Pacific, Vol. 84, 718 - 720 (1972).

HR 6684 has been found to vary in blue light with the period of $0^d.13989$. The light range is $0^m.035$. The star appears to be a β Cephei variable. If confirmed, it would have the shortest period and lowest luminosity of any previously known β Cephei star.

122.098 Photoelectric monitoring of BL Lacertae.

E. F. Milone.
Publ. Astron. Soc. Pacific, Vol. 84, 723 - 731 = Rothney Astrophys. Obs. Publ., No. 2 (1972).

Photoelectric V observations of BL Lac were obtained with the 50-inch telescope at KPNO during eight nights of June 1970. The light variation generally followed previously reported types of fluctuations, with one exception. Previously observed night-to-night variations are confirmed, intra-night variations strongly suspected, but no firm confirmation of the flickering reported by Racine (1970) is possible with the present data.

122.099 High resolution radio observations of AP Lib.
R. G. Conway, D. Stannard.
Monthly Notices Roy. Astron. Soc., Vol. 160, 31P - 33P (1972).

Observations of PKS 1514−24 suggest a radio structure consisting of a compact component coincident with AP Lib, and an extended companion some 20″ arc away. The latter accounts for the displacement of the radio centroid from the optical position of AP Lib.

122.100 On the location of pulsational blue edges and estimates of the luminosity and helium content of RR Lyrae stars. R. S. Tuggle, I. Iben, Jr.
Astrophys. Journ., Vol. 178, 455 - 465 (1972).

Blue edges of the instability region for pulsation in the fundamental and first harmonic modes have been constructed with Cox-Stewart opacities treated by spline interpolation. Estimates of the helium abundance in the envelopes of RR Lyrae stars are reduced by $\Delta Y \sim 0.06$.

122.101 Photometric behaviour of field RR Lyrae stars.
K. Stępień.
Acta Astron., Vol. 22, 175 - 226 (1972).

Three-colour observations of the rising branches of 25 field RR Lyrae stars are presented. This photometry supplemented with the observations by Sturch (1966) allowed to find mean colour indices of a number of single-period variables. Observations of additional single-period variables with well determined light curves were included into the discussion.

122.102 On the chemical composition of classical cepheids of the Galaxy and Magellanic Clouds. N. N. Yakimova.
Astron. Zhurn. Akad. Nauk SSSR, Vol. 49, 1175 - 1183 (1972). In Russian. English translation in Soviet Astron. AJ, Vol. 16, No. 6.

Stobie's theoretical calculations allowed to estimate preliminarily the helium abundance in shells of cepheids of the Galaxy and Magellanic Clouds.

122.103 Flare activity of Gl 669 A.
N. I. Shakhovskaya, W. Sofina.
Inform. Bull. Variable Stars, (IAU Commission 27), Konkoly Obs., Budapest, No. 730 (1972).

122.104 The periods of six RR Lyrae type stars.
D. Hoffleit.
Inform. Bull. Variable Stars, (IAU Commission 27), Konkoly Obs., Budapest, No. 735 (1972).

122.105 Observations of UV Ceti during the 1972 October 1−15 international patrol.
T. J. Moffett, B. W. Bopp.
Inform. Bull. Variable Stars, (IAU Commission 27), Konkoly Obs., Budapest, No. 736, 8 pp. (1972).

122.106 The secondary period of RV Capricorni.
S. Kanyó.
Inform. Bull. Variable Stars, (IAU Commission 27), Konkoly Obs., Budapest, No. 737 (1972).

122.107 Visual observations of EV Lacertae. J. E. Isles.
Inform. Bull. Variable Stars, (IAU Commission 27),
Konkoly Obs., Budapest, No. 738 (1972).

122.108 On the variability of the Mira star UX Cygni.
T. J. Moffett, T. G. Barnes, III.
Inform. Bull. Variable Stars, (IAU Commission 27), Konkoly
Obs., Budapest, No. 739 (1972).

122.109 Polarimetric observations of R Coronae Borealis.
M. Ya. Orlov, M. H. Rodriguez.
Inform. Bull. Variable Stars, (IAU Commission 27), Konkoly
Obs., Budapest, No. 742 (1972).

122.110 UV Ceti. K. Osawa, K. Ichimura, Y. Shimizu.
Inform. Bull. Variable Stars, (IAU Commission 27),
Konkoly Obs., Budapest, No. 747 (1972). – Observations of
Okayama Station, 1972 October 9 - 15.

122.111 Observations of southern flare stars.
W. E. Kunkel.
Inform. Bull. Variable Stars, (IAU Commission 27), Konkoly
Obs., Budapest, No. 748, 5 pp. (1972).

**122.112 Photoelectric surveillance of the flare stars YZ CMi,
AD Leo and EV Lac.** R. C. Kapoor, S. D. Sinvhal.
Inform. Bull. Variable Stars, (IAU Commission 27), Konkoly
Obs., Budapest, No. 750, 5 pp. (1972).

122.113 Light variations of 59 Piscium.
S. K. Gupta, A. K. Bhatnagar.
Inform. Bull. Variable Stars, (IAU Commission 27), Konkoly
Obs., Budapest, No. 751 (1972).

**122.114 Infrared radiation from RV Tauri stars. I. An infrared
survey of RV Tauri stars and related objects.**
R. D. Gehrz.
Astrophys. Journ., Vol. 178, 715 - 725 (1972).
Broad-band photometric measurements from 2.2 to 22 μ
have been made for 25 RV Tauri stars and 50 comparison objects. All 11 RV Tauri stars for which the [3.6]−[11.3] color
was measured showed anomalous excess 11.3-μ radiation. Six
of 8 RV Tauri stars for which upper limits to the [3.6]−[11.3]
color were set may have large infrared excesses. Most RV Tauri
stars are embedded in extended cool dust envelopes. Metallic
silicates are present in the dust surrounding oxygen-rich RV
Tauri stars while an unidentified substance has condensed
around the carbon-rich RV Tauri star AC Her.

**122.115 Infrared measurements of R Coronae Borealis
through its 1972 March−June minimum.**
W. J. Forrest, F. C. Gillett, W. A. Stein.
Astrophys. Journ., (*Letters*), Vol. 178, L129 - L132 (1972).
Infrared (3.5−11 μ) observations of R CrB through the
recent 1972 March−June visual minimum are presented, and
an interpretation in terms of dust clouds surrounding the star
is suggested.

122.116 Pulsation of rotating RR Lyrae stars. N. Lungu.
Stud. Cerc. Astron. Vol. 17, 227 - 233 (1972).
In Romanian.
The paper presents a theoretical interpretation of the
pulsation of variable RR Lyr-type stars for a model with a
"surface" envelope, and a comparison with observations. The
contribution of the rotation to the period of pulsation is
given. As special case the period of XZ Dra is determined.

122.117 Photometry of cepheids in the Magellanic Clouds.
C. J. Butler.
Proc. Third Colloquium on astrophysics. Supergiant stars,
Trieste 1971, (see 012.020), p. 128 - 143 (1972).

**122.118 Revision of the conversion factor 24/17 and the
Wesselink radii of classical cepheids.**
S. B. Parsons.
Proc. Third Colloquium on astrophysics. Supergiant stars,
Trieste 1971, (see 012.020), p. 178 - 183 (1972).

**122.119 Some statistical properties and evolutionary consid-
erations on the S-type stars.** M. Motteran.
Proc. Third Colloquium on astrophysics. Supergiant stars,
Trieste 1971, (see 012.020), p. 292 - 312 (1972).

122.120 CC Eridani. R. E. Nather, J. Harwood.
IAU Circ., No. 2434 (1972).

122.121 BH Crucis−Photoelectric observations.
W. S. G. Walker, B. F. Marino, R. G. Welch.
Roy. Astron. Soc. New Zealand, Variable Star Section, Circ.
No. 188, 7 pp. (1972).
Photoelectric observations of the red variable BH Cru,
discovered by Welch in 1969, are presented. A period and
epoch are given. BH Cru is classified as one of the small group
of R Cen-like variables within the Mira Cet type stars.

122.122 The short-period variable HDE 302013 = V 753 Cen.
R. D. Cannon.
Observatory, Vol. 92, 234 - 236 (1972).
The star HDE 302013 was reported to be variable by
Cannon and Eggen. Further observations have been made to
check the period, to confirm a suggested bump on the V light
curve just before maximum light, and to examine the variabili-
ty in B−V and U−B.

**122.123 Continual photoelectric monitoring of flare stars,
VII. YZ CMi, AD Leo, EV Lac and UV Cet 1971.**
K. Ichimura, Y. Shimizu.
Tokyo Astron. Bull., Second Ser., No. 222, p. 2589 - 2596
(1972).
Flare stars YZ CMi, AD Leo, EV Lac and UV Cet were
monitored photoelectrically with the 91-cm reflector of the
Okayama Station during the periods of international coopera-
tive observations in 1971.

**122.124 *UBV* observations of the RR Lyrae variable HD
176387 (MT Telescopii).** J. B. Alexander.
Royal Obs. Bull., [Herstmonceux: Royal Greenwich Obs.], No.
174, p. 107 - 114 (1972).
UBV observations of HD 176387 made in 1970 at the
Cape Observatory are described. The significance of the anti-
clockwise loop performed by the star in the (U−B, B−V) plane
is discussed.

122.125 Infrared observations of southern RV Tauri stars.
R. D. Gehrz, E. P. Ney.
Publ. Astron. Soc. Pacific, Vol. 84, 768 - 771 (1972).
Photometric measurements from 2.2 μ to 18 μ are report-
ed for six southern RV Tauri stars. Four of the six have anoma-
lous infrared excess radiation similar to that observed in other
RV Tauri stars. The infrared spectra of IW Car, AR Pup, and
SX Cen resemble that of oxygen-rich RV Tau. RU Cen is
similar to AC Her.

**122.126 Über das Sonneberger Programm zur Bestimmung
von Spektraltypen veränderlicher Sterne auf Platten
mit geringfügig verbreitertenSpektrogrammen.**
W. Götz, W. Wenzel.
MVS, *Sonneberg*, Vol. 6, 44 - 47 (1972).
Experiences and results from the determination of spectral types of variable stars on plates with spectrograms which
are little-widened are given. From 222 variables 278 spectral
types are estimated.

122.127 **Scheinperioden bei 10 RRab-Sternen.** H. Geßner.
MVS, *Sonneberg,* Vol. 6, 48 (1972).

122.128 **Flare stars in the Pleiades region observed during the fall and winter 1971–1972.**
L. Pigatto, L. Rosino.
Mem. Soc. Astron. Italiana, Nuova Ser., Vol. 43, 455 - 473 (1972).

In the course of the systematic survey of the Pleiades field carried out at Asiago during the fall and winter 1971–1972, twenty-nine stars have been caught in the phase of flaring. Positions, identification charts, light curves and circumstances of the flares are reported. The light curves of the variables No. 98 and 107 are peculiar, showing a double peak.

122.129 **Maxima von Mirasternen.** H.-J. Blasberg.
MVS, *Sonneberg,* Vol. 6, 56 (1972).

122.130 **Motions and absolute magnitudes of RR Lyrae stars.**
N. K. M. Hemenway.
Thesis, Univ. Virginia, Charlottesville. [Available from Univ. Microfilms, Ann Arbor, Mich., USA. Order No. 72-7187], 89 pp. (1971).

New proper motions of 65 RR Lyrae field stars are derived using McCormick plate pairs with an average time base of 39 years. Solar motion solutions are made using published radial velocities. Using all stars for which accurate photoelectric photometry, radial velocities, and proper motions are known, statistical parallaxes are calculated.

122.131 **Polarimetric observations of nonstable stars and extragalactic objects. II. EV Lac flare polarization.**
Yu. S. Efimov, N. M. Shakhovskoy.
Izv. Krymskoj Astrofiz. Obs., Vol. 45, 111 - 117 (1972).
In Russian.
Polarimetric observations of the flare star EV Lac were carried out in four different spectral regions. The flare of EV Lac on August 17, 1969, was measured polarimetrically in the blue region. It was found that polarization of the EV Lac radiation at the different phases of the flare is the same as at the undisturbed state of the star. The wavelength dependence of the polarization has been found.

122.132 **Spectrographic investigation of the flare star YZ CMi in January 1969.** R. E. Gershberg.
Izv. Krymskoj Astrofiz. Obs., Vol. 45, 118 - 123 (1972).
In Russian.
Results of spectral observations of the star YZ CMi carried out during its international photometric patrol are given. The quantitative spectral characteristics of the star were determined at quiet state and during 5 flares. Equivalent widths and relative intensities of emission lines, depths of the absorption line λ 4227 and the values of intensity jumps at the limits of TiO molecular bands are obtained.

122.133 **Spectral and photoelectric observations of the flare star AC + 39° 1214–608.** N. I. Shakhovskaya.
Izv. Krymskoj Astrofiz. Obs., Vol. 45, 124 - 129 (1972).
In Russian.
Photoelectric observations of this star in a photometric system close to B showed flare activity of UV Cet type. Seven weak flares were observed during the whole monitoring time of 47 hours. The equivalent widths of Ca II and H emission lines have been obtained using six spectrograms with dispersion of 160 Å/mm. The relative intensities of these lines have also been determined. Variations of the emission which are not caused by flares during the exposure time have been found. Two possible causes of these variations are suggested.

122.134 **Photometric peculiarities of AC Her.**
N. L. Magalashvili, J. I. Kumsishvili.
Byull. Abastumansk. Astrofiz. Obs., No. 43, p. 3 - 18 (1972). In Russian.
The variable AC Her has been observed photoelectrically by means of a stellar electrophotometer of the Abastumani Astrophysical Observatory in 1960–1967. The results are given in tables and diagrams.

122.135 **Two-color electrophotometric observations of BD Dra.** I. F. Alania.
Byull. Abastumansk. Astrofiz. Obs., No. 43, p. 19-28 (1972). In Russian.
More than 230 observations of BD Dra in B and V have been carried out with the AZT-14 telescope during 10 nights. The results are given in tables and figures.

122.136 **A re-examination of the Pleiades Tonantzintla photographic material: 1963 - 1970. II.**
G. Haro, G. González.
Bol. Obs. Tonantzintla y Tacubaya, Vol. 6, (No. 38), 149 - 154 (1972).
From the 148 outbursts detected, 12 correspond to "new" flare stars. The rest are flare-up repetitions of previously known flare stars discovered at the Tonantzintla, Asiago, Byurakan and Konkoly Observatories.

122.137 **Flare stars in the Pleiades region (1971–1972). III.**
G. Haro, E. Chavira.
Bol. Obs. Tonantzintla y Tacubaya, Vol. 6, (No. 38), 155 - 160 (1972).
During the months of October, November and December, 1971 and January, 1972 we obtained 115 ultraviolet multiple exposure plates centered in Alcyone and covering an area of approximately 16 square degrees. There were 725 different exposures with a total time of effective observation of $135^h 45^m$. In this material we found 22 objects that can provisionally be considered as "new" flare stars and 50 outburst repetitions in previously known flare stars.

122.138 **Flare stars and rapid irregular variables in the southern Coalsack.** A. D. Andrews.
Bol. Obs. Tonantzintla y Tacubaya, Vol. 6, (No. 38), 179 - 195 (1972).
A series of multiple-exposure ultraviolet plates in the southern Coalsack has revealed 140 stars showing rapid variations in brightness on a time scale of 10 to 40 minutes. Although flare-like outbursts were observed in the majority of these objects, a rapid diminution in brightness sometimes takes place more suggestive of rapid irregular variability. The objects are between U = 14 and 17 with estimated amplitudes of 0.5 to 2 magnitudes. Five stars have shown repeated activity. The detection rate is high, 2 to 3 events per hour per square degree for multiple 10 minute exposures reaching limiting magnitude U = 17.

122.139 **Light variation of an R CrB star in Sculptor.**
P. Pişmiş.
Bol. Obs. Tonantzintla y Tacubaya, Vol. 6, (No. 38), 197 - 200 (1972).
The variation of the luminosity of the star SPC 4 estimated from photographic plates are here presented. The observational material covers an interval of about 60 years. The results confirm the previous suggestion that this blue, hot variable is of the R CrB type. A cycle of 600 days in the light variation is shown by the four last minima. The minima seem to be double. The luminosity at the maxima shows an erratic variation which may reach to 0.70 magnitudes.

122.140 **Photometrische Untersuchungen an sehr schnell variablen Objekten.** R. Schoembs.
Diss. Nat. Fak., Ludwig - Maximilians - Univ. München, 5 + 74 pp. (1971).

122.141 An analysis of the spectrum of the irregular variable
CY Cygni in the wavelength region 5000–6700 Å.
R. B. Culver.
Thesis, Ohio State Univ., Columbus. [Available from Univ.
Microfilms, Ann Arbor, Mich., USA. Order No. 72-4460],
103 pp. (1971).

The spectrum of the irregular variable CY Cygni was analyzed in detail over the wavelength range 5000–6700 Å and
element abundances relative to the solar composition star 63
Cygni were calculated.

122.142 Flare stars in the Pleiades. III.
V. A. Ambartsumyan, L. V. Mirzoyan, É. S. Parsa-
myan, O. S. Chavushyan, L. K. Erastov, E. S. Kazaryan, G.
B. Oganyan.
Byurakanskaya Astrofiz. observ., Preprint No. 5. Erevan. 30 pp.
(1972). In Russian. – Abstr. in Referativ. Zhurn. 51. Astron.,
2.51.669 (1973).

122.143 On the Blazhko effect for ST Bootis. G. A. Lange.
Astron. Tsirk., No. 720, p. 6 - 7 (1972). In Russian.

122.144 Elements of light variation of the RR Lyrae type
variables V 166 and SVS 1374 in the globular cluster
M 3. P. N. Kholopov.
Astron. Tsirk., No. 721, p. 8 (1972). In Russian.

122.145 New RR Lyrae variable SVS 1850. V. P. Goranskij.
Astron. Tsirk., No. 724, p. 8 (1972). In Russian.

122.146 Photoelectric spectrophotometry of R Lyrae.
Yu. A. Medvedev.
Astron. Tsirk., No. 725, p. 3 - 5 (1972). In Russian.

122.147 Monochromatic photometry of R Lyrae.
Yu. A. Medvedev.
Astron. Tsirk., No. 725, p. 5 - 7 (1972). In Russian.

122.148 Outburst of DH Aquilae.
G. V. Zhukov, V. Ya. Solovjev.
Astron. Tsirk., No. 729, p. 8 (1972). In Russian.

122.149 Electrophotometric observations of BD and BK
Draconis. I. F. Alaniya.
Astron. Tsirk., No. 732, p. 7 - 8 (1972). In Russian.

122.150 Flare star SVS 1849. V. P. Tsesevich.
Astron. Tsirk., No. 733, p. 7 (1972). In Russian.

122.151 Elements of seven short period variable stars.
V. P. Tsesevich.
Astron. Tsirk., No. 736, p. 7 - 8 (1972). In Russian.

122.152 Elements of four short period variable stars.
V. P. Tsesevich.
Astron. Tsirk., No. 736, p. 8 (1972). In Russian.

122.153 Cepheid variable stars. R. Lincoln.
Journ. Astron. Soc. Western Australia, 1972 October
p. 2 - 4.

On the induced amplification of synchrotron radiation in cosmic sources. See Abstr. 061.054.

Schwingungsphänomene in der Astrophysik.
See Abstr. 061.056.

Fine structure of shock waves in an RR Lyrae model
atmosphere. See Abstr. 064.020.

Hydrodynamic and radiative-transfer effects on an

RR Lyrae atmosphere. See Abstr. 064.050.

Perturbation of the radial and non-radial oscillations
of a star by a magnetic field. See Abstr. 065.036.

Oscillations of a polytrope with a toroidal magnetic
field. See Abstr. 065.038.

Resonance effects in polytropes.
See Abstr. 065.039.

Iterative nonlinear pulsations in massive stars. I. The
iterative approach. See Abstr. 065.040.

Iterative nonlinear pulsations in massive stars. II.
Terms up to second order. See Abstr. 065.041.

Extrem junge Sterne. See Abstr. 065.094.

Cepheids, presupernovae, and the $^{12}C (\alpha, \gamma)\ ^{16}O$
reaction. See Abstr. 065.097.

Comments on a PLC relationship for cepheids and
on the comparison between pulsation and evolution masses
for cepheids. See Abstr. 065.098.

Miscellaneous UBV observations of symbiotic stars.
See Abstr. 113.020.

Infrared excesses in supergiant stars: Evidence for
silicates. See Abstr. 113.039.

UBV photometry of some selected stars in the
Hyades. See Abstr. 113.047.

Comparative UBVR photometry of Orion flare stars
and H α emission-line stars. See Abstr. 113.062.

Photometric observations of AG Pegasi.
See Abstr. 113.063.

Recent changes in the nature of V1057 Cygni.
See Abstr. 114.102.

Is HR 5491 really a pulsating Am star?
See Abstr. 114.003.

Über die Veränderlichkeit der Sterne des Spektraltyps A. See Abstr. 114.106.

Titanium isotopes in Omicron Ceti.
See Abstr. 114.132.

The angular diameter of X Cancri.
See Abstr. 115.012.

The surface gravities of F and G supergiants and
classical cepheids. See Abstr. 115.021.

Eruptive binaries. Part II. See Abstr. 117.006.

The Delta Scuti stars. An annotated catalogue and
bibliography. See Abstr. 120.019.

Variable stars and the photon rest mass.
See Abstr. 121.056.

Observations of the supernova SVS 1854 and
Roberts-Altizer's variable in the vicinity of NGC 3147.
See Abstr. 125.100.

Observations of rapid blue variables—X. G61—29.
See Abstr. 126.008.

A young stellar group surrounding RCrA.
See Abstr. 131.042.

Variability of polarization in the infrared radiation
of IRC+10216 (CW Leo).   See Abstr. 131.046.

The periodical components of variations in the flux
density of the radio source VRO 42.22.01 (BL Lacertae).
See Abstr. 141.018.

Infrared observations of variable radio objects.
See Abstr. 141.032.

Photographic UBV observations of OJ 287, 3C 345,
and BL Lac.   See Abstr. 141.036.

Radio and optical activity of the peculiar object
BL Lac.   See Abstr. 141.037.

3C 120, BL Lac, and OJ 287: Coordinated multi-
wavelength observations of intraday variability.
See Abstr. 141.039.

3C 120, BL Lacertae, and OJ 287: Coordinated
optical,infrared, and radio observations of intraday variability.
See Abstr. 141.105.

Period-luminosity relation for cepheids in globular
clusters.   See Abstr. 154.004.

Errata

122.901   Erratum: "On the polarization of light of some
           variable stars in the Orion nebula" [Peremennye
Zvezdy, Vol. 18, 131 - 139 (1971). In Russian.]
V. S. Shevchenko, V. I. Kardopolov.
Peremennye Zvezdy, Vol. 18, 317 (1972).

123 Variable Stars: Lists of Observations, Individual Observations

123.001 **Photographic brightness estimates of variables SVS 1564 and SVS 1730 in Taurus.** O. N. Deynichenko.
Peremennye Zvezdy, Prilozhenie, Vol. 1, 151 - 155 (1971). In Russian.

123.002 **SVS 1635 Tauri.** A. N. Kulapova.
Peremennye Zvezdy, Prilozhenie, Vol. 1, 157 - 160 (1971). In Russian.

123.003 **SS Andromedae.** B. L. Shaganyan.
Peremennye Zvezdy, Prilozhenie, Vol. 1, 161 - 164 (1971). In Russian.

123.004 **BE Andromedae.** B. L. Shaganyan, E. P. Stanika.
Peremennye Zvezdy, Prilozhenie, Vol. 1, 165 - 168 (1971). In Russian.

123.005 **Elements of light variations of three variables.** P. N. Kholopov.
Peremennye Zvezdy, Prilozhenie, Vol. 1, 173 - 174 (1971). In Russian.

123.006 **GR Cassiopeiae.** V. P. Ryadchenko.
Peremennye Zvezdy, Prilozhenie, Vol. 1, 191 - 195 (1971). In Russian.

123.007 **V866 Aquilae.** V. P. Tsesevich.
Peremennye Zvezdy, Prilozhenie, Vol. 1, 223 - 225 (1971). In Russian.

123.008 **Investigation of variable stars in Aquila. I.** V. P. Tsesevich.
Peremenny Zvezdy, Prilozhenie, Vol. 1, 235 - 248 (1971). In Russian.

123.009 **Studio sulle stelle variabili in due campi stellari nelle costellazioni della Lacerta e della Lira.** G. Romano, M. Perissinotto.
Mem. Soc. Astron. Italiana, Nuova Ser., Vol. 43, 319 - 338 (1972).
This paper gives the results of the photographic observations of 34 variable stars in two fields around 1 Lac and $18^h 50^m + 36°$ (Lyra). Seven new variable stars have been discovered during this research, one of these (FV Lac) has been already reported in the General Catalogue of Variable Stars.

123.010 **Amateurs observe CH Cygni.** J. E. Isles.
Sky Telescope, Vol. 44, 204 (1972).

123.011 **R Coronae Borealis.** G. Zaffi.
Coelum, Vol. 40, 165 (1972).

123.012 **Photoelectric observations of Wr 16.** G. V. Zhukov.
Astron. Tsirk., No. 668, p. 7 - 8 (1972). In Russian.

123.013 **The variable V154 in the globular cluster M3.** P. N. Kholopov.
Astron. Tsirk., No. 676, p. 7 - 8 (1972). In Russian.

123.014 **Observations of FG Sagittae in 1971.** V. P. Arhipova.
Astron. Tsirk., No. 679, p. 1 - 3 (1972). In Russian.

123.015 **Eruptive variable or intergalactic supernova observed.**
Astron. Tsirk., No. 690, p. 1 (1972). In Russian.

123.016 **Possible new flare star SVS 1753.** A. S. Sharov, A. K. Alksnis.
Astron. Tsirk., No. 694, p. 8 (1972). In Russian.

123.017 **New eruptive variable star in Ursa Major, SVS 1755.** V. P. Goranskij.
Astron. Tsirk., No. 696, p. 1 - 2 (1972). In Russian.

123.018 **Observations of RY Sgr.**
Astron. Tsirk., No. 697, p. 1 (1972). In Russian.

123.019 **VZ Tauri — an RW Aurigae type star.** V. P. Tsesevich, B. A. Dragomiretskaya.
Astron. Tsirk., No. 697, p. 8 (1972). In Russian.

123.020 **New eruptive object.**
Astron. Tsirk., No. 700, p. 2 (1972). In Russian.

123.021 **Interesting variable star in the surrounding of the globular cluster M92 (NGC 6341).** B. V. Kukarkin.
Astron. Tsirk., No. 709, p. 7 - 8 (1972). In Russian.

123.022 **Investigation of variables WR-96, GR-29 and WR-96(2).** G. U. Kovalchuk.
Astrometriya i Astrofiz., *Kiev*, No. 15, (see 003.001), p. 39 - 46 (1972). In Russian.
m_{pg} and m_{pv} magnitudes for three variables were obtained from the plates of a sky survey at the Main Astronomical Observatory of the Ukrainian Academy of Sciences. Variability of WR-96(2) was detected for the first time. Distribution functions of brightness are considered.

123.023 **Photographic observations of variable stars in the vicinity of NGC 6830.** B. L. Shaganian.
Astrometriya i Astrofiz., *Kiev*, No. 15, (see 003.001), p. 46 - 55 (1972). In Russian.
Photographic observations of stars OW, OR, V 433, V 768 and V 1086 Aquilae are given. New elements are obtained for OW and V 1086 Aql. Cycles of light changes in OR Aql are determined more accurately. V 433 Aql is found to have two oscillation cycles. Charts of the variables, photographic magnitudes of comparison stars and epochs of light maxima are given. Photographic light curves are obtained.

123.024 **Observations of variable stars, January – June 1972. Report No. 22.** L. Plaut, H. Feijth.
Nederlandse Vereniging voor Weer- en Sterrenkunde. Kapteyn Astron. Lab., Groningen — Netherlands. 3 + 5 pp. (1972).
This report gives 1810 visual observations of 138 variable stars, 1972 January – June.

123.025 **MU Cas — not an RRs star.** W. Wenzel.
Inform. Bull. Variable Stars (I.A.U. Commission 27), Konkoly Obs., Budapest, No. 701 (1972).

123.026 **New southern variable stars.** D. H. Martins.
Inform. Bull. Variable Stars (I.A.U. Commission 27), Konkoly Obs., Budapest, No. 705, 3 pp. = Veröff. Remeis-Sternw. Bamberg, Astron. Inst. Univ. Erlangen-Nürnberg, Vol. 10, No. 103 (1972).

123.027 **Photoelectric secondary minima of AK Her.** D. J. Killian, T. W. Edwards.
Inform. Bull. Variable Stars (I.A.U. Commission 27), Konkoly Obs., Budapest, No. 710 (1972).

123.028 **Note on three new red variables.**
P. N. J. Wisse, M. Wisse.
Inform. Bull. Variable Stars (I.A.U. Commission 27), Konkoly Obs., Budapest, No. 718 (1972).

123.029 **Photoelectric observations of V 1057 and V 1329 Cyg.** H. Bossen.
Inform. Bull. Variable Stars (I.A.U. Commission 27), Konkoly Obs., Budapest, No. 722 (1972).

123.030 **Variability of BD −10°4662 confirmed.**
D. Hoffleit.
Inform. Bull. Variable Stars (I.A.U. Commission 27), Konkoly Obs., Budapest, No. 724 (1972).

123.031 **New results on known variables in Sagittarius.**
D. Hoffleit.
Inform. Bull. Variable Stars (I.A.U. Commission 27), Konkoly Obs., Budapest, No. 726, 4 pp. (1972).

123.032 **Eight new long period variables in Sagittarius.**
D. Hoffleit.
Inform. Bull. Variable Stars (I.A.U. Commission 27), Konkoly Obs., Budapest, No. 729 (1972).

123.033 **Het minimum van R Coronae Borealis in de eerste helft van 1972.** H. Feijth.
Hemel en Dampkring, Vol. 70, 265 - 266 (1972).

123.034 **SS Cygni 1971 – og en opfordring.** O. Klinting.
Astron. Tidssk., Årg. 5, p. 142 (1972).

123.035 **SY Aurigae.** O. Klinting.
Astron. Tidssk., Årg. 5, p. 142 - 144 (1972).

123.036 **Oversigt over Astronomisk Selskabs observationer af variable stjerner 1970.0–1972.0.** O. Klinting.
Astron. Tidssk., Årg. 5, p. 145 - 146 (1972).

123.037 **V 465 (Proxima) Centauri.**
B. F. Marino, W. S. G. Walker.
Southern Stars, Vol. 24, 130 = Auckland Obs. Observational Note No. 5 (1972).

123.038 **New variable stars in the Large Magellanic Cloud.**
A. D. Andrews, D. J. Mullan.
Irish Astron. Journ., Vol. 10, 149 - 150 (1971).

123.039 **VV 281 - 427, variable stars in a Cepheus–Lacerta field of the Milky Way.**
W. J. Miller, A. A. Wachmann.
Ric. Astron., Specola Vaticana, *Castel Gandolfo*, Vol. 8,(No. 12), 211 - 309 (1971).

Nearly 150 variable stars, mostly new variables discovered by Miller on his Castel Gandolfo plates some twenty years ago, have been processed on these same plates by Wachmann at the Hamburg Observatory. Eleven tables of data on thirty-one pages, twenty-six pages of identification charts, thirteen pages of light curves, and notes on each variable star summarize the results.

123.040 **New variable stars in Scorpius.** V. Satyvaldiev.
Astron. Tsirk., No. 711, p. 7 (1972). In Russian.

123.041 **BS Tauri.** Yu. A. Fadejev.
Astron. Tsirk., No. 714, p. 8 (1972). In Russian.

123.042 **Photographic observations of HZ Her.**
N. E. Kurochkin.
Astron. Tsirk., No. 717, p. 1 - 3 (1972). In Russian.

123.043 **R Leonis 094212.** G. Zaffi.
Coelum, Vol. 40, 238 (1972).

123.044 **γ Cassiopeiae 005561 (1972).** G. Zaffi.
Coelum, Vol. 40, 238 - 239 (1972).

123.045 **Variable star notes.** M. W. Mayall.
Journ. Roy. Astron. Soc. Canada, Vol. 66, 233 - 236, 285 - 288, 325 - 328 (1972).

123.046 **The photometric activity of RU Cam during the years 1970−72.**
P. Broglia, G. Guerrero.
Inform. Bull. Variable Stars, (IAU Commission 27), Konkoly Obs., Budapest, No. 732 (1972).

123.047 **S 10764 – a slowly variable object in the globular cluster M3 with $U−B \approx -1^m0$.** L. Meinunger.
Inform. Bull. Variable Stars, (IAU Commission 27), Konkoly Obs., Budapest, No. 738 (1972).

123.048 **A non-existent suspected variable.**
W. P. Bidelman, W. F. van Altena.
Inform. Bull. Variable Stars, (IAU Commission 27), Konkoly Obs., Budapest, No. 744 (1972).

123.049 **Notes on BV 1481 = AA Ceti.** R. H. Bloomer.
Inform. Bull. Variable Stars, (IAU Commission 27), Konkoly Obs., Budapest, No. 745, 3 pp. = Rosemary Hill Obs., Univ. Florida, Gainesville, Florida, Contr. No. 34 (1972).

123.050 **HBV 479 - 495, variables in a field around SA 18.**
A. A. Wachmann.
Inform. Bull. Variable Stars, (IAU Commission 27), Konkoly Obs., Budapest, No. 749, 12 pp. (1972).

123.051 **12 new variable stars in NGC 6402.**
A.Wehlau, N. Potts.
Inform. Bull. Variable Stars, (IAU Commission 27), Konkoly Obs., Budapest, No. 752 (1972).

123.052 **Observations of variable stars in 1967.**
Mem. Japan Astron. Study Ass., No. 17, Vol. 5, 9 - 41 (1971).

123.053 **Observations of variable stars in 1968.**
Mem. Japan Astron. Study Ass., No. 18, Vol. 5, 109 - 146 (1972). In Japanese.

123.054 **Observations of miscellaneous types of variables, 22 October 1961 - 21 September 1963, J.D. 2,437,595 - 2,438,294.** M. W. Mayall.
American Ass. Variable Star Observers, *Cambridge, Mass.*, AAVSO Rep. 29, 84 pp. (1972).

This report is based on approximately 61,000 observations made by 396 observers. The variables contained in the report are listed with pertinent information about each. An index by constellation concludes the volume. Individual observations are preserved on data cards and on a computer print-out. The means and number of observations used in each are also on a print-out. Copies are available at cost.

123.055 **Eruptive object in Ursa Major.** C. Bertaud.
IAU Circ., No. 2423 (1972).

123.056 **R CrB variables.**
V. L. Matchett, M. Mayall, J. E. Isles, W. E. Pennell.
IAU Circ., No. 2460 (1972).

123.057 **SS Cygni.** C. Scovil, W. Lowder.
IAU Circ., No. 2462 (1972).

123.058 **SU Tauri.**   J. E. Isles, W. E. Pennell, P. A. Moore.
IAU Circ., No. 2467 (1972).

123.059 **Photoelectric observations at Auckland Observatory.**
B. F. Marino, W. S. G. Walker.
Roy. Astron. Soc. New Zealand, Variable Star Section, Circ.
No. 184, 7 pp. (1971).
  Three-colour photoelectric observations made at the
Auckland Observatory of selected variables and suspected
variables are given.

123.060 **UW Centauri.**   F. M. Bateson, A. F. Jones.
Roy. Astron. Soc. New Zealand, Variable Star Section, Circ. No. 185, 4 pp. (1972).
  A discussion of visual observations of UW Cen from J. D.
2,435,106 to 2,440,832 shows that it is a typical R CrB type
variable with deep minima distributed at entirely random intervals.

123.061 **Observations of southern variable stars R. A. 00hrs
to 06hrs,** 1967 July 1 — 1970 December 31, J.D.
2,439,673 — 2,440,952.   F. M. Bateson.
Roy. Astron. Soc. New Zealand, Variable Star Section, Circ.
No. 186, 14 pp. (1972).

123.062 **Veränderliche in einem Feld um M 3 auf Tautenbur-
ger Schmidt-Aufnahmen.**   L. Meinunger.
MVS, *Sonneberg,* Vol. 6, 37 - 43 (1972).
  On plates of M 3 taken with the Tautenburg 135/200/400
cm Schmidt telescope 8 new variables (S 10758 — S 10765),
namely 4 RR Lyrae stars and 4 slowly variable uncoloured
objects, have been found. The slowly variable object S 10764
probably belongs to M 3. The period of XX CVn has been determined.

123.063 **UX Canum Venaticorum.**   W. Wenzel.
MVS, *Sonneberg,* Vol. 6, 43 (1972).

123.064 **9 Mira-Sterne vom Feld β Apodis.**   H. Geßner.
MVS, *Sonneberg,* Vol. 6, 49 - 51 (1972).

123.065 **S 5784 Normae und S 10766 Normae.**
C. Thänert.
MVS, *Sonneberg,* Vol. 6, 51 (1972).

123.066 **S 9313 Lyrae — ein veränderliches extragalaktisches
Objekt?**   C. Thänert.
MVS, *Sonneberg,* Vol. 6, 52 (1972).
  The light curve of S 9313 resembles that of BL Lac. Spec-
trographic observations are recommended.

123.067 **Beobachtungen von Mirasternen im Feld η Ara.**
I. Meinunger.
MVS, *Sonneberg,* Vol. 6, 52 (1972).

123.068 **Photographische Beobachtungen von Veränderlichen
auf Platten der Sonneberger Himmelsüberwachung.**
E. Splittgerber.
MVS, *Sonneberg,* Vol. 6, 53 - 55 (1972).

123.069 **R Coronae Borealis.**   H.-J. Blasberg.
MVS, *Sonneberg,* Vol. 6, 56 (1972).

123.070 **AC Herculis.**   H.-J. Blasberg.
MVS, *Sonneberg,* Vol. 6, 57 (1972).
  Dates of 18 primary minima and the mean light curve
for 1958 to 1969 are given.

123.071 **AG Canis Minoris.**   W. Zschocke.
MVS, *Sonneberg,* Vol. 6, 58 - 59 (1972).
  New mean elements with $P = 1\overset{d}{.}6645567$ were derived
from observations on Sonneberg patrol plates; the variability
of the period was confirmed.

123.072 **RU Camelopardalis 1967 bis 1971.**   W. Zschocke.
MVS, *Sonneberg,* Vol. 6, 59 - 60 (1972).

123.073 **Maxima des Mirasterns R UMa und des RR-Lyrae-
Sterns AA CMi.**   S. Bratner.
MVS, *Sonneberg,* Vol. 6, 60 (1972).

123.074 **ε Pegasi.**   R. J. Wood, B. G. Marsden.
British Astron. Ass., Circ. No. 544 (1972).

123.075 **CVS 2662 – a unique star?**   V. Satyvaldiev,
V. M. Grigorevsky, S. S. Vykhrestyuk.
Astron. Tsirk., No. 723, p. 3 - 6 (1972). In Russian.

123.076 **Photoelectric observations of FG Sge.**
G. V. Zhukov.
Astron. Tsirk., No. 728, p. 7 - 8, with a correction, No. 733,
p. 8 (1972). In Russian.

123.077 **UBV observations of the star V 1057 Cyg.**
N. N. Kiselev.
Astron. Tsirk., No. 742, p. 7 - 8 (1972). In Russian.

123.078 **Visual observations of YY Ursae Majoris.**
H. Honda.
Yamamoto Circ., No. 1751, p. 3 - 4 (1972). In Japanese.

123.079 **Prediction of minima of RV Tau type variables.**
Japan Astron. Study Ass., Circ. 276 (1972). In
Japanese.

Identification of the CSV 6150 with a galaxy.
See Abstr. 158.116.

Errata

123.901 **Erratum: "Photoelectric observations of Y Cygni
near minima"** [Astron. Tsirk., No. 662, p. 1 - 5
(1971). In Russian].   G. V. Zaitseva, V. M. Lyutyj, D. Ya.
Martynov.
Astron. Tsirk., No. 679, p. 8 (1972).

123.902 **Errata: "Photographic brightness estimates of
variables SVS 1564 and SVS 1730 in Taurus"**
[Peremennye Zvezdy, Prilozhenie, Vol. 1, 151 - 156 (1971)].
In Russian.   O. N. Deynichenko.
Astron. Tsirk., No. 733, p. 8 (1972). In Russian. — See Abstr.
08.123.001.

124 Novae

124.001 Analysis of the chemical composition of the envelopes of novae. E. R. Mustel, L. I. Antipova.
Nauchn. Informatsii, vyp. (No.) 19, p. 32 - 40 (1971).
In Russian.

The results of investigations of the chemical composition of the envelope of DQ Her using the method of the curve of growth for different moments are summarized. The conclusion is drawn that the envelope is very inhomogeneous. The possibility of applying the method of the curve of growth for the envelopes of novae is discussed; it is shown, that its application is justified in our case.

124.002 The non-spherical nebulae of nova Delphini 1967, nova Vulpeculae 1968 (I), and nova Serpentis 1970.
J. B. Hutchings.
Monthly Notices Roy. Astron. Soc., Vol. 158, 177 - 198 (1972).

The non-spherical distribution of ejected matter in the nebular stages of novae is discussed in terms of the assumed binary nature of the central stars. An hypothesis is proposed, by which the non-spherical distribution may be formed, and which may explain most of the general features of the nova phenomenon. The nature of line profiles formed by non-spherical ejecta is discussed and illustrated. New spectrophotometric observations are presented of the later stages of three recent novae, and models are derived for their postmaximum and nebular stages, in terms of the foregoing discussion.

124.003 CNO abundances and hydrodynamic models of the nova outburst.
S. Starrfield, J. W. Truran, W. M. Sparks, G. S. Kutter.
Astrophys. Journ., Vol. 176, 169 - 176 (1972).

We have used a fully implicit, Lagrangian, hydrodynamic computer code incorporating a nuclear reaction network to follow thermonuclear runaways in the hydrogen-rich envelopes of white dwarfs in order to produce a nova outburst. Our models have ejected $1.7 \times 10^{-4} M_{\odot}$ with kinetic energies of 8×10^{44} ergs a value that agrees quite closely with the observed values for novae. We have varied the initial abundance of the CNO nuclei and find that the actual ejection of material and production of an outburst requires these nuclei to be strongly enhanced and that observed features of the outburst depend on the degree of CNO enhancement.

124.004 On the novae S 10735 and S 10753 in the region of the Andromeda nebula.
A. K. Alksnis, A. S. Sharov.
Astron. Tsirk., No. 692, p. 7 - 8 (1972).   In Russian.

124.005 The state of ionization in nova shells. I. The ionization equation for hydrogen shells.   P. B. Bosma.
Astron. Astrophys., Vol. 21, 223 - 229 (1972).

The ionization equation for a spherically symmetric hydrogen shell illuminated by a central star is derived. An approximate solution of this equation quantitatively shows that for optically thick shells nearly all photons arising from recombinations are reabsorbed, whereas for optically thin shells nearly all these photons escape. For shells of intermediate optical thickness the fraction of photons arising from recombinations escaping from the shell can be calculated relatively easily with the help of the correction functions introduced in this paper.

124.006 Nova in Large Magellanic Cloud.
F. M. Bateson.
Inform. Bull. Variable Stars (I.A.U. Commission 27), Konkoly Obs., Budapest, No. 725 (1972).

124.007 On the outburst of cataclysmic variables.
B. Warner.
Monthly Notices Roy. Astron. Soc., Vol. 160, 35P - 36P (1972).

It is proposed that outbursts of classical, recurrent and dwarf novae originate in rotating white dwarf stars. The consequent excitation of non-radial oscillations leads to a satisfactory understanding of the non-spherical ejection of nova shells.

124.008 Probable nova in Large Magellanic Cloud.
J. A. Graham.
IAU Circ., No. 2441 (1972).

124.009 Nova in Large Magellanic Cloud.
J. A. Graham, G. Araya.
IAU Circ., No. 2444 (1972).

124.010 Nova in Large Magellanic Cloud.
A. C. Gilmore, P. J. Andrews.
IAU Circ., No. 2445 (1972).

124.011 Nova in Large Magellanic Cloud.
A. C. Gilmore, R. E. Millington, G. A. Chapman.
IAU Circ., No. 2449 (1972).

124.012 Nova in Large Magellanic Cloud.
A. C. Gilmore, P. M. Kilmartin.
IAU Circ., No. 2451 (1972).

124.013 Nova in Large Magellanic Cloud.   J. A. Graham.
IAU Circ., No. 2461 (1972).

On thermal waves in stars.   See Abstr. 065.037.

Observations of rapid blue variables—XI. DQ Herculis.
See Abstr. 121.033.

Ultrashort-period binaries. II. HZ 29 (= AM CVn): a double-white-dwarf semidetached postcataclysmic nova?
See Abstr. 126.002.

Optical studies of Uhuru sources. II. IM Normae, a possible X-ray source.   See Abstr. 142.003.

124.100 Nova Cephei 1971

Nova Cephei 1971.   A. Alksnis, L. Dunzans.
Astron. Tsirk., No. 681, p. 4 - 6 (1972). In Russian.

Photoelectric UBV measures of nova Cephei 1971.
D. J. MacConnell, J. C. Thomas.
Inform. Bull. Variable Stars (I.A.U. Commission 27), Konkoly Obs., Budapest, No. 706 (1972).

Spectrophotometry of nova Cephei 1971.
J. D. R. Bahng.
Monthly Notices Roy. Astron. Soc., Vol. 158, 151 - 158 (1972).

Photoelectric scanner measurements were made on nova Cephei 1971 on eight nights from 1971 July 29 to November 1. The results are presented in detail.

124.101 Nova Vulpeculae 1968 No. 1

**Photoelectric observations of nova Vulpeculae 1968
(1) = LV Vul.**   T. Z. Dworak, M. Winiarski.
Acta Astron., Vol. 22, 33 - 40 (1972).

The brightness of the star nova Vul 1968 (1) = LV Vul
was observed photoelectrically in two colours by means of a
20-cm refractor from April 27 to September 4, 1968. 128
measurements were made in yellow and 126 in blue colour.
The results were reduced to the BV system, and compared with
observations obtained by other authors. The exact position of
nova Vul 1968 (1) was also determined.

**Photoelectric light curve of nova Vulpeculae 1968
N. 1.**   P. Tempesti.
Astron. Astrophys., Vol. 20, 63 - 68 = Contr. Oss. Astron.
Teramo, Ser. 2, No. 2 (1972).

115 photoelectric observations in V light performed at
the Teramo Observatory together with the observations so far
published by other observers yield the light curve of nova
Vulpeculae 1968 N. 1 from the outburst to August 1970.

LV Vulpeculae (nova 1968 No. 1) in 1968–71.
J. E. Isles.
Journ. British Astron. Ass., Vol. 83, 44 - 49 (1972). – Report
of the Variable Star Section of the British Astron. Ass.

124.102 Nova DQ Herculis

**On the origin of a very close similarity between the
spectra of the type I supernova in NGC 3198 and the absorption spectrum of DQ Her.**   See Abstr. 125.101.

124.103 Nova Serpentis 1970

**Preliminary report on the infrared spectrum of
nova Serpentis 1970.**   F. Ciatti, A. Mammano.
Atti Accad. Nazionale Lincei, Ser. Ottava, Rend. Cl. Sci. fis.,
mat., nat., Vol. 52, 62 - 71 = Contr. Oss. Astrofis. Asiago,
No. 260 (1972).

124.104 Nova VW Hydri

An outburst of VW Hydri.
B. F. Marino, W. S. G. Walker.
Southern Stars, Vol. 24, 126 - 129 = Auckland Obs. Observa-
tional Note No. 4 (1972).

124.105 Nova Herculis 1960

Photometric features of N Her 1960.
Sh. G. Gordeladze, L. A. Kudryavtseva.
Problems of cosmic physics. Vyp. (No.) 7, (see 003.028), p.
123 - 128 (1972). In Russian.

125 Supernovae, Supernova Remnants

125.001 Absorption spectra of type II supernovae.
B. Patchett, D. Branch.
Monthly Notices Roy. Astron. Soc., Vol. 158, 375 - 382
(1972).

On the bases of wavelength coincidence and comparison
of observed and synthetic spectra, intensity minima in type II
supernova spectra are attributed to blueshifted absorption
lines of hydrogen and Fe II.

125.002 Supernova remnants in the Large Magellanic Cloud.
D. K. Milne.
Astrophys. Letters, Vol. 11, 167 - 171 (1972).

Confirmatory radio observations are given for 11 possible
supernova remnants in the Large Magellanic Cloud. These
sources are over-luminous compared with remnants in the
Galaxy and it is suggested that they have evolved from the
early-type supergiants so abundant in the Large Cloud.

125.003 Observations of the Cygnus Loop at 6-cm wavelength.
M. R. Kundu, R. H. Becker.
Astron. Journ., Vol. 77, 459 - 463, 527 (1972).

Brightness and polarization distributions of the Cygnus
Loop have been measured at 6-cm wavelength. The derived
magnetic-field distribution is found to be tangential to the
southern boundary of the source, in agreement with previously
published results.

125.004 Spectra of type I supernovae. I. Generalities.
C. Gordon (Pecker-Wimel).
Astron. Astrophys., Vol. 20, 79 - 85 (1972).

Preliminary study of the physical conditions existing in a
supernova shell: mechanism of formation of the emission and
absorption lines, construction of a density model and determination of the electronic density at the time of the maximum
of luminosity. Review of the possible mechanisms of heating
and ionization in the shell with a description of Sobolev's
method used to compute the line intensities. A tentative identification of the lines present in the spectra 20 days, 79 days
and 275 days after the maximum of luminosity is proposed.

125.005 Spectra of type I supernovae. II. Interpretation of
the spectra of SN 1960 in NGC 4496 from 8 to 20
days after the maximum of luminosity.
C. Gordon (Pecker-Wimel).
Astron. Astrophys., Vol. 20, 87 - 98 (1972).

Study of the spectra of type I supernovae in an early
stage of evolution. Identification of the absorption lines and
construction of an empirical model showing the stratification
of ionization. Identification of some emission lines as forbidden coronal lines. Computations of their intensities in the two
cases of collisional or radiative ionization in the shell. Study of
the radiative ionization of the shell and approximative determination of T_e, of some abundances and of the intensity of
the ionizing radiation in the case of a bremsstrahlung or synchrotron radiation.

125.006 Brightness and polarization structure of four super-
nova remnants 3C58, IC443, W28, and W44 at
2.8 centimeter wavelength.   M. R. Kundu, T. Velusamy.
Astron. Astrophys., Vol. 20, 237 - 244 (1972).

Distributions of brightness and polarization over four supernova remnants have been measured at 2.8 centimeter wavelength with a beam of 3'. Distributions of rotation measure over these sources have been computed by comparing the present data with the data at longer wavelengths. Significantly large variations of rotation measure over all the four sources have been observed. The rotation measure varies systematically over the sources 3C58 and W44. Possible origin of the Faraday rotation entirely in the interstellar medium outside the remnants is discussed.

125.007 Observational evidence against supernovae being the source of the universal X-ray background.
J. E. Mack, D. E. Robbins.
Astrophys. Journ., Vol. 176, 99 - 101 (1972).

An analysis has been made of the theory that the universal X-ray background originates in supernovae. Observational evidence relating to such theories has been critically examined. It is concluded that evidence presently exists which can be interpreted to exclude such theories of the origin of the background radiation, regardless of the timescale of the X-ray emission from a supernova.

125.008 A study of galactic supernova remnants. II. Supernova rate, galactic radio emission and pulsars.
S. A. Ilovaisky, J. Lequeux.
Astron. Astrophys., Vol. 20, 347 - 356 (1972).

Using the radio luminosity function at 1 GHz for galactic supernova remnants (SNR) derived in a previous article together with independent age estimates for six SNR, we obtain a mean interval between supernova outbursts in the Galaxy of 50 yr. The total number of discrete remnants in our Galaxy is presented as a function of the linear diameter, D. The surface density of supernova remnants is compared to that of radio pulsars. We find that the surface density of pulsars, within a factor of two, is equal to that of SNR. This, plus the fact that the z-distributions of the two types of objects are quite similar, shows that it is possible that most supernova explosions have given birth to pulsars. A new model is presented for the non-thermal galactic radio emission, based on the 150 MHz sky map of Landecker and Wielebinski (1970).

125.009 New method for estimating the Hubble constant.
S. van den Bergh.
Astron. Astrophys., Vol. 20, 469 - 470 (1972).

Radio observations of galactic supernova remnants indicate that the mean interval between supernova outbursts in the Galaxy is 50 ± 25 y. Comparison of this value with Tammann's determination of the frequency of supernovae in distant spirals yields $H = 95 ± 36$ km s^{-1} Mpc^{-1}.

125.010 New supernova in an anomalous galaxy.
Astron. Tsirk., No. 676, p. 1 (1972). In Russian.

125.011 Polarized radio emission from five supernova remnants. D. K. Milne.
Australian Journ. Phys., Vol. 25, 307 - 313 (1972).

Maps are presented of the polarization and total power emission at 2700 MHz from five southern supernova remnants: MSH 09−32, 10−53, 14−63, 15−52, and 15−56. Observations of 14−63 and 15−56 at 1410 MHz are also presented.

125.012 High resolution observations of supernova remnants at 80 MHz. J. R. Dickel.
Bull. American Astron. Soc., Vol. 4, 319 (1972). − Abstr. AAS.

125.013 Recent high resolution observations of supernova remnants at 11 cm wavelength.
A. G. Willis, J. R. Dickel.
Bull. American Astron. Soc., Vol. 4, 319 (1972). − Abstr. AAS.

125.014 Supernovae and sources of X-radiation.
I. S. Shklovsky.
Astron. Zhurn. Akad. Nauk. SSSR, Vol. 49, 913 - 920 (1972). In Russian. English translation in Soviet Astron. AJ, Vol. 16, No. 5.

It is shown that the main part of X-ray radiation of Cas A is thermal radiation of a hot non-isothermal plasma situated behind the shock wave. We draw attention to the "isotropical" component of optical and X-radiation of pulsar NP0532. The X radiation of the Tycho supernova remnant can be explained as isotropical component of pulsar radiation. We derive approximate formulae for the evolution of pulsar synchrotron radiation. By means of these formulae the possibility of optical observations of Tycho's pulsar is discussed. It is shown that X-ray radiation from young pulsars can explain the light curve of type I supernovae.

125.015 Type I supernovae and the Hubble constant.
D. Branch, B. Patchett.
Bull. American Astron. Soc., Vol. 4, 340 (1972). − Abstr. AAS.

125.016 Supernova in an anonymous galaxy. L. Detre.
Astron. Tsirk., No. 706, p. 1 (1972). In Russian.

125.017 The emission-line spectrum of N 49, a supernova remnant in the Large Magellanic Cloud.
D. E. Osterbrock, R. J. Dufour.
Bull. American Astron. Soc., Vol. 4, 398 (1972). − Abstr. AAS.

125.018 The absorption by the interstellar medium of 80 MHz radio emission from galactic supernova remnants.
G. A. Dulk, O. B. Slee.
Australian Journ. Phys., Vol. 25, 429 - 441 (1972).

High resolution 80 MHz observations of 20 galactic supernova remnants have been made with the Culgoora radioheliograph. More than half of the sources have an unexpectedly low flux density at 80 MHz, probably as a result of free−free absorption taking place in the inner arms of the Galaxy. The large value for the electron density seems to require a greater ionizing flux than that which occurs in the solar vicinity.

125.019 Observations of soft X-rays: Two supernova remnants in the constellation Lupus and the diffuse background. T. M. Palmieri, G. A. Burginyon, R. W. Hill, J. K. Scudder, F. D. Seward.
Astrophys. Journ., Vol. 177, 387 - 393 (1972).

X-rays in the energy range 0.6−1.6 keV have been detected from the Lupus loop and possibly from the remnant of the supernova of 1006 A.D. At lower energies, 0.2−0.6 keV, the sources are not evident, but spatial variations in the diffuse background are such that the flux varies by a factor of 2 within 8° on the sky.

125.020 Supernova remnants. L. Woltjer.
Annual Rev. Astron. Astrophys., Vol. 10, (see 003.005), 129 - 158 (1972). − Contents: (1) Introduction; (2) Catalogs of SNR; (3) Description of individual remnants; (4) The $\Sigma - R$ diagram; (5) Hydrodynamic evolution of the supernova remnants; (6) The filamentary shells; (7) Radio evolution; (8) X-ray emission; (9) Galactic effects.

125.021 Classification of supernova remnants and HII regions from their recombination line emission.
J. R. Dickel, D. K. Milne.
Australian Journ. Phys., Vol. 25, 539 - 544 = Separate print Div. Radiophys., C.S.I.R.O., Sydney (1972).

H109α recombination line observations are used in an attempt to classify 46 galactic radio sources as either supernova remnants or HII regions. Long integrations at the H109α line frequency on two well-known supernova remnants (IC

443 and 3C391) provide improved upper limits on the line emission from these objects. From these results the electron temperature in IC 443 is estimated to be in excess of 1.6×10^4 K.

125.022 **Methods to estimate the contribution of supernovae to the activity of quasars and nuclei of galaxies.**
L. M. Ozernoy.
Astron. Tsirk., No. 712, p. 1 - 3 (1972). In Russian.

125.023 **Cooling and evolution of a supernova remnant.**
D. P. Cox.
Astrophys. Journ., Vol. 178, 159 - 168 (1972).

The structure, evolution, and cooling of an old supernova remnant are reviewed and elaborated in order to provide a theoretical framework to relate remnants of different ages but with similar energies and environments.

125.024 **Pulsars and the evolution of supernova remnants.**
G. Setti, L. Woltjer.
Astrophys. Journ., (*Letters*), Vol. 178, L17 - L19 (1972).

If pulsars are responsible for the particle acceleration in the remnants and if the main energy loss is due to the expansion, then the radio flux of young supernova remnants is expected to be well below that obtained by extrapolating the flux-radius relation of older remnants to small radii. The periods of the pulsars would be comparable to those found in the Crab and Vela.

125.025 **Rotation and the occurrence of carbon detonation in presupernova models of intermediate masses.**
I.-J. Sackmann, V. Weidemann.
Astrophys. Journ., Vol. 178, 427 - 431 (1972).

Degenerate C/O cores of stars between 3.5 and $7 M_\odot$ evolve qualitatively different if they retain a considerable fraction of their original main-sequence angular momentum J. Carbon detonation either is prevented or, with smaller J, will occur at core masses above the Chandrasekhar limit. The correspondingly increasing luminosity will terminate this phase of evolution by mass loss.

125.026 **Ejection of the envelopes of supernovae by means of magnetic pumping.**
P. R. Amnuel, O. H. Guseinov, F. K. Kasumov.
Astron. Zhurn. Akad. Nauk SSSR, Vol. 49, 1139 - 1147 (1972). In Russian. English translation in Soviet Astron. AJ, Vol. 16, No. 6.

A possibility of explaining explosions of supernovae of type I on account of rotational energy of a formed neutron star is examined. The envelope acquires the kinetic energy of expansion as a result of the pressure of a multiply turned magnetic field that originates between the neutron star and the envelope.

125.027 **A supernova remnant in the Small Magellanic Cloud.**
D. S. Mathewson, J. N. Clarke.
Astrophys. Journ., (*Letters*), Vol. 178, L105 - L107 (1972).

A combination of radio and optical techniques has been used to identify a supernova remnant in the emission region N 19 in the bar of the Small Magellanic Cloud. This is the first supernova remnant to be discovered in that galaxy.

125.028 **Production of p-process nuclei during explosive carbon and oxygen burning.**
W. M. Howard, S. E. Woosley.
Bull. American Astron. Soc., Vol. 4, 412 (1972). – Abstr. AAS.

125.029 **X-ray observations of Tycho's supernova remnant.**
D. McCammon, A. N. Bunner, P. L. Coleman,
W. L. Kraushaar, F. O. Williamson.
Bull. American Astron. Soc., Vol. 4, 413 (1972). – Abstr. AAS.

125.030 **Supernova in anonymous galaxy.**   J. P. Huchra.
IAU Circ., No. 2447 (1972).

125.031 **Probable supernova in anonymous galaxy.**
L. Pigatto.
IAU Circ., No. 2453 (1972).

125.032 **Probable supernova in anonymous galaxy.**
D. Deming, J. C. Webber, K. Yoss.
IAU Circ., No. 2458 (1972).

125.033 **Probable supernova in anonymous galaxy.**
L. Rosino, L. Pigatto.
IAU Circ., No. 2464 (1972).

125.034 **The 1971 Palomar supernova search.**
C. T. Kowal, W. L. W. Sargent, L. Searle, F. Zwicky.
Publ. Astron. Soc. Pacific, Vol. 84, 844 - 849 (1972).

125.035 **A high sensitivity search for X-rays from supernova remnants in Aquila.**   D. A. Schwartz, D. A. Bleach,
E. A. Boldt, S. S. Holt, P. J. Serlemitsos.
Rep. NASA-TM-X-65825, NASA, Greenbelt, Md. [Available from NTIS, Springfield, Va.], 16 pp. (1972).

125.036 **On the nature of the Tycho X-ray source.**
I. S. Shklovsky.
Nature, Vol. 238, 144 (1972).

Of the five X-ray sources which are identified with remnants of supernovae, the Tycho 1572 source is of particular interest. It seems probable that this source should be identified with a young pulsar which originated after the explosion of the Tycho supernova. Because of "unfavourable" orientation of the rotation axis of this pulsar, coherent radio emission is not observed.

Systems with negative specific heat.
See Abstr. 061.027.

On the synthesis of neutron-rich iron-peak nuclei.
See Abstr. 061.041.

On the termination of the r-process and the synthesis of superheavy elements from supernovae.
See Abstr. 061.069.

Cepheids, presupernovae, and the $^{12}C(\alpha, \gamma)^{16}O$ reaction.   See Abstr. 065.097.

Convectively driven Urca neutrino losses and the carbon-detonation supernova.   See Abstr. 065.121.

Neutron stars, pulsar radiation and supernova remnants.   See Abstr. 065.130.

An experiment to detect related optical and radio pulses of astrophysical origin.   See Abstr. 066.085.

Eruptive variable or intergalactic supernova observed.
See Abstr. 123.015.

Anionic species of Fe in interstellar dust.
See Abstr. 131.086.

Der Gum-Nebel.   See Abstr. 132.006.

Spectrophotometric investigations of diffuse nebulae. III. Emission lines in the northern part of the nebula M 20. Supernova remnant in the direction of M 20.
See Abstr. 132.008.

Optical observations of the Simeis 59 nebula — a possible supernova remnant.   See Abstr. 132.017.

A model of the Crab nebula derived from dual-frequency radio measurements.   See Abstr. 134.010.

Pulsar associated with the supernova remnant IC 443.   See Abstr. 141.524.

Spatial coincidences of X-ray sources and supernova remnants.   See Abstr. 142.007.

Distances and absolute luminosities of galactic X-ray sources.   See Abstr. 142.099.

A soft X-ray survey of the galactic plane from Cygnus to Norma.   See Abstr. 142.108.

X-ray sources and supernova remnants.
See Abstr. 142.152.

A survey of linear polarization at 1415 MHz. IV. Discussion of the results for the galactic spurs.
See Abstr. 155.017.

125.100   Supernova in NGC 3147

New supernova in NGC 3147.
Astron. Tsirk., No. 670, p. 1 (1972). In Russian.

Supernova Altizer in NGC 3147.
Astron. Tsirk., No. 700, p. 1 (1972). In Russian.

Discovery of a supernova in NGC 3147.
Astron. Tsirk., No. 716, p. 1 (1972). In Russian.

Observations of the supernova SVS 1854 and Roberts-Altizer's variable in the vicinity of NGC 3147.
V. P. Goranskij.
Astron. Tsirk., No. 723, p. 6 - 8 (1972). In Russian.

Supernova in NGC 3147.   V. P. Goranskij.
IAU Circ., No. 2431 (1972).

Supernova in NGC 3147.
R. Wood, M. Penston, B. Jones, S. Tritton.
IAU Circ., No. 2434 (1972).

Supernova in NGC 3147.
R. Wallis, D. Thomas, R. Selmes.
IAU Circ., No. 2452 (1972).

125.101   Supernova in NGC 3198

On the origin of a very close similarity between the spectra of the type I supernova in NGC 3198 and the absorption spectrum of DQ Her.   E. R. Mustel.
Astron. Tsirk., No. 674, p. 1 - 4 (1972). In Russian.

125.102   Supernova in NGC 5253

Supernova in NGC 5253 discovered.
Astron. Tsirk., No. 706, p. 1 (1972). In Russian.

Extragalactic Ca II absorption lines in the spectra

of the supernova in NGC 5253.
R. F. Sisteró, M. E. Castore de Sisteró.
Astrophys. Journ., (Letters), Vol. 176, L123 - L126 (1972).
Spectroscopic observations at 39 Å mm^{-1} of the supernova in NGC 5253 (1972) are reported. The spectrum shows characteristic emission bands of type I supernovae. Interstellar galactic Ca II lines are observed; also a redshifted H and K absorption pair is present(cz = +428 ± 8 km s^{-1}). It is probably formed in the interstellar gas of NGC 5253.

Photometry of supernova 1972 in NGC 5253.
T. A. Lee, W. Wamsteker, W. Z. Wisniewski, T. J. Wdowiak.
Astrophys. Journ., (Letters), Vol. 177, L59 - L62 (1972).
Results are reported for $UBVRIJK$ photometry of the bright supernova 1972 in galaxy NGC 5253. The observations obtained over a span of 43 days show the UBV measurements to be of similar character as those made on previous supernovae classed as type I. Infrared observations made at 1.25 and 2.2 microns during two nights separated by 38 days show nearly identical magnitudes. The measurements show the decline in all bands to have definite irregularities.

Interstellar lines in the spectrum of the supernova in NGC 5253.
G. Wallerstein, P. S. Conti, J. L. Greenstein.
Astrophys. Letters, Vol. 12, 101 - 102 (1972).
Radial velocities and equivalent widths are presented for interstellar Ca II and Na I lines in the spectrum of the supernova in NGC 5253.

Supernova in NGC 5253.   R. R. D. Austin.
IAU Circ., No. 2421 (1972).

Supernova in NGC 5253.   A. Przybylski.
IAU Circ., No. 2434 (1972).

Supernova in NGC 5253.   A. W. J. Cousins.
Inform. Bull. Variable Stars (I.A.U. Commission 27), Konkoly Obs., Budapest, No. 700 (1972).

The supernova in NGC 5253.   S. E. Williams.
Journ. Astron. Soc. Western Australia, 1972 June, p. 5 - 6.

Optical polarization of the supernova in NGC 5253.
R. D. Wolstencroft, J. C. Kemp.
Nature, Vol. 238, 452 (1972).
The possibility of large magnetic fields in the recent supernova in NGC 5253 was investigated by polarimetry. Circular and linear polarization values in the visible were found to be $q \leq 0.8 \times 10^{-4}$ and $p \leq 3.5 \times 10^{-3}$, on 1972 May 13. The conclusion is that no field as large as 10^5 gauss was present.

Hellste Supernova seit 35 Jahren.   K. Locher.
Orion, 30. Jahrgang, p. 152 (1972).

125.103   Supernova in NGC 5457

Observations of a supernova in M 101.
R. G. Mnatsakanian.
Astron. Tsirk., No. 679, p. 3 - 5 (1972). In Russian.

125.104   Supernova in NGC 5055

Search for periodic light pulsations in a supernova outburst: NGC 5055.   R. H. Miller.
Bull. American Astron. Soc., Vol. 4, 320 (1972). — Abstr. AAS.

125.105 Supernova in NGC 7634

Supernova in NGC 7634.   L. Pigatto.
IAU Circ., No. 2437 (1972).

125.106 Supernova in NGC 735

Supernova in NGC 735.   J. P. Huchra.
IAU Circ., No. 2448 (1972).

125.107 Supernova in NGC 4254

Supernova in NGC 4254.   L. Rosino.
IAU Circ., No. 2472 (1972).

125.108 Supernova in NGC 493

Supernova in NGC 493.
Yamamoto Circ., No. 1751, p. 2 (1972). In Japanese.

126   Low-luminosity Stars, Subdwarfs, White Dwarfs

**126.001 A non-LTE analysis of the O-type subdwarf,
HD 49798.**   P. L. Dufton.
Monthly Notices Roy. Astron. Soc., Vol. 159, 79 - 93 (1972).

Equivalent width and line profile data are presented for the O-type subdwarf, HD 49798. These observations cannot be explained on the assumption of LTE.

126.002 Ultrashort-period binaries. II. HZ 29 (= AM CVn): A double-white-dwarf semidetached postcataclysmic nova?
J. Faulkner, B. P. Flannery, B. Warner.
Astrophys. Journ., (*Letters*), Vol. 175, L79 - L83 = Contr. Lick Obs., No. 367 (1972).

The peculiar white dwarf HZ 29 is in fact an eclipsing binary system with $P \sim 1051^s.05$ ($\sim 17^m.5$). A hydrogen-containing secondary is most unlikely. A satisfactory model has a degenerate helium secondary of mass $\sim 0.041\ M_\odot$. We suggest this is the remnant core of a nova secondary which has suffered complete envelope mass loss as a result of gravitational radiation angular momentum losses.

126.003 New measurements of circular polarization and an ephemeris for the variable white dwarf G195−19.
J. R. P. Angel, R. M. E. Illing, J. D. Landstreet.
Astrophys. Journ., (*Letters*), Vol. 175, L85 - L87 (1972).

Observations of the white dwarf G195−19 made for over a year show that the periodic variation in circular polarization continues with constant frequency and amplitude; an accurate value for the period, 1.3309 ± 0.0004 days, is obtained. While the variation in blue-green light is sinusoidal, extensive measurements in red light show an asymmetric curve, reaching a minimum about 6 hours earlier than in the blue-green.

126.004 Model atmospheres of white dwarfs with convection.   T. C. Grenfell.
Astron. Astrophys., Vol. 20, 293 - 297 (1972).

The Feautrier method with full linearization and variable Eddington factors is employed to construct model atmospheres of white dwarfs including simultaneously energy transport by radiation and convection. The procedure converges rapidly giving flux constancy to high precision. Models have been computed for van Maanen 2 and Ross 640 in order to test the procedure against calculations made by Wegner. Agreement is found to be quite good for the emergent flux and temperature stratification although the layers where convection sets in differ for the two methods.

126.005 Non-radial pulsations in white dwarf stars.
B. Warner, E. L. Robinson.
Nature, Phys. Sci., Vol. 239, 2 - 7 (1972).

Five more oscillating white dwarf stars have been disco-

vered. The data now available from these and other sources support an explanation of observed oscillations in dwarf novae in terms of g-mode pulsations.

126.006 Low dispersion spectroscopy of Lowell white dwarf suspects.   K. Szathmary.
Journ. Roy. Astron. Soc. Canada, Vol. 66, 219 (1972).
Abstr. Canadian Astron. Soc.

126.007 Ox+25°6725 − a white dwarf with periodically varying circular polarization.
O. S. Shulov, E. T. Belokon.
Astron. Tsirk., No. 681, p. 6 - 8 (1972). In Russian.

126.008 Observations of rapid blue variables−X. G61−29.
B. Warner.
Monthly Notices Roy. Astron. Soc., Vol. 159, 315 - 319 (1972).

Photoelectric observations of the helium emission line white dwarf G61−29 are presented. The presence of rapid flickering is established. The light curve contains features resembling that of the binary VV Pup, including effects which can be ascribed to an eclipse by the secondary and obscuration by an optically thick ring. A tentative orbital period of $6^h\ 16^m$ is determined for G61−29. Both components are probably the helium cores of evolved stars.

126.009 Quasi-radial pulsations and the stability of rotating white dwarfs with $A/Z = 2$.
G. G. Arutyunyan, É. V. Chubaryan.
Uch. zap. Erevan. un-t. Estestv. n., Vol. 3 (118), 143 - 145 (1971). In Russian. − Abstr. in Referativ. Zhurn. 51. Astron., 9.51.499 (1972).

126.010 Low-luminosity stars.   P. F. Chugajnov.
Zemlya i Vselennaya, 1972, No. 4, p. 32 - 35 .
In Russian.

126.011 Multiple periodicities in the white dwarf HL Tau-76.
C. G. Page.
Monthly Notices Roy. Astron. Soc., Vol. 159, 25P - 27P (1972).

Power spectrum analysis applied to the light curve of HL Tau-76 given by Warner and Nather has revealed at least three independent periodicities which entirely account for the phase deviations in pulse arrival times noted hitherto.

126.012 Pulsed and continuous radiation from white dwarfs.
I. Lerche.
Astrophys. Space Sci., Vol. 17, 117 - 125 (1972).

We demonstrate that the detection of steady kV X-ray emission from the vicinity of a white dwarf star possessing a magnetic field of the order of 10^7 G will provide strong evidence that the white dwarf is rotating with a period of about one minute. We also show that detection of pulsed radiation at about 1 mm wavelength would confirm this. Also some of the interesting dynamical consequences for the interstellar medium due to such white dwarfs are outlined.

126.013 Ferromagnetism in white dwarfs.
H. Pohl, J. Schmid-Burgk.
Nature, Phys. Sci., Vol. 238, 56 - 57 (1972).

Landau Orbital Ferromagnetism as a possible source of magnetic fields in white dwarfs is investigated. Temperature dependence of this mechanism requires a free electron gas to cool below ca. 200°K at white dwarf densities for LOFER to operate. Electron–electron interactions increase limiting temperatures but also lead to selfconsistent field strengths orders of magnitude above what is observed. Several possibilities to reconcile theory with observations are indicated.

126.014 The Einstein redshift in white dwarfs. III.
V. Trimble, J. L. Greenstein.
Astrophys. Journ., Vol. 177, 441 - 452 (1972).

We present radial velocities for 74 white dwarfs measured since 1966. Of these velocities, 51 are classed as reliable and have an average value of +53 km s^{-1} (after the normal solar motion has been removed). Colors and luminosities have been used to obtain radii for the stars, using two different temperature scales. On a temperature scale based on Shipman's, the median radius is 0.0089 $R_\odot$. This and the average redshift give a mass of 0.65–0.87 $M_\odot$, depending on composition, using the Hamada-Salpeter mass-radius relation, and allowing for some uncertainty in both radius and redshift.

126.015 On the effective temperatures of DB white dwarfs.
D. T. Wickramasinghe.
Observatory, Vol. 92, 186 - 187 (1972). – Letter.

126.016 A spectroscopic survey of southern hemisphere white dwarfs. I. The nature of CD −42°14462.
G. Wegner.
Astrophys. Letters, Vol. 12, 219 - 225 (1972).

New observations of the peculiar star CD −42°14462 are presented. The spectrum is composite. The possibilities that the star is a main-sequence star, an old nova, a close short-period binary of two white dwarfs, or a rapidly-rotating white dwarf are explored.

126.017 Masses and radii of white dwarfs.
H. L. Shipman.
Astrophys. Journ., Vol. 177, 723 - 743 (1972).

Additional model atmospheres representing the surface layers of hydrogen-rich and helium-rich white dwarfs have been calculated and are presented in tables. These model atmospheres and multichannel photoelectric scans by Oke have been used to determine the masses of 19 and the radii of 26 stars in Oke's observing program. The program stars are then used to calibrate a $U-V$ versus T_{eff} relation, and radii are determined for all white dwarfs with known distances. The above analyses were applied for the purpose of testing the convection-accretion hypothesis of Strittmatter and Wickramasinghe, which purports to explain the existence of the DB stars. The present results contradict their prediction that there should be a shortage of DA stars with effective temperatures between 15,000° and 18,000°K. I propose a modification of their hypothesis which explains the existence of DB and DC stars in the observed temperature ranges.

126.018 Discovery of circular polarization in the red degenerate star G99–47.
J. R. P. Angel, J. D. Landstreet.
Astrophys. Journ., (*Letters*), Vol. 178, L21 - L22 = Columbia

Astrophys. Lab., Columbia Univ., *New York,* Contr. No. 73 (1972).

Circular polarization is found in the light of the cool DC white dwarf G99−47 = GR 289, varying from 0.45 percent in the ultraviolet to 0.30 percent in the near-infrared. No evidence for time variability of the polarization is found.

126.019 Photometric systems for quantitative studies of white dwarfs.
D. T. Wickramasinghe, P. A. Strittmatter.
Monthly Notices Roy. Astron. Soc., Vol. 160, 421 - 434 (1972).

The suitability of the Strömgren *uvby* system for quantitative studies of white dwarfs is examined. For the cooler DA's the system yields useful indicators of the atmospheric parameters but is inadequate for gravity determinations for $T_{eff} \gtrsim 15000$ K. An alternative intermediate band system is suggested which allows determinations of T_{eff} and g over the observed colour range. The hydrogen deficient DB white dwarfs are also discussed. The *uvby* system is found to be unsuitable for quantitative studies of these stars and alternative methods are suggested. Masses and radii of cool DA white dwarfs for which reliable absolute magnitudes are available are derived from Graham's *uvby* photometry.

126.020 On the radii of white dwarf stars. B. Warner.
Monthly Notices Roy. Astron. Soc., Vol. 160, 435 - 440 (1972).

From a calibration of the surface brightness in the V band as a function of $(b-y)$ colour in the Strömgren system we derive radii of white dwarfs with known distances. There is good agreement between the location of white dwarfs in the M_V versus $(b-y)$ diagram and cooling curves for stars of constant radius.

126.021 On the theory of thermal runaway in the hydrogen envelope of a white dwarf. Yu. N. Redkoborody.
Astrometriya i Astrofiz., *Kiev,* Vyp. (No.) 17, (see 003.012), p. 3 - 49 (1972). In Russian.

An evolutionary sequence was calculated for a white dwarf on which a rich hydrogen envelope is assumed to increase with time. The accretion of matter was assumed to be quasi-static. Near the point of the chemical discontinuity the temperature maximum is shown to arise, then hydrogen ignites, a thin shell energy source being formed. The new shell source is thermally unstable. The resulting thermal runaway was investigated numerically and by means of a simple approximation based on the energy balance in the hydrogen burning shell source.

126.022 Quasi-radial pulsations of rotating white dwarfs and neutron stars in general relativity.
G. G. Arutjunian, D. M. Sedrakian, E. V. Chubarian.
Astron. Zhurn. Akad. Nauk SSSR, Vol. 49, 1216 - 1220 (1972). In Russian. English translation in Soviet Astron. AJ, Vol. 16, No. 6.

In the frame of relativity theory the quasi-radial pulsations of white dwarfs and neutron stars are calculated. It is shown that for white dwarfs the results are similar to the Newtonian case. For the neutron stars the relativistic corrections in rotation compensate the positive input in σ^2 of the Newtonian rotation in the sense of dynamical instability.

126.023 Molecular helium bands in Grw+70°8247.
D. T. Wickramasinghe, R. I. Thompson, P. A. Strittmatter.
Astrophys. Journ., Vol. 178, 763 - 769 (1972).

The possibility that molecular helium bands are present in the spectrum of Grw+70°8247 is discussed in terms of model atmospheres. Problems associated with pressure broadening at the required densities ($10^{22}-10^{23}$ cm^{-3}) are also discussed.

126.024 Observation of the Zeeman effect in the magnetic white dwarf G 99–37.
J. R. P. Angel, J. D. Landstreet.
Bull. American Astron. Soc., Vol. 4, 409 (1972). – Abstr. AAS.

126.025 Theoretical and laboratory studies of circular polarization produced by magnetic fields in the CH spectrum. J. R. P. Angel.
Bull. American Astron. Soc., Vol. 4, 409 (1972). – Abstr. AAS.

Model atmospheres for DA and DB white dwarfs.
See Abstr. 064.002.

Blanketed model atmospheres for cool hydrogen-rich white dwarfs. See Abstr. 064.005.

Analytic approximations to the mass-radius relation and energy of zero-temperature stars. See Abstr. 065.012.

Extended horizontal branch loci.
See Abstr. 065.033.

Ionization energies of hydrogen in magnetic white dwarfs. See Abstr. 065.090.

Polarized radiation from magnetic white dwarfs. II. Solution of Kemp's model at all temperatures.
See Abstr. 065.091.

Reduced-proper-motion diagrams. II. Luyten's white-dwarf catalog. See Abstr. 112.006.

Blaue Objekte in der Umgebung von M 31. I. Entdeckungwahrscheinlichkeit neu aufgefundener Blauer Objekte. See Abstr. 113.032.

Helium spectra in large magnetic fields.
See Abstr. 114.155.

The luminosities of late-type stars of differing metal abundance. See Abstr. 115.015.

On the pointlessness of stellar masses: (i) AM CVn (= HZ 29), a semi-detached double white dwarf binary system, (ii) the general relativistic Roche model.
See Abstr. 117.012.

Observations of rapid blue variables – XII. UX Ursae Majoris. See Abstr. 121.043.

Observations of rapid blue variables–IX. AM CVn (HZ 29). See Abstr. 122.023.

On the outburst of cataclysmic variables.
See Abstr. 124.007.

The variable X-ray source Cen X-3 as a radially pulsating white dwarf. See Abstr. 142.021.

Are cosmic rays produced by white dwarfs?
See Abstr. 143.001.

Possible photoionization of the intergalactic medium by hot white dwarfs. See Abstr. 161.002.

Interstellar Matter, Gaseous Nebulae, Planetary Nebulae

131 Interstellar Space, Interstellar Matter, Polarization of Starlight

131.001 Momentum exchange between small particles and light. P. G. Martin.
Monthly Notices Roy. Astron. Soc., Vol. 158, 63 - 78 (1972).

The transfer of linear and angular momentum between small particles and a radiation field is discussed, with a view to evaluating the importance of the exchange of angular momentum to the alignment of interstellar grains. To this end expressions for the two diffusion coefficients in the Fokker–Planck equation governing the steady-state momentum distribution are derived.

131.002 Observations of recombination lines at decimetre wavelengths. A. Pedlar, R. D. Davies.
Monthly Notices Roy. Astron. Soc., Vol. 159, 129 - 153 (1972).

Recombination lines from a total of 15 H II regions have been measured in the 166α (1424.734 MHz), 192α (921.897 MHz) and 220α (613.405 MHz) transitions. A description is given of the observational and reduction techniques; the spectra of all the observed lines are given. The data from these observations taken with published results of other transitions have been used to derive some of the physical parameters of the H II regions studied.

131.003 Interstellar circular polarization. P. G. Martin.
Monthly Notices Roy. Astron. Soc., Vol. 159, 179 - 190 (1972).

This paper shows that optical observations of circular polarization produced by aligned interstellar grains could yield valuable information about the grain material. Calculations are presented to demonstrate that the wavelength of the circular polarization is sensitive to the imaginary part of the complex refractive index of the grain material.

131.004 Discovery of interstellar circular polarization in the direction of the Crab nebula.
P. G. Martin, R. Illing, J. R. P. Angel.
Monthly Notices Roy. Astron. Soc., Vol. 159, 191 - 201 (1972).

A search in many small regions of the Crab nebula has resulted in the detection of a small component of circular polarization. The variation of the sign and magnitude with position in the nebula indicates that the polarization is of interstellar origin. On the basis of the polarity, strength, and colour dependence, it is concluded that the composition of the aligned grains causing this polarization is dielectric.

131.005 Interferometric positions of the water-vapor emission sources in H II regions.
R. Hills, M. A. Janssen, D. D. Thornton, W. J. Welch.
Astrophys. Journ., (*Letters*), Vol. 175, L59 - L64 (1972).

An interferometer has been constructed to operate at wavelengths as short as 1 cm. We have measured the positions of sources of strong water-vapor emission in the H II regions W3, W49, W51, and Orion A. The water sources lie close to, but not always coincident with, the OH and other molecular sources and infrared objects. In some cases the water sources are clearly separated from the compact H II regions.

131.006 Cosmic-ray heating of low-density interstellar H II regions. R. McCray, J. Buff.
Astrophys. Journ., (*Letters*), Vol. 175, L65 - L68 (1972).

If the interstellar medium is heated by a steady flux of low-energy cosmic rays, then H II regions of density $\lesssim 10^{-2}$ cm^{-3} and temperature $\gtrsim 3 \times 10^5$ °K are likely to occur in and above the galactic disk.

131.007 Formaldehyde absorption in the southern Coal Sack. M. W. Sinclair, J. W. Brooks.
Astrophys. Letters, Vol. 11, 207 - 210 (1972).

Absorption of the 2.7 K background by the $1_{11} \rightarrow 1_{10}$ transition of formaldehyde has been observed in the southern Coal Sack with an antenna beam of 4.4 arc min and a frequency resolution of 10 kHz. The absorbing region appears to be confined to a small dense cloud of diameter 0.3 pc and having a minimum optical depth of 0.17. The radial velocity of the absorbing cloud as given by the absorption profile is −6.5 km sec^{-1}, corresponding to a kinematic distance of 170 pc to the cloud, in agreement with optical estimates.

131.008 Microwave celestial water-vapor sources.
K. J. Johnston, S. H. Knowles, P. R. Schwartz.
Sky Telescope, Vol. 44, 88 - 90 (1972).

131.009 4830 MHz observations of the formaldehyde molecule in the direction of discrete radio sources.
T. L. Wilson.
Astron. Astrophys., Vol. 19, 354 - 368 (1972).

Observations have been made of the formaldehyde (H_2CO) molecule against the peaks of 74 continuum sources using the NRAO 140 foot radio telescope. Sixty-three of the sources have been selected from the H 109α line survey of Reifenstein et al. (1970). The results of the survey are used to assist in resolving the twofold kinematic distance ambiguity for many of the H II regions observed. A summary is made of comparisons of H_2CO, H I and OH. Relations between the H_2CO clouds and the H II regions are briefly discussed. Distributions of the H II regions and the H_2CO in the Galaxy using the Schmidt (1965) model are presented.

131.010 On the λ 4686 He II line intensity in H II regions and the cosmic ray flux.
M. Peimbert, D. W. Goldsmith.
Astron. Astrophys., Vol. 19, 398 - 404 (1972).

To study the presence of dust inside H II regions, the $N(He)/N(H)$ abundance ratio, and the cosmic-ray flux, we have carried out observations of the λ 4686 line in several objects. We discuss the continuum emission from the Orion nebula; we obtain the $N(He)/N(H)$ ratio; we study the implications of the observed $N(He^{++})/N(H^+)$ ratio on the value of the cosmic-ray flux.

131.011 On the abundances of interstellar molecules.
W. D. Watson, E. E. Salpeter.
Astrophys. Journ., Vol. 175, 659 - 671 (1972).

Equilibrium abundances are calculated for various molecules in interstellar H I gas clouds using rates from a previous paper for molecule formation on grain surfaces, and rates from other authors for gas-phase reactions and for photodissociation.

131.012 Formation of clouds in a cooling interstellar medium.
J. Schwarz, R. McCray, R. F. Stein.
Astrophys. Journ., Vol. 175, 673 - 686 (1972).

We explore the growth of thermal condensations in a

cooling region of the interstellar medium. We set forth the equations of fluid dynamics that govern the development of condensations and discuss the initial conditions that we expect to encounter, and the physical mechanisms that will be important. We discuss the results of a one-dimensional numeric al hydrodynamic calculation, and compare these results with observations of the interstellar medium.

131.013 **Polarization of optical radiation of stars with non-spherical atmospheres and envelopes.**
Yu. N. Gnedin, A. Z. Dolginov, N. A. Silant'ev.
Astron. Zhurn. Akad. Nauk SSSR, Vol. 49, 689 - 699 (1972).
In Russian. English translation in Soviet Astron. AJ, Vol. 16, No. 4.

Polarization of radiation scattered by freely oriented anisotropic molecules or grains in stellar nonspherical atmospheres or envelopes is considered. The influence of a magnetic field on the scattering of radiation by electrons is taken into account. The results of the polarimetric observations of red giants, magnetic and X-ray stars are discussed. The magnetic field of the X-ray star Sco X-1 is estimated.

131.014 **On the origin of high velocities of H_2O sources in W49.** V. S. Strelnitsky, R. A. Syunaev.
Astron. Zhurn. Akad. Nauk SSSR, Vol. 49, 704 - 711 (1972).
In Russian. English translation in Soviet Astron. AJ, Vol. 16, No. 4.

The most likely model of the H_2O sources in W49 is a flying away system of light dense clouds of neutral gas around a hot supergiant or a massive protostar. Possible mechanisms of acceleration of the clouds are discussed: stellar wind, "rocket" mechanism, and radiation pressure from the central object. Possible connection between the mechanism of acceleration and the pumping of the maser is pointed out.

131.015 **Space maser with feedback.** V. S. Letokhov.
Astron. Zhurn. Akad. Nauk SSSR, Vol. 49, 737 - 743 (1972). In Russian. English translation in Soviet Astron. AJ, Vol. 16, No. 4.

The model of a space maser with resonant scattering feedback is investigated theoretically. Unlike a model of travelling wave maser amplifier the amplification of such a system occurs during a great number of passes. An estimation for OH and H_2O amplifying clouds is presented.

131.016 **Properties of H I regions heated by X-rays and subcosmic rays.** N. G. Bochkarev.
Astron. Zhurn. Akad. Nauk SSSR, Vol. 49, 756 - 767 (1972).
In Russian. English translation in Soviet Astron. AJ, Vol. 16, No. 4.

Equilibrium conditions in the interstellar medium heated either by soft X-rays having a power-law spectrum or by subcosmic rays having an energy of 2 MeV/nucleon are compared on the basis of thermal balance, and the hydrogen/helium ionization equilibrium equations are solved numerically.

131.017 **A model of the interstellar medium. II. Interpretation of the Na^0/Ca^+ ratio.** S. R. Pottasch.
Astron. Astrophys., Vol. 20, 245 - 258 (1972).

An attempt is made to explain the observed variation of the ratio Na^0/Ca^+ in interstellar "clouds", assuming the abundances of sodium and calcium are everywhere the same. The results of this analysis are applied to about thirty observed "clouds", and it is shown that it is possible to derive values of electron temperature, electron density, and state of ionization for each "cloud". It is shown that at least two and possibly four different groups of clouds exist. These results are compared with other observations of the interstellar medium, especially the pulsar dispersion measures and the low frequency radio absorption measurements.

131.018 **The density of H_2 molecules in dark interstellar clouds.** T. de Jong.
Astron. Astrophys., Vol. 20, 263 - 274 (1972).

We calculate the density of atoms (H), positive ions (H^+), negative ions (H^-), molecules (H_2), positive molecular ions (H_2^+, H_3^+) and electrons (e) in a dark cloud with kinetic temperature T_k = 50° K. The calculation is made for clouds with total hydrogen density n = 100 cm^{-3} and 1000 cm^{-3} and we let the low-energy cosmic ray flux inside the cloud vary over three decades. The results are compared with available observations and discussed.

131.019 **The Cygnus X-region. VII. Radio continuum search for a ring of filaments around the area.**
A. v. Kap-herr, H. J. Wendker.
Astron. Astrophys., Vol. 20, 313 - 315 (1972).

A ring of optically visible filaments around the Cygnus X-region was searched for radio continuum radiation at 1420 MHz. The optical filaments could possibly belong to a Fossil Strömgren Sphere. Although several of the filaments were detected, the search for the ring as a whole was negative.

131.020 **Heating of interstellar H I clouds by ultraviolet photoelectron emission from grains.** W. D. Watson.
Astrophys. Journ., Vol. 176, 103 - 110, with an addendum, p. 271 (1972).

The physical processes and consequences of photoemission from interstellar grains in H I regions are investigated.

131.021 **An attempt to detect the 3-centimeter fine-structure transition of hydrogen in H II regions.**
P. C. Myers, A. H. Barrett.
Astrophys. Journ., Vol. 176, 111 - 126 (1972).

A search for the $2\ ^2P_{3/2}-2\ ^2S_{1/2}$ transition of hydrogen at 9.9 GHz failed to detect this spectral line in six galactic H II regions: W3, W10, W49, W51, DR 21, and Sgr B2. Expressions for line brightness temperature are derived from a model H II region of uniform density, which emits and absorbs both line and continuum radiation. Upper limits on the integrated $2\ ^2P_{3/2}$ density and on the ground-state density are derived from this model and from the results of the observations. Estimates are made of the likelihood of detecting two other microwave lines of hydrogen in H II regions: $3\ ^2P_{3/2}-3\ ^2S_{1/2}$ at 2.94 GHz, and $2\ ^2S_{1/2}-2\ ^2P_{1/2}$ at 1.06 GHz.

131.022 **Models for OH radiation from the interstellar medium.** G. I. Peters, L. Allen.
Astrophys. Journ., (Letters), Vol. 176, L23 - L25 (1972).

Some recently proposed models for OH radiation from the interstellar medium are shown to be in error. A theory for radiation from an inverted medium amplifying both its own spontaneous emission and an external signal allows comment to be made on the possible emission mechanisms occurring in the medium.

131.023 **Hydrogen recombination radio lines.**
M. Kubiak.
Postępy Astron., Vol. 20, 219 - 240 (1972). In Polish.

This article gives a review of the recent interpretations of radio lines data for hydrogen in terms of the non-LTE theory. Brief account of some results is also given.

131.024 **Optical and chemical studies on simulated interstellar grain materials.**
J. D. McCullough, G. R. Floyd, R. H. Prince, W. W. Duley.
Journ. Roy. Astron. Soc. Canada, Vol. 66, 223 (1972).
Abstr. Canadian Astron. Soc.

131.025 **Angular distribution of interstellar atomic hydrogen.**
E. Daltabuit, S. Meyer.
Astron. Astrophys., Vol. 20, 415 - 424 (1972).

A summary of recent values of the columnar density of galactic atomic hydrogen appearing in the literature is presented in the form of plots of N_H vs · l^{II} for every 5° in galactic longitude.

131.026 Studies of the 21-cm line near dark dust clouds.
M. J. Mahoney.
Astrophys. Letters, Vol. 12, 43 - 48 (1972).
This letter reports observations of the 21-cm line on and near twelve dark dust clouds. Four of these objects show significant deficits in their hydrogen emission compared with nearby surrounding regions of the sky. The results of these measurements have been compared with other work and the problem of whether the hydrogen in the clouds is atomic or molecular has been considered.

131.027 On parametric down-conversion in astrophysical masers. P. Goldreich, J. Y. Kwan.
Astrophys. Journ., Vol. 176, 345 - 351 (1972).
The mechanism of parametric down-conversion proposed by Litvak cannot explain the observed preference for circular polarization in OH maser emission because the nonlinear interaction between oppositely circularly polarized microwaves is too weak.

131.028 Fine structure in H II regions: Synthesis observations at 11 and 3.7 centimeters. B. Balick.
Astrophys. Journ., Vol. 176, 353 - 365 (1972).
We report new high-resolution observations of several H II regions. Improved values for the physical parameters describing the structure are deduced from their continuum spectra. The results are compared with estimates obtained by other means, and surprising agreement is found with the results of Andrews et al. (1971). In addition, the unknown, but very important, relation between the structure and the parent H II region is investigated.

131.029 Interstellar hydrogen sulfide. P. Thaddeus, M. L. Kutner, A. A. Penzias, R. W. Wilson, K. B. Jefferts.
Astrophys. Journ., (Letters), Vol. 176, L73 - L76 (1972).
The $1_{10} \rightarrow 1_{01}$ transition of H$_2$S at 168.7 GHz has been detected in seven galactic sources. Hydrogen sulfide is about as abundant as H$_2$CO, having column densities in the range 0.4 to 5×10^{14} cm $^{-2}$.

131.030 On the injection of grains into interstellar clouds.
N. C. Wickramasinghe.
Monthly Notices Roy. Astron. Soc., Vol. 159, 269 - 287 (1972).
Dust grains expelled from cool stars are injected into interstellar clouds with typical velocities of several thousand km s^{-1}. The interaction of such high speed grains with interstellar gas is discussed.

131.031 The formaldehyde absorption associated with W44.
J. B. Whiteoak, F. F. Gardner.
Astron. Astrophys., Vol. 21, 159 - 161 (1972).– Research note.

131.032 Molecular astronomy. D. Buhl.
Bull. American Astron. Soc., Vol. 4, 307 (1972). Abstr. AAS.

131.033 Formaldehyde in dark nebulae: absorption of the isotropic background radiation at 2 cm wavelength.
N. J. Evans II, G. Morris, T. Sato, B. Zuckerman.
Bull. American Astron. Soc., Vol. 4, 307 (1972). – Abstr. AAS.

131.034 Interferometric studies of interstellar CH$^+$ molecules.
L. M. Hobbs.
Bull. American Astron. Soc., Vol. 4, 307 (1972). – Abstr. AAS.

131.035 Observations of thermal and maser emission from OH excited rotational states.
L. J. Rickard, B. Zuckerman, P. Palmer.
Bull. American Astron. Soc., Vol. 4, 307 - 308 (1972). Abstr. AAS.

131.036 $^{12}C^{32}S$, $^{12}C^{34}S$, $^{13}C^{32}S$, and $^{12}C^{33}S$ in the interstellar medium.
M. Morris, B. E. Turner, B. Zuckerman, P. Palmer.
Bull. American Astron. Soc., Vol. 4, 308 (1972). – Abstr. AAS.

131.037 Millimeter wavelength molecular line observations of infrared sources and dust clouds.
P. Palmer, B. Zuckerman, B. E. Turner, M. Morris.
Bull. American Astron. Soc., Vol. 4, 308 (1972). – Abstr. AAS.

131.038 Interferometer observations of the $^2\pi_{3/2}$, J = 5/2 microwave transition of OH. J. A. Ball,
K. J. Johnston, S. H. Knowles, J. M. Moran.
Bull. American Astron. Soc., Vol. 4, 308 - 309 (1972). Abstr. AAS.

131.039 Interstellar molecules and dust in the Ophiuchus cloud. F. H. Chaffee, Jr., S. E. Strom, K. M. Strom, B. L. Lutz, J. G. Cohen.
Bull. American Astron. Soc., Vol. 4, 317 (1972). – Abstr. AAS.

131.040 The C248α transition in W51. V. Pankonin.
Bull. American Astron. Soc., Vol. 4, 317 - 318 (1972). – Abstr. AAS.

131.041 Microwave spectroscopic mapping of NGC 7538.
C. J. Lada, E. J. Chaisson.
Bull. American Astron. Soc., Vol. 4, 319 (1972). – Abstr. AAS.

131.042 A young stellar group surrounding RCrA.
K. M. Strom, S. E. Strom, R. F. Knacke, E. Young, W. Kunkel.
Bull. American Astron. Soc, Vol. 4, 325 (1972). – Abstr. AAS.

131.043 Photoelectric observations of interstellar Ca I.
R. E. White.
Bull. American Astron. Soc., Vol. 4, 327 (1972). – Abstr. AAS.

131.044 On a third thermally stable phase of the interstellar gas. R. Giovanelli.
Bull. American Astron. Soc., Vol. 4, 328 (1972). – Abstr. AAS.

131.045 Structure and stability of shocks in a two component interstellar medium. S. L. Mufson.
Bull. American Astron. Soc., Vol. 4, 328 (1972). – Abstr. AAS.

131.046 Variability of polarization in the infrared radiation of IRC+10216 (CW Leo). G. V. Khozov.
Astron. Tsirk., No. 709, p. 3 - 6 (1972). In Russian.

131.047 Interstellarer Staub. C. Schalén.
Bild der Wiss., 9. Jahrgang, p. 494 - 504 (1972).
Interstellar dust can be recognized by eclipsing effects and colour shiftings of penetrating starlight. These properties along with polarizing effects and the dust's ability of reflection, have led to theories which by now permit certain statements on the nature of interstellar dust. The author gives a comprehensive outline on this research work.

131.048 Ionization structure of interstellar carbon.
J. C. Weisheit, A. Dalgarno.
Astrophys. Letters, Vol. 12, 103 - 106 (1972).
The ionization equilibrium of interstellar carbon is determined for several model H I regions. The importance both of radiationless Auger transitions following K-shell ionizations of

CI, CII, and CIII and of ionizations by the measured isotropic X-ray background is established. The predicted carbon absorption-line strengths are significantly larger than earlier calculations had suggested.

131.049 Microwave observations of a partially ionized interstellar cloud. E. J. Chaisson.
Nature, Phys. Sci., Vol. 239, 83 - 85 (1972).

High resolution spectroscopy has enabled detection of recombination lines of excited hydrogen, carbon, and elements of intermediate mass from a cool cloud in the direction of W3.

131.050 Collisional excitation of carbon monoxide in interstellar clouds. P. F. Goldsmith.
Astrophys. Journ., Vol. 176, 597 - 610 (1972).

Different models for the collisional excitation of carbon monoxide (CO) have been investigated by solving the rate equations for the 10 lowest energy levels, including stimulated emission and absorption by $3°K$ isotropic radiation, spontaneous emission, and collisions. The results are qualitatively applicable to other linear molecules. They show that a wide range of rotational excitation temperatures, including population inversions, can be produced by the various kinetic temperatures and particle densities which are likely to occur in the interstellar regions containing molecules.

131.051 Interstellar circular polarization: Data for six stars and the wavelength dependence.
J. C. Kemp, R. D. Wolstencroft.
Astrophys. Journ., (*Letters*), Vol. 176, L115 - L118 (1972).

Circular polarization attributed to twisted grain alignment along the line of sight has now been detected in light from six early-type stars, including the cases o Sco and σ Sco A reported earlier. Detailed $q(\lambda)$ data for the latter two change sign in the red as predicted by a simple model.

131.052 Interstellar reddening near the north galactic pole.
K. A. Feltz, Jr.
Publ. Astron. Soc. Pacific, Vol. 84, 497 - 514 (1972).

Color excesses are obtained for 182 B-, A-, and F-type stars within about 20 degrees of the north galactic pole with the aid of BV, $uvby\beta$, and (khg) photometry. The average value of the reddening in the vicinity of the pole is found to be $0^m.000 \pm 0.002$.

131.053 A search for elliptical polarization in starlight.
G. W. Wolf.
Astron. Journ., Vol. 77, 576 - 583 (1972).

A survey search for elliptical polarization was undertaken on objects observable from the northern hemisphere. The objects observed include O and B stars with a low ratio of polarization to interstellar absorption, and stars with intrinsic polarization, peculiar spectra, high linear polarization, wavelength dependence of position angles of polarization, and high magnetic fields. Several galaxies, planetary nebulae, reflection nebulae, the Crab nebula, and comet Tago-Sato-Kosaka were also included. The final results list the linear polarization and position angle and the ellipticity for all objects observed.

131.054 On the polarization of magnetic stars radiation.
L. B. Demkina, V. N. Obridko.
Astron. Zhurn. Akad. Nauk SSSR, Vol. 49, 1046 - 1048 (1972). In Russian. English translation in Soviet Astron. AJ, Vol. 16, No. 5.

It is shown that polarization of magnetic stars radiation due to Zeeman effect cannot exceed 1 %. Dependence of polarization on wavelength in the range 4152–4262 Å has been computed for ϵ UMa.

131.055 Absorption effects in six H II regions.
H. F. Gianotti, R. J. Quiroga, C. M. Varsavsky.

Astrophys. Space Sci., Vol. 17, 126 - 133 (1972).

We have measured the profile of the 21-cm line of neutral hydrogen in the direction of six H II regions (RCW 38, RCW 49, RCW 57(b), RCW 74, W22, and W37). By comparison with line profiles outside the nebulae we have deduced the absorption profiles corresponding to each H II region. We compare our results with those of other authors in the 21-cm line, in the line of OH, and in the recombination lines of hydrogen.

131.056 The influence of the surface roughness of grains upon the law of interstellar extinction.
J. Svatoš.
Astrophys. Space Sci., Vol. 17, 238 - 240 (1972). – Research note.

131.057 Hydrogen recombination line and continuum observations at 5000 MHz of 13 southern H II regions.
J. L. Caswell.
Australian Journ. Phys., Vol. 25, 443 - 450 (1972).

Hydrogen 109 α recombination line observations are presented for 13 southern H II regions. For four of these and for one additional H II region new continuum observations at 5000 MHz are also given. Four of the H II regions had previously been incorrectly categorized as supernova remnants. For a number of the other H II regions comparisons are made between the H II velocities and those of associated OH detected in either emission or absorption.

131.058 New OH sources associated with IR-late-type stars.
R. Fillit, M. Gheudin, Nguyen-Quang-Rieu, M. Paschenko, V. Slysh.
Astron. Astrophys., Vol. 21, 317 - 319 (1972).

A search of OH emission from IR-late-type stars has resulted in the detection of four new sources. Their radio characteristics are similar to those of the class of the OH main-line emitters.

131.059 An upper limit on the OH abundance in the intercloud medium. G. R. Knapp, F. J. Kerr.
Astron. Journ., Vol. 77, 649 - 651 (1972).

A search for weak OH emission at 1665 and 1667 MHz was carried out in three high-latitude directions containing no compact clouds of H I or dust. No OH emission was detected down to low limits, and the comparison of our upper limits with the known H I and dust-column densities reinforces the suggestion that OH tends to be concentrated to regions of high dust or gas density.

131.060 Light molecules and dark clouds – Part I.
D. Buhl.
Mercury, (Journ. Astron. Soc. Pacific), Vol. 1, No. 5, p. 4 - 7, 18 (1972).

131.061 Backscatter of solar resonance radiation – II.
H. E. Johnson.
Planet. Space Sci., Vol. 20, 1784 - 1785 (1972). – Research note.

131.062 Infra-red sources in the H II region W3.
C. G. Wynn-Williams, E. E. Becklin, G. Neugebauer.
Monthly Notices Roy. Astron. Soc., Vol. 160, 1 - 14 (1972).

High resolution mapping and photometric observations in the wavelength range $1.65-20 \mu$ have led to the discovery of nine distinct infra-red objects in W3. The main purposes of the observations were (a) to search for the exciting stars of the H II condensations, which are obscured at optical wavelengths; (b) to examine the distribution of heated dust in the nebula; and (c) to see if there exist correlations between OH/H_2O and infra-red emission in H II regions.

131.063 **The mass spectrum of interstellar clouds and the assumption of total coalescence.**
L. G. Taff, M. P. Savedoff.
Monthly Notices Roy. Astron. Soc., Vol. 160, 89 - 97 (1972).

We have investigated some modifications of the statistical mechanical model for the formation and disruption of interstellar clouds originally proposed by Field and Saslaw (1965). We also describe some preliminary results of more detailed model calculations.

131.064 **Molekülwolken und Sternentstehung im interstellaren Raum.** J. Solf.
SuW, Vol. 11, 302 - 306 (1972).

131.065 **Observations of the $^2\Pi_{3/2}$, $J = 5/2$ state of interstellar OH.**
B. Zuckerman, J. L. Yen, C. A. Gottlieb, P. Palmer.
Astrophys. Journ., Vol. 177, 59 - 78 (1972).

With the 140-foot (4267-cm) telescope of the National Radio Astronomy Observatory, OH Λ-doublet radiation at 5-cm wavelength has been observed in the direction of six sources: W3, W75N, NML Cyg, W49, Sgr B2, and NGC 6334N. The $F = 3 \rightarrow 3$ transition was detected in all six sources and the $F = 2 \rightarrow 2$ transition in W3 and NGC 6334N. We briefly describe the equipment, then describe our spectra and discuss the results in terms of the variety of pumping models that have been proposed to explain the OH maser phenomenon.

131.066 **The evolution of interstellar clouds. I.**
P. Mészáros.
Astrophys. Journ., Vol. 177, 79 - 92 (1972).

The consequences of the depletion of heavy elements on grains are examined in the two-phase model of the interstellar medium. It is found that, if depletion is important in the cloud phase, then clouds left to evolve in the absence of external dynamical perturbations must heat up gradually, and eventually destroy themselves on timescales of the order of 3×10^7 years.

131.067 **Rocket infrared observations of H II regions.**
B. T. Soifer, J. L. Pipher, J. R. Houck.
Astrophys. Journ., Vol. 177, 315 - 323 (1972).

Infrared observations of three H II regions, made with a rocket telescope cooled by liquid helium are reported. For two of the H II regions multicolor photometry yields very accurate ($\pm 2°$K) color temperatures. From these temperatures, we are able to rule out grains with infrared absorptivities $\epsilon \propto 1/\lambda^2$ as responsible for the infrared emission in these two nebulae. We also find good agreement between the infrared luminosities for these three H II regions and the ultraviolet luminosities required to ionize the regions.

131.068 **21-micron observations of H II regions.**
D. Lemke, F. J. Low.
Astrophys. Journ., (*Letters*), Vol. 177, L53 - L57 (1972).

The 21-μ contour map of M17 is similar to radio maps with three components. Measured flux densities give color temperatures of both 240° and 75°K for the two principal components of M17 and 44°–63°K for Sgr B2, NGC 2024, and DR–21.

131.069 **A simple analytic approximation for dusty Strömgren spheres.**
V. Petrosian, J. Silk, G. B. Field.
Astrophys. Journ., (*Letters*), Vol. 177, L69 - L73 (1972).

We interpret recent far-infrared observations of H II regions in terms of true absorption by internal dust of a significant fraction of the Lyman-continuum photons. We present approximate analytic expressions describing the effects of internal dust on the ionization structure of H II regions, and outline a procedure for deducing the properties of this dust from optical and infrared observations.

131.070 **Formation of filaments in fossil H II regions.**
R. McCray, R. F. Stein, J. Schwarz.
Astrophys. Journ., (*Letters*), Vol. 177, L75 - L77 (1972).

Ionized filaments of temperature $\sim 10^4$ °K and density contrast $\sim 10{:}1$ are formed by thermal instability in a low-density optically thin medium which cools radiatively from an initial temperature $\sim 10^5$ °K. Typical scale lengths are $\sim 0.1/n_0$ pc. The outlying filaments of the Gum nebula may result from this mechanism.

131.071 **Pumping of interstellar OH molecule by polarized infrared radiation.** A. Kudo.
Sci. Rep. Tôhoku Univ., First Ser., Vol. 54, 120 - 126 (1971).

It is shown that the polarization properties of a class I OH emitter may be related to the pumping by polarized infrared radiation. The population inversion in the ground Λ-doublet of OH molecule due to the pumping is roughly estimated.

131.072 **Heating and ionization of HI regions.**
A. Dalgarno, R. A. McCray.
Annual Rev. Astron. Astrophys., Vol. 10, (see 003.005), 375 - 426 (1972). – Contents: (1)Introduction; (2) Cooling processes; (3) Heating and ionization processes; (4) Observations of interstellar HI regions; (5) Energetic requirements for heating and ionizing the interstellar gas; (6) Time-independent models; (7) Time-dependent models.

131.073 **Between the stars, hot gases... .** S. Souffrin.
Sci. Progrès Découverte, 99e année, No. 3432, p. 9 - 16 (1971). In French.

131.074 **Detection of several new interstellar molecules.**
L. E. Snyder, D. Buhl.
Ann. New York Acad. Sci., Vol. 194, 17 - 24 = National Radio Astron. Obs., *Green Bank*, Repr. Ser. B, No. 317 (1972).

We have detected radio emission from several new interstellar molecules and the preliminary results of our observations are discussed. We have identified the new interstellar molecules on the basis of pure rotational transitions as isocyanic acid (HNCO), methylacetylene (CH_3C_2H), and hydrogen isocyanide (HNC).

131.075 **OGO 5 determination of the local interstellar wind parameters.**
J. L. Bertaux, A. Ammar, J. E. Blamont.
Space Research XII, (see 012.016), Vol. 2, 1559 - 1567 (1972).

131.076 **New interpretations of extraterrestrial Lyman-alpha observations.** P. W. Blum, H. J. Fahr.
Space Research XII, (see 012.016), Vol. 2, 1569 - 1577 (1972).

131.077 **Influence of interstellar hydrogen on the location of the heliospheric shock front.** H. J. Fahr.
Space Research XII, (see 012.016), Vol. 2, 1579 - 1587 (1972).

131.078 **Are the chemical abundances in the cosmic masers abnormal?**
V. S. Strelnitsky, R. A. Sunyaev, D. A. Varshalovich.
Comments Astrophys. Space Phys., Vol. 4, 155 - 159 (1972).

The analysis of the physical processes leading to the phenomenon of the cosmic OH and H_2O masers is made under assumption of normal chemical abundances in the amplifying medium: like those in the stellar atmospheres and gaseous nebulae. This point of view follows from the widespread interpretation of the maser sources in HII regions as protostars. It is shown that the observational data are easier explained, if one assumes a great deficiency of hydrogen in the maser sources.

131.079 **High resolution spectra of some strong galactic OH emission sources.** B. O. Rönnäng.
Res. Lab. Electronics, Onsala Space Obs., Chalmers Univ. Technology, Gothenburg, Sweden, Res. Rep. No. 101, 1 + 44 pp. (1972).

The 84-foot (25.6 m) radio telescope of the Onsala Space Observatory has been used for detailed examinations with 250 Hz resolution of the 18-cm spectral line emission from eight well-known strong OH sources. Models obtained by incoherent superposition of circularly polarized Gaussian features have been fitted to the spectra. The observations were compared with earlier measurements made since 1969 with the same receiver system for evidence of variabilities and of possible correlation between changes in intensity and linewidth. An intensity change by a factor of seven over a week was observed in one component of the right circularly polarized emission of W75B at 1665 MHz.

131.080 **On the determination of the optical depth of 21-cm emission line profiles.**
K. Rohlfs, E. Braunsfurth, U. Mebold.
Astron. Journ., Vol. 77, 711 - 717 (1972).

The method for determining the optical depth τ of cold hydrogen clouds from saturation effects of emission-line spectra is investigated quantitatively. It is shown that the rms noise on the profile introduces a bias towards larger τ and lower spin temperature. From Monte Carlo model computations, estimates are derived for this bias and for the uncertainty of the gas parameters which can be determined from the profiles. The measurements of Verschuur and Knapp (1971) are rediscussed taking these considerations into account.

131.081 **H I clouds with spin temperatures less than 25°K. II. Physical properties of two neutral hydrogen clouds.** G. R. Knapp, G. L. Verschuur.
Astron. Journ., Vol. 77, 717 - 725 (1972).

Narrow-band observations of two neutral hydrogen clouds near $l = 230°$, $b = +45°$ have been used to study the variation of velocity, temperature, and density in the clouds. One of the clouds has a fairly constant spin temperature of 24 K and is very elongated; the other has a more variable temperature with a mean value of about 17 K, and is slowly rotating. If the clouds are in pressure equilibrium with the interstellar medium, their masses are very low, about 0.2 $M_\odot$, and their distances are about 20 pc.

131.082 **Searches for microwave spectral line radiation from some molecules in the interstellar medium.**
T. Cato, J. Elldér, B. Höglund, O. E. H. Rydbeck, B. Rönnäng, A. Sume.
Astron. Astrophys., Vol. 21, 435 - 440 (1972).

The Onsala 25.6 m radio telescope equipped with travelling wave maser radiometers at the appropriate frequencies has been used in searches for absorption or emission lines from a number of different organic and inorganic molecules in the direction of several radio sources. The molecules include OH (the $^2\Pi_{3/2}$, $J = 3/2$, $v = 1$ vibrational state; and the second and third harmonics of some ground state transitions), CH (the ground state $^2\Pi_{1/2}$, $J = 1/2$, and the $^2\Pi_{3/2}$, $J = 3/2$ excited state), SO_2, HCN, H_2CS, NH_2CHO, and CH_3CHO. All searches gave negative results, with possible exception for H_2CS. Details of the observations are tabulated.

131.083 **Ultraviolet pumping of H_2O cosmic masers?**
V. S. Strelnitskij, A. F. Yukin.
Astron. Tsirk., No. 714, p. 1 - 3 (1972). In Russian.

131.084 **The results of the Fabry-Perot Hα-spectrometer observations of faint H II regions of the Milky Way.**
V. F. Zhidkov.
Astron. Tsirk., No. 718, p. 6 - 8 (1972). In Russian.

131.085 **Interstellar magnesium abundances and electron density in the direction of Orion and Cassiopeia.**
A. Boksenberg, B. Kirkham, W. A. Towlson, T. E. Venis, B. Bates, G. R. Courts, P. P. D. Carson.
Nature, Phys. Sci., Vol. 240, 127 - 130 (1972).

The interstellar lines of MgII and MgI in the spectra of γ Cas and γ, β, δ, ϵ and ζ Ori have been observed with a balloon-borne objective grating spectrograph. The column density of Mg relative to H is close to the cosmic value for the Orion stars, but is somewhat lower for γ Cas. Excluding β Ori, the average value of electron density obtained for the cool clouds in the lines of sight to these stars is $(2.8^{+1.6}_{-1.0})$ 10^{-3} cm^{-3}, assuming a cloud temperature of 60 K.

131.086 **Anionic species of Fe in interstellar dust.**
P. G. Manning.
Nature, Vol. 240, 547 (1972).

It is pointed out that interstellar dust may contain anionic species of Fe, namely FeO_4^{2-} and FeO_2^-, which are thermally degradable to a likely interstellar material α-Fe_2O_3. Octahedral-Fe^{3+} bands at 6180 Å and 4430 Å are the strongest diffuse interstellar bands, whereas in supernovae spectra the 5100 Å bands are often at least as intense. It seems that the grains responsible for supernova absorption, in contrast to the diffuse interstellar grains, have not been heated sufficiently in a reducing atmosphere to convert FeO_4^{2-} to α-Fe_2O_3.

131.087 **A new interstellar line: The 5_1-4_0 (E_2) transition in methyl alcohol.**
B. Zuckerman, B. E. Turner, D. R. Johnson, P. Palmer, M. Morris.
Astrophys. Journ., Vol. 177, 601 - 607 (1972).

An emission line has been detected at 84.5 GHz in the direction of the radio continuum source Sgr B2. The new line frequency has been found to coincide with a laboratory measurement of the 5_1-4_0 (E_2) transition of methyl alcohol. Calculations of the spontaneous emission rates from all energy levels of astrophysical interest have been performed. Additional searches for some other molecules were carried out in the 80−92 GHz region without success.

131.088 **Detection of the 4_1-3_0 (E_2) line of interstellar methyl alcohol.**
B. E. Turner, M. A. Gordon, G. T. Wrixon.
Astrophys. Journ., Vol. 177, 609 - 617 (1972).

The $4_1 \rightarrow 3_0$ (E_2) transition of methyl alcohol (CH_3OH) at 36,169 MHz has been detected with a total strength of $\gtrsim 120$ flux units (f.u.) in two extended clouds in Sgr B2. The line is not present to a limit of ~30 f.u. in Ori A or in several other sources. We give limits to other molecular lines searched for but not detected in the range 28−40 GHz.

131.089 **Interstellar isocyanic acid.**
L. E. Snyder, D. Buhl.
Astrophys. Journ., Vol. 177, 619 - 623 (1972).

Interstellar isocyanic acid (HNCO) has been detected in emission through the $4_{04}-3_{03}$ groundstate rotational transition at 3.4 mm. Out of nine galactic sources surveyed, HNCO has been observed only in the direction of Sgr B2 and possibly W51. The Sgr B2 emission pattern is fairly extensive, and our observations indicate that the peak HNCO emission is ~2′ north of the OH position. A search for the 7−6 transition of interstellar OCS yielded negative results.

131.090 **An interstellar emission line from isocyanic acid at 1.4 centimeters.**
D. Buhl, L. E. Snyder, J. Edrich.
Astrophys. Journ., Vol. 177, 625 - 628 (1972).

We have confirmed our earlier identification of interstellar isocyanic acid (HNCO) by detection of a second transition at 1.4 cm in the galactic-center source Sgr B2. An excitation temperature of $12.8°$ K is obtained from the ratio of the two lines. A search for the $1_{01}-0_{00}$ transition of formic acid (HCOOH) was negative; hence we are unable to confirm the interstellar HCOOH detection report of Zuckerman et al.

131.091 A search for interstellar ^{14}CO.
P. R. Schwartz, W. J. Wilson.
Astrophys. Journ., (Letters), Vol. 177, L129 - L130 (1972).
A search for interstellar ^{14}CO was made at 105 871.11 MHz in seven galactic ^{12}CO sources with negative results.

131.092 Polarization of light by circumstellar material.
B. H. Zellner, K. Serkowski.
Publ. Astron. Soc. Pacific, Vol. 84, 619 - 626 (1972). — Invited symposium paper presented at the Santa Cruz meeting of the Astronomical Society of the Pacific, 26—29 June 1972.
Our purpose is to give and overview of the state of the art in the study of circumstellar polarizations, with emphasis on a few of the best-studied and most representative objects. A simple scattering theory for polarization by circumstellar material is described. This theory finds applicability to early-type stars with emission lines in their spectra, to novae, and to red and yellow variables. Finally highly polarized objects are discussed.

131.093 Circumstellar infrared emission.   W. A. Stein.
Publ. Astron. Soc. Pacific, Vol. 84, 627 - 632 (1972)
Invited symposium paper presented at the Santa Cruz meeting of the Astronomical Society of the Pacific, 26—29 June 1972.

131.094 A spectroscopic study of the molecular constituents of dark clouds.
F. H. Chaffee, Jr., S. E. Strom, B. Lutz, J. G. Cohen.
Publ. Astron. Soc. Pacific, Vol. 84, 639 - 640 (1972). — Abstr. Astron. Soc. Pacific.

131.095 Interstellare und intergalaktische Materie im Licht der Kosmogonie.   K.-H. Schmidt.
Sterne, 48. Jahrgang, p. 207 - 212 (1972).

131.096 Light molecules and dark clouds. Part II.
D. Buhl.
Mercury, (Journ. Astron. Soc. Pacific), Vol. 1, No. 6, p. 4 - 8 (1972).

131.097 Theoretical structure and spectrum of a shock wave in the interstellar medium: The Cygnus Loop.
D. P. Cox.
Astrophys. Journ., Vol. 178, 143 - 157 (1972).
The flow structure, spectrum, and luminosity are calculated for a model of a plane shock wave followed by a region in which the shock-heated gas cools to a low temperature and is compressed to very high densities. These calculations are compared with observations of the Cygnus Loop filaments, assuming that they are such regions viewed edge-on. The agreement is very good for reasonable choices of model parameters.

131.098 $^{13}C^{16}O/^{12}C^{18}O$ ratios in nine H II regions.
A. A. Penzias, K. B. Jefferts, R. W. Wilson, H. S. Liszt, P. M. Solomon.
Astrophys. Journ., (Letters), Vol. 178, L35 - L38 (1972).
Observations of the $J = 1$ to 0 transitions in $^{13}C^{16}O$ and $^{12}C^{18}O$ were made in nine H II regions. In each source the intensity ratio of these lines was found to be within a factor of 2 of the terrestrial ratio of 5.5.

131.099 Interstellar molecular clouds.   N. Kaifu.
Kagaku, (Japan), Vol. 42, No. 1, p. 2 - 10 (1972).
Review article on the following topics: Origin of space radiowave spectroscopy; Interstellar molecular clouds; Inter-stellar space chemistry; Anomalous behavior of the interstellar molecules and the future perspective of space radiowave chemistry.

131.100 The formation of diatomic molecules in interstellar clouds.   P. M. Solomon, W. Klemperer.
Astrophys. Journ., Vol. 178, 389 - 421 (1972).
The rates of a number of homogeneous gas-phase chemical reactions likely to occur in clouds composed primarily of atomic hydrogen, with $10 < n_H < 1000$, are estimated. It is shown that the primary mechanism for producing diatomic molecules from atoms is the radiative association of C + H and $C^+ + H$ yielding CH and CH^+. Dielectronic recombination of CH^+ is important in producing CH, destroying CH^+ and increasing the ratio CH/CH^+ relative to C/C^+. Other important processes lead to the production of C_2, CN, and CO, but not NO, N_2, or O_2. The steady-state abundance of electrons, atoms, ions, and diatomic molecules composed of H, C, N, and O is determined as a function of the H density, the kinetic temperature, and the ultraviolet optical depth, which governs the photodissociation and ionization rates. A comparison of the theory with optical observations of the interstellar medium, particularly the well-studied features in the ζ Oph spectrum, shows good qualitative and quantitative agreement.

131.101 Laboratory observation of the $1_{01} \leftarrow 0_{00}$ transitions for the HCO and DCO free radicals by microwave spectroscopy.   S. Saito.
Astrophys. Journ., (Letters), Vol. 178, L95 - L97 (1972).
The frequencies of the hyperfine components of the $1_{01} \leftarrow 0_{00}$ transitions for the HCO and DCO radicals have been measured with laboratory microwave spectroscopy.

131.102 Restrictions on cosmic maser intensity and the possibility of detecting new OH and H_2O radio sources in a short-term sky survey.
G. M. Rudnitskij, V. S. Strel'nitskij.
Astron. Zhurn. Akad. Nauk SSSR, Vol. 49, 1323 - 1325 (1972). In Russian. English translation in Soviet Astron. AJ, Vol. 16, No. 6.
Some restrictions on the intensity of OH and H_2O line radio sources are considered. It is shown that maser sources having flux densities comparable to those of observed bright ones can be detected in a short-term sky survey with a relatively simple technique.

131.103 Studies of small H II regions. I. Infrared photometry of Sharpless 138, 152, and 270.
J. A. Frogel, S. E. Persson.
Astrophys. Journ., Vol. 178, 667 - 672 (1972).
Multiaperture photometric observations from 1.6 to 3.5 μ of three small H II regions, Sharpless 138, 152, and 270, are presented. These three objects are shown to be representative of a distinct class of H II regions. The spatial and spectral distributions of the flux are similar for the three objects.

131.104 A study of the unidentified interstellar diffuse features.   C.-C. Wu.
Astrophys. Journ., Vol. 178, 681 - 699 (1972).
The interstellar diffuse lines $\lambda\lambda$ 5780, 5797 have been observed in 66 stars using the Washburn Observatory echelle spectrograph coupled with an image tube. This investigation proposes that $\lambda\lambda$ 5780, 5797 are produced by the pure electronic transitions of the impurity centers in solid grains, with the broad, shallow absorption feature shortward of λ 5780 as the absorption sideband associated with the parent pure electronic line. Also, evidences are presented to support the idea that the diffuse band λ 4430 may be produced by the preionization of H^-.

131.105 Anomalous hyperfine lines in formaldehyde in a

dust cloud.   N. H. Dieter.
Astrophys. Journ., (*Letters*), Vol. 178, L133 - L137 (1972).

Abnormal intensities have been found in the hyperfine lines of the 6-cm formaldehyde transition observed in dust clouds. The $0 \to 1$ line is enhanced in at least one cloud by an order of magnitude.

131.106   Searches for interstellar molecules and studies of recombination lines from the diffuse matter in the Galaxy.   B. T. Cato.
Res. Lab. Electronics, Onsala Space Obs., Chalmers Univ. Techn., Gothenburg, Sweden, Res. Rep. No. 108, p. 1 - 34 = Publ. Onsala Space Obs., No. 63 (1972).

Contents: Rotational transition frequencies for some molecules of astrophysical interest; Vibrational effects on the OH Λ-doubling state; Radio recombination lines from the diffuse matter in the interstellar space.

131.107   Collisional and radiative processes in interstellar molecules. Part I.
M. M. Litvak, edited and appended by A. Sume.
Res. Lab. Electronics, Onsala Space Obs., Chalmers Univ. Techn., Gothenburg, Sweden, Lecture Note No. 5, 220 pp. = Publ. Onsala Space Obs., No. 73 (1972).

The purpose of these lectures is to elucidate the likely causes of the non-equilibrium effects that appear so strikingly in the OH and water maser (anomalous emitters) and the OH and formaldehyde antimasers (anomalous absorbers), and that appear more subtley in the recently discovered cyanoacetylene and methanol molecules, for example. A table lists those interstellar molecules known to us at this time (June, 1972) by their microwave lines, their date of discovery, the quantum numbers for the levels involved in the microwave transitions, the line rest frequencies, the astronomical objects they were observed in, the estimated maximum projected densities and the Einstein coefficient for the spontaneous radiative transition.

131.108   Compact non-thermal source from collision of two gas clouds.   D. P. Cox.
Bull. American Astron. Soc., Vol. 4, 411 (1972). – Abstr. AAS.

131.109   Temperatures of H II regions from the [N II] forbidden line intensity ratio.
T. J. Bohuski, R. J. Dufour.
Bull. American Astron. Soc., Vol. 4, 424 (1972). – Abstr. AAS.

131.110   Trivalent transition-metal ions in interstellar dust.   P. G. Manning.
Nature, Phys. Sci., Vol. 239, 87 - 88 (1972).

The intensities of the crystal-field absorption bands of Fe^{3+} and Mn^{3+} in oxide and silicate minerals are discussed in light of recent suggestions that type I supernova absorption bands and some diffuse interstellar bands are caused by trivalent transition-metal ions.

131.111   Modelos de nubes cósmicas en contracción gravitatoria.   W. G. L. Pöppel.
Contr. Inst. Argentino Radioastron., Buenos Aires, No. 34, 9 + 162 pp. (1971). – Thesis Fac. Ciencias Exact. y Nat., Univ. Nacional, Buenos Aires.

131.112   First detection of extragalactic carbon monoxide.   W. J. Wilson, E. E. Epstein, P. R. Schwartz.
IAU Circ., No. 2447 (1972).

131.113   Untersuchungen zur Aufheizung des interstellaren Gases durch kosmische Strahlung.   H. Billing.
Separate print Max-Planck-Inst. Phys. Astrophys., Inst. Extraterr. Phys., München. MPI–PAE/Extraterr. 72, 1 + 132 pp. (1972).

131.114   A radical, its radiation and the maser.
Nguyen-Quang-Rieu.
Sci. Progr. Découverte, No. 3442, p. 24 - 32 (1972).
In French.

131.115   A photon rest mass and the absorption of longitudinal electric waves in interstellar space.
R. Burman.
Journ. Phys. A, General Phys., Vol. 5, L78 - L80 (1972).

The author treats the absorption of longitudinal electric waves in a cold plasma; upper limits for absorption in propagation from the galactic centre to the earth are estimated.

131.116   Microwave radiation of water vapor in galactic sources.   W. T. Sullivan, III.
Thesis, Univ. Maryland, College Park. [Available from Univ. Microfilms, Ann Arbor, Mich., USA. Order No. 72-12857], 125 pp. (1971). – See Phys. Abstr., Vol. 75, No. 69016 (1972).

131.117   Gain parameters of interstellar masers.   A. C. Selden.
Phys. Letters A, (*Netherlands*), Vol. 40A, 355 - 356 (1972).

The OH and H_2O masers at the boundaries of HII regions are only partially saturated even for large solid angle. The conditions on H_2O are more stringent than for OH emission.

131.118   Observations of recombination lines at Ku-band.   G. D. Papadopoulos, K. Y. Lo, P. Rosenkranz, E. J. Chaisson.
Quarterly Progr. Rep. Res. Lab. Electron. Mass. Inst. Technology, No. 104, p. 74 - 76 (1972).

Measurements of excited hydrogen 73α recombination lines in the galactic HII regions of Orion A, W3, W49, and W51 have been performed at 16.56 GHz. The purpose of these short-wavelength observations was to help distinguish between local thermodynamic equilibrium (LTE) and non-LTE models of atomic energy $n \lesssim 80$.

131.119   Statistics of the radiation from astronomical masers.   N. J. Evans, II., R. E. Hills, O. E. H. Rydbeck, E. Kollberg.
Phys. Rev. A, General Phys., Vol. 6, 1643 - 1647 (1972).

The results of an experimental determination of the statistical properties of radiation from OH maser sources are reported and interpreted. The radiation is found to have Gaussian statistics with no deviations greater than 1%.

131.120   Molecules in space.   C. Henderson.
Contemporary Phys., (*GB*), Vol. 13, 479 - 499 (1972).

This review tells how the emission and absorption of spectral lines, revealing the presence of molecules, can furnish estimates of their numbers, densities and isotopic abundances. Departures from thermal equilibrium, including maser and reverse maser action are mentioned. The radio-telescopes and their spectrometers are described.

131.121   Alignment of interstellar grains.   M. P. Goldstein.
Thesis, California Inst. Technology, Pasadena. [Available from Univ. Microfilms, Ann Arbor, Mich., USA. Order No. 71-17379], 134 pp. (1971).

Considers the alignment of interstellar dust grains with respect to the magnetic field of our Galaxy. The alignment is found for several values of magnetic field strength internal grain temperature, and grain shape. The following processes which affect the alignment are considered in detail: (i) a dissipative magnetic torque; (ii) the collisions of the grain with interstellar hydrogen; (iii) the non-zero internal temperature of the grain.

131.122 **Polarization observations of nonstable stars and extragalactic objects. I. Equipment, method of observation and reduction.**   N. M. Shakhovskoy, Yu. S. Efimov.
Izv. Krymskoj Astrofiz. Obs., Vol. 45, 90 - 110 (1972).
In Russian.

The new photon-counting polarimeter of the Crimean Astrophysical Observatory is intended for linear polarization measurements of extremely faint and rapidly changing objects. A computer is used to reduce the observations. Linear transformation of Stokes parameters is applied to reduce the observed instrumental parameters to their true values. Observations of standard stars are used to obtain the coefficients of transformation formulae separately for different colours (U, B, V, R) and observational periods. From the analysis of the reduction coefficients the basic characteristics of the instrumental effects were estimated. Their wavelength dependences have been obtained too.

131.123 **Interstellar absorption of light and the distribution of stars around the star cluster NGC 6834.**
V. I. Voroshilov, N. B. Kalandadze, V. I. Kuznetsov.
Byull. Abastumansk. Astrofiz. Obs., No. 43, p. 55 - 66 (1972). In Russian.

A field of 1 square degree in the Milky Way with its center coinciding with the open cluster NGC 6834 has been studied. The interstellar absorption and spatial distribution of stars of different spectral classes have been investigated on the basis of B, V magnitudes and of spectral classes for 645 faint stars (up to V = 14^{m}5).

131.124 **Interstellar absorption of light and the distribution of stars around the star cluster NGC 7654.**
V. I. Voroshilov, N. B. Kalandadze, V. I. Kuznetsov.
Byull. Abastumansk. Astrofiz. Obs., No. 43, p. 67 - 78 (1972). In Russian.

In an area of 1 square degree in the Milky Way the center of which coincides with the open cluster NGC 7654 the interstellar absorption of light and the spatial distribution of stars of different spectral types have been investigated on the basis of B, V magnitudes and spectral classes for 730 faint stars (up to V = 14^{m}5).

131.125 **Interstellar absorption of light in the area of Orion.**
M. S. Kazanasmas.
Byull. Abastumansk. Astrofiz. Obs., No. 43, p. 79 - 94 (1972). In Russian.

Interstellar absorption of light in the area of Orion has been studied on the basis of the author's catalogue of magnitudes, color-indices, spectral and luminosity classes of stars. (In print).

131.126 **Radio observations of massive molecular clouds.**
N. Z. Scoville.
Thesis Fac. Pure Sci., Columbia Univ., 147 pp. (1972).

Deals with formaldehyde, carbon monoxide and OH in the directions of the galactic centre. Sources W49, W51 and dark clouds are discussed. – *RXM*

131.127 **Ring structure of a neutral gas cloud studied in a one-dimensional expansion into space.**
R. E. Davidson.
National Aeronautics Space Adminstration, Greenbelt, NASA Techn. Note, TN D-6760, 24 pp. (1972).

131.128 **The degree of ionisation in interstellar space.**
M. Grewing, U. Mebold, K. Rohlfs.
Accad. Nazionale Lincei, Anno 369, Quaderno No. 162, p. 317 - 321 = Mitt. Astron. Inst. Bonn No. 131 (1972).

131.129 **Two new compact H II regions.**   Yu. N. Parijskij.
Astron. Tsirk., No. 721, p. 1 - 3 (1972). In Russian.

131.130 **18-cm OH emission from W 49 A at negative velocities.**   M. I. Pashchenko.
Astron. Tsirk., No. 721, p. 3 - 5 (1972). In Russian.

131.131 **On the capture of interstellar dust by the solar system.**   N. I. Komarnitskaya.
Astron. Tsirk., No. 722, p. 5 - 7 (1972). In Russian.

131.132 **On the capture of interstellar dust by Saturn and Neptune.**   N. I. Komarnitskaya.
Astron. Tsirk., No. 722, p. 7 - 8 (1972). In Russian.

131.133 **18-cm OH absorption in the direction of Cygnus X.**   M. I. Pashchenko.
Astron. Tsirk., No. 733, p. 1 - 3 (1972). In Russian.

131.134 **Distribution of molecular hydroxyl in the DR 21 region.**   M. I. Pashchenko.
Astron. Tsirk., No. 733, p. 3 - 4 (1972). In Russian.

131.135 **Determination of interstellar extinction in the vicinity of the sun by means of spectrophotometry of late-type stars.**   V. M. Tereshchenko, A. V. Kharitonov.
Astron. Tsirk., No. 737, p. 7 - 8 (1972). In Russian.

Spin alignment of OH molecules by directed infrared radiation.   See Abstr. 022.006.

The e.p.r. spectrum of vibrationally excited hydroxyl radicals.   See Abstr. 022.058.

Populations of excited atoms: Sensitivity to low-energy cross-sections.   See Abstr. 022.067.

Molecular calculations concerning a new candidate for the unidentified emission line at 89.190 GHz. See Abstr. 022.068.

Low-energy elastic and fine-structure excitation scattering of ground-state C^{+} ions by hydrogen atoms. See Abstr. 022.093.

Radiative-lifetime studies of the emission continua of the hydrogen and deuterium molecules. See Abstr. 022.111.

CH and CH^{+} formation in ion-molecule reactions. See Abstr. 022.114.

The condensation of H$_{2}$ and D$_{2}$: Astrophysics and vacuum technology.   See Abstr. 022.115.

Statistics of the radiation from astronomical masers. See Abstr. 034.107.

The abundance of helium in the cosmos—I. See Abstr. 061.005.

Effects of erosion and fragmentation on the mass distribution of colliding particles.   See Abstr. 061.047.

A photon rest mass and the dispersion of longitudinal electric waves in interstellar space. See Abstr. 062.067.

A search for diffuse interstellar features in stars with circumstellar dust shells.   See Abstr. 064.024.

On the coupling of grains to the gas in circumstellar envelopes.   See Abstr. 064.049.

Possible influence on the interstellar Lyman-α absorption measurements from hot star circumstellar envelopes. See Abstr. 064.060.

Ultraviolet stars and the interstellar gas. See Abstr. 065.042.

Formation of protostars by thermal instability. See Abstr. 065.092.

Hier entstehen Sterne.   See Abstr. 065.151.

Accretion onto black holes: The emergent radiation spectrum.   See Abstr. 066.016.

Interaction of the solar wind with the neutral component of the interstellar gas.   See Abstr. 074.059.

Review of the heliosphere. II.   See Abstr. 074.105.

Annual variation of the interplanetary He$^+$ velocity distribution at 1 AU.   See Abstr. 106.021.

Electromagnetic instabilities produced by neutral-particle ionization in interplanetary space. See Abstr. 106.022.

Radial velocities of southern B stars determined at the Radcliffe Observatory – VI. Stars in H II regions. See Abstr. 112.002.

Individual reddening laws from O type stars. I. Computation method, first results.   See Abstr. 113.009.

Interstellar absorptions and colour-excesses in the Vilnius photometric system.   See Abstr. 113.048.

On the ratio of total to selective absorption for carbon stars.   See Abstr. 113.061.

Ten-micron spectroscopy of circumstellar shells. See Abstr. 114.019.

The measurement of polarized 10-micron radiation from cool stars with circumstellar shells. See Abstr. 114.020.

Equivalent width of interstellar molecular lines. II. CO, SiO, CS, H$_2$ and HD in the interstellar spectrum of ξ Persei.   See Abstr. 114.037.

Interstellar lines in the ultraviolet spectrum of Zeta Ophiuchi.   See Abstr. 114.043.

OH in the Hoffman infrared sources. See Abstr. 114.065.

A measurement of the interstellar $^{12}C/^{13}C$ ratio. See Abstr. 114.075.

Rocket-ultraviolet spectra of eight stars in Ophiuchus and Scorpius.   See Abstr. 114.089.

Rocket-ultraviolet spectra of six stars in Perseus. See Abstr. 114.090.

Infrared stars with strong 1665/1667-MHz OH microwave emission.   See Abstr. 114.099.

UV interstellar absorption lines in the spectrum of ζ Pup.   See Abstr. 114.116.

Circular polarimetry of fifteen interesting objects. See Abstr. 116.015.

The heating of the interstellar medium by Mira stars. See Abstr. 122.091.

Polarimetric observations of nonstable stars and extragalactic objects. II. EV Lac flare polarization. See Abstr. 122.131.

Classification of supernova remnants and H II regions from their recombination line emission. See Abstr. 125.021.

Interstellar nitrogen-15 and U169.3–possibly a new methanol line.   See Abstr. 132.015.

Spectrophotometric studies of gaseous nebulae. See Abstr. 132.016.

Spectrophotometric investigations of galactic nebulae. VII. Compact H II regions in NGC 2024 and NGC 7538. See Abstr. 132.023.

Spectrophotometric investigations of galactic nebulae. VIII. Compact H II regions in NGC 7635, IC 1470, Sh 257, NGC 1931 and NGC 2175.   See Abstr. 132.024.

A recombination-line study of the Sagittarius B$_2$ radio complex.   See Abstr. 141.042.

The formaldehyde absorption of W33. See Abstr. 141.053.

Detection of cyanoacetylene at 18 GHz. See Abstr. 141.087.

Observations of methanol in Sagittarius B2 at 48 GHz.   See Abstr. 141.103.

The interstellar scintillation pattern of PSR 0329 + 54.   See Abstr. 141.503.

Interstellar cloud properties revealed by pulsars. See Abstr. 141.545.

The interaction of Sco X-1 with its environment. See Abstr. 142.012.

OH and formaldehyde absorption in the direction of Cygnus X-3.   See Abstr. 142.073.

On the nature of X Persei; evidence from the 1957 outburst.   See Abstr. 142.144.

The role of gaseous dissipation in density waves of finite amplitude.   See Abstr. 151.006.

A new solution to the accretion problem. See Abstr. 151.026.

Recombination line emission from the galactic ridge.   See Abstr. 155.002.

A study of the interstellar extinction in the Carina-Centaurus region.   See Abstr. 155.007.

Diffuse galactic FUV radiation and interstellar dust grains.   See Abstr. 155.011.

The distribution of stars and obscuring matter in a

Monoceros field.   See Abstr. 155.013.

A survey of linear polarization at 1415 MHz. IV. Discussion of the results for the galactic spurs. See Abstr. 155.017.

On the possible existence of different interstellar extinction laws in the spiral arms and in the field regions. See Abstr. 155.018.

An H I velocity—longitude diagram for the southern Milky Way.   See Abstr. 155.019.

A kinematic determination of the ratio of total to selective absorption.   See Abstr. 155.047.

Some comments on "The recombination line emission from the galactic ridge".   See Abstr. 155.048.

Molecular clouds in the galactic center region: Carbon monoxide observations at 2.6 millimeters. See Abstr. 155.052.

Observations of the outer spiral structure of the Milky Way and its relation to the high velocity clouds. See Abstr. 155.054.

A longitude survey of radio recombination lines from the diffuse interstellar medium. See Abstr. 157.001.

A search for H I in elliptical galaxies. See Abstr. 158.064.

Dark nebulae in the Large Magellanic Cloud. See Abstr. 159.001.

132   Emission Nebulae, Reflection Nebulae

132.001   Spectrophotometric investigation of the cometary nebula NGC 2261.
M. A. Kazarian, E. Ye. Khachikian.
Astrofizika, Vol. 8, 17 - 31 (1972). In Russian. – English translation in Astrophysics, Vol. 8, No. 1.
The results of a detailed spectrophotometry of the cometary nebula NGC 2261 and its nucleus R Mon are presented. The spectra obtained with the 200"-, 84"-, 120"- and 36"-telescopes of the Hale, Kitt Peak and Lick Observatories have been used.

132.002   Two astrophotography projects.   J. L. Matteson.
Sky Telescope, Vol. 44, 200 - 201 (1972).

132.003   Nebulae of the southern Milky Way. An atlas.
G. Lyngå, N. Hansson.
Astron. Astrophys., Suppl. Ser., Vol. 6, 327 - 414 (1972).
An atlas of the southern Milky Way in wavelengths around Hα is presented together with a catalogue of identified stars, nebulae and open clusters.

132.004   Bright nebulae near concentrations of high-velocity gas.   R. Minkowski, J. Silk, R. S. Siluk.
Astrophys. Journ., (*Letters*), Vol. 175, L123 - L125 (1972).
We report some coincidences in position on the sky between apparent emission nebulae at $|b^{II}| > 10°$ and concentrations of high-velocity gas ($|v| > 100$ km s^{-1}).

132.005   The dynamical effects of stellar mass loss on diffuse nebulae.   J. E. Dyson, J. de Vries.
Astron. Astrophys., Vol. 20, 223 - 232 (1972).
The dynamical effect of mass loss from an early type star on a diffuse nebula is examined using a similarity method. The maintenance of velocities greatly in excess of the sound speed in the nebular gas for periods $\gtrsim 10^4$ years requires very high mass loss rates. It is therefore very unlikely that the high velocities measured in diffuse nebulae by Sheglov (1968) and

Meaburn (1970, 1971) can be attributed to the operation of this mechanism.

132.006   Der Gum-Nebel.   T. Schmidt-Kaler.
SuW, Vol. 11, 220 - 223 (1972).

132.007   New results of observations of the nebula A21 (YM 29, Medusa).   T. A. Lozinskaja.
Astron. Tsirk., No. 668, p. 1 - 2 (1972). In Russian.

132.008   Spectrophotometric investigations of diffuse nebulae. III. Emission lines in the northern part of the nebula M 20. Supernova remnant in the direction of M 20.
Yu. I. Glushkov, E. S. Eroshevich, Z. V. Karyagina.
Astron. Tsirk., No. 676, p. 1 - 4 (1972). In Russian.

132.009   Spectrophotometric investigations of diffuse nebulae. IV. The electron density in the central regions of NGC 6618.   Yu. I. Glushkov, Z. V. Karyagina.
Astron. Tsirk., No. 676, p. 4 - 7 (1972). In Russian.

132.010   Interferometric observations of NGC 2359 – the nebula around a Wolf-Rayet star.
T. A. Lozinskaya.
Astron. Tsirk., No. 678, p. 7 - 8 (1972). In Russian.

132.011   He 2—10, an extragalactic object.
L. N. Kondratjeva.
Astron. Tsirk., No. 683, p. 7 - 8 (1972). In Russian.

132.012   Spectrophotometric investigations of galactic nebulae. V. The lines of [O III] in the spectrum of the cometary nebula NGC 2245.   Yu. I. Glushkov.
Astron. Tsirk., No. 692, p. 1 - 2 (1972).   In Russian.

132.013   Spectrophotometric investigations of galactic nebulae. VI. NGC 2359 and NGC 1514.

Yu. I. Glushkov.
Astron. Tsirk., No. 692, p. 2 - 4 (1972). In Russian.

132.014 Optical observations of an infrared source in NGC 2264. E. G. Schmidt.
Astrophys. Journ., (Letters), Vol. 176, L69 - L71 (1972).

The presence of a strong infrared source in NGC 2264 has previously been reported. Spectrophotometric observations at optical wavelengths were obtained of a knot of nebulosity which appears to be associated with the infrared object. The total flux of the knot was found along with the slope of its energy distribution at visual wavelengths. These measurements seem to fit well onto the infrared energy distribution.

132.015 Interstellar nitrogen-15 and U169.3–possibly a new methanol line. R. W. Wilson, A. A. Penzias, K. B. Jefferts, P. Thaddeus, M. L. Kutner.
Astrophys. Journ., (Letters), Vol. 176, L77 - L79 (1972).

The $J = 2 \to 1$ lines of $H^{13}C^{14}N$ and $H^{12}C^{15}N$ have been detected in the Ori A infrared nebula. A new line has also been found in the infrared nebula whose rest frequency is 169336.1 ± 0.7 MHz; it is possibly the $J = 10$, $K = 1$–0 transition of the E_1 symmetry species of methanol.

132.016 Spectrophotometric studies of gaseous nebulae. C. T. Hua.
Astron. Astrophys., Vol. 21, 105 - 109 (1972). In French.

The energy distribution in the continuum of the Orion nebula has been determined spectrophotometrically with respect to several standards of the Barbier-Chalonge-Divan classification. On the assumptions of an optically thin nebula for the range $\lambda\lambda$ 3000–5000, an attempt is made to calculate the contribution of scattered light in the central part of the nebula.

132.017 Optical observations of the Simeis 59 nebula – a possible supernova remnant.
V. F. Esipov, T. A. Lozinskaya, V. I. Shenavrin.
Astron. Tsirk., No. 702, p. 1 - 3 (1972). In Russian.

132.018 On the association of the nebula M8 and the cluster NGC 6530. P. S. Hoover.
Bull. American Astron. Soc., Vol. 4, 318 (1972). – Abstr. AAS.

132.019 The albedo of interstellar particles as determined from reflection nebulae. W. F. Rush.
Bull. American Astron. Soc., Vol. 4, 318 - 319 (1972). Abstr. AAS.

132.020 An 85α recombination line survey of the Orion nebula.
L. H. Doherty, L. A. Higgs, J. M. MacLeod.
Astrophys. Letters, Vol. 12, 91 - 98 (1972).

Observations of the 85α (10,522 MHz) recombination lines of hydrogen, helium, and carbon have been obtained at a grid of 49 points in the Orion nebula. The variations across the nebula of radial velocity, electron temperature, hydrogen line width, and ionized helium abundance are presented in the form of contour diagrams. The distribution of ionized carbon is discussed.

132.021 An unusual nebula associated with HD 87634.
S. van den Bergh.
Publ. Astron. Soc. Pacific, Vol. 84, 594 - 595 (1972).

The star HD 87634, which exhibits a nova-like spectrum, is found to be embedded in a nebula of unusual structure.

132.022 Collision strengths for the 10.5 μ transition of S^{3+}.
M. Brocklehurst.
Monthly Notices Roy. Astron. Soc., Vol. 160, 19P - 21P (1972).

In many gaseous nebulae, the observed intensities of the infra-red transition of S^{3+} differ widely from the predicted values. A re-evaluation of the collision strengths using distorted-wave and close-coupling techniques is in reasonable agreement with previous estimates.

132.023 Spectrophotometric investigations of galactic nebulae. VII. Compact H II regions in NGC 2024 and NGC 7538. Yu. I. Glushkov, Z. V. Karyagina.
Astron. Tsirk., No. 711, p. 1 - 4 (1972). In Russian.

132.024 Spectrophotometric investigations of galactic nebulae. VIII. Compact H II regions in NGC 7635, IC 1470, Sh 257, NGC 1931 and NGC 2175.
Yu. I. Glushkov, Z. V. Karyagina.
Astron. Tsirk., No. 711, p. 4 - 7 (1972). In Russian.

132.025 An evolutionary thermal model for the Cygnus Loop. D. P. Cox.
Astrophys. Journ., Vol. 178, 169 - 173 (1972).

A rather approximate but self-consistent model which can explain the density, spectrum, luminosity, and velocity of the optical filaments of the Cygnus Loop as well as the spectrum, luminosity, and shell structure of its X-ray source is derived. Assuming the distance to the Loop is 770 pc, the initial blast energy is found to be $E_0 \sim 4 \times 10^{50}$ ergs, the average interstellar density $n_0 \sim 1$ cm^{-3}, and the apparent age 45,000 years. Comparison is made with the alternative adiabatic blast-wave model which explains only the X-ray source.

132.026 Very-low-excitation compact nebulae.
N. Sanduleak, C. B. Stephenson.
Astrophys. Journ., Vol. 178, 183 - 187 (1972).

A finding list is given for 23 emission-line objects which show an extremely-low-excitation nebular component in their spectra. Some of these may be planetary nebulae in an early stage of development. Such objects are rare, comprising only about five percent of the planetary-like nebulae detected in an objective-prism survey of the southern Milky Way.

132.027 High-velocity gas near the Gum nebula.
A. D. Thackeray, P. R. Warren.
Monthly Notices Roy. Astron. Soc., Vol. 160, 23P - 25P (1972).

Radcliffe coudé spectra of HR 3462 and 3527 confirm Wallerstein and Silk's observation of high velocity Ca II components but show three additional components which complicate the interpretation.

132.028 Radial velocities from the radio recombination lines of Orion A.
A. S. J. Batchelor, M. Brocklehurst.
Monthly Notices Roy. Astron. Soc., Vol. 160, 27P - 30P (1972).

The increase in the values of the radial velocities of Orion A radio recombination lines with n may be explained in terms of masing and a radial expansion of the nebula.

132.029 On the polarization of emission lines in diffuse nebulae with the spectrum C+E.
D. A. Rozhkovskij, K. G. Dzhakusheva.
Trudy Astrofiz. Inst., Alma-Ata, Vol. 19, 72 - 78 (1972). In Russian.

Light scattering in Strömgren's dust zone is considered. In terms of the idealized model of a diffuse nebula with the spectrum C+E, the polarimetric analysis of the emission line Hα emitted by the periphery of the nebula is given.

132.030 Spectrophotometric survey of diffuse galactic nebulae.
Yu. I. Glushkov, E. S. Eroshevich, Z. V. Karyagina.
Trudy Astrofiz. Inst., Alma-Ata, Vol. 19, 79 - 93 (1972). In

Russian.

Spectra of 98 diffuse galactic nebulae in the 6200–7000 Å and 4500–5500 Å wavelength ranges are obtained. For some of these nebulae relative intensities of lines $H\alpha/6584$, $6717/6731$, $H\beta/5007$, and absolute intensities of the $H\alpha$ line are found.

132.031 Absolute intensity of the $H\alpha$ line in the spectrum of the nebulae NGC 2068 and S-57.
Z. V. Karyagina, Yu. I. Glushkov.
Trudy Astrofiz. Inst., *Alma-Ata*, Vol. 19, 94 - 97 (1972). In Russian.

132.032 Interferometric investigations of the nebula A 21 (YM 29, Medusa). T. A. Lozinskaya.
Astron. Zhurn. Akad. Nauk SSSR, Vol. 49, 1158 - 1163 (1972). In Russian. English translation in Soviet Astron. AJ, Vol. 16, No. 6.

Observations of the nebula A 21 with a high contrast Fabry-Perot etalon and contact image converter in 6584 Å and 6563 Å lines were carried out. A satisfactory model of the nebula of a thin shell prolate ellipsoid inclined to the plane of the sky by 50° ± 15° was found to agree with observations. The axis ratio of the ellipsoid is 0.4 ± 0.2, the width of the shell is less than 0.3 of the mean radius. The mean radial velocity of the object is +24 ± 5 km/sec relative to the local standard of rest. An expansion of the shell with a velocity of 53 ± 10 km/sec was discovered.

132.033 Search for coronal line emission from the Cygnus Loop.
D. W. Kurtz, P. A. Vanden Bout, J. R. P. Angel.
Astrophys. Journ., Vol. 178, 701 - 706 (1972).

The flux from the edges of the Cygnus Loop in the coronal line [Fe XIV] λ 5303 is measured to be less than 5×10^{-9} ergs cm^{-2} sterad^{-1} s^{-1} (0.017 R) in a 3 Å band centered on the line. This upper limit is an order of magnitude lower than the predicted total flux in the line.

132.034 Relative emission-line intensities in the Vela X nebula. D. E. Osterbrock, R. Costero.
Bull. American Astron. Soc., Vol. 4, 423 (1972). – Abstr. AAS.

132.035 [O III] line ratios in gaseous nebulae. J. D. R. Bahng.
Observatory, Vol. 92, 237 - 238 (1972). – Letter.

132.036 On the problem of connection of diffuse nebulae with stars of early spectral classes.
Sh. G. Gordeladze, A. G. Belyj.
Problems of cosmic physics. Vyp. (No.) 7, (see 003.028), p. 129 - 136 (1972). In Russian.

The connection of gas nebulae with stars of early spectral classes is studied, data of 1475 stars and of 1200 nebulae being used.

132.037 Pressure broadening of radio recombination lines.
G. Peach, M. J. Seaton.
Comments Atomic and Molecular Phys., (*GB*), Vol. 3, 107 - 112 (1972).

The authors discuss the physical processes which determine the profiles of radio lines. This comment is an extension of an earlier one and is related to recombination spectra of gaseous nebulae.

132.038 Observations of radio recombination lines.
M. Brocklehurst, M. J. Seaton.

Comments Atomic and Molecular Phys., (*GB*), Vol. 3, 113 - 120 (1972).

The authors present some recent work on the interpretation of radio recombination line observations. This is an extension of another comment (see Abstr. 132.037) and is related to the recombination spectra of gaseous nebulae.

Excitation of metastable levels in low density nebular plasmas: [SII] and [ArIV]. See Abstr. 062.013.

Internal dust in nebulae. III. Nonisotropic scattering. See Abstr. 063.014.

Transfer of resonance-line radiation in differentially expanding atmospheres. I. General considerations and Monte Carlo calculations. See Abstr. 064.014.

Z-dependence of the level intervals in $2s^2 2p^2$, $2s^2 2p^3$ and $2s^2 2p^4$. See Abstr. 074.009.

Observations of planets, nebulae, and galaxies at 350 microns. See Abstr. 091.033.

Detection of far-infrared astronomical sources. See Abstr. 113.018.

The peculiar O6f star HD 148937 and the symmetrically surrounding nebulae. See Abstr. 117.014.

Observations of the Cygnus Loop at 6-cm wavelength. See Abstr. 125.003.

Microwave celestial water-vapor sources. See Abstr. 131.008.

On the λ 4686 He II line intensity in H II regions and the cosmic ray flux. See Abstr. 131.010.

Absorption effects in six H II regions. See Abstr. 131.055.

Molekülwolken und Sternentstehung im interstellaren Raum. See Abstr. 131.064.

Formation of filaments in fossil H II regions. See Abstr. 131.070.

Theoretical structure and spectrum of a shock wave in the interstellar medium: The Cygnus Loop. See Abstr. 131.097.

Interstellar absorption of light in the area of Orion. See Abstr. 131.125.

Infrared spectroscopy of M 42, NGC 7027 and IC 418. See Abstr. 133.014.

Errata

132.901 Erratum: "Spectrophotometric investigations of diffuse nebulae. II. Mi I-19 – a compact H II region"
[Astron. Tsirk., No. 632, p. 3 - 6 (1971). In Russian].
Yu. I. Glushkov, S. V. Karyagina.
Astron. Tsirk., No. 679, p. 7 (1972).

133 Planetary Nebulae

133.001 Spatial spectroscopic diagnostic of planetary nebulae. I. Formulation of the synthetic problem for optically thin lines. J. Hekela.
Bull. Astron. Inst. Czechoslovakia, Vol. 23, 197 - 206 (1972).

The formation of optical thin lines is used for spatial spectroscopic diagnostic of planetary nebulae in a synthetic approach. The observed quantities are treated as functions of the position on the disk. The observed values are compared with theoretical predictions obtained by means of coarse analysis, ionization structure and fine analysis. Illustrative examples of both local values are given for the N_1 line (OIII − λ5007 Å) of a hypothetical planetary nebula.

133.002 Spatial spectroscopic diagnostic of planetary nebulae. II. Formulation of the analytical problem for optically thin lines. J. Hekela.
Bull. Astron. Inst. Czechoslovakia, Vol. 23, 207 - 218 (1972).

The present paper describes and discusses the general analytical method of spatial spectroscopic diagnostic. The method is based on the inversion of the integral equation of the local absolute monochromatic intensity, which expresses the line profile of the emergent radiation from 1 cm² column along the line of sight through an optically thin medium. Illustrative examples of constraint solutions are given for the N_1 line (O III − λ 5007 Å) of a hypothetic planetary nebula, in order to infer the variation of the population of the upper level along the line of sight. Spherically symmetric and inhomogeneous models are treated.

133.003 Identification of the 100-micron source No. 15. H. M. Johnson.
Astrophys. Journ., (*Letters*), Vol. 175, L105 - L106 (1972).

The source coincides with the planetary nebula He 2-53 (285 + 1°.2). The 100-μ flux density exceeds the optically predicted bremsstrahlung flux density by a factor of 10^6.

133.004 Photometry of NGC 3587, "the Owl nebula", in Hα light, and its structure. P. Proisy.
Astron. Astrophys., Vol. 20, 115 - 119 (1972). In French.

Isophotic contours of NGC 3587 in Hα light are derived from measurements on direct photographs secured with the 120-cm reflector of the Observatory of Haute Provence.

133.005 Observations of planetary nebulae at 1.65 to 3.4 microns.
S. P. Willner, E. E. Becklin, N. Visvanathan.
Astrophys. Journ., Vol. 175, 699 - 706 (1972).

Photometric measurements at 2.2 μ of 15 planetary nebulae are presented, along with measurements of 12 at 1.65 μ and seven at 3.4μ. The measurements agree with the predicted thermal emission from ionized hydrogen and helium except for four nebulae. The excesses are interpreted in terms of emission from dust distributed throughout or surrounding the nebulae.

133.006 The Bowen fluorescence mechanism in planetary nebulae. J. P. Harrington.
Astrophys. Journ., Vol. 176, 127 - 137 (1972).

Solutions have been obtained for the transfer of He II Lα radiation in model planetary nebulae. The available observations are reviewed and are shown to be in good agreement with the predictions of theory.

133.007 A consistent model of the planetary nebula NGC 7662. R. C. Kirkpatrick.
Astrophys. Journ., Vol. 176, 381 - 393 (1972).

A numerical model of NGC 7662 has been computed which very closely matches the observations. It is found that in order to obtain a model which is consistent with all of the observed line strengths and line ratios, one must take into account not only the double-shell structure but also the contribution of condensations in the nebula and the departure of the central-star flux distribution from that of a blackbody.

133.008 The spectrum of the planetary nebula NGC 7027 in the near infrared.
E. A. Kolotilov, R. I. Noskova.
Astron. Tsirk., No. 697, p. 4 - 6 (1972). In Russian.

133.009 Monochromatic observations of some planetary nebulae. C. T. Hua, R. Louise.
Astron. Astrophys., Vol. 21, 193 - 198 (1972). In French.

Monochromatic images were obtained by using the «Réducteur Focal» apparatus (Courtès, 1960) combined with narrow pass-band interference filters respectively centered on Hα (λ = 6563 Å), [N II] (λ = 6584 Å) and [O III] (λ = 5007 Å) Several plates were taken on eight planetary nebulae with different exposure times. The results show that the [N II] image is generally greater than (or at least equal to) the Hα one. Furthermore, the [N II] line is not observed in two nebulae (NGC 1514 and NGC 7008).

133.010 Spatial spectroscopic diagnostic of planetary nebulae. III. Numerical investigation of local absolute monochromatic energies and local absolute energies in spherically symmetric models. J. Hekela, I. Hubený.
Bull. Astron. Inst. Czechoslovakia, Vol. 23, 331 - 341 (1972).

Changes in line profiles and total intensities of emission lines have been studied in dependence on an arbitrary rearrangement of ion abundance, turbulent and macroscopic velocities on the line of sight through the investigated gas. Due to the easy comparison of the individual line profiles and total line intensities, from different points of the disk, all computations were performed on spherically symmetric models of planetary nebulae. As the line profiles are formed only by Doppler effects, it was sufficient to study the shape of one selected line. The forbidden N_1 line (O III − λ 5007 Å) was chosen. The purpose of this paper is to clarify the individual effects caused by the structural functions, pointed out above, on emergent line profiles.

133.011 Six new planetary nebulae.
B. A. Vorontsov-Velyaminov, E. B. Kostjakova, O. D. Dokuchaeva, V. P. Arkhipova.
Astron. Tsirk., No. 716, p. 7 - 8 (1972). In Russian.

133.012 Kinematics of planetary nebulae in the Large Magellanic Cloud. M. G. Smith, D. W. Weedman.
Astrophys. Journ., Vol. 177, 595 - 600 (1972).

Radial velocities of 27 planetary nebulae in the Large Magellanic Cloud have been measured with a pressure-scanned, Fabry-Perot interferometer. The results are combined with data from previous observers, and it is found that the planetary nebulae have a velocity dispersion perpendicular to the plane of the LMC which is twice the dispersion of the H II regions, but the systemic velocity of the planetaries agrees with that of the H II regions. This confirms previous suggestions that the LMC is a flattened, collapsed galaxy.

133.013 Comparison of angular expansions of planetary nebulae measured at Pulkovo and Harvard Observatories. O. N. Orlova.
Astron. Zhurn. Akad. Nauk SSSR, Vol. 49, 1164 - 1167 (1972). In Russian. English translation in Soviet Astron. AJ,

Vol. 16, No. 6.

The angular expansions of planetary nebulae NGC 6720, NGC 6853 and NGC 7662 measured at Pulkovo have been compared with the results obtained by Liller at Harvard Observatory. The discrepancy of these results can be explained by a systematic error in Liller's observations due to the change from silver to aluminium coating.

133.014   Infrared spectroscopy of M 42, NGC 7027 and
              IC 418.   T. Hilgeman.
Bull. American Astron. Soc., Vol. 4, 424 (1972). – Abstr. AAS.

133.015   On the formation of planetary nebulae.
              D. R. Alexander.
Thesis, Indiana Univ., Bloomington. [Available from Univ. Microfilms, Ann Arbor, Mich., USA. Order No. 72-15907], 105 pp. (1972).

The purpose of this investigation is to study the possible mechanisms responsible for the formation of planetary nebulae. The arguments that planetary nebulae are formed by non-catastrophic events in low mass red supergiants are reviewed.

133.016   Observations of interstellar molecular hydrogen and
              measurements of hydrogen and helium fluxes from
planetary nebulae.   T. R. Gull.
Thesis, Cornell Univ., Ithaca, N.Y. [Available from Univ. Microfilms, Ann Arbor, Mich., USA. Order No. 71-17642], 301 pp. (1971).

An automated Fabry-Perot interferometer was developed for telescope use and was utilized in observations for several astrophysical problems. The search for interstellar molecular hydrogen begun by Werner (1968) was continued; the He λ 10830 Å (2^3P-2^3S) line fluxes and the P12 λ 8750.36 Å (n = 12 to n = 3) hydrogen line fluxes were measured in six planetary nebulae; upper limits to the sky background continuum in the near infrared were established.

133.017   The life spans of condensations in planetary nebulae.
              A. V. Holm.

crofilms, Ann Arbor, Mich., USA. Order No. 71-14144], 92 pp. (1971).

Shows that, under certain conditions, neutral condensations formed in a young planetary nebula and not held together by gravitational or magnetic forces can survive for a time span comparable with the ages of old planetaries in which condensations are observed.

Thermal pulses in helium shell-burning stars.
See Abstr. 065.095.

Peculiar southern emission-line objects with strong [O III] λ 4363.   See Abstr. 114.165.

Circular polarimetry of fifteen interesting objects.
See Abstr. 116.015.

The peculiar O6f star HD 148937 and the symmetrically surrounding nebulae.   See Abstr. 117.014.

Very-low-excitation compact nebulae.
See Abstr. 132.026.

Relative emission-line intensities in the Vela X
nebula.   See Abstr. 132.034.

NGC 2818, an open cluster containing a planetary
nebula.   See Abstr. 153.004.

Errata

133.901   Erratum: 'Infrared photometry of the H II region
              Sharpless 266' [Astrophys. Letters, Vol. 11, 95 -
97 (1972)].   J. A. Frogel, S. E. Persson, D. E. Kleinmann.

and the position angle $155 \pm 3°$.

134.006 Neutron star in Crab nebula — really?
F. C. Michel.
Comments Astrophys. Space Phys., Vol. 4, 101 - 104 (1972).

The history of searching for a neutron star in the Crab nebula is outlined. The arguments for the detection of such a star are examined.

134.007 A search for isolated radio pulses from the Crab nebula at 151.5 MHz.
W. P. S. Meikle, R. W. P. Drever, R. F. Haynes, J. R. Shakeshaft, W. N. Charman, J. V. Jelley.
Monthly Notices Roy. Astron. Soc., Vol. 160, 5P - 8P (1972).

A search has been made for large bursts of radio emission at 151.5 MHz from the direction of the Crab nebula. In 605 hr of observation, no events exceeding a flux of 1.4×10^{-22} W m^{-2} Hz^{-1} were detected. Implications of the results with regard to 'strong pulses' and phase fluctuations in the periodic emissions from the pulsar NP 0532 are also examined.

134.008 The radio polarization of the Crab nebula at 21-cm wavelength. R. M. Duin, H. van der Laan.
Astrophys. Letters, Vol. 12, 177 - 180 (1972).

Polarization and total intensity maps of the Crab nebula, obtained at a wavelength of 21 cm by the Westerbork Synthesis Radio Telescope, are presented. The relation between the filamentary structure and the radio polarization is discussed.

134.009 Nonlinear inverse Compton radiation and the circular polarization of diffuse radiation from the Crab nebula. J. Arons.
Astrophys. Journ., Vol. 177, 395 - 410 (1972).

A detailed calculation is given of the high-frequency radiation from very relativistic particles moving in a strong, circularly polarized electromagnetic wave. The results are used to show that the circular polarization of this "nonlinear inverse Compton" mechanism, when combined with the vacuum oblique-rotator approximation for pulsar environments, may be in disagreement with observational upper limits on the fractional circular polarization of the diffuse emission from the Crab nebula.

134.010 A model of the Crab nebula derived from dual-frequency radio measurements.
K. W. Weiler, G. A. Seielstad.
Astron. Astrophys., Vol. 21, 393 - 400 (1972).

The total intensity and linearly polarized emission from the Crab nebula were synthesized to a resolution of approximately 1 arc-minute at both 1420 and 2880 MHz. From these data were calculated the spectral index, rotation measure, intrinsic position angle, and depolarization ratio distributions. Then, combined with a source model, the physical conditions within the supernova remnant were established. The strength and orientation of both the homogeneous and random components of the magnetic field were determined and a measurement of the thermal electron plasma distribution obtained.

134.011 The structure of the Crab nebula—II. The spatial distribution of the relativistic electrons.
A. S. Wilson.
Monthly Notices Roy. Astron. Soc., Vol. 160, 355 - 371 (1972).

A diffusion-loss model is proposed for the Crab nebula. Relativistic electrons are produced near the centre of the nebula, diffuse outwards and lose energy by synchrotron radiation. Many of the observed features of the continuum emission may be explained in this way.

134.012 The structure of the Crab nebula—III. The radio filamentary radiation. A. S. Wilson.
Monthly Notices Roy. Astron. Soc., Vol. 160, 373 - 379 (1972).

It is proposed that the 5 GHz continuum radiation from the optical filaments in the Crab nebula is synchrotron radiation from an increased magnetic field. The field is circular (or spiral) about the filaments in agreement with Woltjer's suggestion that the filaments carry a current along their length.

134.013 Ionization and relative abundance of hydrogen and helium atoms in the gaseous filaments of the Crab nebula. V. V. Golovatyi, V. I. Pronik.
Izv. Krymskoj Astrofiz. Obs., Vol. 45, 152 - 161 (1972).
In Russian.

Ionization of hydrogen and helium atoms for a gaseous filament embedded in the L_c-radiation field is considered. Calculations have been carried out for several models of filaments having different electron density, abundance of helium and optical depth beyond the Lyman limit. Different values of flux density and spectral index of ionizing radiation have been considered too.

A northern California pictograph that may be another record of the Crab nebula supernova explosion. See Abstr. 004.021.

The polarisation of synchro-Compton radiation. See Abstr. 061.007.

Lorentz transformation properties of the Stokes parameters. See Abstr. 061.051.

On the possibility of observation of circular polarization in the optical emission of some cosmic sources. See Abstr. 061.055.

Discovery of interstellar circular polarization in the direction of the Crab nebula. See Abstr. 131.004.

Very long baseline interferometer observations of Taurus A and other sources at 121.6 MHz. See Abstr. 141.073.

VLBI observations of the Crab nebula pulsar. See Abstr. 141.507.

Detection of 10–100 MeV γ-rays from the Crab nebula pulsar NP 0532. See Abstr. 141.523.

The pulse-height distribution for NP 0532. See Abstr. 141.527.

Mechanisms of optical, X-ray and γ-radiation from Crab pulsar. See Abstr. 141.528.

The pulse shape of the Crab nebula pulsar NP 0532 as a function of color. See Abstr. 141.532.

Upper limit on the gravitational flux reaching the earth from the Crab pulsar. See Abstr. 141.557.

Low energy X-ray survey from the Crab nebula to Cygnus. See Abstr. 142.025.

Radio Sources, Quasars, Pulsars, X Ray-, Gamma Ray-Sources, Cosmic Radiation

141 Radio Sources, Quasars, Pulsars

Radio Sources, Quasars

141.001 Optical variations of the radio sources ON 231 = W Comae and B2 1215 + 30. G. Romano.
Mem. Soc. Astron. Italiana, Nuova Ser., Vol. 43, 309 - 312 (1972).

This paper gives the results of the observations of the peculiar radio sources ON 231 = W Com and B2 1215 + 30 on films obtained with the 40/50/100 cm Schmidt telescope of Asiago from 1962 to 1971. W Com presents a slow irregular variation and an activity consisting in the occurrence of narrow minima of small amplitude. The optical variation of B2 1215 + 30 is slow and irregular with small amplitude.

141.002 Accurate flux densities at 5009 MHz of 1007 radio sources. A. J. Shimmins, J. G. Bolton.
Australian Journ. Phys. Astrophys. Suppl. No. 23, 41 pp. = Separate print Division Radiophys. C.S.I.R.O. Sydney.

Accurate flux densities at 5009 MHz for 1007 radio sources are presented here together with the measured positions at this frequency. The sources were selected between declinations +27° and −90° from the Parkes 408 and 2700 MHz catalogues, but are predominantly south of declination −33°. Because of the small beamwidth of 4.'05 arc, some of the sources are partly resolved, and size correction factors have been calculated from either known source structure or measured beamwidths. The estimated errors in the positions are approximately 15''arc in both coordinates for sources stronger than 1 f.u., increasing to 20'' arc for the weaker sources.

141.003 High resolution observations of 3C 390.3 at 2.7 and 5 GHz. A. Harris.
Monthly Notices Roy. Astron. Soc., Vol. 158, 1 - 11 (1972).

3C 390.3 has been mapped at 2.7 and 5 GHz with the Cambridge One-Mile telescope. It is found to consist of three compact components and an extended region of emission. The central component, not previously detected, is coincident with the associated N-galaxy and has a spectral cut-off below about 3 GHz. Maps of the polarized emission have also been obtained and provide information on physical conditions in the source.

141.004 The emission-line spectrum of Cygnus A. S. Mitton, J. Mitton.
Monthly Notices Roy. Astron. Soc., Vol. 158, 245 - 254 (1972).

Spectra of Cygnus A taken at the Hale Observatories have been analyzed. A line list of the nebular emission is presented. The results can be accounted for by a two-component model of the nucleus in which condensations at $T_e = 10^4$ °K produce most of the emission lines; these condensations are embedded in a tenuous plasma at $T_e \approx 10^6$ °K. A continuing source of excitation probably exists in the optical nucleus of this radio galaxy.

141.005 A study of the lunar occultations of eleven radio sources. A. G. Lyne.
Monthly Notices Roy. Astron. Soc., Vol. 158, 431 - 462 (1972).

Observations of the lunar occultations of 11 radio sources have been made at Jodrell Bank, using an interferometer to surmount some technical difficulties. This paper provides accurate positions and brightness distributions for these sources. Some of the observations have been made at several frequencies, allowing a study of the variations of spectral index within the sources. In particular, 3C 2, 3C 245 and 3C 279 are quasars which bear a striking resemblance to 3C 273. For 3C 2, the data suggest an identification of the radio jet with an optical one.

141.006 The cosmological evolution of radio sources of large angular extent. B. L. Fanaroff, M. S. Longair.
Monthly Notices Roy. Astron. Soc., Vol. 159, 119 - 128 (1972).

The numbers of radio sources of large angular size ($\theta \geqslant 60''$) found in a survey of sources having $0.5 < S_{408} < 5 \times 10^{-26}$ W m^{-2} Hz^{-1} by Windram and Kenderdine is compared with the number expected in different world models. An excess of such sources is found in comparison with the predictions of uniform models but their number is consistent with the postulate that such sources exhibit a moderate rate of evolution. In a complete sample of 5C sources at much lower flux densities, $S_{1400} \geqslant 0.01 \times 10^{-26}$ W m^{-2} Hz^{-1}, there is an excess of sources having steep spectra and a correlation is found between these sources and those of large angular size. A model is proposed in which these sources are associated with relatively weak radio galaxies with long lifetimes.

141.007 Quasars as images of Seyfert nuclei. L. N. K. de Silva.
Monthly Notices Roy. Astron. Soc., Vol. 159, 219 - 231 (1972).

Treating a galaxy as a concentrated mass deflector (gravitational lens) acting in the manner discussed by previous writers, an expression is derived for the probability of finding a deflecting galaxy between the observer and a radiating source that produces at least intensification of the apparent brightness of the source. Barnothy and Barnothy have suggested that QSOs might be images of Seyfert nuclei intensified in this way. It is shown that the gravitational lens mechanism is inadequate to account for the observed number of quasars.

141.008 Evidence for spatially independent outbursts in compact radio sources. W. A. Dent.
Astrophys. Journ., (Letters), Vol. 175, L55 - L58 = Contr. Five College Obs., Amherst, No. 142 (1972).

Observations of the 7.8-GHz flux-density variations of 3C 120 suggest that the apparent super-relativistic expansion of this source and others can be explained as being due to independent spatially separated radio outbursts.

141.009 Meter-wavelength observations of galactic radio sources and the galactic disc radiation. A. deS. Parrish.
Thesis, Cornell Univ., Ithaca, N. Y. [Available from Univ. Microfilms, Ann Arbor, Mich., USA. Order No. 71-17119], 149 pp. (1971).

The 1000 foot diameter radio telescope at the Arecibo Observatory has been used to observe the continuum emission of galactic sources at low radio frequencies. The following sources were observed: W49, W51, W56, NGC1499, IC410, NGC2175, NGC2244, and Orion A. Flux densities of the observed sources are presented.

141.010   Positions and flux densities at 1415 MHz of 5C3 sources near M31.
P. C. van der Kruit, P. Katgert.
Astrophys. Letters, Vol. 11, 181 - 185 (1972).

Observations of part of M31 and NGC 205 with the Westerbork Synthesis Radio Telescope provide positions, flux densities, and spectral indices between 408 and 1415 MHz for 30 5C3 radio sources and upper limits to flux densities and lower limits to spectral indices for 14 sources.

141.011   The optical identification of 3C 220.2.
D. Wills, R. Lynds.
Astrophys. Letters, Vol. 11, 189 - 190 (1972).

The object 3C 220.2, previously thought to be a galactic star, is a quasi-stellar object.

141.012   Flux densities, positions, and structures for a complete sample of intense radio sources at 1400 MHz.
A. H. Bridle, M. M. Davis, E. B. Fomalont, J. Lequeux.
Astron. Journ., Vol. 77, 405 - 443 (1972).

Accurate flux densities, precise positions of unresolved sources, and structures of resolved sources have been derived from full-beam and interferometric observations of intense sources at 1400 MHz. Results are given for 424 sources in the area of sky $-5°<\delta<+70°$, $|b|>5°$ whose 1400-MHz integrated flux densities S_{1400} exceed 1.70 f. u. The 234 sources with $S_{1400} \geqslant 2.00$ f. u., equivalent diameters < 10 arc min, and $|b|>20°$ form a 98±2% complete sample comparable in number to the 178-MHz Revised Third Cambridge Catalogue in this 4.30-sr area of sky, but selected at 1400 MHz. This sample is suitable for statistical studies of the properties of extragalactic radio sources. To facilitate its use, and that of other samples which may be drawn from these data, references to other studies of the positions, fine and extended structure, polarization, and variability of the sources have been assembled in the principal table of this paper. A comparison is made with other 1400-MHz flux-density data, and the spectral content of the complete sample is discussed.

141.013   Optical positions of forty-eight quasistellar objects.
C. Barbieri, M. Capaccioli, R. Ganz, G. Pinto.
Astron. Journ., Vol. 77, 444 - 447 (1972).

Optical positions are presented for 48 quasistellar objects. The positions have been derived in comparison to AGK3 stars (or to SAO stars for fields having declination south of $-2°$ and also for 3C345) from Schmidt-telescope plates taken at Asiago. The results are believed to be accurate at the $0.''3-0.''4$ level.

141.014   On the ability of the luminosity-volume test to reveal the statistical evolution of the luminosity of quasi-stellar sources.   R. Lynds, V. Petrosian.
Astrophys. Journ., Vol. 175, 591 - 599 (1972).

We review the principles and capabilities of the luminosity-volume test and source-count analysis as procedures for studying the statistical evolution of the radio and optical luminosities of quasi-stellar sources. In contrast to the conclusions expressed by Longair and Scheuer, a comparison of the two methods reveals that the luminosity-volume test is substantially more powerful and lucid and that a knowledge of the individual redshifts of objects is required for the most comprehensive determination of the evolutionary history of quasi-stellar sources.

141.015   Physical associations between quasi-stellar objects and galaxies.

G. R. Burbidge, S. L. O'Dell, P. A. Strittmatter.
Astrophys. Journ., Vol. 175, 601 - 611 (1972).

Further evidence is presented relevant to apparent associations between galaxies and quasi-stellar objects. It is shown that for the five 3C QSOs which lie very close to bright galaxies the QSO—galaxy angular separations are inversely proportional to the redshifts of the galaxies. This lends additional support to the association hypothesis. On the assumption that the 3C QSO—galaxy associations are real, the relationship of QSOs to galaxies is discussed.

141.016   Observations of compact objects of cosmic radio emission at 3.55 cm wavelength with maximum angular resolution.
B. G. Clark, J. J. Broderick, V. A. Efanov, K. I. Kellermann, M. H. Cohen, L. R. Kogan, V. I. Kostenko, L. I. Matveyenko, I. G. Moiseev, M. M. Mukhina, V. B. Steinschleier, D. L. Jauncey.
Astron. Zhurn. Akad. Nauk SSSR, Vol. 49, 700 - 703 (1972). In Russian. English translation in Soviet Astron. AJ, Vol. 16, No. 4.

The sizes of compact radio sources have been measured at 3.55 cm wavelength with radio interferometers having Simeis — Goldstone — Green Bank as baselines. The first results are reported.

141.017   Autocorrelation function of "rapid" light variations of the quasar 3C 273.
L. M. Ozernoy, V. E. Chertoprud.
Astron. Zhurn. Akad. Nauk SSSR, Vol. 49, 712 - 721 (1972). In Russian. English translation in Soviet Astron. AJ, Vol. 16, No. 4.

Using Smith's data on light variations of the quasar 3C 273 during the past 80 years, the autocorrelation function of «rapid» light variations was constructed after subtraction of the 9-year cyclic component. The autocorrelation function shows two regions of decrease where the characteristic time is equal to $\tau_1 \lesssim 0.1-0.2$ day and $\tau_2 \approx 1$ year, respectively. The results of the analysis are compared with existing theoretical ideas about the nature of the optical variability of quasars. The difficulties to interpret the autocorrelation function in the framework of supernova flares or an accreting «black hole» are indicated. The possible sources of observable «rapid» light variations are discussed.

141.018   The periodical components of variations in the flux density of the radio source VRO 42.22.01 (BL Lacertae).   A. G. Gorshkov, M. V. Popov.
Astron. Zhurn. Akad. Nauk SSSR, Vol. 49, 722 - 726 (1972). In Russian. English translation in Soviet Astron. AJ, Vol. 16, No. 4.

The periodical components of variations in the flux density of the radio source VRO 42.22.01 at 2.8 cm, 3.75 cm and 4.5 cm are discovered by the Fourier method. A model of the pulsating magnetic rotator is suggested for the explanation of the phenomenon. The analysis of the radio spectrum has made it possible to conclude on the dual nature of VRO 42.22.01, one component being of constant flux density.

141.019   On the use of lunar occultations of radio sources for the investigation of their angular structure. II.
G. L. Abramyan.
Astron. Zhurn. Akad. Nauk SSSR, Vol. 49, 897 - 899 (1972). In Russian. English translation in Soviet Astron. AJ, Vol. 16, No. 4. – Short note.

141.020   The distribution of linear polarization in 3C 270 and 3C 452 at 21-centimeter wavelength.
P. P. Kronberg.
Astrophys. Journ., Vol. 176, 47 - 55 (1972).

Two-dimensional radio distributions for 3C 452 and 3C

270 (NGC 4261) are obtained with a resolution of approximately 1′ arc for both the total radiation and the linearly polarized component at λ 21.1 cm. The results are compared with existing data at other wavelengths, and some physical parameters for both sources are estimated.

141.021 VRS for VRO.   G. W. Swenson.
Nature, Phys. Sci., Vol. 238, 37 (1972).

It is suggested by the author that VRS for "variable radio source" would be a better designation for VRO = variable radio objects.

141.022 Interpretation of rotation measures of radio sources. II.   S. Mitton, M. Reinhardt.
Astron. Astrophys., Vol. 20, 337 - 340 (1972).

The systematic behaviour of the rotation measures of radio galaxies, quasars and unidentified sources is discussed. A redshift dependence of the rotation measures of quasars, after subtraction of the regular galactic contribution, is indicated. The data are compatible with the assumption that the intrinsic rotation measures of quasars on the average do not differ from those of radio galaxies, and that they are independent of redshift.

141.023 Optical positions of radiosources.   M. P. Véron.
Astron. Astrophys., Vol. 20, 471 - 473 (1972).

We have measured optical positions of quasars and radiogalaxies with an accuracy better than one arc second. We have used for this the method of Schlesinger (1926) and usually for reference, stars from the AGK3 catalogue.

141.024 Faraday depolarization of extragalactic radio sources.   R. G. Strom.
Nature, Phys. Sci., Vol. 239, 19 - 21 (1972).

Polarization observations indicate that the Faraday depth decreases with increasing size and supports the view that quasars have strictly cosmological redshifts.

141.025 Radio observations of early-type stars.   B. Balick.
Astrophys. Letters, Vol. 12, 21 - 23 (1972).

A survey of 32 O, B, and Wolf-Rayet stars with the NRAO interferometer failed to detect unresolved radio sources coincident with any of the stellar positions. The detection limits range from 10 to 50×10^{-29} W m^{-2} Hz^{-1}. Consequently, the physical processes that produce the radio radiation observed in other types of stars are not as important for early-type stars.

141.026 The B2 catalogue of radio sources—second part.   G. Colla, C. Fanti, R. Fanti, A. Ficarra, L. Formiggini, E. Gandolfi, C. Lari, B. Marano, L. Padrielli, P. Tomasi.
Astron. Astrophys., Suppl. Ser., Vol. 7, 1 - 34 (1972).

The catalogue lists 3013 radio sources observed at 408 MHz with the Bologna northern cross telescope. It covers an area of 0.53 ster. between 24°02′ and 29°30′ down to 0.25 f.u. Results are given for the radio spectra of the 4C sources and for the log N–log S relationship.

141.027 Brightness changes of the radio source OJ 287.
Astron. Tsirk., No. 679, p. 1 (1972). In Russian.

141.028 On the variation of light of the radio source OJ 287.   V. P. Tsesevich.
Astron. Tsirk., No. 688, p. 7 - 8 (1972). In Russian.

141.029 Statistical studies of the evolution of extragalactic radio sources. I. Quasars.   M. Schmidt.
Astrophys. Journ., Vol. 176, 273 - 287 (1972).

We first discuss evidence concerning quasar evolution, the dependence of radio luminosity function on optical luminosi-ty, and the range of absolute optical luminosity. We find that pure optical-luminosity evolution of quasars is incompatible with the optical counts. We adopt pure density evolution and use two alternative density laws. Redshift-magnitude tables are presented. The consequent derivation of the quasar radio counts is explained. The distribution function of the ratio of radio to optical luminosity is derived. Quasar source counts, redshift–flux density tables and a magnitude–flux density table are presented.

141.030 Statistical studies of the evolution of extragalactic radio sources. II. Radio galaxies.   M. Schmidt.
Astrophys. Journ., Vol. 176, 289 - 301 (1972).

We derive radio source counts attributable to radio galax-ies and ordinary galaxies. Since 5 percent of the giant elliptical galaxies are radio galaxies, we impose a limit of a factor of 20 on the overall density increase of the latter with redshift. Nei-ther this type of restricted density evolution nor luminosity evolution can explain the radio source counts attributed to radio galaxies. We present two alternative luminosity-depend-ent density laws that are compatible with the source counts. We make extensive use in the present paper of tables of red-shift versus flux density that allow insight in the effect of luminosity function, density law, redshift cutoff, etc. on the source counts.

141.031 Statistical studies of the evolution of extragalactic radio sources. III. Interpretation of source counts and discussion.   M. Schmidt.
Astrophys. Journ., Vol. 176, 303 - 314 (1972).

Radio source counts predicted on the basis of evolution models of quasars (Paper I), radio galaxies (Paper II), and or-dinary galaxies are compared to observed total source counts at various radio frequencies. Optical and radio luminosity functions are presented for quasars and radio galaxies. It is shown that certain arguments based on general source counts against the need for evolution or against the large distances of the sources are not supported by detailed evolution models. Finally we emphasize the many uncertainties in the evolution models, some of which can be resolved eventually by exten-sive optical and radio studies at low flux levels.

141.032 Infrared observations of variable radio objects.   G. H. Rieke.
Astrophys. Journ., (*Letters*), Vol. 176, L61 - L63 (1972).

BL Lacertae, OJ 287, AP Lib, B2 1212+30, ON 231 have been observed in the infrared; the first three have been detected. BL Lacertae and OJ 287 show infrared variations which, at least for OJ 287, correlate with the long-term varia-tions observed in the optical and radio regions.

141.033 Precise positions of radio sources measured at 2695 MHz.
R. L. Adgie, J. H. Crowther, H. Gent.
Monthly Notices Roy. Astron. Soc., Vol. 159, 233 - 251 (1972).

The positions of 159 radio sources have been measured at 2695 MHz with the interferometer at the Royal Radar Establishment, Malvern. The positions were calibrated with the positions of the optical counterparts of 36 identified radio sources. The standard error of measurement is believed to be 0.4 arc sec in both coordinates.

141.034 The ratio of two flux density scales at 408 MHz.   R. G. Conway, R. E. B. Munro.
Monthly Notices Roy. Astron. Soc., Vol. 159, 21P - 24P (1972).

A comparison between two sets of measurements of the flux density of radio sources at 408 MHz confirms that the CKL scale (Conway, Kellermann and Long, 1963) of flux den-sity at this frequency differs significantly from the absolute scale due to Wyllie. To be consistent with the Wyllie scale, CKL flux densities should be increased by 1.087±0.028.

141.035 Unusual behaviour of quasar 3C 345 in 1971.
M. K. Babadzhanjanz, V. A. Hagen-Thorn, E. V. Semjonova.
Astron. Tsirk., No. 701, p. 1 - 3 (1972). In Russian.

141.036 Photographic UBV observations of OJ 287, 3C 345, and BL Lac. K. R. Hackney, R. L. Hackney, A. G. Smith, G. H. Folsom, R. J. Leacock, R. L. Scott.
Bull. American Astron. Soc., Vol. 4, 314 (1972). – Abstr. AAS.

141.037 Radio and optical activity of the peculiar object BL Lac. R. L. Hackney, K. R. Hackney, A. G. Smith, G. H. Folsom, R. J. Leacock, R. L. Scott, E. E. Epstein.
Bull. American Astron. Soc., Vol. 4, 314 (1972). – Abstr. AAS.

141.038 Flux density and polarization variations of OJ 287 at 8 GHz. H. D. Aller, R. D. Carpenter.
Bull. American Astron. Soc., Vol. 4, 314 (1972). – Abstr. AAS.

141.039 3C 120, BL Lac, and OJ 287: Coordinated multi-wavelength observations of intraday variability.
E. E. Epstein, E. E. Becklin, G. G. Wynn-Williams, G. Neugebauer, W. G. Fogarty, R. L. Hackney, K. R. Hackney, R. J. Leacock, R. B. Pomphrey, R. L. Scott, A. G. Smith, W. A. Stein, B. Gary, R. W. Hawkins, R. C. Roeder, M. Penston, K. Tritton, G. Wlerick, J. H. Bigay, U. Barnard, C. Bertand, A. Durand, U. Merlin.
Bull. American Astron. Soc., Vol. 4, 314 - 315 (1972). Abstr. AAS.

141.040 Further observations of brightness variation in the small scale structure of 3C 273 and 3C 279.
G. E. Marandino, G. M. Resch, N. R. Vandenberg, T. A. Clark, H. Hinteregger, C. A. Knight, D. S. Robertson, A. E. E. Rogers, I. I. Shapiro, A. R. Whitney, R. M. Goldstein, D. Spitzmesser.
Bull. American Astron. Soc., Vol. 4, 315 (1972). – Abstr. AAS.

141.041 Optical variations of four Ohio radio sources.
E. R. Craine, J. W. Warner.
Bull. American Astron. Soc., Vol. 4, 315 (1972). – Abstr. AAS.

141.042 A recombination-line study of the Sagittarius B_2 radio complex. E. J. Chaisson.
Bull. American Astron. Soc., Vol. 4, 317 (1972). – Abstr. AAS

141.043 An improved expanding source model for variable extragalactic radio sources.
F. W. Peterson, W. A. Dent.
Bull. American Astron. Soc., Vol. 4, 339 (1972). – Abstr. AAS.

141.044 A composite spectrum of quasars. A. Yahil.
Bull. American Astron. Soc., Vol. 4, 339 (1972).
Abstr. AAS.

141.045 The cutoff of quasars beyond $z = 2$.
M. F. Barnothy, J. M. Barnothy.
Bull. American Astron. Soc., Vol. 4, 339 - 340 (1972).
Abstr. AAS.

141.046 UBV observations of the radio source OJ 287.
V. M. Lyutyj.
Astron. Tsirk., No. 708, p. 3 - 4 (1972). In Russian.

141.047 Tentative optical identification of the radio sources VRO 22.17.01 and VRO 24.02.02. L. Pataki.
Bull. American Astron. Soc., Vol. 4, 398 (1972). – Abstr. AAS.

141.048 A flux density scale for microwave frequencies.

W. A. Dent.
Bull. American Astron. Soc., Vol. 4, 398 (1972). – Abstr. AAS.

141.049 The linear polarization of Cassiopeia A at wavelengths of 9.8 and 11.1 cm.
G. S. Downs, A. R. Thompson.
Bull. American Astron. Soc., Vol. 4, 398 (1972). – Abstr. AAS.

141.050 A possible mechanism for energy release in quasars and Seyfert nuclei. J. D. G. Rather.
Bull. American Astron. Soc., Vol. 4, 398 (1972). – Abstr. AAS.

141.051 A physical model of line formation in QSO's.
J. D. Scargle, C. B. Tarter.
Bull. American Astron. Soc., Vol. 4, 398 - 399 (1972).
Abstr. AAS.

141.052 Accurate positions for radio sources.
M. H. Cohen.
Astrophys. Letters, Vol. 12, 81 - 85 (1972).
Fringe-rate residuals from observations with the Goldstack interferometer have been analyzed. The positions for eleven sources, and the baseline, are obtained to an accuracy of about 0.2 arc sec. Five sources are coincident with their associated optical objects to about 0.2 arc sec.

141.053 The formaldehyde absorption of W33.
F. F. Gardner, J. B. Whiteoak.
Astrophys. Letters, Vol. 12, 107 - 112 (1972).
An investigation of the H_2CO absorption across the radio source W33 shows features at the velocities 12, 35, 52 and 61 km/sec. The results are interpreted in terms either of an expanding continuum source with surrounding molecular cloud, or of several cloud-continuum complexes at different distances.

141.054 Cartes d'intensité totale du quasar 3C 249.1 aux fréquences 2695 MHz et 8085 MHz.
C. Faubert, P. P. Kronberg.
Journ. Roy. Astron. Soc. Canada, Vol. 66, 221 (1972).
Abstr. Canadian Astron. Soc.

141.055 Optical identification of Ohio survey radio sources.
M. R. Gearhart, J. M. Lund, D. J. Frantz, J. D. Kraus.
Astron. Journ., Vol. 77, 557 - 559, 617 - 620 (1972).
Tentative optical identifications are made for 47 Ohio survey sources, many of which have flat or unusual radio spectra. Finding charts are provided for the identifications. Twenty eight of the identifications are with stellar objects, 18 with galaxies, and 1 is a blank field.

141.056 Measurements of the flux density and spectra of discrete radio sources at centimeter wavelengths.
III. Observations of weak sources at 2.7 and 5 GHz.
I. I. K. Pauliny-Toth, K. I. Kellermann.
Astron. Journ., Vol. 77, 560 - 568 (1972).
Flux-density measurements at 2.7 and 5 GHz of 283 weak radio sources selected from the 178-MHz Ryle–Neville survey and the 408-MHz Bologna surveys, respectively, are given. The spectral-index distributions are compared with those for the stronger sources, and no relation is found up to source densities of 10^4 source sr^{-1}.

141.057 Results of observations of radio sources at short millimeter wavelengths.
V. F. Zabolotny, I. G. Moiseev, A. V. Pavlov, V. I. Slysh, V. A. Soglasnova, G. B. Sholomitsky, M. B. Shcherbina-Samoylova.
Astron. Zhurn. Akad. Nauk SSSR, Vol. 49, 971 - 981 (1972). In Russian. English translation in Soviet Astron. AJ, Vol. 16,

No. 5.

Astronomical measurements in the short-wave region of millimeter wavelength range·were made in 1969–1970 using a wide-band radiometer with an indium antimonide detector cooled by liquid helium, and the 22-*m* radio telescope of the Crimean Astrophysical Observatory. The beamwidth of the radiotelescope was 2 arc min in diameter at half power level. The values or upper limits of the fluxes for 12 galactic and extragalactic sources were obtained.

141.058 Observations of quasi-stellar objects at frequencies of 7700 MHz.   N. M. Lipovka.
Astron. Zhurn. Akad. Nauk SSSR, Vol. 49, 982 - 985 (1972). In Russian. English translation in Soviet Astron. AJ, Vol. 16, No. 5.

Observations of radio sources with flat spectrum are carried out. Optical identifications are made which allow to suppose that the majority of identified objects is connected with quasars.

141.059 Estimate of macro-parameters of quasar nuclei from the thermal component of their continuum.
L. M. Ozernoy.
Astron. Zhurn. Akad. Nauk SSSR, Vol. 49, 1123 - 1125 (1972). In Russian. English translation in Soviet Astron. AJ, Vol. 16, No. 5.

The quasi-Planckian component of the optical continuum of quasars, which remains after subtracting a power-law (synchrotron) component is apparently caused by the thermal radiation of a supermassive body. A crude estimate of macro-parameters of this body does not contradict the expected parameters of the magnetoid.

141.060 A preferred orientation of extragalactic double radio sources.   M. A. F. Thiel.
Astrophys. Space Sci., Vol. 17, 39 - 47 (1972).

The distribution of the main axes of double radio sources is used to test isotropy of that part of the universe which is accessible to radio investigations. Data for 274 double and/or extended sources have been taken from the literature to compute the fit to several simple models of global orientation by means of a χ^2-test. The best fit has been found on a 3 %-significance level for the preferred orientation along a right-handed helix with pitch angle 82° in the direction $\alpha = 95°$, $\delta = -38°$ ($l \cong 65°$, $b = \cong 20°$). This preferential alignment of radio sources is assumed to be caused by a large-scale magnetic field.

141.061 Improved positions and some identifications for 108 radio sources between declinations −33° and +27°.
J. K. Merkelijn.
Australian Journ. Phys., Vol. 25, 451 - 460 (1972).

Positions of improved accuracy have been obtained for 108 sources from the Parkes catalogue between declinations −33° and +27°. The estimated errors in the positions are less than 15″ arc in both coordinates. Identifications are suggested for 17 sources, 12 with possible quasi-stellar objects and 5 with galaxies.

141.062 Australian east-west baseline interferometer observations at 2.3 GHz.
J. S. Gubbay, A. J. Legg, D. S. Robertson.
Australian Journ. Phys., Vol. 25, 461 - 463 (1972).
The results on 17 radio sources are listed in a table.

141.063 A new identification for PKS 2204−54.
K. P. Tritton, G. D. Nicolson.
Astrophys. Letters, Vol. 11, 187 - 188 (1972).

An object with an ultraviolet excess is suggested as the correct identification for PKS 2204−54, on the basis of intensity variations detected at 13 cm. The possible existence of a galaxy cluster in the field of the object is noted.

141.064 Identification of 4C sources with galaxies.
C. Hazard, D. L. Jauncey.
Astron. Journ., Vol. 77, 621 - 624, 695 - 697 (1972).

We suggest identification with galaxies for 59 radio sources from the 4C catalogue.

141.065 Observations at 750, 1400, and 2700 MHz of radio sources in the Vermilion River Observatory survey.
J. C. Webber, A. G. Willis.
Astron. Journ., Vol. 77, 625 - 636 (1972).

For a sample of radio sources selected at 610.5 MHz, higher frequency measurements are given. 494 sources are listed, of which 410 have been observed at 2700 MHz and have positions accurate to ±20″. A spectral index and other catalog numbers are also listed.

141.066 Ton 155 and 156: a double quasar?
A. N. Stockton.
Nature, Phys. Sci., Vol. 238, 37 (1972).

Ton 155 and 156, separated by 35″arc, are both quasars and have redshifts $z = 1.703$ and $z = 0.549$, respectively.

141.067 Selection effects in spectroscopic observations of QSOs.   D. Wills.
Nature, Phys. Sci., Vol. 238, 70 - 71 (1972).

The importance of selection effects in determining redshift of QSOs has perhaps been overstated. Data available at the time of this paper can be represented by a smooth redshift distribution if the "radio-quiet" QSOs are excluded.

141.068 Brightness distribution and polarization of 3C 273.
R. G. Conway, D. Stannard.
Nature, Phys. Sci., Vol. 239, 22 - 23 (1972).

Interferometric observations have been made of the quasar 3C 273 at two wavelengths, 11 cm and 73 cm, in each case with a resolution of 7 arc sec. The polarization of the radio radiation is determined in the form of strip distributions in position angle 43°, along the direction of the optical jet. From the observations the direction of the magnetic field may be deduced, which is found to lie along the jet at its root, but across the jet at its tip. Such a structure is predicted both by the magnetic loop theory and by the ram-pressure model, but has not hitherto been confirmed observationally.

141.069 Doppler effects and hypervelocities of 3C 279 sources.   C. Gregory.
Nature, Phys. Sci., Vol. 239, 56 - 57 (1972).

Doppler effects at hypervelocities and their origin in the light of observations and interpretations of radio interferometric experiments on quasar 3C 279 as due to a symmetric double source apparently separating at a speed of 10c are discussed. An expression for the red shift is obtained for hypervelocities. Extra-dimensionality and a generalization of Robertson-Walker metric yield an expression for time dependence of angular separation. Possible transformation of optical spectra to X-ray or γ-ray region is noted.

141.070 Observations of extragalactic variable sources at 2.8 and 4.5 cm wavelength.
W. J. Medd, B. H. Andrew, G. A. Harvey, J. L. Locke.
Mem. Roy. Astron. Soc., Vol. 77, 109 - 158 (1972).

The paper describes a programme of observations of 84 variable extragalactic radio sources at centimetre wavelengths. Details are given of the equipment, the observing techniques and the calibration procedures. The results of 5 yr of observations are presented and briefly discussed. It is shown that there is no consistent agreement between the observations and the various theories of variable radio sources. However, there is a strong correlation between variability and spectral type, and the rate of the variations shows the expected dependence on redshift.

141.071 Cosmological inferences from the angular diameters of quasars. M. Reinhardt.
Astrophys. Letters, Vol. 12, 135 - 138 (1972).

The distribution function of linear diameters of quasars is investigated for sources whose largest angular separation is not smaller than 7 arc sec in Miley's compilation. The distribution function is consistent with the assumption of no cosmological evolution in low-density Friedmann universes. The existence of a class of objects with a well-defined linear diameter between 0.33 and 0.43 Mpc is indicated.

141.072 A flux-density scale for microwave frequencies. W. A. Dent.
Astrophys. Journ., Vol. 177, 93 - 99 = Contr. Five College Obs., *Amherst, Mass.*, No. 137 (1972).

Accurate flux-density measurements of the thermal radio source DR 21 have been made at centimeter wavelengths relative to the KPW absolute flux-density scale based on Cas A and at millimeter wavelengths relative to absolute brightness-temperature measurements of Jupiter and Saturn. The absolute spectrum of DR 21 is given. It defines a flux-density scale that can be used to calibrate antennas having beamwidths between 1 and 6 minutes of arc.

141.073 Very long baseline interferometer observations of Taurus A and other sources at 121.6 MHz.
W. C. Erickson, T. B. H. Kuiper, T. A. Clark, S. H. Knowles, J. J. Broderick.
Astrophys. Journ., Vol. 177, 101 - 114 (1972).

VLBI observations with an antenna spacing of 92,000 λ (2″.2 lobe separation) were made on a number of small-angular-diameter sources at a frequency of 121.6 MHz. Through positional and spectral coincidence, these observations confirm the physical association of the compact source in Tau A with the pulsar NP 0532; in the east-west direction, the two objects agree in position to an accuracy of ±0″.1. The fluxes of the small-angular-diameter components of 3C 48, 3C 84, 3C 144, 3C 147, 3C 273, 3C 274, 3C 298, 3C 405, 3C 459, and 3C 461 are estimated.

141.074 Observations of quasars with small Schmidt telescopes. P. Véron.
The role of Schmidt telescopes in astronomy. Conference Hamburg 1972, (see 012.014), p. 77 - 79 (1972).

141.075 Schmidt telescopes and radio astronomy. J. G. Bolton.
The role of Schmidt telescopes in astronomy. Conference Hamburg 1972, (see 012.014), p. 81 - 85 (1972).

141.076 Spectra of some Ohio radio sources: List III.
E. K. Conklin, B. H. Andrew, B. J. Wills, J. D. Kraus.
Astrophys. Journ., Vol. 177, 303 - 307 (1972).

Spectra and positions ($\sim 15''$) are presented for 32 Ohio radio sources. Flux densities have been measured at nine frequencies between 0.6 and 85.3 GHz. Several of the sources are variable.

141.077 The redshifts of quasi-stellar objects and associated galaxies. D. F. Falla.
Observatory, Vol. 92, 179 - 181 (1972).

The association of QSOs with individual galaxies, and with clusters of galaxies, has been studied in several recent attempts to elucidate the origin of the QSO redshifts. We show that it may be possible to resolve their apparently contradictory conclusions, in terms of a simple hypothesis.

141.078 Optical positions for 21 3C objects.
A. N. Argue, C. M. Kenworthy.
Monthly Notices Roy. Astron. Soc., Vol. 160, 197 - 211 (1972).

Optical positions relative to AGK have been obtained for 21 3C objects. The external standard errors have been estimated by comparison with the most accurate radio positions resulting in: for R.A. ± 0.15 arc sec, and for Dec a value varying between ± 0.12 arc sec at $\delta = +40°$ and ± 0.20 arc sec at $\delta = +77°$ and +4°.

141.079 MWC 349, a new radio star.
L. L. E. Braes, H. J. Habing, A. A. Schoenmaker.
Nature, Vol. 240, 230 (1972).

Further examination of the Westerbork 1,415 MHz observations of the Cygnus X-3 region revealed the presence of a radio source at the position of the peculiar emission-line star MWC 349.

141.080 A feasible interpretation of non-exponential cosmic radio source spectra.
Yu. N. Gnedin, A. Z. Dolginov, V. N. Fedorenko.
Pis'ma v ZhEhTF, Vol. 16, 45 - 47 (1972). In Russian.
Abstr. in Referativ. Zhurn. 51. Astron., 11.51.628 (1972).

141.081 The phenomenon of QSS. N. E. Kurochkin.
Trudy Gos. Astron. Inst. Shternberga, Vol. 43, 89 - 118 (1972). In Russian.

The principal observational data on QSS are critically reviewed. Close interrelations between QSS and galaxies properties suggest their common nature. We suggest a conception of QSS as the nuclei of young galaxies – superassociations. The phenomenon of QSS and compact galaxies is explained by the higher luminosity and high density of the nuclear regions of younger galaxies.

141.082 Modell der Variabilität von Quasaren und Seyfert-Galaxien. J. Dorschner, C. Friedemann,
J. Gürtler, H. Oleak, K.-H. Schmidt.
Astron. Nachr., Vol. 294, 65 - 78 (1972).

The observed long-time optical variations of quasi-stellar objects and Seyfert galaxies are interpreted by extinction variations of grains in intervening clouds in the quasar envelopes moving against the central cores of the quasars. Assuming a density ratio between gas and dust in the clouds similar to the value observed in the interstellar medium of our Galaxy the diameters of the clouds are restricted to values between 10^{14} and 10^{18} cm depending on the gas densities in the quasar envelopes obtained by Bahcall and Kozlovsky.

141.083 An interferometric survey of the areas surrounding four intense radio sources.
R. J. Peckham, H. P. Palmer.
Nature, Phys. Sci., Vol. 240, 76 - 77 (1972).

We report here the results of an interferometric survey of the areas surrounding the intense radio sources Cassiopeia A, Cygnus A, Taurus A and Virgo A.

141.084 Further remarks on selection effects in spectroscopic observations of QSOs.
R. C. Roeder, C. C. Dyer.
Nature, Phys. Sci., Vol. 240, 104 - 105 (1972).

Wills has criticized our suggestion that the distribution of quasar redshifts may well be significantly affected by observational selection effects, and that consequently no conclusions should be drawn from the raw distribution. We wish that Wills's presentation had convinced us that our concern is unjustified; this, however, is not the case.

141.085 Radio components with a circumferential magnetic field configuration in 3C 219 and 3C 353.
E. B. Fomalont.
Astrophys. Letters, Vol. 12, 187 - 192 (1972).

The linearly polarized distribution for the radio sources 3C 219 and 3C 353 at 2695 MHz are presented. Both sources show evidence for a circumferential magnetic field associated with one of their components. These components are extended perpendicular to the major axis of the radio source. Such a magnetic field configuration may have some bearing on the confinement and evolution of the source components.

141.086 Four variable radio sources at 408 MHz.
R. W. Hunstead.
Astrophys. Letters, Vol. 12, 193 - 200 (1972).

Evidence is presented for large fractional variations in the 408-MHz emission from two quasi-stellar objects (CTA 102 and 3C454.3) and two radio galaxies (PKS 1504−16.7 and 1524−13). In each source significant changes in flux density have occurred on a time scale of a few months. If the variations observed at 408 MHz are intrinsic to the radiating regions, they raise some important astrophysical questions concerning the physical processes operating in non-thermal radio sources.

141.087 Detection of cyanoacetylene at 18 GHz.
D. F. Dickinson.
Astrophys. Letters, Vol. 12, 235 - 236 (1972).

The principal hyperfine components of the $J = 2 \to 1$ rotational transition of cyanoacetylene have been observed in Sgr B2. This result, coupled with the $J = 1 \to 0$ observations of Turner (1971), suggests that cyanoacetylene cannot be described by a simple LTE model.

141.088 Observations of radio sources with an interferometer of 24-km baseline–I. The angular structures at 408 MHz of 106 sources from the Parkes catalogue.
J. Critchley, H. P. Palmer, B. Rowson.
Monthly Notices Roy. Astron. Soc., Vol. 160, 271 - 282 (1972).

The basic details of the Jodrell Bank–Mk III telescope interferometer system are described. Angular size and structure information at 408 MHz on 106 sources from the Parkes catalogue in the range $-13° > \delta > -17°$ is presented. The data show that there is a larger fraction of double sources consisting of compact widely separated emitting regions than is apparent from published work.

141.089 Observations of radio sources with an interferometer of 24-km baseline–II. The angular structures at 151 and 408 MHz of 46 unidentified radio sources from the revised 3C catalogue.   P. K. Wraith.
Monthly Notices Roy. Astron. Soc., Vol. 160, 283 - 303 (1972).

The angular structures of 46 unidentified radio sources from the revised 3C catalogue have been determined with the Mk I–Mk III interferometer based at Jodrell Bank. The observations were made at 151 and 408 MHz simultaneously giving maximum resolving powers of ≈ 4 arcsec and ≈ 1.2 arcsec respectively. Many of the sources have a double structure, the components of these double sources often having higher brightness regions toward their outer edges and lower brightness 'tails' usually–but not always–pointing toward the centroid of the source.

141.090 Observations of radio sources with an interferometer of 24-km baseline–III. The angular structures at 408 and 1423 MHz of 44 relatively intense radio sources.
P. N. Wilkinson.
Monthly Notices Roy. Astron. Soc., Vol. 160, 305 - 319 (1972).

The results of long baseline radio interferometer studies at 408 and 1423 MHz of the angular structures of a sample of relatively intense radio sources, mainly from the 3CR catalogue, are presented. In this paper the results of the model fitting analysis of the visibility data are given and discussed for each source, and some of the more interesting points arising from

them noted. A detailed discussion of the results in comparison with other angular structure observations will be given in a later paper.

141.091 The spectra of sources in the ±4° declination zone of the Parkes 2700 MHz survey. I. Spectra of individual sources.   J. V. Wall.
Australian Journ. Phys., Astrophys. Suppl., No. 24, p. 1 - 47 (1972).

A series of observations to determine the spectra of sources in the ±4° declination zone of the Parkes 2700 MHz survey is described. The observations have yielded flux densities for 370 sources at 5009 MHz, 300 sources at 1403 and 468 MHz, and 450 sources at 635 MHz, while additional flux densities for many of the sources have been obtained from published measurements from other observatories. Spectral data have also been tabulated for some of the sources found in the Parkes 2700 MHz deep surveys of selected regions. The spectra of a number of the sources are illustrated and discussed. Several of the sources appear to be variable at high frequencies.

141.092 The spectra of sources in the ±4° declination zone of the Parkes 2700 MHz survey. II. Statistical analysis.   J. V. Wall.
Australian Journ. Phys., Astrophys. Suppl., No. 24, p. 49 - 63 (1972).

An analysis is presented of the spectral data for sources in the +4° to −4° declination zone of the Parkes 2700 MHz catalogue. The proportions of the different types of radio spectra in the sample are derived. Spectral differences between quasi-stellar objects and radio galaxies are discussed and the results are used to estimate the proportion of the unidentified sources which belong to each class. The existence of a correlation between spectral index and flux density is established.

141.093 Observations of the radio source PKS 0123−01 at 5000, 408, and 80 MHz.
R. T. Schilizzi, I. A. Lockhart, J. V. Wall.
Australian Journ. Phys., Vol. 25, 545 - 558 = Separate print Div. Radiophys., C.S.I.R.O., Sydney (1972).

Observations of PKS 0123−01 (3C40) have been made at 5000, 408, and 80 MHz with telescopes of comparable beamwidth. The value of the spectral index for the frequency intervals 5000−408 and 408−80 MHz has been calculated for a grid of points covering the source spaced at approximately half–beamwidth intervals. For both frequency intervals the spectrum of emission is found to be significantly steeper for the regions of the source furthest from the associated galaxies NGC 545−547 than for the regions nearest the galaxies. Some implications of this result are discussed. The age and energy requirements of the source have also been estimated.

141.094 Positions and some identifications for 111 sources of about 1 flux unit at 408 MHz.
D. G. Hoskins, H. S. Murdoch, C. Hazard, D. L. Jauncey.
Australian Journ. Phys., Vol. 25, 559 - 579 (1972).

Radio positions with an accuracy of ~5 sec arc have been measured for a sample of 111 sources that have flux densities in the range 0.6 to 5.5 f.u. at 408 MHz but which are not listed in the 4C catalogue. An investigation of the random background rates for both galaxies and BSO's shows that the expected chance identification rate is small for galaxies brighter than 19^m and for BSO's to the plate limit. A comparison of optical and radio positions shows that radio−optical separations of up to 15 sec arc are common for galaxies of 18^m or brighter, but rare for BSO's. The identification content, corrected for chance associations, is 16% BSO's and 20% galaxies.

141.095 Television photometry of the radio source OJ 287.

P. P. Petrov.
Astron. Tsirk., No. 710, p. 1 - 2 (1972). In Russian.

141.096 Rapid variability of the compact object OJ 287.
V. A. Hagen-Thorn.
Astron. Tsirk., No. 714, p. 5 - 8 (1972). In Russian.

141.097 Radio source counts and redshifts in steady state cosmology. M. Schmidt.
Nature, Vol. 240, 399 - 400 (1972).

There appears to be continuing confusion as to whether or not counts of extragalactic radio sources are compatible with the steady state theory. I present here an explicit derivation of source counts expected in the steady state on the basis of identifications and redshifts of a complete sample of radio galaxies and quasi-stellar sources.

141.098 Maiņzvaigžņu radiostarojums. A. Alksnis.
Zvaigžņotā debess, 1971. gada rudens, p. 9 - 16.

141.099 The distribution of radiation from relativistically expanding radio sources. D. S. De Young.
Astrophys. Journ., Vol. 177, 573 - 583 (1972).

The distribution of synchrotron radiation in a relativistically expanding source is considered as seen by a distant observer. A variety of initial conditions, optical depths, and geometries are examined, including sources of constant energy input, blast waves, and jets. It is found that double or core-and-halo sources with apparent expansion velocities in excess of the velocity of light can result from natural geometries and actual expansion rates less than c. The results are compared with observations of 3C 279, 3C 273, and 3C 120, all of which exhibit possible "superrelativistic" expansions.

141.100 Observed anisotropy in the distribution of radio sources. A. Yahil.
Astrophys. Journ., Vol. 178, 45 - 55 (1972).

The anomalous number-flux relation of the stronger radio sources, which has been interpreted as due to evolution, is shown to appear predominantly and perhaps only in the northern galactic hemisphere. Furthermore, it is due to ~40 percent of the sources which have not been identified optically, and whose distance is not known. It is therefore also possible that these unidentified sources are less powerful ones close by, and the irregularities in their flux density distribution is due to local inhomogeneities. It should be possible to rule out this hypothesis by pushing optical identification to $V \lesssim 22.5$, and by discovering that the presently unidentified sources indeed are giant elliptical galaxies at cosmological distances.

141.101 Digicon spectrophotometry of the quasi-stellar object PHL 957.
E. A. Beaver, E. M. Burbidge, C. E. McIlwain, H. W. Epps, P. A. Strittmatter.
Astrophys. Journ., Vol. 178, 95 - 103 (1972).

Measurements of the spectral region centered on the broad Lα absorption in the QSO PHL 957 have been made with the UCSD Digicon (digital image tube) at the Cassegrain focus of the Lick 12-inch telescope, and photographic spectra have also been obtained over a wider spectral range. The broad absorption is very deep, with an intensity less than 2 percent of the continuum over the central 30 Å. The profile of the broad absorption and the relationship between this feature and the other multiple-redshift systems found in PHL 957 by Lowrance et al. is discussed.

141.102 The effects of radiation pressure from resonance scattering in a quasar cloud. R. E. Williams.
Astrophys. Journ., Vol. 178, 105 - 112 (1972).

The suggestion that those absorption-line systems in quasi-stellar objects for which $z_{abs} \ll z_{em}$ are caused by the accel-

eration of clouds by radiation pressure is investigated. The transfer of resonance-line radiation in a static cloud has been solved in order to determine the distribution of the force within the cloud due to radiation pressure. It is shown that radiation pressure will accelerate a discrete cloud coherently only under a restrictive range of conditions. Specifically, if the continuum radiation incident upon a cloud ultimately causes appreciable photon creation within the cloud, e.g., through ionization, the gas will tend to be disrupted by the outward diffusion of the internally created photons as it is accelerated.

141.103 Observations of methanol in Sagittarius B2 at 48 GHz.
A. H. Barrett, R. N. Martin, P. C. Myers, P. R. Schwartz.
Astrophys. Journ., (Letters), Vol. 178, L23 - L27 (1972).

The $J, K = 1,0 \rightarrow 0,0$ lines of methanol at 48.3 GHz have been detected in both Sgr A and Sgr B2 OH. The Sgr B2 emission is found to extend over an area at least $5' \times 10'$. The indicated column density of methanol is $\sim 2 \times 10^{16} \, cm^{-2}$, and the total mass of the neutral cloud is estimated to be $\sim 10^6 M_\odot$.

141.104 The distribution of optical spectral indices for QSOs from U, B, V data. A. Evans.
Monthly Notices Roy. Astron. Soc., Vol. 160, 407 - 420 (1972).

A term is derived to correct the U, B, V measurements for the quasi-stellar objects for the presence of emission lines. The corrected colours for 102 objects are calculated using different values of emission line equivalent widths. From the corrected colours, the spectral indices for the optical continua are obtained. The spectral indices so derived are compared with those calculated from scanner measurements, and the colour spectral indices that give the best agreement with the available scanner spectral indices are used to find the distribution of optical spectral indices for quasi-stellar objects.

141.105 3C 120, BL Lacertae, and OJ 287: Coordinated optical, infrared, and radio observations of intraday variability. E. E. Epstein, W. G. Fogarty, K. R. Hackney, R. L. Hackney, R. J. Leacock, R. B. Pomphrey, R. L. Scott, A. G. Smith, R. W. Hawkins, R. C. Roeder, B. L. Gary, M. V. Penston, K. P. Tritton, C. Bertaud, M. P. Véron, G. Wlérick, A. Bernard, J. H. Bigay, P. Merlin, A. Durand, G. Sause, E. E. Becklin, G. Neugebauer, C. G. Wynn-Williams.
Astrophys. Journ., (Letters), Vol. 178, L51 - L59 (1972).

Simultaneous optical, infrared, and radio observations were made to search for intraday variability of the radiation from 3C 120, BL Lac, and OJ 287, sources known to be active at both radio and optical wavelengths on time scales of days or longer. Optical interday variability was found for all three sources.

141.106 A Green Bank Sky Survey in search of radio sources at 1400 MHz. III. Positions and flux densities of the GB radio sources. J. Maslowski.
Acta Astron., Vol. 22, 227 - 260 (1972).

A continuum 1400 MHz survey with the NRAO 300-foot radiotelescope has been made between right ascensions $7^h 17^m$ to $16^h 23^m$, fully covering the declination range $+45°.8$ to $+51°.7$. Results are presented by a list of 1086 radio sources down to a limiting flux density of $0.09 \times 10^{-26} \, W \, m^{-2} \, Hz^{-1}$ at 1400 MHz. Among these sources, 821 were previously uncatalogued. A study has been made of the spectral index distribution for sources common to the present survey and the BP catalogue (Bailey and Pooley 1968). Differential counts of radio sources at 1400 MHz are presented in the form of a graph and compared with those at 1415 MHz derived from the Ohio Survey (Harris and Kraus 1970). Observational and reduction procedures are also discussed.

141.107 The influence of disperion of velocities of inhomo-

geneities on temporal spectra of radio sources scintillations. V. I. Shishov.
Astron. Zhurn. Akad. Nauk SSSR, Vol. 49, 1258 - 1266 (1972). In Russian. English translation in Soviet Astron. AJ, Vol. 16, No. 6.

Temporal spectra of scintillations of radio sources situated close to the sun are considered. It is shown that the dependence of indices of scintillations of the sources CTA 21 and 3C 279 on the distance to the sun in the region of saturated fluctuations can be explained only taking into account the size of the source and the pass-band of the receiver.

141.108 The distribution of redshifts of quasi-stellar objects and related emission-line objects.
G. R. Burbidge, S. L. O'Dell.
Astrophys. Journ., Vol. 178, 583 - 605 (1972).

A list of all published redshifts of quasi-stellar objects and related compact objects with strong emission-line spectra is given in a table. The distribution of all the redshifts and those of various subsets thereof have been subjected to several statistical tests in order to determine whether they are consistent with the hypothesis of randomness on a scale $\Delta z < 0.1$. A power-spectrum analysis of all the redshifts yields a spectral maximum, corresponding to a wavelength of 0.0705 in z, which may be considered significant at a 97.5 percent confidence level.

141.109 Absorption lines in the spectrum of the quasar Ton 1530. W. A. Morton, D. C. Morton.
Astrophys. Journ., Vol. 178, 607 - 615 (1972).

Two spectra of the quasar Ton 1530 (z_{em} = 2.047) were obtained with the Princeton integrating television system at the coudé focus of the 200-inch (508-cm) telescope. The principal features are two sets of C IV resonance doublets in absorption. We found each member of the shorter-wavelength doublet split into three components with redshifts z = 1.9358, 1.9371, 1.9384.

141.110 Decimeter-wavelength observations of radio recombination lines in W51.
A. Parrish, V. Pankonin, C. E. Heiles, J. M. Rankin, Y. Terzian.
Astrophys. Journ., Vol. 178, 673 - 680 (1972).

The 1000-foot (305-m) Arecibo telescope has been used to detect the 221 and 248α hydrogen and the 248α anomalous (carbon) recombination lines from the galactic radio source W51. The half-power widths of the detected lines are proportional to their radio frequencies; they exhibit no pressure broadening.

141.111 Radio behavior of β Persei.
R. M. Hjellming, E. Webster, B. Balick.
Astrophys. Journ., (Letters), Vol. 178, L139 - L144 (1972).

Radio observations of β Per at 2695 and 8085 MHz during 1972 April–July are summarized, and the details on several unusual events are given. The implications of the data, which show dominant thermal behavior with very rare nonthermal events, are related to a model of a variable thermal source energized by shock waves and/or suprathermal particles ejected when a star adjusts its structure discontinuously by means of "starquakes".

141.112 The NRAO 5-GHz radio source survey. III. The 140-ft "strong" source survey.
I. I. K. Pauliny-Toth, K. I. Kellermann.
Astron. Journ., Vol. 77, 797 - 809 (1972).

The 140-ft telescope at the National Radio Astronomy Observatory has been used to extend the 5-GHz strong source (S) survey over an area of 1.14 ster. Some 240 sources have been detected in·this region, of which 135 are above the completeness limit of 0.6 f.u. The distribution of spectral indices, the dependence of the spectral index on the flux density, and

the number – flux density relation for various groups of sources have been analysed.

141.113 Measurements of the integrated Stokes parameters of compact radio sources.
G. L. Berge, G. A. Seielstad.
Astron. Journ., Vol. 77, 810 - 818 (1972).

The complete state of the integrated radiation from several compact radio sources was measured at three frequencies and at several epochs: 1420 MHz (1971.0), 1602 MHz (1971.1), and 5000 MHz (1971.5, 1971.8, and 1972.0). All four Stokes parameters were obtained, and all of the results are presented.

141.114 7.8-GHz flux density measurements of variable radio sources. W. A. Dent, G. Kojoian.
Astron. Journ., Vol. 77, 819 - 828 = Contr. Five College Obs., Univ. Massachusetts, Amherst, Mass., No. 137 (1972).

Measurements of the variation in the 7.8-GHz flux density of 22 extragalactic radio sources between 1969.0 and 1971.7 are presented.

141.115 Optical monitoring of quasistellar objects. I.
P. K. Lü.
Astron. Journ., Vol. 77, 829 - 844 (1972).

Observational results of 25 quasistellar objects (QSO's) from the Yale Observatory optical patrol program of QSO's are given. Light curves and magnitudes are presented for all objects, generally brigher than 17th magnitude. Fine-structure light curves are given for 3C 345 and 3C 454.3. There is a good correlation of optical variation with the radio variability at 2-cm to 6-cm wavelength.

141.116 On the interpretation of the absorption spectra of quasars. C. F. McKee, C. B. Tarter, J. C. Weisheit.
Bull. American Astron. Soc., Vol. 4, 413 (1972). – Abstr. AAS.

141.117 Transcontinental interferometry at 606 MHz.
G. H. Purcell, Jr.
Bull. American Astron. Soc., Vol. 4, 414 (1972). – Abstr. AAS.

141.118 Blue stellar radio sources without ultraviolet excess.
I. W. A. Browne, N. J. McEwan.
Nature, Phys. Sci., Vol. 239, 101 - 102 (1972).

The identification of eleven radio-sources with blue stellar objects is reported. These objects are unusual in that they do not show the characteristic ultra-violet excess normally associated with quasars. It is tentatively suggested that some of these objects may be quasars with redshifts greater than 2.3.

141.119 The red-shifts and the patterns of electromagnetic waves in quasi-stellar objects. T. L. Chou.
Ann. New York Acad. Sci., Vol. 187, 213 - 233 (1972).

141.120 Radio emission from MWC 349. L. L. E. Braes, H. J. Habing, A. A. Schoenmaker.
IAU Circ., No. 2450 (1972).

141.121 OJ 287. M. Mattei, J. Elliot.
IAU Circ., No. 2453 (1972).

141.122 Quasistellar objects. K. Aizu, H. Tabara, M. Taketani.
Progr. Theor. Phys. Suppl., (Japan), No. 49, p. 228 - 247 (1971).

141.123 UBV observations of 3C 273.V.
M. S. Burkhead, T. W. Rettig.
Publ. Astron. Soc. Pacific, Vol. 84, 850 = Publ. Goethe Link Obs., Indiana Univ., Bloomington, No. 143 (1972).

141.124 Flux densities and spectra of extragalactic radio sources. A. E. Niell.
Thesis, Cornell Univ., Ithaca, New York. [Available from Univ. Microfilms, Ann Arbor, Mich., USA. Order No. 72-13171], 312 pp. (1971).

141.125 The scintillation of extended radio sources when the receiver has a finite bandwidth. III. Further methods. K. G. Budden, B. J. Uscinski.
Proc. Roy. Soc. London, Ser. A, Vol. 330, 65 - 77 (1972).
New methods of solution of the problem are presented which avoid some of the limitations of the solutions used earlier. The method is also extended to deal with the case where the scattering medium varies along the line of sight.

141.126 Extragalactic radio sources. D. M. Mills.
Thesis, Stanford Univ., Stanford, Calif. [Available from Univ. Microfilms, Ann Arbor, Mich., USA. Order No. 72-5953], 81 pp. (1971).

141.127 Quasar redshifts, true or false? N. Sanitt.
New Scient., (GB), Vol. 55, No. 812, p. 494 - 496 (1972).
The most heated debate in physical cosmology during the past decade has concerned the large redshifts of the quasars. Are these objects really at the observable limit of the expanding universe? This interpretation has been challenged several times recently but remains relatively unscathed.

141.128 OJ 287. V. P. Tsesevich.
Yamamoto Circ., No. 1751, p. 2 (1972). In Japanese.

141.129 Observations of variable radio sources at 8.2 mm. V. A. Efanov, V. I. Zagatin, I. G. Moiseev, H. M. Tovmassian, V. B. Shteinshleger.
Izv. Krymskoj Astrofiz. Obs., Vol. 45, 172 - 175 (1972). In Russian.
The results of observations of radio sources 3C 84, 3C 273 and 3C 345 at 8.2 mm made in March 1970 at the Crimean Astrophysical Observatory with the 22-m radio telescope in conjunction with a maser are presented.

141.130 Spectral flux densities of radio emission from discrete sources at 3.5 cm wavelength. II.
A. G. Gorshkov, I. G. Moiseev, V. A. Soglasnov.
Izv. Krymskoj Astrofiz. Obs., Vol. 45, 176 - 181 (1972). In Russian.
Results of measurements of flux densities of radio emission from 99 discrete sources at 3.5 cm wavelength are given. Data on the variability of radio emission from discrete sources and NGC 4278 are given.

141.131 The Parkes 2700 MHz Survey (Fourth Part). Catalogue for the south polar cap zone, declinations −75° to −90°. A. J. Shimmins, J. G. Bolton.
Australian Journ. Phys., Astrophys. Suppl. No. 26, 25 pp. = Separate print Division Radiophys., C.S.I.R.O., Sydney (1972).
This paper presents a catalogue of 454 extragalactic radio sources obtained from a sky survey at 2700 MHz. The area of 0.214 sr covers the south polar cap from declination −75° to −90°. The catalogue is complete to a limiting flux density of 0.26 f.u. (640 sources per steradian) at 2700 MHz and is thought to be at least 90% complete at a flux density of 0.15 f.u. (approximately 1400 sources per steradian). The positions are accurate to $16''$ arc in both coordinates for sources of 0.26 f.u. and to $13''$ arc for the stronger sources; the flux densities are accurate to 0.02 f.u. for the weaker sources and to 3% of the flux density for the stronger sources.

141.132 Occultation of Sgr A.
G. Krishna, G. Swarup, N. V. G. Sarma, M. N. Joshi.

Nature, Vol. 239, 91 - 93 (1972).
Lunar occultation observations made at Ootacamund on September 9, 1970, have provided detailed information on the structure of the source of 327 MHz and have revealed the presence of an extended component 10 X 4 arc min in size and centred approximately on Sgr A. The extended source has an inclination of $45° \pm 15°$ to the galactic plane, similar to that of the 20 cm continuum ridges.

141.133 The proper motion of the object OJ 287. D. K. Karimova.
Astron. Tsirk., No. 720, p. 1 (1972). In Russian.

141.134 A new galactic population of radio sources. A. G. Gorshkov, M. V. Popov.
Astron. Tsirk., No. 720, p. 2 - 4 (1972). In Russian.

141.135 On the apparent brightness temperatures of variable radio structures. V. N. Kurilchik.
Astron. Tsirk., No. 732, p. 1 - 3 (1972). In Russian.

Radiative lifetimes for some resonance transitions of Fe I and Fe II in the region between 2300 Å and 3050 Å, and the application to iron abundance determinations in the sun and in the QSO PHL 938. See Abstr. 022.009.

Intercontinental radio astronomy. See Abstr. 033.035.

Depolarization of synchrotron radiation in various sources. See Abstr. 061.025.

Possible sites of star formation in Sgr B 2. See Abstr. 065.015.

Rotazione in oggetti stellari e quasi-stellari relativistici. See Abstr. 065.063.

Analysis of three-station interplanetary scintillation. See Abstr. 074.047.

The recurrent solar wind streams observed by interplanetary scintillation of 3C 48. See Abstr. 074.063.

Observations of the interplanetary medium and of the structure of radio sources using higher moments of interplanetary scintillations. See Abstr. 106.010.

Blaue Objekte in der Umgebung von M 31. I. Entdeckungwahrscheinlichkeit neu aufgefundener Blauer Objekte. See Abstr. 113.032.

UBV photoelectric observations: I. Stars within 25 parsecs of the sun; II. Stars in quasar and galaxy fields; III. Stars in Kapteyn Selected Areas; IV. Miscellaneous stars. See Abstr. 113.060.

Discovery of infrared emission from the radio source near Cygnus X-3. See Abstr. 114.086.

Radio flare on Beta Persei. See Abstr. 121.024.

Radio flares from Algol. See Abstr. 121.040.

β Persei: Radio star and probable X-ray star. See Abstr. 121.107.

On the interpretation of the photometry of irregular variable stars observed intermittently. See Abstr. 122.032.

Optical and radio flares in BL Lacerta (VRO 42.22.01). See Abstr. 122.078.

I Zw 1727+50: a new lacertid? See Abstr. 122.087.

Observations of linear polarization of optical radiation of BL Lac-type objects. See Abstr. 122.089.

Photoelectric monitoring of BL Lacertae. See Abstr. 122.098.

High resolution radio observations of AP Lib. See Abstr. 122.099.

Brightness and polarization structure of four supernova remnants 3C58, IC443, W28, and W44 at 2.8 centimeter wavelength. See Abstr. 125.006.

Methods to estimate the contribution of supernovae to the activity of quasars and nuclei of galaxies. See Abstr. 125.022.

Infra-red sources in the H II region W3. See Abstr. 131.062.

Restrictions on cosmic maser intensity and the possibility of detecting new OH and H_2O radio sources in a short-term sky survey. See Abstr. 131.102.

Radio search for the pulsing X-ray source in Hercules. See Abstr. 142.014.

The nature of the first Cygnus X-3 radio outburst. See Abstr. 142.040.

Observations at 408 MHz of the Cyg X-3 radio outburst. See Abstr. 142.041.

Large outburst of Cygnus X-3. See Abstr. 142.055.

Discovery of giant radio outburst from Cygnus X-3. See Abstr. 142.056.

Unusual radio events in Cygnus X-3. See Abstr. 142.057.

Search for Hα emission from the companion radio sources of Sco X-1. See Abstr. 142.059.

Determination of the distance of Cygnus X-3 by 21-cm absorption. See Abstr. 142.060.

Observations of Cygnus X-3 at 2.8 cm with a 17 arc s beam. See Abstr. 142.061.

Limits to the 2,695 MHz circular polarization of the September Cyg X-3 flare. See Abstr. 142.062.

The large outburst in Cygnus X-3 at 8 GHz. See Abstr. 142.063.

Observations of Cygnus X-3 by Uhuru. See Abstr. 142.064.

X-ray observations of Cyg X-3 near the time of radio outbursts. See Abstr. 142.065.

Cygnus X-3: 3.3 mm observations. See Abstr. 142.066.

An attempt to detect recombination line emission from the Cygnus X-3 outburst. See Abstr. 142.067.

15.5 GHz observations at the Haystack Observatory of the Cygnus X-3 outburst. See Abstr. 142.068.

Spectrum and polarization of the Cygnus X-3 outburst. See Abstr. 142.069.

Absence of linear polarization in Cygnus X-3. See Abstr. 142.070.

Radio observations of Cygnus X-3 and the surrounding region. See Abstr. 142.071.

21 cm and 18 cm observations of the Cygnus X-3 radio outburst. See Abstr. 142.072.

Observations of Cygnus X-3 at the Mullard Radio Astronomy Observatory. See Abstr. 142.074.

Search for a visible counterpart of the September 2, 1972 radio outburst in Cygnus. See Abstr. 142.075.

X-ray observations of Cygnus X-3 by Copernicus. See Abstr. 142.076.

More unusual radio events in Cygnus X-3. See Abstr. 142.077.

Observation of a correlated X-ray—radio transition in Cygnus X-1. See Abstr. 142.079.

High frequency observations of the second radio flare in Cygnus X-3. See Abstr. 142.090.

Cygnus X-3 radio source: Lower limit on size and upper limit on distance. See Abstr. 142.091.

Giant radio outburst of Cygnus X-3. See Abstr. 142.131.

A low-latitude galactic survey from $l^{II} = 46°$ to 61° and 190° to 290° at 2700 MHz. See Abstr. 157.008.

New redshifts of radio galaxies. See Abstr. 158.005.

Activity in galaxies and quasars. See Abstr. 158.015.

The optical spectrum of the Seyfert galaxy 3C 120. See Abstr. 158.017.

Extragalactic explosive phenomena. See Abstr. 158.049.

A search for circular polarization in extragalactic objects. See Abstr. 158.052.

Observations of bright galaxies at 80 MHz. See Abstr. 158.053.

Infrared photometry of extragalactic sources. See Abstr. 158.057.

Radio galaxies, quasars, and cosmology. See Abstr. 158.063.

The role of Schmidt telescopes in the study of external galaxies. See Abstr. 158.088.

A high-frequency study of Cygnus A.
See Abstr. 158.091.

Counts of galaxies in the fields surrounding quasars.
See Abstr. 158.094.

NGC 2992 and the blue stellar object Weedman No. 2. See Abstr. 158.109.

On the luminosity function of radiogalaxies.
See Abstr. 158.112.

On "pairing" of galaxies and quasars.
See Abstr. 158.115.

On the optical variability of the N-galaxy 3C 371.
See Abstr. 158.136.

The radio continuum of the Large Magellanic Cloud. IV. Spectra of sources. See Abstr. 159.016.

3 C 323.1: A QSO in a rich cluster of galaxies.
See Abstr. 160.010.

Relativistic non-zero pressure cosmology.
See Abstr. 162.026.

The redshift-distance relation. III. Photometry and the Hubble diagram for radio sources and the possible turn-on time for QSOs. See Abstr. 162.051.

Structures of magnetic fields in the universe and galaxies. See Abstr. 162.060.

Pulsars

141.501 Amplitude—time and polarization characteristics of CP 1133 pulsar subpulses.
Iu. I. Alexeev, V. V. Vitkevich, V. M. Malofeev.
Dokl. Akad. Nauk SSSR, Ser. Mat. Fiz., Vol. 205, 307 - 309 (1972). In Russian.

141.502 Hamiltonian analysis of charged particle motion in the pulsar rotating magnetic field. V. G. Endean.
Monthly Notices Roy. Astron. Soc., Vol. 158, 13 - 22 (1972).

A relativistic Hamiltonian analysis of the charged particle motion in the approximate 'cylindrical' pulsar rotating magnetic field is given. The predicted features of the pulses of Doppler-shifted cyclotron radiation emitted from a region near the light cylinder and the leading edge of the $E>cB$ zone show good agreement with the observations.

141.503 The interstellar scintillation pattern of PSR 0329 + 54. J. A. Galt, A. G. Lyne.
Monthly Notices Roy. Astron. Soc., Vol. 158, 281 - 290 (1972).

The pulsar PSR 0329 + 54 was observed at 408 MHz simultaneously for an entire day at Jodrell Bank and Penticton. The separation of these stations is sufficient to permit detection of motions of the diffraction pattern produced by irregularities in the interstellar medium. The present observations suggest that the scintillation pattern is stable.

141.504 Pulsars and pair production in electric fields.
L. Parker, J. Tiomno.
Nature, Phys. Sci., Vol. 238, 57 - 58 (1972).

Electron—positron pair production by a strong radial electric field surrounding a central body gives rise to a radially oscillating plasma. If it is assumed that an energy balance is reached, in which pulses of radiation are emitted in step with the intervals of pair creation, then the predicted pulse width, rate of increase of oscillation period, and power output are in the range observed for pulsars, indicating a possible connection.

141.505 The nature of radio emission from pulsars.
T. N. Rengarajan.
Astrophys. Space Sci., Vol. 17, 65 - 68 (1972).

It is shown that radio emission from pulsars is unlikely to be of coherent synchrotron origin if the surface magnetic field of the central neutron star is greater than 10^8 G.

141.506 Further searches for the pulsar in Vela.
B. M. Lasker, S. B. Bracker, O. Saá.
Astrophys. Journ., (*Letters*), Vol. 176, L65 - L67 (1972).

Several stars and apparently empty fields near the 1971 NRAO interferometric position for the Vela pulsar (PSR 0833−45) were searched for optical pulsations. The results, all negative, are presented.

141.507 VLBI observations of the Crab nebula pulsar.
N. R. Vandenberg, W. C. Erickson, G. M. Resch, T. A. Clark, J. J. Broderick.
Bull. American Astron. Soc., Vol. 4, 320 (1972). − Abstr. AAS.

141.508 On the search for an optical identification for PSR 0833−45. B. M. Lasker, S. B. Bracker, O. Saá.
Bull. American Astron. Soc., Vol. 4, 320 (1972). − Abstr. AAS.

141.509 SEC vidicon system. H. Y. Chiu.
Bull. American Astron. Soc., Vol. 4, 397 (1972).
Abstr. AAS.

141.510 Digicon observations of PHL 957.
E. A. Beaver, E. M. Burbidge, C. E. McIlwain, H. Epps, P. A. Strittmatter.
Bull. American Astron. Soc., Vol. 4, 397 (1972). − Abstr. AAS.

141.511 Whence the pulsars — rotation and carbon detonation. I.-J. Sackmann, V. Weidemann.
Bull. American Astron. Soc., Vol. 4, 399 (1972). − Abstr. AAS.

141.512 Vier Jahre Pulsarforschung. III. K.-H. Schmidt.
Sterne, 48. Jahrgang, p. 144 - 145 (1972).

141.513 Cooling of pulsars. S. Tsuruta, V. Canuto, J. Lodenquai, M. Ruderman.
Astrophys. Journ., Vol. 176, 739 - 744 (1972).

Cooling rates are calculated for superfluid neutron stars of about 1 $M_\odot$ and 10 km radius, with magnetic fields from zero to about 10^{14} gauss, when possible internal friction effects are neglected. Our results show that most old pulsars are so cold that thermal ionization of surface atoms would be negligible. At an age of 10^6 years and with canonical magnetic fields of 10^{12} gauss, the estimated stellar surface temperature is several thousand to a hundred thousand degrees. However, if we neglect magnetic fields and superfluid states of nucleons, the same surfaces would be about $10^6\,°K$.

141.514 **Accurate position of PSR 1749−28 from lunar oc-cultations.** A. P. Rao, S. Krishnamohan.
Nature, Phys. Sci., Vol. 238, 69 - 70 (1972).
From five occultations of the pulsar the position derived is RA $(1950.0) = 17^h\,49^m\,49^s.22 \pm 0^s.03$; Dec $(1950.0) = -28°06'00".2 \pm 0".5$. No narrow continuum source with flux greater than $5 \times 10^{-28}\,Wm^{-2}\,Hz^{-1}$ was found associated with the pulsar. No optical object seems to be seen within 10 arc sec of the pulsar.

141.515 **Interpretation of the microstructure in pulsar pulses.** T. W. Cole.
Astrophys. Letters, Vol. 12, 181 - 183 (1972).
The output from a narrow-band receiver used to study fine detail in a pulsar pulse is governed by the equations of diffraction. Study of the details of the received pulse can yield details of the true pulse shape which are beyond the currently accepted time-resolution for such a receiver.

141.516 **Periodicities in seismic response caused by pulsar CP1133.** D. Sadeh, M. Meidav.
Nature, Vol. 240, 136 - 138 (1972).
A seismometer/computer combination has detected seismic vibrations corresponding to the half-period of pulsar CP1133. A hitherto unknown pulsar may have been detected by the same technique.

141.517 **Search for seismic signals from gravitational radiation of pulsar CP1133.** T. S. Mast, J. E. Nelson, J. Saarloos, R. A. Muller, B. A. Bolt.
Nature, Vol. 240, 140 - 142 (1972).
Attempts to derive seismic signals from CP1133 were unsuccessful. Possible experimental errors in previous work are discussed.

141.518 **Pulsar glitches and the metastability of the super-fluid core.** S. Banerji, R. Chanda.
Nature, Phys. Sci., Vol. 239, 139 - 140 (1972).

141.519 **The pulsar phenomenon.** F. Pacini.
Atti XV Riunione Soc. Astron. Italiana, Bologna 1971, (see 012.013), p. 137 - 147 (1972).
We review the observational aspects of pulsars and the present status of theoretical interpretation.

141.520 **Short-timescale structure in two pulsars.** T. H. Hankins.
Astrophys. Journ., (Letters), Vol. 177, L11 - L15 (1972).
Frequency-independent characteristic timescales of 175 and 575 μs have been found, using a new predetection dispersion-removal technique, in the intensity structure of meter-wavelength radio signals from the pulsars CP 0950 and CP 1133.

141.521 **Pulsars: Structure and dynamics.** M. Ruderman.
Annual Rev. Astron. Astrophys., Vol. 10, (see 003. 005), 427 - 476 (1972). − Contents: (1) Introduction; (2) Some general properties; (3) Outside the neutron star; (4) The neutron star; (5) Dynamics; (6) Evolution in a pulsar; (7) Glitches; (8) Fluctuations in pulsar periods; (9) Pulses.

141.522 **Thermal conductivity and hot magnetic poles of pulsars.** R. Smoluchowski.

Nature, Phys. Sci., Vol. 240, 54 - 56 (1972).
It is usually assumed that the solid crust of pulsars has a very high and isotropic electrical as well as thermal conductivity. This rather simple description is very likely not valid nearer to the surface of pulsars where the temperature, the density and the Debye temperature drop off rapidly and the surface itself is non-degenerate. It is interesting to inquire into the presence of anisotropic electrical and thermal magneto-resistance caused by the very high magnetic fields.

141.523 **Detection of 10−100 MeV γ-rays from the Crab nebula pulsar NP 0532.**
P. Albats, G. M. Frye, Jr., A. D. Zych, O. B. Mace, V. D. Hopper, J. A. Thomas.
Nature, Vol. 240, 221 - 224 (1972).
Pulsed γ-ray emission from NP 0532 has been detected up to 100 MeV. For the main pulse from 10 to 30 MeV the signal to noise is $(5-6\sigma)$. The energy spectrum of the main pulse from 1 keV to 100 MeV can be fitted by a single power law of index (-0.9).

141.524 **Pulsar associated with the supernova remnant IC 443.**
J. G. Davies, A. G. Lyne, J. H. Seiradakis.
Nature, Vol. 240, 229 - 230 (1972).
PSR 0611 + 22 lies about $0°.6$ from the centre of the supernova remnant IC 443, whose diameter is $0°.7$. The period has increased by 0.33 μs in 45 days, giving a value of $P/\dot{P}$ of 125,000 yr. This can be compared with the age of IC 443 estimated by Minkowski of 65,000 yr. The distance inferred from the dispersion measure is not discordant with that given for IC 443, and in view of the period measurements, the association of the pulsar with the supernova remnant seems clear.

141.525 **New and improved parameters for twenty-two pulsars.**
R. N. Manchester, J. H. Taylor, G. R. Huguenin.
Nature, Phys. Sci., Vol. 240, 74 - 75 (1972).
Observations made with the 92 m telescope of the National Radio Astronomy Observatory have resulted in improved positions, periods, period derivatives, dispersion measures, and pulse widths for a number of recently discovered or weak pulsars. The new data are presented in tables.

141.526 **21-cm line absorption profiles of four pulsars.** J. Gomez Gonzalez, E. Falgarone, P. Encrenaz, M. Guélin.
Astrophys. Letters, Vol. 12, 207 - 209 (1972).
21-cm line spectra have been observed in front of the pulsars PSR 0628−28, 1706−16, 1929 + 10, and 2021 + 51. Absorption is detected in the directions of PSR 1706−16 and 1929 + 10 near to the local-standard-of-rest velocity. No lower limit can be derived for the distances of these sources. Tentative upper limits of 1 kpc are set for PSR 1929 + 10 and 2021 + 51.

141.527 **The pulse-height distribution for NP 0532.** E. Argyle, J. F. R. Gower.
Astrophys. Journ., (Letters), Vol. 175, L89 - L91 (1972).
The main pulses from the Crab pulsar at 146 MHz are distributed according to a power law with an exponent of -2.5.

141.528 **Mechanisms of optical, X-ray and γ-radiation from Crab pulsar.**
V. V. Zheleznyakov, V. E. Shaposhnikov.
Astrophys. Space Sci., Vol. 18, 141 - 165, 166 - 189 (1972). In Russian and English.
The synchrotron mechanism of radiation from the Crab pulsar has been investigated on the assumption that the mechanism acts in a source moving with relativistic velocity round

a neutron star. A detailed matching has been made of the theoretical spectra of synchrotron radiation from relativistic electrons with the results of measurements of the radiation flux from the Crab pulsar in the infrared, optical and X-ray ranges. The parameters of the radiating region (intensity of the magnetic field, source dimensions, density and lifetime of radiating electrons) have been found. The level of Compton γ-radiation in this region is estimated. An estimate is presented for the surface magnetic field of the neutron star which does not contradict those obtained from considerations of the magnetic flux conservation when compressing the object up to the neutron star dimensions.

141.529 Influence of interstellar inhomogeneities on radio pulse form of pulsars. L. M. Erukhimov.
Izv. vyssh. ucheb. zavedenij. Radiofizika, Vol. 15, 821 - 825 (1972). In Russian. − Abstr. in Referativ. Zhurn. 51. Astron., 12.51.466 (1972).

141.530 Seismical signals of cosmic origin? L. P. Vinnik.
Astron. Tsirk., No. 710, p. 5 - 7 (1972). In Russian.

141.531 Detection of six pulsars at 2.8 cm.
R. Wielebinski, W. Sieber, D. A. Graham, H. Hesse, R. E. Schönhardt.
Nature, Phys. Sci., Vol. 240, 131 - 132 (1972).
The MPIfR 100 m radio telescope was designed to operate down to low centimetre wavelengths, Since May 1972 first tests, then astronomical observations at 11 cm wavelength have been made. More recently a 2.8 cm wavelength receiver was put into service. The measurement of pulsars reported here is the first result of the 100 m telescope at 2.8 cm wavelength.

141.532 The pulse shape of the Crab nebula pulsar NP 0532 as a function of color.
G. W. Muncaster, W. J. Cocke.
Astrophys. Journ., (*Letters*), Vol. 178, L13 - L15 (1972).
We find differences in the shape of the secondary pulse of NP 0532 in U, B, and V. The leading edge and peak are stronger at higher frequencies, relative to the trailing edge. The peak difference is about 3 percent between the U and the V.

141.533 Pulsar magnetospheres, braking index, polar caps and period-pulse-width distribution.
D. H. Roberts, P. A. Sturrock.
Bull. American Astron. Soc., Vol. 4, 414 - 415 (1972). − Abstr. AAS.

141.534 A model of pulsar magnetospheres. I.
D. H. Roberts, P. A. Sturrock.
Inst. Plasma Res., Stanford Univ., Stanford, California, SUIPR Rep. No. 473, 38 pp. (1972).
Recent studies have suggested that pulsar magnetospheres may contain large amounts of non-relativistic material. This article discusses where such material could collect, and the effect of this collection on the magnetic-field configuration. Gas flow is considered in the simple approximation of non-interacting particles constrained to move on magnetic field lines. The results are discussed for arbitrary relative orientation of the dipole and rotation axes, and explicit solutions are given for the aligned and orthogonal cases. The period-pulse-width distribution is calculated, using recent neutron-star models, and is found to be compatible with that of an updated sample of pulsars.

141.535 Two new pulsars.
J. H. Taylor, G. R. Huguenin, R. N. Manchester.
IAU Circ., No. 2435 (1972).

141.536 Eleven new pulsars.

J. G. Davies, A. G. Lyne, J. H. Seiradakis.
IAU Circ., No. 2436 (1972).

141.537 PSR 0740-28. M. M. Komesaroff, P. M. McCulloch, P. A. Hamilton, J. M. Rankin.
IAU Circ., No. 2461 (1972).

141.538 On the role of pulsars as cosmic ray accelerators.
J. Trümper.
Atomkernenergie, Vol. 19, No. 2, p. 149 - 151 (1972).
By discussing the energetics of the Crab nebula and pulsar as well as the chemical composition of cosmic radiation it is shown that the observational facts can hardly be reconciled with some current ideas on a direct acceleration of cosmic rays by pulsars.

141.539 Problems and pleasures of pulsars. S. P. Maran.
The physics of pulsars, (see 003.022), p. 1 - 7 (1972)

141.540 Radio observations of pulsars. G. R. Huguenin.
The physics of pulsars, (see 003.022), p. 9 - 19 (1972).

141.541 Optical observations of pulsars. E. J. Wampler.
The physics of pulsars, (see 003.022), p. 21 - 32 (1972).

141.542 X-ray observations of pulsars. H. V. Bradt.
The physics of pulsars, (see 003.022), p. 33 - 43 (1972).

141.543 The measurement of pulsar periods. G. S. Downs.
The physics of pulsars, (see 003.022), p. 45 - 55 (1972).

141.544 Searching for pulsars − Results and techniques.
D. H. Staelin.
The physics of pulsars, (see 003.022), p. 57 - 67 (1972).

141.545 Interstellar cloud properties revealed by pulsars.
R. M. Hjellming.
The physics of pulsars, (see 003.022), p. 69 - 83 (1972).

141.546 Dispersion measures and distances of pulsars.
Y. Terzian.
The physics of pulsars, (see 003.022), p. 85 - 99 (1972).

141.547 Rotating neutron stars and the nature of pulsars.
T. Gold.
The physics of pulsars, (see 003.022), p. 101 - 109 (1972).

141.548 Pulsars—rotating magnetic neutron stars.
H. Y. Chiu.
The physics of pulsars, (see 003.022), p. 135 - 150 (1972).

141.549 Magnetosphere theory of pulsar electrodynamics.
P. Goldreich.
The physics of pulsars, (see 003.022), p. 151 - 158 (1972).

141.550 Pulsars and the origin of cosmic rays.
J. P. Ostriker.
The physics of pulsars, (see 003.022), p. 159 - 170 (1972).

141.551 Approximate parameters of 55 pulsars.
S. B. Modali, S. P. Maran.
The physics of pulsars, (see 003.022), p. 171 - 173 (1972).

141.552 Pulsars. D. S. Milan.
Vasiona, Vol. 20, 67 - 74 (1972). In Serbo-Croatian.

141.553 A new method for astronomical observation.

B. Kaplan.
Nuovo Cimento Lettere, Ser. 2, Vol. 4, 985 - 987 (1972).

The author suggests a new method for the observation of pulsars. It involves the reception of an electromagnetic radiation at a very low frequency. It is believed that the frequency of this radiation is related directly to the period of pulsation or rotation (depending on the nature of the model) of a pulsar. The feasibility of this observation is discussed.

141.554 **Search for long-period X-ray pulsations in the NP 0527 region.**
F. Frontera, F. Fuligni, D. Brini, C. Cavani.
Nuovo Cimento Lettere, Ser. 2, Vol. 5, 131 - 134 (1972).

Evidence for a new source of pulsating X-rays in the energy interval between 30 and 100 keV has been recently reported. Association of this source, having a period of 3.8266 s, with the well-known radio pulsar at 3.74549 s NP 0527 has also been suggested.

141.555 **Radio intensity fluctuations in pulsars.**
D. C. Backer.
Thesis, Cornell Univ., Ithaca, N.Y. [Available from Univ. Microfilms, Ann Arbor, Mich., USA. Order No. 72-7537], 254 pp. (1971).

141.556 **Relaxation phenomena at acceleration of rotation of a spherical vessel with helium II and relaxation in pulsars.** J. S. Tsakadze, S. J. Tsakadze.
Phys. Letters A, (*Netherlands*), Vol. 41A, 197 - 199 (1972).

Measurements of the rotational speed of the PSR 0833 pulsar in Vela and PSR 0531 in the Crab nebula have shown that after a sudden speed-up of their rotation a considerably long relaxation process is observed. The present paper is devoted to an attempt of modeling these processes observing the relaxation motions of liquid helium after sharp changes in the rotational speed of a spherical vessel filled with superfluid liquid.

141.557 **Upper limit on the gravitational flux reaching the earth from the Crab pulsar.**
J. Levine, R. Stebbins.
Phys. Rev. D, Particles and Fields, Vol. 6, 1465 - 1468 (1972).

A 30-m laser interferometer has been used in a search for gravitational radiation from the Crab pulsar. The minimum detectable signal would be produced by an incident gravitational flux of 10^9 ergs/sec cm^2 and the authors find no effect at this level.

141.558 **Radio observations of the pulse profiles and dispersion measures of twelve pulsars.** H. D. Craft, Jr.
Thesis Cornell Univ., Center Radiophys. Space Res., CRSR 395, 337 pp. (1970).

141.559 **Radio intensity fluctuations in pulsars.**
D. C. Backer.
Thesis Cornell Univ. National Astron. Ionosph. Center (NATC) 246 pp. (1971).

141.560 **Linear polarization and spectrum of PSR 0833−45 and the effects of scattering.**
M. M. Komesaroff, P. A. Hamilton, J. G. Ables.
Australian Journ. Phys., Vol. 25, 759 - 777 (1972).

Linear polarization characteristics as well as pulse shape have been measured at Parkes for the average of many pulses from PSR 0833−45 at several frequencies between 300 and 1410 MHz. These confirm a previous conclusion, which was based on pulse shape measurements alone, that the change of pulse character with decreasing frequency can be explained in terms of interstellar scattering.

141.561 **Discovery of a new-type period of the pulsar CP**

0808. V. A. Izvekova, V. M. Malofeev.
Astron. Tsirk., No. 721, p. 5 - 7 (1972). In Russian.

141.562 **Investigation of short-time fluctuations of amplitudes of the pulsars CP 1919, CP 0808, CP 0950.**
B. N. Panovkin, D. G. Buyanova.
Astron. Tsirk., No. 722, p. 1 - 3 (1972). In Russian.

141.563 **Spectral observations of pulsars at 18 cm.**
V. I. Slysh.
Astron. Tsirk., No. 731, p. 1 - 3 (1972). In Russian.

141.564 **Intrinsic fine structure of the spectrum of radio emission of pulsars.**
V. V. Vitkevich, Yu. P. Shitov.
Astron. Tsirk., No. 740, p. 1 - 4 (1972). In Russian.

On the induced amplification of synchrotron radiation in cosmic sources. See Abstr. 061.054.

Pair-producing electric fields and pulsars. See Abstr. 062.061.

Time-dependent radiation transfer and a possible explanation of the interpulse in CP 0950. See Abstr. 063.029.

Concerning the atmosphere of magnetic neutron stars (pulsars). See Abstr. 064.073.

Neutron stars, pulsar radiation and supernova remnants. See Abstr. 065.130.

Detection of neutron stars. See Abstr. 065.139.

Rotational energy, pulsars, and active nebulae. See Abstr. 065.149.

The elastic energy and character of quakes in solid stars and planets. See Abstr. 091.054.

Possible sidereal period for the seismic lunar activity. See Abstr. 094.136.

Photometrische Untersuchungen an sehr schnell variablen Objekten. See Abstr. 122.140.

A study of galactic supernova remnants. II. Supernova rate, galactic radio emission and pulsars. See Abstr. 125.008.

Pulsars and the evolution of supernova remnants. See Abstr. 125.024.

Untersuchungen zur Aufheizung des interstellaren Gases durch kosmische Strahlung. See Abstr. 131.113.

The degree of ionisation in interstellar space. See Abstr. 131.128.

Nonlinear inverse Compton radiation and the circular polarization of diffuse radiation from the Crab nebula. See Abstr. 134.009.

Errata

141.901 **Corrigenda: 'Observations at 408 MHz of radio sources from the 4C catalogue', I, II** [Australian Journ. Phys., Vol. 24, 263 - 291, 617 - 630 (1971)].
R. E. B. Munro.
Australian Journ. Phys., Vol. 25, 337 (1972).

142   X Ray-, Gamma Ray-Sources

142.001   On the reality of the ~165-s 'period' in the variation of Sco X-1.   J. Gribbin.
Monthly Notices Roy. Astron. Soc., Vol. 159, 1P - 2P (1972).

The peak found at ~165 s in power spectra of the optical flickering of Sco X-1 by both Gribbin et al. and by Robinson & Warner may after all represent a real, repeatable physical phenomenon.

142.002   On the contribution of transition radiation from dust grains to the diffuse X-ray background.
I. Lerche.
Astrophys. Journ., Vol. 175, 373 - 377 (1972).

We show that the transition radiation produced in the band 0.1 keV—~100 keV by cosmic-ray electrons bombarding interstellar dust grains produces a significant fraction of the observed diffuse X-ray background radiation.

142.003   Optical studies of *Uhuru* sources. II. IM Normae, a possible X-ray source.   J. L. Elliot, W. Liller.
Astrophys. Journ., (*Letters*), Vol. 175, L69 - L72 (1972).

We have discovered that the ex-nova IM Normae lies within the 90 percent confidence error-box surrounding the X-ray source 2U 1536−52. Following a discussion of the light curve of IM Nor, we present several lines of evidence suggesting that at least one type of nova phenomenon can give rise to observable X-ray emission.

142.004   Soft X-rays from Cygnus X-2 and from Cygnus X-1 (in eclipse?).
J. C. Stevens, G. P. Garmire, G. R. Riegler.
Astrophys. Journ., (*Letters*), Vol. 175, L73 - L77 (1972).

Cygnus X-1 and Cygnus X-2 were observed on 1971 October 23. The intensity of Cyg X-1 was in a low state at the time of observation which corresponded to the time of possible eclipse of the companion of BD + 34°3815 (HDE 226868). The intensity of Cyg X-2 was measured to be comparable to previous observations, with an interstellar cut-off corresponding to $0.8 \pm 0.4 \times 10^{21}$ H atoms per cm^2.

142.005   Some observational distinctions among models of pulsing X-ray binaries.   J. Arons, J. N. Bahcall.
Astrophys. Letters, Vol. 11, 191 - 194 (1972).

Observational distinctions between various models for the environment of pulsing X-ray binaries are described. It is pointed out that all observations available so far are consistent with X-ray binaries being systems containing only 'familiar' objects, i.e., a rotating neutron star and a massive normal star such as an A or B supergiant.

142.006   Possible models for the extended X-ray source at the galactic center.   L. Maraschi, A. Treves.
Astrophys. Letters, Vol. 11, 211 - 215 (1972).

X- and γ-ray observations from the direction of the galactic center are discussed.

142.007   Spatial coincidences of X-ray sources and supernova remnants.   A. P. Cowley, D. J. MacConnell.
Astrophys. Letters, Vol. 11, 217 - 218 (1972).

The X-ray source 2U 1509−58 may be identified with the radio supernova remnant Milne 33 (= Downes 38 = Kesteven 23) as well as with an Hα nebulosity RCW 89. The proximity of six ancient novae or supernovae to X-ray sources in the Uhuru Catalogue is pointed out.

142.008   Techniques in balloon X-ray astronomy.
L. E. Peterson, R. M. Pelling, J. L. Matteson.
Space Science Rev., Vol. 13, 320 - 336 (1972). — Conference paper (see 012.002).

142.009   Extragalactic origin of the transient X-ray sources.
S. Sofia.
Astrophys. Journ., (*Letters*), Vol. 175, L113 - L115 (1972).

It is shown that the current assumption that the transient X-ray source 2U 1543−47 is located within our Galaxy leads to serious energetic difficulties. However, all the presently available observations are consistent with the interpretation that the source was produced by a supernova event in a nearby galaxy. On the assumption that the nature of this source is similar to that of the other known transient sources, observational tests are proposed to verify their extragalactic origin.

142.010   On the optical search for Centaurus X-3.
R. J. Brucato, J. Kristian, J. A. Westphal.
Astrophys. Journ., (*Letters*), Vol. 175, L137 - L139 (1972).

The optical eclipsing binary LR Cen has been eliminated as a candidate for Cen X-3 on the basis of a real discrepancy of orbital periods. We believe that the position coincidence of Wray 795 with Cen X-3 is not statistically significant.

142.011   The association of X-ray sources with bright stars.
H. Gursky.
Astrophys. Journ., (*Letters*), Vol. 175, L141 - L144 (1972).

A comparison of selected *Uhuru* X-ray sources with the SAO star catalog reveals 10 stars within the positional error of the X-ray sources where only about two are expected on the basis of chance. Three of these are already under study as optical candidates for X-ray sources. The new results lend confidence to these candidates as being correct and yield candidate stars for seven other X-ray sources.

142.012   The interaction of Sco X-1 with its environment.
J. Silk, D. W. Goldsmith, G. B. Field, L. Carrasco.
Astron. Astrophys., Vol. 20, 287 - 291 (1972).

X-ray emission from Sco X-1 may be a significant source of ionization for the surrounding interstellar gas, and can enable us to interpret, in particular, the possible recent detection by Johnson (1971) of enhanced Hβ emission in the vicinity of Sco X-1. We have assumed a constant temperature in the medium surrounding Sco X-1 and have calculated simple analytic and numerical solutions for the variation in electron density and emission measure with distance from the X-ray source. The Hβ observations set limits on the parameters of a possible variable soft component of Sco X-1 which has been suggested by recent low energy X-ray observations.

142.013   Angular momentum and energy loss of a compact star rotating in a thermal plasma.
R. H. Cohen, A. Treves.
Astron. Astrophys., Vol. 20, 305 - 308 (1972).

Models of thermal X-ray sources in which the energy is produced by the rotation of a compact star are considered. The angular momentum loss is assumed to be caused by a Poynting-Robertson torque. The dependence of the radius of the region of emission on the velocity field of the plasma surrounding the star is examined. It is shown that in the absence of stellar winds this radius must be greater than or equal to the radius of the speed-of-light cylinder.

142.014   Radio search for the pulsing X-ray source in Hercules.   R. Doxsey, G. T. Murthy, S. Rappaport, J. Spencer, W. Zaumen.
Astrophys. Journ., (*Letters*), Vol. 176, L15 - L18 (1972).

The region of the celestial sphere near the pulsing X-ray source in Hercules (2U 1705+34) has been searched for radio emission with the NRAO three-element interferometer. Four weak radio sources, which may be considered as candidates for the radio counterpart of this X-ray source, were detected.

142.015 On the mass limit of Centaurus X-3.
D. W. Weedman, D. S. Hall.
Astrophys. Journ.,(*Letters*), Vol.176, L19 - L21 (1972).

It is possible to construct a model in which the X-ray source in Cen X-3 has a mass of 1 $M_\odot$. Calculations, based on a mass of 1 $M_\odot$ for the X-ray source and relative radius of 0.74 for the larger component, show that the mass which must be lost from the larger component should flow toward the X-ray source.

142.016 The X-ray sky. H. W. Schnopper, J. P. Delvaille.
Sci. American, Vol. 227, No. 1, p. 26 - 37 (1972).

More than 120 celestial X-ray sources are now known; the list includes at least one neutron star, one quasar, two galaxies, one double source and perhaps even a black hole.

142.017 Photometric observations of BD +34°3815 (Cyg X-1) in the interval 3600–22000 Å.
V. M. Lyutyj, V. I. Moroz, G. S. Khromov.
Astron. Tsirk., No. 675, p. 1 - 4 (1972). In Russian.

142.018 Spectrophotometric investigations of BD +34°3815 (Cyg X-1).
I. N. Glushneva, V. T. Doroshenko, E. A. Kolotilov.
Astron. Tsirk., No. 675, p. 4 - 6 (1972). In Russian.

142.019 Spectrum of BD +34°3815 (Cyg X-1) in the λλ6000–9100 Å region. E. A. Kolotilov.
Astron. Tsirk., No. 675, p. 6 - 7 (1972). In Russian.

142.020 Spectrum of BD +34°3815 (Cyg X-1) in the λλ3600–6000 Å region. G. V. Zajtseva.
Astron. Tsirk., No. 675, p. 7 - 8 (1972). In Russian.

142.021 The variable X-ray source Cen X-3 as a radially pulsating white dwarf. G. Vauclair.
Astrophys. Letters, Vol. 12, 17 - 19 (1972).

On the basis of a study of the vibrational stability of white dwarf models, it is argued that the pulsation period of the Cen X-3 variable X-ray source is the first harmonic of radial pulsation of a 0.6 $M_\odot$ white dwarf rather than the fundamental of a 1.1 $M_\odot$ white dwarf, as proposed in a recent model of this source.

142.022 The characteristics of the X-ray source Cen X-3 obtained on the supposition of its identity with the Algol-type variable LR Cen.
I. S. Shklovskij, A. M. Cherepashchuk, Yu. N. Efremov.
Astron. Tsirk., No. 678, p. 1 - 4 (1972). In Russian.

142.023 Accretion disc models for compact X-ray sources.
J. E. Pringle, M. J. Rees.
Astron. Astrophys., Vol. 21, 1 - 9 (1972).

The accretion process is considered, in cases when the infalling matter possesses angular momentum and forms a disc spinning around a central compact mass. This situation occurs when gas falls onto a black hole or neutron star from a binary companion, and may be relevant to galactic X-ray sources. The spectrum of the radiation emitted by gas spiralling down into a black hole is calculated and compared with the data on Cygnus X-1. An interpretation of Centaurus X-3 involving a rotating magnetized neutron star is proposed. Tentative interpretation of some other phenomena are also proposed.

142.024 A further high-resolution search for Fe XXV line emission from Scorpius X-1. R. E. Griffiths.
Astron. Astrophys., Vol. 21, 97 - 103 (1972).

An improved LiF crystal spectrometer has been flown on a sounding rocket to search for emission lines from the helium-like ion Fe XXV in Sco X-1. A 3σ upper limit of 25 eV is set on the E.W. of narrow line emission from this ion. From proportional counters on the same rocket experiment, a 5σ upper limit of 100 eV is set on the total Fe XXV-XXVI line emission. This result is discussed and compared with those of similar proportional counter experiments on Sco X-1.

142.025 Low energy X-ray survey from the Crab nebula to Cygnus. G. Garmire, G. R. Riegler.
Astron. Astrophys., Vol. 21, 131 - 138 (1972).

Observations of cosmic X-rays in the energy band 0.2 - 3.0 keV were made from an experiment on board an Aerobee 170 rocket launched from White Sands N.M. on 19 December 1970. A survey along the galactic plane between the Crab nebula and Cygnus provided data on the interstellar absorption and X-ray flux for the Crab nebula, NP 0532, Cas A, Cyg X-2, and upper limits to the flux from a number of supernova remnants. The 1/4 keV X-ray background was measured over a limited range of galactic latitude and longitude.

142.026 Transfer effects on X-ray lines in optically thick celestial sources.
J. E. Felten, M. J. Rees, T. F. Adams.
Astron. 21, 139 - 150 (1972).

Uniform static isothermal spherical models of Sco X-1 are considered, having temperatures T= 5, 7 and 10×10^7°K, and Thomson-scattering optical thicknesses along a radius of 5, 10 and 20. Expected fluxes and equivalent widths in line cores and in the total lines (including the wings) are calculated approximately for several lines of sulfur and iron. The density of ionizing photons in the source is obtained by diffusion theory. The effects of Thomson and resonance scattering are estimated by an approximate skin-depth method involving functions given by Hummer and Rybicki.

142.027 On the identification of X-ray cosmic sources.
P. R. Amnuel, O. H. Guseinov.
Astron. Tsirk., No. 697, p. 6 - 8 (1972). In Russian.

142.028 A new X-ray binary associated with an O-type star.
C. Jones, W. Forman, W. Liller, E. Schreier, H. Tananbaum, E. Kellogg, H. Gursky, R. Giacconi.
Bull. American Astron. Soc., Vol. 4, 329 (1972). – Abstr. AAS.

142.029 X-ray spectra of binary sources. E. Schreier, R. Giacconi, H. Gursky, E. Kellogg, H. Tananbaum.
Bull. American Astron. Soc., Vol. 4, 329 (1972). – Abstr. AAS.

142.030 The number-intensity distribution of X-ray sources observed by UHURU. T. A. Matilsky, H. Gursky, E. M. Kellogg, S. S. Murray, H. D. Tananbaum, R. Giacconi.
Bull. American Astron. Soc., Vol. 4, 335 (1972). – Abstr. AAS.

142.031 Correlation analysis of X-ray emission from Cygnus X-1. A. C. Brinkman, D. R. Parsignault, E. Schreier, H. Gursky, E. M. Kellogg, H. Tananbaum, R. Giacconi.
Bull. American Astron. Soc., Vol. 4, 336 (1972). – Abstr. AAS.

142.032 Models of extragalactic X-ray sources.
A. Solinger, W. Tucker, A. Cavaliere.
Bull. American Astron. Soc., Vol. 4, 336 (1972). – Abstr. AAS.

142.033 X-ray sources in clusters of galaxies versus isolated extragalactic sources. E. Kellogg, E. Schreier, H. Tananbaum, H. Gursky, R. Giacconi.
Bull. American Astron. Soc., Vol. 4, 336 (1972). – Abstr. AAS.

142.034 X-ray absorption in the spectra of some X-ray

sources.   M. S. Burgin.
Astron. Tsirk., No. 707, p. 5 - 7 (1972). In Russian.

142.035   Possible identification of the X-ray source Her X-1 (2U 1705+34) with HR 6351.
I. S. Shklovsky, Yu. N. Efremov.
Astron. Tsirk., No. 708, p. 1 - 3 (1972). In Russian.

142.036   Centaurus X-3, possible reactivation of an old neutron star by mass exchange in a close binary.
E. P. J. van den Heuvel, J. Heise.
Nature, Phys. Sci., Vol. 239, 67 - 69 (1972).
    An upper limit of $0.7 \pm 0.14\, M_\odot$ is obtained for the mass of the pulsar component of Cen X-3. The system is likely to be a massive close binary in the second stage of mass exchange.

142.037   Observation of several X-ray sources in 1970 September.   R. E. Price, F. D. Seward, C. D. Swift.
Astrophys. Journ., Vol. 176, 611 - 616 (1972).
    Several X-ray sources were detected in a survey of the sky on 1970 September 24. Intensities were measured for Tau XR-1, Cyg XR-1, Cyg XR-2, NGC 1275, Cas A, Cyg XR-3, 2U 0613+9,2U 1908+0, and 2U 1912−5. Locations are reported for the latter four sources, which are thus far unidentified with optical or radio objects.

142.038   Scorpius X-1: A search for optical circular polarization.   R. M. E. Illing, P. G. Martin.
Astrophys. Journ., (*Letters*), Vol. 176, L113 - L114 (1972).
    Upper limits of 0.1 percent for the circular polarization of Sco X-1 have been obtained in several wavelength bands covering the range 3700–8800 Å.

142.039   Observation of cosmic soft X-rays.
    S. Hayakawa, T. Kato, T. Kohno, K. Nishimura, Y. Tanaka, K. Yamashita.
Astrophys. Space Sci., Vol. 17, 30 - 38 (1972).
    Cosmic X-rays in the energy range between 0.2∼10 keV were observed with polypropylene window proportional counters on board a sounding rocket. The field of view crossed the galactic plane in the Sgr region and reached galactic latitudes of 50° and −90°. A new soft X-ray source was found in the Aries-Taurus region. The soft X-ray flux from the direction of NGC 1275 was conspicuous, whereas that of Sgr region source were very weak. The distribution of the intensity of diffuse soft X-rays over the scanned region indicates the galactic emission of soft X-rays.

142.040   The nature of the first Cygnus X-3 radio outburst.
    P. C. Gregory, P. P. Kronberg, E. R. Seaquist, V. A. Hughes, A. Woodsworth, M. R. Viner, D. Retallack, R. M. Hjellming, B. Balick.
Nature, Phys. Sci., Vol. 239, 114 - 117 (1972).
    The data for the event observed in Cyg X-3 early in September at many wavelengths indicate that the radio outburst originated in an expanding cloud of synchrotron emitting electrons. Details of the radiation mechanism and limits on models are discussed.

142.041   Observations at 408 MHz of the Cyg X-3 radio outburst.   B. Anderson, R. G. Conway, R. J. Davis, R. J. Peckham, P. J. Richards, R. E. Spencer, P. N. Wilkinson.
Nature, Phys. Sci., Vol. 239, 117 - 118 (1972).
    Observations of the Cygnus X-3 outburst have been made with interferometers operating at 408 MHz at Jodrell Bank. The data are consistent with a model in which the source has a true angular diameter of ∼0.1 arc s which is increased to the apparent value by interstellar scattering. The true size of the source is thus ∼2 light day at ∼4 kpc.

142.042   On the nature of the optical variations of HZ Her =

Her X1.   A. M. Cherepashchuk, Yu. N. Efremov, N. E. Kurochkin, N. I. Shakura, R. A. Sunyaev.
Inform. Bull. Variable Stars (I.A.U. Commission 27), Konkoly Obs., Budapest, No. 720 (1972).

142.043   Binary stars as X-ray sources.   G. Burbidge.
    Comments Astrophys. Space Phys., Vol. 4, 105 - 111 (1972).
    The original suggestion that binary stars might turn out to be detectable X-ray sources was made in 1964 by Hayakawa and Matsuoka. They included it among many possible sources because they had realized that the dissipation of the kinetic energy in gas streams might give rise to high temperatures and hence to thermal X-ray emission. It therefore seems worthwhile both to summarize the history of this idea so far as I have seen it and then to discuss some of the fruitful lines of investigation stemming from it and also some of the unsolved problems.

142.044   GX 340+0 as a hot neutron star.   G. Greenstein.
    Nature, Phys. Sci., Vol. 238, 71 (1972).
    We derive a cooling curve (temperature vs. time) for superfluid neutron stars within the theory of frictional heating of such objects. Applying it to the X-ray source GX 340+0 yields estimates of its age, period and luminosity.

142.045   On identifying new kinds of astronomical objects.
    J. N. Bahcall, A. Yahil, B. Margon, S. Bowyer, M. Lampton, R. Cruddace.
Nature, Phys. Sci., Vol. 238, 92 - 93 (1972).
    Statistical problems in identifying new kinds of astronomical objects are discussed.

142.046   Search for high frequency optical variability in X-ray sources.   H. B. Richer, J. R. Auman, B. C. Isherwood, J. P. Steele, T. J. Ulrych.
Nature, Phys. Sci., Vol. 238, 131 - 132 (1972).
    High frequency optical variability has been searched for in the optical candidates of the X-ray sources Cyg X-1, Sco X-1, and X Per. Upper limits of about .005 magnitudes were set for periodic fluctuations on a time scale of .020 seconds through 10 minutes for Cyg X-1 and X Per and from .020 through 10 seconds for Sco X-1.

142.047   Her X-1: A precessing binary pulsar?
    K. Brecher.
Nature, Vol. 239, 325 - 326 (1972).
    It is proposed that the pulsating variable X-ray source Her X-1 is a precessing binary pulsar. If the pulsation period $T_s \cong 1.24$ sec is identified with the spin time of a neutron star of density ρ, then the free precession time $T_p \cong T_s^3 G\rho$, can account for the observed 35 day on off period if a beam of radiation precesses in and out of our line of sight. Optical variations of the companion star at the binary orbital period are predicted to continue even when the object is "off" at X-ray wavelengths. Other reasons for, and consequences of the model are discussed.

142.048   *B* and *V* photometry of Cygnus X-1.
    E. N. Walker.
Monthly Notices Roy. Astron. Soc., Vol. 160, 9P - 12P (1972).
    B and *V* photometry of HDE 226868 (= Cygnus X-1) is presented. The star is variable with two maxima and two minima per radial velocity period. The variations indicate that at least one of the stars present in the binary system is not a normal star.

142.049   Pulserende røntgenkilder.   B. Thomsen.
    Astron. Tidssk., Årg. 5, p. 129 - 131 (1972).

142.050   A search for absorption of the soft X-ray diffuse

flux by the Small Magellanic Cloud.
D. McCammon.
Thesis, Univ. Wisconsin, Madison. [Available from Univ. Microfilms, Ann Arbor, Mich., USA. Order No. 72-1042], 134 pp. (1971).

142.051 The direct reduction of astronomical X-ray spectra.
J. F. Dolan.
Astrophys. Space Sci., Vol. 17, 472 - 481 (1972).
A matrix method is outlined for the reduction of astronomical X-ray spectral data which includes the effects of detector resolution and fluorescent escape phenomena. The differences between this method and the 'backward reduction' or multiple grid methods presently employed are discussed.

142.052 A balloon-borne observation of the X-ray source Cygnus XR-1. A. E. Metzger, J. F. Dolan.
Astrophys. Space Sci., Vol. 17, 482 - 488 (1972).
An observation of the variable X-ray source Cygnus XR-1 was made with a balloon-borne proportional counter in the energy region 22–70 keV. The resultant spectrum indicates that the source was in a state of maximum flux during the time period 1915–2005 UT, 15 January, 1968.

142.053 The *UBV* colors of Sco X-1 at the time of an X-ray observation. W. Osborn.
Astron. Astrophys., Suppl. Ser., Vol. 7, 393 - 394 (1972).
The *UBV* colors of Sco X-1 measured during a simultaneous X-ray observation are presented.

142.054 The X-ray absorption measure of Sco X-1.
A. N. Bunner, P. L. Coleman, W. L. Kraushaar, D. McCammon.
Astrophys. Letters, Vol. 12, 165 - 168 (1972).
A soft X-ray spectrum of Sco X-1 was recorded by a rocket-borne proportional-counter payload flown in May 1970. No X-rays were detected below 0.28 keV. The observed pulse-height distribution is compared with predictions based on various source-function models.

142.055 Large outburst of Cygnus X-3. P. C. Gregory.
Nature, Vol. 239, 439 - 440 (1972).
The recent radio outburst of Cygnus X-3 is described and the overall picture of the source which has been gleaned from observations at many wavelengths is presented.

142.056 Discovery of giant radio outburst from Cygnus X-3.
P. C. Gregory, P. P. Kronberg, E. R. Seaquist, V. A. Hughes, A. Woodsworth, M. R. Viner, D. Retallack.
Nature, Vol. 239, 440 - 443 (1972).
A giant radio outburst was observed from the X-ray source Cygnus X-3 on September 2, 1972. Flux density measurements were obtained at 10,522, 6,630 and 3,240 MHz as well as linear polarization measurements at 10,522 MHz. The evidence suggests a model of synchrotron radiation from an expanding cloud of relativistic particles. A computed upper limit for the source distance of ≤ 400 kpc implies that the source of the radio emission is galactic.

142.057 Unusual radio events in Cygnus X-3.
R. M. Hjellming, B. Balick.
Nature, Vol. 239, 443 - 446 (1972).
Observations made at Green Bank confirm the findings of Gregory et al.

142.058 Possible identification of an Uhuru X-ray source in the southern sky. B. N. G. Guthrie.
Nature, Phys. Sci., Vol. 240, 43 (1972).

142.059 Search for Hα emission from the companion radio sources of Sco X-1.
P. A. Vanden Bout, G. Grupsmith, B. W. Bopp.
Nature, Phys. Sci., Vol. 240, 43 - 44 (1972).

142.060 Determination of the distance of Cygnus X-3 by 21-cm absorption.
R. Lauqué, J. Lequeux, Nguyen-Quang-Rieu.
Nature, Phys. Sci., Vol. 239, 119 - 120 (1972).
The recent radio outburst of the X-ray source Cygnus X-3 has offered a unique opportunity for measuring its distance, using hydrogen-line absorption in front of its radio counterpart. By this method, we have found that the distance to Cyg X-3 is between 8 and 11 kpc.

142.061 Observations of Cygnus X-3 at 2.8 cm with a 17 arc s beam. L. R. D'Addario, M. A. Stull.
Nature, Phys. Sci., Vol. 239, 120 - 121 (1972).
Cygnus X-3 was observed at 2.8 cm (10,690 MHz) on each of four days during its early September 1972 outburst using the new five-element minimum redundancy array at Stanford University. All observations were corrected for known instrumental effects, and the average complex ratios of the visibilities of Cygnus X-3 to those of PKS 2134 + 00 were computed for each of the nine different baselines. From these we deduce our estimates of the structure of Cygnus X-3 and the ratio of its flux density to that of PKS 2134 + 00. We estimate our errors from the observed scatter in the visibility samples.

142.062 Limits to the 2,695 MHz circular polarization of the September Cyg X-3 flare.
A. H. Bridle, M. J. L. Kesteven, A. E. Niell.
Nature, Phys. Sci., Vol. 239, 121 (1972).
The decay of the first September Cyg X-3 radio outburst was observed at 2,695 MHz with the resurfaced 300-foot meridian transit telescope of the National Radio Astronomy Observatory. An on-axis feed system provided both left and right-handed circular polarizations.

142.063 The large outburst in Cygnus X-3 at 8 GHz.
H. D. Aller, W. A. Dent.
Nature, Phys. Sci., Vol. 239, 121 - 123 (1972).
Here we present results of observations at 8 GHz during the decay phase of the recent outburst of Cygnus X-3. We also compare our results with the flux density variation observed with the Haystack 120-foot telescope at 15.5 GHz and describe a possible model for the source.

142.064 Observations of Cygnus X-3 by Uhuru.
D. R. Parsignault, H. Gursky, E. M. Kellogg, T. Matilsky, S. Murray, E. Schreier, H. Tananbaum, R. Giacconi, A. C. Brinkman.
Nature, Phys. Sci., Vol. 239, 123 - 125 (1972).
We have analysed 32 transits of the source obtained during the period August 19 to September 6. The observed intensities are within the range of intensities for Cyg X-3 as seen by Uhuru since its launch in December 1970. Thus, the X-ray intensity of Cyg X-3 does not seem to be anomalous during this time period, nor is there any indication of a systematic intensity change coincident with the radio flare within the stated limit. Assuming the radio and X-ray source to be the same object, we conclude that whatever precipitated the radio event did not have a significant impact on the X-ray emitting region within a time period of the order of several days.

142.065 X-ray observations of Cyg X-3 near the time of radio outbursts. J. P. Conner, W. D. Evans, D. E. Mook.
Nature, Phys. Sci., Vol. 239, 125 (1972).
We have made a search through all data available for dates near the time of radio outbursts from the Vela 5B spacecraft to determine if any associated increase in X-ray flux could be detected.

142.066 Cygnus X-3: 3.3 mm observations.
R. B. Pomphrey, E. E. Epstein.
Nature, Phys. Sci., Vol. 239, 125 - 126 (1972).

Observations of the early September 1972 flare of the radio counterpart of Cyg X-3 were made at 3.3 mm (90 GHz) with the 4.6-m antenna of the Aerospace Corporation. Observing and data reduction procedures identical to those described by Fogarty *et al.* yielded the 3.3 mm total fluxes.

142.067 An attempt to detect recombination line emission from the Cygnus X-3 outburst.
B. G. Leslie, W. A. Dent.
Nature, Phys. Sci., Vol. 239, 126 (1972).

Between 0708 and 0737 UT on September 3, 1972, shortly after the Cygnus X-3 outburst had reached its maximum, attempts were made to detect possible recombination line emission in the source. The observations were made at the frequency of the hydrogen 92α line with a 100 K system noise temperature parametric amplifier receiver on the 120-foot Haystack antenna. We have found no recombination line having an amplitude greater than 0.03 K in antenna temperature or 0.2×10^{-26} W m^{-2} Hz^{-1} in flux density.

142.068 15.5 GHz observations at the Haystack Observatory of the Cygnus X-3 outburst.
W. A. Dent, J. E. Kapitzky, B. G. Leslie, G. Kojoian, M. L. Meeks, H. H. Danforth, J. J. Kollasch, E. J. Chaisson, D. F. Dickinson, L. E. Goad, C. J. Lada.
Nature, Phys. Sci., Vol. 239, 126 - 127 (1972).

The recent major outburst in Cygnus X-3 was observed at 15.5 GHz with the 120-foot antenna of the NEROC Haystack Observatory. The flux density of the continuum emission from Cyg X-3 was first measured at 0437 UT on September 3, 1972, when the outburst was near maximum. The flux density measurements were continued during the 6 day period following the maximum during which time the source decayed rapidly. From other observations at the Haystack Observatory it is also possible to place an upper limit to the angular diameter of the radio source associated with Cyg X-3 at the time of the outburst maximum.

142.069 Spectrum and polarization of the Cygnus X-3 outburst. W. A. Dent.
Nature, Phys. Sci., Vol. 239, 127 - 128 (1972).

The large outburst of Cygnus X-3 was observed at 3 h 58 min UT on September 3, 1972, with the 300-foot antenna of the National Radio Astronomy Observatory at a frequency of 2.7 GHz. The flux density and linear polarization were measured as Cygnus X-3 successively transited through three antenna beams aligned in an east—west plane. The observed spectrum of Cygnus X-3 is characteristic of that observed in many compact radio sources in that it shows a normal non-thermal spectrum at the higher frequencies and a sharp low frequency cutoff. An estimate of the radio luminosity of the outburst at maximum can be obtained by integrating the spectrum over frequencies below 10^{11} Hz.

142.070 Absence of linear polarization in Cygnus X-3.
H. D. Aller.
Nature, Phys. Sci., Vol. 239, 128 (1972).

Here I describe the results of linear polarization measurements of the Cyg X-3 outburst made with the University of Michigan 85-foot telescope at 8 GHz between 0 and 8 hours UT on September 3, 4, and 5. Linear polarization was not detected at any time during the three days and the scatter in the measurements was consistent with the measurement uncertainties.

142.071 Radio observations of Cygnus X-3 and the surrounding region.
B. Gary, E. T. Olsen, P. W. Rosenkranz.
Nature, Phys. Sci., Vol. 239, 128 - 130 (1972).

On September 7, 1972, Cygnus X-3 was observed with the Goldstone 64-m antenna at wavelengths of 13.1 cm, 3.55 cm, and 1.95 cm. At 13.1 cm a map of received flux density has been constructed for a region 0.5×0.6 arc degree with a resolution of 8.15 arc min. This map reveals the presence of several partially resolved features having integrated flux densities in excess of one flux unit.

142.072 21 cm and 18 cm observations of the Cygnus X-3 radio outburst.
D. B. Shaffer, G. A. Shields, B. Schupler.
Nature, Phys. Sci., Vol. 239, 131 - 132 (1972).

We have observed the outburst of Cyg X-3 using the Owens Valley Radio Observatory interferometer at 21 cm, and with the 130-foot antenna at 18 cm. The most salient features in our data are the rapid rise in intensity at 21 cm at the beginning of the outburst and the apparently irregular decay at 18 cm.

142.073 OH and formaldehyde absorption in the direction of Cygnus X-3. B. E. Turner.
Nature, Phys. Sci., Vol. 239, 132 - 133 (1972).

Observations of all four 18 cm lines of OH were made between the hours 0620 and 0650 UT on September 4 and September 5, 1972, during the latter stages of the first outburst of Cygnus X-3. OH was detected in 8 of 11 positions searched within 1 arc deg of Cygnus X-3. The observations of OH and formaldehyde absorption suggest the possibility of a much smaller distance for Cygnus X-3.

142.074 Observations of Cygnus X-3 at the Mullard Radio Astronomy Observatory.
N. J. B. A. Branson, A. H. M. Martin, G. G. Pooley, A. C. S. Readhead, J. R. Shakeshaft, A. Slingo, P. J. Warner.
Nature, Phys. Sci., Vol. 239, 133 - 134 (1972).

The radio outbursts of Cygnus X-3 have been observed at several frequencies since September 4, 1972. We have made measurements of the flux density variations at 5.0, 2.7 and 1.4 GHz, and the position and an upper limit to the angular size at 5 GHz; in addition the HI absorption spectrum has been determined and upper limits have been placed on the flux densities at 150 and 81.5 MHz.

142.075 Search for a visible counterpart of the September 2, 1972 radio outburst in Cygnus.
J. A. Westphal, J. Kristian, J. P. Huchra, S. A. Shectman, R. J. Brucato.
Nature, Phys. Sci., Vol. 239, 134 - 135 (1972).

Following notification by P. C. Gregory of the discovery of the radio outburst in Cygnus, we attempted to detect a visible object at the radio position over 6 nights, at several wavelengths from the blue to the near infrared. No visible object was detected at the radio position in any of our data.

142.076 X-ray observations of Cygnus X-3 by Copernicus.
P. W. Sandford, F. H. Hawkins.
Nature, Phys. Sci., Vol. 239, 135 (1972).

Cygnus X-3 was observed by the X-ray instrument on the fourth Orbiting Astronomical Observatory, named Copernicus, for three periods following the radio enhancement. Two features are apparent in the data from Copernicus: there is clear evidence of a regular drop in intensity by a factor of 2 and a preliminary analysis gives a principal period in the region of 288 minutes.

142.077 More unusual radio events in Cygnus X-3.
R. M. Hjellming, B. Balick.
Nature, Phys. Sci., Vol. 239, 135 - 136 (1972).

During September 2–11, 1972, an unprecedented radio

event was observed in the X-ray source Cyg X-3. We now report observations of additional outbursts of similar nature during September 18–25, 1972.

142.078   Asymmetry of soft X-ray emission near M87.
R. C. Catura, P. C. Fisher, H. M. Johnson, A. J. Meyerott.
Astrophys. Journ., (*Letters*), Vol. 177, L1 - L4 (1972).

The angular distribution of X-ray emission near M87 has been measured at energies from 0.5 to 4 keV by a rocketborne optical system. The observed distribution has an angular extent of $0°\!.4 \pm 0°\!.1$ and contains a compact component, coincident with M87.

142.079   Observation of a correlated X-ray–radio transition in Cygnus X-1.   H. Tananbaum, H. Gursky,
E. Kellogg, R. Giacconi, C. Jones.
Astrophys. Journ., (*Letters*), Vol. 177, L5 - L10 (1972).

Analysis of 16 months of Uhuru data on Cyg X-1 has shown a remarkable transition in the source which occurred during 1971 March and April. The average X-ray intensity in the 2–6-keV energy range decreased by about a factor of 4, the average X-ray intensity in the 10–20-keV band increased by a factor of 2, and a weak radio source suddenly appeared. The data were analyzed for an effect due to a binary system.

142.080   Cosmic X-ray spectra.   K. A. Pounds.
Space Sci. Rev., Vol. 13, 871 - 889 (1972). – Invited paper IAU Colloquium No. 14 (see 012.012).

142.081   High-energy cosmic gamma-ray observations from the OSO-3 satellite.   W. L. Kraushaar, G. W. Clark,
G. P. Garmire, R. Borken, P. Higbie, C. Leong, T. Thorsos.
Astrophys. Journ., Vol. 177, 341 - 363 (1972).

Final results from observations carried out in 1967–1968 with an instrument on the OSO-3 satellite confirm the discovery of cosmic γ-rays with energies above ~50 MeV. The celestial distribution demonstrates the existence of a galactic component which is concentrated in a band of directions around the galactic equator with a broad maximum toward the galactic center. It also shows the existence of an isotropic component with a softer energy spectrum which is probably of extragalactic origin.

142.082   Positive detection of an excess of low-energy diffuse X-rays at high galactic latitude.
D. J. Yentis, R. Novick, P. Vanden Bout.
Astrophys. Journ., Vol. 177, 365 - 373 (1972).

We have obtained data in the energy range 0.10–0.28 keV which confirms the existence of a soft diffuse X-ray flux exceeding in intensity an extrapolation to low energy of the spectrum determined at energies above 1 keV. We show that these data are completely free from contamination by ultraviolet and charged-particle events.

142.083   Galactic-latitude dependence of low-energy diffuse X-rays.   D. J. Yentis, R. Novick, P. Vanden Bout.
Astrophys. Journ., Vol. 177, 375 - 386 (1972).

A flux of low-energy diffuse X-rays has been observed which is qualitatively well correlated with 21-cm column densities of hydrogen from intermediate to high galactic latitude. The intensity exceeds that predicted by an extrapolation to low energies of the power-law spectrum observed above 1 keV. The observed X-ray intensity and the apparent correlation with interstellar hydrogen is accounted for by an additional component of extragalactic radiation plus a local component. In addition, we have observed a large flux of X-rays near the galactic plane and a prominent feature in the constellation Gemini.

142.084   The hydrogen lines in the spectrum of Scorpius X-1.
D. E. Mook, S. Edwards, W. A. Hiltner.

Astrophys. Journ., (*Letters*), Vol. 177, L63 - L67 (1972).

Systematic variations in the equivalent widths and profiles of Hα, Hβ, and Hγ and in the Balmer decrement with B magnitude of Sco X-1 are presented.

142.085   Optical identification of GX 17 + 2.
M. Tarenghi, C. Reina.
Nature, Phys. Sci., Vol. 240, 53 - 54 (1972).

A systematic search is being made for the optical counterparts of the better localized X-ray sources. Here we report a preliminary result of the study of GX 17 + 2 (2U 1813–14).

142.086   On the circular polarization of Sco X-1 and the adjacent sky.   J. C. Kemp.
Nature, Phys. Sci., Vol. 240, 103 - 104 (1972).

The unusually strong circular polarization in the region of Sco X-1 may be only coincidence, a matter of the nearness to the ecliptic.

142.087   Turbulent heating of plasma and thermal X-ray sources.   B. Koppi, A. Trievs.
Trudy Mezhdunar. seminara po probl. "Uskorenie chastits v kosmich. prostranstve (okolozem. i mezhplanet. kosmich. prostranstve), Galaktike i Metagalaktike". Moskva, 1972, p. 146 - 159. – Abstr. in Referativ. Zhurn. 51. Astron., 12.51.472 (1972).

142.088   Dimensions of the binary system HDE 226868 = Cygnus X-1.   C. T. Bolton.
Nature, Phys. Sci., Vol. 240, 124 - 127 (1972).

Recent observations of the binary star HDE 226868, which is believed to be associated with Cygnus X-1, are here combined with data already published to obtain improved orbital elements for the binary system. The data imply that the secondary is a black hole.

142.089   Possibility of continuous monitoring of celestial X-ray sources through their ionization effects in the nocturnal D-region ionosphere.   D. P. Sharma, A. K. Jain,
S. C. Chakravarty, K. Kasturirangan, K. R. Ramanathan, U. R. Rao.
Astrophys. Space Sci., Vol. 17, 409 - 425 (1972).

The electron production rates in the night-time D-region arising from the transit of strong celestial X-ray sources Sco X-1, Tau X-1 and the galactic center are estimated and compared with the ambient electron production rates resulting from other known stable agencies. Using the experimentally measured values of the night-time electron densities, the number of additional electrons/cc expected from the passage of these sources is computed. For the 164 kHz transmission from Tashkent, received at Ahmedabad, the associated enhancement in the attenuation is calculated using the full wave admittance technique of Barron and Budden.

142.090   High frequency observations of the second radio flare in Cygnus X-3.
K. Y. Lo, J. Spencer, P. C. Crane, B. F. Burke.
Nature, Phys. Sci., Vol. 240, 158 - 159 (1972).

The radio source identified with Cygnus X-3 was observed throughout the second great flare from 0100 UT on September 20, 1972, until 0800 UT on September 22, 1972, at 15.5 GHz using the 120-foot antenna of the Haystack Observatory in Westford, Massachusetts. In addition, two measurements were made at 22.235 GHz on September 21 and one was made at 8.105 GHz on September 22. The beginning of a third large flare is also present in our data.

142.091   Cygnus X-3 radio source: Lower limit on size and upper limit on distance.
H. F. Hinteregger, G. W. Catuna, C. C. Counselman III, R. A Ergas, R. W. King, C. A. Knight, D. S. Robertson, A. E. E.

Rogers, I. I. Shapiro, A. R. Whitney, T. A. Clark, L. K. Hutton, G. E. Marandino, R. A. Perley, G. Resch, N. R. Vandenberg.
Nature, Phys. Sci., Vol. 240, 159 - 160 (1972).

The sudden increase in radio flux observed from the direction of Cygnus X-3 on September 22–23, 1972, occurred just as we began a four-antenna very-long-baseline interferometry experiment that involved a pair of radio telescopes in Green Bank, West Virginia, and another pair in Massachusetts. We were able to take advantage of this opportunity to observe Cygnus X-3 interferometrically on September 24 and report here the negative results of these observations.

142.092 **Identification of 2U0525–06 with ϑ^2 Orionis.**
R. Barbon, P. L. Bernacca, M. Tarenghi, A. Treves.
Nature, Phys. Sci., Vol. 240, 182 - 183 (1972).

We have compared the Trimble and Thorne lists of collapsed star candidates with the Uhuru catalogue of X-ray sources. Trimble and Thorne considered only systems with unseen secondary stars of mass $M_2 > 1.4\,M_\odot$ in the catalogue examined. We have found that four of these objects fall within 3 arc deg of X-ray sources. ϑ^2 Orionis is the only one of the four which is inside the error box of an X-ray source (2U0525–06).

142.093 **Optical candidate for GX 5-1.** B. L. Webster.
Nature, Phys. Sci., Vol. 240, 183 (1972).

A faint radio source has been detected within the error box of the X-ray source GX 5-1 by Braes et al. Braes et. al. suggest that the system may be similar to Cygnus X-1 (= HD 226868). The possibility was investigated, and it was found that this star is not similar to HD 226868 and has no property to make it an obvious candidate for the X-ray source.

142.094 **Optical candidate for SMC X-1.**
B. L. Webster, W. L. Martin, M. W. Feast, P. J. Andrews.
Nature, Phys. Sci., Vol. 240, 183 (1972).

The one known discrete X-ray source in the Small Magellanic Cloud, SMC X-1, has been shown to be a binary system with eclipses of the X-ray component occurring every 3.8927 day. We wish to point out that there are two relatively bright stars within the X-ray 90 % region. Positions from the Mount Stromlo SMC charts and photoelectric magnitudes measured at Radcliffe for the two stars are given.

142.095 **Comment on inverse Compton models for the isotropic X-ray background and possible thermal emission from a hot intergalactic gas.**
R. Cowsik, E. J. Kobetich.
Astrophys. Journ., Vol. 177, 585 - 593 (1972).

We have calculated the spectrum of the isotropic X-ray background under the assumption that it arises through Compton scattering of the universal microwave photons by relativistic electrons leaking out of the galaxies. Inverse Compton models for the origin of background X-rays are not adequate to explain all the spectral features in detail. In particular, there is evidence for a thermal component in the X-ray background arising possibly from a hot intergalactic gas at $\sim 3 \times 10^8\,^\circ$K.

142.096 **Identification of the X-ray pulsar in Hercules: a new optical pulsar.**
A. Davidsen, J. P. Henry, J. Middleditch, H. E. Smith.
Astrophys. Journ.,(Letters), Vol. 177, L97 - L102 (1972).

A series of photographic, photoelectric, and spectroscopic observations beginning 1972 June 1 has led to the optical identification of Her X-1, a pulsed X-ray source in an eclipsing binary system, with the thirteenth-magnitude blue variable star HZ Herculis. The detection of optical pulses at the frequency of the X-ray pulsar on three nights makes the identification conclusive and establishes HZ Her as the second

known optical pulsar. The strength of the optical pulses may be correlated with the orbital phase but is not obviously related to the high- or low-intensity states of the X-ray source.

142.097 **Optical studies of UHURU sources. III. Optical variations of the X-ray eclipsing system HZ Herculis.**
W. Forman, C. A. Jones, W. Liller.
Astrophys. Journ.,(Letters), Vol. 177, L103 - L107 (1972).

Photographic observations reveal several unusual properties of the X-ray eclipsing system HZ Herculis and suggest that the source of X-rays lies on the surface of a rapidly rotating collapsed object which revolves about an A- or F-type subgiant. The resultant tidal distortions and reflection-excitation effects of the collapsed object on the subgiant appear to explain satisfactorily the optical observations.

142.098 **Measurement of the position and spectrum of Hercules X-1 from the OSO-7 satellite.**
G. W. Clark, H. V. Bradt, W. H. G. Lewin, T. H. Markert, H. W. Schnopper, G. F. Sprott.
Astrophys. Journ., (Letters), Vol. 177, L109 - L113 (1972).

In 1971 December, the periodically variable binary X-ray source Her X-1 was scanned twice by the 1–60 keV X-ray detectors on the OSO-7. The observations provide an improved position and, in addition, spectral information over the energy range from 1 to 60 keV.

142.099 **Distances and absolute luminosities of galactic X-ray sources.**
F. D. Seward, G. A. Burginyon, R. J. Grader, R. W. Hill, T. M. Palmieri.
Astrophys. Journ., Vol. 178, 131 - 142 (1972).

Data from two soft X-ray surveys are combined to propose a distance scale for 20 X-ray sources. These are the bright sources that lie close to the galactic plane. A relationship between measured absorption and distance is derived by using supernova remnants with known distances for nearby objects and by requiring a rough symmetry about the galactic center for the distant sources. Based on this distance scale, the calculated absolute luminosities of the sources range from 10^{35} to 10^{39} ergs s^{-1}. The sources that have not been identified with supernova remnants tend to fall into two groups.

142.100 **On the color-magnitude locus of Scorpius X-1.**
D. E. Mook, V. Blanco, J. Hesser, W. Kunkel, B. Lasker.
Astrophys. Journ., (Letters), Vol. 178, L11 - L12 (1972).

Recent simultaneous observations of Sco X-1 in V and B are shown to define a color-magnitude relation in agreement with those obtained in previous observing seasons.

142.101 **The *Uhuru* catalog of X-ray sources.**
R. Giacconi, S. Murray, H. Gursky, E. Kellogg, E. Schreier, H. Tananbaum.
Astrophys. Journ., Vol. 178, 281 - 308 (1972).

A catalog of X-ray sources observed with the *Uhuru* satellite is presented. About 70 days of data have been analyzed for this catalog resulting in 125 sources. Approximately two-thirds of the sources are located within ± 20° of the galactic plane. Some of the sources at higher galactic latitudes are identified with known extragalactic objects. Most of the strong sources near the galactic plane are found to be variable.

142.102 **Soft X-ray spectra of the Cygnus Loop and Cygnus X-2 in the energy range 0.16–6.7 keV.**
J. A. M. Bleeker, A. J. M. Deerenberg, K. Yamashita, S. Hayakawa, Y. Tanaka.
Astrophys. Journ., Vol. 178, 377 - 387 (1972).

Soft X-ray spectra of the Cygnus Loop and Cyg X-2 were obtained during flight of a rocket borne soft X-ray experiment on 1971 May 26. The spectrum of the Cygnus Loop is best represented by thermal bremsstrahlung of a plasma at $2.7 \times 10^6 \,^\circ$K with interstellar absorption of 5.5×10^{20} H atoms cm^{-2}. The soft X-ray spectrum of Cyg X-2 indicates a column density which is compatible with that expected for its optical counterpart.

142.103 Hard X-ray observations of Hercules X-1 by OSO-7.
M. P. Ulmer, W. A. Baity, W. A. Wheaton, L. E. Peterson.
Astrophys. Journ., (Letters), Vol. 178, L61 - L64 (1972).

The UCSD X-ray telescope on the OSO-7 scanned the periodic variable X-ray source Her X-1 from 1972 May 30 to 1972 June 1. Analysis of the quick-look data in the 7–26 keV range shows low intensity at $\sim 12 \pm 2$ hours UT, 1972 May 31, which is expected from the previously determined 1.7-day period. The source was observed in its high-intensity state ~ 30 hours earlier than predicted from the *Uhuru* determination of the 35.7-day period and its phase.

142.104 Spectroscopic observations of HZ Herculis and a model for Hercules X-1.
D. Crampton, J. B. Hutchings.
Astrophys. Journ., (Letters), Vol. 178, L65 - L69 (1972).

Ten spectrograms of HZ Her have been obtained, through its 1.7-day period. The spectrum is peculiar, varying in mean spectral type from early B at maximum to late A at minimum light. An ultraviolet continuum disappears during X-ray eclipse. A mean velocity variation is indicated through the 1.7-day period, out of phase with the X-ray source. The optical light curve can be synthesised by a model consisting of a small hot object and a normal A-type star at its inner Roche limiting surface.

142.105 Discovery of the binary nature of SMC X-1 from *Uhuru*.   E. Schreier, R. Giacconi, H. Gursky, E. Kellogg, H. Tananbaum.
Astrophys. Journ., (Letters), Vol. 178, L71 - L75 (1972).

The X-ray source SMC X-1 in the Small Magellanic Cloud has been found to undergo periodic intensity changes which have been interpreted as eclipses in a binary system. No regular pulsations of the source on timescales of seconds have been detected with an upper limit on the percentage pulsed of 10 percent. Intensity fluctuations on timescales of hours and possibly minutes have been observed. Extended nonperiodic intervals of low intensity have also been noted. The energy spectrum is flat and shows low-energy absorption, similar to the behavior of other binary X-ray sources. A possible correlation between low-energy cutoff and orbital inclination in binaries is discussed.

142.106 On the variability of X-ray radiation from black holes at disk accretion.   R. A. Sunyaev.
Astron. Zhurn. Akad. Nauk SSSR, Vol. 49, 1153 - 1157 (1972). In Russian. English translation in Soviet Astron. AJ, Vol. 16, No. 6.

The existence of hot spots on the surface of the disk formed by matter accreting on the collapsar must lead to specific quasi-periodic variability of X-ray radiation of the disk.

142.107 Optical variability of Her X-1 (HZ Her).
W. Wenzel, H. Gessner.
Inform. Bull. Variable Stars, (IAU Commission 27), Konkoly Obs., Budapest, No. 733 (1972).

142.108 A soft X-ray survey of the galactic plane from Cygnus to Norma.
R. Borken, R. Doxsey, S. Rappaport.
Astrophys. Journ., (Letters), Vol. 178, L115 - L120 (1972).
A low-energy X-ray survey ($E \gtrsim 150$ eV) of the Cygnus

region and the galactic plane from Cygnus to Norma is described. The soft X-ray emission from the Cygnus Loop was resolved into two distinct regions which contain several of the major optical filaments. Upper limits for soft X-ray emission from the supernova remnants HB 21, CTB 72, CTB 63, and W 63 are given.

142.109 Observations of Vela XR-1 by the UCSD X-ray telescope on OSO-7.
M. P. Ulmer, W. A. Baity, W. A. Wheaton, L. E. Peterson.
Astrophys. Journ., (Letters), Vol. 178, L121 - L126 (1972).

The UCSD X-ray telescope on the OSO-7 scanned Vel XR-1 for 37 days (1971 December 17 to 1972 January 22) in the ~ 7–500 keV range. From an analysis of the data in the 7–26 keV range, we find evidence for an ~ 8.7-day periodicity and we show that the average spectral shape is similar to that of the binary X-ray sources Her X-1 and Cen X-3.

142.110 The parallax and proper motion of Scorpius X-1.
W. F. van Altena.
Astrophys. Journ., (Letters), Vol. 178, L127 - L128 (1972).

The absolute trigonometric parallax of the optical object identified with Sco X-1 is found to be $\pi = -0\rlap{.}''001 \pm 0\rlap{.}''006$ (s.e.) as determined from plates taken with the Yerkes Observatory 40-inch (102-cm) refractor. The absolute proper motion determined from these plates is $\mu_\alpha = -0\rlap{.}''011 \pm 0\rlap{.}''006$ per year and $\mu_\delta = -0\rlap{.}''018 \pm 0\rlap{.}''003$ per year (s.e.). This motion is consistent with reflex solar motion or membership in the Sco-Cen association and agrees with that determined by Sofia, Eichhorn, and Gatewood.

142.111 X-ray astronomy.   J. Mergentaler.
Urania Kraków, Vol. 43, 226 - 229 (1972).
In Polish.

142.112 Observations of Cyg X-1 and Cyg X-3 above 7 keV from OSO-7.
W. A. Baity, L. E. Peterson, M. P. Ulmer, W. A. Wheaton.
Bull. American Astron. Soc., Vol. 4, 410 (1972). – Abstr. AAS.

142.113 Correlated X-ray and optical measurements of Her X-1.
H. Bradt, R. Doxsey, G. Murthy, S. Rappaport, G. Spada.
Bull. American Astron. Soc., Vol. 4, 411 (1972). – Abstr. AAS.

142.114 Identification of the X-ray pulsar in Hercules: A new optical pulsar.
A. Davidsen, J. P. Henry, J. Middleditch, H. E. Smith.
Bull. American Astron. Soc., Vol. 4, 411 (1972). – Abstr. AAS.

142.115 Implications of the X-ray spectrum for models of Sco X-1.   J. E. Felten.
Bull. American Astron. Soc., Vol. 4, 412 (1972). – Abstr. AAS.

142.116 Spectroscopic observations of the Cygnus X-1 optical candidate.
H. E. Smith, B. Margon, P. Conti.
Bull. American Astron. Soc., Vol. 4, 415 (1972). – Abstr. AAS.

142.117 The X-ray spectrum of Scorpius X-1.   H. Stockman, Jr., J. R. P. Angel, R. Novick, B. E. Woodgate.
Bull. American Astron. Soc., Vol. 4, 415 (1972). – Abstr. AAS.

142.118 Hard X-ray observation of Her X-1 and Vela XR-1 from OSO-7.
M. P. Ulmer, W. A. Baity, L. E. Peterson, W. A. Wheaton.
Bull. American Astron. Soc., Vol. 4, 416 (1972). – Abstr. AAS.

142.119 A high-sensitivity curved crystal spectrometer for cosmic X-rays.   B. E. Woodgate.
Bull. American Astron. Soc., Vol. 4, 417 (1972). – Abstr. AAS.

142.120 Point sources of gamma-rays of energy $> 10^{12}$ eV in Cassiopeia and Cygnus regions.
A. A. Stepanian, B. M. Vladimirsky, V. P. Fomin.
Nature, Phys. Sci., Vol. 239, 40 - 41 (1972).

A burst of high-energy gamma-rays ($E_\gamma \geqslant 2 \times 10^{12}$ eV) was revealed using extensive air shower Cerenkov flash detectors of the Crimean Astrophysical Observatory. Excess in counting rate was observed at September 29, October 14 and 15, 1971 in the direction with coordinates $\alpha = 01^h 11^m \pm 6^m$, $\delta = 62° \pm 1°$ on 3.9 σ level. The gamma-ray flux was about 10^{-10} quanta $cm^{-2}\,s^{-1}$. The sources in Cygnus region $\alpha = 20^h 15^m$, $\delta = 35°$, which was observed in 1970, was not seen in 1971. It was concluded that these sources are variable.

142.121 BD +34°3815. R. J. Brucato, J. Kristian.
IAU Circ., No. 2421 (1972).

142.122 HZ Her as a possible optical pulsar.
D. Q. Lamb, J. M. Sorvari.
IAU Circ., No. 2422 (1972).

142.123 Possible identifications of X-ray sources.
B. W. Bopp, G. Grupsmith, P. Vanden Bout, C. T. Bolton, H. Gursky.
IAU Circ., No. 2424 (1972).

142.124 HZ Herculis. J. N. Bahcall, N. A. Bahcall, W. Liller, C. Jones, W. Forman, C. Y. Shao.
IAU Circ., No. 2427 (1972).

142.125 HZ Herculis. J. N. Bahcall, N. A. Bahcall, D. Crampton, C. L. Morbey.
IAU Circ., No. 2428 (1972).

142.126 HZ Herculis.
P. Boynton, R. Canterna, D. Gerend.
IAU Circ., No. 2430 (1972).

142.127 HZ Herculis. H. B. Richer, B. C. Isherwood, M. Fletcher, C. Morbey.
IAU Circ., No. 2431 (1972).

142.128 HZ Herculis. P. Murdin, A. Davidson, J. P. Henry, J. Middleditch, H. E. Smith.
IAU Circ., No. 2433 (1972).

142.129 HZ Herculis. B. W. Bopp, G. Grupsmith, P. Vanden Bout, V. A. Hughes, A. Woodsworth.
IAU Circ., No. 2434 (1972).

142.130 HZ Herculis. A. Frohlich, N. E. Kurochkin.
IAU Circ., No. 2436 (1972).

142.131 Giant radio outburst of Cygnus X-3.
P. C. Gregory, E. R. Seaquist, P. P. Kronberg, V. A. Hughes, A. Woodsworth, M. R. Viner, D. Retallack, R. M. Hjellming, B. Balick.
IAU Circ., No. 2440 (1972).

142.132 HZ Herculis. D. C. Koo.
IAU Circ., No. 2441 (1972).

142.133 Cygnus X-3. T. Matilsky, E. Kellogg, H. Tananbaum, H. Gursky, R. Giacconi.
IAU Circ., No. 2442 (1972).

142.134 Cygnus X-3.
A. N. Argue, C. M. Kenworthy, P. Stewart.
IAU Circ., No. 2444 (1972).

142.135 Cygnus X-3. A. Brinkman, D. Parsignault,
R. Giacconi, H. Gursky, E. Kellogg, E. Schreier, H. Tananbaum.
IAU Circ., Nos. 2446, 2447 (1972).

142.136 HZ Herculis.
J. L. Elliot, R. E. Murphy, D. P. Cruikshank.
IAU Circ., Nos. 2454, 2462 (1972).

142.137 Two X-ray sources from OSO-7 observations.
C. Heinz, T. Markert, G. W. Clark, W. H. G. Lewin, H. W. Schnopper, G. F. Sprott.
IAU Circ., No. 2466 (1972).

142.138 Identification of SMC X-1. W. Liller.
IAU Circ., No. 2469 (1972).

142.139 Cosmic background X-rays.
S. Hayakawa, D. Sugimoto.
Progr. Theor. Phys. Suppl., (*Japan*), No. 49, p. 148 - 180 (1971).

Observations of the diffuse component of cosmic X-rays are reviewed.

142.140 The variable X-ray source Scorpius X-1.
R. Staubert.
Atomkernenergie, Vol. 19, No. 2, p. 153 - 156 (1972).
In German.

142.141 The attenuation length for high-energy γ-rays in the relict radiation. M. C. Allcock, J. Wdowczyk.
Nuovo Cimento B, Ser. 11, Vol. 9B, 315 - 320 (1972).

Detailed calculations of the interaction of electrons and photons with the relict radiation (inverse Compton effect and pair production process, respectively) have been made. The attenuation length for energy and intensity of the photons and electrons has thereby been established.

142.142 Observation of soft X-ray emission from Perseus X-1.
P. C. Agrawal, G. R. Riegler, J. C. Stevens, G. P. Garmire.
Bull. American Astron. Soc., Vol. 4, 409 (1972). – Abstr. AAS.

142.143 Attenuation of X-rays in interstellar space.
K. Schocken.
Journ. Applied Phys., (*USA*), Vol. 43, 3575 - 3577 (1972).

142.144 On the nature of X Persei; evidence from the 1957 outburst. L. R. Wackerling.
Publ. Astron. Soc. Pacific, Vol. 84, 827 - 833 (1972).

In mid-1957 there occurred in the spectrum of the O star X Per, near the X-ray source 2U0352 + 30, abrupt changes in the emission-line radial velocities and V/R ratios. The irregular variation of the V/R ratio suggests the occasional interruption of a more or less steady infall of material through the rotating emission region by outbursts of material from the star. The space velocity of X Per has been rederived, using a revised radial velocity, and compared to the space velocity of the runaway O star ξ Per.

142.145 Spectroscopic changes in the suspected X-ray source X Persei.
A. P. Cowley, D. B. McLaughlin, J. Toney, D. J. MacConnell.
Publ. Astron. Soc. Pacific, Vol. 84, 834 - 838 (1972).

Spectrograms from 1913 to the present show X Per to have bright H lines that are variable in structure, velocity, and intensity. At times the spectrum is veiled and very few features can be distinguished. During some phases emission of Fe II, Ti II, He I, and He II are seen. Very weak, broad absorption lines suggest the underlying star is near O9. No periodicities are suggested by our data.

142.146 Primary cosmic γ rays above 10^{12} eV.
J. Wdowczyk, W. Tkaczyk, A. W. Wolfendale.
Journ. Phys. A, General Phys., Vol. 5, 1419 - 1432 (1972).

An analysis is made of the interactions of very energetic cosmic ray primaries with the universal black body radiation in extragalactic space.

142.147 An X-ray survey of the Cygnus region in the 20 to 300 keV energy range. J. L. Matteson.
Thesis, Univ. California, San Diego. [Available from Univ. Microfilms, Ann Arbor, Mich., USA. Order No. 72-12290], 83 pp. (1971).

142.148 Röntgen-Quelle Her X-1 = HZ Her.
W. Wenzel, H. Geßner.
MVS, *Sonneberg*, Vol. 6, 61 - 62 (1972).

142.149 A pulsation mechanism of stellar X-ray sources.
S. Hayakawa.
Progr. Theor. Phys., (*Japan*), Vol. 47, 1452 - 1453 (1972).

It has been recently discovered for galactic X-ray sources, Sco X-1 and Cyg X-1, that X-rays are emitted as pulse trains of varying periods. The author points out that the phenomena can be explained by extending the acoustic vibration model.

142.150 Electron bremsstrahlung from hot plasma in the presence of strong magnetic field.
A. El-Gowhari, J. Arponen.
Nuovo Cimento B, Ser. 11, Vol. 11B, 201 - 214 (1972).

Relativistic expressions are given for the high-frequency electromagnetic radiation due to bremsstrahlung in the presence of a strong magnetic field ($\lesssim 10^{11}$ G), and the nonrelativistic limit is also considered. The results are applied to the case of a hot plasma in a magnetic field. It is suggested that the results can explain the pulsed X-ray and γ-ray radiation from X-ray pulsars like PSR 0532.

142.151 Time variations of X-ray spectrum and optical luminosity of Sco X-1. U. R. Rao, E. V. Chitnis,
U. B. Jayanthi, A. S. Prakasarao, S. M. Bhandari.
Cosmic rays. 12th International Conference, Hobart, Tasmania, Australia, 1971, Vol. 1, 2 - 6 (1972).

142.152 X-ray sources and supernova remnants.
P. R. Amnuel, O. H. Gusejnov.
Astron. Tsirk., No. 729, p. 6 - 7 (1972). In Russian.

142.153 Circular polarization of Cyg X-1.
O. S. Shulov, E. N. Kopatskaya.
Astron. Tsirk., No. 741, p. 5 - 6 (1972). In Russian.

Gamma-ray astronomy: Gearing up for SAS–B.
Nature, Vol. 239, 303 - 304 (1972).

The sixteenth Herstmonceux conference, 1972 April 5 and 6. Cosmic X-ray sources. See Abstr. 011.045.

Accretion vortices and X-ray sources.
See Abstr. 065.087.

X-ray flux from HR 465. See Abstr. 114.030.

Discovery of infrared emission from the radio source near Cygnus X-3. See Abstr. 114.086.

HD 152667 and Sco X-2. See Abstr. 119.003.

The primary spectrum of the eclipsing binary LR Centauri. See Abstr. 121.017.

The period and light curve of HZ Herculis.
See Abstr. 121.078.

Spectroscopic observations of HZ Herculis.
See Abstr. 121.079.

Optical observations of HZ Herculis.
See Abstr. 121.091.

β Persei: Radio star and probable X-ray star.
See Abstr. 121.107.

Observations of rapid blue variables − VII. EX Hydrae. See Abstr. 122.020.

Observational evidence against supernovae being the source of the universal X-ray background.
See Abstr. 125.007.

Supernovae and sources of X-radiation.
See Abstr. 125.014.

Observations of soft X-rays: Two supernova remnants in the constellation Lupus and the diffuse background.
See Abstr. 125.019.

On the nature of the Tycho X-ray source.
See Abstr. 125.036.

An evolutionary thermal model for the Cygnus Loop. See Abstr. 132.025.

Detection of high-energy gamma rays from the Crab nebula. See Abstr. 134.001.

Detection of 10−100 MeV γ-rays from the Crab nebula pulsar NP 0532. See Abstr. 141.523.

X-ray observations of pulsars. See Abstr. 141.542.

Search for long-period X-ray pulsations in the NP 0527 region. See Abstr. 141.554.

Gamma astronomy and cosmic rays. I.
See Abstr. 143.071.

Distribution of the diffuse component of cosmic soft X-rays. See Abstr. 155.012.

On a galactic origin for the soft X-ray background.
See Abstr. 155.015.

X-ray galaxies? See Abstr. 158.044.

The interpretation of the X-ray emission detected from some nearby radio galaxies. See Abstr. 158.098.

Observations of the extended X-ray sources in the Perseus and Coma clusters from *Uhuru*.
See Abstr. 160.019.

The observation of relic radiation as a test of the nature of X-ray radiation from the clusters of galaxies.
See Abstr. 160.027.

Errata

142.901 Errata:'The low-energy gamma-ray spectrum of Scorpius X-1' [Astrophys. Journ., (*Letters*), Vol. 172, L47 - L49 (1972)]. R. C. Haymes, F. R. Harnden, Jr., W. N. Johnson III, H. M. Prichard, H. E. Bosch.
Astrophys. Journ., (*Letters*), Vol. 176, L93 (1972).

143 Cosmic Radiation

143.001 **Are cosmic rays produced by white dwarfs?**
V. Schönfelder, J. Trümper.
Astrophys. Letters, Vol. 11, 203 - 206 (1972).

The suggestion that magnetic white dwarfs might be the main source of cosmic rays in the Galaxy is discussed. It is shown that such a picture is incompatible with observations of the magnetic fields of white dwarfs and current theories on magnetic field decay.

143.002 **Cosmic-ray electron spectrum and its modulation near solar maximum.** P. J. Schmidt.
Journ. Geophys. Res., Vol. 77, 3295 - 3310 (1972).

We have used a balloon-borne ionization spectrometer to measure the energy spectrum of cosmic-ray electrons in the range 20 Mev to 15 Gev. Balloon flights were made from Fort Churchill, Manitoba, Canada, in the summers of 1968, 1969, and 1970. Primary electron spectra were obtained for these three periods that show the effect of differing levels of solar modulation. It is shown that the data can be explained quantitatively by using the force-field approximation to the cosmic-ray transport equation if the rigidity dependence of the diffusion coefficient is generalized to be time-dependent.

143.003 **Rigidity dependence of cosmic-ray modulation at rigidities >2 Gv in August 1969.**
P. H. Stoker, B. C. Raubenheimer, A. J. van der Walt.
Journ. Geophys. Res., Vol. 77, 3575 - 3582 (1972). − Brief report.

143.004 **The future of balloons in cosmic-ray research.**
B. Peters.
Space Science Rev., Vol. 13, 313 - 318 (1972). − Conference paper (see 012.002).

143.005 **Observations of the radial gradient of galactic cosmic radiation over a solar cycle.**
J. J. O'Gallagher.
Rev. Geophys. Space Phys., Vol. 10, 821 - 835 (1972).

It is the purpose of this review to provide a coherent compilation of current observations of the radial gradient, presented in such a way that the magnitude and importance of existing discrepancies can be clearly interpreted. In addition, it is intended to give the reader a perspective from which to understand the possible reasons for the as yet unresolved discrepancies.

143.006 **Equilibrium energy spectrum for the galactic cosmic electrons.** P. K. Suh.
Astron. Astrophys., Vol. 20, 375 - 382 (1972).

The equilibrium cosmic electron diffusion equation is examined, and perturbational analysis developed. The effect of diffusion across the galactic disk slab is predominantly important, and up to a considerably high energy domain, the cosmic electron spectral behavior is sensitively affected by the leakage probability at the disk-halo boundaries. One dimensional diffusion equation is thus examined in detail.

143.007 **Geomagnetic cutoffs for cosmic-ray protons for seven energy intervals between 1.2 and 39 Mev.**
J. L. Fanselow, E. C. Stone.
Journ. Geophys. Res., Vol. 77, 3999 - 4009 (1972).

The vertical geomagnetic cutoffs for cosmic-ray protons are presented for seven different energy intervals between 1.2 and 39 Mev. These data, representing approximately 160 passes through the cutoff, were taken during 1967 and 1968, between 408- and 912-km altitude, during times of $K_p < 1^+$. We find that the measured invariant latitudes for the cutoffs are 3° to 5° below previous calculations. The data indicate that even during 'quiet' times there are temporal changes in the geomagnetic field that cause the cutoff to fluctuate by 1° to 2°.

143.008 **Further evidence for a cosmic ray selection mechanism.** K. Kristiansson.
Astrophys. Space Sci., Vol. 16, 405 - 412 (1972).

A comparison has been made between the overabundance of heavy elements in the primordial cosmic radiation and the cross-section for ionization by charged particle impacts. It is found that there is evidence that this type of ionization is important for the selection of cosmic ray particles.

143.009 **About the variation of the cosmic ray gradient during the cycle of solar activity.**
A. K. Lavrukhina, G. K. Ustinova.
Geomagn. Aeronom., Vol. 12, 744 - 746 (1972). In Russian. Brief information.

143.010 **The electron component of cosmic rays. I. Spatial distribution and energy spectrum.**
S. V. Bulanov, V. A. Dogel', S. I. Syrovatskij.
Kosmich. Issled., Vol. 10, 532 - 544 (1972). In Russian.

143.011 **Results of measurements of cosmic ray intensity with Venera 7.**
S. N. Vernov, P. V. Vakulov, I. V. Getselev, E. V. Gorchakov, N. L. Grigorov, P. P. Ignat'ev, N. N. Kontor, S. N. Kuznetsov, Yu. I. Logachev, G. P. Lyubimov, A. G. Nikolaev, N. V. Pereslegina, N. F. Pisarenko, I. A. Savenko, V. I. Tkachenko, E. A. Chuchkov, V. A. Yakovlev.
Kosmich. Issled., Vol. 10, 561 - 568 (1972). In Russian.

143.012 **Antimatter and ball lightning.** J. F. Crawford.
Nature, Vol. 239, 395 (1972).

Evidence is presented that fluxes of photons lasting ≤1 s at an energy corresponding to positron annihilation, in connexion with the possibility that "ball lightning" may be associated with the existence in the terrestrial environment of small quantities of antimatter, at about the observed frequency, can be expected from extensive air showers caused by primary cosmic rays of energy $E_0 \sim 10^{16}$ e. V.

143.013 **Cosmic-ray proton and helium spectra above 50 GeV.**
M. J. Ryan, J. F. Ormes, V. K. Balasubrahmanyan.
Phys. Rev. Letters, Vol. 28, 985 - 988, with a correction, p. 1497 (1972).

A program of experiments to determine the charge composition of galactic cosmic rays in the energy range 10^{10} to 10^{14} eV was initiated at Goddard Space Flight Center. In this Letter, we report measurements on the differential spectra of proton and helium nuclei up to 2000 GeV. The results on heavier nuclei ($Z > 2$) will be reported later.

143.014 **Composition of cosmic-ray nuclei at high energies.**
E. Juliusson, P. Meyer, D. Müller.
Phys. Rev. Letters, Vol. 29, 445 - 448 (1972).

We have measured the charge composition of cosmic-ray nuclei from Li to Fe with energies up to about 100 GeV/nucleon. A balloon-borne counter telescope with gas Cherenkov counters for energy determination was used for this experiment.

143.015 **A search for antihelium in primary cosmic radiation.**
P. Evenson.
Astrophys. Journ., Vol. 176, 797 - 808 (1972).

A search for anti-α-particles in the primary cosmic radiation has been carried out, and a new upper limit for these particles in the range 0.2–4.3 GeV per nucleon has been obtained. At the 95 percent confidence level the upper limit is found to be 0.14 percent of the α-particle flux. The instrument used for this purpose is a magnetic spectrometer employing spark chambers for determining particle trajectories and time-of-flight measurement for the rejection of upward-moving particles. Implications of these results for various models of the sources of cosmic radiation are discussed.

143.016 Change in the eleven-year modulation at the time of the June 8, 1969, Forbush decrease.
J. A. Lockwood, J. A. Lezniak, W. R. Webber.
Journ. Geophys. Res., Vol. 77, 4839 - 4844 (1972).

The functional form of the modulation for the 11-year variation changed abruptly after the Forbush decrease of June 8, 1969. This change in the 11-year modulation suggests that this Forbush decrease was an integral part of the 11-year variation.

143.017 Time and space variations of the increases of cosmic ray intensity before Forbush-decreases.
N. S. Kaminer, A. E. Kuzmicheva, L. I. Dorman.
Geomagn. Aeronom., Vol. 12, 814 - 822 (1972). In Russian.

143.018 Intercosmos satellites exploring cosmic rays.
N. L. Grigorov, P. V. Vakulov.
Priroda, No. 10.72, p. 29 - 36 (1972). In Russian.

143.019 Cosmic ray intensity decrease of 23 September, 1966. U. D. Desai.
Canadian Journ. Phys., Vol. 49, 265 - 269 (1971).

In the present study, particle and magnetic field data from satellite-borne detectors and ground-based neutron neutron monitors clearly show the onset of the Forbush decrease coincident with the SSC magnetic storm. It is pointed out that the Forbush decrease arises from a corotating shock front approaching from the east of the sun–earth line and is not associated with any solar flare effect. Further, the increases observed by the various neutron monitors 9 h after the onset of the Forbush decrease are interpreted to be an enhancement of the diurnal anisotropy.

143.020 Cosmic-ray production of deuterium, He³, lithium, beryllium, and boron in the Galaxy. H. E. Mitler.
Astrophys. Space Sci., Vol. 17, 186 - 218 (1972).

The production of deuterium, He³, lithium, beryllium, and boron by galactic cosmic rays in the interstellar medium, over the life of the Galaxy, is calculated. It is found that high-energy $\alpha-\alpha$ reactions contribute in an essential way to the observed lithium. The observed abundances of Li⁶, Be⁹ and boron can be explained. But deuterium, He³ and Li⁷ are not produced in significant amounts. An incidental result is that the mean luminosity of the Galaxy over its lifetime has been about 3 times its present luminosity.

143.021 Forbush microdecreases in the cosmic-ray intensity.
D. Blănariu, A. Moldovanu, E.-B. Bradu.
Gerlands Beiträge Geophys., Vol. 79, 321 - 326 (1970).

The two-years registration of the neutron supermonitor at Alert revealed a great number of short-lived Forbush decreases. The detailed study of these "Forbush microdecreases" (FMD-events) in connection with the geomagnetic perturbations at Surlari, Rumania, is undertaken. The main statistical features of the FMD-events are discussed in terms of their influence on the daily variation of the cosmic-ray intensity, as well as of the "geoefficiency" of the geomagnetic perturbations.

143.022 Absolute calculation of secondary cosmic gamma rays at the top of the atmosphere.
A. Jabs, G. Wibberenz.
Zeitschr. Physik, Vol. 236, 101 - 129 (1970).

Starting from the primary nucleon and the primary electron spectra the vertical secondary gamma spectra in 0–30 g/cm² atmospheric depth and the energy range 1–1000 GeV are calculated and compared with experimental data. The calculated spectra largely depend on the properties of the primary-nucleon–air-nucleus collision. The collision model is considered in detail. The influence of the relevant collision parameters, which are not very well known, on the gamma spectra is studied. In addition the influence of the solar cycle and of the geomagnetic cutoff is shown.

143.023 Nature of the long-term and short-term modulations of cosmic-ray intensity. R. P. Kane.
Journ. Geophys. Res., Vol. 77, 5573 - 5579 (1972). – Brief report.

143.024 On lifetime of cosmic rays in the Galaxy in presence of acceleration. P. Velinov.
Comptes Rendus.Acad. Bulgar. Sci., (Dokl. Bolg. Akad. Nauk), Vol. 24, 431 - 434 (1971).

143.025 Galactic cosmic ray modulation from 1965–1970.
I. H. Urch, L. J. Gleeson.
Astrophys. Space Sci., Vol. 17, 426 - 446 (1972).

Numerical solutions of the cosmic-ray equation of transport within the solar cavity and including the effects of diffusion, convection, and energy losses due to adiabatic deceleration, have been used to reproduce the modulation of galactic electrons, protons and helium nuclei observed during the period 1965–1970. Kinetic energies between 10 and 10⁴ MeV/nucleon are considered. Computed and observed spectra (where data is available) are given for the years 1965, 1968, 1969 and 1970 together with the diffusion coefficients. The force-field solutions are given for these diffusion coefficients and galactic spectra and compared with the numerical solutions. For each of the above years we have (1) determined the radial density gradients near earth; (2) found the mean energy losses suffered by galactic particles as they diffuse to the vicinity of the earth's orbit; (3) shown quantitatively the exclusion of low-energy galactic protons and helium nuclei from near earth by convective effects; and (4), for nuclei of a given energy near earth, obtained their distribution in energy before entering the solar cavity.

143.026 The Fermi mechanism and the source spectrum of cosmic ray nuclei. S. Ramadurai, S. Biswas.
Astrophys. Space Sci., Vol. 17, 467 - 471 (1972).

It is shown that the velocity term, occurring in the expression for the rate of energy gain by the Fermi mechanism of acceleration, is to be taken into account in case of acceleration of non-relativistic particles. A spectral form of accelerated particles is derived on this basis. Using this form of source spectrum of cosmic ray nuclei, satisfactory agreement can be obtained between the calculated values and the observed ones of the ratios of H²/He⁴ and He³/He⁴, and the energy spectra of protons and helium nuclei near the earth.

143.027 Diffusion of cosmic rays and their source composition. M. M. Shapiro, R. Silberberg, C.-H. Tsao.
G. Gamow Memorial Volume, (see 003.003), p. 124 - 149 (1972).

143.028 Cosmic rays near the centre of the Galaxy.
V. L. Ginzburg, Ya. M. Khazan.
Astrophys. Letters, Vol. 12, 155 - 158 (1972).

The question of the proton-nuclear component of cosmic rays near the centre of the Galaxy is discussed. It follows from γ-ray astronomy data, that the cosmic ray intensity in this region is very likely to exceed considerably the corresponding value near the earth.

143.029 Adiabatic propagation of cosmic rays in the Galaxy.
S. C. Barrowes.
Astrophys. Journ., Vol. 177, 45 - 57 (1972).

An analysis is presented for the propagation of charged cosmic rays away from a source in a relatively smooth galactic magnetic field. The expected intensity of the resulting narrow-angle anisotropy is related to the particle flux produced by the source. The probable number of such anisotropies is derived, based on the lifetime of cosmic rays and the number of sources in the Galaxy. The agreement with experimental evidence and future possibilities of the model are discussed.

143.030 The propagation of cosmic rays with $Z \geqslant 74$.
D. N. Schramm.
Astrophys. Journ., Vol. 177, 325 - 339 (1972).

The propagation of relativistic (energy $\gtrsim 1$ GeV per nucleon), $Z \geqslant 74$ cosmic rays has been investigated. An r-process source was assumed, and effects due to spallation, fission, and radioactive decay (including time dilation) were included. In addition to the above-mentioned nuclear effects, ionization losses were explicitly put into the calculation. Cosmic rays in this region fall into four discrete groups: Pt group ($74 \leqslant Z \leqslant 80$), Pb group ($81 \leqslant Z \leqslant 83$), actinides ($90 \leqslant Z \leqslant 100$), and the SH (superheavy) group ($Z \gtrsim 106$). The relative abundances of these groups have been calculated as a function of the cosmic-ray leakage time.

143.031 Observations of galactic cosmic-ray intensity at heliocentric radial distances of from 1.0 to 2.0 astronomical units.   J. A. Van Allen.
Astrophys. Journ., (Letters), Vol. 177, L49 - L51 (1972).

From observations with the Jupiter-bound spacecraft Pioneer 10 it is found that the integral intensity of the galactic cosmic radiation ($E_p > 80$ MeV) varied by less than 3 percent (either plus or minus) over the heliocentric radial range 1.0–2.0 a.u. during the epoch 1972 March–July.

143.032 High-energy cosmic rays in the expanding universe.
K. Arai.
Sci. Rep. Tôhoku Univ., First Ser., Vol. 55, 9 - 14 (1972).

Energy spectra of extragalactic neutrinos and γ-rays are calculated. The calculations are made both for blackbody photons and for graybody photons, using experimental data on photomeson production. The effective spectrum of shower-producing neutrinos are estimated under the assumption of a linear rise of neutrino-nucleon cross-sections up to an energy of 10^{21} eV.

143.033 The residual cosmic ray modulation at the 1954 solar minimum.   D. M. Thomson.
Planet. Space Sci., Vol. 20, 2196 - 2197 (1972).

Analysis of the anomalous solar diurnal variation in cosmic rays at the 1954 solar minimum yields an estimate of the residual solar modulation present at that time. The data are

143.034 Composition of relativistic cosmic rays near the earth and at the sources.
M. M. Shapiro, R. Silberberg, C. H. Tsao.
Space Research XII, (see 012.016), Vol. 2, 1609 - 1615 (1972).

143.035 Study of energy spectra of primary cosmic rays at very high energies on the Proton series of satellites.
N. L. Grigorov, V. E. Nesterov, I. D. Rapoport, I. A. Savenko.
Space Research XII, (see 012.016), Vol. 2, 1617 - 1622 (1972).

143.036 Anomalous recurrent diurnal anisotropy in cosmic ray intensity with maximum along the garden hose direction.   H. Razdan, M. M. Bemalkhedkar.
Cosmic Electrodynamics, Vol. 3, 297 - 315 (1972).

During the recovery of a number of cosmic ray intensity decreases of Forbush or gradual commencement type, cosmic ray diurnal variation exhibits large amplitudes for 2—4 days with maximum along the garden hose direction in the interplanetary space. The results are explained in terms of the important role played by the enhanced outward convection of cosmic ray particles and an inward diffusion along the garden hose direction in the specialized geometry of a transient modulating M region stream or a plasma cloud from an east limb flare.

143.037 Interpretation of the cosmic ray anisotropy.
H. R. Allan.
Astrophys. Letters, Vol. 12, 237 - 241 (1972).

The model of 'compound diffusion' put forward by Lingenfelter et al. to describe the motion of cosmic rays in the Galaxy is criticized on the grounds that their calculated anisotropy refers to the flux across an area large compared with the scale size of magnetic irregularities, whereas the measured values refer to a single tube of force.

143.038 Pulsed galactic nuclei and the origin of cosmic rays.
J. R. Wayland.
Astrophys. Space Sci., Vol. 18, 89 - 93 (1972).

The possibility that a series of explosions of the galactic nuclei every 5×10^6 yr can cause a substantial flux of cosmic ray particles at the vicinity of the earth is investigated. The steady flux of cosmic radiation forces the conclusion that there have been explosions back to 10^9 yr if this is a dominant source of cosmic rays.

143.039 Deuterons and He³ formation and destruction in proton induced spallation of light nuclei ($Z \leqslant 8$).
J. P. Meyer.
Astron. Astrophys., Suppl. Ser., Vol. 7, 417 - 467 (1972).

With the aim of interpreting the deuteron and He³ abundances in galactic cosmic radiation, we survey: (i) their production cross sections and reaction kinematics in p-p, p-He³, p-He⁴ and p-(CNO) interactions, with special emphasis on the p-He⁴ reactions, (ii) their destruction cross sections in p-d and p-He³ reactions, and (iii) the p-p, p-d and p-He⁴ elastic scattering cross sections. The experimental data are thoroughly discussed and the "neutron data" reinterpreted in terms of the mirror proton induced reaction.

143.040 Low energy cosmic rays.
S. N. Vernov, N. N. Kontor, G. P. Lyubimov, N. V. Pereslegina, E. A. Chuchkov.
Trudy Mezhdunar. seminara po probl. "Uskorenie chastits v kosmich. prostranstve (okolozem. i mezhplanet. kosmich. prostranstve), Galaktike i Metagalaktike". Moskva, 1972, p. 193 - 214. In Russian. – Abstr. in Referativ. Zhurn. 51. Astron., 12.51.526 (1972).

143.041 The spectrum of galactic cosmic rays in interstellar space beyond the modulation region.
A. G. Zusmanovich, E. V. Kolomeets.
Trudy Mezhdunar. seminara po probl. "Uskorenie chastits v kosmich. prostranstve (okolozem. i mezhplanet. kosmich. prostranstve), Galaktike i Metagalaktike". Moskva, 1972, p. 215 - 229. In Russian. – Abstr. in Referativ. Zhurn. 51. Astron., 12.51.527 (1972).

143.042 Propagation of galactic cosmic rays.
L. E. Gajnova, G. A. Gonchar, A. G. Zusmanovich, E. V. Kolomeets, R. A. Chumbalova, Yu. A. Shakhova.
Trudy Mezhdunar. seminara po probl. "Uskorenie chastits v kosmich. prostranstve (okolozem. i mezhplanet. kosmich. prostranstve), Galaktike i Metagalaktike". Moskva, 1972, p. 243 - 251. In Russian. – Abstr. in Referativ. Zhurn. 51. Astron., 12.51.528 (1972).

143.043 On the spectrum of relativistic particles accelerated by plasma turbulence. V. N. Tsytovich.
Trudy Mezhdunar. seminara po probl. "Uskorenie chastits v kosmich. prostranstve (okolozem. i mezhplanet. kosmich. prostranstve), Galaktike i Metagalaktike". Moskva, 1972, p. 314 - 315. In Russian. – Abstr. in Referativ. Zhurn. 51. Astron., 12.51.530 (1972).

143.044 Spectrum and composition of primary cosmic rays of high and super-high energy. N. L. Grigorov.
Trudy Mezhdunar. seminara po probl. "Uskorenie chastits v kosmich. prostranstve (okolozem. i mezhplanet. kosmich. prostranstve), Galaktike i Metagalaktike". Moskva, 1972, p. 230 - 242. In Russian. – Abstr. in Referativ. Zhurn. 62. Issled. kosmich. prostranstva, 12.62.161 (1972).

143.045 Rigidity spectrum of helium nuclei above 17 GV and a search for high energy anti-nuclei in primary cosmic rays.
R. P. Verma, T. N. Rengarajan, S. N. Tandon, S. V. Damle, Yash Pal.
Nature, Phys. Sci., Vol. 240, 135 - 137 (1972).
We present here some results obtained with a magnet spectrograph flown from Hyderabad (geomagnetic cutoff ~17 GV) in May 1970. The two principal results are: there is no anomaly in the differential spectrum of helium nuclei close to the geomagnetic cutoff; and no anti-nuclei have been seen up to a rigidity of ~100 GV.

143.046 Off-ecliptic control of cosmic ray modulation.
P. C. Hedgecock, J. J. Quenby, S. Webb.
Nature, Phys. Sci., Vol. 240, 173 - 175 (1972).
In this article the suggestion is made that phenomena happening well away from the ecliptic plane may be important controlling factors in the modulation of cosmic rays received at the earth.

143.047 Cosmic-ray scintillations. 1. Inside the magnetosphere. A. J. Owens, J. R. Jokipii.
Journ. Geophys. Res., Vol. 77, 6639 - 6655 (1972).
Evidence is presented for the existence of statistically significant broad-band fluctuations, far above noise level, in the cosmic-ray flux observed inside the magnetosphere. The observed intensities at the polar cap at both low and high energies were analyzed, and power spectra of the cosmic-ray scintillations are presented and discussed.

143.048 Modulation of low-energy galactic cosmic rays over solar maximum (cycle 20).
M. A. I. Van Hollebeke, J. R. Wang, F. B. McDonald.
Journ. Geophys. Res., Vol. 77, 6881 - 6885 (1972). – Letter.

143.049 Nuclear gamma rays from ^{7}Li in the galactic cosmic radiation. G. J. Fishman, D. D. Clayton.
Astrophys. Journ., Vol. 178, 337 - 340 (1972).
The observation of a γ-ray line feature from the direction of the galactic center by Johnson, Harnden, and Haymes is interpreted as the 478-keV nuclear de-excitation of low-energy ^{7}Li cosmic rays as they inelastically scatter from the interstellar gas. The prediction of an associated line at 432 keV is proposed as a definitive test of this idea.

143.050 Peculiarities of cosmic ray variations near the plane of the solar equator.
L. I. Dorman, N. S. Kaminer, A. E. Kuzmicheva.
Geomagn. Aeronom., Vol. 12, 1112 - 1113 (1972).
In Russian. – Brief information.

143.051 Modulation of cosmic-ray electrons. J. A. Earl.
Astrophys. Journ., Vol. 178, 857 - 862 (1972).
The origin and time variations of the steep spectrum of electrons observed below 20 MeV may be explained by a simple model in which the spectrum of interplanetary cosmic rays is decomposed, at low energies, into two independently varying components.

143.052 Mean path length of high energy cosmic rays in the galactic disk. J. M. Audouze, C. J. Cesarsky.
Bull. American Astron. Soc., Vol. 4, 410 (1972). – Abstr. AAS.

143.053 Nuclear gamma rays from ^{7}Li in the galactic cosmic radiation. D. Clayton, G. J. Fishman.
Bull. American Astron. Soc., Vol. 4, 411 (1972). – Abstr. AAS.

143.054 Energetic neutrons leaking from the top of the atmosphere. M. Merker.
Phys. Rev. Letters, Vol. 29, 1531 - 1534 (1972).
The energy, angular, and latitude distributions of energetic neutrons (up to 2 GeV) leaking from the top of the atmosphere have been calculated. The results agree with the recent measurement of Preszler, Simnett, and White.

143.055 Low-energy cosmic rays.
S. N. Vernov, N. N. Kontor, G. P. Lyubimov, N. V. Pereslegina, E. A. Chuchkov.
Trudy Mezhdunar. seminara po probl. "Uskorenie chastits v kosmich. prostranstve (okolzem. i mezhplanet. kosmich. prostranstve), Galaktike i Metagalaktike". Moskva, 1972, p. 193 - 214. In Russian. – Abstr. in Referativ. Zhurn. 62. Issled. kosmich. prostranstva, 1.62.187 (1973).

143.056 Energy spectrum and composition of pulsar-accelerated cosmic rays. K. Sitte, L. Briatore, M. Dardo.
Nuovo Cimento B, Ser. 11, Vol. 10B, 498 - 510 (1972).
In traversing the expanding shell of the supernova, pulsar-accelerated heavy nuclei will suffer collisions with the shell gas which lead to attenuation, and to fragmentation. Since the attainable energy decreases in time as the pulsar slows down, and the shell thins out, cosmic-ray particles of the highest energy will be most strongly affected.

143.057 Hot universe cosmic rays of ultrahigh energy and absolute reference system. H. Sato, T. Tati.
Progr. Theor. Phys., (*Japan*), Vol. 47, 1788 - 1790 (1972).
The authors discuss the non-existence of the expected cut-off of cosmic rays in the vicinity of 10^{20} eV due to thermal radiation.

143.058 Theory and search for detailed cosmic ray anisotropies. S. C. Barrowes.
Thesis, Univ. Utah, Salt Lake City. [Available from Univ. Microfilms, Ann Arbor, Mich., USA. Order No. 72-1140], 145 pp. (1971).

143.059 The cosmic-ray spectral modulation above 2 GV. IV. The influence on the attenuation coefficient of the nucleonic component.
F. Bachelet, N. Iucci, G. Villoresi, N. Zangrilli.
Nuovo Cimento B, Ser. 11, Vol. 11B, 1 - 12 (1972).
The authors study the influence of the primary cosmic-ray modulation on the attenuation coefficient of the nucleonic component at different latitudes and altitudes.

143.060 Diffusion processes of cosmic rays with energies between 2 and 20 GV during Forbush decreases: The diurnal effect. E. Bussoletti, A. Geranios.
Nuovo Cimento B, Ser. 11, Vol. 11B, 53 - 67 (1972).
A study of the morphology of the diurnal variation of the cosmic-ray intensity during the Forbush decreases in the periods 15 October 1965–30 June 1966 and 21 January 1967–17 December 1968 is presented.

143.061 Cut-off problem in the integral spectrum of ultra-high energy cosmic rays and urbaryons.
K. Terasaki-Okada.
Progr. Theor. Phys., (Japan), Vol. 48, 349 - 350 (1972).

143.062 Quiet-time electron increase. A measure of conditions in the outer solar system.
L. A. Fisk, M. Van Hollebeke.
Rep. NASA-TM-X-65812, NASA, Greenbelt, Md. [Available from NTIS, Springfield, Va.], 47 pp. (1972).

143.063 Three-dimensional cosmic ray anisotropy in interplanetary space. III. Origin of cosmic ray solar semidiurnal variation.
K. Nagashima, H. Ueno, K. Fujimoto, Z. Fujii, I. Kondo.
Rep. Ionosph. Space Res. Japan, Vol. 26, 1 - 30 (1972).

143.064 Three-dimensional cosmic ray anisotropy in interplanetary space. IV. Origin of solar semi-diurnal variation.
K. Nagashima, K. Fujimoto, Z. Fujii, H. Ueno, I. Kondo.
Rep. Ionosph. Space Res. Japan, Vol. 26, 31 - 68 (1972).

143.065 Current problems of high energy cosmic ray primaries. J. G. Wilson.
Acta Phys. Acad. Sci. Hungaricae, Vol. 32, 83 - 93 (1972).
Problems of measurement of the total energy spectrum of primary cosmic ray particles beyond 10^{17} eV are discussed, with the most recent conclusion that no significant change in the slope of the spectrum at higher energies has yet been established.

143.066 A 'restricted diffusion' model of cosmic-ray propagation. K. Sitte.
Nuovo Cimento Lettere, Ser. 2, Vol. 5, 483 - 489 (1972).
The shape of the energy spectrum of primary cosmic radiation presents a puzzling feature, the abrupt change of slope at about 10^{15} eV. Several explanations have been proposed where the authors put forward a theory of propagation by diffusion by magnetic clouds, in which an energy-dependent disc effect is derived.

143.067 Variations of cosmic ray intensity in consequence of the corotation effect. A. J. Somogyi.
Acta Phys. Acad. Sci. Hungaricae, Vol. 32, 261 - 274 (1972).

143.068 Anisotropy and diffusion of cosmic ray electrons.
C. Y. Mao, C. S. Shen.
Chinese Journ. Phys., Vol. 10, No. 1, p. 16 - 28 (1972). — See Phys. Abstr., Vol. 76, No. 3277 (1973).

143.069 Hard rock cosmic ray archaeology. D. Lal.
Space Science Rev., Vol. 14, 3 - 102 (1972).
The studies are entirely based on the natural detector method which utilises two principal cosmogenic effects observed in rocks, (1) isotopic changes and (2) changes in the crystalline structure of rock constituents, due to cosmogenic interactions. The information available to date in the field of hard rock cosmic ray archaeology refers to meteorites and lunar rocks/soil. Additional information based on study of cosmogenic effects in man-made materials exposed to cosmic radiation in space is also discussed.

143.070 Primaries of extensive air showers of cosmic radiation. B. V. Sreekantan.
Space Science Rev., Vol. 14, 103 - 174 (1972).
The experimental situation concerning the primary cosmic radiation on aspects relating to the primary spectrum, composition, and directional distribution and the evidence for discrete sources as derived from the study of extensive air showers and as it stands after the International Conference on Cosmic Rays held at Hobart in August 1971 are reviewed.

143.071 Gamma astronomy and cosmic rays. I.
V. L. Ginzburg.
Comments Astrophys. Space Phys., Vol. 4, 167 - 172 (1972).
Cosmic γ-ray studies not only add essentially to X-ray astronomy data but also enable us to obtain new information, fundamental for the whole of high-energy astrophysics. Energies of cosmic rays far from the earth are estimated and collisions of the protons and nuclei of cosmic rays with those of the intergalactic or interstellar gas are considered.

143.072 Polarization of extensive air shower emission at 6 MHz. D. G. Felgate, T. J. Stubbs.
Nature, Vol. 239, 151 - 152 (1972).

143.073 A possible cosmic ray primary particle energy spectrum above 10^4 GeV and its astrophysical implications. A. Subramanian.
Proc. Indian Acad. Sci. Sect. A, Vol. 76, No. 3, p. 121 - 128 (1972).

On the transport properties of charged particles in one dimension in random electric fields.
See Abstr. 062.062.

Cosmic-ray evolution due to interactions with self-excited plasma waves. See Abstr. 062.063.

Statistical analysis of Forbush-decreases and foregoing increases of cosmic ray intensity.
See Abstr. 078.007.

Resultados de mediciones con globos estratosfericos. I. (Balloon flights data. I.). See Abstr. 082.232.

Raketenexperiment zur Untersuchung von Nordlichtern. Meßergebnisse des Protonendetektors EI 101.
See Abstr. 084.025.

Daily variation of electron and proton geomagnetic cutoffs calculated for Fort Churchill, Canada.
See Abstr. 084.242.

Argon 37/argon 39 activity ratios in meteorites and the spatial constancy of the cosmic radiation.
See Abstr. 105.002.

Cosmic-ray diffusion coefficient in interplanetary space. See Abstr. 106.007.

Review of results in infrared space astronomy.
See Abstr. 114.044.

On the λ 4686 He II line intensity in H II regions and the cosmic ray flux. See Abstr. 131.010.

On the role of pulsars as cosmic ray accelerators.
See Abstr. 141.538.

Pulsars and the origin of cosmic rays.
See Abstr. 141.550.

The electron component of cosmic rays. II. Radio radiation of relativistic electrons in the Galaxy.
See Abstr. 157.003.

Gamma radiation of Magellanic Clouds and metagalactic origin of cosmic rays. See Abstr. 159.018.

Stellar Systems

151 Kinematics and Dynamics of Stellar Systems

151.001 Phase mixing of the second kind in stellar systems. I. L. P. Ossipkov.
Astrofizika, Vol. 8, 139 - 147 (1972). In Russian. – English translation in Astrophysics, Vol. 8, No. 1.

Phase mixing of the second kind presents a phase analogy with differential rotation. It is manifested in the differences between mean phase velocities of stars for various integral hypersurfaces.

151.002 On the simulation of field stars in numerical experiments. M. Hénon.
Astron. Astrophys., Vol. 19, 488 - 490 (1972).

In some recent numerical experiments, field stars have been simulated inside a sphere as follows: first an initial position is selected on the surface of the sphere, and then the direction of the velocity is chosen randomly, with an isotropic probability distribution. We show that this method is incorrect and produces systematic errors in the results.

151.003 Stability of a one dimensional stellar system. J. P. Dorémus, M. R. Feix.
Astron. Astrophys., Vol. 20, 259 - 262 (1972).

The stability of a double Water-Bag is investigated for plane one dimensional systems. For each Bag, the eigenfrequencies and the eigenfunctions are computed and are used as orthonormal bases for the expansion of the perturbed field. The eigenfrequencies for the total system are subsequently computed and it is found that the system is stable as long as the stars on the inner contour rotate in a shorter time than the stars on the outer contour.

151.004 Galactic evolution: Program and initial results. B. M. Tinsley.
Astron. Astrophys., Vol. 20, 383 - 396 (1972).

Models discussed here were not planned to match any galaxy precisely, but the types of birthrate required for various typical galactic populations, and their overall evolution, are clearly indicated. In the models, stars are formed from interstellar gas, and the evolution of each star is followed from the main sequence to death. Semi-empirical tracks based on old open clusters and old-disk moving groups are used for the later stages of low-mass stars. The models studied to date are homogeneous and contain solar-composition stars only. Metal-poor and super-metal-rich stars can be added to a computed population before its photometric properties are found. Stellar birthrates used include a variety of initial luminosity functions and functions of time or gas mass. These functions provide a set of models with which to survey the potentialities of the method.

151.005 A computer study of galactic spiral arms. N. A. Barricelli, O. Havnes, J. Hemphil, E. Bölviken.
Astrophys. Letters, Vol. 12, 37 - 41 (1972).

We present results obtained in a computer study of the formation of galactic spiral arms. We have adopted the explosion theory, according to which the material that forms the arms is supported to have been ejected from the galactic nucleus.

151.006 The role of gaseous dissipation in density waves of finite amplitude. W. W. Roberts, Jr., F. H. Shu.
Astrophys. Letters, Vol. 12, 49 - 52 (1972).

This letter gives a correct derivation of the dissipative effects of galactic shocks based on the application of Bernoulli's theorem to the problem of the angular momentum interaction. In an approximation which ignores the effects of the convective transport and of the interactions with the basic state, we estimate a value for the 'damping time', $\sim 10^9$ yr, which is comparable to the 'propagation time' for a group of spiral waves.

151.007 The collapse of a massless, non-rotating cloud around a point nucleus. A. E. Wright, K. A. Innanen.
Astron. Astrophys., Vol. 21, 151 - 153 (1972).

The limiting inhomogeneous contraction of a spherical cloud of gas may be represented by the collapse of a massless spheroid of gas around a massive point nucleus. Some relevant possible interpretations of observations in the galactic system are briefly considered.

151.008 Non-linear spiral structure in flat galaxies. A. F. Saaf.
Bull. American Astron. Soc., Vol. 4, 315 - 316 (1972). Abstr. AAS.

151.009 Resonant stellar orbits in spiral galaxies. P. O. Vandervoort.
Bull. American Astron. Soc., Vol. 4, 316 (1972). – Abstr. AAS.

151.010 On a singularity of three-dimensional orbits of stars. G. A. Malasidze.
Soobshch. AN GruzSSR, Vol. 66, 309 - 312 (1972). In Russian. – Abstr. in Referativ. Zhurn. 51. Astron., 9.51.740 (1972).

151.011 Model of a stationary star cluster with high binding energy. G. S. Bisnovatyj-Kogan.
Zhurn. ehksperim. i teor. fiz., Vol. 62, 1593 - 1597 (1972). In Russian. – Abstr. in Referativ. Zhurn. 51. Astron., 10.51.572 (1972).

151.012 On the determination of the type of motion in spherically symmetric systems of galaxies.
I. D. Karachentsev, W. Zonn, A. L. Shcherbanovsky.
Astron. Zhurn. Akad. Nauk SSSR, Vol. 49, 998 - 1009 (1972). In Russian. English translation in Soviet Astron. AJ, Vol. 16, No. 5.

Regression analyses of the radial velocities of galaxies in a spherically symmetric system have been carried out for several models. Comparison of the theoretical regression curves with the empirical ones for the Virgo and Coma clusters and a synthetic group, shows that the suggestion on the instationarity of these systems does not contradict the observational data.

151.013 The solution of the problem of the third integral of motion. II. T. A. Agekjan.
Astron. Zhurn. Akad. Nauk SSSR, Vol. 49, 1127 - 1130 (1972). In Russian. English translation in Soviet Astron. AJ, Vol. 16, No. 5.

151.014 The dynamical evolution of a stellar cluster with initial subclustering. S. J. Aarseth, J. G. Hills.
Astron. Astrophys., Vol. 21, 255 - 263 (1972).

Considerations based on star-formation theory suggest that stellar clusters should initially be composed of a hierarchy of subclusters. To test their influence we computer simulated the evolution of a 120-star cluster having initially such a hierarchy of subclusters.

151.015 Kinetic theory of inhomogeneous rotating self-gravitating systems: Local energy production from binary interactions. M. J. Haggerty, G. Severne.
Physica, Vol. 51, 461 - 476 (1971).

A non-markoffian weak-coupling kinetic theory of binary gravitational interactions in a rotating system is presented. Inhomogeneities along the axis of rotation are included. Irreversible increases in local kinetic energy are found, together with a corresponding growth of local correlations. Additional terms of opposite sign can occur when initial correlations or macroscopic inhomogeneities are included.

151.016 Kinetic theory of inhomogeneous rotating self-gravitating systems: Enhanced collisional diffusion and dissipation. M. J. Haggerty.
Physica, Vol. 51, 477 - 488 (1971).

Kinetic equations are presented for binary interactions in strongly inhomogeneous self-gravitating systems that rotate with uniform angular velocities.

151.017 How are intergalactic filaments made?
M. Clutton-Brock.
Astrophys. Space Sci., Vol. 17, 292 - 324 (1972).

Computer simulations are performed which suggest that narrow intergalactic filaments can be produced by tidal forces. The gas clouds are simulated by a disk of points of zero mass, zero velocity dispersion superimposed on the stellar disk. The 'gas clouds' form remarkably long, narrow filaments; but not so narrow as the narrowest filaments observed.

151.018 The relative merits of galactic density functions: An orbit computational viewpoint.
K. A. Innanen, A. G. Ryman.
Astrophys. Space Sci., Vol. 17, 447 - 452 (1972).

Certain analytic functions used in galactic mass models are reviewed from a somewhat biased orbit-computational viewpoint. On this basis, it is argued that at the present time, an optimal galactic mass model should be a superposition of spheroids with two kinds of density laws.

151.019 Depletion of low-mass stars in clusters.
S. J. Aarseth, N. J. Woolf.
Astrophys. Letters, Vol. 12, 159 - 164 (1972).

Analysis of the luminosity function of Hyades members indicates that the population of faint stars falls below the field star distribution by at most a factor of four in the range $M_v = +5$ to $+15$ mag. The hypothesis of dynamical depletion is tested by numerical integrations of N-body systems which include tidal effects and supernova events. Calculated cluster models containing 250 members show no preferential evaporation of light particles on time-scales comparable to the half-life. Most supernova remnants do not escape until about 80 per cent of the initial members are lost.

151.020 Une méthode de résolution de l'équation de Vlasov. Application à une théorie globale, non-linéaire, de la rotation galactique et de la structure spirale galactique.
J.-P. Petit.
Comptes Rendus Acad. Sci., Paris, Sér. B, Vol. 275, 755 - 758 (1972).

Une méthode particulière de résolution de l'équation de Vlasov est proposée, où la solution se présente comme une somme de fonctions elliptiques couplées par l'intermédiaire du champ gravitationnel. Le modèle de galaxie construit à l'aide de cette solution semble offrir de nombreuses concordances avec les résultats d'observation.

151.021 The kinetics of gravitational clustering.
W. C. Saslaw.
Astrophys. Journ., Vol. 177, 17 - 29 (1972).

Starting from Liouville's theorem, and using the Born-Bogolyubov-Green-Kirkwood-Yvon hierarchy, general expressions are developed for the growth of correlations in isotropic systems of gravitating point particles. These are applied to investigate the early evolution of clusters and density fluctuations in an expanding universe. The results suggest a simple explanation for the observed secondary maxima of the density distributions within clusters of galaxies.

151.022 Shock waves in spiral arms and star formation.
S. B. Pikel'ner.
Comments Astrophys. Space Phys., Vol. 4, 129 - 136 (1972).

Indicators of the spiral structure are early type stars, bright HII regions and dark lanes, but lanes define a regular two-arm spiral best of all. It is possible to divide each of the Sa and Sb galaxies into three distinct zones: the central zone, which has lanes but has no HII; the intermediate belt, which has both lanes and HII; and the external low-luminosity region, which has lanes but again has no HII. This means that a shock wave does not always lead to star formation. One of the reasons may be the effect of the stellar background, which is much denser than the gas in the central regions of galaxies. Stars stabilize gravitational instability.

151.023 Approximate calculation of the magnitude of impulse at a material point's passing by a prolate homogeneous ellipsoid of revolution. K. F. Ogorodnikov.
Vestn. Leningr. un-ta, 1972, No. 7, p. 146 - 149. In Russian. Abstr. in Referativ. Zhurn. 51. Astron., 11.51.596 (1972).

151.024 On the disintegration of star clusters. P. Bouvier.
Astron. Astrophys., Vol. 21, 441 - 442 (1972).

By fitting a simple analytical law to some former numerically computed results for a 25-solar mass stellar cluster, a value is obtained for the disruption time of the cluster submitted to three disintegration causes: its inner evolution, the galactic tidal field and the influence of passing interstellar clouds.

151.025 The stability of certain model binary stellar systems in galactic gravitational fields.
K. A. Innanen, A. E. Wright, F. C. House, D. Keenan.
Monthly Notices Roy. Astron. Soc., Vol. 160, 249 - 253 (1972).

Binary stars and relatively massive binary galaxies are well established in the hierarchies of objects found in the sidereal Universe. In this note we wish to consider the questions 'Why apparently, are there no well-documented binary globular clusters (BGC's) in the Galaxy or external systems? ' and 'Can systems such as the Magellanic Clouds and NGC 185 and NGC 147 exist as physical binaries in proximity to more massive galaxies? '

151.026 A new solution to the accretion problem.
R. A. Lyttleton.
Monthly Notices Roy. Astron. Soc., Vol. 160, 255 - 270 (1972).

The Bondi-Hoyle theory of line-accretion is briefly reviewed, and it is shown that there are an infinity of steady-state solutions satisfying the requirements originally imposed as boundary conditions. Questions are raised as to the applicability of these solutions to physical reality. The existence is demonstrated of an entirely different type of steady-state solution that implies slow velocity in the accretion-stream beyond the neutral point. Accurate numerical values are obtained of this solution for a number of cases. The braking-action given by this solution, for any selected cut-off distance, is far stronger than on the Bondi–Hoyle solution. With stronger braking-action, the velocity of stars relative to interstellar material will be reduced in correspondingly shorter times, and more rapid accretion than on the Bondi–Hoyle theory will thereby come about. This result combined with recent measures of gas-densities at various regions of the Galaxy suggests that the problem posed by the brightest stars may well be resolvable on the basis of the accretion-process.

151.027 On the stability of strongly non-homogeneous self-gravitating equilibria.   S. Cuperman, A. Harten.
Astrophys. Space Sci., Vol. 18, 207 - 222 (1972).

The stability analysis of several strongly non-homogeneous, self-gravitating, one-dimensional unstable equilibrium systems is carried out with the help of numerical techniques. The evolution of the 'perturbed' unstable equilibria is studied by following the motion of the boundary curves of 'water bag' configurations defining the systems. It is found that initial perturbations drive the unstable equilibrium states out of equilibrium at rates depending on the typical scale length of the perturbations λ: the instability rates increase with λ.

151.028 Computer experiments on the structure and dynamics of spiral galaxies.   F. Hohl.
National Aeronautics and Space Administration, Washington, D.C. NASA Techn. Note, TN D–6630, 50 pp. (1972).

The evolution of an initially balanced rotating disk of stars with an initial velocity dispersion given by Toomre's local criterion is investigated by means of a computer model for isolated disks of stars. It is found that the disk is unstable against very large-scale modes. After about two rotations the central portion of the disk tends to assume a bar-shaped structure. A stable axisymmetric disk with a velocity dispersion much larger than that given by Toomre's criterion is generated. The final mass distribution for the disk gives a high-density central core and a disk population of stars that is closely approximated by an exponential variation. Various methods and rates of cooling the hot axisymmetric disks were investigated. It was found that the cooling resulted in the development of two-arm spiral structures which persisted as long as the cooling continued. An experiment was performed to induce spiral structure in a galaxy by means of the close passage of a companion galaxy. Parameters similar to those expected for M 51 and its companion were used.

151.029 Effects of evolution on the diameter-redshift relation.   B. M. Tinsley.
Astrophys. Journ., (Letters), Vol. 178, L39 - L42 (1972).

The relation between redshift and metric angular diameters is unaffected by evolution of the light of galaxies, but the relation using isophotal diameters is subject to an evolutionary correction. At small redshifts, the required correction to the indicated value of q_0 is identical to that required in use of the magnitude-redshift relation. An apparent discrepancy, approximately equal to that found between Sandage's preliminary value ($q_0 \sim 1$) from isophotal diameters and Baum's value ($q_0 = 0.3$) from metric diameters, is predicted.

151.030 Star distribution near a collapsed object.
P. J. E. Peebles.
Astrophys. Journ., Vol. 178, 371 - 375 (1972).

The purpose of this paper is to give a derivation of the steady-state distribution of stars near a massive collapsed object (such as a black hole) in a star cluster.

151.031 On the third integral of motion in stellar dynamics.
I.   J. S. Stodółkiewicz.
Acta Astron., Vol. 22, 375 - 386 (1972).

The third integral of motion in dynamical systems with axial symmetry is treated as a function of position variables and an inclination angle. For further analysis the Fourier expansion of this integral is used. The potentials with a third integral quadratic in the velocities are discussed. A rich class of potentials having such an integral for at least one family of orbits is found.

151.032 On the correctness of models of stellar systems.
I. L. Genkin, L. M. Genkina.
Trudy Astrofiz. Inst., Alma-Ata, Vol. 19, 3 - 7 (1972). In Russian.

The problem of correctness of models of stellar systems when integral equations of the first kind are applied is considered. Examples of the instability of solutions with regard to the initial conditions are given.

151.033 On the theory of open spirals.   I. L. Genkin.
Trudy Astrofiz. Inst., Alma-Ata, Vol. 19, 17 - 23 (1972). In Russian.

Dynamics of wave propagation in an axisymmetrical non-infinitesimally thin disk is considered. It is shown that such waves can exist in Lindblad's resonance region only and that only arms having a logarithmic spiral form are stable in the radial direction.

151.034 On the stability of very oblate galaxies.
O. V. Chumak.
Trudy Astrofiz. Inst., Alma-Ata, Vol. 19, 24 - 30 (1972). In Russian.

In first approximation the stability of a model having fundamental properties of real stellar systems is considered. This model was found to be unstable in intermediate frequencies; in a limit case necessary conditions of stability are fulfilled.

151.035 On some problems of particle dynamics integrated by means of Stäckel's theorem.
Yu. I. Ivanov, V. K. Kajsin.
Trudy Astrofiz. Inst., Alma-Ata, Vol. 19, 110 - 114 (1972). In Russian.

151.036 The origin of rotation of galaxies according to non-linear theory of gravitational instability.
A. G. Doroshkevich.
Astron. Zhurn. Akad. Nauk SSSR, Vol. 49, 1221 - 1228 (1972). In Russian. English translation in Soviet Astron. AJ, Vol. 16, No. 6.

The problem of the formation of vortex velocities and the origin of the rotation of cosmic objects is considered according to the non-linear theory of gravitational instability.

151.037 The spectrum of eigenfrequencies of a gravitating cylinder of free particles with finite radius.
G. S. Bisnovaty-Kogan.
Astron. Zhurn. Akad. Nauk SSSR, Vol. 49, 1238 - 1248 (1972). In Russian. English translation in Soviet Astron. AJ, Vol. 16, No. 6.

The spectrum of oscillations of an infinitely long, gravitating homogeneous cylinder of free particles with finite radius is calculated. The regions of instabilities for stream and cyclotronic modes and the eigenfrequencies and increments for different values of parameters are found.

151.038 About a model of star clusters with axial symmetry and homogeneous stellar composition.
V. M. Bagin.
Astron. Zhurn. Akad. Nauk SSSR, Vol. 49, 1249 - 1257 (1972). In Russian. English translation in Soviet Astron. AJ, Vol. 16, No. 6.

A model of star clusters with axial symmetry and homogeneous stellar composition in which in general the average circular velocity is different from zero is investigated. Two types of continuous series of solutions are possible.

151.039 Galactic bridges and tails.
   A. Toomre, J. Toomre.
Astrophys. Journ., Vol. 178, 623 - 666 (1972).
   This paper argues that the bridges and tails seen in some multiple galaxies are just tidal relics of close encounters. These consequences of the brief but violent tidal forces are here studied in a deliberately simple-minded fashion: Each encounter is considered to involve only two galaxies and to be roughly parabolic; each galaxy is idealized as just a disk of noninteracting test particles which initially orbit a central mass point. Besides extensive pictorial surveys of tidal damage, this paper offers reconstructions of the orbits and outer shapes of four specific interacting pairs: Arp 295, M51 + NGC 5195, NGC 4676, and NGC 4038/9. Also discussed are some closely related issues of eccentric bound orbits, orbital decay, accretion, and forced spiral waves.

151.040 Regularization in the N-body problem.
   S. J. Aarseth.
Bull. American Astron. Soc., Vol. 4, 417 (1972). – Abstr. AAS.

151.041 Numerical integration of the N-body problem.
   A. Ahmad, L. Cohen.
Bull. American Astron. Soc., Vol. 4, 417 (1972). – Abstr. AAS.

151.042 Random force in gravitational systems.
   L. Cohen, A. Ahmad.
Bull. American Astron. Soc., Vol. 4, 418 (1972). – Abstr. AAS.

151.043 The method of doubly individual step for N-body computations.   A. Hayli.
Bull. American Astron. Soc., Vol. 4, 420 (1972). – Abstr. AAS.

151.044 A variable order method for the numerical integration of the gravitational N-body problem.
G. Janin.
Bull. American Astron. Soc., Vol. 4, 420 (1972). – Abstr. AAS.

151.045 Numerical difficulties with the gravitational N-body problem.   R. H. Miller.
Bull. American Astron. Soc., Vol. 4, 421 (1972). – Abstr. AAS.

151.046 Integration errors and their effects on macroscopic properties of N-body systems.   H. Smith, Jr.
Bull. American Astron. Soc., Vol. 4, 422 (1972). – Abstr. AAS.

151.047 On the numerical integration of the N-body problem for star clusters.   R. Wielen.
Bull. American Astron. Soc., Vol. 4, 423 (1972). – Abstr. AAS.

151.048 About a possible mechanism of disintegration of gravitating systems.   T. B. Omarov.
Vestn. AN KazSSR, 1972, No. 7, p. 55 - 61. In Russian.
Abstr. in Referativ. Zhurn. 51. Astron., 1.51.674 (1973).

151.049 The investigation of the galactic spiral structure.
   B. Barbanis.
Techn. Chronika, No. 2, p. 63 - 69 (1972).   In Greek.
   The optical and radio methods used for the study of the spiral arms, and the observational data are presented. The problems concerning the spiral structure, the main theories proposed and, finally, some computer experiments simulating the evolution of a system containing many thousands of particles, are examined.

151.050 Tidal interactions in multiple galaxies.
   M. Clutten-Brock.

Thesis, Univ. Washington, Seattle. [Available from Univ. Microfilms, Ann Arbor, Mich., USA. Order No. 72-7330], 239 pp. (1971).

151.051 Self-gravitating gaseous disks.   C. Hunter.
   Annual Rev. Fluid Mechanics, Vol. 4, 219 - 242 (1972).
   This review is largely devoted to various dynamical investigations inspired by astronomical phenomena associated with spiral galaxies.

151.052 An investigation of stellar orbits of a star cluster with allowance for the disturbing force of the Galaxy.   R. M. Dzigvashvili.
Byull. Abastumansk. Astrofiz. Obs., No. 43, p. 223 - 246 (1972). In Russian.
   The present paper deals with the investigation of stellar orbits of the open cluster NGC 6067. This cluster is assumed to be situated in the galactic plane moving with circular velocity around the galactic center. On the basis of an analysis of the computed orbits and the first integral of motion some conclusions are suggested on the characteristics of motions of stars in clusters.

151.053 Some observational consequences of the model of galaxy formation from adiabatic perturbations.
A. D. Narits.
Astron. Tsirk., No. 732, p. 5 - 7 (1972). In Russian.

151.054 On the third integral of motion in stellar dynamics.
   I.   J. S. Stodółkiewicz.
Polish Acad. Sci., Inst. Astron. Warsaw, Preprint No. 15, 19 pp. (1972).

Single close encounters in the planetary problem.
See Abstr. 042.039.

Application of von Zeipel's method to the stellar three-body problem.   See Abstr. 042.068.

The stability of a self-gravitating, nonrotating gas layer with stellar, magnetic, and cosmic-ray components. I. See Abstr. 061.003.

Thermal equilibrium states of a classical system with gravitation.   See Abstr. 061.038.

Gravitational plasmas. II.   See Abstr. 062.070.

The main sequences of synthetic clusters with finite formation times.   See Abstr. 153.018.

Evolution of galactic clusters due to dissipation. See Abstr. 153.019.

The origin and form of the galactic magnetic field. II. The primordial-field model.   See Abstr. 156.001.

On the gas content of galaxies.
See Abstr. 158.021.

The radio emission of NGC 4258 and the possible origin of spiral structure.   See Abstr. 158.067.

On the wave nature of rotation curves of galaxies.
See Abstr. 158.114.

Formation of clusters of galaxies; protocluster fragmentation and intergalactic gas heating.   See Abstr. 162.005.

152 Stellar Associations

152.001 Possible new members of a subgroup of the Scorpio-Centaurus association. J. W. Glaspey.
Astron. Journ., Vol. 77, 474 - 485 (1972).

Photoelectric *uvby* and Hβ photometric observations are presented for 117 A- and F-type stars in the region of the sky covered by the Scorpio-Centaurus association. The data are combined with previously published observations of B-type stars in the same region to discuss the possible membership of fainter stars in two subgroups of the association.

152.002 Metallicism in border regions of the Am domain. I. Extremely young Am stars in the Orion 1 c association. M. A. Smith.
Astrophys. Journ., Vol. 175, 765 - 777 = Contr. Lick Obs., No. 365 (1972).

A search for metallic-line stars by means of 127 and 63 Å mm^{-1} spectrograms and uvby photometry has resulted in the discovery of five such stars in the extremely young Orion 1 c association. A number of arguments in favor of their membership are made. It appears that Am stars in the pre-main-sequence-contraction phase occupy the same domain of the H-R diagram as do evolving Am stars. Evidence is also presented that in open clusters the characteristic rotation of stars, and not age directly, is important in determining the probability that an A star will become metallic lined.

152.003 On some peculiarities of faint stars in the Orion I association. V. N. Sincheskul.
Astron. Tsirk., No. 677, p. 5 - 7 (1972). In Russian.

152.004 Internal motions in the association Cep OB 3. C. D. Garmany.
Bull. American Astron. Soc., Vol. 4, 316 (1972). – Abstr. AAS.

152.005 Note on the Aquila stellar ring. B. L. Webster.
Observatory, Vol. 92, 143 - 145 (1972). – Note.

152.006 An investigation of A-type star groupings in Perseus and Cassiopeia. G. F. Kevanishvili.
Soobshch. AN GruzSSR, Vol. 67, No. 1, p. 65 - 68 (1972). In Russian. – Abstr. in Referativ. Zhurn. 51. Astron., 10.51.569 (1972).

152.007 Stellar kinematics in the Sco OB 1 association. A. Laval.
Astron. Astrophys., Vol. 21, 271 - 278 (1972). In French.

From an earlier paper (H II regions in the Sco OB 1 association) we could deduce that the gas belonging to this association is clumping into two subgroups. We show that the same subgroups appear among the stars: two stellar groups can be distinguished which behave differently in kinematics. These subgroups coincide with the gaseous subgroups.

152.008 Observations of the extremely young stellar group Lk Hα 224 and 225. W. Wenzel.
Inform. Bull. Variable Stars (I.A.U. Commission 27), Konkoly Obs., Budapest, No. 713 (1972).

152.009 On the reality of a group of carbon stars in Auriga. H. B. Richer, S. Sharpless, B. Campbell.
Astrophys. Journ., Vol. 177, 515 - 518 (1972).

The properties of a group of seven carbon stars whose projected positions make them appear clumped together are examined. It is concluded that the seven do not form a physically real group. The possibility is discussed that there are two pairs of carbon stars within this group.

152.010 Three remarkably similar groups of stars in the southern sky. N. J. Rumsey.
Southern Stars, Vol. 24, 130 - 132 (1972).

152.011 A T association in Chamaeleon. E. E. Mendoza V.
Publ. Astron. Soc. Pacific, Vol. 84, 641 - 642 (1972). Abstr. Astron. Soc. Pacific.

152.012 Photometric observations of stellar ring No. 284. T. W. Rettig.
Publ. Astron. Soc. Pacific, Vol. 84, 673 - 676 = Publ. Goethe Link Obs., Indiana Univ., *Bloomington*, No. 140 (1972).

UBV photometry and objective-prism spectral classifications of stars in stellar ring No. 284 indicate the possibility of the member stars being at a common distance and age, but not at a distance consistent with the stellar ring theory.

152.013 A spectroscopic study of the OB association III Cepheus. C. D. Garmany.
Thesis, Univ. Virginia, Charlottesville. [Available from Univ. Microfilms, Ann Arbor, Mich., USA. Order No. 72-7186], 86 pp. (1971).

A spectroscopic and astrometric study of the OB association III Cepheus, one of the youngest within one kiloparsec of the sun, has been carried out, and the results are compared with those associations studied by Blaauw, Van Albada and others.

152.014 On the variability of Hα emission stars in the Scorpius-Ophiuchus association. V. Satyvaldiev.
Astron. Tsirk., No. 728, p. 5 - 7 (1972). In Russian.

Photographic R magnitudes of 228 stars in Orion. See Abstr. 113.033.

Photographic photometry of T Tauri stars and related objects in Orion. See Abstr. 113.038.

Comparative UBVR photometry of Orion flare stars and Hα emission-line stars. See Abstr. 113.062.

A search for He-weak stars in very young clusters. See Abstr. 114.016.

On circumstellar gas emission among pre-main-sequence stars in Ic Orionis and NGC 2264. See Abstr. 114.071.

A young stellar group surrounding RCrA. See Abstr. 131.042.

Zur Struktur der näheren Sonnenumgebung. Untersuchungen über Sterntrupps, Sternfamilien und Sternströme. See Abstr. 155.042.

Errata

152.901 Errata:'Infrared and optical observations of a young stellar group surrounding BD+40°4124' [Astrophys. Journ., (*Letters*), Vol. 173, L65 - L70 (1972)].
K. M. Strom, S. E. Strom, M. Breger, A. L. Brooke, J. Yost, G. Grasdalen, L. Carrasco.
Astrophys. Journ., (*Letters*), Vol. 176, L93 (1972).

153 Galactic Clusters

153.001 The red giants in the Hyades group. O. J. Eggen.
Publ. Astron. Soc. Pacific, Vol. 84, 406 - 419
(1972).

The present note presents the results of an examination of all the late-type giants in the Catalogue of Bright Stars with the aim of obtaining a census of all possible members of the Hyades group near the sun.

153.002 A search for short-period light variability among stars in h and χ Persei. J. R. Percy.
Publ. Astron. Soc. Pacific, Vol. 84, 420 - 423 = Commun. David Dunlap Obs., Univ. Toronto, *Richmond Hill,* No 329 (1972).

Thirty-nine stars in the nuclei of the galactic clusters h and χ Per have been tested for short-period light variability.

153.003 Three-colour photometry of NGC 1647.
S. M. Hassan.
Mem. Soc. Astron. Italiana, Nuova Ser., Vol. 43, 279 - 289 (1972).

A new photometric investigation of the open cluster NGC 1647 in three colours is presented using plates of the Kottamia 74"-reflector (Egypt). Arguments about membership in the cluster and comparisons with the Naval Observatory results are included.

153.004 NGC 2818, an open cluster containing a planetary nebula.
W. G. Tifft, L. P. Connolly, D. F. Webb.
Monthly Notices Roy. Astron. Soc., Vol. 158, 47 - 62 (1972).

Three-colour UBV photometry of the cluster NGC 2818 shows it to lie at a distance of 3.2 kpc. The cluster has a colour excess $E_{B-V} = 0.22$ and an ultraviolet excess of 0.09. The evolutionary turnoff from the main sequence occurs in mid A. The faint main sequence in the cluster shows a distinct bend or discontinuity. Preliminary velocity measurements indicate that the high excitation planetary nebula NGC 2818 is probably associated with the cluster.

153.005 The cluster NGC 330 in the SMC (Paper II): Hα emission in main sequence stars. M. W. Feast.
Monthly Notices Roy. Astron. Soc., Vol. 159, 113 - 118 (1972).

An Hα survey of the cluster NGC 330 in the Small Magellanic Cloud has been carried out. Eighteen stars which are main sequence objects, according to Arp's photometry, were observed and ten found to be Be stars. Three other stars without photometry are probably Be stars. All eight main sequence stars investigated with M_v in the range −3.6 to −4.2 are Be stars. The significance of these results is discussed.

153.006 On the possible existence of different populations in the double cluster h and χ Persei.
P. Galeotti, G. Silvestro, E. Trussoni.
Atti Accad. Nazionale Lincei, Ser. 8, Rend. Cl. Sci. fis., mat., nat., Vol. 51, 228 - 233 (1971).

We discuss the absolute motion of the individual supergiants and the stellar content of the cluster, taking into account new observational data.

153.007 On circumstellar molecules in the Pleiades.
L. M. Hobbs.
Astrophys. Journ., (*Letters*), Vol. 175, L145 - L147 (1972).

Both old and new observations of the interstellar λ4232 line of CH⁺ are considered for the brightest members of the Pleiades. They suggest that the molecules are circumstellar in some sense, perhaps resembling in this respect the micron-

sized grains inferred to be present in this region.

153.008 A study of the galactic cluster NGC 6866.
L. S. Koroleva.
Astron. Zhurn. Akad. Nauk SSSR, Vol. 49, 786 - 795 (1972). In Russian. English translation in Soviet Astron. AJ, Vol. 16, No. 4.

Problems of the existence of a corona around the open stellar cluster NGC 6866 and also the superposition of background stars over the cluster are investigated. Ninety-three members of this cluster, located in its outer space, were surely detected according to two criteria. This may speak in favour of the existence of a corona in this stellar area. Lists of 92 additional cluster members within the cluster region and 93 members of the cluster in its outer areas are given. The colour-magnitude and the vector diagrams for these stars are shown on figures.

153.009 The moderately old open cluster, NGC 752.
R. F. Garrison.
Journ. Roy. Astron. Soc. Canada, Vol. 66, 216 (1972). Abstr. Canadian Astron. Soc.

153.010 Membership in the extremely young open cluster NGC 6530 (M8). W. F. van Altena, B. F. Jones.
Astron. Astrophys., Vol. 20, 425 - 436 = Lick Obs. Bull., No. 623 (1972).

Relative proper motions for 363 stars in the vicinity of the extremely young open cluster NGC 6530 (M8) have been determined from plates taken with the Yerkes Observatory 40-inch refractor. Based on these relative proper motions probabilities of membership are determined with the result that of the 363 stars measured there are 76 expected members. Herschel 36, the exciting star for the "hour glass" nebula, is possibly a cluster member. A distance of 1780 pc and an age of 2×10^6 yr has been determined.

153.011 Possible near-by open clusters. 1.
I. N. Latyshev.
Astron. Tsirk., No. 681, p. 2 - 4 (1972). In Russian.

153.012 On the metal abundance for members of stellar clusters. A. E. Vasilevskij.
Astron. Tsirk., No. 693, p. 1 - 3 (1972). In Russian.

153.013 Infrared studies of the young clusters IC 2944, NGC 6611 and Orion 1.
R. F. Knacke, S. E. Strom, K. M. Strom, E. Young.
Bull. American Astron. Soc., Vol. 4, 325 (1972). – Abstr. AAS.

153.014 On the metallicity of M67.
D. C. Barry, R. H. Cromwell.
Bull. American Astron. Soc., Vol. 4, 326 (1972). – Abstr. AAS.

153.015 Note sur les vitesses radiales des étoiles d'un amas galactique en direction du Grand Nuage de Magellan.
C. Fehrenbach, M. Duflot.
Observatory, Vol. 92, 145 - 146 (1972). – Note.

153.016 Southern open star clusters I. UBV–Hβ photometry of 15 clusters between galactic longitudes 231° and 256°. N. Vogt, A. F. J. Moffat.
Astron. Astrophys., Suppl. Ser., Vol. 7, 133 - 167 (1972).

Fifteen clusters in this region were selected from the Catalogue of Star Clusters and Associations (Ruprecht, Alter and Vanýsek 1970) with special emphasis placed on obtaining distant young clusters. Two-color and color-magnitude diagrams from the photoelectric *UBV* photometry for each cluster are discussed. The cluster parameters are summarized and reveal that seven young clusters are probable spiral arm tracers out to 4.2 kpc. Possible member super-giants were found in each of four of the clusters. Interstellar reddening is very small in most of the region and nearly independent of distance from one to four kpc.

153.017 An investigation of four southern open clusters. U. Lindoff.
Astron. Astrophys., Suppl. Ser., Vol. 7, 231 - 256 (1972).

Magnitudes and colours have been determined on the *UBV* system for stars in the southern open clusters NGC 5138, NGC 6134, NGC 6208, and IC 4651. The investigation has been based on a combination of photoelectric and photographic photometry. Spectral classes for some of the brighter stars have been determined from objective-prism plates. Slit spectra for MK classification have been available for a few stars. Dimensions, foreground absorptions, distances, and ages have been determined. The main results are shown in a table.

153.018 The main sequences of synthetic clusters with finite formation times. B. M. Schlesinger.
Astron. Journ., Vol. 77, 584 - 589 = Publ. Goethe Link Obs., Indiana Univ., *Bloomington*, No. 141 (1972).

Color-magnitude diagrams have been synthesized for clusters whose stars formed in a period of 15 million years. The average cluster ages range from 12 to 600 million years. In addition, H−R diagrams have been synthesized for the two youngest of these clusters.

153.019 Evolution of galactic clusters due to dissipation. V. S. Kaliberda.
Astron. Zhurn. Akad. Nauk. SSSR, Vol. 49, 1026 - 1032 (1972). In Russian. English translation in Soviet Astron. AJ, Vol. 16, No. 5.

The equations describing the evolution of star clusters due to dissipation have been deduced. The parameters of a cluster are given in terms of time. The proposed method has been applied to the Pleiades.

153.020 Photometric observations of the star cluster NGC 2141.
M. S. Burkhead, R. D. Burgess, B. M. Haisch.
Astron. Journ., Vol. 77, 661 - 665, 701 - 702 = Publ. Goethe Link Obs., Indiana Univ. *Bloomington*, No. 142 (1972).

Photoelectric and photographic observations are presented for the open cluster NGC 2141. The data indicate the cluster is of late intermediate age with a main-sequence turnoff at M_v = 3.5, $(B-V)$=0.40. A mean color excess of $E_{(B-V)}$ = 0.30 and a distance modulus of $m-M$ = 14.1 yield (R = 3.0) a distance of 4.4 kpc.

153.021 Photometry of eleven young open star clusters. A. F. J. Moffat.
Astron. Astrophys., Suppl. Ser., Vol. 7, 355 - 383 (1972).

Photographic *UBV* magnitudes to V~17ᵐ were obtained for all stars contained in regions defined by the open clusters: NGC 6823, NGC 6830, NGC 6834, NGC 7235, NGC 457, NGC 581, NGC 654, NGC 663, IC 1805, IC 1848 and NGC 1893. Magnitudes of field stars were measured in additional regions of similar area located outside and symmetrically surrounding each cluster. Comparison of the cluster and field areas made it possible to select with high probability individual bright cluster members of spectral type earlier than B 7−9. A derivation of accurate distances and the recalibration of the absolute magnitudes and colours of the probable member supergiants were made.

153.022 Untersuchungen über die Struktur junger Sternhaufen und die Entwicklungsphasen ihrer Mitglieder.

II. NGC 6530. W. Götz.
Astron. Nachr., Vol. 294, 9 - 22 (1972).

Statistical investigations of the open star cluster NGC 6530 serve to complete, enlarge and test the cosmogonic and genetic relations found in the cluster NGC 2264. The comparison of the results from both clusters offers a possibility to decide which of the statements are generally or specifically valid for each cluster respectively. The structure of the star cluster and the behaviour of the astrophysical parameters in the system were investigated in detail.

153.023 Untersuchungen über die Struktur junger Sternhaufen und die Entwicklungsphasen ihrer Mitglieder.
III. NGC 6611. W. Götz.
Astron. Nachr., Vol. 294, 23 - 28 (1972).

Investigations on the structure of the open cluster and the behaviour of the astrophysical parameters in the system show that the structure of the cluster is closely connected with the continuous formation and the evolution of its members. A time scale allowing statements on the age of certain regions and on the evolutionary phases of the stars was made up.

153.024 The determination of distance, absorption, probable physical members and age for the open clusters Haffner 8, Haffner 6, Basel 11 and NGC 2374.
R. P. Fenkart, R. Buser, H. Ritter, H. Schmitt, H. Steppe, R. Wagner, D. Wiedemann.
Astron. Astrophys., Suppl. Ser., Vol. 7, 487 - 496 (1972).

We determined distance, absorption, probable physical members and age of four open clusters with the help of three-colour photometric methods in the *UBV* system. Long-wave $(V, B - V)$ and short-wave $(V, U - B)$ colour-magnitude diagrams yield the photometric data of the clusters. No interstellar absorption has been observed in this directions in the range of distances covered by these clusters (1.13 to 1.67 kpc). None of these clusters can be considered as a spiralarm indicator since their earliest spectral-types are all later than b2. They yield "photometric" ages between 2.0 and 7.9×10^8 years.

153.025 The old open cluster NGC 6819. U. Lindoff.
Astron. Astrophys., Suppl. Ser., Vol. 7, 497 - 513 (1972).

An investigation of the old open cluster NGC 6819 has been made on the *UBV*-system, using a combination of photographic and photoelectric photometry. The colour excess, E_{B-V}, has been determined to 0ᵐ3 and the distance to 2200 pc. The cluster contains about 60 giant members, the brightest of which is found to be variable. NGC 6819 is one of the oldest open clusters known. The age has been estimated to 2×10^9 years. Comparisons have been made with other old clusters. About 20 stars, situated in the surroundings of the cluster and with spectral classes around A0, have been measured in order to get some information about the colour excesses in front of the cluster.

153.026 Some characteristics of the moderately old open cluster NGC 752. R. F. Garrison.
Astrophys. Journ., Vol. 177, 653 - 656 (1972).

MK classifications are presented for 35 of the F-type stars in the cluster NGC 752. One star, Heinemann 193, has spectral peculiarities similar to those in some δ Scuti stars. Differential comparisons of line strength between NGC 752 and the Hyades are described, and a rediscussion of previous photometric results is presented. No convincing evidence is found, from either the photometry or the classifications, for line weakening.

153.027 Photographic photometry of NGC 6866. B. Hidajat, W. Sutantyo.
Proc. Inst. Teknologi Bandung, Vol. 6, 89 - 103 = Contr.

Bosscha Obs., *Lembang*, No. 44 (1972).

Magnitudes and colors in the *UBV* photometric system have been determined for 575 stars in and around NGC 6866. The color-color relation of the cluster stars provides an estimate of interstellar reddening of 0.16 mag. The distance of cluster is found to be 1336 pc. The age, as determined from the evolutionary deviation curve, is approximately 2.5×10^8 years.

153.028 On the evidence of a gap in the main sequences of open clusters.   I. Mazzitelli.
Mem. Soc. Astron. Italiana, Nuova Ser., Vol. 43, 541 - 543 (1972). – Letter.

153.029 An old open cluster containing a BaII star.
R. D. McClure.
Bull. American Astron. Soc., Vol. 4, 424 (1972). – Abstr. AAS.

153.030 Galactic cluster NGC 7762.   P. E. Zakharova.
Astron. Tsirk., No. 740, p. 6 - 8 (1972). In Russian.

Metal abundances in the atmospheres of red giants, in open clusters and dynamical groups.   See Abstr. 064.071.

Construction of isochrones by interpolation. Age determination for some open clusters.
See Abstr. 065.078.

Experimental trigonometric parallaxes in the Hyades region using a Schmidt telescope of moderate size.
See Abstr. 111.001.

Radial velocities of some stars in NGC 1893.
See Abstr. 112.001.

Proper motions of field M stars in the region of the Pleiades.   See Abstr. 112.003.

Infrared photometry of the Pleiades lower main sequence.   See Abstr. 113.043.

A direct photometric comparison of M 67 and Hyades stars.   See Abstr. 113.044.

UBV photometry of some selected stars in the Hyades.   See Abstr. 113.047.

Photographic photometry of stars in the region of the open cluster NGC 6871 in the Vilnius photometric system. Part I.   See Abstr. 113.049.

Four-color and Hβ photometry for open clusters. VIII. IC 4665.   See Abstr. 113.053.

UBV and Hβ photometry of the galactic cluster NGC 2343.   See Abstr. 113.054.

A search for He-weak stars in very young clusters.
See Abstr. 114.016.

The metal deficiency of two members of M67.
See Abstr. 114.076.

Über die Veränderlichkeit der Sterne des Spektraltyps A.   See Abstr. 114.106.

A spectroscopic study of the peculiar stars in the open cluster NGC 2516.   See Abstr. 114.113.

Preliminary study on early-type supergiants in h and χ Per.   See Abstr. 114.151.

Red supergiants and neutrino emission. II.
See Abstr. 115.003.

The mass-luminosity relationship of the Hyades and other stars.   See Abstr. 115.023.

Pulsating variables in the Pleiades cluster.
See Abstr. 122.042.

Main-sequence pulsation in open clusters.
See Abstr. 122.043.

New flare stars in the Pleiades region (a re-examination of the Tonantzintla photographic material: 1963–1970). See Abstr. 122.070.

New flare stars in the Pleiades region (1971–1972). See Abstr. 122.071.

Flare stars in the Pleiades region observed during the fall and winter 1971–1972.   See Abstr. 122.128.

A re-examination of the Pleiades Tonantzintla photographic material: 1963 - 1970. II.   See Abstr. 122.136.

Flare stars in the Pleiades region (1971–1972). III.
See Abstr. 122.137.

Flare stars in the Pleiades. III.
See Abstr. 122.142.

Elements of light variation of the RR Lyrae type variables V 166 and SVS 1374 in the globular cluster M3.
See Abstr. 122.144.

Interstellar absorption of light and the distribution of stars around the star cluster NGC 6834.
See Abstr. 131.123.

Interstellar absorption of light and the distribution of stars around the star cluster NGC 7654.
See Abstr. 131.124.

Nebulae of the southern Milky Way. An atlas.
See Abstr. 132.003.

On the association of the nebula M8 and the cluster NGC 6530.   See Abstr. 132.018.

Depletion of low-mass stars in clusters.
See Abstr. 151.019.

The stability of certain model binary stellar systems in galactic gravitational fields.   See Abstr. 151.025.

An investigation of stellar orbits of a star cluster with allowance for the disturbing force of the Galaxy.
See Abstr. 151.052.

The distribution of stars and obscuring matter in a Monoceros field.   See Abstr. 155.013.

The variation of the K3–M2 main sequence with space motion.   See Abstr. 155.023.

Theoretical isochrones for disk population stars.
See Abstr. 155.024.

Zur Struktur der näheren Sonnenumgebung. Untersuchungen über Sterntrupps, Sternfamilien und Sternströme.
See Abstr. 155.042.

Electronographic photometry of star clusters in the   NGC 419.   See Abstr. 159.005.
Magellanic Clouds—IV. The colour—magnitude diagram of

154 Globular Clusters

154.001 **On the evolution of globular clusters.**
J. P. Ostriker, L. Spitzer, Jr., R. A. Chevalier.
Astrophys. Journ., (*Letters*), Vol. 176, L51 - L56 (1972).
The evolution of globular clusters is dominated by dynam- ical relaxation, produced by two-body encounters, and by gravitational shocks, which occur whenever a cluster passes through the galactic plane. Observational data are used to eval- uate these two effects.

154.002 **Variation of the nitrogen abundance among popula- tion II objects.**   F. D. A. Hartwick, R. D. McClure.
Astrophys. Journ., (*Letters*), Vol. 176, L57 - L59 (1972).
Intermediate-band photometry, on the DDO system, of individual giant stars in the anomalous globular cluster NGC 7006 confirm the weak-metal-lined character exhibited by integrated light observations but indicate anomalously strong cyanogen bands.

154.003 **Four-color photometry of blue horizontal-branch stars in NGC 3201.**   A. G. D. Philip.
Bull. American Astron. Soc., Vol. 4, 325 (1972). – Abstr. AAS

154.004 **Period-luminosity relation for cepheids in globular clusters.**   B. V. Kukarkin, A. S. Rastorguev.
Astron. Tsirk., No. 707, p. 7 - 8 (1972). In Russian.

154.005 **A search for ionized hydrogen in globular clusters.**
A. G. D. Philip, J. W. Erkes.
Bull. American Astron. Soc., Vol. 4, 399 (1972). – Abstr. AAS.

154.006 **Upper limits on the atomic hydrogen abundance in 12 globular clusters.**   F. J. Kerr, G. R. Knapp.
Astron. Journ., Vol. 77, 573 - 576 (1972).
A search was made with the NRAO 140-ft telescope for 21-cm line emission from 12 globular clusters, with negative results. These observations yielded limits on the atomic hydro- gen content ranging from 11 $M_\odot$ for M 4 to 85 $M_\odot$ for M 15, and gave an average value for the galactic H I gas-to-dust ratio in the directions of the clusters of 1.6×10^{21} atom cm^{-2} mag^{-1}.

154.007 **The axial ratio and position angle of the major axis of the globular cluster M 92.**
W. Högner, Z. I. Kadla, N. Richter, A. A. Strugatskaya.
Astron. Zhurn. Akad. Nauk. SSSR, Vol. 49, 1033 - 1036 (1972). In Russian. English translation in Soviet Astron. AJ, Vol. 16, No. 5.
The method of equidensity curves is used for determin- ing the axial ratio and position angle of the major axis at various angular distances from the center of the globular clus- ter M 92.

154.008 **On the structure of the globular cluster M 4.**

V. Castellani, F. A. D'Antona, A. Natta.
Astrophys. Space Sci., Vol. 17, 23 - 29 (1972).
On the basis of the work by Greenstein on the globular cluster M4, the spatial distribution for groups of cluster stars in different stages of evolution is investigated. The comparison between the various samples is made using the apparent distri- butions. The results are discussed.

154.009 *UBV* **photometry of the metal-rich globular cluster NGC 6171.**   R. J. Dickens, A. Rolland.
Monthly Notices Roy. Astron. Soc., Vol. 160, 37 - 62 (1972).
Three-colour (*UBV*) photographic photometry of stars in the globular cluster NGC 6171 is reported. The present study of the C—M and two-colour ($U-B, B-V$) diagrams for the brighter stars in the cluster was undertaken to improve the quality of the earlier C—M diagram and to investigate the ultra-violet excesses of stars in various regions of the C—M diagram. Points of particular interest are the intrinsic widths of various sequences in the C—M diagram, the existence or otherwise of an asymptotic giant branch and the question of whether these stars in a metal-rich cluster show a smaller ultra- violet excess than the subgiants, as is found in metal-poorer clusters.

154.010 **Search for globular clusters near the centre of the Andromeda nebula.**   A. S. Sharov.
Astron. Tsirk., No. 714, p. 4 - 5 (1972). In Russian.

154.011 **On the cosmological origin of globular clusters.**
T. V. Ruzmaikina.
Astron. Zhurn. Akad. Nauk SSSR, Vol. 49, 1229 - 1237 (1972). In Russian. English translation in Soviet Astron. AJ, Vol. 16, No. 6.
A hypothesis of the cosmological origin of globular clusters is developed. The possibility of a fragmentation of initial clouds into stars due to thermal instability is investigated for a wide spectrum of initial conditions.

154.012 **The R-method for determining the helium abun- dance of globular clusters.**
P. Demarque, A. V. Sweigart, P. G. Gross.
Nature, Phys. Sci., Vol. 239, 85 - 87 (1972).
Theory predicts that the ratio R of the number of hori- zontal branch stars to the number of red giants above the hori- zontal branch luminosity level is a function of helium abun- dance. In this letter, it is pointed out that including the effects of semi-convection on the horizontal branch affects the cali- bration of Y vs R in the sense of decreasing the helium abun- dance of the globular clusters to Y $\simeq$ 0.13. Other methods yield Y $\simeq$ 0.30. It is suggested that the discrepancy could be caused by: 1) incorrect star counts; 2) an underestimate of the critical core mass at the helium flash due to faulty neutrino

loss rates or due to the neglect of core rotation; 3) erroneous opacities.

154.013 An apparent discontinuity in the properties of giants of the disk population.
F. D. A. Hartwick, J. E. Hesser.
Publ. Astron. Soc. Pacific, Vol. 84, 813 - 815 (1972).

By comparing the color-magnitude diagram of the metal-rich globular cluster 47 Tuc with that of bright field giants with the same $\delta(U - B)$, we conclude that there are at least two different groups of giants that have the same line blanketing in the UV. We suggest that a parameter other than age or abundance of elements heavier than the CNO group may be responsible for the observed discontinuity.

154.014 Red giants in NGC 6752. O. J. Eggen.
Astrophys. Journ., (*Letters*), Vol. 178, L109 - L110 (1972).

The previously published observations of red giants in this cluster refer, for the fainter stars, mainly to asymptotic-branch stars. New data have been obtained for stars on the subgiant sequence.

The application of objective gratings to faint star photometry. See Abstr. 113.025.

A second CH star in Omega Centauri and the C^{12}/C^{13} ratio. See Abstr. 114.045.

Carbon stars in the globular cluster Omega Centauri. See Abstr. 114.054.

On the luminosity functions of giant stars in globular clusters. See Abstr. 115.014.

RR Lyrae type variables in the globular cluster NGC 5466. See Abstr. 122.007.

On five variable stars in the vicinity of the globular cluster NGC 5466. See Abstr. 122.008.

Photometry of RR Lyrae variables in the globular cluster NGC 6981. See Abstr. 122.017.

Photometry and spectroscopy of red variables in Omega Centauri. See Abstr. 122.044.

S 10764 – a slowly variable object in the globular cluster M3 with $U-B \approx -1\overset{m}{.}0$. See Abstr. 123.047.

On the metal abundance for members of stellar clusters. See Abstr. 153.012.

Galactic kinematical parameters from star clusters. See Abstr. 155.030.

A spectroscopic study of the blue globular clusters in the Large Magellanic Cloud. See Abstr. 159.006.

155 Structure and Evolution of the Galaxy

155.001 High energy gamma radiation from the galactic centre region.
R. Browning, D. Ramsden, P. J. Wright.
Nature, Vol. 238, 138 - 139 (1972).

Balloon observations of gamma-ray sources near the galactic centre are described. These show good correlation with known X-ray sources. No diffuse emission was observed; gamma-rays apparently originate from point sources.

155.002 Recombination line emission from the galactic ridge.
R. D. Davies, H. E. Matthews, A. Pedlar.
Nature, Phys. Sci., Vol. 238, 101 - 103 (1972).

An excess of ionized hydrogen is found to lie in the spiral structure between the sun and the galactic centre. Its electron temperature is 6,000 K, in contradiction to previous determinations.

155.003 The equilibrium configuration of the gaseous component of the Galaxy.   S. A. Kellman.
Astrophys. Journ., Vol. 175, 353 - 362 (1972).

This paper discusses, among other topics, two observational findings relating to the equilibrium state of the galactic gas disk: (i) the run of gas density ρ_g with distance above and below the galactic plane, and (ii) the rapid increase in half-thickness of the gas layer at distances from the galactic center greater than the solar distance, and its relatively uniform thickness at distances between about 4 kpc and 10 kpc. To compute the equilibrium state of the gas disk, the Poisson equation and the hydrostatic-equilibrium equations for the gaseous and stellar components of a two-fluid, static, plane-parallel mixture are used. Each component is assumed to be isothermal, but each has its own characteristic "temperature". The analysis includes one-dimensional magnetic and cosmic-ray components, each constrained to vary directly with $\rho_g(z)$.

155.004 UBV and Hβ photometry of OB stars in a Milky Way region in Norma.   J. S. Drilling.
Astron. Journ., Vol. 77, 463 - 474, 529 = Contr. Louisiana State Univ. Obs., *Baton Rouge*, No. 67 (1972).

Ninety-nine OB stars, 10 of which show Hα in emission, have been identified by means of an objective-prism survey in a region of 20 square degrees centered near $l=330°$, $b=-2°$. Hβ and UBV photometry has been obtained for these stars and the color excesses and distance moduli of 96 of the stars determined therefrom.

155.005 A balloon search for extraterrestrial high energy γ-rays in the northern hemisphere.
M. Niel, G. Vedrenne, A. Claverie, R. Bouigue.
Astron. Astrophys., Vol. 20, 1 - 8 (1972).

We present the results of 5 balloon flights intended for the research of possible sources of γ-rays above 60 MeV in the northern hemisphere, and of a possible anisotropy of the γ-ray flux detected towards the galactic plane.

155.006 Structure and motions in the Carina spiral feature.
R. M. Humphreys.
Astron. Astrophys., Vol. 20, 29 - 48 (1972).

The distribution and kinematics of the luminous young stars in the Carina spiral feature ($l = 282° - 305°$) are investigated and compared with the available data for the neutral hydrogen gas in the same region. The space distribution of these luminous stars reveals a well-defined spiral feature from 1.5 to 6 kpc from the sun.

155.007 A study of the interstellar extinction in the Carina-Centaurus region.   L. O. Lodén, A. Sundman.
Astron. Astrophys., Vol. 20, 49 - 53 (1972).

An attempt is made to estimate the general distribution of interstellar obscuring matter, particularly the behaviour of the colour excess, in the Carina-Centaurus region of the Milky Way from photoelectric UBV observations of a selection of 1000 stars and to find some applicable value of the coefficient R for some particular regions.

155.008 Kinematics of molecular clouds near the galactic center.   N. Z. Scoville.
Astrophys. Journ., (*Letters*), Vol. 175, L127 - L132 (1972).

A model is proposed in which most of the molecular clouds near the galactic center are situated in a radially moving ring of radius 250 pc.

155.009 Distribution of O to B9 stars and dust in a region in Puppis ($l = 245°$, $b = -0.2°$).
W. J. F. Wilson, M. P. FitzGerald.
Journ. Roy. Astron. Soc. Canada, Vol. 66, 217 (1972).
Abstr. Canadian Astron. Soc.

155.010 Was wissen wir über die Entstehung der Milchstraße?   L. Martinet.
Umschau, 72. Jahrgang, p. 595 - 597 (1972).

Investigations of the motions of stars serve as a clue for the structure and the development of the Galaxy. Interrelationship between the chemical composition and the motion at one hand and the age of stars on the other hand have been detected.

155.011 Diffuse galactic FUV radiation and interstellar dust grains.   S. Yoshioka.
Astrophys. Space Sci., Vol. 16, 421 - 431 (1972).

A re-analysis of the diffuse far UV radiation ($\lambda\lambda$ 1350–1480 Å) observed in the sky region of $l^{II} \approx 180°$ and $0° \lesssim b^{II} \lesssim 40°$ is presented, as a revised version of a paper by Hayakawa *et al.* (1969). In comparison with the previous one, the value of the half optical depth of the Galaxy τ in our wavelength region is reduced, and the values of the albedo coefficient γ and the forward phase function g are not well determined. If, however, we combine our results with the theoretical model of interstellar grains by Gilra, the value of τ is given by $0.13(5) \lesssim \tau \lesssim 0.18(5)$.

155.012 Distribution of the diffuse component of cosmic soft X-rays.   T. Kato.
Astrophys. Space Sci., Vol. 16, 478 - 498 (1972).

The celestial distribution of the intensity of diffuse soft X-rays is mapped by reference to a critical review of observations available. The soft X-rays of extragalactic origin are subtracted, taking the interstellar absorption into account. The distribution of the galactic component thus obtained shows irregularities, but the general behaviour can be accounted for in terms of a cylindrical distribution of the X-ray emissivity with the scale height of $100 \sim 300$ pc in the direction perpendicular to the galactic plane.

155.013 The distribution of stars and obscuring matter in a Monoceros field.   B. Karlsson.
Astron. Astrophys., Suppl. Ser., Vol. 7, 35 - 81 (1972).

Space densities for O–F7 type stars are derived for a Monoceros region ($l^{II} = 203°$, $b^{II} = +2°$) and the behaviour of the interstellar extinction is obtained. The present results together with those by McCuskey and Bok and collaborators form the basis for a discussion of the distribution of stars and interstellar matter in the longitude section 185° to 215°. The open cluster NGC 2264, which is situated within the region, is shortly discussed.

155.014 **Space density of stars.**
L. Kh. Esojan, Z. F. Seidov.
Astron. Tsirk., No. 687, p. 6 - 8 (1972). In Russian.

155.015 **On a galactic origin for the soft X-ray background.**
P. Gorenstein, W. H. Tucker.
Astrophys. Journ., Vol. 176, 333 - 344 (1972).
The results of a new observation of the soft X-ray background in the Virgo region with high angular resolution are reported. Using results from this and previous observations, we derive the following observational features for the background: (a) the intrinsic emission spectrum of the background is quite soft, for an exponential spectrum $KT \lesssim 0.1$ keV, (b) the ratio of the pole to plane flux decreases with energy below 0.28 keV, and (c) the diffuse flux in the Virgo region has a granularity of ~ 14 percent in regions of 4.5 square degrees.

155.016 **An approximate form of the third integral in the Galaxy.** L. Perek.
Bull. Astron. Inst. Czechoslovakia, Vol. 23, 246 - 262 (1972).
An approximate form of the third integral, based on a potential separable in eccentric elliptic coordinates, is proposed for the Galaxy at the distance of the sun and for velocity components up to 30 km/sec. The approximation is accurate to about 10^{-3}. Formulae are given for the third integral, for the perigalactic and apogalactic distances, for the maximum height of the orbit above the plane of symmetry, for the periods in z, in R, and in ϑ. No resonances seem to appear. The influence of the value of the integral of areas is discussed.

155.017 **A survey of linear polarization at 1415 MHz. IV. Discussion of the results for the galactic spurs.**
T. A. T. Spoelstra.
Astron. Astrophys., Vol. 21, 61 - 84 (1972).
The results of a survey of linear polarization at 1415 MHz in the galactic spurs are discussed. The investigation is a continuation of that on the lower part of the North Polar Spur given in paper I (Spoelstra, 1971). Van der Laan's model of a shell expanding into the interstellar medium gives a reasonable explanation of the observed continuum features and polarization characteristics.

155.018 **On the possible existence of different interstellar extinction laws in the spiral arms and in the field regions.** M. Créze.
Astron. Astrophys., Vol. 21, 85 - 89 (1972).
The ratio of total to selective extinction is derived from a comparison of spectroscopic and photometric data with the kinematical parallaxes computed from radial velocities, through a model of galactic differential rotation. H II regions, population I cepheids and open clusters are used for this purpose.

155.019 **An H I velocity–longitude diagram for the southern Milky Way.** G. R. Knapp.
Astron. Astrophys., Vol. 21, 163 - 165 (1972).
An observational velocity–longitude diagram derived from 21-cm hydrogen-line observations from longitude 190° to 12°, taken with the Parkes radio telescope, is presented.

155.020 **A simple interpretation of the "rolling" motions of the spiral arms in the outer parts of the Galaxy.**
C. Yuan, L. Wallace.
Bull. American Astron. Soc., Vol. 4, 316 (1972). – Abstr. AAS.

155.021 **Is the mass of the galactic halo greater than the mass of the disc?** J. P. Ostriker, P. J. E. Peebles.
Bull. American Astron. Soc., Vol. 4, 316 - 317 (1972). Abstr. AAS.

155.022 **A model of the local region of the Galaxy.**
J. C. Novaco, N. R. Vandenberg.
Bull. American Astron. Soc., Vol. 4, 318 (1972). – Abstr. AAS.

155.023 **The variation of the K3–M2 main sequence with space motion.** A. R. Upgren.
Bull. American Astron. Soc., Vol. 4, 325 - 326 (1972). Abstr. AAS.

155.024 **Theoretical isochrones for disk population stars.**
P. Demarque, G. Gisler.
Bull. American Astron. Soc., Vol. 4, 326 - 327 (1972). Abstr. AAS.

155.025 **A new calculation of the average interstellar radiation density in the 1250 to 4250 Ångstrom region.**
A. N. Witt, M. W. Johnson.
Bull. American Astron. Soc., Vol. 4, 327 - 328 (1972). Abstr. AAS.

155.026 **A high-resolution high-sensitivity map of the galactic center region.**
J. E. Kapitzky, W. A. Dent.
Bull. American Astron. Soc., Vol. 4, 328 (1972). – Abstr. AAS.

155.027 **Distribution of soft X-rays over a selected region of the sky.** F. O. Williamson, A. N. Bunner, P. L. Coleman, W. L. Kraushaar, D. McCammon.
Bull. American Astron. Soc., Vol. 4, 335 (1972). – Abstr. AAS.

155.028 **Observation of the soft X-ray structure of the Vela-Puppis region with a focusing collector.**
P. Gorenstein, P. Bjorkholm, B. Harris, F. R. Harnden, Jr., H. Gursky.
Bull. American Astron. Soc., Vol. 4, 336 (1972). – Abstr. AAS.

155.029 **The X-ray structure of the Vela Puppis region observed from UHURU.** E. Kellogg, F. R. Harnden, Jr., H. Tananbaum, H. Gursky, S. Murray, R. Giacconi, J. Grindlay.
Bull. American Astron. Soc., Vol. 4, 336 (1972). – Abstr. AAS.

155.030 **Galactic kinematical parameters from star clusters.**
W. Buscombe.
Observatory, Vol. 92, 141 - 142 (1972). – Note.

155.031 **Proton synchrotron emission as an explanation of the electromagnetic spectrum of the galactic center.**
V. De Sabbata, P. Fortini, C. Gualdi.
Astrophys. Letters, Vol. 12, 87 - 89 (1972).
A model is proposed to explain the main features of γ, infrared, and radio emission from the galactic center, on the basis of hydrogen ionization in a very strong magnetic field and the resulting proton synchrotron radiation.

155.032 **A survey of linear polarization at 1415 MHz. III. Method of reduction and results for the galactic spurs.** T. A. T. Spoelstra.
Astron. Astrophys., Suppl. Ser., Vol. 7, 169 - 230 (1972).
In this paper 1415 MHz observations are presented of linear polarization of the galactic background radiation in the direction of the North Polar Spur, the Cetus Arc and Loop III. The reduction of the observations is discussed; the discussion includes a comparison of the instrumental effects of two feedhorns. The results are presented in tabular form.

155.033 **The spiral wave of our Galaxy near inner Lindblad resonance.** J. W-K. Mark.

Proc. National Acad. Sci., *U.S.A.*, Vol. 68, 2095 - 2098 (1971).

The dispersion relationship for short-wavelength spiral density waves in our Galaxy has been refined to remove the divergences that occurred in wave number and in amplitude as inner Lindblad resonance is approached. The wave is found to be evanescent in an annular region near 4 kpc. By 3 kpc, the inward propagating trailing wave is completely absorbed. The outgoing leading wave is suppressed compared to the trailing one because it begins in the evanescent state. Throughout this region of inner Lindblad resonance, a smooth wave amplitude has been obtained, and it has a sharp peak correlating well with the observed density of ionized hydrogen.

155.034　**The distribution of early type stars in Circinus (l = 316°9; 327°9. b = 2°8; −5°9) along the galactic equator.**　F. Oyen.
Astron. Astrophys., Vol. 21, 315 - 316 (1972).

In four regions of Circinus 249 OB-stars were selected. For one of these regions the density distribution as a function of the distance was drafted based upon photographic magnitudes of 59 OB-stars. It points to the existence of a spiral arm.

155.035　**About Kerr's galactic expansion and Weber's gravitation signals.**　P. Jordan.
Zeitschr. Physik, Vol. 233, 84 - 88 (1970). In German.

The expansion of the Galaxy seeming to be detected by Kerr can probably not be explained as resulting from a rapid diminution of the central mass of the Galaxy. But it follows as a consequence from the scalar-tensor theory of gravitation (as developed by Jordan, Thiry, Brans-Dicke) if the value $\dot{f}/f = -\epsilon$, as taken from the difference between ephemeris time and inertial time, has the correct order of magnitude. The events in the nucleus of the Galaxy, triggering the Weber signals, must probably be collisions between neutron stars and normal stars.

155.036　**Distribution of O to B9 stars and dust in a region in Puppis (l = 245°, b = −0°2).**
W. J. F. Wilson, M. P. FitzGerald.
Journ. Roy. Astron. Soc. Canada, Vol. 66, 254 - 260 = Contr. Univ. Waterloo Obs., No. 16 (1972).

An 8.43 square degree region of the sky located in Puppis has been studied to determine the space distribution of O to B9 stars and dust. It is found that the region is relatively dust free, and that the space density of O to B5 stars is lower than found at neighbouring galactic longitudes. The OB associations Puppis OB1 (II Puppis) and Puppis OB2 (I Puppis) are clearly visible in the O to B5 star data. The space density of B7 to B9 stars remains approximately constant, at about 40 stars per 10^6 pc³ out to 2500 pcs from the sun, with a probable decrease beyond this.

155.037　**Velocity structures in hydrogen profiles.**
M. A. Tuve, S. Lundsager.
Astron. Journ., Vol. 77, 652 - 660 (1972).

An atlas based on more than 5000 hydrogen Doppler profiles has been made ready for publication as a Carnegie Institution Monograph. Discussion of this material reaffirms our reluctant earlier conclusion that interpreting conspicuous Doppler peaks and trends as hydrogen concentrations in spiral arms is not justified or tenable, because it assumes an extremely regular rotation curve. We feel that Doppler-velocity structures usually referred to as the Scutum, Sagittarius, Perseus, and other arms should not be interpreted as confirmed concentrations of neutral hydrogen in our Galaxy.

155.038　**Observations of the general background Hα emission from the galactic plane.**　J. Meaburn.
Astrophys. Space Sci., Vol. 17, 499 - 502 (1972).

A mosaic of photographs, very sensitive to variations in brightness of the diffuse Hα emission 10° fields has now been compiled that reveals particularly well the structure of the

many large areas of this background emission from the less confused region of the anticentre of the galactic plane.

155.039　**The role of Schmidt telescopes in the study of galactic structure. (Photometric methods).**
W. Becker.
The role of Schmidt telescopes in astronomy. Conference Hamburg 1972, (see 012.014), p. 9 - 29 (1972).

155.040　**Galactic structure at high galactic latitudes.**
A. G. D. Philip.
The role of Schmidt telescopes in astronomy. Conference Hamburg 1972, (see 012.014), p. 117 - 126 (1972).

155.041　**Wat zijn de Galactische Spurs?**　T. A. T. Spoelstra.
Hemel en Dampkring, Vol. 70, 281 - 286 (1972).

155.042　**Zur Struktur der näheren Sonnenumgebung. Untersuchungen über Sterntrupps, Sternfamilien und Sternströme.**　J. Hopmann.
Sitzungsber. Österreich. Akad. Wiss., Math.-naturwiss. Kl., Abt II, Vol. 180, 247 - 297 (1971) = Astron. Mitt. Wien, No. 9 (1972).

1.) Introduction, earlier work on wide pairs, multiple stars, star-troops and star-streams; 2.) The two troops near 107 Aqr, the Schütte-family IV a 4, the R.A. Fisher-criterion; 3.) The moving cluster in PsA and the family IIIi3; 4.) Revision of the work of K. Schütte; 5.) The reality of the 74 families of K. Schütte; 6.) The family of the sun; 7.) Extreme stars, eight hyperbolic orbits relative to the Galaxy; 8.) Statistical studies on all Schütte-stars; 9.) The twelve star-streams; 10.) Discussion, a model of the near surroundings of the sun.

155.043　**Galactic models and stellar orbits.**　J. Einasto.
Tartu Astron. Obs. Teated, No. 40, 47 pp. (1972).
Invited report, IAU Regional Meeting, Athens, September 1972.

Methods of determination of composite models of galaxies are outlined. The available observational information renders it possible to distinguish the following populations in nearby galaxies: nucleus, kernel, core, bulge, halo, disc, young population. Parameters of models of six galaxies are presented (our Galaxy, M31, M32, M87, and dwarf galaxies in Fornax and Sculptor). Recent developments in derivation of stellar orbits and kinematical characteristics of various stellar populations in galaxies are summarized. Models of physical and dynamical evolution of galaxies are discussed.

155.044　**Spiral structure and kinematics of the Galaxy from a study of the H II regions. Fabry–Perot interference methods applied to ionized hydrogen.**　G. Courtès.
Vistas in astronomy, Vol. 14, (see 003.008), 81 - 161 (1972).

155.045　**Kinematics of a sample of very red stars at the south galactic pole.**　W. Gliese.
Astron. Astrophys., Vol. 21, 431 - 433 (1972).

From the proper motion distribution of 75 very red stars near the south galactic pole a mean group parallax of about 0."04 is derived. The objects either form a cloud in that region or they are representative of the star density of red dwarfs in the solar neighbourhood. This density is considerably larger than was previously thought. The latter conclusion agrees with similar observations near the north galactic pole.

155.046　**Galactic rotation and second order terms.**
L. G. Taff, J. E. Littleton.
Astron. Astrophys., Vol. 21, 443 - 446 (1972).

An analysis of the motion of objects in the Galaxy to second order in r/R_0 yields a mathematically complete set of equations for the galactic rotation parameters. We have applied this second order analysis to the galactic clusters in the catalogue of Hagen (1970). We also made a reexamination of Humphreys' (1970) work on galactic rotation using supergiants. After analyzing the supergiants in the same manner as the galactic clusters we find that her first order constant, Oort's A, is inconsistent with the value of A obtained from the galactic clusters and other classes of objects.

155.047 A kinematic determination of the ratio of total to selective absorption.   R. A. Bell, M. P. FitzGerald.
Proc. Third Colloquium on Astrophysics. Supergiant Stars, Trieste 1971, p. 168 - 177 = Contr. Univ. Waterloo Obs., No. 11 (1972).

A kinematic method for the determination of the ratio of total to selective absorption, R, Oort's constant, A, and the solar motion, including second order terms, is presented. The method has been applied to 130 Cepheids, 60 galactic clusters, and 454 supergiants. The results do not support values of R much in excess of 3.5.

155.048 Some comments on "The recombination line emission from the galactic ridge".
S. T. Gottesman, M. A. Gordon, with a reply by R. D. Davies, A. Pedlar, H. E. Matthews.
Nature, Phys. Sci., Vol. 240, 160 - 161 (1972).

Davies et al. have reported observations of radio recombination lines from regions of the galactic plane chosen to be free of discrete radio sources. These observations support and considerably extend measurements made earlier with other instruments. Here we comment upon the analysis of the Jodrell Bank measurements. In particular, we point out possible difficulties in assumptions underlying the conclusion that the lines are generated by a gas of electron temperature of 6,200 K and suggest that the temperature of the line-emitting gas might still be quite uncertain.

155.049 On a possible interstellar galactic chromosphere.   D. W. Sciama.
Nature, Vol. 240, 456 - 457 (1972).

I suggest here that the disk of the Galaxy is surrounded by a shell of gas at a temperature intermediate between the 10^2 - 10^4 K of the disk and the 10^6 K of the corona, and that this "chromosphere" may be detectable. This suggestion is related to the hypothesis that the Galaxy is accreting hot gas from the intergalactic space of the Local Group.

155.050 Observations of the soft X-ray background.   A. Davidsen, S. Shulman, G. Fritz, J. F. Meekins, R. C. Henry, H. Friedman.
Astrophys. Journ., Vol. 177, 629 - 642 (1972).

Observations of the diffuse X-ray background over a hemisphere of the sky in the galactic anticenter direction provide clear evidence for an excess intensity below 1 keV in all directions. The correlation of this low-energy flux with galactic latitude and with hydrogen column density implies that the radiation originates partially within and partially outside the absorbing layers of the Galaxy. The data are consistent with a model combining an isotropic extragalactic flux with a component due to emitters distributed in the same way as the galactic hydrogen. If discrete sources are responsible for the intensity observed near the galactic plane, then a high density of low-luminosity objects is consistent with our data.

155.051 Il nucleo galattico.   L. Rosino.
Accad. Nazionale Lincei, Quaderno No. 164, 15 pp. = Oss. Astron. Padova, Comun. Rassegne, No. 90 (1972). – Presented at the ordinary meeting 1971 March 13.

155.052 Molecular clouds in the galactic center region: Carbon monoxide observations at 2.6 millimeters.
P. M. Solomon, N. Z. Scoville, K. B. Jefferts, A. A. Penzias, R. W. Wilson.
Astrophys. Journ., Vol. 178, 125 - 130 (1972).

A preliminary CO emission line survey covering a strip at $b = -2'$ from $l = 359°.7$ to $l = 2°.8$ is presented which shows a continuous band of emission connecting the region between Sgr A and Sgr B. A high-resolution map of the Sgr A cloud near the galactic center shows that there are at least two clouds centered within 3' of each other with a velocity difference of 35 km s^{-1}. Measurement of the $^{13}C^{16}O$ and $^{12}C^{18}O$ emission indicates isotopic abundances similar to those of the solar system.

155.053 The far-infrared and submillimeter background.
J. R. Houck, B. T. Soifer, M. Harwit, J. L. Pipher.
Astrophys. Journ., (*Letters*), Vol. 178, L29 - L33 (1972).

We have repeated our earlier observations of the infrared and submillimeter background radiation. While the measured values of the infrared background radiation remain unchanged, we have failed to observe the high flux previously reported for the 0.4 to 1.3 mm range. This indicates that the flux cannot have been galactic or cosmic, but further observations are needed to rule out a solar-cycle-dependent geocoronal origin.

155.054 Observations of the outer spiral structure of the Milky Way and its relation to the high velocity clouds.   R. D. Davies.
Monthly Notices Roy. Astron. Soc., Vol. 160, 381 - 406 (1972).

New observations have been made of the neutral hydrogen emission from the outer spiral structure of the Milky Way. A number of new arms with densities about 1 per cent of the well-known spiral arms have been discovered at distances of 15 - 20 kpc from the galactic centre. It is suggested that the tilt in the galactic plane and the asymmetrical distribution of the high velocity clouds is produced by a close passage of the Large Magellanic Cloud. A few high velocity clouds having velocities incompatible with galactic rotation are also believed to be a consequence of this interaction.

155.055 The dynamics of M stars in the Southern Galactic Cap.   D. H. P. Jones.
Astrophys. Journ., Vol. 178, 467 - 476 (1972).

The radial velocities of HD M stars in the Southern Galactic Cap have been measured with a photoelectric speedometer. Their velocity dispersion is 7 km s^{-1} in the plane, rises to 28 km s^{-1}, and thereafter remains steady. Using conventional dynamical theory, the density variation of these stars with distance below the plane implies a density of gravitating matter of 0.21 $M_\odot$ pc^{-3} in the solar neighborhood. The distribution of total energies of motion perpendicular to the plane accords with the collapse picture of the Galaxy.

155.056 On the activity of the galactic nucleus.
G. M. Idlis, V. A. Semenenya.
Trudy Astrofiz. Inst., *Alma-Ata*, Vol. 19, 31 - 35 (1972). In Russian.

Taking into consideration the limitation of material and energetic resources of the galactic nucleus, we hold that the observed intense escape of hydrogen from the central zone is not constant and has a recurrent character.

155.057 Diffuse glow of a two-layer galaxy with carbon-silicon particles.   D. A. Rozhkovskij, V. S. Matyagin.
Trudy Astrofiz. Inst., *Alma-Ata*, Vol. 19, 52 - 63 (1972). In Russian.

The transfer of diffuse radiation in the Galaxy is considered. The galactic model consists of a stellar population and dust which form two homogeneous plane-parallel layers with different linear thickness. Equations of diffuse radiative transfer in the far UV and in the visual long-wave regions are solved numerically. The parameters of the models are chosen on the assumption that the particles have a graphite-silicon structure, and their properties in the spectral regions considered conform with the data by Wickramasinghe (1970). The obtained numerical values of the diffuse radiation intensity are compared with the calculations of van de Hulst and de Jong (1969) for finding differences which are brought into the intensity calculation results by choosing one or another shape of the indicatrix.

155.058 **A numerical solution for the equation of transfer of diffuse galactic radiation in two-layer planeparallel models.** V. S. Matyagin, D. A. Rozhkovskij.
Trudy Astrofiz. Inst., *Alma-Ata*, Vol. 19, 64 - 71 (1972).
In Russian.

A method for the numerical solution for the equation of transfer of diffuse radiation in two-layer plane-parallel models of the Galaxy is considered. A block-diagram of the program for the computer BESM-3M and results for models of the Galaxy with different parameters are given.

155.059 **Distribution of O to A0 stars in a region in Vela** $(l = 268°5, b = -0°3)$.
J. E. Stegman, M. P. FitzGerald.
Journ. Roy. Astron. Soc. Canada, Vol. 66, 303 - 309 = Contr. Univ. Waterloo Obs., No. 20 (1972).

Spectral types, V magnitudes, and B—V colour indices have been obtained for 470 stars of types O to A0 in a 15.24 square degree region in Vela. Analysis of this material has confirmed the existence of a region of high interstellar absorption centered about 1000 pcs from the sun. Space densities of O to B3 stars, of B4 to B5 stars, and of B8 to A0 stars are given.

155.060 **Epicyclic motion of stars at different regions of the Galaxy.** A. K. Ray.
Stud. Cerc. Astron., Vol. 17, 235 - 243 (1972).

Sizes of the epicycles at different distances from the galactic centre have been discussed. It is found that the sizes increase with the distance from the centre indicating that mixing of stars of larger volume takes place as we go away from the centre.

155.061 **Statistical population indices of *M* type dwarfs.**
W. Iwanowska.
Stud. Soc. Sci. Torunensis, Sectio F (Astron.), Vol. 5, 1 - 7 = Bull. Astron. Obs. Toruń No. 49/I (1972).

Statistical population indices for 202 *dM* stars and 105 *dMe* stars were calculated from the *U, V, W* galactic velocity components taken from the Catalogue of Nearby Stars by Gliese (1969). The parameter values for the two population types are given, the resulting statistical population indices and population class designations are presented, and the frequencies of population classes among *dM* and *dMe* stars are shown.

155.062 **Detection of a gamma ray spectral line from the galactic center region.**
W. N. Johnson, III, R. C. Haymes.
Bull. American Astron. Soc., Vol. 4, 412 (1972). — Abstr. AAS.

155.063 **Anomalous velocity hydrogen near the galactic center.** I. F. Mirabel, K. C. Turner.
Bull. American Astron. Soc., Vol. 4, 413 - 414 (1972). — Abstr. AAS.

155.064 **Further evidence for a radial composition gradient in the Galaxy.** T. E. Lutz, K. M. Yoss.
Bull. American Astron. Soc., Vol. 4, 424 (1972). — Abstr. AAS.

155.065 **Continuum radio structure of the Galaxy.**
R. M. Price.
Bull. American Astron. Soc., Vol. 4, 424 (1972). — Abstr. AAS.

155.066 **270 pc expanding ring at the galactic centre.**

N. Kaifu, T. Kato, T. Iguchi.
Nature, Phys. Sci., Vol. 238, 105 - 107 (1972).

An expanding and rotating ring of radius of about 270 pc is found by analyses of OH, H_2CO and NH_3 radio lines in the galactic center region. Expansion and rotation velocities are 130 km/sec and 50 km/sec, respectively. Total mass of the ring is estimated to be 10^8 to 10^9 $M_{\odot}$ and it gives the kinetic energy of expansion of 10^{55} to 10^{56} ergs. Existence of a contracting and rotating ring with a radius of 140 pc and contracting velocity of 40 km/sec is also shown. These rings are probably caused by a strong shock wave propagating in the galactic disk outward, formed by an explosion of the galactic center about 10^6 years ago.

155.067 **A search for optical pulses from the galactic centre.**
G. A. Baird, T. J. Delaney, B. G. Lawless.
Observatory, Vol. 92, 233 - 234 (1972). — Note.

155.068 **Radiations from the center of the Galaxy.**
G. Vedrenne.
Sci. Progr. Découverte, No. 3443, p. 33 - 41 (1972). In French.

155.069 **Continuum radio structure of the galactic disk.**
R. M. Price.
Quarterly Progr. Rep. Res. Lab. Electron. Mass. Inst. Technology, No. 104, p. 80 - 82 (1972).

155.070 **The galactic structure in Cepheus near NGC 7235.**
W. R. Kubineo.
Thesis, Case Western Reserve Univ., Cleveland, Ohio. [Available from Univ. Microfilms, Ann Arbor, Mich., USA. Order No. 72-6311], 168 pp. (1971).

Data from a catalogue of objective prism spectral types, photographic V magnitudes and B-V color indices for approximately 1500 stars brighter than V = 13.0 are used to study the galactic structure in an eight square degree region, centered at l = 103°, b = 0° (new galactic coordinates are used throughout).

155.071 **Studies in galactic structure.** Y. K. Minn.
Thesis, Rensselaer Polytechn. Inst., Troy, N.Y.
[Available from Univ. Microfilms, Ann Arbor, Mich., USA. Order No. 72-1206], 193 pp. (1971).

The original data of 'The Maryland-Green Bank Galactic 21-cm Line Survey' are analyzed. Neutral hydrogen distributions in our Galaxy are examined and best fit models are derived.

155.072 **Gravitational waves from the center of the Galaxy.**
S. Persides.
Technika Chronika, (Greece), No. 8, p. 751 - 755 (1972).
In Greek.

One of the main consequences of Einstein's general theory of relativity, the existence of gravitational waves, is examined from a theoretical as well as experimental viewpoint. The linearized version of the field equations is used to support the claim that gravitational waves exist.

155.073 **The equilibrium and stability of the gaseous component of the Galaxy. 4.** S. A. Kellman.
National Aeronautics Space Administration, Goddard Space Flight Center, Greenbelt, Rep. X-661-71-458, 13 pp. (1971).

Galactic rotation and the precession constant.
See Abstr. 043.002.

Gamma-ray production from proton-proton reactions in the galactic disk. See Abstr. 061.047.

Polarization of optical radiation in an absorbing medium in case of scattering. See Abstr. 063.017.

Possible sites of star formation in Sgr B2.
See Abstr. 065.015.

On the stability of the Oort cloud.
See Abstr. 102.037.

Determination of the form of the Oort cometary cloud as the Hill surface in the galactic field.
See Abstr. 102.038.

Secular parallaxes of stars and solar motion determined from absolute proper motions with reference to galaxies of 14600 stars.   See Abstr. 111.002.

UBVRI photometry of North Galactic Pole K giants. II.   See Abstr. 113.034.

UBVRI photometry of North Galactic Pole K giants. III. Search for halo stars.   See Abstr. 113.035.

Photometric standards for the southern hemisphere. II.   See Abstr. 113.036.

Dwarf K and M stars in the southern hemisphere.
See Abstr. 114.013.

Review of results in infrared space astronomy.
See Abstr. 114.044.

Infrared surveys of the southern Milky Way. II. Suspected supergiant M stars.   See Abstr. 114.072.

Studies of heavy-element synthesis in the Galaxy. I. Separation of *r*- and *s*-process abundances.
See Abstr. 114.074.

Spectra of Upgren's unclassified stars.
See Abstr. 114.081.

A finding list of stars of spectral type A 7 and earlier in regions at high galactic latitudes. VI.   See Abstr. 114.173.

The luminosity function and density distribution of disk population stars.   See Abstr. 115.017.

Luminosities and motions of A0 to A2 stars.
See Abstr. 115.025.

The kinematics of semi-regular red variables in the solar neighbourhood.   See Abstr. 122.016.

Flare stars and rapid irregular variables in the southern Coalsack.   See Abstr. 122.138.

Observations of the Cygnus Loop at 6-cm wavelength.
See Abstr. 125.003.

A study of galactic supernova remnants. II. Supernova rate, galactic radio emission and pulsars.
See Abstr. 125.008.

The Cygnus X-region. VII. Radio continuum search for a ring of filaments around the area.   See Abstr. 131.019.

Angular distribution of interstellar atomic hydrogen.
See Abstr. 131.025.

Interstellar reddening near the north galactic pole.
See Abstr. 131.052.

Nebulae of the southern Milky Way. An atlas.

See Abstr. 132.003.

Search for coronal line emission from the Cygnus Loop.   See Abstr. 132.033.

Possible models for the extended X-ray source at the galactic center.   See Abstr. 142.006.

Observation of cosmic soft X-rays.
See Abstr. 142.039.

Distances and absolute luminosities of galactic X-ray sources.   See Abstr. 142.099.

Soft X-ray spectra of the Cygnus Loop and Cygnus X-2 in the energy range 0.16–6.7 keV.
See Abstr. 142.102.

A soft X-ray survey of the galactic plane from Cygnus to Norma.   See Abstr. 142.108.

Equilibrium energy spectrum for the galactic cosmic electrons.   See Abstr. 143.006.

Cosmic-ray production of deuterium, He^3, lithium, beryllium, and boron in the Galaxy.   See Abstr. 143.020.

Adiabatic propagation of cosmic rays in the Galaxy.
See Abstr. 143.029.

Galactic evolution: Program and initial results.
See Abstr. 151.004.

A computer study of galactic spiral arms.
See Abstr. 151.005.

A new solution to the accretion problem.
See Abstr. 151.026.

Southern open star clusters I. *UBV*–Hβ photometry of 15 clusters between galactic longitudes 231° and 256°.
See Abstr. 153.016.

A longitude survey of radio recombination lines from the diffuse interstellar medium.
See Abstr. 157.001.

Lineaire polarisatie in de galactische spurs op 21 cm golflengte.   See Abstr. 157.002.

The latitude extent of diffuse ionization in the Galaxy.   See Abstr. 157.004.

Radiation-driven efflux and circulation of dust in spiral galaxies.   See Abstr. 158.054.

The tilt of the Large Magellanic Cloud and the galactical binary (sextuplet) Andromeda nebula and Milky Way.
See Abstr. 159.004.

Is the existence of a galaxy evidence for a black hole at its center?   See Abstr. 162.038.

Errata

155.901   Erratum: "Gamma-ray observations of the galactic center and some possible point sources" [Astrophys. Journ., Vol. 171, 31 - 40 (1972)].
C. E. Fichtel, R. C. Hartman, D. A. Kniffen, M. Sommer.
Astrophys. Journ., Vol. 176, 271 (1972).

156 Galactic Magnetic Field

156.001 **The origin and form of the galactic magnetic field. II. The primordial-field model.** J. H. Piddington.
Cosmic Electrodynamics, Vol. 3, 129 - 146 (1972).

In Part 1 of this investigation (Piddington, 1971) we discussed the dynamo theory of the origin and form of the magnetic field of our Galaxy and those of other galaxies. This investigation confirmed the idea that the galactic magnetic field is frozen into the gas on a time scale greater than the lifetime of the Galaxy, in which case the field is likely to be primordial in origin. Further support is found in the difficulty of constructing a satisfactory model of the galactic field confined to the thin gas disk. However, if the hypothesis is to be developed into an acceptable theory, then other possible effects of an intergalactic field must be determined and compared with observations. This is the objective of the present paper.

156.002 **Het galactische magneetveld.** T. A. T. Spoelstra.
Hemel en Dampkring, Vol. 70, 315 - 320 (1972).

156.003 **A photon rest mass and magnetic fields in the Galaxy.** J. C. Byrne, R. R. Burman.
Journ. Phys. A, General Phys., Vol. 5, L109 - L111 (1972).

The effect of a photon rest mass on the dissipation of magnetic fields in the Galaxy is discussed and an upper limit on the rest mass is obtained: an earlier treatment is corrected and refined.

On the transport properties of charged particles in one dimension in random electric fields.
See Abstr. 062.062.

157 Galactic Radio Radiation

157.001 **A longitude survey of radio recombination lines from the diffuse interstellar medium.**
M. A. Gordon, T. Cato.
Astrophys. Journ., Vol. 176, 587 - 596 (1972).

Observations of the H157α recombination line (1683.2 MHz), from regions of the galactic plane free of discrete radio sources, show the line radiation to come from a region spatially coincident with the zone of giant H II regions — a zone bounded by galactic radii of approximately 3 and 9 kpc.

157.002 **Lineaire polarisatie in de galactische spurs op 21 cm golflengte.** T. A. T. Spoelstra.
Hemel en Dampkring, Vol. 70, 245 - 252 (1972).

157.003 **The electron component of cosmic rays. II. Radio radiation of relativistic electrons in the Galaxy.**
S. V. Bulanov, V. A. Dogel', S. I. Syrovatskij.
Kosm. Issled., Vol. 10, 721 - 731 (1972). In Russian.

157.004 **The latitude extent of diffuse ionization in the Galaxy.**
M. A. Gordon, R. L. Brown, S. T. Gottesman.
Astrophys. Journ., Vol. 178, 119 - 124 (1972).

Observations of radio recombination lines (H157α) at λ18 cm, from regions free of discrete radio sources, show the line emission to be confined to within $0°.5$ of the galactic equator at $l^{II} = 33°$. This angle implies the largest values of the measured physical parameter $\langle N_e N_{H\,II}/T_e^{1.5}\rangle$ to be entirely within 70 pc of the plane. Such a distribution appears to be consistent with that expected from a thermal gas in hydrostatic pressure equilibrium with a temperature of approximately $3600°$K.

157.005 **Brightness temperatures in the southern sky at 408 MHz.** R. M. Price.
Astron. Journ., Vol. 77, 845 - 848 (1972).

Absolute brightness temperatures for a number of regions of the southern sky have been measured on an l–b grid with $15°$ spacing. The measurements were taken at a nearly constant zenith angle and have been related to the brightness temperature observed at the south celestial pole, at the same zenith angle.

157.006 **Absolute radio brightness temperatures in the southern sky.** R. M. Price.
Quarterly Progr. Rep. Res. Lab. Electron. Mass. Inst. Technology, No. 104, p. 82 - 84 (1972).

The analysis of observations of the southern sky at 408 MHz using the 210ft antenna of the CSIRO in Australia has been completed.

157.007 **New measurements of the galactic radiation spectrum between 200 kHz and 2.6 MHz.** L. W. Brown.
Rep. NASA-TM-X-65820, NASA Greenbelt, Md. [Available from NTIS, Springfield, Va.], 10 pp. (1972).

The Goddard Space Flight Center radio astronomy experiment aboard the IMP-6 satellite has provided new data on the absolute spectrum of the galactic background radiation at frequencies between 0.2 and 2.6 MHz.

157.008 **A low-latitude galactic survey from $l^{II} = 46°$ to $61°$ and $190°$ to $290°$ at 2700 MHz.**
G. A. Day, J. L. Caswell, D. J. Cooke.
Australian Journ. Phys., Astrophys. Suppl. No. 25, 19 pp. = Separate print Division Radiophys., C.S.I.R.O., Sydney (1972).

The results of observations undertaken with the 64 m (210 ft) radio telescope of the Parkes Observatory are given in the form of contour maps and a list of 343 radio sources. Some individual sources are discussed in detail. The present results together with those already published complete the Parkes 11 cm survey of the galactic plane.

A survey of linear polarization at 1415 MHz. III. Method of reduction and results for the galactic spurs.
See Abstr. 155.032.

Continuum radio structure of the Galaxy.
See Abstr. 155.065.

Continuum radio structure of the galactic disk.
See Abstr. 155.069.

158   Single und Multiple Galaxies

158.001   The spectra of Markarian galaxies. IV.
M. A. Arakelian, E. A. Dibay, V. F. Yesipov.
Astrofizika, Vol. 8, 33 - 42 (1972). In Russian. – English
translation in Astrophysics, Vol. 8, No. 1.

The results of spectral observations of eighty-five objects
from Markarian's lists of galaxies with ultraviolet continuum
are presented. Emission lines are detected in the spectra of
seventy objects.

**158.002   Brightness variations in the nucleus of the Seyfert
galaxy NGC 6814.**   G. J. MacPherson.
Publ. Astron. Soc. Pacific, Vol. 84, 392 - 393 = Lick Obs. Bull.,
No. 625 (1972).

**158.003   Energy distribution in the near infrared from nuclei
of galaxies. I. M31, M32, NGC 3115, NGC 4151,
NGC 4406.**   R. Barbon, S. D'Odorico.
Mem. Soc. Astron. Italiana, Nuova Ser., Vol. 43, 263 - 267
(1972).

Photographic spectrophotometry from 0.7 to 1.04 μ for
the nuclear regions of M31, M32, NGC 3115, NGC 4151, and
NGC 4406 is given. Except in the case of NGC 4151, all of the
galaxies show similar energy distributions. The observations
are compared with data obtained by photoelectric techniques.

158.004   The spectrum of the compact galaxy III Zw43.
R. Barbon.
Mem. Soc. Astron. Italiana, Nuova Ser., Vol. 43, 313 - 317
(1972).

New spectroscopic observations of the compact galaxy
III Zw43 confirm earlier conclusions about the overall appearance of its spectrum. The value of the recession velocity previously found is also substantiated. It is concluded that the different set of data recently published by some authors do not
refer to this object.

158.005   New redshifts of radio galaxies.   K. P. Tritton.
Monthly Notices Roy. Astron. Soc., Vol. 158, 277 -
280 (1972).

Redshifts have been obtained for 20 radio galaxies, including 0620–52, a new identification.

**158.006   A comparison between the emission-line galaxies
NGC 5253 and NGC 5408.**
T. J. Bohuski, E. M. Burbidge, G. R. Burbidge, M. G. Smith.
Astrophys. Journ., Vol. 175, 329 - 333 (1972).

It is shown that the nuclear region of the irregular galaxy
NGC 5408 is made up of large H II regions which have similar
structures, electron densities, and temperatures to those found
in the nucleus of the peculiar galaxy NGC 5253. Interferometric measurements give most probable velocities (velocity dispersions) of 23 and 30 km s^{-1} for the gas in the nuclei of
NGC 5408 and NGC 5253, respectively. The heliocentric radial velocities are found to be 505 and 396 km s^{-1} for NGC 5408
and NGC 5253, respectively. The nitrogen abundance relative
to hydrogen is probably lower in NGC 5408 than it is in
NGC 5253. A brief discussion is given of possible ways of interpreting these results.

158.007   On the internal kinematics of two compact galaxies.
R. W. O'Connell, R. P. Kraft.
Astrophys. Journ., Vol. 175, 335 - 345 = Contr. Lick Obs.,
No. 364 (1972).

Spectrograms of two compact galaxies, I Zw 129 and
II Zw 70, taken in several position angles have been obtained
with a Varo-type image tube at the coudé focus of the

120-inch (305 cm) reflector. The observations indicate that the
initial luminosity function for star formation in I Zw 129 does
not resemble that in the solar neighborhood. Our data for
II Zw 70 are not as extensive, but they suggest that its mass-to-
light ratio and mean mass density are similar to those of
I Zw 129.

**158.008   High-velocity neutral hydrogen in the central region
of the Andromeda galaxy.**
R. N. Whitehurst, M. S. Roberts.
Astrophys. Journ., Vol. 175, 347 - 352 (1972).

Hydrogen-line radiation reaching a radial velocity of
-540 km s^{-1} is observed from the central region of Andromeda. This corresponds to -230 km s^{-1} with respect to the
systemic radial velocity. To explain the large velocity range,
models involving a rapidly rotating disk, z-motion, or planar
radial streaming are considered. The present data do not allow
differentiation among these models, although broad constraints
can be placed on them.

**158.009   High resolution observations of the nucleus, spiral
structure and elliptical companions of M31 at 1415
MHz.**   P. C. van der Kruit.
Astrophys. Letters, Vol. 11, 173 - 180 (1972).

Observations of M31 and its companions with the Westerbork Synthesis Radio Telescope at 1415 MHz are reported.

158.010   Radial velocities of galaxies. II.
G. Chincarini, H. J. Rood.
Astron. Journ., Vol. 77, 448 - 450 (1972).

We present a list of radial velocities of galaxies derived
from spectrograms obtained in 1971 with the Carnegie image-
tube Cassegrain spectrograph attached to the KPNO 84-inch
telescope. The list includes data for a $7° \times 4°$ region centered
near R.A. = 12^{h}18^m, Dec. = 29°. It contains members of the
NGC 4274 group, the NGC 4131 group, and the Coma cluster.

**158.011   A radio map of the spiral galaxy Maffei 2 at 1415
MHz.**   R. J. Allen, E. Raimond.
Astron. Astrophys., Vol. 19, 317 - 325 (1972).

A field of one degree square centered near the infrared
object Maffei 2 has been observed with the Westerbork synthesis telescope in the radio continuum at 1415 MHz. The angular resolution of 26″ × 31″ is sufficient to reveal a nuclear
source 19″ ± 4″ in diameter with a flux of 0.21 ± 0.01 flux
units located near the optical center of Maffei 2. The radio
morphology is consistent with the inference from previous infrared and radio H I observations that Maffei 2 is a spiral galaxy. A list of incidental radio sources in the field is given, as
well as the positions of eleven faint stars in the immediate
neighbourhood of the galaxy.

158.012   Study of the continuum of the nuclei of galaxies. II.
Y. Andrillat, S. Souffrin, D. Alloin.
Astron. Astrophys., Vol. 19, 405 - 416 (1972).

We apply a method previously described (Alloin et al.,
1971), based on the study of equivalent widths of absorption
lines, to determine the stellar population of the galactic nuclei,
and we compute a "synthetic" continuum that we compare
with observation. We study here nuclei of four galaxies: NGC
1052, NGC 2655, NGC 2903, NGC 4569.

**158.013   Radial cyanogen-band gradients near the nuclei of
galaxies.**   H. Spinrad, H. E. Smith, D. J. Taylor.
Astrophys. Journ., Vol. 175, 649 - 657 (1972).

Very high spatial resolution area scans of the nuclear

regions of eight nearby galaxies, of types E0–E5, S0, Sa, and Sb were obtained in 1970–71 with 33 Å spectral purity at the blue λ4200 CN band. The interpretations of the CN gradients are briefly discussed in terms of differences in chemical composition relating to the dynamical histories of simple spherical and flattened systems.

158.014 Hα emission in the central regions of the nearest galaxies.   I. I. Pronik.
Astron. Zhurn. Akad. Nauk SSSR, Vol. 49, 768 - 776 (1972). In Russian. English translation in Soviet Astron. AJ, Vol. 16, No. 4.

Spectra of 16 central regions of the nearest normal galaxies have been obtained. Equivalent widths of Hα emission have been measured. The correlation between these and blue colours of Tifft may be explained as the result of excitation of Hα emission by ultraviolet radiation of blue stars. It is supposed that the spectral peculiarity of the nuclei of Seyfert galaxies follows the normal evolution when red giants are evolving to the state of planetary nebulae.

158.015 Activity in galaxies and quasars.
        P. A. Sturrock, C. Barnes.
Astrophys. Journ., Vol. 176, 31 - 45 (1972).

Activity in galaxies and quasars is interpreted in terms of plasma processes occurring in the magnetosphere of a certain magnetoid model. This magnetoid comprises a core and an annulus rotating about a common axis with different angular velocities.

158.016 Linear polarization of the Hα emission line in the halo of M82 and the radiation mechanism of the filaments.   N. Visvanathan, A. Sandage.
Astrophys. Journ., Vol. 176, 57 - 74 (1972).

New polarization measurements have been made of five faint outer regions in the M82 halo to test if the electric vectors are perpendicular to the filaments, or to the radius vector from the center. The measurements agree to within 2 σ with a scattering model, but differ by 4 to 9 σ from the prediction for the synchrotron mechanism. But we are unconvinced that a simple scattering model explains the system. The galaxy remains a mystery to us.

158.017 The optical spectrum of the Seyfert galaxy 3C 120.
        G. A. Shields, J. B. Oke, W. L. W. Sargent.
Astrophys. Journ., Vol. 176, 75 - 89 (1972).

We have studied the spectrum of the nucleus of the Seyfert galaxy 3C 120 with the aid of spectrophotometric scans and photographic spectrograms. The scans cover the wavelength range λλ 3330–10620, and the spectrograms cover the wavelength range λλ 3200–6900. These data have been used to study the emission lines and the optical continuum of 3C 120.

158.018 A new determination of the mass of M32.
        D. Richstone, W. L. W. Sargent.
Astrophys. Journ., Vol. 176, 91 - 98 (1972).

The radial-velocity dispersion of the stars in the nucleus of M32 has been determined from two Palomar coudé spectra to be $\sigma_r = 60 \pm 8$ km s^{-1}. This value of σ_r is smaller than the values ($\sigma_r \sim 100$ km s^{-1}) obtained by Minkowski and by Burbidge, Burbidge, and Fish; possible reasons for the discrepancy are discussed. The mass of M32 calculated from the derived value of σ_r is $1.30 \pm 0.50 \times 10^9 \, M_\odot$, and the mass-to-light ratio is $M/L_p = 6.9 \pm 2.2$; these are also smaller than previous values.

158.019 The absorption-line spectrum of Markarian 231.
        T. F. Adams.
Astrophys. Journ., (Letters), Vol. 176, L1 - L3 (1972).

The faint absorption lines observed by Adams and Weedman in the absorption-line Seyfert galaxy Markarian 231 are shown to be a composite A-type stellar spectrum originating in the nucleus with the same redshift as the emission system. The presence of strong interstellar absorption lines blueshifted with respect to the nucleus is again confirmed. The similarity between Markarian 231 and the QSO 3C 273 is discussed briefly.

158.020 On the evidence for a physical connection between Markarian 205 and NGC 4319.
        R. Lynds, A. G. Millikan.
Astrophys. Journ., (Letters), Vol. 176, L5 - L8 (1972).

New direct photographs and spectroscopic material are discussed in relation to the finding by Arp that there is a filament connecting Markarian 205 and NGC 4319, objects of greatly differing redshift. The new material appears to raise doubts concerning the nature of the feature reported by Arp and tends, to a small degree, to support a more conventional interpretation of Markarian 205 as a Seyfert galaxy located at its indicated Hubble distance.

158.021 On the gas content of galaxies.   W. J. Quirk.
        Astrophys. Journ., (Letters), Vol. 176, L9 - L14 (1972).

The following hypothesis is suggested: If the gas in a galaxy is Jeans unstable according to the criterion of Goldreich and Lynden-Bell, efficient star formation results. No attempt is made to describe how star formation occurs.

158.022 Dimensions, absolute magnitudes and colour characteristics of spiral arm patches in the galaxies NGC 628, NGC 4254 and NGC 5194.
I. I. Pronik, K. K. Chuvaev.
Izv. Krymskoj Astrofiz. Obs., Vol. 44, 40 - 44 (1972). In Russian.

Dimensions, absolute magnitudes and colour characteristics ($K_1 = I\,(\lambda3600)/I\,(\lambda4350)$ and $K_2 = I\,(\lambda4350)/I\,(\lambda5550)$) for dozens of spiral arm patches belonging to the Sc galaxies NGC 628, NGC 4254 and NGC 5194 are given.

158.023 Photometric investigation of the dwarf galaxy Fornax with G.A.L.A.X.Y.   S. Demers.
Journ. Roy. Astron. Soc. Canada, Vol. 66, 217 (1972). Abstr. Canadian Astron. Soc.

158.024 Quadratic programming applied to the problem of galaxy population synthesis.   S. M. Faber.
Astron. Astrophys., Vol. 20, 361 - 374 (1972).

The technique of quadratic programming as applied to the problem of galaxy population synthesis is described. The method offers significant advantages over the trial-and-error approach usually employed. This technique is applied to 38-color data on the nuclei of M31, M32, and M81 and to integrated 10-color photometry for elliptical galaxies. The results indicate that estimates of mean line strengths in external galaxies by means of population synthesis are well determined. Ages based on the main-sequence turnoff point are uncertain by a factor of two.

158.025 The age and structure of II Zwicky 40.
        W. J. Jaffe.
Astron. Astrophys., Vol. 20, 461 - 464 (1972).

The dwarf galaxy II Zw 40 was detected at 1415 MHz continuum with the WSRT. The measured flux density, combined with previous H I measurements, indicates a diameter of about 1 kpc. The kinetic energy of the galaxy seems primarily non-rotational, and the system is probably less than 10^7 years old.

158.026 Comparison of electrographic, photographic and photoelectric photometry of NGC 4881.
H. D. Ables, P. G. Ables.

AAS Photo-Bull., 1972, No. 2, (see 011.003), p. 8 - 11.

The luminosity profile of the Coma cluster galaxy, NGC 4881, has been derived from electrographic, photographic and photoelectric observations. This study was undertaken for two reasons: (1) to compare the three different techniques for photometry of galaxies and (2) to resolve differences between the photographic photometry by Rood and Baum (1968) and our electrographic photometry of NGC 4881.

158.027 Variations in the nuclear luminosities of the Seyfert galaxies NGC 3516 and 5548.
G. de Vaucouleurs, A. de Vaucouleurs.
Astrophys. Letters, Vol. 12, 1 - 4 (1972).

Variations in the nuclear luminosities of the Seyfert galaxies NGC 3516 and 5548 are demonstrated by U, B, V photometry at McDonald Observatory between 1968 and 1972 supplemented by earlier data. Estimates of the variable magnitude B_N of the quasi-stellar nucleus and of the constant total magnitude B_T^* of the stellar component of the galaxy (excluding the nucleus) are, for NGC 3516, $B_T^* \simeq 12.4$, $B_N = 14.2$ (1958) to 15.0 (1968), with a flare up to $B_N \simeq 13.8$ in March 1972 accompanied by a considerable increase in the ultraviolet flux; and, for NGC 5548, $B_T^* \simeq 13.1$, $B_N = 14.7$ (1968) to 15.7 (1971–72).

158.028 Interesting object. B. E. Markarian.
Astron. Tsirk., No. 678, p. 4 - 6 (1972). In Russian.

158.029 Composite photography of Stephan's quintet.
V. G. Christich.
Astron. Tsirk., No. 684, p. 1 - 3 (1972). In Russian.

158.030 Narrow-band photoelectric observations of Hα-line variability in the nuclei of Seyfert galaxies NGC 4151, 3516, 1068. V. M. Lyutyj, A. M. Cherepashchuk.
Astron. Tsirk., No. 688, p. 1 - 4 (1972). In Russian.

158.031 Real flattening of S galaxies for subtypes.
B. A. Vorontsov-Velyaminov, R. I. Noskova.
Astron. Tsirk., No. 690, p. 1 - 3 (1972). In Russian.

158.032 Gross properties of five Scd galaxies as determined from 21-centimeter observations.
D. H. Rogstad, G. S. Shostak.
Astrophys. Journ., Vol. 176, 315 - 321 (1972).

Neutral-hydrogen studies of the five Scd galaxies M33, NGC 2403, IC 342, M101, and NGC 6946 have revealed substantial similarities in their hydrogen distributions and rotation curves. We find that all the H I global properties and ratios are nearly constant or scale approximately with Holmberg radius or total mass. An observed correlation of maximal rotation velocity with Holmberg radius allows characterization of the total mass with only one parameter.

158.033 The "census" of the Local Group of galaxies.
Yu. P. Pskovsky.
Priroda, No. 8.72, p. 98 - 100 (1972). In Russian.

158.034 High frequency radio observations of optically interacting galaxies. C. R. Purton, A. E. Wright.
Monthly Notices Roy. Astron. Soc., Vol. 159, 15P - 20P (1972).

Radio observations of four systems of gravitationally interacting galaxies are reported. It is concluded that no evidence of the interactions appears in the radio characteristics of the systems.

158.035 21 cm observations of NGC 45. B. M. Lewis.
Australian Journ. Phys., Vol. 25, 315 - 328 (1972).

Observations of NGC 45 have been made at a wavelength of 21 cm using the radio telescope at Parkes. Line profiles have been measured for a grid of 44 points spaced at intervals of 6' arc. From the measurements the mass of neutral hydrogen is calculated to be $8.2 \times 10^8 \, M_\odot$, assuming a distance of 3 Mpc. A simple self-consistent model is determined by computing line profiles to compare directly with the observations. This enables the rotation curve to be corrected for the first-order effects of beam smoothing and results in a total limiting mass of $2.5 \times 10^{10} \, M_\odot$.

158.036 Obtaining the redshift–distance dependence for galaxies. S. I. Urbanovich.
Vestsi AN BSSR. Ser. fiz.-mat. n., Izv. AN BSSR. Ser. fiz.–mat. n., 1972, No. 2, p. 118 - 120. In Russian. – Abstr. in Referativ. Zhurn. 51. Astron., 8.51.650 (1972).

158.037 Integrated magnitudes and colours of bright galaxies in the UBV system.
G. de Vaucouleurs, A. de Vaucouleurs.
Mem. Roy. Astron. Soc., Vol. 77, 1 - 53 (1972).

Integrated V magnitudes and $B-V$, $U-B$ colour indices are presented for 461 bright galaxies ($\delta > -50°$) observed between 1960 and 1968 mainly with the McDonald Observatory 91-cm reflector. The catalogue includes over 1000 observations by 13 observers in 219 nights. The mean error of one observation is ±0.04 mag in V, ±0.030 in $B-V$, and ±0.045 in $U-B$.

158.038 Determination of the redshift of the irregular galaxy NGC 4433. E. Ye. Khachikian, G. A. Panosyan.
Astron. Tsirk., No. 698, p. 1 - 2 (1972). In Russian.

158.039 The velocity dispersion of stars in the nucleus of M31. T. X. Thuan, D. C. Morton.
Bull. American Astron. Soc., Vol. 4, 315 (1972). – Abstr. AAS.

158.040 UBV photoelectric area scans of the southern spiral arm of M33. V. Lee.
Bull. American Astron. Soc., Vol. 4, 332 (1972). – Abstr. AAS.

158.041 Preliminary results of a photoelectric study of supergiant (cD) galaxies. G. A. Welch.
Bull. American Astron. Soc., Vol. 4, 332 (1972). – Abstr. AAS.

158.042 Emission line strengths in the nuclei of galaxies.
J. W. Warner.
Bull. American Astron. Soc., Vol. 4, 332 (1972). – Abstr. AAS.

158.043 The non-nuclear regions of the Seyfert galaxy NGC 4151. S. M. Simkin.
Bull. American Astron. Soc., Vol. 4, 332 (1972). – Abstr. AAS.

158.044 X-ray galaxies? S. Murray, E. Kellogg, H. Gursky, H. Tananbaum, R. Giacconi.
Bull. American Astron. Soc., Vol. 4, 336 - 337 (1972). Abstr. AAS.

158.045 The diameter-redshift relation for distant galaxies.
W. A. Baum.
Bull. American Astron. Soc., Vol. 4, 340 (1972). – Abstr. AAS.

158.046 Catalogue of galaxies brighter than 15^m on the magnetic tape of the computer M-220.
N. G. Kogoshvili.
Astron. Tsirk., No. 706, p. 1 - 3 (1972). In Russian.

158.047 Morphology and redshifts of galaxies. H. Arp.
Bull. American Astron. Soc., Vol. 4, 397 (1972). Abstr. AAS.

158.048 Observation de M82 dans le proche infrarouge.
M. Duchesne, Y. Andrillat.

L'Astronomie, 86ᵉ année, p. 381 - 383 (1972).

158.049 Extragalactic explosive phenomena. M. J. Rees.
    Astronaut. Acta, Vol. 17, 315 - 320 (1972).
Presented at the international colloquium on gasdynamics of explosions, Marseille, September 12 - 17, 1971.

158.050 A photoelectric study of Messier 81.
    J. C. Brandt, J. K. Kalinowski, R. G. Roosen.
Astrophys. Journ.,Suppl. Ser., No. 210, Vol. 24, 421 - 448 (1972).

Photoelectric observations in B and V have been made at over 1200 locations in Messier 81 with the 82-inch (208-cm) Struve reflector at the McDonald Observatory. The observations cover an area of approximately 169 square minutes of arc and include the galaxy's nucleus. Isophote maps in V and B based on these observations have been drawn. $B - V$ and (integrated B) − (integrated V) are given as functions of distance from the center of the galaxy along the semi-major axis. Thirty observations of the nucleus, plotted to show the distribution of brightness and color out to a distance of 13″, indicate considerable small-scale structure. A mass model is derived using a rotation curve based on the velocities measured by Münch. The local mass and light surface densities and other quantities are given as functions of distance from the nucleus.

**158.051 A neutral-hydrogen study of the compact galaxy II
    Zw 40.** S. T. Gottesman, L. Weliachew.
Astrophys. Letters, Vol. 12, 63 - 68 (1972).

We have made interferometric observations in the 21-cm line of neutral hydrogen to gain detailed information about the structure and dynamics of II Zw 40.

**158.052 A search for circular polarization in extragalactic
    objects.** K. H. Nordsieck.
Astrophys. Letters, Vol. 12, 69 - 74 (1972).

No significant circular polarization (less than 0.15−0.30 per cent) has been found in the ultraviolet radiation of the nuclei of the Seyfert galaxies NGC 1068 and NGC 4151, and the QSO 3C 273. The variable object OJ 287 has shown linear polarization of up to 14 per cent, variable on a time scale of one day, with a possible (2.5 σ) circular polarization of −0.9 per cent on one day.

158.053 Observations of bright galaxies at 80 MHz.
    O. B. Slee.
Astrophys. Letters, Vol. 12, 75 - 79 (1972).

The 80-MHz radioheliograph with 3.7 arc min beamwidth has been used to measure the flux densities, positions, and angular sizes of radio sources associated with 30 bright galaxies. The 80-MHz sources associated with 12 of the 24 galaxies detected have angular dimensions smaller than the optical diameters and are identified with nuclear or disk components. Radio brightness contours are presented for NGC 253 and used to separate the 80-MHz nuclear and disk sources. Separate spectra are constructed for the components.

**158.054 Radiation-driven efflux and circulation of dust in
    spiral galaxies.**
R. Y. Chiao, N. C. Wickramasinghe.
Monthly Notices Roy. Astron. Soc., Vol. 159, 361 - 373 (1972).

Dust grains, which carry a net electric charge, may be driven out of the galactic disk along magnetic field lines by radiation pressure of starlight. Large scale mixing or escape of the dust occurs along closed loops or open field lines, respectively. Mixing and escape also occur by means of instabilities of the galactic disk caused partially by radiation pressure on the grains.

**158.055 Inclination corrections to the optical luminosities
    and diameters of galaxies.** R. B. Tully.

Monthly Notices Roy. Astron. Soc., Vol. 159, 35P - 40P (1972).

The results of an investigation by Heidmann, Heidmann and de Vaucouleurs on the effects of inclination on observed magnitudes and diameters of galaxies are questioned. It is concluded that corrections must be made for internal absorption in external galaxies and that diameter corrections do not have to be made.

**158.056 The estimation of masses of individual galaxies in
    clusters of galaxies.** R. A. Wolf, J. N. Bahcall.
Astrophys. Journ., Vol. 176, 559 - 580 (1972).

In this paper, we discuss how one can estimate masses for individual galaxies that are members of groups or clusters of galaxies. We discuss three different methods: (I) the "density method," based on the analysis of the density distribution of galaxies around the object whose mass we want to find; (II) the "bound-galaxy method," which gives estimates of the mass of a double, triple, quadruple, ... system from analysis of the orbital motion of the components; and (III) the "virial method," which utilizes the formulae derived for method (II) to obtain estimates of the virial-theorem masses of whole clusters, and thus to obtain upper limits on the mass of any individual galaxy in a cluster.

158.057 Infrared photometry of extragalactic sources.
    G. H. Rieke, F. J. Low.
Astrophys. Journ., (*Letters*), Vol. 176, L95 - L100 (1972).

Observations of 57 extragalactic sources at 10μ are presented. Several of the brightest sources were observed from 2 to 25μ. The observed range of lower limits to the infrared luminosities for galaxies extends from 10^{32} to 10^{38} watts. The infrared and radio spectra of the nuclei in Seyfert and related galaxies appear to be causally related.

158.058 The structure of small sources in M51 at 2695 MHz.
    J. H. Spencer, B. F. Burke.
Astrophys. Journ., (*Letters*), Vol. 176, L101 - L104 (1972).

M51 has been mapped at 2695 MHz using one configuration of the NRAO three-element interferometer. Fluxes, spectral index, and structure of four discrete sources were obtained. The nucleus of M51 is resolved into a complex that strongly resembles the nuclear regions of our own Galaxy.

158.059 Photometry of Markarian 205.
    R. W. O'Connell, W. C. Saslaw.
Astrophys. Journ., (*Letters*), Vol. 176, L109 - L111 (1972).

Strömgren photometry of Markarian 205 shows no evidence for variability larger than 0.05 mag over periods on the order of minutes, hours, and days. The spectral energy distribution of Markarian 205 resembles those of quasi-stellar objects or compact galaxies with broad emission lines.

158.060 Blue emission-line galaxies in rich clusters.
    G. Chincarini, H. J. Rood.
Publ. Astron. Soc. Pacific, Vol. 84, 589 - 591 (1972).

Blue galaxies selected from lists of color indices by Philip and Sanduleak contain emission lines in their spectra. Several are spiral and irregular members of rich regular clusters.

158.061 Two galaxies dominated by bright ultraviolet knots.
    A. Braccesi, L. Formiggini, I. Gioia, W. L. W.
Sargent.
Publ. Astron. Soc. Pacific, Vol. 84, 592 - 593 (1972).

**158.062 Investigation of the process of explosive decay and
    simultaneous collision of n material gravitating
points.** E. M. Nezhinskii.
Dokl. Akad. Nauk SSSR, Ser. Mat. Fiz., Vol. 206, 566 - 567 (1972). In Russian.

158.063 Radio galaxies, quasars, and cosmology.

K. I. Kellermann.
Astron. Journ., Vol. 77, 531 - 542 (1972).

This paper discusses the role of radio observations of galaxies and quasars in observational cosmology. Particular emphasis is given to the interpretation of radio-source counts, and it is concluded that when properly analyzed, the counts are not a function of wavelength, and that the claims for a large observed excess of weak sources have been greatly exaggerated. Evidence for a nonuniform spatial distribution of radio sources is inconclusive.

158.064 A search for H I in elliptical galaxies.
J. S. Gallagher, III.
Astron. Journ., Vol. 77, 568 - 572 (1972).

A search has been made for 21-cm emission from neutral hydrogen in the normal elliptical galaxies NGC 2768, NGC 4125, NGC 4278, and NGC 4472 using the NRAO 140-ft telescope. None of the galaxies were definitely detected and upper limits on the order of several times $10^8 M_\odot$ are set for the mass of optically thin neutral hydrogen. The implications of these upper-limit masses are briefly discussed in terms of the mass balance of the interstellar medium in elliptical galaxies.

158.065 Optical variability of nuclei of Seyfert galaxies.
V. M. Lyutyj.
Astron. Zhurn. Akad. Nauk SSSR, Vol. 49, 930 - 942 (1972). In Russian. English translation in Soviet Astron. AJ, Vol. 16, No. 5.

UBV observations of nuclei of Seyfert galaxies NGC 1068, 1275, 3516, 4151, 5548, 7469, Markarian 10 and N-galaxies 3C 120 and II Zw 136 are given. The data of UBV photometry of the variable radio galaxies NGC 1275 (3C 84) and 3C 120 are compared with radio observations at 1.9 cm. There is a correlation between radio and optical variability of NGC 1275 and 3C 120, and radio, optical and infrared observations of NGC 1275.

158.066 Photometric parameters of flat galaxies.
B. A. Vorontsov-Velyaminov, R. I. Noskova.
Astron. Zhurn. Akad. Nauk SSSR, Vol. 49, 1010 - 1016 (1972). In Russian. English translation in Soviet Astron. AJ, Vol. 16, No. 5.

The photometric parameters of the flat component and of the spherical component were determined for 130 S0 and S galaxies and studied statistically.

158.067 The radio emission of NGC 4258 and the possible origin of spiral structure.
P. C. van der Kruit, J. H. Oort, D. S. Mathewson.
Astron. Astrophys., Vol. 21, 169 - 184 (1972).

The radio structure of the spiral galaxy NGC 4258 has been investigated with a beam of 24″ × 32″ at 1415 MHz. The radio structure shows two curved "ridges" of considerable intensity which differ in position, shape and continuity from the optical arms. The radio ridges and the accompanying filamentary H_α-arms are interpreted as the present location of "clouds" expelled from the nucleus in two opposite directions in the equatorial plane about 18 million years ago, at velocities ranging from about 800−1600 km/s.

158.068 21-cm neutral hydrogen line study of early type galaxies.
C. Balkowski, L. Bottinelli, L. Gouguenheim, J. Heidmann.
Astron. Astrophys., Vol. 21, 303 - 310 (1972).

Eighteen more lenticular to Sab galaxies have been observed with the Nançay radio telescope in the 21-cm line of neutral hydrogen. Adding these data to those previously obtained at Nançay for 15 other galaxies of these types, we show that the neutral hydrogen content of early type galaxies is relatively large and that the mass-to-light ratio is practically constant from lenticular to irregular galaxies. The hydrogen

diameter has been determined for three of them, including the peculiar galaxy NGC 1808.

158.069 Variations d'éclat de 3C 120.
C. Bertaud, M.-P. Véron, C. Pollas.
Inform. Bull. Variable Stars (I.A.U. Commission 27), Konkoly Obs., Budapest, No. 703 (1972).

158.070 Variable compact galaxy or supernova in a compact galaxy. F. Zwicky.
Inform. Bull. Variable Stars (I.A.U. Commission 27), Konkoly Obs., Budapest, No. 709 (1972).

158.071 The evolution of galaxies − a heretical view.
S. van den Bergh.
Journ. Roy. Astron. Soc. Canada, Vol. 66, 237 - 248 = Commun. David Dunlap Obs., Univ. Toronto, *Richmond Hill*, No. 337 (1972).

Evidence is presented which suggests that bursts of star formation were triggered by the explosions that recently took place in the giant elliptical galaxies NGC 1275 (= Per A) and NGC 5128 (= Cen A). It is speculated that galaxies age by gradual evolution and by rapid mutations that are triggered by explosive events.

158.072 Low frequency radio observations of the Andromeda galaxy. J. M. Durdin, Y. Terzian.
Astron. Journ., Vol. 77, 637 - 641, 699 (1972).

Radio observations of M31 and its halo are presented at 73.8, 111.5, and 196.5 MHz. These observations were performed with the NAIC 1000-ft radio telescope. Brightness-temperature contour maps are presented and the spectrum of M31 and its halo are discussed. It may be possible that the spectrum of the observed region flattens out at frequencies $\lesssim 200$ MHz.

158.073 Electrographic photometry of NGC 4881.
H. D. Ables, P. G. Ables.
Astron. Journ., Vol. 77, 642 - 648 (1972).

Luminosity profiles, integrated magnitudes, effective diameters, and other photometric parameters are derived for the Coma cluster galaxy, NGC 4881, from V and B electrographic plates taken with a Kron electronic camera attached to the 40-inch Ritchey-Chrétien telescope. The luminosity profiles indicate the presence of a weak lens which suggests that NGC 4881 is probably a lenticular S0 galaxy rather than an E0 as previously classified.

158.074 Extremely compact galaxy CGCG 1439 + 5344.
M. Takada, K. Kodaira.
Publ. Astron. Soc. Japan, Vol. 24, 525 - 531 (1972).

The extremely compact galaxy CGCG 1439 + 5344 was observed with an image-intensifier spectrograph ($\lambda\lambda$ 3900 − 6600 Å, 220 Å mm^{-1}). The results are presented and discussed. Due to the spectroscopic characteristics of this galaxy one might interpret it as being a dense aggregate of ionized gas and normal stars.

158.075 Are galaxies formed by material flowing out of singularities? J. N. Bahcall, P. C. Joss.
Comments Astrophys. Space Phys., Vol. 4, 95 - 99 (1972).

In the present communication, we describe plausible observational consequences of the "outflow hypothesis" of galaxy formation. Our examination leads us to propose that some compact galaxies may be in the process of being born. Because of the small angular dimensions of known quasistellar objects, it is not feasible to carry out our proposed observational tests on quasars.

158.076 Infrared photometry of Markarian 231.
E. T. Young, R. F. Knacke, R. R. Joyce.
Nature, Vol. 238, 263 (1972).

Observations of the Seyfert galaxy Markarian 231 at 3.5 and 10.8 μ give a lower limit of 10^{45} ergs sec^{-1} for the infrared luminosity. Thus Markarian 231 is the most luminous galaxy known.

158.077 Small "non-velocity" redshifts in galaxies.
S. M. Simkin.
Nature, Vol. 239, 43 - 44 (1972).

Recent determinations of some galaxy redshifts, made using impersonal, numerical techniques, show that the small, excess redshifts of E and S0 galaxies relative to neighboring spirals and irregulars reported in the literature are probably the result of measurement errors.

158.078 The Andromeda galaxy.
Ya. E. Einasto, M. M. Iyeveer.
Priroda, No. 11.72, p. 70 - 80 (1972). In Russian.

158.079 Observations of NGC 4319 and Markarian 205.
H. C. Ford, H. W. Epps.
Astrophys. Letters, Vol. 12, 139 - 141 (1972).

A narrow-band, sky-limited, Hα photograph taken at the prime focus of the Lick 120-in telescope during excellent seeing does not show a connection between NGC 4319, whose redshift is 0.006, and Markarian 205, whose redshift is 0.070. A wide-band, sky-limited, green-light photograph taken during good seeing does not show a connection at a photographic density that can be considered significant. We suggest that Markarian 205 is an N-type galaxy at cosmological distance.

158.080 Limits on Hα emission in NGC 4319 associated with Markarian 205. T. F. Adams, R. J. Weymann.
Astrophys. Letters, Vol. 12, 143 - 146 (1972).

Photographic plates of Markarian 205 taken through a narrow-band filter centered on Hα at the redshift of the nearby galaxy NGC 4319 fail to reveal the 'connection' reported by Arp. The observational limits on the emission measure are used to rule out a model in which Markarian 205 was ejected in the plane of NGC 4319. Other models in which Markarian 205 is not in the plane of NGC 4319, but still at the same distance, cannot be rejected without a substantial improvement in the observations.

158.081 The theoretical gas contents of galaxies.
J. G. Hills.
Astrophys. Letters, Vol. 12, 153 - 154 (1972).

The observation that the neutral gas contents of spiral or irregular galaxies of a given morphological class are directly proportional to their luminosity is shown to follow from energy conservation if their gas was cumulated from the mass shed by evolving stars. The increase in the constant of proportionality in this relation from morphological class S0 to I is shown quantitatively to reflect a systematic change in these galaxies of that part of their stellar populations primarily responsible for their integrated luminosities.

158.082 Infall of matter into galaxies. G. de Vaucouleurs.
Nature, Phys. Sci., Vol. 239, 139 (1972).

158.083 Cinématique de l'émission Hα du filament anormal de la galaxie NGC 4258. G. Courtès.
Comptes Rendus Acad. Sci., Paris, Sér. B, Vol. 275, 759 - 762 (1972).

Le filament d'hydrogène ionisé issu du noyau de NGC 4258 participe à la rotation générale de cette galaxie. Cette observation confirme l'interprétation radio de Van der Kruit, Oort et Mathewson, assimilant le filament à un bras spiral en formation dans le plan même de rotation de NGC 4258.

158.084 The process of galaxy formation according to the universal turbulence hypothesis.
N. Dallaporta.
Atti XV Riunione Soc. Astron. Italiana, Bologna 1971, (see 012.013), p. 149 - 178 (1972).

After a short presentation of the main principles of the theory of galaxy formation through gravitational instability from primordial matter and of its shortcomings in trying to explain the high value of angular momentum of spiral galaxies, the assumption of universal turbulence of primeval matter is introduced. The principal theories rooted on this assumption for interpreting the main properties of spiral galaxies are then successively discussed. Moreover, some of the yet unpublished results of Jones's approach, in which dissipation phenomena are also included, are briefly quoted; finally the possible signification of elliptical galaxies and clusters of galaxies in the frame of the turbulence theory of primeval matter is shortly outlined.

158.085 The rotation of galaxies. F. Bertola.
Atti XV Riunione Soc. Astron. Italiana, Bologna 1971, (see 012.013), p. 199 - 203 (1972).

After a brief review of the methods of observing and analysing the rotation of the galaxies, preliminary observational data on the rotation of four early type galaxies are presented.

158.086 Variation of emission-line strengths across M 31.
V. C. Rubin, C. K. Kumar, W. K. Ford, Jr.
Astrophys. Journ., Vol. 177, 31 - 44 (1972).

Intensities of [N II] λ 6584, [O III] λ 5007, Hβ, and [S II] $\lambda\lambda$ 6717 and 6731 emission lines, relative to Hα, have been measured for 53 H II regions between 3 and 24 kpc from the nucleus of M 31. When these results are combined with earlier studies of the nucleus, it is shown that the [N II]/Hα line intensity ratio decreases systematically from a value of about 2 at the nucleus to about 0.2 at 15 kpc, beyond which distance there is no certain trend. A systematic increase in the [O III]/Hβ ratio from 0.4 to 2 is observed in going from 3 kpc to 15 kpc.

158.087 On the investigation of errors in a detailed photometry of galaxies. A. I. Shapovalova.
Vestn. Kiev. Un-ta, Ser. Astron., No. 14, p. 103 - 106 (1972). In Russian.

The necessity of applying Student – Fisher's method for the calculation of the errors of the measurements in a detailed photometry of galaxies is shown. This method is used for the calculation of the errors in U, B, V colours of the irregular galaxy NGC 4449.

158.088 The role of Schmidt telescopes in the study of external galaxies. S. van den Bergh.
The role of Schmidt telescopes in astronomy. Conference Hamburg 1972, (see 012.014), p. 67 - 76 (1972).

158.089 Surface colour photometry of galaxies with Schmidt telescopes. J. D. Wray.
The role of Schmidt telescopes in astronomy. Conference Hamburg 1972, (see 012.014), p. 87 - 96 (1972).

158.090 A comparison of spectral scans of some E and S0 galaxies. H. Spinrad.
Astrophys. Journ., Vol. 177, 285 - 289 (1972).

Scans of the nuclear regions of four E and S0 galaxies covering a luminosity range of about 5 mag are presented. They confirm and extend the trends of increasing blueness and decreasing metal-line strength with decreasing luminosity. An attempt at a stellar population synthesis for the center of NGC 3379 is presented to illustrate the difficulty encountered by the simultaneous presence of strong metallic absorption lines with a relatively blue continuum produced by hotter stars.

158.091 A high-frequency study of Cygnus A.
K. B. W. Yip, G. A. Seielstad.
Astrophys. Journ., Vol. 177, 291 - 301 (1972).

We have combined measurements at 8300 and 9600 MHz to map the distribution of total radiation across Cygnus A with a beam of half-power width ~20″EW × 30″NS. We have also mapped the distributions of linearly polarized radiation at both 8300 and 9600 MHz. The Sf component of Cygnus A is probably more distant from us than the Np. Within each component the density of the thermal plasma and the magnetic field strength are greatest at the outer edges.

158.092 The curious cases of the colliding galaxies and the rapidly expanding universe. S. Mitton.
Observatory, Vol. 92, 183 - 185 (1972). – Letter.

158.093 Nogmaals Maffei 2. F. P. Israel.
Hemel en Dampkring, Vol. 70, 294 - 297 (1972).

158.094 Counts of galaxies in the fields surrounding quasars.
M. Różyczka.
Acta Astron., Vol. 22, 93 - 102 (1972).

The counts of galaxies on the Palomar Atlas Prints show that there is a slight excess of galaxies around the quasars with redshifts smaller than 0.5.

158.095 Zur Spiralarmstruktur am südwestlichen Rand von M 31. F. Börngen.
Astron. Nachr., Vol. 294, 79 - 81 = Mitt. Karl-Schwarzschild-Obs. Tautenburg, No. 59 (1972).

The spiral structure on the south-western border of M 31 is discussed using the distribution of there well known OB associations. Starting from four clearly defined spiral-arm segments the course of the spiral-arms is asked. The most distant spiral-arm on the northern end of the major axis of M 31 has a considerably smaller distance from the nucleus than the corresponding arm on the southern end of it. Therefore on the north-eastern border of M 31 some OB associations till now unknown may be suspected.

158.096 Structure and dynamics of barred spiral galaxies, in particular of the Magellanic type.
G. de Vaucouleurs, K. C. Freeman.
Vistas in astronomy, Vol. 14, (see 003.008), 163 - 294 (1972).

158.097 Observations of some Markarian galaxies at 9.5-mm wavelength. H. M. Tovmassian.
Astron. Journ., Vol. 77, 705 - 706 (1972).

Only one out of 29 observed Markarian galaxies is suspected to have radio emission at 9.5-mm wavelength.

158.098 The interpretation of the X-ray emission detected from some nearby radio galaxies. S. E. Okoye.
Monthly Notices Roy. Astron. Soc., Vol. 160, 339 - 348 (1972).

Values of the X-ray emission expected from nearby radio galaxies on the basis of the inverse Compton effect are derived. The theoretical results are discussed in the context of three nearby radio galaxies, for which recent observational data are available and from which X-rays have reportedly been detected.

158.099 Nota sobre la fotometría de NGC 3783.
J. L. Sersic.
Obs. Astron. Municipal Rosario, (*Argentina*), Bol. No. 2, p. 5 - 10 (1972).

We discuss the photometric characteristics of the brightness distribution of this Seyfert galaxy and establish the existence of three subsystems. We give the total photographic magnitude and that of the nuclear region.

158.100 The infrared variability of a dust model for Seyfert galaxies. N. Kaneko, K. Toyama, M. Nishimura.
Astrophys. Space Sci., Vol. 18, 121 - 127 (1972).

Assuming a dust model of Rees et al. (1969), we study the time dependence of the infrared emission from grains which absorb variable ultraviolet radiation. A time scale for the variability is shown to be fairly small compared with the light travel-time across the infrared emitting region, of which size depends on the wavelength strongly. Comparison of calculated infrared light curves with observations of NGC 4151 leads to the conclusion that the variability at 2.2 μ is consistent with interpretation by the infrared emission on the basis of thermal process.

158.101 A neutral hydrogen survey of the galaxy M 33. I. Observations. W. Huchtmeier.
Astron. Astrophys., Suppl. Ser., Vol. 7, 397 - 416 (1972).

A survey of the neutral hydrogen of the nearby galaxy M 33 has been performed with the 200 m transit telescope of the Nançay field station of Meudon Observatory. A bandwidth of 60 kHz and a telescope beamwidth at half power of 4 min of arc in right ascension and 24 min of arc in declination were used to get detailed information on the H I in this galaxy. This paper describes the equipment and the methods of observation and data reduction, and presents the observations as drift scans, which give the most detailed information on the observational material.

158.102 On the radio emission of the galaxy Markarian 6.
H. M. Tovmassian, R. Šramek.
Astron. Tsirk., No. 715, 2 pp. (1972). In Russian.

158.103 On the spectrum of the Markarian galaxy 388.
M. A. Arakelian, E. A. Dibaj, V. F. Esipov.
Astron. Tsirk., No. 717, p. 7 (1972). In Russian.

158.104 On the thermal component of radiation from the nuclei of galaxies in "excited" and "ground" state. I. "Excited" nuclei. L. M. Ozernoy.
Astron. Tsirk., No. 718, p. 1 - 3 (1972). In Russian.

158.105 On the thermal component of radiation from the nuclei of galaxies in "excited" and "ground" state. II. "Quiescent" nuclei. L. M. Ozernoy.
Astron. Tsirk., No. 718, p. 3 - 6 (1972). In Russian.

158.106 Eksplodējošās galaktikas. U. Dzērvitis.
Zvaigžņotā debess, 1972. gada vasara, p. 1 - 13.

158.107 Variability of extragalactic sources at 10 microns.
G. H. Rieke, F. J. Low.
Astrophys. Journ.,(*Letters*), Vol. 177, L115 - L119 (1972).

Variations have been detected at 10 μ in the fluxes from NGC 1068, NGC 4151, and 3C 273. The observations virtually eliminate the possibility that the infrared fluxes from these objects are generated thermally.

158.108 The compact central region of the galaxy NGC 1614.
M.-H. Ulrich.
Astrophys. Journ., Vol. 178, 113 - 118 (1972).

Spectrographic investigations of the peculiar galaxy NGC 1614 (or II Zw 15) with a dispersion of 28 Å mm⁻¹, and scales perpendicular to the dispersion of 24″ and 13.″5 mm⁻¹, have given the following results: (a) The inclination of the emission lines on spectra taken in P.A. 0° and 30° is interpreted as due to rotation of the galaxy and gives a mass $M \simeq 1.5 \times 10^9 M_\odot$ for the central region, 570 pc in radius. (b) A wing of Hα extending 10 Å to the blue is interpreted as evidence for an outflow of ionized gas from the nucleus of the galaxy and toward the observer with a radial velocity up to 450 km s⁻¹ with respect to the nucleus.

158.109   NGC 2992 and the blue stellar object Weedman No. 2.
E. M. Burbidge, P. A. Strittmatter, H. E. Smith, H. Spinrad.
Astrophys. Journ., *(Letters)*, Vol. 178, L43 - L46 (1972).

Observations of the possible triple system consisting of NGC 2992, NGC 2993, and the blue stellar object Weedman 2 are reported. It is shown that, although Weedman 2 has no detectable emission lines, its continuous energy distribution is similar to those found in quasi-stellar objects. A luminous bridge extends from NGC 2992 toward Weedman 2, and the possibility that this could be due to excitation by Weedman 2 is discussed. Both galaxies have strong emission-line spectra. The system may provide further evidence for the association of QSOs with bright galaxies.

158.110   The radio emission and the nuclei of spiral galaxies.
H. M. Tovmassian.
Astrophys. Journ., *(Letters)*, Vol. 178, L47 - L49 (1972).

It is shown that radio emission occurs mainly in galaxies that have starlike nuclei or other evidences of nuclear structure.

158.111   IC 3576: an unusual spiral galaxy in Virgo.
B. Margon, H. Spinrad, C. Heiles, H. Tovmassian, E. Harlan. S. Bowyer, M. Lampton.
Astrophys. Journ., *(Letters)*, Vol. 178, L77 - L80 (1972).

We have detected a weak, soft X-ray source in Virgo. If the positional information from our experiments is combined with that for the *Uhuru* source 2U 1231+7, the resulting error box of size 0.4 square degrees is centered on IC 3576, an unusual spiral galaxy. Several direct photographs of this galaxy have been obtained. We have observed neutral hydrogen emission at 21 cm from IC 3576, and have obtained upper limits on emission at 5, 1.4, and 0.43 GHz. The galaxy has an extraordinarily large ratio of H I to optical luminosity. Regardless of the identification of the X-ray source, the optical and radio data are difficult to reconcile.

158.112   On the luminosity function of radiogalaxies.
A. Sołtan.
Acta Astron., Vol. 22, 261 - 265 (1972).

The luminosity function of radiogalaxies is determined over the range of radio power $22 < \log P_{178} < 26$ at frequency 178 MHz on the basis of the complete sample of radiogalaxies identified with objects from 3C catalogue down to limiting visual magnitude $m_v = 15$.

158.113   Do elliptical galaxies rotate?
I. L. Genkin, L. M. Genkina.
Trudy Astrofiz. Inst., *Alma-Ata*, Vol. 19, 8 - 12 (1972). In Russian.

Indirect evidence of the rotation of elliptical galaxies is considered.

158.114   On the wave nature of rotation curves of galaxies.
N. G. Makarenko.
Trudy Astrofiz. Inst., *Alma-Ata*, Vol. 19, 13 - 16 (1972). In Russian.

Analyzing the rotation curves of galaxies, regular deviations from smoothness (velocity waves) are noted. This paper indicates how to construct a simple stochastic model based upon weak interactions in galaxies that gives rise to an analogous pattern. By means of the stochastic integrals the frequency spectrum associated with that model is obtained and discussed.

158.115   On "pairing" of galaxies and quasars.
L. M. Ozernoy.
Astron. Zhurn. Akad. Nauk SSSR, Vol. 49, 1148 - 1152 (1972). In Russian. English translation in Soviet Astron. AJ, Vol. 16, No. 6.

The relation between the angular distance of a quasar from a nearby galaxy and the redshift of this galaxy, found by G. Burbidge, O'Dell and Strittmatter for five galaxy-quasar pairs, disappears if we consider a more complete sample of "pairing" systems.

158.116   Identification of the CSV 6150 with a galaxy.
N. E. Kurochkin.
Inform. Bull. Variable Stars, (IAU Commission 27), Konkoly Obs., Budapest, No. 743 (1972).

158.117   The energy distribution of NGC 1068 and of other stellar systems.   R. E. Schild.
Astrophys. Journ., Vol. 178, 617 - 621 (1972).

The energy distribution of the nonnuclear portions of Seyfert galaxy NGC 1068 is compared with those of M81 and M51; a good correspondence between stellar population and form type on the Morgan system is found. Energy distributions for NGC 1068, M3, M67, M87, M11, and NGC 188 are intercompared in a further discussion of the stellar population of the nonnuclear portion of NGC 1068. An unexpected depression in the ultraviolet continuum of M81 relative to M87 is noted.

158.118   Resolution of one of the companions to M31.
S. van den Bergh.
Astrophys. Journ., *(Letters)*, Vol. 178, L99 (1972).

The dwarf spheroidal galaxy And III, which appears to be a companion to the Andromeda galaxy, has been resolved into stars with the 200-inch (5-m) Hale telescope.

158.119   Faint surface brightness features between NGC 7331 and Stephan's Quintet.
H. Arp. J. Kormendy.
Astrophys. Journ., *(Letters)*, Vol. 178, L101 - L103 (1972).

Limiting exposures with the 48-inch Schmidt telescope have been taken in order to search for optical connections between NGC 7331 and the region of Stephan's Quintet. The superposition print of six such plates reveals faint nebulous material extending southwestward from NGC 7331 toward Stephan's Quintet, and also material reaching several galaxy diameters to the west.

158.120   Preliminary evidence for bursts of star formation in giant elliptical galaxies.   S. van den Bergh.
Bull. American Astron. Soc., Vol. 4, 409 (1972). – Abstr. AAS.

158.121   IC3576: an unusual spiral galaxy in Virgo.
B. Margon, H. Spinrad, C. Heiles, S. Bowyer, M. Lampton, G. Tovmassion, E. Harlan.
Bull. American Astron. Soc., Vol. 4, 413 (1972). – Abstr. AAS.

158.122   Redshifts of a BSO and galaxies in the vicinity of the radio source RN 8.
J. S. Miller, L. Robinson, E. J. Wampler.
Bull. American Astron. Soc., Vol. 4, 413 (1972). – Abstr. AAS.

158.123   Spectroscopic observations of some faint radio galaxies.   H. Spinrad, H. E. Smith.
Bull. American Astron. Soc., Vol. 4, 415 (1972). – Abstr. AAS.

158.124   On the evolution of evolutionary corrections.
B. M. Tinsley.
Bull. American Astron. soc., Vol. 4, 416 (1972). – Abstr. AAS.

158.125   Radio continuum emission at 21 cm near Stephan's Quintet.   R. J. Allen, J. W. Hartsuiker.
Nature, Vol. 239, 324 - 325 (1972).

A report is given of observations made in the region of Stephan's Quintet using the 24″ beam of the Westerbork Synthesis Radio Telescope. The observations reveal a small-diameter source centered on the nucleus of NGC 7319 and a peculiar arc of emission located between NGC 7319 and NGC 7318. Several possible explanations are discussed. The most favourable model could be that of an intergalactic bow shock occurring during the accretion of cluster gas by NGC 7319.

158.126 Compass in the world of galaxies. V. Komarov.
Nauka i zhizn', 1972, No. 8, p. 9 - 12. In Russian.

158.127 Radial velocities in the tail of NGC 4676 A.
J. C. Theys, E. A. Spiegel, J. Toomre.
Publ. Astron. Soc. Pacific, Vol. 84, 851 - 853 (1972).
A spectrum of the long tail of NGC 4676 A shows [O II] $\lambda 3727$ in emission. The radial velocity measured from one plate varies by about 400 km sec^{-1} along the length of the tail.

158.128 Photometry of elliptical galaxies in multiple systems.
S. M. Faber.
Thesis, Harvard Univ., Cambridge, Mass. [Available from Univ. Microfilms, Ann Arbor, Mich., USA. Order No. 72-15056], 206 pp. (1972).
Elliptical galaxies, globular clusters and stars have been observed on a 10-colour intermediate-band photometric system designed for the study of the integrated spectra of old stellar populations. Observations of 31 elliptical galaxies, largely members of double galaxies or small groups, are presented. The technique of quadratic programming as directed to the problem of galaxy synthesis is described.

158.129 On the mass-to-light ratio of M 87.
F. Bertola, M. Capaccioli.
Mem. Soc. Astron. Italiana, Nuova Ser., Vol. 43, 539 (1972). Letter.

158.130 Multi-colour photometry of five SBc galaxies: NGC 925, NGC 1073, NGC 3359, NGC 4088, and NGC 7741. I. I. Pronik.
Izv. Krymskoj Astrofiz. Obs., Vol. 45, 162 - 171 (1972). In Russian.
Multi-colour photometry of bars and spiral arm patches for five SBc galaxies (NGC 925, NGC 1073, NGC 3359, NGC 4088 and NGC 7741) has been carried out. The spectral energy distribution of bars was found to be similar to that of the central parts of ordinary type S and I galaxies. It is possible that there is an ultraviolet excess in some of them. All bars are redder than the spiral arm patches.

158.131 U, B, V photometry of irregular galaxies NGC 5363, NGC 5360. A. I. Shapovalova.
Problems of cosmic physics. Vyp. (No.) 7, (see 003.028), p. 137 - 149 (1972). In Russian.
The results of a detailed three-color photometry of two M 82 type galaxies — NGC 5363, NGC 5360 are presented. The integral magnitudes and $U-B$, $B-V$ color indices are determined; the absolute magnitude of NGC 5363 is $M_B = -19^m.65$. The distribution of brightness and colors along the axes of the galaxies are given. $U-B$, $B-V$ diagrams are constructed. The dependences of the relative intensities and the mean surface brightnesses on the colors are investigated. A similarity of NGC 5363 with M 82 and NGC 5360 with NGC 3077 is marked.

158.132 General results of an investigation of Sculptor-type dwarf galaxies. V. E. Karachentseva.
Problems of cosmic physics. Vyp. (No.) 7, (see 003.028), p. 150 - 172 (1972). In Russian.
The following questions are considered: 1) the Sculptor-type dwarfs in the Local Group (common description, distribution of surface brightness and structure); 2) problems of terminology and determination of type; 3) the Sculptor-type dwarfs outside the Local Group (superclusters, clusters, groups of galaxies); 4) the detection of physical systems as normal

galaxy—a Sculptor-type dwarf galaxy; 5) some statistical properties of the Sculptor-type dwarf galaxies.

158.133 New Seyfert-type objects from the IIIrd and IVth lists of galaxies with UV-continuum.
M. A. Arakelyan, E. A. Dibaj, V. F. Esipov.
Astron. Tsirk., No. 722, p. 3 - 5 (1972). In Russian.

158.134 The Sculptor-type dwarf galaxies detected in the zone $\delta = -30°$ of the Palomar Sky Survey.
V. E. Karachentseva.
Astron. Tsirk., No. 723, p. 1 - 2 (1972). In Russian.

158.135 Ring galaxies without nucleus.
B. A. Vorontsov-Velyaminov.
Astron. Tsirk., No. 731, p. 3 - 5 (1972). In Russian.

158.136 On the optical variability of the N-galaxy 3C 371.
V. B. Nebelitskij.
Astron. Tsirk., No. 732, p. 3 - 5 (1972). In Russian.

158.137 Stellar component in the radiation of N-galaxies?
B. V. Komberg, V. M. Lyutyj.
Astron. Tsirk., No. 735, p. 1 - 4 (1972). In Russian.

158.138 Hubble's object N 9 in the region of the Andromeda nebula — a probable galaxy.
A. S. Sharov, V. F. Esipov.
Astron. Tsirk., No. 741, p. 7 - 8 (1972). In Russian.

Catalogue of selected compact galaxies and of posteruptive galaxies.   See Abstr. 003.014.

Galaxies.   See Abstr. 003.132.

Resolution enhancement of astronomical spectra. See Abstr. 031.015.

Morphology of rapid cosmic processes. See Abstr. 061.018.

Electron scattering in spherically expanding envelopes.   See Abstr. 064.048.

Hydrodynamic model calculations for supermassive stars II. The collapse and explosion of a nonrotating $5.2 \times 10^5 M_\odot$ star.   See Abstr. 065.057.

Stellar evolution in elliptical galaxies. See Abstr. 065.093.

Stellar evolution in elliptical galaxies. See Abstr. 065.096.

The evolution of stars and galaxies; condensation or expansion?   See Abstr. 065.150.

Observations of planets, nebulae, and galaxies at 350 microns.   See Abstr. 091.033.

UBV photoelectric observations: I. Stars within 25 parsecs of the sun; II. Stars in quasar and galaxy fields; III. Stars in Kapteyn Selected Areas; IV. Miscellaneous stars. See Abstr. 113.060.

Die blauen Objekte.   See Abstr. 113.064.

Review of results in infrared space astronomy. See Abstr. 114.044.

Methods to estimate the contribution of supernovae

to the activity of quasars and nuclei of galaxies. See Abstr. 125.022.

High resolution observations of 3C 390.3 at 2.7 and 5 GHz.   See Abstr. 141.003.

The emission-line spectrum of Cygnus A. See Abstr. 141.004.

Quasars as images of Seyfert nuclei. See Abstr. 141.007.

Positions and flux densities at 1415 MHz of 5C3 sources near M31.   See Abstr. 141.010.

Physical associations between quasi-stellar objects and galaxies.   See Abstr. 141.015.

Interpretation of rotation measures of radio sources. II.   See Abstr. 141.022.

Statistical studies of the evolution of extragalactic radio sources. II. Radio galaxies.   See Abstr. 141.030.

Statistical studies of the evolution of extragalactic radio sources. III. Interpretation of source counts and discussion.   See Abstr. 141.031.

A possible mechanism for energy release in quasars and Seyfert nuclei.   See Abstr. 141.050.

Identification of 4C sources with galaxies. See Abstr. 141.064.

The redshifts of quasi-stellar objects and associated galaxies.   See Abstr. 141.077.

The phenomenon of QSS.   See Abstr. 141.081.

Modell der Variabilität von Quasaren und Seyfert-Galaxien.   See Abstr. 141.082.

Observations of the radio source PKS 0123–01 at 5000, 408, and 80 MHz.   See Abstr. 141.093.

Positions and some identifications for 111 sources of about 1 flux unit at 408 MHz.   See Abstr. 141.094.

Observed anisotropy in the distribution of radio sources.   See Abstr. 141.100.

The distribution of redshifts of quasi-stellar objects and related emission-line objects.   See Abstr. 141.108.

Transcontinental interferometry at 606 MHz. See Abstr. 141.117.

Spectral flux densities of radio emission from discrete sources at 3.5 cm wavelength. II.   See Abstr. 141.130

Asymmetry of soft X-ray emission near M87. See Abstr. 142.078.

Galactic bridges and tails.   See Abstr. 151.039.

Search for globular clusters near the centre of the Andromeda nebula.   See Abstr. 154.010.

Spiral structure and kinematics of the Galaxy from a study of the H II regions. Fabry—Perot interference methods applied to ionized hydrogen.   See Abstr. 155.044.

Il nucleo galattico.   See Abstr. 155.051.

The tilt of the Large Magellanic Cloud and the galactical binary (sextuplet) Andromeda nebula and Milky Way. See Abstr. 159.004.

The cluster of compact galaxies Zw Cl 0152+33. See Abstr. 160.012.

Coherence in the distribution of Abell clusters. See Abstr. 160.026.

The redshift-distance relation. III. Photometry and the Hubble diagram for radio sources and the possible turn-on time for QSOs.   See Abstr. 162.051.

Structures of magnetic fields in the universe and galaxies.   See Abstr. 162.060.

Errata

158.901   Corrections to the Reference Catalogue of Bright Galaxies (second list).
G. de Vaucouleurs, A. de Vaucouleurs.
Publ. Astron. Soc. Pacific, Vol. 84, 461 (1972).

158.902   Corrections to the Reference Catalogue of Bright Galaxies (third list).   H. G. Corwin, Jr.
Publ. Astron. Soc. Pacific, Vol. 84, 462 - 463 (1972).

158.903   Erratum: "A new outstanding chain of galaxies" [Astron. Tsirk., No. 654, p. 1 - 2 (1971). In Russian]
B. A. Vorontsov-Velyaminov.
Astron. Tsirk., No. 679, p. 8 (1972).

159 Magellanic Clouds

159.001 Dark nebulae in the Large Magellanic Cloud.
P. W. Hodge.
Publ. Astron. Soc. Pacific, Vol. 84, 365 - 372 (1972).

Schmidt plates in three colors from the Curtis and ADH Schmidt telescopes were searched for dark nebulae. Positions, dimensions, and photometric properties are given for 68 identified dark nebulae. They are generally larger than the well-known examples in the Milky Way. Many show alignment with the bar, and there is a strong suggestion of an elongation of the dust clouds parallel to the magnetic field.

159.002 On the extended van Wijk sequence in the Large Magellanic Cloud. A. Ardeberg.
Astron. Astrophys.,Vol. 19, 384 - 387 (1972).

Ten stars in the van Wijk sequence in the Large Magellanic Cloud have been observed extensively in the UBV system using 5 E regions for standards. In V and $(B-V)$ the results obtained are in very good agreement with previous results. However, for the $(U-B)$ colour indices the differences are considerable.

159.003 Far-ultraviolet observations of the Large Magellanic Cloud. T. L. Page, G. R. Carruthers.
Bull. American Astron. Soc., Vol. 4, 331 (1972). – Abstr. AAS.

159.004 The tilt of the Large Magellanic Cloud and the galactical binary (sextuplet) Andromeda nebula and Milky Way. S. I. Gaposhkin.
Bull. American Astron. Soc., Vol. 4, 331 - 332 (1972). Abstr. AAS.

159.005 Electronographic photometry of star clusters in the Magellanic Clouds–IV. The colour–magnitude diagram of NGC 419. M. F. Walker.
Monthly Notices Roy. Astron. Soc., Vol. 159, 379 - 388 = Contr. Lick Obs., No. 371 (1972).

Electronographic magnitudes and colours of 146 stars in and near the cluster NGC 419 in the Small Magellanic Cloud have been measured to $V = 20.9$ on electrographs taken with a Spectracon image-converter attached to the f/7.5 focus of the Cerro Tololo 60-in. reflector. The results obtained are discussed.

159.006 A spectroscopic study of the blue globular clusters in the Large Magellanic Cloud.
P. J. Andrews, T. Lloyd Evans.
Monthly Notices Roy. Astron. Soc., Vol. 159, 445 - 452 (1972).

Spectra have been obtained for 15 blue globular clusters in the LMC. Spectral types on an arbitrary system correlate well with Johnson's reddening free parameter Q. The velocities follow the same rotation curve as that defined by the objects of extreme Population I.

159.007 Remarks on the comparison between the Sanduleak and the Fehrenbach-Duflot catalogs of LMC member stars. C. Fehrenbach, M. Duflot.
Astron. Astrophys., Vol. 21, 321 - 326 (1972). In French.

Some thirty stars for which the spectral classification gives a luminosity class between giants and supergiants are, actually, LMC members of absolute magnitude −6. Their spectra and colours set problems but no difference can be made between them and proven LMC members of the same magnitude.

159.008 Light variations of high luminosity O and B stars in the Large Magellanic Cloud. I. Appenzeller.
Publ. Astron. Soc. Japan, Vol. 24, 483 - 494 (1972).

A differential UBV photometry was carried out for 18 bolometrically very bright O and B stars in the Large Magellanic Cloud. Almost all observed stars were found to be definitely variable on a time scale of several days to several weeks. Only for the two hottest stars (O5 and O6) and for one star with a very peculiar spectrum no variations could be detected.

159.009 Remarque sur la structure du Petit Nuage de Magellan. A. Florsch.
Comptes Rendus Acad. Sci., Paris, Sér. B, Vol. 275, 763 - 764 (1972).

Dans une publication récente, nous avons émis l'hypothèse que le Petit Nuage de Magellan possédait une profondeur importante le long de la visée. L'étude des Céphéides fournit un argument supplémentaire en faveur de cette hypothèse.

159.010 Photometric standards in the Magellanic Clouds. C. J. Butler.
Commun. Dublin Inst. Advanced Studies, Ser. C. Dunsink Obs. Publ., Vol. 1, (No. 6), 133 - 192 (1972).

Photoelectric $UBVR$ and photographic UBV magnitudes are presented for standards in two regions in the LMC and one region in the SMC. Some tentative conclusions are drawn concerning the differences in the stellar population of the regions. Photometric spectral classification has been made for early type stars and the degree of interstellar reddening derived. From the derived spectral types and measured V-R colours it has been possible to evaluate the ratio of colour excesses E_{V-R}/E_{B-V}. All three regions show a higher value for this ratio than is normal within the Galaxy and this may indicate an unusual composition for the interstellar medium in the direction of the Magellanic Clouds. The ratio of E_{B-V} to the number of hydrogen atoms in the line of sight is evaluated and discussed.

159.011 Des étoiles supergéantes à très fortes raies d'hydrogène dans le Grand Nuage de Magellan.
C. Fehrenbach, M. Duflot, R. Burnage.
Comptes Rendus Acad. Sci. Paris, Sér. B, Vol. 275, 833 - 836 (1972).

Une trentaine d'étoiles supergéantes du Grand Nuage de Magellan, de magnitude absolue − 6,2 et de type spectral A0–F0 ont des raies d'hydrogène très fortes et une discontinuité de Balmer importante. Ces étoiles n'ont pas d'équivalent connu dans notre Galaxie.

159.012 Extension of the objective-prism techniques to the search of faint stars in the Large Magellanic Cloud.
N. Martin, E. Rebeirot.
Astron. Astrophys., Vol. 21, 329 - 335 (1972). In French.

The present objective-prism technique has been adapted to solve two problems. The first is to detect LMC OB stars of fourteenth magnitude. The second is to have a greater number of radial velocities especially in this very crowded region which is the bar.

159.013 The radio continuum of the Large Magellanic Cloud. I. The sources at 6 cm wavelength.
R. X. McGee, J. W. Brooks, R. A. Batchelor.
Australian Journ. Phys., Vol. 25, 581 - 597 = Separate print Div. Radiophys., C.S.I.R.O., Sydney (1972).

The Large Magellanic Cloud has been surveyed in the radio continuum at a wavelength of 6 cm with the 4' arc beam of the Parkes radio telescope. A catalogue of 95 sources and diagrams of their brightness contours are presented here.

159.014 The radio continuum of the Large Magellanic Cloud.

II. Continuum observations at 11 cm wavelength.
N. W. Broten.
Australian Journ. Phys., Vol. 25, 599 - 612 = Separate print
Div. Radiophys., C.S.I.R.O., Sydney (1972).
Contour diagrams are presented for the radio continuum brightness of the Large and Small Magellanic Clouds observed at a wavelength of 11 cm. The observations were made with the Parkes 64 m radio telescope, whose half-power beamwidth was 7'.35 arc.

159.015 **The radio continuum of the Large Magellanic Cloud. III. The sources at 11 cm wavelength.**
R. X. McGee, J. W. Brooks, R. A. Batchelor.
Australian Journ. Phys., Vol. 25, 613 - 617 = Separate print
Div. Radiophys., C.S.I.R.O., Sydney (1972).
A catalogue is given of radio sources observed in the Large Magellanic Cloud at 11 cm wavelength. A map of the radio brightness in the 30 Doradus region is included. The observations were made with a low-noise correlation radiometer and the Parkes 64 m radio telescope.

159.016 **The radio continuum of the Large Magellanic Cloud. IV. Spectra of sources.**
R. X. McGee, L. M. Newton.
Australian Journ. Phys., Vol. 25, 619 - 635 = Separate print
Div. Radiophys., C.S.I.R.O., Sydney (1972).
In this paper the sources of the LMC are classified on the basis of estimates of their spectral indices over as large a range as practicable. The sources with well-determined spectra fall into three classes: sources with spectral indices $\geqslant -0.20$, most of which are identified with HII regions from the Henize (1956) catalogue; sources with spectral indices < -0.20 and which have been identified with entries in the Henize catalog; and finally other sources with spectral indices < -0.20, which are likely to be external to the LMC.

159.017 **A catalogue of spectral classifications, radial velocities and UBV photometry for stars in the Large Magellanic Cloud.**
A. Ardeberg, J. P. Brunet, E. Maurice, L. Prévot.
Proc. Third Colloquium on astrophysics. Supergiant stars, Trieste 1971, (see 012.020), p. 144 - 146 (1972).

159.018 **Gamma radiation of Magellanic Clouds and metagalactic origin of cosmic rays.** V. L. Ginzburg.
Nature, Phys. Sci., Vol. 239, 8 - 9 (1972).
If the main part of cosmic rays observed near the earth have metagalactic origin the flux of gamma-rays (with energy $E_\gamma > 100$ MeV) from the Magellanic Clouds must be equal to 3×10^{-7} photons/cm^2 sec. On the author's opinion if the observed flux is considerably lower this would be the definite disproof of all metagalactic origin models.

Model atmosphere analysis of the A 3Ia–O supergiant HD 33579 in the Large Magellanic Cloud.
See Abstr. 064.007.

Radial velocities measurement with the coudé spectrograph of the 152-cm telescope (Observatoire de Haute Provence). See Abstr. 112.004.

Photométrie photoélectrique U.B.V. de 160 étoiles supergéantes O–B membres probables du Grand Nuage de Magellan. See Abstr. 113.055.

Photométrie en six couleurs de 12 supergéantes F et G du Grand Nuage de Magellan. See Abstr. 113.056.

The analysis of the Small Magellanic Cloud supergiant HD 7583. See Abstr. 114.008.

Astronomical observations with the Apollo 16 far-ultraviolet camera/spectrograph.
See Abstr. 114.056.

The $M_{bol} - \log T_e$ diagram of the brightest supergiants in the Large Magellanic Cloud. See Abstr. 115.020.

The R Coronae Borealis variables in the Large Magellanic Cloud. See Abstr. 122.019.

The influence of metal lines on the colors of cepheids in the Small Magellanic Cloud. See Abstr. 122.034.

Red supergiant variables of large amplitude in the SMC. See Abstr. 122.083.

Photometry of cepheids in the Magellanic Clouds. See Abstr. 122.084.

Further observations of HV 13055. See Abstr. 122.088.

Photometry of cepheids in the Magellanic Clouds. See Abstr. 122.117.

New variable stars in the Large Magellanic Cloud. See Abstr. 123.038.

Nova in Large Magellanic Cloud. See Abstr. 124.006.

Supernova remnants in the Large Magellanic Cloud. See Abstr. 125.002.

A supernova remnant in the Small Magellanic Cloud. See Abstr. 125.027.

Kinematics of planetary nebulae in the Large Magellanic Cloud. See Abstr. 133.012.

Optical candidate for SMC X-1. See Abstr. 142.094.

Discovery of the binary nature of SMC X-1 from *Uhuru.* See Abstr. 142.105.

The cluster NGC 330 in the SMC (Paper II): Hα emission in main sequence stars. See Abstr. 153.005.

Structure and dynamics of barred spiral galaxies, in particular of the Magellanic type. See Abstr. 158.096.

Errata

159.901 **Erratum: 'Spectrographic and photometric observations of supergiants and foreground stars, in the direction of the Large Magellanic Cloud'** [Astron. Astrophys., Suppl. Ser., Vol. 6, 249 - 309 (1972)].
A. Ardeberg, J. P. Brunet, E. Maurice, L. Prévot.
Astron. Astrophys., Suppl. Ser., Vol. 7, 395 (1972).

160 Clusters of Galaxies

160.001 Morphology of galaxies in clusters. I. The cluster A262. A. T. Kalloghlian.
Astrofizika, Vol. 8, 43 - 51 (1972). In Russian. — English translation in Astrophysics, Vol. 8, No.1.

On plates obtained with the 2-m telescope of the Tautenburg Observatory morphological types of galaxies in the cluster Abell 262 have been determined. Plates of the 1-m Schmidt telescope of the Byurakan Observatory and Palomar charts were also used.

160.002 Models and mass estimates for ionized intergalactic gas in the Coma cluster.
G. A. Welch, G. N. Sastry.
Astrophys. Journ., Vol. 175, 323 - 328 (1972).

Comparison is made of the ability of thermal, synchrotron, and inverse Compton mechanisms to match currently available observations of diffuse radiation (apparent dimension $\sim 45'$) in the central region of the Coma cluster.

160.003 Static properties of galaxies in the cluster Abell 2199. H. J. Rood, G. N. Sastry.
Astron. Journ., Vol. 77, 451 - 458, 525 (1972).

For 170 galaxies in a 2-square-degree area of Abell 2199, a catalogue is presented of radial and angular coordinates, major axes, ellipticities, position angles of major axis, color, and morphological type. The data are analyzed statistically.

160.004 The correlation of redshift with magnitude and morphology in the Coma cluster. W. G. Tifft.
Astrophys. Journ., Vol. 175, 613 - 625 (1972).

In this paper the data on the galaxies in the Coma cluster are considerably extended and the presence of nondynamical patterns in redshift-morphology and redshift-magnitude is demonstrated.

160.005 The structure of the Coma cluster of galaxies. H. J. Rood, T. L. Page, E. C. Kintner, I. R. King.
Astrophys. Journ., Vol. 175, 627 - 647 (1972).

Radial velocities in the Coma cluster, many of them new, are analyzed in conjunction with previous data on the density distribution. The cluster extends more than $200'$ in radius, with bright and faint galaxies distributed nearly the same. Some spirals are members. Several different determinations of M/L give values around 250. The cluster is stabilized by a "missing mass" some 7 times the mass of the galaxies, which must be distributed in the same way as the galaxies.

160.006 Relationship between X-ray luminosity and velocity dispersion in clusters of galaxies.
A. B. Solinger, W. H. Tucker.
Astrophys. Journ., (Letters), Vol. 175, L107 - L111 (1972).

The connection between the X-ray luminosity L_X and the dynamics of clusters of galaxies is considered. L_X is found to be strongly correlated with the velocity dispersion ΔV. The observed relationship can be explained most simply within the framework of a thermal-bremsstrahlung model with the X-ray luminous mass being a fraction of the cluster binding mass. The X-ray intensities or velocity dispersions are predicted for several clusters.

160.007 On the infall of matter into clusters of galaxies and some effects on their evolution.
J. E. Gunn, J. R. Gott III.
Astrophys. Journ., Vol. 176, 1 - 19 (1972).

A theory of infall of material into clusters of galaxies is developed and applied to the Coma cluster. It is suggested that the infall phenomenon is responsible for the growth of cluster galaxies. The generation of a hot intracluster medium is discussed and its relation to the observed absence of normal spirals in rich clusters investigated. The inference made earlier by Gott and Gunn that the observed X-ray luminosity of Coma puts severe constraints on the deceleration parameter q_0 is further elucidated. We discuss the relation of these phenomena to the morphology of clusters, and find that some observed regularities in their observed properties can be explained.

160.008 Absolute magnitudes of E and S0 galaxies in the Virgo and Coma clusters as a function of $U-B$ color.
A. Sandage.
Astrophys. Journ., Vol. 176, 21 - 30 (1972).

Three-color photoelectric data are listed for 25 galaxies in the Coma cluster and six faint dwarf ellipticals in the Virgo cluster. The correlation between standard isophotal apparent magnitude of the galaxies and their $U-B$ values has a correlation coefficient of $r = 0.94$, and extends over 8.5 magnitudes. K-corrections are computed and listed for $U-B$ and Q indices for $z \lesssim 0.03$. Applying the corrections to the Coma cluster permits the color-magnitude effect in Coma and Virgo to be compared, so as to test the zero-point difference between the clusters.

160.009 On the preference orientation of clusters of galaxies. A. V. Mandzhos, V. V. Telnjuk-Adamchuk.
Astron. Tsirk., No. 681, p. 1 - 2 (1972). In Russian.

160.010 3C 323.1: A QSO in a rich cluster of galaxies. A. Oemler, Jr., J. E. Gunn, J. B. Oke.
Astrophys. Journ., (Letters), Vol. 176, L47 - L50 (1972).

The QSO 3C 323.1 is at a projected distance of $6\rlap{.}'5$ from the center of a compact Zwicky cluster and has the same redshift as the cluster. The cluster is investigated and found to be a typical cluster of richness 1 on Abell's scale. The probability of a chance coincidence is small.

160.011 Approximate solutions of the relativistic gravitational field equations to describe clusters of galaxies.
M. W. Cook.
Australian Journ. Phys., Vol. 25, 299 - 305 (1972).

Approximate solutions to the Einstein field equations are found which describe a spherically symmetric inhomogeneity in a general Robertson-Walker model, i.e. one with an arbitrary equation of state. Reference is made to observed data to determine which categories of stellar objects may be described by the results.

160.012 The cluster of compact galaxies Zw Cl 0152+33. W. L. W. Sargent.
Astrophys. Journ., Vol. 176, 581 - 586 (1972).

Redshifts have been determined for 14 seventeenth-magnitude, mostly red, compact galaxies which form an elongated system at the center of the cluster Zw Cl 0152+33. The mean redshift is 26,304 km s^{-1} and the apparent distance modulus of the cluster is $(m - M) = 38.2$. It is argued that the galaxies in the elongated system do not form a linear chain, but are in a flattened system seen nearly edge-on. The galaxies on the outer parts of the region considered have a higher velocity dispersion σ_r than those in the center. As a result we find that the center of the cluster is probably a bound system while the outer members are escaping from the cluster on a timescale of about 10^9 years.

160.013 A reinterpretation of Abell's counts of clusters. M. Rowan-Robinson.
Astron. Journ., Vol. 77, 543 - 549 (1972).

A reexamination of Abell's counts of clusters indicates that they show the same evolutionary effect as quasars and radio galaxies. Possible interpretations in terms of the breakup of clusters, or of the extinction of galaxies, are discussed.

160.014 The distribution of galaxies in the cluster Abell 31.
N. A. Bahcall.
Astron. Journ., Vol. 77, 550 - 556 (1972).

The distribution of galaxies in the cluster Abell 31 ($\equiv$ HMS 0025 + 2223; $z = 0.16$) is derived from galaxy counts on plates taken with the Palomar 48-inch Schmidt, the Mount Wilson 100-inch, and the Hale 200-inch telescopes. The distributions are given for two different limiting magnitudes; a total of 790 galaxies were counted in the photo-red magnitude range $R \simeq 16\overset{m}{.}6$ to 19^m. It is shown that the observed distributions of galaxies are in good quantitative agreement with the distribution appropriate to an Emden isothermal gas sphere.

160.015 On the cosmological influence on insular systems.
A. V. Mandjos.
Vestn. Kiev. Un-ta, Ser. Astron., No. 14, p. 83 - 88 (1972). In Russian.

The cosmological influence on an extended astronomical system is investigated in terms of general relativity. The approximate expression of the cosmological force for an arbitrary cosmological model is obtained in the coordinates measured in principle. The problem of galaxy cluster stationarity, in particular of Coma, is discussed, considering a possible cosmological influence.

160.016 The problem of the stability of clusters of galaxies.
M. Karpowicz.
Postępy Astron., Vol. 20, 297 - 305 (1972). In Polish.

A number of attempts made during the last eighteen years at the solution of the problem of the stability of clusters of galaxies are considered. Two important approaches to it, the hypothesis of expansion and the assumption of the presence of invisible intergalactic matter, are shown to involve serious difficulties of observational and theoretical nature.

160.017 Orientation of galaxies and the Local Supercluster.
M. Reinhardt, M. S. Roberts.
Astrophys. Letters, Vol. 12, 201 - 206 (1972).

The indication that the flattening of a system of galaxies is correlated with the flattening of its constituent galaxies is tested for the Local Supercluster. Two independent sets of data are used. Both sets yield a positive, but modest, correlation that is suggestive of a true physical relation between the orientation of the angular momentum vector of member galaxies and the flattening of the Local Supercluster.

160.018 Remarks on the radial velocities of galaxies in the Virgo cluster.   G. A. Tammann.
Astron. Astrophys., Vol. 21, 355 - 359 (1972).

Redshifts for 122 probable members of the Virgo cluster are compiled and, where possible, corrected for systematic errors. The mean radial velocity of the cluster relativ to the galactic center is determined to 1141 ± 60 km s^{-1} and the velocity dispersion to $\sigma = 666$ km s^{-1}. There is no significant radial velocity difference between elliptical-lenticular galaxies and spiral galaxies. The velocity distribution of elliptical-lenticular galaxies as well as of spiral galaxies can reasonably well be approximated by a gaussian distribution.

160.019 Observations of the extended X-ray sources in the Perseus and Coma clusters from *Uhuru*.
W. Forman, E. Kellogg, H. Gursky, H. Tananbaum, R. Giacconi.
Astrophys. Journ., Vol. 178, 309 - 316 (1972).

The X-ray source in Perseus identified as NGC 1275 is found to have a finite angular extent of about 35 arc minutes. The improved location of the center of the emission is con-

sistent with NGC 1275. An improved location for the center of the Coma X-1 source is consistent both with the kinematic center of the cluster and with NGC 4874. These extended sources—Coma, Perseus, and the source in the Virgo cluster—may be a new class of X-ray objects associated with active galaxies in rich clusters.

160.020 The central part of the Coma cluster of galaxies. III.
R. Kh. Gainullina.
Trudy Astrofiz. Inst., *Alma-Ata*, Vol. 19, 36 - 45 (1972). In Russian.

This paper deals with the study of positions, flattenings and position angles of 497 spheroidal galaxies of the central part of the Coma cluster ($r \leqslant 100'$).

160.021 On the frequency function of galaxy flattenings for clusters of galaxies.   R. Kh. Gainullina.
Trudy Astrofiz. Inst., *Alma-Ata*, Vol. 19, 46 - 51 (1972). In Russian.

The necessity of rejecting the generally accepted suggestion of the random distribution of angles between the tangent plane and the equatorial planes of members of clusters of galaxies is pointed out. The integral equation for the frequency function of galaxy flattenings for clusters of galaxies in which the effect of favourable orientation of the equatorial planes of cluster members manifests itself is deduced.

160.022 Discovery of hot gas in the Coma cluster.
Yu. N. Parijsky.
Astron. Zhurn. Akad. Nauk SSSR, Vol. 49, 1322 - 1323 (1972). In Russian. English translation in Soviet Astron. AJ, Vol. 16, No. 6.

As a result of scattering of cold relict quanta on hot gas in clusters of galaxies dark spots in the background radiation are appeared. This phenomenon is observed in the direction of the Coma cluster.

160.023 The problem of galaxy clusters stability.
M. Karpowicz.
Urania Kraków, Vol. 43, 194 - 203 (1972). In Polish.

160.024 The Coma cluster is not bound by an intracluster gas.   S. Bowyer, J. Holberg, M. Lampton.
Bull. American Astron. Soc., Vol. 4, 410 (1972). – Abstr. AAS.

160.025 Soft X-ray flux of the Coma cluster of galaxies.
P. Gorenstein, P. Bjorkholm, B. Harris, F. R. Harnden, Jr.
Bull. American Astron. Soc., Vol. 4, 412 (1972). – Abstr. AAS.

160.026 Coherence in the distribution of Abell clusters.
M. G. Hauser, P. J. E. Peebles.
Bull. American Astron. Soc. Vol. 4, 412 (1972). – Abstr. AAS.

160.027 The observation of relic radiation as a test of the nature of X-ray radiation from the clusters of galaxies.   R. A. Sunyaev, Ya. B. Zeldovich.
Comments Astrophys. Space Phys., Vol. 4, 173 - 178 (1972).

It is assumed that clusters of galaxies form an important class of powerful X-ray sources, possibly giving the main contribution to the X-ray background radiation of the universe. What is the nature of these sources? What physical mechanisms give the observed X-ray radiation? The observations of small perturbation in angular distribution of relic radiation can give an answer to these questions.

A new identification for PKS 2204—54.
See Abstr. 141.063.

The kinetics of gravitational clustering.
See Abstr. 151.021.

The estimation of masses of individual galaxies in clusters of galaxies.   See Abstr. 158.056.

Blue emission-line galaxies in rich clusters. See Abstr. 158.060.

Indication of an intergalactic extinction effect connected with the largest clusters of galaxies.

See Abstr. 161.003.

Formation of clusters of galaxies; protocluster fragmentation and intergalactic gas heating.   See Abstr. 162.005.

The redshift-distance relation. II. The Hubble diagram and its scatter for first-ranked cluster galaxies: A formal value for q_0.   See Abstr. 162.050.

161   Intergalactic Matter

161.001   X-ray emission from intergalactic gas in the neighbourhood of galaxies.   R. Hunt, D. W. Sciama.
Nature, Vol. 238, 320 - 323 (1972).
Massive galaxies may have X-ray coronas associated with accretion of hot intergalactic gas. X-ray observations of Virgo A fit this hypothesis. Andromeda may also have an observable X-ray corona.

161.002   Possible photoionization of the intergalactic medium by hot white dwarfs.   J. G. Hills.
Astrophys. Letters, Vol. 12, 9 - 11 (1972).
Ionizing radiation from hot pre-white dwarfs (called UV stars) may profoundly affect the interstellar medium. It was noted that the scale height of these stars is several times that of the gaseous disk, which must allow much of their radiation to escape the galaxy. This preliminary note points out some of the effects of the integrated ionizing radiation from all galaxies on the intergalactic medium.

161.003   Indications of an intergalactic extinction effect connected with the largest clusters of galaxies.
I. Toborek.
Acta Astron., Vol. 22, 67 - 71 (1972).

161.004   Infall of intergalactic matter into the Galaxy.
D. P. Cox, B. Smith.
Bull. American Astron. Soc., Vol. 4, 316 (1972). — Abstr. AAS.

161.005   Theoretical models of photoionized intergalactic hydrogen.   J. Arons, D. W. Wingert.
Astrophys. Journ., Vol. 177, 1 - 15 (1972).
The effects of ionizing radiation emitted by quasi-stellar objects on intergalactic hydrogen are studied. The hydrogen is assumed to expand with the universe, and the amount of ionizing radiation is estimated from observations of the QSO luminosity function. The amount of ionizing radiation is shown to be sufficient to allow the existence of a universal medium with a temperature less than $10^4 \, °K$ and a density several times the smeared-out density of luminous matter in galaxies. Such a medium would escape detection with presently available techniques.

161.006   Intergalactic matter.   G. B. Field.
Annual Rev. Astron. Astrophys., Vol. 10, (see 003. 005), 227 - 260 (1972). — Contents: (1) Introduction; (2) Search for predominantly neutral gas; (3) Search for predominantly ionized gas; (4) Clustering and clumpiness; (5) Physical problems connected with a hot intergalactic medium; (6) Summary.

161.007   The possibility that nongaseous hydrogen supplies the missing cosmological mass.   T. W. Noonan.
Astrophys. Journ., Vol. 178, 317 - 318 (1972).
Considerations of intergalactic extinction and thermal equilibrium with intergalactic radiation fields indicate that intergalactic particles with radii larger than 2 mm and a temperature of order 3°K may exist in sufficient quantity to supply the missing cosmological mass. However, the vapor pressure of solid hydrogen imposes an upper limit of 1.7°K. Although there is the possibility of "asteroids" of solid hydrogen, the existence of particles of solid or liquid hydrogen must be regarded as an impossibility.

Interstellare und intergalaktische Materie im Licht der Kosmogonie.   See Abstr. 131.095.

The origin and form of the galactic magnetic field. II. The primordial-field model.   See Abstr. 156.001.

Radiation-driven efflux and circulation of dust in spiral galaxies.   See Abstr. 158.054.

Infall of matter into galaxies.   See Abstr. 158.082.

Faint surface brightness features between NGC 7331 and Stephan's Quintet.   See Abstr. 158.119.

Models and mass estimates for ionized intergalactic gas in the Coma cluster.   See Abstr. 160.002.

Formation of clusters of galaxies; protocluster fragmentation and intergalactic gas heating.   See Abstr. 162.005.

162 Structure and Evolution of the Universe, Cosmology

162.001 **On the cosmological equations in a universe with small scale condensations.** A. H. Nelson.
Monthly Notices Roy. Astron. Soc., Vol. 158, 159 - 175 (1972).

The metric describing a universe containing small scale condensations is divided into two terms. One term represents the large scale, global development of space-time, while the other represents the small scale effects of local condensations. The global term is approximated by an average of the metric over a region containing many condensations. Equations to be satisfied by the individual terms of the metric are then derived from a combination of the Einstein equations and the averaged Einstein equations.

162.002 **Exact expressions for the properties of the zero-pressure Friedmann models.** D. Edwards.
Monthly Notices Roy. Astron. Soc., Vol. 159, 51 - 66 (1972).

Exact expressions in terms of Jacobian elliptic functions are obtained for the properties of the expanding singular cosmological models of the Friedmann kind, assuming zero pressure and a general cosmological constant.

162.003 **Quarks as a thermometer for cosmologies.**
S. Frautschi, G. Steigman, J. Bahcall.
Astrophys. Journ., Vol. 175, 307 - 322 (1972).

We derive the consequences for conventional cosmologies of the Hagedorn type of hadron spectrum which implies a "warm" limiting temperature of order $kT \approx 160$ MeV. With initial temperatures of this order, the predicted quark density relative to ordinary baryons is consistent with present experimental limits if the quark mass exceeds 9 GeV. We also discuss predictions for the quark density in several less conventional cosmologies (mixmaster, Brans-Dicke, Lemaître, steady state, and Klein-Alfvén) and with forms of the hadron density of states that differ from Hagedorn's proposed form.

162.004 **Particles of antimatter.** Yu. D. Prokoshkin.
Naturwissenschaften, 59. Jahrgang, p. 281 - 284 (1972).

New antimatter particles have been produced. The experimental work is summarized. Future experiments may elucidate the role that antimatter plays in our universe.

162.005 **Formation of clusters of galaxies; protocluster fragmentation and intergalactic gas heating.**
R. A. Sunyaev, Ya. B. Zeldovich.
Astron. Astrophys., Vol. 20, 189 - 200 (1972).

We are trying to give a consecutive theoretical picture based on modern cosmology of galaxies formation. The paper deals with dynamics of shock waves, heat processes in the gas, its radiation and contribution to the X-ray background as well as the process of fragmentation of the cold gas to form the above mentioned objects, starting from the cosmological model of the hot universe with small density perturbations.

162.006 **Robertson-Walker cosmology and Friedmann cosmology.** M. Heller.
Postępy Astron., Vol. 20, 241 - 250 (1972). In Polish.

One introduces a differentiation between the Robertson-Walker and Friedmann cosmologies. A classification of the R-W cosmological models is given, according to the possible isometry groups.

162.007 **Scalar-tensor cosmology and the classical tests.**
S. K. Luke, G. Szamosi.
Astron. Astrophys., Vol. 20, 397 - 405 (1972).

In this note the relativistic theory of gravitation as pro-
posed by Brans and Dicke (1961) is compared with cosmological observations. The theory known also as scalar-tensor theory admits cosmological solutions of the Friedman type. The three classical tests of cosmology are calculated for a range of the parameters ρ_0 and $(d\phi/dt)_0$. Comparing observational data with theory suggests that the absolute visual magnitude of the first rank galaxy is -22, the cosmic density is of the order 10^{-30} g cm^{-3} and the average linear diameter of the galaxies is 4.3 kpc. Finally, the possibility of deciding for or against the scalar-tensor theory is considered.

162.008 **Perturbations in an expanding universe of free particles.** J. M. Stewart.
Astrophys. Journ., Vol. 176, 323 - 332 (1972).

In the canonical big-bang cosmology, one of the major energy sources in the temperature range $10^{10}°> T > 10^5$ °K would be collision-free neutrinos. These might be expected to interact with gravitational radiation in a radically different way from the usual cosmological models filled with collision-dominated radiation. The Jeans swindle, Newtonian cosmology, and relativistic cosmology are discussed using a kinetic-theory approach, and the same qualitative results are found in all three cases.

162.009 **Anisotropic cosmology and cosmic structure.**
A. D. Chernin, A. N. Shvarts.
Dokl. Akad. Nauk SSSR, Ser. Mat. Fiz., Vol. 205, 1057 - 1058 (1972). In Russian.

162.010 **A non-uniform relativistic cosmological model.**
W. B. Bonnor.
Monthly Notices Roy. Astron. Soc., Vol. 159, 261 - 268 (1972).

The model is constructed to accommodate the variable cosmic density claimed by de Vaucouleurs. It is an exact solution of Einstein's equations for spherically symmetric dust flows, and is a special case of a class of solutions given by Tolman. The model has an imploding big bang, and evolves from extreme inhomogeneity to an eventual Einstein–de Sitter universe. It seems to be inconsistent with the observations of red shift versus luminosity distance.

162.011 **The interaction of primordial gravitational waves with groups of galaxies.** M. J. Rees.
Monthly Notices Roy. Astron. Soc., Vol. 159, 11P - 14P (1972).

Two processes whereby very long wavelength gravitational waves may cumulatively feed energy into a cluster of galaxies during its lifetime are considered, and found to be unimportant. The only potentially observable manifestation of primordial gravitational radiation is the velocity dispersion induced by waves currently interacting with groups of galaxies, as discussed in an earlier paper by the present author.

162.012 **The generalized Taub solution.** J. Horský, J. Novotný.
Bull. Astron. Inst. Czechoslovakia, Vol. 23, 266 - 267 (1972).

162.013 **Observation of relict radio radiation fluctuations as a method to distinguish adiabatic from other forms** of mass density disturbances in the universe leading to formation of galaxies. Ya. B. Zel'dovich, A. Kh. Rakhmatulina, R. A. Syunyaev.
Izv. vyssh. ucheb. zavedenij. Radiofizika, Vol. 15, 161 - 171 (1972). In Russian. – Abstr. in Referativ. Zhurn. 51. Astron., 8.51.704 (1972).

162.014 **The Cauchy problem for the Dirac equation in a de Sitter space and the anticommutator.**
N. S. Shavokhina.
Teor. i mat. fiz., Vol. 10, 412 - 423 (1972). In Russian.
Abstr. in Referativ. Zhurn. 51. Astron., 8.51.760 (1972).

162.015 **The possibility that non-gaseous hydrogen supplies the missing cosmological mass.** T. W. Noonan.
Bull. American Astron. Soc., Vol. 4, 340 (1972). – Abstr. AAS.

162.016 **Evidence of the relict nature of the black-body radiation of the Metagalaxy.** V. N. Kurilchik.
Astron. Tsirk., No. 709, p. 1 - 3 (1972). In Russian.

162.017 **On the physical nature of cosmic electromagnetic absorption. V: The Einstein–de Sitter cosmology with plasma coupled to radiation at non-relativistic temperature.** R. Burman.
Observatory, Vol. 92, 86 - 89 (1972).

This paper deals with a stage of an Einstein–de Sitter universe during which the cosmic medium is fully ionized hydrogen strongly coupled to the background radiation, with the electron gas having non-relativistic temperature. An approximation is obtained for the refractive index of electromagnetic waves, and the effects of collisions and radiation reaction on both retarded and advanced waves are investigated.

162.018 **On the physical nature of cosmic electromagnetic absorption. VI: The Einstein–de Sitter cosmology with plasma coupled to radiation at relativistic temperature.** R. Burman.
Observatory, Vol. 92, 90 - 93 (1972).

Whereas Part V dealt with the epoch in which the temperature has fallen below the level at which the thermal motions of the plasma electrons are relativistic, this part treats the earlier epoch in which these motions are highly relativistic.

162.019 **On the physical nature of cosmic neutrino absorption. I: Cosmological models with continuous creation.** R. Burman.
Observatory, Vol. 92, 128 - 131 (1972).

This paper deals with certain conformally flat cosmological models of zero spatial curvature, with matter density kept constant by continuous creation; the steady-state universe is included as a special case. The absorption process for retarded neutrino fields is studied in both the current–current and photon–neutrino weak interaction theories.

162.020 **On the physical nature of cosmic neutrino absorption. II: Cosmological models without continuous creation.** R. Burman.
Observatory, Vol. 92, 131 - 135 (1972).

This paper deals with certain conformally flat cosmological models of zero spatial curvature, without continuous creation of matter. The models include the spatially flat form of the Brans–Dicke universe and, its particular case, the Einstein–de Sitter universe. The absorption process for retarded neutrino fields is studied in both the current–current and photon–neutrino weak interaction theories.

162.021 **The effect of energy separation on the emission spectrum in a hot universe.**
Ya. B. Zeldovich, A. F. Illarionov, R. A. Syunyaev.
Zhurn. ehksperim. i teor. fiz., Vol. 62, 1217 - 1227 (1972). In Russian. – Abstr. in Referativ. Zhurn. 51. Astron., 9.51.786 (1972).

162.022 **Hydrodynamic motions and vacuum stage in an anisotropic cosmological model.**
A. D. Chernin.
Dokl. Akad. Nauk SSSR, Ser. Mat. Fiz., Vol. 206, 62 - 63

(1972). In Russian.

162.023 **On some properties of cosmological models.**
S. P. Novikov.
Zhurn. ehksperim. i teor. fiz., Vol. 62, 1977 - 1989 (1972). In Russian. – Abstr. in Referativ. Zhurn. 51. Astron., 10.51.617 (1972).

162.024 **About constructing a general cosmological solution of Einstein's equations which possess a time singularity.** V. A. Belinskij, E. M. Lifshits, I. M. Khalatnikov.
Zhurn. ehksperim. i teor. fiz., Vol. 62, 1606 - 1613 (1972). In Russian. – Abstr. in Referativ. Zhurn. 51. Astron., 10.51.620 (1972).

162.025 **Extrapolation in cosmological models.**
V. P. Lebedev.
Nauch. dokl. vyssh. shkoly. Filos. n., 1972, No. 3, p. 64 - 71. In Russian. – Abstr. in Referativ. Zhurn. 51. Astron., 10.51.625 (1972).

162.026 **Relativistic non-zero pressure cosmology.**
R. F. Sisteró.
Astrophys. Space Sci., Vol. 17, 150 - 160 (1972).

A further extension of the theory of interacting matter-radiation cosmological models is presented. The neutrino contribution to the radiation field is explicitly included. A discussion and interpretation of the observables within the theory is given. Mean evolutionary corrections for galaxies are shown to be implied by these models. Finally, as an example we present a cosmological interpretation of quasars.

162.027 **A cosmological basis for hyperbolic velocity space.**
S. J. Prokhovnik.
Proc. Cambridge Philos. Soc., Vol. 67, 391 - 395 (1970).

It is shown that a cosmological model of light propagation associated with a uniformly expanding universe provides a physical significance to the hyperbolic velocity space by which this model can be described. Some implications of this result are discussed.

162.028 **On the collisional absorption of radio waves in cosmology.** R. R. Burman.
Publ. Astron. Soc. Japan, Vol. 24, 533 - 535 (1972).

The absorption of radio waves through electron–proton collisions is discussed for the steady-state and Einstein–de Sitter cosmologies.

162.029 **Chaos in cosmology.**
B. T. J. Jones, P. J. E. Peebles.
Comments Astrophys. Space Phys., Vol. 4, 121 - 128 (1972).

The idea that the structure in the universe originated from some primordial vortical motion has had a remarkably long history and central role in cosmological speculation. In the sixty years since then, astronomers have made a number of discoveries, but the vortex cosmogonies in the form of turbulence theories are still with us and still are the center of controversy. A survey on the turbulence theories is given and the comment is concluded with a brief survey of some of the simple constraints that we think are cause for serious concern with primeval turbulence in a hot big bang model.

162.030 **A hypothesis, unifying the structure and the entropy of the universe.** Ya. B. Zeldovich.
Monthly Notices Roy. Astron. Soc., Vol. 160, 1P - 3P (1972).

A hypothesis about the averaged initial state and its perturbations is put forward, describing the entropy of the hot universe (due to damping of short waves) and its structure (clusters of galaxies due to long wave perturbations).

162.031 **Reflections on "big bang" cosmology.**

R. A. Alpher, R. Herman.
G. Gamow Memorial Volume, (see 003.003), p. 1 - 14 (1972).

162.032 Conformal invariance in physics and cosmology.
F. Hoyle, J. V. Narlikar.
G. Gamow Memorial Volume, (see 003.003), p. 15 - 28 (1972).

162.033 Cosmology and microwave astronomy.
A. A. Penzias.
G. Gamow Memorial Volume, (see 003.003), p. 29 - 47 (1972).

162.034 On a model of the expanding universe.
G. Wataghin.
G. Gamow Memorial Volume, (see 003.003), p. 48 - 55 (1972).

162.035 Are the constants constant? E. Teller.
G. Gamow Memorial Volume, (see 003.003), p. 60 - 66 (1972).

162.036 New observations & old nucleocosmochronologies.
W. A. Fowler.
G. Gamow Memorial Volume, (see 003.003), p. 67 - 123 (1972).

162.037 L'aspect mécanique de l'expansion de l'Univers.
O. Onicescu.
Comptes Rendus Acad. Sci., Paris, Sér. A, Vol. 275, 1015 - 1018 (1972).

162.038 Is the existence of a galaxy evidence for a black hole
at its center? M. P. Ryan, Jr.
Astrophys. Journ., (Letters), Vol. 177, L79 - L83 (1972).
We present a Newtonian analysis of a cosmological perturbation with a central black hole. We are able to show that
a central mass of $\sim 10^7 M_\odot$ would have caused our Galaxy to
accrete from a small initial perturbation.

162.039 Evolution of uniform cosmological models containing both matter and radiation. M. Kubo.
Sci. Rep. Tôhoku Univ., First Ser., Vol. 54, 113 - 119 (1971).
The evolution, the horizon and the age of uniform cosmological models containing both matter and radiation without cosmological constant are investigated by means of the
matter and the radiation energy density parameters. The analytical expressions of the particle horizon and the age are given.
The particle horizon is redefined to avoid the discontinuity
without loss of its intrinsic meaning. It is also pointed out that
the radiation lessens the particle horizon and the age.

162.040 Classification of uniform cosmological models.
M. Kubo.
Sci. Rep. Tôhoku Univ., First Ser., Vol. 55, 1 - 8 (1972).
A classification of uniform cosmological models with
cosmological constant containing both matter and radiation
without interaction is presented. A model is specified by the
present values of the deceleration parameter and the matter
and radiation energy density parameters.

162.041 Vainberg's model and the "hot" universe.
D. A. Kirzhnits.
Pis'ma v ZhEhTF, Vol. 15, 745 - 748 (1972). In Russian.
Abstr. in Referativ. Zhurn. 51. Astron., 11.51.647 (1972).

162.042 The metrics of an open model "universe" in an
centrally symmetric frame of reference.
O. Sharshekeev.
Trudy Kirg. un-ta. Ser. fiz. n., 1972, vyp. (No.) 1, p. 87 - 90.
In Russian. – Abstr. in Referativ. Zhurn. 51. Astron.,
11.51.654 (1972).

162.043 Investigations of an open model "universe" in a

centrally symmetric frame of reference.
O. Sharshekeev.
Trudy Kirg. un-ta. Ser. fiz. n., 1972, vyp. (No.) 1, p. 90 - 93.
In Russian. – Abstr. in Referativ. Zhurn. 51. Astron.,
11.51.655 (1972).

162.044 Quantum models for the lowest-order velocity-
dominated solutions of irrotational dust cosmologies.
E. P. T. Liang.
Phys. Rev. D, Particles and Fields, Vol. 5, 2458 - 2466 (1972).
The lowest-order velocity-dominated solutions to the
Einstein dust equations of Eardley, Liang, and Sachs are quantized using the canonical methods of DeWitt and of Arnowitt,
Deser, and Misner. The quantum dynamics of these models is
shown to be governed by the Einstein-Klein-Gordon (EKG)
equation. Exact solutions of the decoupled EKG equations in
the discrete limit are obtained, which have the striking feature
that the state amplitude vanishes at the singularity for anisotropic models. The geometry of the manifold of the classical
3-metrics is studied and it turns out to be composed of conformally flat geodesic submanifolds. Other difficulties related
to the quantum theory such as factor ordering, divergence,
interpretation of the volume measure, etc. are also discussed.

162.045 The universe as a black hole. R. K. Pathria.
Nature, Vol. 240, 298 - 299 (1972).
It is demonstrated that the universe may not only be a
closed structure (as perceived by its inhabitants at the present
epoch) but may also be a black hole, confined to a localized
region of space which cannot expand without limit.

162.046 A comparative study of Brans-Dicke and general
relativistic cosmologies in terms of observationally
measurable quantities. R. C. Barnes, R. Prondzinski.
Astrophys. Space Sci., Vol. 18, 34 - 39 (1972).
A comparison between general relativistic and Brans-
Dicke cosmologies is made in terms of quantities measurable
by an observational astronomer. Numerical integration of the
Brans-Dicke field equations was employed to find the relationships of the mean density of cosmic matter, the age, and the
time derivative of the gravitational constant to the Hubble
constant and deceleration parameter. The difference between
general relativistic and Brans-Dicke apparent magnitude-redshift diagrams was found to be negligible even at large redshifts under the assumption of no galactic evolution in absolute magnitude.

162.047 The evolution of Friedmann models with radiation.
D. Edwards.
Astrophys. Space Sci., Vol. 18, 40 - 48 (1972).
The evolution of uniform cosmological models containing both matter and radiation and with non-zero cosmological
constant is examined. Models are specified by means of three
fundamental parameters and are represented by points in a
three dimensional space.

162.048 The decline of the Hubble constant: A new age for
the universe. W. D. Metz.
Science, Vol. 178, 600 - 601 (1972). – Report on the Henry
Norris Russell lecture, held by A. Sandage during the 138th
meeting of the American Astron. Soc., 1972 August 15 - 18
at Michigan State Univ.

162.049 The cosmic numbers. E. R. Harrison.
Phys. Today, Vol. 25, No. 12, p. 30 - 34 (1972).
The large dimensionless numbers in cosmology have led
to fascinating questions about the possible significance of their
puzzling coincidence.

162.050 The redshift-distance relation. II. The Hubble diagram and its scatter for first-ranked cluster galaxies:

A formal value for q_0.   A. Sandage.
Astrophys. Journ., Vol. 178, 1 - 24 (1972).

The Hubble diagram for first-ranked cluster galaxies is discussed on the basis of new photoelectric measurements of 41 clusters. Additional data by Westerlund and Wall and by Peterson, reduced by the same corrections, increase the sample to 84 clusters. The diagram has small scatter about a line of slope 5. Analysis of the scatter gives upper limits to the dispersions in redshift and in apparent magnitude, respectively. If a strictly homogeneous Friedmann model universe is adopted, the deceleration parameter is calculated to be $q_0 = 0.96 \pm \sim 0.4$ (p.e.). With $q_0 = +1$ and $H_0 = 50$ km s^{-1} Mpc^{-1}, the time to the Friedmann singularity is 11×10^9 years, which agrees with the age of globular clusters in our own Galaxy.

162.051   The redshift-distance relation. III. Photometry and the Hubble diagram for radio sources and the possible turn-on time for QSOs.   A. Sandage.
Astrophys. Journ., Vol. 178, 25 - 44 (1972).

UBV photometry is given for 59 radio galaxies, and is summarized for 103 radio and 25 radio-quiet quasars. The Hubble diagram for the radio galaxies is similar to that for first-ranked cluster galaxies, but is displaced faintward by 0.3 mag in the mean. The Hubble diagram for quasars is scattered, but no quasars lie to the right (fainter) of the radio galaxy distribution. Previously unreported photometry for 22 quasars is listed in an appendix.

162.052   Galaxy formation from annihilation-generated supersonic turbulence in the baryon-symmetric big-bang cosmology and the gamma-ray background spectrum.
F. W. Stecker, J. L. Puget.
Astrophys. Journ., Vol. 178, 57 - 76 (1972).

Following the big-bang baryon-symmetric cosmology of Omnès where an initial phase separation of matter and antimatter leads to regions of pure matter and pure antimatter containing masses of the size of galaxy clusters by a redshift which we calculate to be of the order of 500−600, we show that at these redshifts, annihilation pressure at the boundaries between the regions of matter and antimatter drives large-scale supersonic turbulence which can trigger galaxy formation. This picture is consistent with the γ-ray background observations discussed previously by Stecker, Morgan, and Bredekamp. Gravitational binding of galaxies then occurs at a redshift of ~ 70 at which time vortical turbulent velocities of $\sim 3 \times 10^7$ cm s^{-1} lead to angular momenta for galaxies comparable with measured values.

162.053   Primordial turbulence and the formation of galaxies.
J. Silk, S. Ames.
Astrophys. Journ., Vol. 178, 77 - 93 (1972).

The purpose of this paper is to study the development of turbulence at early epochs in a Friedmann universe. We calculate the density perturbations that would arise on the scale of galaxies through incompressible turbulence, using a post-Newtonian approximation and retaining terms which are nonlinear in the velocity. This theory can also account for the observed angular momentum of spiral galaxies.

162.054   The cosmological implications of counts of galaxies.
M. Rowan-Robinson.
Astrophys. Journ., (*Letters*), Vol. 178, L81 - L83 (1972).

An error in a recent paper by Sandage, Tammann, and Hardy, while in no way invalidating their argument against hierarchical models of the universe, does admit the possibility of a strong evolutionary effect in the interpretation of counts of galaxies.

162.055   Some remarks on Hubble's law.   C. Popovici.
Stud. Cerc. Astron., Vol. 17, 171 - 175 (1972).
In Romanian.

The author deduces Hubble's law in an elementary manner in the hypothesis of a simple cosmological principle for an Euclidean space and constant velocity of light, giving the expression of the redshift z, taking into account not only the expansion, but also the Einstein gravitational shift. The importance of the time delay of the light propagation on Hubble's law is stressed.

162.056   The hadron era in cosmology.   G. Steigman.
Bull. American Astron. Soc., Vol. 4, 409 (1972).
Abstr. AAS.

162.057   Development of correlation in an expanding universe.
P. J. E. Peebles.
Bull. American Astron. Soc., Vol. 4, 414 (1972). – Abstr. AAS.

162.058   Perturbations on the mixmaster universe.
B. L. Hu, T. Regge.
Phys. Rev. Letters, Vol. 29, 1616 - 1620 (1972).

Mathematical formalisms for the separation and solution of the tensor perturbation equations in an empty, diagonal, type-IX space is developed, based upon group-symmetry properties of homogeneous spaces. Numerical results in sampling solutions of the "mixmaster universe" show damping amplitudes of perturbations as the universe expands, a behavior in qualitative accordance with earlier results on the Friedmann universe.

162.059   Qualitative magnetic cosmology.   C. B. Collins.
Commun. Math. Phys., Vol. 27, 37 - 43 (1972).

The technique of phase plane analysis which was used in a previous paper (1971) to study the behaviour of a class of perfect-fluid anisotropic cosmological models, is applied to some simple anisotropic models that contain a uniform magnetic field. A formal correspondence is established between these magnetic models (Bianchi type I) and certain perfect fluid models (Bianchi type II), and new exact solutions are consequently discovered.

162.060   Structures of magnetic fields in the universe and galaxies.
M. Fujimoto, K. Kawabata, Y. Sofue.
Progr. Theor. Phys. Suppl., (*Japan*), No. 49, p. 181 - 227 (1971).

162.061   Proton-neutron concentration ratio in the expanding universe at the stages preceding the formation of the elements.   C. Hayashi.
Progr. Theor. Phys. Suppl., (*Japan*), No. 49, p. 248 - 260 (1971).

162.062   Evolution of the expanding hot universe [galaxy formation].   H. Sato, T. Matsuda, H. Takeda.
Progr. Theor. Phys., Suppl., (*Japan*), No. 49, p. 11 - 82 (1971).

Evolution of the expanding hot universe is discussed from the point of view of astrophysical cosmology. The main effort is devoted to the theory of galaxy formation in connection with the physical state of matter and radiation in the early stage of the hot universe.

162.063   Formation of proto-galaxies in the expanding universe. Gravitational instability.
H. Nariai, K. Tomita.
Progr. Theor. Phys. Suppl., (*Japan*), No. 49, p. 83 - 119 (1971).

162.064   Thermal instability in the expanding universe [galaxy formation].   M. Kondo, Y. Sofue, W. Unno.
Progr. Theor. Phys. Suppl., (*Japan*), No. 49, p. 120 - 147 (1971).

In the expanding universe, the thermal instability, if it occurs, can develop much more quickly than the gravitational instability and, therefore, may provide a possible mechanism to initiate the formation of galaxies. The general characteristics of thermal instability are discussed for non-equilibrium media with due regard to the ionization change and the optical depth effect of fluctuations.

162.065 Hooke's symmetries and nonrelativstic cosmological kinematics. I. J.-R. Derome, J.-G. Cubois.
Nuovo Cimento B, Ser. 11, Vol. 9B, 351 - 376 (1972).

162.066 Hamiltonian approach to the dynamics of expanding homogeneous universes in the Brans-Dicke cosmology. H. Nariai.
Progr. Theor. Phys. (*Japan*), Vol. 47, 1824 - 1843 (1972).

In view of a grave importance of the problem of initial singularity in theoretical cosmology, the dynamical behavior of expanding homogeneous universes (without rotation) in the Brans-Dicke cosmology is studied by means of extending suitably the canonical formalism due to Arnowitt, Deser, and Misner.

162.067 The magnitude—redshift relation in Hoyle-Narlikar cosmology. J. M. Barnothy, B. M. Tinsley.
Bull. American Astron. Soc., Vol. 4, 410 (1972).—Abstr. AAS.

162.068 Mach's principle in the light of modern cosmology. S. J. Prokhovnik.
Structure of matter. Rutherford Centennial Symposium, Christchurch 1971, [Univ. Canterbury, Christchurch, New Zealand], p. 320 - 329 (1972).

The author discusses the consequences of a uniformly expanding universe in order to provide evidence for the unity of the universe and the principle of Mach which reigns over this unity.

162.069 Remarks on some symmetry problems in cosmology and in the theory of particles and fields.
G. Wataghin.
Nuovo Cimento Lettere, Ser. 2, Vol. 4, 608 - 610 (1972).

162.070 Faraday rotation in the Brans-Dicke cosmology. R. Burman.
Nuovo Cimento Lettere, Ser. 2, Vol. 4, 643 - 644 (1972).

Previous authors have presented evidence that Faraday rotation occurs in extra-galactic space. It has been suggested that observations of Faraday rotation could be used to select a model for the universe. The present author has discussed Faraday rotation in a cosmology based on the scalar-tensor theory of gravitation.

162.071 Electromagnetic dispersion in the Brans-Dicke cosmology. R. Burman.
Nuovo Cimento Lettere, Ser. 2, Vol. 4, 645 - 646 (1972).

The dispersion of an electromagnetic wave in an ionized cosmic medium is calculated for the spatially flat form of the Brans-Dicke cosmology.

162.072 Joining of two semiclosed worlds and a cosmological model of matter-antimatter asymmetry.
N. Hokkyo.
Progr. Theor. Phys., (*Japan*), Vol. 48, 104 - 109 (1972).

It is pointed out that the solutions of general relativity field equations allow a model of the closed universe in which two semiclosed worlds, one filled with the dust of ordinary matter and the other filled with the dust of antimatter, are joined through a narrow Schwarzschild throat. The two semiclosed worlds do not exchange matter in a finite world time but are gravitationally connected to form a single closed world possessing zero baryonic charge.

162.073 Cosmology and information. D. H. Gudehus.

Publ. Astron. Soc. Pacific, Vol. 84, 818 - 822 (1972).

The dependence of information derived from a photographic plate on the expansion of the universe is discussed and the accuracy with which the deceleration parameter, q_0, can be obtained from the redshift-magnitude relation is estimated for various telescopes.

162.074 Surface-of-revolution cosmology. M. P. Ryan, Jr.
Ann. Physics, Vol. 72, 584 - 604 (1972).

162.075 On criteria of cosmological spatial homogeneity. M. A. H. MacCallum.
Phys. Letters A, (*Netherlands*), Vol. 40A, 325 - 326 (1972).

162.076 On 'diagonal' Bianchi cosmologies. M. A. H. MacCallum.
Phys. Letters A, (*Netherlands*), Vol. 40A, 385 - 386 (1972).

A list of the possible 'diagonal' Bianchi cosmologies is given, with a proof that it is complete.

162.077 Static cosmological solutions of generalized field equations. B. O. J. Tupper.
Progr. Theor. Phys., (*Japan*), Vol. 48, 678 - 684 (1972).

The generalized field equations of general relativity derived previously are used to find static Robertson-Walker models. It is shown that in the context of these field equations the negative curvature model has physically acceptable properties.

162.078 Redshift magnitude relations. S. E. Kaufman.
Thesis, New York Univ., New York. [Available from Univ. Microfilms, Ann Arbor, Mich., USA. Order No. 72-13375], 71 pp. (1971).

A closed formula for the relation between luminosity and redshift in an expanding Friedmann universe is derived for the case of positive cosmological constant, positive space curvature, and vanishing pressure. This formula is then generalized so that it is valid for all values of the cosmological constant and the three possible space curvatures. The general formula is tested with the well known special cases. A completely similar procedure is followed for radiation filled world models.

162.079 Galaxy formation in anisotropic cosmologies. T. E. Perko, R. A. Matzner, L. C. Shepley.
Phys. Rev. D, Particles and Fields, Vol. 6, 969 - 983 (1972).

The authors analyze the growth rates of perturbations of the generic dust-filled Bianchi type-I cosmology (which exhibits anisotropy but not rotation). Anisotropy induces coupling between gravitational wave and density modes and can enhance the power-law rate of growth of the density perturbations. A maximum growth rate for pregalaxy perturbations is $t^{8/3}$ (where t is cosmic time), so that no conclusive solution to the galaxy formation problem is found.

162.080 The electrodynamic Green functions in a closed universe. H. Stephani.
Acta Phys. Polonica B, Vol. B3, 427 - 436 (1972).
In German.

In a closed Einstein universe Green's functions of electrostatics and magnetostatics are given. An integral representation of the retarded potentials is constructed which is valid for a large class of timedependent sources; the exceptional cases are discussed. The results are generalized for closed universes with Robertson-Walker metric.

162.081 Friedmann perfect-fluid cosmologies for Deser scalar-tensor theory. C. Aragone, A. Restuccia.
Nuovo Cimento Lettere, Ser. 2, Vol. 4, 962 - 964 (1972).

Deser (1970) formulated a new scalar-tensor theory where the cosmological term has been introduced in order to break the conformal invariance of the field equations. The present authors examine Friedmann solutions having a perfect fluid as their source in the context of this theory.

162.082 Power spectrum of the neutrino sea.
M. Ruderfer.
Nuovo Cimento Lettere, Ser. 2, Vol. 5, 86 - 88 (1972).

The neutrino sea refers to the totality of unbound neutrinos and antineutrinos in the universe. The authors discuss the fact that since there are essentially no limits presently applicable to the low energy portion of the power spectrum, the total energy in the neutrino sea may feasibly exceed the total observable energy in the universe by some unknown amount.

162.083 Perturbations in anisotropic Euclidean-homogeneous cosmologies. T. K. Perko.
Thesis, Univ. Texas, Austin. [Available from Univ. Microfilms, Ann Arbor, Mich., USA. Order No. 72-15813], 90 pp. (1971).

162.084 Unit transformations and cosmology. E. A. Lord.
Nuovo Cimento B, Ser. 11, Vol. 11B, 185 - 200 (1972).

The mathematical consequences of the hypothesis that physical laws should be invariant under space-time-dependent changes in the unit of length are investigated. The theory obtained does not satisfy energy-momentum conservation and therefore allows creation of matter. The relationship between the conformally covariant theory and the cosmological theories of Hoyle and Narlikar (1971) and of Brans and Dicke (1961) is discussed.

162.085 Closed anisotropic cosmological models.
N. Batakis.
Ann. Physics, Vol. 73, 578 - 588 (1972).

The authors give a new exact solution to Einstein's equations representing homogeneous nonisotropic cosmological models of a closed universe containing electro-magnetic and solar fields.

162.086 Generalization of the Taub-Kazner cosmological metric in the scalar-tensor gravitation theory.
V. A. Ruban, A. M. Finkelstein.
Nuovo Cimento Lettere, Ser. 2, Vol. 5, 289 - 293 (1972).

162.087 On irrotational Bianchi-type universes in the Brans-Dicke cosmology. H. Nariai.
Progr. Theor. Phys., (Japan), Vol. 48, 703 - 705 (1972).

In a previous paper, the Hamiltonian approach to the dynamics of expanding homogeneous universes in the Brans-Dicke cosmology was proposed in order to see how their dynamical behavior is different from that in relativistic cosmology. In this note, the author examines the situation in an irrotational Bianchi-type universe.

162.088 Temperature maximum of the early (hadron) universe. D. Stauffer.
Phys. Rev. D, Particles and Fields, Vol. 6, 1797 - 1798 (1972).

162.089 Is the universe transparent or opaque?
P. C. W. Davies.
Journ. Phys. A, General Phys., Vol. 5, 1722 - 1737 (1972).

The author determines the maximal class of conformally flat cosmological models which eventually absorb all the electromagnetic radiation they contain. This is the requirement of the Wheeler-Feynman absorber theory of radiation. The condition for complete absorption is more or less written down immediately, and the various cosmological models tested by inspection.

162.090 A finite expanding universe with matter injection.
F. M. Gomide.
Nuovo Cimento B, Ser. 11, Vol. 12B, 11 - 19 (1972).

A fluid homogeneous cosmological model is proposed with a Robertson-Walker metric and positive curvature. The model affords an interpretation to the matter production process in the following terms: the matter injection into the observable universe is due to the work done along cosmic expansion by a negative-pressure substratum and to the variation of a negative field energy. The properties of the expansion function and the vector field demand a negative cosmological constant.

162.091 The cosmological transformation. H. Hafner.
Revue Roumaine Phys., Vol. 17, 933 - 946 (1972).

162.092 Cosmological significance of e^2/Gm^2 and related "large numbers". M. J. Rees.
Comments Astrophys. Space Phys., Vol. 4, 179 - 185 (1972).

162.093 Particle barriers in cosmology. E. R. Harrison.
Comments Astrophys. Space Phys., Vol. 4, 187 - 192 (1972).

In this discussion it has been argued, in admittedly a rather tendentious fashion, that the Bahcall-Frautschi hadron barrier probably does not exist.

162.094 Das Alter des Universums. G. A. Tammann.
Separate print from Neue Zürcher Zeitung, No. 353, 8 pp. (1972).

162.095 On the total electric charge and mass of an elliptic universe. R. A. Asanov.
Theor. Math. Phys., Vol. 6, 242 - 243 (1971).

162.096 On the effect of scalar and vector fields on the nature of the cosmological singularity.
V. A. Belinskij, I. M. Khalatnikov.
Zhurn. ehksperim. i teor. fiz., Vol. 63, 1121 - 1134 (1972). In Russian. – Abstr. in Referativ. Zhurn. 51. Astron., 2.51.748 (1973).

Problems of modern cosmogony.
See Abstr. 003.035.

Problèmes de cosmogonie contemporaine.
See Abstr. 003.036.

Gravitation and cosmology: Principles and applications of the general theory of relativity. See Abstr. 003.124.

An upper limit on the neutrino rest mass.
See Abstr. 061.022.

On the cosmic abundance of helium.
See Abstr. 065.062.

Backscattering caused by the expansion of the universe. See Abstr. 066.070.

Gravitation, strong interactions, and the creation of the universe. See Abstr. 066.073.

On gravitational aberrations in stellar images.
See Abstr. 066.170.

The sun's rotation and the Brans-Dicke cosmology.
See Abstr. 080.029.

Absence of rapid fluctuations in the ground level γ-ray background. See Abstr. 082.103.

New method for estimating the Hubble constant.
See Abstr. 125.009.

Type I supernovae and the Hubble constant. See Abstr. 125.015.

The cosmological evolution of radio sources of large angular extent.   See Abstr. 141.006.

Faraday depolarization of extragalactic radio sources. See Abstr. 141.024.

Radio source counts and redshifts in steady state cosmology.   See Abstr. 141.097.

The distribution of redshifts of quasi-stellar objects and related emission-line objects.   See Abstr. 141.108.

Comment on inverse Compton models for the isotropic X-ray background and possible thermal emission from a hot intergalactic gas.   See Abstr. 142.095.

Effects of evolution on the diameter-redshift relation.   See Abstr. 151.029.

Radio galaxies, quasars, and cosmology. See Abstr. 158.063.

Are galaxies formed by material flowing out of singularities?   See Abstr. 158.075.

The process of galaxy formation according to the universal turbulence hypothesis.   See Abstr. 158.084.

The curious cases of the colliding galaxies and the rapidly expanding universe.   See Abstr. 158.092.

On the cosmological influence on insular systems. See Abstr. 160.015.

The possibility that nongaseous hydrogen supplies the missing cosmological mass.   See Abstr. 161.007.

Author Index

BOTTON, C.
099.026
BOUDON, Y.
052.009
BOUIGUE, R.
155.005
BOURGOIS, G.
106.010 .011
BOURNE, S. R.
094.055
BOUSKA, J.
102.075
103.123
BOUSKA, V.
105.017
BOUSTEAD, J.
105.094
BOUT, P. VANDEN
SEE VANDEN BOUT, P.
BOUVIER, P.
151.024
BOUW, G. D.
115.019
BOWERS, R. L.
066.021
BOWHILL, S. A.
012.016
083.037
BOWMAN, M. R.
082.089
BOWYER, S.
142.045
158.111 .121
160.024
BOYARCHUK, M. E.
114.028
BOYCE, P. B.
097.043
099.017 .040
BOYD, F. R.
094.229
BOYD, R. L. F.
031.078
BOYLE, R. P.
071.069
BOYNTON, P.
142.126
BOZIS, G.
042.016
BOZOKI, G.
080.057
BRACCESI, A.
158.061
BRACE, L. H.
083.020 .060
BRACKER, S. B.
141.506 .508
BRACKMAN, R. T.
022.103
BRADLEY, C. C.
022.139
BRADT, H.
010.002
142.113
BRADT, H. V.
141.542
142.098
BRADU, E.-B.
143.021
BRADY, J. L.
103.104

BRAES, L. L. E.
141.079 .120
BRAGIN, YU. A.
082.043
BRAGINSKIJ, V. B.
066.079
BRAGINSKY, V. B.
061.026
BRAHDE, R.
072.052
BRAILOVSKAYA, I. Y.
073.055
BRAMANTI, D.
066.126
BRAMER, B.
033.124
BRANCH, D.
125.001 .015
BRANDI, E.
114.023
BRANDT, J. C.
004.021
034.047
074.008 .067
102.063
158.050
BRANDT, L.
041.006
BRANDT, P.
073.089
BRANDT, V. E.
034.012
BRANNER, G. R.
033.088
BRANSON, N, J. B. A.
142.074
BRATNER, S.
123.073
BRATOLJUBOVA, L. S.
041.039
BRAUDE, B. V.
033.003
BRAUNE, W.
011.039
120.006 .016
BRAUNSFURTH, E.
131.080
BRAVO, S.
084.037
BRECHER, K.
142.047
BRECKENRIDGE JR., R. W.
034.027
BRECKINRIDGE, J. B.
034.117
BREGER, M.
114.003
122.042 .043
152.901
BREIDO, I. I.
113.022
BREIG, E. L.
022.097
BREIHAN, E.
034.115
BREINHORST, R.
121.056
BREKKE, A.
084.039
BRETT, R.
094.007

BREUER, R. A.
066.020 .128
BREVES FILHO, J. A.
042.033
BRIATORE, L.
143.056
BRICARD, J.
082.035
BRIDLE, A. H.
141.012
142.062
BRIGGS, F. H.
097.098
BRIGGS, G. A.
097.023 .086
BRILL, D. R.
066.020 .062
BRINI, D.
008.013
141.554
BRINKMAN, A.
142.135
BRINKMAN, A. C.
142.031 .064
BRINKMANN, R. T.
091.021
BRINTON, H. C.
083.020
BRISTOW, F. E.
097.007
BRITTEN, W. E.
003.050
BRKE, B. F.
033.065
BROCKLEHURST, M.
132.022 .028 .038
BRODERICK, J. J.
141.016 .073 .507
BRODKORB, E.
011.022
036.007
BRODZINSKI, R. L.
094.126
BROENSTAD, K.
084.055
BROGLIA, P.
123.046
BROMANDER, J.
022.071 .135 .136
BROMWELL, L. G.
094.122
BRONSHTEHN, V. A.
010.033
098.034
BROOKE, A. L.
152.901
BROOKES, C. J.
101.002
BROOKS, E. M.
079.101 .106
BROOKS, J. W.
131.007
159.013 .015
BROSCHE, P.
081.016
BROTEN, N. W.
159.014
BROUCKE, R.
042.052
BROVAR, V. V.
081.022

BROWN, D. R.
073.015 .016
BROWN, G. M.
011.006 .007
BROWN, G. W.
099.083
BROWN, J. C.
011.045
076.009 .038
BROWN, L. W.
157.007
BROWN, P. LANCASTER
003.051
BROWN, R. C.
022.150
BROWN, R. L.
157.004
BROWN, R. R.
084.055
BROWN, R. T.
022.011
BROWN, R. W.
094.007 .108 .192
BROWN JR., W. P.
063.020
BROWNE, I. W. A.
141.118
BROWNING, R.
155.001
BROWNLEE, D. E.
105.050
BRU, P.
041.018
BRUBAK, H.
096.009
BRUCATO, R. J.
142.010 .075 .121
BRUCKNER, H.-P.
105.107
BRUECK, H. A.
008.032
BRUECK, M. T.
113.027 .033
BRUECKNER, G. E.
071.022
073.108
074.028
076.036
BRUENN, S. W.
065.067
BRUIJN, P. J.
079.101
BRUIN, F.
075.033
BRUMBERG, V. A.
003.052
042.003
BRUMBERG, V. L.
042.901
BRUN, A.
120.001
BRUNET, J. P.
115.020
159.017 .901
BRUNING, D. H.
072.064
BRUWER, J. A.
103.112
BRUZEK, A.
071.046

BRYANT, D. A.
084.031
BRYDEN, D. J.
004.005
BRZOSTKIEWICZ, S. R.
005.022
097.105
BRZOZOWSKI, J.
022.070
BUCHANCOWA, N.
122.074
BUCHAR, E.
032.045
BUCHAU, J.
084.007 .015 .069
BUCHER, W.
105.050
BUCHET, J. P.
022.073
BUCHLER, J.-R.
065.051
BUCHROEDER, R.
032.026
BUCHROEDER, R. A.
032.016 .039
BUCHTA, R.
022.070 .135 .136
BUCHWALD, V. F.
105.024 .097
BUCK, R. M.
084.297
BUCKBESCH, F.
044.031
BUCKINGHAM, M. J.
033.100
BUCKMASTER, H. A.
033.038 .039
BUDDEN, K. G.
141.125
BUDENKOV, N. A.
046.029
BUD'KO, N. I.
084.209
BUDNIKOVA, N. A.
042.004
BUEHLER, F.
084.071
BUEREN, H. G. VAN
033.068
034.106
BUERGER, P. F.
064.040
BUFF, J.
131.006
BUFTON, J. L.
082.085 .222
BUGAENKO, O. I.
100.003
BUGAEVSKIJ, A. V.
094.006
BUHAGIAR, M.
104.065
BUHL, D.
093.004
131.032 .060 .074 .089
    .090 .096
BUKATA, R. P.
078.027
BUKOW, H. H.
022.066

BULANOV, S. V.
143.010
157.003
BULIRSCH, R.
021.011
BULLARD, E.
011.007
BULLEN, K. E.
091.055
BUN, F. O. VON
054.002
BUNCH, T. E.
010.018
094.105 .232 .252
105.009
BUNNER, A. N.
022.126
125.029
142.054
155.027
BUONOCORE, B.
045.043 .045
BURBIDGE, E. M.
141.101 .510
158.006 .109
BURBIDGE, G.
013.003
142.043
BURBIDGE, G. R.
141.015 .108
158.006
BURCH, J. L.
084.291
106.030
BURDJUZHA, V. V.
022.006 .122
BURGER, J. J.
106.007
BURGER, M.
076.007
BURGESS, D. D.
062.043
BURGESS, R. D.
153.020
BURGIN, M. S.
142.034
BURGINYON, G. A.
125.019
142.099
BURHOP, E. H. S.
003.053
BURKARD, O. M.
044.022
BURKE, B. F.
033.106
046.023
142.090
158.058
BURKE, J. A.
065.082
BURKE, J. D.
053.017
BURKE, J. J.
031.031
BURKE, J. R.
066.136
BURKE, T.
097.089
BURKE, T. E.
097.028 .031

FRAGOSO, N.
080.006
094.031
FRANCESE, G.
032.037
079.100
FRANCIS, D.
097.056
FRANCMANIS, J.
009.021
011.033 .034
FRANK, L. A.
034.078
078.046
084.002 .013
FRANK, P.
120.007 .009
FRANTZ, D. J.
141.055
FRANZ, O. G.
122.052
FRANZINI, M.
105.093
FRASER, D. R. E.
011.007
FRASER-SMITH, A. C.
084.220
FRAUTSCHI, S.
162.003
FRAZHO, D. B.
021.017
FRAZIER, E. N.
071.047
073.047 .080
080.062
FREDRICK, L.
117.011
FREDRICKS, R. W.
074.004
084.280
FREEMAN, K. C.
099.020
158.096
FREEMAN, N. C.
064.064
FREITAS MOURAO, R. R. DE
118.013
FRENCH, B. M.
105.010
FRENCH, C. E.
053.028
FRENCH, V. A.
114.093
FREW, N. M.
094.238
FREY, N.
022.040
FRICKE, K.
065.057
FRICKE, K. J.
065.069 .116
FRICKE, W.
002.037
041.022
043.001
FRICKER, P. E.
094.087
FRIDMAN, V. M.
077.015
079.102

FRIEDEMANN, C.
141.082
FRIEDEN, B. R.
031.031
FRIEDLANDER, A. L.
051.011
FRIEDMAN, H.
155.050
FRIEDMAN, J. L.
066.005 .024 .082
FRIEDMAN, M.
061.036
FRIEDMAN, V. E.
073.101
FRIIS-CHRISTENSEN, E.
106.002
FRIMOUT, D.
082.034 .189
FRISCH, C. VON
003.070
FRISCH, H.
073.003
FRISHBERG, L. D.
114.041 .161
FRITZ, G.
155.050
FRITZE, K.
065.124
FROGEL, J. A.
113.021
131.103
133.901
FROHLICH, A.
142.130
FROLOV, M. S.
122.030
FROLOV, V. V.
104.015 .022
FRONTERA, F.
141.554
FROST, K. J.
076.017
FRYE, R. L.
114.002 .013
FRYE JR., G. M.
141.523
FUCHS, E.
011.053
FUCHS, L. H.
105.120
FUCHS, R. W.
051.013
FUERST, E.
074.014
FUERSTENBERG, F.
075.024
FUJII, Z.
143.063 .064
FUJIMOTO, K.
143.063 .064
FUJIMOTO, M.
066.155
080.053
162.060
FUJIWARA, K.
034.112
FUKUSHIMA, N.
084.224
FULCHIGNONI, M.
094.071

FULIGNI, F.
141.554
FULLER, J. C.
082.200
FULLER, M.
094.121
FULLERTON, L. W.
102.053
FUNK, H.
105.116
FURENLID, I.
034.048
FURNISS, I.
113.018
FURUTA, T.
103.107 .120 .124
FUTRELLE, R. P.
022.110
FYMAT, A. L.
034.074
063.023

GABRIEL, A. H.
011.045
012.012
022.061
062.071
076.028
GABRIEL, M.
065.035
GADSDEN, M.
011.006
082.017
GADZHIEV, M. S.
034.159
097.011
GAEBERT, H.-W.
003.161
GAFFEY, M.
097.055
GAHM, G. F.
114.142 .153
GAIGNEBET, J.
034.088
046.021 .022
GAINULLINA, R. KH.
160.020 .021
GAJDUK, A. R.
097.011 .099
GAJLANS, A. G.
033.061
GAJNOVA, L. E.
143.042
GALAKTIONOV, V. N.
097.097
GALAKTIONOVA, YU. F.
078.024
GALATOLA, A.
121.016
GALE, N. H.
094.208
105.044
GALE, W. A.
093.003 .004
GALEHOUSE, J. S.
081.018
GALEOTTI, P.
153.006
GALIBINA, I. V.
098.011

IWANOWSKA, W.
155.061
IYEVEER, M. M.
158.078
IZAKOV, M. N.
082.204
IZVEKOV, V. A.
041.031
093.025 .049
IZVEKOVA, A. A.
041.065
IZVEKOVA, V. A.
141.561

JABS, A.
143.022
JACKISCH, G.
114.106
JACKSON, D. W.
082.202
JACKSON, E. D.
094.158 .237
JACKSON, J. C.
011.045
061.030
066.111
JACKSON, W. M.
102.061
JACOBS, J. A.
081.010
JAEGER, F. W.
011.035
031.091
JAFFE, J.
061.046
JAFFE, L. D.
012.016
094.146
JAFFE, W. J.
158.025
JAGER, C. DE
007.000
064.039
071.013 .037 .050
JAGODZINSKI, H.
094.054
JAIN, A.
022.147
JAIN, A. K.
142.089
JAIN, A. V.
105.110 .131
JAIN, S. K.
022.055
JAKES, P.
094.007
JAKI, S. L.
003.156
004.004
091.047
JAKI, S. T.
091.040
JAKUBCOVA, I.
081.045
JAMAR, C.
114.029
JAMBOR, B. J.
103.115
JAMES, A. N.
082.103

JAMESON, R. F.
034.025
JAMIESON, H. D.
010.003
094.104
JAMIESON, T. H.
003.080
JANES, A. F.
034.003
051.026
JANICZEK, P. M.
098.007 .033
JANIN, G.
151.044
JANIN, L.
004.040 .042
JANOSSY, L.
117.039
JANSEN, J. K. M.
033.102
JANSSEN, M. A.
131.005
JANSSENS, T. J.
073.036 .105
080.004
JAROSEWICH, E.
003.057
105.011 .102 .113
JARRETT, A. H.
008.011
010.007
082.045
122.035 .067
JASCHEK, M.
114.023
JASEVICIUS, V.
113.050
JASTREBOV, A. A.
083.028
JAUNCEY, D. L.
141.016 .064 .094
JAYANTHI, U. B.
142.151
JEANSAUME, G.
044.015
JEFFERIES, J. T.
009.019
JEFFERTS, K. B.
131.029 .098
132.015
155.052
JEFFERYS, W. H.
021.005
JEFFREYS, B.
012.004
JEFFREYS, H.
045.007
081.009 .054
094.057
JEGIBEKOV, P.
106.034
JELLEY, J. V.
134.007
JENKINS, E. B.
114.089 .090
JENKINS, R. E.
004.022
JENNINGS, M. C.
064.029 .067
JENNINGS, R. E.
008.104

JENNINGS, R. E.
113.018
JEPSEN, P. L.
034.124
JERZYKIEWICZ, M.
122.058 .097
JEUKEN, M.
033.102
JEUKEN, M. E. J.
033.070
JOCKERS, K.
103.121
JODINSKIENE, E.
113.050
JOERGENSEN, H. E.
014.010
JOHANNESSON, G.-A.
022.140
JOHANSON, A. E.
082.125
JOHANSSON, U. R.
004.051
JOHNSON, C. E.
022.100
JOHNSON, D. R.
131.087
JOHNSON, F. S.
094.170
JOHNSON, H. E.
061.071
131.061
JOHNSON, H. L.
114.163 .164
JOHNSON, H. M.
117.014
133.003
142.078
JOHNSON, H. R.
064.017 .023
073.103
114.053
JOHNSON, J. H.
105.051 .053
JOHNSON, L. R.
081.063
JOHNSON, M.
066.103
JOHNSON, M. A.
034.122
JOHNSON, M. W.
155.025
JOHNSON, N. P.
083.022
JOHNSON, R. G.
034.075
084.054 .223
JOHNSON, R. S.
064.064
JOHNSON, T. S.
045.033
JOHNSON, T. V.
091.019 .022
094.019
JOHNSON III, W. N.
142.901
155.062
JOHNSTON, D. H.
094.901
JOHNSTON, K.
121.064

MAGALASHVILI, N. L.
122.134
MAGALINSKY, V. B.
066.033
MAGEE, N. H.
076.043
MAGEE JR., N. H.
065.064
MAGNAN, C.
064.037
MAGNI, G.
094.071
MAGUIRE, W.
097.089
MAGUIRE, W. C.
097.031
MAHONEY, M. J.
131.026
MAISCHBERGER, K.
066.126
MAITRE, V.
007.000
041.035
MAITZEN, H. M.
114.003
MAJOR, S. P.
003.069
045.003
MAK, M. K.
091.028
MAKALKIN, A. B.
101.017
MAKARENKO, N. G.
158.114
MAKARENKO, N. L.
094.213 .262
MAKAROV, E. S.
094.175
MAKAROV, V. I.
073.024 .107
074.024
MAKAROVA, E. A.
034.011 .118
074.110
MAKAROVA, V. V.
073.107
MAKROYANNIS, T. J.
082.240
MALAFEEV, L. I.
097.097
MALAISE, D.
114.029
MALAKHOVA, O. F.
052.032
081.036
MALASIDZE, G. A.
151.010
MALAVIYA, V.
022.059
MALEY, P. D.
104.051
MALIN, S. R. C.
011.007
084.262 .304
085.004
MALKUS, W. V. R.
081.065
084.359
MALL, A. P.
105.009

MALLA, Y. B.
033.080
MALLIA, E. A.
072.049
MALOFEEV, V. M.
141.501 .561
MALOMYZHEV, L. M.
032.055
MALOV, N. N.
091.025
MALTBY, P.
072.045 .066
MALTZEVA, O. A.
083.015
MALYSHEV, M. I.
034.012
MALYSHEVA, T. V.
094.173
105.078 .087
MAMADOV, O.
103.100 .121
MAMAKOV, A. S.
034.157
MAMEDOV, M. A.
102.051
MAMMANO, A.
124.103
MAMYRIN, B. A.
078.029
MANASSAH, J. T.
065.144
MANCHANDA, R. K.
085.002
MANCHESTER, R. N.
141.525 .535
MANCUSO, S.
041.005
MANDEL'SHTAM, S.
061.078
MANDEL'SHTAM, S. L.
073.085 .086 .091 .092
MANDELSTAM, S. L.
034.065
073.013
076.035
MANDJOS, A. V.
160.015
MANDUJANO O., F. J.
075.002
106.033
MANDZHOS, A. V.
160.009
MANGANIELLO, E. J.
051.027
MANGO, S. A.
121.064
MANN, G. R.
073.015
MANN, H. M.
051.004
MANNING, P. G.
131.086 .110
MANNO, V.
012.005
MANOCHINA, A. V.
105.049
MANSFIELD, M. W. D.
022.151
MANSILLA, L. A.
075.035

MANSON, A. J.
094.123
MANSON, J. E.
071.064
MANSOURI, R.
066.118
MANTZ, A. W.
099.015 .045
MANUEL, O. K.
105.001 .021 .055 .057
MANUKIN, A. B.
061.026
066.079
MAO, C. Y.
143.068
MARAN, S. P.
004.021
034.047 .051
141.539 .551
MARANDINO, G. E.
141.040
142.091
MARANO, B.
141.026
MARASCHI, L.
142.006
MARCH, N. H.
022.150
MARCHESINI, F.
010.027
MARCOLUNGO, P.
061.075
MARCUS, E.
041.007 .079
MARDUS, F.
015.003
MAREK, K.-H.
055.019 .023
MARENIN, I.
064.042
MARGOLIS, J. S.
022.028
099.036
MARGON, B.
142.045 .116
158.111 .121
MARGRAVE JR., T. E.
071.079
MARGULIS, L.
003.097
MARIANI, F.
034.084
084.322 .901
MARINESCU, A.
052.022
MARINO, B. F.
122.085 .121
123.037 .059
124.104
MARINOV, S.
066.152
MARIS, G.
075.030
MARISKA, J. T.
071.048
084.239
MARK, J. W-K.
155.033
MARKARIAN, B. E.
158.028

MEDD, W. J.
141.070

MEDNIKOV, V. I.
105.087

MEDNIKOVA, N. G.
105.087

MEDVEDEV, V. I.
055.011

MEDVEDEV, YU. A.
064.074
082.210
122.146 .147

MEEKINS, J. F.
073.063
076.015
155.050

MEEKS, M. L.
134.005
142.068

MEEUS, J.
041.036
047.001
054.005 .008
079.106
091.032
092.003
095.002
096.006 .007 .010
098.024 .038
102.004

MEFFROY, J.
052.039

MEGRELISHVILI, T. G.
082.163 .176

MEHLMAN-BALLOFFET, G.
022.018
062.045

MEIDAV, M.
141.516

MEIER, P. J.
033.091

MEIKLE, W. P. S.
134.007

MEIN, P.
034.160

MEINIG, M.
032.046

MEINUNGER, I.
113.032
123.067

MEINUNGER, L.
123.047 .062

MEISEL, D. D.
103.100

MEISTAS, E.
113.050

MELCHIOR, P.
002.038
012.004
044.002
045.004 .038
081.057

MELCHIORRI, B.
034.032

MELLIN, J. R.
053.029

MELNIKOV, O. A.
071.019 .062

MEL'NIKOVA, N. S.
064.062

MELROSE, D. B.
062.017 .033

MENATH, A.
004.048

MENDE, S. B.
084.034 .248

MENDOZA V, E. E.
082.203 .233
113.063
152.011

MENG, C.-I.
084.246

MENGEL, J. G.
064.022
065.122

MENON, M. G. K.
061.028 .029

MENSHUTINA, I. N.
084.017

MENTALL, J. E.
022.049

MENTEK, J. S.
051.013

MENZEL, H.
101.024

MENZEL, K.
004.076

MENZIES, B.
120.020

MENZIES, J. W.
122.018

MEREDITH, B. L.
093.001

MEREZHIN, V. P.
121.032

MERGENTALER, J.
015.015
142.111

MERKELIJN, J. K.
141.061

MERKER, M.
143.054

MERLIN, P.
141.105

MERLIN, U.
141.039

MERMAN, N. V.
041.080

MERTS, A. L.
065.064
076.043

MESROBIAN, W. S.
111.005

MESSAGE, P. J.
091.051

MESSELL, K.
097.073

MESTIASHVILI, Z. D.
034.144 .145

MESZAROS, P.
131.066

METZ, W. D.
013.002
066.012
162.048

METZGER, A. E.
142.052

MEURERS, J.
003.098

MEWE, R.
022.007

MEWE, R.
071.044

MEYER, B.
083.042

MEYER, E. R.
033.088

MEYER, J. P.
143.039

MEYER, K.
003.076

MEYER, P.
143.014

MEYER, R. X.
073.040

MEYER, S.
131.025

MEYER-HOFMEISTER, E.
065.111

MEYEROTT, A. J.
074.070 .072
142.078

MEZGER, P. G.
008.014

MIANES, P.
113.056

MICHARD, R.
074.066
075.007

MICHAUD, G.
064.010
065.024

MICHEL, F. C.
062.065
084.355
134.006

MICHEL, G.
031.066

MICHELINI, R. D.
046.026

MICHELS, D. J.
073.108

MICHLOVIC, J.
034.057

MIDDLEDITCH, J.
142.096 .114 .128

MIDDLEHURST, B. M.
094.094

MIEGHEM, J. VAN
003.091

MIGAL', N. K.
081.023

MIGUNOV, V. M.
083.041

MIHALAS, D.
064.008 .027 .075
114.012 .021 .057

MIHALOV, J. D.
084.246
099.006
106.032

MIJIC, M.
011.047

MIKERINA, N. V.
074.025

MIKHAILOVA, O. M.
113.022

MIKHAJLOV, M. V.
105.073

MILAN, D. S.
141.552

NEFED'EV, A. A.
094.255 .261
NEFED'EVA, A. I.
041.060
NEFF, J.
003.084
NEFF, J. S.
101.008
114.061
NEGUS, C. R.
073.068
NEJMAN, YU. M.
091.058
NELEN, J.
003.057
105.113
NELSON, A. H.
162.001
NELSON, H. F.
062.056
NELSON, J. E.
141.517
NELSON, M. R.
121.091
NELSON, R. W.
063.039
NEMETH, J.
065.125 .132
NEMIRO, A. A.
041.052
NEO, Y. P.
082.026
NESMYANOVICH, A. T.
033.047
073.113
079.101 .102
106.041
NESMYANOVICH, E. I.
073.113
106.041
NESS, N. F.
084.271 .322 .901
106.032
NESTEROV, V. E.
143.035
NESTEROV, V. V.
045.034 .036 .037
NESTOROV, G. T.
083.039 .074
NEUBAUER, F. M.
062.021
NEUBERT, R.
055.022
NEUGEBAUER, G.
114.086 .099
131.062
141.039 .105
NEUGEBAUER, M.
074.004 .064
NEUGEBAUER, O.
003.108
NEUPERT, W. M.
034.052
074.036
NEUSS, H.
083.042
NEVEN, L.
071.037
NEWBURN, R. L.
053.012

NEWELL, E. B.
036.003
NEWELL, H. E.
094.065
NEWKIRK, G. A.
074.033
NEWKIRK JR., G.
074.010 .039
NEWTON, J. B.
009.003 .006
NEWTON, L. M.
159.016
NEWTON, R. R.
003.109
004.022
044.005
NEY, E. P.
034.062
113.021 .039
122.125
NEZHINSKII, E. M.
158.062
NEZHINSKIJ, E. M.
102.037
NGUEN-NGAN
072.014
NGUEN BIK LAN
083.069
NGUYEN-HUU-DOAN
084.044
NGUYEN-QUANG-RIEU
131.058 .114
142.060
NGUYEN XUAN VINH
042.069
NI, W.-T.
066.025
NICELL, D.
104.026
NICHOLLS, R. W.
022.021 .027
079.101
082.211
NICOGHOSSIAN, A. G.
063.002
NICOLAIDES, C.
022.101
NICOLAS, K.
071.022
NICOLESCU, S.
075.030
NICOLET, M.
082.095
NICOLET, M.-A.
031.016
NICOLLIER, C.
113.057
NICOLSON, G. D.
008.023
141.063
NIEHOFF, J. C.
051.011
NIEL, M.
155.005
NIELL, A.
079.107
NIELL, A. E.
141.124
142.062
NIEMANN, H. B.
053.005

NIEMELA, V. S.
114.005
NIEMELAE, V. S.
114.903
NIER, A. O.
082.150
NIETO, M. M.
003.110
NIEWENHUIJZEN, N.
034.133
NIIMI, Y.
044.023
NIKITIN, A. A.
114.177
NIKOGOSYAN, A. G.
063.027
NIKOLAEV, A. G.
143.011
NIKOLAEV, S. I.
042.063
NIKOLAEV, V. D.
083.040 .070
NIKOLAYEV, G. B.
053.021 .022
NIKOLOV, N.
013.015
122.033
NIKOLOV, N. S.
014.003
NIKOLSKAYA, T. K.
021.003
042.064
NIKOL'SKII, G. M.
003.079
NIKOLSKY, G. M.
034.045
NIKONOV, V. B.
113.050
NIKONOVA, L. E.
102.022
NIKSCH, M. A.
094.241
NIKULIN, I. F.
072.016
NILL, K. W.
082.036
NILSSON, S. G.
061.069
NISBET, J. S.
082.064
083.063
NISHIDA, A.
084.219 .285
NISHIMURA, H.
105.118
NISHIMURA, K.
142.039
NISHIMURA, M.
158.100
NISHIMURA, S.
031.059 .061
NISHINO, Y.
008.106
NITA, I.
075.030
NITSCH, J.
065.058
NOBILI, L.
065.107 .109
NOBIS, H. M.
004.050

NOCI, G.
074.029 .082

NOE, J. DE LA
077.002

NOEL, F.
041.032
044.004

NOELS, A.
065.035 .128

NOERDLINGER, P. D.
064.014 .015
066.096

NOLAND, M.
097.087

NOLLEZ, G.
062.018

NOLT, I. G.
082.057

NOMOTO, K.
064.063

NOONAN, T. W.
161.007
162.015

NOONKESTER, V. R.
083.059

NORDSIECK, K. H.
158.052

NORDTVEDT JR., K.
066.083 .084 .142

NORDTVEDT JR., K. L.
066.159

NORRIS, J.
114.114

NORTON, R. B.
082.007
083.064

NOSKOV, B. N.
042.074

NOSKOVA, R. I.
133.008
158.031 .066

NOVACO, J. C.
155.022

NOVAK, B. L.
054.012

NOVAK, M.
034.043

NOVAK, Z. L.
099.071

NOVELLO, M.
066.130

NOVICK, R.
022.005
034.120
142.082 .083 .117

NOVIKOV, G. G.
102.071
103.100

NOVIKOV, I. D.
066.077 .149

NOVIKOV, S. B.
031.012
082.032

NOVIKOV, S. P.
162.023

NOVIKOV, V. V.
081.017
094.181
097.017

NOVIKOVA, G. V.
082.033

NOVOPASHENNYJ, B. V.
034.150

NOVOSELOV, V. S.
003.111

NOVOTNY, J.
162.012

NOVOTNY, V.
095.004

NOVRUZOVA, H. I.
115.006

NOWAK, T.
103.121

NOWATZKI, E. A.
094.017

NOXON, J. F.
082.109 .125

NOYES, R. W.
071.018 .030
072.002 .030
073.042 .093 .099
074.068
076.021
080.032 .060

NUCKOLLS, J.
022.023

NUSINOV, A. A.
084.250 .337
102.001 .003 .006

NUTKU, Y.
066.068

NYGAARD, K. J.
022.090

NYSTROEM, G.
061.037

OBAYASHI, T.
074.104 .105

OBENSON, G.
081.033

OBERBECK, V. R.
097.113 .118

OBERNDORFER, H.
003.112
009.007

OBREGON, O.
066.043

OBRIDKO, V. N.
072.006 .007 .022
131.054

O'BRIEN, K. C.
033.089

OBUKHOV, G. G.
094.100

OBURKA, O.
009.024

OCCHIONERO, F.
065.063
066.003

O'CONNELL, D. J. K.
005.027
121.072

O'CONNELL, R. F.
065.090 .091
066.121 .137

O'CONNELL, R. W.
158.007 .059

ODABASI, H.
003.050

ODELL, A. P.
119.013

O'DELL, C. R.
051.017
103.121

O'DELL, S. L.
141.015 .108

O'DONNELL, E. J.
121.097

OEMLER JR., A.
160.010

OEPIK, E.
005.027

OEPIK, E. J.
002.017

OERTEL, G. K.
034.105

OESTBERG, K.
062.068

OFFERMANN, D.
082.158

O'GALLAGHER, J. J.
143.005

OGANESIAN, R. S.
061.039

OGANYAN, G. B.
122.142

OGATA, H.
121.090 .094

OGILVIE, K. W.
074.019 .101

OGIR, M. B.
073.055

OGORODNIKOV, K. F.
151.023

OGRINS, M. P.
031.038 .039

OGUMA, I.
034.113

O'HANDLEY, D. A.
031.026

OHANESYAN, J. B.
034.091
114.035 .092 .178

O'HARA, M. J.
094.069 .133 .157

OHKI, K.
073.035

OINES, R. K.
114.063

OJA, H.
042.062
052.005

OJHA, S. N.
065.134

OKADA, A.
105.048

OKANO, J.
105.118

OKAZAKI, S.
044.003
045.011

OKE, J. B.
114.159
158.017
160.010

OKEAN, H. C.
033.090 .091

O'KEEFE, J. A.
094.004

READE, V.
009.005
READER, J.
022.041 .042 .043
READHEAD, A. C. S.
142.074
READMAN, P. W.
094.123
REAGAN, J. B.
084.406
REAMES, D. V.
078.002 .010
REASONBERG, R. D.
097.037
REASONER, D. L.
094.203
REAY, N. K.
082.195
REBEIROT, E.
159.012
REBER, E. E.
082.010
REBRISTYJ, V. T.
114.161
REDDISH, V. C.
032.027
REDKOBORODY, YU. N.
126.021
REDMAN, R. O.
008.020
REED, S. J. B.
105.032 .124
REED JR., G. W.
094.111
REES, D. E.
073.016
REES, M.
003.115
REES, M. J.
142.023 .026
158.049
162.011 .092
REESE, D. E.
053.004
REESE, E. J.
093.031
099.018 .074 .086
REEVES, E. M.
073.099
080.060
REEVES, H.
091.043
099.062
REFSDAL, S.
065.020 .110
REGAS, J. L.
093.014
REGENER, V. H.
121.096
REGGE, T.
162.058
REGGIANI VIANI, E.
082.024
REGO, M. E.
114.103
REHFUSS, D. E.
094.182
REID, A. M.
094.007 .108 .192
REID, G. C.
083.064

REINA, C.
142.085
REINES, F.
003.003
061.033
REINHARD, P.
008.005
REINHARDT, G. W.
082.223
REINHARDT, M.
121.056
141.022 .071
160.017
REISS, P.
082.035
REISZ, A. C.
094.127
REITMEYER, W. L.
034.142
REMO, J.
072.047
RENARD, M.
034.024
RENGARAJAN, T. N.
141.505
143.045
RENNILSON, J. J.
094.212
RENSE, W. A.
073.098
RENSON, P.
082.047
116.002
122.060
RENY, H. R.
084.351
RENZINI, A.
065.055
REPAPIS, C. C.
077.054
RESCH, G.
142.091
RESCH, G. M.
141.040 .507
RESHETNYAK, L. N.
072.069
RESTUCCIA, A.
162.081
RETALLACK, D.
142.040 .056 .131
RETTIG, T. W.
141.123
152.012
REYNOLDS, R.
099.095
REYNOLDS, R. T.
094.087
REZACOVA, V.
034.043
RHODES, J. M.
094.196
RHODES, M.
034.115
RIABCHIKOV, E.
003.116
RICCIARDI, O.
114.154
RICE, J. R.
081.007
RICHARD, J.-P.
066.122

RICHARDS, D. W.
077.031 .068
RICHARDS, M. L.
011.007
RICHARDS, P. J.
142.041
RICHARDS-JONES, P.
014.025
RICHARDSON, F. F.
034.081
RICHARDSON, R. W.
065.079
RICHARDSON, W. W.
036.005
RICHER, H. B.
142.046 .127
152.009
RICHOU, J.
022.019
RICHSTONE, D.
158.018
RICHTER, G. A.
113.032
RICHTER, N.
004.064
013.011
113.064
114.080
154.007
RICHTER, P. H.
031.017
RICKARD, J. J.
114.055
RICKARD, L. J.
131.035
RICKER JR., G. R.
031.032
RICKMAN, H.
098.079
RIDGELEY, A.
071.077
RIDGWAY, S. T.
099.081
RIDLEY, W. I.
094.007 .108 .192
RIEDLER, W.
084.030
RIEGER, E.
083.042
RIEGLER, G. R.
142.004 .025 .142
RIEKE, G. H.
091.033
099.060
141.032
158.057 .107
RIGBY, A.
079.106
RIGHI, A.
033.072
RIGHINI, A.
082.004 .028 .227
RIGHINI, G.
012.013
RIGHINI, G. M.
034.032
RIGTERINK, P. V.
121.105
RIGUTTI, M.
073.078 .079

SORVARI, J. M.
142.122
SOSA, C. F.
008.089
075.034
SOSNOVA, A. K.
104.010 .060
SOSNOVETS, E. N.
084.419
SOSNOVETZ, E. N.
084.411
SOTER, S.
097.041
SOUFFRIN, S.
131.073
158.012
SOUTH, R. H.
103.016 .107 .117
SOUTH, R. H. S.
103.107 .120
SOUTHWOOD, D. J.
084.074 .278
SOWARD, A. M.
064.025
SOWERBY, P. L.
011.012
SPADA, G.
142.113
SPAENKUCH, D.
063.036
082.077
SPALL, H.
084.331
SPANNAGEL, G.
105.101
SPARKS, P. R.
097.009
SPARKS, W. M.
064.006
065.011
124.003
SPARROW, J. G.
034.062
SPEER, R. J.
034.126
SPEISER, D.
066.133
SPENCER, J.
142.014 .090
SPENCER, J. H.
158.058
SPENCER, N. W.
053.005
097.021
SPENCER, R. E.
142.041
SPERAUSKAS, J.
113.050
SPERLING, H. J.
052.018
SPEYBROECK, L. VAN
034.092
076.019 .020
SPIEGEL, E. A.
062.078
065.070
158.127
SPINRAD, H.
158.013 .090 .109 .111
.121 .123

SPITKOVSKIJ, V. M.
033.007 .029 .030
SPITZ, A. L.
084.009
SPITZER JR., L.
154.001
SPITZMESSER, D.
141.040
SPLITTGERBER, E.
114.143
123.068
SPOELSTRA, T. A. T.
121.026
155.017 .032 .041
156.002
157.002
SPREITER, J. R.
074.045
SPRENGER, K.
083.006
SPROTT, G. F.
142.098 .137
SPRUNG, D. W. L.
065.145
SPRYSAK, S. J.
084.350
SPYROU, N. K.
066.108
SQUEREN, A. M. LE
122.087
SRAMEK, R.
158.102
SREEKANTAN, B. V.
143.070
SRINIVASAN, B.
105.001
SRIVASTAVA, A. N.
022.055
SRIVASTAVA, D. C.
066.165
SRIVASTAVA, S.
061.053
STABELL, R.
065.020
066.162
STADNIKOVA, N. P.
094.254
STADT, H. VAN DE
034.106 .132
STAEHLE, V.
105.061
STAELIN, D. H.
141.544
STAFEEV, A. M.
034.153
041.072
STAFFORD, E. G.
034.082
STAGAT, R. W.
082.067
STAKHEYEV, YU. I.
094.169 .172
STALIO, R.
064.051
114.151
STALLING, D. L.
094.230
STAN, A. S.
094.163
STANDISH JR., E. M.
042.051

STANDISH JR., E. M.
117.020
STANEK, W.
072.057
STANGE, L.
055.016
STANIFORTH, A.
035.001
STANIKA, E. P.
123.004
STANNARD, D.
099.065
122.099
141.068
STARIKOVA, G. A.
118.001
STARK, B.
082.137
STARK, H.
033.033
STARKOV, G. V.
084.067 .252
STARODUBTSEVA, O. M.
097.122
STARR, W. L.
022.031
STARRFIELD, S.
124.003
STAUBERT, R.
142.140
STAUDE, J.
014.006
080.002
106.003
STAUFFER, D.
162.088
STAVELAND, L.
071.075
072.048 .066
STEBBINS, R.
141.557
STECHER, T. P.
022.114
STECKER, F. W.
162.052
STEELE, I. M.
094.131
STEELE, J. P.
142.046
STEENSTRUP, F.
083.001
STEFANOV, A. P.
072.043
STEFANOVITCH, D.
034.073
STEGMAN, J. E.
155.059
STEIGMAN, G.
162.003 .056
STEIN, A.
072.034 .053
STEIN, R. F.
065.092
080.042
131.012 .070
STEIN, W. A.
010.006
122.115
131.093
141.039

STROM, S. E.
114.901
122.092
131.039 .042 .094
152.901
153.013
STRONG, I. B.
084.299
STROUD, D. B.
065.099
STRUGATSKAYA, A. A.
154.007
STRUNNIKOVA, L. V.
084.340
STUB, H.
121.014
STUBBE, P.
082.059
083.009
STUBBS, T. J.
143.072
STUDNICKA, J.
034.043
STULL, M. A.
142.061
STUMPFF, P.
103.105
STURCH, C.
113.044
STURCH, C. R.
113.034 .035
114.081
STURIALE, M. L.
075.011 .013
STURROCK, P. A.
072.024
141.533 .534
158.015
STYAZHKIN, V. A.
077.060
SUBE, R.
003.076
SUBRAMANIAN, A.
143.073
SUDZIUS, J.
113.050
SUENDERMANN, J.
081.016
SUFFOLK, G.
116.003
SUFFOLK, G. C. J.
071.033
SUGAWA, C.
044.010
SUGI, N.
045.027
SUGIMOTO, D.
064.063
066.151
080.053
142.139
SUGIURA, M.
084.299 .324
SUH, P. K.
143.006
SULIDI-KONDRATIEV, E. D.
094.099
SULLIVAN, J. D.
073.004
078.016

SULLIVAN, S.
094.222
SULLIVAN III, W. T.
131.116
SULTANOV, D. D.
094.073
SULTANOV, G. F.
008.096
SUMARUK, P. V.
106.018
SUME, A.
131.082 .107
SUMMERS, A.
099.095
SUMMERS, A. L.
094.087
SUMMERS, H. P.
071.001
SUNDBERG, L. L.
094.199
SUNDMAN, A.
155.007
SUNG, C. C.
022.146
SUNYAEV, R. A.
131.078
142.042 .106
160.027
162.005
SUOMI, V. E.
031.047
SURKOV, E. P.
072.020
SURKOV, YU. A.
022.123
094.163 .206 .207
SURKOVA, L. P.
121.004 .005
SURMELIAN, G. L.
065.090
SUSHCHINSKY, M. M.
094.168
SUTANTYO, W.
061.064
153.027
SUTHERLAND, R.
022.040
SUZUKI, K.
103.124
SUZUKI, Y.
047.005
SVALGAARD, L.
106.006 .012
SVANBERG, S.
022.063
SVATOS, J.
131.056
SVECHNIKOV, M. A.
121.004 .005 .031
SVEEN, O. P.
015.009
079.004
SVENSSON, L. A.
022.072
SVESHNIKOV, M. L.
042.065
SVESTKA, Z.
073.074 .083 .087
SVIDERSKIENE, Z.
113.050

SVIRIDOV, A. M.
033.047
SVOLOPOULOS, S. N.
066.146
121.088
SWAMY, K. S. K.
114.037
SWANENBURG, P. N.
106.007
SWARTZ, W. E.
082.064
083.063
SWARUP, G.
141.132
SWEDLUND, J. B.
097.052
100.008 .020
116.015
SWEET, P. A.
008.037
SWEIGART, A. V.
065.034
154.012
SWENSON, G. W.
141.021
SWENSON JR., L. S.
003.139
SWIDER, W.
083.019
SWIFT, C. D.
142.037
SWINGS, J. P.
103.130
113.006 .019
SWISHER, R. L.
034.078
SY, W. N.
062.017
SYKES, D.
103.107
SYKES, M. J.
097.034 .094 .095
SYKORA, J.
079.106
SYMMS, L. S. T.
007.000
SYNEK, I.
031.082
032.049
SYNGE, J. L.
066.156
SYNITSYN, V. M.
082.141
SYROVATSKIJ, S. I.
062.050
143.010
157.003
SYUNAEV, R. A.
131.014
SYUNYAEV, R. A.
162.013 .021
SZAFRANIEC, R.
121.081
SZAMOSI, G.
162.007
SZATHMARY, K.
126.006
SZEBEHELY, V.
042.032
117.038

WALLERSTEIN, G.
064.024
122.094
125.102
WALLIS, M. K.
102.070
WALLIS, R.
125.100
WALSH, W. J.
083.065
WALT, A. J. VAN DER
143.003
WALT, M.
084.417 .422
WALTER, K.
117.035
WALTER, L. S.
105.038
WALTON, J. R.
084.297
WAMPLER, E. J.
141.541
158.122
WAMSTEKER, W.
125.102
WANG, A. P.
052.003
WANG, A. P.-I.
052.016
WANG, C. G.
065.086 .117
134.004
WANG, C. Y.
062.056
WANG, C.-Y.
081.008
WANG, J.
078.013
WANG, J. R.
078.018
143.048
WANG, T. I.
063.022
WANG, Y.-M.
073.116
WARD, P. L.
105.010
WARD, W. R.
101.021
WARD JR., F. W.
032.014
WARE, N.
094.210
WARE, N. G.
094.251
WARES, G. W.
072.036
WARMAN, I. M.
011.006
WARNER, B.
115.002
117.022
121.033 .043 .065
122.020 .022 .023
124.007
126.002 .005 .008 .020
WARNER, J.
094.007 .108 .192
WARNER, J. W.
114.125
141.041

WARNER, J. W.
158.042
WARNER, P. J.
142.074
WARNOCK, W. W.
093.038 .045
WARNOW, J. N.
003.127
WARREN, P. R.
132.027
WARREN JR., W. H.
113.012
WASSERBURG, G. J.
094.117 .164 .197 .198
WASSERMAN, L.
099.068
WASSON, J. T.
094.199
105.015 .024 .115
WATAGHIN, G.
162.034 .069
WATANABE, A.
022.035
WATANABE, E.
031.059
WATANABE, T.
004.052 .055
074.063
WATERFIELD, R. L.
103.016 .107 .116 .117
.120
WATERS, J. I.
051.011
WATSON, D. E.
105.033
WATSON, M. D.
084.009
WATSON, P. A.
033.125
WATSON, W. D.
131.011 .020
WATTENBERG, D.
004.014
010.038
WATTS JR., R. N.
053.001 .018
054.001 .006
093.021
WAUCHOP, T. S.
022.107
WAWRUKIEWICZ, A. S.
121.050
WAYLAND, J. R.
143.038
WAYLEN, P. C.
066.039 .112
WAYMAN, P. A.
004.031
005.027
113.028
122.088
WDOWCZYK, J.
142.141 .146
WDOWIAK, T. J.
125.102
WEAGANT, R. A.
034.072
WEART, S.
073.064

WEAVER, H.
006.000
WEAVER, W. B.
114.167
WEBB, C. J.
071.012
WEBB, D. F.
153.004
WEBB, S.
143.046
WEBBER, J. C.
125.032
141.065
WEBBER, W. R.
076.018
078.011 .034
143.016
WEBER, E.
103.130
WEBER, J.
066.134 .166
WEBER, R. R.
051.018
WEBER, W.
003.167
WEBROVA, L.
044.039
WEBSTER, B. L.
011.045
142.093 .094
152.005
WEBSTER, E.
121.040
141.111
WEBSTER JR., W. J.
077.019
WEEDMAN, D. W.
133.012
142.015
WEEKES, T. C.
003.129
134.001
WEGNER, G.
126.016
WEGNER, W.
082.214
WEHINGER, P.
114.132
122.014 .093
WEHLAU, A.
123.051
WEHLAU, W. H.
114.031
WEHMEYER, R.
121.015
WEHRSE, R.
064.005
WEIBLEN, P. W.
094.033
WEIDEMANN, V.
125.025
141.511
WEIDNER, D. K.
082.135
WEIGERT, A.
065.019
WEILER, K. W.
134.010
WEILL, G.
082.034

WEINBERG, J. L.
051.004
WEINBERG, S.
003.124
WEINER, C.
003.127
WEINSTEIN, A.
042.087
WEISHEIT, J. C.
022.014 .087 .093
131.048
141.116
WEISS, W.
031.023
WEISSKOPF, M. C.
034.120
WEISSLER, G. L.
011.017
WEISTROP, D.
115.017
WELCH, G. A.
158.041
160.002
WELCH, R. G.
122.121
WELCH, W. J.
131.005
WELIACHEW, L.
158.051
WELIN, G.
114.142
WELLER, W. G.
034.008
WELLINGTON, K. J.
033.070
WELLS, C. A.
022.099
WELLS, D. C.
115.012
WELLS, F. J.
045.028
WELLS, J. S.
022.129
WELLS, M. B.
063.035
WELLS, R. A.
097.053 .059 .072
WELSH, H. L.
005.001
022.035
WELTER, H.
084.321
WEMPE, J.
111.007
WENDKER, H. J.
131.019
WENGLER, P.
004.070
WENTZEL, D. G.
010.002
062.007
065.042
WENZEL, W.
065.094 .151
122.086 .126
123.025 .063
142.107 .148
152.008
WESSELINK, A. J.
115.009

WEST, F. R.
119.004
WEST, M. L.
072.065
WEST, R. M.
114.172
WEST JR., H. I.
084.297
WESTERLUND, B. E.
082.088
WESTFALL, J. E.
094.104 .266
WESTHAUS, P.
022.101
WESTON, E. B.
072.036
115.016
118.015
WESTPHAL, J. A.
091.012
099.043
103.122
142.010 .075
WETHERELL, W. B.
032.038
WEYMANN, R. J.
032.001
158.080
WHALEN, J. A.
084.069
WHANG, Y. C.
074.090
106.032
WHEATON, W. A.
142.103 .109 .112 .118
WHEELER, J.
003.115
WHEELER, J. C.
065.121
WHELAN, J.
121.036
WHELAN, J. A. J.
117.023
WHIPPLE, F. L.
005.027
102.056
103.103
107.005
WHITAKER, A. J. T.
033.093
WHITE, J. A.
022.106 .127
WHITE, J. L.
007.000
WHITE, K. S.
085.006
WHITE, M. L.
073.053
107.018
WHITE, N. M.
113.058
WHITE, O. R.
071.063
WHITE, R. E.
131.043
WHITE, R. S.
084.410
WHITE III, K. P.
077.032
WHITEHILL, L.
094.038

WHITEHURST, R. N.
158.008
WHITEOAK, J. B.
131.031
141.053
WHITFORD, C. H.
034.003
WHITING, E. E.
082.054
WHITMIRE, D. P.
066.094
WHITNEY, A. R.
046.023
055.007
141.040
142.091
WHITNEY, C. K.
063.038
WHITNEY, H. E.
083.001
WHITROW, G. J.
004.027
WHITTEKER, J. H.
083.060
WHITWORTH, D. P. D.
101.007
WIANT, J. R.
115.012
WIBBERENZ, G.
143.022
WICKRAMASINGHE, D. T.
064.002
126.015 .019 .023
WICKRAMASINGHE, N. C.
131.030
158.054
WICKWAR, V. B.
082.188
WIDING, K. G.
073.007
076.013 .029
WIEDEMANN, D.
153.024
WIEDEMANN, E.
007.000
031.056
103.104
WIELEBINSKI, R.
008.014
141.531
WIELEN, R.
151.047
WIEMER, W.
113.030
114.024
WIERCIOCH, K.
104.049
WIESENFARTH, H. J.
033.120
WIESMANN, H.
094.196
WIJBENGA, J. W.
121.077
WIJNBERGEN, J. J.
034.004 .033 .056
WILCOX, J. M.
071.055
080.017 .038
106.002 .013 .020 .028
WILD, J. P.
077.041

Subject Index

ABSOLUTE MAGNITUDES
115.000
ABSOLUTE MAGNITUDES
RR LYRAE STARS
122.061 .130
ABSOLUTE MAGNITUDES
SUPERGIANTS
115.001
ABSORPTION
H II REGIONS
131.055
ABSORPTION
INTERSTELLAR MATTER
113.048 .061
131.123 .124 .125
ABSORPTION
RADIO SOURCES
141.053
ABSORPTION
STELLAR ATMOSPHERES
114.035
ACCRETION
MAGNETIC STARS
065.087
ACCRETION
NEUTRON STARS
117.003
ACHONDRITES
105.006
ACOUSTIC WAVES
SOLAR ATMOSPHERE
080.042
AIRGLOW
012.019
082.000
AIRGLOW
MARS ATMOSPHERE
097.091
ALBEDO
MOON
094.025
ALBEDO
PLANETARY ATMOSPHERES
091.003

ALFVEN WAVES
PLASMA
062.007
ALFVEN WAVES
SOLAR ATMOSPHERE
080.030
ALGOL
PHOTOMETRY
121.064
ALGOL SYSTEMS
121.101
ALMANACS
047.000
ANDROMEDA NEBULA
141.010
158.008 .009 .078
.086 .095 .118
ANDROMEDA NEBULA
GLOBULAR CLUSTERS
154.010
ANDROMEDA NEBULA
RADIO RADIATION
158.072
ANTIMATTER
143.012
162.004 .072
ARTIFICIAL SATELLITES
054.000
ARTIFICIAL SATELLITES
OBSERVATIONS
055.000
ASSOCIATIONS
010.000
ASSOCIATIONS STELLAR
152.000
A STARS
FINDING LISTS
114.173
A STARS
GALACTIC DISTRIBUTION
155.059
A STARS
LUMINOSITIES
115.025

A STARS
METAL ABUNDANCES
152.002
A STARS
MK TYPES
116.011
A STARS
PECULIAR
114.023 .133 .138
.171
116.003 .006
A STARS
ROTATION
116.003 .011
A STARS
SPACE MOTIONS
115.025
A STARS
SPECTRA
114.071
ASTEROID BELT
098.023
ASTRODYNAMICS
052.000
ASTROMETRY
003.012
012.022
ASTRONOMICAL
ACCESSORIES
012.016
034.000
ASTRONOMICAL CONSTANTS
043.000
ASTRONOMICAL
INSTRUMENTS
012.016
032.000
ASTRONOMICAL UNIT
043.004
ATLASES
041.000
ATMOSPHERE
EARTH
082.000

DUE TO A MASCHINE ERROR DURING THE SORTING PROCESS THE
FOLLOWING NAMES ARE MISSING IN THE AUTHOR INDEX OF VOLUME 6.

PLEASE INSERT THIS PAGE IN VOL. 6.